C. Oliver Kappe, Alexander Stadler, and Doris Dallinger

Microwaves in Organic and Medicinal Chemistry

Methods and Principles in Medicinal Chemistry

Edited by R. Mannhold, H. Kubinyi, G. Folkers

Previous Volumes of this Series:

Smith, Dennis A. / Allerton, Charlotte / Kalgutkar, Amit S. / van de Waterbeemd, Han / Walker, Don K.

Pharmacokinetics and Metabolism in Drug Design

Third, Revised and Updated Edition

2012

ISBN: 978-3-527-32954-0

Vol. 51

De Clercq, Erik (Ed.)

Antiviral Drug Strategies

2011

ISBN: 978-3-527-32696-9

Vol. 50

Klebl, Bert / Müller, Gerhard / Hamacher, Michael (Eds.)

Protein Kinases as Drug Targets

2011

ISBN: 978-3-527-31790-5

Vol. 49

Sotriffer, Christoph (Ed.)

Virtual Screening

Principles, Challenges, and Practical Guidelines

2011

ISBN: 978-3-527-32636-5

Vol. 48

Rautio, Jarkko (Ed.)

Prodrugs and Targeted Delivery

Towards Better ADME Properties

2011

ISBN: 978-3-527-32603-7

Vol. 47

Smit, Martine J. / Lira, Sergio A. / Leurs, Rob (Eds.)

Chemokine Receptors as Drug Targets

2011

ISBN: 978-3-527-32118-6

Vol. 46

Ghosh, Arun K. (Ed.)

Aspartic Acid Proteases as Therapeutic Targets

2010

ISBN: 978-3-527-31811-7

Vol. 45

Ecker, Gerhard F. / Chiba, Peter (Eds.)

Transporters as Drug Carriers

Structure, Function, Substrates

2009

ISBN: 978-3-527-31661-8

Vol. 44

Faller, Bernhard / Urban, Laszlo (Eds.)

Hit and Lead Profiling

Identification and Optimization of Drug-like Molecules

2009

ISBN: 978-3-527-32331-9

Vol. 43

Sippl, Wolfgang / Jung, Manfred (Eds.)

Epigenetic Targets in Drug Discovery

2009

ISBN: 978-3-527-32355-5

Vol. 42

C. Oliver Kappe, Alexander Stadler, and Doris Dallinger

Microwaves in Organic and Medicinal Chemistry

Second, Completely Revised and Enlarged Edition

WILEY-VCH

WILEY-VCH Verlag GmbH & Co. KGaA

Series Editors

Prof. Dr. Raimund Mannhold
Molecular Drug Research Group
Heinrich-Heine-Universität
Universitätsstrasse 1
40225 Düsseldorf
Germany
mannhold@uni-duesseldorf.de

Prof. Dr. Hugo Kubinyi
Donnersbergstrasse 9
67256 Weisenheim am Sand
Germany
kubinyi@t-online.de

Prof. Dr. Gerd Folkers
Collegium Helveticum
STW/ETH Zurich
8092 Zurich
Switzerland
folkers@collegium.ethz.ch

The Authors

Prof. Dr. C. Oliver Kappe
Karl-Franzens-Universität Graz
Christian Doppler Laboratory
Heinrichstrasse 28
8010 Graz
Austria

Dr. Alexander Stadler
Anton Paar GmbH
Anton-Paar Str. 20
8054 Graz
Austria

Dr. Doris Dallinger
Karl-Franzens-University Graz
Christian Doppler Laboratory
Heinrichstrasse 28
8010 Graz
Austria

Library of Congress Card No.: applied for

British Library Cataloguing-in-Publication Data
A catalogue record for this book is available from the British Library.

Bibliographic information published by the Deutsche Nationalbibliothek
The Deutsche Nationalbibliothek lists this publication in the Deutsche Nationalbibliografie; detailed bibliographic data are available on the Internet at http://dnb.d-nb.de.

Print ISBN: 978-3-527-33185-7
ePDF ISBN: 978-3-527-64785-9
ePub ISBN: 978-3-527-64784-2
mobi ISBN: 978-3-527-64783-5
oBook ISBN: 978-3-527-64782-8

Cover Design Grafik-Design, Schulz Fußgönheim
Typesetting Thomson Digital, Noida, India
Printing and Binding betz-druck GmbH, Darmstadt, Germany

Printed on acid-free paper.

Contents

Preface

The application of microwaves marks a real revolution in synthetic organic chemistry. Although it was more or less a curiosity, only a few decades ago, the rapid development within this field made it necessary to come up with a second, completely revised edition of the standard monograph, *Microwaves in Organic and Medicinal Chemistry*, by Oliver Kappe and Alexander Stadler, published in this book series in 2005. Indeed, the current edition is not just an updated version, but a completely new monograph as one can see from the increase in size, from originally 409 pages to almost 700 pages! An enormous amount of recent literature has been considered and included, making these two volumes now the new "gold standard" of microwave chemistry.

Especially in medicinal chemistry, yield and elegance of the synthesis of a new compound are no issue – only a minor amount of pure material is needed to screen for biological properties. Only later and only for a negligibly small number of potential candidates, better synthetic strategies have to be developed. Thus, microwave-supported synthesis is the first choice to quickly (and simply) create a multitude of test compounds.

We, the editors of the book series Methods and Principles in Medicinal Chemistry, are very grateful to Oliver Kappe, Alexander Stadler, and Doris Dallinger for having undertaken this enormous effort. We are also grateful to Frank Weinreich for his ongoing engagement in our book series and to Heike Noethe, both at Wiley-VCH Verlag GmbH, for her editorial support.

January 2012

Düsseldorf	*Raimund Mannhold*
Weisenheim am Sand	*Hugo Kubinyi*
Zurich	*Gerd Folkers*

Personal Foreword to the First Edition

We are currently witnessing an explosive growth in the general field of "microwave chemistry." The increase of interest in this technology stems from the realization that microwave-assisted synthesis, apart from many other enabling technologies, actually provides significant practical and economic advantages. Although microwave chemistry is currently used in both academic and industrial contexts, the impact on the pharmaceutical industry especially has developed microwave-assisted organic synthesis (MAOS) from a laboratory curiosity in the 1980s and 1990s to a fully accepted technology today. The field has grown such that nearly every pharmaceutical company and more and more academic laboratories now actively utilize this technology for their research.

One of the main barriers facing a synthetic chemist contemplating to use microwave synthesis today is – apart from access to suitable equipment – obtaining education and information on the fundamental principles and possible applications of this new technology. Thus, the aim of this book is to give the reader a well-structured, up-to-date, and exhaustive overview of known synthetic procedures involving the use of microwave technology and to illuminate the "black box" stigma that microwave chemistry still has.

Our main motivation for writing *Microwaves in Organic and Medicinal Chemistry* derived from our experience in teaching microwave chemistry in the form of short courses and workshops to researchers from the pharmaceutical industry. In fact, the structure of this book closely follows a course developed for the American Chemical Society and can be seen as a compendium for this course. It is hoped that some of the chapters of this book are sufficiently convincing as to encourage scientists not only to use microwave synthesis in their research but also to offer training for their students or coworkers.

We would like to thank Hugo Kubinyi for his encouragement and motivation to write this book. Thanks are also due to Mats Larhed, Nicholas E. Leadbeater, Erik Van der Eycken, and scientists from Anton Paar GmbH, Biotage AB, CEM Corp., and Milestone srl, who have been kind enough to read various sections of this book and to provide valuable suggestions. First and foremost, we would like to thank Doris Dallinger, Bimbisar Desai, Toma Glasnov, Jenny Kremsner, and other members of the Kappe research group for spending their time searching the "microwave literature" and for tolerating this distraction. We are particularly indebted to Doris Dallinger for carefully proofreading the complete text and to Jenny Wheedby for providing the cover art. We are very grateful to Dr. Frank Weinreich and other editors at Wiley-VCH Verlag GmbH for their assistance in bringing out this book.

This book is dedicated to Rajender S. Varma, a pioneer in the field of microwave synthesis, who inspired us to enter this exciting research area in the 1990s.

Graz, Austria
December 2004

C. Oliver Kappe
Alexander Stadler

Personal Foreword to the Second Edition

In more than 6 years since the manuscript submission for the first edition of *Microwaves in Organic and Medicinal Chemistry*, many things have changed. In contrast to 2004, microwave chemistry now is truly an established technology, especially in the pharmaceutical industry. Most medicinal chemists are now so accustomed to this nonclassical form of heating that taking their microwave reactors away from them would probably cause significant chaos in the laboratory. To a somewhat smaller extent, dedicated microwave instruments are however slowly replacing oil baths and heating mantles in many academic labs. Importantly, the speculation and confusion about "microwave effects" that persisted for many years have now subsided and most scientists today accept the fact that microwave chemistry is a great way to heat reaction mixtures in sealed tubes with very accurate control of the reaction parameters and to do synthesis in general.

Based on these facts, we now present the second, extensively updated, edition of *Microwaves in Organic and Medicinal Chemistry*. This edition covers the literature till early 2011, which has led to a significant increase in the number of references and examples in most chapters. We have tried not to greatly increase the page numbers of the introductory Chapters 1–4, but rather to selectively update the fundamental and more technical information on the concept of microwave chemistry contained therein (removing some outdated content). Having the practicing organic and medicinal chemist in mind, most of the changes and additions have occurred in the chapters (now Chapters 5–8) describing the examples of microwave chemistry. Close to 1000 additional references have been included in these chapters. We hope that this revised version will become an indispensable reference work for all chemists interested in microwave chemistry.

Graz, Austria
July 2011

C. Oliver Kappe
Alexander Stadler
Doris Dallinger

1
Introduction: Microwave Synthesis in Perspective

1.1
Microwave Synthesis and Medicinal Chemistry

Improving research and development (R&D) productivity is one of the biggest tasks facing the pharmaceutical industry. In a few years, the pharmaceutical industry will see many patents of drugs expire. In order to remain competitive, pharma companies need to pursue strategies that will offset the sales decline and see robust growth and improved shareholder value. The impact of genomics and proteomics is creating an explosion in the number of drug targets. Today's drug therapies are solely based on approximately 500 biological targets; in a few years' time, it is expected that the number of targets will well reach 10 000. In order to identify more potential drug candidates for all these targets, pharmaceutical companies have made major investments in high-throughput technologies for genomic and proteomic research, automated/parallel chemistry, and biological screening. However, lead compound optimization and medicinal chemistry remain one of the bottlenecks in the drug discovery process. Developing chemical compounds with the desired biological properties is time-consuming and expensive. Consequently, increasing interest is being directed toward technologies that allow more rapid synthesis and screening of chemical substances to identify compounds with functional qualities.

Medicinal chemistry has benefited tremendously from the technological advances in the field of combinatorial chemistry and high-throughput synthesis. This discipline has been an innovative machine for the development of methods and technologies that accelerate the design, synthesis, purification, and analysis of compound libraries. These new tools have had a significant impact on both lead identification and lead optimization in the pharmaceutical industry. Large compound libraries can now be designed and synthesized to provide valuable leads for new therapeutic targets. Once a chemist develops a suitable high-speed synthesis of a lead, it becomes possible to synthesize and purify hundreds of molecules in parallel

Microwaves in Organic and Medicinal Chemistry, Second Edition.
C. Oliver Kappe, Alexander Stadler, and Doris Dallinger.

to discover new leads and/or derive structure–activity relationships (SAR) in unprecedented timeframes.

The bottleneck of conventional parallel/combinatorial synthesis is typically optimization of reaction conditions to afford the desired products in high yields and with suitable purities. Since many reaction sequences require at least one or more heating steps for extended time periods, these optimizations are often difficult and time-consuming. Microwave-assisted heating under controlled conditions has been shown to be an invaluable technology for medicinal chemistry and drug discovery applications since it often dramatically reduces reaction times, typically from days or hours to minutes or even seconds. Many reaction parameters can be evaluated in a few hours to optimize the desired chemistry. Compound libraries can then be rapidly synthesized in either a parallel or (automated) sequential format using this new, enabling technology. In addition, microwave synthesis allows the discovery of novel reaction pathways that serve to expand "chemical space" in general and "biologically relevant, medicinal chemistry space" in particular.

Specifically, microwave synthesis has the potential to impact upon medicinal chemistry efforts in at least three major phases of the drug discovery process: lead generation, hit-to-lead efforts, and lead optimization. Medicinal chemistry addresses what are fundamentally biological and clinical problems. Focusing first on the preparation of suitable molecular tools for mechanistic validation, efforts ultimately turn to the optimization of biochemical, pharmacokinetic, pharmacological, clinical, and competitive properties of drug candidates. A common theme throughout this drug discovery and development process is speed. Speed equals competitive advantage, more efficient use of expensive and limited resources, faster exploration of structure–activity relationship, enhanced delineation of intellectual property, more timely delivery of critically needed medicines, and ultimately determines positioning in the marketplace. To the pharmaceutical industry and the medicinal chemist, time truly does equal money, and microwave chemistry has become a central tool in this fast-paced, time-sensitive field.

Chemistry, like all sciences, consists of never-ending iterations of hypotheses and experiments, with results guiding the progress and development of projects. The short reaction times provided by microwave synthesis make it ideal for rapid reaction scouting and optimization, allowing very rapid progress through the "hypotheses–experiment–results" iterations, resulting in more decision points per time unit. In order to fully benefit from microwave synthesis, one has to "be prepared to fail in order to succeed." While failure could cost a few minutes, success would gain many hours or even days. The speed at which multiple variations of reaction conditions can be performed allows a morning discussion of "What should we try?" to become an after lunch discussion of "What were the results?" (the "let's talk after lunch" mantra) [1]. Not surprisingly, therefore, most pharmaceutical, agrochemical, and biotechnology companies are already heavily using microwave synthesis as frontline methodology in their chemistry programs, both for library synthesis and for lead optimization, as they realize the ability of this enabling technology to speed chemical reactions and therefore the drug discovery process.

1.2
Microwave-Assisted Organic Synthesis (MAOS): A Brief History

While fire is now rarely used in synthetic chemistry, it was not until Robert Bunsen invented the burner in 1855 that the energy from this heat source could be applied to a reaction vessel in a focused manner. The Bunsen burner was later superseded by the isomantle, the oil bath, or the hot plate as a means of applying heat to a chemical reaction. In the past few years, heating and driving chemical reactions by microwave energy has been an increasingly popular theme in the scientific community [1, 2].

Microwave energy, originally applied for heating foodstuff by Percy Spencer in the 1940s, has found a variety of technical applications in the chemical and related industries since the 1950s, in particular in food processing, drying, and polymer industries. Other applications range from analytical chemistry (microwave digestion, ashing, and extraction) [3] to biochemistry (protein hydrolysis and sterilization) [3], pathology (histoprocessing and tissue fixation) [4], to medical treatments (diathermy) [5]. Somewhat surprisingly, microwave heating has only been implemented in organic synthesis since the mid-1980s. The first reports on the use of microwave heating to accelerate organic chemical transformations (MAOS) were published 25 years ago by the groups of Gedye *et al.* (Scheme 1.1) [6] and Giguere *et al.* [7] in 1986. In those early days, experiments were typically carried out in sealed Teflon or glass vessels in a domestic household microwave oven without any temperature or pressure measurements. The results were often violent explosions due to the rapid uncontrolled heating of organic solvents under closed-vessel conditions. In the 1990s, several groups started to experiment with solvent-free microwave chemistry (so-called dry media reactions), which eliminated the danger of explosions [8]. Here, the reagents were preadsorbed onto either a more or less microwave-transparent (i.e., silica, alumina, or clay) or strongly absorbing (i.e., graphite) inorganic support that additionally may have been doped with a catalyst or reagent. Particularly in the early days of MAOS, the solvent-free approach was very popular since it allowed the safe use of domestic microwave ovens and standard open-vessel technology. While a large number of interesting transformations using "dry media" reactions have been published in the literature [8], technical difficulties relating to nonuniform heating, mixing, and the precise determination of the reaction temperature remained unsolved, in particular when scale-up issues needed to be addressed.

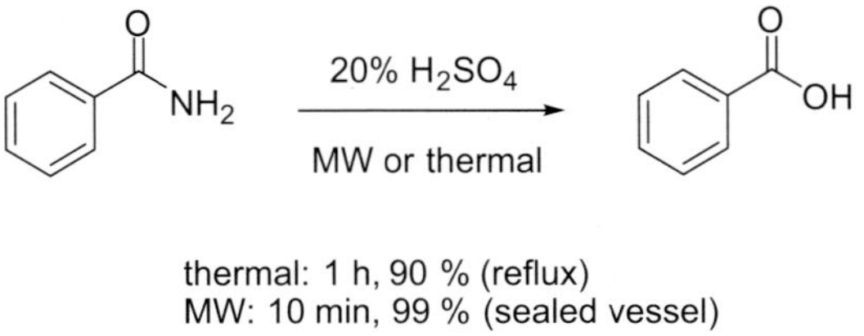

Scheme 1.1 Hydrolysis of benzamide. The first published example (1986) of microwave-assisted organic synthesis.

Alternatively, microwave-assisted synthesis has been carried out using standard organic solvents under open-vessel conditions. If solvents are heated by microwave irradiation at atmospheric pressure in an open vessel, the boiling point of the solvent typically limits the reaction temperature that can be achieved. Nonetheless, in order to achieve high reaction rates, high-boiling microwave-absorbing solvents have been frequently used in an open-vessel microwave synthesis [9]. However, the use of these solvents presented serious challenges in relation to product isolation and recycling of the solvent. Because of the recent availability of modern microwave reactors with online monitoring of both temperature and pressure, MAOS in dedicated sealed vessels using standard solvents – a technique pioneered by Christopher R. Strauss in the mid-1990s [10] – has been celebrating a comeback in recent years. This is clearly evident surveying the recently published (since 2001) literature in the area of controlled microwave-assisted organic synthesis (Figure 1.1). In addition to the primary and patent literature, many review articles, several books, special issues of journals, feature articles, online databases, information on the World Wide Web, and educational publications provide extensive coverage of the subject (see Section 5.1 for a comprehensive survey). Among the approximately 1000 original publications that appeared in 2010 describing microwave-assisted reactions under controlled conditions, a careful analysis demonstrates that in about 90% of all cases, sealed-vessel processing (autoclave technology) in dedicated single-mode microwave instruments has been employed. A 2007 survey has however found that as many as 30% of all published MAOS papers still employ kitchen microwave ovens [11], a practice

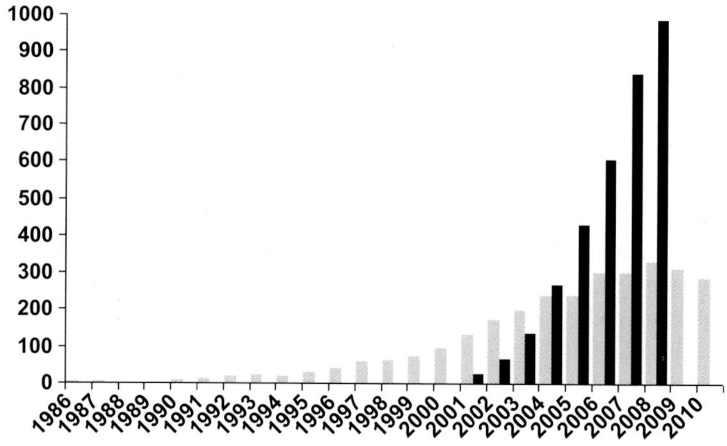

Figure 1.1 Publications on microwave-assisted organic synthesis (1986–2010). Gray graphs: Number of articles involving MAOS for seven selected synthetic organic chemistry journals (*Journal of Organic Chemistry*, *Organic Letters*, *Tetrahedron*, *Tetrahedron Letters*, *Synthetic Communications*, *Synthesis*, and *Synlett*; SciFinder scholar search, keyword: "microwave"). The black graphs represent the number of publications (2001–2008) reporting MAOS experiments in dedicated reactors with adequate process control (about 50 journals, full text search: microwave). Data for 2009 and 2010 are not available, but are estimated to be in the 1000–1200 publications per year range.

banned by most of the respected scientific journals today. For example, the American Chemical Society (ACS) organic chemistry journals will typically not consider manuscripts describing the use of kitchen microwave ovens or the absence of a reaction temperature as specified in the relevant author guidelines [12].

Since the early days of microwave synthesis, the observed rate accelerations and sometimes altered product distributions compared to oil bath experiments have led to speculation on the existence of so-called "specific" or "nonthermal" microwave effects [13]. Historically, such effects were claimed when the outcome of a synthesis performed under microwave conditions was different from that of the conventionally heated counterpart at the same apparent temperature. Reviewing the present literature [14, 15], it appears that today most scientists agree that in the majority of cases the observed rate enhancement is a purely thermal/kinetic effect, that is, a consequence of the high reaction temperatures that can rapidly be attained when irradiating polar materials in a microwave field, although effects that are caused by the unique nature of the microwave dielectric heating mechanism (specific microwave effects) also need to be considered. While for the medicinal chemist in industry, this discussion may seem futile, the debate on "microwave effects" is undoubtedly going to continue for a few years in the academic world. Regardless of the nature of the observed rate enhancements (for further details on microwave effects, see Section 2.5), microwave synthesis has now truly matured and has moved from a laboratory curiosity in the late 1980s to an established technique in organic synthesis, heavily used in both academia and industry.

The initially slow uptake of the technology in the late 1980s and 1990s has been attributed to its lack of controllability and reproducibility, coupled with a general lack of understanding of the basics of microwave dielectric heating. The risks associated with the flammability of organic solvents in a microwave field and the lack of available dedicated microwave reactors allowing adequate temperature and pressure control were major concerns. Important instrument innovations (see Chapter 3) now allow a careful control of time, temperature, and pressure profiles, paving the way for reproducible protocol development, scale-up, and transfer from laboratory to laboratory and scientist to scientist. Today, microwave chemistry is as reliable as the vast arsenal of synthetic methods that preceded it. Since 2001, therefore, the number of publications related to MAOS has increased dramatically (Figure 1.1) to such a level that it might be assumed that in a few years, many more chemists than today will probably use microwave energy to heat chemical reactions on a laboratory scale [1, 2]. However, it should be emphasized that the potential for growth is still very large as a recent survey has found that less than 10% of all publications in synthetic organic chemistry currently make use of microwave technology [15].

Recent innovations in microwave reactor technology now allow controlled parallel and automated sequential processing under sealed-vessel conditions and the use of continuous or stop-flow reactors for scale-up purposes. In addition, dedicated vessels for solid-phase synthesis, for performing transformations using pre-pressurized conditions and for a variety of other special applications, have been developed. Today, there are four major instrument vendors that produce microwave instrumentation dedicated toward organic synthesis. All those instruments offer temperature and pressure sensors, built-in magnetic stirring, power control, software operation, and sophisticated safety controls. The number of users of dedicated microwave reactors is therefore growing at a rapid rate, and it appears only to be a question of time until most laboratories will be equipped with suitable microwave instrumentation.

In the past, microwave chemistry was often used only when all other options to perform a particular reaction failed or when exceedingly long reaction times or high temperatures were required to complete a reaction. This practice is now slowly changing and due to the growing availability of microwave reactors in many laboratories, routine synthetic transformations are also now being carried out by microwave heating. One of the major drawbacks of this relatively new technology still is equipment cost. While prices for dedicated microwave reactors for organic synthesis have come down considerably since their first introduction in the late 1990s, the current price range for microwave reactors is still many times higher than that of conventional heating equipment. As with any new technology, the current situation is bound to change over the next several years and less expensive equipment should become available. By then, microwave reactors will have truly become the "Bunsen burners of the twenty first century" and will be a standard equipment in every chemical laboratory.

1.3
Scope and Organization of the Book

Today, a large body of work on microwave-assisted synthesis exists in the published and patent literature. Many review articles, several books, and information on the World Wide Web already provide extensive coverage of the subject (see Section 5.1). The goal of the present book is to present carefully scrutinized, useful, and practical information for advanced practitioners of microwave-assisted organic synthesis. Special emphasis is placed on concepts and chemical transformations that are of importance to medicinal chemists, and that have been reported in the most recent literature (2002–2010). The extensive literature survey is limited to reactions that have been performed using controlled microwave heating conditions, that is, where dedicated microwave reactors for synthetic applications with adequate temperature and pressure measurements have been employed. After a discussion of microwave dielectric heating theory and microwave effects (Chapter 2), a review of the existing equipment for performing MAOS will be presented (Chapter 3). This is followed by a chapter outlining the different processing techniques in a microwave-heated experiment (Chapter 4). Finally, a literature survey with more than 1500 references will be presented in Chapters 5–8.

Beginners in the field of microwave-assisted organic synthesis are referred to a recent book containing a chapter with useful practical tips ("How To Get Started") and an additional section with carefully selected and documented microwave experiments that may be used by scientists in academia to design a course on microwave-assisted organic synthesis [16].

References

1 Leadbeater, N. (2004) *Chemistry World*, **1**, 38–41.

2 (a) Adam, D. (2003) *Nature*, **421**, 571–572; (b) Marx, V. (2004) *Chemical and Engineering News*, **82** (50), 14–19; (c) Yarnell, A. (2007) *Chemical and Engineering News*, **85** (21), 32–33.

3 Kingston, H.M. and Haswell, S.J. (eds) (1997) *Microwave-Enhanced Chemistry: Fundamentals, Sample Preparation and Applications*, American Chemical Society, Washington.

4 Giberson, R.T. and Demaree, R.S. (eds) (2001) *Microwave Techniques and Protocols*, Humana Press, Totowa, NJ.

5 Prentice, W.E., (2002) *Therapeutic Modalities for Physical Therapists*, McGraw-Hill, New York.

6 Gedye, R., Smith, F., Westaway, K., Ali, H., Baldisera, L., Laberge, L., and Rousell, J. (1986) *Tetrahedron Letters*, **27**, 279–282.

7 Giguere, R.J., Bray, T.L., Duncan, S.M., and Majetich, G. (1986) *Tetrahedron Letters*, **27**, 4945–4958.

8 (a) Loupy, A., Petit, A., Hamelin, J., Texier-Boullet, F., Jacquault, P., and Mathé, D. (1998) *Synthesis*, 1213–1234; (b) Varma, R.S. (1999) *Green Chemistry*, 43–55.

9 (a) Bose, A.K., Banik, B.K., Lavlinskaia, N., Jayaraman, M., and Manhas, M.S. (1997) *Chemtech*, **27**, 18–24; (b) Bose, A.K., Manhas, M.S., Ganguly, S.N., Sharma, A.H., and Banik, B.K. (2002) *Synthesis*, 1578–1591.

10 (a) Strauss, C.R. and Trainor, R.W. (1995) *Australian Journal of Chemistry*, **48**, 1665–1692; (b) Strauss, C.R. (1999) *Australian Journal of Chemistry*, **52**, 83–96.

11 Moseley, J.D., Lenden, P., Thomson, A.D., and Gilday, J.P. (2007) *Tetrahedron Letters*, **48**, 6084–6087 (Ref. 13).

12 (2011) *The Journal of Organic Chemistry*, **76** (1), Author Guidelines.

13 (a) Perreux, L. and Loupy, A. (2001) *Tetrahedron*, **57**, 9199–9223; (b) Perreux, L. and Loupy, A. (2006) Chapter 4, in *Microwaves in Organic Synthesis*, 2nd edn (ed. A. Loupy), Wiley-VCH Verlag GmbH, Weinheim, pp. 134–218; (c) de la Hoz, A., Díaz-Ortiz, A., and Moreno, A. (2005) *Chemical Society Reviews*, **34**, 164–178; (d) de la Hoz, A., Diaz-Ortiz, A., and Moreno, A. (2006) Chapter 5, in *Microwaves in Organic Synthesis*, 2nd edn (ed. A. Loupy), Wiley-VCH Verlag GmbH, Weinheim, pp. 219–277.

14 (a) Caddick, S. and Fitzmaurice, R. (2009) *Tetrahedron*, **65**, 3325–3355; (b) Kappe, C.O. and Dallinger, D. (2009) *Molecular Diversity*, **13**, 71–193.

15 Leadbeater, N.E. (ed.) (2011) *Microwave Heating as a Tool for Sustainable Chemistry*, CRC Press, Boca Raton.

16 Kappe, C.O., Dallinger, D., and Murphree, S.S. (2009) *Practical Microwave Synthesis for Organic Chemists*, Wiley-VCH Verlag GmbH, Weinheim.

2
Microwave Theory

The physical principles behind and the factors determining the successful application of microwaves in organic synthesis are not widely familiar to chemists. Nevertheless, it is essential for the synthetic chemist involved in microwave-assisted organic synthesis to have at least a basic knowledge of the underlying principles of microwave–matter interactions and of the nature of microwave effects. The basic understanding of macroscopic microwave interactions with matter was formulated by von Hippel in the mid-1950s [1]. In this chapter, a brief summary of the current understanding of microwaves and their interactions with matter is given. For more in-depth discussion on this quite complex field, the reader is referred to recent review articles [2–5].

2.1
Microwave Radiation

Microwave irradiation is an electromagnetic irradiation in the frequency range of 0.3–300 GHz, corresponding to wavelengths of 1 mm–1 m. The microwave region of the electromagnetic spectrum (Figure 2.1) therefore lies between infrared (IR) and radio frequencies. The major use of microwaves is either for transmission of information (telecommunication) or for transmission of energy. Wavelengths between 1 mm and 25 cm are extensively used for RADAR transmissions and the remaining wavelength range is used for telecommunications. All domestic "kitchen" microwave ovens and all dedicated microwave reactors for chemical synthesis that are commercially available today operate at a frequency of 2.45 GHz (corresponding to a wavelength of 12.25 cm) in order to avoid interference with telecommunication, wireless networks, and cellular phone frequencies. There are other frequency allocations for microwave heating applications (ISM (industrial, scientific, and medical) frequencies (see Table 2.1) [6], but these are generally not employed in dedicated reactors for synthetic chemistry. Indeed, published examples of organic synthesis carried out with microwave heating at frequencies other than 2.45 GHz are extremely rare [7].

Microwaves in Organic and Medicinal Chemistry, Second Edition.
C. Oliver Kappe, Alexander Stadler, and Doris Dallinger.

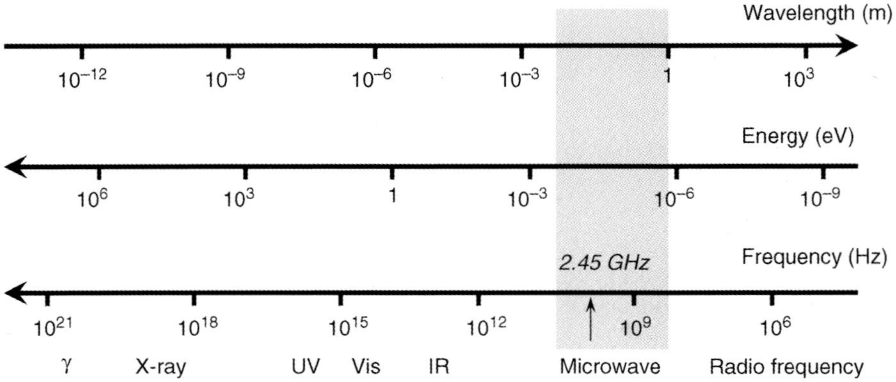

Figure 2.1 The electromagnetic spectrum.

From comparison of the data presented in Table 2.2 [8], it is obvious that the energy of the microwave photon at a frequency of 2.45 GHz (about 10^{-5} eV) is too low to cleave molecular bonds and is also lower than Brownian motion. It is therefore clear that microwaves cannot "induce" chemical reactions by direct absorption of electromagnetic energy, as opposed to ultraviolet and visible radiation (photochemistry).

Table 2.1 ISM microwave frequencies.

Frequency (MHz)	Wavelength (cm)
433.92 ± 0.2%	69.14
915 ± 13	32.75
2450 ± 50	12.24
5800 ± 75	5.17
24 125 ± 125	1.36

Data from Ref. [6].

Table 2.2 Comparison of radiation types and bond energies.

Radiation type	Frequency (MHz)	Quantum energy (eV)	Bond type	Bond energy (eV)
Gamma rays	3.0×10^{14}	1.24×10^{6}	C–C	3.61
X-Rays	3.0×10^{13}	1.24×10^{5}	C=C	6.35
Ultraviolet	1.0×10^{9}	4.1	C–O	3.74
Visible light	6.0×10^{8}	2.5	C=O	7.71
Infrared light	3.0×10^{6}	0.012	C–H	4.28
Microwaves	2450	1.01×10^{-5}	O–H	4.80
Radio frequencies	1	4.0×10^{-9}	Hydrogen bond	0.04–0.44

Data from Refs [6, 8].

2.2
Microwave Dielectric Heating

Microwave chemistry is based on the efficient heating of materials by "microwave dielectric heating" effects [4, 5]. Microwave dielectric heating depends on the ability of a specific material (e.g., a solvent or reagent) to absorb microwave energy and convert it into heat. Microwaves are electromagnetic waves that consist of an electric and a magnetic field component (Figure 2.2). For most practical purposes related to microwave synthesis, it is the electric component of the electromagnetic field that is of importance for wave–material interactions, although in some instances magnetic field interactions (e.g., with metals or metal oxides) can also be of relevance [9, 10].

The electric component of an electromagnetic field causes heating by two main mechanisms: dipolar polarization and ionic conduction. The interaction of the electric field component with the matrix is called the dipolar polarization mechanism (Figure 2.3a) [4, 5]. For a substance to be able to generate heat when irradiated with microwaves, it must possess a dipole moment. When exposed to microwave frequencies, the dipoles of the sample align with the applied electric field. As the field oscillates, the dipole field attempts to realign itself with the alternating electric field and, in the process, energy is lost in the form of heat through molecular friction and dielectric loss. The amount of heat generated by this process is directly related to the ability of the matrix to align itself with the frequency of the applied field. If the dipole does not have enough time to realign (high-frequency irradiation) or it reorients too quickly (low-frequency irradiation) with the applied field, no heating occurs. The allocated frequency of 2.45 GHz, used in all commercial systems, lies between these two extremes and gives the molecular dipole time to align in the field but not to follow the alternating field precisely. Therefore, as the dipole reorients to align itself with the electric field, the field is already changing and generates a phase difference between the orientation of the field and that of the dipole. This phase difference causes energy to be lost from the dipole by molecular friction and collisions, giving rise to dielectric heating. In summary, field energy is transferred to the medium and electrical energy is converted into kinetic or thermal energy and ultimately into heat. It should be emphasized that the interaction between microwave

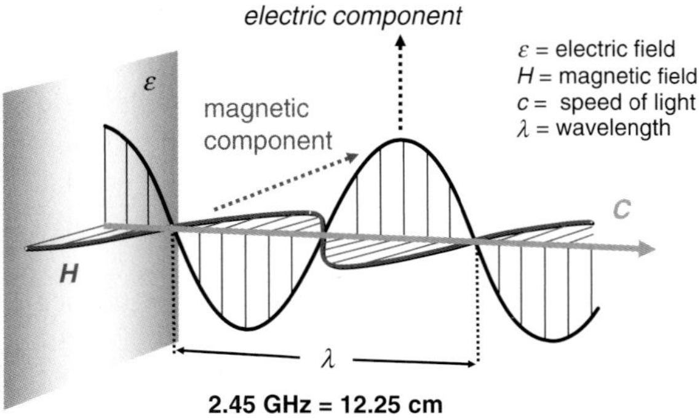

Figure 2.2 Electric and magnetic field components in microwaves.

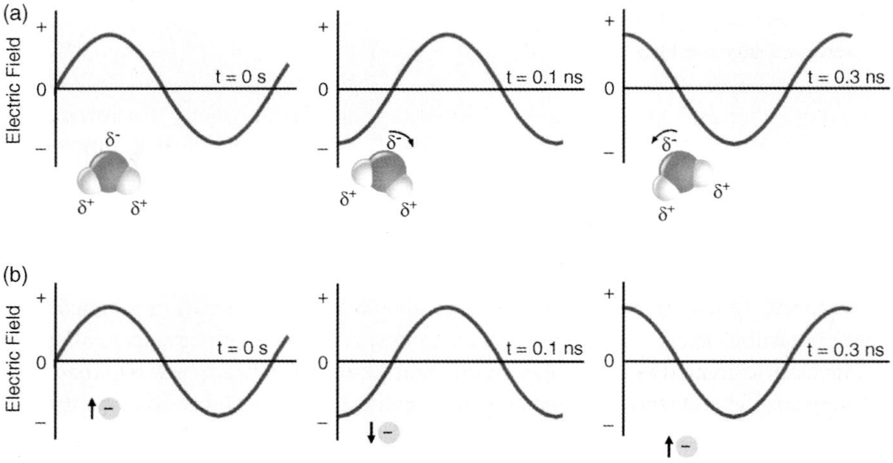

Figure 2.3 (a) Dipolar polarization mechanism. Dipolar molecules try to align with an oscillating electric field. (b) Ionic conduction mechanism. Ions in solution will move in the electric field.

radiation and the polar solvent, which occurs when the frequency of the radiation approximately matches the frequency of the rotational relaxation process, is not a quantum mechanical resonance phenomenon. Transitions between quantized rotational bands are not involved and the energy transfer is not a property of a specific molecule but the result of a collective phenomenon involving the bulk [4, 5]. The heat is generated by frictional forces occurring between the polar molecules whose rotational velocity has been increased by the coupling with the microwave irradiation. It should also be noted that gases cannot be heated under microwave irradiation, since the distance between the rotating molecules is too far. Similarly, ice is also (nearly) microwave transparent, since the water dipoles are constrained in a crystal lattice and cannot move as freely as in the liquid state.

The second major heating mechanism is the ionic conduction mechanism (Figure 2.3b) [4, 5]. During ionic conduction, as the dissolved charged particles in a sample (usually ions) oscillate back and forth under the influence of the microwave field, they collide with their neighboring molecules or atoms. These collisions cause agitation or motion, creating heat. Thus, if two samples containing equal amounts of distilled water and tap water, respectively, are heated by microwave irradiation at a fixed radiation power, more rapid heating will occur for the tap water sample due to its ionic content. Such ionic conduction effects are particularly important when considering the heating behavior of ionic liquids in a microwave field (see Section 4.5.2). The conductivity principle is a much stronger effect than the dipolar rotation mechanism with regard to the heat-generating capacity.

A related heating mechanism exists for strongly conducting or semiconducting materials such as metals, where microwave irradiation can induce a flow of electrons on the surface. This flow of electrons can heat the material through resistance (ohmic) heating mechanisms [11]. In the context of organic synthesis, this becomes important for heating strongly microwave-absorbing materials, such as thin metal films (Pd and Au), graphite supports (see Section 4.1), or so-called passive heating elements made of silicon carbide (see Section 4.6).

2.3 Dielectric Properties

The heating characteristics of a particular material (e.g., a solvent) under microwave irradiation conditions depend on the dielectric properties of the material. The ability of a specific substance to convert electromagnetic energy into heat at a given frequency and temperature is determined by the so-called loss tangent, tan δ. The loss factor is expressed as the quotient, $\tan\delta = \varepsilon''/\varepsilon'$, where ε'' is the dielectric loss, indicative of the efficiency with which electromagnetic radiation is converted into heat, and ε' is the dielectric constant describing the polarizability of molecules in the electric field. A reaction medium with a high tan δ is required for efficient absorption and, consequently, for rapid heating. Materials with a high dielectric constant, such as water (ε' at 25 °C = 80.4), may not necessarily have a high tan δ value. In fact, ethanol has a significantly lower dielectric constant (ε' at 25 °C = 24.3), but heats much more rapidly than water in a microwave field due to its higher loss tangent (tan δ: ethanol = 0.941, water = 0.123). The loss tangents for some common organic solvents are summarized in Table 2.3 [12]. In general, solvents can be classified as high (tan $\delta > 0.5$), medium (tan δ 0.1–0.5), and low microwave-absorbing (tan $\delta < 0.1$) solvents. Other common solvents without a permanent dipole moment, such as carbon tetrachloride, benzene and dioxane, are more or less microwave transparent. It has to be emphasized that a low tan δ value does not preclude a particular solvent from being used in a microwave-heated reaction. Since either the substrates or some of the reagents/catalysts are likely to be polar, the overall dielectric properties of the reaction medium will, in most cases, allow sufficient heating by microwaves.

Table 2.3 Loss tangents (tan δ) of different solvents (2.45 GHz, 20 °C).

Solvent	tan δ	Solvent	tan δ
Ethylene glycol	1.350	*N,N*-Dimethylformamide	0.161
Ethanol	0.941	1,2-Dichloroethane	0.127
Dimethylsulfoxide	0.825	Water	0.123
2-Propanol	0.799	Chlorobenzene	0.101
Formic acid	0.722	Chloroform	0.091
Methanol	0.659	Acetonitrile	0.062
Nitrobenzene	0.589	Ethyl acetate	0.059
1-Butanol	0.571	Acetone	0.054
2-Butanol	0.447	Tetrahydrofuran	0.047
1,2-Dichlorobenzene	0.280	Dichloromethane	0.042
1-Methyl-2-pyrrolidone	0.275	Toluene	0.040
Acetic acid	0.174	Hexane	0.020

Data from Ref. [12].

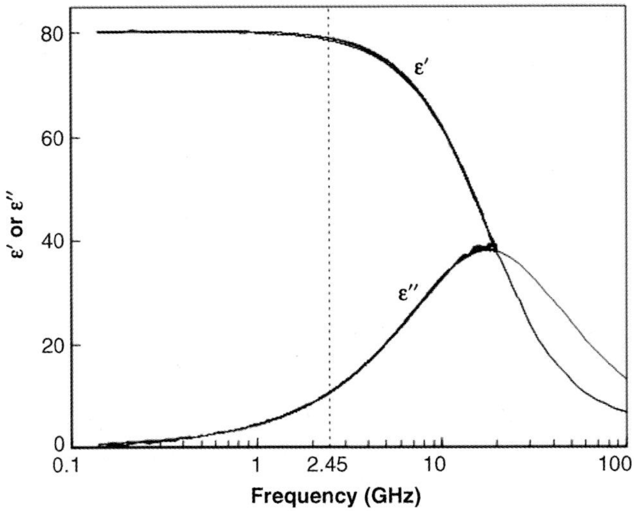

Figure 2.4 Dielectric properties of water as a function of frequency at 25 °C [13].

Furthermore, polar additives (such as alcohols or ionic liquids) or passive heating elements can be added to otherwise low-absorbing reaction mixtures in order to increase the absorbance level of the medium (see Sections 4.5.2 and 4.6).

The loss tangent values are both frequency and temperature dependent. Figure 2.4 shows the dielectric properties of distilled water as a function of frequency at 25 °C [1, 4, 5]. It is apparent that appreciable values of the dielectric loss ε'' exist over a wide frequency range. The dielectric loss ε'' goes through a maximum as the dielectric constant ε' falls. The heating, as measured by ε'', reaches its maximum around 18 GHz, while all domestic microwave ovens and dedicated reactors for chemical synthesis operate at a much lower frequency of 2.45 GHz. The practical reason for the lower frequency is the necessity to heat food efficiently throughout its interior. If the frequency is optimal for a maximum heating rate, the microwaves are absorbed in the outer regions of the food and penetrate only a short distance (skin effect) [4].

According to definition, the penetration depth is the point where 37% (1/e) of the initially irradiated microwave power is still present [6]. The penetration depth is inversely proportional to tan δ and, therefore, critically depends on factors such as temperature and irradiation frequency. Materials with relatively high tan δ values are thus characterized by low values of penetration depth and, therefore, microwave irradiation may be totally absorbed within the outer layers of these materials. For a solvent such as water (tan $\delta = 0.123$ at 25 °C and 2.45 GHz), the penetration depth at room temperature is only on the order of a few centimeters (Table 2.4). Beyond this penetration depth, volumetric heating due to absorption of microwave energy becomes negligible. This means that during microwave experiments on a larger scale, only the outer layers of the reaction mixture may be directly heated by microwave irradiation via dielectric heating mechanisms. The inner part of the reaction mixture will, to a large extent, be heated by conventional heat convection

Table 2.4 Penetration depth of some common materials.

Material	Temperature (°C)	Penetration depth (cm)
Water	25	1.4
Water	95	5.7
Ice	−12	1100
Polyvinylchloride	20	210
Glass	25	35
Teflon	25	9200
Quartz glass	25	16 000

Data from Ref. [11].

and/or conduction mechanisms. Issues relating to the penetration depth are therefore critically important when considering the scale-up of MAOS (see Section 4.8).

The dielectric loss and loss tangent of pure water and most other organic solvents decrease with increasing temperature (Figure 2.5). The absorption of microwave radiation in water therefore decreases at higher temperatures. While it is relatively easy to heat water from room temperature to 100 °C by 2.45 GHz microwave irradiation, it is significantly more difficult to heat water further to 200 °C and beyond in a sealed vessel. In fact, supercritical water ($T >$ 374 °C) is transparent to microwave irradiation (see Section 4.5.1).

Most organic materials and solvents behave like that of water, in the sense that the dielectric loss ε'' will decrease with increasing temperature [2–5]. From the practical point of view, this may be somewhat inconvenient, since microwave heating at higher temperatures may often be compromised. On the other hand, from the standpoint of

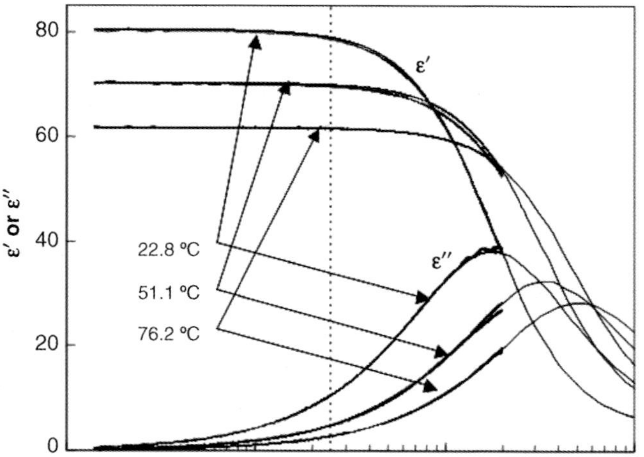

Figure 2.5 Dielectric properties of water as a function of temperature and frequency [13].

safety, it should be stressed that the opposite situation may lead to a scenario where a material will become a stronger microwave absorber with increasing temperature. This is the case for some inorganic/polymeric materials [4], and will lead to the danger of a thermal runaway during microwave heating. Another notable exception of more practical relevance to synthetic chemistry is ionic liquid, which is heated via the ionic conduction mechanism rather than by dipolar polarization. As the temperature increases, the dielectric loss ε'' sometimes increases dramatically [14].

In summary, the interaction of microwave irradiation with matter is characterized by three different processes: absorption, transmission, and reflection (Figure 2.6). Highly dielectric materials, like polar organic solvents, lead to a strong absorption of microwaves and consequently to a rapid heating of the medium (tan δ 0.05–1) (Table 2.3). Nonpolar microwave-transparent materials exhibit only small interactions with penetrating microwaves (tan $\delta < 0.01$) (Table 2.5) and can thus be used as construction materials (insulators) for reactors because of their high penetration depth values (Table 2.4). If microwave radiation is reflected by the material surface, there is no, or only small, coupling of energy into the system. The temperature increases in the material only marginally. This holds true especially for metals with high conductivity, although in some cases resistance heating for these materials can occur [10].

2.4
Microwave versus Conventional Thermal Heating

Traditionally, organic synthesis is carried out by conductive heating with an external heat source (e.g., an oil bath or heating mantle). This is a comparatively slow and inefficient method for transferring energy into the system since it depends on convection currents and on the thermal conductivity of the various materials that

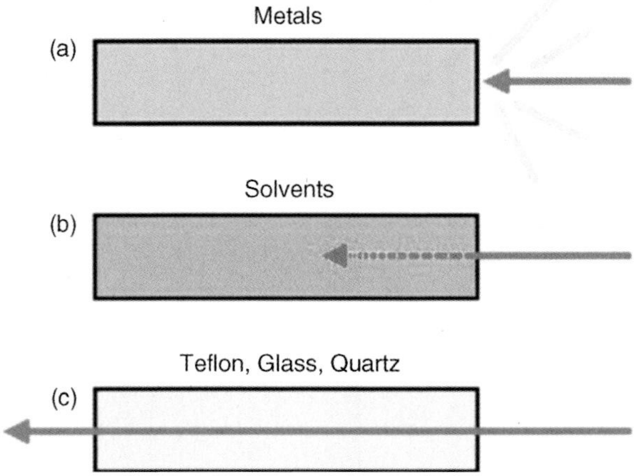

Figure 2.6 Interaction of microwaves with different materials. (a) Electrical conductors. (b) Absorbing materials (tan δ 0.05–1). (c) Insulators (tan $\delta < 0.01$).

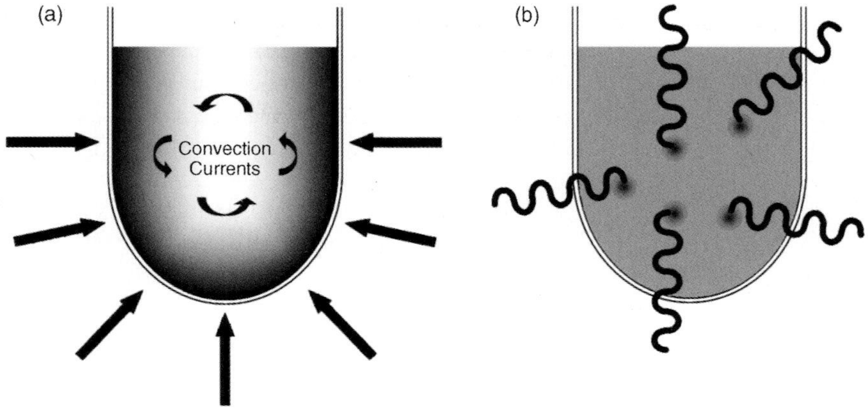

Figure 2.7 Comparison of conventional (a) and microwave heating (b).

must be penetrated, and generally results in the temperature of the reaction vessel being higher than that of the reaction mixture (Figure 2.7). This is particularly true if reactions are performed under reflux conditions, whereby the temperature of the bath fluid is typically kept at 10–30 °C above the boiling point of the reaction mixture in order to ensure an efficient reflux. In addition, a temperature gradient can develop within the sample and local overheating can lead to product, substrate, or reagent decomposition.

In contrast, microwave irradiation produces efficient internal heating (in core volumetric heating) by direct coupling of microwave energy with the molecules (solvents, reagents, and catalysts) that are present in the reaction mixture. Microwave irradiation, therefore, raises the temperature of the whole volume simultaneously (bulk heating), whereas in the conventionally heated vessel, the reaction mixture in contact with the vessel wall is heated first (Figure 2.7a). Since the reaction vessels employed in modern microwave reactors are typically made of (nearly) microwave-transparent materials such as borosilicate glass, quartz, or Teflon (Table 2.5), the radiation passes through the walls of the vessel and an inverted temperature gradient compared to conventional thermal heating results. If the microwave cavity is well designed, the temperature increase will be uniform throughout the sample (see Section 2.5.1). The very efficient internal heat transfer results in minimized wall

Table 2.5 Loss tangents (tan δ) of low-absorbing materials (2.45 GHz, 25 °C).

Material	tan δ ($\times 10^{-4}$)	Material	tan δ ($\times 10^{-4}$)
Quartz	0.6	Plexiglass	57
Ceramic	5.5	Polyester	28
Porcelain	11	Polyethylene	31
Phosphate glass	46	Polystyrene	3.3
Borosilicate glass	10	Teflon	1.5

Data from Ref. [11].

effects (no hot vessel surface) that may in principle lead to the observation of so-called specific microwave effects (see Section 2.5.3), for example, in the context of diminished catalyst deactivation. It should be emphasized that microwave dielectric heating and thermal heating by convection are totally different processes, and that any comparison between the two is inherently difficult.

2.5
Microwave Effects

Despite the relatively large body of published work on microwave-assisted chemistry (Figure 1.1) and the basic understanding of high-frequency electromagnetic irradiation and microwave–matter interactions, the exact reasons why and how microwaves enhance chemical processes are still a matter of debate. Since the early days of microwave synthesis, the observed rate accelerations and sometimes altered product distributions compared to conventionally heated experiments have led to speculations on the existence of so-called specific or nonthermal microwave effects [15, 16]. Such effects have been claimed when the outcome of a synthesis performed under microwave conditions was different from the conventionally heated counterpart at the same measured reaction temperature. Today it is generally agreed that in most standard cases, the observed enhancements in microwave-heated reactions are in fact the result of purely thermal/kinetic effects; in other words, they are a consequence of the high reaction temperatures that can rapidly be attained when irradiating polar materials/reaction mixtures under closed-vessel conditions in a microwave field (see Section 2.5.2). Similarly, the possible existence of so-called specific microwave effects that cannot be duplicated by conventional heating and result from the uniqueness of the microwave dielectric heating phenomenon is largely undisputed [15, 16]. In this category fall, for example (i) the superheating effect of solvents at atmospheric pressure, (ii) the selective heating of, for example, strongly microwave-absorbing heterogeneous catalysts or reagents in a less polar reaction medium, and (iii) the elimination of wall effects caused by inverted temperature gradients (see Section 2.5.3).

In contrast, the subject of “nonthermal microwave effects” (also referred to as athermal effects) is highly controversial and has led to heated debates in the scientific community [17]. Essentially, nonthermal effects have been postulated to result from a proposed direct interaction of the electric field with specific molecules in the reaction medium that is not related to a macroscopic temperature effect (see Section 2.5.4) [15, 16]. It has been argued, for example, that the presence of an electric field leads to orientation effects of dipolar molecules or intermediates and hence changes the preexponential factor A or the activation energy (entropy term) in the Arrhenius equation for certain types of reactions. Furthermore, a similar effect has been proposed for polar reaction mechanisms, where the polarity is increased going from the ground state to the transition state, resulting in an enhancement of reactivity by lowering of the activation energy. Significant nonthermal microwave effects have been suggested for a wide variety of synthetic transformations [15, 16].

It should be obvious from a scientific standpoint that the question of nonthermal microwave effects needs to be addressed in a serious manner, given the rapid increase in the use of microwave technology in chemical sciences, in particular organic synthesis. There is an urgent need to provide a scientific rationalization for the observed effects and to investigate the general influence of the electric field (and therefore of the microwave power) on chemical transformations. This is even more important if one considers engineering and safety aspects once this technology moves from the small-scale laboratory work to pilot or production-scale instrumentation. Although the detailed discussion on microwave effects lies outside the scope of this book, this chapter provides a short summary of the basic concepts of relevance to the microwave chemistry practitioner.

Historically, microwave effects were claimed when the outcome of a synthesis performed under microwave conditions was different from the conventionally heated counterpart at the same apparent temperature. An extreme example is highlighted in Scheme 2.1. Here, Soufiaoui and coworkers [18] have synthesized a series of 1,5-aryldiazepin-2-ones in high yield in only 10 min by the condensation of *ortho*-aryldiamines with β-ketoesters in xylene under microwave irradiation in an open vessel at reflux temperature, utilizing a conventional domestic microwave oven. Surprisingly, they observed that no reaction occurred when the same reactions were heated conventionally for 10 min at the same temperature. In their publication, the authors specifically point to the involvement of "specific effects" (which are not necessarily thermal) in rationalizing the observed product yields. These results could be taken as clear evidence for a specific microwave effect. Interestingly, Gedye and Wei have later reinvestigated the exact same reaction under thermal and microwave conditions and found that there is virtually no difference in the rate of the microwave and the conventionally heated reactions, leading to similar product yields [7, 19]. The literature is full of examples like the one highlighted above, with conflicting reports on the involvement or noninvolvement of "specific" or "nonthermal" microwave effects for a wide variety of different types of chemical reactions [15–17]. Microwave effects are the subject of considerable current debate and controversy and it is evident that extensive research efforts are necessary in order to truly understand these and related phenomena.

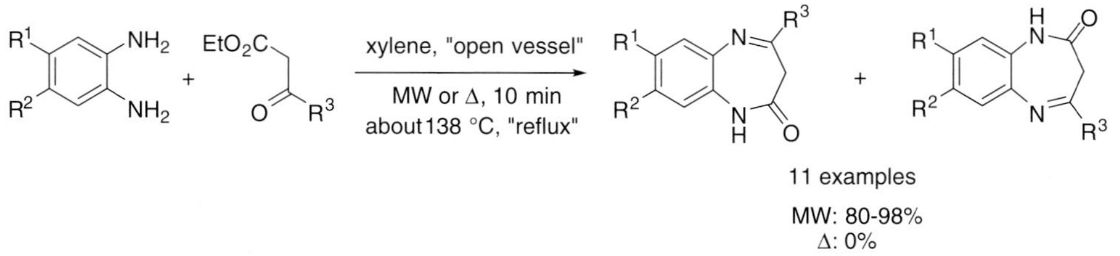

Scheme 2.1 Molecular magic with microwaves.

Essentially, one can envision three different possibilities for rationalizing rate enhancements observed in a microwave-assisted chemical reaction [20]:

- Thermal effects (kinetics)
- Specific microwave effects
- Nonthermal (athermal) microwave effects.

Clearly, a combination of two or all three contributions may be responsible for the observed phenomena, which makes the investigation of microwave effects an extremely complex subject. Before discussing the above-mentioned effects in detail, it is important to have an understanding of how the reaction temperature in a microwave-heated reaction can be adequately determined. In order to obtain reproducible and reliable results from a microwave-assisted reaction, it is absolutely essential to have an accurate way of directly measuring the temperature of the reaction mixture online during the irradiation process. This is even more important if a comparison with conventionally heated experiments is performed.

2.5.1
Temperature Monitoring in Microwave Chemistry

Dedicated microwave reactors for organic synthesis are in most cases operated in "temperature control" mode, which means that the desired reaction temperature is selected by the user (see Chapter 3). By coupling the feedback from a suitable temperature probe to the modulation of magnetron output power, the reaction mixture is heated and kept at the preselected value (see, for example, Figures 2.12 and 2.13). This process requires a reliable way of rapidly monitoring the reaction temperature online during the microwave irradiation process. The correct temperature measurement in microwave-assisted reactions, however, often presents a problem since classical temperature sensors such as thermometers or metal-based thermocouples will fail as they will couple with the electromagnetic field [6]. In the most popular single-mode microwave reactors (Biotage Initiator, CEM Discover, see Section 3.4), the reaction temperature is generally determined by a calibrated external infrared sensor, integrated into the cavity, that detects the surface temperature of the reaction vessel from a predefined distance. It is assumed that the measured temperature on the outside of the reaction vessel will correspond more or less to the temperature of the reaction mixture contained inside. Unfortunately, this is not always the case and extreme care must be taken relying on these data [6, 21–26]. The reactor wall is typically the coldest spot of the reaction system due to the inverted heat flux in comparison to conventional heating as the energy conversion using microwave irradiation takes place directly in the reaction mixture (Figure 2.7) [6].

A more accurate way is to determine the temperature of the reaction mixture directly by an internal probe such as a fiber-optic sensor [21–26], as implemented in the Anton Paar Monowave 300 reactor, that allows both external temperature measurement by an IR sensor and internal temperature monitoring by a ruby-based immersing fiber-optic probe (see Section 3.4) [26]. Fiber-optic probes are more accurate than IR sensors, but are also more expensive. Another disadvantage, compared to other temperature measurement systems, is the generally more narrow operating range of 0–300 °C. In addition, for some types of probes, permanent aging phenomena can already be observed above 250 °C after a few hours [6]. These probes are also very sensitive toward mechanical stress and one reason for the lower temperature resistance is the unavoidable use of polymers during their fabrication, for example, for gluing the sensor crystal to the optical fiber. Until recently, the routine use of fiber-optic probes in microwave-assisted synthesis was therefore often not practical. Fiber-optic probes are also available to monitor internal reaction temperatures in the CEM Discover system and are used in some of the multimode reactors discussed in Chapter 3. In certain instances, it can also be of interest to investigate the temperature of a microwave-heated reaction mixture or vessel surface with the aid of a thermovision camera [27–29].

For routine synthetic applications in single-mode microwave reactors, the use of standard IR probes is often acceptable, mainly because of the convenience, the robust nature, and the low cost of these types of probes. However, the user should be aware of the limitations of these devices and should recognize situations where the use of these external probes is not appropriate. In general, external IR sensors will only represent the internal reaction temperature properly if efficient agitation of the homogeneous reaction mixture is ensured. Inefficient agitation can lead to temperature gradients within the reaction mixture due to field inhomogeneities in the high-density single-mode microwave cavities [25, 26, 30]. Extreme care must therefore be taken with heterogeneous reactions, such as solvent-free, dry media, or highly viscous systems (see Section 4.1).

In addition, it has to be emphasized that in the three most popular single-mode microwave reactors, the temperature is measured at different positions of the otherwise more or less identical microwave vessels (Figure 2.8). Taking into account inherent field inhomogeneities that likely exist in all these cavities [25], this fact in itself can lead to discrepancies when comparing the results obtained from running the exact same chemical reaction in these systems [30]. It has to be noted that in the

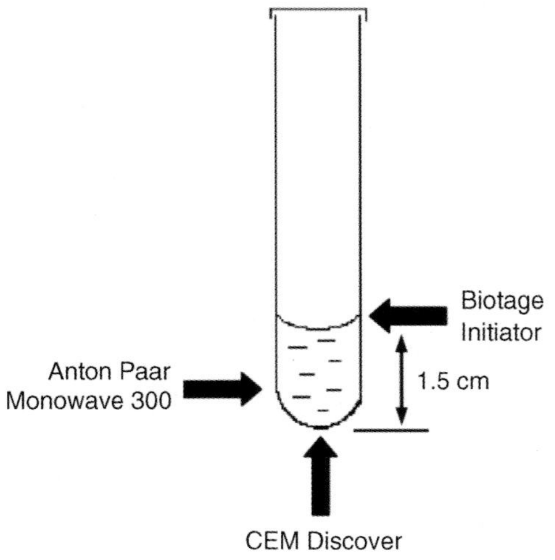

Figure 2.8 Position of infrared temperature sensors in single-mode microwave cavities from Anton Paar, Biotage, and CEM (10 mL reaction vessel).

Biotage microwave systems, a certain minimum filling volume must be used in order to ensure a proper temperature reading. These differences are aggravated when biphasic mixtures are concerned, where one of the phases is strongly microwave absorbing and the other phase is only weakly absorbing. A case in point are, for example, unstirred biphasic mixtures of ionic liquids and nonpolar organic solvents where a strong differential heating (see Section 2.5.3) of the ionic liquid phase will occur [23]. Depending on the microwave system used, either the temperature of the very hot ionic liquid phase (IR from the bottom) or the temperature of the cooler organic layer (IR from the side) will be recorded.

Importantly, external IR sensors should never be used in conjunction with simultaneous external cooling of the reaction vessel. Using this patented technique, the reaction vessel is cooled from the outside by compressed air while being irradiated by microwaves [31]. This allows a higher level of microwave power to be directly administered to the reaction mixture, but will prevent overheating by continuously removing latent heat [32]. It has been demonstrated by several research groups that by using this technique the internal reaction temperatures will be significantly higher than recorded by the IR sensor on the outside [6, 21, 22, 24, 25]. When using simultaneous external cooling, an internal fiber-optic probe device must therefore be employed. Even without using external cooling, one should be aware of the fact that the IR sensor will need some time until it reflects the actual internal reaction temperature. This is because it will take a certain time for the reaction vessel, made of glass, to be warmed "from the inside" by microwave dielectric heating of its polar contents. Although this delay is typically only on the order of a few seconds, it may suffice to lead to an undetected small overshooting of the internal reaction temperature, in particular in case of strongly microwave-absorbing reaction mixtures that are rapidly heated by microwave irradiation [25, 26, 33].

In case of low-absorbing or nearly microwave-transparent reaction mixtures, the opposite phenomenon may occur. Since the glass used for making the comparatively low-cost microwave process vials used in single-mode reactors is not completely microwave transparent (for loss tangents of different types of glasses, see Table 2.5), significant heating of the reaction vessel, rather than of the reaction mixtures, will occur under these circumstances (Figure 2.9). In contrast, no detectable heating of the microwave-transparent reaction mixture is seen when a custom-made reaction vessel made of high-purity quartz is employed (Figure 2.9). Heating of the microwave-transparent solvent, when using the standard glass vessel, is the result of indirect heating by conduction and convection phenomena via the hot surface of the self-absorbing glass. Since an IR sensor directly monitors the surface temperature of the glass (rather than of its contents), the observed effects are more pronounced using this type of monitoring method [23]. It is important to note, however, that in case of medium or strongly microwave-absorbing reaction mixtures, the heating of the glass reaction vessel can be considered negligible and is therefore of little practical concern in microwave synthesis [23].

From a practical point of view, it should be highlighted that IR sensors need to be re-calibrated from time to time against internal probes, and that the path between the

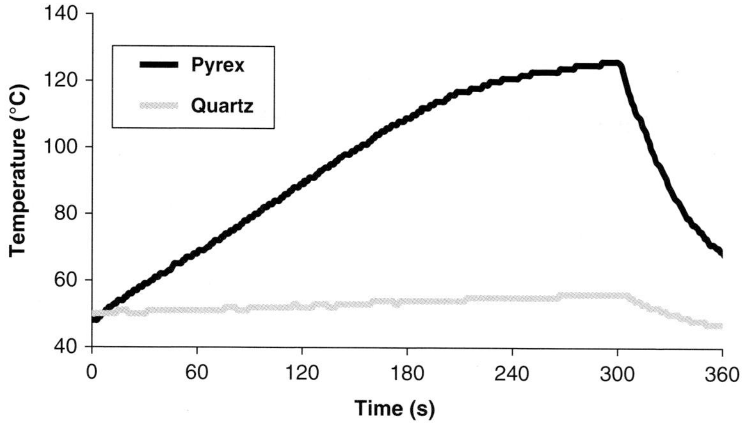

Figure 2.9 Heating profiles for microwave-transparent CCl_4 in Pyrex and quartz reaction vessels at constant 150 W magnetron output power (CEM Discover, IR sensor). Reproduced with permission from Ref. [23].

actual sensor and the reaction vessel must be unobstructed in order to ensure a proper temperature measurement. This is particularly important when the IR sensor is housed at the bottom of the microwave cavity where debris can more easily accumulate (Figure 2.8).

Based on the information provided above, it is evident that more accurate temperature measurements in conjunction with microwave-assisted reactions can be obtained using internal fiber-optic probes. In contrast to thermocouples, fiber-optic sensors are immune to electromagnetic interference and high voltage, do not require shielding, and do not spark or transmit current. Although different types of sensing technologies exist, most microwave reactor manufacturers that provide fiber-optic temperature sensors rely on probes that use semiconductor bandgap technology (CEM, Milestone) or ruby-based probes (Anton Paar). These devices typically have an accuracy of $\pm 1.5\,°C$.

Although internal fiber-optic temperature probes are more accurate than external IR sensors, their use is also not without complications. This is, in part, because the mechanically sensitive sensor crystal needs to be protected, requiring the use of appropriate protective immersion wells for the fiber-optic probes. In some fiber-optic probes, the actual sensor crystal (GaAs) is in addition protected by a polymer coating. This increases the lifetime of the probe, but slows down the response time. Delay times of up to 13 s have been measured for some commercially available fiber-optic probes/immersion wells [25]. In other, "faster," probes, the GaAs crystal is unprotected and can in fact be seen at the tip of the probe, but at the same time it is more prone to destruction. In some commercial systems, a very fast probe is used in combination with an inert immersion well that slows down the response time significantly. Care must therefore be taken in selecting a fiber-optic probe with a short response time for a particular measurement problem [25, 33].

Recent evidence suggests that in fact the use of one single fiber-optic probe may not suffice to represent the temperature profile of a microwave-heated reaction

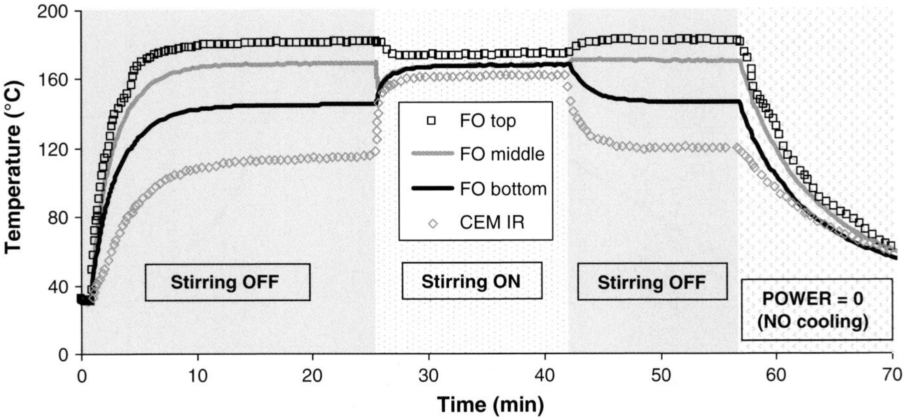

Figure 2.10 Temperature profiles for a sample of 5 mL of NMP contained in a 10 mL quartz vessel equipped with three internal fiber-optic sensors positioned at different heights. The sample was irradiated with constant 50 W magnetron output power (CEM Discover). Shown are the profiles for the three internal fiber-optic probes and the external IR sensor. Magnetic stirring reduces the temperature differences between the individual fiber-optic probes from max 36 °C to less than 6 °C. The temperature of the IR sensor deviates by 12 °C from the top fiber-optic probe with stirring. Adapted from Ref. [25].

mixture [25]. If efficient stirring/agitation cannot be ensured, temperature gradients may develop as a consequence of inherent field inhomogeneities inside a single-mode microwave cavity (Figure 2.10). In contrast to an oil bath experiment, even completely homogeneous solutions, therefore, need to be stirred when using single-mode microwave reactors. The formation of temperature gradients will therefore be a particular problem in case of, for example, solvent-free or dry media reactions (see Section 4.1) and for very viscous or biphasic reaction systems where standard magnetic stirring is not effective, as in the synthesis of polymers.

The temperature monitoring studies shown in Figure 2.10 demonstrate that microwave heating in high field density single-mode cavities is in fact not as homogeneous as often portrayed, and that extreme care must be taken in determining the proper reaction temperature in these experiments, especially in those cases where adequate mixing cannot be ensured.

It should be stressed that when studying differences between microwave heating and conventional heating (microwave effects), it is particularly important to use highly accurate and fast responding temperature monitoring devices. In order to accurately compare the results obtained by direct microwave heating with the outcome of a conventionally heated reaction, a reactor system should be used that allows to perform both types of transformations *in the identical reaction vessel* and to monitor the internal reaction temperature in both experiments directly with the same fiber-optic probe device. Such a reactor setup, originally introduced by Maes and coworkers for the CEM Discover reactor (Figure 2.11) [34], can be immersed either into the cavity of the microwave reactor or into a preheated and temperature-equilibrated oil or metal bath placed on a magnetic stirrer/hot plate. In both cases,

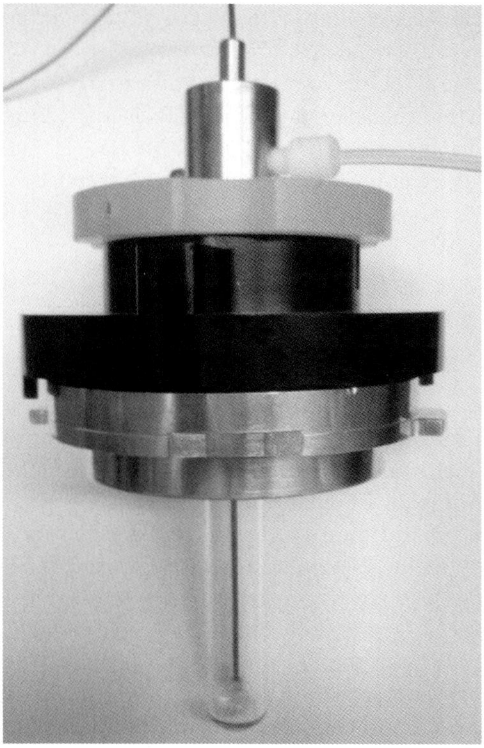

Figure 2.11 Setup for monitoring internal reaction temperatures with fiber-optic probes in microwave and oil bath experiments (CEM Discover).

the software of the microwave instrument is recording the internal temperature. Such a system has the advantage that the same reaction vessel and the same method of temperature measurement are used. In this way, all parameters apart from the mode of heating are identical and, therefore, a fair comparison between microwave heating and thermal heating can generally be made [24, 25, 34, 35].

A more recent and much simpler concept to carefully simulate a microwave-heated experiment using conventional heating conditions is to employ a microwave reaction vessel made of strongly microwave-absorbing silicon carbide (SiC) (see Figure 3.5) [36, 37]. Microwave irradiation will induce a flow of electrons in the semiconducting SiC ceramic that heats the material very efficiently through resistance heating mechanisms. The use of SiC reaction vessels in combination with a single-mode microwave reactor (Anton Paar Monowave 300) provides an almost complete shielding of the contents inside from the electromagnetic field. Therefore, these "microwave" experiments do not involve electromagnetic field effects on the chemistry since the semiconducting ceramic vial is effectively preventing microwave irradiation to penetrate into the reaction mixture. The involvement of electromagnetic field effects (specific/nonthermal microwave effects) on a number of chemical transformations was evaluated by comparing the results obtained in microwave-transparent Pyrex vials with experiments performed in SiC vials at the same reaction temperature (see Section 2.5.4) [36, 37].

Unfortunately, at the time when most of the early work on microwave effects was published, many of the complications and subtleties of accurate online temperature measurement under microwave irradiation conditions were not known [15, 16]. The conclusions of these studies should therefore be treated with extreme skepticism. The following sections provide an overview of the currently existing hypotheses on different types of microwave effects.

2.5.2
Thermal Effects (Kinetics)

Reviewing the present literature, it appears that today many scientists would agree that in the majority of cases, the reason for the observed rate enhancements seen in microwave chemistry is a purely thermal/kinetic effect, that is, a consequence of the high reaction temperatures that can rapidly be attained when irradiating polar materials in a microwave field. As shown in Figure 2.12, even a moderately strong microwave-absorbing solvent such as 1-methyl-2-pyrrolidone (NMP, bp 202–204 °C, tan $\delta = 0.275$) (Table 2.3) can be heated very rapidly (microwave flash heating) in a microwave cavity. As indicated in Figure 2.12, a sample of NMP can be heated to 200 °C within about 40 s, depending on the maximum output power of the magnetron [38].

Today, most of the published microwave-assisted reactions are performed under sealed-vessel conditions in the relatively small, so-called single-mode microwave reactors with high power density (see Chapter 3). Under these autoclave-type

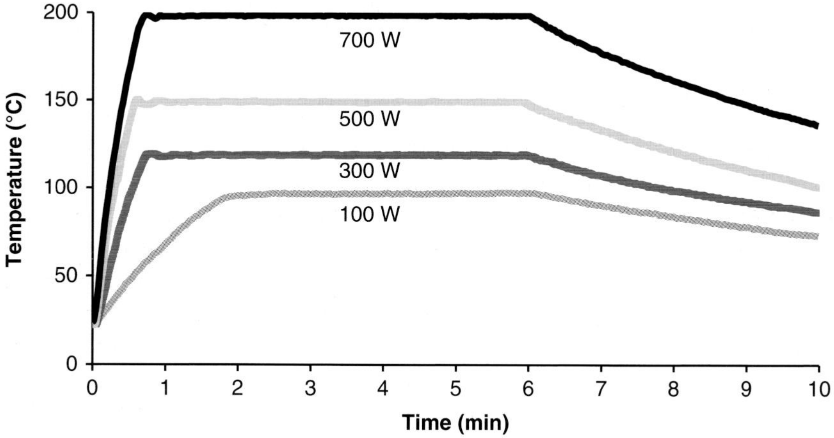

Figure 2.12 Temperature profiles for a 30 mL sample of 1-methyl-2-pyrrolidone heated under open-vessel microwave irradiation conditions [38]. Multimode microwave heating (MicroSYNTH, Milestone) at different maximum power levels for 6 min with temperature control using the feedback from a fiber-optic probe. After the set temperatures of 200 °C (700 W), 150 °C (500 W), 120 °C (300 W), and 100 °C (100 W) are reached, the power regulates itself down to an appropriate level (not shown). Reproduced with permission from Ref. [38].

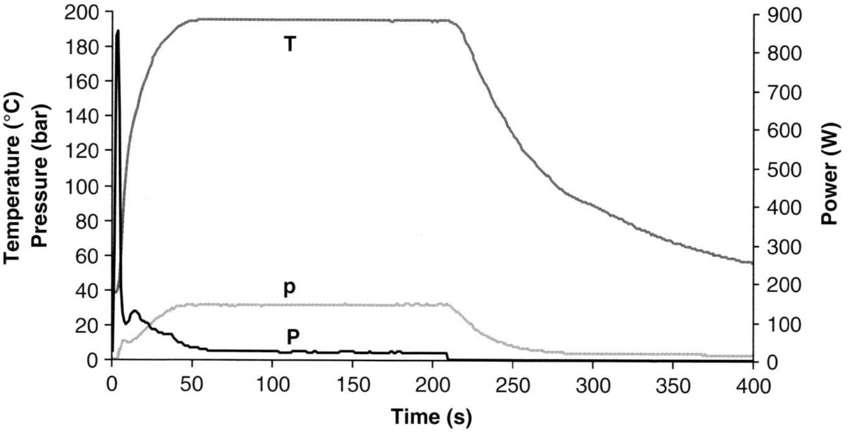

Figure 2.13 Temperature (*T*), pressure (*p*), and power (*P*) profile for a 3 mL sample of methanol heated under sealed-vessel microwave irradiation conditions. Anton Paar Monowave 300 (850 W maximum power).

conditions, microwave-absorbing solvents with a comparatively low boiling point such as methanol (bp 65 °C, tan $\delta = 0.659$) can be rapidly superheated to temperatures more than 130 °C in excess of their boiling points when irradiated under microwave conditions (Figure 2.13). The rapid increase in temperature can be even more pronounced for media with extreme loss tangents such as ionic liquids (see Section 4.5.2), where temperature jumps of 200 °C within a few seconds are not uncommon [26]. Naturally, such temperature profiles are very difficult if not impossible to reproduce by standard thermal heating. Therefore, comparisons with conventionally heated processes are inherently troublesome.

The temperature, power, and pressure profiles shown in Figure 2.13 nicely illustrate the operating principles of a modern dedicated microwave reactor such as the Anton Paar Monowave 300 (see Section 3.4.1.1). After the set temperature of 195 °C is reached (about 40 s), the microwave magnetron power regulates itself down to about 30 W, which is all that is needed to keep the sample of methanol superheated at 195 °C, 130 °C above its boiling point at atmospheric pressure. Depending on the initially selected heating mode (as fast as possible or ramp heating), the temperature control algorithm properly adjusts the magnetron power to ensure that the desired set temperature is reached as fast as possible, at the same time trying not to overheat the sample. Since the reaction is performed in a sealed microwave process vial, an autogenic internal pressure of about 29 bar develops in the vessel, which is controlled and monitored by the pressure measurement system of the instrument (see Section 3.4.1.1). At the end of the microwave irradiation period (after 210 s), the reaction mixture is rapidly cooled to a temperature of typically about 50 °C by a stream of compressed air (active gas jet cooling). This allows the user to remove the processed microwave reaction vial from the cavity in a reasonably short time period, which is particularly important when processing several reactions in sequence using robotic vial handling (see Section 4.7.1).

It appears obvious that a specific reaction performed under the conditions depicted in Figure 2.13 utilizing superheated methanol as solvent at 195 °C will occur at a much faster rate than when carried out in refluxing methanol at 65 °C. Dramatic rate enhancements when comparing reactions that are performed under standard oil bath conditions (heating under reflux) with high-temperature microwave-heated processes are therefore not uncommon. As Baghurst and Mingos [4] have pointed out, based on simply applying the Arrhenius law [$k = A\ \exp(-E_a/RT)$], a transformation that requires 68 days to reach 90% conversion at 27 °C will show the same degree of conversion within 1.61 s (!) when performed at 227 °C (Table 2.6). Due to the very rapid heating and extreme temperatures observable in microwave chemistry, it appears obvious that many of the reported rate enhancements can be rationalized by purely thermal/kinetic effects. In the absence of any "specific" or "nonthermal" effects, however, one would also expect the reactions carried out under open-vessel, reflux conditions to proceed at the same reaction rate, regardless of whether they are heated by microwaves or in a thermal process (see Section 4.3). It should be emphasized that for these strictly thermal effects, the preexponential factor A and the energy term (activation energy E_a) in the Arrhenius equation are not affected, only the temperature term changes (see Section 2.5.4).

It should also be noted that the rapid heating and cooling typical of small-scale microwave-assisted transformations (see Figure 2.13) may lead to altered product distributions compared to a conventional oil bath reflux experiment, where heating (and cooling) typically is not as fast and the reaction temperature is generally lower. It has been argued that the very different heating profiles experienced in microwave and conventional heating can actually lead to different reaction products if the reaction product distribution is controlled by complex temperature-dependent kinetic profiles [3]. This may be the reason why in many cases microwave-assisted reactions have been found to be cleaner, leading to less by-products compared to the conventionally heated processes [30].

At the same time, it is obvious that microwave heating will not always favor the desired reaction pathway and that there may be cases where, because of the higher reaction temperatures, unwanted reaction products (e.g., isomers), not seen during a conventionally heated experiment performed at a lower temperature, will be formed.

Table 2.6 Relationship between temperature and time for a typical first-order reaction ($A = 4 \times 10^{10}\ \mathrm{mol^{-1}\ s^{-1}}$, $E_a = 100\ \mathrm{kJ\ mol^{-1}}$).

	$k = Ae^{-E_a/RT}$	
Temperature (°C)	**Rate constant, k (s^{-1})**	**Time (90% conversion)**
27	1.55×10^{-7}	68 d
77	4.76×10^{-5}	13.4 h
127	3.49×10^{-3}	11.4 min
177	9.86×10^{-2}	23.4 s
227	1.43	1.61 s

Data from Ref. [4].

2.5.3 Specific Microwave Effects

In addition to the above-mentioned thermal/kinetic effects, microwave effects that are caused by the uniqueness of the microwave dielectric heating mechanisms (see Section 2.2) must also be considered. These effects should be termed "specific microwave effects" and shall be defined as *changes in a chemical transformation in a microwave field that cannot be achieved or duplicated by conventional heating, but essentially are still thermal effects.* In this category falls, for example, the superheating effect of solvents at atmospheric pressure [39–42]. The question of the boiling point of a liquid undergoing microwave irradiation is one of the basic problems facing microwave heating. Several groups have established that the enthalpy of vaporization is the same under both microwave and conventional heating [43]. These studies have also shown that the rate of evaporation, as well as the temperature of both vapor and liquid at the interface, strongly depends on the experimental conditions. Initial studies of boiling phenomena related to microwave chemistry appeared in 1992 by Baghurst and Mingos [40], and later by Saillard *et al.* [41]. It was established that microwave-heated liquids boil at temperatures above the equilibrium boiling point at atmospheric pressure. For several solvents, the superheating temperature can be up to 40 °C above the classical boiling point [42]. Therefore, in a microwave-heated reactor, the average temperature of the solvent can be significantly higher than the atmospheric boiling point. This is because the microwave power is dissipated over the whole volume of the solvent. The most significant way to lose excess thermal energy is by boiling. However, this will only occur at the existing liquid–gas interfaces, in contrast to a thermally heated solvent where boiling typically occurs at nucleation points (cavities, pits, and scratches) on the glass reactor surface [40]. The bulk temperature of a microwave-irradiated solvent under boiling depends on many factors, such as the physical properties of the solvent, reactor geometry, mass flow, heat flow, and electric field distribution. It should be emphasized that practically all superheating can be removed by adding boiling chips or stirring [42]. Importantly, however, the kinetics of homogeneous organic reactions shows an extension of Arrhenius behavior into the superheated temperature region [42]. Therefore, 10–100-fold reaction rate enhancements can be achieved, which is normally only possible under pressure. Since all dedicated microwave reactors offer a stirring option (see Chapter 3), and most of the current microwave chemistry is performed under sealed-vessel conditions (for exceptions, see Section 4.3), the microwave superheating effect under atmospheric pressure conditions is of little practical relevance and concern.

Closely related to the superheating effect under atmospheric pressure are wall effects, more specifically the elimination of wall effects caused by inverted temperature gradients (Figure 2.7). With microwave heating, the surface of the wall is generally not heated since the energy is dissipated inside the bulk liquid. Therefore, the temperature at the inner surface of the reactor wall is lower than that of the bulk liquid. It can be assumed that in a conventional oil bath experiment (hot vessel surface, Figure 2.7a), the temperature-sensitive species, for example, catalysts, may decompose at the hot reactor surface (wall effects). The elimination of such a hot surface would increase the lifetime of the catalyst and would, therefore, lead to better conversions in a microwave-heated than in a conventionally heated process. A recent dedicated study on the potential elimination of wall effects in microwave-assisted ring-closing metathesis transformations using ruthenium-based catalysts has, however, shown that in all cases the results obtained with microwave irradiation could also be reproduced using conventional heating in a hot oil bath [44]. The often suggested ability of a hot reactor wall to trigger unwanted thermal side reactions for well-agitated small-scale organic reaction mixtures of low viscosity needs to be reconsidered. Correspondingly, the notion that microwave heating will have a significant advantage over conventional heating in minimizing or eliminating these wall effects needs to be reevaluated.

Another phenomenon characteristic to microwave dielectric heating is mass heating, that is, the rapid and even heating of the whole reaction mixture by microwaves (volumetric heating). An example to illustrate this effect, involving the decomposition of urea to cyanuric acid (Scheme 2.2), was studied by Berlan [45]. Cyanuric acid is obtained by heating urea to temperatures around 250 °C. Under conventional heating, the reaction is sluggish and chemical yields are low due to the formation of various side products. The reason for this is that cyanuric acid, which is first formed as a solid at the walls of the reactor, is a poorly heat-conductive material (it decomposes without melting at 300 °C), and it forms an insulating crust that prevents heat transfer to the rest of the reaction mixture. Increasing the temperature of the wall (i.e., the oil bath temperature) results in partial decomposition and does not improve the chemical yield of cyanuric acid significantly. In contrast, very good yields of cyanuric acid (83%) can be obtained under volumetric microwave heating on a 2 g scale for 2 min without any urea or biuret side product being detected at the end of the reaction. Based on the discussion on penetration depth issues in conjunction with microwave heating (Section 2.3), it should be stressed that these effects will be seen only on a comparatively small scale.

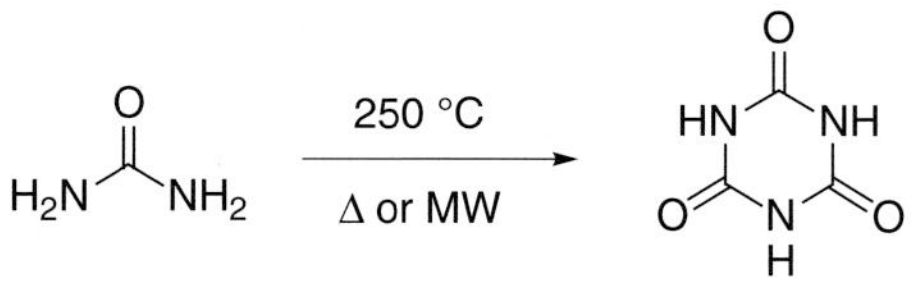

Scheme 2.2 Decomposition of urea to cyanuric acid.

The same concept of volumetric *in situ* heating by microwaves was also exploited by Larhed and coworkers in the context of scaling-up a biochemical process such as the polymerase chain reaction (PCR) [46]. In PCR technology, strict control of temperature in the heating cycles is essential in order to not to deactivate the enzymes involved. With classic heating of a milliliter-scale sample, the time required for heat transfer through the wall of the reaction tube and to obtain an even temperature in the whole sample is still substantial. In practice, the slow distribution of heat (temperature gradients), together with the importance of short processing times and reproducibility, limits the volume for most PCR transformations in conventional thermocyclers to 0.2 mL. With microwave heating, the thermal gradients are eliminated since the full volume is heated simultaneously. Therefore, microwave heating under strict temperature control has been shown to be an extremely valuable tool for carrying out large-scale PCR processing up to a 15 mL scale [46].

Probably one of the most important "specific microwave effects" results from the selective heating of strongly microwave-absorbing heterogeneous catalysts or reagents in a less polar reaction medium [47]. Selective heating generally means that in a sample containing more than one component, only those components that couple with microwaves are selectively heated. The nonabsorbing components are thus not heated directly, but only by heat transfer from the heated component. For heterogeneous mixtures, in particular for gas/solid systems involving heterogeneous gas-phase catalysis [47, 48], selective heating of the catalyst bed is of importance and here the sometimes observed rate enhancements and changes in selectivities have been attributed to the formation of localized (macroscopic) hot spots having temperatures of 100–150 °C above the measured bulk temperature [49]. The measurement and estimation of temperature distributions induced by microwave heating in solid materials are, however, very difficult. Consequently, most local temperature fluctuations are greater than those measured. Under stronger microwave irradiation, it is, therefore, very easy to obtain local temperature gradients. Temperature measurements usually yield an average temperature, because temperature gradients induce convective motions. Despite these difficulties, some methods, for example, IR thermography, can reveal surface temperature distribution without any contact with the sample under study.

Of greater importance for the organic chemist are microwave-assisted transformations in organic solvents catalyzed by a heterogeneous catalyst (liquid/solid systems) such as palladium-on-charcoal (Pd/C). Since here the catalyst is a very strong absorber of microwave energy, it can be assumed that the reaction temperature on the catalyst surface is significantly higher than the bulk temperature of the solvent, in particular when a solvent with a low tan δ value is chosen (Table 2.3). The potential selective heating/activation by microwave irradiation of a Pd/C catalyst was exploited by Vanier in the Pd-catalyzed hydrogenation of various carbon–carbon double bond systems [35]. In the example shown in Scheme 2.3, hydrogenation of the butadiene under single-mode microwave irradiation conditions at 80 °C gave complete conversion after 5 min of microwave irradiation. When the same process was performed in an oil bath at the same measured bulk temperature under strictly comparable conditions (fiber-optic temperature measurement) (Figure 2.11), the conversion was only 55% [35]. Similar observations were also made by Holzgrabe and coworkers in related microwave-assisted hydrogenation reactions [50]. However,

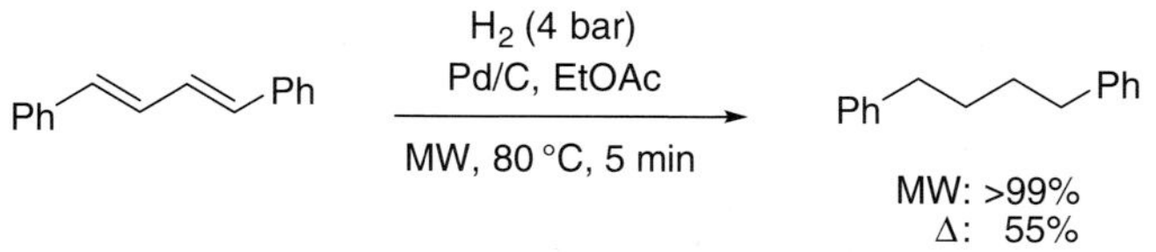

Scheme 2.3 Hydrogenation reactions using a heterogeneous palladium catalyst.

subsequent carefully executed control experiments have revealed that the previously observed differences found in the hydrogenation studies shown in Scheme 2.3 were in fact likely due to differences in stirring efficiency between the experiments carried out under microwave irradiation and the ones performed in an oil bath [51]. No evidence for selective catalyst heating could be obtained.

The selective absorption of microwave energy by a heterogeneous encapsulated palladium catalyst (PdEnCat) was also suggested to be responsible for very efficient Suzuki couplings of aryl bromides performed by the Ley group under both batch and flow microwave conditions, although no control experiments using conventional heating were presented [52]. Subsequent attempts to confirm selective catalyst heating (Pd/C, Ni/C, and Cu/C) in Mizoroki–Heck, Suzuki–Miyaura, Negishi, and Ullmann couplings using microwave heating were unsuccessful. In all cases, the observed rate accelerations could be linked to a purely thermal effect by performing control experiments in conventionally heated autoclave environments at the same temperature [51, 53].

The direct interaction of the microwave field with magnesium metal turnings was recently reported for the formation of Grignard reagents under microwave conditions [54]. Irradiating a solution of an aryl halide in dry tetrahydrofuran together with magnesium turnings led to strong arcing of the magnesium turnings, as observed through the front glass door of the multimode microwave reactor. The formation of the Grignard reagent using microwave heating was significantly faster than by conventional heating at the same temperature of 65 °C (refluxing tetrahydrofuran) [54, 55]. This could be due to a cleansing effect (electrostatic etching) of the magnesium from a layer of magnesium oxide initiated by microwave irradiation [55]. Subsequent investigations have demonstrated that applying high electric field strength conditions, the same transformation experiencing an identical 65 °C macroscopic bulk temperature is almost completely retarded. Apparently, exposure of the reaction mixture to high field density conditions results in more intense electrostatic discharges resulting in the creation of carbonaceous material by solvent decomposition. This disintegration of the THF solvent into carbonaceous material produces a passivating layer of graphitized core/shell MgO/carbon nanoparticles covering the Mg metal, effectively preventing access of the organohalogen reagent to the Mg metal surface and thus shutting down the formation of the organomagnesium (Grignard) reagent [56]. Both pathways are probably best categorized as examples of "specific" microwave effects as the macroscopic reaction temperature in both cases compared to the conventionally heated experiment is the same, and clearly both the activation and deactivation seen under microwave irradiation cannot be duplicated by conventional conductive heat transfer methods.

Recent evidence suggests that heterogeneity itself plays a major role in the enhancement of chemical processes by microwave irradiation. It has been demonstrated that microwave irradiation can change the energies and/or the "effective temperatures" of individual species at interfaces as the result of Maxwell–Wagner interfacial microwave polarization [57].

The above-mentioned studies provide clear evidence for the existence of specific microwave effects in MAOS for very few selected cases, which most probably are linked to heterogeneity and the use of metals or metal oxides leading to arcing phenomena. It should be stressed that the standard methods for determining the temperature in microwave-heated reactions, namely, with an IR pyrometer from the outside of the reaction vessel or with a fiber-optic probe on the inside, would here allow measurement of only the average bulk temperature of the solvent, and not the "true" reaction temperature on the surface of the solid substrate or on an interface.

For homogeneous mixtures, for example, polar reagents in a microwave-transparent solvent (liquid/liquid systems), in principle the same arguments about selective heating can be made. However, the existence of such "molecular radiators" [58] is experimentally difficult to prove and it would have to be assumed that the energy of these "hot" molecules would be instantaneously dissipated to the surrounding "cooler" solvent molecules [2–5]. It should also be stressed that it is not possible to selectively "activate" polar functional groups (so-called antenna groups [59]) within a larger molecule by microwave irradiation. It is tempting for a chemist to give a chemical significance to the fact that localized rotations of such antenna groups are indeed possible [5] and to speculate that microwave dielectric heating of molecules containing these groups may result in an enhancement of reaction rates specifically at these groups. However, the dielectric heating process involves the rapid energy transfer from these groups to neighboring molecules and it is not possible to store the energy in a specific part of the molecule [5].

Another specific microwave effect not easily duplicated by conventional heating is the differential heating of bi- or multiphasic liquid/liquid systems. This type of selective heating was exploited by Strauss and coworkers in a Hofmann elimination reaction using a two-phase water/chloroform system (Figure 2.14) [60]. The temperatures of the aqueous and organic phases under microwave irradiation

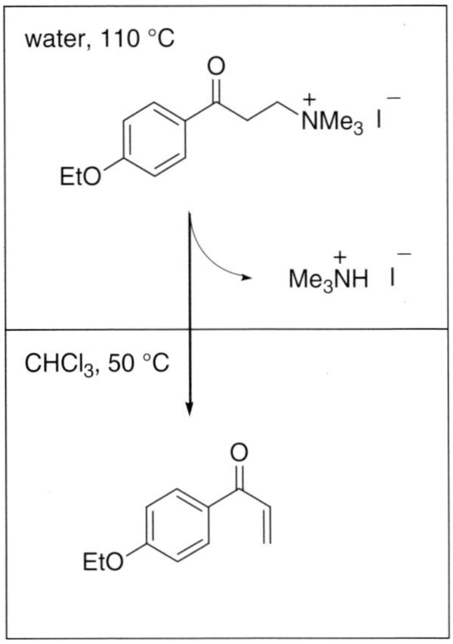

Figure 2.14 Selective dielectric heating of water/chloroform mixtures.

were 110 and 50 °C, respectively, due to differences in the dielectric properties of the solvents (Table 2.3). This difference avoids decomposition of the final product that is soluble in the cooler organic chloroform phase. Comparable conditions would be difficult to obtain using traditional heating methods. A similar effect has been observed by Hallberg and coworkers in the preparation of β,β-diarylated aldehydes by hydrolysis of enol ethers in a two-phase toluene/aqueous hydrochloric acid system [61].

Differential heating phenomena will almost always be observed when heterogeneous liquid/liquid systems are irradiated by microwaves, since there is likely to be a difference in loss tangents between the two phases. The effects can be extreme, as in case of an ionic liquid/hexane mixture, or more moderate in nature. The user, however, should always be aware of the possibility of differential heating when dealing with nonhomogeneous reaction mixtures under microwave irradiation conditions. Because of the potentially different temperatures in the phases, mass and heat transfer across the phase boundaries may be altered compared to conventional heating where both phases have the same temperature. Naturally, particular care must be given to the temperature measurement in these cases, as it will be critically important in which phase the temperature is measured. In any event, intensive stirring should always be applied when dealing with heterogeneous mixtures [25]. It is perhaps no coincidence that a recent study has found that microwave effects can often be eliminated when heterogeneity of the reaction system is reduced by adding appropriate solvents [54].

In summary, all potential rate enhancements discussed above falling under the category of "specific microwave effects," such as the superheating effect of solvents at atmospheric pressure, the selective heating of strongly microwave-absorbing heterogeneous catalysts or reagents in a less polar reaction medium, differential heating of liquid/liquid biphasic mixtures, and the potential elimination of wall effects caused by inverted temperature gradients, are essentially still a result of a *thermal* effect (i.e., a change in temperature compared to heating by standard convection methods), although it may be difficult to determine the exact reaction temperature experimentally.

2.5.4 Nonthermal (Athermal) Microwave Effects

In contrast to so-called specific microwave effects described in Section 2.5.3, some authors have suggested the possibility of "nonthermal microwave effects" (also referred to as athermal effects). These should be classified as *changes in chemical transformations in a microwave field that cannot be rationalized by either purely thermal/kinetic or specific microwave effects* [20]. Essentially, nonthermal effects should result from a proposed direct interaction of the electric field with specific molecules in the reaction medium. It has been argued, for example, that the presence of an electric field leads to orientation effects of dipolar molecules and, hence, changes the preexponential factor A [62] or the activation energy (entropy term) [63] in the Arrhenius equation. Furthermore, it has been argued that a similar effect should be observed for polar reaction mechanisms, where the polarity is increased on going from the ground state to the transition state, resulting in an enhancement of reactivity by lowering of the activation energy.

Several publications in the literature have used arguments like this to explain the outcome of a chemical reaction carried out under microwave irradiation conditions [15, 16]. A 2004 study by Loupy *et al.* provides a representative example [64]: as shown in Scheme 2.4, two irreversible Diels–Alder cycloaddition processes were compared. In the first example (Scheme 2.4a), no difference in either yield or selectivity was observed between the conventionally and microwave-heated reactions. Detailed *ab initio* calculations on the cycloaddition process revealed that here a synchronous, isopolar (concerted) mechanism is operational, where no charges on going from the ground state to the transition state are developed. On the contrary, in the second example (Scheme 2.4b), a significant difference in product yield was observed comparing the thermally and microwave-heated runs. Here, *ab initio* calculations on transition-state geometries and dipole moments revealed a significant development of charges on going from the ground state to the transition state. The authors have taken these experimental results as clear evidence for the involvement of nonthermal microwave effects via electrostatic interactions of polar molecules with the electric field, that is, for the stabilization of the transition state and thereby a decrease in the activation energy [64]. It should be noted, however, that a careful recent reinvestigation of the Diels–Alder process shown in Scheme 2.4b using a fiber-optic temperature setup similar to that shown in Figure 2.11 has revealed no differences for the oil bath and microwave experiments when performed at the exact same temperature [25].

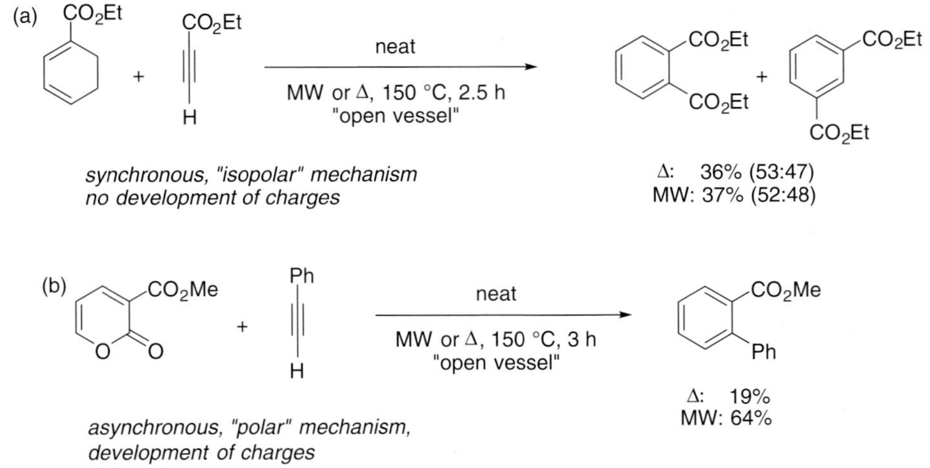

Scheme 2.4 Proposed involvement of nonthermal microwave effects in Diels–Alder cycloaddition reactions.

Along similar lines, it has been argued that reactions that occur via a late, product-like, transition state and, therefore, have a large enthalpy of activation compared to reactions that involve early transition states (Hammond postulate) would be prone to show large microwave effects [15, 16]. Several authors have used quantum mechanical methods to study the interaction between reaction molecules and the electromagnetic field [65, 66].

As already mentioned above, the issue of nonthermal microwave effects is highly controversial. Many scientists denounce the existence of dipolar orientation effects in electric fields on the grounds of overriding disorientation phenomena (thermal agitation) that should prevent any statistically significant orientation (alignment) of dipoles [2]. In this context, it may be noted that it is probably no coincidence that nonthermal microwave effects have, in many cases, been claimed for processes involving solvent-free/dry media reactions and/or for transformations involving polar reaction intermediates or products (which will strongly absorb microwave energy) [15, 16]. Based on the difficulties of exact online temperature measurement under these experimental conditions (see Section 2.5.1), it can be argued that in many of the published cases, the observed differences between microwave and conventional heating may be rationalized by inaccurate temperature measurements, often using external IR temperature probes, rather than being the consequence of a genuine nonthermal effect [6].

Using the silicon carbide (SiC) technology described in Section 2.5.1, it has recently been possible to study a wide range of microwave-assisted processes ranging from synthetic organic reactions [36, 37] to the acid-mediated hydrolysis of proteins [67] and the preparation of inorganic nanocrystals [68]. In all cases, the results indicate that the observed rate accelerations and/or changes in product properties observed using sealed-vessel microwave heating were the result of a purely thermal effect. In combination with other recently published studies [69, 70], these investigations give support to the notion that nonthermal microwave effects probably do not exist, and that even specific microwave effects are difficult to trace and probably are of little relevance to the synthetic organic and medicinal chemist. There appears to be a growing consensus in the microwave chemistry community that, as pointed out in a recent book, "heating is just that – heating" [71].

References

1 von Hippel, A.R. (1954) *Dielectric Materials and Applications*, MIT Press, Cambridge, MA.
2 Stuerga, D. (2006) Chapter 1, in *Microwaves in Organic Synthesis*, 2nd edn (ed. A. Loupy), Wiley-VCH Verlag GmbH, Weinheim, pp. 1–34.
3 Mingos, D.M.P. (2005) Chapter 1, in *Microwave-Assisted Organic Synthesis* (eds P. Lidström and J.P. Tierney), Blackwell Publishing, Oxford, pp. 1–22.
4 Baghurst, D.R. and Mingos, D.M.P. (1991) *Chemical Society Reviews*, **20**, 1–47.
5 Gabriel, C., Gabriel, S., Grant, E.H., Halstead, B.S., and Mingos, D.M.P. (1998) *Chemical Society Reviews*, **27**, 213–223.
6 Nüchter, M., Ondruschka, B., Bonrath, W., and Gum, A. (2004) *Green Chemistry*, **6**, 128–141.
7 (a) Gedye, R.N. and Wei, J.B. (1998) *Canadian Journal of Chemistry*, **76**, 525–532; (b) Horikoshi, S., Iida, S., Kajitani, M., Sato, S., and Serpone, N. (2008) *Organic Process Research & Development*, **12**, 257–263.

8 Neas, E. and Collins, M. (1988) *Introduction to Microwave Sample Preparation: Theory and Practice* (eds H.M. Kingston and L.B. Jassie), American Chemical Society, Washington, DC.

9 (a) Stass, D.V., Woodward, J.R., Timmel, C.R., Hore, P.J., and McLauchlan, K.A. (2000) *Chemical Physics Letters*, **329**, 15–22; (b) Timmel, C.R. and Hore, P.J. (1996) *Chemical Physics Letters*, **257**, 401–408(c)Woodward, J.R., Jackson, R.J., Timmel, C.R., Hore, P.J., and McLauchlan, K.A. (1997) *Chemical Physics Letters*, **272**, 376–382.

10 Gupta, M. and Wong, E.W.W. (2007) *Microwaves and Metals*, John Wiley & Sons (Asia) Pte Ltd, Singapore.

11 Bogdal, D. and Prociak, A. (2007) *Microwave-Enhanced Polymer Chemistry and Technology*, Blackwell Publishing, Oxford.

12 Hayes, B.L. (2002) *Microwave Synthesis: Chemistry at the Speed of Light*, CEM Publishing, Matthews, NC.

13 Kappe, C.O., Dallinger, D., and Murphree, S.S. (2009) *Practical Microwave Synthesis for Organic Chemists*, Wiley-VCH Verlag GmbH, Weinheim.

14 Robinson, J., Kingman, S., Irvine, D., Licence, P., Smith, A., Dimitrakis, G., Obermayer, D., and Kappe, C.O. (2010) *Physical Chemistry Chemical Physics*, **12**, 4750–4758.

15 (a) Perreux, L. and Loupy, A. (2001) *Tetrahedron*, **57**, 9199–9223; (b) Perreux, L. and Loupy, A. (2002) Chapter 3, in *Microwaves in Organic Synthesis* (ed. A. Loupy), Wiley-VCH Verlag GmbH, Weinheim, pp. 61–114; (c) Perreux, L. and Loupy, A. (2006) Chapter 4, in *Microwaves in Organic Synthesis*, 2nd edn (ed. A. Loupy), Wiley-VCH Verlag GmbH, Weinheim, pp. 134–218.

16 (a) Langa, F., de la Cruz, P., de la Hoz, A., Díaz-Ortiz, A., and Díez-Barra, E. (1997) *Contemporary Organic Synthesis*, **4**, 373–386; (b) de la Hoz, A., Díaz-Ortiz, A., and Moreno, A. (2004) *Current Organic Chemistry*, **8**, 903–918(c)de la Hoz, A., Díaz-Ortiz, A., and Moreno, A. (2005) *Chemical Society Reviews*, **34**, 164–178; (d) de la Hoz, A., Díaz-Ortiz, A., and Moreno, A. (2006) Chapter 5, in *Microwaves in Organic Synthesis*, 2nd edn (ed. A. Loupy), Wiley-VCH Verlag GmbH, Weinheim, pp. 219–277.

17 (a) Kuhnert, N. (2002) *Angewandte Chemie International Edition*, **41**, 1863–1866; (b) Strauss, C.R. (2002) *Angewandte Chemie International Edition*, **41**, 3589–3590; (c) Panunzio, M., Campana, E., Martelli, G., Vicennati, P., and Tamanini, E. (2004) *Materials Research Innovations*, **8**, 27–31.

18 Bougrin, K., Bannani, A.K., Tetouani, S.F., and Soufiaoui, M. (1994) *Tetrahedron Letters*, **35**, 8373–8376.

19 Koizumi, H., Itoh, Y., and Ichikawa, T. (2006) *Chemistry Letters*, **35**, 1350–1351.

20 Kappe, C.O. (2004) *Angewandte Chemie International Edition*, **43**, 6250–6284.

21 Nüchter, M., Ondruschka, B., Weiß, D., Bonrath, W., and Gum, A. (2005) *Chemical Engineering & Technology*, **28**, 871–881.

22 Leadbeater, N.E., Pillsbury, S.J., Shanahan, E., and Williams, V.A. (2005) *Tetrahedron*, **61**, 3565–3585.

23 Kremsner, J.M. and Kappe, C.O. (2006) *The Journal of Organic Chemistry*, **71**, 4651–4658.

24 Hosseini, M., Stiasni, N., Barbieri, V., and Kappe, C.O. (2007) *The Journal of Organic Chemistry*, **72**, 1417–1424.

25 Herrero, M.A., Kremsner, J.M., and Kappe, C.O. (2008) *The Journal of Organic Chemistry*, **73**, 36–47.

26 Obermayer, D. and Kappe, C.O. (2010) *Organic & Biomolecular Chemistry*, **8**, 114–121.

27 (a) Bogdal, D., Lukasiewicz, M., Pielichowski, J., Miciak, A., and Bednarz, Sz. (2003) *Tetrahedron*, **59**, 649–653; (b) Lukasiewicz, M., Bogdal, D., and Pielichowski, J. (2003) *Advanced Synthesis and Catalysis*, **345**, 1269–1272.

28 Bogdal, D., Bednarz, S., and Lukasiewicz, M. (2006) *Tetrahedron*, **62**, 9440–9445.

29 Damm, M. and Kappe, C.O. (2009) *Molecular Diversity*, **13**, 529–543.

30 Moseley, J.D., Lenden, P., Thomson, A.D., and Gilday, J.P. (2007) *Tetrahedron Letters*, **48**, 6084–6087.

31 Hayes, B.L. and Collins, M.J., Jr. (2004) World Patent WO 04002617.

32 Hayes, B.L. (2004) *Aldrichim Acta*, **37**, 66–77.

33 Bacsa, B., Horváti, K., Bo7pt2pt´sze, S., Andreae, F., and Kappe, C.O. (2008) *The Journal of Organic Chemistry*, **73**, 7532–7542.

34 Hostyn, S., Maes, B.U.W., Van Baelen, G., Gulevskaya, A., Meyers, C., and Smits, K. (2006) *Tetrahedron*, **62**, 4676–4684.

35 Vanier, G.S. (2007) *Synlett*, 131–135.

36 Obermayer, D., Gutmann, B., and Kappe, C.O. (2009) *Angewandte Chemie International Edition*, **48**, 8321–8342.

37 Gutmann, B., Obermayer, D., Reichart, B., Prekodravac, B., Irfan, M., Kremsner, J.M., and Kappe, C.O. (2010) *Chemistry: A European Journal*, **16**, 12182–12194.

38 Stadler, A. and Kappe, C.O. (2001) *European Journal of Organic Chemistry*, 919–925.

39 Kappe, C.O. (2008) *Chemical Society Reviews*, **37**, 1127–1139.

40 Baghurst, D.R. and Mingos, D.M.P. (1992) *Journal of the Chemical Society, Chemical Communications*, 674–677.

41 Saillard, R., Poux, M., Berlan, J., and Audhuy-Peaudecerf, M. (1995) *Tetrahedron*, **51**, 4033–4042.

42 Chemat, F. and Esveld, E. (2001) *Chemical Engineering & Technology*, **24**, 735–744.

43 (a) Abtal, E., Lallemant, M., Bertrand, G., and Watelle, G. (1985) *Journal of Chemical Physics*, **82**, 381–399; (b) Roussy, G., Thibaut, J.M., and Collin, P. (1986) *Thermochimica Acta*, **98**, 57–62.

44 Dallinger, D., Irfan, M., Suljanovic, A., and Kappe, C.O. (2010) *The Journal of Organic Chemistry*, **75**, 5278–5288.

45 Berlan, J. (1995) *Radiation Physics and Chemistry*, **45**, 581–589.

46 Orrling, K., Nilsson, P., Gullberg, M., and Larhed, M. (2004) *Chemical Communications*, 790–791.

47 Hajek, M. (2006) Chapter 13, in *Microwaves in Organic Synthesis*, 2nd edn (ed. A. Loupy), Wiley-VCH Verlag GmbH, Weinheim, pp. 615–652.

48 Will, H., Scholz, P., and Ondruschka, B. (2002) *Chemie Ingenieur Technik*, **74**, 1057–1067.

49 Zhang, X., Hayward, D.O., and Mingos, D.M.P. (2003) *Catalysis Letters*, **88**, 33–38.

50 Heller, E., Lautenschläger, W., and Holzgrabe, U. (2005) *Tetrahedron Letters*, **46**, 1247–1249.

51 Irfan, M., Fuchs, M., Glasnov, T.N., and Kappe, C.O. (2009) *Chemistry: A European Journal*, **15**, 11608–11618.

52 Baxendale, I.R., Griffith-Jones, C.M., Ley, S.V., and Tranmer, G.K. (2006) *Chemistry: A European Journal*, **12**, 4407–4416.

53 Glasnov, T.N., Findenig, S., and Kappe, C.O. (2009) *Chemistry: A European Journal*, **15**, 1001–1015.

54 Dressen, M.H.C.L., van de Kruijs, B.H.P., Meuldijk, J., Vekemans, J.A.J.M., and Hulshof, L.A. (2007) *Organic Process Research & Development*, **11**, 865–869.

55 van de Kruijs, B.H.P., Dressen, M.H.C.L., Meuldijk, J., Vekemansa, J.A.J.M., and Hulshof, L.A. (2010) *Organic & Biomolecular Chemistry*, **8**, 1688–1694.

56 Gutmann, B., Schwan, A.M., Reichart, B., Gspan, C., Hofer, F., and Kappe, C.O. (2011) *Angewandte Chemie International Edition*, **50**, 7636–7640.

57 Conner, W.C. and Tompsett, G.A. (2008) *The Journal of Physical Chemistry B*, **112**, 2110–2118.

58 Kaiser, N.-F.K., Bremberg, U., Larhed, M., Moberg, C., and Hallberg, A. (2000) *Angewandte Chemie International Edition*, **39**, 3596–3598.

59 Laurent, R., Laporterie, A., Dubac, J., Berlan, J., Lefeuvre, S., and Audhuy, M. (1992) *The Journal of Organic Chemistry*, **57**, 7099–7102.

60 Raner, K.D., Strauss, C.R., Trainor, R.W., and Thorn, J.S. (1995) *The Journal of Organic Chemistry*, **60**, 2456–2460.

61 Nilsson, P., Larhed, M., and Hallberg, A. (2001) *Journal of the American Chemical Society*, **123**, 8217–8225.

62 (a) Jacob, J., Chia, L.H.L., and Boey, F.Y.C. (1995) *Journal of Materials Science*, **30**, 5321–5327; (b) Binner, J.G.P., Hassine, N.A., and Cross, T.E. (1995) *Journal of Materials Science*, **30**, 5389–5393; (c) Shibata, C., Kashima, T., and Ohuchi, K. (1996) *Japanese Journal of Applied Physics*, **35**, 316–319.

63 (a) Berlan, J., Giboreau, P., Lefeuvre, S., and Marchand, C. (1991) *Tetrahedron*

Letters, **32**, 2363–2366; (b) Lewis, D.A., Summers, J.D., Ward, T.C., and McGrath, J.E. (1992) *Journal of Polymer Science Part A: Polymer Chemistry*, **30**, 1647–1653.
64 Loupy, A., Maurel, F., and Sabatié-Gogová, A. (2004) *Tetrahedon*, **60**, 1683–1691.
65 Kalhori, S., Minaev, B., Stone-Elander, S., and Elander, N. (2002) *Journal of Physical Chemistry A*, **106**, 8516–8524.
66 Miklavc, A. (2001) *ChemPhysChem.*, 553–555.
67 Damm, M., Holzer, M., Radspieler, G., Marsche, G., and Kappe, C.O. (2010) *Journal of Chromatography A*, **1217**, 7826–7832.
68 Pein, A., Baghbanzadeh, M., Rath, T., Haas, W., Maier, E., Amenitsch, H., Hofer, F., Kappe, C.O., and Trimmel, G. (2011) *Inorganic Chemistry*, **50**, 193–200.
69 (a) Gilday, J.P., Lenden, P., Moseley, J.D., and Cox, B.G. (2008) *The Journal of Organic Chemistry*, **73**, 3130–3134; (b) Schmink, J.R. and Leadbeater, N.E. (2009) *Organic & Biomolecular Chemistry*, **7**, 3842–3846; (c) Lombard, C.K., Myers, K.L., Platt, Z.H., and Holland, A.W. (2009) *Organometallics*, **28**, 3303–3306; (d) Dressen, M.H.C.L., Stumpel, J.E., van de Kruijs, B.H.P., Meuldijk, J., Vekemans, J.A.J.M., and Hulshof, L.A. (2009) *Green Chemistry*, **11**, 60–64; (e) Burns, M., Lloyd-Jones, G.C., Moseley, J.D., and Renny, J.S. (2010) *The Journal of Organic Chemistry*, **75**, 6347–6353.
70 (a) Strauss, C.R. (2009) *Organic Process Research & Development*, **13**, 915–923; (b) Strauss, C.R. and Rooney, D.W. (2010) *Green Chemistry*, **12**, 1330–1339.
71 Leadbeater, N.E. (ed.) (2011) *Microwave Heating as a Tool for Sustainable Chemistry*, CRC Press, Boca Raton, FL.

3
Equipment Review

3.1
Introduction

Although many of the early pioneering experiments in microwave-assisted organic synthesis were carried out in domestic microwave ovens, the current trend undoubtedly is to use dedicated instruments for chemical synthesis (Figure 1.1). In a domestic microwave oven, the irradiation power is generally controlled by on–off cycles of the magnetron (pulsed irradiation), and it is typically not possible to monitor the reaction temperature reliably. Combined with the inhomogeneous field produced by the low-cost multimode designs and the lack of safety controls, the use of such equipment cannot be recommended for scientific purposes. In contrast, all of today's commercially available dedicated microwave reactors for synthesis feature built-in magnetic stirrers or alternative agitation devices, direct temperature control of the reaction mixture with the aid of IR sensors, fiber-optic probes or other appropriate thermometers, and software that enables online temperature/pressure control by regulation of microwave power output. Currently, two different philosophies with respect to microwave reactor design have been established: multimode and monomode (also referred to as single-mode). In so-called multimode instruments (conceptually similar to a domestic microwave oven), the microwaves that enter the cavity are reflected by the walls and the load over the typically large cavity. In many multimode instruments, a mode stirrer ensures that the field distribution is as homogeneous as possible. In the much smaller monomode cavities, only one mode is present and the electromagnetic irradiation is directed through a precision-designed rectangular or circular waveguide onto the reaction vessel mounted at a fixed distance from the radiation source, creating a standing wave. The key difference between the reactor systems is that whereas in typical multimode cavities, several reaction vessels can be irradiated simultaneously in multivessel rotors (parallel synthesis), in the monomode systems, typically only one vessel can be irradiated at a time. In the latter case, high-throughput can be achieved by integrated robotics that move individual reaction vessels in and out of the microwave cavity. Most instrument manufacturers offer a variety of diverse reactor platforms with different degrees of sophistication with

Microwaves in Organic and Medicinal Chemistry, Second Edition.
C. Oliver Kappe, Alexander Stadler, and Doris Dallinger.

respect to automation, database capabilities, safety features, temperature and pressure monitoring, and vessel design. Importantly, single-mode reactors processing comparatively small volumes also have an efficient built-in cooling feature that allows rapid cooling of the reaction mixture by compressed air after completion of the irradiation period (see Figure 2.13). The dedicated single-mode instruments available today can process volumes ranging from about 200 μL to 50 mL under sealed-vessel conditions (up to 300 °C, 30 bar), and somewhat higher volumes (125 mL) under open-vessel reflux conditions. In the much larger multimode instruments, liter quantities can be processed under both open- and closed-vessel conditions. For both single- and multimode cavities, flow reactors are nowadays available allowing the preparation of kilograms of materials using microwave technology.

This chapter provides a detailed description of the various currently commercially available microwave reactors that are dedicated for microwave-assisted organic synthesis. Publications describing the use of such equipment are discussed in Chapters 5–8. A comprehensive coverage of microwave reactor design, applicator theory, description of waveguides, magnetrons, and microwave cavities can be found elsewhere [1–5]. An overview of experimental, noncommercial microwave reactors has been presented by Ondruschka *et al.* [4] and more recently by Moseley [6].

3.2
Domestic Microwave Ovens

At its beginning in the mid-1980s, microwave-assisted organic synthesis was carried out exclusively in conventional multimode domestic microwave ovens [7, 8]. The main drawback of these household appliances is the lack of control systems. In general, it is not possible to determine the reaction temperature accurately using domestic microwave ovens. The lack of pressure control and no possibility to stir the reaction mixture additionally makes performing chemical syntheses in domestic microwave ovens troublesome. Furthermore, the pulsed irradiation (on–off duty cycles) and the resulting inhomogeneity of the microwave field may lead to problems of reproducibility. Due to the nonhomogeneous distribution of the energy intensity throughout the cavity, some areas receive higher amounts of energy (so-called hot spots), whereas others receive less energy (so-called cold spots).

Another concern is safety. Heating organic solvents in open vessels in a microwave oven can lead to violent explosions induced by electric arcs inside the cavity or by sparking resulting from the switching of the magnetrons. On the other hand, working with sealed vessels under pressure without real-time monitoring of pressure can lead to unexpected vessel failures in case of a thermal runaway and to serious accidents. Early on, simple modifications of the available ovens with self-made accessories like mechanical stirrers or reflux condensers mounted through holes in the cavity were attempted in order to generate instrumentation useful for chemical synthesis (Figure 3.1). However, some safety risks remained as these instruments were not explosion proof and leakage of microwaves harmful to the operator can occur. Clearly, the use of any microwave equipment not specifically designed for

Figure 3.1 Modified domestic household microwave oven. Inlets for temperature measurement by IR pyrometer (left side) and for attaching reflux condensers (top) are visible. A magnetic stirrer is situated below the instrument.

organic synthesis cannot be recommended and is generally banned from most industrial and academic laboratories today.

3.3
Dedicated Microwave Reactors for Organic Synthesis

With growing interest in MAOS during the mid-1990s, the demand for more sophisticated microwave instrumentation, offering, for example, stirring of the reaction mixture, temperature measurement, and power control features, increased. For scientifically valuable, safe, and reproducible work, the utilized microwave reactors should offer the following features:

- Built-in magnetic or mechanical stirring
- Accurate temperature measurement
- Pressure control
- Continuous power regulation
- Efficient postreaction cooling
- Computer-aided method programming
- Explosion proof cavities

A particularly difficult problem in microwave processing is the correct measurement of the reaction temperature during the irradiation phase. Classical temperature sensors (thermometers and thermocouples) will fail since they will couple with the electromagnetic field. Temperature measurement can be achieved either by an immersed temperature probe (fiber-optic, gas balloon thermometer, rising Pt100 sensor) or, on the outer surface of the reaction vessels, by a remote IR sensor. Due to the volumetric character of microwave heating, the surface temperature of the reaction vessel will not always reflect the actual temperature inside the reaction vessel (see Section 2.5.1 for details) [9]. Recent publications clearly point out the importance of accurate internal temperature monitoring in microwave synthesis, in particular for solvent-free or highly viscous reaction mixtures [10].

In general, a microwave reactor consists of a microwave power source (magnetron), a transmission line (waveguide) that delivers microwaves from the magnetron into an antenna or applicator, and a microwave applicator (cavity). According to the geometry and dimensions of the cavities, multimode or single-mode reactors can be distinguished. A state-of-the-art microwave reactor design should ensure that all the incident power is absorbed by the load since the magnetron may be damaged when too much energy is reflected back. Several techniques to overcome this problem are incorporated in dedicated reactors (see Sections 3.4 and 3.5).

Since the early applications in microwave-assisted synthesis were based on the use of domestic multimode microwave ovens, the initial focus in the development of dedicated microwave instruments was on the improvement of multimode reactors. In multimode instruments, typically one or two magnetrons create the microwave irradiation, which is normally directed into the cavity through a waveguide and distributed by a mode stirrer (Figure 3.2). The microwaves are reflected from the walls of the cavity, thus interacting with the sample in a chaotic manner. Multimode cavities may show multiple energy pockets with different levels of energy intensity, thus resulting in hot and cold spots. To provide an equal energy distribution, the samples are continuously rotated within the cavity. Therefore, multimode instruments offer convenient platforms for the increase of reaction throughput by utilizing multivessel rotors for parallel synthesis or scale-up. A general problem for multimode instruments is the weak performance for small-scale experiments (<3 mL). While the generated microwave power is high (800–1700 W), the power density of the field is generally rather low, making heating of small individual samples rather difficult – a major drawback, especially for research and development purposes.

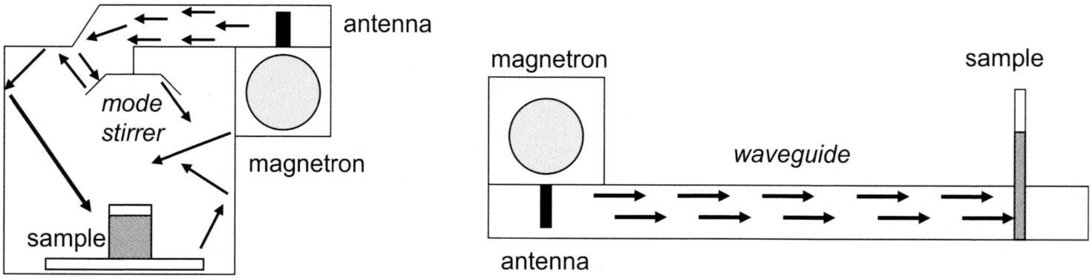

Figure 3.2 Multimode (left) versus single-mode cavities (right).

Although some vendors provide special equipment for parallel reaction screening and method development, the general use of multimode instruments for small-scale synthetic organic chemistry is not so extensive compared to the much more popular single-mode cavities.

In contrast, single-mode (monomode) instruments generate a single, comparatively homogeneous energy field of high power intensity. Thus, these systems couple efficiently with small samples and the maximum output power is limited to 300, 400, or 850 W. The microwave energy is created by a single magnetron and is typically directed through a rectangular waveguide to the sample, which is positioned at a maximized energy point (Figures 3.2 and 3.3). To create optimum conditions for variations in microwave absorptivity, systems are equipped either with an optimized designed applicator (Anton Paar) or with a tuning device that provides proper adjustment of the microwave field (Biotage). The homogenous field generally enables excellent reproducibility, although field inhomogeneities also do exist with these devices (see Section 2.5.1).

In addition to the rectangular waveguide applicator, instruments with a self-tuning circular waveguide are also available (Figure 3.3). This single-mode cavity (CEM) features multiple entry points for the microwave energy to be delivered into the cavity (slots), compensating for variations in the coupling characteristics of the sample. Due to the cavity design, it is suitable for different vessel types and sizes such as sealed 10–80 mL vials or up to 125 mL round-bottom flasks.

Recent advances and further improvements have led to a broad variety of applications for single-mode microwave instruments, offering flow-through systems as well as special features such as solid-phase peptide synthesis. The use of single-mode reactors has therefore increased tremendously since the year 2000 and these types of instruments have become very popular in many synthetic laboratories, both in industry and academia. However, it should be pointed out that the type of instrumentation used is based on the desired application and scale, rather than on

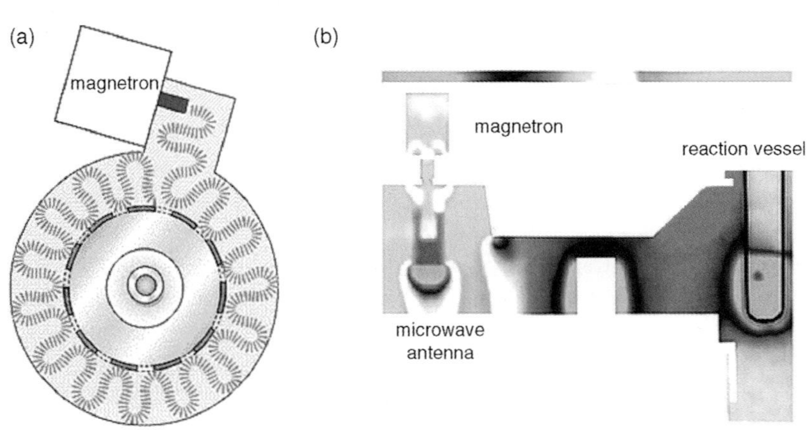

Figure 3.3 (a) Circular waveguide (CEM Discover), (b) Rectangular waveguide (Biotage Initiator).

the kind of chemistry to be performed. Both multimode and single-mode reactors are able to carry out chemical reactions efficiently and to improve classic heating protocols.

The following sections give a comprehensive description of all commercial multimode and single-mode microwave reactors globally available as of 2011, including various accessories and special application tools. Essentially, there are currently four major instrument manufacturers that produce microwave reactors for laboratory-scale organic synthesis: Anton Paar GmbH (Graz, Austria) [11], Biotage AB (Uppsala, Sweden) [12], CEM Corporation (Matthews, NC) [13], and Milestone s.r.l./MLS GmbH (Sorisole, Italy and Leutkirch, Germany) [14].

3.4 Single-Mode Instruments

The first microwave instrument company offering single-mode cavities was the French company Prolabo [15]. In the early 1990s, the Synthewave 402 was released, later followed by the Synthewave 1000 [16]. The instruments featured cylindrical glass or quartz tubes for reactions restricted to atmospheric pressure conditions. Although the company does not exist anymore, some instruments are still in use, mainly throughout the French scientific community. Another pioneer for microwave synthesis equipment was the Swedish company PersonalChemistry. In 2000, the Emrys monomode reactor series was introduced, comprising the stand-alone Creator, the Optimizer with robotic vial handling, and the Liberator with fully automated vial handling and liquid dispensing. After the merger with Biotage, this instrument line has been ceased and was succeeded by the Initiator instruments (see Section 3.4.2.1). Although not commercially available anymore, many Emrys-type reactors are still in use. This chapter provides a description of the currently commercially available instrumentation of the main globally active vendors.

3.4.1 Anton Paar GmbH

3.4.1.1 Monowave 300

This microwave reactor represents the most recent contribution to monomode instrumentation, launched in 2009. The Monowave 300 (Figure 3.4a) provides the highest applicable field density, offering 850 W continuous microwave output power from a single magnetron. It serves reaction conditions up to 300 °C and 30 bar. The instrument is equipped with a touch screen controller for method programming and reaction control. For high-throughput purpose, the basic instrument can be upgraded with the autosampler unit MAS24 hosting carousels for unattended sequential processing of 24 reaction vessels (Figure 3.4b). The autosampler unit is equipped with a pneumatic gripper and provides a software-guided loading strategy.

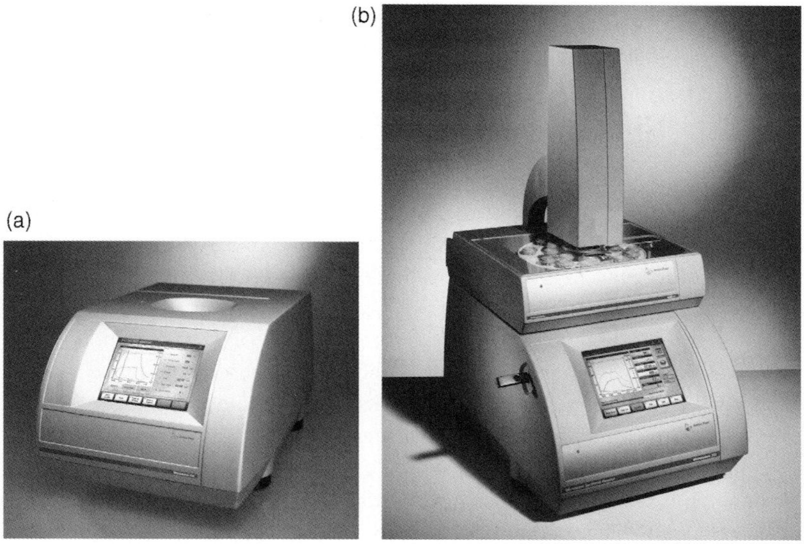

Figure 3.4 Anton Paar Monowave 300 (a) and with autosampler MAS24 upgrade (b).

Apart from three glass vessel types (4, 10, and 30 mL) for reaction volumes from 0.5 to 20 mL, a special 10 mL silicon carbide vessel is also available (Figure 3.5). This vessel enables the processing of fully microwave-transparent mixtures, allows employing aggressive reaction media in a high-temperature regime, and facilitates investigations toward specific/nonthermal microwave effects (see Sections 2.5 and 4.6). For each vessel type, appropriate stir bars for proper mixing are available. All

Figure 3.5 Monowave 300 vessel types: glass vials 6.0–20, 2.0–6.0, and 0.5–2.0 mL; and silicon carbide vessel 2.0–6.0 mL (left to right).

vessels are sealed with a reusable single-size snap cap made of PEEK comprising a PTFE-coated silicone septum. The pneumatic driven swiveling cover of the instrument tightly seals the vessels inside the cavity to withstand pressures up to 30 bar. A sophisticated opening strategy ensures release of remaining overpressure to avoid handling of pressurized vessels outside the cavity.

The Monowave 300 microwave unit consists of a rectangular waveguide tube and a specially designed self-tuning applicator. Depending on the employed volume of the reaction mixture, the optimized quantity of microwave energy is delivered at any time. Thus, uniform and high-density heating is ensured providing high reproducibility and scale-up from 0.5 to 20 mL.

The installed single magnetron delivers a maximum of 850 W, sufficient for rapid heating of various reaction mixtures in any applicable scale. Each Monowave 300 unit is equipped with a built-in magnetic stirrer device with variable stirring speed from 0 to 1200 rpm. Temperature measurement is achieved by an IR sensor perpendicular to the position of the vial in the applicator, working in a measuring range of 30–310 °C. This arrangement requires a specific minimum filling height in each vessel type in order to obtain reliable temperature values when monitoring on the outer surface of the reaction vessels. For accurate measurement of the reaction temperature, an optional ruby-based immersing fiber-optic probe is available, applicable to all vessel types. Pressure control is achieved by means of a noninvasive hydraulic sensor integrated into the swiveling cover of the cavity, which senses the deformation of the seal due to the internal pressure buildup. Efficient cooling is accomplished by means of a pressurized air supply with a rate of approximately 50 L min^{-1}, enabling cooling from 300 to 55 °C within approximately 3 min, depending on the heat capacity of the employed solvent.

Reaction control is typically temperature based and can be performed either by achieving the adjusted target temperature as fast as possible by using the appropriate microwave magnetron output power up to 850 W or by approaching the reaction temperature in a certain period of time when programming a defined ramp time. The latter is beneficial for scale-up in terms of accuracy and reproducibility since the microwave output power will be set individually for different reaction scale, but the relevant processing time (heating and hold time) will be identical. A corresponding software algorithm reduces the microwave power output when nearing the target temperature to prevent overheating of the sample.

In addition, reactions can be performed under "constant power," that is, the microwave power can be adjusted to a defined value until the target temperature is reached. As a special feature, the so-called cooling-while-heating can be performed, introducing more microwave energy into the reaction as the cooling is activated during the irradiation (see Section 2.5.1 for details).

In terms of user's convenience, the touch screen interface allows modifying the performed reaction protocols by "one-click editing." Changes in temperature, time, and stirring speed can be done immediately without terminating the experiment. For data management, the Monowave 300 is equipped with two USB ports for method transfer and data export. Furthermore, *in situ* generated reaction reports can be printed on local printers or network printers connected via the LAN port.

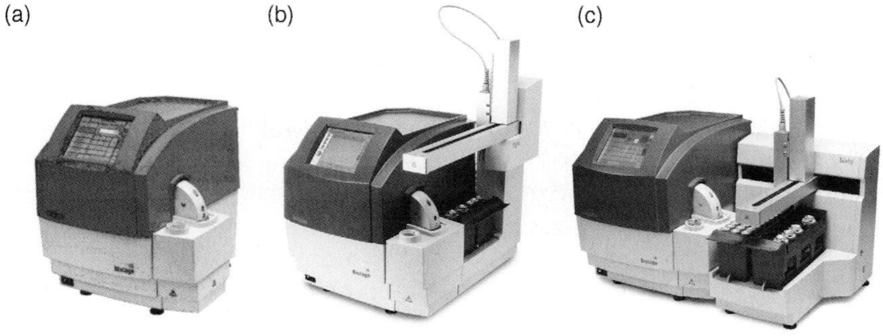

Figure 3.6 (a) Biotage Initiator, (b) Initiator Eight, (c) Initiator Sixty.

3.4.2 Biotage AB

3.4.2.1 Initiator Platform

The standard instrumentation from Biotage is the Initiator reactor for small-scale reactions in a single-mode cavity. This instrument, launched in 2004, is closely related to the former Emrys platform (2000–2004), but 45% smaller in footprint and now equipped with a touch screen so that no external PC is needed. The Initiator is upgradeable from the single-sample manual format to the automated systems Initiator Eight or Initiator Sixty (Figure 3.6). Both are equipped with a vial rack and a robotic gripper, featuring sequential processing of up to 8 reactions for the Initiator Eight and up to 60 for the Initiator Sixty.

For the basic setup, two different vial types with reaction volumes from 0.5 to 5 mL can be used (Figure 3.7). Additional vessel sizes from a very small-scale (0.2–0.5 mL) to a low-level scale-up (20 mL vessel) are available with the EXP upgrade that offers a larger microwave cavity. In this way, a direct scale-up from the milligram to gram range using different vessels from 0.2 to 20 mL operating volume (Figure 3.7) can be achieved without any system modifications or reoptimizations. The reaction vials, made of borosilicate glass, are permanently sealed by a cap and can withstand operating pressures up to 20 bar. For each vial type, appropriate stir bars are provided.

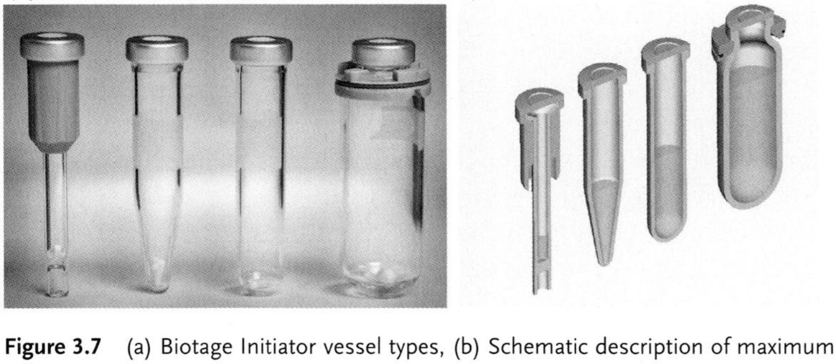

Figure 3.7 (a) Biotage Initiator vessel types, (b) Schematic description of maximum volume: 0.2–0.5, 0.5–2.0, 2.0–5.0, and 10.0–20.0 mL (left to right).

The vial cap consists of an aluminum crimp in which a Teflon septum is inserted. The advantage of this septum is the Reseal design that allows sample withdrawal for analysis or addition of reagents, albeit not when the reaction vessel is contained inside the microwave cavity. The sealing ring located in the cavity lid reseals the septum after penetration while the cavity is closed.

The Initiator microwave unit consists of a closed rectangular waveguide tube (see Figure 3.3) combined with a deflector device, which via a power sensor physically maximizes the energy absorption by the reaction mixture (Dynamic Field Tuning). Dynamic Field Tuning allows the system to detect the absorbance characteristics of the reaction mixture and to optimize the coupling and quantity of microwave energy delivered.

Currently, the Initiator 2.5 (referring to the actual software version) is available and all the following specifications refer to this model. The output power is maximized to continuously deliver 400 W, sufficient for rapid heating of most reaction mixtures. The Initiator series is equipped with built-in magnetic stirring with stirring speed variable from 300 to 900 rpm. Temperature measurement is achieved by an IR sensor perpendicular to the position of the vial in the waveguide, working in a measuring range of 40–250 °C. This arrangement requires a specific minimum filling height in each vessel type in order to obtain accurate temperature values. On the other hand, if the maximum filling volume of the vial (Figure 3.7) is exceeded, then insufficient space for pressure buildup is left. The temperature is measured on the outer surface of the reaction vessels. The pressure limit for the Initiator instruments is 20 bar, imposed by the sealing mechanism that utilizes Teflon-coated silicon seals in aluminum crimp tops. Pressure control is achieved by means of a noninvasive sensor integrated into the closing lid of the cavity, which senses the deformation of the seal due to the pressure buildup. Efficient cooling is accomplished by means of a pressurized air supply with a rate of approximately 60 L min^{-1}, enabling cooling from 250 to 50 °C within approximately 1 min, depending on the heat capacity of the used solvent.

Reaction control is temperature based, with the system trying to attain the adjusted maximum temperature as fast as possible by using the appropriate microwave magnetron output power up to 400 W. In order to account for differences in the microwave absorptivity of the reaction mixture, four different power level settings (absorption level: very high, high, normal, and low) can be specified by the user. The better the absorptivity of the solvent/reaction mixture, the higher the user-selected absorption level and the lower the initial power value chosen by the software algorithm in order to guarantee that the set temperature is reached or to avoid an overshoot of the temperature.

Via the touch screen, the user is able to change the performed reaction protocols "on-the-fly" without aborting the experiment – temperature and time changes can be made immediately. Furthermore, reactions can be performed under "power control," that is the power can be adjusted to a defined value over the process time. In addition, the so-called cooling-while-heating can be applied, introducing more microwave energy into the reaction system as the cooling is activated during the irradiation process (see Section 2.5.1 for details).

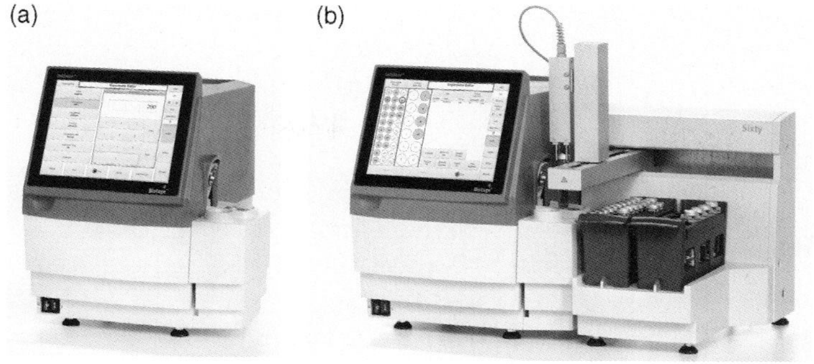

Figure 3.8 (a) Biotage Initiator$^+$, (b) Initiator$^+$ Sixty.

For data management purposes, the Initiator 2.5 is equipped with an USB port for saving and transferring methods and results. Instruments that are connected to a Biotage HUB can share data, methods, and user profiles; furthermore, electronic lab books can be directly interfaced. For instruments that are connected to a network, online monitoring of the processing reaction via the office PC is possible. In addition, the result files can be sent via e-mail.

In 2011, a revised version of the Initiator series was launched. The new Initiator$^+$ modular platform (Figure 3.8) features a 10.4″ touch screen controller and the updated software version 4.0 provides extended operation limits of 300 °C and 30 bar for the smaller vials (Figure 3.7). The 20 mL vial can be operated at the usual 250 °C and 20 bar limit as in the standard Initiator series. As its predecessor, the Initiator$^+$ is upgradeable from the single-sample manual format to automated systems with Biotage's Robot Eight or Robot Sixty (Figure 3.8b) for unattended sequential operation. In addition, it can be equipped with the SP Wave module resulting in a semiautomated solution for peptide chemistry, complete with vortex mixer and liquid handling (see Section 3.4.2.4, Figure 3.9). For these peptide applications a fiber-optic probe is applied to measure the reaction temperature directly inside the vial, not externally by the IR sensor.

3.4.2.2 **Chemspeed SWAVE**

The SWAVE unattended microwave synthesizer station from Chemspeed (Figure 3.10) incorporates a Biotage Initiator with all the features that are necessary to enable a fully automated synthesis workflow platform [17]. Automation is given, starting from reaction preparation up to purification, via the robotic platform that consists of a XYZ robotic arm and 8 racks of up to 30 samples each. Addition of reagents under an inert atmosphere, solid weighing, and direct dispensing, liquid handling is covered, as well as an automated single tool for capping, crimping, and gripping of the microwave vials. For multistep synthesis, reagents can be added directly through the septum, while the vial is located in the microwave cavity. Solid-phase extraction (SPE), automated filtration, liquid–liquid extraction, and online chromatographic analysis are some of the available features for the workup process.

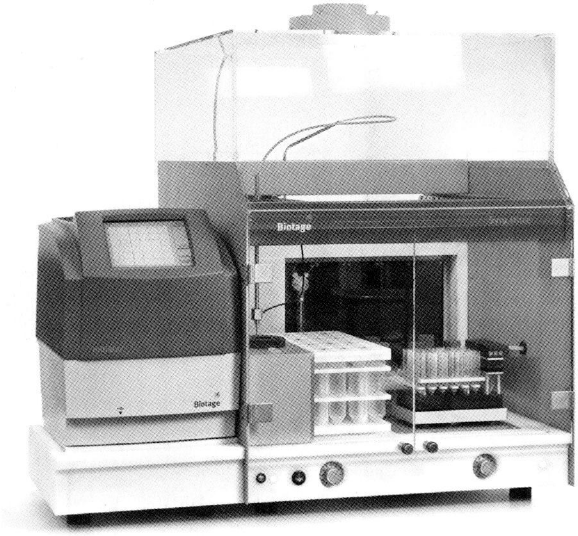

Figure 3.9 Biotage Syro *Wave*: fully automated microwave and parallel peptide synthesizer.

To increase the throughput, a second Initiator microwave reactor can be incorporated on the right-hand side of the platform (Figure 3.10).

3.4.2.3 **Peptide Synthesizers**

In 2010, Biotage launched the Syro *Wave* parallel peptide synthesizer. This programmable workstation connects the amenities of unattended parallel solid-phase peptide synthesis with the benefits of direct microwave heating (Figure 3.9). It is a combination of the MultiSynTech Syro I with a slightly modified Biotage Initiator, featuring vortexing in the microwave cavity instead of stirring. The microwave unit is limited to a maximum of 200 W microwave output power (capped at 60 W during the

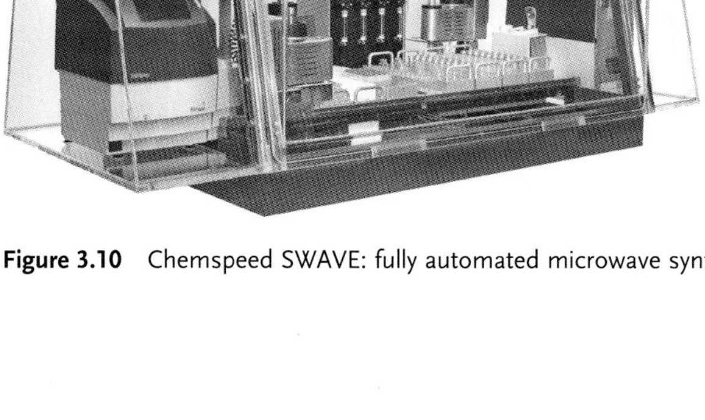

Figure 3.10 Chemspeed SWAVE: fully automated microwave synthesizer station.

hold time), serving atmospheric pressure reactions at 40–80 °C. Three polypropylene vial types (2, 5, and 10 mL) are applicable providing an operation range of 0.8–1.1 mL, 1.6–3.2 mL, and 3.2–6.4 mL, respectively, to prepare peptide sequences in 1–300 μmol scale. The peptide synthesizer part consists of a type U reactor block (24 or 48 positions) with vortex mixer, a robotic arm, a syringe pump, a vacuum pump, an amino acid rack (32 × 50 mL), a corresponding reagent bottle rack (2 × 500 mL, 3 × 200 mL), and a waste bottle (10 L). The included desktop PC controls both units with the Syro XP software.

As an entry system to microwave-assisted peptide synthesis, the Initiator Peptide Workstation was introduced in 2011 (Figure 3.11a). This system enables manual peptide synthesis under atmospheric conditions utilizing 2–5 mL and a 10–20 mL microwave peptide glass vials with disposable HDPE frits and reusable snap caps. In combination with the corresponding wash station that consists of a specially designed height-adjustable vacuum head on a stand, the basic Initiator EXP reactor can be efficiently employed for peptide synthesis up to 500 μmol scale without further system modifications.

More recently, with the presentation of the new Initiator$^+$ platform (see Section 3.4.2.1, Figure 3.8), the corresponding Initiator$^+$ SP *Wave* was introduced. This semiautomated microwave peptide synthesizer (Figure 3.11b) is equipped with robotics for automated deprotection and washing steps, whereas building blocks are added manually. The same 2, 5, and 10 mL polypropylene vials as described for the Syro *Wave* can be employed for peptide synthesis. The system can be switched between peptide synthesis, utilizing vortex mixing, and standard microwave synthesis, employing magnetic stirring. In the peptide synthesis mode, the microwave power output is limited to a maximum of 200 W (capped at 60 W during the hold

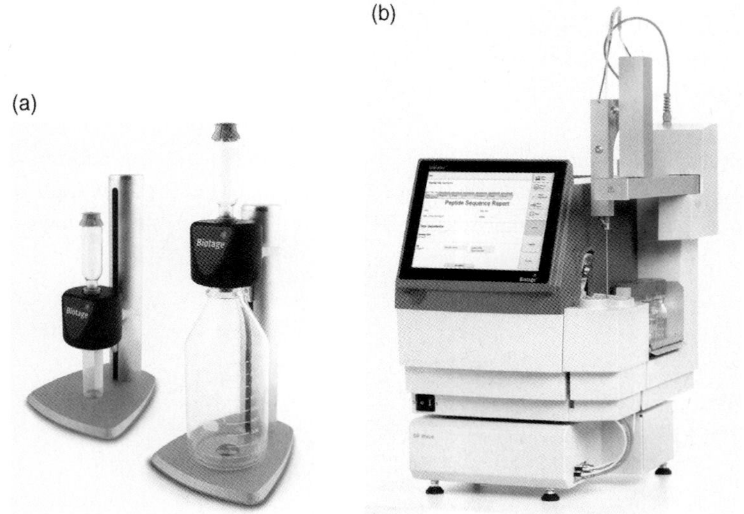

Figure 3.11 (a) Biotage Initiator Peptide Workstation for manual microwave peptide synthesis, (b) Initiator$^+$ SP *Wave* for automated peptide synthesis.

time), serving atmospheric pressure reactions at 40–100 °C. In the MAOS mode, the common limitations of the Initiator$^+$ platform apply (see Section 3.4.2.1).

3.4.3 CEM Corporation

3.4.3.1 Discover Platform

The CEM Discover system, introduced in 2001, is a single-mode instrument based on the self-tuning circular waveguide technique (see Figure 3.3). This circular cavity automatically adjusts to ensure that the reaction receives the optimum amount of energy, regardless of the reaction volume. This concept provides modularity for automation; scale-up under closed-vessel, open-vessel, and flow-through conditions; low-temperature chemistries; and application in biosciences. All Discover instruments are equipped with a built-in keypad for programming the reaction procedures and allowing on-the-fly changes. The manual Discover reactor (Figure 3.12, Table 3.1) covers a variety of reaction conditions in open-vessel (up to 125 mL round-bottom flasks) and closed-vessel systems (up to 50 mL filling volume).

The output power is maximized to continuously deliver 300 W (power can be set between 0 and 300 W), sufficient for rapid heating of most reaction mixtures. Routine temperature measurement within the Discover series is achieved by an IR sensor positioned at the bottom of the cavity, below the vessel. This allows temperature control of the reaction while using minimum amounts of materials (200 μL for the 10 mL vials). The platform can also be equipped with an optional fiber-optic temperature sensor for internal temperature measurement.

Offering an economical choice in terms of footprint, the Discover BenchMate provides an entry-level system to sealed-vessel reactions with basic reaction temperature and pressure management (Figure 3.12a). The pressure management device does not measure pressure, but due to the "snap-on" cap design, automatic venting is feasible when the internal pressure exceeds 20 bar (IntelliVent technology). With the BenchMate system, 10 mL reaction vials with a maximum filling volume of 7 mL can be used.

The Discover LabMate (Figure 3.12b) is the next level, possessing additional features like the pressure measurement device with the IntelliVent Pressure Control System. If the pressure in the vial exceeds 20 bar, the IntelliVent sensor allows a

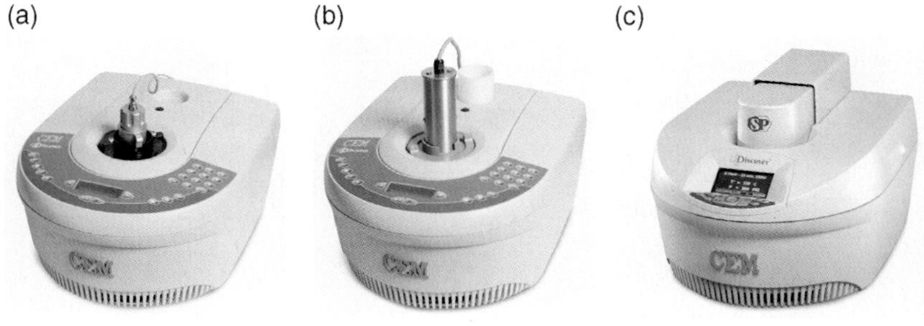

Figure 3.12 Available CEM Discover systems, (a) BenchMate, (b) LabMate, (c) SP.

Table 3.1 Comparison of standard monomode instruments.

Features	Anton Paar Monowave 300	Biotage Initiator^{+}	CEM Discover SP
Waveguide	Rectangular	Rectangular	Circular
Max. output power	850 W	400 W	300 W
Operation temperature	30–300 °C	40–300 °C	rt–300 °C
Max. pressure	30 bar	30 bar	20 bar 15 bar (80 mL vessel)
Vessel sizes	4–30 mL	0.5–20 mL	10–80 mL; max. 125 mL round-bottom flask
Sealing mechanism	Pneumatic supported snap caps	Permanently with crimped caps	"Snap-on" IntelliVent caps
IR sensor	From the side at a defined height	From the side at a defined height	From the bottom
Fiber-optic	✓	✓	✓
Simultaneous cooling	✓	✓	✓
Closed vessel	✓	✓	✓
Open vessel	✓	✓	✓
Magnetic stirring	0–1200 rpm	300–900 rpm	Three different speeds
Method programming	Touch screen	Touch screen	Touch pad or PC

controlled venting of the pressure and subsequently automatically reseals to maintain optimum safety. The same 10 mL vial as for the BenchMate can be employed with this system.

In 2006, the newest Discover system, the Discover S-Class was introduced. Supplementary to the features of the LabMate, the instrument incorporates a fully automated pressure control and an USB interface. In addition to the 10 mL vial, a 35 mL reaction vial with working volumes of 2.0–25 mL is available, allowing a low-level scale-up. As another option for further scale–up, an 80 mL reaction vessel with working volumes up to 50 mL is also available.

An interesting feature is the optional integrated CCD camera for *in situ* reaction monitoring. In 2009, the Discover S-Class was slightly enhanced with the so-called ActiVent Technology for improved control of pressure reactions. Since then the current microwave system is traded as Discover SP (Figure 3.12c). To facilitate reactions requiring inert atmosphere or gaseous reagents, an appropriate gas addition kit is available. This tool provides the ability to purge 10 mL vials up to 14 bar and allows reaction control by an immersed fiber-optic probe (see Figure 3.13a).

All Discover systems work with the Synergy software that is included in the base package for the LabMate and SP. The systems can be run and programmed via PC with the Synergy software that allows documentation, automated data handling, and parameter control capabilities (e.g., variable stirring speed: off, low, medium, and high). For achieving simultaneous cooling (see Section 2.5.1 for details), the patented enhanced cooling system (PowerMAX) can be used during irradiation.

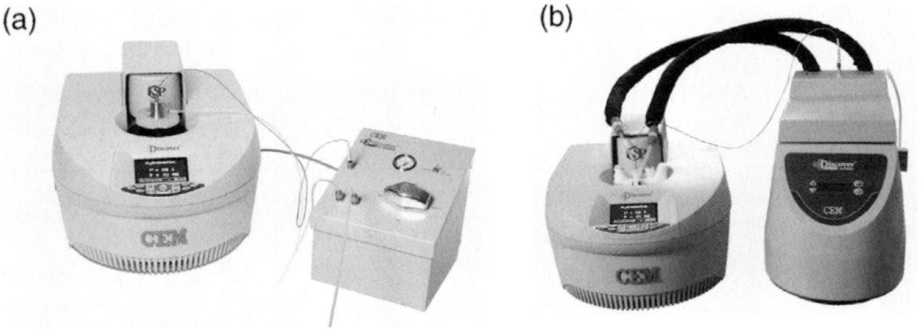

Figure 3.13 CEM Discover special features: (a) Gas loading kit, (b) low temperature cooling module CoolMate.

An extension to the common Discover reactors is the Discover CoolMate (Figure 3.13b), a subzero cooling module that can be used on all Discover systems and which is designed to perform subambient temperature chemistry. The reactor is equipped with a jacketed low-temperature vessel and the system's microwave-transparent cooling media and chilling technology to keep the bulk temperature low (-80 to $+65\,^{\circ}$C). Temperature measurement is performed by internal fiber-optic temperature control. Thus, thermal degradation of materials in temperature-sensitive reactions may be prevented while microwave energy is introduced to the reaction mixture (see Section 2.5.1 for more details).

3.4.3.2 **Explorer Systems**

The Explorer series provides an automation upgrade for the Discover platform by introducing sample racks with a robotic gripper on the top. The Discover BenchMate and LabMate can be converted to the Explorer-24, a 24-position system that can be used only with the 10 mL vials.

For the Discover SP, four other different autosampler rack sizes are optional. The standard vial types (10 and 35 mL) can be employed for the Explorer-12, Explorer-48, Explorer-72, and Explorer-96 instruments (Figure 3.14). Due to the rack design and

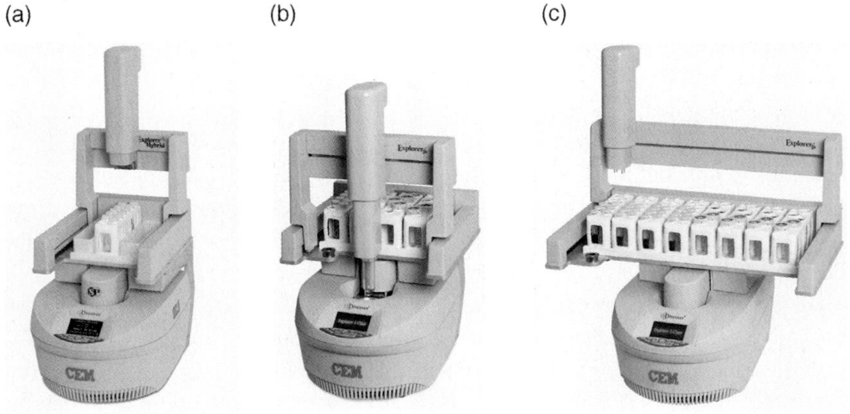

Figure 3.14 Available CEM Explorer systems: (a) Explorer-12 Hybrid, (b) Explorer-48, (c) Explorer-96.

software, the vial size and position are recognized automatically by the autosampler. The most recent automation unit is the Explorer-12 Hybrid (Figure 3.14a), which holds a single rack (12×10 mL or 6×35 mL vials). It represents the most economic entry in sequential, unattended processing, especially addressing academic groups. For the other, earlier introduced variations, the Explorer module holds multiples of the respective rack to serve high-throughput synthesis and library production.

3.4.3.3 **Voyager System**

The Voyager System converts the Discover BenchMate and LabMate reactors into an automated flow system designed to allow the scale-up of reactions from milligram quantities to a daily output of about 1 kg, while still maintaining the advantages of single-mode energy transfer. While the technology accommodates both continuous and stop-flow formats, the stop-flow technique better serves the majority of scale-up applications encountered in today's synthesis laboratory, since it combines the advantages of a batch reactor with those of a continuous flow reactor. The Voyager_{SF} system in stop-flow mode is operated with a special 80 mL vessel (see Figure 3.15), where the reaction mixture is pumped into and out by peristaltic pumps. Reaction limits are 250 °C and 18 bar and the system is also applicable for heterogeneous mixtures, slurries, and solid-phase reactions. Uniform mixing is ensured by dynamic stirring. With this system, a direct scale-up of reactions that were performed on the Discover or Explorer units is possible.

A fiber-optic temperature control module is standard for internal temperature measurement. The initially available continuous flow reactor Voyager_{CF} is no longer advertised. It was recommended to be used only for homogeneous solution-phase chemistry, as slurried mixtures likely caused problems with the pumping system.

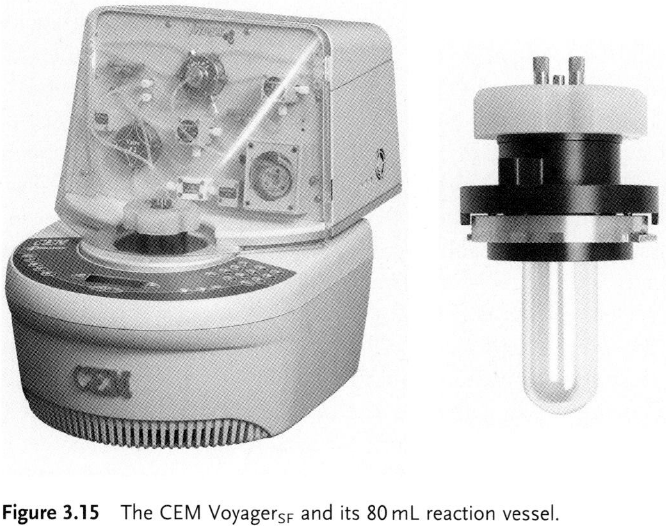

Figure 3.15 The CEM Voyager_{SF} and its 80 mL reaction vessel.

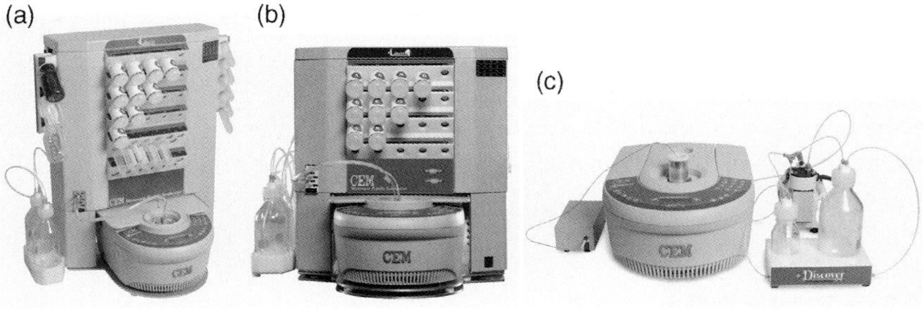

Figure 3.16 CEM peptide synthesizers: (a) Liberty, (b) Liberty 1, (c) Discover SPS.

3.4.3.4 **Peptide Synthesizers**

For solid-phase peptide synthesis, the Liberty, a fully automated microwave peptide synthesizer, is available (Figure 3.16a). This instrument, based on the same Discover reactor core, enables the sequential synthesis of up to 12 peptides, unattended, using a fluidics module to allow the controlled addition of resins, amino acids, coupling, deprotection, and washing reagents, as well as cleavage cocktails. The system is equipped with up to 25 amino acid reservoirs with 125 or 250 mL capacity and 7 (upgradable to 12) external bottle positions. Vessels with 30 and 125 mL volumes allow the synthesis of peptides in a 0.025–5 mmol scale applying either Fmoc or Boc strategies. A spraying system for top–down washing contained in the vessel ensures that the resin beads are properly washed from the vessel wall and covered with the reaction mixture. Appropriate mixing of the resin with reagents is ensured by low-pressure nitrogen bubbling. Typical cycle times with this system are 15 min, including all washings, for each residue addition. The Liberty also allows programmable cleavage from the resin, either immediately after synthesis or at a later programmed time. Potential issues with racemization have been addressed via internal temperature control by a fiber-optic probe.

Recently, the Liberty1 was introduced as an entry-level system for automated single-channel peptide synthesis (Figure 3.16b). It accommodates the same beneficial processing as the multichannel synthesizer conveniently generating single peptides. Up to 20 amino acid reservoirs are provided, as well as 4 (1 additional optional) external bottle positions. Besides the standard 30 and 125 mL vessels, a 10 mL variation is optionally available for peptide synthesis in a 0.05–5 mmol scale by Fmoc strategies.

In addition to the automated peptide synthesizers, a manual microwave reactor, the Discover SPS system, is also available (Figure 3.16c). Reactions are performed at a 0.025–1 mmol scale in a 25 mL polypropylene vessel that can be considered as a modified SPE cartridge with a filtration device at the bottom. Leakage is prevented by a special sealing technology: a Teflon seal ball at the bottom stops reagents from dripping out of the vessel. For washing and filtration steps, a vacuum manifold station, consisting of one waste container and one for product collection after cleavage, is used in combination with the Discover SPS. As for the Liberty system, agitation can be performed via inert gas bubbling and internal temperature measurement via a fiber-optic sensor.

In Table 3.1, some of the most important features of the presented basic systems are summarized as the majority of all published microwave chemistry today is performed in one of these three systems (see Chapter 1). Each instrument provides a variety of different features and it is for the user to decide which instrument is appropriate for individual demands.

3.5 Multimode Instruments

The development of multimode reactors for organic synthesis occurred mainly from already available microwave acid digestion/solvent extraction systems. Instruments for this purpose were first designed in the 1980s and with the growing demand for synthesis systems, these reactors were subsequently adapted for organic synthesis applications. Today, a broad spectrum of instruments with a large range of different accessories for organic synthesis is available. In addition, the reactors are designed in such a way that a direct scale-up from small scales performed in single-mode platforms to larger scales in multimode instruments is possible without any change in already optimized reaction conditions.

3.5.1 Anton Paar GmbH

Two different types of multimode instruments for scale-out and scale-up purposes are available from Anton Paar. The oven-like Synthos 3000 features standard microwave technology with a large cubic cavity for parallel scale-up, whereas the recently introduced Masterwave BTR follows a new technology concept featuring a radial taper surrounding a compact microwave applicator hosting a single 1 L reaction vessel.

3.5.1.1 Synthos 3000

The Anton Paar Synthos 3000 microwave reaction platform (Figure 3.17, Table 3.2) is based on the related digestion instrument Multiwave 3000 and was originally dedicated for parallel scale-up synthesis in quantities of up to 1 L reaction volume per run. The instrument is designed for chemistry under high-pressure and high-temperature conditions. To extend the scope toward combinatorial and library synthesis under more moderate conditions and at smaller scale, appropriate accessories have also been introduced.

The use of two magnetrons (1400 W continuously delivered output power) allows mimicking of small-scale runs to produce large amounts of the desired compounds within a similar time frame. The homogeneous microwave field guarantees identical conditions at every position of the rotors, resulting in good reproducibility of experiments. Offering high operating limits (80 bar at 300 °C), the instrument facilitates the investigation of new reaction methods like near-critical water chemistry. The instrument can be operated with various rotor types ranging from

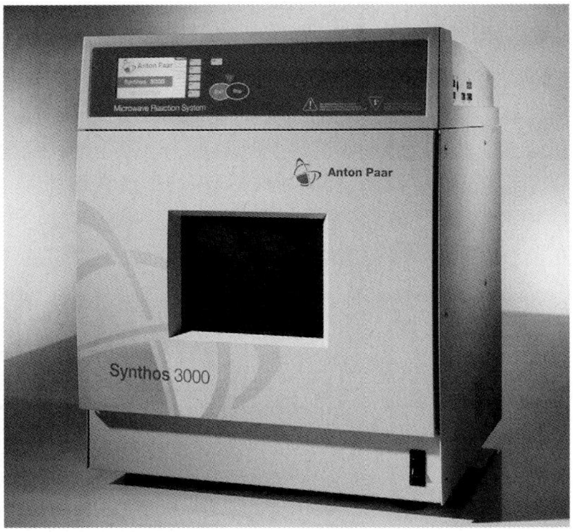

Figure 3.17 Anton Paar Synthos 3000.

an 8- to a 96-position rotor, equipped with a choice of vessel types for different pressure and temperature conditions. For a combinatorial approach, a rotor system that holds four 48-well microtiter plates made of silicon carbide is also available. A gas loading accessory allows creation of inert/reactive gas atmosphere or reactions in pre-pressurized vessels.

Temperature measurement is achieved by a remote IR sensor from the bottom on the outer surface of the vessels. The software-controlled operating limit of the IR sensor is 280 °C. For additional control, temperature measurement in one reference vessel of the 8-, 16-, or 48-position rotor by an immersed gas balloon thermometer is available. The operating limit of this temperature probe is 300 °C, suitable for reactions at extreme temperature and pressure conditions.

The pressure is measured by a hydraulic system, either in one reference vessel of the 16- and 48-vessel rotor or simultaneously for all positions of the 8-vessel rotor. The

Table 3.2 Anton Paar Synthos 3000 features.

Feature	Description
Cavity size (volume)	42 × 57 × 62 cm (66 L)
Installed power	1700 W (two magnetrons)
Max. output power	1400 W
Temperature control	Immersed gas balloon thermometer (max. 300 °C); outside IR remote sensor (max. 280 °C)
Pressure control	Hydraulic system (max. 80 bar)
Cooling system	190 $m^3 h^{-1}$ (forced airflow through rotor, four steps adjustable)
Magnetic stirring	0–600 rpm (four steps software adjustable)
External PC	Optional, not required as key panel + keyboard is standard equipment

operating limit is 80 bar, sufficient for most synthetic applications. In addition, a pressure rate limit is set to 2.0 bar/s by the included control software. Protection against sudden pressure peaks is achieved by metal safety disks, integrated in the vessel caps (safety limit 70 or 120 bar, respectively), and by software settings, depending on the used rotor and vessel type.

All parameters are transmitted wirelessly by IR data transfer from the sensors to the system control computer of the instrument to eliminate disturbing cables and hoses from inside the cavity.

Seven individual rotor types are available for the Synthos 3000 platform. The spectrum comprises a combichem rotor with sealable silicon carbide microtiter plates, high-throughput rotors employing various small glass vials in silicon carbide plates, and multivessel rotors for multigram-scale parallel reactions. Depending on the vessel or the pressure jacket material, different temperatures and pressures can be achieved (see Tables 3.3 and 3.4).

- **8-Vessel Rotor (8SXF100, 8SXQ80)** (Figure 3.18b, Table 3.3) This rotor system was specially designed for high-temperature and high-pressure reaction conditions. Two different vessel types are provided, PTFE–TFM liners or quartz vessels, which enable performing reactions at 260 °C and 60 bar for the former and at 300 °C and 80 bar for the latter. Optional for this rotor system is a gas loading system that allows reactions to be performed under inert or reactive gas atmosphere. Pre-pressurization up to 20 bar is possible.
- **16-Vessel Rotor (16MF100, 16HF100)** (Figure 3.18c, Table 3.3) This tool is dedicated for standard synthetic reactions up to 240 °C and offers 100 mL screw cap vessels with PTFE–TFM liners. Applying different pressure jackets allows continuous operation of these vessels at a maximum of 20 or 40 bar and temperatures up to 200 or 240 °C, respectively.
- **48-Vessel Rotor (48MF50)** (Figure 3.18a, Table 3.3) For library generation on the gram scale, this rotor with its 50 mL vessels can be used. The vessels are arranged in 3 circles of 16 vessels each, and temperature measurement is performed at the center circle via IR in addition to internal measurement in one reference vessel. For this rotor system, screw cap vessels with conical seals for proper sealing are employed. Operation limits are 200 °C and 20 bar.

Table 3.3 Synthos 3000: features of high-performance rotor systems.

	48MF50	16MF100	16HF100	8SXF100	8SXQ80
No. of vessels	48	16	16	8	8
Volume (ml)	50	100	100	100	80
Operating volume (ml)	6–25	6–60	6–60	6–60	6–60
Max. temperature (°C)	200	200	240	260	300
Max. pressure (bar)	20	20	40	60	80
Liner material	PFA	PTFE–TFM	PTFE–TFM	PTFE–TFM	Quartz
Pressure jacket	PEEK	PEEK	Ceramics	Ceramics	×
Pre-pressurizing	×	10 bar	10 bar	20 bar	20 bar

Table 3.4 Synthos 3000: features of high-throughput rotor systems.

	64MG5	4 × 24MG5	4 × 20MGC	4 × 48MC
No. of vessels/wells	64	96	80	192
Volume (mL)	5	5	1.5	0.2/0.4
Operating volume (mL)	0.3–3	0.3–3	0.1–1	0.02–0.15 0.1–0.3
Max. temperature (°C)	200	200	200	200
Max. pressure (bar)	20	20	8/20	20
Vessel material	Glass	Glass	Glass	SiC

- **64-Vial Rotor (64MG5)** (Figure 3.19, Table 3.4) Higher throughput on a smaller scale can be performed with this rotor type featuring 64 disposable 5 mL standard glass vials with screw cap sealing. The vials are arranged in 16 groups of 4 and allow reaction conditions up to 200 °C and 20 bar.
- **96-Vial Rotor (4 × 24MG5)** (Figure 3.19, Table 3.4) This high-throughput rotor is specially designed for parallel method optimization in microwave synthesis. The same glass vials as used with the 64-vessel rotor are here arranged in silicon carbide (SiC) blocks within a 6 × 4 matrix. Silicon carbide ensures maximum temperature homogeneity and, therefore, even allows the use of low-absorbing solvents (see Section 4.6). Reaction conditions are again defined to 200 °C and 20 bar.
- **80-Vial Rotor (4 × 20MGC)** (Figure 3.19, Table 3.4) For microwave-assisted derivatizations and proteomics reactions in parallel manner, this high-throughput rotor was developed. This rotor type accommodates standard 1.5 mL HPLC/GC vials (screw cap or crimp top applicable) in silicon carbide (SiC) blocks within a

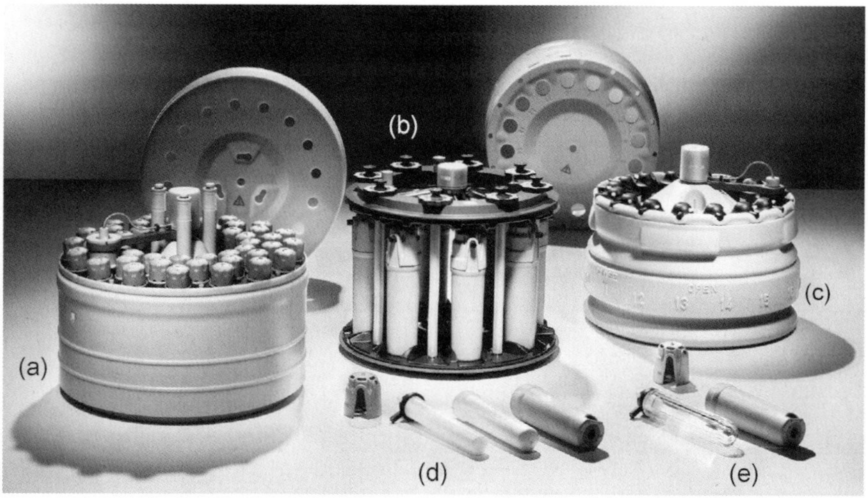

Figure 3.18 Anton Paar 48- (a), 8- (b), and 16-vessel rotors (c) utilizing PTFE–TFM liners with ceramic pressure jackets (d) and quartz vessels (e).

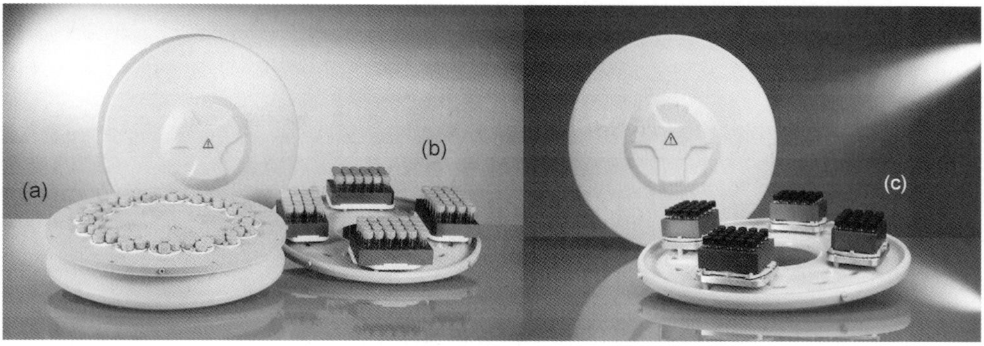

Figure 3.19 Anton Paar 64- (a) and 96-position rotors (b) and 80-position rotor for HPLC/GC vials (c).

5 × 4 matrix. Reaction conditions are defined to 200 °C and 8 bar (limited by the tightness of the HPLC/GC vials). By employing a sealing system with a supporting aluminum top plate, the accessible pressure range can be extended to 20 bar.

- **Combichem Rotor (4 × 48MC Well Plate)** (Figure 3.20, Table 3.4) Here, four SiC microtiter well plates with a standard 6 × 8 matrix are arranged on the rotor enabling to perform 192 reactions in parallel. The well plates are available in two variations, either with round-bottom wells (100–300 μL capacity per well) or with conical wells (20–150 μL capacity per well) for combinatorial approach in ultralow volumes. The plates are covered with a PFA foil and sealed with an aluminum top plate with corresponding bore holes, which allow sample withdrawal. Due to the advanced sealing mechanism, temperatures up to 200 °C and pressures up to 20 bar can be achieved.

3.5.1.2 **Masterwave Benchtop Reactor**

Launched in 2010, Masterwave BTR (Figure 3.21a, Table 3.5) is the most recent multimode reactor on the market. The instrument was specially designed for efficiently processing 1 L batches to simplify high-pressure microwave-assisted scale-up in industrial kilolab applications. The compact applicator hosts a single

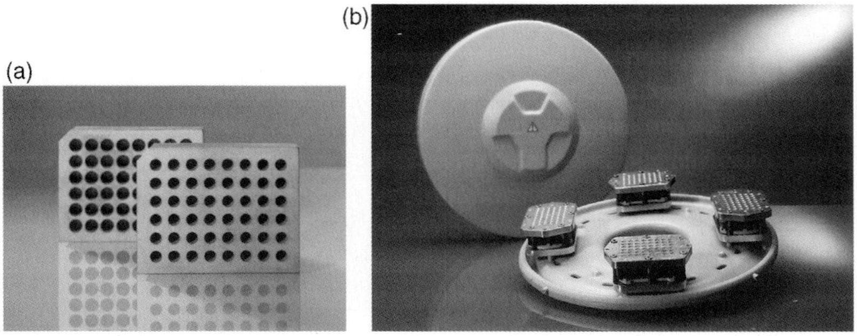

Figure 3.20 Anton Paar SiC 48-well microtiter plates (a) and rotor setup (b).

Figure 3.21 Anton Paar Masterwave BTR (a) and its 1 l PTFE reaction vessel (b).

1 L PTFE reaction vessel with bayonet-locked screw cap and integrated paddle stirrer (Figure 3.21b). The compact dimensions of the instrument (see Table 3.5) allow installation in any standard laboratory fume hood.

An operating volume of 250–750 mL at a maximum of 30 bar enables the preparation of approximately 200 g product within one run. With a typical cycle time of 30–45 min, the reactor could provide a daily productivity on the kilogram scale. Homogeneous heating is ensured by a novel radial design of the waveguide and the applicator itself in combination with thorough agitation, which is monitored and optimized by the software. Utilizing the identical touch screen user interface as with the Monowave 300, a seamless transfer of the optimized reaction conditions to larger scale is provided. An optional remote control tool (VNC open source) allows reaction monitoring and user interaction from the office desk.

The maximum output power of the Masterwave BTR is 1700 W delivered from two stacked magnetrons. Due to the high field density provided in the entire applicator, significant heating rates can be achieved to reach the maximum temperature of

Table 3.5 Masterwave Benchtop Reactor: general features.

Feature	Description
Size	50 × 72.6 × 48 cm (W × D × H)
Delivered power	1700 W
Max. output power	1700 W
Temperature control	Rising Pt100 sensor (max. 250 °C)
Pressure control	Hydraulic sensor (max. 30 bar)
Cooling system	Integrated closed cooling circuit with microwave-transparent fluid
Agitation	Magnetic-driven paddle-stirrer (0–700 rpm)
Reaction control	Integrated touch screen user interface

250 °C for 750 mL reaction volume in an acceptable time frame. The temperature is controlled via a rising Pt100 temperature sensor. An inverted nozzle at the bottom of the reaction vessel allows direct positioning of the vessel onto the sensor to measure directly inside the reaction mixture and giving immediate feedback of the current temperature. The sliding cover of the applicator comprises a hydraulic pressure sensor to monitor the reaction pressure buildup.

Cooling is achieved by an internal closed cooling circuit utilizing a microwave-transparent cooling fluid. The fluid is permanently circulating around the vessel withdrawing the heat effectively by the aid of a heat exchanger fan in the rear of the instrument.

As protection against overpressure or thermal runaways during the run, the vessel cap is equipped with a metal rupture disk. Depending on the reaction temperature, the rupture disk will open at 46–54 bar providing sufficient overpressure tolerance during the entire operation range. In case of a venting action, the vessel content is emptied via a standard-sized (Ø 10 mm) stainless steel tubing into a corresponding expansion tank.

3.5.2
Biotage AB

For scale-up applications, Biotage offers the Advancer batch reactor (see Figure 3.22), serving a multimode cavity for operations with one 350 mL Teflon reaction vessel at high-pressure conditions. An operating volume of 50–300 mL at a maximum pressure of 20 bar enables the production of 10–100 g product within one run. Homogeneous heating is ensured by a precise field tuning mechanism and vigorous overhead stirring (up to 1000 rpm) of the reaction mixture. Direct scalability allows

(a) (b)

Figure 3.22 Biotage Advancer scale-up instrument (a) with automated/tilting cavity lid (Advancer Kilobatch) (b).

translation of the optimized reaction conditions from the Initiator/Emrys systems (see Section 3.4.1) to a larger scale.

The maximum output power of the Emrys Advancer is 1200 W, generating a heating rate of 0.5–4 °C s^{-1} to reach the maximum temperature of 250 °C for 300 mL reaction volume in comparable times to the monomode experiments. The temperature is controlled via an internal fiber-optic probe. Several connection ports in the chamber head enable adding of reagents during irradiation, sample removal for analysis, *in situ* monitoring by real-time spectroscopy, or creation of inert/reactant gas atmosphere. Cooling is achieved by an effective gas expansion mechanism (adiabatic flash cooling) to ensure drastically shortened cooling periods (200 mL EtOH within 30 s from 180 to 65 °C).

In 2008, the Advancer Kilobatch was launched. Based on the standard Advancer, this microwave reactor provides a kilogram scale-up of both homogeneous and heterogeneous reactions in a sequential batch format. With the liquid loading device, liquid components are delivered unattended and homogeneous reaction mixtures can be processed up to 12 h. In addition, by using the automated solid loading carrousel, heterogeneous reactions can be performed in four sequential cycles of 250 mL each to obtain a 1 L batch. For an appropriate mixing, a mechanical overhead stirrer is featured (see Figure 3.21b).

Due to the dimensions of the instrument (140 × 65 × 185 cm) extra lab space is required to make operations comfortable.

3.5.3
CEM Corporation

The MARS S microwave synthesis system (Figure 3.23, Table 3.6) is based on the related MARS 5 digestion instrument and offers different sets of rotor systems with

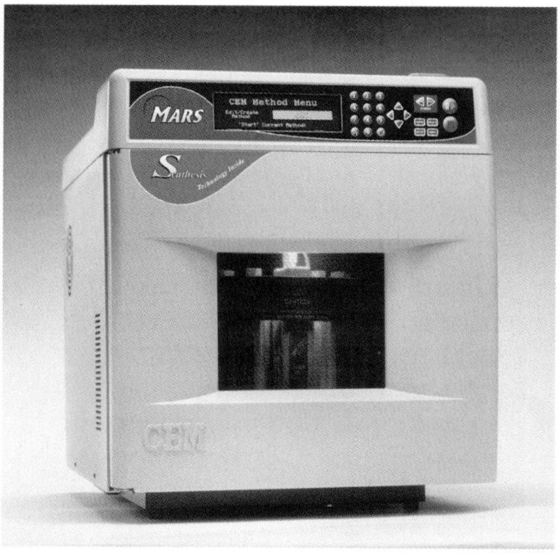

Figure 3.23 CEM MARS S synthesis system.

Table 3.6 MARS S synthesis system: general features.

Feature	Description
Cavity size (volume)	48 L
Delivered power	1600 W
Max. output power	1600 W
Temperature control	Outside dual IR remote sensor
	Immersed fiber-optic probe (optional)
Pressure control	Pneumatic pressure sensor (optional)
Cooling system	Airflow through cavity 210 $m^3 h^{-1}$
Stirring	Magnetic stirring at variable speed overhead stirring (optional for open vessel)
External PC	Optional, not required as integrated key panel is standard equipment

several vessel designs and sizes for various synthesis applications under open- and closed-vessel conditions.

Temperature measurement in the rotor systems is conducted by an immersed fiber-optic probe in one reference vessel or by two IR sensors on the surface of the vessels from the bottom of the cavity. Pressure measurement in HP and XP rotors is achieved by an electronic sensor in one reference vessel. Correct temperature and pressure measurement via the sensors is ensured up to 300 °C and 100 bar. The simultaneous use of the fiber-optic probe and dual IR sensors provides a temperature measurement in all vessels of the turntable, for example, for the MARSXpress option, the dual IR sensors allow temperature measurement in up to 40 vessels simultaneously. For reactions at high pressures, the HP and XP rotor vessels offer a choice of seals and covers, fully sealed or self-venting. The temperature and pressure feedback control of the MARS system monitors and regulates the amount of power being applied to the reactions to provide optimum reaction control. The system will automatically shut the microwave power down if the temperature in the control vessel rises too high or if the vessel starts to overpressurize. In addition, a built-in pressure limit control for all vessels is offered where the magnetron is turned off when the sensor detects excessive venting in the cavity. For the MARSXpress and the GlassChem rotors, no internal pressure measurement is available, as the vessels are "self-regulating" to prevent overpressure. The MARSXpress maintains reaction control via the self-regulating pressure vessels and the temperature feedback control. The temperature control sensor monitors the temperature in each vessel and adjusts the power output accordingly to maintain the user-defined temperature set point. The MARSXpress offers real-time display of the temperature in each vessel. All of CEMs high-pressure vessels have an open-architecture design that allows airflow within the cavity to cool the vessels quickly.

The general maximum output power of the instrument is 1600 W, but the MARS control panel offers two additional low-energy levels with unpulsed microwave output power of 400 and 800 W, respectively. This feature avoids overheating of the reaction mixture and the unit itself when small amounts of reagents are used.

The MARS comes with a software package, operated from the integrated spill proof keypad. The instrument can be connected to an external PC, but this is not required for most common operations. Methods and reaction protocols can be designed as temperature/time profiles or with precise control of constant power during the reaction.

3.5.3.1 **MARS Scale-Up System Accessories**

For reactions at atmospheric pressure, standard laboratory glassware such as round-bottom flasks from 250 mL up to 5 L can be used (see Figure 3.24a). An inlet/outlet port on the top of the cavity allows connection of a reflux condenser or distillation equipment as well as addition of reagents, sample withdrawal, or overhead stirring.

In addition, reactions can be performed in continuous flow mode with appropriate 2 or 4 L flow cells. For this system, the port on the top of the cavity provides the entrance and exit for the inlet and outlet lines. For both described systems, fiber-optic temperature control is available. In addition to the open-vessel systems, a 300 mL closed vessel for performing larger pressurized reactions up to a volume of 150 mL is also available (Figure 3.24b). Operating limits for this vessel are 250 °C and 35 bar.

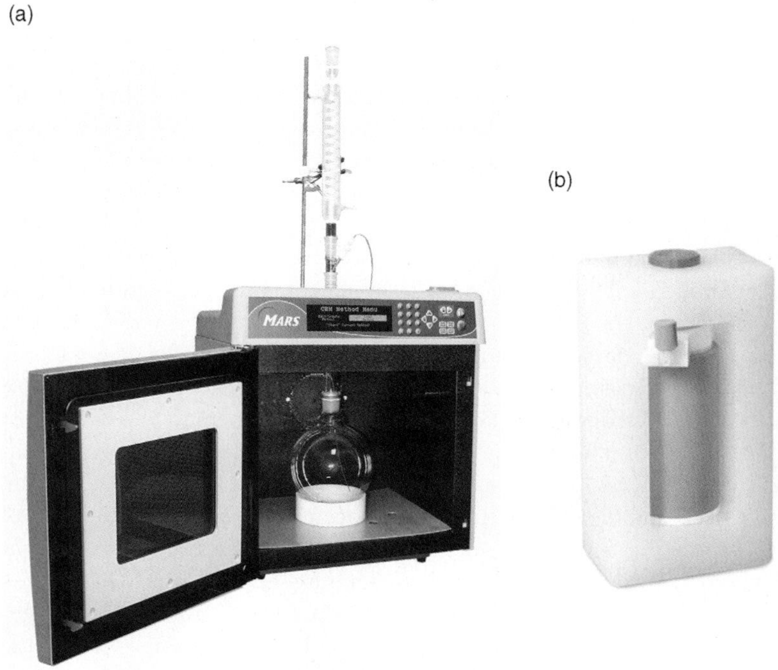

Figure 3.24 CEM MARS scale-up systems: (a) Open-vessel round-bottom flask with reflux condenser, (b) High-pressure 300 mL vessel.

3.5.3.2 MARS Parallel System Accessories

Four individual rotor systems with different reaction vessels for performing reactions under closed-vessel and moderate- to high-pressure conditions are available. Inlet and outlet ports on the side of the cavity allow the introduction of an inert atmosphere into the reaction vial or addition and withdrawal of reagents.

For open-vessel high-throughput parallel synthesis, a rotor system that holds microtiter plates is also available.

- **Microplate Rotor** (Figure 3.25a) Up to three 96-well microtiter plates can be used in combination with a turntable for a combinatorial approach under open-vessel conditions at up to 150 °C. A Teflon stand holds the fiber-optic temperature probe that is immersed into one reference well.
- **MARSXpress Rotor** (Figure 3.25b, Table 3.7) In this 40-position rotor system, reaction volumes of up to 2 L per run can be accommodated. The vessels consist of a liner that holds the reaction mixture, a sealing plug, and a screw cap and are available in four different sizes (10, 25, 55, and 75 mL). Temperatures up to 260 °C for PFA vessels and up to 300 °C for vessels made of TFM can be reached. Here, an all-vessel temperature control via IR monitoring is provided.
- **GlassChem Rotor** (Table 3.7) This rotor system similar to the Xpress enables reactions in up to 24 glass vessels at conditions up to 200 °C and 14 bar. The 20 mL vessels use the same screw cap design as the MARSXpress. Temperature measurement is provided via a fiber-optic probe.
- **XP-1500 Plus Rotor** (Figure 3.26a, Table 3.7) With this 12-position rotor, reactions at high pressures and temperatures (100 bar and 300 °C) can be performed in Teflon, Pyrex, or quartz vessels. Temperature and pressure measurement is possible via immersed sensors in one reference vessel. Dual IR sensors in conjunction with a fiber-optic probe provide an all-vessel temperature control, the so-called DuoTemp.
- **HP-500 Plus Rotor** (Figure 3.26b, Table 3.7) In this rotor type, 14 Teflon, Pyrex, or quartz vessels with operating limits of 33 bar and 260 °C are accommodated. Temperature and pressure measurement is identical to the XP-1500 Plus system.

(a) (b)

Figure 3.25 (a) CEM Microplate rotor, (b) MARSXpress rotor.

Table 3.7 MARS-S: features of parallel rotors.

	GlassChem	MARSXpress		XP-1500+	HP-500+
No. of vessels	24	40	40	12	14
Vessel volume (mL)	20	55	10–75	100	100
Operating volume (mL)	3–14	6–35	1–50	10–70	10–70
Max. temperature (°C)	200	300	260	300	260
Max. pressure (bar)	14	35	35	100	34
Vessel material	Glass	TFM	PFA	Teflon, Pyrex, quartz	
Temperature control	Fiber-optic	IR	IR	Fiber-optic + optional IR DuoTemp	

3.5.4 Milestone s.r.l

Milestone offers a large repertoire of multimode instruments with appropriate accessories to perform reactions at volumes up to about 3.5 L under closed-vessel conditions. For open-vessel conditions, round-bottom flasks up to a size of 4 L can be utilized. For both applications, several rotor systems as well as single-vessel modules or microtiter plates are available. A continuous flow system is also provided when a larger scale-up is considered.

3.5.4.1 MultiSYNTH System

The MultiSYNTH instrument, a so-called hybrid instrument, was introduced in 2006 (see Figure 3.27, Table 3.8). It has the unique feature of being able to merge both single-mode and multimode technologies in a single unit. The benefits of single-mode reactor features like fast heating, full control of single vessels, and fast cooling are combined with the main advantage of multimode technologies of performing reactions in parallel using rotor systems. In the single-mode setup, one vessel is

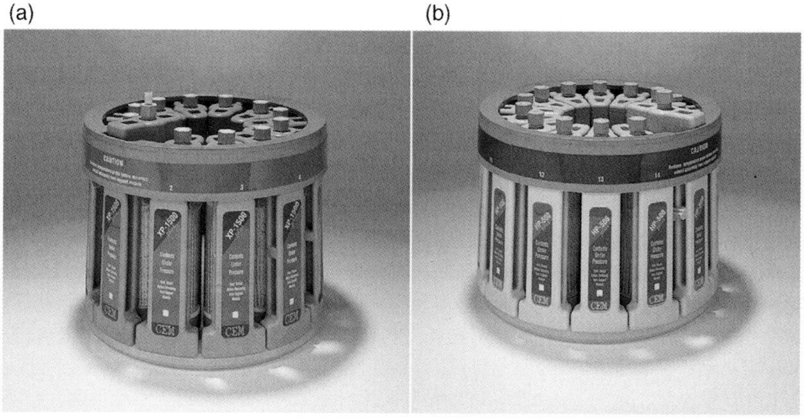

Figure 3.26 CEM MARS high-pressure rotors: (a) XP-1500 Plus, (b) HP-500 Plus.

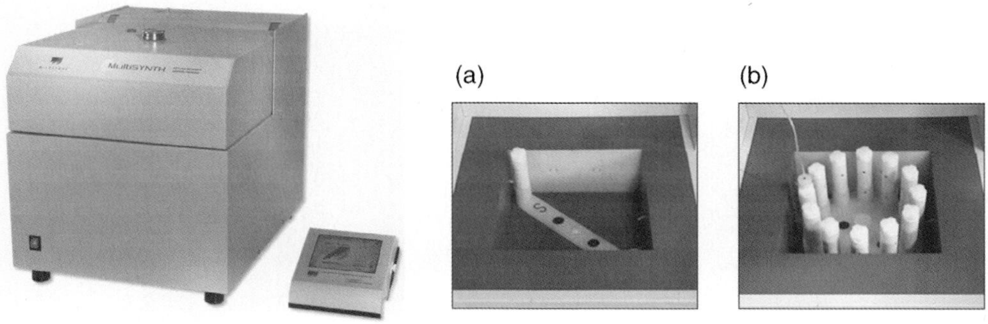

Figure 3.27 Milestone MultiSYNTH with single-mode (a) and multimode (b) setup.

located at a defined position where the microwave energy intensity is the highest; whereas in the multimode setup, a classical rotor system is employed (see Figure 3.27).

A single magnetron with homogeneous microwave distribution in the cavity delivers 800 W output power. The full 800 W can be employed for the multimode configuration only, while for the single-mode setup, 400 W is the maximum power. In both cases, the user can select between continuous and pulsed delivery of the microwave power.

Temperature control is performed by internal measurement via a fiber-optic probe in one reference vessel plus measurement by an IR sensor that is located on the sidewall at a defined height that allows temperature recording for all vials. Both sensors are interfaced with a microprocessor-controlled rotor-positioning system that permits vessel recognition and thus profile tracking for all vessels. The MultiSYNTH features an indirect pressure control through a spring-type valve preloaded at 20 bar. This special valve is included in the TFM safety shield in which the standard glass vials are inserted and allows safe pressure release in case of overpressurization, with subsequent resealing of the vessel. An additional built-in sensor is available where the magnetron reduces the power when a set vapor concentration in the cavity is

Table 3.8 MultiSYNTH System – general features.

Feature	Description
Cavity size (volume)	26.5 × 24.5 × 20 cm (13 L)
Delivered power	800 W
Max. output power	800 W (multimode)
	400 W (single-mode)
Temperature control	Immersed fiber-optic probe (max. 250 °C)
	Outside IR remote sensor (max. 300 °C)
Pressure control	Indirect pressure control
Cooling system	Compressed air
Agitation	Magnetic stirring (0–400 rpm)
	Shaking (0–100%)
External PC	External touch screen terminal

(a) (b)

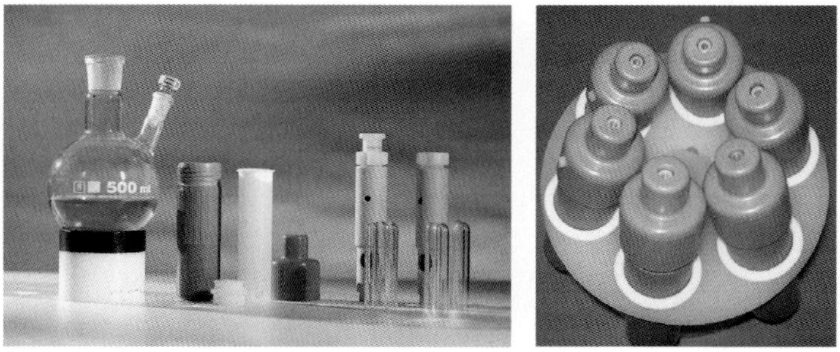

Figure 3.28 Milestone MultiSYNTH accessories: (a) Teflon/Weflon stand for round-bottom flasks: 70, 2.5, and 10 mL vessels with corresponding safety shields, (b) 6-position rotor.

exceeded. In addition to the magnetic stirring, a vessel "vibration" system is offered where the vessel is mechanically shaken to prevent hot and cold spot formation, thus ensuring reaction and temperature homogeneity. In the multimode format, the MultiSYNTH labstation oscillates/rotates each vial to achieve a similar mixing effect. Effective cooling is achieved via compressed air.

For the MultiSYNTH system, reaction monitoring is achieved via an external control terminal with a touch screen display utilizing the EasyCONTROL software package. The system can be operated either with a temperature/time mode where the microwave power is automatically adjusted to maintain the set temperature or with a power/time mode where a fixed power value is set.

For this system, two different vial types in three different sizes are available permitting a scale-up from 0.25 up to 300 mL under closed-vessel conditions (see Figure 3.28). For the single-mode format, 2.5 and 10 mL glass vials are provided with an operating volume of 0.25–1.2 mL or 1–7 mL, respectively, and operation limits of 250 °C and 20 bar. The same vials can be used for a 12-position rotor system in the multimode setup (see Figure 3.27b). In addition, a 6-position rotor for 70 mL vessels (Figure 3.28) made of TFM with a PEEK protecting shield and limits of 250 °C and 30 bar is offered. Furthermore, standard round-bottom flasks up to 1 L can be positioned in the cavity and refluxed up to 250 °C under multimode conditions.

Due to the unique feature of the MultiSYNTH to provide single and parallel techniques in a single instrument unit, the user is able to carry out optimization studies via the single-mode setup and directly scale up the reactions by employing one of the rotor systems in the multimode format.

3.5.4.2 MicroSYNTH Labstation

The MicroSYNTH multimode instrument (also known as ETHOS series, Figure 3.29, Table 3.9) is available with a broad range of accessories. Tools are offered starting from combichem via medchem to parallel synthesis and large-scale synthesis in batch and continuous flow mode. Two magnetrons deliver up to 1000 W microwave output power and a patented pyramid-shaped diffuser ensures homogeneous microwave distribution within the cavity.

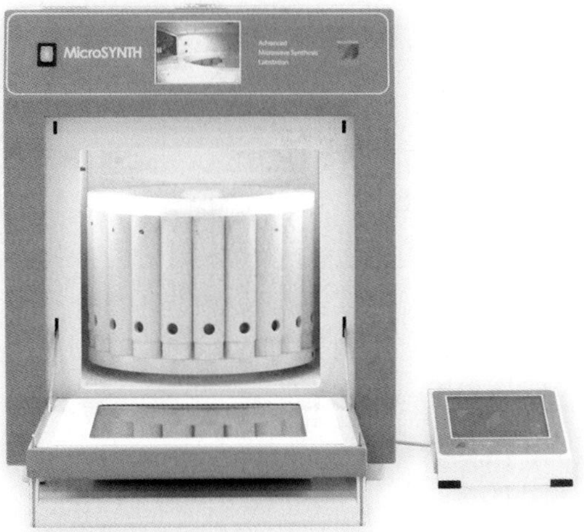

Figure 3.29 Milestone MicroSYNTH labstation with SafeVIEW system.

In case of parallel synthesis in multivessel rotors, temperature measurement is achieved by using a fiber-optic probe, immersed in one reference vessel. Also available is an IR sensor for monitoring the outside surface temperature of each vessel, mounted on the sidewall of the cavity, about 5 cm above the bottom. The reaction pressure is measured by a pneumatic sensor connected to the reference vessel. An additional sensor monitors the vapor concentration in the microwave cavity and thus controls all vessels simultaneously. This technology switches off the applied microwave power until the vapors have been cleared from the cavity by the exhaust module.

Postreaction cooling of the reaction mixture is achieved by a constant airflow through the cavity and a stream of compressed air (see Table 3.9). Due to the design of the thick polymer/composite segments, the cooling of the high-pressure rotors is not very efficient, external cooling by immersing the rotor in a water bath is recommended. Thus, vessels under pressure have to be handled outside the cavity, but a special cooling rack is available for this purpose.

Table 3.9 MicroSYNTH platform: general features.

Feature	Description
Cavity size (volume)	35 × 35 × 35 cm (43 L)
Delivered power	1600 W (two magnetrons)
Max. output power	1000 W
Temperature control	Immersed fiber-optic probe (max. 250 °C) Outside IR remote sensor (max. 300 °C)
Pressure control	Pneumatic pressure sensor (max. 55 bar) Indirect pressure control
Cooling system	Built-in exhaust airflow 1.8 m^3 min^{-1}, compressed air
Magnetic stirring	0–400 rpm
External PC	External touch screen terminal

For all MicroSYNTH systems, reaction monitoring is achieved via an external control terminal with a touch screen display utilizing the EasyCONTROL software package. The runs can be controlled either by temperature, pressure, or microwave power output. The software enables online modifications of any method parameter and the reaction process is monitored by an appropriate graphical interface. An included solvent library and electronic lab journal feature simplifies the experimental documentation.

Optionally, the newest generation of the MicroSYNTH labstation can be equipped with the SafeVIEW system (Figure 3.29). This add-on comprises a high-definition digital camera combined with a 5.6″ TFT–LCD module and allows to visually monitor the course of the reaction inside the cavity. In addition, a video of the entire run can be recorded and saved as MPEG file on a Windows-compatible memory card.

The MicroSYNTH platform offers a broad range of different rotor and vessel systems, enabling reactions from 3 mL up to 4 L under open- and sealed-vessel conditions in batch and parallel manner up to 55 bar. The spectrum includes single reaction vessels operational at different pressures and rotor systems to large volume batch reactors.

- **SV Module** (Figure 3.30a) This tool is designed for single optimization runs at elevated pressure on a comparatively small scale. It includes a 45 mL quartz vessel QV50, a TFM cover, and a safety shield with built-in safety valve. This module is suitable for volumes from 3 to 30 mL at operating limits of 250 °C and 40 bar. The airflow device allows rapid cooling of the reaction mixture using a stream of compressed air. Temperature measurement is conducted via both an internal fiber-optic probe and a surface IR sensor. A special cover that allows working under inert atmosphere or pre-pressurization of the system with a reactive gas is also available.
- **PRO-16/24 Rotor** (Figure 3.30b, Table 3.10) This is a rotor for high-throughput purposes at elevated conditions utilizing 16 or 24 screw cap reaction containers. Each 75 mL PTFE–TFM vessel offers up to 50 mL working volume at 250 °C up to

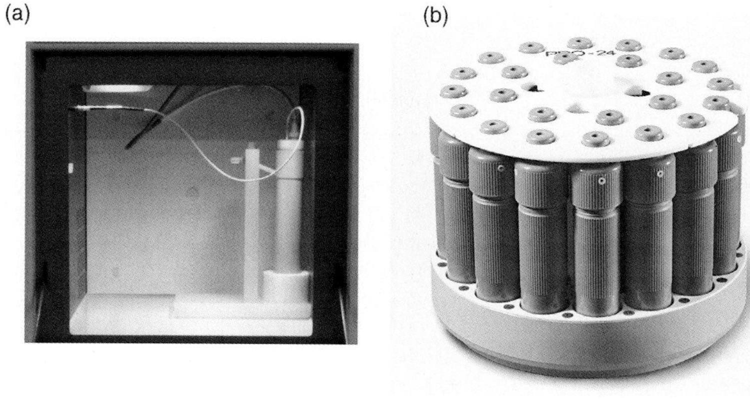

Figure 3.30 (a) Milestone SV module, (b) Rotor PRO 16/24.

Table 3.10 MicroSYNTH: features of parallel rotors.

	PRO-16/24	Q20	High Pressure	Large Volume
No. of vessels	16/24	20	10	6
Vessel volume (mL)	75	45	100	270
Operating volume (mL)	10–50	3–30	10–60	50–180
Vessel material	TFM	Quartz	TFM	TFM
Safety shield material	PEEK	PEEK	PEEK	PEEK
Max. pressure (bar)	30	40	55	10
Max. temperature (°C)	250	250	250	250
Internal T control	Fiber-optic in one reference vessel			
External T control	IR for all vessels			

30 bar. The special pressure release valve in the vessels vents in case of overpressure and reseals automatically.

- **High-Pressure Rotor** (Figure 3.31a, Table 3.10) This segmented 10-position rotor accommodates 100 mL TFM pressure reactors, which are suitable for reactions up to 250 °C at 55 bar.
- **Large-Volume Rotor** (Figure 3.31b, Table 3.10) Dedicated mainly for parallel scale-up at relatively low pressures, this rotor comes with six segments of a 270 mL TFM vessel. Thus, volumes up to 1080 mL can be processed in a single run up to 250 °C and 10 bar.
- **Q20 Rotor** (Figure 3.32, Table 3.10) This rotor type accommodates 20 QV50 vessels in parallel. Volumes up to 600 mL can be processed at conditions up to 250 °C and 40 bar.
- **Scale-Up at Normal Pressure** (Figure 3.32) Standard laboratory glassware such as round-bottom flasks from 50 mL to 4 L can be used for reactions at atmospheric pressure. A protective mount in the ceiling of the cavity enables connection of standard reflux condensers or distillation equipment. Additional mounts on the sidewall allow sample withdrawal or flushing with gas to create inert atmospheres.

(a) (b)

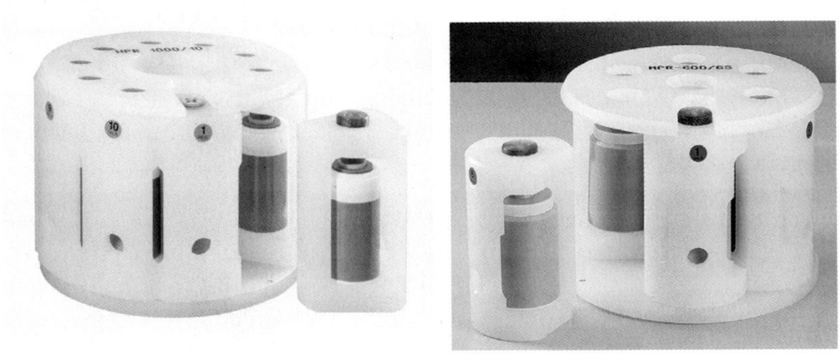

Figure 3.31 Milestone rotor systems: (a) High-pressure rotor, (b) Large-volume rotor.

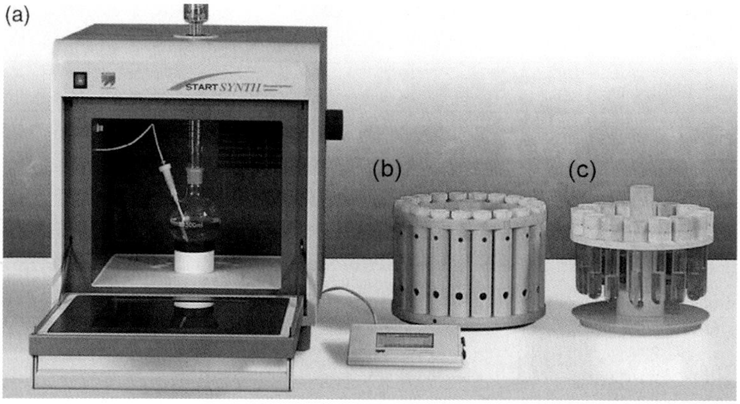

Figure 3.32 (a) Milestone StartSYNTH with atmospheric pressure setup, (b) Rotor Q-20, (c) Teaching Lab Kit.

3.5.4.3 **StartSYNTH**

This microwave platform (see Figure 3.32) is specifically designed for the academic lab with proper safety features that incorporate all the safety elements of the MicroSYNTH system. Here, a single magnetron system with a rotating diffuser provides a homogeneous microwave distribution in the cavity and delivers an output power up to 1200 W. Magnetic stirring of the reaction mixtures further ensures homogeneous temperature distribution. Temperature and pressure control is achieved via the same sensor systems as described above for the MicroSYNTH (direct temperature and pressure control in one reference vessel and contactless for all vessels). Reaction monitoring is achieved via an external control terminal with a touch screen display utilizing the EasyCONTROL software package. The runs can be controlled either by temperature or by microwave power.

- **Teaching Lab Kit** (Figure 3.32) This is a basic rotor for standard organic reactions allowing an initial approach toward microwave-mediated chemistry in teaching laboratories. It is designed for 32 × 25 mL glass vessels and for temperatures up to 200 °C. Reactions can either be performed at ambient pressure, employing a special valve that uses a reflux-style pressure venting mechanism, or at pressures up to 1.5 bar, using weighted valves that work with a gravity-based venting mechanism that vents when overpressure occurs and reseals afterward.

  Additionally available is the same SV module for a single QV50 vessel that is offered for the MicroSYNTH with the required holder, cover, stir bars, and Weflon buttons.

For reaction scale-up under normal pressure open-vessel conditions, standard laboratory glassware like round-bottom flasks from 50 mL up to 4 L can be employed. The same assembly as for the MicroSYNTH with a reflux condenser attached through the opening in the ceiling of the cavity is also utilized for the StartSYNTH. Optionally, the standard rotors Q-20 and high-pressure rotors can also be applied for parallel reactions at elevated conditions.

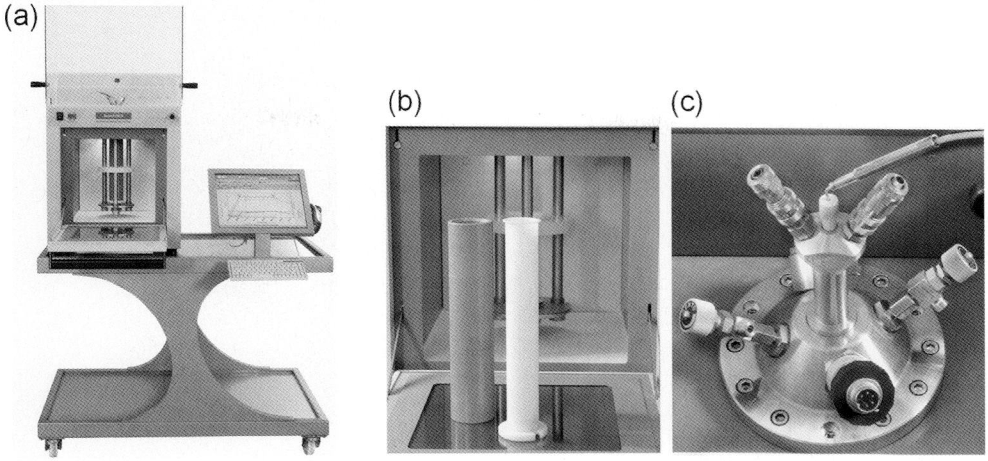

Figure 3.33 Milestone scale-up reactor: (a) BatchSYNTH, (b) Liner and safety shield, (c) Stainless steel flange with sensors and inlet/outlet ports.

3.5.4.4 **Scale-Up Systems**

In many industrial laboratories, the scale-up of microwave-assisted reactions from the gram to the kilogram region is of specific interest. For this purpose, Milestone offers specific batch and continuous flow reactors that are based on the MicroSYNTH^{plus} unit that incorporates all the features of the MicroSYNTH reactor. Parallel, batch, and continuous flow reactions can thus be performed using a single microwave system.

- **BatchSYNTH** (Figure 3.33) To address the needs for scaling up reactions in a batch format under high temperatures and pressures, the BatchSYNTH was developed. It is equipped with a vertically mounted 300 mLTFM reactor (working volume 250 mL) that is inserted into a high-temperature compound (HTC) safety shield and operates up to 230 °C and 30 bar. The temperature is measured internally via a K-type thermocouple and pressure monitoring is also carried out via an internal probe. Efficient cooling is provided by an internal cooling finger. Inlet and outlet ports for creating an inert atmosphere inside the vessel, reagent addition, and reaction mixture sampling are also included (see Figure 3.33).
- **FlowSYNTH** (Figure 3.34) FlowSYNTH is a continuous flow system where gram to kilogram scale-up is accomplished. The reagents are pumped through the microwave field from the bottom of the TFM reactor (200 mL) to the top, at maximum operating conditions of 200 °C and 30 bar. The reaction mixture is then cooled by flowing through a water-cooled heat exchanger that is also located on the mobile platform. Temperature and pressure control during the entire course of the process is achieved by built-in sensors. Temperature and reaction homogeneity along the entire length are ensured by a paddle-stirrer homogenizer. Flow rates from 12 to 100 mL min^{-1}, reaction times from 120 s to 16 min, and a throughput from 0.8 to 6 L h^{-1} can be achieved with this system.

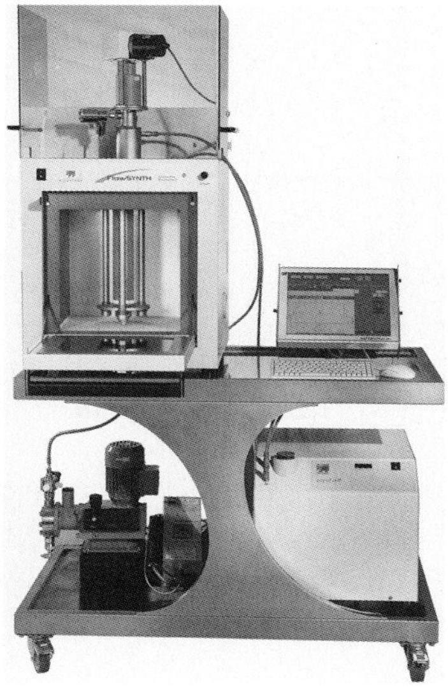

Figure 3.34 Milestone scale-up reactor FlowSYNTH.

- **RotoSYNTH** (Figure 3.35) This rotative solid-phase microwave reactor system was initially designed to perform reactions under solventless conditions, for example, where reagents are adsorbed on solid supports like silica or alumina. A glass vessel that is fitted in a tilted position in the microwave cavity rotates to obtain proper mixing of the reactants, thus ensuring a homogeneous temperature distribution, even for larger amounts of solids. Furthermore, the glass vessels

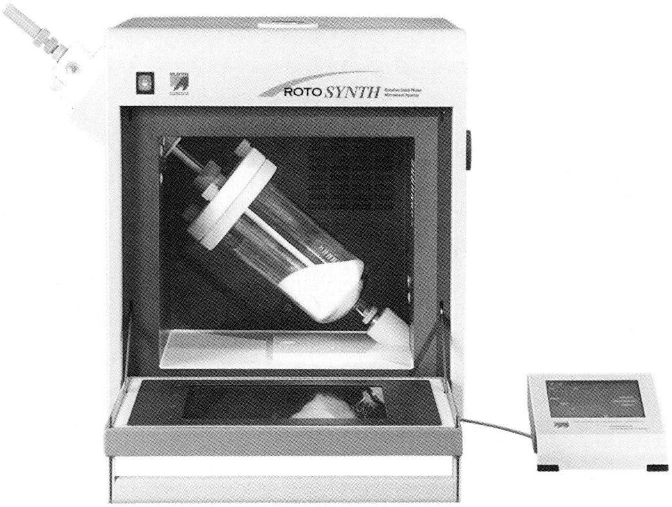

Figure 3.35 Milestone RotoSYNTH: rotative solid-phase microwave reactor.

have special paddles built in for more effective mixing of reaction mixtures. Vessels in different sizes are available, from 300 mL up to 4 L with working volumes of 150 mL up to 2 L. However, the RotoSYNTH can also be used for liquid-phase reactions. A special application in this case would be product or by-product distillation out of the reaction mixture that can be realized by the attachment of a vacuum pump to an opening on the outside of the microwave unit on the left-hand side (see Section 4.3).

A single magnetron delivers up to 1200 W microwave output power ensuring a homogeneous microwave distribution in the cavity via a rotating diffuser. Temperature measurement can be conducted either by an IR sensor located in the right-hand wall of the microwave unit controlling the temperature at the bottom of the vessel or by a fiber-optic probe inserted directly into the reaction mixture. Reactions can be performed at temperatures up to 250 °C at atmospheric pressure or under dynamic vacuum. The instrument is controlled with a touch screen terminal running the EasyCONTROL software.

Further application versatility can be achieved by using the complete series of single vessels available for the MicroSYNTH unit.

3.5.4.5 **Microwave-Heated Autoclave Systems**

An addition to the spectrum of available reactors are Milestone's microwave autoclave systems. The UltraCLAVE serves large-scale batch processing up to 3.5 L as well as parallel reactions, whereas the recently introduced UltraWAVE represents a desktop version of this technology with a 1 L cavity.

- **UltraCLAVE** (Figure 3.36a) The UltraCLAVE batch reactor is a large-scale microwave autoclave initially designed for high-temperature/high-pressure digestions, but the technology can also be applied toward synthesis applications.

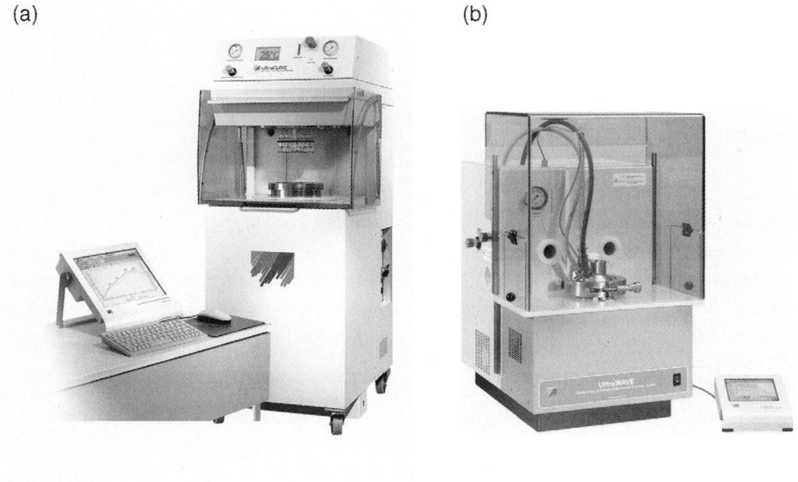

Figure 3.36 Milestone microwave-heated autoclaves: (a) UltraCLAVE, (b) UltraWAVE.

Continuous unpulsed microwave energy (0–1000 W) from the system's 1200 W magnetron is introduced into the vessel through a special microwave-transparent port. The internal geometry of the vessel is optimized for direct microwave coupling with zero reflectance, ensuring maximum sample heating efficiency. A unique feature of this system is the pressurization of the reaction chamber with nitrogen prior to heating so that boiling of the reaction mixture is prevented. In this way, reactions can be conducted at temperatures up to 300 °C and pressures up to 200 bar. The 4.2 L chamber of the UltraCLAVE allows processing of either multiple reactions by employing rotors with as many as 77 positions or a single large reaction mixture of 3.5 L utilizing the installed PTFE vessel. For the rotor systems, standard glass, quartz, or PTFE vessels can be employed. All functions (raising/lowering the reaction chamber, inert gas pressurization, venting, etc.) are performed automatically under computer control. Reaction monitoring and programming is achieved via a touch screen control terminal and the EasyCLAVE software. Due to the dimensions of the instrument (64 × 100 × 164 cm), extra lab space is required for appropriate operation.

- **UltraWAVE** (Figure 3.36b) The UltraWAVE single reaction chamber (SRC) reactor is the miniaturized version of the UltraCLAVE fitting on any bench in a standard laboratory fume hood. Continuous unpulsed microwave energy (0–1500 W) from the system's 1500 W magnetron is introduced into the vessel through a special microwave-transparent port. Pressurization of the reaction chamber prior to heating with up to 40 bar nitrogen allows performing reactions at up to 300 °C and pressures up to 199 bar. The 1 L stainless steel chamber of the UltraWAVE can be used for single batch processing in the installed 1 L PTFE vessel or for parallel processing employing 15 glass vessels with 10 mL operation volume. Cooling is achieved by an external water circuit for acceptable process cycles. Similar to the UltraCLAVE reaction, monitoring and programming is realized via an external touch screen control terminal and the EasyCONTROL software.

References

1 Decareau, R.V. and Peterson, R.A. (1986) *Microwave Processing and Engineering*, Wiley-VCH Verlag GmbH, Weinheim.

2 Metaxas, A.C. and Meredith, R.J. (1983) *Industrial Microwave Heating*, Peter Peregrinus Ltd., London, UK.

3 Balanis, C.A. (1989) *Advanced Engineering Electromagnetics*, John Wiley & Sons, Inc., New York.

4 Ondruschka, B., Bonrath, W., and Stuerga, D. (2006) Chapter 2, in *Microwaves in Organic Synthesis*, 2nd edn (ed. A. Loupy), Wiley-VCH Verlag GmbH, Weinheim, pp. 62–103.

5 Bogdal, D. (2005) Chapter 3, in *Microwave-Assisted Organic Synthesis: One Hundred Reaction Procedures*, Elsevier, Oxford, UK, pp. 23–32.

6 Moseley, J.D. (2011) *Microwave Heating as a Tool for Sustainable Chemistry* (ed. N.E. Leadbeater), CRC Press, Taylor & Francis Group, Boca Raton, FL, pp. 105–147.

7 Gedye, F.R., Smith, F., Westaway, K., Ali, H., Baldisera, L., Laberge, L., and

Rousell, J. (1986) *Tetrahedron Letters*, **27**, 279–282.

8 Giguere, R.J., Bray, T.L., Duncan, S.M., and Majetich, G. (1986) *Tetrahedron Letters*, **27**, 4945–4948.

9 Nüchter, M., Ondruschka, B., Bonrath, W., and Gum, A. (2004) *Green Chemistry*, **6**, 128–141.

10 Obermayer, D. and Kappe, C.O. (2010) *Organic & Biomolecular Chemistry*, **8**, 114–121.

11 Anton Paar GmbH, Anton-Paar-Str. 20, A-8054 Graz, Austria; http://www.anton-paar.com.

12 Biotage AB, Kungsgatan 76, SE-753 18 Uppsala Sweden; http://www.biotage.com.

13 CEM Corporation, P.O Box 200, Matthews, NC 28106, USA; www.cem.com (main website), www.cem.de (German website).

14 Milestone s.r.l., Via Fatebenefratelli, 1/5, I-24010 Sorisole, BG, Italy; MLS GmbH, Auenweg 37, D-88299 Leutkirch, Germany; Milestone Inc., 25 Controls Dr., Shelton, CT 06484, USA; http://www.milestonesci.com (US website), www.milestonesrl.com (Italian website), www.mls-mikrowellen.de (German website).

15 Commarmont, R., Didenot, R., and Gardais, J.F. (1986) Prolabo, French Patent 84/03496.

16 Cleophax, J., Liagre, M., Loupy, A., and Petit, A. (2000) *Organic Process Research & Development*, **4**, 498–504.

17 Chemspeed Technologies AG, Rheinstrasse 32, 4302 Augst, Switzerland; http://www.chemspeed.com.

4
Microwave Processing Techniques

In modern microwave synthesis, a variety of processing techniques can be utilized, supported by the availability of different types of dedicated single-mode and multimode microwave reactors. While in the past much interest has been focused on solvent-free reactions under open-vessel conditions [1], today the large majority of the published examples in the area of controlled microwave-assisted organic synthesis (MAOS) involve the use of organic solvents under sealed-vessel conditions [2]. In this chapter, general processing techniques in MAOS as well as processing techniques used in drug discovery and high-throughput synthesis are discussed.

4.1
Solvent-Free Reactions

A frequently used processing technique employed in microwave-assisted organic synthesis since the early 1990s involves solvent-less (dry media) procedures [1] where the reagents are reacted neat in the absence of a solvent. Alternatively, reagents can be pre-adsorbed onto either an essentially microwave-transparent (silica, alumina, or clay) or a strongly absorbing (graphite) inorganic support, which can additionally be doped with a catalyst or reagent.

Particularly in the early days of MAOS, the solvent-free approach was very popular since it allowed the safe use of domestic household microwave ovens and standard open-vessel technology. Even today, many of the solvent-free chemistries reported in the literature are performed in domestic microwave ovens taking advantage of the benefits related to this method. The solvent-free technique is claimed to be environmentally benign and contributes to the green chemistry philosophy. In some cases, workup procedures are simplified since the products can be obtained by simple extraction, distillation, or sublimation techniques [3].

One of the simplest methods involves the mixing of the neat reagents and subsequent irradiation by microwaves. In general, pure, dry solid organic substances do not absorb microwave energy; therefore, almost no heating will occur. If none of the reagents is a microwave-absorbing liquid, small amounts of a polar solvent

Microwaves in Organic and Medicinal Chemistry, Second Edition.
C. Oliver Kappe, Alexander Stadler, and Doris Dallinger.

(e.g., *N,N*-dimethylformamide and water) can be added to the reaction mixture in order to allow dielectric heating by microwave irradiation.

An alternative technique utilizes microwave-transparent or only weakly absorbing inorganic supports such as silica, alumina, or clay materials [1]. These reactions are effected by the reagents/substrates immobilized on the porous solid supports and have advantages over the conventional solution-phase reactions because of their good dispersion of active reagent sites, associated selectivity, and easier workup. The recyclability of some of these solid supports also renders these processes into eco-friendly "green" protocols. In general, the substrates are pre-adsorbed onto the surface of the solid support and then exposed to microwave irradiation.

Apart from examples where the inorganic support merely acts as a catalyst, there are many instances where a solid-supported reagent can be used very effectively in the process. This is particularly true for oxidation reactions with metal-based reagents. For example, Varma and Dahiya have developed a method where Montmorillonite K 10 clay-supported iron(III) nitrate (the so-called clayfen) is used under solvent-free conditions for the oxidation of alcohols to carbonyl compounds in less than 1 min [4]. Related microwave-assisted solvent-free oxidations were also carried out with manganese [5], copper [6], and chromium-based [7] oxidation reagents adsorbed on suitable inorganic supports.

In addition to cases where the inorganic support itself acts as a catalyst or where a reagent has been impregnated on the solid support, it is also possible to additionally dope the support material with metal catalysts. For instance, a Sonogashira coupling was performed on a strongly basic potassium fluoride/alumina support, doped with a palladium/copper(I) iodide/triphenylphosphine mixture. The resulting aryl alkynes were synthesized in very high yields (67–97%) [8]. Many more examples of solvent-free microwave-assisted transformations can be found in review articles [1] and recent books on microwave synthesis [3, 9].

In contrast to solvent-free (dry media) microwave processing involving (when properly dried) weakly microwave-absorbing supports such as silica or alumina, an alternative is to use strongly microwave-absorbing supports such as graphite. For reactions that require high temperatures, the idea of using a reaction support that takes advantage of both strong microwave coupling and strong adsorption of organic molecules has been considered in recent years [10]. Since many organic compounds do not interact appreciably with microwave irradiation, such a support could be an ideal "sensitizer," able to absorb, convert, and transfer energy provided by a microwave source to the reaction mixture.

Most forms of carbon interact strongly with microwaves. Amorphous carbon and graphite, in their powdered form, irradiated at 2.45 GHz, rapidly reach about 1000 °C within 1 min of irradiation. The main published work employing graphite as sensitizer is focused on heterocyclic synthesis [10]. The amount of graphite can be varied, either an excess is employed (graphite-supported reaction) or a graphite is used in "catalytic amounts" ($\leq$10% by weight), which often suffices to reach high temperatures in a short time. More recently, Johnson and coworkers have reported on graphite-sensitized microwave reactions at temperatures of 100–300 °C which otherwise are typically performed by flash vacuum pyrolysis at temperatures >500 °C [11].

Since the measured bulk temperature of the graphite is in the range of the reaction temperature, the authors suggested the formation of surface "hot spots" as a reason for their results. Due to the high temperature, rapid heating, and frequent ejection of material from the irradiation zone, this method is termed as microwave flash pyrolysis (MFP).

In addition to the graphite being used as a "sensitizer" (energy converter), there are several examples in the literature where the catalytic activity of metal inclusions in graphite has been exploited (graphimets). The Friedel–Crafts acylation of anisole is a case in point, where the presence of catalytic amounts of iron oxide magnetite crystallites (Fe_3O_4) allowed efficient acylation with benzoyl chloride within 5 min of irradiation [10].

Since graphite is a very strong absorber of microwave heating, the temperature must be carefully controlled to avoid melting of the reactor. The use of a quartz reaction vessel is highly preferable.

4.2 Phase-Transfer Catalysis

In addition to the solvent-free processing, phase-transfer catalytic conditions (PTC) have also been widely employed as a processing technique in MAOS [12]. In phase-transfer catalysis, the reactants are situated in two separate phases, for example, liquid–liquid or solid–liquid. In liquid–liquid PTC, the phases are mutually insoluble, ionic reagents are typically dissolved in the aqueous phase, while the substrate remains in the organic phase. In solid–liquid PTC, on the other hand, ionic reagents may be used in their solid state as a suspension in the organic medium. Transport of the anions from the aqueous or solid phase to the organic phase, where the reaction takes place, is facilitated by phase-transfer catalysts, typically quaternary onium salts (e.g., tetrabutylammonium bromide, Aliquat 336) or cation-complexing agents. At least one liquid component in solid–liquid PTC is necessary: usually the electrophilic reagent acts both as substrate and liquid phase. Phase-transfer catalytic reactions are perfectly tailored for microwave irradiation, and the combination of solid–liquid PTC and microwave irradiation typically gives the best results in this area. Numerous transformations in organic synthesis have been achieved under solid–liquid PTC and microwave irradiation in the absence of solvent, generally under atmospheric pressure in open vessels [12].

Although solid–liquid PTC is best suited for microwave irradiation, in recent years, liquid–liquid PTC has also become a popular technique. It has found a widespread application in palladium-catalyzed carbon–carbon cross-coupling reactions (Heck, Suzuki, and Sonogashira) [12]. Here, TBAB is again the preferred phase-transfer catalyst and water is used as a solvent that renders the process environmentally benign and "green" (for reactions in water as solvent, see also Section 4.5.1).

While a large number of interesting transformations using "dry media" reactions, involving classical solvent-free synthesis as well as phase-transfer catalysis under solvent-free conditions, have been published in the literature [9, 10, 12], technical difficulties relating to nonuniform heating leading to localized superheating and the formation of macroscopic hot spots, mixing (stirring), and the precise determination of the reaction temperature remain unsolved [13], in particular when scale-up issues need to be addressed.

Recent evidence suggests that a correct temperature measurement in microwave chemistry relies on the use of internal fiber-optic probes (especially when comparing conventional oil bath with microwave heating). In addition, it becomes now evident that proper stirring during the reaction is also an important factor (see also Section 2.5.1). Particularly in solvent-free synthesis, effective stirring is essential in order to avoid temperature gradients that can lead to irreproducible results.

Bogdal *et al.* have investigated the correlation between stirring and temperature gradients in the solvent-free reaction of chloroacetic ethyl ester and salicylaldehyde (Scheme 4.1) since different distributions of intermediate **1** and final product **2** were obtained when performing the reaction at the identical temperature under both conventional and microwave heating [14]. Under conventional conditions, it was revealed that at temperatures as high as 150 °C, the final product **2** could be obtained with about 90% selectivity; whereas at lower temperatures, intermediate **1** is favored (110 °C, 74 : 26). In the microwave experiment using a single-mode reactor (Synthewave 402, Prolabo), the internal temperature of the reaction mixture was recorded with a fiber-optic probe, whereas the surface temperature was recorded with a thermovision camera. If the reaction mixture is not stirred, temperature differences of up to 130 °C (P1: 70 °C, P2: 125 °C, P3: 200 °C, see Figure 4.1) were observed on the surface and a product distribution of 33 : 67 **1**/**2** was obtained. The existence of the products could also be detected visually since intermediate **1** has a light yellow color and product **2** is dark brown when adsorbed on K_2CO_3 (Figure 4.1). The rotation of the reaction vessel in the microwave reactor decreases the temperature gradient and the product distribution was closer to that under conventional conditions at 110 °C. When the reaction mixture was additionally stirred with a quartz spatula, the thermal homogeneity was greatly improved, which was also detected via the surface homogeneity (Figure 4.1), and the results were now comparable to those obtained under conventional conditions at 110 °C.

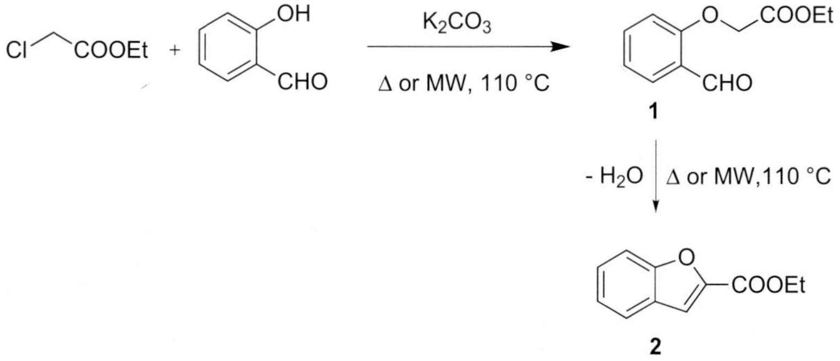

Scheme 4.1 Synthesis of benzofuran-2-carboxylic acid ethyl ester.

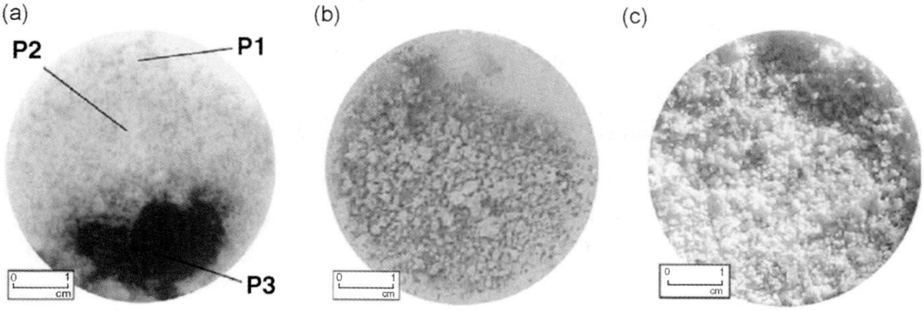

Figure 4.1 Surface images of the reaction mixture: (a) No stirring, (b) Vessel rotation, (c) Vessel rotation and stirring. Reproduced with permission from Ref. [14].

As demonstrated in these experiments, in order to maintain good temperature homogeneity, effective stirring has to be provided that is sometimes difficult to achieve for heterogeneous reaction mixtures or under solvent-free conditions. Even temperature monitoring using a fiber-optic probe can lead to incorrect measurements (see also Section 2.5.1). By employing a multiple fiber-optic probe device, temperature inhomogeneities in a sample of Montmorillonite K10 clay were observed when the reaction vessel was heated in a standard single-mode microwave reactor [15]. The temperature was measured with an external IR sensor and in addition with three fiber-optic sensors that were located at different heights inside the reaction vessel. It was revealed that the temperature gradients could not be eliminated by magnetic stirring since the stirring proved to be inefficient.

4.3 Open- versus Closed-Vessel Conditions

Microwave-assisted syntheses can be carried out using standard organic solvents either under open- or sealed-vessel conditions. If solvents are heated by microwave irradiation at atmospheric pressure in an open vessel, the boiling point of the solvent (as in an oil bath experiment) typically limits the reaction temperature that can be achieved. In the absence of any specific or nonthermal microwave effects (such as the superheating effect at atmospheric pressure, see Section 2.5.3), the expected rate enhancements would be comparatively small. In order to nonetheless achieve high reaction rates, high-boiling microwave-absorbing solvents such as dimethylsulfoxide, 1-methyl-2-pyrrolidone, 1,2-dichlorobenzene, or ethylene glycol (see Table 2.3) have frequently been used in open-vessel microwave synthesis [16, 17]. However, the use of these solvents presents serious challenges during product isolation. Because of the recent availability of modern microwave reactors with online monitoring of both temperature and pressure (see Chapter 3), MAOS in sealed vessels is the preferred technique today. The main advantage of closed-vessel conditions is that solvents can be heated far above their boiling points, eliminating the need for high-boiling solvents to reach high temperatures and, therefore, significant rate enhancements may be achieved compared to reactions that are performed under conventional reflux conditions.

The main advantage of using open-vessel microwave processing is the reduced safety risk, since pressurized reaction vessels can be avoided. This is particularly important when the scale-up of microwave-assisted reactions is concerned (see Section 4.8). Apart from that, microwave heating using open vessels can allow a more rapid heating compared to conventional techniques, provided that the reaction mixture is highly microwave absorbing.

An area where high-boiling solvents are frequently used is microwave-assisted solid-phase organic synthesis (SPOS). In solid-phase synthesis, the synthesized compounds are attached to an insoluble polymer support, which can easily be removed by filtration. Therefore, the high-boiling solvents do not represent a problem during workup. An example of solid-phase microwave synthesis where the use of open-vessel technology is essential is shown in Scheme 4.2. The transesterification of β-keto esters with a supported alcohol (Wang resin) is carried out in 1,2-dichlorobenzene (DCB) as a solvent under controlled microwave heating conditions [18]. The temperature is kept constant at 170 °C, about 10 °C below the boiling point of the solvent, thereby allowing safe processing in the microwave cavity. In order to achieve full conversion to the desired resin-bound β-keto esters **3**, it is essential that the alcohol formed can be removed from the equilibrium. The second step, a Knoevenagel condensation, is also performed under open-vessel conditions. Here, a lower temperature of 125 °C is necessary to prevent cleavage from the resin, but is still sufficient to remove the water from the reaction mixture and to obtain the resulting polymer-bound enones **4** in quantitative conversion [18].

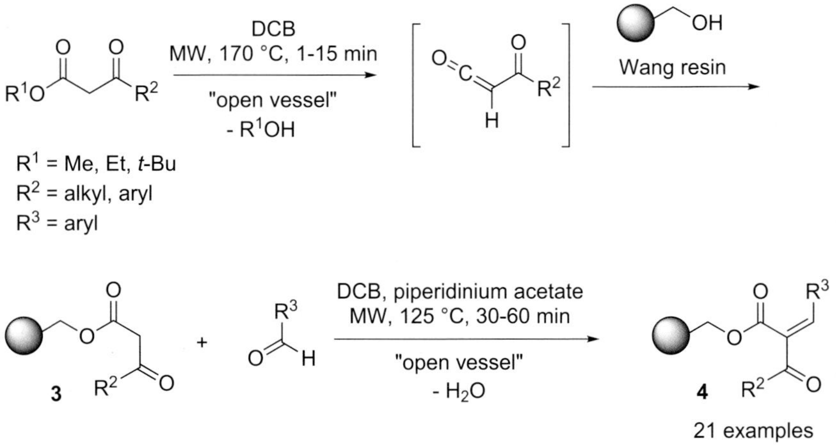

Scheme 4.2 Microwave-assisted solid-phase enone synthesis.

Another example where open-vessel microwave processing is required is outlined in Scheme 4.3, namely, the formation of pyranoquinolone **5** from *N*-methylaniline and diethyl malonate in diphenyl ether as high-boiling solvent (bp 259 °C) [19].

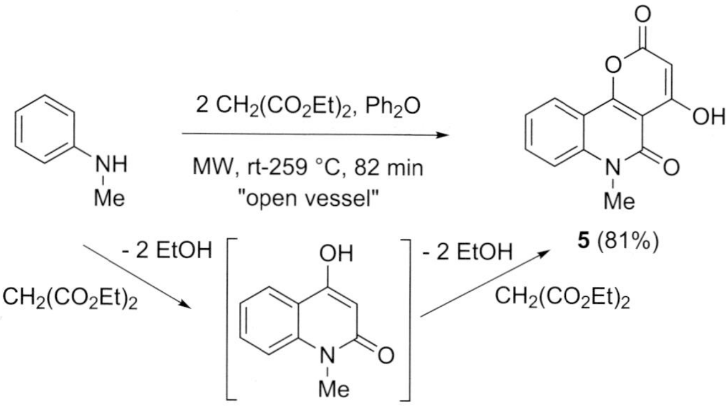

Scheme 4.3 Formation of pyrano[3,2-*c*]quinolone.

This double condensation process requires the use of open-vessel technology, since 4 equiv of the volatile ethanol by-product are formed that are continuously removed from the reaction mixture by fractional distillation from the reagents. The reaction has been performed in a 0.2 mol scale in a 500 mL round-bottom flask in the MicroSYNTH, a multimode microwave reactor (Figure 3.29) equipped with a Vigreux column through the mount in the ceiling of the instrument and a standard distillation kit. After rapid heating to about 200 °C, the ethanol is distilled off and the temperature is slowly ramped to the boiling point of diphenyl ether at which point the double condensation was completed (Scheme 4.3). The conversion of this transformation can be monitored by the amount of formed ethanol in the receiver. Importantly, the speed of the distillation is controlled by the modulation of the microwave power (too rapid heating leads to codistillation of the reagents). Under standard closed-vessel conditions, no reaction to the desired product takes place.

Another related experiment is the synthesis of 4-hydroxyquinolinone from anilines and malonic esters [20]. Again, it was essential to use open-vessel technology here, since the formed 2 equiv of ethanol need to be removed from the equilibrium. Preventing the removal of ethanol from the reaction mixture, for example, by using a standard closed-vessel microwave system, results in a pressure buildup in the reaction vial and leads to significantly lower yields (Table 4.1, entries 1–3). Comparable yields could only be achieved when the reaction under closed-vessel conditions is conducted on a 1 mmol scale with 0.5 mL of solvent (Table 4.1, entry 4). In this case, the pressure buildup is relatively small since enough headspace in the vial is guaranteed due to the low filling volume. Scale-up of this synthesis would clearly be feasible by only using open-vessel technology [21].

A similar situation was encountered by Amore and Leadbeater when conducting esterification reactions [22]. Typically, either the ester product or the generated water needs to be removed from the equilibrium in order to drive these reactions to completion. Since under standard sealed-vessel conditions this was not feasible, a multimode microwave instrument was utilized where the water could be distilled out from the reaction mixture by connecting a vacuum pump on the outside of the reactor

Table 4.1 Dependence of the yield of hydroxyquinolinone product on the use of closed- or open-vessel microwave heating.[a)]

Entry	Reagents (mmol)	Solvent (mL)	Yield (%)	Pressure (bar)
1	1	2	76	3.6
2	2	2	67	5.3
3	4	2	60	7.4
4	1	0.5	91	2.0
5	1	Neat	90	
6[b)]	2	Neat	92	
7[b)]	4	Neat	90	

Data from Ref. [20].
a) Microwave heating, 250 °C, 10 min, 1,2-dichlorobenzene or neat.
b) Open vessel.

(RotoSYNTH, Figure 3.35). Excellent yields (82–94%) were obtained after 15 min heating to 100–120 °C for primary alcohols on a 0.3 or 3 mol scale in a 0.3 or 2 L vessel, respectively [22].

The employment of a Dean–Stark setup for distillation and thus azeotropic removal of water was demonstrated by Lukács *et al.* for the synthesis of benzophenone and acetophenone ethylene ketals [23]. Excellent yields of the ethylene ketal products could be obtained by reacting the ketones with ethylene glycol, *p*-toluenesulfonic acid (PTSA), and toluene under reflux conditions for 2–3 h at constant 650 W (Scheme 4.4).

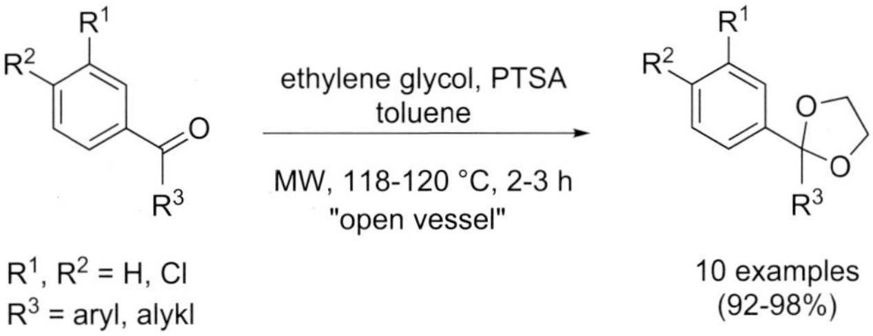

Scheme 4.4 Synthesis of ethylene ketals with azeotropic removal of water.

Not only the liquid by-products can be removed from the reaction mixture, but also the gaseous by-products have been successfully purged off to shift the equilibrium to the product. For the ring-closing metathesis (RCM) of diene **6** employing Grubbs II catalyst, it was crucial to remove the developing ethylene during the reaction in order to obtain the tricyclic product **7** in high yields (Scheme 4.5) [24]. This was possible by passing an argon stream through the reaction solution (gas sparging). Importantly, under closed-vessel conditions, starting material **6** was recovered almost quantitatively. Similar observations were made by Kappe and coworkers toward investigations

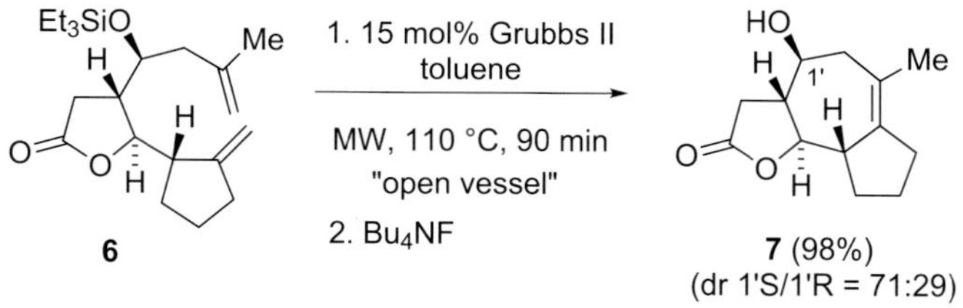

Scheme 4.5 RCM with concurrent removal of ethylene by gas sparging.

of wall effects in MAOS in the ring-closing metathesis reaction of 1,2-bis(allyloxy) benzene to provide the eight-membered 2,5-dihydro-1,6-benzodioxocin [25]. Here, better results were also achieved using an open-vessel gas sparging protocol in 1,2-dichloroethane (DCE) at reflux temperature (83 °C) compared to closed-vessel experiments at the same temperature measured internally via a fiber-optic probe (85% versus 53% conversion after 5 min and 1 mol% Grubbs I catalyst).

All the experiments discussed in this section highlight the importance of choosing appropriate experimental conditions when using microwave heating technology. In particular, the user should be aware of the consequences of performing reactions in a sealed vessel.

4.4 Pre-pressurized Reaction Vessels

Relatively little work has been performed with gaseous reagents in sealed-vessel microwave experiments [26]. Although several publications describe this technique in the context of heterogeneous gas-phase catalytic reactions important for industrial processes [27], the use of pre-pressurized reaction vessels in conventional microwave-assisted organic synthesis involving solvents is rare. Several authors have earlier described the use of reactive gases in such experiments and the experimental techniques as to how to apply a slight overpressure (2–3 bar) [28]. However, in the last few years, multimode and single-mode reactors have been developed where pre-pressurization typically up to 20 bar with the appropriate accessories is possible (see Chapter 3).

For example, the Diels–Alder cycloaddition reaction of the pyrazinone heterodiene **8** with ethene (Table 4.2) led to the corresponding bicyclic cycloadduct **9** [29]. Under conventional conditions, these cycloaddition reactions have to be carried out in an autoclave applying an ethene pressure of 25 bar before the setup is heated at 110 °C for 12 h. When the solution of pyrazinone **8** in DCB was saturated with gaseous ethene prior to sealing (1 bar), the cycloaddition was completed (89% yield) after irradiation for 140 min at 190 °C in a single-mode reactor (Biotage). It was, however, not possible to further increase the reaction rate by raising the temperature. At temperatures above 200 °C, an equilibrium between the cycloaddition and the competing retro-Diels–Alder fragmentation process was observed [29]. Only by

Table 4.2 Diels–Alder reaction of pyrazinone **8** with ethene under pre-pressurized conditions.

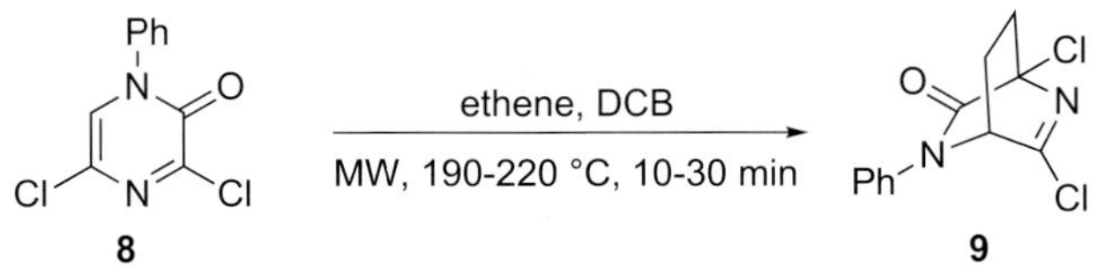

Entry	Pressure (bar)	Temperature (°C)	Time (min)	Yield (%)
1	1	190	100	12
2	5	190	30	87
3	10	190	20	85
4	10	220	10	85

Data from Ref. [30].

using a multimode microwave reactor that allowed pre-pressurization of the reaction vessel with 10 bar of ethene (Anton Paar, Synthos 3000, see Figures 3.17 and 3.18), could the Diels–Alder addition be carried out much more efficiently at 190 °C within 20 min (Table 4.2, entry 3) [30]. The reaction time could be even reduced to 10 min when the reaction temperature was raised to 220 °C (Table 4.2, entry 4), with the cycloaddition product being still stable at this higher temperature.

Leadbeater and Kormos have performed palladium-catalyzed carbonylation reactions employing gaseous carbon monoxide (CO) using the same multimode instrument with the gas loading interface as described above. The vessels were pre-pressurized either with 14 bar carbon monoxide for hydroxycarbonylation [31] or with 10 bar carbon monoxide for alkoxycarbonylation [32]. For alkoxycarbonylations (Scheme 4.6b), a lower temperature of 125 °C was sufficient to reach high conversions since carbon monoxide is better soluble in alcohol than in water that is used as solvent for the hydroxycarbonylations (Scheme 4.6a). In both cases, only aryl iodides could be converted to the corresponding acids **10** or esters **11**, aryl bromides were unreactive.

Furthermore, the same authors have shown that alkoxycarbonylations can be conducted successfully when near-stoichiometric amounts of carbon monoxide are employed [33]. These reactions were performed in a single-mode instrument (CEM Discover, see Figure 3.13a) with a gas loading interface where the exact loading pressure can be monitored since the pressure sensor is directly connected to the reaction vessel. The reactions were run in an 80 mL glass vessel on a 2 mmol scale. First, the reaction mixture was loaded with about 1 bar of carbon monoxide that corresponds to 2.5 mmol, and subsequently about 9 bar of nitrogen were additionally loaded into the vial to reach an initial pressure of about 10 bar. Interestingly, the total pressure had a significant effect on the reaction outcome, and 10 bar proved to give the best results, although the yields of the corresponding esters were somewhat lower compared to those described in Scheme 4.6b [32]. The scale-up of

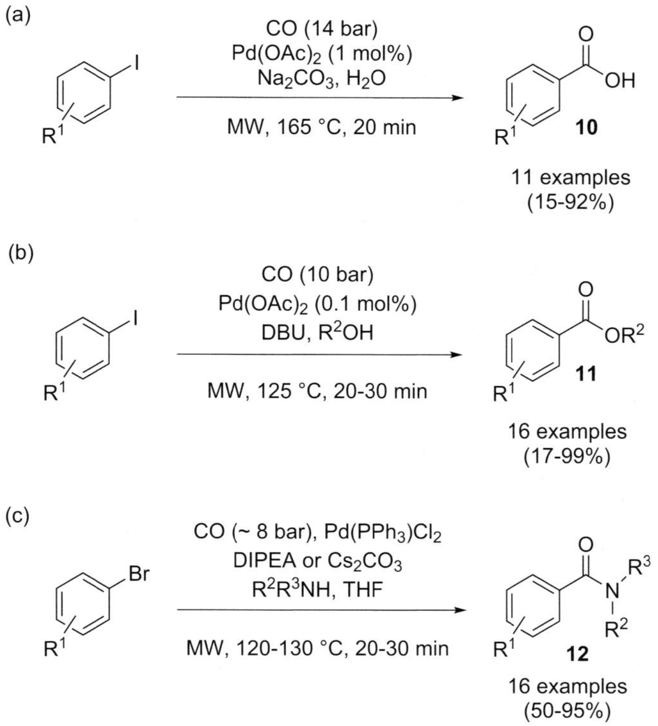

Scheme 4.6 Hydroxy-, alkoxy-, and aminocarbonylations using gaseous carbon monoxide and pre-pressurized reaction vessels.

ethoxycarbonylation of iodobenzene with near-stoichiometric amounts of carbon monoxide (1.08 equiv) on the 1 mol scale was demonstrated in a multimode batch reactor (Milestone UltraCLAVE, see Figure 3.36a) by applying similar reaction conditions, as shown in Scheme 4.6b [34]. Effective sealing of the 3 L reaction vessel was ensured by pre-pressurization with 11 bar of nitrogen, followed by 27 bar of carbon monoxide and a further 12 bar of nitrogen, leading to a total pressure of 50 bar. Ethyl benzoate was obtained in 80% isolated yield after heating for 30 min at 125 °C, hence being comparable to the small-scale experiment at 1 mmol. In a further experiment, ethoxycarbonylation of six different aryl iodides in parallel on a 50 mmol scale employing 150 mL glass vials was studied [34]. Excellent conversions for all six reactions were observed under the same reaction conditions as before, indicating that no cross-contamination of the reaction mixtures occurred.

Organometallic ruthenium and osmium complexes were prepared by the same group by using the identical gas loading accessory for the CEM Discover as mentioned above [35]. $Ru_3(CO)_{12}$, $H_4Ru_4(CO)_{12}$, and $H_2Os_3(CO)_{10}$ could be synthesized by pre-pressurizing the reaction mixture with about 3.5 bar of carbon monoxide or hydrogen gas, respectively. In addition to this approach, the synthesis of Zeise's salt [$KPtCl_3(C_2H_4)$] was developed using K_2PtCl_4 as starting material, a 1 : 1 : 1 ratio of water/ethanol/HCl_{conc} as solvent, a loading of about 3.5 bar of ethene, and heating at 130 °C for 15 min [36].

An analogous gas inlet device for the CEM Discover system, where the 10 mL vial is provided with a tube connection to an external pressure control system, was employed by Petricci and coworkers for aminocarbonylations of aryl bromides at low carbon monoxide pressure of about 8 bar [37]. Amides **12** were prepared from primary and secondary amines using 5 mol% of $Pd(PPh_3)Cl_2$ as catalyst and 3 equiv of DIPEA as base in THF at 130 °C within 20 min in good yields (Scheme 4.6c). For less nucleophilic aromatic and heteroaromatic amines, a switch to Cs_2CO_3 as base, a slightly lower temperature of 120 °C, and a prolonged reaction time of 30 min is necessary in order to obtain the corresponding amides in high yields (Scheme 4.6c).

Similarly, hydroformylations of alkenes can be carried out by using a 1 : 1 mixture of carbon monoxide and hydrogen (syngas), a rhodium catalyst, and a suitable ligand. Under standard conditions, high pressures (50–80 bar) of syngas and long reaction times from 1 to 3 days in an autoclave are required. When a similar setup as described above (real-time pressure monitoring) in the CEM Discover is employed, 80 mL vial including the reaction mixture can be filled with about 3 bar of syngas [38]. After irradiation for 4–6 min (2 min cycles) at 110 °C, the corresponding aldehydes were obtained in high yields (Scheme 4.7a). Importantly, the ionic liquid is crucial for the reaction, since without $bmimBF_4$ the reaction mixture could not be heated up to 110 °C. Only 60 °C could be reached otherwise, due to the low microwave-absorbing characteristics of toluene (see also Section 4.5.2) and complete conversion could not be achieved even after 30 min.

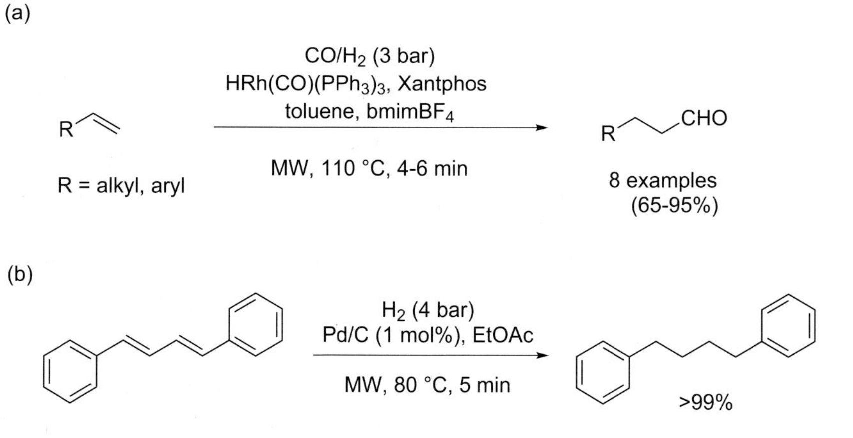

Scheme 4.7 Hydroformylations (a) and hydrogenation (b) reactions performed in a single-mode instrument.

In order to perform hydrogenations with gaseous hydrogen, Vanier utilized a comparable gas inlet device for the CEM Discover, where the 10 mL fiber-optic accessory was additionally equipped with a gas inlet [39]. The reaction vessel, containing the substrate, 1 mol% of palladium-on-carbon (Pd/C), and ethyl acetate as solvent, was charged with about 4 bar of hydrogen (Scheme 4.7b). Different types of substrates were reduced nearly quantitatively after irradiation at 80–100 °C for 3–20 min. It should be noted that toluene as solvent should be avoided in combination with Pd/C, since it is not able to suspend the catalyst properly and explosions can occur due to deposition of Pd/C on the vessel wall. In addition, a selective catalyst heating effect (see Section 2.5.3) was postulated, since significant differences in conversions between conventionally heated (3–55% conversion) and microwave-irradiated (full conversion) hydrogenations were reported by applying otherwise identical reaction conditions. After a careful reinvestigation of the reported hydrogenation reactions, it was discovered that the stirring speed is a crucial factor for the outcome in both microwave and conventionally heated hydrogenations, because it plays a major role in controlling mass transfer from the gas to the liquid phase [40]. When the stirring speed was set to "high" in the microwave and to 1200 rpm in the conventionally heated experiments, full conversions were observed for both heating modes.

Hydrogenation reactions in a specifically designed reactor for a multimode instrument (based on MicroSYNTH, see Figure 3.29) were reported by Holzgrabe and coworkers [41]. Hydrogen gas of up to 25 bar can be charged to the reaction mixture prior to irradiation. Dearomatizations, debenzylations, azide hydrogenation, and double-bond reductions are conducted in shorter reaction times compared to classical conditions.

Instead of employing gaseous reagents like carbon monoxide or hydrogen gas in MAOS, where special equipment is necessary, a popular and more convenient technique is to utilize solid reagents that liberate, for example, carbon monoxide during the reaction upon heating. One of these solid reagents is $Mo(CO)_6$ that is known to liberate carbon monoxide smoothly at higher temperatures [42]. The Larhed group has extensively employed $Mo(CO)_6$ as solid carbon monoxide source for diverse carbonylations (such as alkoxy- and aminocarbonylations) under microwave conditions [42]. One example is shown in Scheme 4.8a, where palladium-catalyzed aminocarbonylations were successfully performed with $Mo(CO)_6$ in water as solvent [43]. Aryl iodides, bromides, and even the otherwise unreactive aryl chlorides could be reacted with diverse primary and secondary amines to the aryl amide products **13** in moderate to excellent yields (Scheme 4.8a). The competing hydroxycarbonylation could be inhibited by fine-tuning of the reaction parameters. In particular, the stoichiometry of aryl halide to amine was crucial for the successful reaction as well as the choice of the proper catalyst. With this general protocol, aryl iodides could be reacted at 110 °C, whereas the bromides and chlorides needed a higher temperature of 170 °C and sometimes even longer reaction times.

Catalytic transfer hydrogenation (CTH) is another possibility for avoiding gaseous reagents, that is, hydrogen gas in this case. In Scheme 4.8b, the reduction of the nitro group in pyrazole **14** by using cyclohexene as the hydrogen source is displayed [44]. Clean and complete reduction of the nitro group was obtained using 1 mol% of Pd/C catalyst, 2 equiv of cyclohexene and ethanol as solvent, at 160 °C within 2 min. The corresponding aniline derivative – a precursor of the canonical transient receptor potential channel inhibitor Pyr3 – was provided in excellent 92% yield.

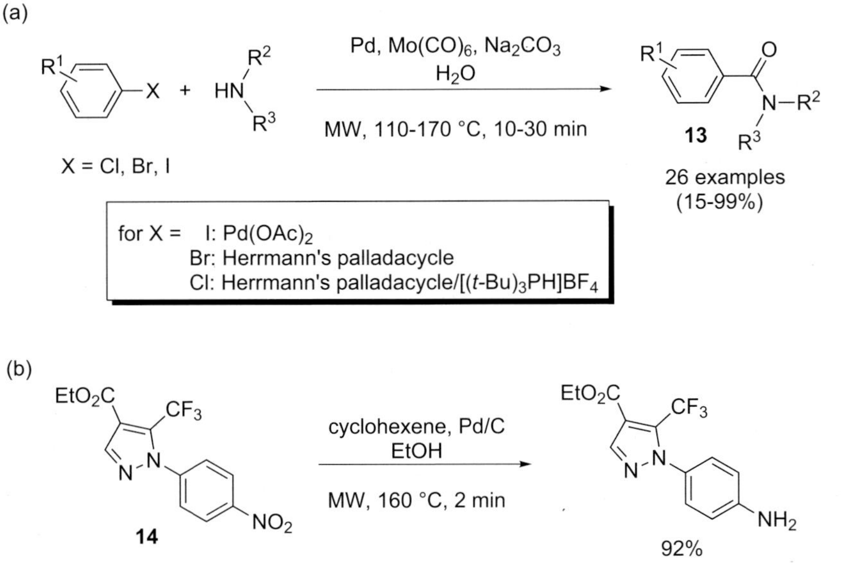

Scheme 4.8 (a) Aminocarbonylation using a solid carbon monoxide source. (b) Catalytic transfer hydrogenation employing cyclohexene as hydrogen donor.

4.5
Nonclassical Solvents

Apart from using standard organic solvents in conjunction with microwave synthesis, the use of either water or so-called ionic liquids as alternative reaction media has become increasingly popular in recent years.

4.5.1
Water as Solvent

Synthetic organic reactions in aqueous media at ambient or slightly elevated temperatures have become of great interest as water as a solvent for organic reactions often displays unique reactivity and selectivity [45], exploiting, for example, so-called hydrophobic effects [46]. Apart from performing reactions in aqueous solutions in a moderate temperature range (0 to 100 °C), chemical processing in water is also possible and is of considerable interest under "superheated conditions" (>100 °C) and so-called near-critical (also termed subcritical) conditions (150–300 °C) in sealed vessels because of the favorable changes that occur in the chemical and physical properties of water at high temperatures and pressures [47, 48].

As the water temperature is increased, it exhibits properties like polar organic solvents and thus organic compounds become soluble in high-temperature water. In addition to the environmental advantages of using water as so-called pseudo-organic solvent, isolation of products is normally facilitated. Once cooled, most organic products are no longer soluble in ambient temperature water, and this allows easy postsynthesis product separation.

Most importantly, the ionic product (dissociation constant) of water is increased by three orders of magnitude on going from room temperature to 250 °C. This means that water becomes both a stronger acid and a stronger base as the temperature increases to the near-critical water (NCW) range and can therefore act as an acid, base, or acid–base bi-catalyst [47, 48]. However, the application of supercritical water (SCW) (>374 °C) for preparative organic synthesis is limited due to its degenerative properties.

As with most organic solvents, the loss tangent (tan δ) for water is strongly influenced by temperature (see Section 2.3). Since the dielectric constant ε' for water drastically decreases with temperature, the loss tangent is also reduced. For this reason, it is not a trivial affair to heat pure water to high temperatures under microwave conditions. While water can be heated rather effectively from room temperature to 100 °C, it is more difficult to superheat water in sealed vessels from 100 to 200 °C and very difficult to reach 300 °C by microwave dielectric heating [49]. In fact, SCW is transparent to microwave radiation.

The loss tangent of a solvent such as water can be significantly increased, for example, by addition of small amounts of inorganic salts that improve microwave absorbance by ionic conduction (see Section 2.2). For most applications in NCW chemistry (200–300 °C), it can be assumed that a comparatively low salt concentration would not significantly influence the reactivity of the water medium [49]. However, pure water can be easily and rapidly heated to temperatures up to 200 °C in the high-temperature water region, if microwave-absorbing substrates or metal catalyst like palladium are present in the reaction mixture [50].

The application of microwave heating in combination with the use of high-temperature water as "green" solvent has received much interest in recent years [50], a field originally pioneered by Strauss and coworkers in the mid-1990s [51–53]. Due to the existing pressure limit of 20–30 bar for most of today's commercially available single-mode microwave reactors, the majority of published chemistry has generally been restricted to reaction temperatures at or below 200 °C [50]. Only a small number of publications exist where reactions are performed under near-critical water conditions at temperatures close to 300 °C in one of the few accessible dedicated instruments with higher pressure limits [50]. By employing the Synthos 3000 multimode instrument with a pressure and temperature limit of 80 bar and 300 °C (see Figures 3.17 and 3.18), respectively, Kremsner and Kappe have conducted benzamide hydrolysis at 295 °C without the addition of an acid catalyst, taking advantage of the increased ionic product of water at this temperature [49]. The Diels–Alder reaction of 2,3-dimethyl butadiene and acrylonitrile produced only trace amounts of product when performed in water at 200 °C, whereas full conversion was achieved at 295 °C within 20 min [49]. For both syntheses, a 0.03 M NaCl solution was employed as solvent because of the improved microwave absorbance. Direct conversion of aryl halides to the corresponding phenols at 300 °C was reported by Kormos and Leadbeater [54]. The reactions could also be carried out at 200 °C, although higher yields, especially for aryl bromide and chloride substrates, were obtained at 300 °C.

Similar to water, microwave irradiation in the above-mentioned multimode reactor makes reactions at near-critical or supercritical conditions of alcohols possible. Catalyst-free transesterifications of triglycerides (rapeseed oil) with butanol ($T_c = 287$ °C at 49 bar) under supercritical conditions at 310 °C and 80 bar were performed by Maes and coworkers, resulting in high conversions to the fatty acid butyl ester [55]. To reach such high temperatures, silicon carbide (SiC) passive heating elements had to be used to improve the heating performance due to the low microwave absorbance of butanol at higher temperatures (see Section 4.6).

In another example, Molteni *et al.* have described the three-component, aqueous one-pot synthesis of fused pyrazoles **17** by reacting cyclic 1,3-diketones **15** with *N,N*-dimethylformamide dimethyl acetal **16** (DMFDMA) and a suitable bidentate nucleophile like a hydrazine derivative (Scheme 4.9) [56]. The reaction proceeds via initial formation of an enaminoketone followed by a tandem addition–elimination/cyclodehydration step. An amount of 2.6 equiv of acetic acid is necessary to ensure a clean conversion at 200 °C within 2 min. Upon cooling and stirring, the desired reaction products crystallized and were isolated in high purity by simple filtration. Pyrimidines and isoxazoles could be synthesized in a similar fashion applying the same protocol employing amidines and hydroxylamine as nucleophiles.

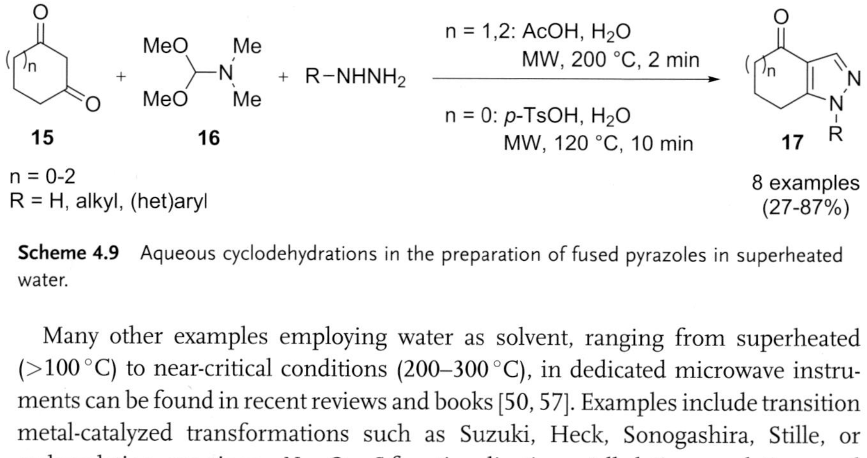

Scheme 4.9 Aqueous cyclodehydrations in the preparation of fused pyrazoles in superheated water.

Many other examples employing water as solvent, ranging from superheated (>100 °C) to near-critical conditions (200–300 °C), in dedicated microwave instruments can be found in recent reviews and books [50, 57]. Examples include transition metal-catalyzed transformations such as Suzuki, Heck, Sonogashira, Stille, or carbonylation reactions, *N*-, *O*-, *S*-functionalizations (alkylation, acylation, and arylation), heterocycle synthesis, epoxide ring-opening reactions, and protection/deprotection reactions, as well as solid-phase organic synthesis, polymer synthesis, enzymatic reactions, and nanomaterial synthesis.

4.5.2
Ionic Liquids

Ionic liquids (ILs) are a new class of solvents that are entirely constituted of ions. They usually consist of an organic cation (mainly quaternary nitrogen) and an inorganic or organic anion and either are liquids at room temperature or have melting points below 100 °C [58]. They generally have negligible vapor pressure and are immiscible with common nonpolar solvents, meaning that organic products can be easily removed by extraction and the ionic liquid can be recycled. In addition, they have a wide accessible temperature range (typically >300 °C), a low toxicity, and are nonflammable. Due to these advantages, they have attracted much recent attention as environmentally benign solvents [58].

From the perspective of microwave chemistry, one of the points of key importance is their high polarity that is variable depending on the cation and anion and hence can effectively be tuned to a particular application. Ionic liquids interact very efficiently with microwaves through the ionic conduction mechanism (see Section 2.2) and are rapidly heated at rates easily exceeding 10 °C per second without any significant pressure buildup [59]. Therefore, safety problems arising from overpressurization of heated sealed reaction vessels can be minimized. The rapid heating of ionic liquids is demonstrated in Figure 4.2, where the heating profile of neat 1-butyl-3-methylimidazolium hexafluorophosphate **18** (bmimPF_6), which is irradiated at different constant microwave power, is shown [60]. The very strong coupling characteristics are demonstrated by the fact that only 2 W of microwave energy suffices to heat IL to 140 °C within about 5 min. Temperatures up to 220 °C can be rapidly reached within less than 1 min when higher power is applied.

The use of ionic liquids in conjunction with MAOS can be classified into three areas [61–63]:

1) Employment of microwave heating in the synthesis of ILs.
2) Use of ILs as solvents, reagents, and catalysts for microwave-assisted synthesis.
3) Application of ILs as "doping agents" for poor microwave-absorbing solvents.

Although promoted as environmentally benign reaction media, the synthesis of ionic liquids is not. Especially for the purification step, large volumes of organic solvents are typically required under conventional conditions. Halide-based ionic

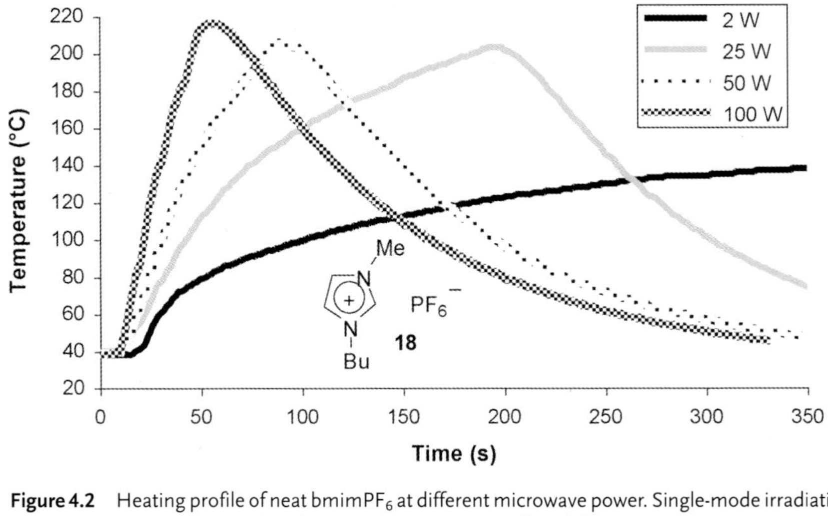

Figure 4.2 Heating profile of neat bmimPF_6 at different microwave power. Single-mode irradiation (CEM Discover), 5 mL sample volume, 10 mL quartz vessel, and fiber-optic temperature measurement. Reproduced with permission from Ref. [60].

liquids can be easily prepared by reacting nitrogen-containing heterocyclic starting materials with an appropriate alkyl halide. To reduce the large excess of alkyl halides needed under traditional conditions, microwave-assisted methods have been developed. Much of the early work in this field was published by Varma and coworkers, applying open-vessel conditions [64]. Later, Khadlikar and Rebeiro demonstrated that the preparation of ionic liquids was also feasible applying closed-vessel microwave conditions, eliminating the dangers of working with toxic and volatile alkyl halides in the open atmosphere [65]. A comprehensive study of the microwave-assisted preparation of ionic liquids was published by Deetlefs and Seddon in 2003. These authors synthesized a large number of ionic liquids based on 1-alkyl-3-methylimidazolium (**19**), 1-alkyl-2-methylpyrazolium (**20**), 3-alkyl-4-methylthiazolium (**21**), and 1-alkylpyridinium (**22**) cations by solvent-free alkylation of the corresponding basic heterocyclic cores (Scheme 4.10) [66]. The published procedures feature dramatically reduced reaction times compared to conventional methods, minimize the generation of organic waste, and also afford the ionic liquid products in excellent yields and with high purity. The syntheses were performed on flexible reaction scales ranging from 50 mmol to 2 mol in either sealed or open vessels, with a significant reduction of the large molar excess of haloalkane.

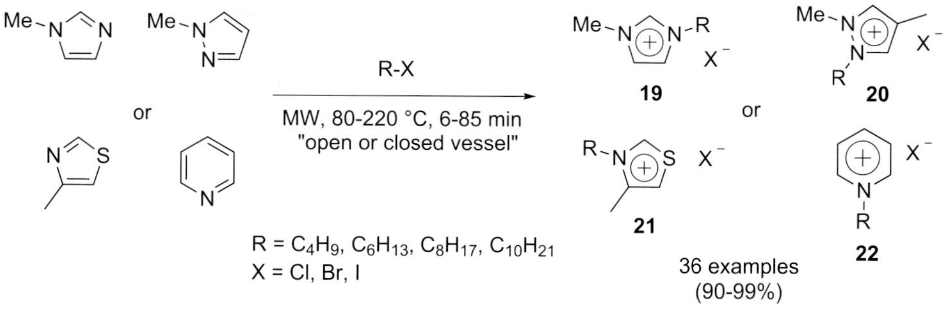

Scheme 4.10 Preparation of ionic liquids under microwave conditions.

Due to the extremely high microwave absorptivity of ILs, one should be aware that in particular, the syntheses of ILs under microwave conditions are notoriously difficult to control since not only these *N*-alkylations are generally exothermic, but the microwave absorptivity during the process also changes significantly from moderate (starting materials) to high (ionic liquid) [67]. To maintain such a high heating rate, another phenomenon plays a significant role. In contrast to other solvents, the loss tangent (tan δ) of ILs increases with increasing temperature [68]. In a case study by Obermayer and Kappe, the importance of dual IR/fiber-optic temperature measurements (see Section 2.5.1) in the preparation of the imidazolium-based IL bmimBr was highlighted [67]. Utilizing a single-mode microwave reactor that allows simultaneous infrared/fiber-optic temperature measurements (see Section 3.4.1.1), significant differences between the two methods of temperature monitoring were revealed, showing an internal temperature overshoot of up to ~50 °C measured by the fiber-optic probe. Even more drastic overshoots of up to 100 °C were detected when other microwave instruments were used. It is therefore apparent that microwave reactors that rely exclusively on external IR temperature probes as lead sensors should not be used for transformations that involve ILs or other strongly absorbing materials as solvents or reagents.

Using a similar *N*-alkylation protocol as described above, the Loupy group has reported the synthesis of chiral ionic liquids based on (1*R*,2*S*)-(−)-ephedrinium salts under microwave irradiation conditions [69]. Importantly, the authors were also able to demonstrate that the desired hexafluorophosphate salts could be prepared in a one-pot protocol by *in situ* anion exchange metathesis. In 2007, Gaertner and coworkers have developed chiral imidazolium ionic liquids bearing a bornyl moiety from (1*S*)-(+)-camphorsulfonic acid and (+)-camphene precursors with the aid of microwave heating in the quaternization step [70]. These ILs were utilized as solvents in diastereoselective Diels–Alder reactions.

In the last few years, the scope of ionic liquids has been expanded by the introduction of additional functional groups in the ionic liquid structure [71]. In these so-called task-specific ionic liquids (TSILs), the functional group is tethered either to the cation or to the anion (or both) and can be utilized as soluble support. In recent publications, both the synthesis of TSILs [72] and further transformations of the introduced functional groups were reported using microwave irradiation [62, 63]. De Kort *et al.* have designed a *N*-methylimidazolium-based ionic support that incorporates an aldehyde functionality and can be seen as ionic analog of the AMEBA solid support [73]. The ionic AMEBA support is synthesized via simple alkylation of *N*-methylimidazole with the corresponding aldehyde-functionalized alkyl chloride and is further converted to supported amines by reductive amination of the aldehyde. Furthermore, heterocycle synthesis has been performed on these ionic supports, like the synthesis of dihydropyrimidines (DHPMs) and dihydropyridines via Biginelli and Hantzsch reactions, respectively, on PEG-ionic liquid matrices [74].

More established than the synthesis of ionic liquids under microwave irradiation is the employment of ionic liquids as solvents in microwave chemistry because of their efficient coupling characteristics (see Figure 4.2). One example is shown in Scheme 4.11, where the Larhed group performed Heck reactions in $bmimPF_6$ as solvent [75]. Aryl bromides and iodides were reacted with butyl acrylate and $PdCl_2/P(o\text{-tolyl})_3$ as catalyst/ligand system. Full conversions were achieved within 5 min at 180 °C (X = I) and 20 min at 220 °C (X = Br). A key feature of this catalytic/ionic liquid system is its recyclability: the phosphine-free ionic catalyst phase $PdCl_2/bmimPF_6$ was recyclable at least five times. After each cycle, the volatile product was directly isolated in high yield by rapid distillation under reduced pressure.

Apart from the application of neat ionic liquids as solvents, they can also act as a combined solvent–reagent system in microwave-assisted reactions. Leadbeater and coworkers have disclosed the synthesis of primary alkyl halides from the corresponding alcohols where the 1,3-dialkylimidazolium halide-based ionic liquids **23** serve as both reagents and solvents (Scheme 4.12) [76]. Depending on the employed ionic liquid, reactions could be completed within 30 s–10 min at 200 °C.

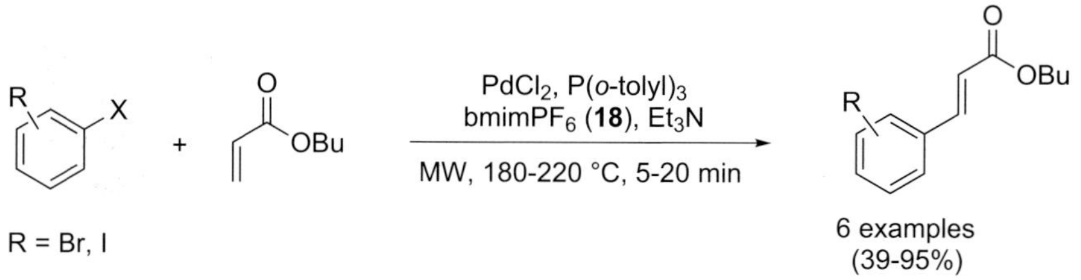

Scheme 4.11 Heck reactions in ionic liquids.

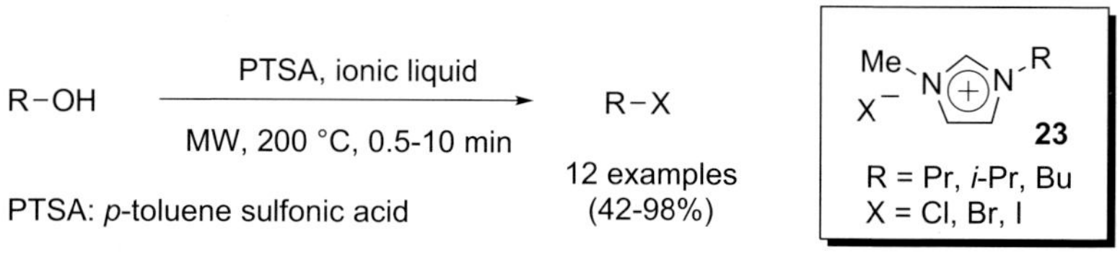

Scheme 4.12 Ionic liquids as reagents.

Ionic liquids can also serve as solvent–catalyst combination. In a recent publication, Friedel–Crafts sulfonylations were carried out using $FeCl_3$-based ionic liquids as both solvent and catalyst systems [77]. On the other hand, only 2 mol% of several imidazolium-based ionic liquids were utilized as catalysts in the solvent-free benzoin condensation [78].

As an alternative to the use of the rather expensive ionic liquids as solvents, several research groups have used ionic liquids as "doping agents" for microwave heating of otherwise nonpolar solvents such as hexane, toluene, tetrahydrofuran, and dioxane. First introduced by Ley and coworkers in 2001 for the conversion of amides to thioamides using a polymer-supported thionating reagent in a weakly microwave-absorbing solvent such as toluene [79], this technique is becoming increasingly popular as demonstrated by the many published examples [61–63]. One representative example is presented in Scheme 4.13. Under conventional reflux conditions in chlorobenzene at 135 °C, the intramolecular hetero-Diels–Alder reaction of alkenyl-tethered pyrazinone **24** required 1–2 days for completion. When performing the cycloaddition in 1,2-dichloroethane as solvent at 170 °C under microwave irradiation, the reaction time could be reduced to about 1 h [29]. To further accelerate the reaction,

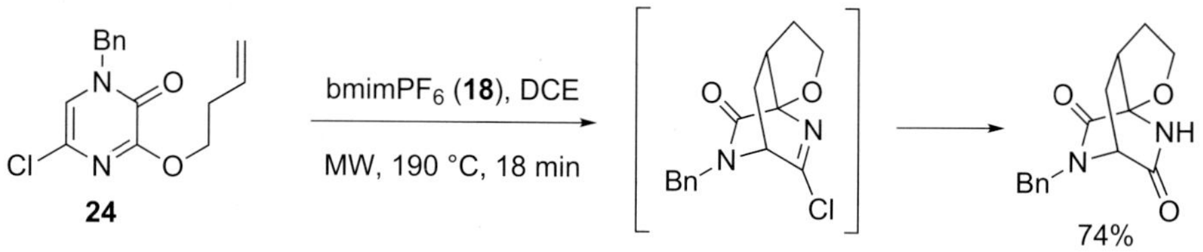

Scheme 4.13 Use of ionic liquid-doped 1,2-dichloroethane.

the authors doped the weakly microwave-absorbing DCE with a small amount of thermally stable bmimPF_6 (0.15 mmol for 2 mL DCE) in order to reach a higher temperature in a shorter time. Under these conditions, a temperature of 190 °C could be attained and the reaction was completed within 18 min (Scheme 4.13).

However, it should be noted, that the development of monomode microwave instrumentation that use higher power levels such as 400 or 850 W (see Section 3.4) has somewhat reduced the need to employ ILs as doping agents for low microwave-absorbing solvents, since higher temperatures can now be achieved for the reaction mixtures due to the increased microwave field density provided by these reactors.

The concept of performing microwave synthesis aided by room-temperature ionic liquids as reaction media has been applied to several different organic transformations, such as 1,3-dipolar cycloaddition reactions, ring-closing metathesis, Knoevenagel reactions, multicomponent reactions, and several others. These have been summarized in recently published reviews or book chapters [61–63, 80].

Systematic studies on temperature profiles and the thermal stability of ionic liquids under microwave irradiation conditions by the Leadbeater [81], Ondruschka [59], and Kappe groups [29, 60] have shown that even the addition of a small amount of an ionic liquid (0.1 mmol mL^{-1} solvent) suffices to obtain dramatic changes in the heating profiles by changing the overall dielectric properties (tan δ value) of the reaction medium (Table 4.3).

Table 4.3 Microwave heating effects of doping organic solvents with ionic liquids (ILs) **25** and **26**.[a)]

25

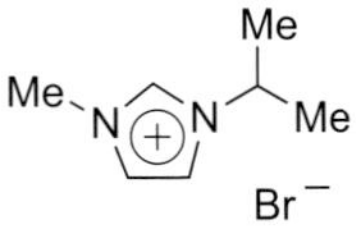

26

Solvent	IL added	Temperature attained (°C)	Time taken (s)	Temperature attained without IL (°C)[b)]
Hexane	25	217	10	46
	26	228	15	
Toluene	25	195	150	109
	26	234	130	
THF	25	268	70	112
	26	242	60	
Dioxane	25	264	90	76
	26	246	60	

Data from Ref. [81].

a) Experiments run using a constant 200 W irradiation power (CEM Discover), with 0.2 mmol IL/2 mL solvent under sealed-vessel conditions.

b) Temperature attained during the same irradiation time but without any IL added.

Despite the unique advantage of ionic liquids being very strong microwave absorbers and thus finding application as doping agents for poor microwave-absorbing solvents, some limitations arise. In particular, the use of ionic liquids is sometimes incompatible with certain reaction types and even small amounts of an ionic liquid may prevent specific reaction pathways. Cavaleiro and coworkers observed the complete decomposition of a specific porphyrin starting material in the presence of small amounts of $bmimPF_6$ as doping reagent in the course of Diels–Alder reactions [82]. Other authors have shown that the thermal stability of alkylimidazolium-based ionic liquids is reduced in the presence of nucleophiles due to reaction of the nucleophile with the alkyl groups of the IL [83]. Similar observations were made by Leadbeater and Torenius on the alkylation of pyrazoles with alkyl halides and, therefore, reactions that use or generate nucleophiles are generally not compatible with ionic liquids [81].

Stability studies of a neat IL at high temperatures have been presented in a 2007 publication [84]. The authors investigated the recyclability of a bicyclic imidazolium IL (*b*-3C-$imNTf_2$) in microwave-assisted Claisen rearrangements at 250 °C. In comparison to other ILs, these bicyclic imidazolium structures proved stable at temperatures up to 250 °C and could be reused several times.

In addition to the thermal stability and chemistry incompatibility problem, a more instrument design-related issue arises when biphasic mixtures (e.g., ILs that are insoluble in nonpolar organic solvents) are heated under microwave irradiation. The immiscibility creates a severe problem related to the accurate temperature measurement and reproducibility since differential heating will occur. Depending on the position of the IR sensor, either the temperature of the very hot ionic liquid phase (IR from the bottom) or the temperature of the cooler organic layer (IR from the side) will be recorded [60]. In order to avoid these problems, it is recommend to use ILs that are soluble only in the particular solvent in question. In general, differential heating of biphasic mixtures by microwave irradiation is likely to occur when effective stirring is not possible [60, 85], especially in solvent-free chemistry or in the presence of heavy slurries and viscous liquids (see also Section 4.1). In these cases, the use of internal temperature measurement as provided with some microwave reactors is highly advisable [67].

4.6
Passive Heating Elements

As already discussed in Section 4.5, microwave synthesis in low-absorbing or microwave-transparent solvents such as dioxane, tetrahydrofuran, toluene, hexane, or carbon tetrachloride (CCl_4) is often not feasible, since the required temperatures for a particular transformation to proceed cannot be reached. For this reason, many nonpolar solvents that are popular in conventional chemistry are potentially precluded from the use as solvents in microwave synthesis. To overcome this problem and to avoid the switch to a polar solvent, it is sometimes sufficient to add a small amount of an ionic liquid (see Table 4.3). As an alternative to ionic liquids, a small quantity of a strongly microwave-absorbing solvent can be added to an otherwise low-absorbing solvent. By adding only 5% of ethanol to 2 mL of CCl_4, the temperature can be raised from about 50 °C to 100 °C at 150 W constant power [60]. Water that is known to be only a moderate microwave-absorbing solvent can be doped with sodium chloride or TBAB [60]. All these above-mentioned methods are so-called invasive methods, having the disadvantage that the polarity of the original solvent system is modified. In particular, severe problems can arise when ionic liquids are employed due to a possible incompatibility of the ionic liquid with certain substrates.

Passive heating elements (PHEs) are noninvasive heating aids since they are chemically inert and therefore avoid the above-mentioned difficulties that arise with invasive heating aids. They are strongly microwave-absorbing materials (see Section 2.2) that transfer the generated heat via conduction phenomena to the reaction mixture, similar to conventional oil bath heating. Furthermore, the use of PHEs is more practical since they can be mechanically removed from the reaction mixture and thus facilitate the purification step.

However, not many publications employing PHEs exist in the field of MAOS. Carboflon (CEM), a fluoropolymer doped with carbon black, was applied in a siloxy-Cope rearrangement in hexane as solvent [86]. Importantly, desilylation of the Cope product could be avoided in contrast to the ionic liquid-doped rearrangement. Weflon (Teflon doped with graphite, Milestone), a related PHE, was employed in aza-Claisen rearrangements of allylic imidates to the corresponding amides [87]. Due to the higher temperatures that could be achieved, higher product yields in shorter reaction times were obtained. Since both types of PHEs are based on an organic polymer, problems regarding deformation or degradation occur when heated to high temperatures and for extended reaction times and their use is limited to about 200 °C bulk temperature [60].

In a 2006 study, silicon carbide (SiC) was introduced as a novel passive heating element for MAOS [60]. SiC is a very strong microwave absorber, has a high thermal conductivity, and a low thermal expansion coefficient [88]. Moreover, it is mechanically, thermally, and chemically resistant up to 1500 °C and is thus compatible with any solvent or reagent, virtually indestructible, and can be reused unlimited without any loss of efficiency. The SiC heating elements are available in different cylindrical shapes and therefore are compatible with the different vessel sizes of both single-mode and multimode instruments (Figure 4.3)

In Table 4.4, the heating performance of SiC is presented. Even microwave-transparent solvents like CCl_4 can be rapidly heated far above the boiling point that could not be reached without the added SiC heating element.

One application of SiC as noninvasive heating element was found in the Claisen rearrangement of allyl phenyl ether (Scheme 4.14) [60]. In general, thermal Claisen rearrangements require high temperatures and proceed quite slowly. When a solution of the allyl ether in toluene in a Pyrex vessel was heated under microwave heating, the attained 160 °C proved not high enough to induce the rearrangement. In contrast, by adding a SiC cylinder, 250 °C was easily realized within 30 s and full conversion to the desired allyl phenol was accomplished after 105 min (Scheme 4.14). In addition, the alkylation of pyrazole with alkyl halides previously reported by

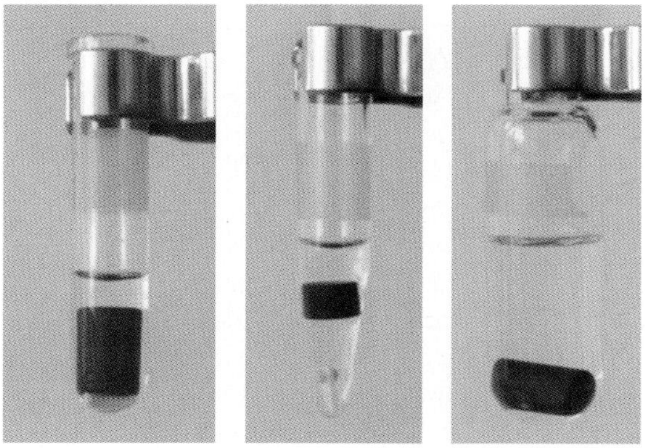

Figure 4.3 Differently shaped SiC cylinders (Anton Paar) in single-mode reaction vessels including magnetic stirrers. Reproduced with permission from Ref. [60].

Table 4.4 Temperatures of nonpolar solvents attained by microwave heating in the absence and presence of SiC heating elements.[a)]

Solvent	bp (°C)	*T* without SiC (°C)[b)]	*T* with SiC (°C)	Time taken (s)[c)]
CCl_4	76	40	172	81
Dioxane	101	41	206	114
Hexane	69	42	158	77
Toluene	111	54	231	145
THF	66	93	151	77

Data from Ref. [60].

a) CEM Discover, single-mode sealed-vessel microwave irradiation, 150 W constant power, 2 mL solvent, sealed 10 mL quartz or Pyrex vessel.
b) After 77–145 s of microwave irradiation (see footnote c).
c) Time until the maximum pressure limit of the instrument (20 bar) was reached and the experiment had to be aborted (with SiC).

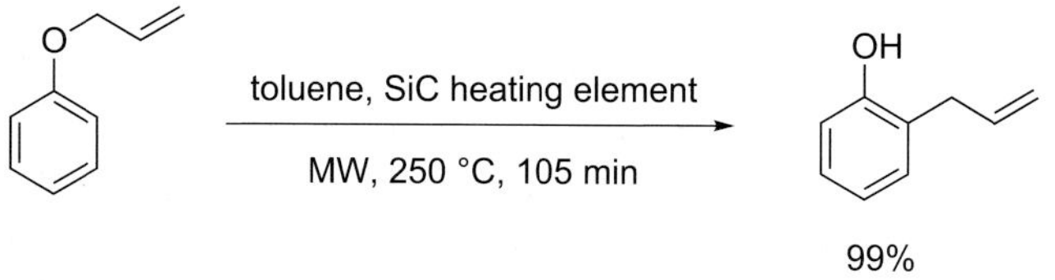

Scheme 4.14 Claisen rearrangement in toluene with the addition of a SiC cylinder.

Leadbeater as unfeasible using ionic liquids as doping agents was also successfully carried out [60, 81].

In a follow-up study, this method has been used to probe the influence of microwave power (electromagnetic field strength) on chemical reactions [89]. Six diverse types of chemical transformations were performed in the presence or absence of a SiC heating element at the same reaction temperature, but at different microwave power levels. In all six cases, the measured conversions/yields were similar regardless of whether a heating element was used or not. The applied microwave power had no influence on the reaction rate, and only the attained temperature governed the outcome of a specific chemical process under microwave conditions.

A further development was the design of a reaction vial entirely made out of sintered SiC ceramic that can be used for single-mode microwave chemistry (see Figure 3.5). This concept was introduced in 2009 and provides several unique opportunities in microwave chemistry [90]. Most importantly, it is a simple method for separating thermal from specific/nonthermal microwave effects since the SiC vial shields the contents inside the vial from microwave irradiation [90]. In addition, it is also perfectly suited for the utilization of aggressive reaction media in a high-temperature regime due to the high chemical resistance. Examples include high-temperature fluorine–chlorine exchange reactions using triethylamine trihydrofluoride, and the hydrolysis of nitriles with aqueous potassium hydroxide that could be otherwise problematic when performed in standard glass vials because of the corrosive character of the reagents [91].

4.7
Processing Techniques in Drug Discovery and High-Throughput Synthesis

The generation of diverse compound libraries has been shown to be a valuable tool for lead structure identification in the drug discovery process [92]. However, lead compound optimization and traditional medicinal chemistry remain the bottlenecks in this high-throughput discipline. Microwave-assisted heating under controlled conditions is an invaluable technology for medicinal chemistry and drug discovery applications because it often dramatically reduces reaction times, typically from days or hours to minutes or even seconds [2, 93]. Many reaction parameters such as reaction temperature and time, variations in solvents, additives and catalysts, or the molar ratios of the substrates can be evaluated in a few hours to optimize the desired chemistry.

As speed is a critical factor in the field of drug discovery and medicinal chemistry, the combination of microwave heating with high-throughput techniques for compound library generation is nowadays a popular and convenient application [93]. Therefore, it is not surprising that most pharmaceutical and biotechnology companies are already heavily using MAOS as frontline methodology in their chemistry programs, both for lead generation and lead optimization as they realize the ability of this technology to speed chemical reactions and, therefore, ultimately the drug discovery process [93].

4.7.1
Automated Sequential versus Parallel Processing

For the preparation of compound libraries, two different high-throughput techniques can be applied using microwave technology, the automated sequential and parallel approaches. Due to the typically short reaction times experienced with MAOS (minutes compared to hours), the concept of automated sequential microwave-assisted library synthesis in the single-mode instruments is a very attractive tool if small focused libraries containing about 20–100 compounds need to be produced. If larger libraries (>200 compounds) need to be generated, the sequential approach can become impractical since the time saving aspect of microwave synthesis is diminished by having to irradiate each reaction mixture individually and the parallel processing technique is favored.

Library synthesis in dedicated single-mode instruments can become as efficient as a parallel approach under conventional heating when robotic vial handling is integrated since it is currently not feasible to have more than one reaction vessel in a single-mode microwave cavity. Even more efficient are instrument setups where a liquid handler additionally allows dispensing of reagents into sealed reaction vials, while a gripper moves each sealed vial in and out of the microwave cavity after irradiation (see Section 3.4). Some equipment can process more than 100 reactions per run with a typical throughput of 12–15 reactions per hour in an unattended fashion. In contrast to the parallel synthesis application in multimode cavities, this approach allows the user to perform a series of optimization or library production reactions with each reaction separately programmed.

In an early case study, a 48-member library of dihydropyrimidines via the Biginelli reaction (Scheme 4.15) – a one-pot, three-component condensation of a CH-acidic building block (**27**), aldehyde (**28**), and (thio)urea (**29**) – was performed employing automated sequential processing in a fully automated microwave unit specifically designed for library production (Biotage Emrys Liberator) [94]. With the incorporated software liquid, dispensing of stock solutions or liquid reagents was possible and each experiment was generated and carried out individually. In summary, of the 3400 possible DHPM derivatives, a subset of 48 compounds was prepared within 12 h in 0.2–1 g scale. Compared to a conventional protocol, the reaction times were reduced from 3–12 h to 10–20 min, with initial reaction optimization being accomplished

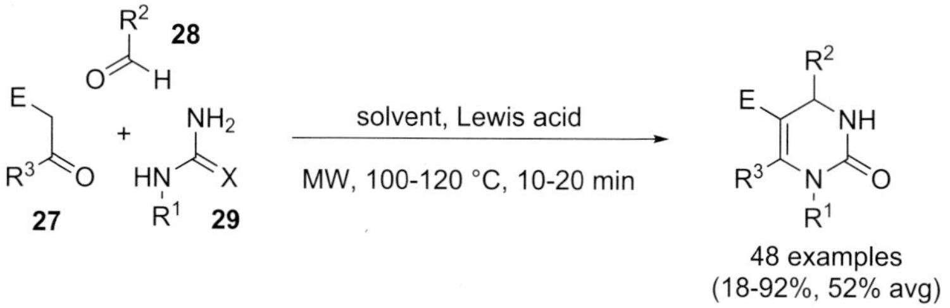

Scheme 4.15 DHPM-library generation employing the automated sequential process technique.

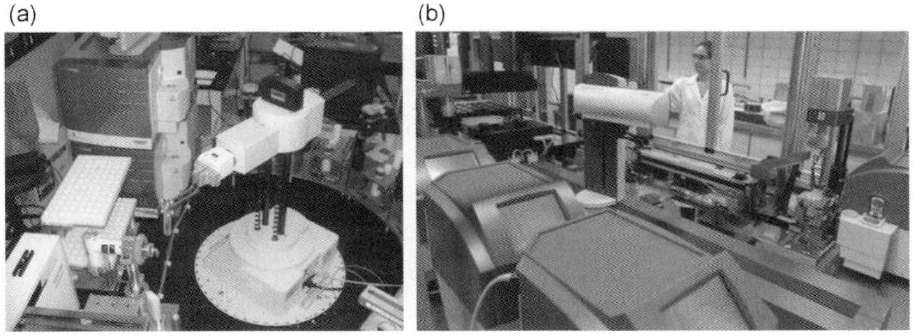

Figure 4.4 Abbott Laboratories (Illinois) robotic microwave facility (a) and Novartis (Basel, Switzerland) high-throughput microwave synthesis factory (b). Courtesy of D. Sauer (Abbott) and S. Chamoin (Novartis).

within a few hours. Importantly, the sequential treatment allowed the use of optimized conditions for specific building block combinations, not possible in parallel processing.

As an alternative to the instrument employed above, fully automated workstations that integrate a Biotage Initiator (Chemspeed SWAVE, see Figure 3.10), a single-mode instrument can be utilized for compound library synthesis. Due to the advantages of microwave heating for library synthesis, it is not surprising that pharmaceutical companies have established microwave facilities in their high-throughput synthesis divisions. As shown in Figure 4.4, at Abbott Laboratories (Illinois), a microwave station combining two single-mode instruments with additionally incorporated liquid reagent addition, automated capping, and solid-phase extraction (SPE) purification tools has been employed for high-throughput microwave synthesis. With this setup, a 480-member library of DHPM-5-carboxamides was generated by reacting 10 different DHPM-acid cores with 48 diverse amines (Scheme 4.16) [95]. A 76% success rate (365 compounds isolated) could be achieved with a 55% average isolated yield and >95% purity after SPE. The whole processing time per compound took 30–45 min. In

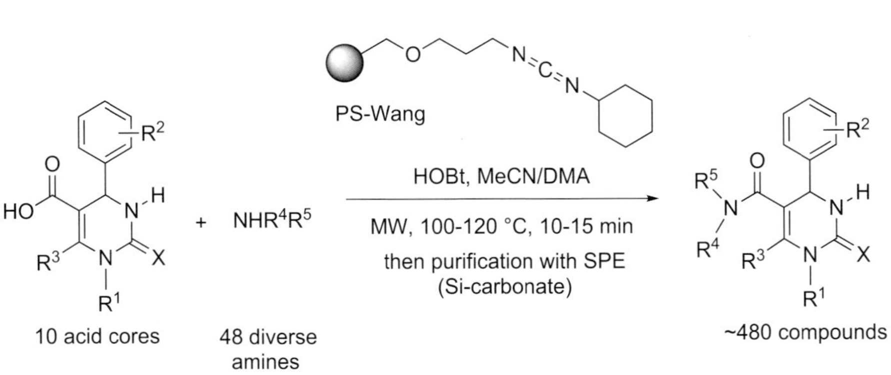

Scheme 4.16 Fully automated DHPM amide synthesis performed with the setup shown in Figure 4.4.

the Novartis high-throughput microwave synthesis factory (Basel, Switzerland), four single-mode instruments are integrated in the robotic station where several hundred reactions can be performed within 24 h (Figure 4.4).

Despite the current trend in the pharmaceutical industry to synthesize smaller, focused libraries, medicinal chemists in a high-throughput synthesis environment often still need to generate large compound libraries using a parallel synthesis approach. Parallel microwave synthesis can be performed in multimode instruments using either dedicated multivessel rotor systems or deep-well microtiter plates that allow higher throughput (see Section 3.5). The first published example of parallel reactions carried out under microwave irradiation conditions involved the nucleophilic substitution of an alkyl iodide with 60 diverse piperidine or piperazine derivatives [96]. Reactions were carried out in a multimode microwave reactor in individually sealed polypropylene vials using acetonitrile as solvent.

An important issue in parallel microwave processing is the homogeneity of the electromagnetic field in the microwave cavity. Inhomogeneities in the field distribution may lead to the formation of so-called hot and cold spots, resulting in different reaction temperatures in individual vessels or wells and thus different product conversions. Conducting microwave-assisted parallel synthesis in a reproducible manner is therefore often a nontrivial affair. In this context, investigations toward the reaction homogeneity in a 36 sealed-vessel rotor system (MicroSYNTH) were conducted [97]. For that purpose, 36 Biginelli condensations using 6 different aldehydes, ethyl acetoacetate, and urea as building blocks (see Scheme 4.15) were performed employing ethanol as solvent and hydrochloric acid as catalyst [97, 98]. Importantly, the yields of isolated products did not vary significantly depending on the position in the rotor, although slightly increased yields were obtained for mixtures that were placed in the inner circle, which would indicate a somewhat higher temperature in these reaction vessels.

Similar results were achieved when Biginelli reactions in acetic acid/ethanol (3 : 1) as solvent (120 °C, 20 min) were run in parallel in an eight-vessel rotor system (Anton Paar, Synthos 3000, see Figure 3.18b) on an 8×80 mmol scale [99]. Here, the temperature in one reference vessel was monitored with the aid of a suitable internal probe, while the surface temperature of all eight quartz reaction vessels was also monitored (deviation less than 10 °C). The product yield in all eight vessels was nearly identical and the same setup was also used for performing a variety of different chemistries in parallel mode [99].

The two-step synthesis of a 21-member library of polymer-bound enones depicted in Scheme 4.6 under open-vessel conditions was conducted in a parallel fashion employing PFA (perfluoroalkoxyethylene) vessels in a 50-position rotor (MicroSYNTH) [18]. Here, the temperature was monitored with the aid of a fiber-optic probe inserted into one of the reaction vessels. It was confirmed by standard temperature measurements performed immediately after the reaction period that the resulting end temperature in each vial was the same to within ± 2 °C.

The same group reported on temperature and reaction homogeneity studies for a 48-vessel rotor system introduced in 2006 (Synthos 3000, Figure 3.18a) by performing

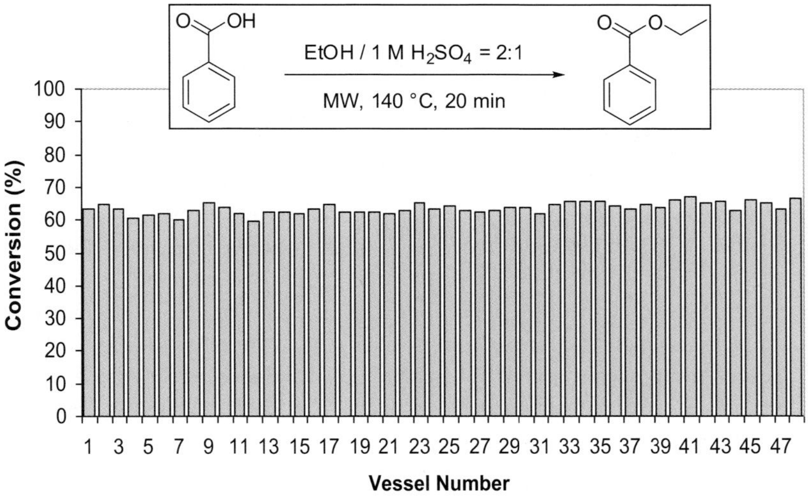

Figure 4.5 Reaction homogeneity for an esterification reaction in a 48-vessel rotor. Adapted from Ref. [100].

the acid-catalyzed esterification of benzoic acid with ethanol [100]. As can be seen in Figure 4.5, good homogeneity for this temperature-sensitive reaction between the individual vessels was guaranteed with only 1.7% standard deviation (60–67% conversion). Similar results were also obtained for a 64-vessel rotor system (Synthos 3000). The utility of the 48-vessel rotor system for library generation was demonstrated by synthesizing a set of 16 5-aroyl-DHPM derivatives by the Liebeskind–Srogl coupling of thiol esters with boronic acids at 130 °C within 1 h (Scheme 4.17) [100]. Only a minimal yield deviation (3–4%) compared to an automated sequential single-mode protocol was reported. Importantly, when performed in a parallel format, a significant time saving is possible for the library synthesis, since it requires 16 h to produce the 16-member library under automated sequential conditions.

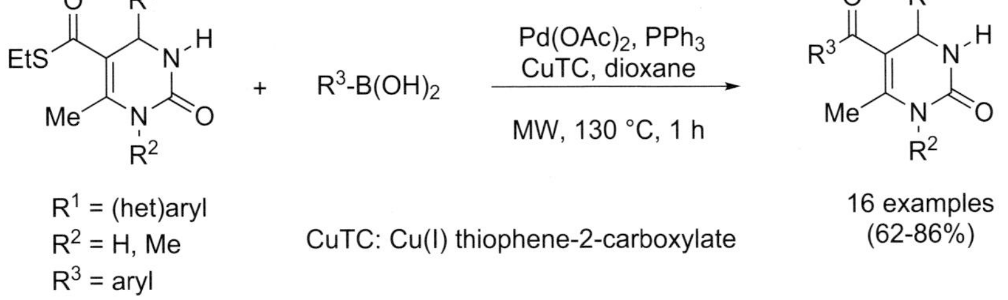

Scheme 4.17 Parallel thiol ester-boronic acid couplings performed in a 48-vessel rotor.

The same 48-vessel rotor system was employed by Kirschning and coworkers to perform multiple Suzuki–Miyaura couplings [101]. Boronic acids and organotrifluoroborates have been utilized with various palladium catalyst systems. The authors found that heterogenized palladium catalysts performed less effective in the multimode approach that was related to diffusion phenomena of the utilized PTFE vessels.

Since all reaction vessels, in parallel microwave synthesis, are exposed to the same irradiation conditions, problems regarding the individual conversions in specific vessels can occur. Investigations toward this phenomenon have been conducted by Leadbeater in a combined multimode and single-mode instrument (Milestone MultiSYNTH, Figure 3.27) [102]. In particular, when low microwave-absorbing solvents are used (or no solvent at all), the heating characteristics of the reaction mixture strongly depend on the microwave absorptivity of the building blocks, thus leading to different temperatures in the reaction vessels, and different conversions compared to a single-mode sequential method can arise. A case in point is the Michael addition of anilines with methyl acrylate, which is best performed under neat conditions at 200 °C, as demonstrated by a single-mode experiment (Scheme 4.18) [103]. Since this reaction is highly temperature dependent, lower temperatures in general give poorer yields, whereas higher temperatures may lead to side product formation and decomposition. When different anilines were heated in parallel, different product yields have been observed compared to the corresponding single-mode experiments. For example, decomposition of the reaction involving *m*-anisidine **30** was observed that indicated a temperature higher than 200 °C, whereas for *N*-ethylaniline **31** a lower conversion was obtained corresponding to a temperature lower than 200 °C (Scheme 4.18) [103]. These observations were additionally confirmed by the evaluation of individual heating profiles recorded with a fiber-optic probe in one reference vessel and an IR sensor that recorded the temperatures for all the rotor vessels. Moreover, the positioning of the fiber-optic sensor is crucial since the vessel with the fiber-optic probe is used as the control. For example, if the fiber-optic probe is placed in a vessel with high-absorbing reagents, a lower microwave

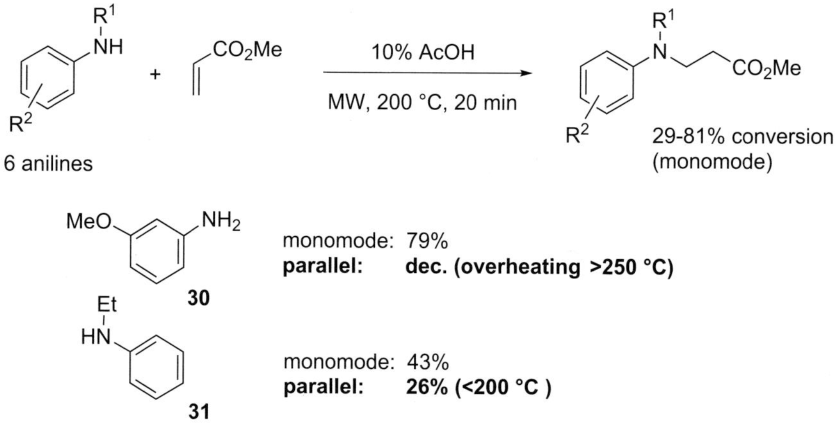

Scheme 4.18 Different microwave absorptivity of anilines in Michael additions performed in parallel.

power will be used to reach and keep the temperature. As a consequence, the set temperature will not be attained for lower absorbing substrates leading to lower yields. These effects could be minimized by either using a high-absorbing solvent or by adding polar additives such as TBAB [103].

In order to achieve an even higher throughput and to address the needs of the combinatorial chemistry community to produce several hundreds of compounds per day, microwave chemistry performed in microtiter plates in multimode instruments has emerged [104].

In a key 1998 publication, the concept of microwave-assisted parallel synthesis in microtiter plates was introduced for the first time [105]. Using the three-component Hantzsch pyridine synthesis as a model reaction, libraries of substituted pyridines were prepared in a high-throughput parallel fashion. Microwave irradiation in a domestic oven was carried out in the standard 96-well polypropylene bottom filtration plates that contained the corresponding 8 1,3-dicarbonyl compounds, 12 aldehyde building blocks, and ammonium nitrate adsorbed on clay. HPLC/MS analysis indicated that the reactions were uniformly successful across the 96-well reactor plate, without any residual starting material remaining.

Since then, several articles in the area of microwave-assisted parallel synthesis have described irradiation of 96-well polypropylene bottom filtration plates in conventional household microwave ovens for high-throughput synthesis. While some authors have not reported any difficulties associated with the use of such equipment [105], others have experienced problems in connection with the thermal instability of the polypropylene material itself [106] and with respect to the creation of temperature gradients between individual wells upon microwave heating [106, 107]. Figure 4.6 shows the temperature gradients after irradiation of a conventional 96-well plate for 1 min in a domestic microwave oven. For the particular chemistry involved, the 20 °C difference between inner and outer wells was, however, not critical. Furthermore, conducting pressurized reactions is troublesome in conventional microtiter plates due to inappropriate sealing devices.

While issues of temperature stability with standard polypropylene microtiter plates can be overcome to some extent by utilizing PTFE (Teflon) or HTPE (high-

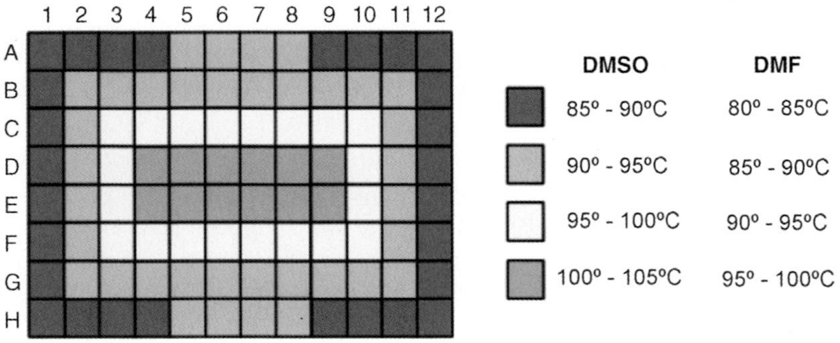

Figure 4.6 Temperature gradients within a microwave-heated microtiter plate; 1 mL per well, heated continuously for 1 min at full power in a conventional microwave oven. Adapted from Ref. [107].

temperature polyethylene) as plate materials [108, 109], dealing with transient and static temperature gradients across a microtiter plate is a nontrivial affair. In particular, the significant lower temperature attained in wells located on the outside region of the plate due to both a radiative heat loss from the plate to the ambient air and a lower microwave coupling (due to the absence of neighboring wells) constitutes a problem regarding lower conversions or product purities compared to reactions performed in wells near to the center of the plate. To address these problems, custom-built variations were designed by scientists from Sanofi-Aventis [108] and Boehringer Ingelheim [109].

To overcome the problems associated with using conventional polypropylene deep well plates in a microwave reactor, specifically designed well plates in combination with rotor systems for the combinatorial chemistry approach are available from the instrument manufacturers. For a detailed description of these rotor systems including microtiter plates, see Section 3.5.

The reaction homogeneity in a 24-well plate that consists of a basis of carbon-doped Teflon (Weflon) for better heat distribution with glass inserts as reaction vessels was investigated by monitoring the esterification of octanoic acid with 1-octanol at 120 °C for 30 min [98]. The difference in conversion between the individual vessels was 3% with a standard deviation of 2.7%. It appears therefore that all individual reactions were irradiated homogeneously in the applied microwave field. It is important to note that with this system, the material used for the preparation of the plates (Weflon) absorbs microwave energy, which means that the sealed glass vials will be heated by microwave irradiation regardless of the dielectric properties of the reactants/solvents.

This Weflon-based microtiter plate system has been used for a variety of library applications [110–112]. Alcázar has recently reported on the synthesis of a 24-member library of tertiary amines via the alkylation of four amines with six alkylation agents, after having confirmed the reaction homogeneity with a model reaction (yields 80–87%) [110]. Importantly, optimized reaction protocols from single-mode microwave synthesis performed in the Emrys Optimizer can be successfully transferred to this microtiter plate system when high boiling solvents are used in order to prevent a significant pressure buildup (2.3% average yield difference).

The preparation of a 96-member hexa-β-peptide library on solid phase employing a standard 96-well polypropylene bottom filtration plate in a dedicated rotor system (Microplate system for MARS reactor, Figure 3.25a) was reported by Murray and Gellman [113]. Prior to the library generation, a temperature homogeneity study was conducted by synthesizing model peptide **32** in 26 different wells distributed across the plate. As can be seen in Figure 4.7, lower purities were obtained in the outer wells. The authors have ascribed this to poor stirring that especially occurred in the outer regions of the plate, but could be overcome by switching to a smaller stir bar. Unfortunately, for the hexa-β-peptide library synthesis, it was necessary to couple one β-amino acid at a time due to the different microwave absorption characteristics that would otherwise lead to different temperatures in individual wells.

The deep-well plate systems described so far are limited to the use of high-boiling solvents under atmospheric pressure or to sealed vessels at low pressures up to 4 bar. With these setups, no direct translation from reactions performed under standard

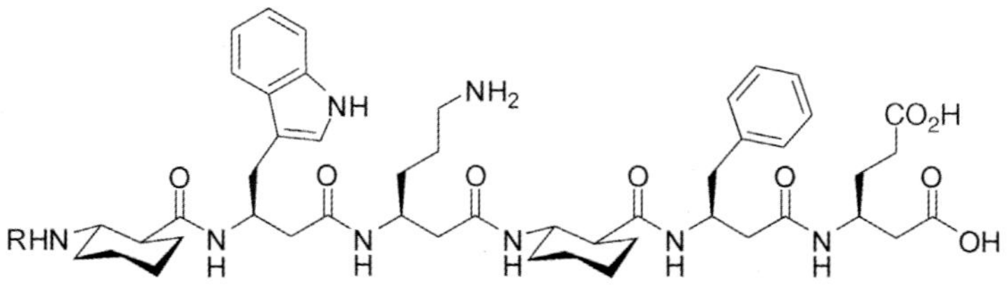

32

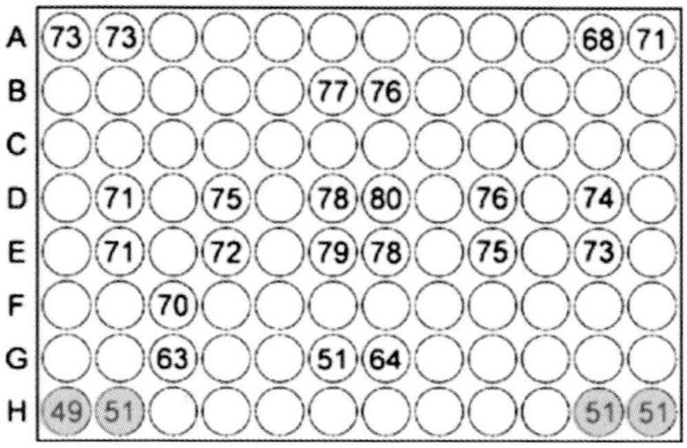

Figure 4.7 Model hexa-β-peptide **32** and its purity distribution across the plate. Reproduced with permission from Ref. [113].

single-mode sealed-vessel conditions (20 bar pressure) is possible. In a 2007 publication, a novel 48-well microtiter plate was introduced that is constructed for performing reactions under sealed-vessel conditions at elevated pressure similar to what can be achieved with a single-mode reactor (4 × 48-well plate system for the Synthos 3000, Figure 3.20) [114]. These special plates are made of strongly microwave-absorbing silicon carbide (SiC) (see Section 4.6) and can hold pressures up to 20 bar. Investigating the reaction homogeneity, it was found that by performing the esterification of benzoic acid with ethanol at 145 °C for 20 min, the conversions in all 48 wells were virtually identical with only 0.6% standard deviation (Figure 4.8). The successful application of the SiC microtiter plate setup was demonstrated by synthesizing a 30-member library of 2-aminopyridines by reacting a set of five 2-sulfonylpyrimidines with six diverse amines (Scheme 4.19) [114]. The generated

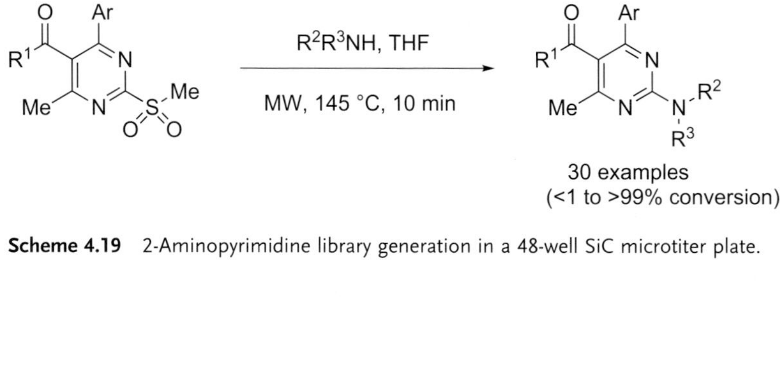

Scheme 4.19 2-Aminopyrimidine library generation in a 48-well SiC microtiter plate.

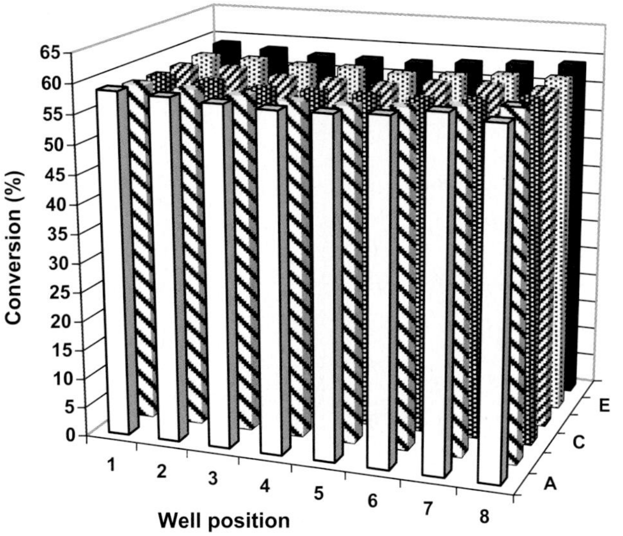

Figure 4.8 Reaction homogeneity for the esterification of benzoic acid in a 48-well SiC microtiter plate (Anton Paar Synthos 3000). Reproduced with permission from Ref. [114].

7 bar pressure was tolerated well by the system and in the majority of cases, high conversions to the corresponding 2-aminopyridines were achieved.

In a further development of this SiC microtiter plate, a 96-position rotor system that consists of four SiC blocks with a standard 6×4 matrix and holds disposable screw-capped 5 mL glass vials was designed (Figure 3.19b) [115]. With a filling volume of 0.3–3 mL, larger product quantities can now be synthesized for the hit-to-lead and lead optimization phases, while still maintaining the 200 °C temperature and 20 bar pressure limit, comparable with standard single-mode microwave instruments. Temperature homogeneity was confirmed again by the esterification of benzoic acid with ethanol at 140 °C for 25 min (see Figure 4.7), resulting in an average conversion to ethyl benzoate of 61.3% with a standard deviation of 2.5%. For the application of this rotor type for parallel library synthesis, the decoration of oxindole scaffolds by a Knoevenagel condensation (Scheme 4.20a) and the synthesis of multiple substituted benzimidazoles (Scheme 4.20b) were chosen [115]. By heating six oxindoles with four benzaldehydes in ethanol with substoichiometric amounts of piperidine at 140 °C for 10 min, the benzylidene oxindoles were obtained in good to high yields (50–98%). Starting from 8 substituted phenylene diamines and 3 carboxylic acids, 22 benzimidazoles could be prepared in 40–95% yield and with high purities after heating for 20 min at 145 °C.

In the same report, a 24-member library of 2-aminothiophenes by one-step Gewald synthesis was shown employing a 64-position rotor (Scheme 4.20c), where the same 5 mL glass vials are arranged in 16 groups of 4 vials [115]. No significant

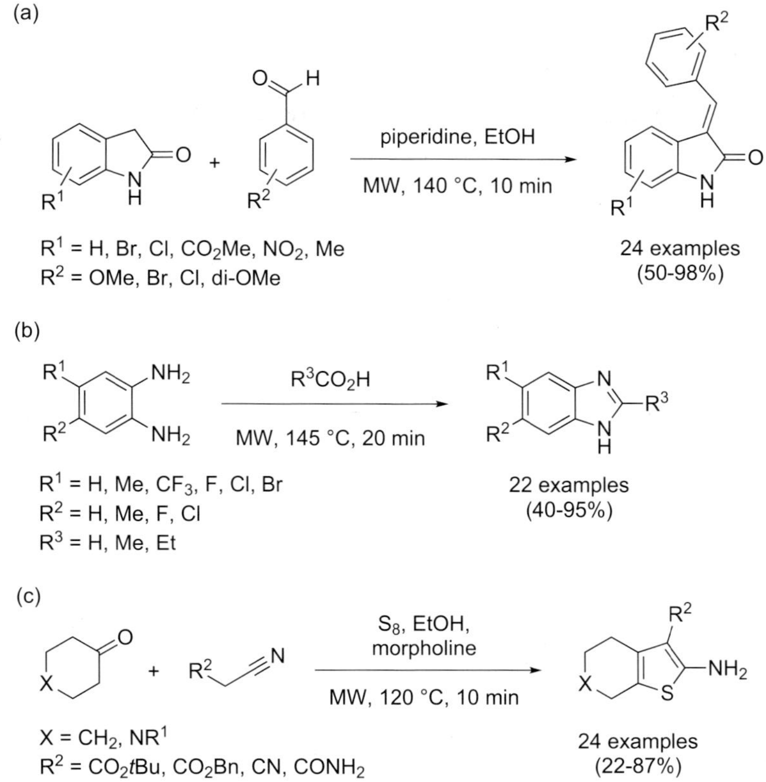

Scheme 4.20 Library generation of benzylidene oxindoles (a) and benzimidazoles (b) using a 6 × 4 SiC microtiter plate in a 96-position rotor and aminothiophenes (c) in a 64-position rotor.

difference in yield and purity was observed when comparing optimized results from a monomode instrument that were transferred to the rotor system of the multimode reactor.

The above-described 6 × 4 SiC microtiter plate was employed by the group of Leadbeater in the library synthesis of biaryls and 1,4-dihydropyridines (Scheme 4.21) [116]. In an initial study, the authors determined the heating characteristics and stirring efficiency of the plate, by performing the Suzuki coupling of 4-bromoanisole with phenylboronic acid (reaction conditions, see Scheme 4.21a) in eight vials placed in representative positions across the plate. Near-complete conversion was obtained in six vessels, in the other two vials – where no stir bar was added – a lower conversion of 72–74% was achieved, indicating the stirring importance. After having confirmed the temperature homogeneity and stirring efficiency within this plate system, a biaryl library was prepared via the Suzuki coupling of four aryl bromides with six arylboronic acids employing 0.002 mol% of a ligandless palladium catalyst, NaOH as base, a 1 : 1 mixture of H_2O/EtOH as solvent, and heating at 140 °C for 20 min (Scheme 4.21a). By merging six aldehydes with four β-dicarbonyl compounds and

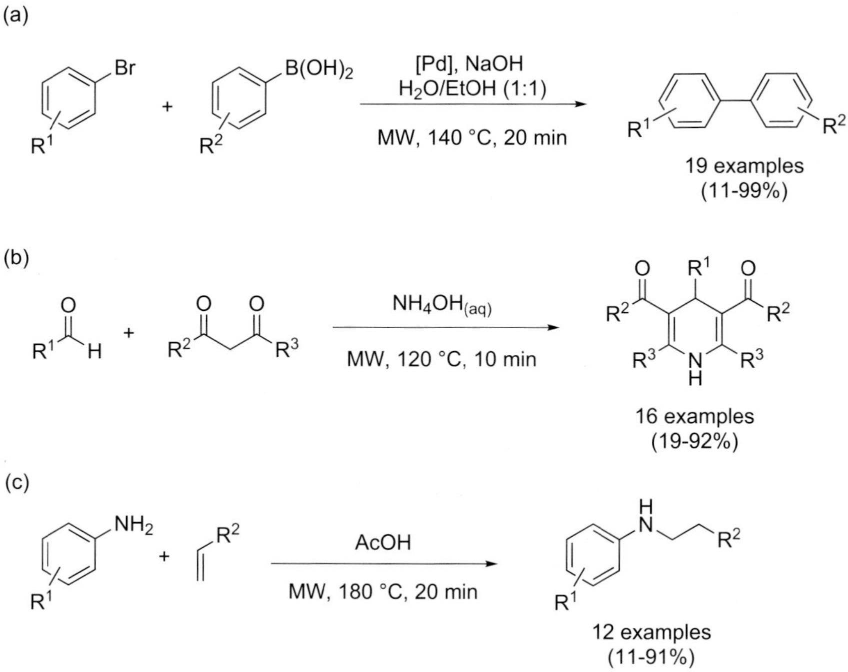

Scheme 4.21 Library generation of biaryls (a) and 1,4-dihydropyridines (b) using a 6 × 4 SiC microtiter plate in a 96-position rotor and *N*-aryl functionalized β-amino esters (c) using the 48-well SiC microtiter plate.

ammonia via the Hantzsch multicomponent reaction, a 1,4-dihydropyridine library was synthesized (Scheme 4.21b). In addition, the 48-well SiC plate was tested for the library synthesis of 12 *N*-aryl functionalized β-amino esters via an aza-Michael reaction of 4 anilines with 3 Michael acceptors (Scheme 4.21c) and for performing the proteolytic digest of insulin chain B by trypsin.

The same authors have reported on the library generation of bis-imidazolium salts and the corresponding Pd complexes with a subsequent screening of these complexes as catalysts in a Suzuki reaction where each step was performed in the 24-position SiC plate [117]. The combination of five imidazole-based substrates with four haloalkanes in ethyl acetate as solvent and heating at 140 °C for 10 min delivered the bis-imidazolium salts (Scheme 4.22). In the next step, these salts were converted to the analogous Pd complexes by reaction with $Pd(OAc)_2$ in THF (Scheme 4.22). After cooling, K_2CO_3 and a THF solution of 4-bromoanisole and phenylboronic acid were added to each vial for the Suzuki screen and the plate was heated for additional 5 min at 110 °C.

Another type of SiC platform that consists of a 5 × 4 matrix, in which standard HPLC/GC autosampler vials (Figure 3.19c) with filling volumes of 0.5–1.5 mL are employed, was introduced by Damm and Kappe in 2009 [118]. In combination with an aluminum sealing plate, a temperature/pressure limit of 250 °C/20 bar can be

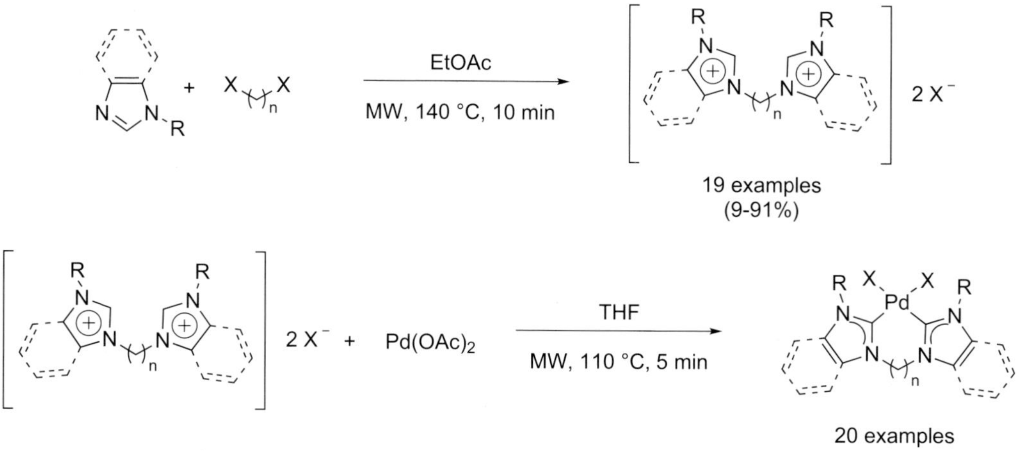

Scheme 4.22 Parallel ligand synthesis and formation of palladium complexes, which were screened in a Suzuki reaction, in a 6 × 4 SiC microtiter plate using a 96-position rotor.

achieved. After microwave processing, the vials can be directly transferred into the appropriate HPLC or GC autosampler racks for direct analysis, eliminating the necessity for reaction mixture transfer and leading to more time efficiency. This plate system has been used for optimization studies involving the parallel screening of catalyst, solvent and substrate reactivity for esterification reactions, and metal-catalyzed dehydrative C–C couplings [118]. The same authors have reported on the library generation of 39 2-styrylquinazolin-4(3*H*)-ones irradiating two SiC plates containing 20 vials each simultaneously [119]. The styrylquinazoline products **35** were synthesized via a two-step/one-pot protocol involving the initial three-component condensation of 4 anthranilic acids **33** with acetic anhydride and ammonium acetate at 250 °C for 30 min and a subsequent catalyst-free condensation of the resulting 2-methylquinazolinones **34** with a selection of 15 aromatic aldehydes (Scheme 4.23). An in-depth investigation, in particular toward heating characteristics of SiC microtiter-type plates, employing online thermoimaging cameras and multiple fiber-optic probe temperature sensors was performed by the same authors [120].

The issue of parallel versus sequential synthesis using multimode or single-mode cavities, respectively, deserves special comment. While the parallel setup allows a considerable higher throughput achievable in the relatively short time frame of a microwave-enhanced chemical reaction, the individual control over each reaction vessel in terms of reaction temperature/pressure is limited. In the parallel mode, all reaction vessels are exposed to the same irradiation conditions. In order to ensure similar temperatures in each vessel, the same amount of the identical solvent should be used in each reaction vessel because of the dielectric properties involved [98, 102]. However, this issue can be overcome when special plates made of strongly microwave-absorbing SiC are employed. It has been argued that because of the strongly microwave-absorbing SiC plate, the microwave absorption characteristics of the individual reaction mixtures contained in the wells/vials will be practically irrelevant,

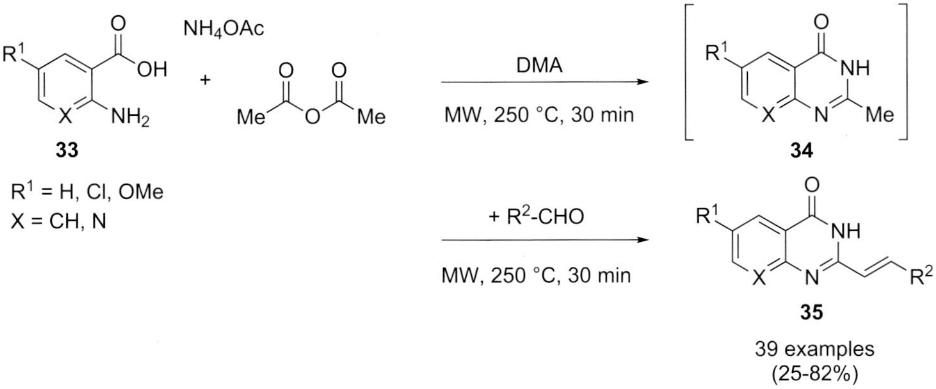

Scheme 4.23 Library generation of 2-styrylquinazolin-4(3*H*)-ones employing 20-position SiC plates containing HPLC/GC vials.

since the semiconducting plate itself absorbs microwave energy much stronger than any organic material contained inside the wells [118–120]. Hence, in temperature-controlled experiments using dedicated multimode reactors, solvents with different microwave absorption characteristics can be heated in parallel in individual wells/vials of the silicon carbide plate.

As an alternative to parallel processing, the automated sequential synthesis of libraries can be a viable strategy if small focused libraries (20–200 compounds) need to be prepared. Irradiating each individual reaction vessel separately gives not only better control over the reaction parameters but also fast iterations in protocol development and individual rapid optimization of reaction conditions is ensured. For the preparation of relatively small libraries, where delicate chemistries are to be performed, the sequential format may be preferable.

4.7.2
High-Throughput Synthesis Methods

Modern drug discovery relies on high-speed organic synthesis and high-throughput chemistry techniques for the rapid generation of compound libraries. Several high-throughput synthesis methods in conjunction with parallel or automated sequential high-speed microwave chemistry have proven to be very efficient for library production. In particular, methods that involve a polymer support are suitable for this approach since they facilitate purification and thus allow automation by using appropriate robotics for filtration and evaporation.

Since an in-depth overview of microwave-assisted transformations employing high-throughput synthesis techniques is presented in Chapter 8, only selected examples that highlight the usefulness of these methods are presented in the following sections. The reader is also referred to several recent reviews and book chapters [121–125].

4.7.2.1 Solid-Phase Synthesis

In solid-phase organic synthesis, a molecule is attached to a solid support and subsequent chemistry is then performed on the molecule until, at the end of the multistep synthesis, the desired product scaffold is released from the support. To accelerate reactions and to drive them to completion, a large excess of reagents can be used since this can easily be removed by filtration. Thus, final purification of the desired product is simplified, as by-products formed in solution do not affect the outcome of the target. Several articles reporting rate enhancements of this otherwise time-consuming technique by applying microwave irradiation have been published in the literature (see Section 8.1) [121].

As far as polymer supports for microwave-assisted SPOS are concerned, the use of cross-linked macroporous or microporous polystyrene (PS) resins has been most prevalent. In contrast to common belief, which states that the use of polystyrene resins limits reaction conditions to temperatures below 130 °C [126], it has been shown that these resins can withstand microwave irradiation for short periods of time, even at 200 °C for 20–30 min in solvents such as 1-methyl-2-pyrrolidone or 1,2-dichlorobenzene [127]. Standard polystyrene Merrifield resin shows thermal stability up to 220 °C without any degradation of the macromolecular structure of the polymer backbone, which allows reactions even at significantly elevated temperatures.

In recent years, there has been intense interest in the use of microwave irradiation in solid-phase peptide synthesis (SPPS) [128]. The first example of microwave-assisted SPPS of a decamer was published in 1992 [129] where a significant improvement in the coupling efficiency (two- to fourfold) was obtained. However, the procedure is not easily reproducible due to the use of a domestic microwave oven. With the availability of dedicated microwave instruments, the synthesis of a small tripeptide containing three of the most hindered natural amino acids was reported in 2002 by Erdélyi and Gogoll [130]. The authors observed enhanced couplings employing a microwave protocol compared to standard conditions without racemization.

Since a standard protocol for one SPPS cycle consists of four steps – deprotection, washing, coupling, and washing – the effort for the synthesis of longer peptide sequences in traditional dedicated microwave instruments using the standard glass vials is rather cumbersome due to the transfer of the resin suspension out of the microwave vial for each washing step. To overcome this problem, a MicroKan reactor that contains the resin beads can be introduced into standard microwave process vials [131]. MicroKans are made of a porous Teflon derivative that is fully penetrable by small molecules in solution but not by the resin particles. By using the MicroKan reactor, physical loss of resin particles is avoided [131]. Alternatively, microwave-assisted solid-phase reactions can be performed on SynPhase Lanterns that are rigid polymeric supports on which reagents can be attached [132].

In order to cope with the labor-intensive handling of resin beads in solid-phase peptide synthesis, automated microwave peptide synthesizers were introduced in 2003 (CEM Liberty, Figure 3.16) and 2009 (Biotage Syro *Wave*, Figure 3.9). These microwave instruments are able to perform all the necessary SPPS cycles in a fully automated fashion. In addition to the entirely automated microwave peptide

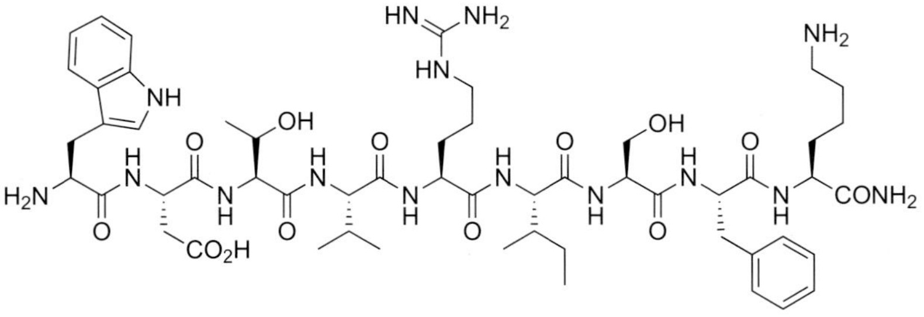

Figure 4.9 Model nonapeptide synthesized in a manual single-mode microwave peptide synthesizer.

synthesizer, manual versions are also available (CEM Discover SPS, Figure 3.16c, and Biotage Initiator Peptide Workstation, Figure 3.11). The reaction vessel in these instruments is designed for solid-phase synthesis, allowing bottom filtration and therefore mimicking the workflow of conventional peptide synthesizers. The preparation of a nonapeptide (Figure 4.9) using conventional Fmoc/*t*-Bu orthogonal protecting strategy has been described by employing the manual CEM instrument [133]. The coupling steps were performed within 5 min at 60 °C and the Fmoc-deprotection steps were completed within 3 min at 60 °C. The authors demonstrated that the model nonapeptide could be synthesized in a shorter time (about 3.5 h) and with higher purity (>95%) under microwave conditions compared to standard room-temperature methods (11 h).

In a 2007 publication, the group of Papini has conducted comparison studies for the synthesis of difficult α-peptide sequences performed either under conventional conditions in a standard peptide synthesizer or under microwave conditions in the automated peptide synthesizer (Liberty, Figure 3.16) [134]. For the hydrophobic antibiotic peptide Gramicidin A (15 mer) and the glycopeptide CSF114(Glc) (21 mer), the microwave SPPS approach was more effective in terms of yield and purity since deletion sequences occurred by the conventional strategy (Table 4.5). In addition, the reaction time for each coupling cycle could be reduced from 2 h to 30 min employing the Liberty system.

Table 4.5 Comparison of peptide yield and purity between conventional and MW-assisted SPPS strategies.

Peptide	SPPS strategy	Yield (%)[a]	HPLC purity (%)
Gramicidin A (15mer)	rt	11	<20
	MW	59	72
CSF114(Glc) (21mer)	rt	10	<20
	MW	46	72

Data from Ref. [134].

a) Yield of crude peptide, desalted.

The positive impact of microwave heating on the coupling and deprotection steps in peptide synthesis in terms of higher purities and enhanced reaction times could potentially be ascribed to a reduction in chain aggregation [128]. It had been originally proposed that the polar *N*-terminal amine group and polar backbone constantly try to align with the oscillating field and that this movement could lead to a deaggregation of the peptide backbones, thus allowing reagents to reach the reaction sites at the end of the growing chains more easily. However, carefully executed control experiments with proper internal temperature measurements have demonstrated that the enhancements seen in microwave-assisted solid-phase peptide synthesis are most likely due to a purely thermal effect [135].

An interesting alternative technique for conducting solid-phase synthesis is the so-called SPOT synthesis on planar supports. This method involves a spatially addressed synthesis on derivatized cellulose membranes (e.g., standard filter paper) to generate arrays of single compounds (1–10 000 spots per array) [136]. The membrane sheets are mechanically robust and moreover are compatible with various "on support" biological screening methods. This technique was applied for the preparation of an 8000-member library of 1,3,5-triazines on a 18 cm × 26 cm cellulose membrane via microwave-assisted nucleophilic substitution of the corresponding monochlorotriazines [137].

In a more recent SPOT synthesis study, the Blackwell group has performed chalcone and dihydropyrimidine syntheses (Scheme 4.24) [136]. In the first step, six diverse hydroxy acetophenones were spotted onto the cellulose support that was functionalized with an acid labile Wang-type linker and subsequently subjected to

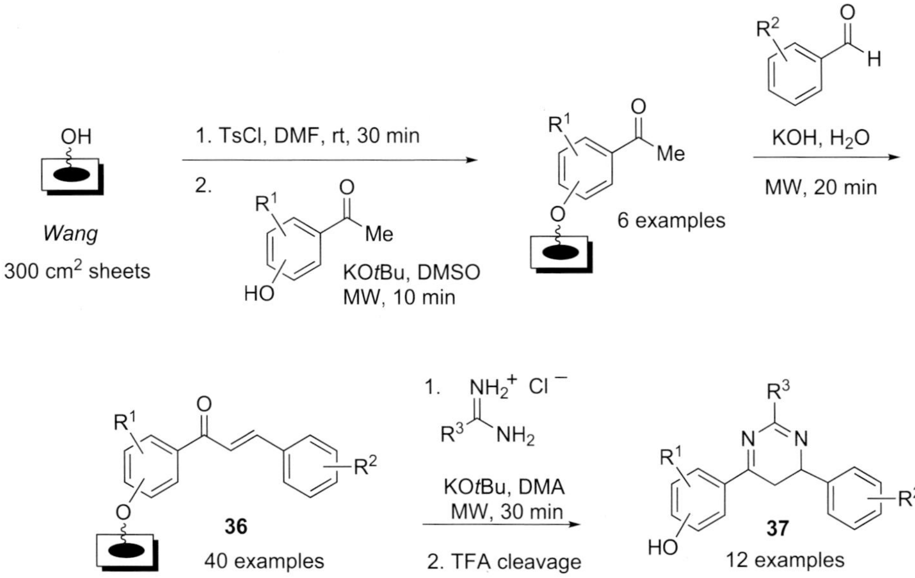

Scheme 4.24 SPOT synthesis of chalcones and dihydropyrimidines.

microwave irradiation in a multimode instrument for 10 min. The attached acetophenones were further reacted with several aryl aldehydes via Claisen–Schmidt condensation to give a set of 40 cellulose-bound chalcones **36**. The successfully generated chalcones could be cleaved by treatment with trifluoroacetic acid vapor or used for the subsequent synthesis of dihydropyrimidines **37**. The SPOT technique proved to be highly compatible with microwave conditions; furthermore, rapid access to compounds in small quantities (nanomolar–micromolar) – enough for characterization and biological screening – is feasible [136].

4.7.2.2 **Soluble Polymer-Supported Synthesis**

A viable alternative to solid-phase organic synthesis (SPOS) is the use of a soluble polymer as support [138]. Soluble polymer-supported synthesis offers several advantages over SPOS: the polymer support is not cross-linked and thus soluble in several organic solvents so that reactions can be carried out under homogeneous conditions, thus allowing standard spectroscopic characterization techniques to be used. Product isolation is typically performed by precipitation of the soluble support by addition of an appropriate solvent in which the support is insoluble (e.g., diethyl ether or hexane) with subsequent cleavage of the product from the support. Separation by membrane filtration or size-exclusion chromatography is an alternative. The most common soluble supports are polyethylene glycol (PEG) and monomethoxypolyethylene glycol (MeOPEG-OH). The synthesis of a set of diversely substituted benzimidazoles **39** was performed by Sun and coworkers starting from polyethylene glycol (PEG 6000) and 4-fluoro-3-nitrobenzoic acid to give the polymer-bound intermediate **38** (Scheme 4.25) [139]. In the first reaction sequence, ipso-fluoro displacement with

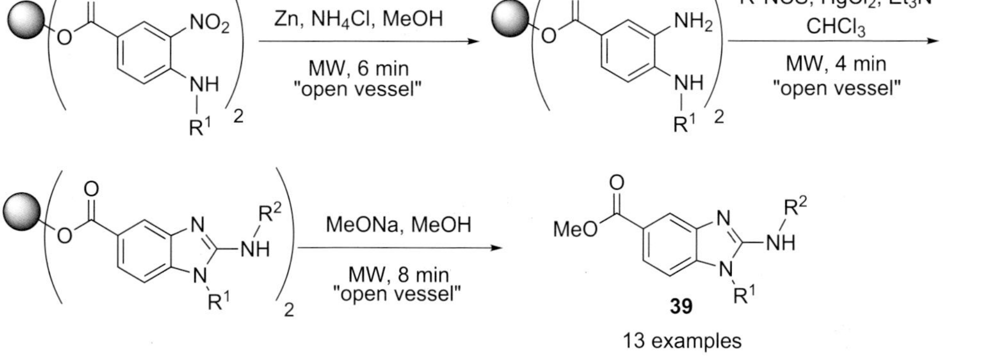

Scheme 4.25 Multistep benzimidazole synthesis on soluble PEG support.

various amines was conducted followed by nitro-group reduction. For the ring-closure step in one pot, the primary amines were converted to thioureas with a variety of isothiocyanates and subsequent intramolecular cyclization was possible under $HgCl_2$ mediation. All reactions were performed under open-vessel conditions in a CEM Discover and after each reaction step, the PEG-bound products were precipitated from a suitable solvent combination of dichloromethane and diethyl ether. The final benzimidazoles were cleaved from the support by using methanolic sodium methoxide and PEG 6000 was removed by precipitation with diethyl ether and filtration. Furthermore, the synthesis of bis-benzimidazoles using a similar protocol was performed by the same group [140] as well as the traceless synthesis of thiohydantoins [141] and hydantoin-fused β-carboline scaffolds [142].

More examples for the use of soluble polymers in MAOS can be found in reviews, book chapters, and in Section 8.2 [121, 133].

4.7.2.3 **Fluorous-Phase Organic Synthesis**

Fluorous-phase organic synthesis is a separation and purification technique for organic synthesis and process development that combines the advantages of solution-phase reaction conditions with the convenient purification of solid-phase synthesis [124]. Perfluorinated (fluorous) chains such as C_6F_{13} and C_8F_{17} instead of resin beads are employed as phase tags that facilitate product separation. Molecules on which a fluorous tag is attached can be easily isolated from the reaction mixture by fluorous separation techniques such as fluorous solid-phase extraction (F-SPE) on fluorous silica gel [143]. Compounds with a fluorous tag are soluble in a range of organic solvents allowing reactions to be conducted under homogeneous conditions and monitored with standard analytical methods such as TLC, HPLC, and NMR. In contrast to SPOS, more than one fluorous tagged compound can be employed in a single reaction.

Zhang *et al.* have reported on fluorous Suzuki couplings applying fluorous aryl sulfonates **40** as precursor that were obtained by the reaction of phenols with perfluorooctylsulfonylfluoride (Scheme 4.26) [144]. After the reaction with diverse boronic acids, the biaryl products were isolated via F-SPE by elution with $MeOH/H_2O$ 80 : 20 in high yields and with purities >90%, while the cleaved fluorous tag remained on the cartridge. Under similar reaction conditions, the perfluorooctylsulfonyl group

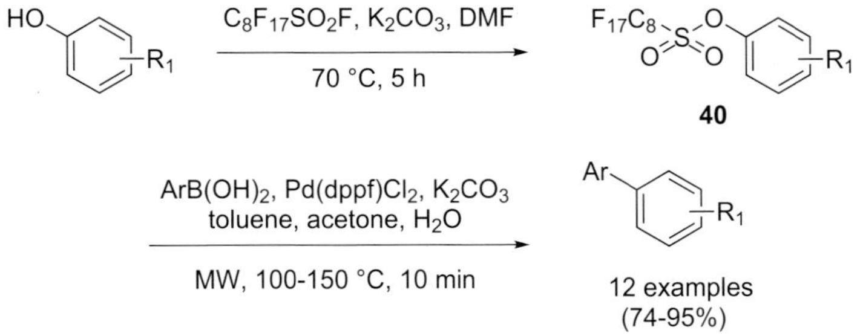

Scheme 4.26 Suzuki coupling of fluorous-tagged aryl sulfonates.

in **40** serves as traceless tag for palladium-catalyzed deoxygenations with formic acid [145].

In addition to fluorous tagged substrates that can be compared with polymer-bound substrates in SPOS, fluorous reagents, catalysts, and scavengers are also available. Although fluorous substrates are more suitable for multistep synthesis, in particular for combinatorial chemistry approaches, fluorous reagents are commonly used for single-step synthesis [124].

Numerous applications of microwave-assisted fluorous transformations have been discussed in the recent literature and are highlighted in Section 8.3 [121, 124, 146].

4.7.2.4 **Polymer-Supported Reagents, Catalysts, and Scavengers**

The use of polymer-supported reagents for solution-phase chemistry has attracted increasing attention in high-throughput organic synthesis [125, 147]. The polymer-assisted solution-phase (PASP) synthesis technique combines the benefits of SPOS in terms of workup with the advantages of solution-phase synthesis. Excess amounts of polymer-supported reagents can be used to drive reactions to completion without affecting the purification step. Considering the characteristics of polymeric material as support for catalysts and/or reagents, the key factors that have made their application increasingly popular are ease of separation after the reaction, recyclability, and the possibility of reinstating the catalytic activity postreaction. Importantly, reactions can be easily monitored in real time by conventional methods such as TLC, HPLC, or NMR. In addition, this technique is highly suitable for automation.

For example, the rapid synthesis of esters – starting from carboxylic acids and alcohols – in solution phase employing polymer-supported Mukaiyama-type reagent **41** was developed by the group of Taddei [148]. The 2-iodo-1-methylpyridinium salt attached to a PS-DVB resin (**41**) was efficient in activating carboxylic acids and thus various esters could be synthesized at 80 °C within 6–18 min (Scheme 4.27). Hindered alcohols such as cyclohexyl alcohol or long chain acids and alcohols reacted in good yields to the corresponding esters that were obtained with high purity after a simple filtration step.

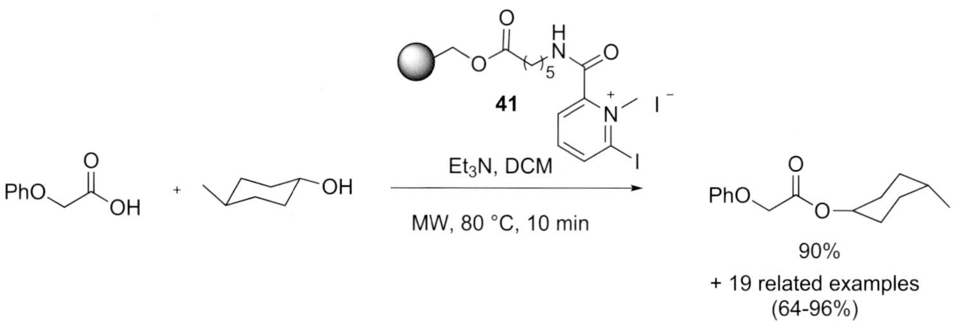

Scheme 4.27 Esterifications using polymer-supported Mukaiyama reagent.

Catalysts immobilized on solid supports (heterogeneous catalysts) have an important advantage over conventional homogeneous catalysts since workup is facilitated – the catalyst is simply filtered upon completion of the reaction [149]. In addition, the catalyst system can often be regenerated and recycled several times without significant loss of activity [149]. Furthermore, transition metal catalysts immobilized on polymer resins have significant benefits in reducing metal contamination of the reaction mixture and the product.

Immobilized palladium catalysts for microwave-assisted Suzuki couplings were introduced by Wang and Sauer (Scheme 4.28) [150]. In the so-called FibreCats, the palladium is coordinated to a phosphine ligand that is covalently bound to a polyethylene support. Compared to reactions that were performed under homogeneous catalysis with $PdCl_2(PPh_3)_2$, the supported palladium reactions were cleaner, and most importantly, no phosphine by-products that are usually difficult to remove were detected. Another feature of this protocol in terms of purification simplification is the use of carbonate-functionalized silica (Si-carbonate) for a solid-phase extraction step after the reaction in order to remove excess boronic acid.

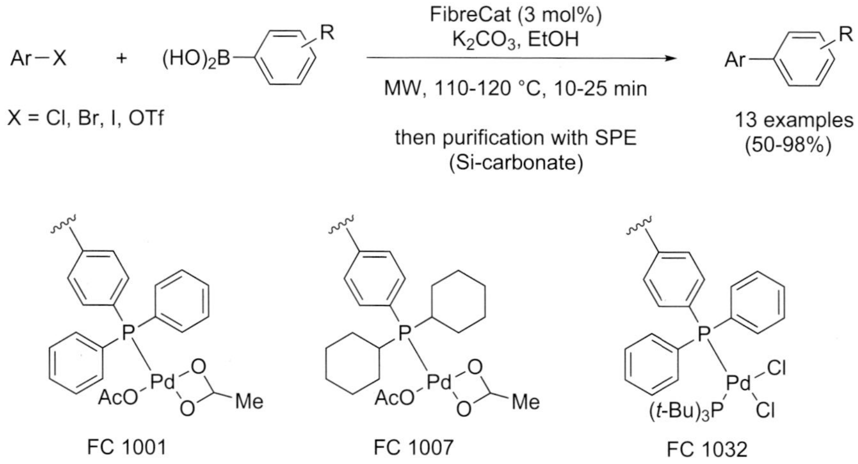

Scheme 4.28 Immobilized palladium catalysts for Suzuki couplings.

A palladium catalyst that is anchored to a glass/polymer composite material (Raschig rings, Figure 4.10) has been successfully applied for Suzuki couplings of aryl bromides and iodides with boronic acids in water as solvent by Dawood and Kirschning [151]. In a recyclability study, the authors demonstrated that under microwave conditions at 160 °C for 3 min, this palladium catalyst can be reused with almost full conversion (97%) up to the seventh run.

Several other supports for transition metals were introduced in the recent literature for a range of microwave-assisted coupling reactions. Ligand-free heterogeneous layered double hydroxide-supported nanopalladium proved to be a highly reactive catalyst system for Heck-, Suzuki-, Sonogashira-, and Stille-type couplings [152].

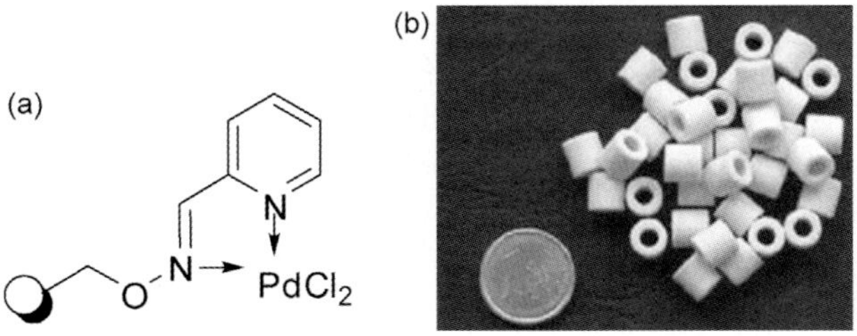

Figure 4.10 Palladium(II) complex (a) that is anchored to a glass/polymer composite material (Raschig rings) (b). Reproduced with permission from Ref. [151].

Suzuki and Heck reactions were catalyzed by palladium that was deposited as a thin film on the inner surface of capillaries (palladium-coated capillaries) [153] and nickel-in-charcoal demonstrated to be a very reactive catalyst for Negishi, Suzuki, and amination reactions under microwave conditions [154].

The isolation of a clean and homogeneous product is an integral part of any synthesis, and microwave heating has also been instrumental in improving many purification techniques. Therefore, polymer-supported scavengers – for removing excess reactants or by-products – play an increasingly important role in solution-phase combinatorial chemistry, although not many studies including the combination with microwave heating exist till date.

The microwave-induced *N*3-acylation of DHPM scaffolds using different anhydrides has been discussed in a comprehensive report [155]. The process included a microwave-assisted scavenging sequence to remove excess anhydride from the reaction mixture. Several polymer-supported sequestration reagents containing amino functionalities (polystyrene and silica supports **42** and **43**, StratoSphere Plugs **44**, and SynPhase Lanterns **45**) were employed for scavenging excess benzoic anhydride (Scheme 4.29). In both synthesis and purification, applying microwave heating reduced reaction times from several hours to minutes.

A very efficient method for the scavenging of metal contaminants from reaction mixtures was presented by Pitts and coworkers [156]. With the strict guidelines limiting metal levels in pharmaceuticals, there is a growing need for practical techniques for the removal of trace metals from reaction products, in particular, since homogeneous metal-catalyzed reactions are very popular reactions in medicinal chemistry. Packed cartridges of QuadraPure metal scavengers have been used to reduce levels of metal contaminants in flow. QuadraPure scavengers are functionalized macroporous polystyrene-based resins, highly cross-linked, and have low-swelling properties in organic solvents. In this study, copper by-products from a Rosemund-von Braun cyanation performed under microwave heating at 250 °C for 30 min were removed by passing the reaction mixture through an imino diacetate-functionalized QuadraPure cartridge. The copper content could be reduced from 345 ppm after the reaction to <1 ppm after the scavenging step. In the same way but with different functionalized scavengers, palladium, copper, iron, and rhodium were efficiently removed up to 99%.

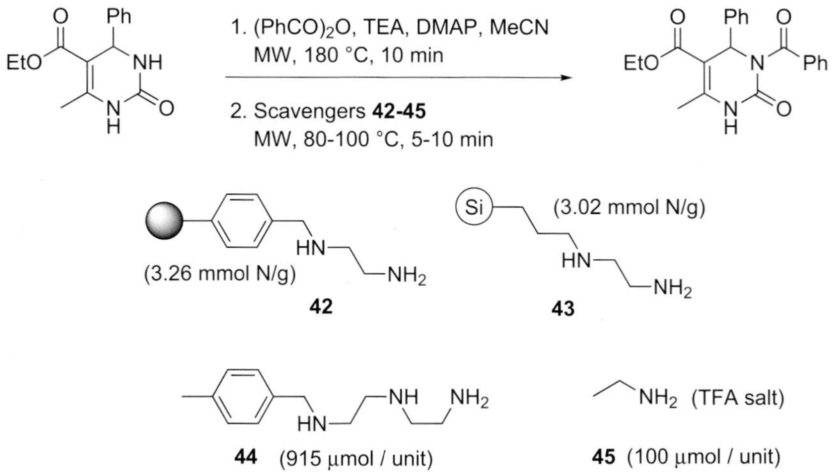

Scheme 4.29 N_3-Acylation of DHPM with subsequent anhydride scavenging employing different scavengers **42–45**.

Instead of scavenging excess reagents or by-products out of the reaction mixture, an alternative approach enables the selective capture of the product. In the course of the parallel synthesis of α-branched amines by microwave-assisted imine formation, the Ellman group applied resin capture of the amine product when the sulfinyl group of **46** was removed via acidic alcoholysis (Scheme 4.30) [157]. Cleavage of the sulfinyl group and concomitant amine capture was induced by macroporous sulfonic acid **47**. The final product was subsequently released with methanolic ammonia.

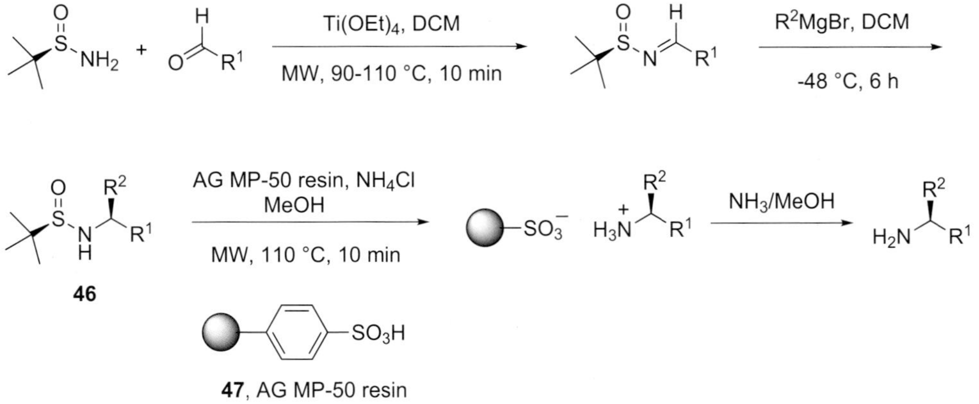

Scheme 4.30 Microwave-assisted resin capture.

An impressive example of applying polymer-assisted solution-phase synthesis was demonstrated by Ley group (Scheme 4.31). In the course of their investigations toward the application of polymer-supported reagents and scavengers in multistep

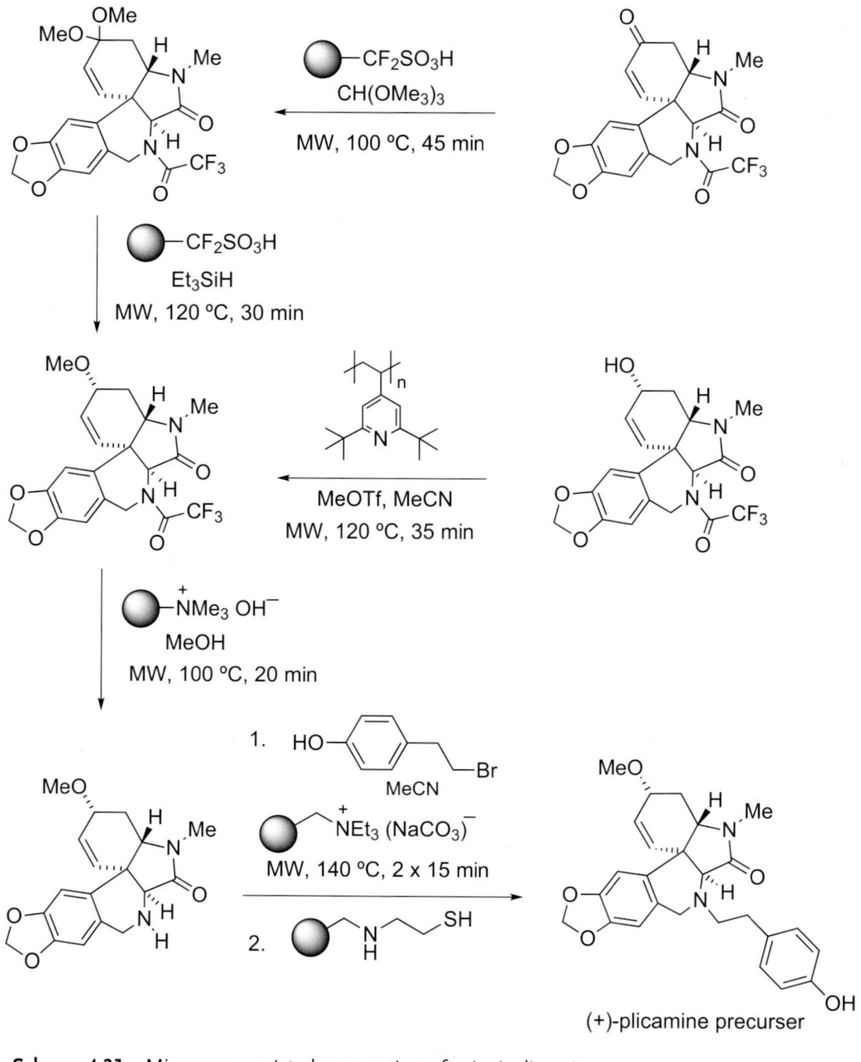

Scheme 4.31 Microwave-assisted preparation of a (+)-plicamine precursor.

synthesis of small compound libraries and more advanced natural products, they reported on the synthesis of the alkaloid (+)-plicamine where immobilized reagents and scavengers are employed in every single step [158]. No conventional purification such as chromatography or crystallization of the intermediates was necessary, only filtration was required and the precursors obtained could be used for further synthesis. In Scheme 4.31, the microwave-assisted sequences toward (+)-plicamine are highlighted.

More information on microwave-assisted polymer-assisted solution phase syntheses can be found in reviews, book chapters, and in Section 8.5 [121, 125].

4.8
Scale-Up in Batch and Continuous Flow

Most examples of microwave-assisted chemistry published to date have been performed on a less than 1 g scale with a typical reaction volume of 1–5 mL. This is in part a consequence of the availability and popularity of single-mode microwave reactors that allow the safe processing of small reaction volumes under sealed-vessel conditions by microwave irradiation (see Section 3.4). Due to limitations in the vessel and microwave cavity size of these single-mode instruments, microwave-assisted synthesis so far has focused predominantly on reaction optimization and method development on small scale (<10 mmol). While these instruments have been very successful in this field, it is clear that for microwave-assisted synthesis to become a fully accepted technology in the future, there is a need to develop larger scale MAOS techniques that can ultimately routinely provide products on a multikilogram (or even higher) scale.

Bearing in mind some of the physical limitations of microwave heating technology such as magnetron power or penetration depth (see Section 2.3), two different approaches for microwave synthesis on larger scale (>100 mL volume) have emerged. While some groups have employed larger batch-type multimode reactors (≤10 L processing volume), others have used continuous flow (CF) or stop-flow (SF) techniques in multimode and single-mode reactors to overcome the inherent problems associated with MAOS scale-up. An additional key point in processing comparatively large volumes under pressure in a microwave field is the safety aspect, as any malfunction or rupture of a large pressurized reaction vessel may have significant consequences. Furthermore, many of the large-scale reactors – due to safety concerns – do not have the same temperature/pressure ratings as modern single-mode instruments (300 °C/20–30 bar) (see Chapter 3). Moreover, solvents, reagents, and products should be stable at the required reaction temperatures, since instability and degradation of the reaction mixture may also lead to safety problems [159].

When heating rates of small volumes that are processed in single-mode instruments are compared with that of the larger volumes in multimode reactors, it becomes evident that in multimode instruments a higher microwave power has to be employed to reach identical reaction temperatures in the same time frame. Otherwise, lower heating rates are achieved or in some cases the set temperatures cannot be reached when performing large-scale experiments [160]. This is true in particular when low-absorbing solvents, such as toluene, are utilized. Maes and coworkers experienced this problem in the scale-up of a Buchwald–Hartwig reaction using toluene as solvent [161]. When the reaction is performed on a 1 mmol scale in a 10 mL vessel in a CEM Discover instrument with 300 W maximum power, the reaction temperature of 150 °C could be reached within 2 min. However, when performing the reaction on a 20 mmol scale in an 80 mL vessel in the same instrument, the set temperature of 150 °C could not be achieved within the 10 min reaction time. With the final maximum temperature being only 128 °C, incomplete conversion to product was observed (38% versus 76% isolated yield). Direct scalability and thus full conversion could only be achieved when switching from toluene to higher absorbing benzotrifluoride (BTF) as solvent, allowing to attain the 150 °C within the specified ramp time.

In addition to the heating profile, a significantly different cooling profile also needs to be considered when going from small-scale single-mode to large-scale multimode instruments. In particular, when batch reactors are employed for synthesis, one has to bear in mind that a longer cooling period is necessary and thus resulting in a longer total processing time [160].

With today's commercially available single-mode cavities having different vessel types available, scale-up in a linear fashion is feasible from 0.5 to 20 mL (Anton Paar Monowave 300), from 0.2 to 20 mL (Biotage Initiator EXP series), or from 0.2 to 50 mL (CEM Discover platform, see Section 3.4). The scale-up of the microwave-assisted reactions can be defined in different ranges depending on the discipline the user is involved with. In case of medicinal chemistry, a scale-up to 50 mL reaction volume – corresponding to a 10- to 100-fold scale-up performed in standard single-mode microwave vials and to multigram quantities of product – is a significant amount. On the other hand, in a preparative laboratory, the synthesis of >100 g compound quantity or the use of at least 1 L reaction volume, respectively, is required. Ideally, reactions should be directly scalable from small to large scale, heterogeneous mixtures should be processable, and the possibility for automation (sequential or continuous) should be feasible [159].

A possibility for further scale-up using the above-mentioned single-mode instruments would be using the "numbering up" approach, where repetitive cycles of small-scale runs are performed employing the automated sequential processing technique (Section 4.7.1). Alternatively, these reactions can also be conducted by parallel synthesis in multivessel rotor systems switching to multimode instruments.

While the use of microwave heating for performing reactions in the milligram to gram region is straightforward, scale-up of microwave synthesis from the laboratory to process and production scale has proven more difficult to achieve and is still a challenging area. Nevertheless, in the last few years, a significant number of microwave protocols have already made their way into kilo and process research laboratories, since microwave technology is increasingly required in early scale-up phases on a ~0.5–5 kg scale, specifically for the synthesis of key intermediates and active compounds.

Scale-up as defined for this section covers batch reactions in open vessels up to 3 L and closed vessels at the ≥50 mL scale, flow systems employing flow cells ≥5 mL, and stop-flow vessels of ≥50 mL volume. For an overview of and a more detailed information on the microwave instruments and accessories described in this section, see Chapter 3 and reviews cited in Ref. [162]. A detailed review on scale-up of microwave chemistry covering all commercially available instruments was published in 2011 by Moseley [163].

4.8.1 Scale-Up in Batch and Parallel

An important issue for the process chemist is the potential of direct scalability of microwave reactions, allowing rapid translation of previously optimized small-scale conditions to a larger scale. Several authors have reported independently the feasibility of directly scaling reaction conditions from small-scale single-mode (typically 0.5–5 mL) to larger scale multimode batch microwave reactors (20–500 mL) without reoptimization of the reaction conditions [20, 99, 164–166].

The successful scale-up of Suzuki couplings under open-vessel conditions using low palladium concentrations (Scheme 4.32) was demonstrated by Leadbeater *et al.* [167]. In order to prepare multigram quantities of biaryls but at the same time keep a high level of safety, a switch from sealed-vessel to open-vessel conditions employing standard round-bottom flasks was performed. Direct scalability without changing any of the reaction condition parameters was possible when going from a 5 mmol to a 1 mol scale. The small-scale reactions were performed in a 100 mL round-bottom flask in a single-mode instrument (CEM Discover), whereas for the scale-up approach, a 3 L reaction vessel in a multimode reactor (MARS, Figure 3.24) was employed. As discussed above, a higher microwave power (600 W initial power) is necessary to reach the reflux temperature in similar time when larger amounts need to be heated. In addition, efficient stirring is a problem on such large scales, but could be overcome by using an overhead paddle stirrer. The same authors have also reported on up to 1000-fold scale-up from the millimoles to mole region for Heck couplings [168] and *N*-heterocyclization reactions [169] under open-vessel conditions with the same instrument setup as described above.

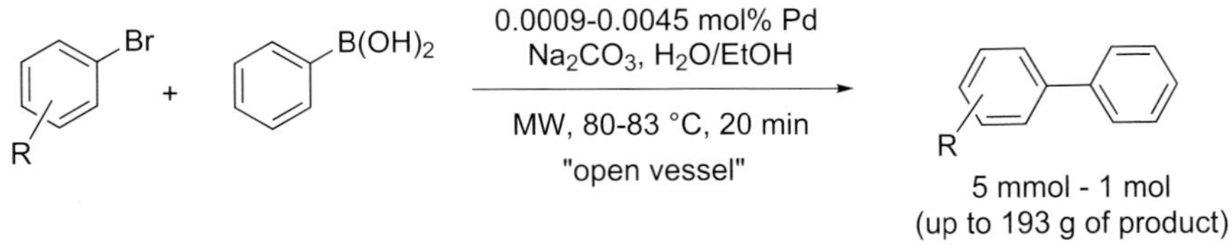

Scheme 4.32 Scale-up of Suzuki couplings under open-vessel conditions.

An early comprehensive study on the scalability of optimized small-scale microwave protocols in single-mode reactors to large-scale experiments in a dedicated multimode instrument utilizing multivessel rotors (Synthos 3000, Figures 3.17 and 3.18) has been presented by the Kappe group [99]. The examples performed included Biginelli dihydropyrimidine synthesis, Kindler thioamide synthesis, Heck and Negishi couplings, solid-phase amination, and Diels–Alder cycloaddition reactions. In all cases, the yields obtained in the small-scale single-mode experiments (1–4 mmol) could be reproduced on a larger scale (40–640 mmol) without the need of reoptimizing the reaction conditions. Despite the somewhat longer heating and cooling period, no appreciable difference in the outcome of the reactions studied was found.

The same instrument (Synthos 3000) was employed for the parallel scale-up of a range of pharmaceutically relevant reactions from 15 mL to 1 L and it was demonstrated that the synthesis of compounds on greater than 100 g scale is feasible in one batch [170]. In a 2010 study, the Kappe group reported on the scale-up of benzimidazole and pyrazole syntheses from 20 mL in single-mode instruments up to 960 mL in the Synthos 3000 with product quantities of ~0.5 kg within less than 1 h of overall processing time [160]. However, heating of a Diels–Alder toluene reaction mixture up to 250 °C proved more or less ineffective with a temperature ~115 °C being reached after 7 min of microwave irradiation with 1400 W magnetron power, which highlights the limitations of parallel scale-up for low-absorbing reaction mixtures (see also above).

Similar scale-up results were obtained utilizing a different multimode batch reactor with a single reaction vessel (Emrys Advancer, Figure 3.22). Mannich reactions (2 mmol → 40 mmol) [165], oxidative Heck processes (1 mmol → 10 mmol) [28], 2,5-diketopiperazine synthesis (7 mmol → 42 mmol) [171], ketone reductions (0.5 mmol → 81 mmol) [172], and aminocarbonylations (0.4 mmol → 25 mmol) [173] could be successfully scaled to larger product quantities. In a more recent report, a number of other reactions have been scaled up in the Advancer, most being performed between 50 and 400 mmol per batch, with the aza-Michael addition performed even on the 1.5 mol scale [174]. One particular example of pharmaceutical interest was the Grandberg synthesis of 2-methyltryptamine [174]. Again, yields were comparable on going from a small-scale single-mode reactor to a larger multimode reactor. Here, rapid cooling after the microwave heating step is possible by a patented expansion cooling process.

For the palladium-catalyzed ethoxycarbonylation of aryl iodides, the Milestone UltraCLAVE multimode reactor (Figure 3.36) has been used [175]. By employing the 3.5 L single vessel (maximum filling volume 2 L), the reaction was performed on a 1 mol scale (1.8 L volume). The unit was pre-pressurized with 27 bar of CO and 23 bar of N_2. However, the cooling time at the end of the reaction was rather long (45 min) due to the vessel size and thickness of the walls. In addition, six different aryl iodide substrates were converted to the corresponding products by using the parallel setup, where six vessels were charged with 50 mmol each (90 mL).

In a recent 2011 study, several microwave-assisted organic reactions, including Newman–Kwart and Diels–Alder reactions, Pd-catalyzed cross-couplings, heterocycle synthesis, aromatic substitution, and a Knoevenagel condensation have been scaled-up in a benchtop microwave batch reactor (Anton Paar Masterwave BTR, Figure 3.21) that uses a single 1 L reaction vessel [176]. A range of different solvents (high and low microwave absorbing) and varying reaction times (4 s–2 h) and temperatures (120 to 250 °C) have been explored in these investigations. For all studied transformations, it was possible to perform a direct scale-up (from 2 to 720 mL reaction volume), obtaining similar isolated product yields. A scalability up to 360-fold, when moving from 3 mmol up to 1.08 mol, was demonstrated and isolated product yields up to 300 g (2.5 mol scale) in a single run could be accomplished.

On the other hand, the fact that direct scalability is not always achievable, in particular when heterogeneous reaction mixtures are present, has been earlier reported by researchers from Merck [177]. During the synthesis of Rasta resins via living free radical polymerization (LFRP), it was discovered that when applying the same reaction conditions previously optimized in a single-mode instrument (180 °C, 10 min) in the multimode reactor (Biotage Emrys Advancer), after 2 min the temperature spiked to above 250 °C and instead of resin beads, a polymeric mass was obtained. One explanation for the spiking could be that the resin polymerized around the fiber-optic temperature probe is used in the multimode instrument for internal temperature measurement, whereas the single-mode instrument is equipped with an external IR sensor. However, when the reaction conditions are changed to 30 min at 160 °C and NMP is added as a spectator co-solvent, a 150-fold larger scale compared to single-mode experiments could be prepared with the same loading level.

4.8.2 Scale-Up Using Continuous Flow Techniques

Mainly because of safety concerns and issues related to the penetration depth of microwaves into absorbing materials such as organic solvents, the preferable option for processing volumes of >1 L under sealed-vessel microwave conditions is a continuous flow technique, although here the number of published examples using dedicated microwave reactors is limited [178, 179]. In such a system, the reaction mixture is passed through a microwave-transparent coil that is positioned in the cavity of a single-mode or multimode microwave reactor. The previously optimized reaction time under batch microwave conditions now needs to be related to a "residence time" (the time for which the sample stays in the microwave-heated coil) at a specific flow rate. While the early pioneering work in this area stems from the group of Strauss [180], others have since made notable contributions to this field, often utilizing custom-built microwave reactors or modified domestic microwave units. In addition, several groups reported on examples of continuous flow organic microwave synthesis utilizing modified dedicated monomode microwave reactors [181–183]. More information on these systems and on MAOS employing continuous flow techniques in general can be found in a 2007 review [178].

Much larger volumes than in single-mode reactors can be processed in continuous flow reactors that are housed inside a multimode microwave system. In a 2001 publication, Shieh *et al.* described the methylation of phenols, indoles, and benzimidazoles with dimethyl carbonate under continuous flow microwave conditions, using a Milestone ETHOS-CFR reactor [184]. The same authors also reported the usefulness of this general method for the esterification of carboxylic acids (up to 100 g product within 20 min) [185]. Similar results were also achieved for benzylations employing dibenzyl carbonate using the same continuous flow instrument [186].

In a 2007 study [187], Moseley and Lawton have reported preliminary results using the FlowSYNTH continuous flow reactor (an improved version of the Milestone ETHOS-CFR reactor (Figure 3.34). For their investigations, the Newman–Kwart rearrangement was chosen as model reaction since the authors have previously performed extensive studies under microwave batch conditions [85, 188]. Moreover, this reaction is completely homogeneous, an important issue in continuous flow synthesis regarding line clogging by solids or slurries. Newman–Kwart rearrangements were performed at 200 °C, which is the limit of the instrument, at a flow rate of 2.1 L h^{-1} that would correspond to a reaction time of 10 min under batch conditions (Scheme 4.33). However, the actual residence time was slightly shorter – about 6 min

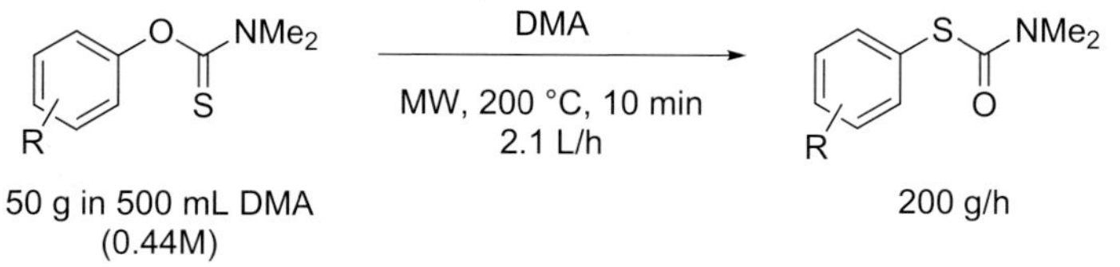

Scheme 4.33 Newman–Kwart rearrangements performed in the FlowSYNTH.

– resulting in reduced conversions, but this problem could be solved by simply passing partially converted reaction mixtures several times through the reactor, thus enabling longer reaction times. Employing the above-mentioned flow rate, 200 g of *S*-thiocarbamate product per hour could be generated [187]. In addition, with the incorporated product cooler, efficient cooling was provided; hence, a heating and cooling profile similar to small-scale runs in single-mode microwave instruments could be achieved. The reported results using microwave flow processing were in good agreement in terms of reaction times, yield, and purity with the previously reported batch microwave data for the Newman–Kwart rearrangement [187].

The FlowSYNTH continuous flow reactor was also used for performing pharmaceutically relevant reactions with production rates between 0.5 and 3.0 $\text{mol}\,\text{h}^{-1}$ ($1-6\,\text{L}\,\text{h}^{-1}$) [189], esterifications and aspirin synthesis [190], the racemization of *N*-acetylamino acids [191], and for the Bohlmann–Rahtz cyclodehydration reaction [192].

The scale-up of polymerization reactions under CF conditions is not a trivial affair since for the synthesis of well-defined polymers, not only a homogeneous heating profile is required but also a homogeneous concentration profile through the entire polymerization mixture to ensure narrow molecular weight distribution. In addition, the higher viscosities that are present might lead to undesired concentration profiles under microwave CF conditions. In order to investigate these issues, the Schubert group performed the previously microwave batch-tested cationic ring-opening polymerization of 2-ethyl-2-oxazoline in a monomode and a multimode reactor suitable for continuous flow conditions [193]. The authors have found that the polymerization results regarding molecular weight distributions and PDI (polydispersity indices) strongly depend on the flow profile (laminar versus tubular flow). In general, all the CF polymerizations resulted in broader molecular weight distributions, which is probably a result of the residence time distributions in the flow reactors.

The reactions reported so far under continuous flow conditions have been conducted in closed systems under pressure using back-pressure regulators. In contrast, the continuous flow synthesis of biodiesel under open-vessel conditions employing the CEM MARS in conjunction with flow cell accessories was demonstrated by Leadbeater and coworkers [194]. By using a 4 L flow vessel, a maximum flow rate of 7.2 $\text{L}\,\text{min}^{-1}$ could be achieved, producing 6.1 L of biodiesel per minute (99% conversion, see Scheme 4.34). The same reaction conditions as for batch synthesis in the same microwave unit could be applied – a 1 : 6 molar ratio of oil/methanol and 1 wt% KOH were heated to 50 °C and held there for 1 min, followed by

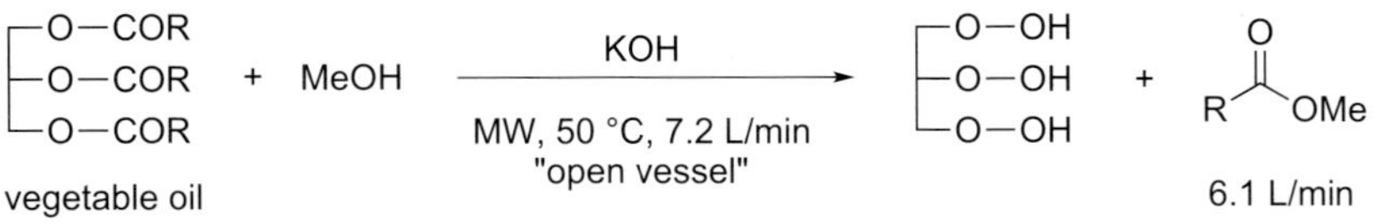

Scheme 4.34 Open-vessel CF preparation of biodiesel using a 4 L flow cell in the CEM MARS.

pumping new material into the flow cell. The preparation of biodiesel derived from butanol was described by the same authors [195].

4.8.3
Scale-Up Using Stop-Flow Techniques

As mentioned in Section 4.8.2, a serious problem with continuous flow reactors is the clogging of the lines and the difficulties in processing heterogeneous mixtures. Since many organic transformations involve some form of insoluble reagent or catalyst, so-called single-mode stop-flow microwave reactors have been developed (CEM $Voyager_{SF}$, Figure 3.15), in which peristaltic pumps – capable of pumping slurries and even solid reagents – are used to fill a batch reaction vessel (80 mL) with the reaction mixture (up to 50 mL). After microwave processing in batch, the product mixture is pumped out of the system that is then ready to receive the next batch of the reaction mixture.

In the course of investigations toward the scale-up of palladium-catalyzed Buchwald–Hartwig aminations of aryl chlorides using dedicated multimode and single-mode instruments, Maes and coworkers have employed the CEM $Voyager_{SF}$ (Figure 3.15) for the batchwise scale-up in the coupling of 4-chloroanisole with morpholine [161]. Two different stock solutions were prepared, one containing the palladium catalyst/ligand system in benzotrifluoride and the other the aryl chloride, with amine and base as well in BTF. Due to the low solubility of NaO*t*Bu in BTF, the second stock solution was not completely homogeneous. Three cycles of 20 mmol 4-chloroanisole were performed under the same conditions as in small scale (the solvent had to be changed from toluene to BTF due to low absorbance, see above) giving the product in nearly identical average yield (78% versus 76%, see Scheme 4.35). The same is true for the other Buchwald–Hartwig aminations performed in this study. Around 261 g (1.35 mol) of product **48** could be synthesized in 1 day since one complete cycle takes 16 min.

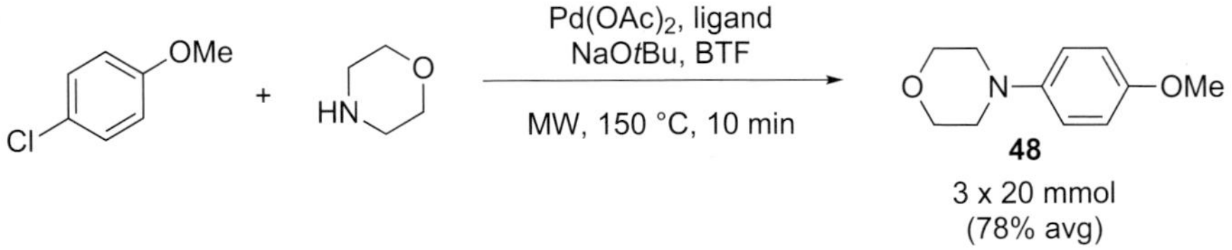

Scheme 4.35 Buchwald–Hartwig aminations under stop-flow conditions.

The same instrument has been employed by Leadbeater and coworkers for the scale-up of Suzuki and Heck couplings in water using ultralow palladium concentrations [196]. When going from a 1 mmol to a 10 mmol scale in the Suzuki reaction of 4-bromoacetophenone and phenylboronic acid using 250 ppb palladium at 150 °C, pumping the reaction out of the 80 mL vessel proved to be problematic since the biaryl product precipitated below 90 °C and blocked the exit tube. This problem could be solved by programming an additional step to the protocol: after cooling to 110 °C, the vessel is vented and ethyl acetate is added to dissolve the biaryl product. A 95% average yield corresponding to 18.6 g of biaryl product over 10 cycles of 10 mmol each – with 15 min processing time per cycle – was obtained. The reaction conditions for the Heck coupling of 4-bromoanisole and styrene at 170 °C had to be slightly modified as well. Here, a small amount of DMF had to be added to the reaction mixture in water to guarantee efficient pumping in and out of the reaction vessel.

The palladium-catalyzed ($Pd_2(dba)_3$/Xantphos) cyanation with $Zn(CN)_2$ as final step in the synthesis of citalopram, an antidepressant drug, could be scaled up under SF conditions as was demonstrated by Pitts and coworkers [197]. The reaction was performed at 160 °C and a cycle time of about 10 min per batch (200 s hold time under microwave irradiation). After four cycles and a total run time of 40 min, 47 g of citalopram was obtained.

Moseley and Woodman reported on SF protocols for a Heck coupling, Claisen and Newman–Kwart rearrangements, heterocycle formation, hydrolysis, and an alkylation yielding between 200 g and 1.4 kg per 24 h period [198].

A comparison study of seven commercially available microwave reactors for scale-up in terms of process chemistry has been conducted by Moseley *et al.* [199]. The scale-up of Newman–Kwart rearrangements (see Scheme 4.34) was evaluated in single batch (Biotage Advancer, Figure 3.22; Milestone UltraCLAVE, Figure 3.36; CEM MARS, Figure 3.23; Milestone MicroSYNTH, Figure 3.29), multibatch (Anton Paar Synthos 3000, Figure 3.17), stop-flow (CEM Voyager$_{SF}$, Figure 3.15), and continuous flow (Milestone FlowSYNTH, Figure 3.34) reactors under both open- and closed-vessel conditions. The authors reported a linear and reliable scalability from small-scale experiments to larger scale experiments (from 2 mL to >1 L) accomplished by each large-scale microwave reactor. In addition, a comparison of daily throughputs for the rearrangement of 2-nitrophenyl-*O*-thiocarbamate performed in the large-scale microwave instruments indicated in Table 4.6 was conducted. As highlighted in Table 4.6, up to 4 kg of product per day could be produced in the FlowSYNTH, but one should bear in mind that these data refer to a highly concentrated, homogeneous reaction. At present, no commercially available scale-up microwave reactor is capable of performing the majority of reactions important for pharmaceutical industry in a >1 kg scale. For these kinds of reactions, scale-up to several 100 g of product is feasible and useful for the top end of medicinal chemistry scale-up, but for genuine process chemistry, further improvements with regard to reaction volume and automation are desirable [159, 199].

Similar comparison studies have also been reported by Leadbeater *et al.* [200], Lehmann [159], and Schubert *et al.* [201] for evaluation of the scale-up of diverse reactions under both batch and continuous flow processing conditions.

Table 4.6 Daily throughput for the rearrangement of 2-nitrophenyl-O-thiocarbamate in different microwave instruments.

Reactor type	Amount of product (g)	Solvent (mL)	Cycle time (min)	Batches/day	Total daily throughput (kg)
Advancer	60	240	30[a]	16	0.96[b]
UltraCLAVE	400	800	100[a]	5	1.92
MARS	500	2000	96[c]	5	2.5
Synthos 3000	200	800	45[c]	10	2.1
$Voyager_{SF}$	10	40	16	30	0.3
FlowSYNTH	500 g h^{-1}	2000	2.1 L h^{-1}	Cont.	4.0

Data from Ref. [199].
a) Good estimate.
b) Assumes automated.
c) Extrapolated from similar reaction conditions.

4.8.4
Microwave Reactor Systems for Production Scale

The concept of performing reactions in a flow format using microchannels (microreactors) is a recent addition to the powerful toolbox of continuous manufacturing technologies [202–204]. Notably, there have been some attempts to integrate microwave heating with microreactor technology, but so far no commercially available solution has been developed. Examples of employing microwave heating in combination with the use of microreactors have been described by Haswell and coworkers [205] and Hessel and coworkers [206]. Another custom-built microreactor-type device employing glass capillaries integrated in a Biotage single-mode microwave unit was developed by Organ and coworkers [207].

In contrast to the merging of microreactor – with microwave technology for performing reactions in the milligram range, the largest microwave reactor for organic synthesis so far is a pilot plant-scale prototype installed at Sairem in France, developed and designed in collaboration with BioEurope and De Dietrich. This custom-built 1 m^3 reactor with a powerful 6 kW microwave generator was used for the production of Laurydone [208]. Running in a batch-type recycling process, the equipment accomplished a 40% power reduction compared to the classical thermal approach. Moreover, the overall processing time could be reduced by 80%. Recently, another microwave flow reactor has been developed by the pharmaceutical supplier Cambrex (GENESIS), working in collaboration with C-Tech Innovation, with maximum operating conditions of 200 °C, 20 bar, and flow rates up to 200 mL min^{-1} www.cambrex.com. A Hantzsch reaction was performed at 150 °C in only 1 min at 20 mL min^{-1}, giving the dihydropyridine product on a 1.8 kg scale. In addition, a Suzuki reaction with low palladium loading (5 ppm) was reported with a processed volume of 140 L and a flow rate of 60–70 mL min^{-1} giving 21 kg of the biaryl product within 32 h continuous run time.

A prototype large batch microwave reactor was introduced in 2010 by AccelBeam Synthesis, allowing the processing of reaction volumes from 2 to 12 L in a single vessel. The reactor employs three magnetrons with an accessible power of 2.5 kW each with a maximum output power of 7.5 kW. The operating limits are 250 °C and 24 bar. The Leadbeater group has reported on the direct scale-up of a range of reactions, including palladium-mediated transformations, condensation reactions, nucleophilic aromatic substitution reactions, and alkylations [209]. In some cases, reactions were scaled over 18 000-fold when moving from small (0.1–1 mmol) to large (1–18 mol) runs. However, this device was not converted into a commercially available solution till date.

Critically evaluating the currently available instrumentation for microwave scale-up in batch and continuous flow (Chapter 3), one may argue that for processing volumes of <1000 mL, a batch process is the preferable option. By carrying out sequential runs in batch mode, kilogram quantities of product can easily be obtained. When larger quantities of a specific product need to be prepared on a regular basis, it may be worthwhile evaluating a continuous flow protocol. Large-scale continuous flow microwave reactors (flow rate 20 L h^{-1}) are currently under development [162, 163, 210]. However, at the present time, apart from the prototype at Sairem, there are no further documented published examples of the use of microwave technology for organic synthesis on a production-scale level (>1000 kg), which is a clear limitation of this otherwise successful technology.

In the context of microwave-assisted reactions on a larger scale, the question of energy efficiency and therefore of the relative greenness and sustainability of microwave heating compared to conventional heating processes needs to be addressed. Recently published data suggest that in general, microwave processing under sealed-vessel conditions (taking advantage of increased reaction rates at higher temperatures) will be significantly more energy efficient than the conventional heating in open vessels at the solvent reflux temperature [211]. While thermal reflux processes often require several hours to reach completion, the same transformations can sometimes be completed within a few minutes using sealed-vessel microwave heating. It is important to note that the energy savings in these cases are mainly the result of the significantly shortened reaction times and are not directly connected to the heating mode. This has been made evident by comparing the consumed energy from open-vessel microwave heating experiments with the corresponding data from conventional thermal reflux heating runs, lasting for the same period of time [211]. In all the investigated cases, the thermal runs were more energy efficient than the microwave experiments, independent of the scale, the absorbance characteristics of the medium, or the particular microwave instrument used (single-mode or multimode). This can be rationalized by considering the moderate energy efficiency (50–65%) of a magnetron, the central component of any microwave reactor transforming electrical energy into electromagnetic irradiation [162].

Similar reports have been published [212, 213], stating that microwave heating is not very energy efficient on small to moderate scale, but becomes more efficient when going to larger scale in multimode instruments – in particular, continuous flow reactors, but still with overall energy efficiencies of only ~30% [212]. As with the other studies, it was shown that energy efficiency fell with longer reaction times, lower temperatures, and low-absorbing solvents [213].

A full energy balance between microwave and conventional heating would also have to take into account the losses through the glassware, the oil bath, and the condenser. In order to evaluate the overall ecobalance of a particular process, the consumed energy for pre- and posttreatment of the reaction mixture should also be considered. The widespread general opinion on the relative "greenness" of microwave heating in chemical processing – at least in terms of energy efficiency – however needs to be critically questioned [214].

References

1 (a) Loupy, A., Petit, A., Hamelin, J., Texier-Boullet, F., Jacquault, P., and Mathé, D. (1998) *Synthesis*, 1213–1234; (b) Varma, R.S. (1999) *Green Chemistry*, **1**, 43–55; (c) Kidawi, M. (2001) *Pure and Applied Chemistry*, **73**, 147–151; (d) Varma, R.S. (2001) *Pure and Applied Chemistry*, **73**, 193–198; (e) Varma, R.S. (2002) *Tetrahedron*, **58**, 1235–1255; (f) Varma, R.S. (2002) *Advances in Green Chemistry: Chemical Syntheses Using Microwave Irradiation*, Kavitha Printers, Bangalore.

2 Kappe, C.O. (2004) *Angewandte Chemie International Edition*, **43**, 6250–6284.

3 Strauss, C.R. and Varma, R.S. (2006) *Topics in Current Chemistry*, **266**, 199–231.

4 Varma, R.S. and Dahiya, R. (1997) *Tetrahedron Letters*, **38**, 2043–2044.

5 Varma, R.S., Saini, R.K., and Dahiya, R. (1997) *Tetrahedron Letters*, **38**, 7823–7824.

6 Varma, R.S. and Dahiya, R. (1998) *Tetrahedron Letters*, **39**, 1307–1308.

7 Varma, R.S. and Saini, R.K. (1998) *Tetrahedron Letters*, **39**, 1481–1482.

8 Kabalka, G.W., Wang, L., Namboodiri, V.N., and Pagni, R.M. (2000) *Tetrahedron Letters*, **41**, 5151–5154.

9 Varma, R.S. and Ju, Y. (2006) Chapter 8, in *Microwaves in Organic Synthesis*, 2nd edn (ed. A. Loupy), Wiley-VCH Verlag GmbH, Weinheim, pp. 362–415.

10 Besson, T., Thiéry, V., and Dubac, J. (2006) Chapter 9, in *Microwaves in Organic Synthesis*, 2nd edn (ed. A. Loupy), Wiley-VCH Verlag GmbH, Weinheim, pp. 416–455.

11 Cho, H.Y., Ajaz, A., Himali, D., Waske, P.A., and Johnson, R.P. (2009) *The Journal of Organic Chemistry*, **74**, 4137–4142.

12 (a) Deshayes, S., Liagre, M., Loupy, A., Luche, J.-L., and Petit, A. (1999) *Tetrahedron*, **55**, 10851–10870; (b) Loupy, A., Petit, A., and Bogdal, D. (2006) Chapter 6, in *Microwaves in Organic Synthesis*, 2nd edn (ed. A. Loupy), Wiley-VCH Verlag GmbH, Weinheim, pp. 278–326; (c) Bogdal, D. and Loupy, A. (2008) *Organic Process Research & Development*, **12**, 710–722.

13 Nüchter, M., Müller, U., Ondruschka, B., Tied, A., and Lautenschläger, W. (2003) *Chemical Engineering & Technology*, **26**, 1207–1216.

14 Bogdal, D., Bednarz, S., and Lukasiewicz, M. (2006) *Tetrahedron*, **62**, 9440–9445.

15 Herrero, M.A., Kremsner, J.M., and Kappe, C.O. (2008) *The Journal of Organic Chemistry*, **73**, 36–47.

16 Bose, A.K., Banik, B.K., Lavlinskaia, N., Jayaraman, M., and Manhas, M.S. (1997) *Chemtech*, **27**, 18–24.

17 Bose, A.K., Manhas, M.S., Ganguly, S.N., Sharma, A.H., and Banik, B.K. (2002) *Synthesis*, 1578–1591.

18 Strohmeier, G.A. and Kappe, C.O. (2002) *Journal of Combinatorial Chemistry*, **4**, 154–161.

19 Razzaq, T. and Kappe, C.O. (2007) *Tetrahedron Letters*, **48**, 2513–2517.

20 Stadler, A., Pichler, S., Horeis, G., and Kappe, C.O. (2002) *Tetrahedron*, **58**, 3177–3183.

21 Lange, J.H.M., Verveer, P.C., Osnabrug, S.J.M., and Visser, G.M. (2001) *Tetrahedron Letters*, **42**, 1367–1369.

22 Amore, K.M. and Leadbeater, N.E. (2007) *Macromolecular Rapid Communications*, **28**, 473–477.

23 Lukács, G., Porcs-Makkay, M., Komáromi, A., and Simig, G. (2008) *ARKIVOC*, **iii**, 17–24.

24 Nosse, B., Schall, A., Jeong, W.B., and Reiser, O. (2005) *Advanced Synthesis and Catalysis*, **347**, 1869–1874.

25 Dallinger, D., Irfan, M., Suljanovic, A., and Kappe, C.O. (2010) *The Journal of Organic Chemistry*, **75**, 5278–5288.

26 Petricci, E. and Taddei, M. (2007) *Chimica Oggi: Chemistry Today*, **25**, 40–45.

27 (a) Will, H., Scholz, P., and Ondruschka, B. (2004) *Topics in Catalysis*, **29**, 175–182; (b) Will, H., Scholz, P., and Ondruschka, B. (2004) *Chemical Engineering & Technology*, **27**, 113–122; (c) Will, H., Scholz, P., Ondruschka, B., and Burckhardt, W. (2003) *Chemical Engineering & Technology*, **26**, 1146–1149; (d) Will, H., Scholz, P., and Ondruschka, B. (2002) *Chemie Ingenieur Technik*, **74**, 1057–1067; (e) Zhang, X., Lee, C.S.-M., Mingos, D.M.P., and Hayward, D.O. (2003) *Catalysis Letters*, **88**, 129–139; (f) Zhang, X., Hayward, D.O., and Mingos, D.M.P. (2003) *Catalysis Letters*, **88** 33–38.

28 (a) Miljanić, O.Š., Vollhardt, K.P.C., and Whitener, G.D. (2003) *Synlett*, 29–34; (b) Andappan, M.M.S., Nilsson, P., von Schenck, H., and Larhed, M. (2004) *The Journal of Organic Chemistry*, **69**, 5212–5218.

29 Van der Eycken, E., Appukkuttan, P., De Borggraeve, W., Dehaen, W., Dallinger, D., and Kappe, C.O. (2002) *The Journal of Organic Chemistry*, **67**, 7904–7909.

30 Kaval, N., Dehaen, W., Kappe, C.O., and Van der Eycken, E. (2004) *Organic & Biomolecular Chemistry*, **2**, 154–156.

31 Kormos, C.M. and Leadbeater, N.E. (2006) *Synlett*, 1663–1666.

32 Kormos, C.M. and Leadbeater, N.E. (2007) *Organic & Biomolecular Chemistry*, **5**, 65–68.

33 Kormos, C.M. and Leadbeater, N.E. (2007) *Organic & Biomolecular Chemistry*, **5**, 65–68.

34 Iannelli, M., Bergamelli, F., Kormos, C.M., Paravisi, S., and Leadbeater, N.E. (2009) *Organic Process Research & Development*, **13**, 634–637.

35 Leadbeater, N.E. and Shoemaker, K.M. (2008) *Organometallics*, **27**, 1254–1258.

36 Shoemaker, K.M. and Leadbeater, N.E. (2009) *Inorganic Chemistry Communications*, **12**, 341–342.

37 Cardullo, F., Donati, D., Merlo, G., Paio, A., Petricci, E., and Taddei, M. (2009) *Synlett*, 47–50.

38 (a) Petricci, E., Mann, A., Schoenfelder, A., Rota, A., and Taddei, M. (2006) *Organic Letters*, **8**, 3725–3727; (b) Piras, L., Genesio, E., Ghiron, C., and Taddei, M. (2008) *Synlett*, 1125–1128.

39 Vanier, G.S. (2007) *Synlett*, 131–135.

40 Irfan, M., Fuchs, M., Glasnov, T.N., and Kappe, C.O. (2009) *Chemistry: A European Journal*, **15**, 11608–11618.

41 Heller, E., Lautenschläger, W., and Holzgrabe, U. (2005) *Tetrahedron Letters*, **46**, 1247–1249.

42 (a) Kaiser, N.-F.K., Hallberg, A., and Larhed, M. (2002) *Journal of Combinatorial Chemistry*, **4**, 109–111; (b) Georgsson, J., Hallberg, A., and Larhed, M. (2003) *Journal of Combinatorial Chemistry*, **5**, 350–352; (c) Wannberg, J. and Larhed, M. (2003) *The Journal of Organic Chemistry*, **68**, 5750–5753; (d) Herrero, M.A., Wannberg, J., and Larhed, M. (2004) *Synlett*, 2335–2338; (e) Wu, X., Roenn, R., Gossas, T., and Larhed, M. (2005) *The Journal of Organic Chemistry*, **70**, 3094–3098; (f) Wannberg, J., Dallinger, D., Kappe, C.O., and Larhed, M. (2005) *Journal of Combinatorial Chemistry*, **7**, 574–583; (g) Wannberg, J., Kaiser, N.-F.K., Vrang, L., Samuelsson, B., Larhed, M.,

and Hallberg, A. (2005) *Journal of Combinatorial Chemistry*, **7**, 611–617; (h)Wu, X., Wannberg, J., and Larhed, M. (2006) *Tetrahedron*, **62**, 4665–4670.

43 Wu, X., Ekegren, J.K., and Larhed, M. (2006) *Organometallics*, **25**, 1434–1439.

44 Glasnov, T.N., Groschner, K., and Kappe, C.O. (2009) *ChemMedChem*, **4**, 1816–1818.

45 (a) Li, C.-J. and Chan, T.-H. (1997) *Organic Reactions in Aqueous Media*, John Wiley & Sons, Inc., New York; (b) Lindström, U.M. (ed.) (2007) *Organic Reactions in Water*, Blackwell Publishing, Oxford; (c) Lindström, U.M. (2002) *Chemical Reviews*, **102**, 2751–2772; (d) Li, C.-J. (2005) *Chemical Reviews*, **105**, 3095–3166; (e) Li, C.-J. and Chen, L. (2006) *Chemical Society Reviews*, **35**, 68–82; (f) Breslow, R. (2004) *Accounts of Chemical Research*, **37**, 471–478; (g) Pirrung, M.C. (2006) *Chemistry: A European Journal*, **12**, 1312–1317.

46 (a) Blokzijl, W. and Engberts, J.B.F.N. (1993) *Angewandte Chemie International Edition in English*, **32**, 1545–1579; (b) Widom, B., Bhimalapuram, P., and Koga, K. (2003) *Physical Chemistry Chemical Physics*, **5**, 3085–3093.

47 (a) Krammer, P. and Vogel, H. (2000) *Journal of Supercritical Fluids*, **16**, 189–206; (b) Krammer, P., Mittelstädt, S., and Vogel, H. (1999) *Chemical Engineering & Technology*, **22**, 126–130.

48 (a) Bröll, D., Kaul, C., Krämer, A., Krammer, P., Richter, T., Jung, M., Vogel, H., and Zehner, P. (1999) *Angewandte Chemie*, **111**, 3180–3196; (b) Savage, P.E. (1999) *Chemical Reviews*, **99**, 603–621; (c) Siskin, M. and Katritzky, A.R. (2001) *Chemical Reviews*, **101**, 825–836; (d) Katritzky, A.R., Nichols, D.A., Siskin, M., Murugan, R., and Balasubramanian, M. (2001) *Chemical Reviews*, **101**, 837–892; (e) Akiya, N. and Savage, P.E. (2002) *Chemical Reviews*, **102**, 2725–2750; (f) Watanabe, M., Sato, T., Inomata, H., Smith, R.L., Jr., Arai, K., Kruse, A., and Dinjus, E. (2004) *Chemical Reviews*, **104**, 5803–5822; (g) Weingärtner, H. and Franck, E.U. (2005) *Angewandte Chemie International Edition*, **44**, 2672–2692.

49 Kremsner, J.M. and Kappe, C.O. (2005) *European Journal of Organic Chemistry*, 3672–3679.

50 Dallinger, D. and Kappe, C.O. (2007) *Chemical Reviews*, **107**, 2563–2591.

51 Strauss, C.R. and Trainor, R.W. (1995) *Australian Journal of Chemistry*, **48**, 1665–1692.

52 Strauss, C.R. (1999) *Australian Journal of Chemistry*, **52**, 83–96.

53 Roberts, B.A. and Strauss, C.R. (2005) *Accounts of Chemical Research*, **38**, 653–661.

54 Kormos, C.M. and Leadbeater, N.E. (2006) *Tetrahedron*, **62**, 4728–4732.

55 Geuens, J., Kremsner, J.M., Nebel, B.A., Schober, S., Dommisse, R.A., Mittelbach, M., Tavernier, S., Kappe, C.O., and Maes, B.U.W. (2008) *Energy Fuels*, **22**, 643–645.

56 Molteni, V., Hamilton, M.M., Mao, L., Crane, C.M., Termin, A.P., and Wilson, D.M. (2002) *Synthesis*, 1669–1674.

57 (a) Polshettiwar, V. and Varma, R.S. (2008) *Chemical Society Reviews*, **37**, 1546–1557; (b) Polshettiwar, V., Decottignies, A., Len, C., and Fihri, A. (2010) *ChemSusChem*, **3**, 502–522; (c) Polshettiwar, V. and Varma, R.S. (eds) (2010) *Aqueous Microwave Assisted Chemistry*, RSC Publishing, Cambridge.

58 Wasserscheid, P. and Welton, T. (eds) (2007) *Ionic Liquids in Synthesis*, 2nd edn, Wiley-VCH Verlag GmbH, Weinheim.

59 Hoffmann, J., Nüchter, M., Ondruschka, B., and Wasserscheid, P. (2003) *Green Chemistry*, **5**, 296–299.

60 Kremsner, J.M. and Kappe, C.O. (2006) *The Journal of Organic Chemistry*, **71**, 4651–4658.

61 Leadbeater, N.E., Torenius, H.M., and Tye, H. (2004) *Combinatorial Chemistry & High Throughput Screening*, **7**, 511–528.

62 Habermann, J., Ponzi, S., and Ley, S.V. (2005) *Mini-Reviews in Organic Chemistry*, **2**, 125–137.

63 Leadbeater, N.E. and Torenius, H.M. (2006) Chapter 7, in *Microwaves in Organic Synthesis*, 2nd edn (ed. A. Loupy),

Wiley-VCH Verlag GmbH, Weinheim, pp. 327–361.

64 (a) Varma, R.S. and Namboodiri, V.V. (2001) *Chemical Communications*, 643–644; (b) Varma, R.S. and Namboodiri, V.V. (2002) *Tetrahedron Letters*, **43**, 5381–5383; (c) Namboodiri, V.V. and Varma, R.S. (2002) *Chemical Communications*, 342–343.

65 Khadilkar, B.M. and Rebeiro, G.L. (2002) *Organic Process Research & Development*, **6**, 826–828.

66 Deetlefs, M. and Seddon, K.R. (2003) *Green Chemistry*, **5**, 181–186.

67 Obermayer, D. and Kappe, C.O. (2010) *Organic & Biomolecular Chemistry*, **8**, 114–121.

68 Robinson, J., Kingman, S., Irvine, D., Licence, P., Smith, A., Dimitrakis, G., Obermayer, D., and Kappe, C.O. (2010) *Physical Chemistry Chemical Physics*, **12**, 4750–4758.

69 Vo Thanh, G., Pegot, B., and Loupy, A. (2004) *European Journal of Organic Chemistry*, 1112–1116.

70 Bica, K., Gmeiner, G., Reichel, C., Lendl, B., and Gaertner, P. (2007) *Synthesis*, 1333–1338.

71 (a) Lee, S. (2006) *Chemical Communications*, 1049–1063; (b) Davis, J.H., Jr. (2004) *Chemistry Letters*, **33**, 1072–1077.

72 (a) Fu, S.-K. and Liu, S.-T. (2006) *Synthetic Communications*, **36**, 2059–2067; (b) Arfan, A. and Bazureau, J.P. (2005) *Organic Process Research & Development*, **9**, 743–748.

73 de Kort, M., Tuin, A.W., Kuiper, S., Overkleeft, H.S., van der Marel, G.A., and Buijsman, R.C. (2004) *Tetrahedron Letters*, **45**, 2171–2175.

74 Legeay, J.-C., Vanden Eynde, J.J., and Bazureau, J.P. (2005) *Tetrahedron*, **61**, 12386–12397.

75 Vallin, K.S.A., Emilsson, P., Larhed, M., and Hallberg, A. (2002) *The Journal of Organic Chemistry*, **67**, 6243–6246.

76 Leadbeater, N.E., Torenius, H.M., and Tye, H. (2003) *Tetrahedron*, **59**, 2253–2258.

77 Alexander, M.V., Khandekar, A.C., and Samant, S.D. (2004) *Journal of Molecular Catalysis A: Chemical*, **223**, 75–83.

78 Estager, J., Lévêque, J.-M., Turgis, R., and Draye, M. (2006) *Journal of Molecular Catalysis A: Chemical*, **256**, 261–264.

79 Ley, S.V., Leach, A.G., and Storer, R.I. (2001) *Journal of the Chemical Society, Perkin Transactions 1*, 358–361.

80 (a) Martínez-Palou, R. (2010) *Molecular Diversity*, **14**, 3–25; (b) Boros, E., Seddon, K.R., and Strauss, C.R. (2008) *Chimica Oggi: Chemistry Today*, **26**, 28–30.

81 Leadbeater, N.E. and Torenius, H.M. (2002) *The Journal of Organic Chemistry*, **67**, 3145–3148.

82 Silva, A.M.G., Tomé, A.C., Neves, M.G.P.M.S., Cavaleiro, J.A.S., and Kappe, C.O. (2005) *Tetrahedron Letters*, **46**, 4723–4726.

83 Glenn, A.G. and Jones, P.B. (2004) *Tetrahedron Letters*, **45**, 6967–6969.

84 Lin, Y.L., Cheng, J.-Y., and Chu, Y.-H. (2007) *Tetrahedron*, **63**, 10949–10957.

85 Moseley, J.D., Lenden, P., Thompson, A.D., and Gilday, J.P. (2007) *Tetrahedron Letters*, **48**, 6084–6087.

86 Davies, H.M. and Beckwith, R.E. (2004) *The Journal of Organic Chemistry*, **69**, 9241–9247.

87 Gonda, J., Martinková, M., Zadrošová, A., Šoteková, M., Raschmanová, J., Čonka, P., Gajdošíková, E., and Kappe, C.O. (2007) *Tetrahedron Letters*, **48**, 6912–6915.

88 (a) Harris, G.L. (ed.) (1995) *Properties of Silicon Carbide*, Institute of Electrical Engineers, London; (b) Choyke, W.J., Matsunami, H., and Pensl, G. (eds) (2004) *Silicon Carbide: Recent Major Advances*, Springer, Berlin; (c) Saddow, S.E. and Agarwal, A. (eds) (2004) *Advances in Silicon Carbide Processing and Applications*, Artech House Inc., Norwood.

89 Razzaq, T., Kremsner, J.M., and Kappe, C.O. (2008) *The Journal of Organic Chemistry*, **73**, 6321–6329.

90 Obermayer, D., Gutmann, B., and Kappe, C.O. (2009) *Angewandte Chemie International Edition*, **48**, 8321–8324.

91 Gutmann, B., Obermayer, D., Reichart, B., Prekodravac, B., Irfan, M., Kremsner, J.M., and Kappe, C.O. (2010) *Chemistry: A European Journal*, **16**, 12182–12194.

92 (a) Thompson, L.A. and Ellman, J.A. (1996) *Chemical Reviews*, **96**, 555–600; (b) Dolle, R.E., Le Bourdonnec, B., Morales, G.A., Moriarty, K.J., and Salvino, J.M. (2006) *Journal of Combinatorial Chemistry*, **8**, 597–635.

93 (a) Kappe, C.O. and Dallinger, D. (2006) *Nature Reviews. Drug Discovery*, **5**, 51–64; (b) Cao, P. and Leadbeater, N.E. (2011) Chapter 4, in *Microwave Heating as a Tool for Sustainable Chemistry* (ed. N.E. Leadbeater), CRC Press, Boca Raton, pp. 73–103.

94 (a) Stadler, A. and Kappe, C.O. (2001) *Journal of Combinatorial Chemistry*, **3**, 624–630; (b) Stadler, A. and Kappe, C.O. (2003) *Methods in Enzymology*, **369**, 197–223; (c) Dallinger, D. and Kappe, C.O. (2007) *Nature Protocols*, **2**, 1713–1721.

95 Desai, B., Dallinger, D., and Kappe, C.O. (2006) *Tetrahedron*, **62**, 4651–4664.

96 Selway, C.N. and Terret, N.K. (1996) *Bioorganic & Medicinal Chemistry*, **4**, 645–654.

97 Nüchter, M., Ondruschka, B., Tied, A., Lautenschläger, W., and Borowski, K.J. (2001) *American Genomic/Proteomic Technology*, **1**, 34–39.

98 Nüchter, M. and Ondruschka, B. (2003) *Molecular Diversity*, **7**, 253–264.

99 Stadler, A., Yousefi, B.H., Dallinger, D., Walla, P., Van der Eycken, E., Kaval, N., and Kappe, C.O. (2003) *Organic Process Research & Development*, **7**, 707–716.

100 Pisani, L., Prokopcová, H., Kremsner, J.M., and Kappe, C.O. (2007) *Journal of Combinatorial Chemistry*, **9**, 415–421.

101 Schön, U., Messinger, J., Eichner, S., and Kirschning, A. (2008) *Tetrahedron Letters*, **49**, 3204–3207.

102 Leadbeater, N.E. and Schmink, J.R. (2007) *Tetrahedron*, **63**, 6764–6773.

103 Amore, K.M., Leadbeater, N.E., Miller, T.A., and Schmink, J.R. (2006) *Tetrahedron Letters*, **47**, 8583–8586.

104 (a) Matloobi, M. and Kappe, C.O. (2007) *Chimica Oggi: Chemistry Today*, **25**, 26–31; (b) Matloobi, M. and Kappe, C.O. (2007) *Combinatorial Chemistry & High Throughput Screening*, **10**, 735–750.

105 Cotterill, I.C., Usyatinsky, A.Ya., Arnold, J.M., Clark, D.S., Dordick, J.S., Michels, P.C., and Khmelnitsky, Y.L. (1998) *Tetrahedron Letters*, **39**, 1117–1120.

106 Glass, B.M. and Combs, A.P. (2001) Chapter 4.6, in *High-Throughput Synthesis. Principles and Practices* (ed. I. Sucholeiki), Marcel Dekker, Inc., New York, pp. 123–128.

107 Glass, B.M. and Combs, A.P. (2001) Article E0027, *Fifth International Electronic Conference on Synthetic Organic Chemistry* CD-ROM edn (eds C.O. Kappe, P. Merino, A. Marzinzik, H. Wennemers, T. Wirth, J.J. Vanden Eynde, and S.-K. Lin), MDPI, Basel, Switzerland.

108 Al-Obeidi, F.A.D. and Austin, R.E. (2002) *Chemical Abstracts*, **138**, 26204.

109 Sarko, C.R. (2005) Chapter 8, in *Microwave Assisted Organic Synthesis* (eds P. Lidström and J.P. Tierney), Blackwell Publishing, Oxford, pp. 222–236.

110 Alcázar, J. (2005) *Journal of Combinatorial Chemistry*, **7**, 353–355.

111 Martinez-Teipel, B., Green, R.C., and Dolle, R.E. (2004) *QSAR & Combinatorial Science*, **23**, 854–858.

112 Macleod, C., Martinez-Teipel, B., Barker, W.M., and Dolle, R.E. (2006) *Journal of Combinatorial Chemistry*, **8**, 132–140.

113 Murray, J.K. and Gellman, S.H. (2006) *Journal of Combinatorial Chemistry*, **8**, 58–65.

114 Kremsner, J.M., Stadler, A., and Kappe, C.O. (2007) *Journal of Combinatorial Chemistry*, **9**, 285–291.

115 Treu, M., Karner, T., Kousek, R., Berger, H., Mayer, M., McConnell, D.B., and Stadler, A. (2008) *Journal of Combinatorial Chemistry*, **9**, 863–868.

116 Stencel, L.M., Kormos, C.M., Avery, K.B., and Leadbeater, N.E. (2009) *Organic & Biomolecular Chemistry*, **7**, 2452–2457.

117 Avery, K.B., Devine, W.G., Kormos, C.M., and Leadbeater, N.E. (2009) *Tetrahedron Letters*, **50**, 2851–2853.

118 Damm, M. and Kappe, C.O. (2009) *Journal of Combinatorial Chemistry*, **11**, 460–468.

119 Baghbanzadeh, M., Molnar, M., Damm, M., Reidlinger, C., Dabiri, M., and Kappe, C.O. (2009) *Journal of Combinatorial Chemistry*, **11**, 676–684.

120 Damm, M. and Kappe, C.O. (2009) *Molecular Diversity*, **13**, 529–543.

121 (a) High-throughput synthesis methods: Stadler, A. and Kappe, C.O. (2006) Chapter 16, in *Microwaves in Organic Synthesis*, 2nd edn (ed. A. Loupy), Wiley-VCH Verlag GmbH, Weinheim, pp. 726–787; (b) Erdélyi, M. (2006) *Topics in Heterocyclic Chemistry*, **1**, 79–128; (c) Lange, T. and Lindell, S. (2005) *Combinatorial Chemistry & High Throughput Screening*, **8**, 595–606.

122 Al-Obeidi, F., Austin, R.E., Okonya, J.F., and Bond, D.R.S. (2003) *Mini-Reviews in Medicinal Chemistry*, **3**, 449–460.

123 Swamy, K.M.K., Yeh, W.-B., Lin, M.-J., and Sun, C.-M. (2003) *Current Medicinal Chemistry*, **10**, 2403–2423.

124 (a) Kadam, A., Zhang, Z., and Zhang, W. (2011) *Current Organic Synthesis*, **8**, 295–309; (b) Zhang, W. (2006) *Topics in Current Chemistry*, **266**, 145–166; (c) Zhang, W. (2004) *Chemical Reviews*, **104**, 2531–2556; (d) Olofsson, K. and Larhed, M. (2004) Chapter 10.19, in *Handbook of Fluorous Chemistry* (eds J.A. Gladysz, D.P. Curran, and I.T. Horvath), Wiley-VCH Verlag GmbH, Weinheim, pp. 359–365.

125 (a) Crosignani, S. and Linclau, B. (2006) *Topics in Heterocyclic Chemistry*, **1**, 129–154; (b) Baxendale, I.R., Lee, A.-L., and Ley, S.V. (2005) Chapter 6, in *Microwave Assisted Organic Synthesis* (eds P. Lidström and J.P. Tierney), Blackwell Publishing Ltd., Oxford, UK, pp. 133–176; (c) Bhattacharyya, S. (2005) *Molecular Diversity*, **9**, 253–257; (d) Parlow, J.J. (2005) *Current Opinion in Drug Discovery & Development*, **8**, 757–775.

126 (a) Terret, N.K. (1998) *Combinatorial Chemistry*, Oxford University Press, New York; (b) Winter, M. and Warrass, R. (2000) Resins and anchors for solid-phase organic synthesis, in *Combinatorial Chemistry* (ed. H. Fenniri), Oxford University Press, New York, pp. 117–138.

127 Pérez, R., Beryozkina, T., Zbruyev, O.I., Haas, W., and Kappe, C.O. (2002) *Journal of Combinatorial Chemistry*, **4**, 501–510.

128 (a) Collins, J.M. and Collins, M.J. (2006) Chapter 20, in *Microwaves in Organic Synthesis*, 2nd edn (ed. A. Loupy), Wiley-VCH Verlag GmbH, Weinheim, pp. 898–930; (b) Collins, J.M. and Leadbeater, N.E. (2007) *Organic & Biomolecular Chemistry*, **5**, 1141–1150; (c) Sabatino, G. and Papini, A.M. (2008) *Current Opinion in Drug Discovery & Development*, **11**, 762–770.

129 Yu, H.-M., Chen, S.-T., and Wang, K.-T. (1992) *The Journal of Organic Chemistry*, **57**, 4781–4784.

130 Erdélyi, M. and Gogoll, A. (2002) *Synthesis*, 1592–1596.

131 Bacsa, B., Desai, B., Dibó, G., and Kappe, C.O. (2006) *Journal of Peptide Science*, **12**, 633–638.

132 (a) Feliu, L., Font, D., Soley, R., Tailhades, J., Martinez, J., and Amblard, M. (2007) *ARKIVOC*, **iv**, 65–72; (b) Monroc, S., Feliu, L., Planas, M., and Bardají, E. (2006) *Synlett*, 1311–1314.

133 Bacsa, B. and Kappe, C.O. (2007) *Nature Protocols*, **2**, 2222–2227.

134 Rizzolo, F., Sabatino, G., Chelli, M., Rovero, P., and Papini, A.M. (2007) *International Journal of Peptide: Research and Therapeutics*, **13**, 203–208.

135 (a) Bacsa, B., Horváti, K., Bősze, S., Andreae, F., and Kappe, C.O. (2008) *The Journal of Organic Chemistry*, **73**, 7532–7542; (b) Bacsa, B., Bősze, S., and Kappe, C.O. (2010) *The Journal of Organic Chemistry*, **75**, 2103–2106.

136 (a) Bowman, M.D., Jeske, R.C., and Blackwell, H.E. (2004) *Organic Letters*, **6**, 2019–2022; (b) Bowman, M.D., Jacobson, M.M., Pujanauski, B.G., and Blackwell, H.E. (2006) *Tetrahedron*, **62**, 4715–4727.

137 Scharn, D., Wenschuh, H., Reineke, U., Schneider-Mergener, J., and

Germeroth, L. (2000) *Journal of Combinatorial Chemistry*, **2**, 361–369.

138 (a) Chen, J., Yang, G., Zhang, H., and Chen, Z. (2006) *Reactive & Functional Polymers*, **66**, 1434–1451; (b) Lee, M.-J. and Sun, C.-M. (2003) *Chinese Pharmaceutical Journal*, **55**, 405–452; (c) Sun, C.-M. (2002) Chapter 10, in *Combinatorial Chemistry Methods and Protocols* (ed. L. Bellavance), Methods in Molecular Biology Series, The Humana Press Inc., New Jersey, pp. 345–371; (d) Sun, C.-M. (1999) *Combinatorial Chemistry & High Throughput Screening*, **2**, 299–318.

139 Su, Y.-S., Lin, M.-J., and Sun, M.-C. (2005) *Tetrahedron Letters*, **46**, 177–180.

140 Wu, C.-H. and Sun, C.-M. (2006) *Tetrahedron Letters*, **47**, 2601–2604.

141 (a) Yeh, W.-B., Lin, M.-J., Lee, M.-J., and Sun, C.-M. (2003) *Molecular Diversity*, **7**, 185–198; (b) Lin, M.-J. and Sun, C.-M. (2003) *Tetrahedron Letters*, **44**, 8739–8742.

142 (a) Chang, W.-J., Kulkarni, M.V., and Sun, C.-M. (2006) *Journal of Combinatorial Chemistry*, **8**, 141–144; (b) Yeh, W.-P., Chang, W.-J., Sun, M.-L., and Sun, C.-M. (2007) *Tetrahedron*, **63**, 11809–11816.

143 Curran, D.P. (2004) Chapter 7, in *Handbook of Fluorous Chemistry* (eds J.A. Gladysz, D.P. Curran, and I.T. Horvath,), Wiley-VCH Verlag GmbH, Weinheim, pp. 101–127.

144 Zhang, W., Chen, C.H.-T., Lu, Y., and Nagashima, T. (2004) *Organic Letters*, **6**, 1473–1476.

145 Zhang, W., Nagashima, T., Lu, Y., and Chen, C.H.-T. (2004) *Tetrahedron Letters*, **45**, 4611–4613.

146 (a) Zhang, W., Lu, Y., Chen, C.H.-T., Curran, D.P., and Geib, S. (2006) *European Journal of Organic Chemistry*, 2055–2059; (b) Zhang, W. and Chen, C.H.-T. (2005) *Tetrahedron Letters*, **46**, 1807–1810; (c) Lu, Y. and Zhang, W. (2004) *QSAR & Combinatorial Science*, **23**, 827–835; (d) Zhang, W. and Tempest, P. (2004) *Tetrahedron Letters*, **45**, 6757–6760; (e) Zhang, W., Lu, Y., and Chen, C.H.-T. (2003) *Molecular Diversity*, **7**, 199–202.

147 (a) Solinas, A. and Taddei, M. (2007) *Synthesis*, 2409–2453; (b) Ley, S.V., Ladlow, M., and Vickerstaffe, E. (2006) Chapter 1, in *Exploiting Chemical Diversity for Drug Discovery* (eds P.A. Bartlett and M. Entzeroth), RSC Publishing, pp. 3–32; (c) Ley, S.V., Baxendale, I.R., and Myers, R.M. (2006) Chapter 6, in *Combinatorial Synthesis of Natural Product-Based Libraries* (ed. A.M. Boldi), CRC Press, Boca Raton, pp. 131–163; (d) Hinzen, B. and Hahn, M.G. (2006) Chapter 6, in *Combinatorial Chemistry* (eds W. Bannwarth and B. Hinzen), Wiley-VCH Verlag GmbH, Weinheim, pp. 457–512; (e) Buchmeiser, M.R. (ed.) (2003) *Polymeric Materials in Organic Synthesis and Catalysis*, Wiley-VCH Verlag GmbH, Weinheim; (f) Ley, S.V. and Baxendale, I.R. (2002) *Nature Reviews. Drug Discovery*, **1**, 573–586.

148 Donati, D., Morelli, C., and Taddei, M. (2005) *Tetrahedron Letters*, **46**, 2817–2819.

149 (a) Kirschning, A., Solodenko, W., and Mennecke, K. (2006) *Chemistry: A European Journal*, **12**, 5972–5990; ; (b) (2004) *Topics in Current Chemistry*, **244**, 1–317.

150 Wang, Y. and Sauer, D.R. (2004) *Organic Letters*, **6**, 2793–2796.

151 (a) Dawood, K.M. and Kirschning, A. (2005) *Tetrahedron*, **61**, 12121–12130; (b) Michrowska, A., Mennecke, K., Kunz, U., Kirschning, A., and Grela, K. (2006) *Journal of the American Chemical Society*, **128**, 13261–13267.

152 Choudary, B.M., Madhi, S., Chowdari, N.S., Kantam, M.L., and Sreedhar, B. (2002) *Journal of the American Chemical Society*, **124**, 14127–14136.

153 Shore, G., Morin, S., and Organ, M.G. (2006) *Angewandte Chemie International Edition*, **45**, 2761–2766.

154 Lipshutz, B.H., Frieman, B.A. Lee, C.-T., Lower, A., Nihan, D.M., and Taft, B.R. (2006) *Chemistry: An Asian Journal*, **1**, 417–429.

155 Dallinger, D., Gorobets, N.Yu., and Kappe, C.O. (2003) *Molecular Diversity*, **7**, 229–245.

156 Hinchcliff, A., Hughes, C., Pears, D.A., and Pitts, M.R. (2007) *Organic Process Research & Development*, **11**, 477–481.

157 Mukade, T., Dragoli, D.R., and Ellman, J.A. (2003) *Journal of Combinatorial Chemistry*, **5**, 590–596.

158 (a) Baxendale, I.R., Ley, S.L., and Piutti, C. (2002) *Angewandte Chemie International Edition*, **41**, 2194–2197; (b) Baxendale, I.R., Ley, S.L., Nessi, M., and Piutti, C. (2002) *Tetrahedron*, **58**, 6285–6304.

159 Lehmann, H. (2007) *Ernst Schering Foundation Symposium Proceedings* (eds P.H. Seeberger and T. Blume), Springer, Berlin, pp. 133–149.

160 Damm, M., Glasnov, T.N., and Kappe, C.O. (2010) *Organic Process Research & Development*, **14**, 215–224.

161 Loones, K.T.J., Maes, B.U.W., Rombouts, G., Hostyn, S., and Diels, G. (2005) *Tetrahedron*, **61**, 10338–10348.

162 (a) Strauss, C.R. (2009) *Organic Process Research & Development*, **13**, 915–923; (b) Kremsner, J.M., Stadler, A., and Kappe, C.O. (2006) *Topics in Current Chemistry*, **266**, 233–278; (c) Ondruschka, B., Bonrath, W., and Stuerga, D. (2006) Chapter 2, in *Microwaves in Organic Synthesis*, 2nd edn (ed. A. Loupy), Wiley-VCH Verlag GmbH, Weinheim, pp. 62–107.

163 Moseley, J.D. (2011) *Microwave Heating as a Tool for Sustainable Chemistry* (ed. N.E. Leadbeater), CRC Press, Boca Raton, pp. 105–147.

164 Iqbal, M., Vyse, N., Dauvergne, J., and Evans, P. (2002) *Tetrahedron Letters*, **43**, 7859–7862.

165 Lehmann, F., Pilotti, Å., and Luthman, K. (2003) *Molecular Diversity*, **7**, 145–152.

166 Shackelford, S.A., Anderson, M.B., Christie, L.C., Goetzen, T., Guzman, M.C., Hananel, M.A., Kornreich, W.D., Li, H., Pathak, V.P., Rabinovich, A.K., Rajapakse, R.J., Truesdale, L.K., Tsank, S.M., and Vazir, H.N. (2003) *The Journal of Organic Chemistry*, **68**, 267–275.

167 Leadbeater, N.E., Williams, V.A., Barnard, T.M., and Collins, M.J., Jr. (2006) *Organic Process Research & Development*, **10**, 833–837.

168 Leadbeater, N.E., Williams, V.A., Barnard, T.M., and Collins, M.J., Jr. (2006) *Synlett*, 2953–2958.

169 Barnard, T.M., Vanier, G.S., and Collins, M.J., Jr. (2006) *Organic Process Research & Development*, **10**, 1233–1237.

170 Lehmann, H. and La Vecchia, L. (2010) *Organic Process Research & Development*, **14**, 650–656.

171 Carlsson, A.-C., Jam, F., Tullberg, M., Pilotti, Å., Ioannidis, P., Luthman, K., and Grøtli, M. (2006) *Tetrahedron Letters*, **47**, 5199–5201.

172 Ekström, J., Wettergren, J., and Adolfsson, H. (2007) *Advanced Synthesis and Catalysis*, **349**, 1609–1613.

173 Appukkuttan, P., Axelsson, L., Van der Eycken, E., and Larhed, M. (2008) *Tetrahedron Letters*, **49**, 5625–5628.

174 Bowman, M.D., Schmink, J.R., McGowan, C.M., Kormos, C.M., and Leadbeater, N.E. (2008) *Organic Process Research & Development*, **12**, 1078–1088.

175 Ianelli, M., Bergamelli, F., Kormos, C.M., Paravisi, S., and Leadbeater, N.E. (2009) *Organic Process Research & Development*, **13**, 634–637.

176 Dallinger, D., Lehmann, H., Moseley, J.D., Stadler, A., and Kappe, C.O. (2011) *Organic Process Research & Development*, **15**, 841–854.

177 Pawluczyk, J.M., McClain, R.T., Denicola, C., Mulhearn, J.J., Jr., Rudd, D.J., and Lindsley, C.W. (2007) *Tetrahedron Letters*, **48**, 1497–1501.

178 Glasnov, T.N. and Kappe, C.O. (2007) *Macromolecular Rapid Communications*, **28**, 395–410.

179 (a) Baxendale, I.R. and Pitts, M.R. (2006) *Chimica Oggi: Chemistry Today*, **24**, 41–45; (b) Baxendale, I.R., Hayward, J.J., and Ley, S.L. (2007) *Combinatorial Chemistry & High Throughput Screening*, **10**, 802–836; (c) Singh, B.K., Kaval, N., Tomar, S., Van der Eycken, E., and Parmar, V.S. (2008) *Organic Process Research & Development*, **12**, 468–474.

180 Cablewski, T., Faux, A.F., and Strauss, C.R. (1994) *The Journal of Organic Chemistry*, **59**, 3408–3412.

181 Glasnov, T.N., Vugts, D.J., Koningstein, M.M., Desai, B.,

Fabian, W.M.F., Orru, R.V.A., and Kappe, C.O. (2006) *QSAR & Combinatorial Science*, **25**, 509–518.
182 Bagley, M.C., Jenkins, R.L., Lubinu, M.C., Mason, C., and Wood, R. (2005) *The Journal of Organic Chemistry*, **70**, 7003–7006.
183 Smith, C.J., Iglesias-Sigüenza, F.J., Baxendale, I.R., and Ley, S.V. (2007) *Organic & Biomolecular Chemistry*, **5**, 2758–2761.
184 Shieh, W.-C., Dell, S., and Repič, O. (2001) *Organic Letters*, **3**, 4279–4281.
185 Shieh, W.-C., Dell, S., and Repič, O. (2002) *Tetrahedron Letters*, **43**, 5607–5609.
186 Shieh, W.-C., Lozanov, M., and Repič, O. (2003) *Tetrahedron Letters*, **44**, 6943–6945.
187 Moseley, J.D. and Lawton, S.J. (2007) *Chimica Oggi: Chemistry Today*, **25**, 16–19.
188 (a) Moseley, J.D., Sankey, R.F., Tang, O.N., and Gilday, J.P. (2006) *Tetrahedron*, **62**, 4685–4689; (b) Moseley, J.D. and Lenden, P. (2007) *Tetrahedron*, **63**, 4120–4125.
189 Bergamelli, F., Ianelli, M., Marafie, J.A., and Moseley, J.D. (2010) *Organic Process Research & Development*, **14**, 926–930.
190 Dressen, M.H.C.L., Van de Kruijs, B.H.P., Meduldijk, J., Vekemans, J.A.J.M., and Hulshof, L.A. (2010) *Organic Process Research & Development*, **14**, 351–361.
191 Dressen, M.H.C.L., Van de Kruijs, B.H.P., Meduldijk, J., Vekemans, J.A.J.M., and Hulshof, L.A. (2009) *Organic Process Research & Development*, **13**, 888–895.
192 Bagley, M.C., Fusillo, V., Jenkins, R.L., Lubinu, M.C., and Mason, C. (2010) *Organic & Biomolecular Chemistry*, **8**, 2245–2251.
193 Paulus, R.M., Erdmenger, T., Becer, C.R., Hoogenboom, R., and Schubert, U.S. (2007) *Macromolecular Rapid Communications*, **28**, 484–491.
194 Barnard, T.M., Leadbeater, N.E., Boucher, M.B., Stencel, L.M., and Wilhite, B.A. (2007) *Energy Fuels*, **21**, 1777–1781.
195 Leadbeater, N.E., Barnard, T.M., and Stencel, L.M. (2008) *Energy Fuels*, **22**, 2005–2008.
196 Arvela, R.K., Leadbeater, N.E., and Collins, M.J., Jr. (2005) *Tetrahedron*, **61**, 9349–9355.
197 Pitts, M.R., McCormack, P., and Whittall, J. (2006) *Tetrahedron*, **62**, 4705–4708.
198 Moseley, J.D. and Woodman, E.K. (2008) *Organic Process Research & Development*, **12**, 967–981.
199 Moseley, J.D., Lenden, P., Lockwood, M., Ruda, K., Sherlock, J.-P., Thomson, A.D., and Gilday, J.P. (2008) *Organic Process Research & Development*, **12**, 30–40.
200 Bowman, M.D., Holcomb, J.L., Kormos, C.M., Leadbeater, N.E., and Williams, V.A. (2008) *Organic Process Research & Development*, **12**, 41–57.
201 Erdmenger, T., Paulus, R.M., Hoogenboom, R., and Schubert, U.S. (2008) *Australian Journal of Chemistry*, **61**, 197–203.
202 (a) Watts, P. (2004) *Current Opinion in Drug Discovery & Development*, **7**, 807–812; (b) Watts, P. and Haswell, S.J. (2003) *Current Opinion in Chemical Biology*, **7**, 380–387.
203 For recent selected reviews on continuous flow/microreactor chemistry, see: (a) Hartman, R.L., McMullen, J.P., and Jensen, K.F. (2011) *Angewandte Chemie International Edition*, **50**, 7502–7519; (b) Willes, C. and Watts, P. (2011) *Chemical Communications*, **47**, 6512–6535; (c) Baumann, M., Baxendale, I.R., and Ley, S.V. (2011) *Molecular Diversity*, **15**, 61–630; (d) Yoshida, J.-I., Kim, H., and Nagaki, A. (2011) *ChemSusChem*, **4**, 331–340; (e) McMullen, J.P. and Jensen, K.F. (2010) *Annual Review of Analytical Chemistry*, **3**, 19–42; (f) Illg, T., Löb, P., and Hessel, V. (2010) *Bioorganic & Medicinal Chemistry*, **18**, 3707–3719; (g) Frost, C.G. and Mutton, L. (2010) *Green Chemistry*, **12**, 1687–1703; (h) Geyer, K., Gustafsson, T., and Seeberger, P.H. (2009) *Synlett*, 2382–2391; (i) Hartman, R.L. and Jensen, K.F. (2009) *Lab on a Chip*, **9**, 2495–2507.
204 For recent books on continuous flow/microreactor chemistry, see: (a) Willes, C. and Watts, P. (2011) *Micro Reaction Technology in Organic Synthesis*, CRC Press, Boca Raton; (b) Luis, S.V. and

Garcia-Verduqo, E. (2010) *Chemical Reactions and Processes under Flow Conditions*, Royal Society of Chemistry, Cambridge; (c) Wirth, T. (2008) *Microreactors in Organic Synthesis and Catalysis*, Wiley-VCH Verlag GmbH, Weinheim; (d) Hessel, V., Schouten, J.C., Renken, A., Wang, Y., and Yoshida, J.-i. (eds) (2009) *Handbook of Micro Reactors*, Wiley-VCH Verlag GmbH, Weinheim; (e) Yoshida, J.-i. (2008) *Flash Chemistry: Fast Organic Synthesis in Microsystems*, Wiley-VCH Verlag GmbH, Weinheim.

205 (a) He, P., Haswell, S.J., and Fletcher, P.D. (2004) *Lab on a Chip*, **4**, 38–41; (b) He, P., Haswell, S.J., and Fletcher, P.D. (2004) *Applied Catalysis A: General*, **274**, 111–114.

206 Benaskar, F., Hessel, V., Krtschil, U., Löb, P., and Stark, A. (2009) *Organic Process Research & Development*, **13**, 970–982.

207 (a) Comer, E. and Organ, M.G. (2005) *Journal of the American Chemical Society*, **127**, 8160–8167; (b) Comer, E. and Organ, M.G. (2005) *Chemistry: A European Journal*, **11**, 7223–7227; (c) Bremner, W.S. and Organ, M.G. (2007) *Journal of Combinatorial Chemistry*, **9**, 14–16; (d) Shore, G., Morin, S., Mallik, D., and Organ, M.G. (2008) *Chemistry: A European Journal*, **14**, 1351–1356; (e) Ullah, F., Samarakoon, T., Rolfe, A., Kurtz, R.D., Hanson, P.R., and Organ, M.G. (2010) *Chemistry: A European Journal*, **16**, 10959–10962.

208 Howarth, P. and Lockwood, M. (2004) *The Chemical Engineer*, **756**, 29–31.

209 Schmink, J.R., Kormos, C.M., Devine, W.G., and Leadbeater, N.E. (2010) *Organic Process Research & Development*, **14**, 205–214.

210 Bierbaum, R., Nüchter, M., and Ondruschka, B. (2004) *Chemie Ingenieur Technik*, **76**, 961–965.

211 Razzaq, T. and Kappe, C.O. (2008) *Chemistry & Sustainability: Energy & Materials*, **1**, 123–132.

212 Hoogenboom, R., Wilms, T.F.A., Erdmenger, T., and Schubert, U.S. (2009) *Australian Journal of Chemistry*, **62**, 236–243.

213 Moseley, J.D. and Woodman, E.K. (2009) *Energy Fuels*, **23**, 5438–5447.

214 For a critical analysis of heating and energy efficiency in microwave chemistry, see: Moseley, J.D. and Kappe, C.O. (2011) *Green Chemistry*, **13**, 794–806.

5
Literature Survey Part A: Transition Metal-Catalyzed Reactions

5.1
General Comments

The literature survey presented in Chapters 5–8 highlights selected examples of controlled microwave heating technology in organic synthesis. The term "controlled" here refers to the use of a dedicated microwave reactor for synthetic chemistry purposes (single-mode or multimode). Therefore, the exact reaction temperature *during the irradiation process* has been adequately determined in the original literature source. In terms of processing techniques (Chapter 4), preference is given to transformations in solution under sealed-vessel conditions, since this reflects the recent trend in the literature. Most of the examples have been taken from the period 2002 to the early 2011. Microwave-assisted solid-phase organic synthesis and related transformations involving immobilized reagents or catalysts are described in Chapter 8. Here, in some instances, results obtained with domestic microwave ovens have been included.

Today, several books [1–12] and an extensive collection of review articles (>240, see Table 5.1) [13–254] cover MAOS from every conceivable angle. The focus of the following microwave literature survey is on modern synthetic methods of interest to organic/medicinal chemists working in industry or academia.

5.2
Carbon–Carbon Bond Formations

Homogeneous transition metal-catalyzed reactions represent one of the most important and most extensively studied reaction types in MAOS. Using traditional heating under reflux conditions, transition metal-catalyzed carbon–carbon and carbon–heteroatom bond-forming reactions typically need hours or days to reach completion and often require an inert atmosphere. Over the last few years, the groups of Hallberg, Larhed, and others have demonstrated that many of these transformations can be significantly enhanced by employing microwave heating under sealed-vessel conditions (microwave flash heating), in most cases without

Microwaves in Organic and Medicinal Chemistry, Second Edition.
C. Oliver Kappe, Alexander Stadler, and Doris Dallinger.

Table 5.1 Review articles in microwave-assisted organic synthesis.

Microwave theory/dielectric heating	[13–17]
Microwave effects	[19–28]
Microwave reactors/equipment	[13, 29]
Technical reviews	[30–36]
General organic synthesis	[30, 31, 37–56]
Solvent-free conditions	[57–66, 211]
Reactions on graphite supports	[67, 68]
Phase-transfer catalysis	[69–71]
Gaseous reagents	[72]
Green chemistry	[59, 73–88]
Water as solvent	[74, 89–93]
Ionic liquids	[77, 94–100]
Medicinal chemistry/drug discovery	[89, 101–118, 123, 129, 166, 168, 211]
Combinatorial chemistry	[119–128, 143]
Parallel processing	[119, 129–131, 136]
Solid-phase synthesis	[132–137]
Polymer-supported reagents	[138–143]
Fluorous synthesis	[144–147]
Peptide synthesis	[148–151]
Enzymatic reactions	[152, 153]
Carbohydrate chemistry	[154–157]
Biosciences	[158, 159]
Organocatalysis	[160]
Heterogeneous catalysis	[161–163]
Transition metal catalysis	[91, 164–182]
Metathesis chemistry	[175, 176]
Organometallics	[183, 184]
Multicomponent reactions	[144, 185–191]
Heterocycles	[58, 74, 134, 139, 169, 192–213, 217]
Cycloadditions	[199, 214–218]
Natural products	[219–221]
Photochemistry	[222–224]
Radiochemistry	[225–229]
Hyphenated methods	[180, 230–232]
Miscellaneous	[233–243]
Scale-up	[244–249]
Continuous flow processing	[250–253]
Energetic aspects	[29, 33, 254]

requiring an inert atmosphere [171, 173, 174]. The use of metal catalysts in conjunction with microwaves may have significant advantages in comparison to traditional heating methods, since the inverted temperature gradients under microwave conditions may lead to an increased lifetime of the catalyst by elimination of wall effects. In fact, the elimination of wall effects and low thermal gradients (bulk heating) in microwave-heated reactions has frequently been invoked to rationalize the outcome of microwave-assisted reactions involving homogeneous transition metal catalysts (see Section 2.5.3).

5.2.1
Heck Reactions

The Heck reaction, a palladium-catalyzed vinylic substitution, is typically conducted employing olefins and organohalides or pseudohalides as reactants. Numerous elegant synthetic transformations based on C–C bond-forming Heck chemistry have been developed both in classical organic synthesis and natural product chemistry [255]. As early as 1996, the solution-phase Heck chemistry has been carried out successfully by MAOS, reducing reaction times from several hours under conventional reflux conditions to sometimes less than 5 min [256]. These early examples of microwave-assisted Heck reactions have been extensively reviewed by the Larhed group and will not be further discussed herein [171, 173, 174]. Scheme 5.1 shows an example of a standard Heck reaction reported by Kappe and coworkers involving aryl bromides and acrylic acid to furnish the corresponding cinnamic acid derivatives [257]. Optimization of reaction conditions under small-scale (2 mmol) single-mode microwave conditions led to a protocol that employed acetonitrile as solvent, 1 mol% palladium acetate/tri(*ortho*-tolyl)phosphine as catalyst/ligand system, triethylamine as base, and 180 °C reaction temperature for 15 min. Interestingly, the authors have discovered that here the rather expensive homogeneous catalyst system can be replaced by 5% palladium-on-charcoal (<0.1 mol% concentration of palladium) without the need to change any of the other reaction parameters [257]. Yields for the Heck reaction providing cinnamic acids (X = H) were very similar using either homogeneous or heterogeneous Pd catalysis. In the same article [257], the authors also demonstrate that it is possible to directly scale-up the 2 mmol Heck chemistries to 80 mmol (about 120 mL total reaction volume) switching from a single-mode to a larger multimode microwave reactor. Importantly, the optimized small-scale reaction conditions could be directly used for the larger scale run, giving rise to very similar product yields (see Section 4.8).

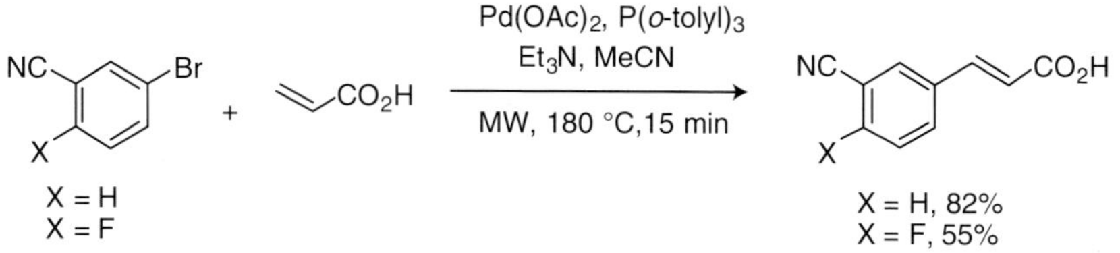

Scheme 5.1 Heck Reactions on a 2 and 80 mmol scale.

In 2002, Larhed and coworkers reported microwave-promoted Heck arylations using the ionic liquid 1-butyl-3-methylimidazolium hexafluorophosphate (bmimPF_6) as reaction medium [258]. This reaction (see Scheme 4.11) and the unique properties of ionic liquids in the context of microwave synthesis were described in detail in Section 4.5.2. More recently, the same group has exploited the combination of bmimPF_6 and dioxane in a Heck coupling of both electron-rich and electron-poor aryl chlorides with butyl acrylate (Scheme 5.2) [259]. Transition

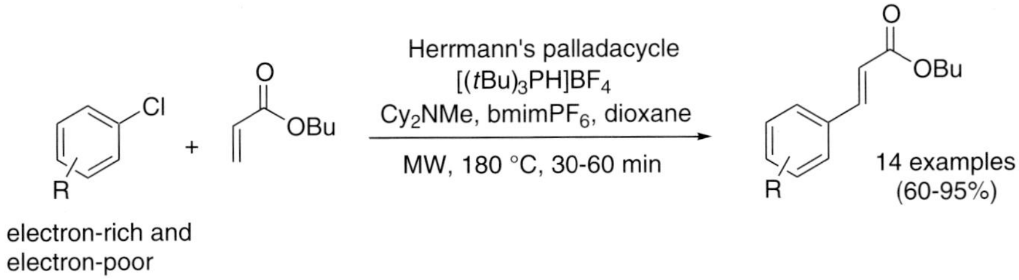

Scheme 5.2 Heck reactions of aryl chlorides involving air-stable phosphonium salts as ligand precursors.

metal-catalyzed carbon–carbon bond-forming reactions involving unreactive aryl chlorides have represented a synthetic challenge for a long time. Due to advances in the development of highly active catalyst/ligand systems, it is only recently that such transformations have been accessible [260]. For the Heck coupling shown in Scheme 5.2, the air-stable but highly reactive tri-*tert*-butylphosphonium tetrafluoroborate described by Netherton and Fu [261] was employed as a ligand precursor using Herrmann's palladacycle [*trans*-di(μ-acetato)bis[*o*-(di-*o*-tolyl-phosphino)benzyl]dipalladium(II)] [262] as palladium precatalyst. Depending on the reactivity of the aryl chloride, 1.5–10 mol% of palladium catalyst (3–20% of ligand), 1.5 equiv of *N,N*-dicyclohexylmethylamine as a base, and 1 equiv of $bmimPF_6$ in dioxane were irradiated at 180 °C with the aryl chloride and butyl acrylate for 30–60 min. Under these optimized conditions, the desired cinnamic esters were obtained in moderate to excellent yields (Scheme 5.2) [259].

In an article by Botella and Nájera, controlled mono and double Heck arylations in water catalyzed by an oxime-derived palladacycle were described [263]. When the reaction was carried out under microwave irradiation at 120 °C in the presence of dicyclohexylmethylamine with only 0.01 mol% palladium catalyst (palladium acetate or palladacycle), monoarylation took place in only 10 min with a very high turnover frequency (TON) of >40 000 (Scheme 5.3). In order to achieve successful diarylation, 1 mol% of the palladacycle catalyst and 2 equiv of iodobenzene had to be utilized to obtain moderate to good yields of diarylated product. Although microwave heating to 120 °C provided a 31% yield after 10 min, a 66% isolated yield of product was observed by heating the reaction mixture under reflux for 13 h at 100 °C.

This protocol could be extended to a range of different α,β-unsaturated carbonyl compounds and activated or deactivated aryl iodides, respectively [263]. An application of related Heck chemistry toward the synthesis of methylated resveratrol (3,4′,5-trihydroxy-(*E*)-stilbene) is shown in Scheme 5.4 [264]. The phytoalexin resveratrol exhibits a variety of interesting biological and therapeutic properties, including activity against several human cancer cell lines. Botella and Nájera have shown that the trimethyl ether of resveratrol (Scheme 5.4) can be rapidly prepared by microwave-assisted Heck reaction of the corresponding aryl iodide and styrene derivatives, using the same oxime-derived palladacycle as described in Scheme 5.3.

Using a different catalytic system, the Larhed group was able to perform regioselective microwave-promoted chelation-controlled double β-arylations of terminal

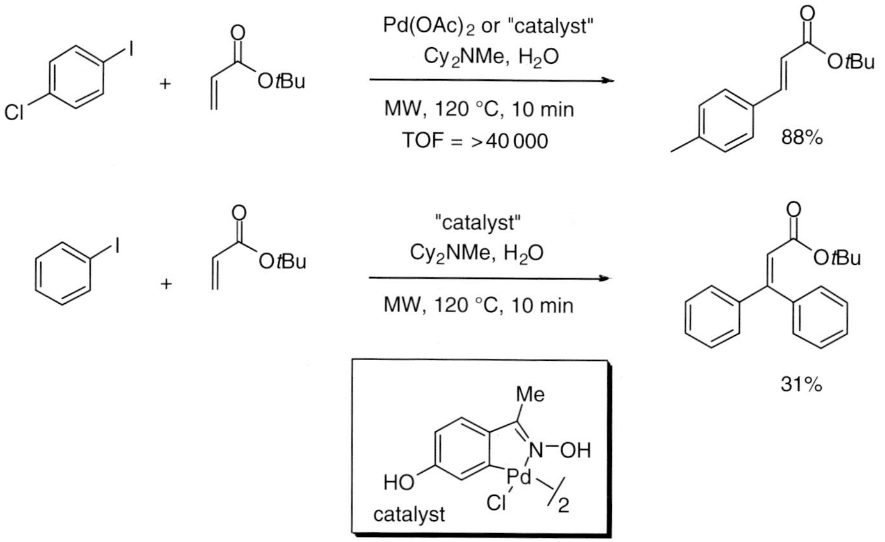

Scheme 5.3 Mono and double Heck arylations in water using oxime-derived palladacycles.

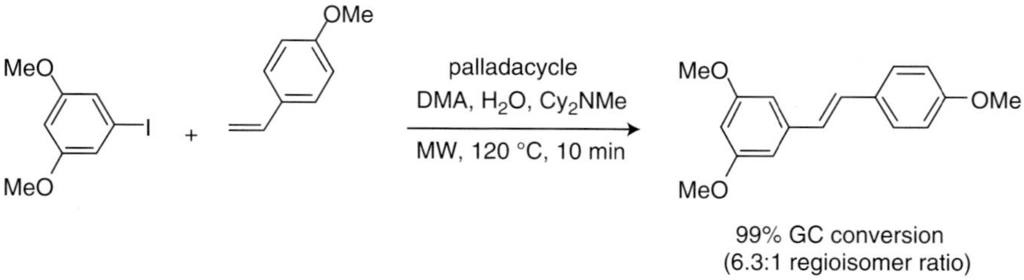

Scheme 5.4 Synthesis of resveratrol trimethylether via Heck arylation.

olefins (Scheme 5.5) [265]. In this Heck approach, the authors used vinyl ethers as chelating olefins and aryl bromides as coupling partners, employing Herrmann's palladacycle as a palladium source. By proper selection of the experimental parameters, it was possible to achieve symmetrical and nonsymmetrical terminal β,β-diarylations with both electron-rich and electron-poor aryl bromides. One-pot microwave-heated symmetric bisarylations were carried out employing an excess of the aryl bromide (Ar^1Br, 3.0–5.0 equiv) with a low amount of Herrmann's palladacycle (0.5 mol%) in 10% aqueous DMF. To increase the stability of the underligated catalytic palladium(0) system, a highly ionic reaction cocktail was preferred using lithium chloride and sodium acetate as additives and potassium carbonate as base. Optimum conditions involved microwave heating at 160–180 °C for 10–55 min. In order to obtain unsymmetrical products, the mono β-arylated olefins were reacted with aryl bromides (Ar^2Br) under almost identical reaction conditions to those employed in the symmetrical examples (Scheme 5.5) [265]. For related Heck vinylations of chelating vinyl ethers, see Refs [266,267].

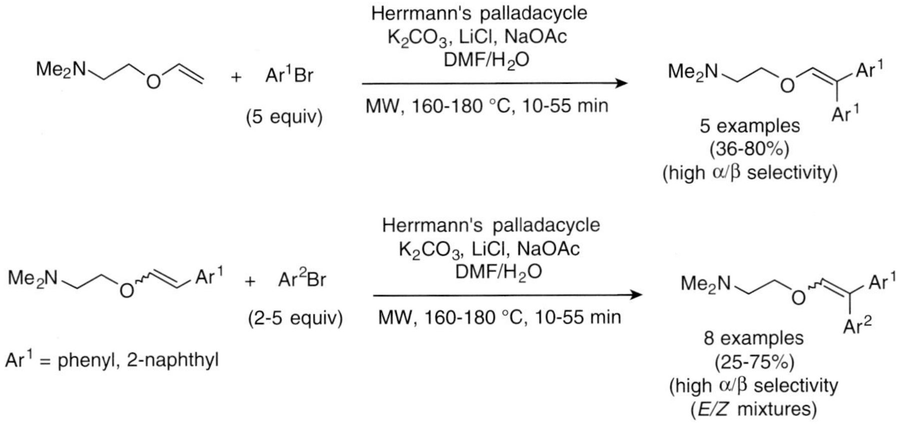

Scheme 5.5 Double Heck arylations of chelating olefins with aryl bromides.

The same group has also reported on chelation-controlled diastereoselective Heck arylations on route to enantiopure 2-aryl-2-methyl cyclopentanones [268].

Leadbeater and Kormos have disclosed a one-pot two-step Mizoroki–Heck coupling strategy for the preparation of asymmetrically substituted stilbenes (Scheme 5.6) [269]. In the first step, styrenes **2** were synthesized selectively by employing ethene as coupling partner using a gas loading device. For aryl iodides (**1**) as coupling partners, 0.02 mol% Pd using an ICP standard and 125 °C are required to obtain the coupled products **2**, whereas for aryl bromides (**1**) 0.5 mol% of the Herrmann's catalyst and 150 °C are necessary. In the next step, styrenes **2** are coupled with aryl bromides using a 1 : 1 molar ratio and Herrmann's catalyst in the same reaction vial to get stilbenes **3**. It has to be noted that the less reactive aryl halide

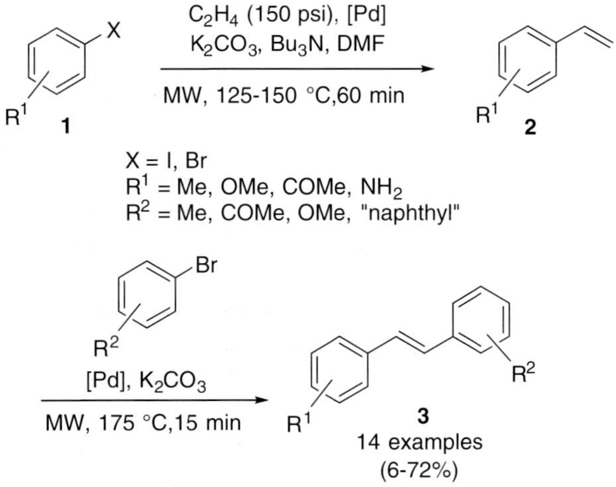

Scheme 5.6 Synthesis of asymmetrically substituted stilbenes.

has to be used in the first coupling step in order to reduce the competitive symmetric stilbene formation and thus increase the product yield.

Similar microwave-assisted Mizoroki–Heck transformations have been reported by the same group, again involving low levels of Pd catalysts (500 ppb) [270].

An addition to the already powerful spectrum of microwave Heck chemistry was the development of a general procedure for carrying out oxidative Heck couplings, that is, the palladium(II)-catalyzed carbon–carbon coupling of arylboronic acids with alkenes using copper(II) acetate as a reoxidant [271]. In a 2003 publication (Scheme 5.7), Larhed and coworkers utilized lithium acetate as a base and the polar and aprotic *N,N*-dimethylformamide as solvent. The coupling reaction could be conducted in ambient atmosphere without any discernible difference in outcome compared to the inert nitrogen atmosphere. Generally, 100–140 °C reaction temperature and 5–30 min of irradiation time produced moderate to good yields of arylated products with predominant (*E*)-configuration. Electron-rich arylboronic acids were found to be the superior coupling partners affording high yields of adducts, devoid of any apparent side reactions.

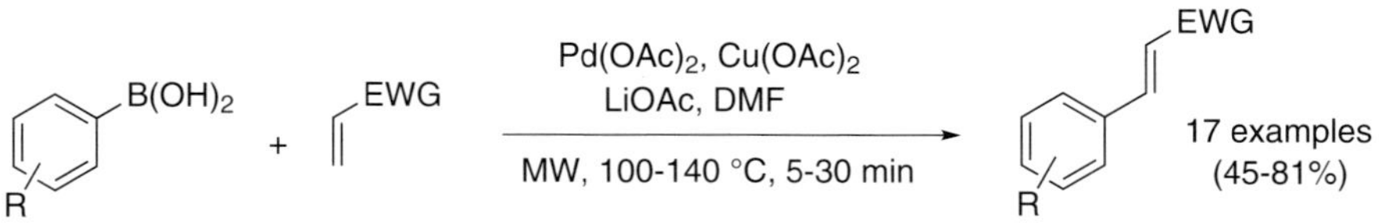

Scheme 5.7 Oxidative Heck coupling of boronic acids and alkenes using copper(II) acetate as a reoxidant.

In 2004, the same group presented a modified method for performing oxidative Heck arylations, exploiting molecular oxygen gas for environmentally benign reoxidation and a stable 1,10-phenanthroline bidentate ligand (dmphen) to promote the palladium(II) regeneration and to control the regioselectivity (Scheme 5.8) [272]. While the conventional thermal reaction often required 18 h to reach completion, microwave irradiation in pre-pressurized vessels (3 bar oxygen pressure) (see also Section 4.4) allowed the oxidative Heck coupling to proceed within 1 h [272].

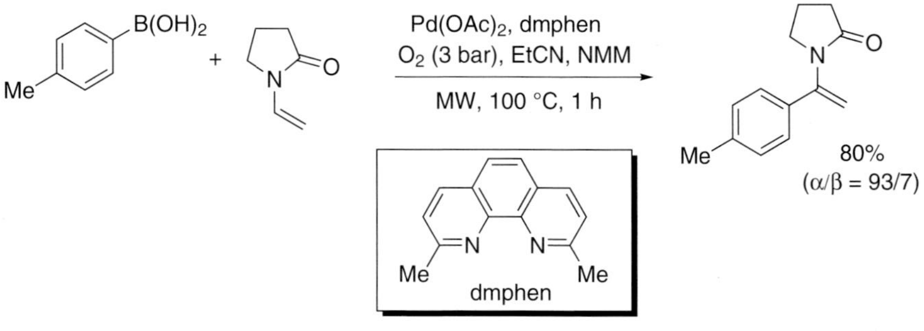

Scheme 5.8 Oxidative Heck coupling of boronic acids and olefins using dioxygen as reoxidant.

In another example of oxidative Heck couplings, the Larhed group presented the vinylation of boronic acids and aryl trifluoroborates to produce the corresponding styrenes [273]. Small-scale single-mode microwave conditions led to an effective protocol employing low-cost vinyl acetate as vinylation agent. Best results were achieved with *N,N*-dimethylformamide as solvent, 2 mol% palladium acetate with 2.2 mol% 1,3-bis(diphenylphosphino)propane (dppp) as catalyst/ligand system, and 10 equiv vinyl acetate at 140 °C for 15 min (Scheme 5.9). Beneficially, the discovered method does neither need inertization nor addition of a base and a reoxidant. The protocol works satisfactory for both electron-rich and electron-poor boronic acids, even vinyl boronic acids can be employed as substrates [273].

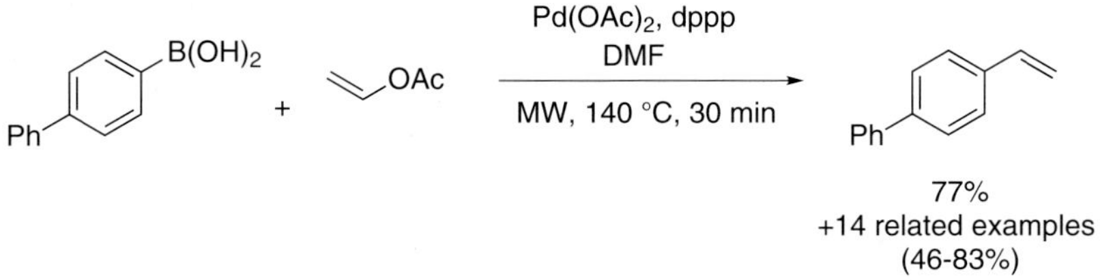

Scheme 5.9 Vinylation of arylboronic acids by oxidative Heck coupling.

Microwave-assisted Heck reactions have also been carried out with triflates as coupling partners involving very complex molecules. Winterfeld and coworkers have reported the multigram synthesis of a complex nonsymmetrical bis-steroidal diene by microwave-promoted coupling of the corresponding olefin and triflate steroidal moieties (Scheme 5.10) [274].

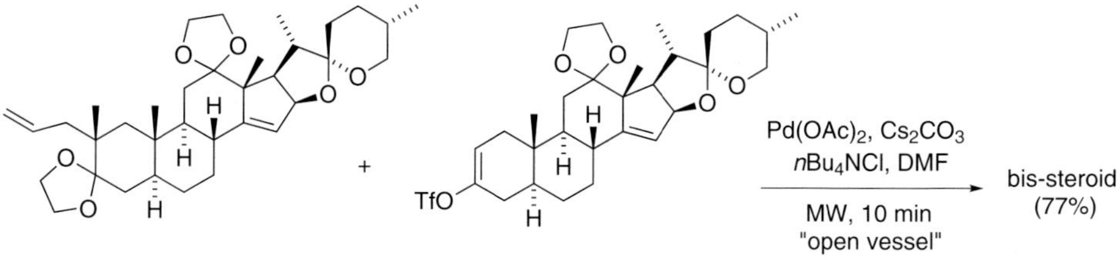

Scheme 5.10 Heck reaction for the synthesis of bis-steroids.

A synthetically useful application of an intramolecular microwave-assisted Heck reaction was described by Gracias *et al.* (Scheme 5.11) [275]. In their approach toward the synthesis of *N*-containing seven-membered heterocycles, the initial product of an Ugi four-component reaction was subjected to an intramolecular Heck cyclization using 5 mol% palladium acetate/triphenylphosphine as the catalytic system. For the example shown in Scheme 5.11, microwave irradiation at 125 °C in acetonitrile for 1 h provided a 98% isolated yield of product. A number

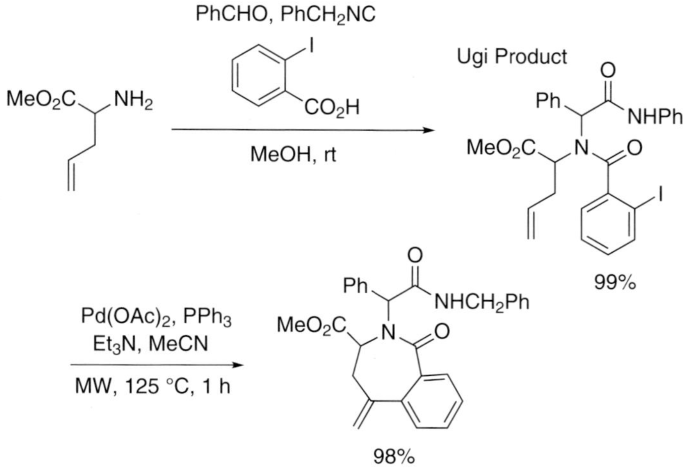

Scheme 5.11 Sequential Ugi/Heck cyclizations for the synthesis of seven-membered *N*-heterocycles.

of related sequential Ugi/Heck cyclizations were reported in the original publication, also involving aryl bromides instead of iodides.

Similarly, Tietze *et al.* have described an intramolecular microwave-promoted Heck reaction for the construction of the B ring in the synthesis of enantiopure B-norestradiol analogs (Scheme 5.12a) [276]. The Heck coupling took place from below, anti to the angular methyl group to form a single diastereoisomer. The best results were obtained using 5 mol% of Herrmann's palladacycle as catalyst and tetrabutylammonium acetate as additive in a mixture of DMF, acetonitrile, and water. Under these conditions, the reaction time could be shortened to 5 min and it also allowed the authors to suppress aromatization of ring C with opening of ring D as

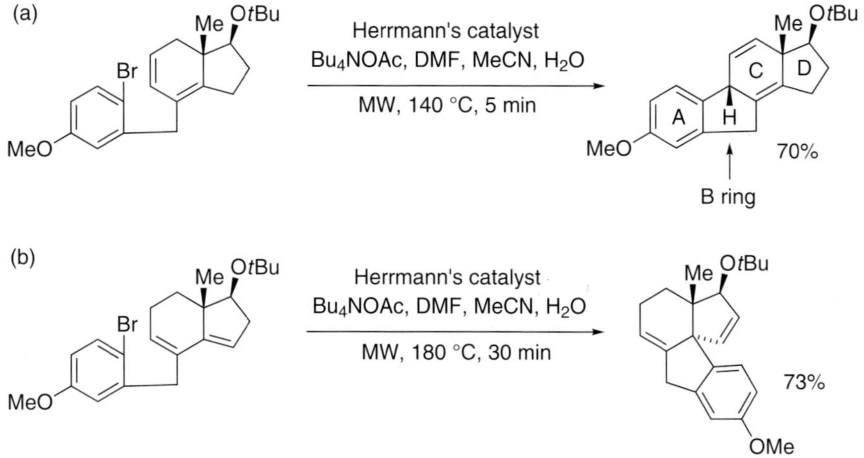

Scheme 5.12 Intramolecular Heck coupling for the construction of B-nor steroids.

unwanted side reaction, which was observed when the reaction was performed under standard conditions (120 °C, 18 h). Treatment of the isomeric seco-B-norsteroid under identical reaction conditions led to a novel spirocyclic ring system as a single diastereomer (Scheme 5.12b) [277]. The isolated product yield was 73% under both the microwave conditions (180 °C, 30 min) and the conventional oil bath heating (120 °C, 18 h).

Another application of an intramolecular Heck reaction using microwave irradiation was disclosed by Sørensen and Pombo-Villar [278]. In the context of synthesizing analogs of a group of seratonin (5-HT_3) receptor antagonists structurally related to the tricyclic carbazolone ring system, the preparation of a cyclopenta[*b*]indole-1-one intermediate via intramolecular Heck reaction of a suitable iodo precursor was envisaged (Scheme 5.13). Optimum conditions were found to involve 5 mol% palladium(II) acetate/tri(*ortho*-tolyl)phosphine as a catalytic system and tetrabutylammonium chloride as an additive. The analogous bromoaniline precursor turned out to be less reactive in the desired cyclization reaction resulting in only 30% conversion under similar microwave-promoted reaction conditions. Surprisingly, however, a 95% yield of the desired product could be obtained when the Heck cyclization was carried out under conventional heating in an oil bath at 120 °C for 16 h.

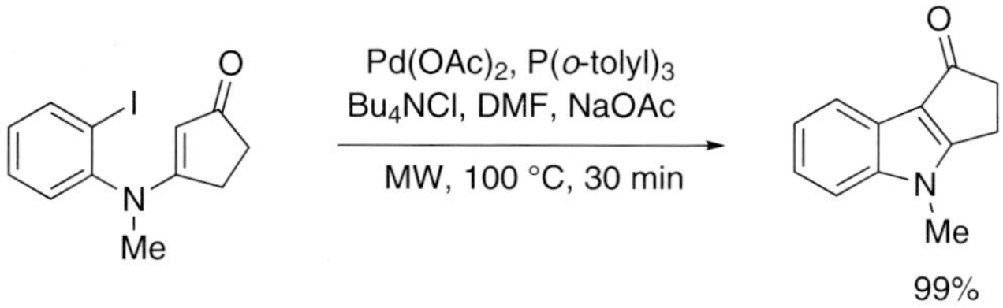

Scheme 5.13 Intramolecular Heck coupling for the preparation of cyclopenta[*b*]indoles.

Lachance *et al.* have shown that imines/enamines **6** can be converted via a microwave-assisted intramolecular Mizoroki–Heck reaction to the corresponding azaindoles **7** in good yields (Scheme 5.14) [279]. Several synthesis routes have been investigated, including the one-step reaction (route **A**) of aminopyridines **4** with ketones **5** or ketals as well as the one-pot two-step reaction (route **B**) leading to azaindoles **7** (Scheme 5.14a). Using either of the two strategies, a small library of 17 examples was synthesized employing chloro-, bromo-, and iodoaminopyridines with toleration of sensitive groups like ketones and esters (Scheme 5.14b).

The Larhed group has disclosed the regioselective synthesis of spiro[cyclohexane-1,1′-isobenzofurane] derivatives via intramolecular Mizoroki–Heck cyclization (Scheme 5.15) [280]. Five *exo*-cyclization of *o*-bromobenzyl cyclohexenyl ethers proceeded at 180 °C within 10 min. When *o*-iodobenzyl cyclohexenyl ethers are employed as substrates, the addition of 5% of water (v/v) to the solvent system

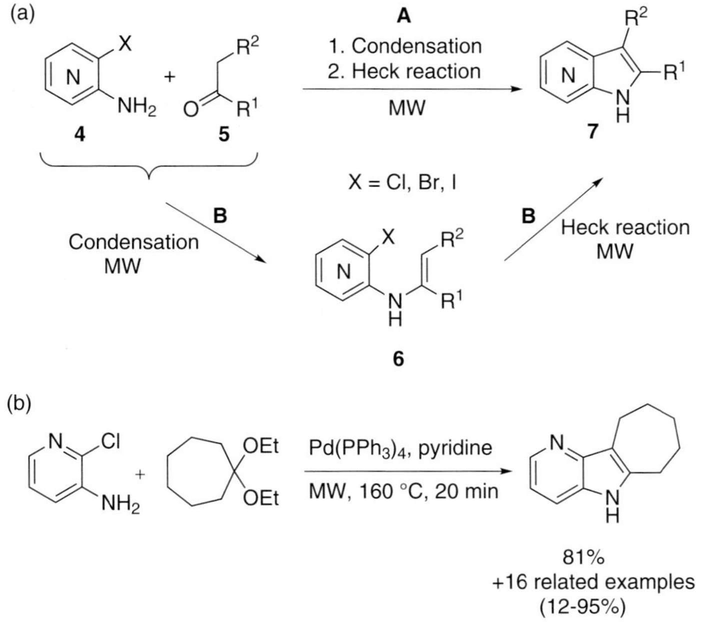

Scheme 5.14 Synthesis of 4-, 5-, 6-, and 7-azaindoles.

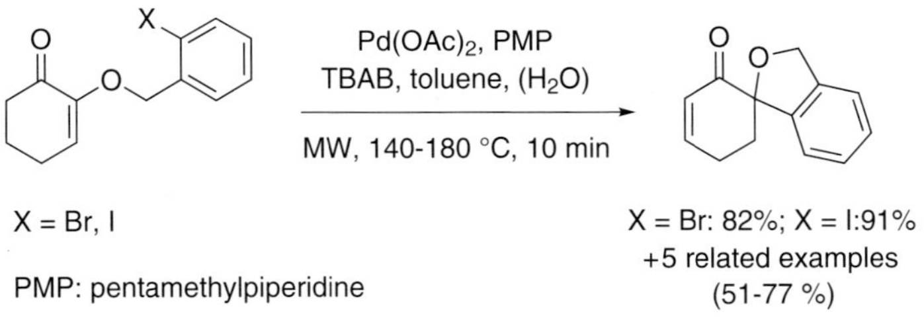

Scheme 5.15 Pd-catalyzed spiro cyclizations.

proved to promote the cyclization within 10 min at 140 °C. Conventional heating at 100 °C provided the products in similar yields, but a reaction time of 18 h is required.

The same group reported on regioselective internal Heck arylations of hydroxyalkyl vinyl ethers [281].

Intramolecular Heck couplings have also been used by Vasudevan *et al.* for the construction of cyclic sulfonamides [282], by Porco and coworkers for the generation of indanes and related polycyclic scaffolds [283], by Donets and Van der Eycken for the synthesis of 3-benzazepines [284], and by Chruma and coworkers for the synthesis of polyfunctionalized 1-aminoindanes [285]. A number of more complex intramolecular examples have also been disclosed [286–288], for example, for the total synthesis of the batrachotoxin ring system [289]. A Cu-catalyzed

Heck-type coupling of aryl iodides and acrylates has been suggested by Lamaty and coworkers [290]. Enantioselective Heck reactions are covered in Section 6.1.7.

5.2.2
Suzuki–Miyaura Reactions

The Suzuki reaction (the palladium-catalyzed cross-coupling of aryl halides with boronic acids) is arguably one of the most versatile and at the same time also one of the most often used cross-coupling reactions in modern organic synthesis [291]. Carrying out the high-speed Suzuki reactions under controlled microwave conditions can today be considered almost a routine synthetic procedure, given the enormous literature precedent for this transformation [7].

A significant advance in Suzuki chemistry has been the observation that Suzuki couplings can be readily carried out using water as solvent in conjunction with microwave heating [292,293]. Water, being cheap, readily available, nontoxic, and nonflammable, has clear advantages as a solvent for use in organic synthesis (see Section 4.5.1). Leadbeater and Marco have described very rapid, ligand-free palladium-catalyzed aqueous Suzuki couplings of aryl halides with arylboronic acids (Scheme 5.16) [292]. Key to the success of this method was the use of 1 equiv of tetrabutylammonium bromide (TBAB) as a phase-transfer catalyst (PTC) additive. The role of the ammonium salt is to facilitate the solubility of the organic substrates and to activate the boronic acid by formation of a $[ArB(OH)_3)]^-[R_4N]^+$ species. By using controlled microwave heating at 150 °C for 5 min with only 0.4 mol% of palladium acetate as catalyst, a wide variety of aryl bromides and iodides were successfully coupled with arylboronic acids (Scheme 5.16) [292]. Aryl chlorides also reacted but required higher temperatures (175 °C).

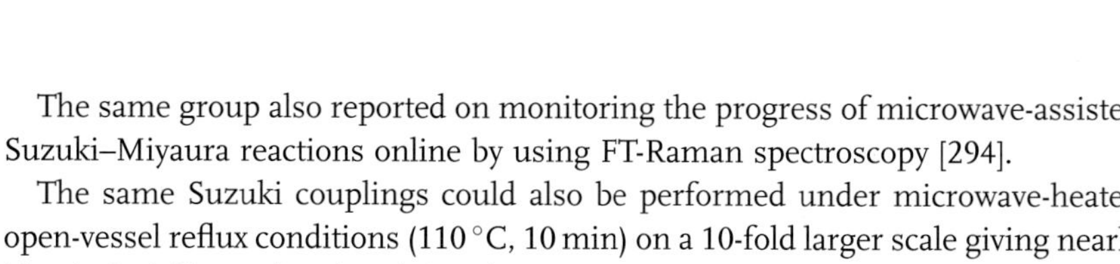

Scheme 5.16 Ligand-free Suzuki reactions using TBAB as additive.

The same group also reported on monitoring the progress of microwave-assisted Suzuki–Miyaura reactions online by using FT-Raman spectroscopy [294].

The same Suzuki couplings could also be performed under microwave-heated open-vessel reflux conditions (110 °C, 10 min) on a 10-fold larger scale giving nearly identical yields to the closed-vessel runs [292,295]. Importantly, nearly the same yields were also obtained when the Suzuki reactions were carried out in a preheated oil bath (150 °C) instead of using microwave heating, clearly indicating the absence of any specific or nonthermal microwave effects [293]. The same group has also reported on a further scalability of these reactions going from a 5 mmol to a 1 mol scale in larger microwave reactors [296].

In another modification, the same authors reported that, somewhat surprisingly, it was also possible to carry out the Suzuki reactions depicted in Scheme 5.16 in the absence of any added palladium catalyst [297–299]. These transition metal-free aqueous Suzuki-type couplings again utilized 1 equiv of tetrabutylammonium bromide as an additive, 3.8 equiv of sodium carbonate as a base, and 1.3 equiv of the corresponding boronic acid (150 °C, 5 min). High yields were obtainable for aryl bromides and iodides, whereas aryl chlorides proved unreactive under the conditions used. The reaction is also limited to electron-poor and electron-neutral boronic acids. It was subsequently discovered that ultralow concentrations of palladium (50 ppb) contained in the sodium carbonate base were responsible for these Suzuki reactions to proceed [300]. Comparatively low catalyst loadings of aqueous Suzuki couplings have also been reported by Nájera *et al.* [301].

Cross-coupling reactions of unactivated aryl chlorides represent one of the most challenging problems in transition metal-catalyzed transformations. A series of air- and moisture-stable (*N*-heterocyclic carbene)palladium(allyl) complexes (NHC) has been shown by Nolan and coworkers to catalyze Suzuki cross-coupling reactions of aryl chlorides with boronic acids (Scheme 5.17) [302]. This catalytic system is compatible to microwave conditions and rapid couplings were observed within 1.5 min at 120 °C employing 2 mol% of the NHC ligand in the presence of a strong base (sodium *tert*-butoxide, 3 equiv). Under these high temperature conditions, the catalyst appeared to be stable as no palladium black was formed. The conventionally heated reactions (60 °C) required several hours to reach

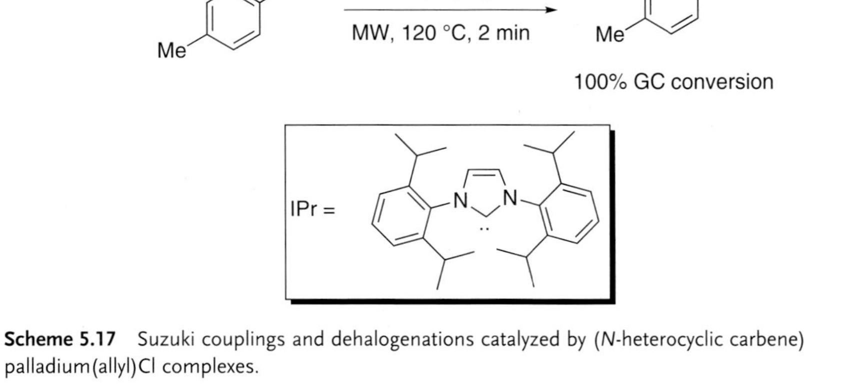

Scheme 5.17 Suzuki couplings and dehalogenations catalyzed by (*N*-heterocyclic carbene) palladium(allyl)Cl complexes.

completion. In the same article, the authors also reported on microwave-assisted dehalogenations of aryl chlorides using the same catalytic system but switching to 2-propanol as a protic solvent (Scheme 5.17) [302]. By using 1.05 equiv of sodium *tert*-butoxide, the amount of catalyst required to dehalogenate 4-chlorotoluene was reduced to 0.025 mol% of the NHC system at 120 °C (2 min). Similar Suzuki couplings of aryl chlorides under microwave conditions were disclosed by Bedford *et al.* [303]. Correspondingly, Clarke *et al.* demonstrated that aryl chlorides couple with boronic acids using Pd complexes formed from the amine–phosphine ligand "dcpmp" [304].

Many examples in the literature demonstrate the versatility of the Suzuki protocol under microwave conditions. In the example shown in Scheme 5.18, Gong and He have described the direct synthesis of unprotected 4-aryl phenylalanines via a microwave-assisted Suzuki protocol, utilizing standard coupling conditions [305]. For most reactions, 5 mol% of palladium(II) bis(triphenylphosphine) dichloride [$Pd(PPh_3)_2Cl_2$] and the use of sodium carbonate as a base allowed the high yielding coupling of 4-borono phenylalanine with a variety of (hetero)aryl halides within 10 min of microwave irradiation (150 °C). A control experiment in a preheated oil bath under otherwise identical conditions showed a significantly lower conversion. The authors have also investigated the effect of switching the positions of boronic acid and halide in the coupling partners. This complementary methodology led to equally high yields. It was also possible to employ enantiomerically pure boron phenylalanine in the Suzuki coupling with a heteroaryl chloride. In addition to the excellent conversion, no significant racemization was observed in the coupling product.

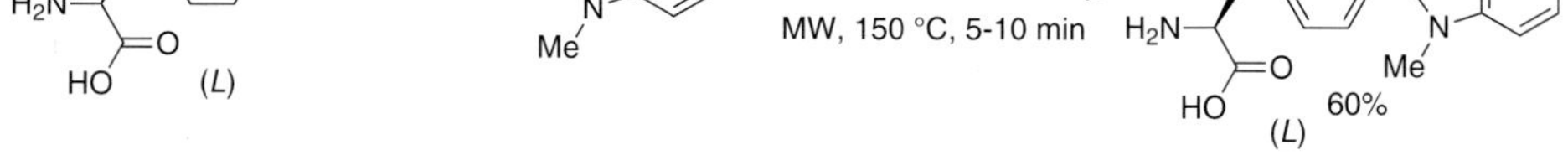

Scheme 5.18 Synthesis of unprotected 4-aryl phenylalanine via Suzuki cross-coupling reactions.

Similar reaction conditions were employed by Savall and Fontimayor for the synthesis of a series of 2-arylbenzimidazoles via Suzuki–Miyaura couplings of unprotected 2-chlorobenzimidazoles (Scheme 5.19) [306]. Higher yields could be achieved when switching from arylboronic acids to aryltrifluoroborate salts as coupling partners. To overcome the problem with homocoupling of pyridylboronic acids, a change in the catalyst system to $PdCl_2(dppf){\cdot}CH_2Cl_2$ and the use of aryltrifluoroborate salts were advantageous.

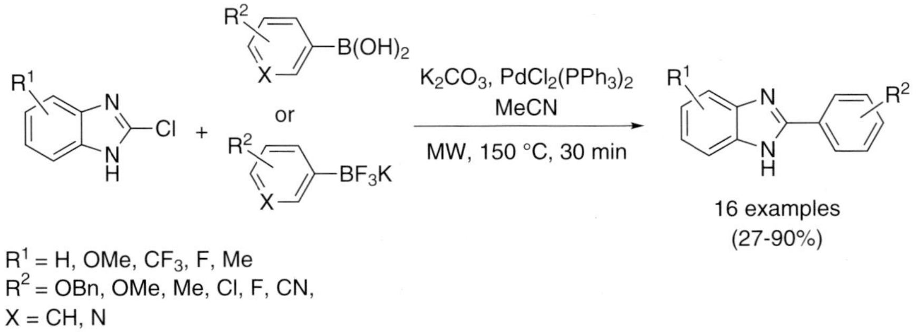

Scheme 5.19 Suzuki–Miyaura coupling of unprotected 2-chlorobenzimidazoles.

The application of a microwave-assisted Suzuki reaction for the synthesis of electron-rich phenethylamines and apogalanthamine analogs was described by Van der Eycken and coworkers (Scheme 5.20) [307]. Here, the use of sodium hydrogen carbonate and 5 mol% tetrakis(triphenylphosphine)palladium(0) as catalyst in a mixture of *N,N*-dimethylformamide and water was found to be the optimum medium for the microwave-assisted Suzuki reactions (120–150 °C, 10–15 min). In order to access the required apogalanthamine analogs, this reaction was also attempted with boronic acids bearing highly electron-withdrawing substituents in the sterically unfavorable *ortho*-position. In order to prevent protodeboronation, to which these systems are prone, the reaction conditions were slightly modified (cesium carbonate) in order to provide an 84% yield of the desired biaryl coupling product. An acid-mediated, microwave-assisted deprotection procedure, followed by a one-pot reductive amination then furnished the desired apogalanthamine analog **8** (Scheme 5.20) [307].

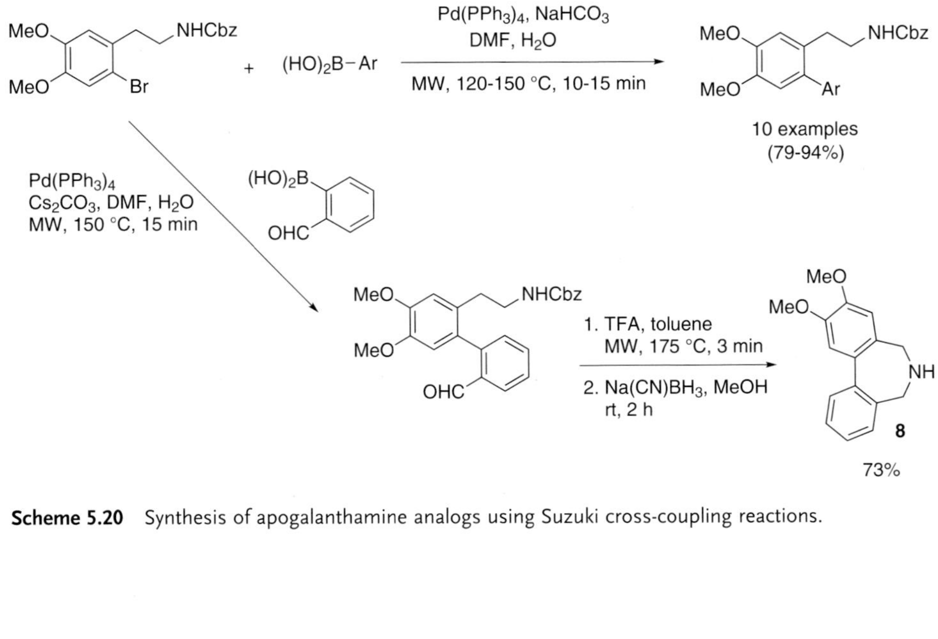

Scheme 5.20 Synthesis of apogalanthamine analogs using Suzuki cross-coupling reactions.

A 2004 publication by Barbarella and coworkers has disclosed the rapid preparation of poorly soluble unsubstituted and modified α-quinque and sexithiophenes by the extensive use of bromination/iodination steps and microwave-assisted Suzuki and Sonogashira cross-couplings (Scheme 5.21) [308]. Suzuki reactions were either carried out under solvent-free conditions on a strongly basic potassium fluoride/alumina support for the synthesis of soluble oligothiophenes or in solution phase for the preparation of the rather insoluble α-quinque and sexithiophenes. In both cases, 5 mol% of [1,1′-bis(diphenylphosphino)ferrocene]dichloropalladium(II) [$PdCl_2$dppf] was used as the catalyst. A particularly noteworthy aspect of this publication is the introduction of a one-pot borylation/Suzuki reaction of 5-bromoterthiophene with commercial bis(pinacolato)diboron. The resulting sexithiophene was obtained within 10 min of microwave irradiation in 84% isolated yield. The same concept was also applied toward the synthesis of modified oligothiophenes (not shown) [308]. Several of the prepared oligothiophenes show liquid–crystalline properties.

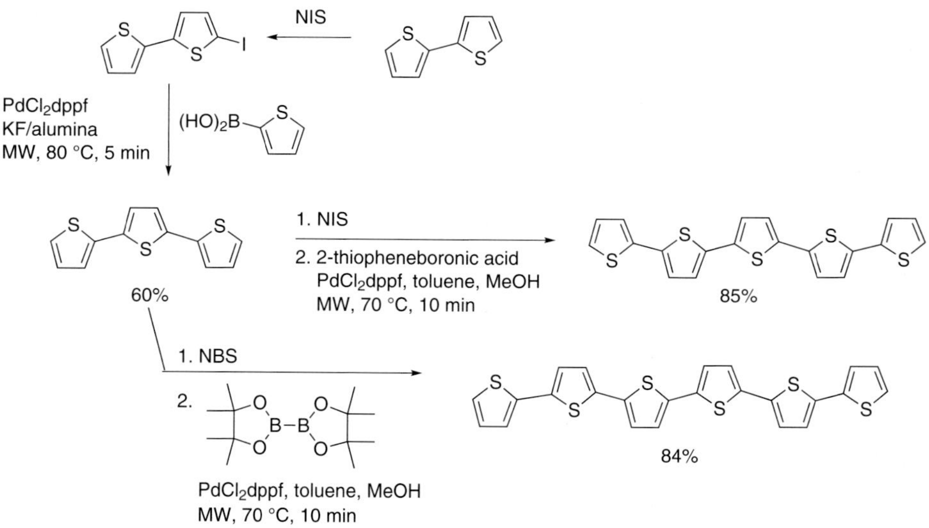

Scheme 5.21 Synthesis of oligothiophenes using Suzuki cross-coupling reactions.

More recently, the same group has revisited the synthesis of oligothiophenes and has presented a Suzuki coupling procedure based on the use of a chitosan-supported Pd catalyst [309].

Similar one-pot borylation/Suzuki sequences were used by Van der Eycken and coworkers for the preparation of bis-2(1*H*)-pyrazinones [310]. Poly-*p*-phenylene conjugated polymers based on spirobifluorene scaffolds were obtained via Suzuki couplings of appropriate bis-boronic acids and *p*-dibromoarene building blocks [311]. Water-soluble polyacetylenes have been obtained from diborylethyne and *p*-diiodoarene synthons [312].

The asymmetric diboration of olefins provides versatile, reactive 1,2-diboron intermediates in a catalytic enantioselective fashion. Unsymmetrical 1,2-bis(boronates), such as those derived from terminal alkenes, engage in selective cross-coupling reactions involving the more accessible, less hindered primary carbon–boron bond [313]. In this context, Morken and coworkers have reported a catalytic asymmetric carbohydroxylation of alkenes by a tandem diboration/Suzuki cross-coupling/oxidation reaction. In the example shown in Scheme 5.22, the sterically encumbered terminal alkene was converted into the desired 1,2-diboron intermediate by a rhodium-catalyzed diboration [5 mol% rhodium(I)-(2,5-norbornadien)-acetylacetonat, (nbd)Rh(acac); 5 mol% (*S*)-(−)-1-(2-diphenylphosphino-1-naphthyl) isoquinoline, (*S*)-quinap] at room temperature. After addition of phenyl triflate as coupling partner and 10 mol% of [1,1′-bis(diphenylphosphino)ferrocene]dichloropalladium(II) [$PdCl_2$dppf] as a catalyst, the selective Suzuki coupling was performed under microwave conditions at 80 °C for 1 h, followed by oxidation of the remaining carbon–boron bond with hydrogen peroxide. The anticipated chiral secondary alcohol, the product of a net carbohydroxylation reaction, was isolated in 70% yield and with 93% enantiomeric excess [313].

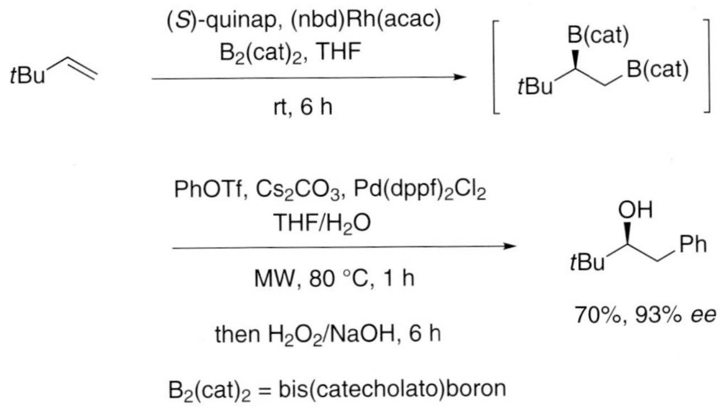

Scheme 5.22 Catalytic asymmetric carbohydroxylation of alkenes.

A 2004 publication by Turner and coworkers has described the microwave-assisted palladium-catalyzed cross-coupling of β-chloroalkylidene/arylidene malonates (vinyl chlorides) with boronic acids in a Suzuki-type fashion (Scheme 5.23) [314]. The Suzuki arylation reaction was found to proceed with a wide range of arylboronic acids, including electron-rich, electron-deficient, and sterically demanding systems. For most reactions, the presence of 1 mol% of the commercially available air-stable palladium catalyst $[(t\text{Bu})_2P(OH)]_2PdCl_2$ provided a high yield of the desired β-aryl/alkylarylidene malonates within 30 min in tetrahydrofuran as solvent. The Suzuki reaction was found to be compatible with amino, hydroxy, and chloro functionalities in the arylboronic acids leading to functionalized products that could be further derivatized (Scheme 5.23). The catalyst was also found active to promote other standard Suzuki reactions of aryl chlorides under microwave conditions [314].

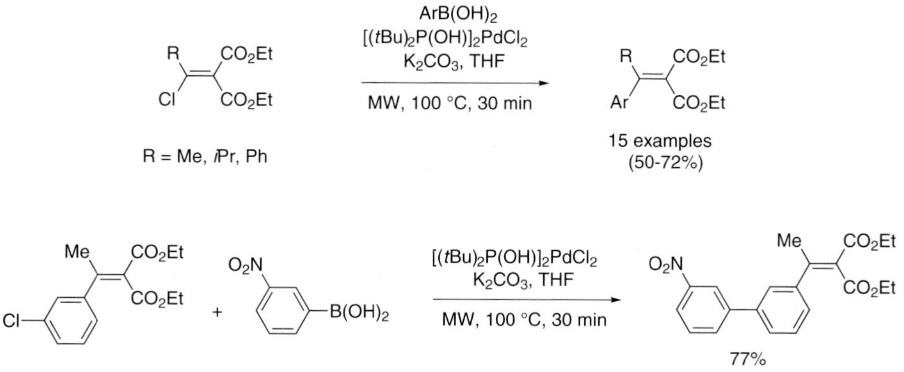

Scheme 5.23 Palladium-catalyzed cross-coupling of β-chloroalkylidene/arylidene malonates.

Greaney and coworkers have described a protocol for the arylation of oxazoles in the 4- or 2-position, respectively, via Suzuki–Miyaura reactions (Scheme 5.24) [315]. 2-Aryl-4-trifloyl oxazoles **9** were coupled with a variety of boronic acids, whereas the coupling at the 2-position was performed with 2-chloro-4-phenyl oxazole as coupling partner. Both couplings were conducted at the same reaction conditions (5 mol% $PdCl_2(PPh_3)_2$, Na_2CO_3, dioxane, 150 °C, and 20 min). The reaction was further extended to the one-pot synthesis of dioxazoles. In the first step, boronic esters were generated *in situ* via the reaction of **9a-9c** with bis(pinacolato)diboron (B_2Pin_2). In the next sequence, 1 equiv of **9a**, **9b**, or **9c** was added to the reaction mixture together with the reagents needed for the Suzuki–Miyaura coupling (catalyst, base). In this way, homo- and heterodimeric dioxazoles **10** were obtained.

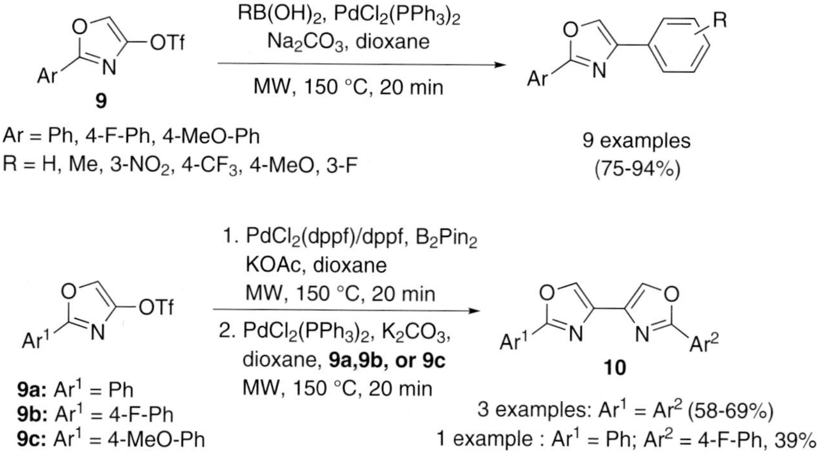

Scheme 5.24 Suzuki–Miyaura couplings of oxazoles.

In 2007, Högermeier and Reißig reported on Suzuki–Miyaura couplings of alkenyl nonaflates with aryl boronic acids [316].

A related approach based on Suzuki or Stille couplings to 2′-chlorobisthiazoles was presented by Stanetty *et al.* [317].

The same authors presented a correlated study of thiazole boronic esters involved in Suzuki–Miyaura cross-couplings. Under microwave conditions, the corresponding 2,4- and 2,5-diarylthiazols **11** can be generated easily [318]. Furthermore, microwave irradiation gave immediate access to the required thiazole-4-boronic ester. Although the preparation of the thiazole-5-boronic esters relies on lithiation strategies at −50 to −80 °C, the thiazole-4-boronic ester can be achieved by treatment of 4-bromo-2-phenylthiazole with bis(pinacolato)diboron under elevated temperatures [318]. In a typical procedure, 4-bromo-2-phenylthiazole and bis(pinacolato) diboron in dioxane with ($Pd_2(dba)_3$) and (PCy_3) as catalyst system together with potassium acetate as a base are heated at 170 °C to quantitatively yield the desired thiazole-4-boronic ester (Scheme 5.25). However, optimizations toward the cross-coupling reaction have been investigated utilizing the thiazole-5-boronic esters, finding that dioxane as a solvent and 3.4 equiv cesium carbonate as a base were the most effective combination. The final protocol was employed with a variety of aryl halides to explore the scope of the reaction (Scheme 5.25). The obtained yields were rather moderate since hydrolysis of the boronic ester, homocoupling of the boronic ester, and biphenyl formation diminish the amounts of the desired diaryl thiazoles. The initially prepared thiazole-4-boronic ester was used to present direct arylation at the 5-position via C−H activation as an alternative two-step route to 2,5-diarylated thiazoles applying the developed microwave protocol.

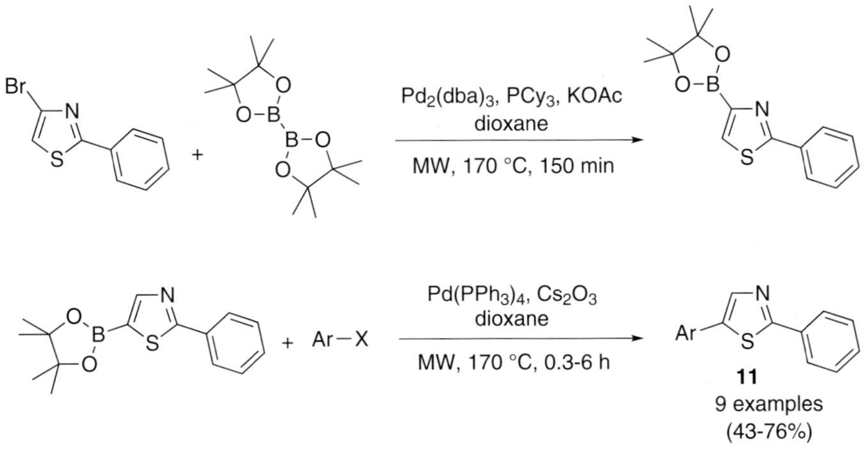

Scheme 5.25 Cross-coupling reactions of 1,3-thiazoles.

Recent literature examples involve the use of the Suzuki protocol for the high-speed decoration of various heterocyclic scaffolds of pharmacological or biological interest, including pyrimidines [319], pyridazines [320], pyrazines [321], chromans [322], and pyrazoles [323] (Scheme 5.26). For an application of Suzuki–Miyaura reactions in the scaffold decoration of quinazoline heterocycles, see Ref. [324]. Along

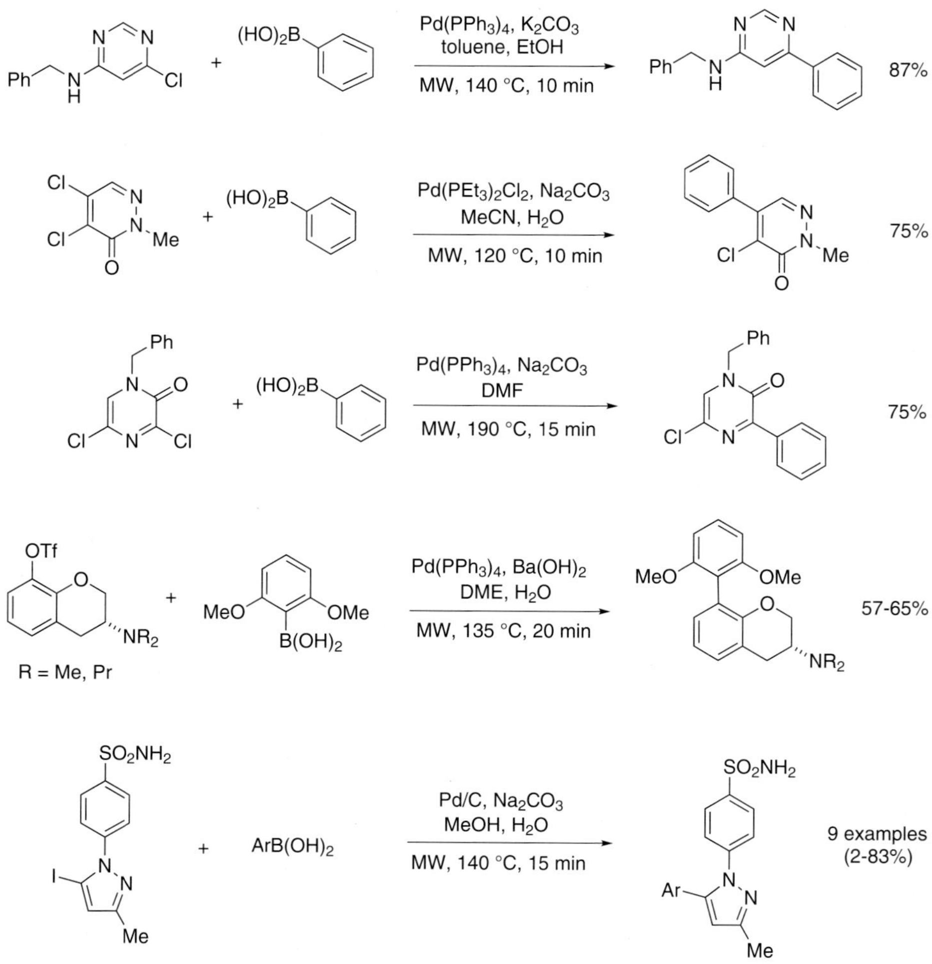

Scheme 5.26 Scaffold decoration of heterocycles using Suzuki cross-coupling chemistry.

similar lines, flavones have been decorated by Caddick and coworkers using Suzuki–Miyaura and Buchwald–Hartwig chemistry [325]. Additional published examples of microwave-assisted Suzuki–Miyaura couplings involve the scaffold decoration of 3-chloropyrazolines [326] and 3-bromoflavones [327] or the preparation of oligophenyl derivatives [328].

In this context, a highly convergent three-step microwave-assisted sequence for the preparation of diversely functionalized pyrazolopyrimidines was reported by Schultz and coworkers (Scheme 5.27) [329]. In the first step, 4-chloropyrazolopyrimidine was brominated on a multigram scale using *N*-bromosuccinimide in acetonitrile under microwave irradiation conditions (100 °C, 10 min). The resulting 4-chloro-5-bromopyrazolopyrimidine was then subjected to microwave-promoted nucleophilic displacement of the C4 chlorine atom by primary and secondary amines (12 examples)

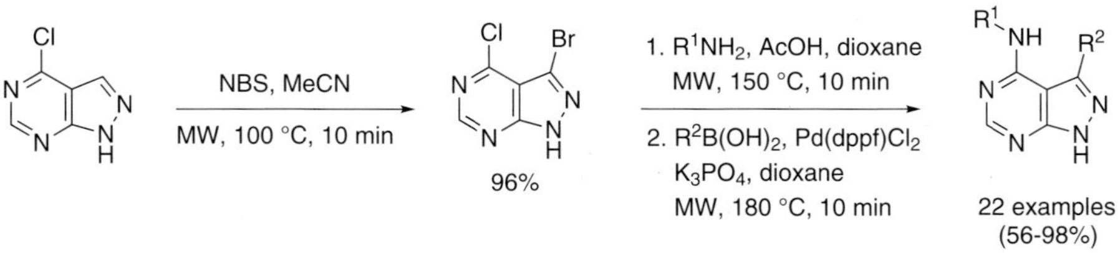

Scheme 5.27 Sequential one-pot nucleophilic aromatic substitutions and Suzuki cross-coupling reactions.

under mild acidic conditions in dioxane as solvent. After adding an appropriate base (potassium phosphate) and palladium catalyst (20 mol% [$PdCl_2$dppf], Suzuki coupling with a variety of boronic acids (12 examples) was performed in the same reaction vessel by microwave heating to 180 °C for 10 min. A diverse set of substituents was utilized in this two-step one-pot sequence providing the desired 4,5-disubstituted pyrazolopyrimidines in excellent overall yields. Boronic acids and amines containing hydroxyl, amino, ketone, amide, and chloro groups were well tolerated in this protocol [329].

An even simpler protocol for performing nucleophilic substitutions (aminations) and Suzuki reactions in one pot was reported by Organ *et al.* for the generation of a 42-member library of styrene-based nicotinic acetylcholine receptor (nAChR) antagonists (Scheme 5.28) [323]. After considerable experimentation, the authors found that the nucleophilic displacement and the Suzuki coupling process can be carried out very effectively simultaneously by charging the microwave process vessel with the palladium catalyst (0.5 mol% palladium-on-charcoal), the boronic acid [$R^1B(OH)_2$], the amine substrate (HNR^2R^3), and an inorganic base required to drive the Suzuki reaction to completion. The desired basic target compounds were isolated as hydrochloride salts in high yield and with excellent purity utilizing a "catch-and-release" strategy.

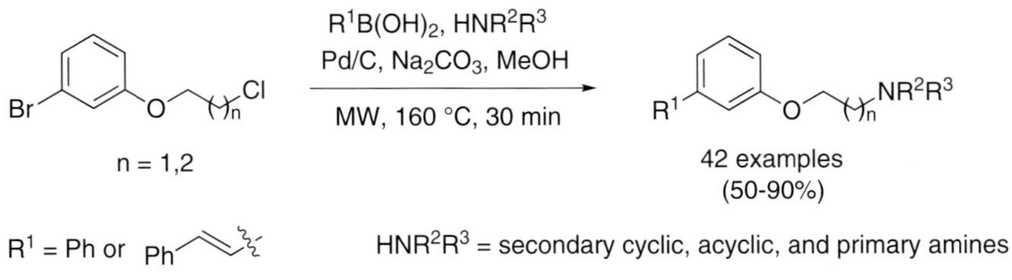

Scheme 5.28 One-pot nucleophilic substitutions and Suzuki cross-coupling reactions for the synthesis of nAChR antagonists.

Recently, Vishnumurthy and Makriyannis presented a novel microwave-mediated one-step synthesis of dibenzopyranones via Suzuki–Miyaura cross-coupling and subsequent spontaneous lactonization [330]. The method was developed and optimized using commercially available methyl 2-bromo-5-methoxybenzoate and methyl 2-bromo-4-fluorobenzoate, respectively, together with *o*-hydroxyphenylboronic acid (Scheme 5.29). A solution of 0.5 mmol corresponding benzoate with 1.3 equiv boronic acid and 4 equiv cesium carbonate in dimethoxyethane/water (20 : 3) was degassed with argon. Then, 10 mol% $Pd(PPh_3)_4$ were added and the mixture was irradiated at 125 °C for 15 min. Finally, the optimized method was employed to generate a small 32-membered library furnishing benzopyranones and heterogeneous analogs in 68–98% yield, demonstrating the suitability of the developed method for a parallel synthesis approach of these pharmaceutically attractive scaffolds [330].

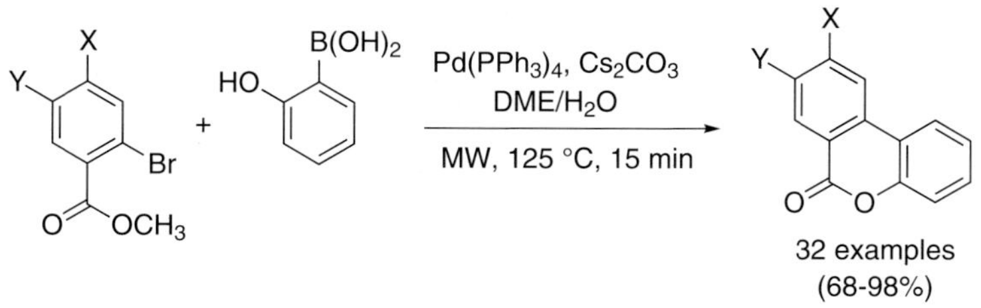

Scheme 5.29 One-step synthesis of dibenzopyranones via Suzuki–Miyaura cross-coupling.

Heo and his group have disclosed a cascade reaction of *o*-substituted aryl halides and *o*-formyl or *o*-acetyl arylboronic acids – involving first a Suzuki–Miyaura coupling and then a subsequent intramolecular aldol condensation – for the synthesis of phenanthrene derivatives (Scheme 5.30) [331]. Crucial for obtaining the phenan-

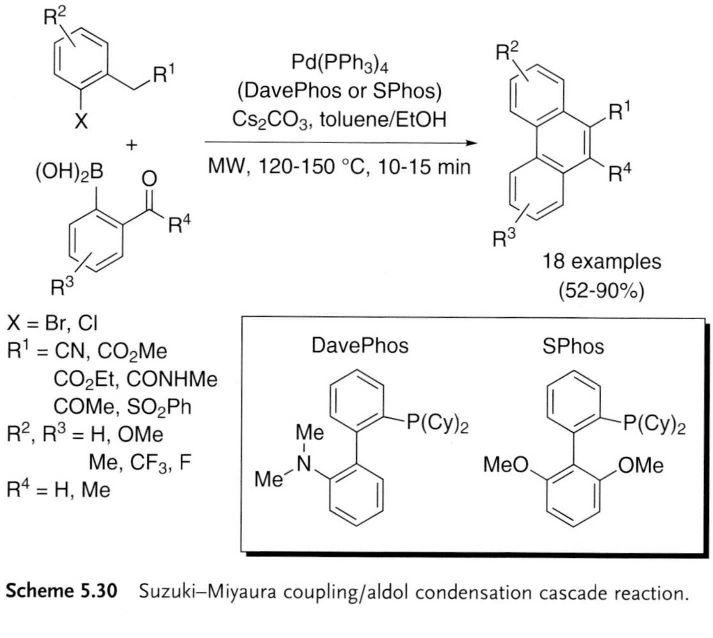

Scheme 5.30 Suzuki–Miyaura coupling/aldol condensation cascade reaction.

threnes in high yields is the ratio of the solvent mixture, with toluene/EtOH 2 : 1 leading to optimum results. When a ketone is employed as aldol acceptor ($R^4 = Me$), DavePhos as ligand has to be used, and a lower temperature of 120 °C for 5 min for the Suzuki coupling was necessary followed by an additional 10 min at 150 °C for the aldol condensation. The same is true for some aryl halide building blocks ($R^1 =$ CO_2Et, CN; X = Cl), where either DavePhos or SPhos has to be applied in order to prevent side reactions and to receive the phenanthrene products in good yields. The reversed approach was also possible where boronate esters are reacted with 2-bromoaryl carboxaldehydes.

The same group has published a similar Suzuki–Miyaura coupling/aldol condensation cascade reaction as the key step in the total synthesis of natural and unnatural aristolactams (Scheme 5.31) [332]. In this one-pot synthesis, 4-bromoisoindolin-1-one is reacted with various 2-formylphenylboronic acids using $Pd(PPh_3)_4$ as catalyst, Cs_2CO_3 as base, and toluene/EtOH as solvent system to construct the core phenanthrene ring system. Only when 3-thienylboronic acid is employed, a lower product yield of 35% was obtained.

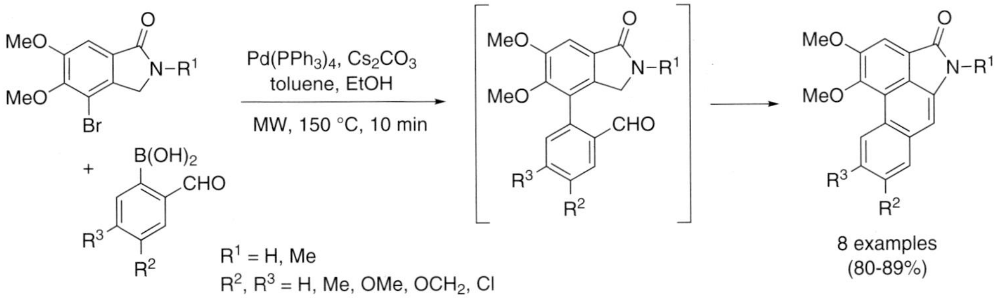

Scheme 5.31 Suzuki–Miyaura coupling/aldol condensation cascade reaction toward the total synthesis of aristolactams.

For a one-pot Hunsdiecker–Suzuki strategy leading to stilbene derivatives, see Ref. [333].

In the course of the synthesis of quinazolinone TGF-β RI inhibitors **16**, Li *et al.* have disclosed the synthesis of key intermediate **14** via a one-pot three-component Suzuki–Miyaura/etherification sequence (Scheme 5.32) [334]. In order to prevent side reactions in the Suzuki coupling of boronic acid **12** and quinazoline **13** (in particular, Suzuki reaction at the 2-chloro position of **13**), ethylene glycol was employed for the etherification step at the 2-position of **13**. In addition, product purification was improved by performing the reaction sequence in one pot. The final products **16** were obtained by amination of scaffold **15**, again using microwave heating.

The group of Larhed has reported the preparation of focused libraries of inhibitors of the malarial proteases plasmepsin I (Plm I) and II [335]. Using high-speed microwave Suzuki reactions, optimization of the initial lead compound

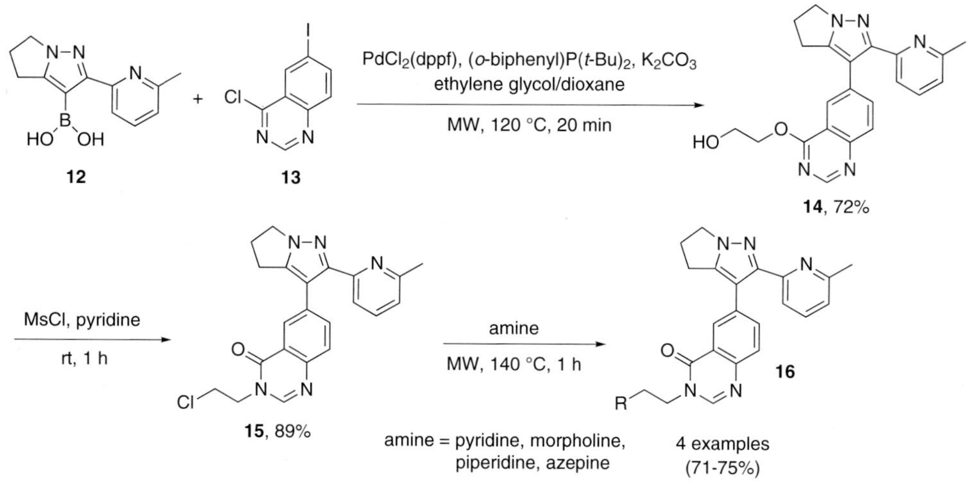

Scheme 5.32 One-pot Suzuki–Miyaura/etherification reaction.

used in the study led to inhibitors with highly improved activity. A combinatorial optimization protocol utilizing an experimental design technique afforded plasmepsin inhibitors with K_i values in the low nanomolar range and with high selectivity versus the human protease cathepsin D. For the decoration of the P1′ subunit, a Suzuki protocol involving 8 mol% of palladium(II)bis(triphenylphosphine) dichloride as catalyst was utilized. Although the yields remained low to moderate, enough material was generated for screening and biological evaluation. One of the most active compounds disclosed in this investigation had a K_i value (Plm I) of 4.1 nM (Scheme 5.33) [335].

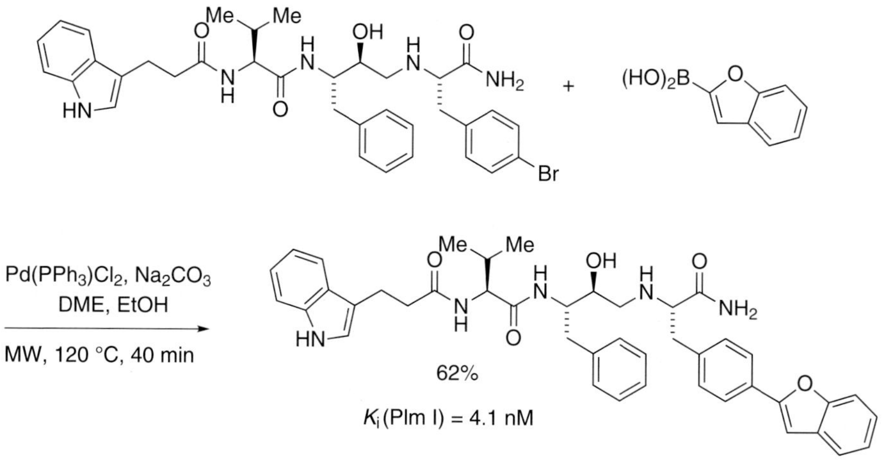

Scheme 5.33 Decoration of the P1′ subunit in plasmepsin I inhibitors.

In a somewhat related study by Hallberg and coworkers, C_2-symmetric compounds encompassing a 1,2-dihydroxyethylene scaffold with elongated P1/P1′ side chains were synthesized using a series of microwave-assisted double Suzuki, Heck, and Sonogashira coupling reactions (Scheme 5.34) [336]. The compounds synthesized exhibited picomolar to nanomolar inhibition constants for plasmepsin I and II with no measurable affinity to the human enzyme cathepsin D. Because of their reliability and robustness, microwave-assisted Suzuki reactions are often used in medicinal chemistry projects [337,338].

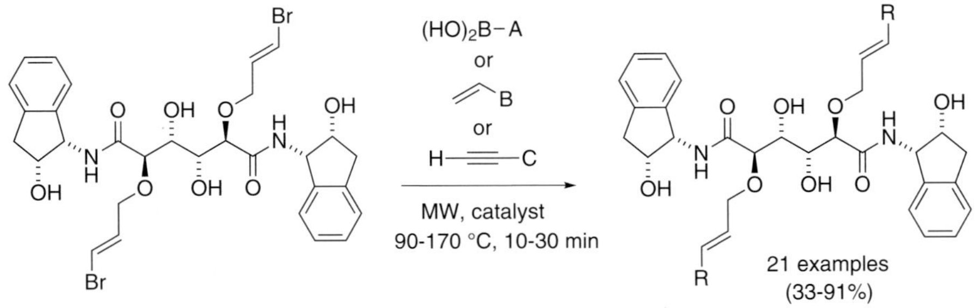

Scheme 5.34 Synthesis of plasmepsin I and II inhibitors with elongated P1/P1′ side chains.

The same group of authors has reported a combination of various palladium and copper-catalyzed Suzuki, cyanation, and Ullmann condensation reactions for the synthesis of thiophene-based selective angiotensin II AT_2 receptor antagonists (Scheme 5.35). [339]

Santagada and his group have reported on the synthesis of biphenyls **19** that incorporate an amino acid moiety with the carboxyl and amino function in *ortho*-position (Scheme 5.36) [340]. These biphenyl scaffolds were synthesized in order to develop new compounds that are able to induce β-hairpin folding in peptides. Biphenyls **19** were obtained in a single step by the Suzuki–Miyaura coupling of boronic acids **17** and aniline ($n = 0$), benzylamine ($n = 1$), or phenethyl amine ($n = 2$) bromides **18** in aqueous DMF as solvent and 30 min microwave heating at 150 °C.

Along similar lines, 2,2′-binaphthyls have been obtained by the Suzuki–Miyaura cross-coupling chemistry. Modest levels of enantioselectivity were obtained by employing chiral ligands [341].

Numerous other applications of microwave-assisted Suzuki couplings applied to medicinal chemistry-oriented target molecules have been reported in the literature [342–346].

The formation of a 15-membered *meta,meta*-cyclophane, and biphenomycin B, featuring a key microwave-induced intramolecular Suzuki–Miyaura reaction step, has been demonstrated by Lépine and Zhu (Scheme 5.37) [347]. Controlled microwave heating has been utilized to increase the cyclization efficiency in the key C–C coupling reaction of the tripeptide precursor. The study of solvent effect, taking into account the advantage of best ligand–base combination, found the toluene–water solvent system in the presence of TBAB as an additive to give best yields (50 %)

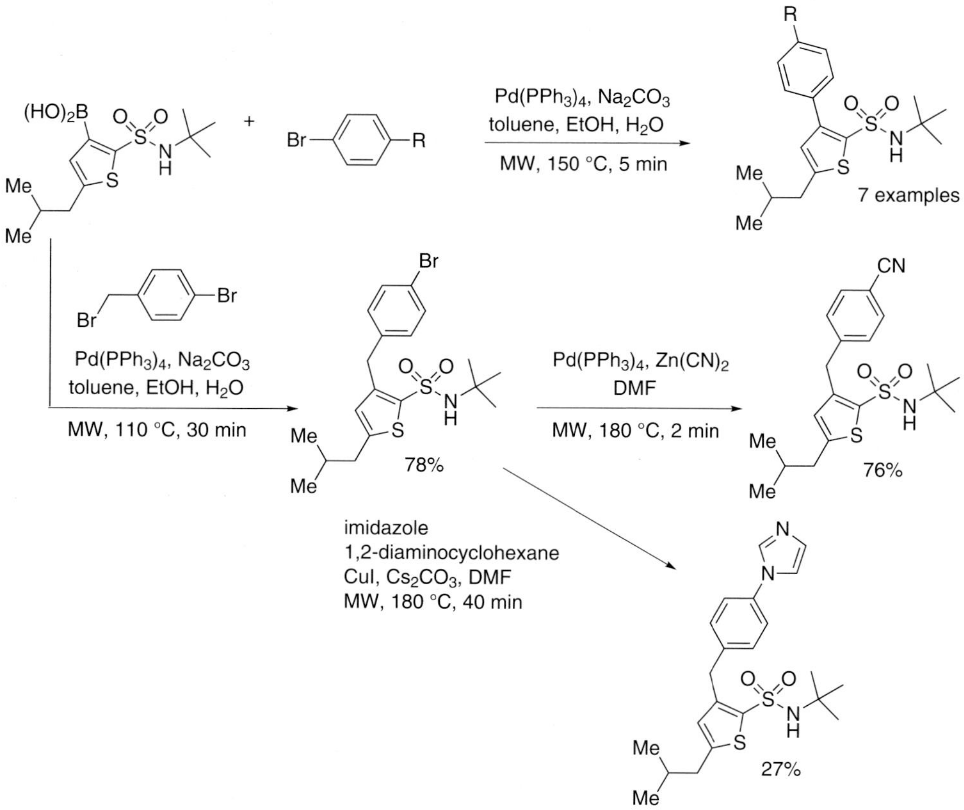

Scheme 5.35 Synthesis of selective nonpeptide AT_2 receptor antagonists.

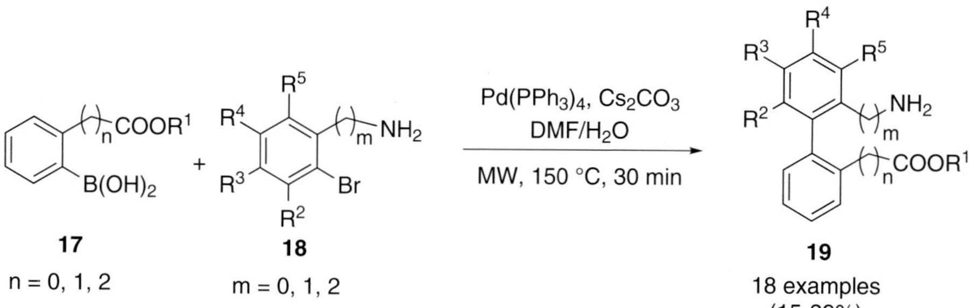

Scheme 5.36 Synthesis of potential β-sheet nucleators via Suzuki–Miyaura coupling.

compared to MeCN and DMSO under identical conditions. The choice of 2-(2′,6′-dimethoxybiphenyl)dicyclohexylphosphine as the ligand, in comparison to ligand-free conditions, has favored higher efficiency of cyclization in the key Suzuki–Miyaura reaction step.

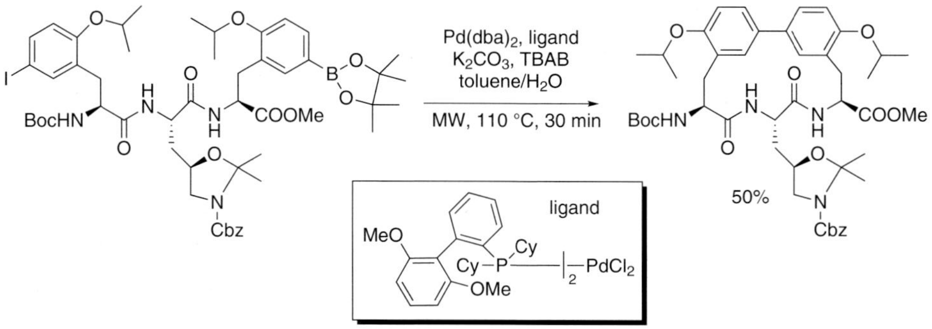

Scheme 5.37 Intramolecular Suzuki–Miyaura reaction leading to macrocycles.

A rather unusual Suzuki-type coupling involving a cyclic boronate was discovered by Zhou *et al.* in 2004 (Scheme 5.38) [348]. Treatment of 2-methoxybiaryls with boron tribromide unexpectedly led to the formation of a 10-hydroxy-10,9-boroxarophenanthrene derivative. This formation most likely proceeds via intramolecular electrophilic aromatization of a reactive dibromoaryloxyborane intermediate. The boroxarene structures are interesting synthetic intermediates. Treatment with an aryl iodide under Suzuki-type reaction conditions (palladium catalyst, base) led to the formation of the expected carbon–carbon coupling product in high yield (Scheme 5.38) [348].

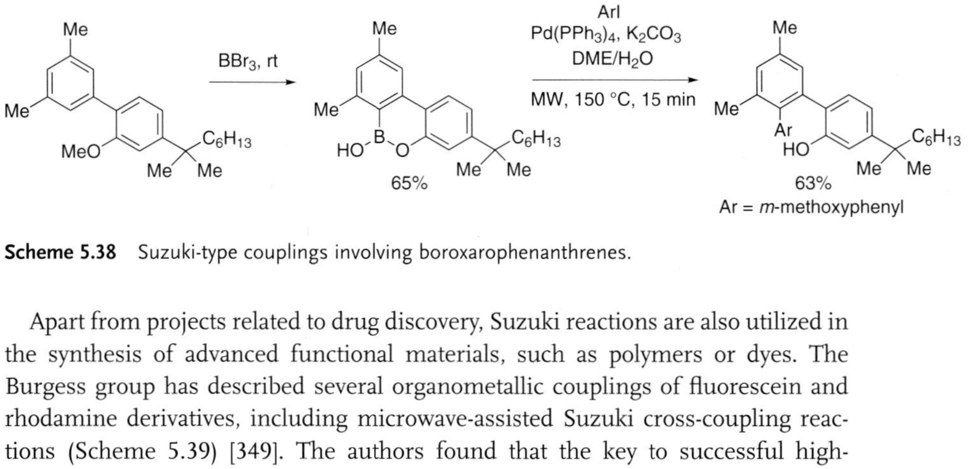

Scheme 5.38 Suzuki-type couplings involving boroxarophenanthrenes.

Apart from projects related to drug discovery, Suzuki reactions are also utilized in the synthesis of advanced functional materials, such as polymers or dyes. The Burgess group has described several organometallic couplings of fluorescein and rhodamine derivatives, including microwave-assisted Suzuki cross-coupling reactions (Scheme 5.39) [349]. The authors found that the key to successful high-yielding Suzuki couplings in this series of molecules was the use of a water-soluble palladium catalyst. A 2 mol% quantity of a sulfonic acid-derived triarylphosphine ligand was used in an acetone/water mixture (100 °C, 15 min) to couple a fluorescein-type boronic acid (prepared by microwave-assisted borylation) (see Section 5.2.3) with a rosamine-derived bromide (Scheme 5.39). The product was subsequently converted to a protic form, followed by counterion metathesis to the *tetra*(3,5-trifluoromethylbenzene)boronate (BARF) anion to allow convenient isolation by chromatography.

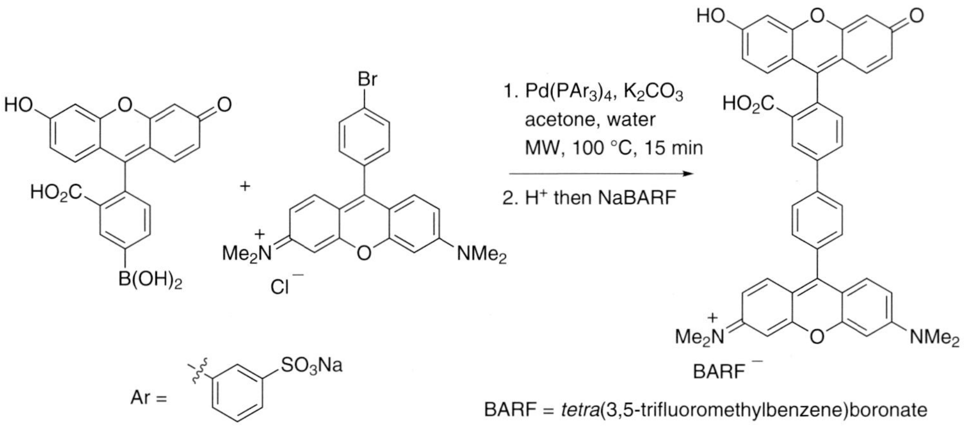

Scheme 5.39 Suzuki-type couplings for the preparation of fluorescent dyes.

The synthesis of fully conjugated semiconducting *para*-phenylene ladder polymers via microwave-assisted palladium-mediated "double" Suzuki- and Stille-type reactions has been demonstrated by Scherf and coworkers (Scheme 5.40) [350]. The procedure yielding polymeric material in about 10 min has no adverse effects on the quality of the polymers and displays a high degree of reproducibility. Compared to the classical thermal protocols, reaction times were reduced from 1–3 days to less than 1 h. Comparing the results of conventional heating with that of microwave-assisted heating experiments, the molecular weights were found in most cases to be of similar magnitudes and the ^1H and ^{13}C NMR data of the polymers were identical.

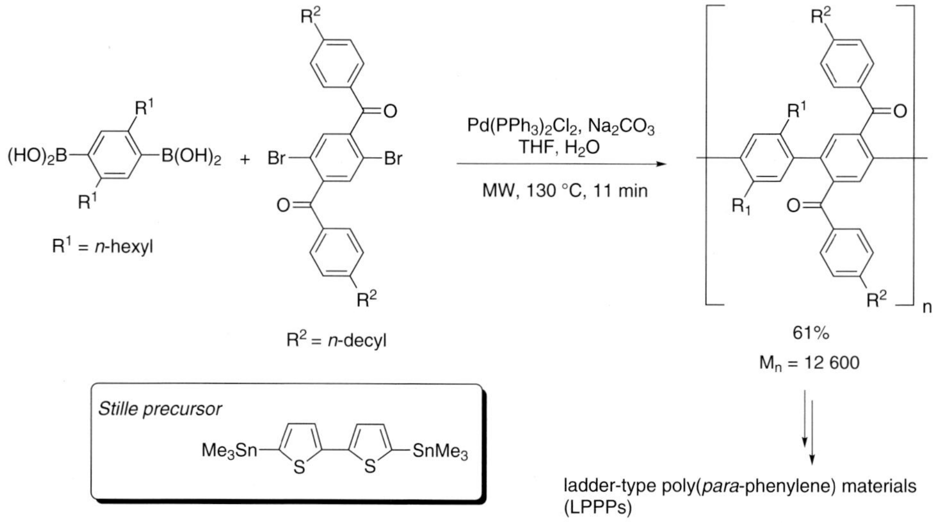

Scheme 5.40 Semiconducting polymers via Suzuki and Stille couplings.

In 2008, Alacid and Nájera introduced a microwave-assisted method for cross-coupling of alkenyl trifluoroborates as an alternative to boronic acids with aryl bromides in aqueous media to prepare styrenes, stilbenoids, and alkenyl benzenes [351]. For the reaction, simple palladium acetate or 4-hydroxyacetophenone oxime-derived palladacycle are employed as precatalyst. In the optimized protocol aryl bromide with 1.5 equiv of the corresponding trifluoroborate, equimolar amount of TBAB, and 1 mol% Pd catalyst were admixed with water and heated at 100 °C or 120 °C for 20–30 min (Scheme 5.41). The microwave protocol led to a significant improvement of reaction times since under thermal heating, refluxing for 2–24 h is required.

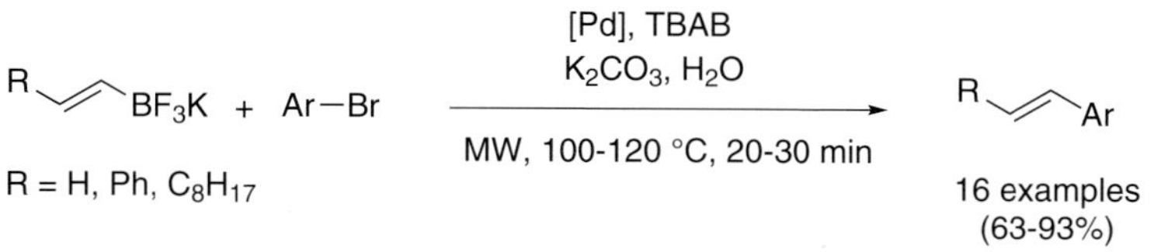

Scheme 5.41 Alkenylation of aryl bromides with trifluoroborates in aqueous media.

The same group reported on the coupling of potassium aryltrifluoroborates with organic chlorides in aqueous media catalyzed by an oxime-derived palladacycle [352].

Wyatt and coworkers studied the efficiency of the Suzuki–Miyaura coupling of sterically hindered substrates under microwave irradiation [353]. Here too, potassium vinyltrifluoroborate was employed as the vinylation agent and benzyl 3,5-bis(benzyloxy)-4-bromobenzoate served as model substrate to investigate the optimum reaction conditions (Scheme 5.42). Stepwise optimization of time, catalyst, compound, and solvent ratio finally led to a protocol that could be applied to a number of substrates. In a typical procedure, benzyl 3,5-bis(benzyloxy)-4-bromobenzoate, 5 equiv potassium vinyltrifluoroborate, 3 equiv cesium carbonate, 5 mol% $PdCl_2$dppf, and degassed tetrahydrofuran/deionized water (9 : 1) as a solvent, the mixture was subjected to microwave irradiation at 150 °C for 30 min. Corresponding workup furnished the desired benzyl 3,5-bis(benzyloxy)-4-vinylbenzoate in excellent yield (93%). Treating other substrates under these conditions showed that the presence of an electron-withdrawing group in *para*-position to the halide is beneficial.

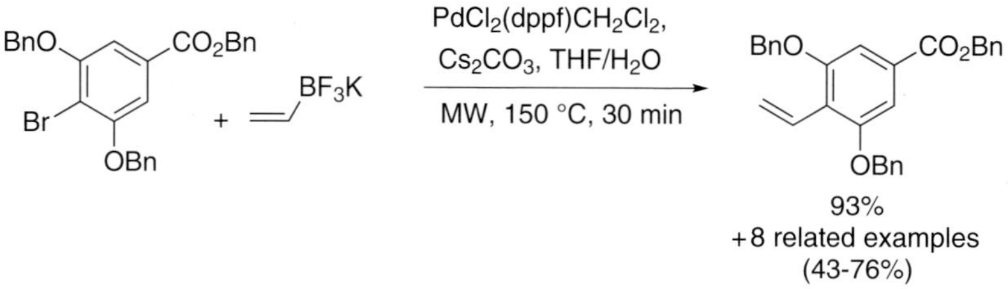

Scheme 5.42 Microwave-enhanced vinylation of benzyl 3,5-bis(benzyloxy)-4-bromobenzoate.

In a 2006 publication, Leadbeater and coworkers have demonstrated the synthesis of functionalized biaryls via Suzuki–Miyaura coupling (Scheme 5.43) [354]. By using potassium organotrifluoroborates and ultralow Pd loading (2.5 ppm), efficient coupling with arylhalides could be performed in 5 min at 150 °C under microwave irradiation.

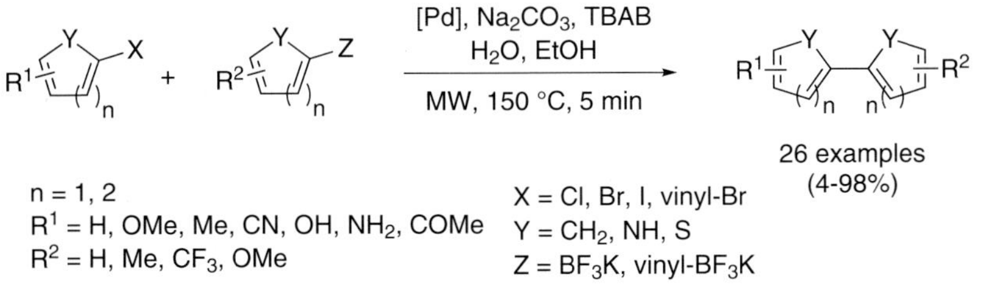

Scheme 5.43 Suzuki–Miyaura couplings with organotrifluoroborates.

Closely related work was published simultaneously by the Kabalka and Al-Masum [355].

A set of 19 symmetrical and unsymmetrical aryl ketones was prepared via Suzuki-type coupling of arylboronic acids with acid chlorides in moderate to high yields under comparatively mild reaction conditions (98 °C, 10 min) by Poláčková *et al.* (Scheme 5.44) [356]. A catalyst screen revealed that $Pd(PPh_3)_4$ and Cs_2CO_3 as catalyst/base system generally provided the aryl ketones in highest yield under the given microwave conditions.

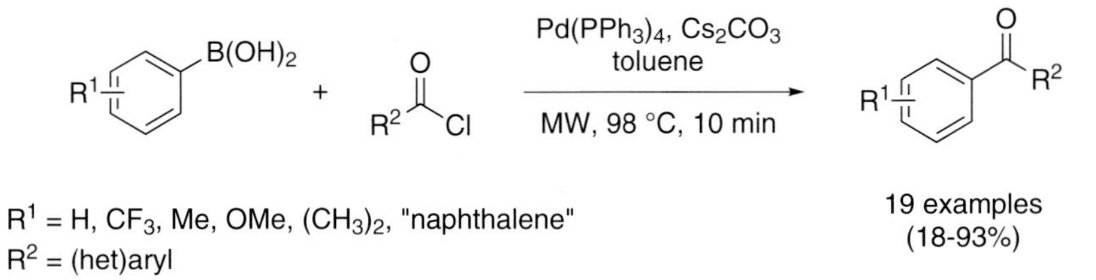

Scheme 5.44 Cross-coupling of arylboronic acids with acid chlorides.

Along the same lines, the group of Wolf has reported on the synthesis of benzophenone and acetophenone derivatives via the Suzuki-type cross-coupling of aromatic and aliphatic acyl chlorides with boronic acids that thus overcomes the drawbacks of Friedel–Crafts acylations such as harsh conditions, untunable regiocontrol, and low substrate scope (Scheme 5.45) [357]. When using Pd–phosphinous acid, POPd, as catalyst system in combination with microwave heating, the ketone products are obtained in good to high yields in 10 min reaction time (versus 1 h at the same temperature under thermal heating). Importantly, aryl halides in both the acyl chloride and boronic acid substrate were tolerated.

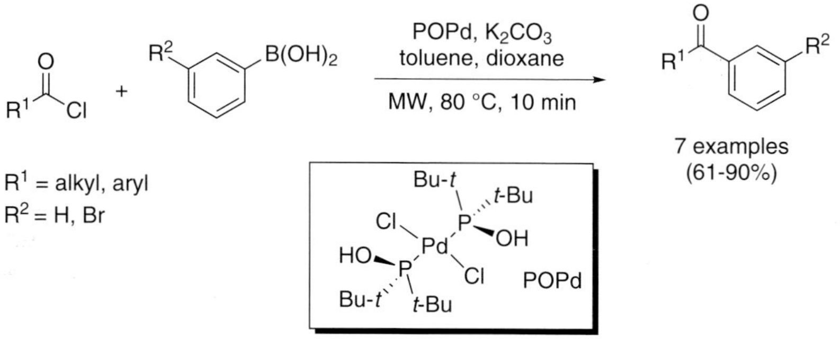

Scheme 5.45 Pd-phosphinous acid-catalyzed cross-coupling of acyl chlorides.

Apart from examples involving Suzuki reactions with standard soluble palladium catalysts, there is a growing number of publications reporting the use of immobilized, recyclable palladium catalysts for carrying out Suzuki and other cross-coupling transformations under microwave conditions [358,359]. Two examples are highlighted in Scheme 5.46. Kirschning and coworkers [358] and a team from Abbott Laboratories [359] have reported very efficient microwave-promoted Suzuki coupling reactions involving immobilized palladium catalysts. Both methods can be used to couple a range of aryl halides and triflates with boronic acids in solvents such as water or ethanol. While the Kirschning group has used a novel type of solid palladium(II) precatalyst that can easily be prepared from 4-pyridine-aldoxime and sodium tetrachloropalladate(II) (catalyst 1), the Abbott researchers employed polyethylene-supported, so-called FibreCat palladium catalysts (catalysts 2–4) under similar conditions. The insoluble catalyst 1 can be recycled at least 14 times without

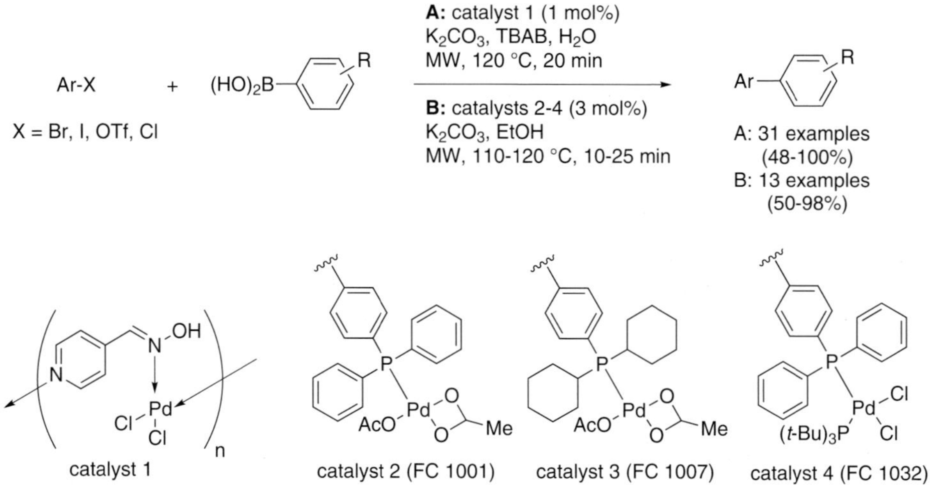

Scheme 5.46 Suzuki couplings involving immobilized palladium catalysts.

any appreciable loss of activity when contained in an IRORI Kan [358]. Using water as solvent and tetrabutylammonium bromide as an additive, a small library of biaryls was prepared. Coupling was observed with aryl bromides, iodides, chlorides, and triflates. Similar coupling behavior was also observed with the FibreCat palladium catalysts [359]. Compared to the homogeneous control reactions, the supported palladium reactions were much cleaner, usually yielding a colorless solution upon completion of the reaction. Here, rapid purification of the reaction mixture was achieved by solid-phase extraction (SPE), passing the crude reaction mixture through a SPE cartridge filled with Si-carbonate that sequestered any excess boronic acid.

Suzuki–Miyaura couplings using a microencapsulated Pd catalyst (Pd EnCat) under microwave heating were investigated by the group of Ley (Scheme 5.47) [360]. For those compounds giving low purities under conventional microwave conditions (**A**), higher purities could be achieved by performing the reactions at 50 W (reaction temperature did not exceed 76 °C) with simultaneous cooling (compressed air, **B**) by preventing otherwise occurring thermal decompositions. Even higher purities were obtained by conducting the reactions via a flow-through approach again in combination with cooling (**C**) due to shorter reaction times (about 1 min) and therefore diminished side reactions.

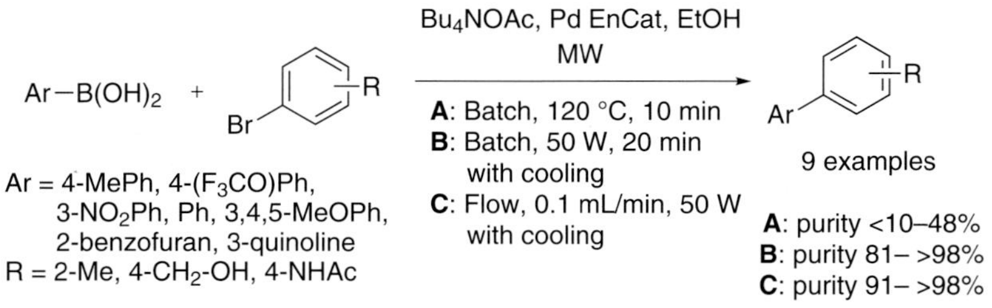

Scheme 5.47 Suzuki–Miyaura couplings catalyzed by microencapsulated Pd.

Sharma *et al.* have described related Suzuki–Miyaura couplings using Pd EnCat catalysts [361]. Other immobilized forms of Pd catalysts have also been used in Suzuki–Miyaura couplings [362,363]. For the use of Pd/C in microwave-assisted Suzuki–Miyaura couplings, see Ref. [364]. Further examples of microwave-assisted Suzuki cross-couplings involving supported substrates/catalysts or fluorous phase reaction conditions are described in Chapter 8.

In a 2011 report, Kappe and coworkers discussed nickel-catalyzed Suzuki–Miyaura cross-couplings under microwave irradiation [365]. The authors employed aryl carbamates and sulfamates as phenol precursors in this widely used transformation. Initial optimization studies using naphthyl carbamate as model compound led to an efficient protocol to be applied with a variety of substrates. The corresponding carbamate or sulfamate with 2.5 equiv boronic acid in toluene, 5 mol% bis(tricyclohexylphosphine)-nickel(II) chloride ($Ni(PCy_3)_2Cl_2$) as catalyst, and 7 equiv potassium phosphate as a base were heated at 180 °C for 10 min to yield the desired biaryls

(Scheme 5.48a). This optimized protocol was also successfully applied toward a protected 2-aminophenol to generate the corresponding 2-aminobipheyl derivatives. Finally, the authors highlight the scalability of their developed protocol by efficient multigram preparation of the Sartan precursor **20** employing a large-scale microwave reactor (Scheme 5.48b). About 1000-fold increase in scale was possible furnishing the desired key intermediate **20** in the same good yield and product quality as in the development scale [365].

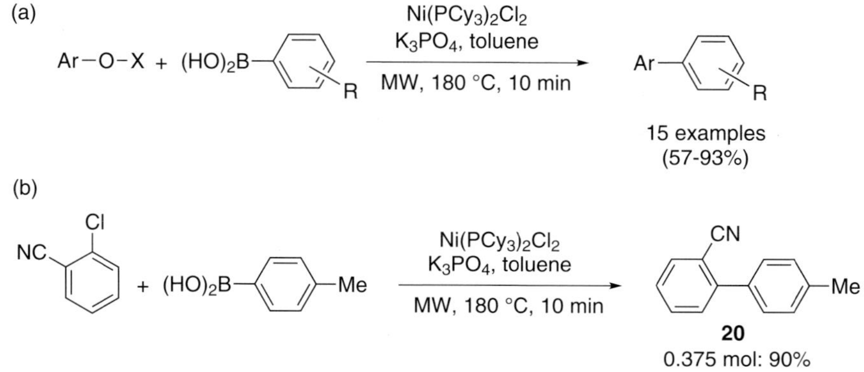

Scheme 5.48 Applying aryl carbamates and sulfamates in nickel-catalyzed Suzuki–Miyaura cross-couplings.

Suzuki-type couplings in water without the use of catalyst and base have been reported in 2006 by Yan *et al.* (Scheme 5.49) [366]. Both symmetrical and unsymmetrical biaryls could be synthesized by coupling of sodium tetraphenylborate with hypervalent iodonium salts or iodanes under mild reaction conditions.

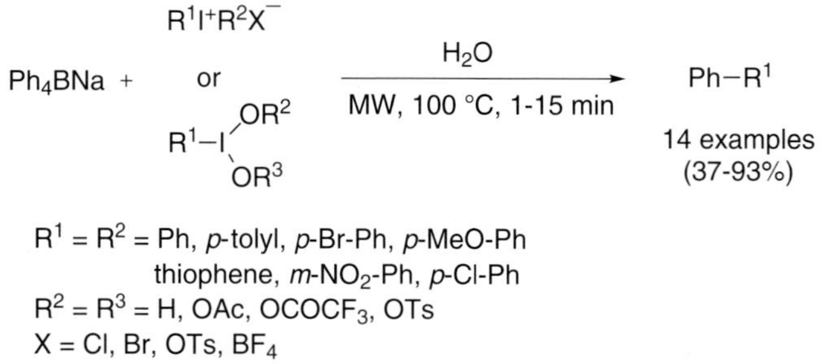

Scheme 5.49 Catalyst- and base-free Suzuki-type couplings.

Somewhat related to the Suzuki cross-coupling reaction involving organoboron reagents is the Hiyama coupling of organosilanes with halides and triflates to form unsymmetrical biaryl compounds. Seganish and DeShong have described rapid, microwave-promoted Hiyama couplings of bis(catechol) silicates with aryl bromides (Scheme 5.50) [367]. Suitable reaction conditions entailed the use of 5 mol% of tris(dibenzylideneacetone)dipalladium(0) as a palladium source, 5 mol% of 2-(dicyclohexylphosphanyl)biphenyl as ligand, and 1.5 equiv of tetrabutylammonium fluoride (TBAF) in tetrahydrofuran as solvent. Exposure of the reaction mixture to microwave irradiation at 120 °C for 10 min typically provided good yields of biaryl coupling products for a wide range of substrates. The only functional group that was found to fail in the coupling studies was the amino group, probably as a result of catalyst poisoning.

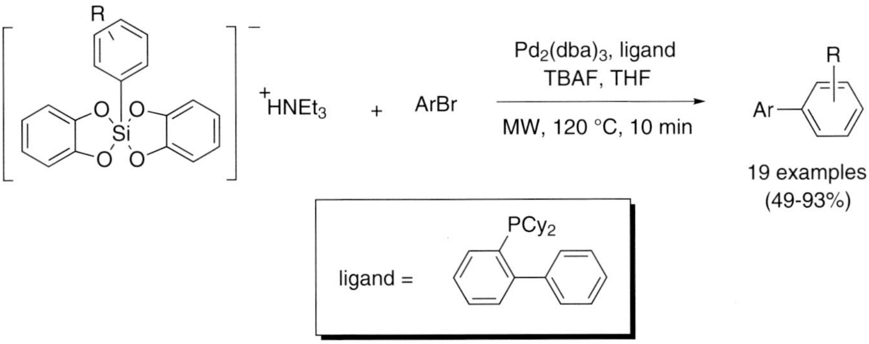

Scheme 5.50 Hiyama couplings of bis(catechol) silicates with aryl bromides.

In addition, Hiyama couplings involving aryl- and vinylsiloxane derivatives have been described by Clarke [368]. Fluoride-free Hiyama couplings of aryl bromides/chlorides with arylsiloxanes and vinylsiloxanes have been reported by Alacid and Nájera [369–371].

5.2.3
Sonogashira Reactions

The Sonogashira reaction (palladium and copper co-catalyzed coupling of terminal acetylenes with aryl and vinyl halides) enjoys considerable popularity as a reliable and general method for the preparation of unsymmetrical alkynes [372]. General protocols for microwave-assisted Sonogashira reactions under controlled conditions were first reported in 2001 by Erdélyi and Gogoll [373]. Typical reaction conditions for the coupling of aryl iodides, bromides, chlorides, and triflates involve *N,N*-dimethylformamide as solvent, diethyl amine as base, and palladium(II)bis(triphenylphosphine) dichloride (2–5 mol%) as a catalyst with copper(I) iodide (5 mol%) as additive [373]. In Ref. [374], such a protocol was utilized in a rapid "domino" Sonogashira sequence in order to synthesize amino esters, as outlined in Scheme 5.51.

Essentially the same experimental protocol was employed by Vollhardt and coworkers in order to synthesize *ortho*-dipropynylated arenes, which served as

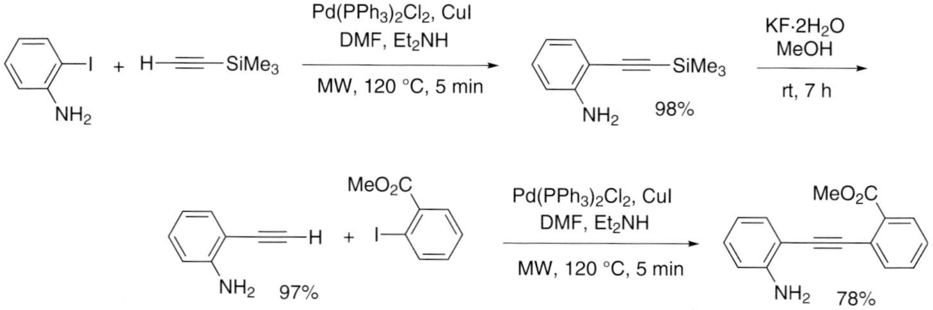

Scheme 5.51 "Domino" Sonogashira sequence for the synthesis of bis(aryl)-acetylenes.

precursor to tribenzocyclynes via an alkyne metathesis reaction (Scheme 5.52) [375]. Here, the Sonogashira reaction was carried out in a pre-pressurized (about 2.5 bar of propyne) sealed microwave vessel (see Section 4.4). Double Sonogashira coupling of the dibromodiiodobenzene was completed within 3.75 min at 110 °C. It is worth mentioning that the authors did not carry out the subsequent tungsten-mediated alkyne metathesis chemistry under microwave conditions to shorten the exceedingly long reaction times and perhaps to improve the low yield (see Scheme 5.160 for a microwave-assisted alkyne metathesis reaction). Related double Sonogashira chemistry has been described by Nielsen and coworkers [376]. Additional examples of Sonogashira [377] and bis-Sonogashira couplings [378] for the generation of polyacetylenes have been reported.

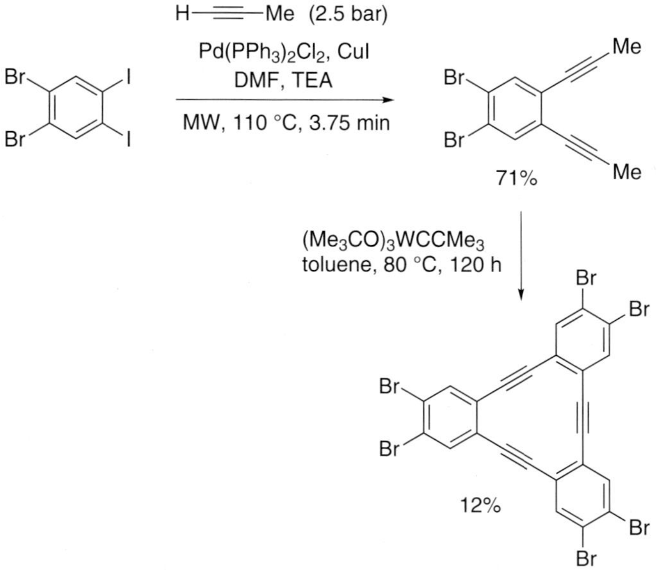

Scheme 5.52 Double Sonogashira reactions in prepressurized vessels.

Microwave-assisted Sonogashira protocols have also been used for the decoration or functionalization of various heterocyclic core structures. Some recent examples involving pyrazines [319], pyrimidines [379], and thiophenes [308] are shown in Scheme 5.53.

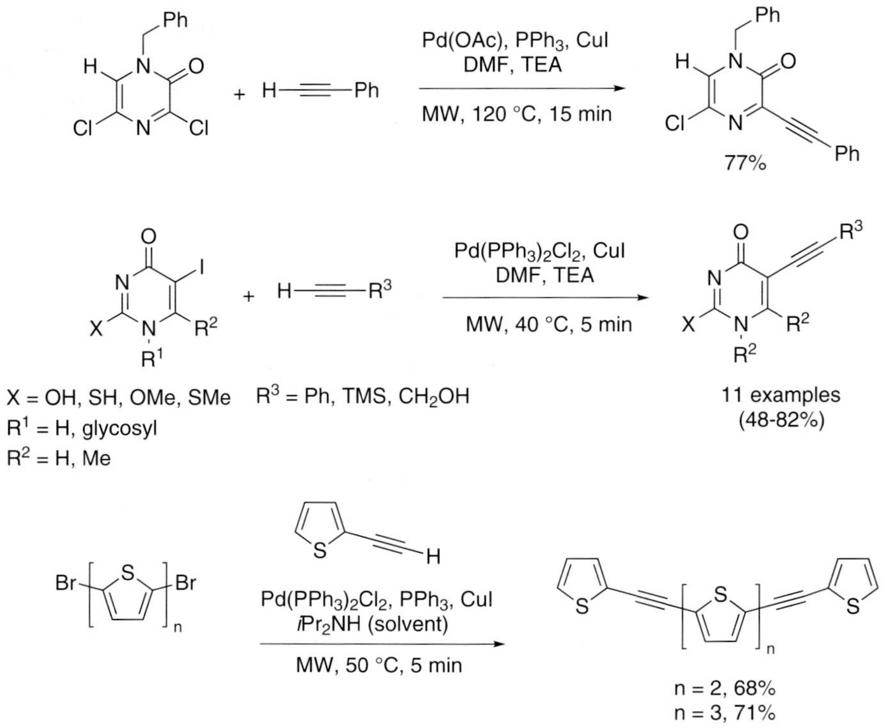

Scheme 5.53 Scaffold decoration of heterocycles using Sonogashira cross-coupling chemistry.

A microwave-driven Sonogashira coupling step is involved in the total synthesis of azaphilones, a structurally diverse family of natural products containing a highly oxygenated bicyclic core and a quaternary center. Porco and his colleagues have described the alkynylation of densely functionalized bromobenzaldehydes with methyl propargyl ether and propargyl cyclohexyl amide under standard Sonogashira coupling conditions (10 mol% palladium(II)bis(triphenylphosphine) dichloride, 20 mol% copper(I) iodide) (Scheme 5.54) [380]. The resulting *ortho*-alkynylbenzaldehydes were obtained in 68% and 65% isolated yield, respectively, and were subsequently converted to the desired bicyclic substrates (unnatural azaphilones) via a gold(III)-catalyzed cycloisomerization–oxidation sequence.

Hopkins and Collar have reported a one-pot Sonogashira/heteroannulation strategy for the synthesis of 6-substituted-5*H*-pyrrolo[2,3-*b*]pyrazines (Scheme 5.55) [381]. The reaction could either be performed by a two-step protocol, by first performing a

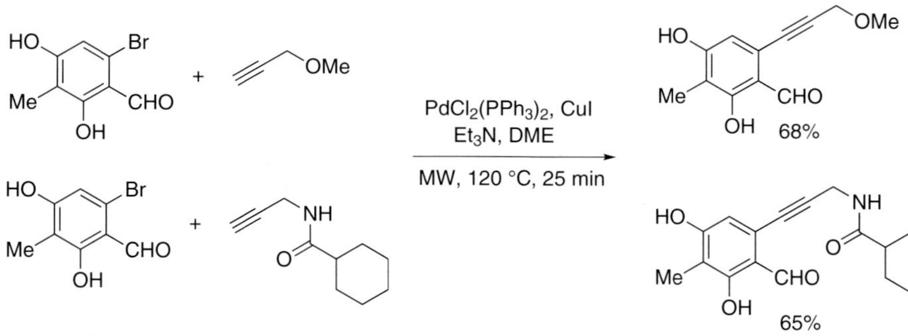

Scheme 5.54 Alkynylation of bromobenzaldehydes.

classical Sonogashira coupling on 2-amino-3-chloropyrazine, followed by base-induced cyclization (Scheme 5.55a), or in a one-step method by reacting the corresponding sulfonamide, *N*-(3-chloropyrazin-2-yl)-methanesulfonamide, directly with terminal acetylenes in the presence of a suitable palladium catalyst (3 mol%) (Scheme 5.55b). The microwave conditions (150 °C, 20 min) tolerated much functional diversity (both electron-withdrawing and electron-donating substituents). Halogens as well as cyano groups were also tolerated, along with silyl protecting groups.

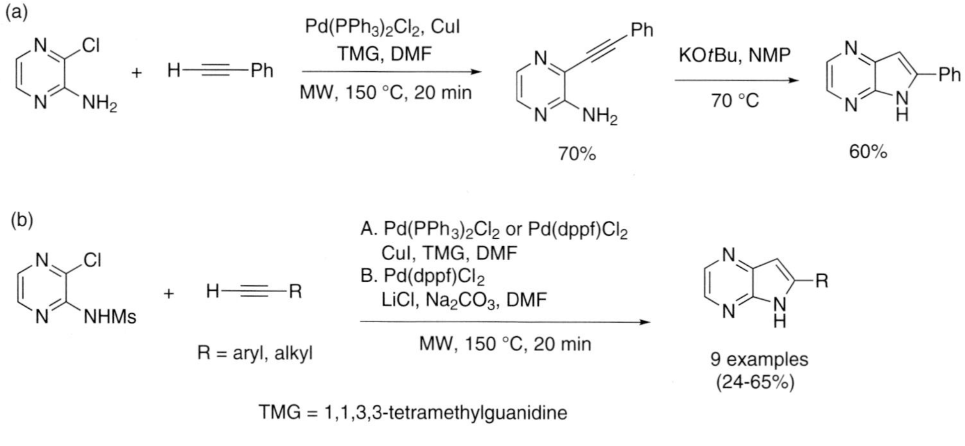

Scheme 5.55 Palladium-catalyzed heteroannulation.

A similar strategy for the synthesis of 2-substituted indoles by a one-pot two-step sequence starting from *o*-iodoanilines and terminal alkynes was described by Sanz *et al.* (Scheme 5.56) [382]. The first step includes a Sonogashira coupling leading to *o*-alkynylanilines that were subsequently cyclized under NaOH mediation to the 2-substituted indoles. Both steps were performed under microwave heating

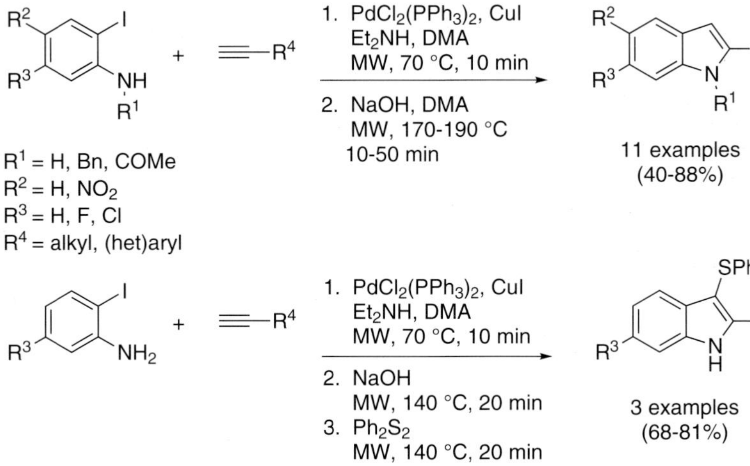

Scheme 5.56 One-pot Sonogashira coupling–NaOH-mediated cyclizations.

that had the advantage of employing less NaOH (4 versus 10 equiv) and reduced reaction times compared to conventional heating (1–2 h at room temperature for step 1 and up to 6 h at 140 °C for step 2). In addition, an arylthio group can be selectively introduced at the C-3 position in a one-pot three-step procedure.

A one-pot two-step microwave synthesis of indoles via Sonogashira coupling was also developed by the group of Larock [383]. Standard Sonogashira conditions (3 mol% $Pd(PPh_3)_2Cl_2$, 2 mol% copper(I) iodide) applied toward substituted iodoanilines and terminal alkynes in triethylamine furnish the corresponding internal alkynes in high yield. Addition of an aryl iodide in a polar solvent like acetonitrile and treatment under slightly elevated temperatures led to subsequent indole formation by palladium-catalyzed cyclization of the Sonogashira coupling product (Scheme 5.57). For the cyclization step, 1.1 equiv aryl iodide and acetonitrile are subsequently added and the resulting mixture is heated additionally 20–50 min at 90 °C. Although electron-rich aryl and aliphatic acetylenes proceed smoothly, somewhat longer reaction times are required for both steps when electron-deficient or steric hindered aryl acetylenes are employed. The applicable 2-iodoanilines and aryl iodides showed no significant electronic effect. The optimized method showed potential for gram-scale synthesis of indole libraries, even in parallel manner.

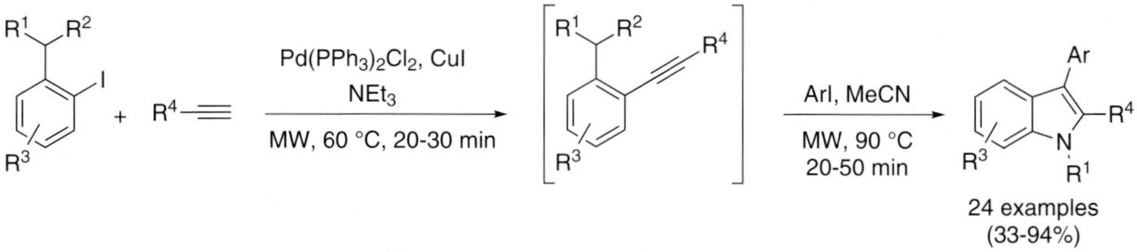

Scheme 5.57 Microwave-assisted one-pot synthesis of indoles under Sonogashira conditions.

Adenosine-based nucleosides **22**, which exhibit fluorescence in the visible range and that are bridged via a 1,2,3-triazole moiety between the purine base and the ribose, were prepared by Grøtli and coworkers (Scheme 5.58) [384]. To incorporate the 8-alkynyl functionality for the cyclization precursor, a Sonogashira reaction between 8-bromoadenosine derivative **21** and various terminal alkynes was performed under microwave heating. After triflation of the 2′-OH group, the 2′-azide functionality was introduced by nucleophilic displacement. The displacement and subsequent intramolecular cyclization via Huisgen [3 + 2] cycloaddition of the alkyne and the 2′-azide was again accomplished under microwave heating. Since these conditions proved to be incompatible with the silyl protecting groups, complete deprotection and subsequent protection with acetic anhydride had to be performed to obtain cyclonucleosides **22** in moderate yields. For both synthesis steps, better results regarding conversion were achieved with microwave heating.

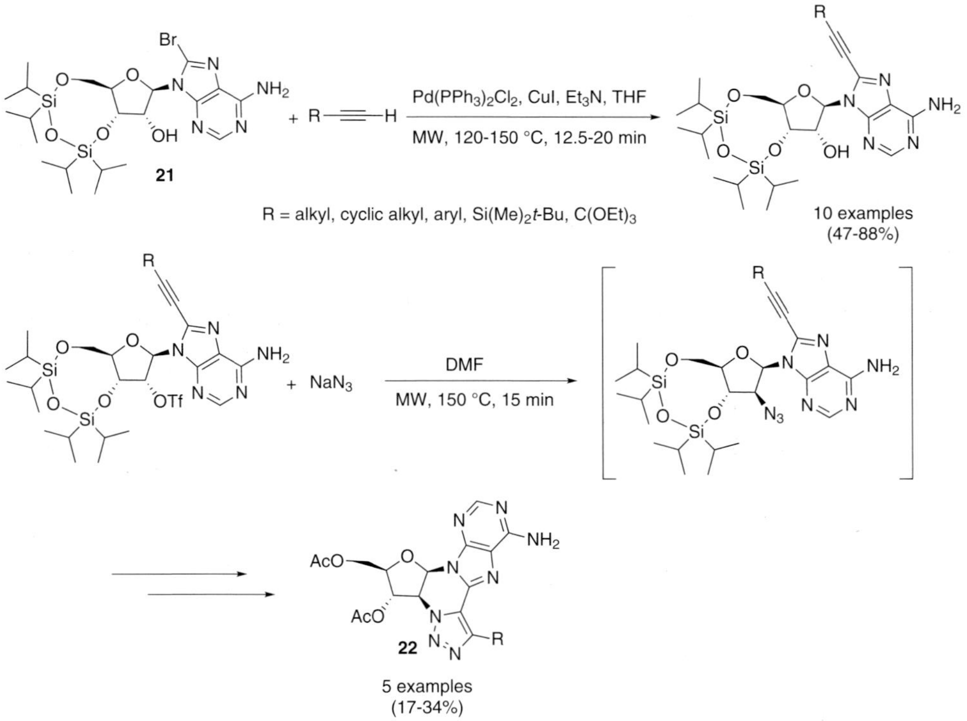

Scheme 5.58 Synthesis of cyclonucleosides.

A coupling–isomerization reaction (CIR) of (hetero)aryl halides with propargyl alcohols **23** under Sonogashira conditions for the synthesis of chalcones **24** was disclosed by Schramm and Müller (Scheme 5.59) [385]. The first step in this one-pot synthesis is a Pd/Cu-catalyzed alkyne coupling followed by base-catalyzed propargyl alcohol to enone isomerization. The reaction times could be reduced from 16–24 h

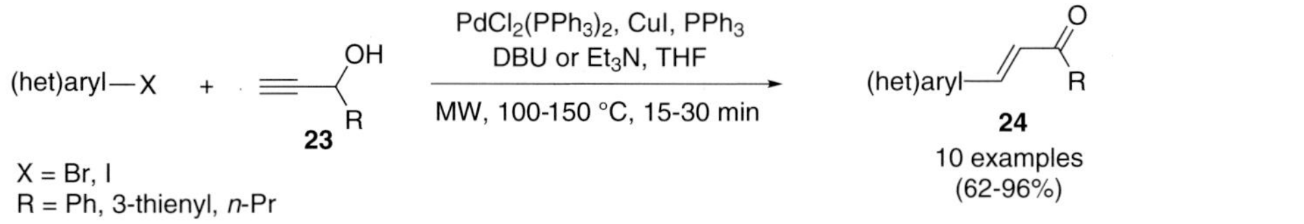

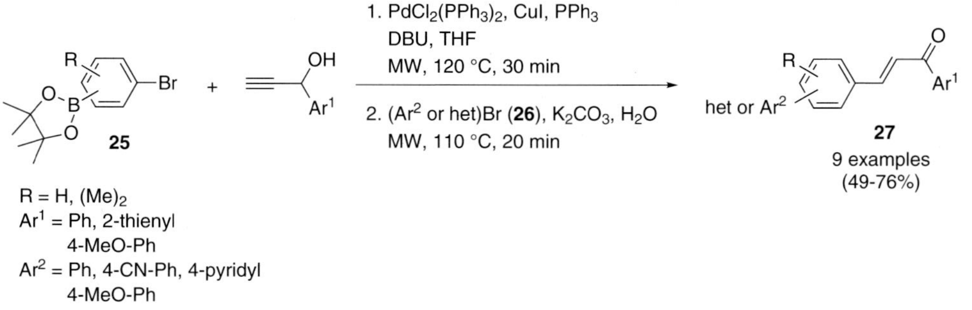

Scheme 5.59 Sequential coupling–isomerization-coupling reactions.

under conventional heating to only 15–30 min by applying microwave irradiation. In a subsequent publication, Liao and Müller extended this protocol to a coupling–isomerization–coupling (CIC) sequence for the synthesis of chalcones **27** [386]. The boronate moiety attached to bromoaryls **25** initially serves as electron-withdrawing substituent for the CIR and after addition of K_2CO_3, (het)aryl bromides **26**, and water, this organometallic group is activated by the alkali base for the Suzuki-coupling sequence.

The same group has also reported on the one-pot reaction of electron-deficient (hetero)aryl bromides **28**, (hetero)aryl propargyl alcohols **29**, and enamino carbonyl compounds **30** or **31**, respectively (Scheme 5.60) [387]. The reaction proceeds via a

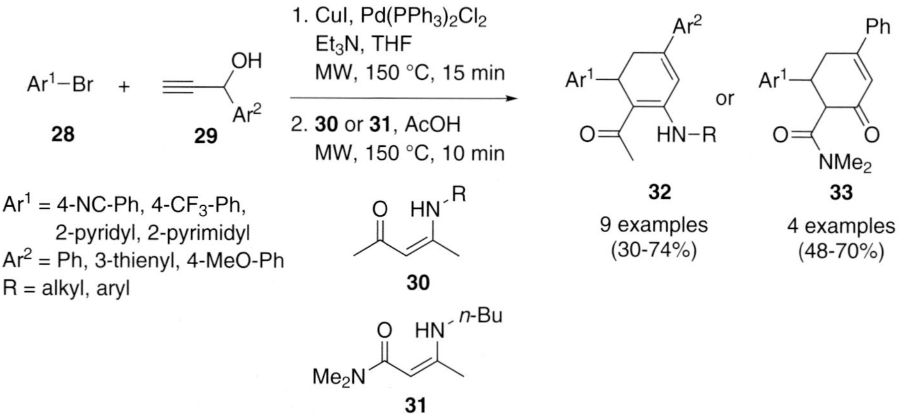

Scheme 5.60 Synthesis of 1-acetyl-2-amino-cyclohexa-1,3-dienes.

coupling–isomerization–enamine addition–aldol condensation sequence and furnishes dienes **32** when enaminones **30** are used or the hydrolyzed products **33** if β-amino crotonamide (**31**) is applied. Compared to thermal heating where dihydropyridines are obtained, an aldol-type condensation is favored in the second step under microwave conditions providing cyclohexa-1,3-dienes **32** and cyclohexanones **33**, respectively.

In a more recent publication, the Müller group described the synthesis of 3,4,5-substituted isoxazoles via a one-pot three-component reaction pathway based on an initial Sonogashira coupling (Scheme 5.61) [388]. In the first sequence, alkynones **34** are obtained by Sonogashira coupling of acid chlorides with terminal alkynes at room temperature. After addition of hydroximinoyl chlorides **35** and Et_3N to the same reaction vessel, alkynones **34** reacted further to the corresponding isoxazoles via a 1,3-dipolar cycloaddition with *in situ* generated aromatic nitrile oxides from **35**. By employing microwave heating for the cycloaddition step, reaction times could be reduced from 3 days to 30 min, while increased yields of the final products and less by-product formation (dimerization of nitrile oxides to furoxan oxides) were observed.

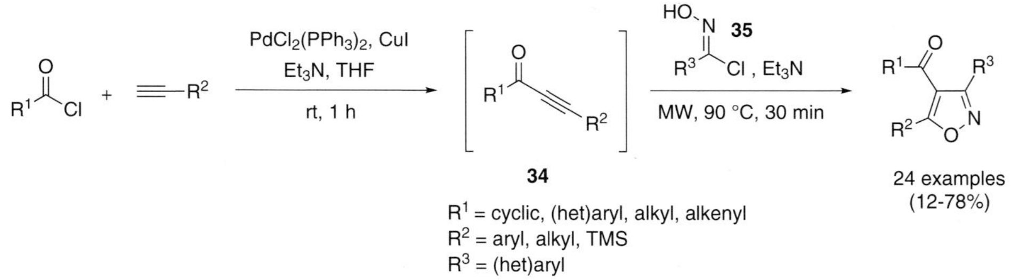

Scheme 5.61 One-pot three-component isoxazole synthesis.

A similar strategy was used by the same authors for the synthesis of highly fluorescent 3,5- and 1,3,5-substituted pyrazoles (Scheme 5.62) [389]. In the first sequence, alkynones are obtained by a Sonogashira coupling of acyl chlorides with terminal alkynes at room temperature. Subsequent addition of hydrazines, MeOH, and acetic acid under microwave heating to 150 °C for 10 min furnished the pyrazoles regioselectively – depending on the hydrazine substituents R^3 – via a Michael addition/cyclocondensation step. Further scaffold decoration for one example was accomplished in a one-pot four-component fashion by performing an additional Suzuki-coupling step employing the catalyst system from step 1, again under microwave heating giving a biphenyl substituted pyrazole.

Cryofluorescent 2,4-disubstituted 3*H*-1,5-benzodiazepines have been prepared by the same group using a similar strategy [390].

An application of the Sonogashira reaction in supramolecular chemistry and chemosensor development is highlighted in Scheme 5.63. Swager and coworkers have employed a microwave-assisted double Sonogashira–Hagihara coupling for

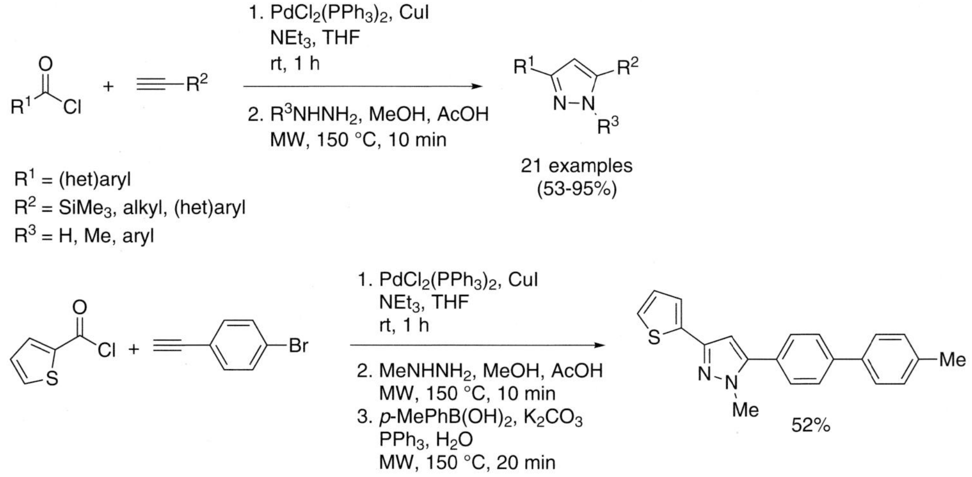

Scheme 5.62 One-pot three-component synthesis of 1,3,5-trisubstituted pyrazoles.

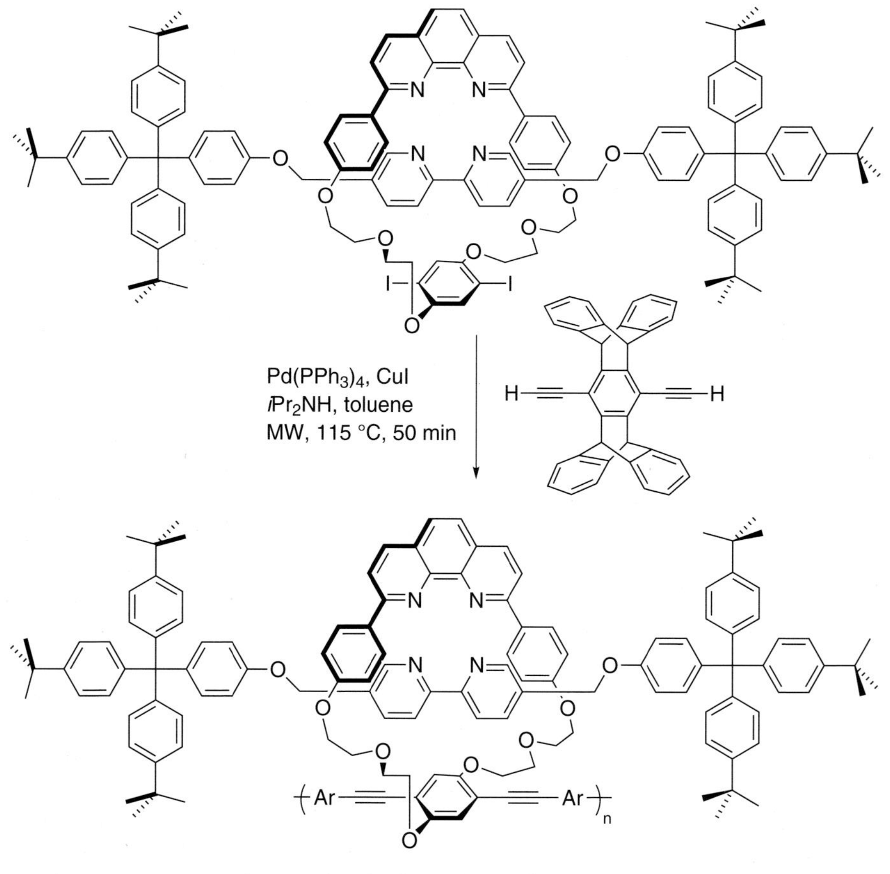

Scheme 5.63 Sonogashira coupling for the synthesis of sensory polymers.

the synthesis of rotaxanated conjugated sensory polymers [391]. Microwave irradiation of a suitable diiodo rotaxane with a corresponding aryl diacetylene under standard microwave Sonogashira coupling conditions [tetrakis(triphenylphosphine)-palladium(0), copper(I) iodide] at 115 °C for 50 min provided the poly(*para*-phenylene ethynylene)s that displayed interesting macromolecular photophysical properties. Using the microwave approach, the reaction time for the synthesis of the rotaxanated conjugated polymer was reduced from 2 days to less than 1 h.

Similarly, functionalized poly(*meta*-phenyleneethynylene)s were obtained by Khan and Hecht using polycondensation routes involving microwave-assisted Sonogashira couplings (Scheme 5.64) [392]. The rate-limiting *in situ* deprotection of the trimethylsilyl (TMS) protecting group (by water) minimized the concentration of free terminal acetylene in the polymerization mixture and therefore limited competing side reactions. Optimum conditions for the one-pot activation/coupling procedure involved the use of 6 mol% each of tetrakis(triphenylphosphine)palladium(0) and copper(I) iodide, in addition to 1,8-diazabicyclo[5.4.0]undec-7-ene (DBU) as a strong base. Utilizing microwave irradiation to 40 °C for 4 h, a 94% yield of diyne defect-free poly(*meta*-phenyleneethynylene) was obtained, showing respectable chain lengths and polydispersity.

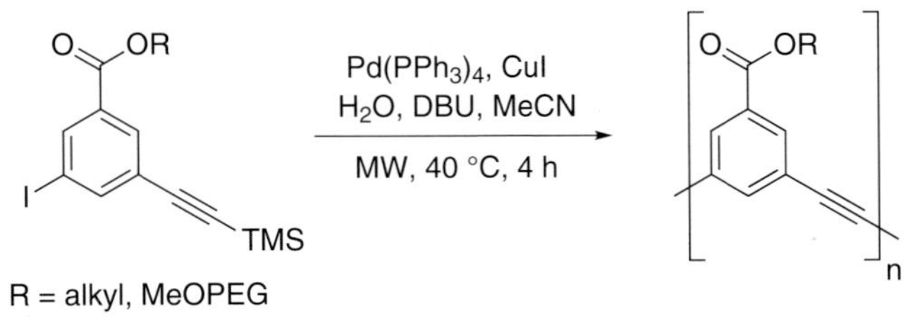

Scheme 5.64 Synthesis of lengthy and defect-free poly(*meta*-phenyleneethynylene)s via polycondensation.

The desulfitative Sonogashira-type cross-coupling of thioether-functionalized pyrazinones **36** and various terminal alkynes was disclosed by Van der Eycken (Scheme 5.65) [393]. The reaction proceeded well for both ethylthio-substituted and resin-linked pyrazinones, although longer reaction times (60 and 90 min) compared to the phenylthio-substituted scaffolds are necessary for resin-bound reactions. In addition, the desulfitative alkynylation was extended to other (het)aryls such as oxazinones, pyrazines, and phenyl thioesters.

The same group has also reported on desulfitative Suzuki-type couplings of pyrazinones **36** with boronic acids under Liebeskind–Srogl conditions [394,395], on Sonogashira couplings starting from the corresponding chloro-substituted pyrazinones [396], and on general C–C and C–N cross-coupling protocols using chloro-substituted pyrazinone precursors [397].

Starting from oxazolinethiones and oxazolidinethiones, desulfitative Sonogashira cross-couplings with terminal alkynes were realized by Tatibouët under similar conditions as described in Scheme 5.65 [398].

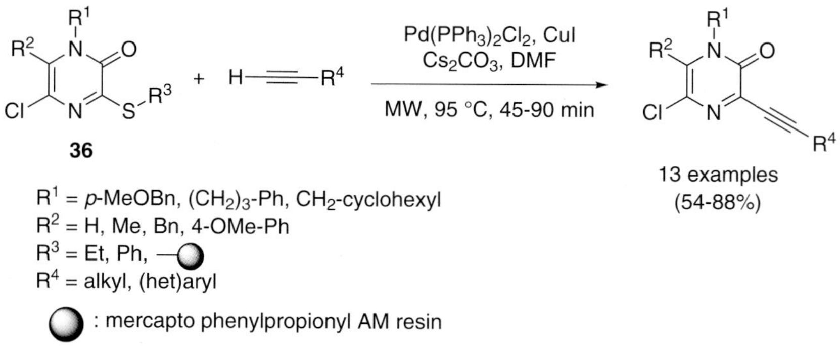

Scheme 5.65 Desulfitative Sonogashira-type cross-couplings.

Another desulfitative Sonogashira coupling of terminal alkynes with heteroaryl methylthioethers (Scheme 5.66) was introduced by Shook *et al.* in 2009 [399]. From a variety of copper salts and palladium catalysts, copper(I) iodide and $PdCl_2$dppf emerged as the most beneficial catalytic system. In the optimized protocol, 2-(methylthio)pyridine, 2 equiv phenylacetylene, and 2 equiv triethylamine in tetrahydrofuran together with 10 mol% $PdCl_2$dppf and 20 mol% CuI were heated at 100 °C for 1 h. Then another 10 mol% $PdCl_2$dppf, 20 mol% CuI, and 2 equiv triethylamine were added and the mixture was additionally heated for 1 h. This two-step protocol also applies for 2-(methylthio)-1,3-thiazole and 2-(methylthio)furan, while for the more reactive 2-(methylthio)pyrazine as a substrate, a single 1 h step is sufficient to achieve maximum yield [399]. In terms of the employed acetylenes, derivatives with electron-withdrawing groups showed somewhat lower efficiency. In general, the procedure proves beneficial for the preparation of heterocycle systems to increase diversity in biological systems.

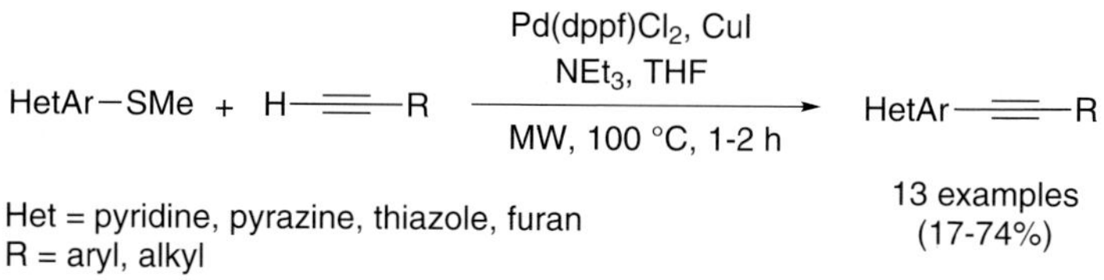

Scheme 5.66 Microwave-assisted Sonogashira coupling of heteroaromatic thiomethylethers.

Liu and coworkers have successfully performed Cu-free Sonogashira cross-couplings of terminal alkynes with a variety of aryl chlorides (Scheme 5.67) [400]. This method proved to be very general and also both sterically hindered and electron-rich aryl chlorides can be coupled very efficiently in addition to electron-neutral and electron-deficient ones giving the products in high yield and in short reaction times. Furthermore, this protocol can be applied for other Pd-catalyzed C−C and C−N bond-forming reactions by switching from $PdCl_2(PPh_3)_2$ to $Pd(OAc)_2$ as catalyst.

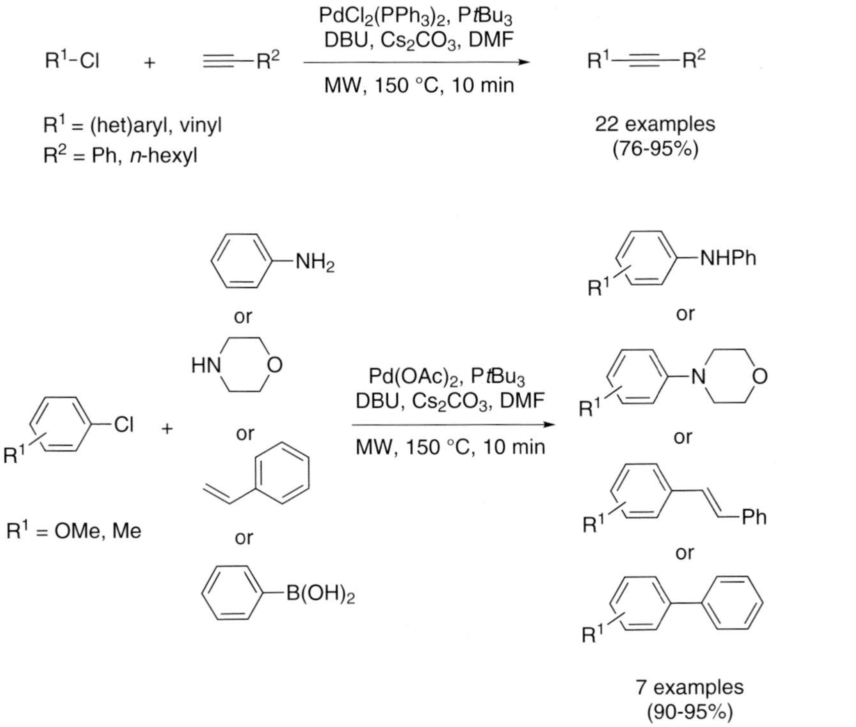

Scheme 5.67 Cu-free Sonogashira and related cross-coupling reactions with aryl chlorides.

Buchwald–Hartwig aminations and Suzuki and Heck coupling were conducted delivering the coupled products in excellent yields (90–95%).

Selective Sonogashira couplings of het(aryl) bromides containing a boronic ester functionality with TMS-acetylene have been described by Wang and coworkers [401].

The Pd-catalyzed synthesis of diaryl acetylenes has been demonstrated by Sørensen and Pombo-Villar (Scheme 5.68) [402]. A direct coupling of activated aryl and heteroaryl bromides and iodides with 1-aryl-2-trimethylsilylacetylenes has been developed for the synthesis of diarylacetylenes, avoiding the use of Cu(I) iodide as a

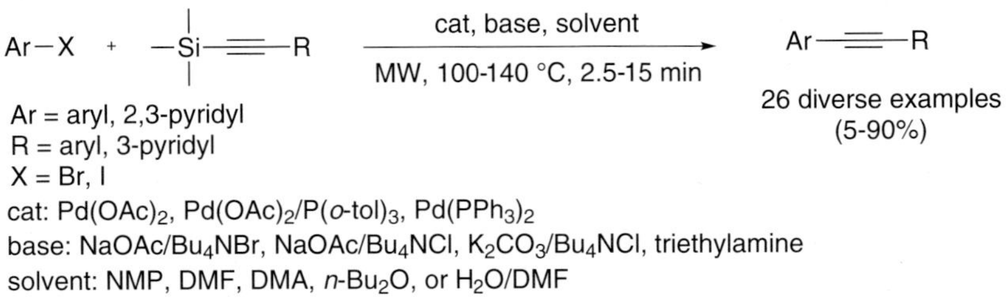

Scheme 5.68 Cu-free, Pd-catalyzed Sonogashira-type couplings.

co-catalyst. Microwave dielectric heating has shown improvement in reaction yields over the conventional oil bath heating.

Lee and coworkers very recently presented a novel method for the synthesis of symmetrical diaryl alkynes from propiolic acid [403]. As model system, the decarboxylative Cu-free Sonogashira coupling of phenyl bromide with propiolic acid in water was investigated; during the optimization, a microwave approach was also involved. Although propiolic acid shows good solubility in water, the addition of a phase-transfer surfactant is required to increase the solubility of the phenyl bromide. In the best working microwave process, 5 mol% $Pd(PPh_3)_2Cl_2$ and 10 mol% 1,4-bis(diphenylphosphino)butane [dppb] as catalyst/ligand system, 5 equiv 1,8-diazabycyclo[5.4.0]undec-7-ene (DBU) as a base, 2 equiv phenyl bromide and octadecyl trimethyl ammonium chloride (OTAC) are combined. Then propiolic acid and distilled water are added and the mixture is irradiated at 100 °C for 15 min yielding 89% of the desired diphenylacetylene (Scheme 5.69). However, since full conversion could not be achieved under microwave conditions, the scope of the aqueous protocol was explored under classical thermal heating

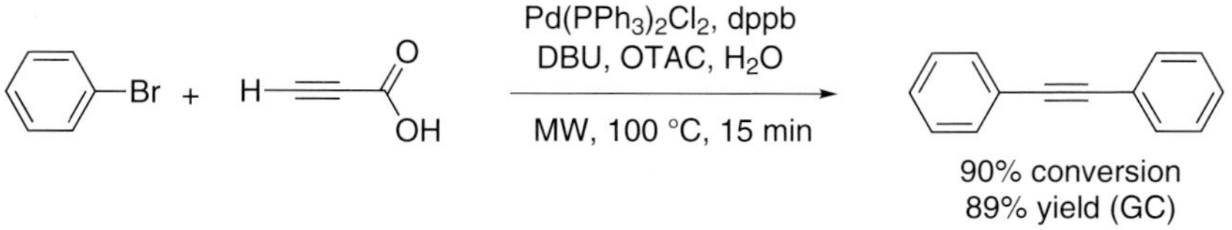

Scheme 5.69 Synthesis of diphenylacetylene from propiolic acid.

In 2009, Awuah and Capretta presented a microwave-assisted one-pot two-step procedure to generate flavones utilizing gaseous carbon monoxide [404]. The process involves a Cu-free Sonogashira coupling subsequently followed by a carbonylation–annulation reaction involving 2-iodophenol derivatives to furnish a variety of the biologically attractive flavone moieties. After preliminary investigations under classical thermal heating to determine the best suitable catalytic system/base/solvent system, the process was transferred to microwave irradiation in order to diminish reaction times and improve overall yields of this two-step reaction. The Sonogashira coupling was performed with trimethylsilylacetylene and aryl iodides or aryl bromides, respectively, *N,N*-dimethylformamide as solvent, 1.5 mol% $Pd_2(dba)_3$ and 3 mol% 1,3,5,7-tetramethyl-2,4,8-trioxa-6-phenyl-6-phosphaadamantane (PA-Ph) as catalyst system, and DBU as a strong base. The mixture is heated for 30 min at 90 °C (120 °C for aryl bromides) to obtain the corresponding arylated alkynes (Scheme 5.70). The products of the first step were then combined with a mixture containing 0.5 equiv substituted 2-iodophenol, fresh Pd/PA-Ph catalyst system, and tetrabutylammonium fluoride, which allows deprotection of the TMS group for smoother and more reproducible annulation reaction (Scheme 5.70). Finally, carbon monoxide gas was bubbled through the mixture before sealing the vessel and irradiating for an additional 30 min at 90 °C. This procedure furnished the desired flavones in moderate to good overall yield.

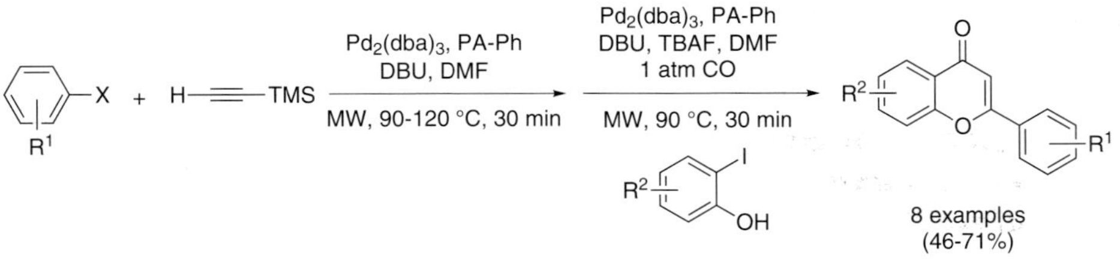

Scheme 5.70 Flavone synthesis via one-pot Sonogashira carbonylation–annulation reaction.

Pd-catalyzed, Cu-free alkenylations and alkynylations of aryl bromides and iodides were also reported by Botta and coworkers [405]. The use of Au nanoparticles supported on inorganic oxides (e.g., silica) for Cu-free Sonogashira reactions was described by Antunes and coworkers [406].

As with the Suzuki reaction, there have been two independent reports by Leadbeater *et al.* [407] and Van der Eycken and coworkers [408] that have shown that it is also possible to perform transition metal-free Sonogashira couplings (Scheme 5.71a). Again, these methods rely on the use of microwave-heated water as solvent, a phase-transfer catalyst (tetrabutylammonium bromide or polyethylene glycol), and a base (sodium hydroxide or sodium carbonate). So far, these metal-free procedures are successful for aryl bromides and iodides, and typical reaction conditions involve heating to about 170 °C for 5–25 min. Employing an organic solvent (1-methyl-2-pyrrolidone), He and Wu demonstrated the ability to perform palladium-free Sonogashira couplings, using 10 mol% of Cu(I) iodide as a catalyst (Scheme 5.71b) [409]. However, the coupling is successful only with aryl iodides and relatively harsh conditions (195 °C, 2–6 h) need to be employed, which limits the practicability of this method.

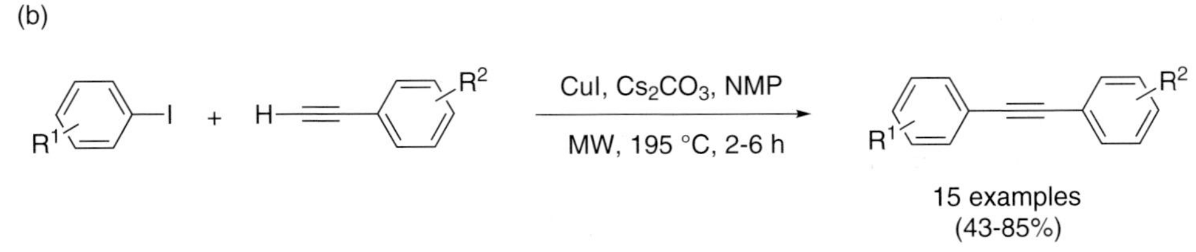

Scheme 5.71 Transition metal-free-catalyzed Sonogashira-type couplings.

5.2.4
Stille Reactions

Compared to the previously described transition metal-catalyzed transformations in this chapter, microwave-assisted Stille reactions [410] involving organotin reagents as coupling partners are comparatively rare. A few examples describing both inter- and intramolecular Stille reactions in heterocyclic systems are summarized in Scheme 5.72 [319,411–413]. Thiophene-based copolymers were also obtained by Stille coupling chemistry [414]. Microwave-assisted Stille cross-couplings for the scaffold decoration of 2(1*H*)-pyrazinones were described by the Van der Eycken group [415]. Several other Stille cross-couplings performed under microwave conditions were reported [416,417].

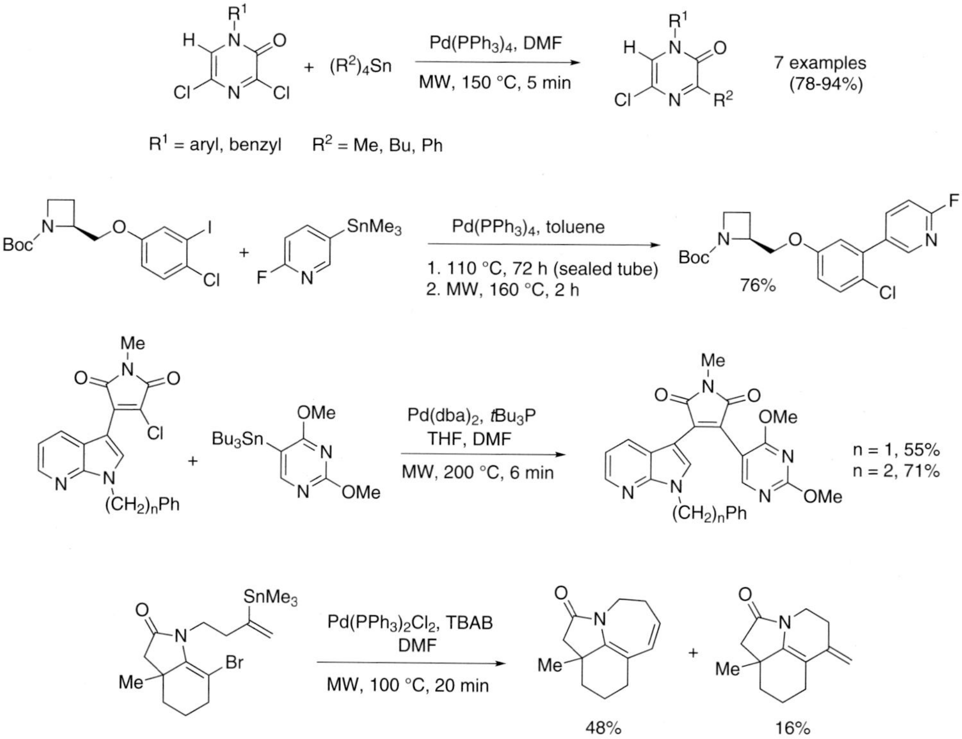

Scheme 5.72 Scaffold decoration and modification of heterocycles using Stille cross-coupling chemistry.

5.2.5
Negishi, Kumada, and Related Reactions

Until recently, very little work has been published on Negishi (organozinc reagents) [418] and Kumada (organomagnesium reagents) [419] cross-coupling chemistry utilizing microwave conditions. There are two examples in the peer-reviewed literature describing Negishi cross-coupling reactions of activated aryl bromides [420] and heteroaryl chlorides [421] with organozinc halides (Scheme 5.73). Öhberg and Westman [422] have reported that microwave-assisted Negishi couplings of activated aryl bromides with arylzinc bromides can be performed within 1 min of microwave heating to 160 °C using 5 mol% of palladium(II)bis(triphenylphosphine) dichloride as the catalyst (Scheme 5.73a). Importantly, this protocol initially carried out on a less than 1 mmol scale was directly scalable to 40 mmol in a larger multimode batch reactor without reoptimization of the reaction conditions [257]. On the other hand, Stanetty *et al.* have demonstrated that activated heteroaryl chlorides can be coupled with heteroarylzinc iodides or chlorides under similar conditions to deliver pyridinyl pyrimidines in high yields (Scheme 5.73b) [421]. Optimized conditions utilized 0.5 mol% tetrakis(triphenylphosphine)palladium(0) as the catalyst and tetrahydrofuran as solvent. At 100 °C reaction temperature, the highest selectivity toward the desired 4-pyrimidinyl coupling product was obtained, with only small amounts of homo- and biscoupling products being formed as by-products.

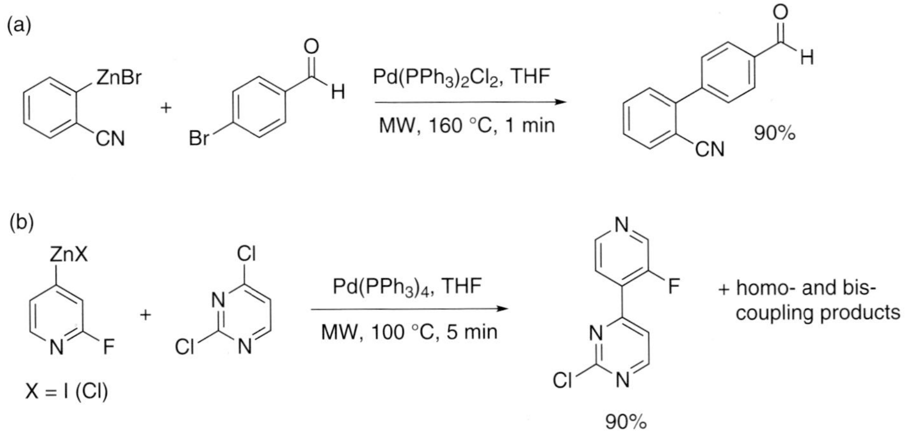

Scheme 5.73 Aryl and heteroaryl cross-coupling using Negishi chemistry.

A more general procedure describing high-speed microwave-assisted Negishi and Kumada couplings of unactivated aryl chlorides was recently reported by Walla and Kappe (Scheme 5.74) [422]. This procedure uses 0.015–2.5 mol% of tris(dibenzylideneacetone)dipalladium(0) as a palladium source and the air-stable tri-*tert*-butylphosphonium tetrafluoroborate as ligand precursor. Successful couplings were observed for both arylorganozinc chlorides and iodides (Scheme 5.74a). Using this methodology, it was also possible to successfully couple aryl chlorides with *alkyl*zinc reagents such as *n*-butylzinc chloride in a very rapid manner without the need for an inert atmosphere. Optimized conditions utilized sealed-vessel microwave irradiation in a

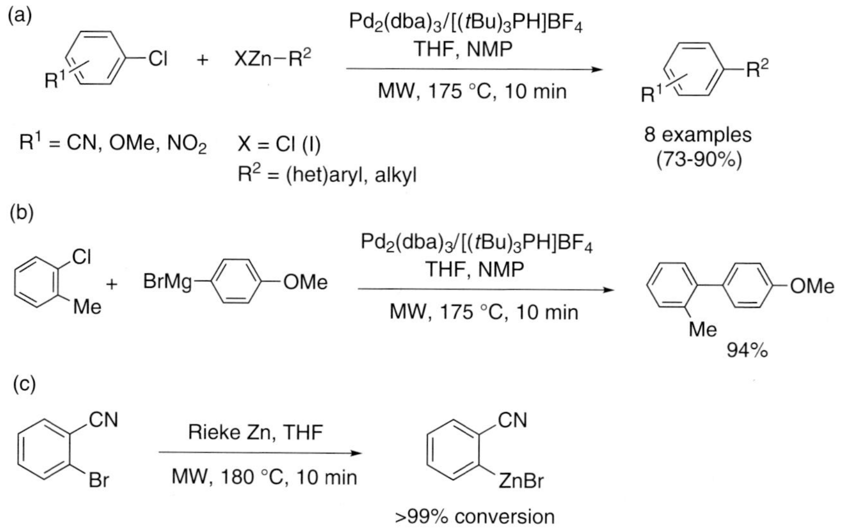

Scheme 5.74 Negishi and Kumada cross-coupling reactions.

THF/1-methyl-2-pyrrolidone mixture at 175 °C for 10 min. Applying the same reaction conditions, Kumada cross-couplings with organomagnesium (Grignard) reagents were also carried out successfully (Scheme 5.74b) [422]. In the same article, the authors have also described rapid microwave-assisted methods for the preparation of the corresponding organozinc and magnesium compounds by insertion of Rieke zinc or magnesium metal, respectively, into aryl halides (Scheme 5.74c) [422].

Ni- and Pd-catalyzed Kumada-type couplings of aryl Grignard reagents with aryl fluorides have been reported by Dankwardt [423].

A similar protocol for the high-speed Negishi cross-coupling transformations was developed by Mutule and Suna (Scheme 5.75) [424]. Here, the required organozinc reagents were readily prepared from aryl iodides using freshly prepared zinc–copper couple. Zinc insertion readily occurred in polar solvents such as *N,N*-dimethylformamide using a sixfold excess of the zinc–copper couple under microwave conditions at 80–125 °C within 2–15 min, depending on the reactivity of the aryl iodide. Subsequent Negishi cross-coupling with 4-bromobenzaldehyde was accomplished by adding the organozinc reagent (supernatant solution after centrifugation) to a

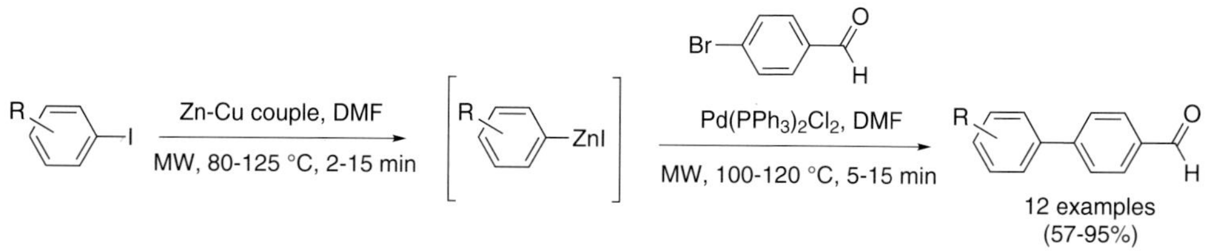

Scheme 5.75 Sequential arylzinc formation and Negishi cross-coupling.

solution of the bromide in *N,N*-dimethylformamide in the presence of 3 mol% palladium(II)bis(triphenylphosphine) dichloride (microwave heating, 100–120 °C, 5–15 min) [424].

An application of this high-speed Negishi coupling methodology for the preparation of enantiopure 2,2′-diarylated 1,1′-binaphthyls was reported by Kappe and coworkers (Scheme 5.76) [425]. Reaction times as short as 40 s in some cases were sufficient to achieve complete conversions in the stereoconservative Negishi coupling of commercially available 2,2′-diiodo-1,1′-binaphthyl (DIBN) with arylzinc chlorides. The catalyst loading for typical runs was 5 mol% of $Pd(PPh_3)_4$, but could be lowered to 0.5 mol% in some instances without appreciable reduction of coupling efficiency. The formed enantiopure 2,2′-diarylated 1,1′-binaphthyls are of interest in material sciences application. In the same article, the corresponding Negishi alkynations using zinc phenylacetylide and zinc trimethylsilylacetylide were also described [425].

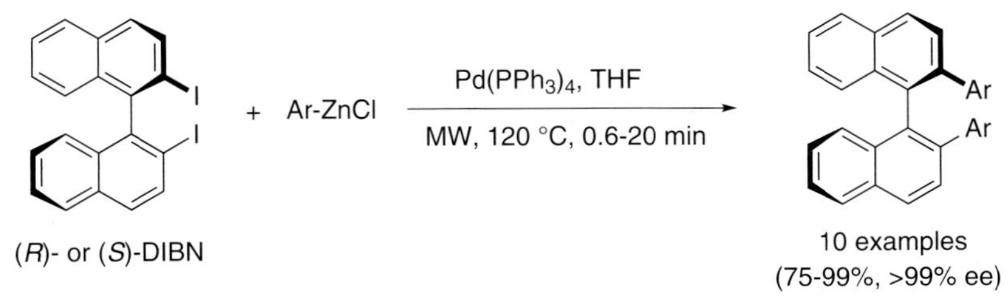

Scheme 5.76 Synthesis of enantiopure 2,2′-diarylated 1,1′-binaphthyls utilizing stereoconservative Negishi cross-coupling reactions.

Through chemistry related to the Negishi reaction, Moloney and coworkers have reported the palladium-catalyzed α-arylation of esters and amides with organozinc reagents in a one-pot fashion (Scheme 5.77) [426]. The required Reformatsky reagents were readily prepared by microwave irradiation of the corresponding

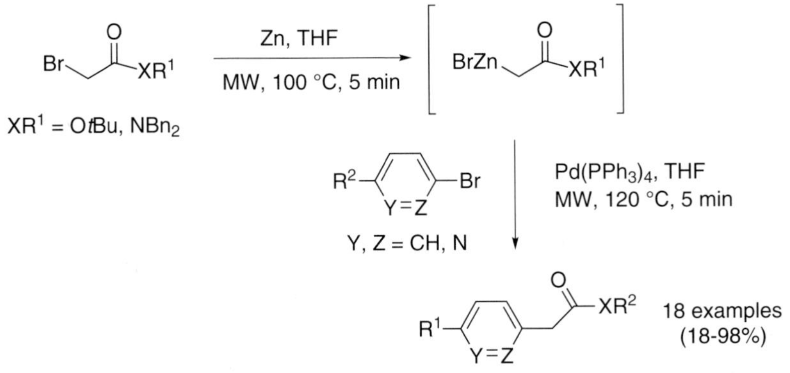

Scheme 5.77 Palladium-catalyzed α-arylation of esters and amides with organozinc reagents.

bromoacetate or bromoacetamide with zinc metal (2 equiv) in THF for 5 min at 100 °C. Addition of this Reformatsky reagent to the coupling partner, an aryl bromide and the relevant catalyst/ligand, in tetrahydrofuran followed by further irradiation for 5–10 min at 120 °C provided the arylacetic esters or amides in good yields. Best results were achieved using 10 mol% of $Pd(PPh_3)_4$ as palladium catalyst.

In addition to the classical Negishi cross-coupling utilizing organozinc reagents, the "zirconium version" involving the coupling of zirconocenes with aryl halides has also been described using sealed-vessel microwave technology. Lipshutz and Frieman have reported the rapid coupling of both vinyl and alkyl zirconocenes (prepared *in situ* by hydrozirconation of alkynes or alkenes, respectively) with aryl iodides, bromides, and chlorides (Scheme 5.78) [427]. While aryl iodides required only 5 mol% nickel-on-carbon (Ni/C) as a ligand-free heterogeneous catalytic system, the presence of triphenylphosphine as a ligand was necessary in order to successfully couple aryl bromides (10 mol%) and chlorides (20 mol% ligand). Under these conditions, full conversion was achieved within 10–40 min at 200 °C using tetrahydrofuran as solvent. The same group subsequently demonstrated that the triphenylphosphine ligands used in the reactions shown in Scheme 5.78 can conveniently be sequestered and subsequently recovered by precipitation of the phosphines with copper(I) chloride [428]. Independently, a publication from the Fu laboratory has described a similar high-yielding palladium-catalyzed process, where alkyl bromides are coupled under ligandless conditions (2.5 mol% palladium(II) acetylacetonate, 2 equiv of lithium bromide, tetrahydrofuran/1-methyl-2-pyrrolidone, 100 °C, 15 min) with alkenyl zirconium reagents [429].

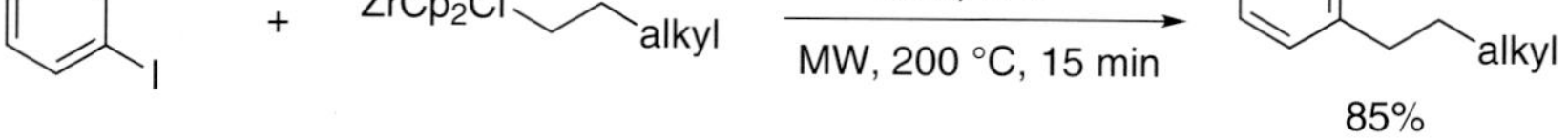

Scheme 5.78 Nickel-catalyzed cross-coupling of zirconocenes and aryl halides.

Hydrozirconation is a mild method for the selective preparation of functionalized organometallics and its compatibility with a range of common protecting groups represents a considerable advantage of these species over traditional organometallic reagents [430]. Wipf and coworkers reported that the hydrozirconation of alkynes with zirconocene hydrochloride can be greatly accelerated by microwave irradiation, as illustrated for the examples depicted in Scheme 5.79a [431]. Under optimum conditions, hydrozirconation of unsymmetrical internal alkynes was accomplished with 2 equiv of Cp_2ZrHCl at 60 °C in 30 min. In this context, a synthetically useful

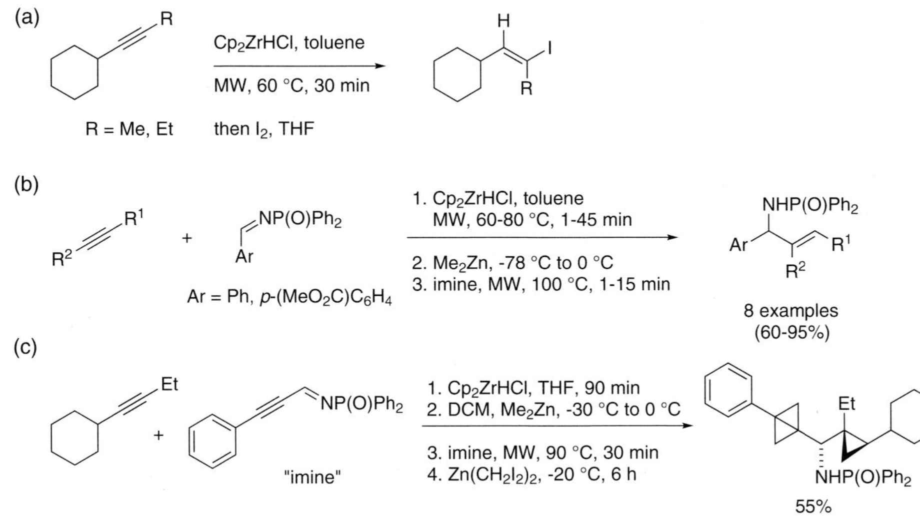

Scheme 5.79 Hydrozirconations and zirconocene-imine additions.

one-pot method for the preparation of allylic amides was elaborated where an alkyne was first hydrozirconated by microwave irradiation, followed by rapid addition of imines in the presence of dimethyl zinc (Scheme 5.79b) [431]. The synthetically useful allylic amides were isolated in good to excellent overall yield (60–95%). A further extension of this protocol involved the synthesis of *C,C*-dicyclopropylmethylamines via multicomponent condensation of alkynes, imines, and zinc carbenoids [432]. For the example shown in Scheme 5.79c, the alkynyl imine–zirconocene addition step was accelerated by microwave heating to 90 °C.

5.2.6
Carbonylation Reactions

Taking advantage of the rapid and controlled heating made possible by microwave irradiation of solvents under sealed-vessel conditions, the group of Larhed has reported a number of valuable palladium-catalyzed carbonylation reactions [433–441]. The key feature of many of these protocols is to use molybdenum hexacarbonyl [$Mo(CO)_6$] as a solid precursor of carbon monoxide (CO), required in carbonylation chemistry utilizing aryl halides as starting materials. At temperatures of 150 °C, $Mo(CO)_6$ liberates enough CO *in situ* for rapid aminocarbonylation reactions to take place (at 210 °C, CO is liberated instantaneously). The initially reported conditions used a combination of Herrmann's palladacycle (7.4 mol% palladium) and 2,2′-bis(diphenylphosphino)-1,1′-binaphthyl (BINAP) as the catalytic system in a diglyme–water mixture, providing the desired secondary and tertiary amides in high yield (Scheme 5.80a) [433]. As in many other cases, an inert atmosphere was not required. Subsequent experimental improvements of the protocol allowed the use of sterically and electronically more demanding amines

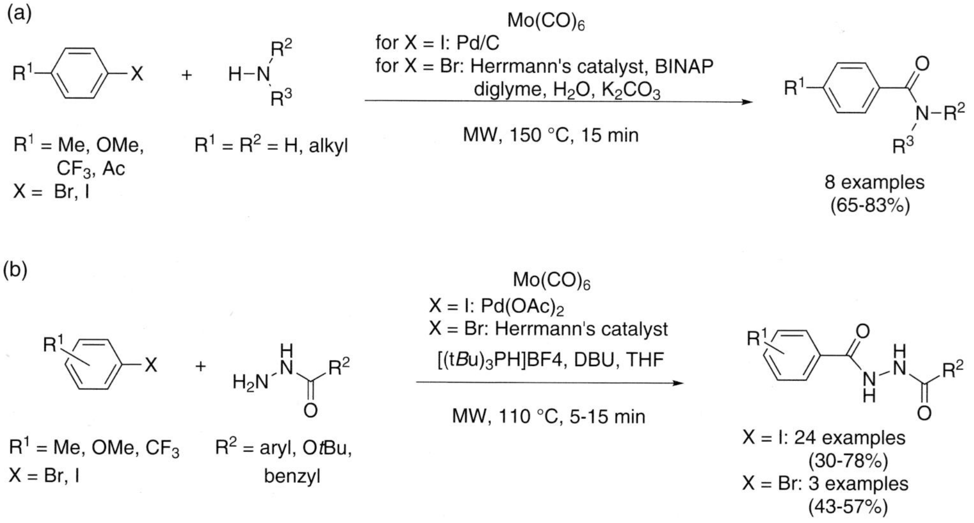

Scheme 5.80 Palladium-catalyzed aminocarbonylations using molybdenum hexacarbonyl as a solid source of carbon monoxide.

(e.g., anilines, unprotected amino acids), by using 1,8-diazabicyclo[5.4.0]undec-7-ene (DBU) as the base and tetrahydrofuran as solvent for both aryl bromides and iodides [434]. The protocol could also be extended to hydrazidocarbonylations employing similar reaction conditions (Scheme 5.80b) [435].

In a further report of the same group, water could be used as solvent in the synthesis of benzamides via aminocarbonylations of aryl chlorides, aryl bromides, and aryl iodides using $Mo(CO)_6$ as solid CO source (Scheme 5.81) [442]. Intensive investigations have been made toward reaction conditions, especially in terms of reagent ratios. By using either aryl halide or amine as a limiting reagent, accompanied with the proper Pd catalyst, hydroxycarbonylations could be suppressed as side reactions and even the otherwise unreactive aryl chlorides could be employed under these conditions.

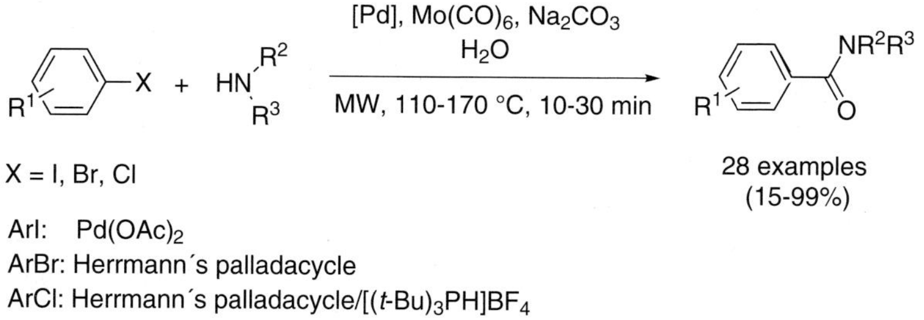

Scheme 5.81 Aminocarbonylations of aryl halides in water.

Larhed and coworkers have also reported on aminocarbonylations using allylamine under a similar set of conditions [443]. Similarly, the Pd-catalyzed hydroxycarbonylation of aryl and vinyl triflates in water using $Mo(CO)_6$ as CO source has been described [444].

The same authors also applied alkenyl phosphates, chlorides, bromides, and triflates as electrophile coupling partners for different amines to generate acrylamide derivatives [445]. For the alkenyl phosphates, previously not used in carbonylation reactions, a slightly modified method for carbonylation of aryl chlorides has been applied. The reactions proceed satisfactory when alkenyl phosphate and the corresponding amines, together with 2.5 mol% Herrmann's palladacycle and 5 mol% Fu-salt [$HP(t\text{-}Bu)_3BF_4$] as catalyst system and DBU as a base in tetrahydrofuran, are heated for 20 min at 170 °C (Scheme 5.82). The corresponding naphthylphosphate required 190 °C to achieve full completion, although yields were only moderate. Alkenyl bromides only needed 140 °C to afford acceptable yields of the desired acrylamides.

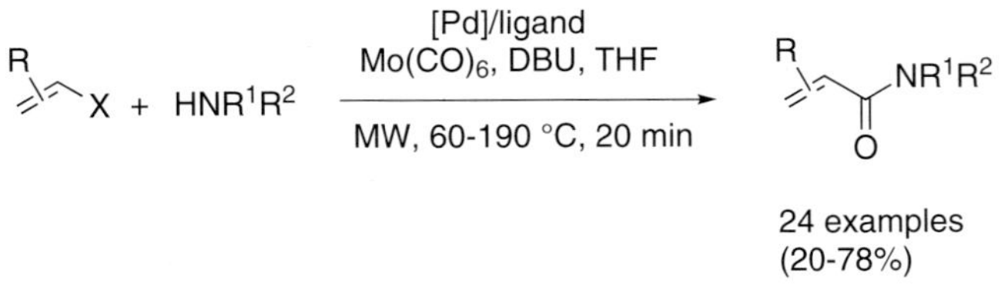

Scheme 5.82 Synthesis of acrylamides from different alkenyl electrophiles.

When turning to alkenyl triflates, the authors discovered that a change of the catalytic system was beneficial. Switching to Xphos instead of the Fu salt as ligand improved the reaction outcome significantly. Moreover, the reaction proceeds already at moderate 60 °C allowing using inexpensive palladium acetate as catalyst. In addition, even secondary amines afforded good yields with alkenyl triflates, which is in sharp contrast to the reaction behavior of such nucleophiles with alkenyl phosphates. Nevertheless, this novel method for aminocarbonylation of alkenyl phosphates certainly expands the scope for the synthesis of acrylamide derivatives.

In another recent example, Larhed and coworkers reported on the novel synthesis of Weinreb and *N*-methylamino pyridyl (MAP) aryl amides utilizing solid wolfram or molybdenum hexacarbonyl as CO source [446]. For the synthesis of Weinreb amides, aryl bromides were coupled with *O,N*-dimethylhydroxylamine hydrochloride using 0.33 equiv $W(CO)_6$, 7.5 mol% palladium acetate and 15 mol% Xantphos as catalyst system, *N,N*-dimethylaminopyridine (DMAP), potassium phosphate as a base, dioxane as a solvent, and heated at 120 °C for 20 min (Scheme 5.83a). The resulting reaction mixture also contains besides the desired Weinreb amides varying amounts of corresponding *N*-methylamides formed by reduction of the employed hydrochloride to undergo aminocarbonylation. Interestingly, although a variety of aryl bromides could be used with this method, corresponding aryl iodides do not tolerate the use of $W(CO)_6$ as carbon monoxide source. Here, $Mo(CO)_6$ could be successfully

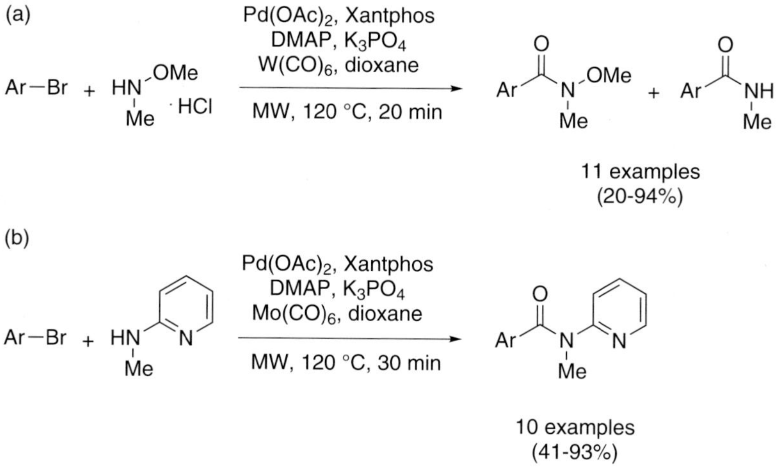

Scheme 5.83 Microwave-assisted synthesis of Weinreb and MEP aryl amides.

applied with the iodides, combined with somewhat lower reaction temperature and shorter time (110 °C, 10 min), which can be attributed to the increased reactivity of these substrates.

Also, when turning toward MAP aryl amides, $Mo(CO)_6$ is the CO source of choice. A set of 10 examples could be prepared by reacting 2-methylaminopyridine with the corresponding aryl bromide, 1 equiv $Mo(CO)_6$, 5 mol% palladium acetate and 10 mol% Xantphos as catalyst system, DMAP, potassium phosphate as a base, dioxane as a solvent, and heating at 120 °C for 30 min (Scheme 5.83b). Using a significant excess (5 equiv) of aryl bromide was required to achieve full conversion of the applied electron-poor nucleophile [446]. Similar to the Weinreb amides, also for MAP amides the use of iodide substrates required slight modifications of the protocol. Here, excess of the aryl halide was not necessary, utilizing the aryl iodide as limiting agent. However, overall yields are somewhat lower compared to the applied bromo substrates. In general, the developed method tolerates a variety of functionalities with potential for further optimization when employing particular substrates.

Larhed and coworkers have also reported on the Pd-catalyzed carbonylation of both aryl iodides and bromides using sulfonamides as nucleophiles (Scheme 5.84) [447].

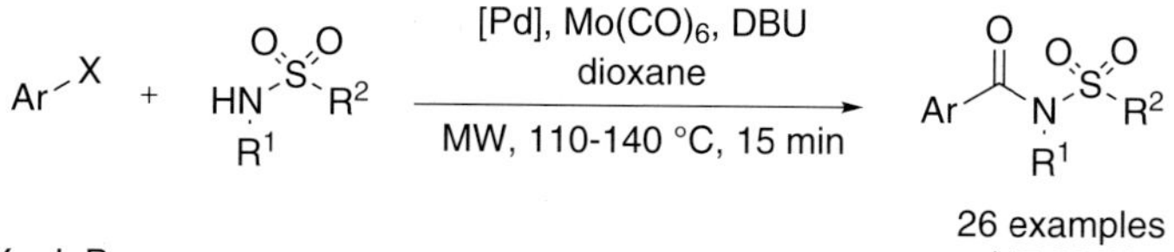

X = I, Br
R^1 = H, Me
R^2 = Ph, 4-tolyl, 4-Br-Ph, Me, CF_3
for X = I: $Pd(OAc)_2$
for X = Br: Herrmann´s palladacycle/Fu salt

Scheme 5.84 Synthesis of acyl sulfonamides via carbonylation reactions.

Good to excellent yields of acyl sulfonamide products were achieved by employing microwave heating for 15 min at 110–140 °C and using $Mo(CO)_6$ as the CO source. Utilizing this protocol, a novel hepatitis C virus NS3 protease inhibitor including acyl sulfonamide elements was also synthesized.

A similar protocol toward aryl acyl sulfamides as potential bioisosteres was also presented by Roberts and coworkers [448]. Palladium acetate/tri-*tert*-butyl phosphine tetrafluoro hydroborate ($P^tBu_3[HBF_4]$) as catalyst/ligand system and DBU as base were used with a small subset of aryl iodides and pyrrolidine sulfamide to furnish the corresponding aryl acyl sulfamides in excellent yields after heating for 2 h at 100 °C (Scheme 5.85a). To further explore the potential of the protocol, it should also be applied to aryl bromides, although it remained unsuccessful. However, slight reoptimization of the catalytic system resulted in a general protocol being beneficial for a variety of aryl and heteroaryl bromides as well as for different sulfamides. Here, $[Pd(OAc)(P(o\text{-tolyl})_3)]_2$ and $P^tBu_3[HBF_4]$ as catalyst/ligand system and microwave heating at 100 °C for 2.5 h are required (Scheme 5.85b). Although most substrates gave good yields, the *ortho*-substituted halides furnished rather poor results, likely due to steric effects in the intermediate acyl palladium species. These yields could be improved by adding 1 equiv DMAP as acyl transfer agent to the reaction mixture.

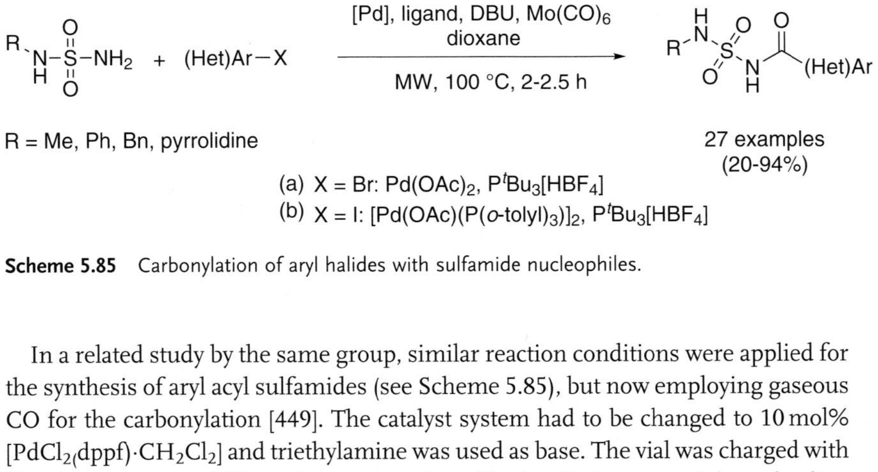

Scheme 5.85 Carbonylation of aryl halides with sulfamide nucleophiles.

In a related study by the same group, similar reaction conditions were applied for the synthesis of aryl acyl sulfamides (see Scheme 5.85), but now employing gaseous CO for the carbonylation [449]. The catalyst system had to be changed to 10 mol% $[PdCl_2(dppf)\cdot CH_2Cl_2]$ and triethylamine was used as base. The vial was charged with CO to a pressure of 65 psi (approximately 4.5 bar) utilizing a special gas loading system. After irradiation at 100 °C for 4 h, the corresponding acyl sulfamides are furnished in good overall yield (36–94%). The method is tolerable to a broad variety of substrates, including sulfonamide with various substitution patterns and aryls and heteroaryls with electron-donating and electron-withdrawing groups, regardless of whether in *ortho*-, *meta*-, or *para*-position. Only aryl chlorides reacted sluggish, even when reaction times are expanded to 20 h, with unactivated 3-methoxychlorobenzene showing no conversion at all.

Another work of Roberts *et al.* dealt with various molybdenum-based sources of CO for carbonylation of aryl halides with nucleophiles [450]. Besides the already established $Mo(CO)_6$, the authors screened a number of Group VI metal carbonyl complexes for their efficiency in the transition metal-free carbonylation of iodobenzene with benzylamine. Not surprisingly, the molybdenum complexes gave the best results, whereas the wolfram analogs afforded only poor yields and the chrome analogs furnished just traces. However, the effective molybdenum complexes required 3–4 h at 130–160 °C to complete the reactions. With such information in hand, the carbamoylation of benzylamine was further examined with a range of aryl and heteroaryl halides. Using $Mo(CO)_6$, aryl iodides were irradiated at 160 °C for 3 h, whereas aryl bromides required 5 h at 200 °C to furnish the corresponding aryl amides in high yields (Scheme 5.86). For the use of the more efficient $Mo(CO)_5Cl{\cdot}NEt_4$, a novel one-pot microwave procedure was envisaged where the required complex is formed *in situ* by heating $Mo(CO)_6$ in the presence of tetraethylammonium chloride in dioxane for 2 min at 140 °C. Subsequently, aryl halide and benzylamine were added and the mixture was again subjected to microwave heating. Aryl iodides were kept at 130 °C for 4 h and aryl bromides at 150 °C for 5 h to obtain the required amides in excellent yields (Scheme 5.86). Finally, this novel approach with *in situ* generation of the carbonylating agent was used to expand the scope of the carbamoylation toward other nucleophiles. Aqueous ammonia, primary and secondary amines, sulfonamides, and even alcohols and water were applied with this effective method [450].

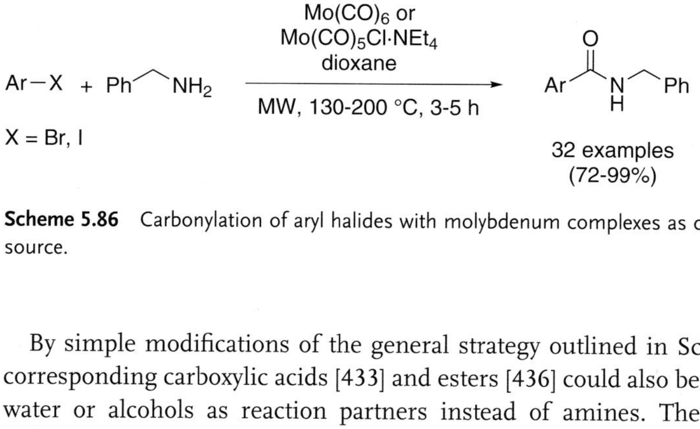

Scheme 5.86 Carbonylation of aryl halides with molybdenum complexes as carbon monoxide source.

By simple modifications of the general strategy outlined in Scheme 5.87, the corresponding carboxylic acids [433] and esters [436] could also be obtained using water or alcohols as reaction partners instead of amines. The ester forming carbonylation was also exploited in a tandem carbonylation–lactone formation reaction sequence for the synthesis of phthalides [437]. Here, again optimum conditions involved the use of $Mo(CO)_6$ as a solid source of carbon monoxide, and palladium acetate/1,1′-bis(diphenylphosphino)ferrocene as a catalyst (5 mol%) at 180 °C. This microwave-assisted carbonylation–cyclization method was also applied for the synthesis of other scaffolds, such as dihydroisocoumarins, dihydroisoindones, and phthalimides (not shown) [437], and could also be applied to the preparation of indanones [438]. Further modifications by Alterman and coworkers have resulted in the use of *N,N*-dimethylformamide as a source of carbon monoxide [439] or in the use of formamide as a combined source of

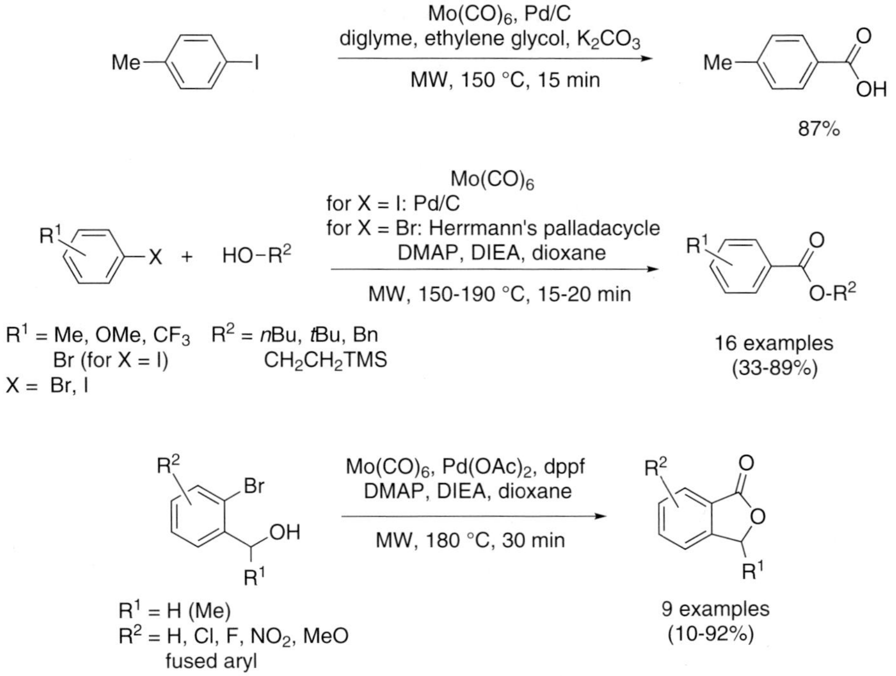

Scheme 5.87 Palladium-catalyzed carbonylation reactions yielding acids, esters, and lactones using molybdenum hexacarbonyl as a solid source of carbon monoxide.

ammonia and carbon monoxide (Scheme 5.88) [440]. The latter method is useful for the preparation of primary aromatic amides from aryl bromides. In both cases, strong bases and temperatures around 180 °C (7–20 min) have to be used to mediate the reaction.

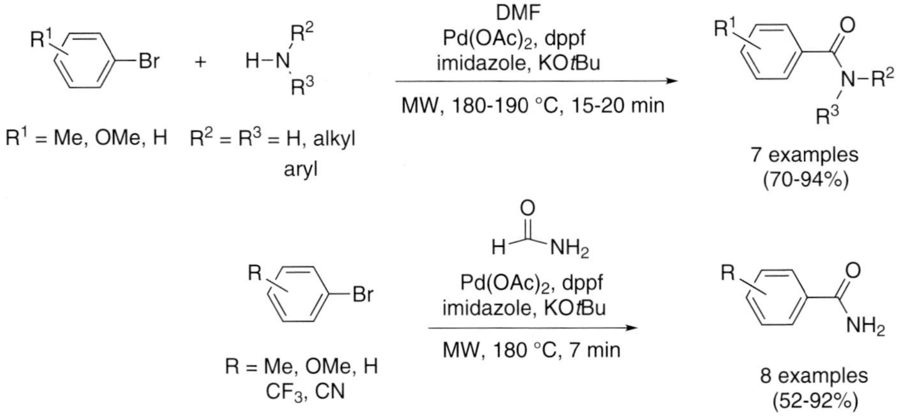

Scheme 5.88 Palladium-catalyzed aminocarbonylations using formamides as sources of carbon monoxide.

Related hydroxycarbonylation [451] and alkoxycarbonylation chemistry [452,453] has been performed by Leadbeater and Kormos using CO gas in a dedicated microwave reactor.

As already described above, gaseous CO has also been applied for microwave-assisted carbonylation chemistry. In 2009, Taddei and coworkers published a method for rapid aminocarbonylation of aryl bromides working at low CO pressures [454]. The corresponding amides were synthesized by coupling aryl bromides with disubstituted amines in anhydrous tetrahydrofuran. DIPEA was used as base and 5 mol% $Pd(PPh_3)_2Cl_2$ was employed as catalyst. After sealing, the vial was charged with CO gas to 120 psi (approximately 8.3 bar) using a special adapter and heated for 20 min at 130 °C to furnish the amides in good yields (Scheme 5.89). When the less nucleophilic anilines were employed, the expected anilides could not be observed; thus, a slight modification of the reaction procedure was required. It turned out that the aromatic amines worked best with 10 mol% of the applied palladium catalyst and 3 equiv of cesium carbonate as the base while lowering the reaction temperature to 120 °C and the reaction time set to 30 min. A small subset of eight aromatic and heteroaromatic anilides in good to excellent yield were obtained [454].

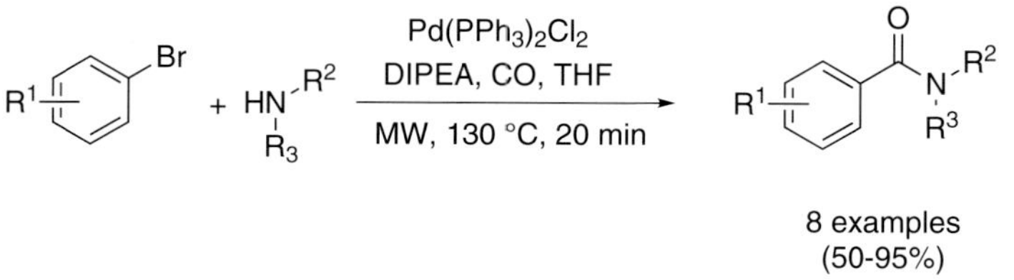

Scheme 5.89 Microwave-assisted aminocarbonylation of aryl bromides with carbon monoxide gas.

Employing the identical instrument set up for controlled introduction of gases, the same group has also reported on the rapid synthesis of linear aldehydes via the hydroformylation of alkenes (Scheme 5.90) [455]. The microwave reaction vial containing the reaction mixture was filled with syngas (CO/H_2 1:1) up to an internal pressure of 3 bar and was subsequently subjected to microwave heating. The reaction protocol (Rh-catalyst, Xantphos, [bmim][BF_4], toluene, 110 °C, 4–6 min) was applicable to differently substituted alkenes to obtain the linear aldehydes without notable formation of the branched isomers. Importantly, the ionic liquid was essential for the reaction; without it only 60 °C could be reached resulting in incomplete conversion of the alkene.

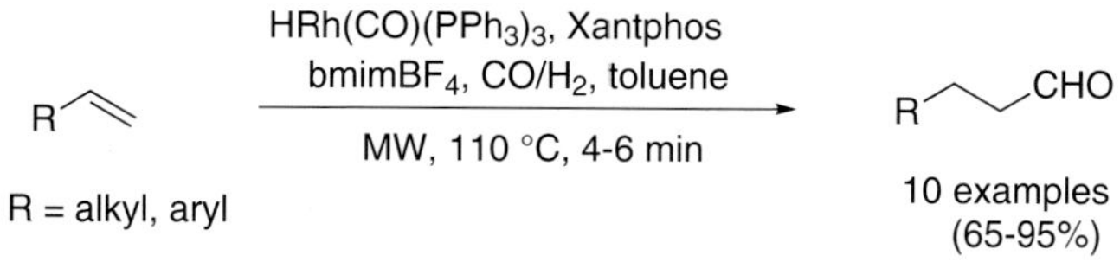

Scheme 5.90 Hydroformylation of alkenes.

The same group has reported that similar catalyst/ligand systems can also be used for the hydroaminomethylation of alkenes [456].

Another report describes the synthesis of aryl *N*-acylureas via microwave-assisted carbonylation using CO gas [457]. Various substituted ureas as nucleophiles were employed in the coupling process to generate pharmaceutically important moieties. The aryl or heteroaryl halide was with urea, 10 mol% $PdCl_2$(dppf)·CH_2Cl_2 as catalyst, and 3 equiv of triethylamine as a base in dioxane. The sealed vial was charged with CO to a pressure of 65 psi (approximately 4.5 bar) and irradiated at 100 °C for 4 h to afford the corresponding acyl ureas in satisfactory overall yield (Scheme 5.91). Although the general acceptance of different substrates for this method is good, *ortho*-substituted and 2,6-disubstituted halides were not well tolerated and nonactivated aryl chlorides could not be converted at all. In contrast to the latter, conversion of the *ortho*-bromides could be increased by adding 1 equiv *N,N*-dimethylaminopyridine (DMAP) to the reaction mixture and enhancing the reaction time to 7 h [457].

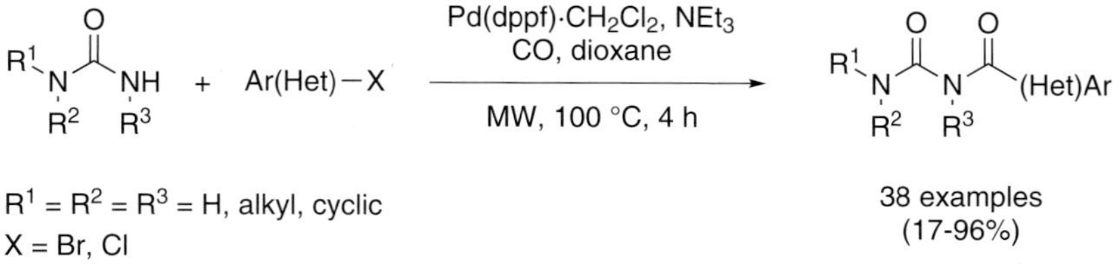

Scheme 5.91 Synthesis of aryl and heteroaryl *N*-acylureas via carbonylation using carbon monoxide gas.

A somewhat related process, the cobalt-mediated synthesis of symmetrical benzophenones from aryl iodides and dicobalt octacarbonyl, is shown in Scheme 5.92 [441]. Here, dicobalt octacarbonyl is used as a combined Ar–I bond activator and CO source. Employing acetonitrile as solvent, a variety of aryl iodides with different steric and electronic properties underwent the carbonylative coupling in excellent yields. Remarkably, in several cases, microwave irradiation for 6 s was sufficient to achieve full conversion! An inert atmosphere, a base, or other additives were all unnecessary. No conversion occurred in the absence of heating regardless of the reaction time. However, equally high yields could also be achieved by heating the reaction mixture in an oil bath for 2 min.

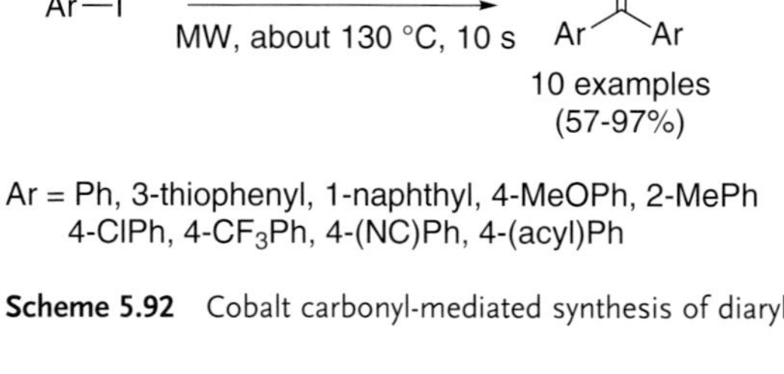

Scheme 5.92 Cobalt carbonyl-mediated synthesis of diaryl ketones.

The same group also reported on a novel and very fast carbonylation method, employing again $Co_2(CO)_8$, for the preparation of ureas starting from primary amines (Scheme 5.93) [458]. Under high-intensity microwave heating with an *in situ* generation of intermediate isocyanates from $Co_2(CO)_8$, amines have been converted to the corresponding ureas within 10 s of reaction times. This method has facilitated the preparation of symmetrical and unsymmetrical ureas in good to moderate yields, respectively.

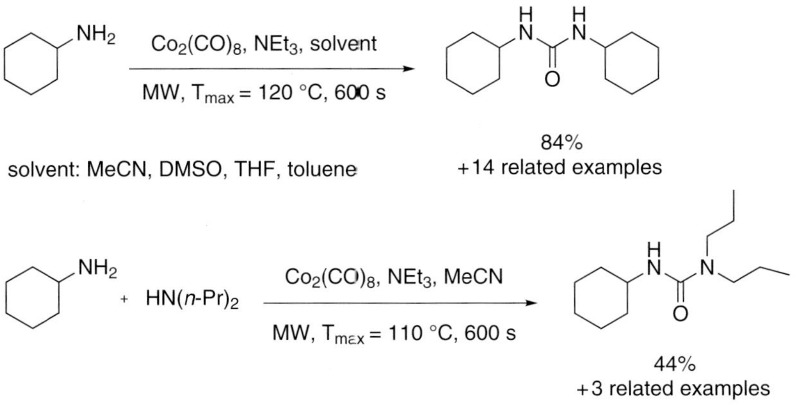

Scheme 5.93 Cobalt carbonyl-mediated synthesis of ureas.

5.2.7
Asymmetric Allylic Alkylations

A frequent criticism of microwave synthesis has been that the typically high reaction temperatures will invariably lead to reduced selectivities. This is perhaps the reason why comparatively few enantioselective processes driven by microwave heating have been reported in the literature. Nevertheless, a number of very impressive enantioselective reactions involving chiral transition metal complexes have been described in the literature.

In 2000, the groups of Moberg, Hallberg, and Larhed have already reported on microwave-mediated palladium- [459,460] and molybdenum-catalyzed [461–464] asymmetric allylic alkylation reactions involving neutral carbon, nitrogen, and oxygen nucleophiles. Both processes, carried out under noninert conditions, yielded the desired products in high chemical yield with typically >98% ee (Scheme 5.94).

A similar example has been presented more recently by the group of Braga who has disclosed Pd-catalyzed allylic alkylations of *rac*-1,3-diphenyl-2-propenyl acetate with malonates (Scheme 5.95) [465]. By applying chiral β-seleno amides as ligands in combination with microwave heating, reaction times of 2–4 min were possible compared to 24 h at room temperature with only a minimal loss of enantioselectivity

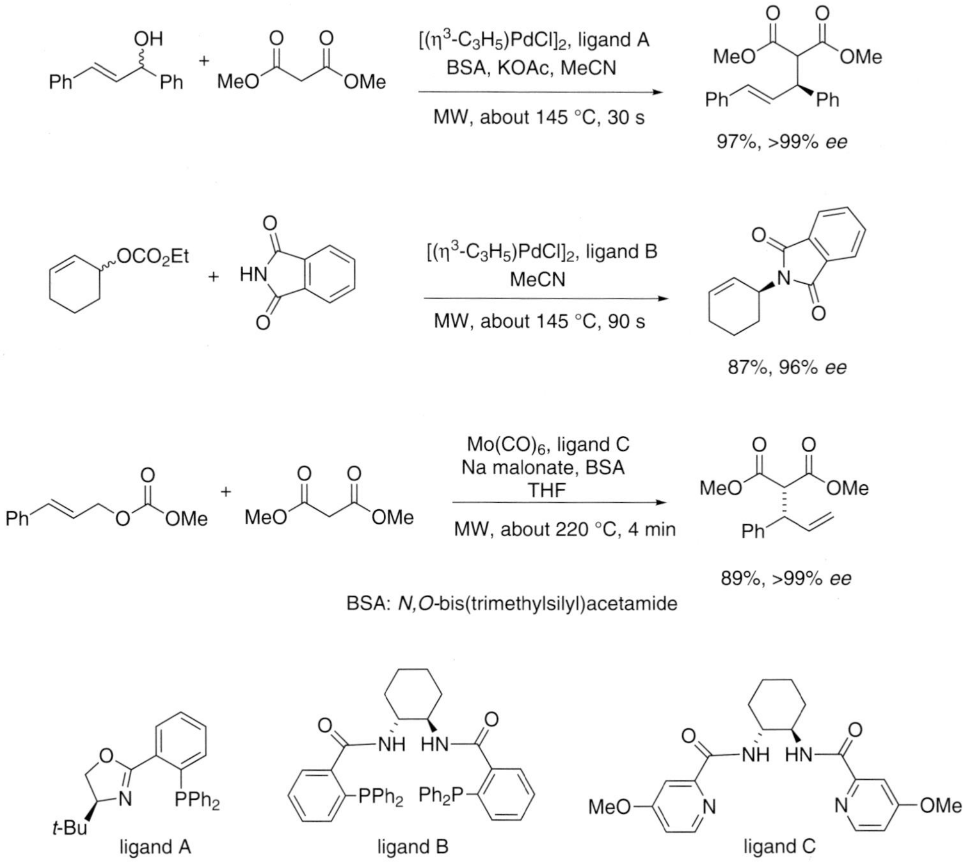

Scheme 5.94 Palladium- and molybdenum-catalyzed asymmetric allylic alkylations.

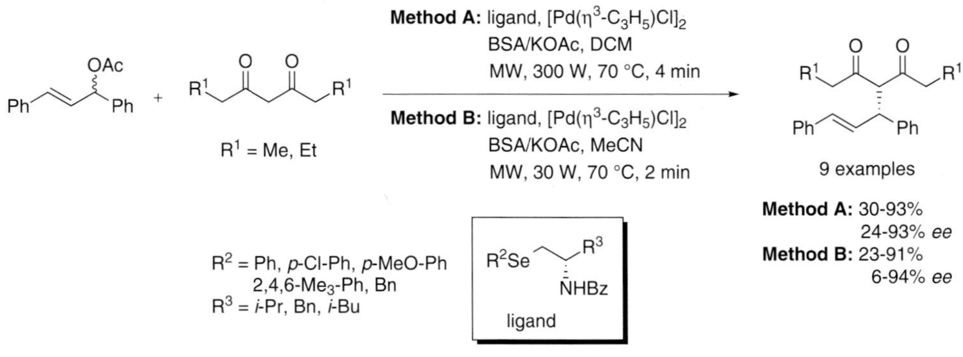

Scheme 5.95 Pd-catalyzed asymmetric allylic alkylations.

(ee values up to 94% were obtained). During optimization studies, two methods were found to give high yields (methods A and B). For ligands with R^2 = Ph and R^3 = Bn, *i*-Bu, it turned out that the method applied has a significant influence on the ee values, better values could be reached with method A.

The groups of Porco have reported on the synthesis of aryl ether *C*-glycoside derivatives **39** via Pd(0)-mediated allylic substitution of biscarbonates **37** with phenols **38** (Scheme 5.96a) [466]. This reaction was both regio- and stereoselective with an overall retention of configuration when the enantiomeric (*S,S*)-Trost ligand was employed. The protected aryl ether *C*-glycosides **40** can be further diversified by Eu(III)-catalyzed Claisen rearrangement where glycosides **40** undergo a [3,3]-sigmatropic rearrangement and furnish the novel phenol derivatives **41** (Scheme 5.96b).

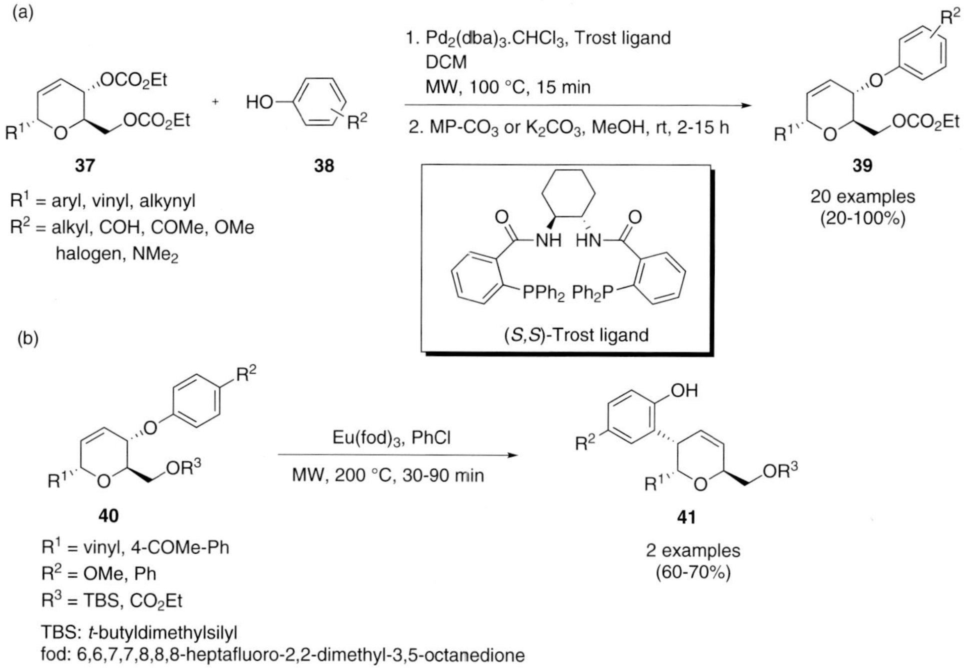

Scheme 5.96 Scaffold decoration of *C*-glycosides.

Trost and Andersen have applied the molybdenum-catalyzed concept in their approach to the orally bioavailable HIV inhibitor Tipranavir (Scheme 5.97) [467]. Synthesis of the key C-3α chiral intermediate was achieved by asymmetric allylic alkylation starting from the corresponding carbonate. Employing 10 mol% of the molybdenum precatalyst and 15 mol% of a suitable chiral ligand with 2 equiv of dimethyl sodiomalonate as additive, a 94% product yield was achieved. The reaction was carried out under sealed-vessel microwave heating at 180 °C for 20 min. Thermal heating under reflux conditions (67 °C) required 24 h and produced the same

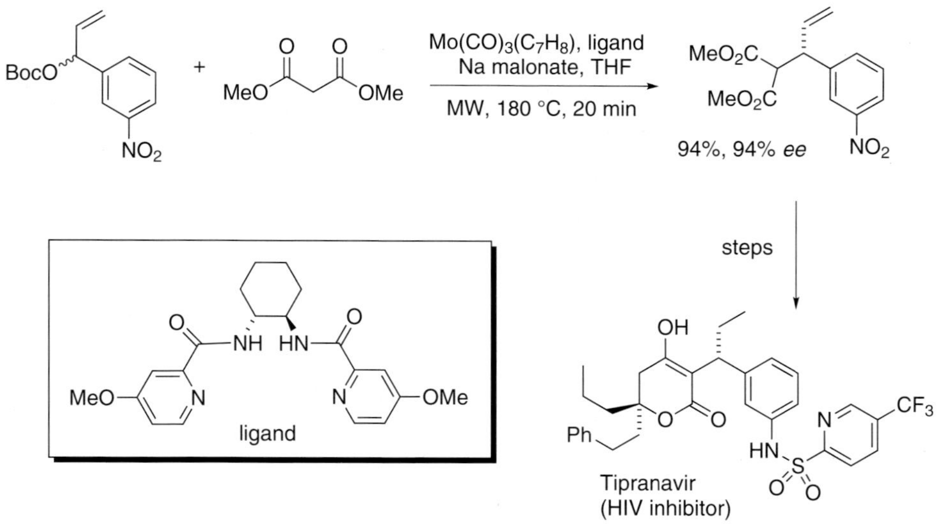

Scheme 5.97 Molybdenum-catalyzed asymmetric allylic alkylation for the synthesis of Tipranavir.

chemical yield of intermediate, albeit with slightly higher enantiomeric purity (96% ee).

A similar pathway involving a microwave-driven molybdenum-catalyzed asymmetric allylic alkylation as the key step was elaborated by Moberg and coworkers for the preparation of the muscle relaxant (*R*)-baclofen (Scheme 5.98) [468]. The racemic form of baclofen is used as a muscle relaxant (antispasmodic) lipophilic derivative

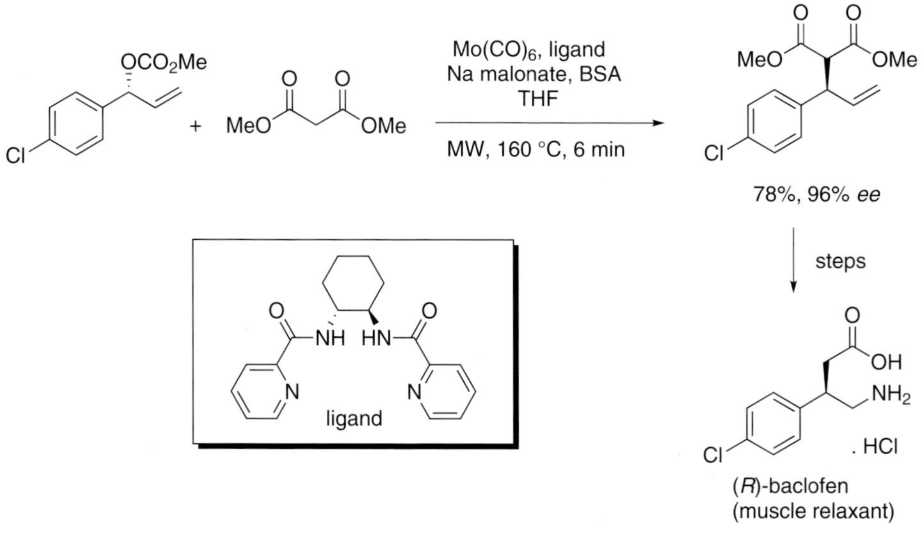

Scheme 5.98 Molybdenum-catalyzed asymmetric allylic alkylation for the synthesis of (*R*)-baclofen.

of γ-aminobutyric acid (GABA). Pharmacological studies have shown that the (*R*)-enantiomer is the therapeutically useful agonist of the $GABA_B$ receptor. Asymmetric alkylation of the allylic carbonate precursor with dimethyl malonate afforded the desired chiral intermediate in high chemical yield and with 96% ee. The reaction was performed in THF as solvent using 4 mol% of $Mo(CO)_6$ as precatalyst at 160 °C. A somewhat lower enantiomeric purity (89% ee) was obtained when an immobilized version of the ligand was employed.

Other enantioselective reactions performed by microwave heating include asymmetric Heck reactions (Scheme 5.99a) [469] and ruthenium-catalyzed asymmetric hydrogen transfer processes (Scheme 5.99b) [470].

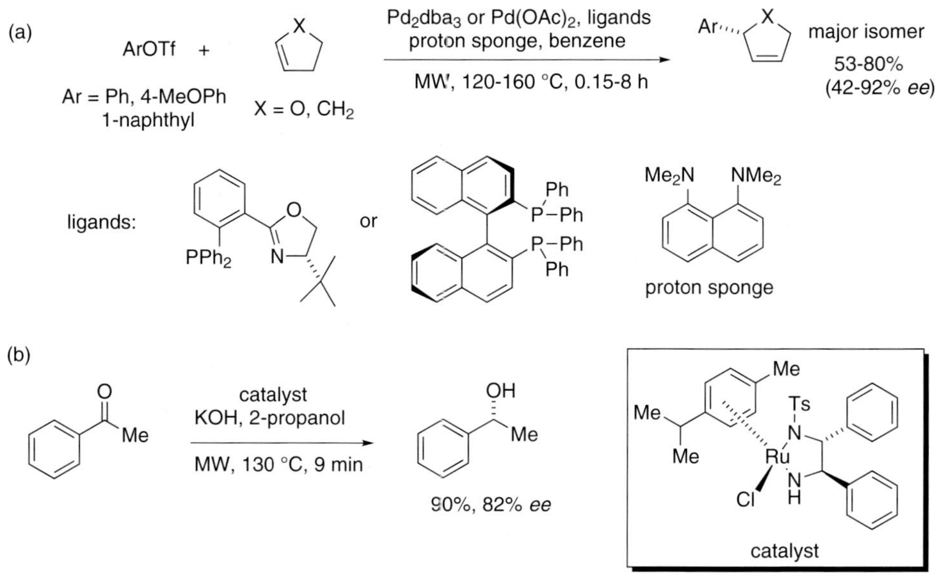

Scheme 5.99 Asymmetric Heck and hydrogen transfer reactions.

More recent examples of asymmetric Heck reactions have been reported by several authors. A library of sugar-based phosphate–oxazoline ligands for use in asymmetric Pd-catalyzed Mizoroki–Heck couplings was synthesized by Diéguez and coworkers (Scheme 5.100) [471]. Performing the Mizoroki–Heck reactions of 2,3-dihydrofurane (**42**, X = O), cyclopentene (**42**, X = CH), and 4,7-dihydro-1,3-dioxepin **43** with phenyl or cyclohexenyl triflate employing either ligand **L1** or ligand **L2** in combination with a Pd catalyst provided the corresponding coupled products in excellent conversions, with regio- (up to 98%) and enantioselectivities. Compared to conventional heating at 50 °C (where excellent conversions and regio- and enantioselectivities were also obtained), applying microwave heating at 70 °C accelerated the reactions from up to 2.5 days to only 10–45 min.

In a more recent study, the same groups have reported the use of a new class of ligands for the same asymmetric Heck reaction described in Scheme 5.100 [472].

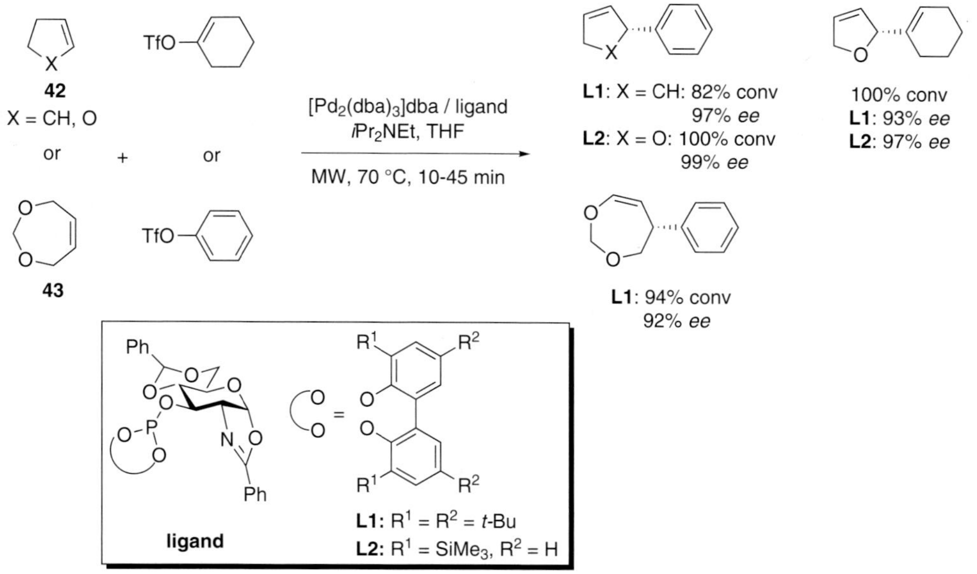

Scheme 5.100 Pd-catalyzed asymmetric Heck reactions.

The employed phosphite–oxazole/imidazole ligands (Figure 5.1) have proven increased stability, less sensitiveness against oxidation, and are easily achieved from readily available alcohols. After studying the potential of the ligands under classic thermal conditions employing $[Pd_2(dba)_3]$·dba as palladium source, the developed system has also been applied under microwave conditions (70 °C, 30 min). The general trends developed under classic thermal heating were followed, including the negative effect of the imidazole moiety (**L5**) on activity, regioselectivity, and enantioselectivity. As expected, reaction times could be significantly reduced under the sealed-vessel microwave conditions.

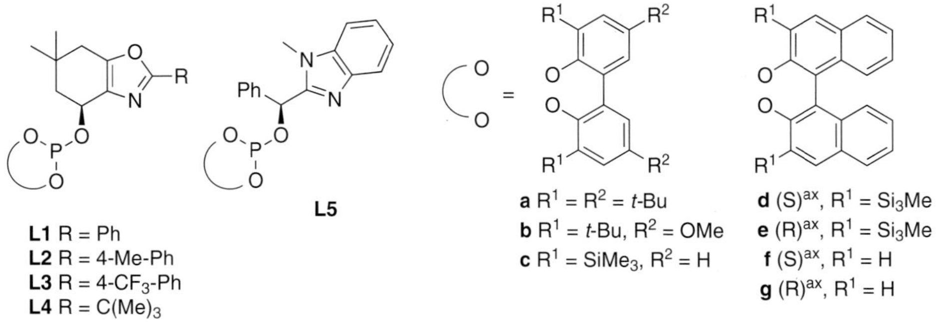

Figure 5.1 Ligands for the enantioselective Heck reaction of 2,3-dihydrofuran.

Similar investigations toward the impact of the ligand structure on the asymmetric Mizoroki–Heck coupling of 2,3-dihydrofuran with various triflates have been made by Andersson and coworkers (Scheme 5.101) [473]. With all employed phosphine-thiazole ligands **44–49**, the regioselectivity was excellent, whereas the enantioselectivity and the reaction rate were dependent on the substitution patterns of the ligands. For the coupling with aryl triflates, bulkier substituents in the 2-position of the ligands enhanced the enantioselectivity and reaction rate: ligand **49** proved to be the best regarding ee values (94–98%) and reaction time (1 h), whereas with ligand **47** low conversions (28–30%) and ee values (77–80%) were obtained. By employing microwave heating, the reactions proceed faster with constant high enantioselectivities.

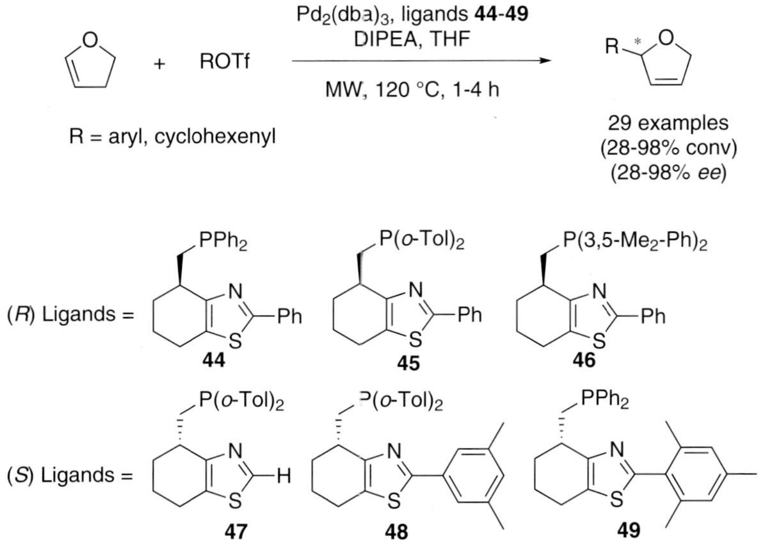

Scheme 5.101 Asymmetric Heck reactions using phosphine-thiazole ligands.

In the course of the multistep synthesis of the *Strychnos* alkaloid minfiensine, Overman and his group have conducted two reaction sequences with microwave irradiation (Scheme 5.102) [474]. As a key step, the intramolecular asymmetric Mizoroki–Heck reaction of precursor **50** was performed at 170 °C within 45 min and afforded the cyclized product **51** in high yield and with excellent 99% ee. Due to the short reaction times, the catalyst loading could be decreased from 20 to 10 mol%. The employed PHOX ligand proved to be optimal in order to prevent double bond migration. In a second-generation total synthesis of (+)-minfiensine, the hydroboration of scaffold **52** proceeded under microwave heating with 9-BBN in THF at 100 °C, whereas in refluxing THF no reaction took place. After oxidation of the crude product with TPAP/NMO, the desired ketone **53** was obtained as major product in 63% yield. (+)-Minfiensine was synthesized in 15 steps and 6.5% overall yield with the second-generation protocol.

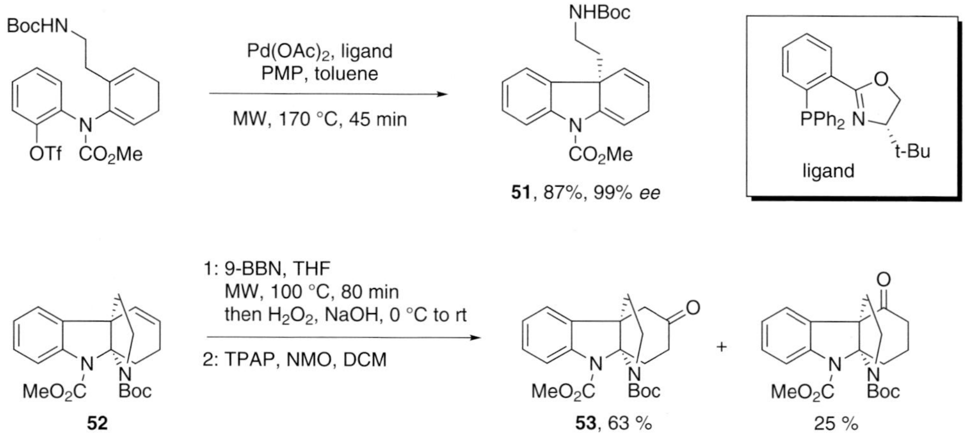

9-BBN: 9-boraybicyclo[3.3.1]nonane
TPAP: tetrapropylammonium perruthenate
NMO: *N*-methylmorpholine-*N*-oxide

Scheme 5.102 Total synthesis of (+)-minfiensine.

The use of Mizoroki–Heck chemistry as one of the key steps in a natural product synthesis has also been highlighted by Qin and coworkers [475].

Scheidt and coworkers employed an asymmetric carbonyl-ene/intramolecular Heck cyclization strategy toward the enantioselective synthesis of substituted indanones [476]. Enantioenriched carbinol precursors **54** were prepared from racemic silyloxyallenes and 2-halo aromatic aldehydes and were further converted to the desired indanones by a rapid microwave-assisted Heck coupling (Scheme 5.103). Preliminary investigations proved 1,2,2,6,6-pentamethylpiperidine (PMP) as best-suited base for complete chirality transfer. In a typical microwave procedure, carbinol in *N,N*-dimethylformamide with 1 mol% $Pd(PPh_3)_2Cl_2$ and PMP is heated at 150 °C for 25 min to furnish the indene derivatives **55** in high yields and with enantioselectivities (79–91% ee). Both electron-rich and electron-deficient substrates performed equally well under these optimized conditions.

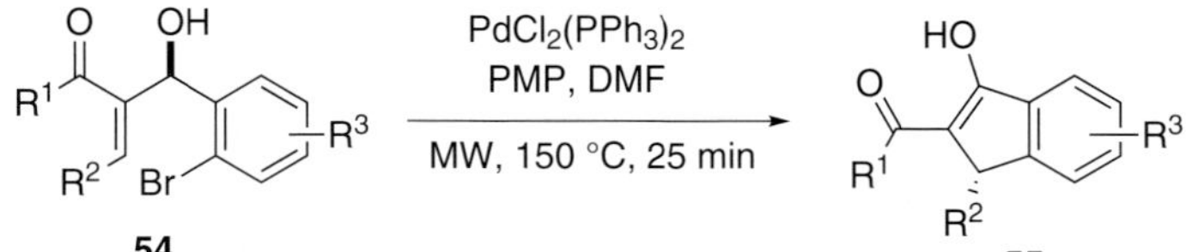

54

R^1 = Me, *n*-Pe
R^2 = aryl, alkyl
R^3 = H, 4-F, "benzodioxole"

55
7 examples
(70-92%)

Scheme 5.103 Enantioselective synthesis of substituted indanones via an asymmetric Heck cyclization.

5.2.8
Miscellaneous Carbon–Carbon Bond-Forming Reactions

A publication by Nájera and coworkers described the palladium-catalyzed acylation of terminal acetylenes with acid chlorides [477]. In general, high yields of the corresponding ynone coupling products were obtained using low loadings of an oxime-derived palladacycle as catalyst at elevated temperatures (Scheme 5.104). Utilizing microwave heating to 80 °C, a 0.2 mol% concentration of the palladacycle catalyst was sufficient to provide 88% conversion within 6 min. Under otherwise identical conditions, a 41% conversion was achieved at room temperature after 25 h (97% conversion at 110 °C for 1 h).

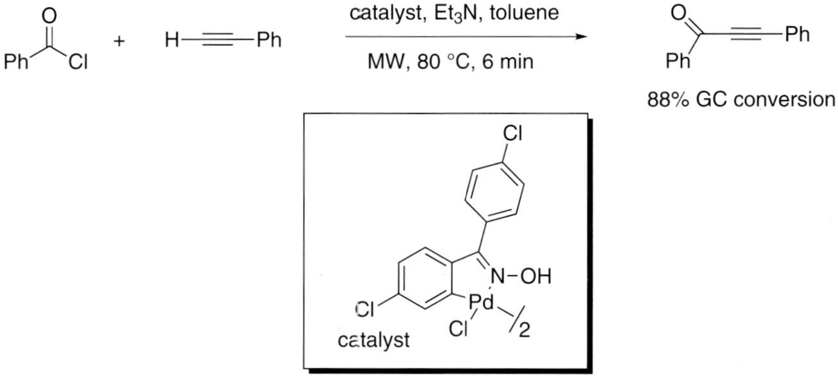

Scheme 5.104 Palladium-catalyzed acylation of terminal acetylenes with acid chlorides.

Recently, Liebeskind and Srogl developed a novel carbon–carbon cross-coupling protocol, involving the palladium(0)-catalyzed, copper(I)-mediated coupling of thioether-type species with boronic acids under neutral conditions [478]. A key feature of these protocols is the requirement of stoichiometric amounts of copper(I) carboxylate (e.g., copper(I) thiophene-2-carboxylate and CuTC) as metal cofactor. Due to the higher thiophilicity of the soft Cu(I) metal, selective sulfide coupling under Liebeskind–Srogl conditions can be performed even in the presence of a Suzuki-active bromide [478]. In this context, Lengar and Kappe have reported a microwave-assisted version of the Liebeskind–Srogl cross-coupling reaction (Scheme 5.105a) [479]. Here, rapid carbon–carbon bond forming was achieved, employing 2-methylthio-1,4-dihydropyrimidine derivative as starting material. Coupling with phenylboronic acid was performed at 130 °C under microwave conditions, providing a 84% isolated yield of the desired product within 25 min. Optimum yields were achieved using 3 mol% of tetrakis(triphenylphosphine)palladium(0) as catalyst and 2.5 equiv of copper(I) thiophene-2-carboxylate as an additive. Interestingly, it was also possible to perform carbon–carbon coupling directly with the corresponding cyclic thiourea structures using similar palladium(0)-catalyzed, copper(I)-mediated coupling conditions.

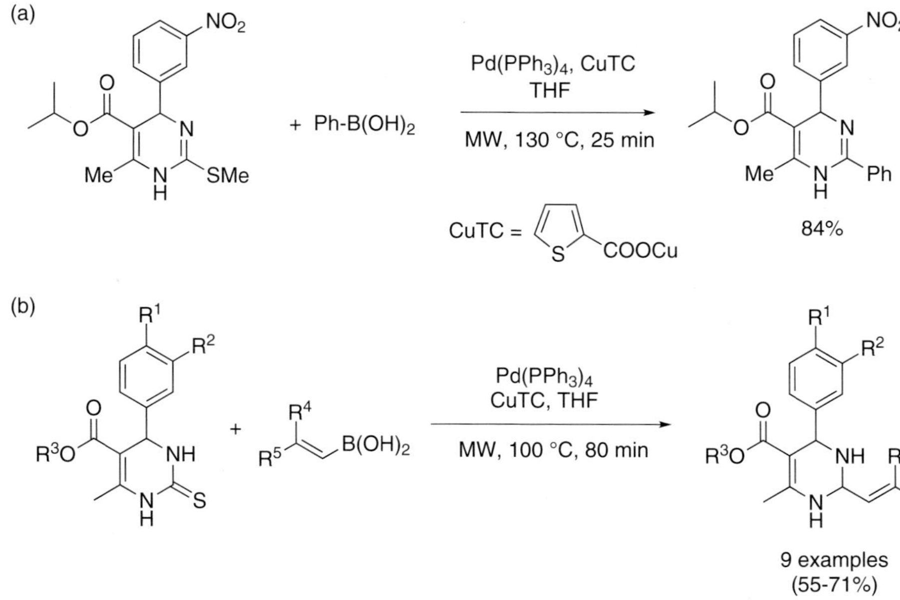

Scheme 5.105 Palladium(0)-catalyzed, copper(I)-mediated Liebeskind–Srogl-type couplings.

The methodology was used to synthesize a small focused library of 2-aryl-1,4-dihydropyrimidines that are highly potent, nonnucleosidic inhibitors of hepatitis B virus replication having *in vitro* and *in vivo* antiviral activity [479].

In a more recent report, the same group described the synthesis of a set of functionalized cyclic thioureas to generate potentially biologically active 2-alkenyl-1,4-dihydropyrimidines applying similar conditions as described above (Scheme 5.105b) [480]. The reaction could also be performed in parallel manner utilizing a corresponding multimode microwave platform to minimize the overall process time. Furthermore, when slightly modified, the protocol could be used in a Stille coupling fashion with organostannanes or in a Hiyama-type coupling using organosilicon compounds to obtain the corresponding 2-functionalized pyrimidines.

The group of Van der Eycken presented in 2011 comprehensive investigations toward the Liebeskind–Srogl cross-coupling for a variety of heterocycles (pyrazines, benzothiazoles, pyrimidines, imidazones, and tetrazoles) bearing a benzylthioether group at the C2-position in order to increase the diversity of pharmaceutically attractive heterocyclic motifs [481]. The thioethers were treated under standard Liebeskind–Srogl conditions (see above) to obtain the corresponding arylated products. After heating the reaction mixture for 30 min at 120 °C, another 1 equiv boronic acid, 5 mol% $Pd(PPh_3)_4$, and 1 equiv CuTC were added and the mixture was irradiated again for 30 min at 120 °C to afford the desired compounds in good yields (Scheme 5.106a). In case of model pyrazines, the remaining chloro-substituent is now prone to common cross-coupling reactions of the Sonogashira type. Treating the 2-chloro pyrazines with triispropylsilylacetylene (TIPSA), 5 mol%

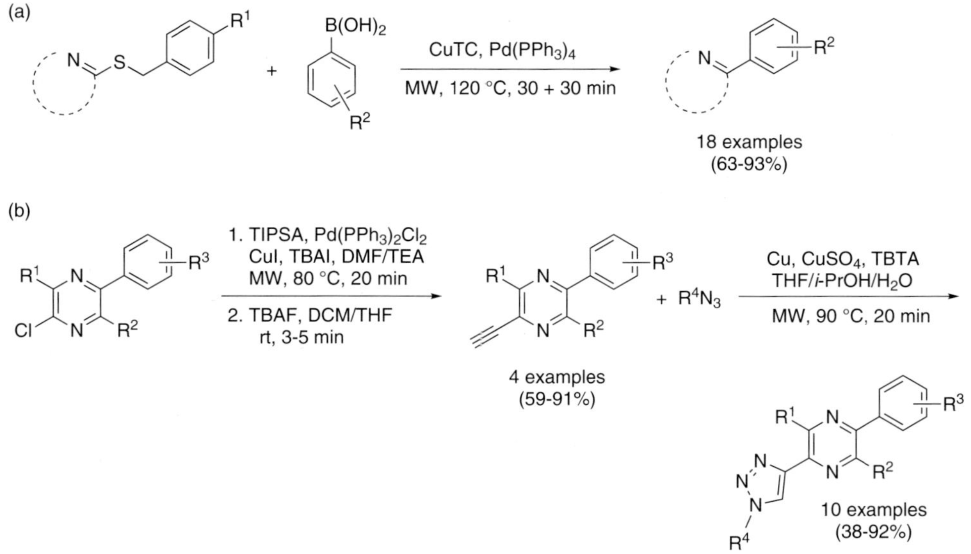

Scheme 5.106 Liebeskind–Srogl cross-coupling procedure for several heterocycles.

$Pd(PPh_3)_2Cl_2$, 10 mol% Cu(I) iodide, and tetrabutylammonium iodide (TBAI) in *N,N*-dimethylformamide/triethylamine (1 : 1) under microwave irradiation for 20 min at 80 °C yields the corresponding ethynyl derivatives (Scheme 5.106b). In the final step, several protected azides can be applied to generate the corresponding triazoles in a copper(I)-catalyzed [3 + 2] dipolar cycloaddition with full regioselectivity in good yields [481].

A method for the microwave-promoted conversion of aryl triflates to the corresponding nitriles was presented by Zhang and Neumeyer (Scheme 5.107a) [482,483]. Based on the pioneering work by Alterman and Hallberg [484], the authors have shown that 3-cyano-3-desoxy-10-ketomorphinans – that are key intermediates in the generation of κ opioid receptor-selective agonists/antagonists – can be readily prepared by palladium-catalyzed cyanation from the corresponding triflates. Optimum results were achieved employing 8 mol% of tetrakis(triphenylphosphine) palladium(0) and 2 equiv of zinc cyanide in *N,N*-dimethylformamide as solvent. After microwave heating to 200 °C for 15 min, the corresponding nitriles were obtained in excellent yields. The authors found that the presence of a keto group adjacent to the phenyl ring in the substrates is important for the efficient conversion of triflates to nitriles. A related method by Arvela and Leadbeater uses aryl bromides or chlorides as starting materials for cyanation reactions (Scheme 5.107b) [485]. The methodology reported for aryl bromides involved either the use of 0.6 equiv of nickel cyanide or a mixture of sodium cyanide (2 equiv) and nickel bromide (1 equiv). With aryl chlorides, a mix of sodium cyanide and nickel bromide was required and the reaction proceeded via *in situ* formation of the corresponding aryl bromides. Typically, complete conversions were achieved after 10 min at 200 °C. Comparable

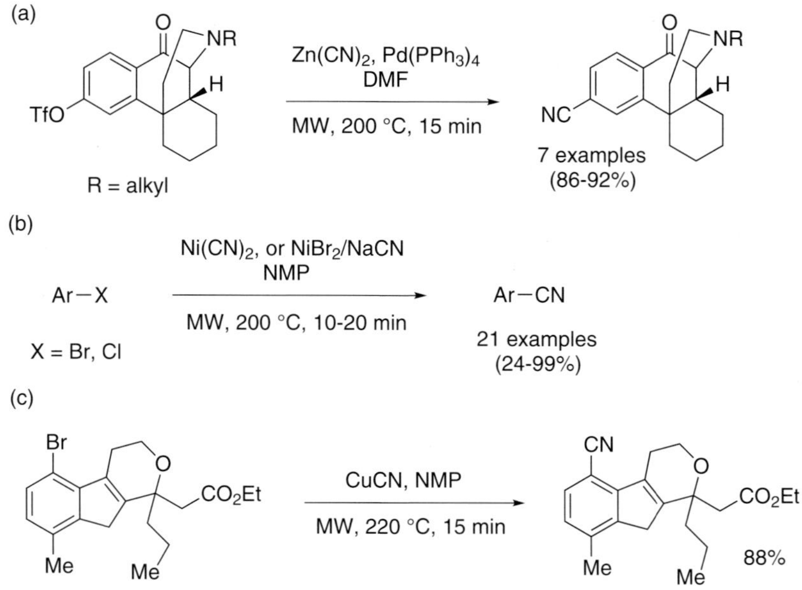

Scheme 5.107 Transition metal-mediated cyanation reactions.

results were described by Gopalsamy *et al.* using copper(I) cyanide under similar reaction conditions (Scheme 5.107c) [486].

Slightly lower temperatures were sufficient for the Cu(I)-catalyzed cyanation of aryl halides in aqueous solution to generate valuable nitriles as synthetic intermediates, reported by DeBlase and Leadbeater [487]. Utilizing simple copper(I) iodide as ligandless catalyst and potassium hexacyanoferrate ($K_4[Fe(CN)_6]$) as efficient cyanide source in aqueous media with tetraethylene glycol (TEG) as co-catalyst resulted in an environmentally benign microwave protocol. Within 30 min at 175 °C, a number of aryl iodides could be converted into the corresponding nitriles in good yields (Scheme 5.108). When 2-iodopyridine was employed, product isolation proved rather sluggish and the obtained yield was only moderate, whereas the cyanation of aryl bromides was not successful at all.

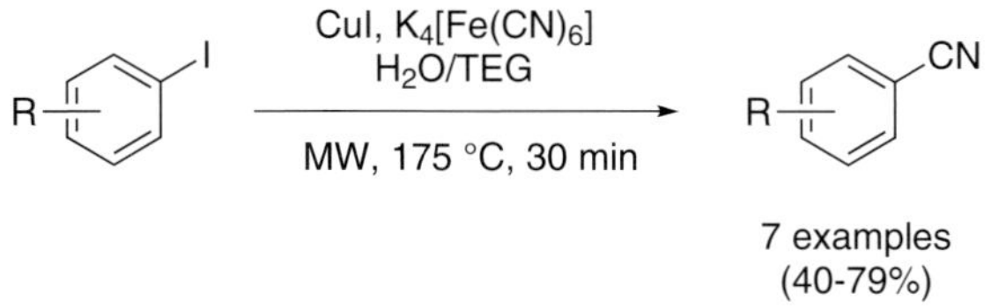

Scheme 5.108 Copper(I)-catalyzed cyanation of aryl iodides.

Similarly, Wan and coworkers also employed nontoxic $K_4[Fe(CN)_6]$ [488], whereas Pd-catalyzed cyanations of aryl bromides using $Zn(CN)_2$ have been disclosed by Pitts *et al.* [489].

Loupy and coworkers have presented a chelation-assisted rhodium(I)-catalyzed *ortho*-alkylation of aromatic imines with olefins (Scheme 5.109) [490]. The use of 2 mol% of Wilkinson's catalyst, $RhCl(PPh_3)_3$, and 5 equiv of the corresponding olefin under solvent-free conditions proved to be optimal, providing the desired *ortho*-alkylated ketones in high yields after acidic hydrolysis. Somewhat lower yields were obtained when the imine preparation and the *ortho*-alkylation were realized in a one-pot procedure.

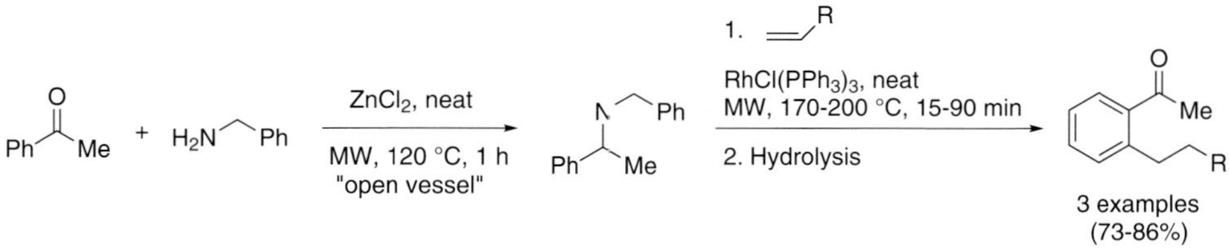

Scheme 5.109 Rhodium-catalyzed *ortho*-alkylation of ketimines.

Lipshutz *et al.* have performed several heterogeneous cross-coupling reactions catalyzed by nickel-on-charcoal (Ni/C) (Scheme 5.110) [491]. In combination with microwave heating, the C–C and C–N cross-coupling reactions ((a) Negishi, (b) Suzuki, (c) aminations, and (d) the coupling of vinyl alanes with benzylic chlorides) could be accelerated from several hours to minutes compared to conventional heating giving the products in excellent yields. A reduction of the Ni^{II}/C precatalyst with *n*-BuLi at room temperature to the active Ni^0/C prior to introduction of the coupling partners is necessary, except for the Negishi coupling. Here, a direct use of Ni^{II}/C gives the same results than with Ni^0/C.

Similar to the chemistry using Ni/C described in Scheme 5.110, carbon–carbon cross-couplings catalyzed by activated Ni(II) mounted on graphite (Ni/C_g) were also recently reported by the Lipshutz group (Scheme 5.111) [492]. In addition to the Suzuki coupling of aryl chlorides and tosylates with boronic acids, vinylalanes and vinylzirconocenes were reacted with aryl halides and tosylates. Compared to the corresponding Ni/C-catalyzed reactions, couplings proceed faster, higher yields are obtained, and cleaner reactions are afforded with Ni/C_g as catalyst. In some cases, different chemoselectivities depending on the used catalyst were obtained.

Lerebours and Wolf have reported on the Pd-phosphinous acid-catalyzed conjugate addition of (het)aryl siloxanes to α,β-unsaturated substrates in water (Scheme 5.112) [493]. Key to the success of a general method is the employment of 10 mol% of a Cu co-catalyst ($Cu(ACN)_4PF_6$) in addition to 5 mol% of the Pd catalyst POPd**1**. An advantage of the Pd-phosphinous acid catalyst POPd**1** is the air and water stability that eliminates an inert atmosphere during the reaction. With this protocol, β-substituted ketones, aldehydes, and nitroalkanes were obtained in high yields. Even otherwise difficult available β-substituted esters and nitriles can be generated smoothly in this way.

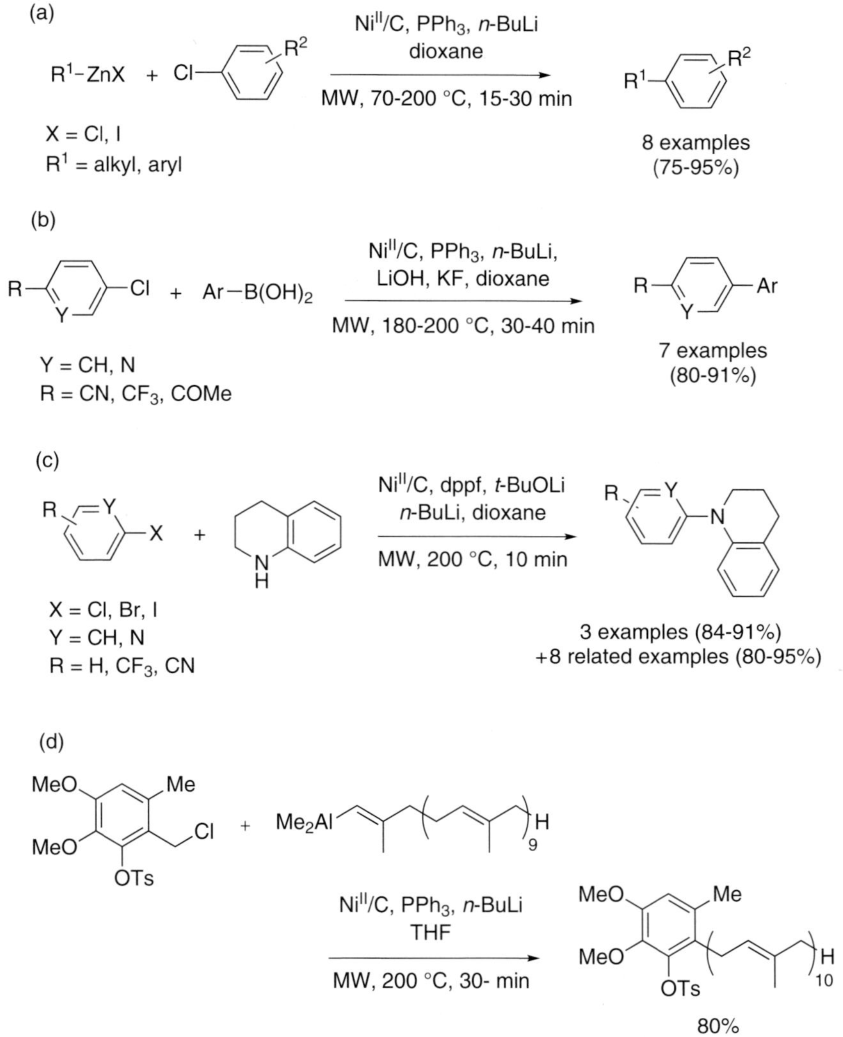

Scheme 5.110 Ni-on-charcoal-catalyzed cross-couplings.

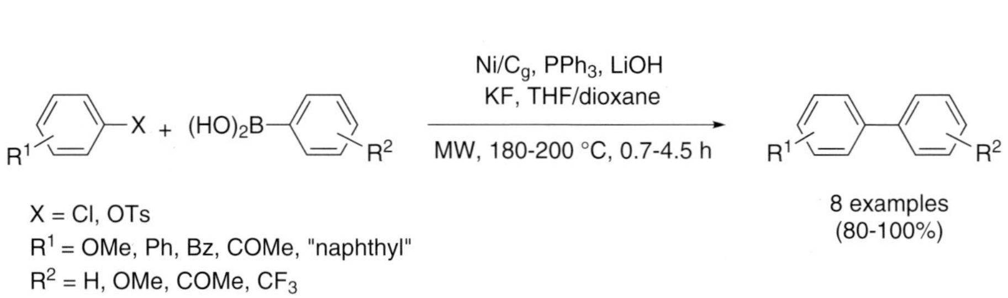

Scheme 5.111 Ni-on-graphite-catalyzed cross-couplings.

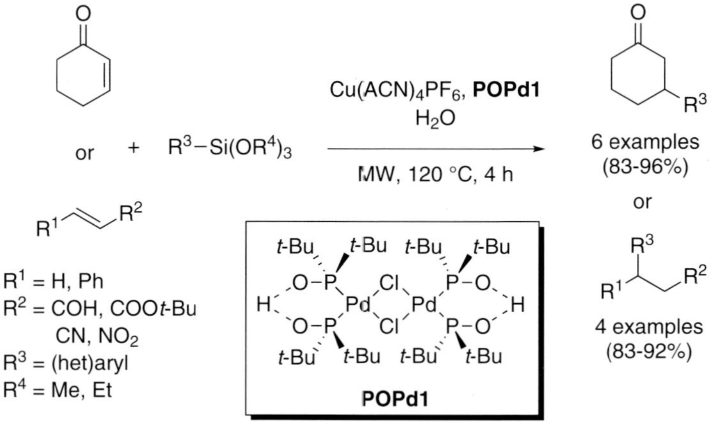

Scheme 5.112 Pd(II)-catalyzed conjugate additions.

The asymmetric synthesis of α,α′-dibenzyl esters **56** from α-benzyl acrylates and diversely substituted boronic acids was performed by Frost *et al.* (Scheme 5.113) [494]. Dibenzyl esters **56** were generally obtained with good enantioselectivities via a tandem Rh-catalyzed conjugate addition–enantioselective protonation protocol. (*S*)-BINAP as ligand and $B(OH)_3$ as proton source proved to be the best combination in terms of yields and enantioselectivities in optimization studies.

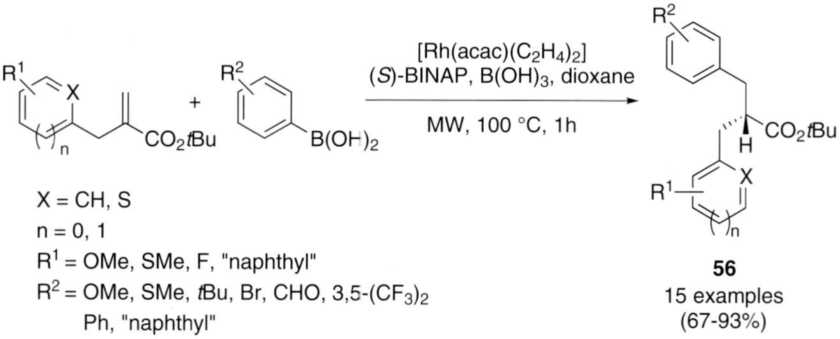

Scheme 5.113 Rh-catalyzed conjugate addition–enantioselective protonation.

The same group has also reported in the Rh-catalyzed addition of organotrialkoysilanes to α-substituted acryl esters leading to a range of useful products, including 2-alkyl succinates and α-amino acid derivatives [495]. For Rh-catalyzed additions of boronic acids to maleimides, see Ref. [496].

Comparison studies of microwave versus conventional heating in the arylation of α-ketones with aryl chlorides have been performed by Nolan and his group (Scheme 5.114) [497]. By employing the versatile *N*-heterocyclic carbene palladacycle **57** as catalyst, full conversion was reached under both microwave and conventional

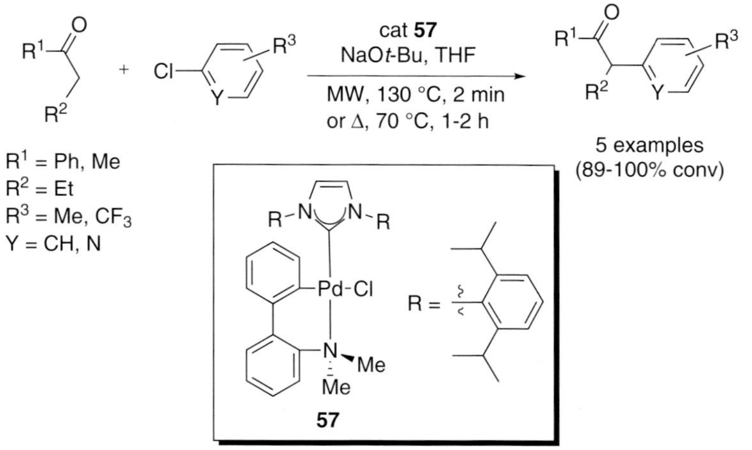

Scheme 5.114 Pd-catalyzed α-ketone arylations.

conditions. Due to the higher temperatures (130 °C versus 70 °C) that can be reached by microwave heating, the reactions proceeded up to 60 times faster.

Along the same lines, Pd-catalyzed arylations of bicyclic lactones with aryl bromides have been reported [498].

In the course of a multidimensional reaction screening of *o*-alkynyl benzaldehydes with a variety of catalysts and reactants in order to identify novel chemical transformations, Beeler *et al.* have discovered two new reaction pathways where microwave heating is employed (Scheme 5.115) [499]. In the first reaction, *o*-alkynyl benzaldehyde **58** is reacted with diethyl malonate in dichloroethane (DCE) under $Au(OAc)_3$ catalysis. Via this cycloisomerization/nucleophilic addition sequence, isochromene **59** was obtained as major product along with the ring-opened benzylidenemalonate **60**. When MeCN is exclusively used as solvent, the ring-opened product **60** is obtained in 90% yield. The second novel reaction employing microwave heating was the cycloisomerization with subsequent Friedel–Crafts addition and phenol annulation. This sequence is catalyzed by $PtCl_2$ and furnished tetracycle **61** as product.

Lautens *et al.* have reported on Pd-catalyzed intramolecular cross-coupling reactions between aryl iodides and allyl moieties (Scheme 5.116) [500]. Various five- to seven-membered carbo- and heterocycles were synthesized applying microwave irradiation at 160 °C for only 1 min. Compared to conventional thermal heating, the yields could be improved by up to 40%.

In an extension of the strategy shown in Scheme 5.116, Alberico and Lautens have investigated the synthesis of tricyclic heterocycles (**64** and **65**) starting from aryl iodide **62** that contains two tethered alkyl bromides and a Heck acceptor **63** or Zn $(CN)_2$, respectively. Products **64** are obtained via a norbornene-mediated Pd-catalyzed intramolecular bis-alkylation/intermolecular alkenylation sequence of **62** and **63** (Scheme 5.117) [501], whereas tricyclic benzonitriles **65** are furnished via tandem intramolecular bis-alkylation/intermolecular cyanation applying the same catalytic system [502].

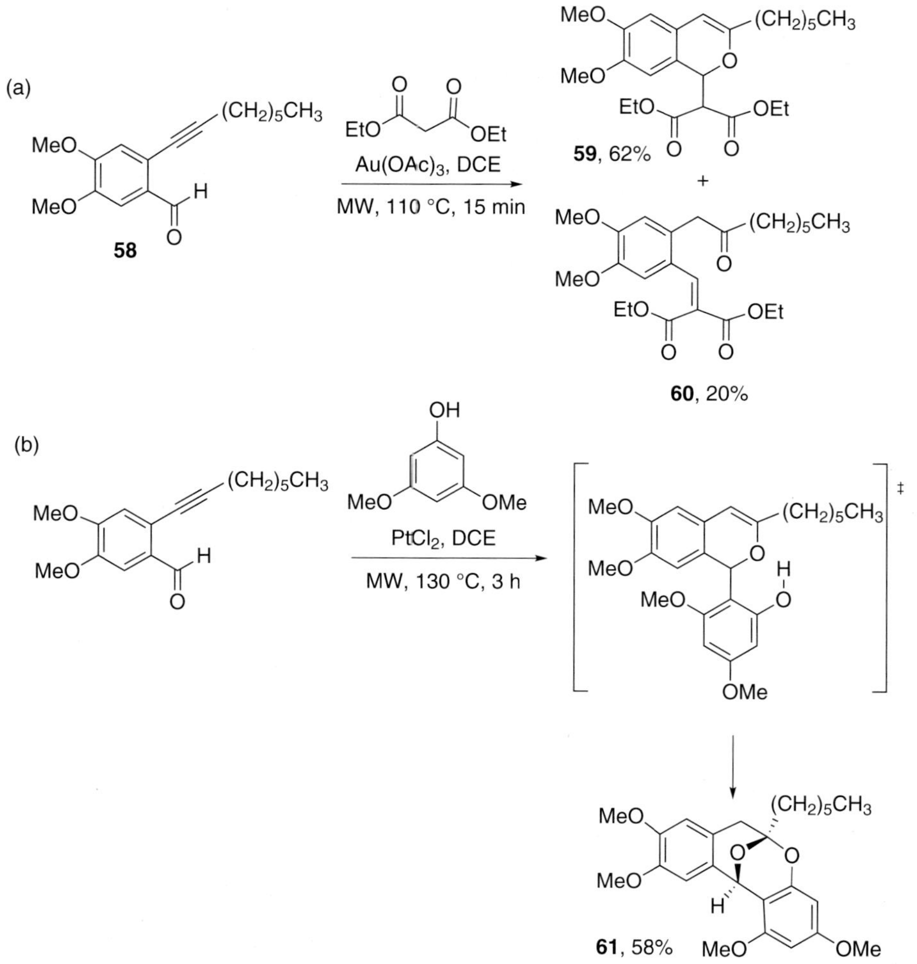

Scheme 5.115 Multidimensional reaction screening of *o*-alkynyl benzaldehydes.

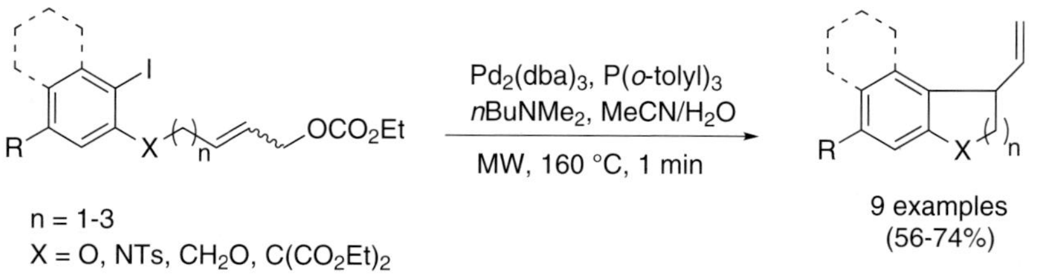

Scheme 5.116 Pd-catalyzed intramolecular cyclization.

A similar approach has been used by the same group for the generation of other tricyclic heterocyclic ring systems [503], highly substituted benzonitriles [504], and polycyclic heterocycles [505].

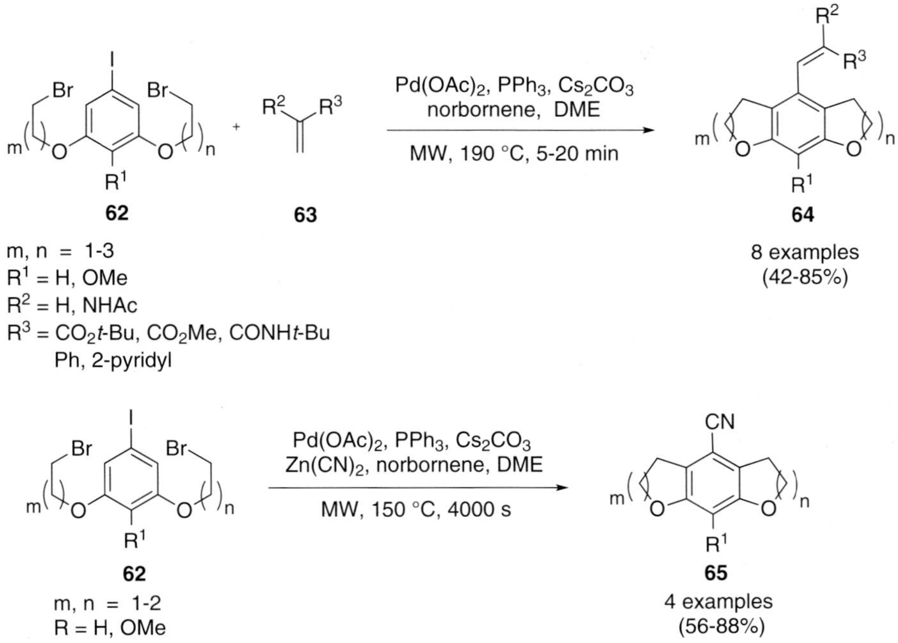

Scheme 5.117 Synthesis of tricyclic symmetrical and unsymmetrical oxo-heterocycles.

In a comprehensive 2009 report, Goossen *et al.* presented their investigations toward palladium-catalyzed decarboxylative coupling of carboxylate salts with aryl triflates (Scheme 5.118a) [506]. Utilizing aryl triflates allows coupling a broader range of carboxylate substrates to increase the diversity of products. Furthermore, the use of microwave heating also extends the scope toward nonactivated carboxylates. Thorough development of the most effective catalyst system involving Cu_2O/1,10-phenanthroline as co-catalyst revealed that the choice of phosphine was the

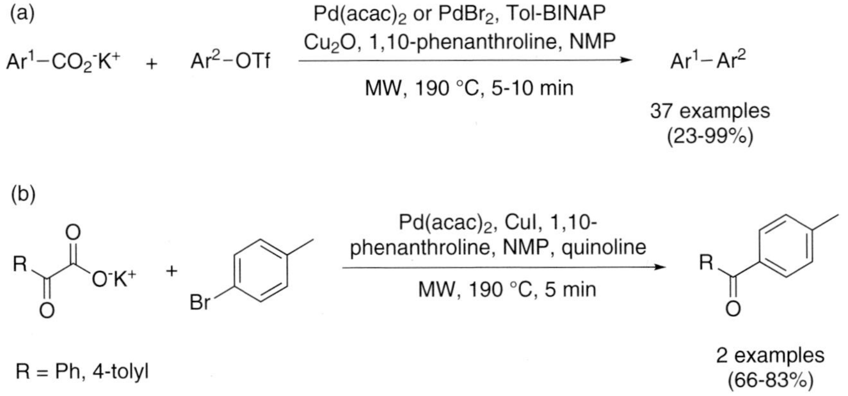

Scheme 5.118 Decarboxylative coupling of carboxylate salts with aryl triflates and halides.

crucial point in order to accomplish stabilizing coordination of the palladium, whereas the utilized palladium precursor was less critical. Interestingly, palladium (II) iodide that was best yielding precursor under classical heating was rather less effective under microwave conditions. The achieved yields could further be optimized when switching to palladium(II) bromide as precursor. Finally, as an example of nonactivated carboxylates, potassium 3-nitrobenzoate was coupled with several aryl triflates. With slightly modified conditions, satisfactory yields for the biaryl products could be achieved [506].

A related work of the same group dealt with the optimization of these decarboxylative couplings [507]. Potassium carboxylates were now coupled with aryl halides employing slightly different conditions as presented above (see Scheme 5.118b). This protocol is also feasible for aryl chloride, if di(*tert*-butyl) biphenylphosphine is added as stabilizing ligand. Even α-oxocarboxylates can be applied for the decarboxylative coupling under these microwave conditions with substantial rate acceleration.

In 2010, Larhed and coworkers extended the scope of aryl ketone synthesis by a decarboxylative addition method involving carboxylic acids and nitriles [508]. In contrast to ketone synthesis from boronic acids and halides, the use of readily available, inexpensive nitriles also allows facile coupling of alkyl groups. With the simple method involving decarboxylation of the carboxylic acid and subsequent *in situ* hydrolysis of the intermediate ketamine, five different nitriles acting as solvent/ reactant could be effectively coupled to a variety of *ortho*-substituted aryl and heteroaryl carboxylic acids. Best conditions were found to employ 10 mol% palladium(II) trifluoroacetate, 12 mol% 6-methyl-2,2′-bipyridine, H_2O, and formic acid to support hydrolyzation of sterically hindered ketimines (Scheme 5.119). Upon heating for 1 h, the desired ketones are obtained in moderate to excellent yields, thus providing a convenient method for synthesis of aryl ketones.

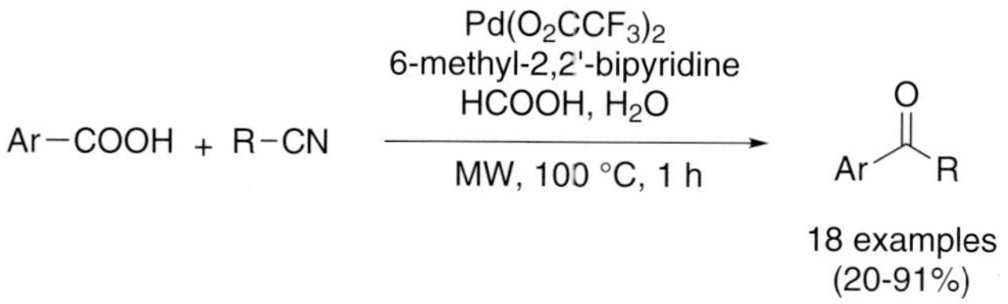

Scheme 5.119 Aryl ketone synthesis by decarboxylative addition from carboxylic acids and nitriles.

Wang and coworkers reported on the palladium-catalyzed cross-coupling of biochemically attractive diazirines with aryl halides [509]. The authors developed the optimum palladium(0) catalyst system supporting carbine migratory insertion and utilized it under microwave conditions with a series of diazirines and aryl halides to generate 21 different olefins in good yields (Scheme 5.120). In the optimized procedure, aryl halide together with aziridine, 2.5 mol% $Pd_2(dba)_3$, 10 mol% Xphos, and 1.5 equiv triethylamine in 1,2-dichloroethane are heated for 10 min at 110 °C.

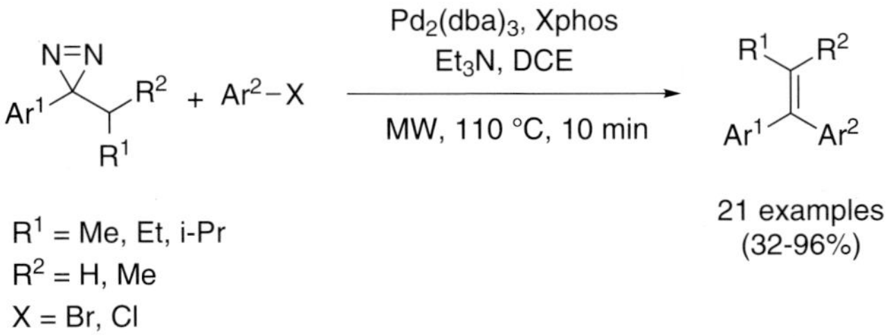

Scheme 5.120 Palladium-catalyzed cross-coupling of diazirines.

The method tolerates a variety of halides with electron-donating and electron-withdrawing groups, as well as sterically hindered derivatives. The stereoselectivity of the resulting products depended on the substrates, with *ortho*-substituted aryl bromides furnishing olefins with predominantly *Z*-isomers.

Tanner and coworkers presented comprehensive investigations toward palladium-catalyzed α-arylation of tetramic acids [510]. Among various optimizations, the authors verified the effectiveness of microwave irradiation toward the coupling of Boc-*py*PHE-OH with 4-chloroanisole (Scheme 5.121). Through a mixture of tetramic acid with 40 mol% 2-di-*tert*-butylphosphino-2′,4′,6′-triisopropylbiphenyl, 2.3 equiv potassium carbonate and 4-chloroanisole in tetrahydrofuran was bubbled with N_2 before adding 0.02 equiv palladium acetate. The vial was flushed again with nitrogen and then subjected to microwave heating. After 5 min at 110 °C, almost full conversion was achieved, whereas thermal heating for 1 h at 80 °C furnished 79% isolated yield.

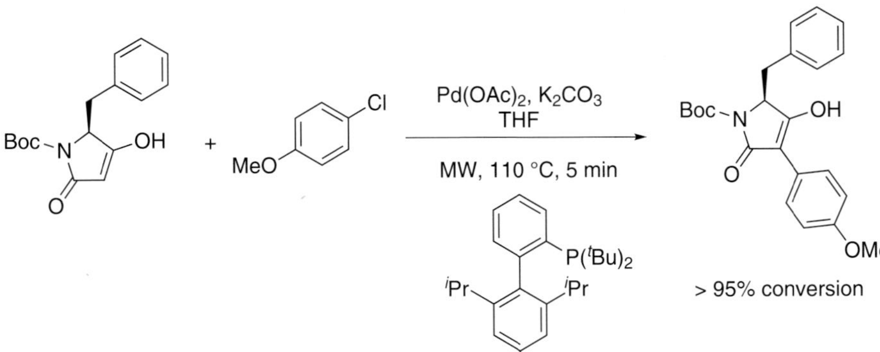

Scheme 5.121 Coupling of tetramic acid with 4-chloroanisole.

Synthetically valuable 1,6-dioxo-2,4-dienes have been obtained by Pd-catalyzed reductive dimerization from *N*-vinylpyridinium tetrafluoroborate salts [511]. For an example of a microwave-assisted Dötz benzannulation, see Ref. [512]. Pd-catalyzed 4-*exo*-dig cyclocarbopalladations have been reported as an approach to cyclobutanediols [513] and strained aromatic ring systems [514]. The

Ru-catalyzed *N*-directed *o*-arylation and heteroarylation of 2-phenylpyridine and 2-phenyloxazolidine with aryl bromides have been described by Oi *et al.* [515]. For examples of Pd-catalyzed phosphonations in the steroid series, see Ref. [516]. For a report on Pd-catalyzed allylations of aryl bromides with homoallyl alcohols via retroallylation, see Ref. [517].

5.3
Carbon–Heteroatom Bond Formations

5.3.1
Buchwald–Hartwig Reactions

The groups of Buchwald [518] and Hartwig [519] have developed a large variety of useful palladium-mediated methods for C–O and C–N bond formation. These arylations have been enormously popular in recent years and indeed the vast amount of published material available describing a wide range of palladium-catalyzed methods, ligands, solvents, temperatures, and substrates have led to a "toolbox" of tunable reaction conditions, the scope of which allows access to most target molecules that incorporate an aryl amine motif. In 2002, Alterman and coworkers described the first high-speed Buchwald–Hartwig aminations using controlled microwave heating (Scheme 5.122) [520,521]. Using *N,N*-dimethylformamide as solvent under noninert conditions, best results were obtained by employing 5 mol% of palladium acetate as precatalyst and 2,2′-bis(diphenylphosphino)-1,1′-binaphthyl (BINAP) as ligand. The procedure proved to be quite general and provided moderate to high yields for both electron-rich and electron-poor aryl bromides within the short time frame of a microwave-heated reaction.

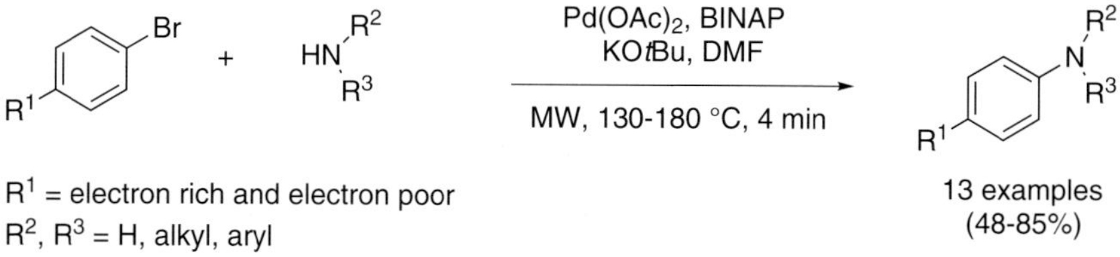

Scheme 5.122 Palladium-catalyzed amination reactions (Buchwald–Hartwig).

Following the pioneering work by Alterman, several microwave-assisted palladium-catalyzed aminations have been reported on a number of different substrates using different types of palladium sources and ligands. The examples shown in Scheme 5.123 include bromoquinolines (a) [522], aryl triflates (b) [523], and intramolecular aminations (c) for the synthesis of benzimidazoles [524]. In all cases, the use of microwave irradiation dramatically reduced the required reaction times and in many cases also improved the yields. Several authors also found that

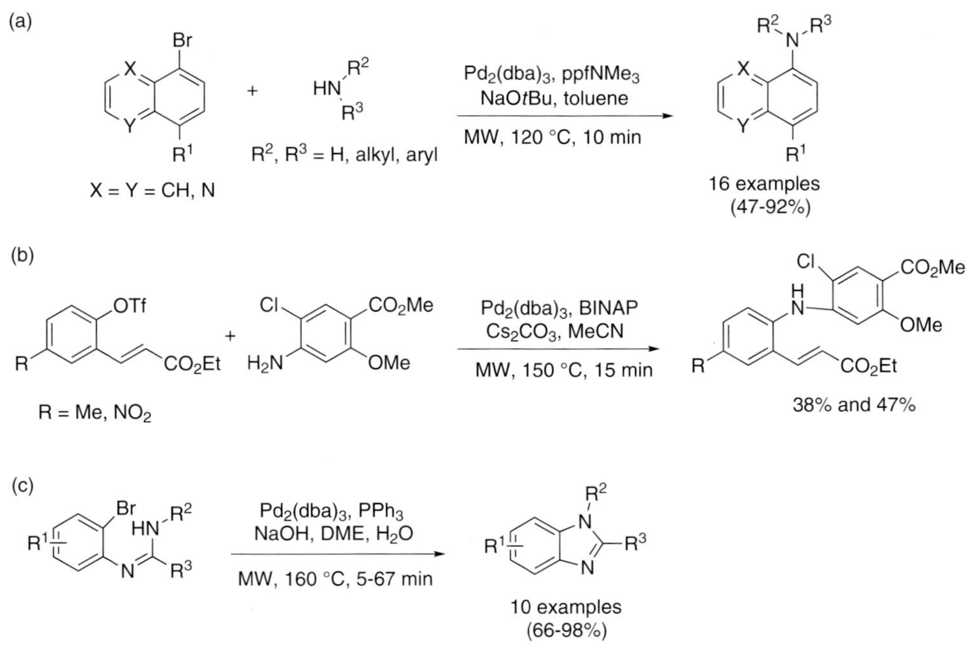

Scheme 5.123 Inter- and intramolecular palladium-catalyzed amination reactions of aryl bromides and triflates.

the microwave-driven reaction required significantly less catalyst than the conventionally heated reactions [524].

An extensive and detailed optimization study of microwave-assisted Buchwald–Hartwig reactions was disclosed by Skjaerbaek and coworkers in the context of elaborating an efficient protocol for the synthesis of aryl aminobenzophenone p38 MAP kinase inhibitors (Scheme 5.124) [525]. Several different strategies involving halide, triflate, and tosylate leaving groups were investigated, in addition to an alternative amination mode (method B). Among the many ligands screened for the palladium-catalyzed amination, the Xphos ligand system (2 mol% palladium acetate, 4 mol% ligand **66**) provided the highest product yields and cleanest reaction profiles. As a base, both sodium *tert*-butoxide and cesium carbonate worked equally well in a 5 : 1 toluene/*tert*-butyl alcohol solvent mixture. Amination of an electronically diverse array of aryl halides with a variety of anilines was realized in good to excellent yields for most cases, without the necessity to work under an inert atmosphere. Depending on the structure and reactivity of the coupling partners, reaction times of 3–30 min at 120–160 °C were required to achieve full conversion.

Buchwald and his group have described a fast protocol for the coupling of (het)aryl nonaflates and amines (Scheme 5.125) [526]. An excellent functional group tolerance was observed for the synthesis of diverse arylamines using the catalyst systems $Pd_2(dba)_3$ with ligands **66**, **67**, or **68**, respectively, and the soluble amine bases DBU and MTBD (7-methyl-1,5,7-triazabicyclo[4.4.0]dec-5-ene).

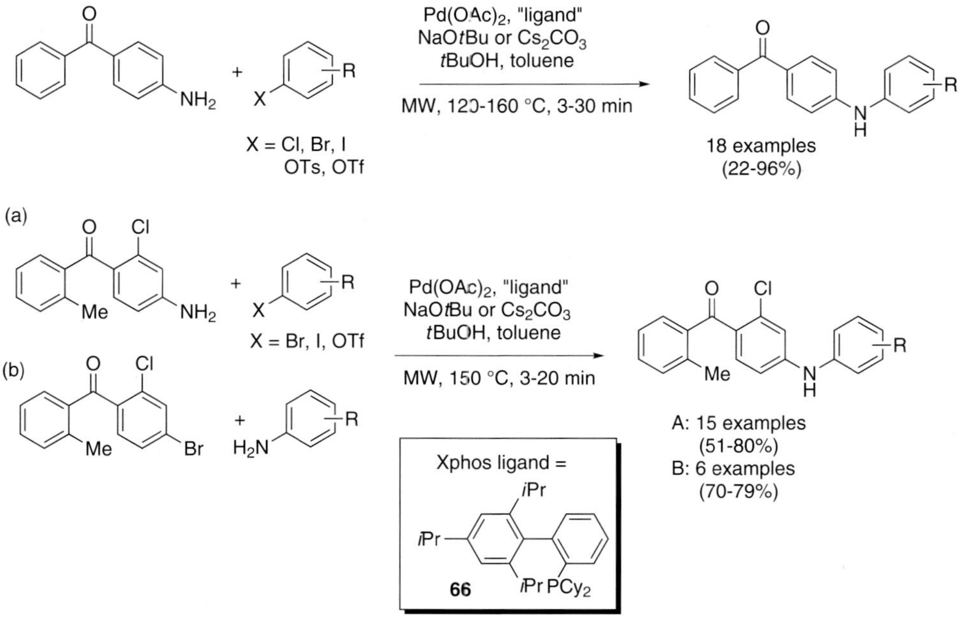

Scheme 5.124 Synthesis of aryl aminobenzophenones using palladium-catalyzed aminations.

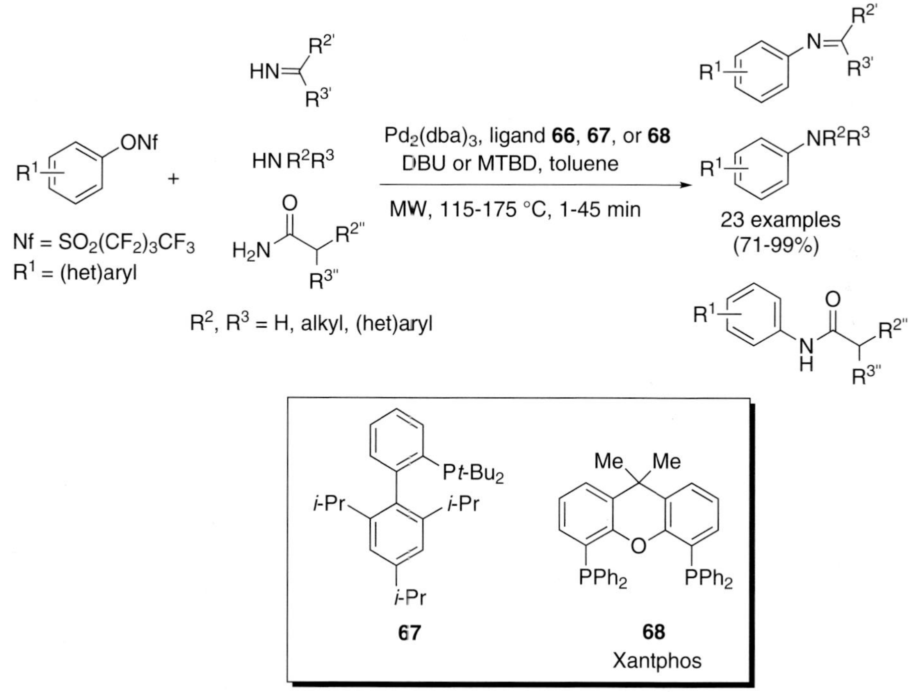

Scheme 5.125 Pd-catalyzed amination of aryl nonaflates.

Zhang *et al.* have shown that 2-chloro pyrimidines **69** can be converted to the corresponding 2-aryl or 2-heteroaryl pyrimidines **70**, respectively, via Pd-mediated aminations (Scheme 5.126) [527]. The reactions were also conducted under both Pd-free conventional and microwave heating, furnishing the 2-aryl pyrimidines in similar yields as under Pd-catalyzed microwave conditions. However, when 2-heteroaryl amines are reacted with **69**, only the Pd-catalyzed microwave protocol afforded the desired products. 2-Bromo pyrimidines were also found to react under these conditions.

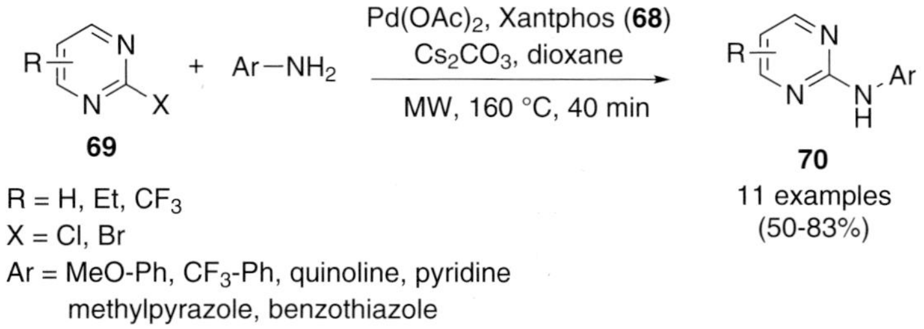

Scheme 5.126 Pd-catalyzed synthesis of 2-arylaminopyrimidines.

Utilizing more reactive discrete palladium-*N*-heterocyclic carbene (NHC) complexes (e.g., $Pd(carb)_2$) or *in situ* generated palladium/imidazolium salt complexes (1 mol% ligand A), Caddick and coworkers were able to extend the rapid amination protocols described above also to electron-rich aryl chlorides (Scheme 5.127) [528].

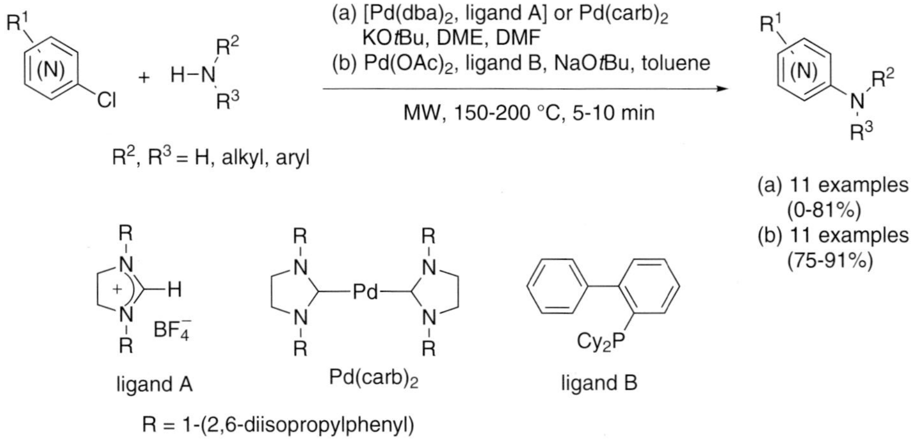

Scheme 5.127 Buchwald–Hartwig amination reactions of aryl chlorides.

In a related work, Nolan and coworkers have reported on Buchwald–Hartwig aminations of unactivated aryl chlorides with amines using NHC containing *N-C*-palladacycles as catalysts [529].

Independent investigations by Maes *et al.* have described the use of commercially available and air-stable 2-(dicyclohexylphosphanyl)biphenyl (ligand B) as ligand system for the successful and rapid coupling of (hetero)aryl chlorides with amines under microwave Buchwald–Hartwig conditions (0.5–2 mol% palladium catalyst) [530,531]. Both methods provide very high yields of products within 10 min of irradiation time.

The same group has developed a protocol that allows selective hydrolysis of nitriles to amides or avoids hydrolysis of esters and nitriles by employing a similar system as described in Scheme 5.127b under microwave conditions (Scheme 5.128) [532]. Selective hydrolysis can be achieved by the use of a phase-transfer catalyst, whereas no hydrolysis occurs in the absence of a PTC under otherwise identical conditions. By employing the reaction conditions discovered in this study for the Pd-catalyzed amination of aryl chlorides bearing sensitive functional groups such as nitriles or esters with aliphatic amines, successful coupling to the corresponding products without the occurrence of hydrolysis could be achieved. When primary amines are employed, JohnPhos proved to be a superior ligand. Anilines can also be coupled, although the yields are rather low. Heteroaromatics are also suitable substrates for this coupling protocol.

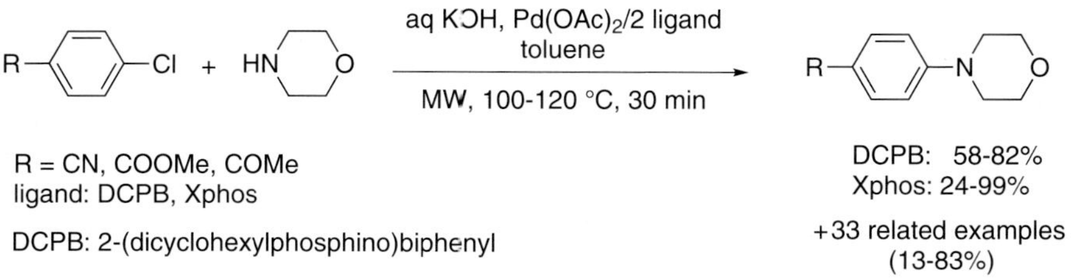

Scheme 5.128 Pd-catalyzed amination of aryl chlorides.

A method to generate *N*-alkylated tacrine (**71**) and huprin (**72**) compounds from chloroquinolones was recently presented by Renard and coworkers [533]. The interest in these compounds is due to their reversible acetylcholinesterase inhibition efficiency. Employing 4 mol% $Pd_2(dba)_3$ and 8 mol% *rac*-BINAP as catalyst/ligand system and microwave heating at 150 °C for 30 min furnished the desired products in good to high yields (Scheme 5.129a). The protocol tolerates various primary and secondary amines as well as diamines. However, the amination turned out to show low selectivity when several chloride functionalities are present in the substrate. The efficiency of the method could be demonstrated to prepare two heterodimeric AChE inhibitors (**73**) in a two-step process starting from 9-chlorotetrahydroacridine and 1,7-diaminoheptane by simply expanding the heating time to 2 h per step (Scheme 5.129b).

An extensive optimization study toward the synthesis of 6-heterocyclic substituted 2-aminoquinolines via Buchwald–Hartwig amination of 6-bromo-2-chloroquinoline with cyclic amines was conducted by the group of Pyke (Scheme 5.130) [534]. Crucial for the selective amination at the 6-position is the choice of the solvent. By changing

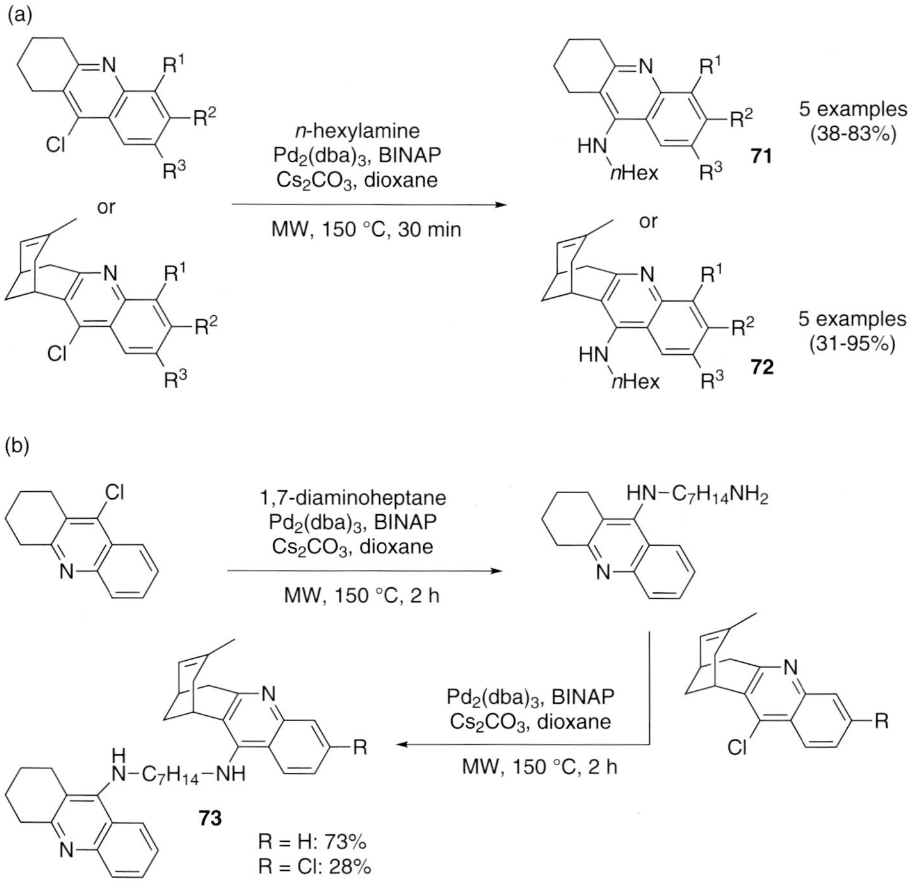

Scheme 5.129 Synthesis of heterodimeric AChE inhibitors.

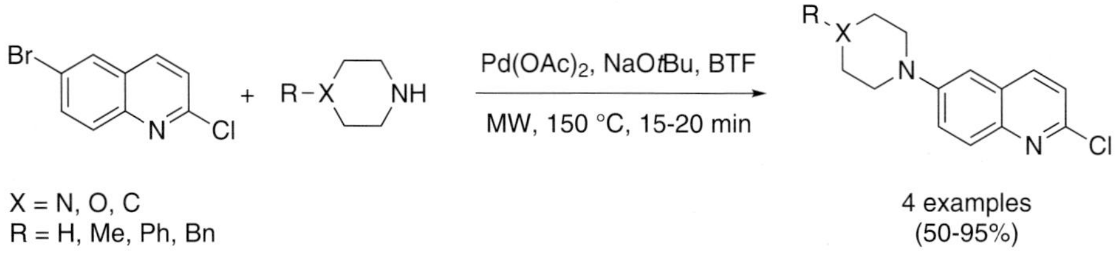

Scheme 5.130 Selective Buchwald–Hartwig aminations.

from toluene to benzotrifluoride (BTF), not only the reaction can be performed in the microwave at 150 °C due to the better microwave absorbance characteristics of BTF compared to toluene (see also Section 4.8) but also higher product yields were obtained. Under thermal heating, BTF also proved to be superior to toluene with respect to product yields.

An application of Buchwald–Hartwig aminations toward the synthesis of *N*-arylsulfonamides is highlighted in Scheme 5.131a. *N*-Arylsulfonamides constitute an important class of therapeutic agents in medicinal chemistry. Over 30 drugs containing this moiety are in clinical use in the areas of antibacterials, nonnucleosidic reverse transcriptase inhibitors, antitumor agents, and HIV-1 protease inhibitors. A group from GlaxoSmithKline has demonstrated that *N*-arylsulfonamides can be readily obtained by palladium-catalyzed intermolecular coupling of heteroaryl chlorides (e.g., 4-chloroquinoline) with sulfonamides under microwave conditions [535]. The reactions proceeded at 180 °C with 2–10 mol% of palladium catalyst in the presence of a hemilabile *N,P* ligand. Similarly, Harmata *et al.* have disclosed an efficient protocol for the palladium-catalyzed *N*-arylation of enantiopure sulfoximines with aryl chlorides (Scheme 5.131b) [536]. Optimal results were achieved by using palladium acetate as palladium source and *rac*-2,2′-bis(diphenylphosphino)-1,1′-binaphthyl (BINAP) or tri-(*tert*-butyl)phosphine as ligands under microwave irradiation conditions. For aryl chlorides bearing *ortho*-carbonyl substituents, the corresponding benzothiazines were obtained.

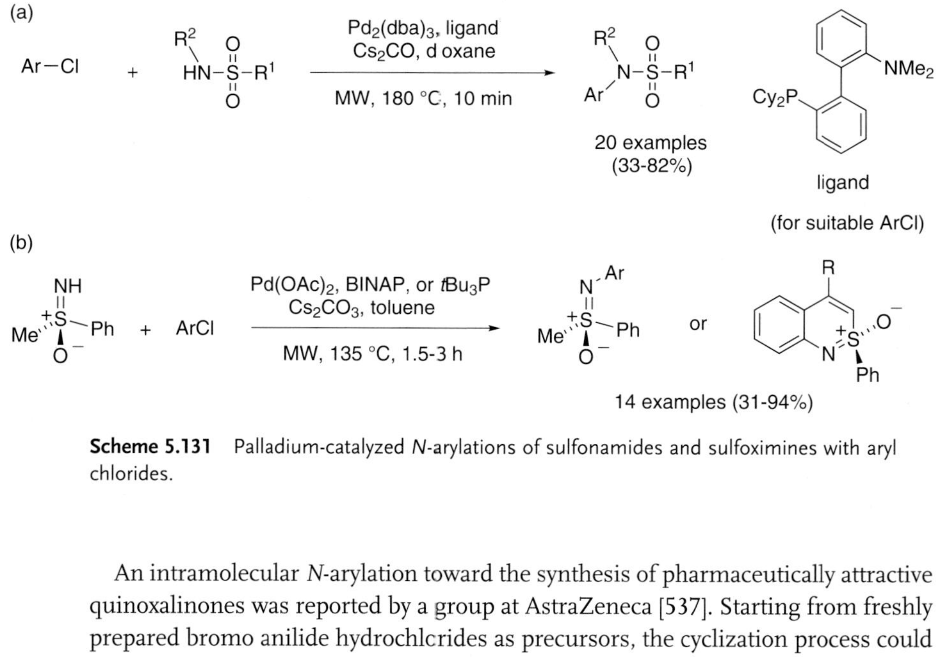

Scheme 5.131 Palladium-catalyzed *N*-arylations of sulfonamides and sulfoximines with aryl chlorides.

An intramolecular *N*-arylation toward the synthesis of pharmaceutically attractive quinoxalinones was reported by a group at AstraZeneca [537]. Starting from freshly prepared bromo anilide hydrochlorides as precursors, the cyclization process could be initiated by microwave-mediated Buchwald–Hartwig conditions. The substrate was admixed in dioxane with 1 mol% $Pd_2(dba)_3$ and 2 mol% ligand **74** as best-suited catalyst system together with potassium *tert*-butoxide as strong base and heated at 160 °C for 10 min to achieve full conversion (Scheme 5.132). The generality of this method has been explored with several aryl-substituted anilides to demonstrate that neutral and electron-donating groups always gave high yields, whereas electron-withdrawing groups only on position 7 afforded quantitative conversions.

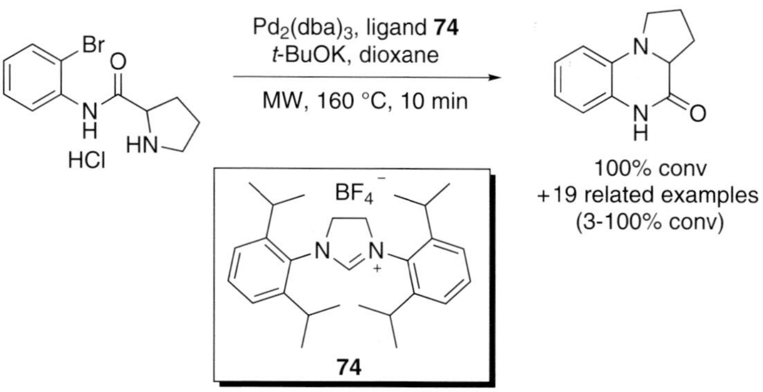

Scheme 5.132 Quinoxalinone synthesis via Pd-catalyzed intramolecular *N*-arylation.

The protocol proved to work also for bicyclic products and employing various cyclic amine precursors successfully furnishes even polycyclic quinoxalinones.

In a 2009 report, Zhu and coworkers discussed the palladium-catalyzed conversion of linear amides into bicyclic heterocycles [538]. Depending on the ligand used, the authors either observed 3,4-dihydroquinoxalin-3-ones **76** or 2-(2-oxoindolin-1-yl) acetamides **77** (Scheme 5.133). Due to their significant biological activities, quinoxalinones **76** are considered as privileged scaffolds in medicinal chemistry, thus the synthesis of these compounds was the focus of the present work. Best reaction conditions employ heating of Ugi adducts **75** for 2 h at 150 °C in dioxane/acetonitrile with 2 equiv cesium carbonate and 5 mol% $Pd(dba)_2$. Using 5 mol% Xphos as a ligand furnished the desired quinoxalinones **76** in mostly high yields. Switching to 5 mol% *rac*-BINAP completely reversed the cyclization selectivity to afford the oxindoles **77** as a mixture of diastereomers. With these results it was demonstrated that ligands can not only influence the catalyst efficiency but also lead to a switch in the reaction manifold when multifunctional substrates are utilized.

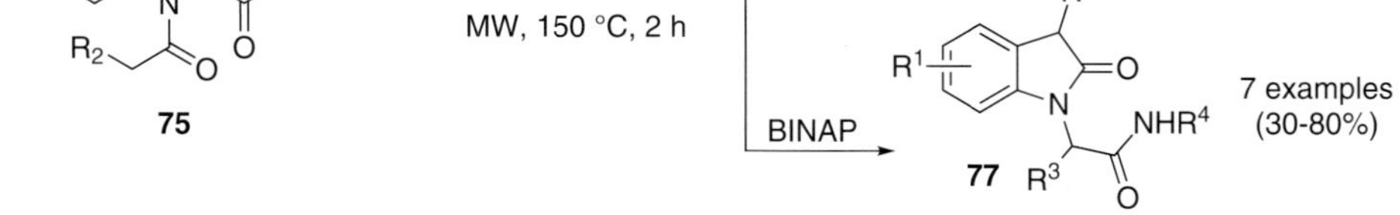

Scheme 5.133 Influence of ligands in the cyclization of linear amides.

A similar strategy has been reported by the same group earlier [539]. Closely related Pd-catalyzed intramolecular amination sequences for the synthesis of oxindoles have been studied by Turner [540].

The Pd-catalyzed *N*-arylation of *o*-nitroaryl bromides with anilines was described by Beifuss and coworkers in 2005 [541]. Pyrazino[1,2-*a*]benzimidazol-1(2*H*)-ones have been obtained in a similar way [542]. A recent report by Liu suggested that *N*-arylations of aryl halides with amines can also be carried out using inexpensive Fe_2O_3 as a catalyst in the presence of L-proline [543].

5.3.2
Ullmann Condensation Reactions

A survey of the literature on the Ullmann and related condensation reactions has highlighted the growing importance and popularity of copper-mediated carbon–nitrogen, carbon–oxygen, and carbon–sulfur bond-forming protocols [544]. In Scheme 5.134, two examples of microwave-assisted Ullmann-type condensations from a group of researchers at Bristol–Myers Squibb are shown. In the first example, (*S*)-1-(3-bromophenyl)-ethylamine was coupled with 11 N–H containing heteroarenes in the presence of 10 mol% of copper(I) iodide and 2 equiv of potassium carbonate base [545]. The comparatively high reaction temperature (1-methyl-2-pyrrolidone, 195 °C) and the long reaction times are noteworthy. For the coupling of 3,5-dimethylpyrazole, for example, microwave heating for 22 h was required to afford a 49% isolated yield of product! The average reaction times were 2–3 h. In the second example, similar conditions were chosen to react mostly aromatic thiols with aryl bromides and iodides to afford aryl sulfides [546]. The same authors have also described the synthesis of diaryl ethers by copper-catalyzed arylation of phenols with aryl halides [547].

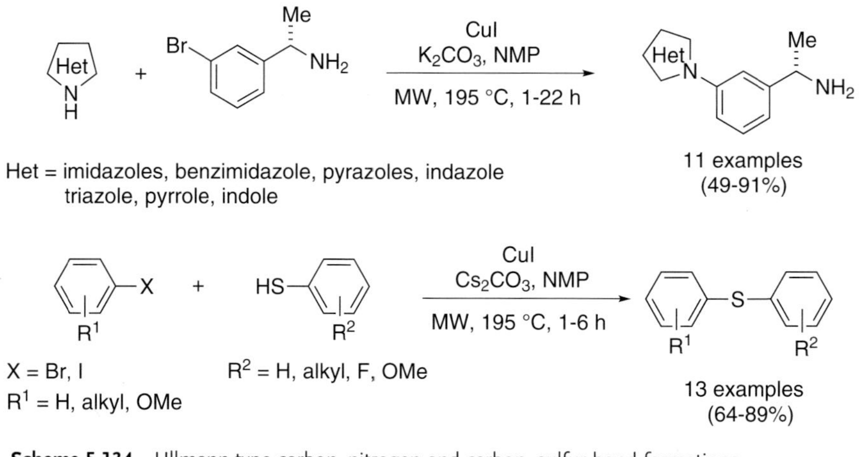

Scheme 5.134 Ullmann-type carbon–nitrogen and carbon–sulfur bond formations.

The amination of various functionalized heterocyclic bromides with aliphatic primary and cyclic amines and pyrazole was disclosed by researchers from Abbott Laboratories (Scheme 5.135) [548]. Optimization studies have shown that high conversions to the desired product could be achieved with the catalyst system CuI/proline that required a much lower temperature of 140 °C compared to the conditions described in Scheme 5.134 and proved to be very general in the further process. However, a rather high catalyst (20%) and ligand (40%) loading was necessary to achieve the aminated products in moderate to good yields.

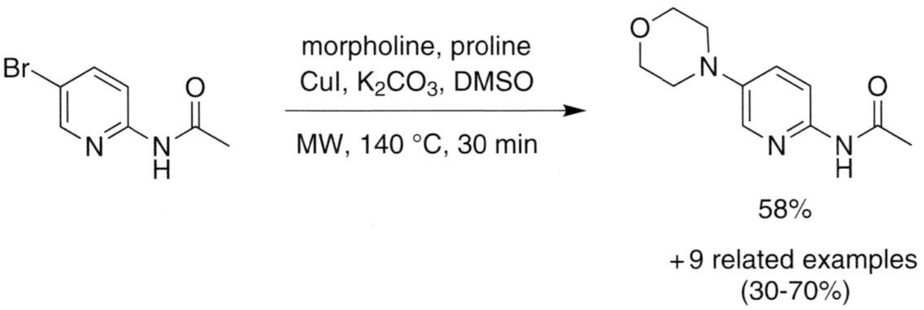

Scheme 5.135 Carbon–nitrogen bond formations employing a CuI/proline system.

The Larhed group has shown that using microwave irradiation, free and protected amino acids can be *N*-arylated by aryl bromides in the presence of CuI using water as solvent [549].

Rolfe and Hanson developed a sequential one-pot protocol for the synthesis of benzothiadiazine-3-one-1,1-dioxides utilizing a microwave-mediated copper-catalyzed *N*-arylation strategy [550]. The sulfonamides were reacted with three different amines using copper(I) iodide/1,10-phenanthroline as catalyst/ligand system and DMF as solvent. After the first heating cycle, triethylamine and carbonyl diimidazole (CDI) were added directly through the septum of the vial. The mixture was heated for another 11 min at 150 °C to furnish the desired product in good overall yield (Scheme 5.136). In particular, DMF as solvent proved to be essential for an effective cyclization step.

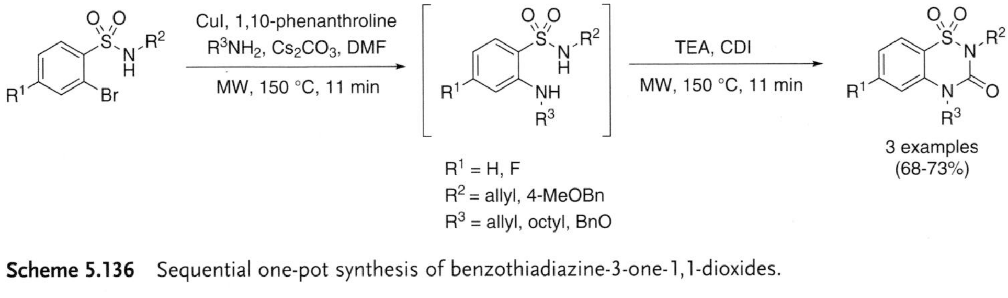

Scheme 5.136 Sequential one-pot synthesis of benzothiadiazine-3-one-1,1-dioxides.

A reaction related to the Ullmann condensation is the Goldberg reaction, that is, the copper-catalyzed amidation of aryl halides. Due to the usually required drastic reaction conditions, the Goldberg reaction has not been recognized as a powerful synthetic methodology in organic synthesis. A team from Solvay Pharmaceuticals has reported that Goldberg reactions can be carried out efficiently under microwave irradiation conditions, applying 10 mol% of copper(I) iodide as a catalyst and small amounts (2 molar equiv) of 1-methyl-2-pyrrolidone as a solvent (Scheme 5.137) [551]. Apart from simple acetamides, the reaction could be extended to cyclic amides, such as 3,4-dihydro-2-quinolones and piperazin-2-ones. The reactions were carried out on a comparatively large scale (25 mmol) under open-vessel conditions under a gentle stream of nitrogen. Yields were in the range of 48–77%.

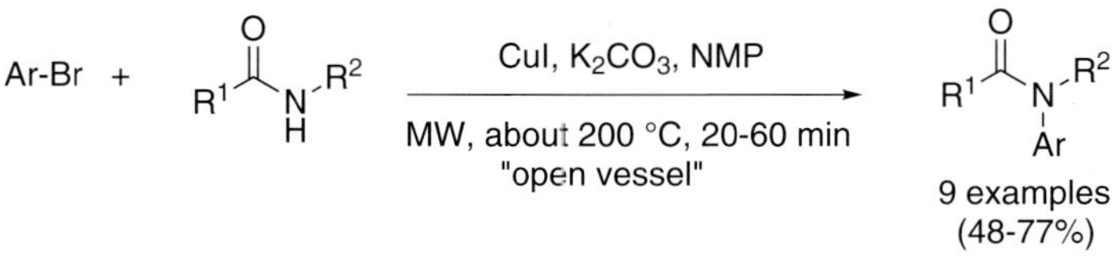

Scheme 5.137 N-Arylation of amides via Goldberg reactions.

Milder conditions can be applied when the Goldberg reaction is performed employing a suitable ligand. A set of asymmetrically substituted *N,N′*-diarylimidazolinones was synthesized by Hafner and Kunz via Cu-catalyzed *N*-arylation of arylimidazolinones with aryl iodides or bromides using a combination of $(CuOTf)_2{\cdot}C_6H_6$ and 1,2-*trans*-diaminocyclohexane (Scheme 5.138) [552]. Similar yields can be achieved by conventional heating at lower temperatures (120 °C), however, a longer reaction time of 15 h has to be applied. The synthesis of symmetrically substituted *N,N′*-diarylimidazolinones via double *N*-arylation was less successful, since substrate decomposition of imidazolinone occurred as a side reaction. Thus, only electron-poor aryl halides can be employed that furnish the products in moderate yield.

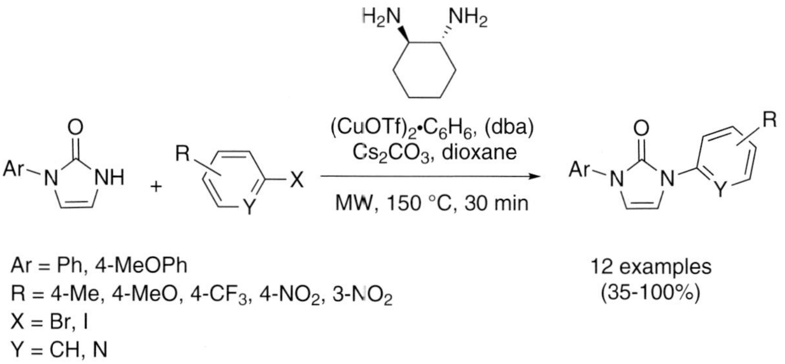

Scheme 5.138 Synthesis of unsymmetrically substituted *N,N′*-diarylimidazolinones.

Recently, interest in copper-catalyzed carbon–heteroatom bond-forming reactions has shifted to the use of boronic acids as reactive coupling partners [544]. One example of carbon–sulfur bond formation is displayed in Scheme 5.139. Lengar and Kappe have reported that in contrast to the palladium(0)/copper(I)-mediated process described in Scheme 5.105 leading to carbon–carbon bond formation, reaction of the same starting materials in the presence of 1 equiv of copper(II) acetate and 2 equiv of phenanthroline ligand furnished the corresponding carbon–sulfur cross-coupled product [479]. Although the reaction at room temperature needed 4 days to reach completion, 45 min of microwave irradiation at 85 °C in 1,2-dichloroethane provided 72% isolated yield of the product.

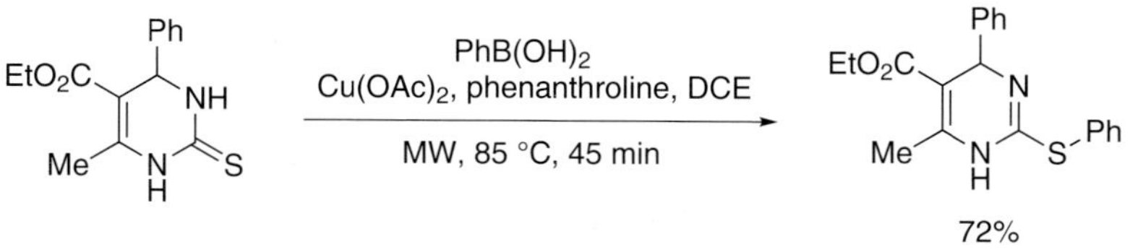

Scheme 5.139 Copper(II)-mediated carbon–sulfur cross-coupling.

Similarly, Liu and coworkers have reported on the Cu-mediated *N*-arylation of amines with boronic acids (Scheme 5.140) [553]. The coupling of various amines with differently substituted boronic acids under mild conditions and in the absence of a Pd catalyst led to a small library of 21 arylated products. Importantly, in addition to anilines, this protocol is applicable to primary and secondary alkyl amines and nitrogen-containing heterocycles where a reduced reaction time of 15 min is sufficient to obtain the coupled products in 48–96% yield.

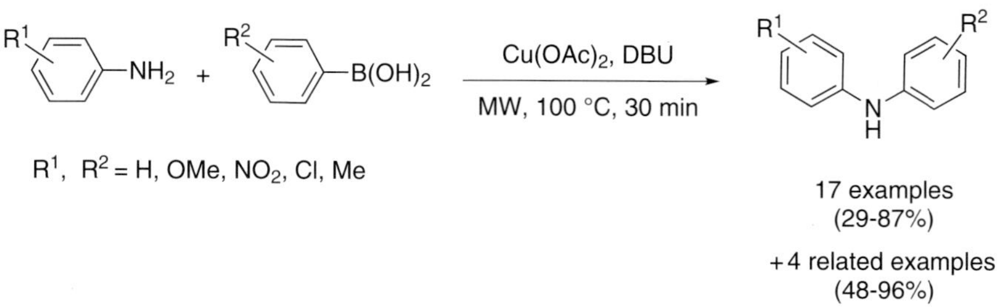

Scheme 5.140 Cu-mediated *N*-arylation of amines with boronic acids.

Van der Eycken and coworkers performed Cu(II)-mediated scaffold decorations on the pyrazinone core (Scheme 5.141) [554]. *N*-Arylations utilizing arylboronic acids applying the Chan–Lam protocol were successfully conducted under microwave heating with intensive simultaneous cooling at 0 °C. Due to prevented product

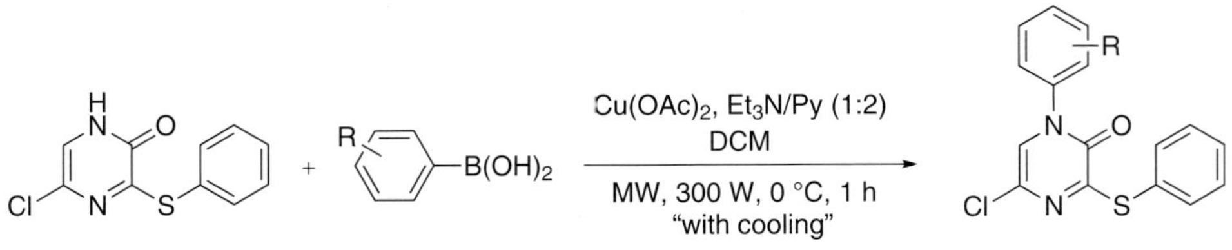

R = H, 3–CF3, 3–Cl, 3–Br, 3–EtO
4-MeO, 4-PhO, 2-naphthyl

8 examples
(83-97%)

Scheme 5.141 Cu(II)-mediated *N*-arylation of amides with boronic acids.

decomposition, excellent yields could be achieved compared to reactions at room temperature or elevated temperatures, respectively. High yields could be obtained by using a mixture of Et_3N/pyridine (1 : 2) as opposed to Et_3N alone: 93% compared to 69% for R = 3-CF_3.

The copper(I)-catalyzed C–S bond formation from thiophenols and aryl halides utilizing (±)-*trans*-cyclohexane-1,2-diol as ligand for the synthesis of aryl sulfides was presented by the Bagley group [555]. Irradiation at 120 °C for 3 × 1 h afforded the desired sulfides in excellent yield. This effective procedure could be successfully applied toward a range of electron-rich and electron-poor aryl halides and a selection of thiols (Scheme 5.142). Although aryl thiols generally furnished high yields, cyclohexanethiol as representative for alkyl thiols proceeded rather sluggish. However, even the simultaneous formation of two carbon–sulfur bonds was successful employing dihalobenzene substrates.

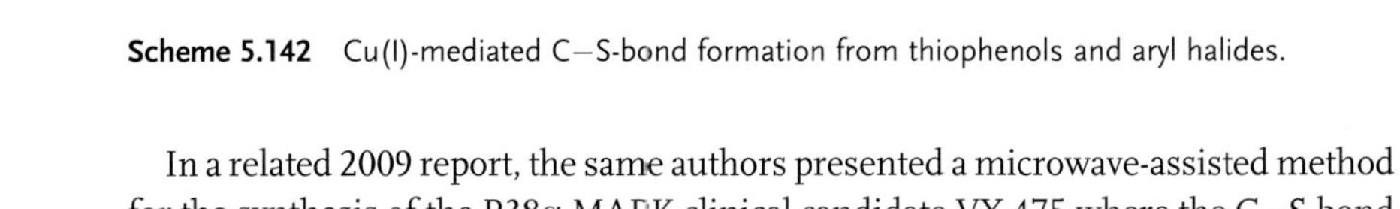

R^1 = H, F
R^2 = MeO, Me, Br, I
X = Cl, Br, I

10 examples
(16-95%)

Scheme 5.142 Cu(I)-mediated C–S-bond formation from thiophenols and aryl halides.

In a related 2009 report, the same authors presented a microwave-assisted method for the synthesis of the P38α MAPK clinical candidate VX-475 where the C–S bond formation is the key transformation in the four-step process [556]. Previously reported methods suffered from poor reproducibility and low yields for the formation of *S*-heteroaryl bond, but the herein established method allows efficient preparation of the biologically relevant heterocyclic motif. Especially the Ullmann-type coupling of the intermediate pyridazine **78** with 2,4-difluorothiophenol (Scheme 5.143) was significantly facilitated by microwave irradiation employing the same conditions as described in Scheme 5.142. Interestingly, almost identical yield could be obtained in

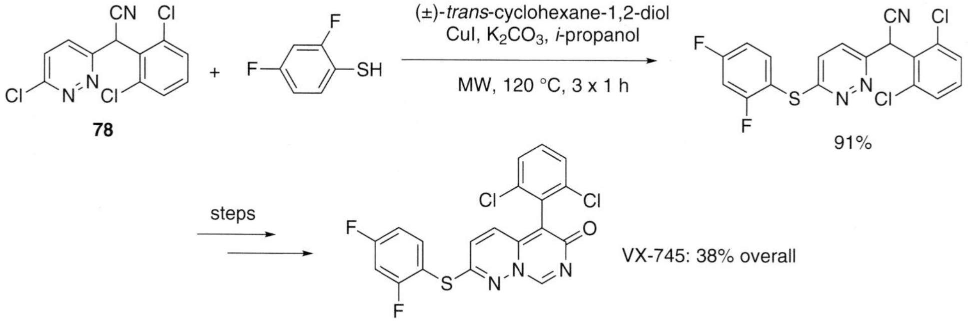

Scheme 5.143 Synthesis of P38α MAPK clinical candidate VX-745.

the absence of the copper catalyst indicating that this transformation is predominantly a S_NAr process. Subsequent hydrolyzation to the amide and further reaction with dimethyl acetal furnished the target compound VX-745 in satisfactory overall yield (Scheme 5.143). Utilizing microwave irradiation in all steps significantly reduces the overall process time and simplifies compound purifications.

A palladium-catalyzed protocol for carbon–sulfur bond formation between an aryl triflated and *para*-methoxybenzylthiol was introduced by Macmillan and Anderson (Scheme 5.144) [557]. Using palladium(II) acetate as a palladium source BINAP as a ligand, microwave heating of the two starting materials in *N,N*-dimethylformamide at 150 °C for 20 min in the presence of triethylamine base led to the formation of the desired sulfide in 85% yield.

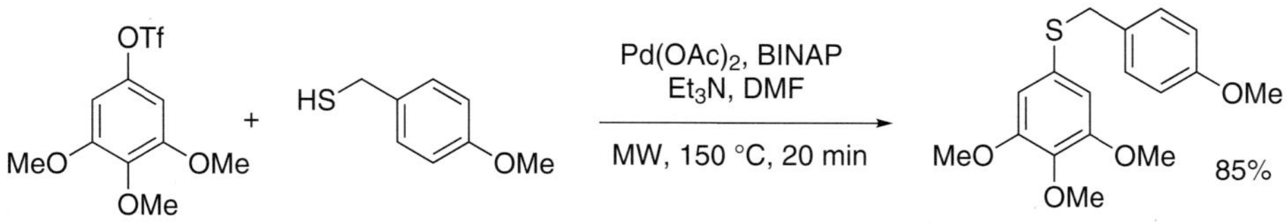

Scheme 5.144 Palladium-catalyzed carbon–sulfur cross-coupling.

Pd-catalyzed reductive Ullmann-type homocouplings of isothiazoles have been studied by Christoforou and Koutentis [558]. Numerous examples of microwave-assisted Ullmann condensations have been additionally reported [559–561].

5.3.3
Miscellaneous Carbon–Heteroatom Bond-Forming Reactions

Tertiary phosphines are a very important class of ligands in transition metal-catalyzed reactions. Among the various methods for the synthesis of such phosphine ligands, direct carbon–phosphorous (C–P) bond formation by transition metal-catalyzed cross-coupling of unprotected secondary phosphines with aryl halides/triflates can be considered as one of the most valuable procedures. All these protocols utilize a homogeneous transition metal catalyst and a base, and generally require many hours of reaction time at high temperatures or otherwise impractical or complex synthetic manipulations. Stadler and Kappe have demonstrated that aryl iodides, bromides, and triflates can be successfully coupled with diphenylposphine under microwave conditions (Scheme 5.145) [562]. Optimized reaction conditions for aryl iodides utilized combinations of 1-methyl-2-pyrrolidone, potassium acetate, and 2.5 mol% of palladium(II) acetate as catalyst. A 50% excess of iodobenzene was found to give the best yields and within 20 min a conversion of >90% was typically achieved at 180–200 °C. At higher temperatures, decomposition of the palladium(II) catalyst was observed, resulting in lower isolated product yields and in the deposition of a very thin film of Pd black on the microwave process vial. The reaction was also successful using palladium-on-charcoal (<0.1 mol% palladium) as a catalyst. For the less reactive aryl bromide and triflate precursors, more active catalytic systems had to be employed (Scheme 5.145).

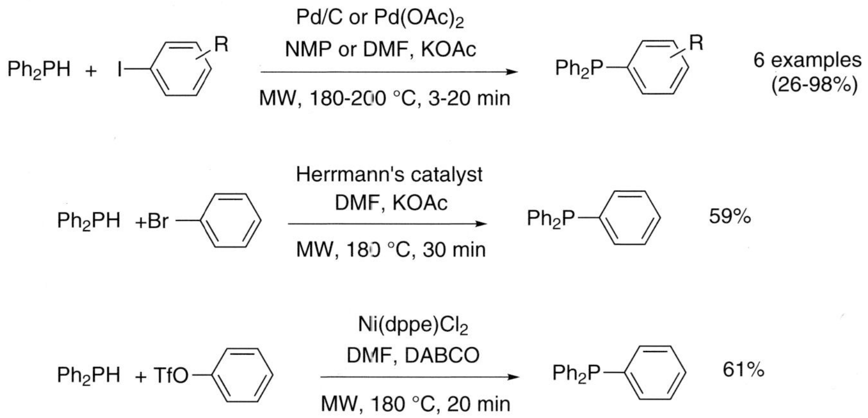

Scheme 5.145 Palladium-catalyzed carbon–phosphorous cross-coupling.

In a 2009 report, the Larhed group presented investigations toward C—P bond formation by coupling arylboronic acids or aryltrifluoroborates with H-phosphonate dialkyl esters [563]. To establish convenient and effective reaction conditions, simple palladium(II) acetate for arylboronic acids and $[Pd(O_2CCF_3)_2]$ for aryltrifluoroborates, respectively, together with the air-stable dmphen bidentate ligand, was chosen as catalytic system to allow noninert reaction conditions. Initial experimentation identified *para*-benzoquinone (*p*-BQ) as the most efficient reoxidant and DMF or MeOH as the suitable solvents (Scheme 5.146). The protocol also tolerates the use of aryltrifluoroborates as boronic coupling substrate to generate a variety of corresponding phosphonate diesters. Utilizing vinyl boronic acid led to a rather unsatisfactory yield of only 37%. Finally, the established protocol for the synthesis of arylphosphonate diesters was successfully applied in the preparation of a *Mycobacterium tuberculosis* glutamine synthase inhibitor.

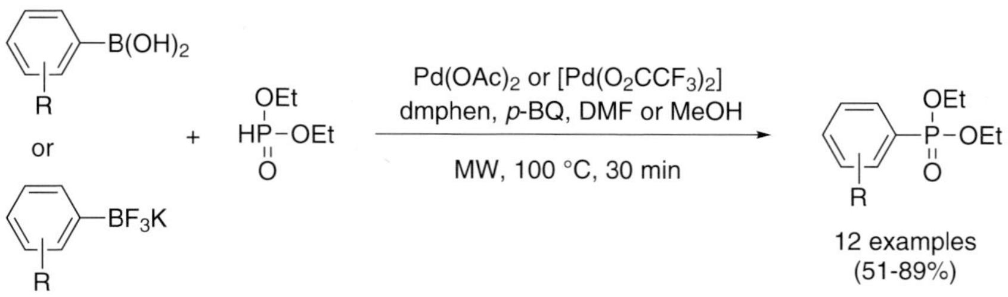

Scheme 5.146 Palladium(II)-catalyzed synthesis of aryl phosphonate esters.

The large number of boronic acids that are commercially available makes the Suzuki and related types of coupling chemistries highly attractive in the context of high-throughput synthesis and scaffold decoration (see Section 5.1.2). In addition, boronic acids are air and moisture stable, of relatively low toxicity, and the boron-derived by-products can easily be removed from the reaction mixture. Therefore, it is not surprising that efficient and rapid microwave-assisted protocols have been developed for their preparation. In 2002, Fürstner and Seidel have outlined the synthesis of pinacol aryl boronates from aryl chlorides bearing electron-withdrawing groups and commercially available bis(pinacol)borane, using a palladium catalyst formed *in situ* from palladium acetate and a corresponding imidazolium chloride (Scheme 5.147, X = Cl) [564]. The very reactive *N*-heterocyclic carbene (NHC) ligand (6–12 mol%) allowed this transformation to proceed to completion within 10–20 min at 110 °C in THF. The conventionally heated process (reflux THF, about 65 °C, argon atmosphere) gave comparable yields, although it required 4–6 h to reach completion. A complementary approach subsequently disclosed by Dehaen and coworkers used electron-rich aryl bromides as substrates (Scheme 5.147, X = Br) and (3 mol%) of Pd(dppf)Cl_2 as a catalyst [565]. Here, a somewhat higher reaction temperature (125–150 °C) was employed producing a variety of different arylboronates in good to excellent yields. Burgess and coworkers have employed a similar procedure for the preparation of fluorescein-derived boronic esters (see Scheme 5.39) [349].

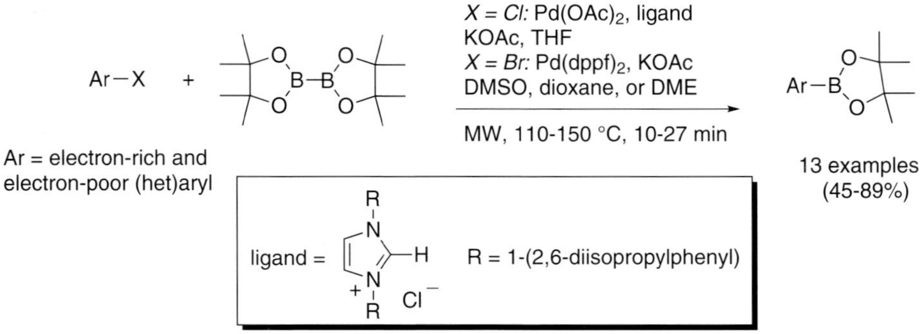

Scheme 5.147 Palladium-catalyzed formation of arylboronates.

The Lipshutz group has reported on the cross-coupling of (het)aryl bromides with phenols to generate C–O bonds that was catalyzed by Cu impregnated into charcoal (Cu/C) (Scheme 5.148) [566]. If the aryl bromide substrate was used as the limiting reagent, it was discovered that active Cu was released from the charcoal into solution. By simply adjusting the molar ratio to phenol being the limiting reagent, this problem could be overcome with the advantage that less Cs_2CO_3 base and thus less solvent could be used. However, longer reaction times were necessary in these cases. When aryl chlorides were employed, the reaction time under microwave heating turned out to be too long (8 h) to be of practical use.

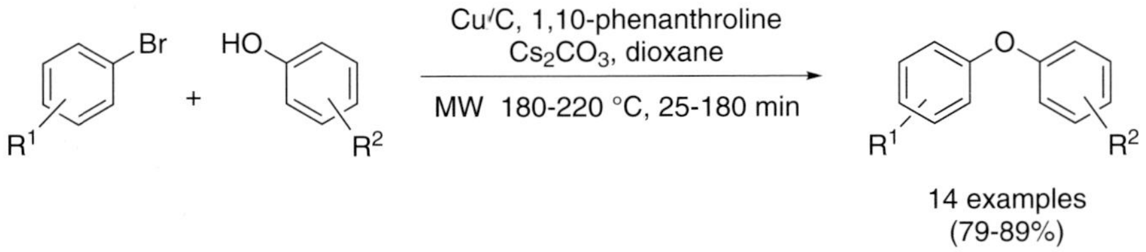

Scheme 5.148 Cu-on-charcoal-catalyzed diaryl ether synthesis.

In a more recent article, Lipshutz presented copper + nickel-on-charcoal (Cu–Ni/C) as a bimetallic heterogeneous catalyst that promotes a variety of cross-couplings normally catalyzed by Pd, Ni, or Cu [567].

Similar cross-couplings of aryl iodides and alcohols employing an octanuclear Cu cluster as catalyst were reported by Guzei [568]. Related Cu-catalyzed couplings of aryl halides with phenols have also been described [569–571]. The Cu-catalyzed conversion of aryl halides to phenols in the presence of L-proline in high-temperature water has been demonstrated by Kormos and Leadbeater [572].

Alternatively, Raders and Verkade have reported on the synthesis of diaryl ethers by the reaction of electron-deficient aryl fluorides with various TBDMS-protected phenols (Scheme 5.149) [573]. Proazaphosphatrane **79** was utilized as strong base in this protocol, and compared to conventional heating the amount could be reduced from 10 to 50 mol% to only 1–10 mol% giving the products in comparable high yields.

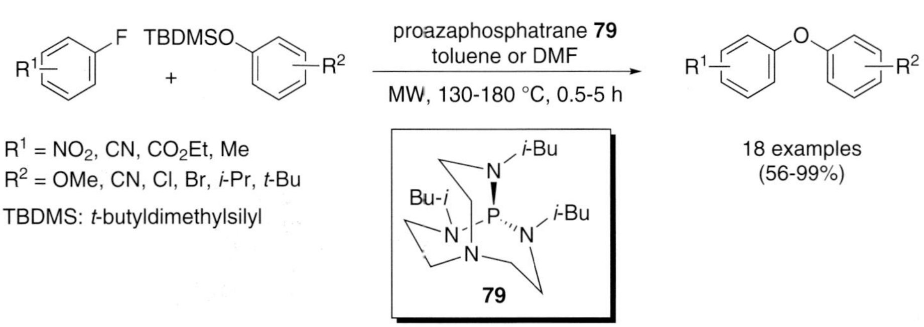

Scheme 5.149 Synthesis of diaryl ethers without metal catalyst.

For NO_2-substituted aryl fluorides, toluene was used as solvent, whereas the other aryl fluorides required DMF.

A 24-member benzoxazole library was synthesized by Batey and coworkers using an automated sequential processing technique (Scheme 5.150) [574]. The reaction proceeds in one pot via initial acylation of the aniline building blocks with acid chlorides giving the 2-haloanilide intermediates **80**, followed by Cu-catalyzed intramolecular cyclization of **80** to form the C–O bond of the benzoxazole products. Shorter reaction times could be achieved by employing microwave heating compared to conventional heating (15 min at 210 °C versus 24 h at 95 °C).

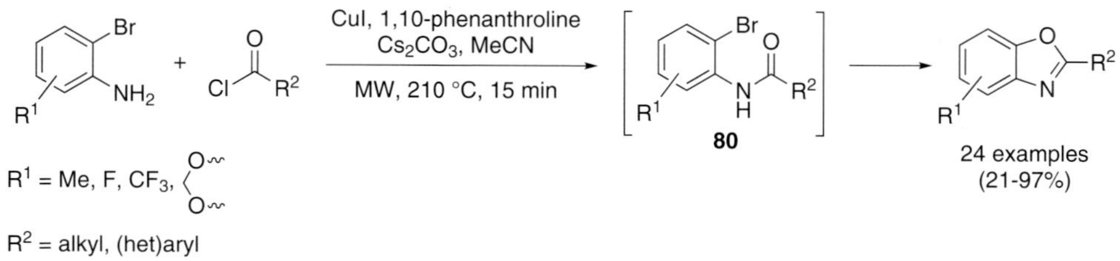

Scheme 5.150 Cu-catalyzed one-pot benzoxazole synthesis.

In a related fashion, intramolecular C–O coupling has been used for the synthesis of benzopyranones and isolamellarin alkaloids [575]. *N*-Substituted benzimidazol-2-ones have been obtained via intramolecular Cu-catalyzed amination from the appropriate *N*′-substituted *N*-(2-halophenyl)ureas [576].

Ranu *et al.* have developed the ligand-free coupling of aryl iodides with thiophenols and alkanethiols under Cu nanoparticle (4–6 nm) catalysis (Scheme 5.151) [577]. The arylsulfide synthesis proceeded in good to excellent yields and high purities. Compared to conventional heating at 120 °C, the reaction times could be accelerated from 12 to 15 h to only 5–7 min. Other Cu sources such as metallic Cu, CuI, or Cu powder lead to lower product yields, indicating the key role of Cu nanoparticles in this C–S coupling reaction.

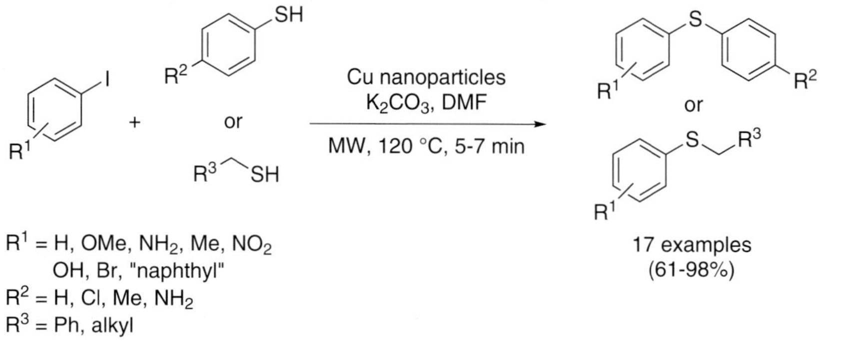

Scheme 5.151 Cu nanoparticle-catalyzed aryl-sulfur bond formation.

In a 2005 publication, Liang and coworkers have reported on optimization studies for the synthesis of aryl azides from aryl halides using Cu(I) catalysis (Scheme 5.152) [578]. In a first screening, several ligands have been tested, resulting in the selection of 1,2-diamino ligands **81** and **82**. In further optimizations, different solvents and catalyst/ligand ratios have been studied, reaching full conversion by using EtOH/H_2O as solvent mixture, 10 mol % of CuI, 15 mol % of ligand **82**, and irradiation at 100 °C for 30 min.

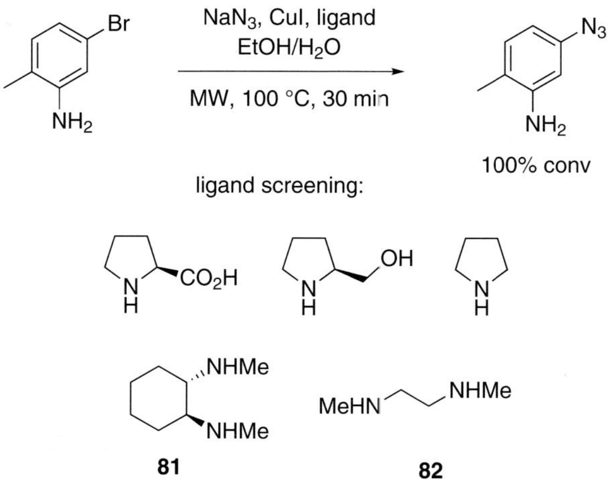

Scheme 5.152 Cu(I)-catalyzed synthesis of aryl azides.

Che and coworkers have reported on the intra- and intermolecular hydroamination of unactivated alkenes (Scheme 5.153) [579]. In a catalyst screening, the phosphine Au(I) catalyst combinations $(PCy_3)AuCl/AgOTf$ and $(PPh_3)AuCl/$

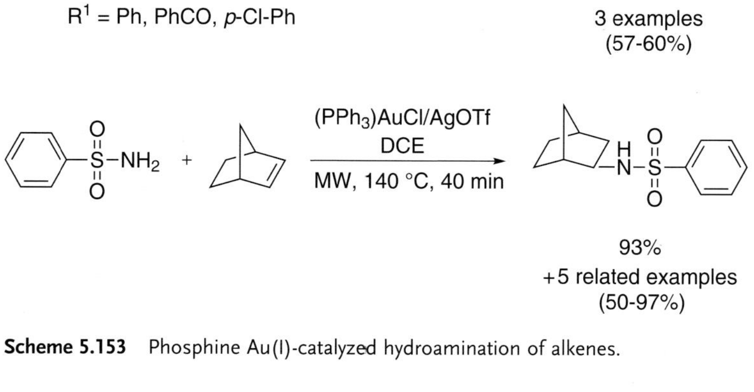

Scheme 5.153 Phosphine Au(I)-catalyzed hydroamination of alkenes.

AgOTf, respectively, showed the best catalytic activity. By applying microwave heating, not only were reaction times reduced (e.g., 30 h → 30 min), but also the amount of catalyst, compared to conventional heating.

5.4 Other Transition Metal-Mediated Processes

5.4.1 Ring-Closing Metathesis and Cross-Metathesis

In recent years, the olefin metathesis reaction has attracted widespread attention as a versatile carbon–carbon bond-forming method [580]. Among the numerous different metathesis methods, the ruthenium-catalyzed ring-closing metathesis (RCM) has emerged as very powerful method for the construction of small, medium, and macrocyclic ring systems. In general, metathesis reactions are carried out at room or at slightly elevated temperatures (e.g., at 40 °C in refluxing dichloromethane), sometimes requiring several hours of reaction time to achieve full conversion. Employing microwave heating, otherwise sluggish RCM protocols have been reported to be completed within minutes or even seconds, as opposed to hours at room temperature [581–584].

One limitation of ring-closing metathesis is that the reaction can be rather sensitive to external substituents. In most cases, only substituents with little or no steric or electronic bias are compatible. External groups with an electronic bias (ester, nitrile, etc.) often result in little to no yield of the RCM product. Wilson and coworkers have described microwave-enhanced ring-closing metathesis transformations with diolefin substrates containing external carboxymethyl substituents (Scheme 5.154a) [585]. The authors report that most of the reactions studied would lead to 40% conversion utilizing 3 mol% of Grubbs II catalyst at room temperature using 1,2-dichloroethane (DCE) as solvent. Heating to 50 °C resulted in conversion to about 75% after 5 h, but proceeded no further. In sharp contrast, microwave irradiation for 5 min at 50 °C gave 85% conversion and at 150 °C for 5 min gave 97% conversion. Independent investigations by Grigg *et al.* with a similar set of diolefin substrates (Scheme 5.154b) arrived at similar conclusions [586]. Again, utilizing controlled microwave heating, the RCM reactions could be run much more efficiently, allowing a significant reduction in catalyst loading and reaction time, comparing microwave-assisted and thermal runs.

An interesting combination of ring-closing metathesis chemistry with the aza Baylis–Hillman reaction was described by Balan and Adolfsson and is shown in Scheme 5.155a [587]. The authors have reported that functionalized 2,5-dihydropyrroles can be obtained by microwave-mediated ruthenium-catalyzed ring-closing metathesis. The required bisolefin precursors were conveniently obtained from aza-Baylis–Hillman adducts. Microwave irradiation for 1–2 min at 100 °C of a dilute solution of the diene with 5 mol% of Grubbs II catalyst in dichloromethane produced the desired pyrroles in high yield. The same conditions were used by the group of

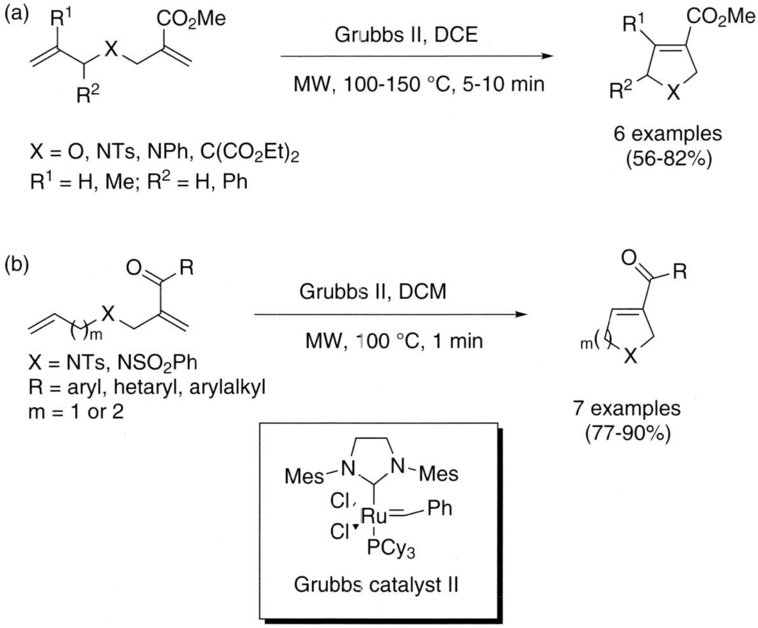

Scheme 5.154 Ring-closing metathesis with external carbonyl substituents.

Lamaty for the ring-closing metathesis of related 2-trimethylsilylethylsulfonyl (SES)-protected substrates (Scheme 5.155b) [588]. The required starting materials were again prepared by aza-Baylis–Hillman chemistry.

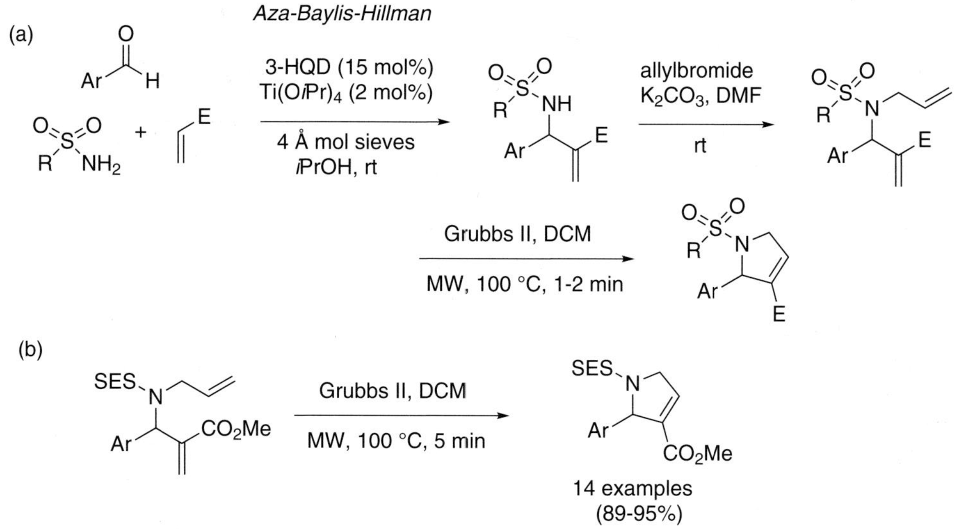

Scheme 5.155 Ring-closing metathesis of aza-Baylis–Hillman adducts.

Related microwave-assisted ring-closing metathesis reactions, leading to seven-membered [589] and eight-membered aza-heterocyclic products [323], are displayed in Scheme 5.156.

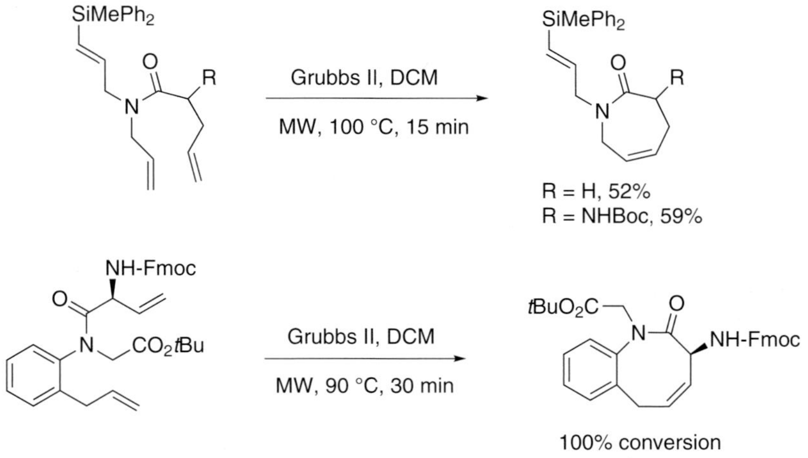

Scheme 5.156 Formation of seven- and eight-membered rings by ring-closing metathesis reactions.

The group of Van der Eycken has employed microwave-enhanced Suzuki–Miyaura cross-coupling and RCM as the key steps in the six-step synthesis of so far unknown *N*-shifted buflavine analogs of type **86** (Scheme 5.157) [590]. Particularly, the generation of the rigid, medium-sized ring system by RCM was enhanced by employing microwave heating for 5 min at 150 °C using 3 mol % of Grubbs II catalyst.

In 2007, the same group has also reported on the synthesis of ring-expanded (in addition to *N*-shifted) buflavine analogs (Scheme 5.157) [591]. Key steps in both reaction pathways are the Suzuki coupling of highly electron-rich aryl halides **83** with *ortho*-substituted boronic acids **84** followed by RCM for the medium-sized ring construction. The *N*-shifted buflavine derivatives **86**, possessing an eight-membered ring, were obtained by allylation of biaryls **85** and subsequent RCM utilizing Grubbs II catalyst. For the preparation of the nine-membered ring buflavine scaffolds **88**, reductive amination was first performed on the biaryl aldehydes **87** followed by RCM. The double bond in both derivatives **86** and **88** was reduced by Pd-catalyzed hydrogenation to obtain the corresponding buflavine analogs (not shown).

The combination of multicomponent chemistry, RCM and Heck coupling chemistry was highlighted by Martin and coworkers for the generation of diverse heterocyclic scaffolds following the concept of diversity-oriented synthesis (DOS) [592].

Reiser and his group have reported on the RCM of a sterically demanding substrate in the context of the total synthesis of tricyclic 5.7.5-sesquiterpene lactones of the

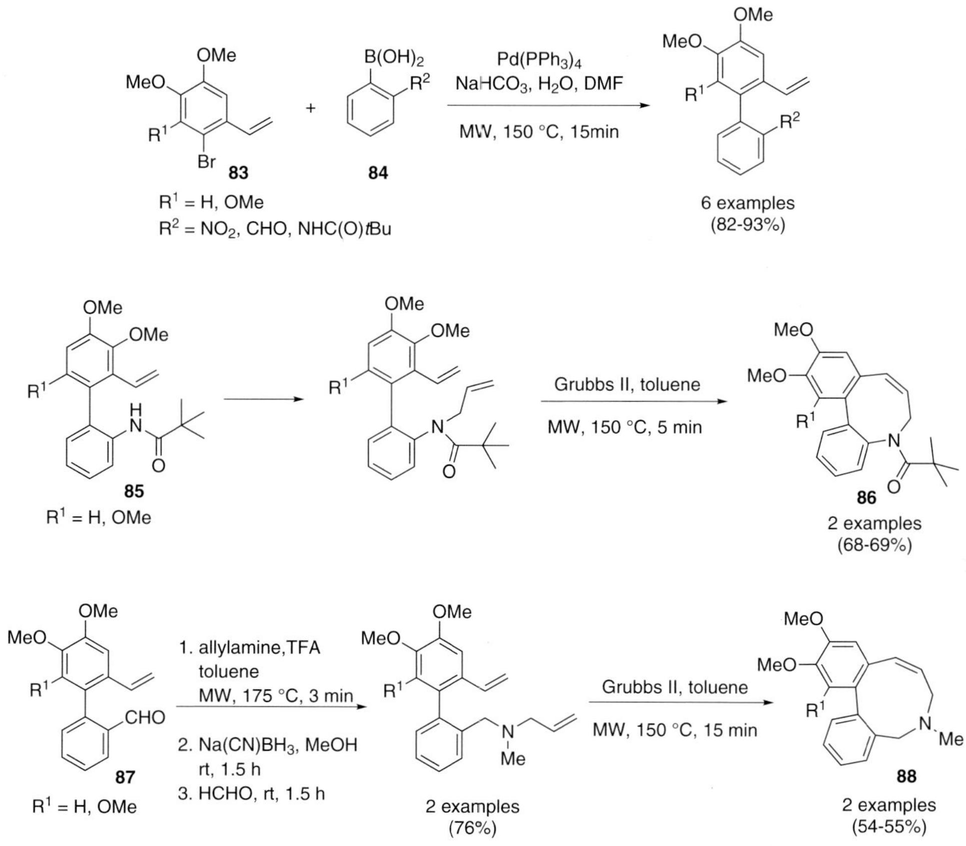

Scheme 5.157 Synthesis of *N*-shifted and ring-expanded buflavine analogs.

guaianolide type (see Scheme 4.5) [593]. Key to the success for the formation of the tetrasubstituted double bond system by RCM was the combination of microwave heating and sparging an inert gas through the solution in order to remove the formed ethylene. An optimized 98% isolated product yield was obtained by irradiating the bisolefin precursor in the presence of 15 mol% of Grubbs II catalyst in toluene for 90 min at reflux temperature (110 °C), while passing a gentle stream of argon through the solution. Importantly, under sealed-vessel microwave irradiation, the RCM failed completely and starting material was recovered almost quantitatively. Similar observations were made by Kappe and coworkers toward investigations of wall effects in MAOS in the ring-closing metathesis reaction of 1,2-bis(allyloxy)benzene to provide the eight-membered 2,5-dihydro-1,6-benzodioxocin [594].

Chapman and Arora were successful in the solid-phase synthesis of hydrogen-bond surrogate (HBS) α-helices (artificial α-helices where the *N*-terminal H-bond is replaced by a covalent C–C bond) including the RCM reaction as key step (Scheme 5.158) [595]. Compared to conventional heating at 60 °C where reaction times up to 72 h are required for maximum conversion, under microwave irradiation

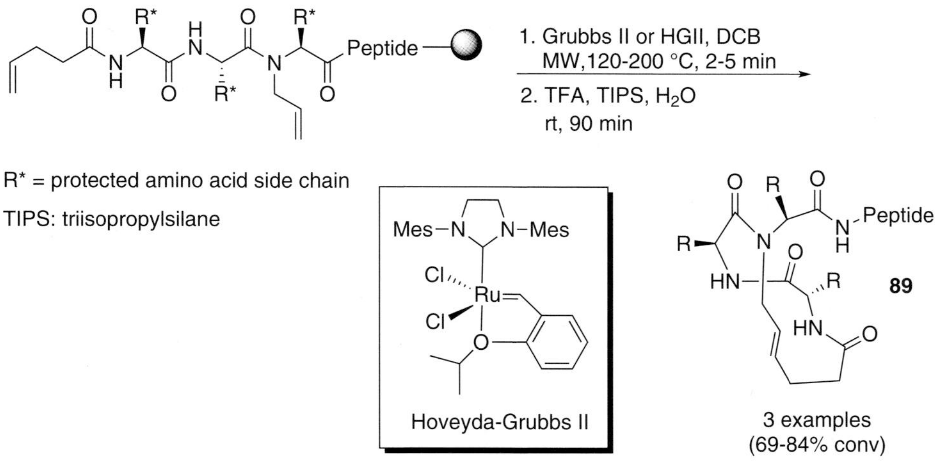

Scheme 5.158 Synthesis of hydrogen bond surrogate helices via ring-closing metathesis.

the RCM could be accelerated to 2–5 min. Three different substituted 13-membered macrocycles (**89**) and one 16-membered macrocycle were obtained in high yields either at 120 °C within 2 min when Grubbs II catalyst is employed or at 200 °C within 5 min for the Hoveyda–Grubbs II (HG II). Importantly, in this HBS approach, Grubbs II produces the macrocycles in high yields compared to conventional heating, where it showed to be inactive. In addition, it was found that a greater variety of amino acid residues is tolerated, which was also one of the limitations using oil bath heating.

For related RCM strategies in the solid-phase synthesis of carbocyclic peptides, see Ref. [596]. An RCM approach was also employed in the generation of helicenes using Grubbs II catalyst [597,598], in the preparation of macrocyclic bispyridinium salts [599] and in the synthesis of complex norbornenes [600].

In 2003, Efskind and Undheim reported dienyne and triyne domino RCMs of appropriately functionalized substrates with Grubbs type I or II catalysts (Scheme 5.159) [601]. While the thermal processes (toluene, 85 °C) required multiple addition of fresh catalyst (3 × 10 mol%) over a period of 9 h to furnish a 92% yield of product, microwave irradiation for 10 min at 160 °C (5 mol% catalyst, toluene) led to full conversion. The authors ascribe the dramatic rate enhancement to the rapid and uniform heating of the reaction mixture and increased catalyst lifetime by the elimination of wall effects. In some instances, the use of Grubbs I catalyst was more efficient than the more common Grubbs II equivalent.

An interesting series of ring-closing alkyne metathesis (RCAM) reactions was recently reported by Fürstner *et al.* (Scheme 5.160) [602]. Treatment of biaryl-derived diynes with 10 mol% of a catalyst prepared *in situ* from molybdenum hexacarbonyl and 4-trifluoromethylphenol at 150 °C for 5 min led to about 70% isolated yield of the desired cycloalkynes, which were further manipulated into a naturally occurring DNA cleaving agent of the turriane family. Conventional heating under reflux conditions in chlorobenzene for 4 h produced about 80% isolated yield of product under otherwise identical conditions.

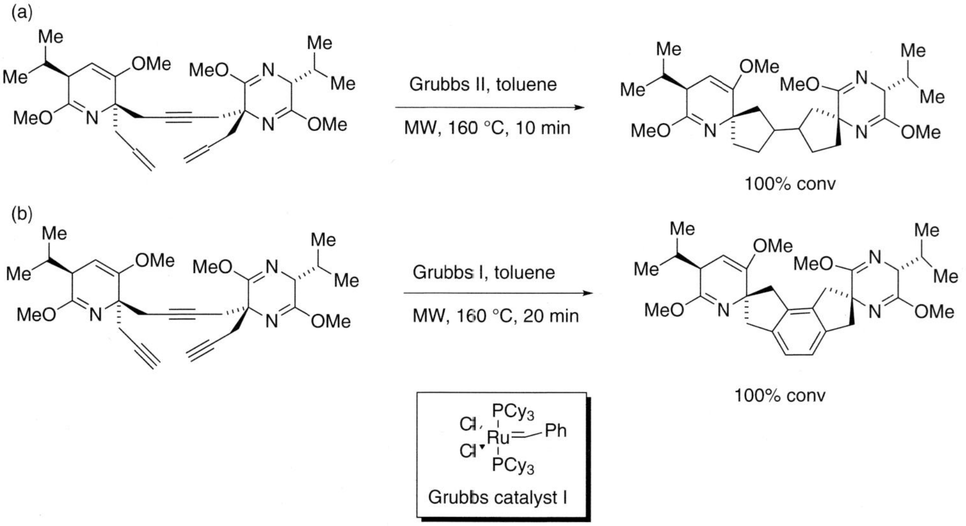

Scheme 5.159 Dienyne (a) and triyne (b) domino ring-closing metathesis reactions.

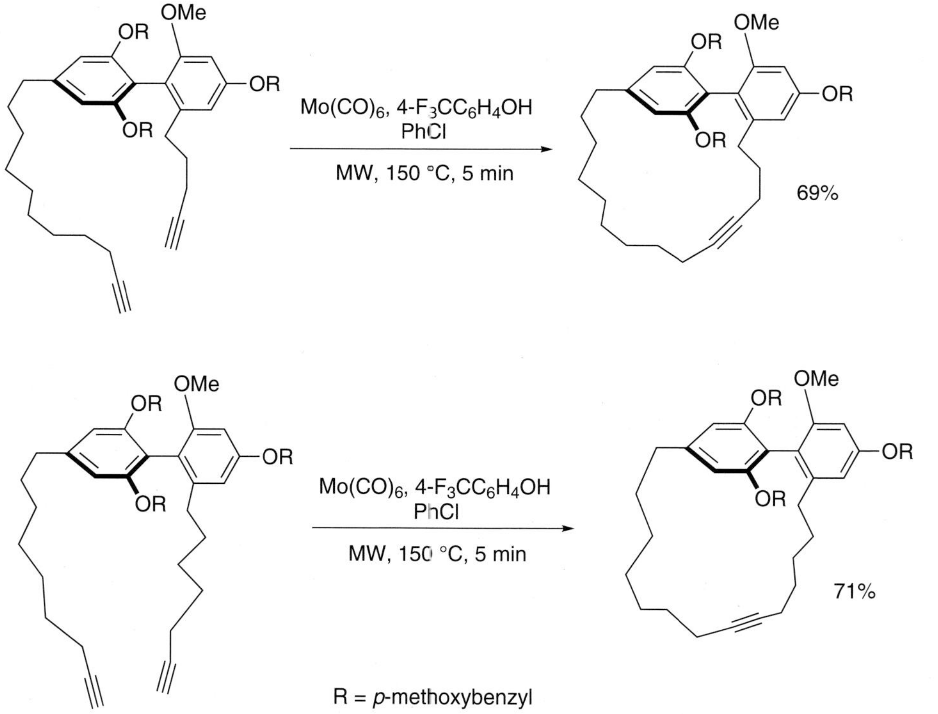

Scheme 5.160 Ring-closing alkyne metathesis reactions.

A series of synthetically valuable RCM/alkene cross-metathesis (CM) transformations starting from sulfamide-linked enynes was described by Brown and coworkers in 2004 (Scheme 5.161) [603]. A range of enyne substrates was subjected to ring-closing metathesis using 3–20 mol% of Grubbs II catalyst. While the reactions of internal alkyne substrates were sluggish at room temperature, they proceeded rapidly by microwave heating to 100 °C within 1 h, furnishing seven-membered cyclic sulfamides in good yields (Scheme 5.161a). For substrates bearing a terminal alkyne group, the authors have developed a one-pot RCM–CM reaction in the presence of 2–3 equiv of olefins such as styrene using 6 mol% of Grubbs II catalyst. As anticipated, the desired enyne RCM–CM products were produced selectively in good yields with the expected *E*-isomer being predominant.

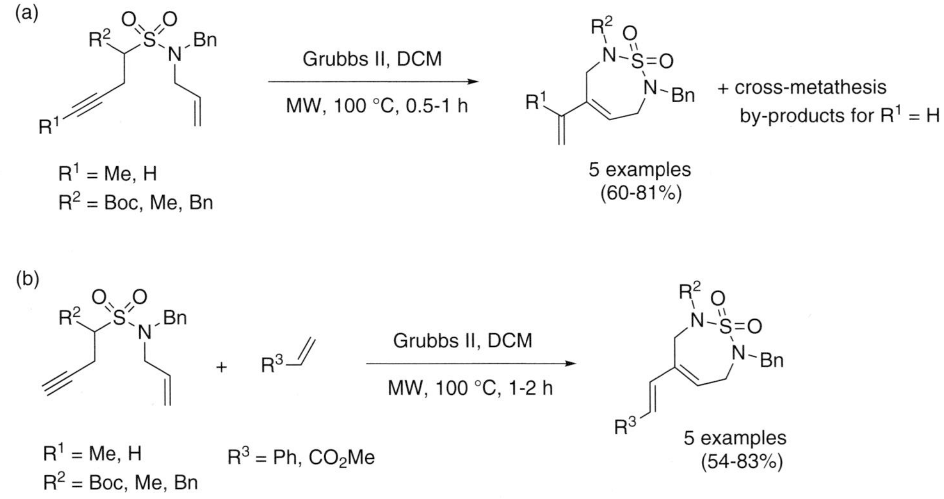

Scheme 5.161 Enyne ring-closing and cross-metathesis reactions.

Successful enyne metathesis reactions of homopropargylic homoallylic alcohols leading to functionalized vinylcyclopentenols have been reported in Ref. [604]. Enyne metathesis was also used by Porco and coworkers for the generation of polycyclic ring systems applying a diversity-oriented synthesis approach [605]. In addition, 1,5-enyne metathesis transformations were extensively studied by Debleds and Campagne for the synthesis of substituted cyclobutenes as valuable 1,3-diene units [606].

In a recent publication, Vincent and Kouklovsky reported on the ring-rearrangement of nitroso Diels–Alder adducts representing the first approach toward ring-rearrangement metathesis of substrates with two contiguous heteroatoms [607]. The obtained fused bicycles are of relevance in the total synthesis of alkaloids. The investigations focused on strained 3,6-dihydro-1,2-oxazines that were initially prepared by conventional nitroso Diels–Alder reaction involving cyclic dienes and acylnitroso reagents. Optimization identified the common Grubbs II catalyst as most efficient and furnished an applicable microwave protocol requiring two heating cycles of 20 min each with fresh catalyst addition in the second cycle. Depending on the chain length of the alkene substituent on the Diels–Alder adduct, tetrahydroisoxazolo[2,3-*a*]pyridine-7-ones **90** (Scheme 5.162), -pyrrol-6-ones, and -azepin-6-ones can be obtained in varying yield, which is influenced by several competing side reactions.

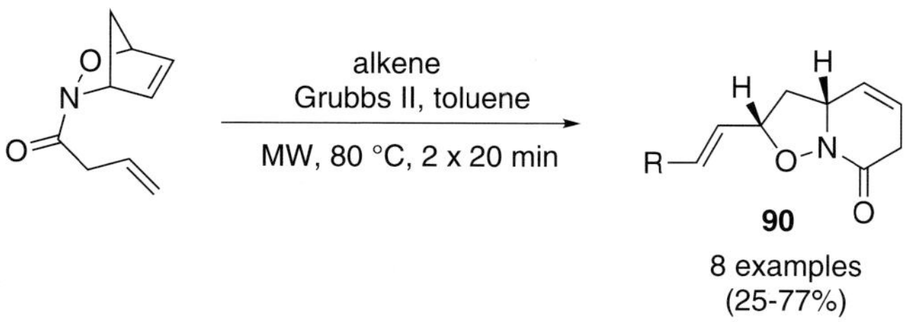

Scheme 5.162 Ring-rearrangement metathesis.

In a 2005 publication, Bargiggia and Murray have reported on cross-coupling metathesis between deactivated olefins (Scheme 5.163) [608]. By applying microwave irradiation, reaction rates could be enhanced compared to classical heating, showing superior yields for the Hoveyda–Grubbs catalyst **91** than for Grubbs II catalyst. No differences in yields or selectivities in homo- or hetero-couplings, respectively, have been observed by comparison studies of microwave and oil bath heating.

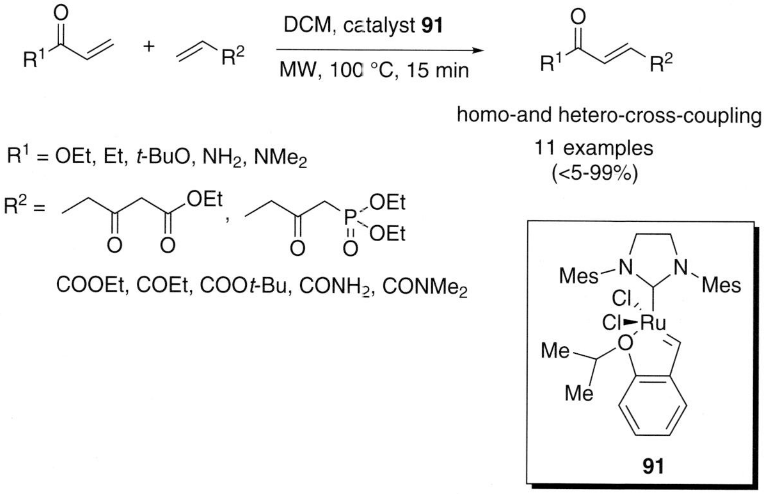

Scheme 5.163 Cross-coupling metathesis between deactivated olefins.

Caddick and coworkers were successful in the generation of peptidomimetics via stereoselective cross-metathesis (Scheme 5.164) [609]. By applying Grubbs II catalyst in the reaction of two different single amino acids, products that still have the length of two amino acids but with a modified main chain (lack of the amide bond) are obtained. This structural feature is of interest with regard to protein folding and protein–protein interactions. Notably, for those products that needed 60 min reaction time, the reaction mixture was degassed after 30 min to remove any dissolved ethylene and subsequently subjected to irradiation for additional 30 min.

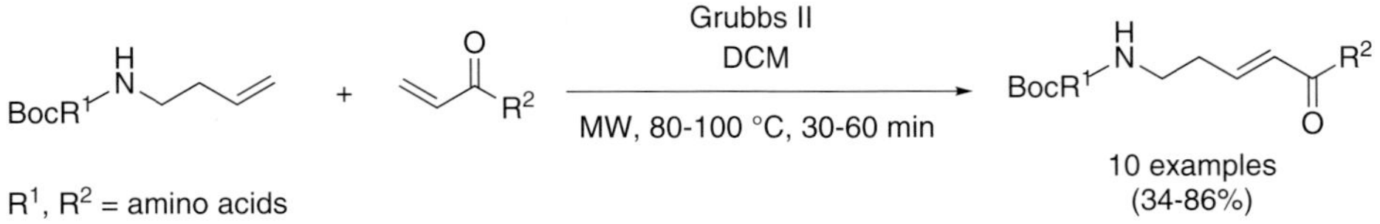

Scheme 5.164 Synthesis of peptidomimetics via cross-metathesis.

For an example of cross-metathesis reactions used for the construction of dicarba analogs of multicysteine-containing peptides, see Ref. [610].

Botta and his group have employed microwave-mediated ethylene–alkyne cross-metathesis using Grubbs II catalyst for the synthesis of enantioenriched 2-substituted butadienes (Scheme 5.165) [611]. Importantly, microwave heating was essential for the fast reaction of enantiomerically enriched alkynes with ethylene (closed vessel saturated with ethylene) under retention of configuration at the propargylic/allylic position. At atmospheric pressure, no reaction was observed [612].

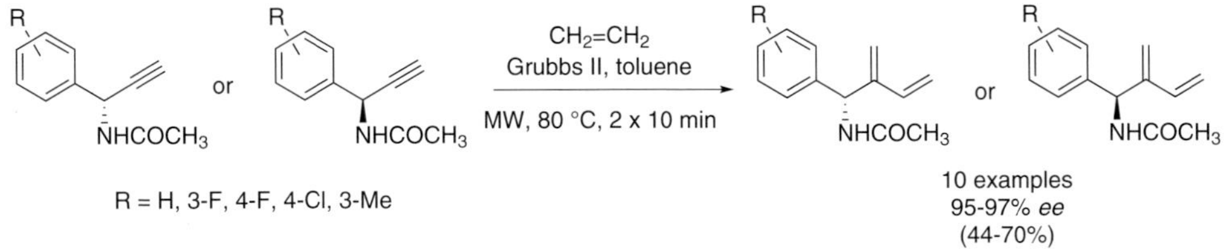

Scheme 5.165 Ethylene–alkyne cross-metathesis.

Cyclic β-amino carbonyl derivatives were synthesized via a cross-metathesis aza-Michael tandem process by Fustero *et al.* (Scheme 5.166) [613]. Vinyl ketones are reacted with Cbz-protected amines to give pyrrolidines ($n = 1$) and piperidines ($n = 2$), respectively, in good to excellent yields. Key to the success of the reaction sequence was the combination of Hoveyda–Grubbs (HG) catalyst and $BF_3.OEt_2$ as Lewis acid that activates the cyclization step. By applying microwave heating, the reaction time could be dramatically reduced from 4 days (45 °C) to only 20 min. Interestingly, under microwave heating, an inversion of the selectivity was observed

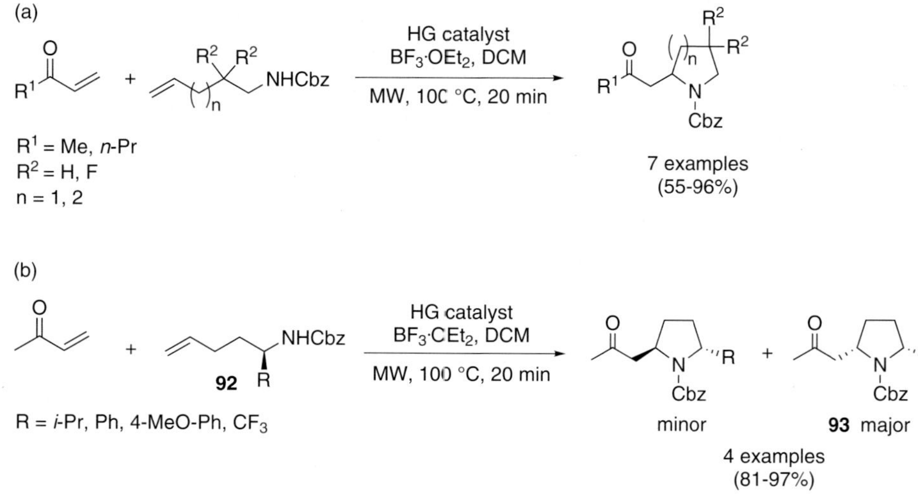

Scheme 5.166 Tandem cross-metathesis–intramolecular aza-Michael reaction.

when α-substituted amines **92** were used, giving pyrrolidine derivatives **93** as major diastereoisomers.

Microwave-assisted cross-metathesis protocols have also been employed in the context of steroid chemistry [614], and in the synthesis of α-substituted prolines [615].

5.4.2
Pauson–Khand Reactions

The [2 + 2 + 1] cycloaddition of alkene, alkyne, and carbon monoxide is known as the Pauson–Khand reaction and is often the method of choice for the preparation of complex cyclopentenones [616]. Groth and coworkers have demonstrated that Pauson–Khand reactions can be carried out very efficiently under microwave heating conditions (Scheme 5.167a) [617]. Taking advantage of the sealed vessel

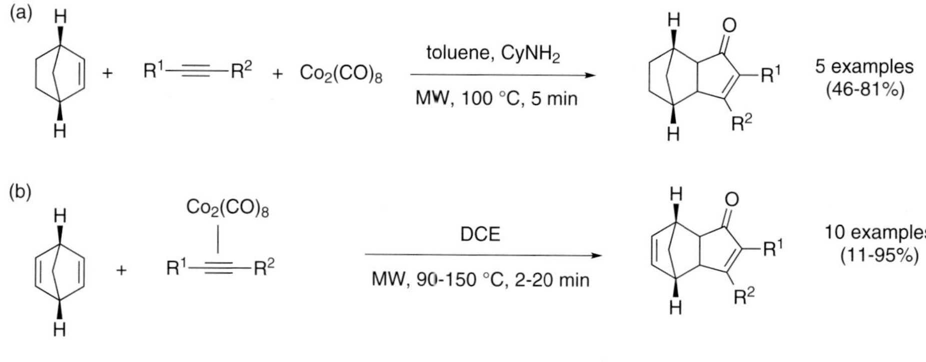

Scheme 5.167 Pauson–Khand [2 + 2 + 1] cycloaddition reactions.

technology, 20 mol% of dicobalt octacarbonyl was found sufficient to drive all the studied Pauson–Khand reactions to completion, without the need for additional carbon monoxide. The carefully optimized reaction conditions utilized 1.2 equiv of cyclohexylamine as additive in toluene as solvent. Microwave heating for 5 min at 100 °C provided good yields of the desired cycloadducts. Similar results were published independently by Evans and coworkers [618,619]. However, here the preformed alkyne–dicobalt hexacarbonyl complexes were used as substrates (Scheme 5.167b). These authors were able to perform the microwave-assisted reactions in different mono- and multimode microwave instruments with equal success [618].

For other microwave-assisted Pauson–Khand-type processes, see Refs [620,621].

5.4.3
Carbon–Hydrogen Bond Activation

Another important reaction principle in modern organic synthesis is carbon–hydrogen bond activation [622]. Bergman and coworkers have introduced a protocol that allows otherwise extremely sluggish inter- and intramolecular rhodium-catalyzed C−H bond activation to occur efficiently under microwave heating conditions. In their investigations, the authors have found that heating of olefin-tethered benzimidazoles in a mixture of 1,2-dichlorobenzene and acetone in the presence of di-μ-chloro-bis-[(cycloocten)-rhodium(I)] (2.5–5 mol%) and tricyclohexylphosphine hydrochloride (5–10 mol%) as the catalyst system provided the desired tricyclic heterocycles in moderate to excellent yields (Scheme 5.168) [623]. Microwave heating to 225–250 °C for 6–12 min proved to be the optimum conditions. The solvents were not degassed or dried before use, but air was excluded by purging the reaction vessel with nitrogen.

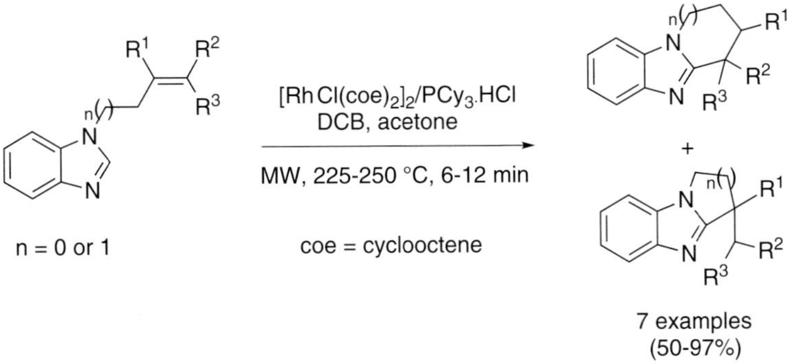

Scheme 5.168 Intramolecular benzimidazole C−H alkene coupling.

In a subsequent study, the same group has reported on the direct coupling of azoles **94** with aryl bromides and iodides, respectively (Scheme 5.169) [624]. Superior yields could be obtained by using a mixture of the bulky trialkylphosphines **95a** and **95b** as

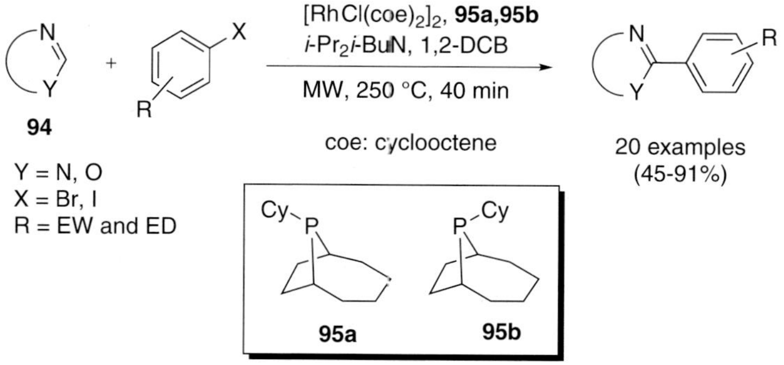

Scheme 5.169 Rh-catalyzed arylation of heterocycles.

ligand compared to PCy_3 due to less hydrodehalogenation. A variety of functional groups in *para*- and *meta*-position was tolerated and the method was also compatible with different heterocycles.

In order to overcome problems – such as substrate scope, functional group tolerance, and practicability – observed in previously published Rh-catalyzed arylations (see Scheme 5.169), the same authors have subsequently developed a new protocol for the direct arylation of a variety of heterocycles with aryl bromides employing phosphepine ligand **96** (Scheme 5.170) [625]. This ligand proved to be the most effective in a ligand screen and coordinates to Rh in a bidentate P-olefin fashion and thus generates an active and a temperature-stable catalyst. For a greater practicability of the protocol, the air-sensitive ligand **96** can be protected as its HBF_4 salt and $[RhCl(cod)]_2$ as air-stable Rh source can be employed. Thus, reaction mixture preparation in a glove box is avoided and only an inert atmosphere in the microwave vial is required. In addition, by using THF as low-boiling solvent product, isolation is simplified.

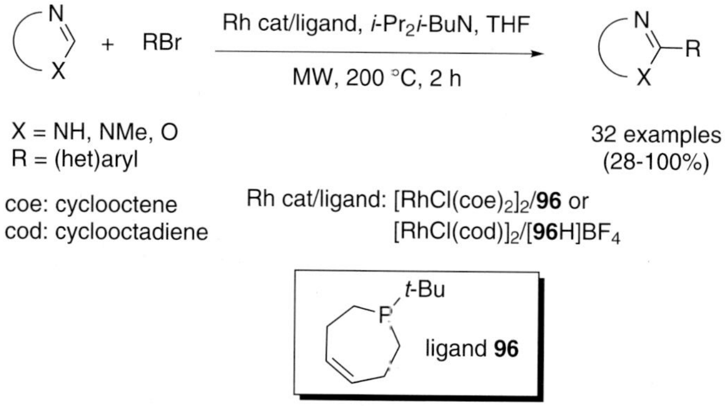

Scheme 5.170 Rh(I)-catalyzed arylation of heterocycles employing a phosphepine ligand.

Itami and coworkers have developed Rh complex **97** as a catalyst precursor for the C–H arylation of heteroarenes with aryl iodides (Scheme 5.171) [626]. Key to the success of the arylation is the use of $P[OCH(CF_3)_2]_3$ as ligand due to its strongly π-accepting character. Various five-membered heteroarenes were selectively arylated either at the 2- (X = S, O) or 3-position (X = N). To show the generality of this method, anisole and 1,3-dimethoxybenzene were reacted with *p*-nitrophenyl iodide. For arylated anisole, a mixture of regioisomers was obtained ($o:p = 29:71$), whereas the latter was arylated regiospecifically at the 4-position in 76% yield. For related examples involving the C–H arylation of heteroarenes involving Rh complexes, see Ref. [627].

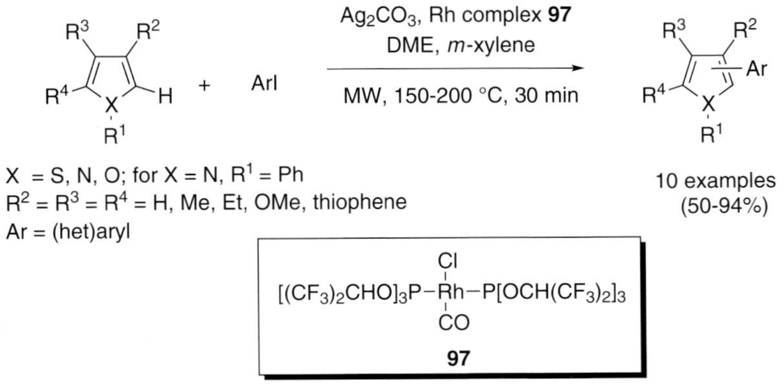

Scheme 5.171 Rh-catalyzed C–H arylations of heteroarenes.

The same group recently reported that electron-deficient *N*-heterocycles coupling with haloarenes can be efficiently promoted by KO^tBu alone, without the addition of any exogenous transition metal species [628].

Marder and coworkers investigated a one-pot procedure for the selective formation of β-aryl ketones **98** or the corresponding alcohols **99** by simple variation of reaction conditions [629]. The first step of the sequence is an iridium-catalyzed C–H borylation featuring methyl-*tert*-butyl ether (MTBE) as best-suited solvent. The required boronation is achieved within 10–60 min at 80 °C utilizing bis(pinacolato-*O,O′*)diboron [B_2pin_2] (Scheme 5.172). The boronated intermediate undergoes simply 1,4-conjugate addition reactions at 100 °C in the presence of $[Rh(COD)Cl]_2$ as catalyst. Interestingly, the choice of solvent leads to different products (Scheme 5.172). The nonoxidizable MTBE in the second step leads to the expected β-aryl ketones **98**, a switch to isopropanol in combination with slightly extended reaction times furnished β-aryl alcohols **99**. Electron-rich as well as electron-poor arenes and heteroarenes can be successfully reacted with a variety of ketones under both reducing and nonreducing conditions in good overall yields.

In a related work of the same group, the initial generation of aryl and heteroaryl boronates under microwave irradiation was investigated [630]. Applying the same reaction conditions as described in Scheme 5.172, the transformations were

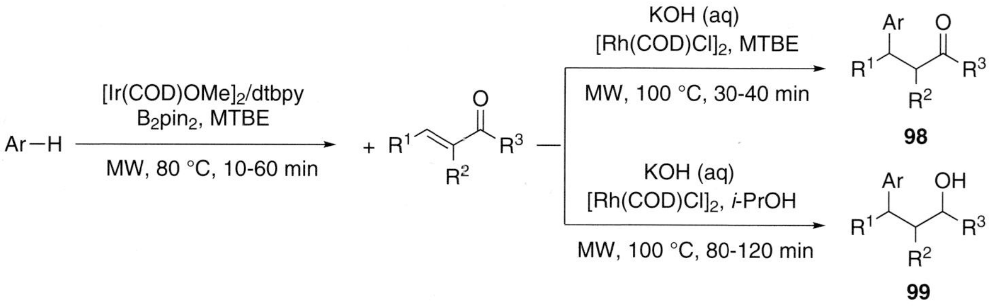

Scheme 5.172 One-pot tandem C—H borylation/1,4-conjugate addition/reduction sequence.

accomplished at 80 °C in 3–60 min (Scheme 5.173). A significant time saving could be achieved since under thermal heating, the reactions required up to 18 h to reach full conversion. A variety of cyclic, polycyclic, and heterocyclic arenes can be successfully converted with this efficient microwave protocol. It even tolerates *N*-protected substrates as demonstrated with *N*-Boc pyrrole (Scheme 5.173). In addition, a one-pot C—H borylation/Suzuki coupling sequence was possible giving the biaryls in high yields.

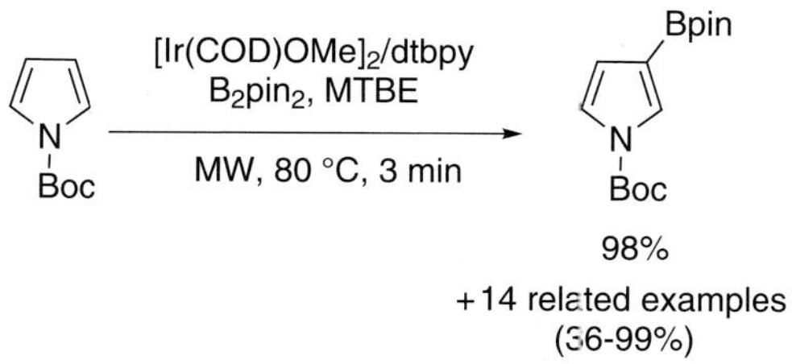

Scheme 5.173 Iridium-catalyzed C—H borylation of *N*-Boc pyrrole.

Alami and coworkers have reported on the direct arylation of adenines under ligandless $Pd(OH)_2/C$ catalysis (Scheme 5.174) [631]. Regioselective coupling at the C8-position was obtained with a variety of aryl iodides, bromides, and chlorides, although for aryl chlorides a longer reaction time (1–2 h) was necessary compared to the corresponding iodide or bromide derivatives. Advantages of this general protocol are that there is no need for protecting free NH_2 substituents and that CuI is used in

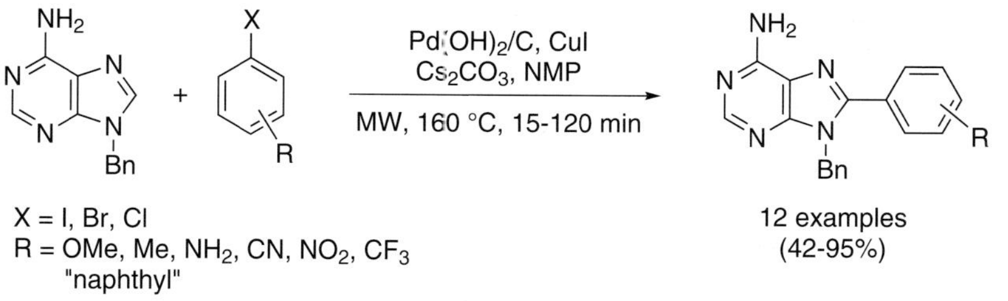

Scheme 5.174 Direct C—H arylation of adenines.

stoichiometric amounts. β-(*E*)-Bromostyrene as an example for vinyl halide coupling was also successfully reacted giving exclusively the *E*-isomer in 55% yield.

Along similar lines, the Van der Eycken group has described the Pd-catalyzed direct C–H arylation of imidazo[1,2-*a*]pyrimidines using aryl bromides [632].

The Pd-catalyzed, Cu-mediated C-2 direct arylation of 5-substituted oxazoles with aryl bromides was disclosed by Piguel and coworkers (Scheme 5.175) [633]. This method shows a high functional group tolerance and in addition the reaction times can be reduced from hours to minutes when employing microwave heating. In particular, electron-poor oxazoles need only very short reaction times of 4–8 min to give the 2,5-diaryloxazoles in higher yields.

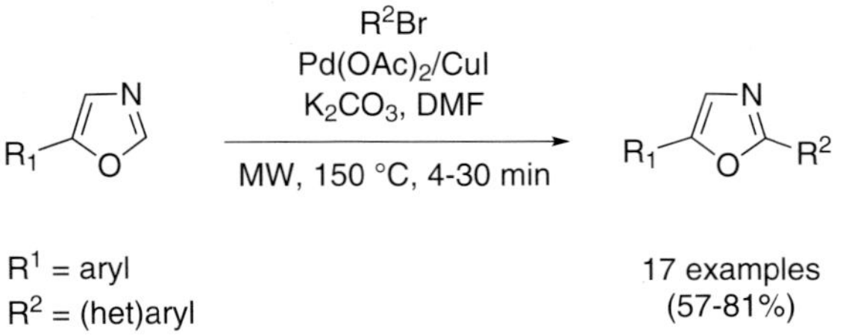

Scheme 5.175 Direct C–H arylation of oxazoles.

Yorimitsu and coworkers have disclosed Pd-catalyzed direct arylations of 1,4-disubstituted triazoles with aryl chlorides (Scheme 5.176) [634]. Slightly longer reaction times (20–30 min) were required in order to achieve high product yields when hexyl-substituted triazoles ($R^1 = n\text{-}C_6H_{13}$) are employed as precursors. Interestingly, when tricyclohexylphosphine [$P(Cy)_3$] is used as ligand, aryl chlorides were superior compared to aryl bromides and iodides. A solvent mixture of toluene:DMF 5/1 is crucial for the outcome of the reaction, since pure toluene cannot be heated to 250 °C, whereas pure DMF afforded the products in lower yields.

Scheme 5.176 Direct C–H arylation of substituted triazoles.

A protocol for the Pd-catalyzed fluorination of C–H bonds was developed by the group of Sanford (Scheme 5.177) [635]. This reaction proceeds via a $Pd(OAc)_2$-catalyzed C–H activation/oxidative fluorination step where an electrophilic

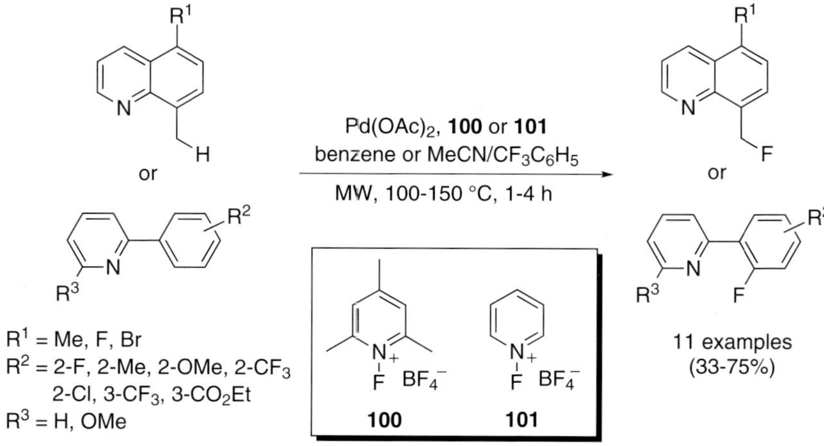

Scheme 5.177 Pd-catalyzed C–F-couplings.

fluorinating reagent (**100** or **101**) is necessary for successful aromatic and benzylic C–F bond formations. Compared to thermal heating, higher yields could be achieved with microwave irradiation due to shorter reaction times.

Bedford and Betham have disclosed a Pd-catalyzed one-pot process for the synthesis of carbazoles from 2-chloroanilines and aryl bromides that relies on consecutive amination and C–H activation chemistry [636,637]. A somewhat related approach was chosen by Maes and coworkers to generate 9*H*-α-carbolines from 2,3-dichloropyridines and substituted anilines [638]. A selective sp^2 and benzylic sp^3 Pd-catalyzed arylation in the picoline *N*-oxide series has been disclosed by Fagnou and coworkers [639].

In 2009, the group of Ellman investigated the synthesis of 5-aryl benzotriazepines by copper-catalyzed direct arylation [640]. Utilizing CuI as catalyst and lithium *tert*-butoxide as base in combination with microwave heating, reaction times can be significantly reduced (from 12 to 1 h) and working under inert conditions is no longer required (Scheme 5.178).

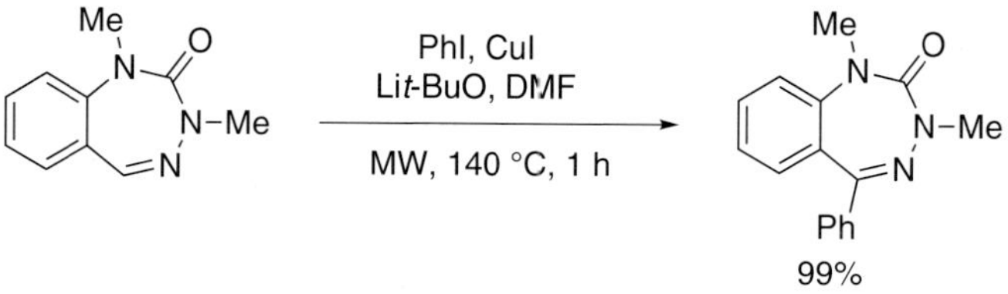

Scheme 5.178 Cu(I)-catalyzed direct arylation of benzotriazepines.

Johansen and Kerr reported on the functionalization of indoles by copper-catalyzed insertions [641]. Utilizing dimethyl diazomalonate as carbene transfer reagent allows regioselective C–H insertion on the pyrrole moiety of the indole scaffold. Especially

2-malonyl indole is an important intermediate in the total synthesis of natural compounds. Dimethyl diazomalonate together with 1.5 equiv of the corresponding indole and 1 mol% copper(II) acetylacetonate in benzene is heated at 100 °C for 2 h. 3-Substituted indoles furnish high yields of the C2−H insertion product, whereas 2-substituted derivatives led to the corresponding C3−H insertion compounds (Scheme 5.179).

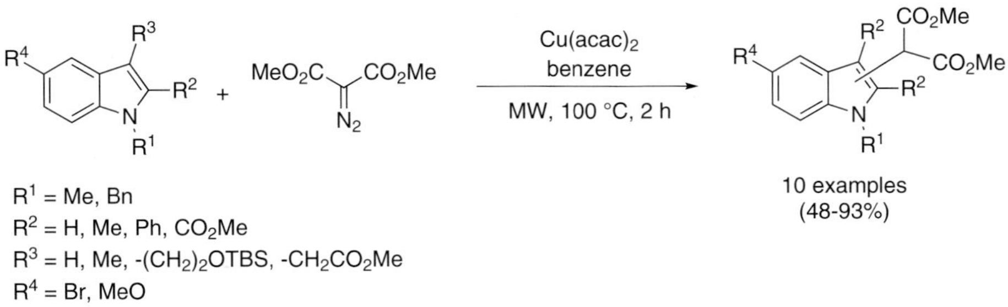

Scheme 5.179 Microwave-promoted malonyl carbenoid insertion.

5.4.4
Copper-Catalyzed Azide–Acetylene Cycloaddition (CuAAC)

First described almost a decade ago [642], "click" reactions such as the Cu(I)-catalyzed azide–alkyne cycloaddition (CuAAC) are widely used today in organic and medicinal chemistry. Kappe and coworkers exploited a microwave-assisted version of the Cu(I)-catalyzed azide–acetylene ligation process (click chemistry) for the preparation of 6-(1,2,3-triazol-1-yl)-dihydropyrimidones (Scheme 5.180) [643]. Here, a suitable heterocyclic azide intermediate (obtained by microwave-assisted azidation) was treated with phenyl acetylene in *N,N*-dimethylformamide employing 2 mol% Cu(II) sulfate/sodium ascorbate as catalyst precursor. After completion of the cycloaddition process, the triazole product can be precipitated in high yield (73%) and with purity by addition to ice/water. For the model reaction displayed in Scheme 5.180, full conversion at room temperature required 1 h. By carrying out the same reaction utilizing controlled microwave heating at 80 °C, complete conversion was achieved within 1 min. A library of 27 6-(1,2,3-triazol-1-yl)-dihydropyrimidones was prepared with 4 points of diversity.

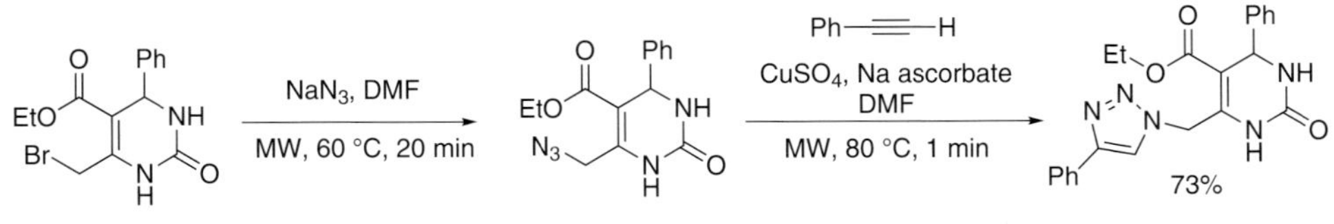

Scheme 5.180 Cu(I)-catalyzed azide–acetylene ligations (click chemistry).

For certain substrates, Fokin, Van der Eycken and coworkers discovered that the azidation and ligation step can be carried out in a one-pot fashion, thereby simplifying the overall protocol [644]. This procedure eliminates the need to handle organic azides, as they are generated *in situ*. Similarly, triazoles can be obtained directly from amines in one pot using Cu(II)-catalyzed diazo transfer with trifluoromethanesulfonyl azide, coupled with the CuAAC reaction [645]. An alternative *in situ* method for azide generation uses the reagent combination *t*-BuONO and TMS-N_3 [646]. The Lipshutz and Taft have shown that copper-on-charcoal (cf. Scheme 5.148) is an effective catalyst for CuAAC chemistry under microwave conditions [647].

The synthesis of 1,2,3-triazoles via Ru-catalyzed aryl azide–alkyne cycloadditions was described by the group of Fokin (Scheme 5.181) [648]. In comparison to the Cu(I)-catalyzed azide–alkyne cycloaddition where 1,4-disubstituted triazoles are obtained (see Scheme 5.180), the Ru-catalyzed version produces 1,5-regioisomers of 1,2,3-triazoles. $[Cp^*RuCl]_4$ proved to be superior in activity than the original $Cp^*RuCl(PPh_3)_2$ catalyst and in combination with microwave heating, higher yields, cleaner products, and shorter reaction times were accomplished. By-product formation and lower yields were noticed under conventional heating conditions in an oil bath.

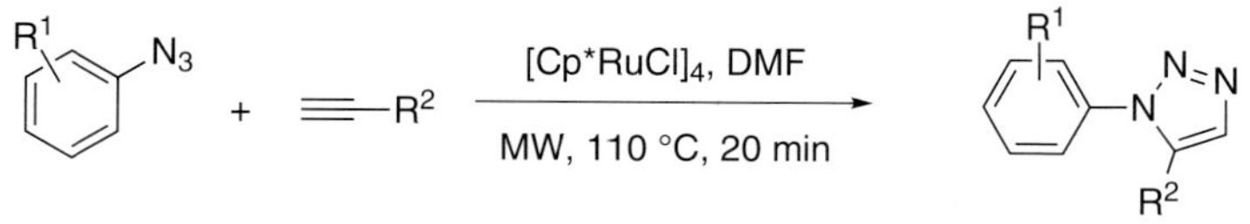

R^1 = H, Me, OMe, Cl, I, COOEt
R^2 = alkyl, hetaryl

15 examples
(43-92%)

$[Cp^*RuCl]_4$: pentamethylcyclopentadienyl ruthenium(II) chloride tetramer

Scheme 5.181 Ru-catalyzed aryl azide–alkyne cycloaddition.

For an application of the Ru-catalyzed azide–alkyne cycloaddition reaction in carbohydrate chemistry, see Ref. [649].

In a subsequent work, Gevorgyan and coworkers have reported on the synthesis of imidazoles via the transannulation of 1-sulfonyl triazoles with nitriles using Rh(II) octanoate as catalyst (Scheme 5.182) [650]. When applying microwave heating, no

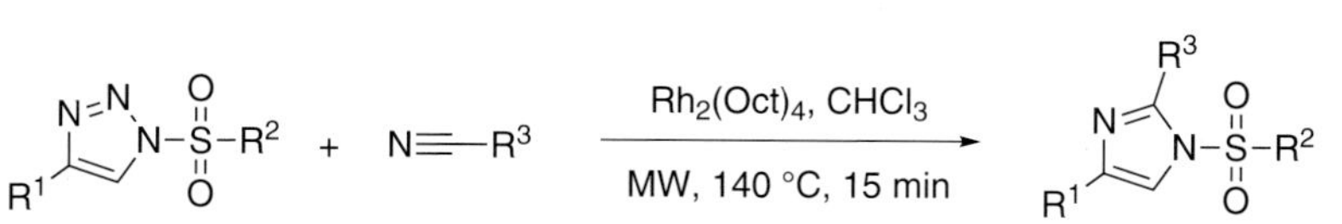

R^1 = aryl, Bn
R^2 = aryl
R^3 = aryl, alkyl, alkenyl

15 examples
(42-88%)

Scheme 5.182 Rh-catalyzed transannulation of 1,2,3-triazoles.

inert atmosphere is necessary compared to performing the reactions under conventional heating at 80 °C where – in addition – longer reaction times of 12–24 h are needed. Furthermore, the authors were successful in combining the synthesis of 1-sulfonyl triazoles – via the Cu(I)-catalyzed cycloaddition of sulfonyl azides with alkynes – and the transannulation in a one-pot two-step sequence.

Applications of CuAAC reaction in the carbohydrate [651,652], glycodendrimer [653], cyclodextrin [654], glycocyclodextrin [655], porphyrine [656], and nucleic acid field [657] have been reported. For the preparation of triazol oligomers using this approach, see Ref. [658]. The use of chiral acetylene building blocks in CuAAC reaction has also been described [659].

5.4.5
Miscellaneous Reactions

In the context of a total synthesis toward the C-1–C-28 ABCD unit of the marine macrolide spongistatin 1, Ley and coworkers have reported an efficient methylenation reaction of an advanced ketone intermediate using the titanium-based Petasis reagent (Scheme 5.183a [660]. Treatment of the ketone with the Petasis reagent in toluene at 120 °C for 3 h proved to be the optimal conditions, generating the desired alkene in 71% yield. The reaction proved to be much more efficient when carried out under sealed-vessel microwave heating at 160 °C, forming the alkene after 10 min in an improved 82% yield. An ionic liquid (1-ethyl-3-methylimidazolium hexafluorophosphate) (emimPF$_6$) was utilized to modify the dielectric properties of the solvent (see Section 4.5.2).

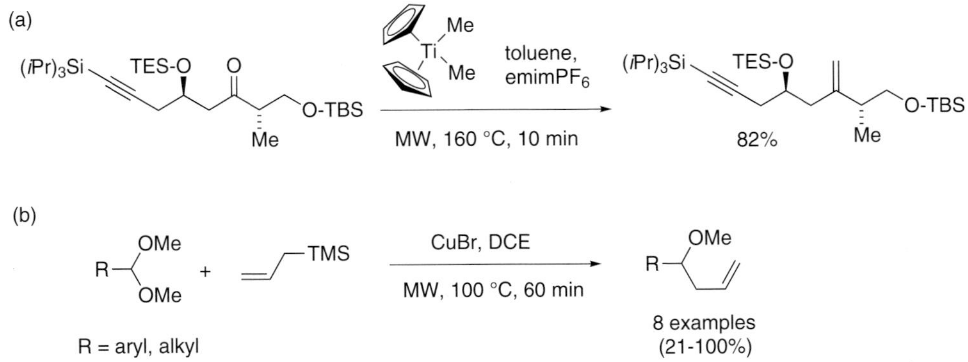

Scheme 5.183 (a) Methylenation of ketones with the Petasis reagent. (b) Copper(I) bromide-mediated allylation of acetals.

Jung and Maderna have reported the microwave-assisted allylation of acetals with allyltrimethylsilane in the presence of copper(I) bromide as promoter (Scheme 5.183b) [661]. Stoichiometric amounts of copper(I) bromide are required and the reaction works best for aromatic acetals in the absence of strong electron-withdrawing substituents on the aromatic ring.

Mejía-Oneto and Padwa have described the rhodium(II) perfluorobutyrate-catalyzed decomposition of an α-diazo ketoamide precursor (Scheme 5.184) [662]. Microwave heating of a benzene solution of the diazo compound with catalytic amounts of the rhodium(II) carboxylate catalyst unexpectedly led to a lactam, formed by a formal insertion of the metal carbene into the carbon–hydrogen bond at the 5-position of the pyridone ring followed by an ethoxy-decarboxylation.

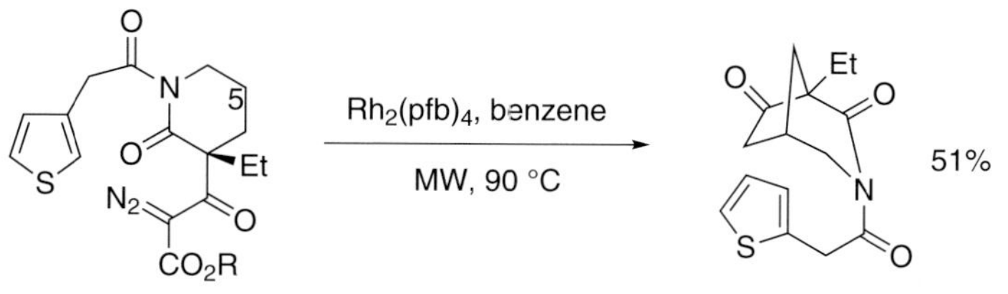

Scheme 5.184 Rhodium(II)-catalyzed carbon–hydrogen insertion.

Snyder and Jones investigated Rh(I)-catalyzed intramolecular [2 + 2 + 2] cyclizations of diynes tethered to enones [663]. After microwave heating of the cyclization precursors using tris(triphenylphosphine) rhodium(I) chloride ([RhCl(PPh$_3$)$_3$], Wilkinson's catalyst), at 150 °C for 15 min, the crude residue needed to be stirred at room temperature for 1 h with 2,3-dichloro-5,6-dicyano-1,4-benzoquinone (DDQ) in order to achieve full aromatization (Scheme 5.185). Both internal and terminal alkynes worked well and the analogous five- and six-membered rings resulting from varying the tether length could be produced. Depending on the location of the carbonyl group, endo- (**102**) or exocyclic products (**103**) could be obtained (Scheme 5.185). In addition, the synthesis of the marine sesquiterpene (−)-alcyopterosin I has been described involving an additional reduction step of the ketone with borohydride.

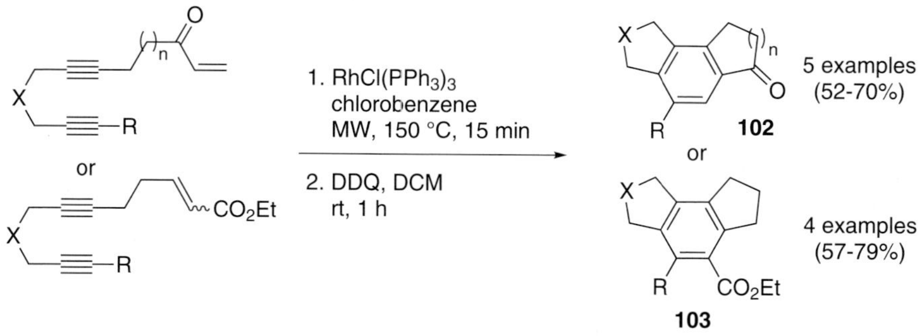

Scheme 5.185 Rh(I)-catalyzed intramolecular [2 + 2 + 2] cyclizations.

Taddei and coworkers investigated domino hydroformylation/cyclization reactions toward terminal olefins containing various C, N, or O nucleophiles [664]. The first step of the sequence to prepare valuable heterocyclic scaffolds is a rhodium-catalyzed hydroformylation in THF employing syngas (H_2/CO) (Scheme 5.186). The formed aldehydes are subsequently cyclized by the aid of corresponding additives. Utilizing sodium acetate/acetic acid, the corresponding heterocyclic acetate is generated, whereas water furnishes acetals. Switching to ethanol as a solvent, consequently, ethyl acetals are formed. The procedure proved tolerable for a variety of unsaturated alcohols, whereas nitrogen-containing nucleophiles proceeded rather sluggish under microwave heating. In a subsequent process, the generated acetates can be transformed into spirodihydrofurans, which serve as versatile intermediates in medicinal chemistry research (Scheme 5.186).

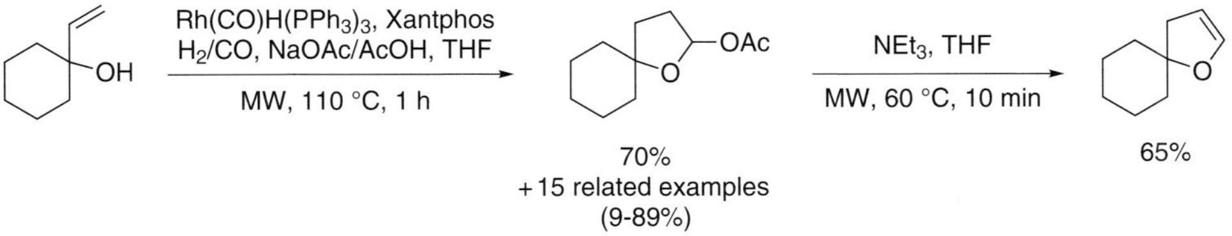

Scheme 5.186 Microwave-assisted tandem hydroformylation/cyclization of unsaturated alcohols.

The group of Einhorn reported on the regiospecific synthesis of 1,3-diarylisobenzofurans by metal-catalyzed couplings of boronic acids with aryl aldehydes (Scheme 5.187) [665]. For the activation of the aryl boronic substrates, either Pd (II) or Rh(III) catalysis turned out to be feasible. Two mol% **104** as *N*-heterocyclic carbene ligand precursor and 2 mol% Rh(III) chloride trihydrate as catalyst/ligand system were employed. Subsequent acidic treatment at room temperature to maximize the yield by converting intermediate lactols to the corresponding isobenzofurans furnished the desired products in moderate to good yields. The alternative Pd-route utilized 10 mol% $PdCl_2$ with 10 mol% *tri*-1-naphthylphospine as catalytic system. Microwave heating and acidic treatment remained unchanged. Within their investigations, the authors identified an interesting case when aryl aldehyde **105** was used as a substrate in the Rh-catalyzed protocol (Scheme 5.187). While classical heating at 90 °C for 48 h almost quantitatively led to the corresponding Suzuki coupling product, microwave heating at 90 °C for 80 min exclusively furnished the desired 5-iodo isobenzofuran. This result was addressed to the formation of different catalytic species from the $RhCl_3$/NHC mixture according to the heating process. Thus, microwave heating generates the hydroxorhodium(I) species (L-Rh-OH) for addition on the carbonyl, whereas classical heating forms the chlororhodium(I) species (L-Rh-Cl) for Suzuki cross-couplings.

The intramolecular [4 + 2] cycloaddition of alkenyl-substituted **106** and aryl-substituted 1,6-enynes **107** was described by the group of Echavarren (Scheme 5.188) [666]. The transformations proceed via catalysis involving cationic Au(I) complexes and furnish bicyclic products from **106** and tricyclic derivatives from **107**. Heating under microwave conditions leads to reduced reaction times (hour versus minute) and improved yields – precatalyst **109** proved to be more reactive than **108** especially

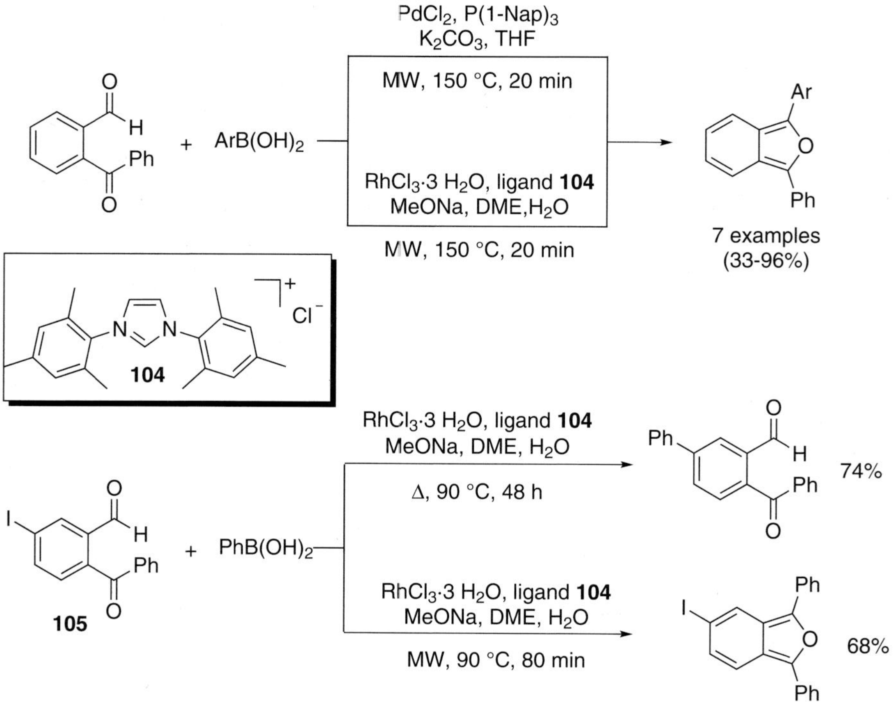

Scheme 5.187 Regiospecific synthesis of functionalized 1,3-diarylisobenzofurans.

under room-temperature conditions. Substrates **107** with *m*-substituents at the aryl moiety gave mixtures of regioisomers, whereas other minor by-product formation was observed via a 6-*endo* cyclization pathway.

The Au-catalyzed rearrangement of allylic acetates was described by Nolan and coworkers from the same institute [667].

Che and coworkers have developed a method for the synthesis of substituted 1,2-dihydroquinolines and quinolines under Au catalysis (Scheme 5.189) [668]. The reaction of anilines **110** with alkynes **111** employing the Au(I) catalyst **112**/AgOTf combination afforded the corresponding dihydroquinolines **113** via a tandem hydroamination–hydroarylation sequence. By utilizing the proper substituted alkynes **111**, products containing multiple alkyne groups can also be obtained. Under similar conditions, quinolines **115** can be synthesized when alkynes are reacted with *o*-acetyl or *o*-benzoyl-substituted anilines **114**.

The same group has also disclosed related inter- and intramolecular hydroaminations of alkenes using Au catalysis [669] and intermolecular hydroarylations of alkenes with indoles [670]. Late transition metals have been shown to be efficient catalysts in the intramolecular hydroamination of *C*-propargyl vinylogous amides into pyrroles [671].

The synthesis of *N*-protected β-amino aldehydes via an alkynylation/anti-Markovnikov alkyne hydration sequence was developed by the groups of Hintermann and coworkers (Scheme 5.190) [672]. By performing the Ru-catalyzed hydration under

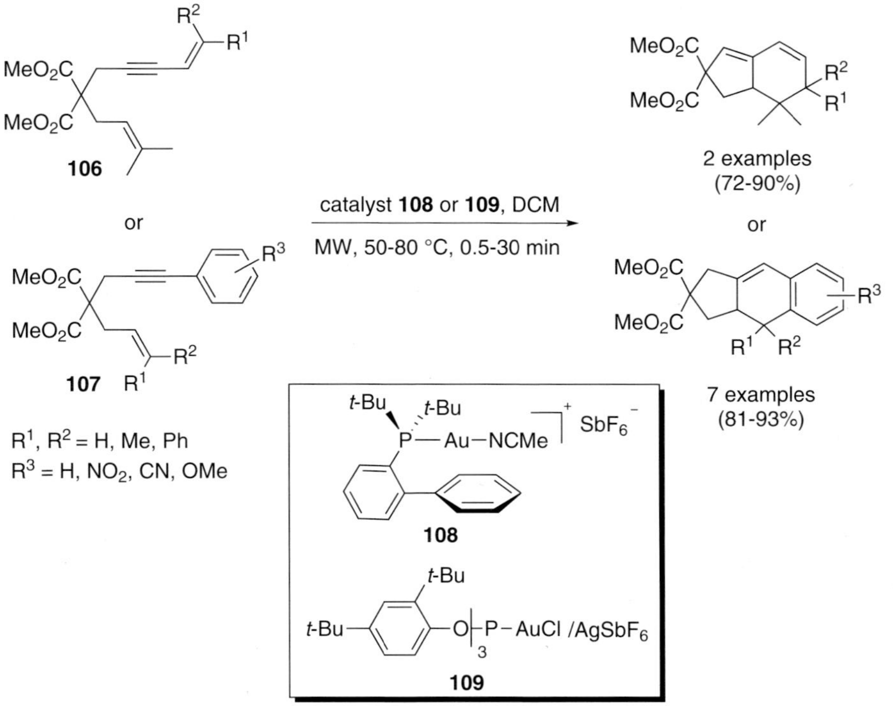

Scheme 5.188 Au(I)-catalyzed intramolecular [4 + 2] cycloadditions.

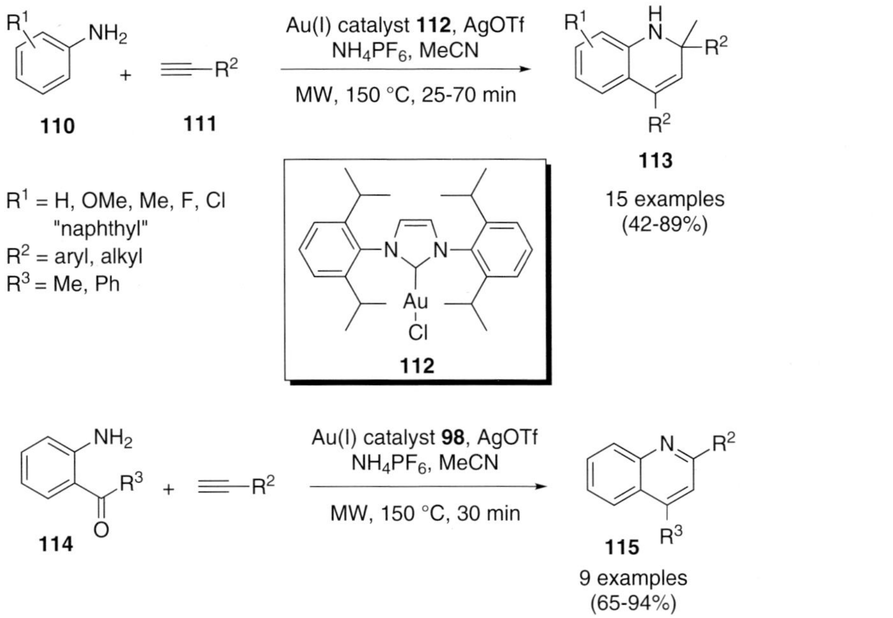

Scheme 5.189 Dihydroquinoline and quinoline synthesis via tandem hydroamination–hydroarylation.

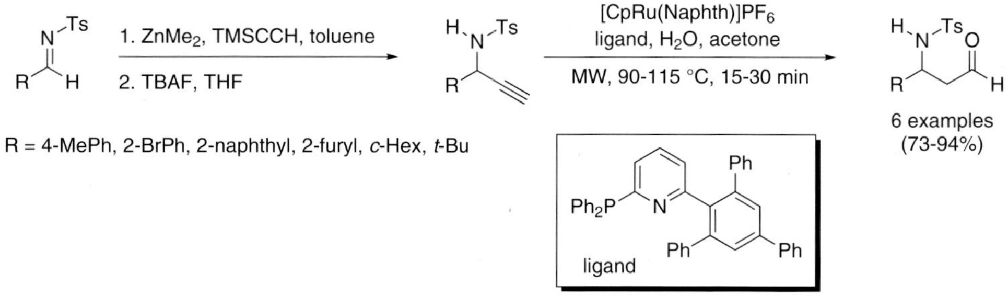

Scheme 5.190 Synthesis of β-amino aldehydes.

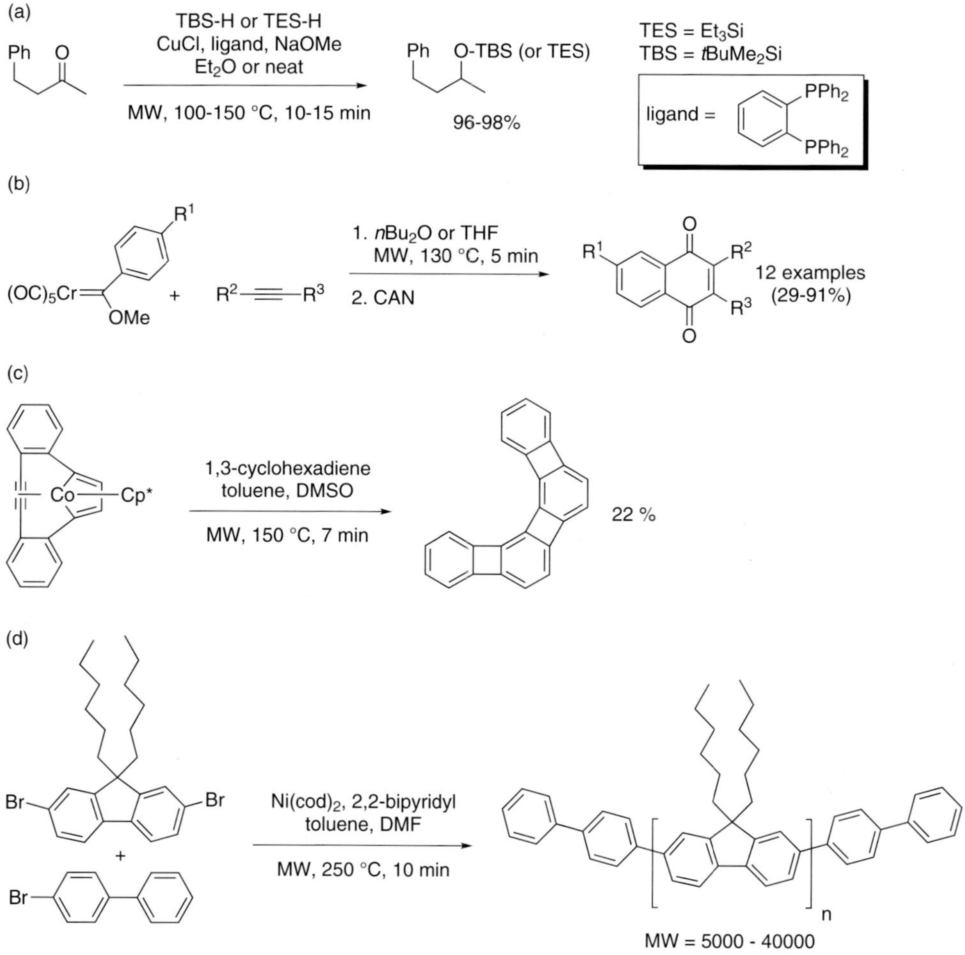

Scheme 5.191 (a) Hydrosilylation of ketones [679], (b) Dötz benzannulation [680], (c) Cobalt-mediated synthesis of angular [4] phenylenes [681], (d) Nickel-mediated coupling polymerizations [682].

microwave heating, the reaction times could be reduced from several hours at 55 °C to a maximum of 30 min. In addition, for most of the derivatives, the catalyst loading could be reduced to 5 mol%, compared to thermal heating, without any decrease in yields.

In a related work, the Rh-catalyzed hydrophosphinylation of ethynyl steroids was disclosed by Stockland *et al.* [673].

For a study on Ru-catalyzed isomerizations of alkenols to alkanones, see Ref. [674]. A Ru-catalyzed cycloisomerization of 1,6-dienes has also been described [675]. The Pd-catalyzed bisdiene cyclizations have been reported by Takacs *et al.* [676]. The Pd-catalyzed regioselective cyanothiolation of alkynes with thiocyanates has been described by Chung and coworkers [677]. For examples of Cu(II)-promoted intramolecular carboaminations of olefins, see Ref. [678].

Other microwave-assisted reactions involving metal catalysts or metal-based reagents are shown in Scheme 5.191 [679–682].

References

1 Loupy, A. (2002) *Microwaves in Organic Synthesis*, Wiley-VCH Verlag GmbH, Weinheim.
2 Loupy, A. (2006) *Microwaves in Organic Synthesis*, 2nd edn, Wiley-VCH Verlag GmbH, Weinheim.
3 Hayes, B.L. (2002) *Microwave Synthesis: Chemistry at the Speed of Light*, CEM Publishing, Matthews, NC.
4 Bogdal, D. (2005) *Microwave-Assisted Organic Synthesis: One Hundred Reaction Procedures*, Elsevier, Oxford.
5 Lidström, P. and Tierney, J.P. (2005) *Microwave-Assisted Organic Synthesis*, Blackwell Publishing, Oxford.
6 Kappe, C.O. and Stadler, A. (2005) *Microwaves in Organic and Medicinal Chemistry*, Wiley-VCH Verlag GmbH, Weinheim.
7 Larhed, M. and Olofsson, K. (2006) *Microwave Methods in Organic Synthesis*, Springer, Berlin.
8 Van der Eycken, E. and Kappe, C.O. (2006) *Microwave-Assisted Synthesis of Heterocycles*, Springer, Berlin.
9 McGowan, C. and Leadbeater, N.E. (2006) *Clean, Fast Organic Chemistry: Microwave-Assisted Laboratory Experiments*, CEM Publishing, Matthews, NC.
10 Kappe, C.O., Dallinger, D., and Murphree, S.S. (2009) *Practical Microwave Synthesis for Organic Chemists: Strategies, Instruments, and Protocols*, Wiley-VCH Verlag GmbH, Weinheim.
11 Polshettiwar, V. and Varma, R.S. (2010) *Aqueous Microwave Assisted Chemistry: Synthesis and Catalysis*, RSC Publishing, Cambridge, UK.
12 Leadbeater, N.E. (2011) *Microwave Heating as a Tool for Sustainable Chemistry*, CRC Press, Boca Raton.
13 Stuerga, D. (2006) *Microwaves in Organic Synthesis*, 2nd edn (ed. A. Loupy), Wiley-VCH Verlag GmbH, Weinheim, pp. 1–61.
14 Michael, D. and Mingos, P. (2005) *Microwave-Assisted Organic Synthesis* (eds P. Lidström and J.P. Tierney), Blackwell Publishing, Oxford, pp. 1–22.
15 Stuerga, D. and Delmotte, M. (2002) *Microwaves in Organic Synthesis* (ed. A. Loupy), Wiley-VCH Verlag GmbH, Weinheim, pp. 1–33.
16 Gabriel, C., Gabriel, S., Grant, E.H., Halstead, B.S.J., and Mingos, D.M.P. (1998) *Chemical Society Reviews*, **27**, 213–224.
17 Mingos, D.M.P. and Baghurst, D.R. (1991) *Chemical Society Reviews*, **20**, 1–47.

18 Perreux, L. and Loupy, A. (2006) *Microwaves in Organic Synthesis*, 2nd edn (ed. A. Loupy), Wiley-VCH Verlag GmbH, Weinheim, pp. 134–218.

19 Loupy, A. and Varma, R.S. (2006) *Chimica Oggi: Chemistry Today*, **24** (3), 36–40.

20 de la Hoz, A., Díaz-Ortiz, A., and Moreno, A. (2006) *Microwaves in Organic Synthesis*, 2nd edn (ed. A. Loupy), Wiley-VCH Verlag GmbH, Weinheim, pp. 219–277.

21 de la Hoz, A., Díaz-Ortiz, A., and Moreno, A. (2005) *Chemical Society Reviews*, **34**, 164–178.

22 de la Hoz, A., Díaz-Ortiz, A., and Moreno, A. (2004) *Current Organic Chemistry*, **8**, 903–918.

23 Panunzio, M., Campana, E., Martelli, G., Vicennati, P., and Tamanini, E. (2004) *Materials Research Innovations*, **8**, 27–31.

24 Perreux, L. and Loupy, A. (2002) *Microwaves in Organic Synthesis* (ed. A. Loupy), Wiley-VCH Verlag GmbH, Weinheim, pp. 61–114.

25 Kuhnert, N. (2002) *Angewandte Chemie International Edition*, **41**, 1863–1866.

26 Strauss, C.R. (2002) *Angewandte Chemie International Edition*, **41**, 3589–3590.

27 Perreux, L. and Loupy, A. (2001) *Tetrahedron*, **57**, 9199–9223.

28 Loupy, A., Perreux, L., Liagre, M., Burle, K., and Moneuse, M. (2001) *Pure and Applied Chemistry*, **73**, 161–166.

29 Ondruschka, B., Bonrath, W., and Stuerga, D. (2006) *Microwaves in Organic Synthesis*, 2nd edn (ed. A. Loupy), Wiley-VCH Verlag GmbH, Weinheim, pp. 62–107.

30 Le Ngoc, T., Roberts, B.A., and Strauss, C.R. (2006) *Microwaves in Organic Synthesis*, 2nd edn (ed. A. Loupy), Wiley-VCH Verlag GmbH, Weinheim, pp. 108–133.

31 Strauss, C.R. (2002) *Microwaves in Organic Synthesis* (ed. A. Loupy), Wiley-VCH Verlag GmbH, Weinheim, pp. 35–60.

32 Ondruschka, B. and Bonrath, W. (2006) *Chimia*, **60**, 326–329.

33 Stankiewicz, A. (2006) *Chemical Engineering Research & Design*, **84**, 511–521.

34 Nüchter, M., Ondruschka, B., Weiß, D., Beckert, R., Bonrath, W., and Gum, A. (2005) *Chemical Engineering & Technology*, **28**, 871–881.

35 Nüchter, M., Ondruschka, B., Bonrath, W., and Gum, A. (2004) *Green Chemistry*, **6**, 128–141.

36 Nüchter, M., Müller, U., Ondruschka, B., Tied, A., and Lautenschläger, W. (2003) *Chemical Engineering & Technology*, **26**, 1207–1216.

37 Kappe, C.O. (2008) *Chemical Society Reviews*, **37**, 1127–1139.

38 Kappe, C.O. (2006) *Comprehensive Medicinal Chemistry II* (eds J.B. Taylor and D.J. Triggle), Elsevier, Oxford, pp. 837–860.

39 Varma, R.S. (2006) *Kirk-Othmer Encyclopedia of Chemical Technology*, 5th edn (ed. A. Seidel) Wiley-VCH Verlag GmbH, Weinheim, pp. 538–594.

40 Kappe, C.O. (2006) *Chimia*, **60**, 308–312.

41 de la Hoz, A., Díaz-Ortiz, A., and Moreno, A. (2005) *Advances in Organic Synthesis*, **1**, 119–171.

42 Romanova, N.N., Gravis, A.G., and Zyk, N.V. (2005) *Russian Chemical Reviews*, **74**, 969–1013.

43 Kuznetsov, D.V., Raev, V.A., Kuranov, G.L., Arapov, O.V., and Kostikov, R.R. (2005) *Russian Journal of Organic Chemistry*, **41**, 1719–1749.

44 Kappe, C.O. (2004) *Angewandte Chemie International Edition*, **43**, 6250–6284.

45 Hayes, B.L. (2004) *Aldrichimica Acta*, **37**, 66–77.

46 Gedye, R.N. (2002) *Microwaves in Organic Synthesis* (ed. A. Loupy), Wiley-VCH Verlag GmbH, Weinheim, pp. 115–146.

47 Lidström, P., Tierney, J., Wathey, B., and Westman, J. (2001) *Tetrahedron*, **57**, 9225–9283.

48 Krstenansky, J.L. and Cotterill, I. (2000) *Current Opinion in Drug Discovery & Development*, **4**, 454–461.

49 Bose, A.K., Banik, B.K., Lavlinskaia, N., Jayaraman, M., and Manhas, M.S. (1997) *Chemtech*, **27**, 18–24.

50 Caddick, S. (1995) *Tetrahedron*, **51**, 10403–10432.

51 Strauss, C.R. and Trainor, R.W. (1995) *Australian Journal of Chemistry*, **48**, 1665–1692.

52 Majetich, G. and Wheless, K. (1997) *Microwave-Enhanced Chemistry: Fundamentals, Sample Preparation, and Applications* (eds H.M. Kingston and S.J. Haswell), American Chemical Society Publication, Washington DC, pp. 455–505.

53 Abramovitch, R.A. (1991) *Organic Preparations and Procedures International*, **23**, 685–711.

54 Kappe, C.O. and Dallinger, D. (2009) *Molecular Diversity*, **13**, 71–193.

55 Stockland, R.A., Jr. (2011) *Microwave Heating as a Tool for Sustainable Chemistry* (ed. N.E. Leadbeater), CRC Press, Boca Raton, pp. 25–51.

56 Caddick, S. and Fitzmaurice, R. (2009) *Tetrahedron*, **65**, 3325–3355.

57 Varma, R.S. and Ju, Y. (2006) *Microwaves in Organic Synthesis*, 2nd edn (ed. A. Loupy), Wiley-VCH Verlag GmbH, Weinheim, pp. 362–415.

58 Bougrin, K., Loupy, A., and Soufiaoui, M. (2005) *Journal of Photochemistry and Photobiology C: Photochemistry Reviews*, **6**, 139–167.

59 Loupy, A. (2004) *Comptes Rendus Chimie*, **7**, 103–112.

60 Varma, R.S. (2002) *Microwaves in Organic Synthesis* (ed. A. Loupy), Wiley-VCH Verlag GmbH, Weinheim, pp. 181–218.

61 Varma, R.S. (2002) *Tetrahedron*, **58**, 1235–1255.

62 Pillai, U.R., Sahle-Demessie, E., and Varma, R.S. (2002) *Journal of Materials Chemistry*, **12**, 3199–3207.

63 Varma, R.S. (2001) *Pure and Applied Chemistry*, **73**, 193–198.

64 Kidwai, M. (2001) *Pure and Applied Chemistry*, **73**, 147–151.

65 Varma, R.S. (1999) *Green Chemistry*, **1**, 43–55.

66 Loupy, A., Petit, A., Hamelin, J. *et al.* (1998) *Synthesis*, 1213–1234.

67 Besson, T., Thiéry, V., and Dubac, J. (2006) *Microwaves in Organic Synthesis*, 2nd edn (ed. A. Loupy), Wiley-VCH Verlag GmbH, Weinheim, pp. 416–455.

68 Laporterie, A., Marquié, J., and Dubac, J. (2002) *Microwaves in Organic Synthesis* (ed. A. Loupy), Wiley-VCH Verlag GmbH, Weinheim, pp. 219–252.

69 Bogdal, D. and Loupy, A. (2008) *Organic Process Research & Development*, **12**, 710–722.

70 Loupy, A., Petit, A., and Bogdal, D. (2006) *Microwaves in Organic Synthesis*, 2nd edn (ed. A. Loupy), Wiley-VCH Verlag GmbH, Weinheim, pp. 278–326.

71 Loupy, A., Petit, A., and Bogdal, D. (2002) *Microwaves in Organic Synthesis* (ed. A. Loupy), Wiley-VCH Verlag GmbH, Weinheim, pp. 147–180.

72 Petricci, E. and Taddei, M. (2007) *Chimica Oggi: Chemistry Today*, **25** (3), 40–45.

73 Polshettiwar, V. and Varma, R.S. (2008) *Accounts of Chemical Research*, **41**, 629–639.

74 Polshettiwar, V. and Varma, R.S. (2007) *Current Opinion in Drug Discovery & Development*, **10**, 723–737.

75 Strauss, C.R. and Varma, R.S. (2006) *Topics in Current Chemistry*, **266**, 199–231.

76 Varma, R.S. (2006) *Indian Journal of Chemistry Section B: Organic Chemistry Including Medicinal Chemistry*, **45**, 2305–2312.

77 Lévêque, J.M. and Cravotto, G. (2006) *Chimia*, **60**, 313–320.

78 Roberts, B.A. and Strauss, C.R. (2005) *Accounts of Chemical Research*, **38**, 653–661.

79 Strauss, C.R. (2005) *Chemistry in Australia*, **72**, 9–11.

80 Carrillo, J.R., Díaz-Ortiz, A., de la Hoz, A., Moreno, A., Gomez, M.V., Prieto, P., Sanchez-Migallon, A., and Vazquez, E. (2003) *Targets in Heterocyclic Systems* (eds O.A. Attanasi and D. Spinelli), Springer, Berlin, pp. 64–85.

81 Bose, A.K., Manhas, M.S., Ganguly, S.N., Sharma, A.H., and Banik, B.K. (2002) *Synthesis*, 1578–1591.

82 Strauss, C.R. (2002) *Handbook of Green Chemistry and Technology* (eds J. Clark and D. Macquarrie), Blackwell Publishing, Oxford, pp. 397–415.

83 Varma, R.S. (2002) Kavitha Printers, Bangalore.

84 Strauss, C.R. (1999) *Australian Journal of Chemistry*, **52**, 83–96.

85 Schmink, J.R. and Leadbeater, N.E. (2011) *Microwave Heating as a Tool for Sustainable Chemistry* (ed. N.E. Leadbeater), CRC Press, Boca Raton, pp. 1–24.

86 Polshettiwar, V., Nadagouda, M.N., and Varma, R.S. (2009) *Australian Journal of Chemistry*, **62**, 16–26.

87 Strauss, C.R. (2009) *Australian Journal of Chemistry*, **62**, 3–15.

88 Polshettiwar, V. and Varma, R.S. (2009) *Eco-Friendly Synthesis of Fine Chemicals* (ed. R. Ballini), RSC Publishing, Cambridge, UK, pp. 275–292.

89 Polshettiwar, V. and Varma, R.S. (2008) *Chemical Society Reviews*, **37**, 1546–1557.

90 Dallinger, D. and Kappe, C.O. (2007) *Chemical Reviews*, **107**, 2563–2591.

91 Leadbeater, N.E. (2005) *Chemical Communications*, **23**, 2881–2902.

92 Polshettiwar, V. and Varma, R.S. (2010) *Aqueous Microwave Assisted Chemistry: Synthesis and Catalysis* (eds V. Polshettiwar, and R.S. Varma), RSC Publishing, Cambridge, UK, pp. 1–24.

93 Polshettiwar, V., Decottignies, A., Len, C., and Fihri, A. (2010) *ChemSusChem*, **3**, 502–522.

94 Lévêque, J.-M., Estager, J., Draye, M., Cravotto, G., Boffa, L., and Bonrath, W. (2007) *Monatshefte fur Chemie*, **138**, 1103–1113.

95 Leadbeater, N.E. and Torenius, H.M. (2006) *Microwaves in Organic Synthesis*, 2nd edn (ed. A. Loupy), Wiley-VCH Verlag GmbH, Weinheim, pp. 327–361.

96 Habermann, J., Ponzi, S., and Ley, S.V. (2005) *Mini-Reviews in Organic Chemistry*, **2**, 125–137.

97 Leadbeater, N.E., Torenius, H.M., and Tye, H. (2004) *Combinatorial Chemistry & High Throughput Screening*, **7**, 511–528.

98 Martinez-Palou, R. (2007) *Journal of Mexican Chemical Society*, **51**, 252–264.

99 Martinez-Palou, R. (2010) *Molecular Diversity*, **14**, 3–25.

100 Boros, E., Seddon, K.R., and Strauss, C.R. (2008) *Chimica Oggi: Chemistry Today*, **26** (6), 28–30.

101 Alcázar, J., Diels, G., and Schoentjes, B. (2007) *Mini-Reviews in Medicinal Chemistry*, **7**, 345–369.

102 Chighine, A., Sechi, G., and Bradley, M. (2007) *Drug Discovery Today*, **12**, 459–464.

103 Mavandadi, F. and Pilotti, A. (2006) *Drug Discovery Today*, **11**, 165–174.

104 Kappe, C.O. and Dallinger, D. (2006) *Nature Reviews. Drug Discovery*, **5**, 51–64.

105 Shipe, W.D., Wolkenberg, S.E., and Lindsley, C.W. (2005) *Drug Discovery Today: Technologies*, **2**, 155–161.

106 Sarko, C.R. (2005) *Microwave-Assisted Organic Synthesis* (eds P. Lidström and J.P. Tierney), Blackwell Publishing, Oxford, pp. 222–236.

107 Ersmark, K., Larhed, M., and Wannberg, J. (2004) *Current Opinion in Drug Discovery & Development*, **7**, 417–427.

108 Mavandadi, F. and Lidström, P. (2004) *Current Topics in Medicinal Chemistry*, **4**, 773–792.

109 Alexandre, F.-R., Domon, L., Frère, S., Testard, A., Thiéry, V., and Besson, T. (2003) *Molecular Diversity*, **7**, 273–280.

110 Wathey, B., Tierney, J., Lidström, P., and Westman, J. (2002) *Drug Discovery Today*, **7**, 373–380.

111 Dzierba, C.D. and Combs, A.P. (2002) *Annual Reports in Medicinal Chemistry*, **37**, 247–256.

112 Santagada, V., Perissutti, E., and Caliendo, G. (2002) *Current Medicinal Chemistry*, **9**, 1251–1283.

113 Wilson, N.S. and Roth, G.P. (2002) *Current Opinion in Drug Discovery & Development*, **5**, 620–629.

114 Collins, M.J. (2001) *American Genomic/Proteomic Technology*, **1**, 40–43.

115 Larhed, M. and Hallberg, A. (2001) *Drug Discovery Today*, **6**, 406–416.

116 Patil, S.A., Patil, R., and Miller, D.D. (2011) *Current Medicinal Chemistry*, **18**, 615–637.

117 Santagada, V., Frecentese, F., Perissutti, E., Fiorino, F., Severino, B., and Caliendo, G. (2009) *Mini-Reviews in Medicinal Chemistry*, **9**, 340–358.

118 Cao, P. and Leadbeater, N.E. (2011) *Microwave Heating as a Tool for Sustainable Chemistry* (ed. N.E. Leadbeater), CRC Press, Boca Raton, pp. 73–103.
119 Matloobi, M. and Kappe, C.O. (2007) *Chimica Oggi: Chemistry Today*, **25** (2), 26–31.
120 Stadler, A. and Kappe, C.O. (2006) *Microwaves in Organic Synthesis*, 2nd edn (ed. A. Loupy), Wiley-VCH Verlag GmbH, Weinheim, pp. 726–787.
121 Martínez-Palou, R. (2006) *Molecular Diversity*, **10**, 435–462.
122 Lange, T. and Lindell, S. (2005) *Combinatorial Chemistry & High Throughput Screening*, **8**, 595–606.
123 Santagada, V., Frecentese, F., Perissutti, E., Favretto, L., and Caliendo, G. (2004) *QSAR & Combinatorial Science*, **23**, 919–944.
124 Blackwell, H.E. (2003) *Organic & Biomolecular Chemistry*, **1**, 1251–1255.
125 Kappe, C.O. and Stadler, A. (2002) *Microwaves in Organic Synthesis* (ed. A. Loupy), Wiley-VCH Verlag GmbH, Weinheim, pp. 405–433.
126 Lidström, P., Westman, J., and Lewis, A. (2002) *Combinatorial Chemistry & High Throughput Screening*, **5**, 441–458.
127 Kappe, C.O. (2002) *Current Opinion in Chemical Biology*, **6**, 314–320.
128 Lew, A., Krutzik, P.O., Hart, M.E., and Chamberlin, A.R. (2002) *Journal of Combinatorial Chemistry*, **4**, 95–105.
129 Alcázar, J., Diels, G., and Schoentjes, B. (2007) *Combinatorial Chemistry & High Throughput Screening*, **10**, 918–932.
130 Kappe, C.O. and Matloobi, M. (2007) *Combinatorial Chemistry & High Throughput Screening*, **10**, 735–750.
131 Nüchter, M., Ondruschka, B., Tied, A., Lautenschläger, W., and Borowski, K.J. (2002) *Chemistry in Australia*, **69**, 28–29.
132 O'Neill, J.C. and Blackwell, H.E. (2007) *Combinatorial Chemistry & High Throughput Screening*, **10**, 857–876.
133 Dai, W.M. and Shi, J. (2007) *Combinatorial Chemistry & High Throughput Screening*, **10**, 837–856.
134 Erdélyi, M. (2006) *Topics in Heterocyclic Chemistry*, **1**, 79–128.
135 Stadler, A. and Kappe, C.O. (2005) *Microwave-Assisted Organic Synthesis* (eds P. Lidström, and J.P. Tierney), Blackwell Publishing, Oxford, pp. 177–221.
136 Al-Obeidi, F., Austin, R.E., Okonya, J.F., and Bond, D.R.S. (2003) *Mini-Reviews in Medicinal Chemistry*, **3**, 449–460.
137 Kappe, C.O. (2001) *American Laboratory*, **33**, 13–19.
138 Baxendale, I.R., Lee, A.L., and Ley, S.V. (2005) *Microwave-Assisted Organic Synthesis* (eds P. Lidström and J.P. Tierney), Blackwell Publishing, Oxford, pp. 133–176.
139 Crosignani, S. and Linclau, B. (2006) *Topics in Heterocyclic Chemistry*, **1**, 129–154.
140 Kirschning, A., Solodenko, W., and Mennecke, K. (2006) *Chemistry: A European Journal*, **12**, 5972–5990.
141 Bhattacharyya, S. (2005) *Molecular Diversity*, **9**, 253–257.
142 Desai, B. and Kappe, C.O. (2004) *Topics in Current Chemistry*, **242**, 177–208.
143 Swamy, K.M.K., Yeh, W.-B., Lin, M.-J., and Sun, C.-M. (2003) *Current Medicinal Chemistry*, **10**, 2403–2423.
144 Zhang, W. (2007) *Combinatorial Chemistry & High Throughput Screening*, **10**, 219–229.
145 Zhang, W. (2006) *Topics in Current Chemistry*, **266**, 145–166.
146 Olofsson, K. and Larhed, M. (2004) *Handbook of Fluorous Chemistry* (eds J.A. Gladysz, D.P. Curran, and I.T. Horvath), Wiley-VCH Verlag GmbH, Weinheim, pp. 359–365.
147 Kadam, A., Zhang, Z., and Zhang, W. (2011) *Current Organic Synthesis*, **8**, 295–309.
148 Collins, J.M. and Collins, M.J. (2006) *Microwaves in Organic Synthesis*, 2nd edn (ed. A. Loupy) Wiley-VCH Verlag GmbH, Weinheim, pp. 898–930.
149 Grieco, P. (2004) *Chimica Oggi: Chemistry Today*, **22** (7/8), 18–20.
150 Collins, J.M. and Leadbeater, N.E. (2007) *Organic & Biomolecular Chemistry*, **5**, 1141–1150.

151 Sabatino, G. and Papini, A.M. (2008) *Current Opinion in Drug Discovery & Development*, **11**, 762–770.
152 Rejasse, B., Lamare, S., Legoy, M.-D., and Besson, T. (2007) *Journal of Enzyme Inhibition and Medicinal Chemistry*, **22**, 518–526.
153 Zhao, H. (2010) *Aqueous Microwave Assisted Chemistry: Synthesis and Catalysis* (eds V. Polshettiwar and R.S. Varma), RSC Publishing, Cambridge, UK, pp. 123–144.
154 Cioffi, E.A. (2008) *Current Topics in Medicinal Chemistry*, **8**, 152–158.
155 Corsaro, A., Chiacchio, U., Pistarà, V., and Romeo, G. (2006) *Microwaves in Organic Synthesis*, 2nd edn (ed. A. Loupy) Wiley-VCH Verlag GmbH, Weinheim, pp. 579–614.
156 Das, S.K. (2004) *Synlett*, 915–932.
157 Corsaro, A., Chiacchio, U., Pistara, V., and Romeo, G. (2004) *Current Organic Chemistry*, **8**, 511–538.
158 Vanier, G.C. (2011) *Microwave Heating as a Tool for Sustainable Chemistry* (ed. N.E. Leadbeater), CRC Press, Boca Raton, pp. 231–269.
159 Paolini, I., Rizzolo, F., and Papini, A.M. (2008) *Chimica Oggi: Chemistry Today*, **26** (Suppl.), 36–38.
160 Bruckmann, A., Krebs, A., and Bolm, C. (2008) *Green Chemistry*, **10**, 1131–1141.
161 Richel, A., Laurent, P., Wathelet, B., Wathelet, J.-P., and Paquot, M. (2011) *Catalysis Today*, **167**, 141–147.
162 Bag, S., Dasgupta, S., and Torok, B. (2011) *Current Organic Synthesis*, **8**, 237–261.
163 Durka, T., Van Gerven, T., and Stankiewicz, A. (2009) *Chemical Engineering & Technology*, **32**, 1301–1312.
164 Appukkuttan, P. and Van der Eycken, E. (2008) *European Journal of Organic Chemistry*, 1133–1155.
165 Singh, B.K., Kaval, N., Tomar, S., Van der Eycken, E., and Parmar, V.S. (2008) *Organic Process Research & Development*, **12**, 468–474.
166 Larhed, M., Wannberg, J., and Hallberg, A. (2007) *QSAR & Combinatorial Science*, **26**, 51–68.
167 Olofsson, K., Nilsson, P., and Larhed, M. (2006) *Microwaves in Organic Synthesis*, 2nd edn (ed. A. Loupy), Wiley-VCH Verlag GmbH, Weinheim, pp. 685–725.
168 Wannberg, J., Ersmark, K., and Larhed, M. (2006) *Topics in Current Chemistry*, **266**, 167–198.
169 Maes, B.U.W. (2006) *Topics in Heterocyclic Chemistry*, **1**, 155–211.
170 Nilsson, P., Olofsson, K., and Larhed, M. (2006) *Topics in Current Chemistry*, **266**, 103–144.
171 Olofsson, K. and Larhed, M. (2005) *Microwave-Assisted Organic Synthesis* (eds P. Lidström and J.P. Tierney), Blackwell Publishing, Oxford, pp. 23–43.
172 Lee, J. and Hlasta, D.J. (2004) *Transition Metals for Organic Synthesis* (eds M. Beller and C. Bolm), Wiley-VCH Verlag GmbH, Weinheim, pp. 597–608.
173 Olofsson, K., Hallberg, A., and Larhed, M. (2002) *Microwaves in Organic Synthesis* (ed. A. Loupy), Wiley-VCH Verlag GmbH, Weinheim, pp. 379–403.
174 Larhed, M., Moberg, C., and Hallberg, A. (2002) *Accounts of Chemical Research*, **35**, 717–727.
175 Coquerel, Y. and Rodriguez, J. (2008) *European Journal of Organic Chemistry*, 1125–1132.
176 Aitken, S.G. and Abell, A.D. (2005) *Australian Journal of Chemistry*, **58**, 3–13.
177 Fihri, A. and Len, C. (2010) *Aqueous Microwave Assisted Chemistry: Synthesis and Catalysis* (eds V. Polshettiwar and R.S. Varma), RSC Publishing, Cambridge, UK, pp. 55–90.
178 Cadierno, V., Crochet, P., and Garcia-Garrido, S.E. (2010) *Aqueous Microwave Assisted Chemistry: Synthesis and Catalysis* (eds V. Polshettiwar and R.S. Varma), RSC Publishing, Cambridge, UK, pp. 10–54.
179 Nicks, F., Borguet, Y., Delfosse, S., Bicchielli, D., Delaude, L., Sauvage, X., and Demonceau, A. (2009) *Australian Journal of Chemistry*, **62**, 184–207.
180 Barge, A., Tagliapietra, S., Tei, L., Cintas, P., and Cravotto, G. (2008) *Current Organic Chemistry*, **12**, 1588–1612.

181 Bonrath, W., Letinois, U., Netscher, T., and Schuetz, J. (2008) *Chimica Oggi: Chemistry Today*, **26** (3), 24–29.

182 Bonrath, W., Letinois, U., Netscher, T., and Schuetz, J. (2008) *Chimica Oggi: Chemistry Today*, **26** (3), 31–36.

183 Powell, G.L. (2011) *Microwave Heating as a Tool for Sustainable Chemistry* (ed. N.E. Leadbeater), CRC Press, Boca Raton, pp. 175–205.

184 Navarro, O. (2010) *Chimica Oggi: Chemistry Today*, **28** (Suppl. 3), 15–17.

185 Bagley, M.C. and Lubinu, M.C. (2006) *Topics in Heterocyclic Chemistry*, **1**, 31–58.

186 de Boer, T., Amore, A., and Orru, R.V.A. (2006) *Microwaves in Organic Synthesis*, 2nd edn (ed. A. Loupy), Wiley-VCH Verlag GmbH, Weinheim, pp. 788–819.

187 Westman, J. (2005) *Microwave-Assisted Organic Synthesis* (eds P. Lidström and J.P. Tierney), Blackwell Publishing, Oxford, pp. 102–132.

188 Bariwal, J.B., Trivedi, J.C., and Van der Eycken, E. (2010) *Topics in Heterocyclic Chemistry*, **25**, 169–230.

189 Kruithof, A., Ruijter, E., and Orru, R.V.A. (2011) *Current Organic Chemistry*, **15**, 204–236.

190 Jiang, B., Shi, F., and Tu, S.-J. (2010) *Current Organic Chemistry*, **14**, 357–378.

191 Hugel, H.M. (2009) *Molecules*, **14**, 4936–4972.

192 Polshettiwar, V. and Varma, R.S. (2008) *Pure and Applied Chemistry*, **80**, 777–790.

193 de la Hoz, A., Díaz-Ortis, A., Moreno, A., Sanchez-Migallon, A., Prieto, P., Carrillo, J.R., Vazquez, E., Gomez, M.V., and Herrero, M.A. (2007) *Combinatorial Chemistry & High Throughput Screening*, **10**, 877–902.

194 Besson, T. and Chosson, E. (2007) *Combinatorial Chemistry & High Throughput Screening*, **10**, 903–917.

195 Bazureau, J.P., Hamelin, J., Mongin, F., and Texier-Boullet, F. (2006) *Microwaves in Organic Synthesis*, 2nd edn (ed. A. Loupy), Wiley-VCH Verlag GmbH, Weinheim, pp. 456–523.

196 Suna, E. and Mutule, I. (2006) *Topics in Current Chemistry*, **266**, 49–101.

197 Pemberton, N., Chorell, E., and Almqvist, F. (2006) *Topics in Heterocyclic Chemistry*, **1**, 1–30.

198 Besson, T. and Thiéry, V. (2006) *Topics in Heterocyclic Chemistry*, **1**, 59–78.

199 Rodriguez, M. and Taddei, M. (2006) *Topics in Heterocyclic Chemistry*, **1**, 213–266.

200 Kaval, N., Appukkuttan, P., and Van der Eycken, E. (2006) *Topics in Heterocyclic Chemistry*, **1**, 267–304.

201 Shipe, W.D., Yang, F., Zhao, Z., Wolkenberg, S.E., Nolt, M.B., and Lindsley, C.W. (2006) *Heterocycles*, **70**, 655–689.

202 Rakhmankulov, D.L., Shavshukova, S.Y., and Latypova, F.N. (2005) *Chemistry of Heterocycle Compounds*, **41**, 951–961.

203 Besson, T. and Brain, C.T. (2005) *Microwave-Assisted Organic Synthesis* (eds P. Lidström and J.P. Tierney), Blackwell Publishing, Oxford, pp. 44–74.

204 Molteni, V. and Ellis, D.A. (2005) *Current Organic Synthesis*, **2**, 333–375.

205 Xu, Y. and Guo, Q.X. (2004) *Heterocycles*, **63**, 903–974.

206 Katritzky, A.R. and Singh, S.K. (2003) *ARKIVOC*, **xiii**, 68–86.

207 Hamelin, J., Bazureau, J.P., and Texier-Boullet, F. (2002) *Microwaves in Organic Synthesis* (ed. A. Loupy), Wiley-VCH Verlag GmbH, Weinheim, pp. 253–293.

208 Polshettiwar, V. and Varma, R.S. (2010) *Aqueous Microwave Assisted Chemistry: Synthesis and Catalysis* (eds V. Polshettiwar and R.S. Varma), RSC Publishing, Cambridge, UK, pp. 91–122.

209 Kranjc, K. and Kocevar, M. (2010) *Current Organic Chemistry*, **14**, 1050–1074.

210 Abid, M., Torok, B., and Huang, X. (2009) *Australian Journal of Chemistry*, **62**, 208–222.

211 Jindal, R. and Bajaj, S. (2008) *Current Organic Chemistry*, **12**, 836–849.

212 Mehta, V.P., Appukkuttan, P., and Van der Eycken, E. (2011) *Current Organic Chemistry*, **15**, 265–283.

213 Kappe, C.O. and Van der Eycken, E. (2010) *Chemical Society Reviews*, **39**, 1280–1290.

214 Bougrin, K., Soufiaoui, M., and Bashiardes, G. (2006) *Microwaves in Organic Synthesis*, 2nd edn (ed. A. Loupy) Wiley-VCH Verlag GmbH, Weinheim, pp. 524–578.

215 de la Hoz, A., Díaz-Ortiz, A., and Langa, F. (2002) *Microwaves in Organic Synthesis* (ed. A. Loupy) Wiley-VCH Verlag GmbH, Weinheim, pp. 295–343.

216 de la Hoz, A., Díaz-Ortis, A., Moreno, A., and Langa, F. (2000) *European Journal of Organic Chemistry*, 3659–3673.

217 Pineiro, M. and Pinho e Melo, T.M.V.D. (2009) *European Journal of Organic Chemistry*, 5287–5307.

218 Appukkuttan, P., Mehta, V.P., and Van der Eycken, E. (2010) *Chemical Society Reviews*, **39**, 1467–1477.

219 Liu, J.F. (2007) *Current Organic Synthesis*, **4**, 223–237.

220 Appukkuttan, P. and Van der Eycken, E. (2006) *Topics in Current Chemistry*, **266**, 1–47.

221 Rajabi, F., Luque, R., and Vargas, C. (2010) *Chimica Oggi: Chemistry Today*, **28** (3), 12–14.

222 Klán, P. and Církva, V. (2006) *Microwaves in Organic Synthesis*, 2nd edn (ed. A. Loupy), Wiley-VCH Verlag GmbH, Weinheim, pp. 860–897.

223 Klán, P. and Církva, V. (2002) *Microwaves in Organic Synthesis* (ed. A. Loupy) Wiley-VCH Verlag GmbH, Weinheim, pp. 463–486.

224 Cirkva, V. and Relich, S. (2011) *Current Organic Chemistry*, **15**, 248–264.

225 Stone-Elander, S., Elander, N., Thorell, J.-O., and Fredriksson, A. (2007) *Ernst Schering Research Foundation Workshop*, **64**, 243–269.

226 Jones, J.R., and Lu, S.Y. (2006) *Microwaves in Organic Synthesis*, 2nd edn (ed. A. Loupy), Wiley-VCH Verlag GmbH, Weinheim, pp. 820–859.

227 Jones, J.R. and Lu, S.Y. (2002) *Microwaves in Organic Synthesis* (ed. A. Loupy), Wiley-VCH Verlag GmbH, Weinheim, pp. 435–462.

228 Stone-Elander, S. and Elander, N. (2002) *Journal of Labelled Compounds & Radiopharmaceuticals*, **45**, 715–746.

229 Elander, N., Jones, J.R., Lu, S.-Y., and Stone-Elander, S. (2000) *Chemical Society Reviews*, **29**, 239–250.

230 Cravotto, G. and Cintas, P. (2007) *Chemistry: A European Journal*, **13**, 1902–1909.

231 Barge, A., Tagliapietra, S., Binello, A., and Cravotto, G. (2011) *Current Organic Chemistry*, **15**, 189–203.

232 Leonelli, C. and Mason, T.J. (2010) *Chemical Engineering and Processing*, **49**, 885–900.

233 Langa, F. and de la Cruz, P. (2007) *Combinatorial Chemistry & High Throughput Screening*, **10**, 766–782.

234 Langa, F. and de la Cruz, P. (2006) *Microwaves in Organic Synthesis*, 2nd edn (ed. A. Loupy), Wiley-VCH Verlag GmbH, Weinheim, pp. 931–958.

235 Vanetsev, A.S. and Tretyakov, Y.D. (2007) *Russian Chemical Reviews*, **76**, 397–413.

236 Millos, C.J., Whittaker, A.G., and Brechin, E.K. (2007) *Polyhedron*, **26**, 1927–1933.

237 Hájek, M. (2006) *Microwaves in Organic Synthesis*, 2nd edn (ed. A. Loupy), Wiley-VCH Verlag GmbH, Weinheim, pp. 615–652.

238 Hájek, M. (2002) *Microwaves in Organic Synthesis* (ed. A. Loupy), Wiley-VCH Verlag GmbH, Weinheim, pp. 345–378.

239 Gellis, A., Rathelot, P., Crozet, M.P., and Vanelle, P. (2002) *Electron Transfer Reactions in Organic Synthesis* (ed. P. Vanelle) Research Signpost, Trivandrum, pp. 113–130.

240 Schmoeger, C., Stolle, A., Bonrath, W., and Ondruschka, B. (2011) *Current Organic Chemistry*, **15**, 151–167.

241 de Souza, R.O.M.A. and Miranda, L.S.M. (2010) *Mini-Reviews in Organic Chemistry*, **7**, 212–220.

242 Ide, D.M., Eastlund, M.P., Jupe, C.L., and Stockland, R.A., Jr. (2008) *Current Organic Chemistry*, **12**, 1258–1278.

243 Tymoshenko, D.O. (2008) *Mini-Reviews in Organic Chemistry*, **5**, 85–95.

244 Mavandadi, F. (2008) *Process Chemistry in the Pharmaceutical Industry* (ed. K. Gadamasetti), CRC Press, Danvers, pp. 403–426.

245 Kremsner, J.M., Stadler, A., and Kappe, C.O. (2006) *Topics in Current Chemistry*, **266**, 233–278.

246 Lehmann, H. (2006) *Ernst Schering Foundation Symposium Proceedings* (eds P.H. Seeberger, and T. Blume), Springer, Berlin, pp. 133–149.

247 Roberts, B.A. and Strauss, C.R. (2005) *Microwave-Assisted Organic Synthesis* (eds P. Lidström, and J.P. Tierney), Blackwell Publishing, Oxford, pp. 237–271.

248 Wolkenberg, S.E., Shipe, W.D., Lindsley, C.W., Guare, J.P., and Pawluczyk, J.M. (2005) *Current Opinion in Drug Discovery & Development*, **8**, 701–708.

249 Moseley, J.D. (2011) *Microwave Heating as a Tool for Sustainable Chemistry* (ed. N.E. Leadbeater), CRC Press, Boca Raton, pp. 105–147.

250 Baxendale, I.R., Hayward, J.J., and Ley, S.V. (2007) *Combinatorial Chemistry & High Throughput Screening*, **10**, 802–836.

251 Glasnov, T.N. and Kappe, C.O. (2007) *Macromolecular Rapid Communications*, **28**, 395–410.

252 Schwalbe, T. and Simons, K. (2006) *Chimica Oggi: Chemistry Today*, **24** (2), 56–61.

253 Baxendale, I.R. and Pitts, M.R. (2006) *Chimica Oggi: Chemistry Today*, **24** (3), 41–45.

254 Moseley, J.D. and Kappe, C.O. (2011) *Green Chemistry*, **13**, 794–806.

255 Beletskaya, I.P. and Cheprakov, A.V. (2000) *Chemical Reviews*, **100**, 3009–3066.

256 Larhed, M. and Hallberg, A. (1996) *The Journal of Organic Chemistry*, **61**, 9582–9584.

257 Stadler, A., Yousefi, B.H., Dallinger, D., Walla, P., Van der Eycken, E., Kaval, N., and Kappe, C.O. (2003) *Organic Process Research & Development*, **7**, 707–716.

258 Vallin, K.S.A., Emilsson, P., Larhed, M., and Hallberg, A. (2002) *The Journal of Organic Chemistry*, **67**, 6243–6246.

259 Datta, G.K., Vallin, K.S.A., and Larhed, M. (2003) *Molecular Diversity*, **7**, 107–114.

260 Littke, A.F. and Fu, G.C. (2002) *Angewandte Chemie International Edition*, **41**, 4176–4211.

261 Netherton, M.R. and Fu, G.C. (2001) *Organic Letters*, **3**, 4295–4298.

262 Herrmann, W.A., Bohm, V.P.H., and Reisinger, C.P. (1999) *Journal of Organometallic Chemistry*, **576**, 23–41.

263 Botella, L. and Nájera, C. (2004) *Tetrahedron Letters*, **45**, 1833–1836.

264 Botella, L. and Nájera, C. (2004) *Tetrahedron*, **60**, 5563–5570.

265 Svennebring, A., Nilsson, P., and Larhed, M. (2004) *The Journal of Organic Chemistry*, **69**, 3345–3349.

266 Stadler, A., von Schenck, H., Vallin, K.S.A., Larhed, M., and Hallberg, A. (2004) *Advanced Synthesis and Catalysis*, **346**, 1773–1781.

267 Datta, G.K., von Schenck, H., Hallberg, H., and Larhed, M. (2006) *The Journal of Organic Chemistry*, **71**, 3896–3903.

268 Datta, G.K., Nordeman, P., Dackenberg, J., Nilsson, P., Hallberg, A., and Larhed, M. (2008) *Tetrahedron Asymmetry*, **19**, 1120–1126.

269 Kormos, C.M. and Leadbeater, N.E. (2008) *The Journal of Organic Chemistry*, **73**, 3854–3858.

270 Arvela, R.K. and Leadbeater, N.E. (2005) *The Journal of Organic Chemistry*, **70**, 1786–1790.

271 Andappan, M.M.S., Nilsson, P., and Larhed, M. (2003) *Molecular Diversity*, **7**, 97–106.

272 Andappan, M.M.S., Nilsson, P., von Schenck, H., and Larhed, M. (2004) *The Journal of Organic Chemistry*, **69**, 5212–5218.

273 Lindh, J., Sävmarker, J., Nilsson, P., Sjöberg, P.J.R., and Larhed, M. (2009) *Chemistry: A European Journal*, **15**, 4630–4636.

274 Flessner, T., Ludwig, V., Siebeneicher, H., and Winterfeld, E. (2002) *Synthesis*, 1373–1378.

275 Gracias, V., Moore, J.D., and Djuric, S.W. (2004) *Tetrahedron Letters*, **45**, 417–420.

276 Tietze, L.F., Wiegand, J.M., and Vock, C. (2003) *Journal of Organometallic Chemistry*, **687**, 346–352.

277 Tietze, L.F., Wiegand, J.M., and Vock, C. (2004) *European Journal of Organic Chemistry*, 4107–4112.

278 Sørensen, U.S. and Pombo-Villar, E. (2004) *Helvetica Chimica Acta*, **87**, 82–89.

279 Lachance, N., April, M., and Joly, M.A. (2005) *Synthesis*, 2571–2577.

280 Svennebring, A., Nilsson, P., and Larhed, M. (2007) *The Journal of Organic Chemistry*, **72**, 5851–5854.

281 Arvela, R.K., Pasquini, S., and Larhed, M. (2007) *The Journal of Organic Chemistry*, **72**, 6390–6396.

282 Vasudevan, A., Tseng, P.S., and Djuric, S.W. (2006) *Tetrahedron Letters*, **47**, 8591–8593.

283 Kesavan, S., Panek, J.S., and Porco, J.A. (2007) *Organic Letters*, **9**, 5203–5206.

284 Donets, P.A. and Van der Eycken, E. (2007) *Organic Letters*, **9**, 3017–3020.

285 Fields, W.H., Khan, A.K., Sabat, M., and Chruma, J.J. (2008) *Organic Letters*, **10**, 5131–5134.

286 Püschl, A., Rudbeck, H.C., Faldt, A., Confante, A., and Kehler, J. (2005) *Synthesis*, 291–295.

287 Cheung, W.S., Patch, R.J., and Player, M.R. (2005) *The Journal of Organic Chemistry*, **70**, 3741–3744.

288 Franck, P., Hostyn, S., Dajka-Halász, B., Polonka-Bálint, Á., Monsieurs, K., Mátyus, P., and Maes, B.U.W. (2008) *Tetrahedron*, **64**, 6030–6037.

289 Lacrouts, P., Parsons, P.J., Penkett, C.S., and Rauf Raza, A. (2005) *Synlett*, 2767–2768.

290 Declerck, V., Martinez, J., and Lamaty, F. (2006) *Synlett*, 3029–3032.

291 Miyaura, N. and Suzuki, A. (1995) *Chemical Reviews*, **95**, 2457–2483.

292 Leadbeater, N.E. and Marco, M. (2002) *Organic Letters*, **4**, 2973–2976.

293 Leadbeater, N.E. and Marco, M. (2003) *The Journal of Organic Chemistry*, **68**, 888–892.

294 Leadbeater, N.E. and Smith, R.J. (2006) *Organic Letters*, **8**, 4589–4591.

295 Bai, L., Wang, J.-X., and Zhang, Y. (2003) *Green Chemistry*, **5**, 615–617.

296 Leadbeater, N.E., Williams, V.A., Barnard, T.M., and Collins, M.J., Jr. (2006) *Organic Process Research & Development*, **10**, 833–837.

297 Leadbeater, N.E. and Marco, M. (2003) *Angewandte Chemie International Edition*, **42**, 1407–1409.

298 Leadbeater, N.E. and Marco, M. (2003) *The Journal of Organic Chemistry*, **68**, 5660–5667.

299 Li, C.-J. (2003) *Angewandte Chemie International Edition*, **42**, 4856–4858.

300 Arvela, R.K., Leadbeater, N.E., Sangi, M.S., Williams, V.A., Granados, P., and Singer, R.D. (2005) *The Journal of Organic Chemistry*, **70**, 161–168.

301 Nájera, C., Gil-Moltó, J., and Karlström, S. (2004) *Advanced Synthesis and Catalysis*, **346**, 1798–1811.

302 Navarro, O., Kaur, H., Mahjoor, P., and Nolan, S.P. (2004) *The Journal of Organic Chemistry*, **69**, 3173–3180.

303 Bedofrd, R.B., Butts, C.P., Hurst, T.E., and Lindström, P. (2004) *Advanced Synthesis and Catalysis*, **346**, 1627–1630.

304 Clarke, M.L., France, M.B., Fuentes, J.A., Milton, E.J., and Roff, G.J. (2007) *Beilstein Journal of Organic Chemistry*, **3**, 18–21.

305 Gong, Y. and He, W. (2002) *Organic Letters*, **4**, 3803–3805.

306 Savall, B.M. and Fontimayor, J.R. (2008) *Tetrahedron Letters*, **49**, 6667–6669.

307 Appukkuttan, P., Orts, A.B., Chandran, R.P., Goeman, J.L., Van der Eycken, J., Dehan, W., and Van der Eycken, E. (2004) *European Journal of Organic Chemistry*, 3277–3285.

308 Melucci, M., Barbarella, G., Zambianchi, M., and Bongini, A. (2004) *The Journal of Organic Chemistry*, **69**, 4821–4828.

309 Alesi, S., Di Maria, F., Melucci, M., Macquarrie, D.J., Luque, R., and Barbarella, G. (2008) *Green Chemistry*, **10**, 517–523.

310 De Borggraeve, W.M., Appukkuttan, P., Azzam, R., Dehaen, W., Compernolle, F., Van der Eycken, E., and Hoornaert, G. (2005) *Synlett*, 777–780.

311 Weber, J. and Thomas, A. (2008) *Journal of the American Chemical Society*, **130**, 6334–6335.

312 Kang, Y.K., Deria, P., Carroll, P.J., and Therien, M.J. (2008) *Organic Letters*, **10**, 1341–1344.

313 Miller, S.P., Morgan, J.B., Nepveux, F.J., and Morken, J.P. (2004) *Organic Letters*, **6**, 131–133.

314 Poondra, R.R., Fischer, P.M., and Turner, N.J. (2004) *The Journal of Organic Chemistry*, **69**, 6920–6922.

315 Flegeau, E.F., Popkin, M.E., and Greaney, M.F. (2006) *Organic Letters*, **8**, 2495–2498.

316 Högermeier, J. and Reißig, H.U. (2007) *Chemistry: A European Journal*, **13**, 2410–2420.

317 Stanetty, P., Schnürch, M., and Mihovilovic, M.D. (2006) *The Journal of Organic Chemistry*, **71**, 3754–3761.

318 Schnürch, M., Hämmerle, J., Mihovilovic, M.D., and Stanetty, P. (2010) *Synthesis*, **5**, 837–843.

319 Luo, G., Chen, L., and Pointdexter, G.S. (2002) *Tetrahedron Letters*, **43**, 5739–5742.

320 Gong, Y. and He, W. (2004) *Heterocycles*, **62**, 851–856.

321 Kaval, N., Bisztray, K., Dehaen, W., Kappe, C.O., and Van der Eycken, E. (2003) *Molecular Diversity*, **7**, 125–134.

322 Holmberg, P., Sohn, D., Leideborg, R., Caldirola, P., Zlatoidsky, P., Hanson, S., Mohell, N., Rosqvist, S., Nordvall, G., Johansson, A.M., and Johansson, R. (2004) *Journal of Medicinal Chemistry*, **47**, 3927–3930.

323 Organ, M.G., Mayer, S., Lepifre, F., N'Zemba, B., and Khatri, J. (2003) *Molecular Diversity*, **7**, 211–227.

324 DiMauro, E.F. and Vitullo, J.R. (2006) *The Journal of Organic Chemistry*, **71**, 3959–3962.

325 Fitzmaurice, R.J., Etheridge, Z.C., Jumel, E., Woolfson, D.N., and Caddick, S. (2006) *Chemical Communications*, 4814–4816.

326 Wang, H.-J., Keilman, J., Pabba, C., Chen, Z.-J., and Gregg, B.T. (2005) *Tetrahedron Letters*, **46**, 2631–2634.

327 Ito, F., Iwasaki, M., Watanabe, T., Ishikawa, T., and Higuchi, Y. (2005) *Organic & Biomolecular Chemistry*, **3**, 674–681.

328 Nehls, B.S., Galbrecht, F., Bilge, A., Brauer, D.J., Lehmann, C.W., Scherf, U., and Farrell, T. (2005) *Organic & Biomolecular Chemistry*, **3**, 3213–3219.

329 Wu, T.Y.H., Schultz, P.G., and Ding, S. (2003) *Organic Letters*, **5**, 3587–3590.

330 Vishnumurthy, K. and Makriyannis, MA. (2010) *Journal of Combinatorial Chemistry*, **12**, 664–669.

331 Kim, Y.H., Lee, H., Kim, Y.J., Kim, B.T., and Heo, J.-N. (2008) *The Journal of Organic Chemistry*, **73**, 495–501.

332 Kim, J.K., Kim, Y.H., Nam, H.T., Kim, B.T., and Heo, J.-N. (2008) *Organic Letters*, **10**, 3543–3546.

333 Bazin, M.-A., El Kihel, L., Lancelot, J.-C., and Rault, S. (2007) *Tetrahedron Letters*, **48**, 4347–4351.

334 Li, H.-y., Wang, Y., McMillen, W.T., Chatterjee, A., Toth, J.E., Mundla, S.R., Voss, M., Boyer, R.D., and Sawyer, J.S. (2007) *Tetrahedron*, **63**, 11763–11770.

335 Nöteberg, D., Schaal, W., Hamelink, E., Vrang, L., and Larhed, M. (2003) *Journal of Combinatorial Chemistry*, **5**, 456–464.

336 Esmark, K., Feierberg, I., Bjelic, S., Hamelink, E., Hackett, F., Blackman, M.J., Hultén, J., Samuelsson, B., Åqvist, J., and Hallberg, A. (2004) *Journal of Medicinal Chemistry*, **47**, 110–112.

337 Kurukulasuriya, R., Sorensen, B.K., Link, J.T., Patel, J.R., Jae, H.-S., Winn, M.X., Rohde, J.R., Grihalde, N.D., Lin, C.W., Ogiela, C.A., Adler, A.L., and Collins, C.A. (2004) *Bioorganic & Medicinal Chemistry Letters*, **14**, 2047–2050.

338 Tromp, R.A., van Ameijde, S., Pütz, C., Sundermann, C., Sundermann, B., von Frijtag Drabe Künzel, J.K., and IJzerman, A.P. (2004) *Journal of Medicinal Chemistry*, **47**, 5441–5450.

339 Wan, Y., Wallinder, C., Plouffe, B., Beaudry, H., Mahalingam, A.K., Wu, X., Johansson, B., Holm, M., Botoros, M., Karlén, A., Pettersson, A., Nyberg, F., Fändriks, L., Gallo-Payet, N., Hallberg, A., and Alterman, M. (2004) *Journal of Medicinal Chemistry*, **47**, 5995–6008.

340 Perissutti, E., Frecentese, F., Lavecchia, A., Fiorino, F., Severino, B., De Angelis, F., Santagada, V., and Caliendo, G. (2007) *Tetrahedron*, **63**, 12779–12785.

341 Genov, M., Almorín, A., and Espinet, P. (2007) *Tetrahedron Asymmetry*, **18**, 625–627.

342 VanAlstine, M.A., Wentland, M.P., Cohen, D.J., and Bidlack, J.M. (2007) *Bioorganic & Medicinal Chemistry Letters*, **17**, 6516–6520.

343 Wilson, D.P., Wan, Z.-K., Xu, W.-X., Kirincich, S.J., Follows, B.C., Joseph-McCarthy, D., Foreman, K., Moretto, A., Wu, J., Zhu, M., Binnun, E., Zhang, Y.-L., Tam, M., Erbe, D.V., Tobin, J., Xu, X., Leung, L., Shilling, A., Tam, S.Y., Mansour, T.S., and Lee, J. (2007) *Journal of Medicinal Chemistry*, **50** 4681–4698.

344 Duncton, M.A.J., Estiarte, M.A., Tan, D., Kaub, C., O'Mahony, D.J.R., Johnson, R.J., Cox, M., Edwards, W.T., Wan, M., Kincaid, J., and Kelly, M.G. (2008) *Organic Letters*, **10**, 3259–3262.

345 Tamayo, N., Liao, H., Stec, M.M., Wang, X., Chakrabarti, P., Retz, D., Doherty, E.M., Surapaneni, S., Tamir, R., Bannon, A.W., Gavva, N.R., and Norman, M.H. (2008) *Journal of Medicinal Chemistry*, **51**, 2744–2757.

346 Benakki, H., Colacino, E., André, C., Guenoun, F., Martinez, J., and Lamaty, F. (2008) *Tetrahedron*, **64**, 5949–5955.

347 Lépine, R. and Zhu, J. (2005) *Organic Letters*, **7**, 2981–2984.

348 Zhou, Q.J., Worm, K., and Dolle, R.E. (2004) *The Journal of Organic Chemistry*, **69**, 5147–5148.

349 Han, J.W., Castro, J.C., and Burgess, K. (2003) *Tetrahedron Letters*, **44**, 9359–9362.

350 Nehls, B.S., Asawapirom, U., Füldner, S., Preis, E., Farrell, T., and Scherf, U. (2004) *Advanced Functional Materials*, **14**, 352–356.

351 Alacid, E. and Nájera, C. (2009) *The Journal of Organic Chemistry*, **74**, 2321–2327.

352 Alacid, E. and Nájera, C. (2008) *Organic Letters*, **10**, 5011–5014.

353 Brooker, M.D., Cooper, S.M., Jr., Hodges, D.R., Carter, R.R., and Wyatt, J.K. (2010) *Tetrahedron Letters*, **51**, 6748–6752.

354 Arvela, R.K., Leadbeater, N.E., Mack, T.L., and Kormos, C.M. (2006) *Tetrahedron Letters*, **47**, 217–220.

355 Kabalka, G.W. and Al-Masum, M. (2006) *Tetrahedron Letters*, **46**, 6329–6331.

356 Poláčková, V., Toma, S., and Augustínová, I. (2006) *Tetrahedron*, **62**, 11675–11678.

357 Ekoue-Kovi, K., Xu, H., and Wolf, C. (2008) *Tetrahedron Letters*, **49**, 5773–5776.

358 Solodenko, W., Schön, U., Messinger, J., Glinschert, A., and Kirschning, A. (2004) *Synlett*, 1699–1702.

359 Wang, Y. and Sauer, D.R. (2004) *Organic Letters*, **6**, 2793–2796.

360 Baxendale, I.R., Griffiths-Jones, C.M., Ley, S.V., and Tranmer, G.K. (2006) *Chemistry: A European Journal*, **12**, 4407–4416.

361 Sharma, A.K., Gowdahalli, K., Krzeminski, J., and Amin, S. (2007) *The Journal of Organic Chemistry*, **72**, 8987–8989.

362 Antonow, D., Cooper, N., Howard, P.W., and Thurston, D.E. (2007) *Journal of Combinatorial Chemistry*, **9**, 437–445.

363 Vickerstaffe, E., Villard, A.L., Ladlow, M., and Ley, S.V. (2007) *Synlett*, 1251–1254.

364 Freundlich, J.S. and Landis, H.E. (2006) *Tetrahedron Letters*, **47**, 4275–4279.

365 Baghbanzadeh, M., Pilger, C., and Kappe, C.O. (2011) *The Journal of Organic Chemistry*, **76**, 1507–1510.

366 Yan, J., Zhu, M., and Zhou, Z. (2006) *European Journal of Organic Chemistry*, 2060–2062.

367 Seganish, W.M. and DeShong, P. (2004) *Organic Letters*, **6**, 4379–4381.

368 Clark, M.L. (2005) *Advanced Synthesis and Catalysis*, **347**, 303–307.

369 Alacid, E. and Nájera, C. (2006) *Advanced Synthesis and Catalysis*, **348**, 945–952.

370 Alacid, E. and Nájera, C. (2006) *Advanced Synthesis and Catalysis*, **348**, 2085–2091.

371 Alacid, E. and Nájera, C. (2008) *The Journal of Organic Chemistry*, **73**, 2315–2322.

372 Sonogashira, K. (2002) *Journal of Organometallic Chemistry*, **653**, 46–49.

373 Erdélyi, M. and Gogoll, A. (2001) *The Journal of Organic Chemistry*, **66**, 4165–4169.

374 Erdélyi, M., Langer, V., Karlén, A., and Gogoll, A. (2002) *New Journal of Chemistry*, **26**, 834–843.

375 Miljanić, O.Š., Vollhardt, K.P.C., and Whitener, G.D. (2003) *Synlett*, 29–34.

376 Ovotrup, K., Andersson, A.S., Mayer, J.-P., Jepsen, A.S., and Nielsen, M.B. (2004) *Synlett*, 2818–2820.

377 Collings, J.C., Parsons, A.C., Porrès, L., Beeby, A., Batsanov, A.S., Howard, J.A.K., Lydon, D.P., Low, P.J., Fairlamb, I.J.S., and Marder, T.B. (2005) *Chemical Communications*, 2666–2668.

378 Stuhr-Hansen, N., Sørensen, J.K., Moth-Poulsen, K., Christensen, J.B., Bjørnholm, T., and Nielsen, M.B. (2005) *Tetrahedron*, **61**, 12288–12295.

379 Petricci, E., Radi, M., Corelli, F., and Botta, M. (2003) *Tetrahedron Letters*, **44**, 9181–9184.

380 Zhu, J., Germain, A., and Porco, J.A. (2004) *Angewandte Chemie International Edition*, **43**, 1239–1243.

381 Hopkins, C.R. and Collar, N. (2004) *Tetrahedron Letters*, **45**, 8631–8633.

382 Sanz, R., Guilarte, V., and Castroviejo, M.P. (2008) *Synlett*, 3006–3010.

383 Chen, Y., Markina, N.A., and Larock, R.C. (2009) *Tetrahedron*, **65**, 8908–8915.

384 O'Mahony, G., Ehrman, E., and Grøtli, M. (2008) *Tetrahedron*, **64**, 7151–7158.

385 Schramm, O.G. and Müller, T.J.J. (2006) *Advanced Synthesis and Catalysis*, **348**, 2565–2570.

386 Liao, W.W. and Müller, T.J.J. (2006) *Synlett*, 3469–3473.

387 Schramm, O.G. and Müller, T.J.J. (2006) *Synlett*, 1841–1846.

388 Willy, B., Rominger, F., and Müller, T.J.J. (2008) *Synthesis*, 293–303.

389 Willy, B. and Müller, T.J.J. (2008) *European Journal of Organic Chemistry*, 4157–4168.

390 Willy, B., Dallos, T., Rominger, F., Schönhaber, J., and Müller, T.J.J. (2008) *European Journal of Organic Chemistry*, 4796–4805.

391 Kwan, P.H., MacLachlan, M.J., and Swager, T.M. (2004) *Journal of the American Chemical Society*, **126**, 8638–8639.

392 Khan, A. and Hecht, S. (2004) *Chemical Communications*, 300–301.

393 Mehta, V.P., Sharma, A., and Van der Eycken, E. (2008) *Organic Letters*, **10**, 1147–1150.

394 Singh, B.K., Mehta, V.P., Parmar, V.S., and Van der Eycken, E. (2007) *Organic & Biomolecular Chemistry*, **5**, 2962–2965.

395 Singh, B.J., Parmar, V.S., and Van der Eycken, E. (2008) *Synlett*, 3021–3025.

396 Ermolat'ev, D., Mehta, V.P., and Van der Eycken, E. (2007) *Synlett*, 3117–3122.

397 Mehta, V.P., Sharma, A., Van Hecke, K., Van Meervelt, L., and Van der Eycken, E. (2008) *The Journal of Organic Chemistry*, **73**, 2382–2388.

398 Silva, S., Sylla, B., Suzenet, F., Tatibouët, A., Rauter, A.P., and Rollin, P. (2008) *Organic Letters*, **10**, 853–856.

399 Shook, B.C., Chakravarty, D., and Jackson, P.F. (2009) *Tetrahedron Letters*, **50**, 1013–1015.

400 Huang, H., Liu, H., Jiang, H., and Chen, K. (2008) *The Journal of Organic Chemistry*, **73**, 6037–6040.

401 Zheng, S.-L., Reid, S., Lin, N., and Wang, B. (2006) *Tetrahedron Letters*, **47**, 2331–2335.

402 Sørensen, U.S. and Pombo-Villar, E. (2005) *Tetrahedron*, **61**, 2697–2703.

403 Park, K., Bae, G., Park, A., Kim, Y., Choe, J., Ho Sung, K., and Lee, S. (2011) *Tetrahedron Letters*, **52**, 576–580.

404 Awuah, E. and Capretta, A. (2009) *Organic Letters*, **11**, 3210–3213.

405 Togninelli, A., Gevariya, H., Alongi, M., and Botta, M. (2007) *Tetrahedron Letters*, **48**, 4801–4803.

406 de Souza, R.O.M.A., Bittar, M.S., Mendes, L.V.P., da Silva, C.M.F., da Silva, V.T., and Antunes, O.A.C. (2008) *Synlett*, 1777–1780.

407 Leadbeater, N.E., Marco, M., and Tominack, B.J. (2003) *Organic Letters*, **5**, 3919–3922.

408 Appukkuttan, P., Dehaen, W., and Van der Eycken, E. (2003) *European Journal of Organic Chemistry*, 4713–4716.

409 He, H. and Wu, Y.-J. (2004) *Tetrahedron Letters*, **45**, 3237–3239.

410 Stille, J.K. (1986) *Angewandte Chemie International Edition in English*, **25**, 508–523.

411 Zhang, Y., Pavlova, O.A., Chefer, S.I., Hall, A.W., Kurian, V., Brown, L.L., Kimes, A.S., Mukhin, A.G., and Horti, A.G. (2004) *Journal of Medicinal Chemistry*, **47**, 2453–2465.

412 O'Neill, D.J., Shen, L., Prouty, C., Conway, B.R., Westover, L., Xu, J.Z., Zhang, H.-C., Maryanoff, B.E., Murray, W.V., Demarest, K.T., and Kuo, G.-H. (2004) *Bioorganic & Medicinal Chemistry*, **12**, 3167–3185.

413 Rashatasakhon, P., Ozdemir, A.D., Willis, J., and Padwa, A. (2004) *Organic Letters*, **6**, 917–920.

414 Usta, H., Facchetti, A., and Marks, T.J. (2008) *Journal of the American Chemical Society*, **130**, 8580–8581.

415 Appukkuttan, P., Husain, M., Gupta, R.K., Parmar, V.S., and Van der Eycken, E. (2006) *Synlett*, 1491–1496.

416 Jeon, S.L., Kim, J.K., Son, J.B., Kim, B.T., and Jeong, I.H. (2007) *Journal of Fluorine Chemistry*, **128**, 153–157.

417 Hodgson, D.M., Chung, Y.K., Nuzzo, I., Freixas, G., Kulikiewicz, K.K., Cleator, E., and Paris, J.-M. (2007) *Journal of the American Chemical Society*, **129**, 4456–4462.

418 Negishi, E.-I. (1998) *Metal-Catalyzed Cross-Coupling Reactions* (eds F. Diederich and P.J. Stang), Wiley-VCH Verlag GmbH, New York.

419 Hassan, J., Sévignon, M., Gozzi, C., and Schulz, E. (2002) *Chemical Reviews*, **102**, 1359–1470.

420 Öhberg, L. and Westman, J. (2001) *Synlett*, 1893–1896.

421 Stanetty, P., Schnürch, M., and Mihovilovic, M.D. (2003) *Synlett*, 1862–1864.

422 Walla, P. and Kappe, C.O. (2004) *Chemical Communications*, 564–565.

423 Dankwardt, J.W. (2005) *Journal of Organometallic Chemistry*, **690**, 932–938.

424 Mutule, I. and Suna, E. (2004) *Tetrahedron Letters*, **45**, 3909–3912.

425 Krascsenicsová, K., Walla, P., Kasáka, P., Uray, G., Kappe, C.O., and Putala, M. (2004) *Chemical Communications*, 2606–2607.

426 Bentz, E., Moloney, M.G., and Westaway, S.M. (2004) *Tetrahedron Letters*, **45**, 7395–7397.

427 Lipshutz, B.H. and Frieman, B. (2004) *Tetrahedron*, **60**, 1309–1316.

428 Lipshutz, B.H., Frieman, B., and Birkedal, H. (2004) *Organic Letters*, **6**, 2305–2308.

429 Wiskur, S.L., Korte, A., and Fu, G.C. (2004) *Journal of the American Chemical Society*, **126**, 82–83.

430 Lipshutz, B.H., Pfeiffer, S.S., Tomioka, T., and Noson, K. (2002) *Titanium and Zirconium in Organic Synthesis* (ed. I. Marek), Wiley-VCH Verlag GmbH, Weinheim, pp. 110–148.

431 Wipf, P., Janjic, J., and Stephenson, C.R.J. (2004) *Organic & Biomolecular Chemistry*, **2**, 443–445.

432 Wipf, P., Stephenson, C.R.J., and Okumura, K. (2003) *Journal of the American Chemical Society*, **125**, 14694–14695.

433 Kaiser, N.-F.K., Hallberg, A., and Larhed, M. (2002) *Journal of Combinatorial Chemistry*, **4**, 109–111.

434 Wannberg, J. and Larhed, M. (2003) *The Journal of Organic Chemistry*, **68**, 5750–5753.

435 Herrero, M.A., Wannberg, J., and Larhed, M. (2004) *Synlett*, 2335–2338.

436 Georgsson, J., Hallberg, A., and Larhed, M. (2003) *Journal of Combinatorial Chemistry*, **5**, 456–458.

437 Wu, X., Mahalingam, A.K., Wan, Y., and Alterman, M. (2004) *Tetrahedron Letters*, **45**, 4635–4638.

438 Wu, X., Nilsson, P., and Larhed, M. (2005) *The Journal of Organic Chemistry*, **70**, 346–349.

439 Wan, Y., Alterman, M., Larhed, M., and Hallberg, A. (2002) *The Journal of Organic Chemistry*, **67**, 6232–6235.

440 Wan, Y., Alterman, M., Larhed, M., and Hallberg, A. (2003) *Journal of Combinatorial Chemistry*, **5**, 82–84.

441 Enquist, P.-A., Nilsson, P., and Larhed, M. (2003) *Organic Letters*, **5**, 4875–4878.

442 Wu, X., Ekegren, J.K., and Larhed, M. (2006) *Organometallics*, **25**, 1434–1439.

443 Appukkuttan, P., Axelsson, L., Van der Eycken, E., and Larhed, M. (2008) *Tetrahedron Letters*, **49**, 5625–5628.

444 Lesma, G., Sacchetti, A., and Silvani, A. (2006) *Synthesis*, 594–596.

445 Lagerlund, O., Mantel, M.L.H., and Larhed, M. (2009) *Tetrahedron*, **65**, 7646–7652.

446 Wieckowska, A., Fransson, R., Odell, L.R., and Larhed, M. (2011) *The Journal of Organic Chemistry*, **76**, 978–981.

447 Wu, X., Rönn, R., Gossas, T., and Larhed, M. (2005) *The Journal of Organic Chemistry*, **70**, 3094–3098.

448 Liptrot, D., Alcaraz, L., and Roberts, B. (2010) *Tetrahedron Letters*, **51**, 5341–5343.

449 Roberts, B., Liptrot, D., and Alcaraz, L. (2010) *Organic Letters*, **12**, 1264–1267.

450 Roberts, B., Liptrot, D., Alcaraz, L., Luker, T., and Stocks, M.J. (2010) *Organic Letters*, **12**, 4280–4283.

451 Kormos, C.M. and Leadbeater, N.E. (2006) *Synlett*, 1663–1666.

452 Kormos, C.M. and Leadbeater, N.E. (2007) *Organic & Biomolecular Chemistry*, **5**, 65–68.

453 Kormos, C.M. and Leadbeater, N.E. (2007) *Synlett*, 2006–2010.

454 Cardullo, F., Donati, D., Merlo, G., Paio, A., Petricci, E., and Taddei, M. (2009) *Synlett*, 47–50.

455 Petricci, E., Mann, A., Schoenfelder, A., Rota, A., and Taddei, M. (2006) *Organic Letters*, **8**, 3725–3727.

456 Petricci, E., Mann, A., and Salvadori, J. (2007) *Tetrahedron Letters*, **48**, 8501–8504.

457 Liptrot, D., Alcaraz, L., and Roberts, B. (2010) *Advanced Synthesis and Catalysis*, **352**, 2183–2188.

458 Enquist, P.A., Nilsson, P., Edin, J., and Larhed, M. (2005) *Tetrahedron Letters*, **46**, 3335–3339.

459 Larhed, M., Moberg, C., and Hallberg, A. (2002) *Accounts of Chemical Research*, **35**, 717–727.

460 Kaiser, N.-F.K., Bremberg, U., Larhed, M., Moberg, C., and Hallberg, A. (2000) *Journal of Organometallic Chemistry*, **603**, 2–5.

461 Bremberg, U., Lutsenko, S., Kaiser, N.-F.K., Larhed, M., Hallberg, A., and Moberg, C. (2000) *Synthesis*, 1004–1008.

462 Kaiser, N.-F.K., Bremberg, U., Larhed, M., Moberg, C., and Hallberg, A. (2000) *Angewandte Chemie International Edition*, **39**, 3596–3598.

463 Belda, O., Kaiser, N.-F., Bremberg, U., Larhed, M., Hallberg, A., and Moberg, C. (2000) *The Journal of Organic Chemistry*, **65**, 5868–5870.

464 Belda, O. and Moberg, C. (2002) *Synthesis*, 1601–1606.

465 Braga, A.L., Vargas, L.F., Sehnem, J.A., and Wessjohann, L.A. (2006) *European Journal of Organic Chemistry*, 4993–4997.

466 Yeager, A.R., Min, G.K., Porco, J.A., Jr., and Schaus, S.E. (2006) *Organic Letters*, **8**, 5065–5068.

467 Trost, B.M. and Andersen, N.G. (2002) *Journal of the American Chemical Society*, **124**, 14320–14321.

468 Belda, O., Lundgren, S., and Moberg, C. (2003) *Organic Letters*, **5**, 2275–2278.

469 Nilsson, P., Gold, H., Larhed, M., and Hallberg, A. (2002) *Synthesis*, 1611–1614.

470 Lutsenko, S. and Moberg, C. (2001) *Tetrahedron: Asymmetry*, **12**, 2529–2532.

471 Mata, Y., Pàmies, O., and Diéguez, M. (2007) *Chemistry: A European Journal*, **13**, 3296–3304.

472 Mazuela, J., Tolstoi, P., Pàmies, O., Andersson, P.G., and Diéguez, M. (2011) *Organic & Biomolecular Chemistry*, **9**, 941–946.

473 Kaukoranta, P., Källström, K., and Andersson, P.G. (2007) *Advanced Synthesis and Catalysis*, **349**, 2595–2602.

474 Dounay, A.B., Humphreys, P.G., Overman, L.E., and Wrobleski, A.D. (2008) *Journal of the American Chemical Society*, **130**, 5368–5377.

475 Yang, J., Wu, H., Shen, L., and Qin, Y. (2007) *Journal of the American Chemical Society*, **129**, 13794–13795.

476 Brekan, J.A., Reynolds, T.E., and Scheidt, K.A. (2010) *Journal of the American Chemical Society*, **132**, 1472–1473.

477 Alonso, D.A., Nájera, C., and Pacheco, M.C. (2004) *The Journal of Organic Chemistry*, **69**, 1615–1619.

478 Liebeskind, L.S. and Srogl, J. (2002) *Organic Letters*, **4**, 979–982.

479 Lengar, A. and Kappe, C.O. (2004) *Organic Letters*, **6**, 771–774.

480 Arshad, N., Hashim, J., and Kappe, C.O. (2009) *The Journal of Organic Chemistry*, **74**, 5118–5121.

481 Modha, S.G., Trivedi, J.C., Mehta, V.P., Ermolat'ev, S., and Van der Eycken, E.V. (2011) *The Journal of Organic Chemistry*, **76**, 846–856.

482 Zhang, A. and Neumeyer, J.L. (2003) *Organic Letters*, **5**, 201–203.

483 Zhang, A., Xiong, W., Bidlack, J.M., Hilbert, J.E., Knapp, B.I., Wentland, M.P., and Neumeyer, J.L. (2004) *Journal of Medicinal Chemistry*, **47**, 165–174.

484 Alterman, M. and Hallberg, A. (2000) *The Journal of Organic Chemistry*, **65**, 7984–7989.

485 Arvela, R.K. and Leadbeater, N.E. (2003) *The Journal of Organic Chemistry*, **68**, 9122–9125.

486 Gopalsamy, A., Lim, K., Ciszewski, G., Park, K., Ellingboe, J.W., Bloom, J., Insaf, S., Upeslacis, J., Mansour, T.S., Krishnamurthy, G., Damarla, M., Pyatski, Y., Ho, D., Howe, A.Y.M., Orlowski, M., Feld, B., and O'Connell, J. (2004) *Journal of Medicinal Chemistry*, **47** 6603–6608.

487 DeBlase, C. and Leadbeater, N.E. (2010) *Tetrahedron*, **66**, 1098–1101.

488 Chen, G., Weng, J., Zheng, Z., Zhu, X., Cai, Y., Cai, J., and Wan, Y. (2008) *European Journal of Organic Chemistry*, 3524–3528.

489 Pitts, M.R., McCormack, P., and Whittall, J. (2006) *Tetrahedron*, **62**, 4705–4708.

490 Vo-Thanh, G., Lahrache, H., Loupy, A., Kim, I.-J., Chang, D.-H., and Jun, C.-H. (2004) *Tetrahedron*, **60**, 5539–5543.

491 Lipshutz, B.H., Frieman, B.A., Lee, C.-T., Lower, A., Nihan, D. M., and Taft, B.R. (2006) *Chemistry: An Asian Journal*, **1**, 417–429.

492 Lipshutz, B.H., Butler, T., and Swift, E. (2008) *Organic Letters*, **10**, 697–700.

493 Lerebours, R. and Wolf, C. (2007) *Organic Letters*, **9**, 2737–2740.

494 Frost, C.G., Penrose, S.D., Lambshead, K., Raithby, P.R., Warren, J.E., and Gleave, R. (2007) *Organic Letters*, **9**, 2119–2122.

495 Hargrave, J.D., Herbert, J., Bish, G., and Frost, C.G. (2006) *Organic & Biomolecular Chemistry*, **4**, 3235–3241.

496 Iyer, P.S., O'Malley, M.M., and Lucas, M.C. (2007) *Tetrahedron Letters*, **48**, 4413–4418.

497 Navarro, O., Marion, N., Oonishi, Y., Kelly, J., III,R.A., and Nolan,S.P. (2006) *The Journal of Organic Chemistry*, **71**, 685–692.

498 Malcolm, S.C., Ribe, S., Wang, F., Hewitt, M.C., Bhongle, N., Bakale, R.P., and Shao, L. (2005) *Tetrahedron Letters*, **46**, 6871–6873.

499 Beeler, A.B., Su, S., Singleton, C.A., and Porco, J.A., Jr. (2007) *Journal of the American Chemical Society*, **129**, 1413–1419.

500 Lautens, M., Tayama, E., and Herse, C. (2005) *Journal of the American Chemical Society*, **127**, 72–73.

501 Alberico, D. and Lautens, M. (2006) *Synlett*, 2969–2972.

502 Mariampillai, B., Alberico, D., Bidau, V., and Lautens, M. (2006) *Journal of the American Chemical Society*, **128**, 14436–14437.

503 Alberico, D., Rudolph, A., and Lautens, M. (2007) *The Journal of Organic Chemistry*, **72**, 775–781.
504 Mariampillai, B., Alliot, J., Li, M., and Lautens, M. (2007) *Journal of the American Chemical Society*, **129**, 15372–15379.
505 Rudolph, A., Rackelmann, N., Turcotte-Savard, M.-O., and Lautens, M. (2009) *The Journal of Organic Chemistry*, **74**, 289–297.
506 Goossen, L.J., Linder, C., Rodriguez, N., and Lange, P.P. (2009) *Chemistry: A European Journal*, **15**, 9336–9349.
507 Goossen, L.J., Zimmermann, B., Linder, C., Rodriguez, N., Lange, P.P., and Hartung, J. (2009) *Advanced Synthesis and Catalysis*, **351**, 2667–2674.
508 Lindh, J., Sjöberg, P.J.R., and Larhed, M. (2010) *Angewandte Chemie International Edition*, **49**, 7733–7737.
509 Zhao, X., Wu, G., Yan, C., Lu, K., Li, H., Zhang, Y., and Wang, J. (2010) *Organic Letters*, **12**, 5580–5583.
510 Storgaard, M., Zaragoza Dörwald, F., Peschke, B., and Tanner, D. (2009) *The Journal of Organic Chemistry*, **74**, 5032–5040.
511 Gao, G., Brown, N., Minatoya, M., and Buszek, K.R. (2008) *Tetrahedron Letters*, **49**, 6491–6494.
512 Shanmugasundaram, M., Garcia-Martinez, I., Li, Q., Estrada, A., Martinez, N.E., and Martinez, L.E. (2005) *Tetrahedron Letters*, **46**, 7545–7548.
513 Bour, C. and Suffert, J. (2006) *European Journal of Organic Chemistry*, 1390–1395.
514 Blond, G., Bour, C., Salem, B., and Suffert, J. (2008) *Organic Letters*, **10**, 1075–1078.
515 Oi, S., Funayama, R., Hattori, T., and Inoue, Y. (2008) *Tetrahedron*, **64**, 6051–6059.
516 Jiang, W., Allan, G., Fiordeliso, J.J., Linton, O., Tannenbaum, P., Xu, J., Zhu, P., Gunnet, J., Demarest, K., Lundeen, S., and Sui, Z. (2006) *Bioorganic & Medicinal Chemistry*, **14**, 6726–6732.
517 Iwasaki, M., Hayashi, S., Hirano, K., Yorimitsu, H., and Oshima, K. (2007) *Tetrahedron*, **63**, 5200–5203.
518 (a) Muci, A.R. and Buchwald, S.L. (2002) *Topics in Current Chemistry*, **219**, 131–209; (b) Wolfe, J.P., Wagaw, S., Marcoux, J.F., and Buchwald, S.L. (1998) *Accounts of Chemical Research*, **31**, 805–818.
519 Hartwig, J.F. (1998) *Angewandte Chemie International Edition in English.*, **37**, 2046–2067.
520 Wan, Y., Alterman, M., and Hallberg, A. (2002) *Synthesis*, 1597–1600.
521 Antane, S. (2003) *Synthetic Communications*, **33**, 2147–2149.
522 Wang, T., Magnin, D.R., and Hamann, L.G. (2003) *Organic Letters*, **5**, 897–900.
523 Ulrich, T. and Giraud, F. (2003) *Tetrahedron Letters*, **44**, 4207–4211.
524 Brain, C.T. and Steer, J.T. (2003) *The Journal of Organic Chemistry*, **68**, 6814–6816.
525 Jensen, T.A., Liang, X., Tanner, D., and Skjaerbaek, N. (2004) *The Journal of Organic Chemistry*, **69**, 4936–4947.
526 Tundel, R.E., Anderson, K.W., and Buchwald, S.L. (2006) *The Journal of Organic Chemistry*, **71**, 430–433.
527 Zhang, H.Q., Xia, Z., Vasudevan, A., and Djuric, S.W. (2006) *Tetrahedron Letters*, **47**, 4881–4884.
528 McCarroll, A.J., Sandham, D.A., Titcomb, L.R., de A.K., Lewis, K., Cloke, F.G.N., Davies, B.P., de Santana, A.P., Hiller, W., and Caddick, S. (2003) *Molecular Diversity*, **7**, 115–123.
529 Broggi, J., Clavier, H., and Nolan, S.P. (2008) *Organometallics*, **27**, 5525–5531.
530 Maes, B.U.W., Loones, K.T.J., Lemiére, G.L.F., and Dommisse, R.A. (2003) *Synlett*, 1822–1825.
531 Maes, B.U.W., Loones, K.T.J., Hostyn, S., and Diels, G. (2004) *Tetrahedron*, **60**, 11559–11564.
532 Van Baelen, G. and Maes, B.U.W. (2008) *Tetrahedron*, **64**, 5604–5619.
533 Ronco, C., Jean, L., Outaabout, H., and Renard, P.-Y. (2011) *Chemistry: A European Journal*, **17**, 302–310.
534 Smith, J.A., Jones, R.K., Booker, G.W., and Pyke, S.M. (2008) *The Journal of Organic Chemistry*, **73**, 8880–8892.

535 Burton, G., Cao, P., Li, G., and Rivero, R. (2003) *Organic Letters*, **5**, 4373–4376.
536 Harmata, M., Hong, X., and Gosh, S.K. (2004) *Tetrahedron Letters*, **45**, 5233–5236.
537 Luo, X., Chenard, E., Martens, P., Cheng, Y.-X., and Tommaszwski, M.J. (2010) *Organic Letters*, **12**, 3574–3577.
538 Erb, W., Neuville, L., and Zhu, J. (2009) *The Journal of Organic Chemistry*, **74**, 3109–3115.
539 Bonnaterre, F., Bois-Choussy, M., and Zhu, J. (2006) *Organic Letters*, **8**, 4351–4354.
540 Poondra, R.R. and Turner, N.J. (2005) *Organic Letters*, **7**, 863–866.
541 Tietze, M., Iglesias, A., Merisor, E., Conrad, J., Klaiber, I., and Beifuss, U. (2005) *Organic Letters*, **7**, 1549–1552.
542 Alen, J., Robeyns, K., De Borggraeve, W.M., Van Meervelt, L., and Compernolle, F. (2008) *Tetrahedron*, **64**, 8128–8133.
543 Guo, D., Huang, H., Xu, J., Jiang, H., and Liu, H. (2008) *Organic Letters*, **10**, 4513–4516.
544 (a) Thomas, A.W. and Ley, S.V. (2003) *Angewandte Chemie International Edition*, **42**, 5400–5449; (b) Kunz, K., Scholz, U., and Ganzer, D. (2003) *Synlett*, 2428–2439.
545 Wu, Y.-J., He, H., and L'Heureux, A. (2003) *Tetrahedron Letters*, **44**, 4217–4218.
546 Wu, Y.-J. and He, H. (2003) *Synlett*, 1789–1790.
547 He, H. and Wu, Y.-J. (2003) *Tetrahedron Letters*, **44**, 3445–3446.
548 Yeh, V.S.C. and Wiedeman, P.E. (2006) *Tetrahedron Letters*, **47**, 6011–6016.
549 Rottger, S., Sjoberg, P.J.R., and Larhed, M. (2007) *Journal of Combinatorial Chemistry*, **9**, 204–209.
550 Rolfe, A. and Hanson, P.R. (2009) *Tetrahedron Letters*, **50**, 6935–6937.
551 Lange, J.H.M., Hofmeyer, L.J.F., Hout, F.A.S., Osnabrug, S.J.M., Verveer, P.C., Kruse, C.G., and Feenstra, R. (2002) *Tetrahedron Letters*, **43**, 1101–1104.
552 Hafner, T. and Kunz, D. (2007) *Synthesis*, 1403–1411.
553 Chen, S., Huang, H., Liu, X., Shen, J., Jiang, H., and Liu, H. (2008) *Journal of Combinatorial Chemistry*, **10**, 358–360.
554 Singh, B.K., Appukkuttan, P., Claerhout, S., Parmar, V.S., and Van der Eycken, E. (2006) *Organic Letters*, **8**, 1863–1866.
555 Bagley, M.C., Dix, M.C., and Fusillo, V. (2009) *Tetrahedron Letters*, **50**, 3661–3664.
556 Bagley, M.C., Davis, T., Dix, M.C., Fusillo, V., Pigeaux, M., Rokicki, M.C., and Kipling, D. (2009) *The Journal of Organic Chemistry*, **74**, 8336–8342.
557 Macmillan, D. and Anderson, D.W. (2004) *Organic Letters*, **6**, 4659–4662.
558 Christoforou, I.C. and Koutentis, P.A. (2006) *Organic & Biomolecular Chemistry*, **4**, 3681–3693.
559 Pabba, C., Wang, H.-J., Mulligan, S.R., Chen, Z.-J., Stark, T.M., and Gregg, B.T. (2005) *Tetrahedron Letters*, **46**, 7553–7557.
560 Yadav, L.D., Yadav, B.S., and Rai, V.K. (2006) *Synthesis*, 1868–1872.
561 Zhu, X., Ma, Y., Su, L., Song, H., Chen, G., Liang, D., and Wan, Y. (2006) *Synthesis*, 3955–3962.
562 Stadler, A. and Kappe, C.O. (2002) *Organic Letters*, **4**, 3541–3544.
563 Andaloussi, M., Lindh, J., Sävmarker, J., Sjöberg, P.J.R., and Larhed, M. (2009) *Chemistry: A European Journal*, **15**, 13069–13074.
564 Fürstner, A. and Seidel, G. (2002) *Organic Letters*, **4**, 541–543.
565 Appukkuttan, P., Van der Eycken, E., and Dehaen, W. (2003) *Synlett*, 1204–1206.
566 Lipshutz, B.H., Unger, J.B., and Taft, B.R. (2007) *Organic Letters*, **9**, 1089–1092.
567 Lipshutz, B.H., Nihan, D.M., Vinogradova, E., Taft, B.R., and Bošković, Ž.V. (2008) *Organic Letters*, **10**, 4279–4282.
568 Manbeck, G.F., Lipman, A.J., Stockland, R.A., Jr., Freidl, A.L., Hasler, A.F., Stone, J.J., and Guzei, I.A. (2005) *The Journal of Organic Chemistry*, **70**, 244–250.

569 D'Angelo, N.D., Peterson, J.J., Booker, S.K., Fellows, I., Dominguez, C., Hungate, R., Reider, P.J., and Kim, T.-S. (2006) *Tetrahedron Letters*, **47**, 5045–5048.

570 Rudolph, J., Esler, W.P., O'Connor, S., Coish, P.D.G., Wickens, P.L., Brands, M., Bierer, D.E., Bloomquist, B.T., Bondar, G., Chen, L., Chuang, C.-Y., Claus, T.H., Fathi, Z., Fu, W., Khire, U.R., Kristie, J.A., Liu, X.-G., Lowe, D.B., McClure, A.C., Michels, M., Ortiz, A.A., Ramsden, P.D., Schoenleber, R.W., Shelekhin, T.E., Vakalopoulos, A., Tang, W., Wang, L., Yi, L., Gardell, S.J., Livingston, J.N., Sweet, L.J., and Bullock, W.H. (2007) *Journal of Medicinal Chemistry*, **50** 5202–5216.

571 Zhu, X.-H., Chen, G., Ma, Y., Song, H.-C., Xu, Z.-L., and Wan, Y.-Q. (2007) *Chinese Journal of Chemistry*, **25**, 546–552.

572 Kormos, C.M. and Leadbeater, N.E. (2006) *Tetrahedron*, **62**, 4728–4732.

573 Raders, S.M. and Verkade, J.G. (2008) *Tetrahedron Letters*, **49**, 3507–3511.

574 Viirre, R.D., Evindar, G., and Batey, R.A. (2008) *The Journal of Organic Chemistry*, **73**, 3452–3459.

575 Thasana, N., Worayuthakarn, R., Kradanrat, P., Hohn, E., Young, L., and Ruchirawat, S. (2007) *The Journal of Organic Chemistry*, **72**, 9379–9382.

576 Li, Z., Sun, H., Jiang, H., and Liu, H. (2008) *Organic Letters*, **10**, 3263–3266.

577 Ranu, B.D., Saha, A., and Jana, R. (2007) *Advanced Synthesis and Catalysis*, **349**, 2690–2696.

578 Andersen, J., Madsen, U., Björkling, F., and Liang, X. (2005) *Synlett*, 2209–2213.

579 Liu, X.Y., Li, C.H., and Che, C.M. (2006) *Organic Letters*, **8**, 2707–2710.

580 (a) Grubbs, R.H. and Chang, S. (1998) *Tetrahedron*, **54**, 4413–4450; (b) Fürstner, A. (2000) *Angewandte Chemie International Edition*, **39**, 3012–3043.

581 Mayo, K.G., Nearhoof, E.H., and Kiddle, J.J. (2002) *Organic Letters*, **4**, 1567–1570.

582 Garbacia, S., Desai, B., Lavastre, O., and Kappe, C.O. (2003) *The Journal of Organic Chemistry*, **68**, 9136–9139.

583 Vo Thanh, G. and Loupy, A. (2003) *Tetrahedron Letters*, **44**, 9091–9094.

584 Varry, S., Gauzy, C., Lamaty, F., Lazaro, R., and Martinez, J. (2000) *The Journal of Organic Chemistry*, **65**, 6787–6790.

585 Yang, C., Murray, W.V., and Wilson, L.J. (2003) *Tetrahedron Letters*, **44**, 1783–1786.

586 Grigg, R., Martin, W., Morris, J., and Sridharan, V. (2003) *Tetrahedron Letters*, **44**, 4899–4901.

587 Balan, D. and Adolfsson, H. (2004) *Tetrahedron Letters*, **45**, 3089–3092.

588 Declerck, V., Ribière, P., Martinez, J., and Lamaty, F. (2004) *The Journal of Organic Chemistry*, **69**, 8372–8381.

589 Schultz-Fademrecht, C., Deshmukh, P.H., Malagu, K., Procopiou, P.A., and Barrett, A.G.M. (2004) *Tetrahedron*, **60**, 7515–7524.

590 Appukkuttan, P., Dehaen, W., and Van der Eycken, E. (2005) *Organic Letters*, **7**, 2723–2726.

591 Appukkuttan, P., Dehaen, W., and Van der Eycken, E. (2007) *Chemistry: A European Journal*, **13**, 6452–6460.

592 Sunderhaus, J.D., Dockendorff, C., and Martin, S.F. (2007) *Organic Letters*, **9**, 4223–4226.

593 Nosse, B., Schall, A., Jeong, W.B., and Reiser, O. (2005) *Advanced Synthesis and Catalysis*, **347**, 1869–1874.

594 Dallinger, D., Irfan, M., Suljanovic, A., and Kappe, C.O. (2010) *The Journal of Organic Chemistry*, **75**, 5278–5288.

595 Chapman, R.N. and Arora, P.S. (2006) *Organic Letters*, **8**, 5825–5828.

596 Robinson, A.J., Elaridi, J., Van Lierop, B.J., Mujcinovic, S., and Jackson, W.R. (2007) *Journal of Peptide Science*, **13**, 280–285.

597 Collins, S.K., Grandbois, A., Vachon, M.P., and Côté, J. (2006) *Angewandte Chemie International Edition*, **45**, 2923–2926.

598 Collins, S.K. (2006) *Journal of Organometallic Chemistry*, **691**, 5122–5128.

599 Perez-Balado, C., Nebbioso, A., Rodriguez-Grana, P. *et al.* (2007) *Journal of Medicinal Chemistry*, **50**, 2497–2505.

600 Spandl, R.J., Rudyk, H., and Spring, D.R. (2008) *Chemical Communications*, 3001–3003.

601 Efskind, J. and Undheim, K. (2003) *Tetrahedron Letters*, **44**, 2837–2839.

602 Fürstner, A., Stelzer, F., Rumbo, A., and Krause, H. (2002) *Chemistry: A European Journal*, **8**, 1856–1871.

603 Salim, S.S., Bellingham, R.K., and Brown, R.C.D. (2004) *European Journal of Organic Chemistry*, 800–806.

604 Wang, L., Maddess, M.L., and Lautens, M. (2007) *The Journal of Organic Chemistry*, **72**, 1822–1825.

605 Comer, E., Rohan, E., Deng, L., and Porco, J.A., Jr. (2007) *Organic Letters*, **9**, 2123–2126.

606 Debleds, O. and Campagne, J.M. (2008) *Journal of the American Chemical Society*, **130**, 1562–1563.

607 Vincent, G. and Kouklovsky, C. (2011) *Chemistry: A European Journal*, **17**, 2972–2980.

608 Bargiggia, F.C. and Murray, W.V. (2005) *The Journal of Organic Chemistry*, **70**, 9636–9639.

609 Morris, T., Sandham, D., and Caddick, S. (2007) *Organic & Biomolecular Chemistry*, **5**, 1025–1027.

610 Elaridi, J., Patel, J., Jackson, W.R., and Robinson, A.J. (2006) *The Journal of Organic Chemistry*, **71**, 7538–7545.

611 Castagnolo, D., Renzulli, M.L., Galletti, E., Corelli, F., and Botta, M. (2005) *Tetrahedron: Asymmetry*, **16**, 2893–2896.

612 Castagnolo, D., Giorgi, Gi., Spinosa, R., Corelli, F., and Botta, M. (2007) *European Journal of Organic Chemistry*, 3676–3686.

613 Fustero, S., Jiménez, D., Sánchez-Roselló, M., and del Pozo, C. (2007) *Journal of the American Chemical Society*, **129**, 6700–6701.

614 Kirschning, A., Harmrolfs, K., Mennecke, K., Messinger, J., Schön, U., and Grela, K. (2008) *Tetrahedron Letters*, **49**, 3019–3022.

615 Lumini, M., Cordero, F.M., Pisaneschi, F., and Brandi, A. (2008) *European Journal of Organic Chemistry*, 2817–2824.

616 Pauson, P.L. (1985) *Tetrahedron*, **41**, 5855–5860.

617 Fischer, S., Groth, U., Jung, M., and Schneider, A. (2002) *Synlett*, 2023–2026.

618 Iqbal, M., Vyse, N., Dauvergne, J., and Evans, P. (2002) *Tetrahedron Letters*, **43**, 7859–7862.

619 Iqbal, M., Li, Y., and Evans, P. (2004) *Tetrahedron*, **60**, 2531–2538.

620 Fager-Jokela, E., Kaasalainen, E., Leppänen, K., Tois, J., and Helaja, J. (2008) *Tetrahedron*, **64**, 10381–10387.

621 Lee, H.W., Kwong, F.Y., and Chan, A. (2008) *Synlett*, 1553–1556.

622 (a) For reviews, see: Dyker, G. (1999) *Angewandte Chemie International Edition*, **38**, 1699–1712; (b) Kakiuchi, F. and Murai, S. (2002) *Accounts of Chemical Research*, **35**, 826–834; (c) Ritleng, V., Sirlin, C., and Pfeffer, M. (2002) *Chemical Reviews*, **102**, 1731–1769.

623 Tan, K.L., Vasudevan, A., Bergman, R.G., Ellman, J.A., and Souers, A.J. (2003) *Organic Letters*, **5**, 2131–2134.

624 Lewis, J.C., Wu, J.Y., Bergman, R.G., and Ellman, J.A. (2006) *Angewandte Chemie International Edition*, **45**, 1589–1591.

625 Lewis, J.C., Berman, A.M., Bergman, R.G., and Ellman, J.A. (2008) *Journal of the American Chemical Society*, **130**, 2493–2500.

626 Yanagisawa, S., Sudo, T., Noyori, R., and Itami, K. (2006) *Journal of the American Chemical Society*, **128**, 11748–11749.

627 Yanagisawa, S., Sudo, T., Noyori, R., and Itami, K. (2008) *Tetrahedron*, **64**, 6073–6081.

628 Yanagisawa, S., Ueda, K., Taniguchi, T., and Itami, K. (2008) *Organic Letters*, **10**, 4673–4676.

629 Tajuddin, H., Shukla, L., Maxwell, A.C., Marder, T.B., and Steel, P.G. (2010) *Organic Letters*, **12**, 5700–5703.

630 Harrisson, P., Morris, J., Marder, T.B., and Steel, P.G. (2009) *Organic Letters*, **11**, 3586–3589.

631 Sahnoun, S., Messaoudi, S., Peyrat, J.-F., Brion, J.-D., and Alami, M. (2008) *Tetrahedron Letters*, **49**, 7279–7283.

632 Ermolat'ev, D.S., Giménez, V.N., Babaev, E.V., and Van der Eycken, E. (2006) *Journal of Combinatorial Chemistry*, **8**, 659–663.

633 Besselièvre, F., Mahuteau-Betzer, F., Grierson, D.S., and Piguel, S. (2008) *The Journal of Organic Chemistry*, **73**, 3278–3280.

634 Iwasaki, M., Yorimitsu, H., and Oshima, K. (2007) *Chemistry: An Asian Journal*, **2**, 1430–1435.

635 Hull, K.L., Anani, W.Q., and Sanford, M.S. (2006) *Journal of the American Chemical Society*, **128**, 7134–7135.

636 Bedford, R.B. and Betham, M. (2006) *The Journal of Organic Chemistry*, **71**, 9403–9410.

637 Bedford, R.B., Betham, M., Charmant, J.P.H., and Weeks, A.L. (2008) *Tetrahedron*, **64**, 6038–6050.

638 Hostyn, S., Van Baelen, G., Lemière, G.L.F., and Maes, B.U.W. (2008) *Advanced Synthesis and Catalysis*, **350**, 2653–2660.

639 Campeau, L.C., Schipper, D.J., and Fagnou, K. (2008) *Journal of the American Chemical Society*, **130**, 3266–3267.

640 Yotphan, S., Bergman, R.G., and Ellman, J.A. (2009) *Organic Letters*, **11**, 1511–1514.

641 Johansen, M.B. and Kerr, M.A. (2010) *Organic Letters*, **12**, 4956–4959.

642 (a) Tornøe, C.W., Christensen, C., and Meldal, M. (2002) *The Journal of Organic Chemistry*, **67**, 3057–3064; (b) Tornøe, C.W., Christensen, C. and Meldal, M. (2002) *The Journal of Organic Chemistry*, **67**, 3057–3064; (c) Rostovtsev, V.V., Green, L.G., Fokin, V.V., and Sharpless, K.B. (2002) *Angewandte Chemie International Edition*, **41**, 2596–2599.

643 Khanetskyy, B., Dallinger, D., and Kappe, C.O. (2004) *Journal of Combinatorial Chemistry*, **6**, 884–892.

644 Appukkuttan, P., Dehaen, W., Fokin, V.V., and Van der Eycken, E. (2004) *Organic Letters*, **6**, 4223–4225.

645 Beckmann, H.S.G. and Wittmann, V. (2007) *Organic Letters*, **9**, 1–4.

646 Moorhouse, A. and Moses, J.E. (2008) *Synlett*, 2089–2092.

647 Lipshutz, B.H. and Taft, B.R. (2006) *Angewandte Chemie International Edition*, **45**, 8235–8238.

648 Rasmussen, L.K., Boren, B.C., and Fokin, V.V. (2007) *Organic Letters*, **9**, 5337–5339.

649 Pradere, U., Roy, V., McBrayer, T.R., Schinazi, R.F., and Agrofoglio, L.A. (2008) *Tetrahedron*, **64**, 9044–9051.

650 Horneff, T., Chuprakov, S., Chernyak, N., Gevorgyan, V., and Fokin, V.V. (2008) *Journal of the American Chemical Society*, **130**, 14972–14974.

651 Pietrzik, N., Schips, C., and Ziegler, T. (2008) *Synthesis*, 519–526.

652 Broggi, J., Díez-González, S., Petersen, J. *et al.* (2008) *Synthesis*, 141–148.

653 Joosten, J.A.F., Tholen, N.T.H., Ait El Maate, F., Brouwer, A.J., van Esse, G.W., Rijkers, D.T.S., Liskamp, R.M.J., and Pieters, R.J. (2005) *European Journal of Organic Chemistry*, 3182–3185.

654 Song, Y., Kohlmeir, E.K., and Meade, T.J. (2008) *Journal of the American Chemical Society*, **130**, 6662–6663.

655 Ortega-Muñoz, M., Morales-Sanfrutos, J., Perez-Balderas, F., Hernandez-Mateo, F., Giron-Gonzalez, M.D., Sevillano-Tripero, Na., l Salto-Gonzalez, R., and Santoyo-Gonzalez, F. (2007) *Organic & Biomolecular Chemistry*, **5**, 2291–2301.

656 Fazio, M.A., Lee, O.P., and Schuster, D.I. (2008) *Organic Letters*, **10**, 4979–4982.

657 Géci, I., Vyacheslav, F.V., and Pedersen, E.B. (2007) *Chemistry: A European Journal*, **13**, 6379–6386.

658 Angelo, N.G. and Arora, P.S. (2007) *The Journal of Organic Chemistry*, **72**, 7963–7967.

659 Castagnolo, D., Dessì, F., Radi, M. *et al.* (2007) *Tetrahedron Asymmetry*, **18**, 1345–1350.

660 Gaunt, M.J., Jessiman, A.S., Orsini, P., Tanner, H.R., Hook, D.F., and Ley, S.V. (2003) *Organic Letters*, **5**, 4819–4822.

661 Jung, M.E. and Maderna, A. (2004) *The Journal of Organic Chemistry*, **69**, 7755–7757.
662 Mejía-Oneto, J.M. and Padwa, A. (2004) *Tetrahedron Letters*, **45**, 9115–9118.
663 Jones, A.L. and Snyder, J.K. (2009) *The Journal of Organic Chemistry*, **74**, 2907–2910.
664 Airiau, E., Chemin, C., Girard, N., Lonzi, G., Mann, A., Petricci, E., Salvadori, J., and Taddei, M. (2010) *Synthesis*, 2901–2914.
665 Jacq, J., Bessiéres, B., Einhorn, C., and Einhorn, J. (2010) *Organic & Biomolecular Chemistry*, **8**, 4927–4933.
666 Nieto-Oberhuber, C., Pérez-Galán, P., Herrero-Gómez, E., Lauterbach, T., Rodríguez, C., López, S., Bour, C., Rosellón, A., Cárdenas, D.J., and Echavarren, A.M. (2008) *Journal of the American Chemical Society*, **130**, 269–279.
667 Marion, N., Gealageas, R., and Nolan, S.P. (2007) *Organic Letters*, **9**, 2653–2656.
668 Liu, X.Y., Ding, P., Huang, J.S., and Che, C.M. (2007) *Organic Letters*, **9**, 2645–2648.
669 Liu, X.Y., Li, C.H., and Che, C.M. (2006) *Organic Letters*, **8**, 2707–2710.
670 Wang, M.Z., Wong, M.K., and Che, C.M. (2008) *Chemistry: A European Journal*, **14**, 8353–8364.
671 Prior, A.M. and Robinson, R.S. (2008) *Tetrahedron Letters*, **49**, 411–414.
672 Labonne, A., Zani, L., Hintermann, L., and Bolm, C. (2007) *The Journal of Organic Chemistry*, **72**, 5704–5708.
673 Stockland, R.A., Jr., Lipman, A.J., Bawiec, J.A., III, Morrison, P.E., Guzei, I.A., Findeis, P.M., and Tamblin, J.F. (2006) *Journal of Organometallic Chemistry*, **691**, 4042–4053.
674 Ishibashi, K., Takahashi, M., Yokota, Y., Oshima, K., and Matsubara, S. (2005) *Chemistry Letters*, **34**, 664–665.
675 Fairlamb, I.J.S., McGlacken, G.P., and Weissberger, F. (2006) *Chemical Communications*, 988–990.
676 Takacs, J.M., Venkataraman, S., Andrews, R.N., and Pelter, L.S.W. (2005) *Journal of Organometallic Chemistry*, **690**, 6205–6209.
677 Lee, Y.T., Choi, S.Y., and Chung, Y.K. (2007) *Tetrahedron Letters*, **48**, 5673–5677.
678 Sherman, E.S., Fuller, P.H., Kasi, D., and Chemler, S.R. (2007) *The Journal of Organic Chemistry*, **72**, 3896–3905.
679 Lipshutz, B.H., Caires, C.C., Kuipers, P., and Chrisman, W. (2003) *Organic Letters*, **5**, 3085–3088.
680 Hutchinson, E.J., Kerr, W.J., and Magennis, E.J. (2002) *Chemical Communications*, 2262–2263.
681 Dosa, P.I., Whitener, G.D., and Vollhardt, K.P.C. (2002) *Organic Letters*, **4**, 2075–2078.
682 Carter, K.R. (2002) *Macromolecules*, **35**, 6757–6759.

6
Literature Survey Part B: Miscellaneous Organic Transformations

6.1
Rearrangement Reactions

6.1.1
Claisen Rearrangements

In their synthesis of the natural product carpanone, Ley and coworkers have described the microwave-assisted Claisen rearrangement of an allyl ether (Scheme 6.1a) [1]. A 97% yield of the rearranged product could be obtained by three successive 15 min irradiations at 220 °C, employing toluene doped with the ionic liquid 1-butyl-3-methylimidazolium hexafluorophosphate ($bmimPF_6$) as solvent (see Section 4.5.2). Interestingly, one single 45 min irradiation event at the same temperature gave a somewhat lower yield (86%). A related Claisen rearrangement, albeit on a much more complex substrate, was reported by the same group, again using "pulsed" microwave irradiation conditions. Heating a solution of a propargylic enol ether (Scheme 6.1b) in 1,2-dichlorobenzene (DCB) at 180 °C for 15 min resulted in a 71% isolated yield of the desired allene as a single diastereomer, which was further elaborated into the skeleton of the triterpenoid natural product azadirachtin [2]. An 88% yield of product was obtained by applying 15 pulses of 1 min irradiation.

Nordmann and Buchwald have reported a diastereoselective Claisen rearrangement of an allyl vinyl ether to an aldehyde (Scheme 6.2) [3]. Using *N,N*-dimethylformamide (DMF) as solvent, an 80% yield with a diastereomeric ratio of 91 : 9 was obtained by microwave heating at 250 °C for 5 min. Conventional heating at 120 °C for 24 h provided somewhat higher yields and selectivities (90% yield, dr = 94 : 6).

A selection of other microwave-assisted Claisen rearrangements is shown in Scheme 6.3. The groups of Wada and Yanagida have discussed the solvent-free double Claisen rearrangement of bis(4-allyloxyphenyl)sulfone into bis(3-allyl-4-hydroxyphenyl)sulfone (Scheme 6.3a), an important color developer for heat- or pressure-sensitive recording [4]. The process was carried out on a 10 g scale, by first melting the solid starting material by conventional heating and then exposing the melt to microwave irradiation at 180 °C for 5 min, leading to a 87% yield of the desired product. Similarly, 2′-allyloxy-acetophenone was heated by microwave irradiation to

Microwaves in Organic and Medicinal Chemistry, Second Edition.
C. Oliver Kappe, Alexander Stadler, and Doris Dallinger.

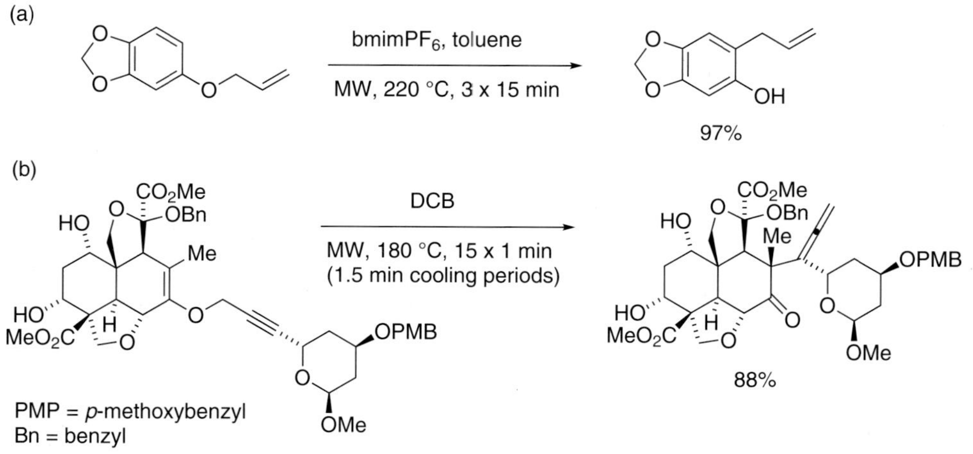

Scheme 6.1 Claisen rearrangements in natural product synthesis.

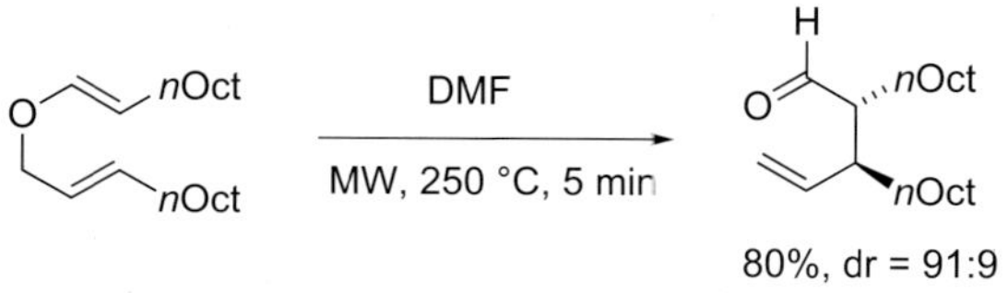

Scheme 6.2 Diastereoselective Claisen rearrangements.

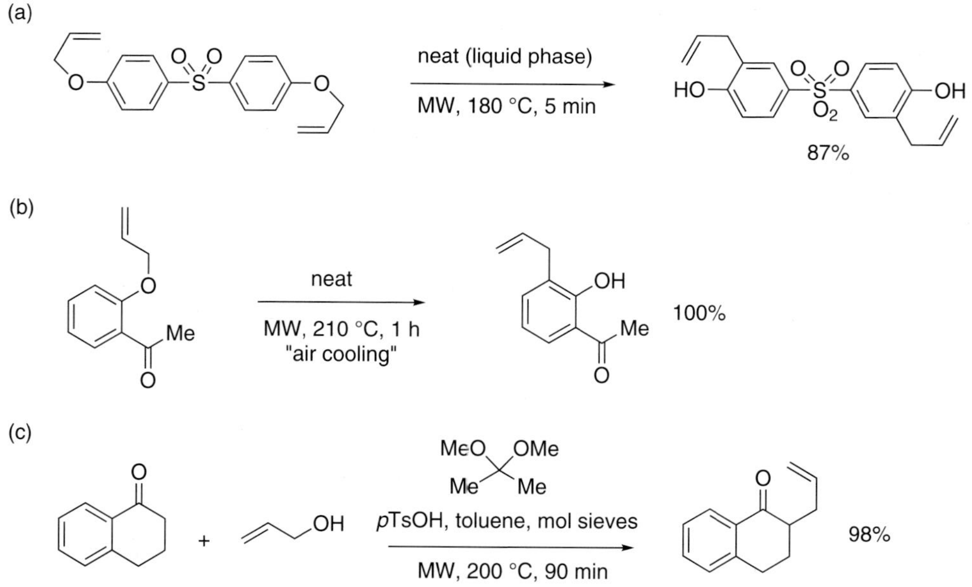

Scheme 6.3 Miscellaneous Claisen rearrangements.

210 °C for 1 h with simultaneous air cooling (see Section 2.5.1) to provide 3′-allyl-2′-hydroxy-acetophenone in quantitative yield (Scheme 6.3b) [5]. The corresponding thermal process (200 °C) took 44 h to reach completion. A method for the selective α-monoalkylation of phenyl ketones (e.g., α-tetralone) with allyl alcohol, involving the *in situ* formation and acid-catalyzed cracking of the corresponding ketone diallyl ketals, was described by Trabanco and coworkers (Scheme 6.3c) [6]. The process relied on the use of 2,2-dimethoxpropane and 3 Å molecular sieves as water scavengers, and allowed the preparation of α-allyl-substituted ketones in moderate to good yields. Optimum conditions required the use of 5 equiv of allyl alcohol, 1.5 equiv of 2,2-dimethoxypropane, and a catalytic amount of *p*-toluenesulfonic acid (*p*-TsOH). In case of α-tetralone, microwave heating to 200 °C for 90 min provided a 98% yield of the desired product, with only minor amounts of diallylated by-product being formed. The optimized protocol was applied to a set of 12 related benzocycloalkanones and phenyl ketones.

α-Chiral aldehydes were obtained by Trost and Zhang with minimum loss of chirality by Claisen rearrangement of allyl vinyl ethers at 140 °C within 15 min [7]. For a description of a decarboxylative Claisen rearrangement, see Ref. [8]. For Claisen rearrangements carried out in ionic liquids, see Ref. [9]. Aza-Claisen rearrangements under microwave conditions have been discussed by Gonda *et al.* [10].

6.1.2
Domino/Tandem Claisen Rearrangements

A series of complex sigmatropic rearrangements of allyl tetronates and allyl tetramates to furnish 3-allyltetronic or allyltetramic acids, respectively, were investigated by Schobert *et al.* (Scheme 6.4) [11]. The authors discovered that allyl tetronates (X = O) and allyl tetramates (X = NH) undergo a microwave-accelerated Claisen rearrangement allowing – in contrast to the conventional procedure – the isolation of the Claisen intermediates. The consecutive (homo)sigmatropic [1, 5] hydrogen shifts such as the oxa-ene (Conia) reaction (Scheme 6.4), leading to 3-(spirocyclopropyl) dihydrofuran-2,4-diones (X = O), are promoted less effectively, which allows the isolation of the Claisen intermediates in these sigmatropic domino sequences.

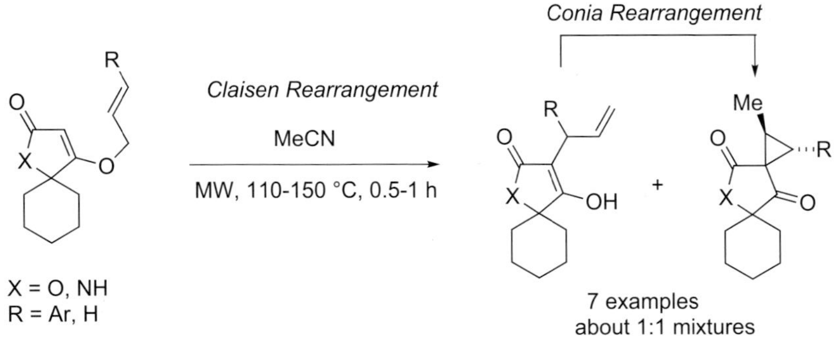

Scheme 6.4 Domino Claisen/Conia rearrangements.

Microwave heating to 110–150 °C for 30–60 min in acetonitrile (MeCN) typically provided mixtures of the desired and readily separable Claisen and Conia products.

Over the last few years, the group of Barriault has reported in a series of publications on microwave-assisted tandem oxy-Cope/Claisen/ene and closely related reactions [12–15]. These pericyclic transformations typically proceed in a highly stereoselective fashion and can be exploited for the synthesis of complex natural products possessing decalin skeletons, such as the abietane diterpene wiedamannic acid [14]. Some of the transformations described by Barriault and coworkers are summarized in Scheme 6.5. Typically, microwave heating in an inert solvent such as

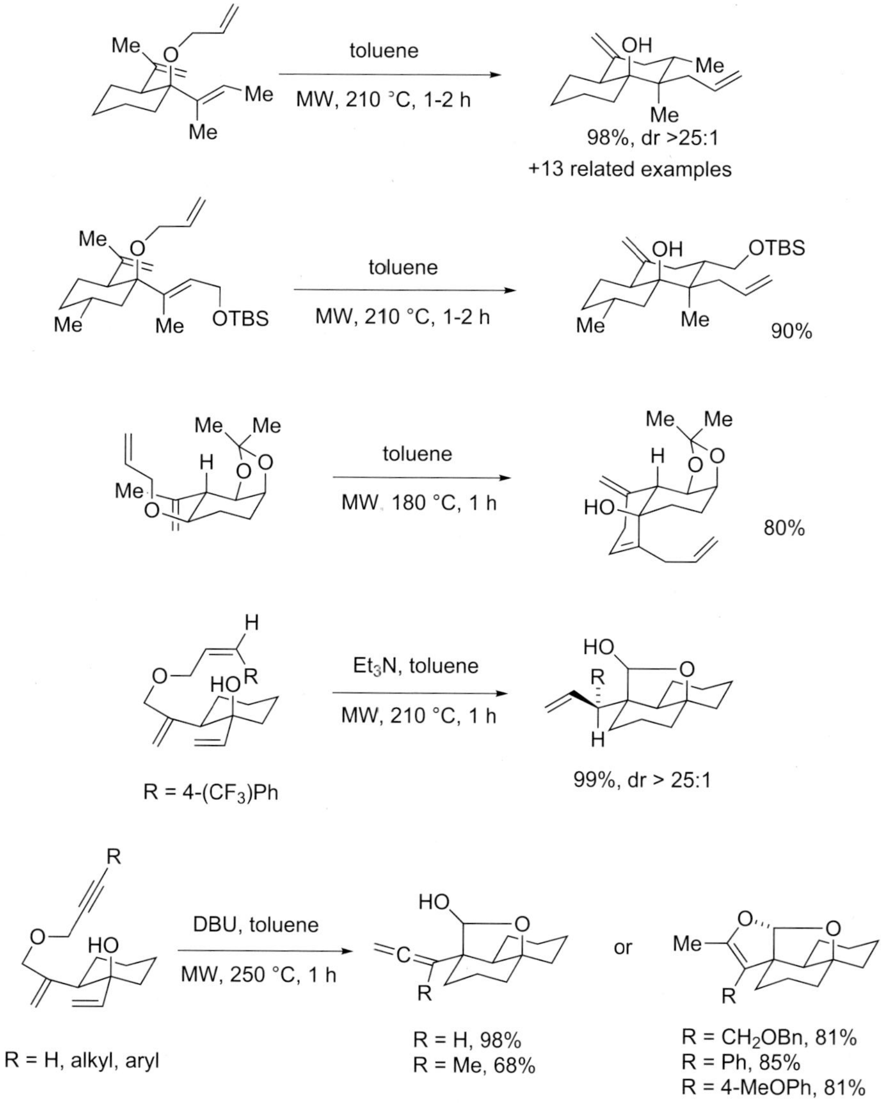

Scheme 6.5 Tandem oxy-Cope/Claisen/ene and related transformation.

toluene to temperatures of 180–250 °C is required for these rearrangements to proceed, often in the presence of a strong base.

The base-catalyzed intramolecular cyclization of appropriately substituted 4-alkyn-1-ols, followed by *in situ* Claisen rearrangement, was investigated by Ovaska and coworkers (Scheme 6.6) [16]. The tandem cyclization–Claisen rearrangements were best carried out in DMF or phenetole as solvent in the presence of 10 mol% of methyllithium base. In most cases, the resulting cycloheptanoid ring systems were produced in high yields in a matter of minutes upon microwave irradiation at 150–200 °C. Some of the reactions were also performed under solvent-free conditions providing similar isolated product yields. Several other bicyclo[5.3.0]decane ring systems were also available from relatively simple acetylenic alcohols using this strategy.

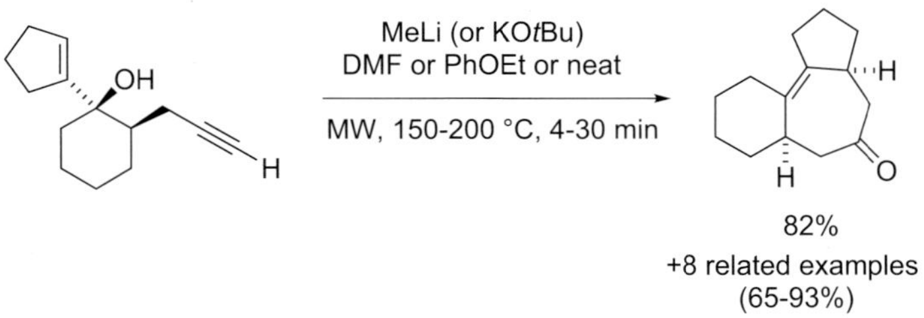

Scheme 6.6 Tandem 5-exo cyclization/Claisen rearrangement.

For related cyclizations published by the same group, see Ref. [17].

Nicolaou and his group have performed the total synthesis of artochamins F and H–J applying microwave heating for the key step (Scheme 6.7) [18]. After synthesizing stilbene **1**, the artochamin skeleton **2** was obtained via a cascade sequence that required catalytic amounts of Ph_3PO. The cascade reaction incorporated two consecutive Claisen rearrangements, collapsing of the Boc groups and a formal [2 + 2] cycloaddition. Further chemoselective protection of **2** with subsequent methylation afforded artochamins H and I and artochamin J by generation of a benzopyran scaffold. If the TBS-protected stilbene **1** is employed under the same conditions but without Ph_3PO, the cascade reaction stops after the Claisen rearrangements, indicating the need of the unprotected hydroxyl groups for the formal [2 + 2] cycloaddition. After debromination and TBS deprotection, artochamin F was obtained in 92% yield.

Pelc and Zakarian have reported on the synthesis of A,G-spiroimine **3** of pinnatoxins (Scheme 6.8) [19]. Key step in this procedure is a cascade sigmatropic route that integrates the Claisen and Mislow–Evans rearrangements. By applying microwave heating at 170 °C, the reaction time could be reduced to 20 min compared to 15 h under conventional heating at 150 °C.

A combined microwave-assisted Mitsunobu reaction/Claisen rearrangement has been reported by Moody and coworkers [20, 21]. A microwave transformation involving a tandem Horner–Wadsworth–Emmons olefination/Claisen rearrangement/hydrolysis

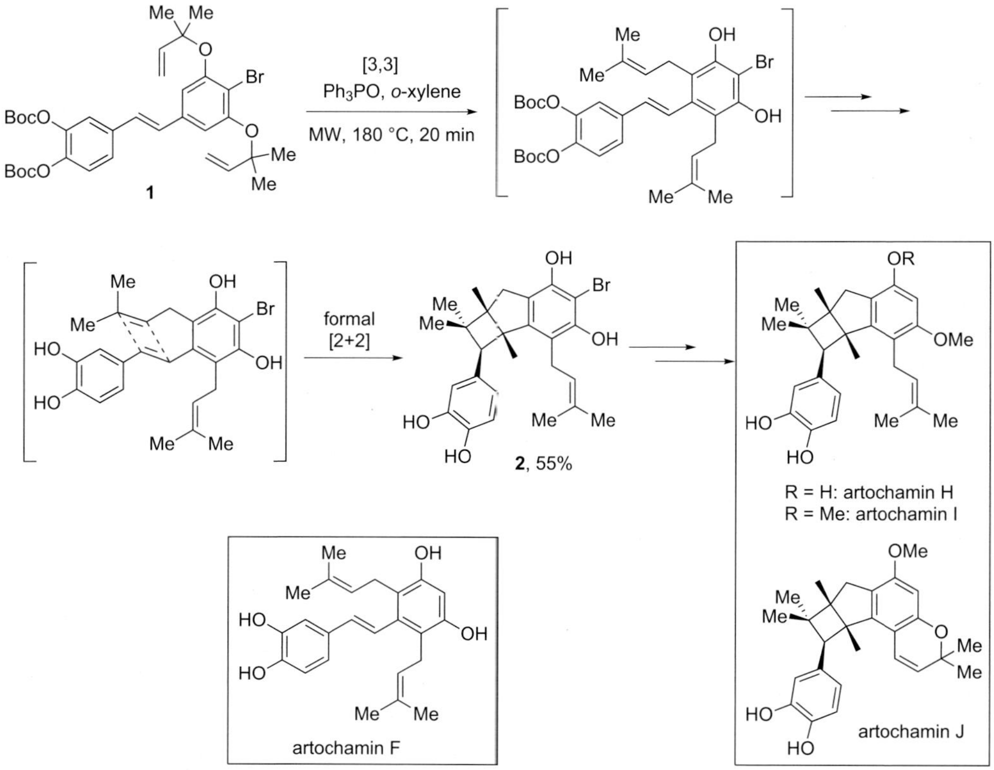

Scheme 6.7 Cascade reactions in the total synthesis of artochamins F, H, I, and J.

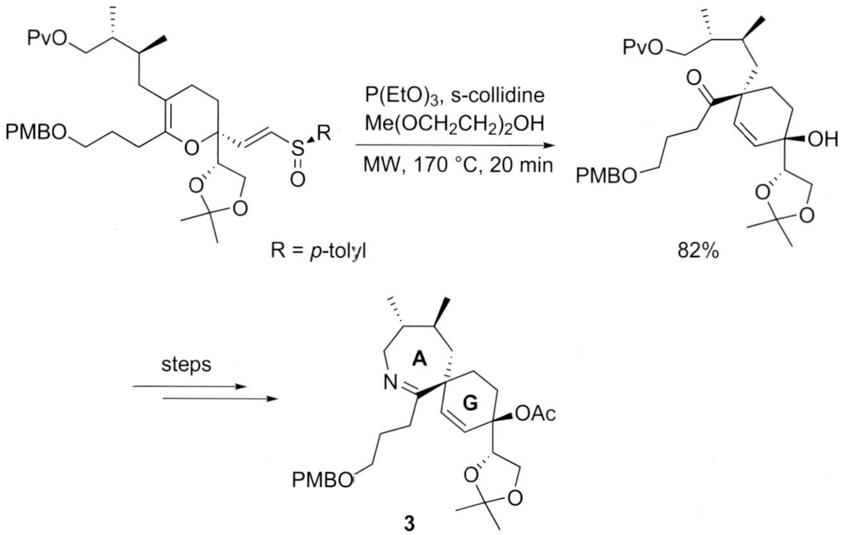

Scheme 6.8 Tandem Claisen–Mislow–Evans rearrangement.

sequence has been disclosed by Taylor and Quesada [22]. Pyrroles have been obtained via an imine condensation/aza-Claisen rearrangement/imine–allene cyclization process [23].

6.1.3
Squaric Acid–Vinylketene Rearrangements

Exploring synthetic routes to analogs of the furaquinocin antibiotics, Trost *et al.* have utilized a microwave-assisted squaric acid–vinylketene rearrangement to synthesize a dimethoxynaphthoquinone, a protected analog of furaquinocin E (Scheme 6.9) [24]. Although successfully applied in closely related series of transformations, use of the conventional rearrangement conditions (toluene, 110 °C) in this case led to incomplete conversion. Thus, the reaction was attempted by microwave heating at 180 °C, which afforded an acceptable yield of 58% of the desired product after oxidation to the naphthoquinone.

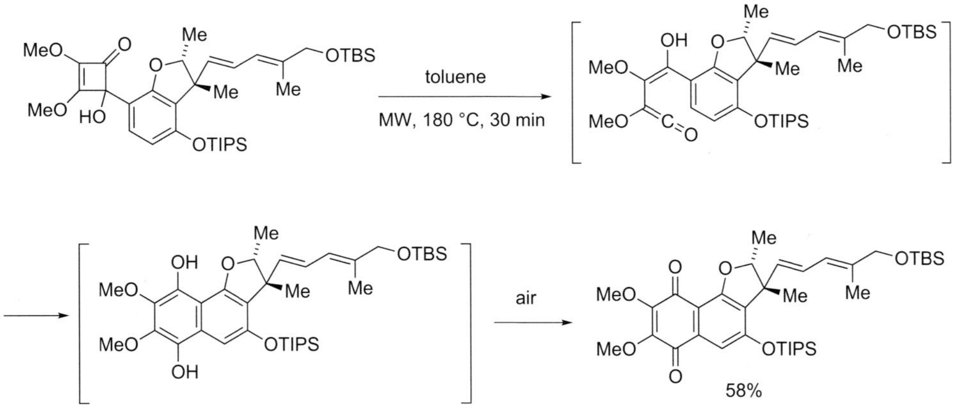

Scheme 6.9 Squaric acid–vinylketene rearrangement.

6.1.4
Vinylcyclobutane–Cyclohexene Rearrangements

A publication by the group of Baran has disclosed the total synthesis of ageliferin, an antiviral agent with interesting molecular architecture [25]. Microwave irradiation of sceptrin, another natural product, for just 1 min at 195 °C in water as solvent under sealed-vessel conditions provided ageliferin in 40% yield, along with 52% of recovered starting material (Scheme 6.10). Remarkably, if the reaction is performed without microwaves at the same temperature, only starting material and decomposition products are observed. Microwave heating of sceptrin in deuteriomethanol for 5 min to 80 °C led exclusively to [D_2]sceptrin in quantitative yield.

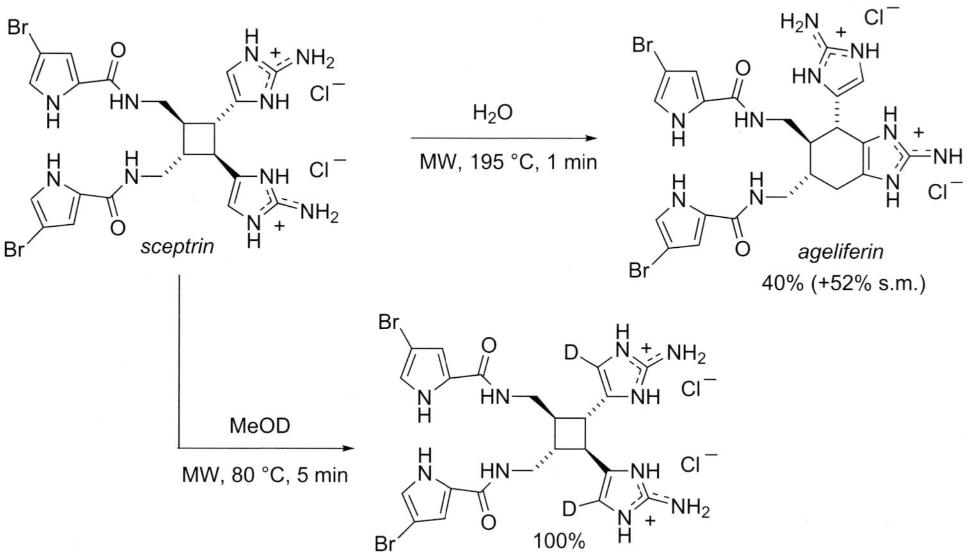

Scheme 6.10 Vinylcyclobutane–cyclohexene rearrangements.

6.1.5
Miscellaneous Rearrangements

The substance 4,12-dibromo[2.2]paracyclophane is the key intermediate *en route* to several functional C_2-symmetric planar-chiral 4,12-disubstituted[2.2]paracyclophanes. Braddock *et al.* have shown that this important intermediate can be obtained by microwave-assisted isomerization of 4,16-dibromo[2.2]paracyclophane, itself readily prepared by bromination of [2.2]paracyclophane (Scheme 6.11) [26]. By performing the isomerization in DMF as a solvent (microwave heating to 180 °C for 6 min), in which the pseudo-*para*-isomer is insoluble at room temperature and the desired pseudo-*ortho*-isomer is soluble, the authors were able to isolate the desired product in 38% yield, with a recovery of 43% of starting material.

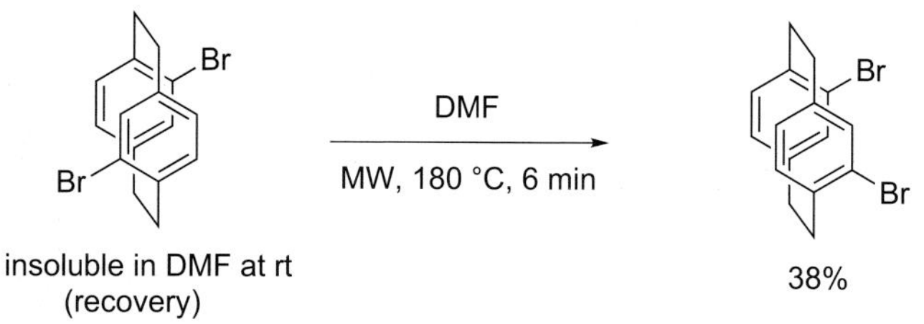

Scheme 6.11 Isomerization of 4,16-dibromo[2.2]paracyclophane.

Vanderwal and his group have discovered a new method for the stereoselective synthesis of *Z*-$\alpha,\beta,\gamma,\delta$-unsaturated amides (Scheme 6.12) [27]. Riecke aldehydes (5-amino-2,4-pentadienals) rearrange thermally to the corresponding $\alpha,\beta,\gamma,\delta$-unsaturated amides via a pericyclic cascade process. Excellent *Z*-selectivity (>20 : 1) is observed in most of the cases. Preliminary experiments indicated that the rearrangement rates could be increased by addition of small amounts of camphorsulfonic acid.

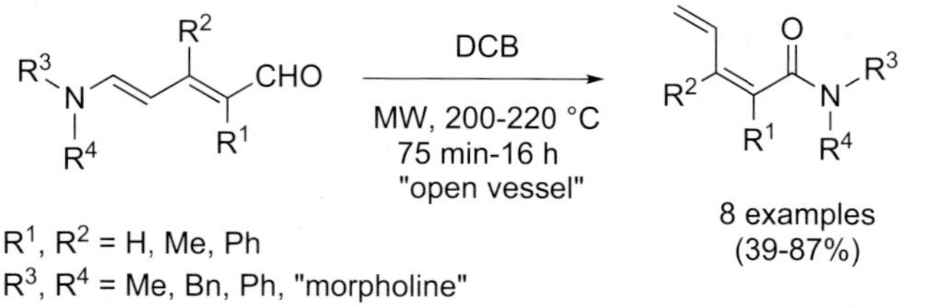

Scheme 6.12 Stereoselective synthesis of *Z*-dienes.

A protocol for the synthesis of substituted pyridines was disclosed by Trost and Gutierrez (Scheme 6.13) [28]. Ru-catalyzed cycloisomerization of primary and secondary propargyl diynols at room temperature afforded unsaturated ketones and aldehydes **4**, respectively. The $\alpha,\beta,\gamma,\delta$-unsaturated aldehydes and ketones **4** are subsequently reacted with hydroxylamine hydrochloride to the corresponding pyridines via 1-azatriene intermediates that undergo an electrocyclization–dehydration sequence. This last step can be either performed in a one-pot format for aryl or unsubstituted aldehydes or ketones (**4**, $R^2 = H$, Ph) applying microwave irradiation or in a two-step protocol for isopropyl-substituted aldehydes and ketones (addition of $NH_2OH{\cdot}HCl$, NaOAc, MeOH, reflux for 1–6 h and then extraction, removal of solvent, and subsequent microwave irradiation of the oxime at 220–230 °C for 40 min–4.25 h).

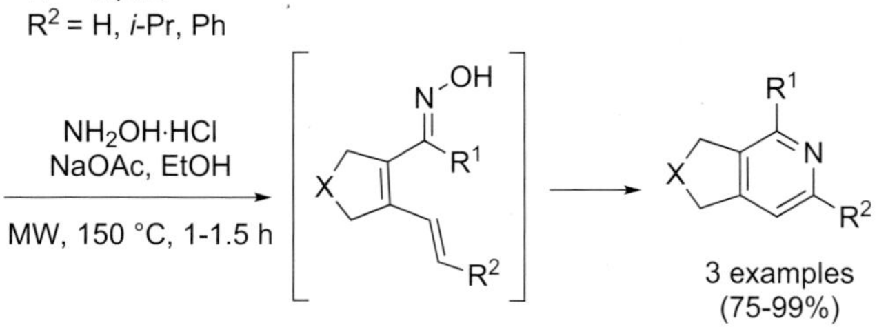

Scheme 6.13 Synthesis of pyridines via a cycloisomerization-6π-cyclization sequence.

The groups of Pardo and Cossy were successful in the stereoselective rearrangement of β-amino alcohols via an aziridinium intermediate under catalytic conditions (Scheme 6.14) [29]. Linear *N,N*-dibenzyl β-amino alcohols **5** and *N*-benzyl prolinol **6** rearranged to β-amino alcohols **7** and 3-hydroxypiperidine **8**, respectively, by treatment with 0.2 equiv of trifluoroacetic anhydride followed by saponification with 0.3 equiv of NaOH in high yields and with excellent *ee* values. Compared to the traditional method where stoichiometric amounts are employed, Et_3N can be omitted from the procedure. See also Ref. [30].

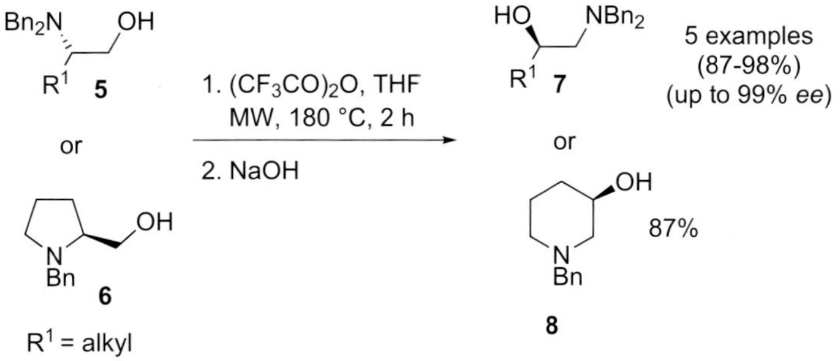

Scheme 6.14 Synthesis of β-amino alcohols.

A reevaluation of the Newman–Kwart rearrangement under microwave conditions was carried out by Moseley *et al.* from AstraZeneca (Scheme 6.15) [31]. Comparisons of results obtained by microwave and thermal heating showed no significant difference that excluded any microwave effects. In addition, a solvent rate effect has been confirmed under both microwave and thermal heating, demonstrating the stabilization of the polar transition-state intermediate **9** by polar solvents like formic acid (78% compared to 23% conversion with NMP after 30 min at 140 °C).

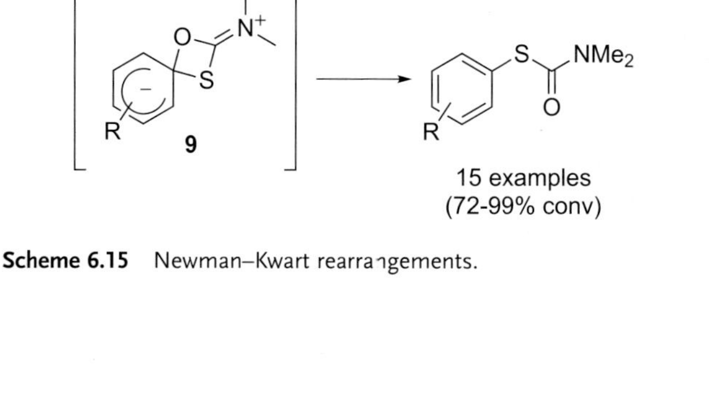

Scheme 6.15 Newman–Kwart rearrangements.

In 2009, Coquerel and coworkers reported on the microwave-assisted Wolff rearrangement of cyclic 2-diazo-1,3-diketones [32]. The developed method tolerates various alcohols, amines, or thiols as nucleophile (NuH) to trap the transient α-oxoketenes after the initial ring contraction to furnish the corresponding α-carbonylated cycloalkanones (Scheme 6.16). Even poorly nucleophilic or sterically hindered alcohols, and secondary amines could be employed efficiently. Moreover, full chemoselectivity was observed when substrates with two nucleophilic nitrogens were used. The reaction was stopped when either 180 °C or 17 bar pressure have been reached (due to liberation of nitrogen) or 3 min of irradiation time have elapsed. In some cases, additional irradiation cycles (3 min each) have been conducted. Since stoichiometric amounts of reagents could be used, no additive was needed and purification was not necessary; thus, the microwave protocol was stated as eco-compatible.

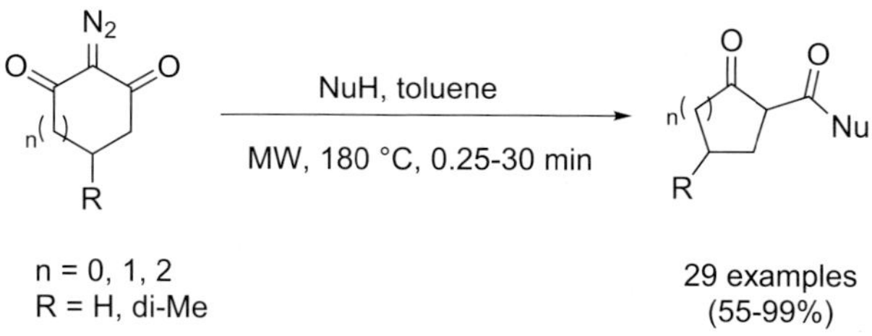

Scheme 6.16 Wolff rearrangement of cyclic 2-azo-1,3-diketones.

Microwave-assisted Wolff rearrangements have also been reported by Sudrik *et al.* [33].

Shen and Dong presented a novel benzofuran synthesis based on the thermal-induced 1,2-silicon-to-oxygen migration of acyl silanes (Brook rearrangement) and subsequent C−H bond insertion [34]. The authors investigated the influence of the solvent on this rearrangement showing that *o*-dichlorobenzene furnished the expected 3-silyloxy-2,3-dihydrobenzofuranes with good stereoselectivity, whereas dimethylsulfoxide (DMSO) led to the corresponding benzofurans by desiloxylation (Scheme 6.17).

The group of Cossy investigated the synthesis of 1,3-enynes from allylic acetates [35]. The authors could demonstrate that under microwave conditions, the 1,3-rearrangement of allylic acetates can be initiated by readily available silica gel instead of rather expensive metal catalysts. The simple protocol involving microwave heating in dichloromethane (DCM) for a couple of minutes proved beneficial for a variety of allylic esters bearing electron-rich substituents (Scheme 6.18a). Substrates with electron-withdrawing groups reacted sluggish and required acidified silica gel to furnish satisfactory results, whereas pyridine derivatives failed at all. Interestingly, even polyfunctional allylic acetates were successfully transformed in good yields by consecutive migration of the acetate group. Finally, the method was applied toward propargylic acetates as well to prepare the corresponding 1,3-enynes (Scheme 6.18b).

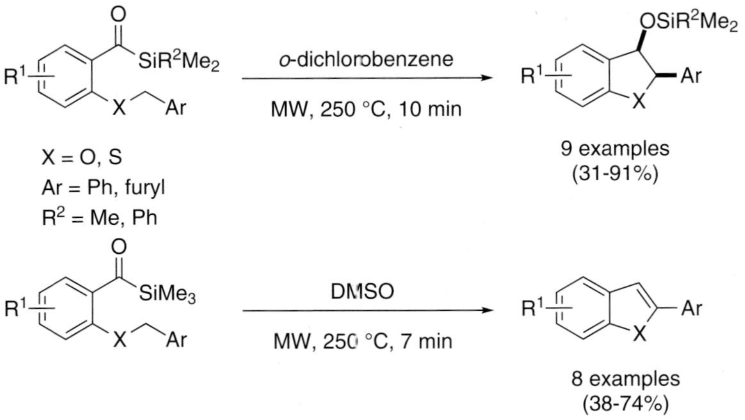

Scheme 6.17 Synthesis of benzofurans from acylsilanes via Brook rearrangement.

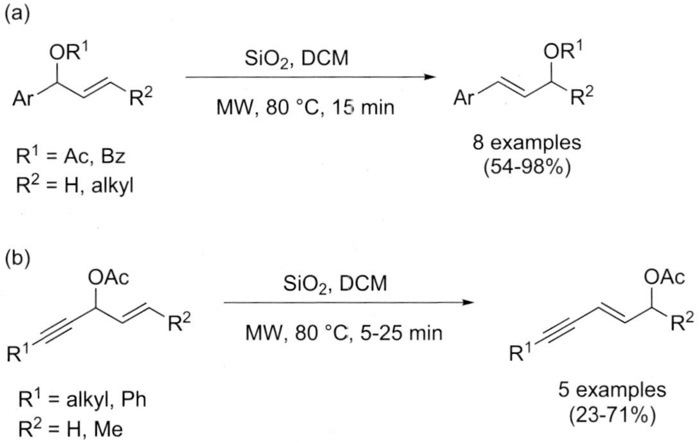

Scheme 6.18 Synthesis of 1,3-enynes by silica gel-mediated rearrangement of propargylic acetates.

If a trimethylsilyl (TMS) substituent is present, somewhat harsher conditions are required: switching to 1,2-dichloroethane and heating at 120 °C for 30 min furnished the desired product in moderate yield.

Microwave irradiation has been used successfully to carry out Cope rearrangements leading to spiroindane frameworks [36], Caroll rearrangements [37], and various 6π-electrocyclizations [38]. Nazarov cyclizations have been carried out in ionic liquids in very short reaction times [39]. For examples of Nazarov cyclizations proceeded by Friedel–Crafts acylations leading to 1-indanones involving arenes and α,β-unsaturated acid chlorides, see Ref. [40]. Alder-ene-type cyclizations of enallenes have also been reported under microwave conditions [41]. Phosphonylated isoindoles have been obtained by rearrangement of o-ethynylbenzyl α-aminophosphonates [42]. Morphine has been rearranged to apomorphine in the presence of $MeSO_3H$ [43].

Microwave-assisted Curtius [44] and Ferrier rearrangements [45] have also been reported.

6.2
Cycloaddition Reactions

6.2.1
Diels–Alder Reactions

Cycloaddition reactions have been performed with great success with the aid of microwave heating, as they require, in many cases, the use of harsh conditions such as high temperatures and long reaction times. Scheme 6.19 shows two examples of Diels–Alder cycloadditions, performed by microwave dielectric heating. In both cases, the diene and dienophile were reacted neat without the addition of solvent. For the transformation in Scheme 6.19a, described by Trost and coworkers, irradiation for 20 min at 165 °C (or for 1 h at 150 °C) gave the cycloadduct in near quantitative yield [46]. In the process reported by the group of de la Hoz (Scheme 6.19b), open-vessel irradiation of 3-styryl chromones with maleimides at 160–200 °C for 30 min furnished the shown tetracyclic adducts along with minor amounts of other diastereoisomers [47].

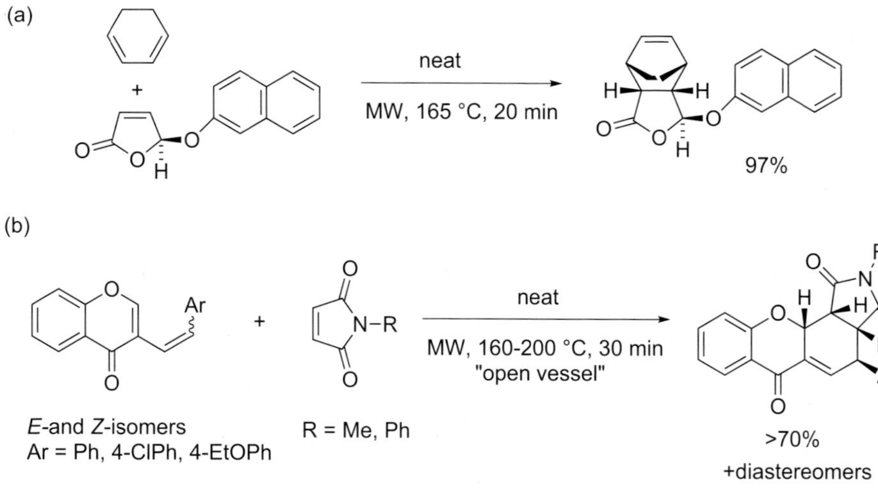

Scheme 6.19 Diels–Alder cycloaddition reactions under solvent-free conditions.

The Diels–Alder cycloaddition of 5-bromo-2-pyrone with the electron-rich *tert*-butyldimethylsilyl (TBS) enol ether of acetaldehyde using superheated DCM as solvent was investigated by Joullié and coworkers (Scheme 6.20) [48]. While the reaction in a sealed tube at 95 °C required 5 days to reach completion, the anticipated

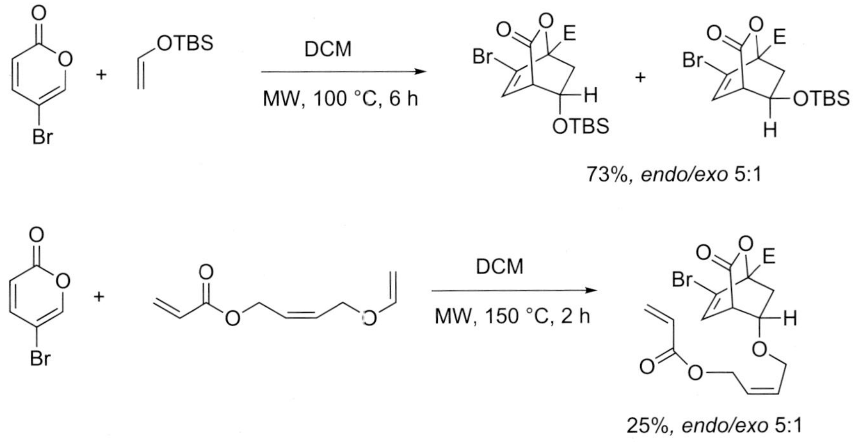

Scheme 6.20 Synthesis of stable oxabicyclo[2.2.2]octenones.

oxabicyclo[2.2.2]octenone core was obtained within 6 h by microwave irradiation at 100 °C. The *endo*-adduct was obtained as the predominant product. Similar results and selectivities were also obtained with a more elaborate bisolefin, albeit providing the desired product in diminished yield.

Pellegrinet and coworkers elaborated a facile Diels–Alder reaction of vinylboronates [49]. Although unsaturated boronates show poor reactivity in Diels–Alder additions, requiring high temperatures and long reaction times, these versatile building blocks were preferred over dialkylboranes because of their stability to air and moisture. Furthermore, convenient microwave heating simplified the desired transformation. Vinylboronate and diene in toluene were heated at 150 °C in three 20 min cycles with an additional 1 equiv of diene each in the repeated cycles. Depending on the substrates, more equiv of the diene, higher temperatures (200 or 220 °C), and more and longer cycles were required to achieve full conversion. In general, all applied vinylboronates furnished the desired bicyclic products in high yields (Scheme 6.21). To demonstrate the synthetic value of the generated bicyclic boronates, the authors subjected these compounds to functionalization of the carbon–boron bond by oxidation to the alcohols with alkaline peroxide. This process could even be performed as a tandem reaction before subjecting to the oxidation step without isolating the cycloadducts. Importantly, *endo* : *exo* stereoselectivity did not change during the oxidation process.

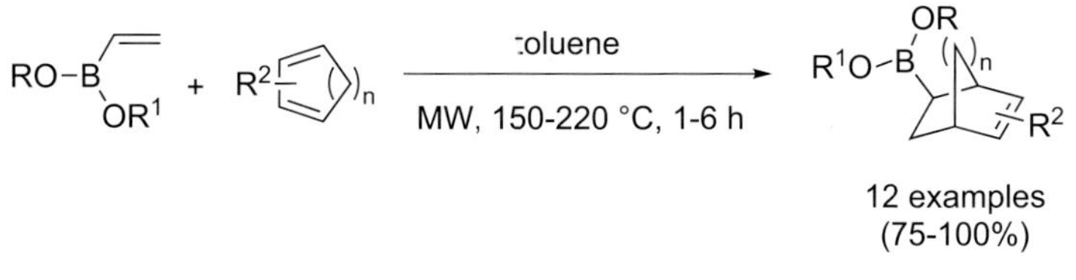

Scheme 6.21 Diels–Alder reaction of vinylboronates.

Maguire and coworkers investigated the reaction conditions for the Diels–Alder cycloaddition of 2-thio-3-chloroacrylamides [50]. Microwave heating proved superior over classical thermal heating since reaction times could be drastically reduced and yields significantly improved. Initially, the reaction behavior with highly reactive cyclopentadiene has been explored. Although classical heating turned out to be time consuming (16 h reaction time) even with the activated, electron-withdrawing sulfoxide species, a quite convenient microwave protocol could be established. Irradiating a solution of the sulfoxide substrate in DCM together with a 10-fold excess of cyclopentadiene at 100 °C for 5 min afforded the corresponding cycloadducts in high yields and with similar *endo* : *exo* selectivity as the thermal attempt (Scheme 6.22a). With aid of microwave irradiation, the far less reactive sulfides could also be reacted successfully. Although thermal heating at reflux required 10 days of reaction time with continuous adding of fresh cyclopentadiene, the microwave protocol furnished the desired product in excellent yield only after 10 min heating at 150 °C. With these results in hand, the authors turned their attention toward Diels–Alder reactions involving 2,3-dimethyl-1,3-butadiene. Slight modifications of the microwave protocol leading to somewhat longer reaction times and higher temperatures resulted in satisfactory conversions of the chloroacrylamides. Heating neat sulfoxide substrates with 10 equiv of the butadiene for 30 min unexpectedly afforded the corresponding benzamides in moderate to high yields (Scheme 6.22b). Aromatization occurs, since the cycloaddition proceeds via the cyclohexene adduct that immediately eliminates both the sulfoxide and the chloride group. When employing the sulfide derivatives, microwave heating for 2 h at 180 °C generates the corresponding cyclohexenecarboxamides as the cycloadduct in almost quantitative yield.

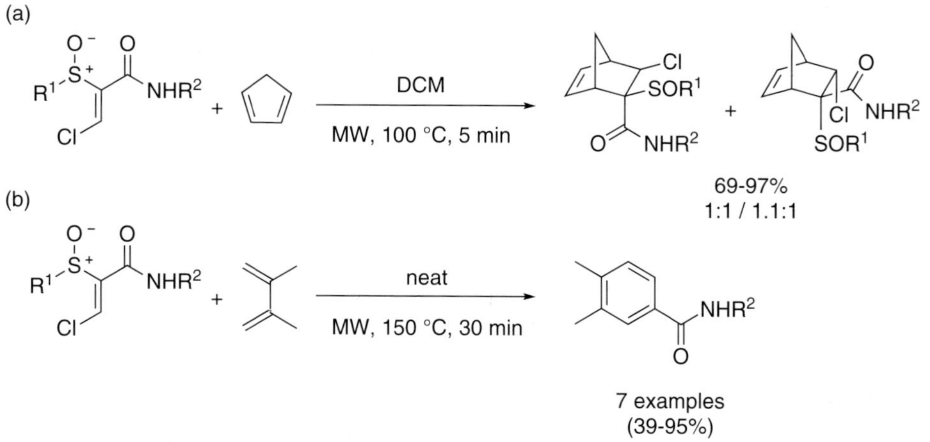

Scheme 6.22 Diels–Alder reaction of 2-thio-chloroacrylamides.

Applying the concept of using solvents doped with ionic liquids in order to allow microwave heating to high temperatures (see Section 4.5.2), Leadbeater and Torenius have studied the Diels–Alder reaction between 2,3-dimethylbutadiene and methyl acrylate (Scheme 6.23) [51]. This reaction is traditionally performed in toluene or xylene and takes 18-24 h to reach completion, giving yields of cycloadduct varying from 9% to 90% depending on the solvent used and the temperature. Using a mixture of toluene and the ionic liquid 1-(2-propyl)-3-methylimidazolium hexafluorophosphate (pmimPF_6), the authors were able to perform the cycloaddition within 5 min to obtain the product in 80% yield, thus offering a significant rate enhancement over the conventional methods.

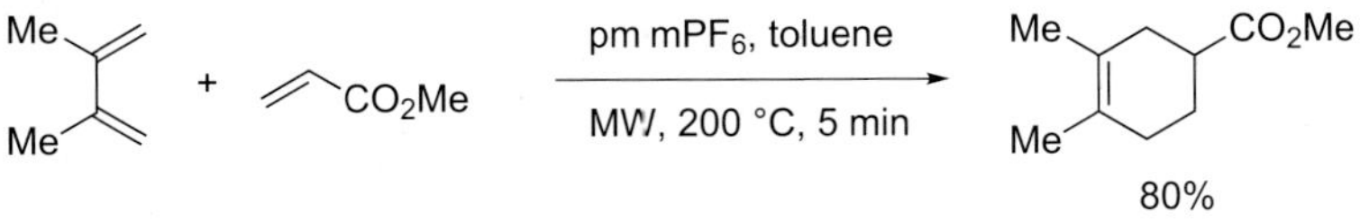

Scheme 6.23 Diels–Alder cycloaddition reactions in ionic liquid-doped solvents.

Hong *et al.* have investigated the cycloaddition chemistry of fulvenes with a large variety of alkenes and alkynes in great detail [52]. As one example, the reaction of 6,6-dimethylfulvene with benzoquinone is shown in Scheme 6.24. Under microwave conditions in DMSO at 120 °C, an unusual hetero-[2 + 3]adduct is formed in 60% yield, the structure of which was determined by X-ray crystallography. The adduct is a structural analog of the natural products aplysin and pannellin and differs completely from the reported thermal (benzene, 80 °C) Diels–Alder cycloaddition product of the fulvene and benzoquinone (Scheme 6.24) [52].

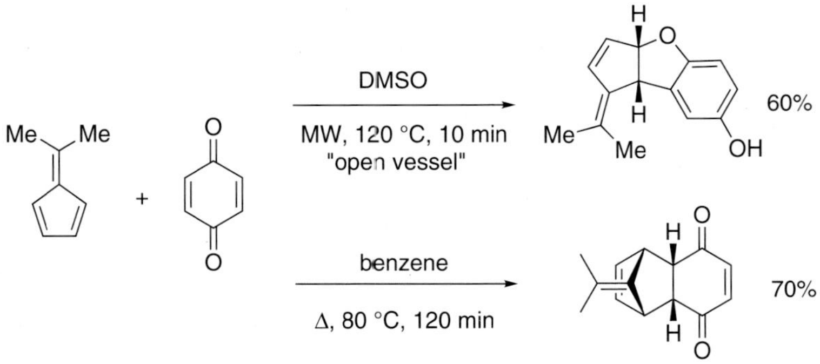

Scheme 6.24 Thermal versus microwave-assisted Diels–Alder cycloaddition reactions of fulvenes with benzoquinones.

The microwave-assisted sidewall functionalization of single-wall carbon nanotubes (SWNTs) via Diels–Alder cycloaddition with *o*-quinodimethane has been demonstrated by Langa and coworkers (Scheme 6.25) [53]. The required *o*-quinodimethane was generated *in situ* from 4,5-benzo-1,2-oxathiin-2-oxide (sultin) by refluxing in 1,2-dichlorobenzene under open-vessel microwave irradiation conditions. In the presence of ester-functionalized SWNTs, cycloadditions take place

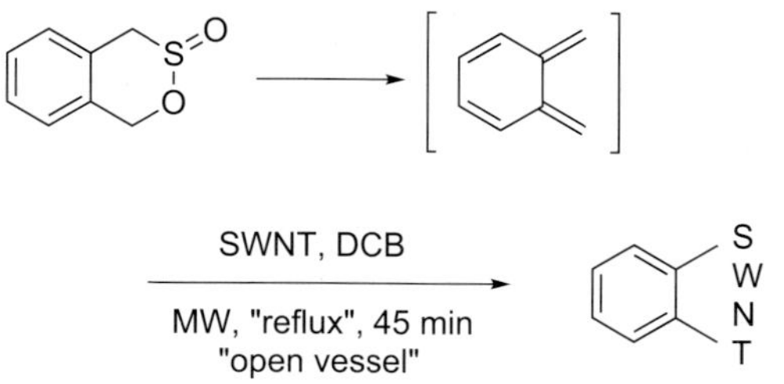

Scheme 6.25 Diels–Alder cycloaddition to single-wall carbon nanotubes.

within 45 min. Conventional refluxing in 1,2-dichlorobenzene for 3 days leads only to a low degree of conversion.

Multiwalled carbon nanotubes (MWNT) have been functionalized using 1,3-dipolar cycloaddition chemistry [54, 55].

Yu and coworkers have reported a recyclable organotungsten Lewis acid as a catalyst for Diels–Alder cycloaddition reactions performed in water or ionic liquids (Scheme 6.26) [56]. Using either of the two solvent systems, all studied cycloaddition reactions were completed in less than 1 min by microwave irradiation at 50 °C employing 3 mol% of the catalyst. An additional advantage of using the ionic liquid 1-butyl-3-methylimidazolium hexafluorophosphate ($bmimPF_6$) as solvent is the facilitated catalyst recycling.

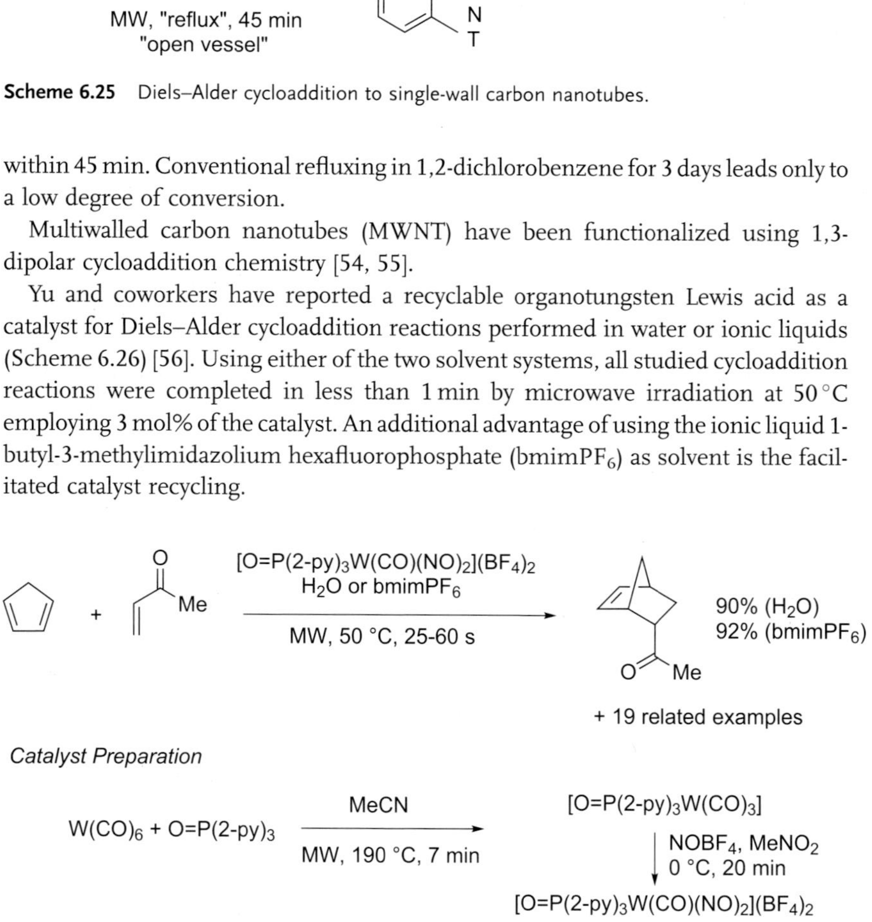

Scheme 6.26 Organotungsten Lewis acid-catalyzed Diels–Alder cycloaddition reactions.

Microwave heating has also been employed for performing retro Diels–Alder cycloaddition reactions, as exemplified in Scheme 6.27. In the context of preparing optically pure cross-conjugated cyclopentadienones as precursors to arachidonic acid derivatives, Evans and Eddolls and their coworkers have performed microwave-mediated Lewis acid-catalyzed retro-Diels–Alder reactions of suitable *exo*-cyclic enone building blocks [57, 58]. The microwave-mediated transformations were performed in DCM at 60–100 °C with 0.5 equiv of methyl aluminum dichloride as catalyst and 5 equiv of maleic anhydride as cyclopentadiene trap. In most cases, the

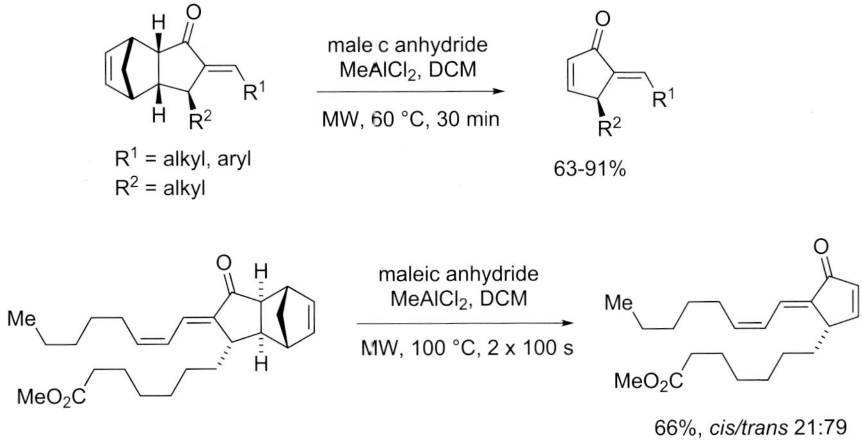

Scheme 6.27 Retro Diels–Alder reactions.

reaction was stopped after 30 min since continued irradiation eroded the yields of products. The use of short bursts of microwave irradiation minimized double-bond isomerization.

Wipf and coworkers presented an innovative route to 4-substituted indoles by intramolecular Diels–Alder reaction of furan derivatives under microwave conditions [59]. For the indole preparation, Boc-protected furanyl aminoenols, initially prepared from α-lithiated 2-alkylaminofuran and α,β-unsaturated carbonyls, were utilized. These compounds undergo *in situ* cycloaddition and subsequent dehydrative aromatization with thermal Boc-deprotection to furnish the pharmaceutically attractive 4-substituted indole moieties (Scheme 6.28). To accomplish the reaction, the substrate in *o*-dichlorobenzene is heated under microwave irradiation at 180 °C for 20 min. This cascade method also gives access to the introduction of alkyl and alkenyl groups in 4-position by slightly increasing the heating time to 25 or 30 min,

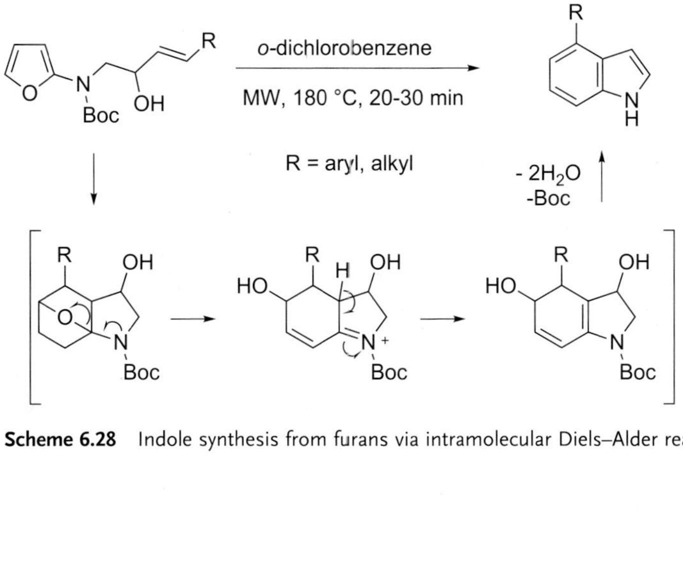

Scheme 6.28 Indole synthesis from furans via intramolecular Diels–Alder reaction.

respectively. When 2-cyclohexen-1-one was employed to prepare the substrate, the resulting tertiary alcohol led to cyclohexanone-annulated indole, which represents the core tricycle of Ergot alkaloids. Therefore, this novel strategy proves a convenient alternative to metal-catalyzed coupling processes in heterocycle synthesis.

Inter- and intramolecular hetero-Diels–Alder cycloaddition reactions in a series of functionalized 2(1*H*)-pyrazinones have been studied in detail by the groups of Van der Eycken and Kappe (Scheme 6.29) [60, 61]. In the intramolecular series, cycloaddition of alkenyl-tethered 2(1*H*)-pyrazinones requires 1–2 days under conventional thermal conditions involving chlorobenzene as solvent under reflux conditions (132 °C). Switching to 1,2-dichloroethane doped with the ionic liquid 1-butyl-3-methylimidazolium hexafluorophosphate (bmimPF_6) and sealed-vessel microwave technology, the same transformations were completed within 8–18 min at 190 °C reaction temperature (Scheme 6.29a) [60]. Without isolating the primary imidoyl chloride cycloadducts, rapid hydrolysis was achieved by addition of small amounts of water and resubjection of the reaction mixture to microwave irradiation (130 °C, 5 min). The isolated overall yields of the desired cycloadducts were in the same range as reported for the conventional thermal protocols.

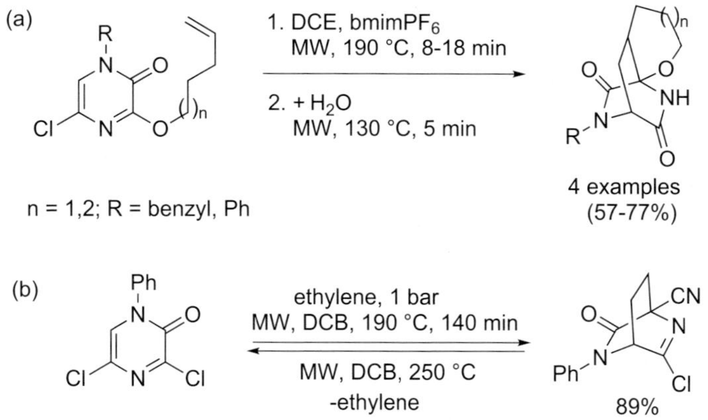

Scheme 6.29 Hetero-Diels–Alder cycloaddition reactions of 2(1*H*)pyrazinones.

In the intermolecular series, Diels–Alder cycloaddition reaction of the pyrazinone heterodiene with ethylene led to the expected bicyclic cycloadduct (Scheme 6.29b) [60]. The details of this transformation utilizing pre-pressurized reaction vessels are given in Section 4.4 [61].

Moody and coworkers have employed a "biomimetic" hetero-Diels–Alder aromatization sequence for the construction of the pyridine ring in amythiamicin D (Scheme 6.30) [62]. The key cycloaddition reaction between the azadiene and enamine component was carried out by microwave irradiation at 120 °C for 12 h and gave the required 2,3,6-tris(thiazolyl)pyridine intermediate in moderate yield. Coupling of the remaining building blocks then completed the first total synthesis of the thiopeptide antibiotic amythiamicin D.

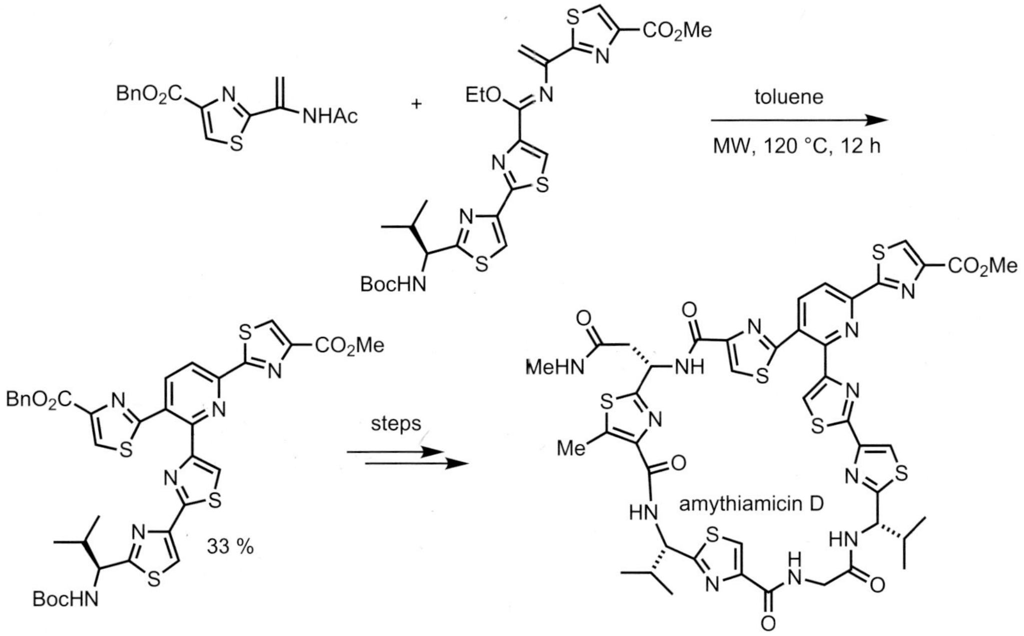

Scheme 6.30 Diels–Alder cycloaddition chemistry in natural product synthesis.

The group of Osborn investigated the rapid synthesis of carbohydrate derivatives via hetero-Diels–Alder reactions [63]. It was demonstrated that various aldehydes and imines could be successfully transformed with *trans*-1-methoxy-3-trimethylsilyloxy-1,3-butadiene (Danishefsky's diene) under mild microwave conditions (Scheme 6.31). The applied imines were initially derived from corresponding aldehydes by treatment with benzylamine under anhydrous conditions. Gratifyingly,

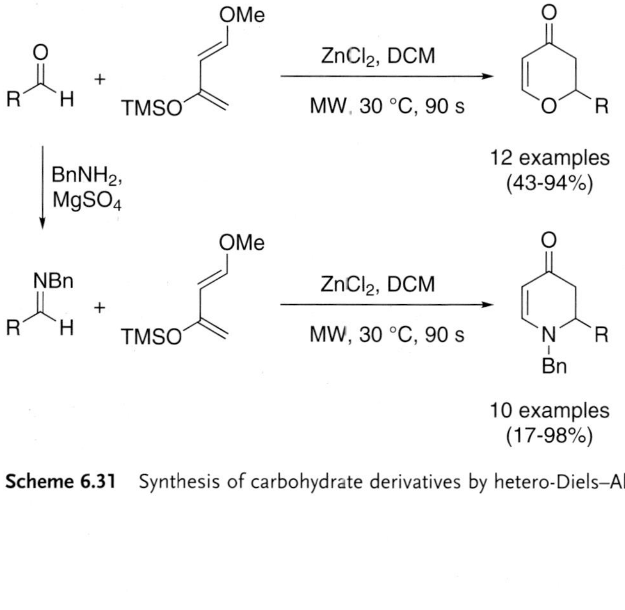

Scheme 6.31 Synthesis of carbohydrate derivatives by hetero-Diels–Alder reaction.

the generated imines could be used in the microwave protocol without prior isolation, thus further simplifying the overall protocol, although with isolated imines the observed yields were slightly higher. The aldehyde or imine, respectively, was admixed with 2 equiv of Danishefsky's diene and 10 mol% of zinc(II) chloride and irradiated at 30 °C for 90 s. Subsequent quenching with 1 equiv of trifluoroacetic acid furnished the corresponding dihydropyranones or tetrahydropyridinones in generally good to high yields. The simple protocol tolerates numerous substrates, however, the authors focused specially on the transformation of complex carbohydrate derivatives to synthesize *C*-linked disaccharide mimetics. Both aldehyde and imine substrates reacted satisfactory furnishing the desired disaccharides as single diastereomers on C-6 position.

A one-pot Diels–Alder/acylation (or vice versa) sequence for the synthesis of isoquinol-1-one-8-carboxylic acids **10** has been developed by the group of Aubé (Scheme 6.32) [64]. A set of six dienes containing an attached secondary amine functionality were reacted with maleic anhydride ($R^2 = H$) or citraconic anhydride ($R^2 = Me$) at 165 °C for 1.5 h to afford the isoquinolone products **10** in one step. In addition, the synthesis of the aminodiene precursors was also performed under microwave conditions by displacement of a mesylate with the corresponding primary amine at 130 °C within 1 h in MeCN. The six carboxylic acid scaffolds obtained by the reaction with maleic anhydride were further reacted with 12 diverse amines to generate a 72-member isoquinolone–amide library, however, this step was performed at room temperature.

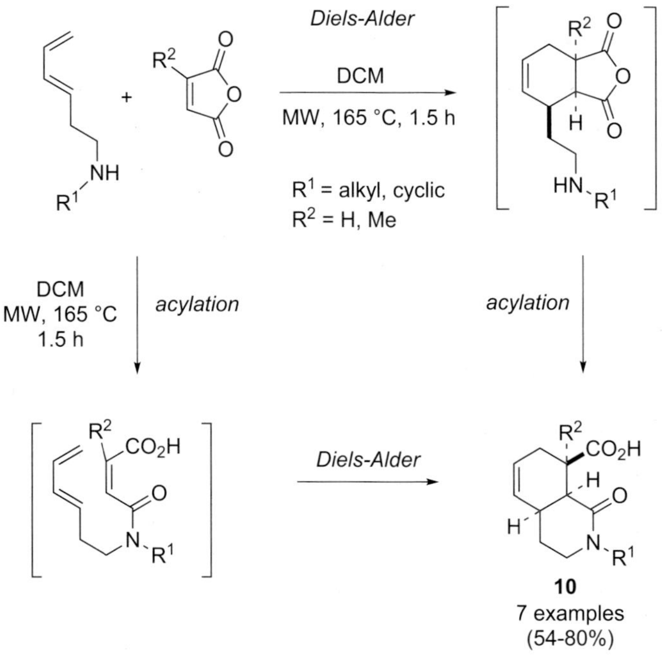

Scheme 6.32 Tandem Diels–Alder/acylation sequence.

The group of Danishefsky has reported on an oxidative dearomatization/transannular Diels–Alder sequence as key step in the total synthesis of (±)-11-*O*-debenzoyltashironin (**14**) (Scheme 6.33) [65]. Upon exposure of **11** to a hypervalent iodine species such as PIDA, a mixture of oxidized intermediate **12** and Diels–Alder adduct **13** was obtained; however, subsequent microwave irradiation of this mixture leads exclusively to the cycloadduct **13** in 65% yield. A following series of deprotections and diastereoselective oxidation steps furnished 11-*O*-debenzoyltashironin (**14**).

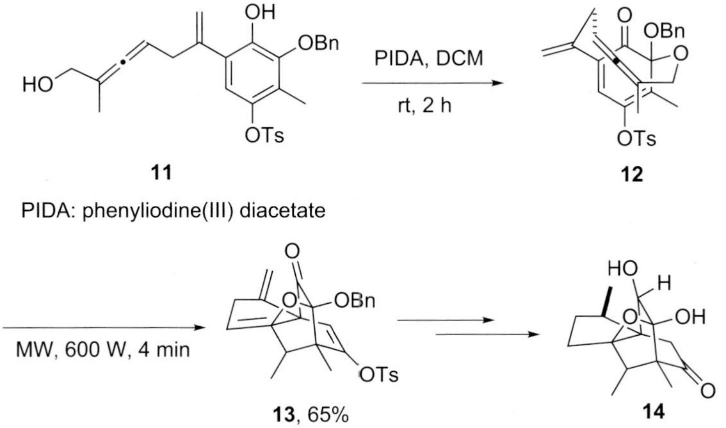

Scheme 6.33 Oxidative dearomatization/transannular Diels–Alder cascade.

Microwave-assisted Diels–Alder cycloaddition reactions using water-soluble aquocobaloxime complexes have been reported by Welker and coworkers [66].

Diels–Alder reactions involving 2-oxazolidinone dienes have been reported by Tamariz and coworkers [67]. Xanthones have been prepared by microwave-assisted Diels–Alder reaction of 3-styrylchromones with suitable dienophiles [68]. Polycyclic oxygen heterocycles have been obtained by intramolecular Diels–Alder cycloaddition of *N*-propargyl-*N*-2-furfurylamines [69] and from 1,3,8-nonatrienes [70]. For examples of aza-Diels–Alder reactions leading to 1,3-oxazines [71] or polyheterocycles, see Ref. [72]. Tandem Wittig intramolecular Diels–Alder cycloadditions have been used by Dai and coworkers for the construction of bicyclic lactones [73]. Tandem domino Knoevenagel Diels–Alder strategies have been used for the preparation of bis-pyrano-1,4-benzoquinones [74]. Anilines have been prepared by Diels–Alder addition of acetylenes to 3-amino-substituted 2*H*-pyran-2-ones [75]. Regioselective cycloaddition reactions between 9-substituted anthracenes and levoglucosenone have been studied in Ref. [76]. Acetylketene can be generated by microwave-assisted retro-Diels–Alder reaction from 2,2,4-trimethyl-4*H*-1,3-dioxin-4-one precursor, and has been trapped *in situ* by a range of alcohols, aldehydes, and ketones [77]. Many more examples of microwave-assisted cycloaddition processes leading to heterocycles are described Chapter 7.

6.2.2
Miscellaneous Cycloadditions

Ley and his group have shown that a variety of oligomeric alkynes cyclotrimerized in a [2 + 2 + 2] manner to the corresponding arenes without the need of any metal catalyst if microwave heating is used (Scheme 6.34) [78]. Furthermore, a flow process was applied for one example, pumping the sample through a glass coil in the microwave cavity.

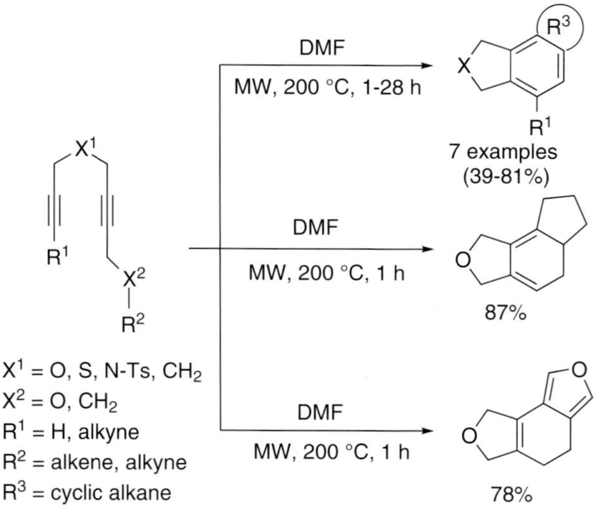

Scheme 6.34 Metal-free intramolecular alkyne trimerizations.

Co, Ni, and Ru-catalyzed [2 + 2 + 2] cyclotrimerization reactions on both solid phase [79, 80] and solution phase [81–83] have been described by Deiters and coworkers. Martinez and coworkers have disclosed similar transformations mediated by an Ir catalyst [84], whereas the Kotora group has applied a Rh catalyst in cyclotrimerizations involving 1-alkynyl-2-deoxyribofuranose [85].

For $Co(CO)_8$-mediated intramolecular cycloadditions of dienynes and related precursors, see Refs [86–88].

Cossio and coworkers have made investigations toward the stereochemical outcome for the [3 + 2] cycloaddition of stabilized azomethine ylides and nitrostyrenes (Scheme 6.35) [89]. By performing the reactions without any solvent, three stereoisomers (*endo*, *exo*, and *endo'*) of the pyrrolidine products are obtained, regardless of whether the cycloaddition is conducted under microwave or conventional heating. Further aromatization of the corresponding pyrrolidines with DDQ in refluxing xylene, again under microwave irradiation, gives pyrrole-2-carboxylate **15** as major and **16** as minor product.

In a related cycloaddition approach, Bergner and Opatz have synthesized 2,3,4,5-tetrasubstituted pyrroles by the reaction of α-(alkylideneamino)nitriles **17** and

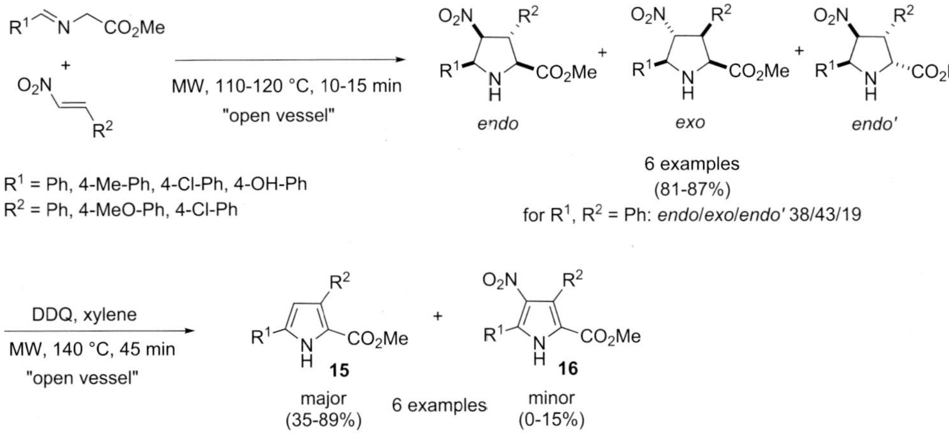

Scheme 6.35 Synthesis of pyrroles via [3 + 2] cycloaddition.

α,β-unsaturated nitro compounds **18** under basic conditions (Scheme 6.36) [90]. The reaction proceeds via cycloaddition of **17** and **18** with subsequent elimination of HCN and HNO_2. Complete regioselectivity was achieved when Cs_2CO_3 was employed as base. The only exception being 4-cyanophenyl-substituted **18** ($R^3 =$ 4-CN-Ph), here, isomeric mixtures of pyrroles were obtained. When microwave heating is applied, the reaction time could be reduced from several hours in refluxing THF to only 2 min at 100 °C.

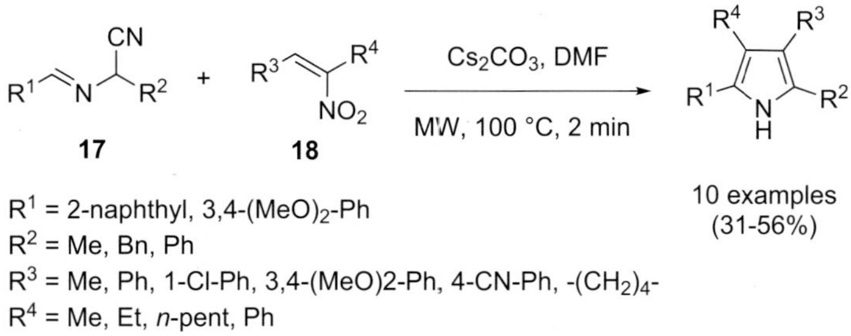

Scheme 6.36 Tetrasubstituted pyrrole synthesis via [3 + 2] cycloaddition.

Another synthesis of tetrasubstituted pyrroles via [3 + 2] cycloaddition under microwave irradiation was reported by the group of Park (Scheme 6.37) [91]. The pyrrole moiety is formed from α,β-unsaturated benzofuran-3(2*H*)-ones and azalactones. The initial cycloadduct undergoes spontaneous decarboxylation to furnish the desired pyrroles. Preliminary investigations showed that silver(I) acetate was the best suited Lewis acid to catalyze the cycloaddition. The substitution pattern on the benzofuran scaffold turned out to have an influence on the reaction time and the regioselectivity. The electron-withdrawing groups on the aromatic ring, preferably on 5-position, led to excellent regioselectivity and yields. However, 2-methoxy benzofurans

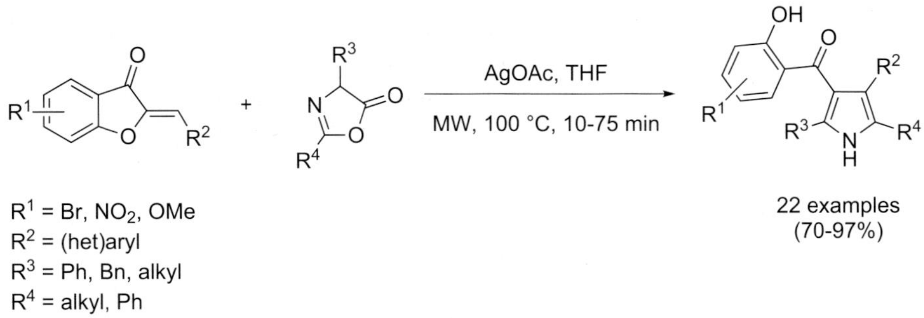

Scheme 6.37 Synthesis of tetrasubstituted pyrroles by cycloaddition and spontaneous decarboxylation.

required longer reaction times, which resulted in only moderate regioselectivity. The microwave procedure proved successful to increase molecular diversity on biologically useful pyrroles by tolerating a broad variety of substrates.

In the course of the preparation of a polycyclic pyrrolidine library, Kurth and coworkers have disclosed an intramolecular azomethine ylide 1,3-dipolar cycloaddition as key transformation (Scheme 6.38) [92]. Fused pyrrolidine scaffolds **21** were obtained with high regio- and stereoselectivity via condensation of benzimidazole-carbaldehydes **18** with a secondary amino ester **19** to form the S-shaped ylide **20** and the subsequent intramolecular 1,3-dipolar cycloaddition of **20**. Importantly, the nature of the secondary amine plays a crucial role on the outcome of the reaction ($R^1 = H$, $R^2 = Ph$: 10%; $R^1 = H$, $R^2 = Me$: 59%).

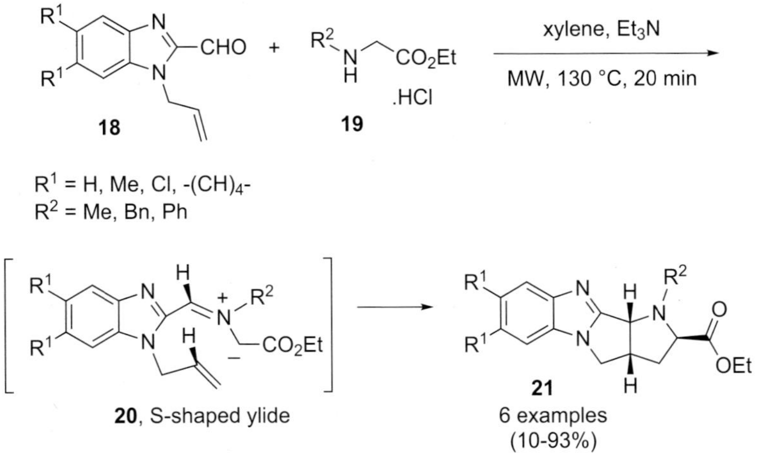

Scheme 6.38 Intramolecular cycloaddition of azomethine ylides.

1,3-Dipolar cycloaddition reactions of cinnamaldehyde with azomethine ylides generated through decarboxylation of iminium or oxazolidinone intermediates formed by the reaction of L-proline with cinnamaldehydes have been reported.

The stereoselective reaction provides an efficient access to the highly functionalized hexahydro-1*H*-pyrrolizines [93]. An intramolecular 1,3-dipolar cycloaddition of an azide and an acetylene functionality was used by Van der Eycken and coworkers as a key step in the synthesis of steganone aza analogs [94]. Bimolecular azide–acetylene cycloadditions have been described by Katritzky *et al.* [95] and Santagada and coworkers [96]. Generation of nitrile sulfide by microwave-assisted decomposition of a suitable precursor followed by trapping of the dipole with a selection of dipolarophiles allows the preparation of five-membered *S*-heterocycles [97, 98]. Intramolecular [2 + 2] allenic cycloaddition reactions carried out in toluene doped with ionic liquids have been published by Brummond and Chen [99].

6.3 Oxidations

The osmium-catalyzed dihydroxylation reaction, that is, the addition of osmium tetroxide to olefins producing a vicinal diol, is one of the most selective and reliable organic transformation. Work by Sharpless and coworkers has uncovered that electron-deficient olefins can be converted to the corresponding diols much more efficiently when the pH of the reaction medium is maintained on the acidic side [100]. One of the most useful additives in this context has been proven to be citric acid (2 equiv), which, in combination with 4-methylmorpholine *N*-oxide (NMO) as reoxidant for osmium(VI) and potassium osmate [$K_2OsO_2(OH)_4$] (0.2 mol%) as a stable, nonvolatile substitute for osmium tetroxide, allows the conversion of many olefinic substrates to their corresponding diols at ambient temperature. In specific cases, such as for extremely electron-deficient olefins (Scheme 6.39), the reaction had to be carried out under microwave irradiation at 120 °C to produce an 81% isolated yield of the pure diol.

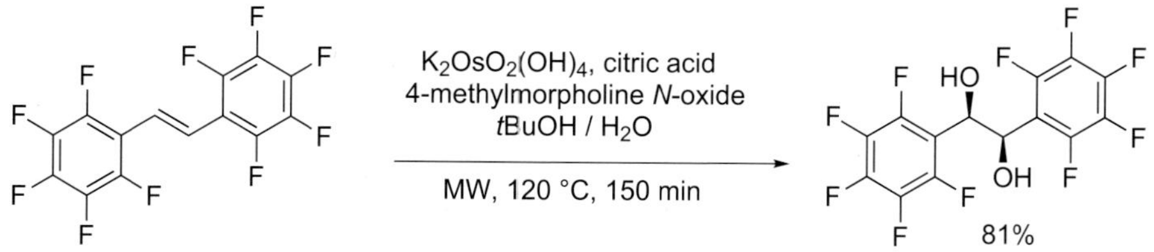

Scheme 6.39 Osmium-catalyzed dihydroxylation of electron-deficient olefins.

The asymmetric allylic oxidation of bridged bicyclic alkenes using a copper-catalyzed symmetrizing–desymmetrizing Kharasch–Sosnovsky reaction was disclosed by Clark and coworkers (Scheme 6.40) [101]. Here, a racemic bridged bicyclic alkene was used as a starting material in the presence of 5 mol% of a copper catalyst and 6 mol% of a chiral pyridyl-bisoxazoline ligand. Microwave heating in MeCN with a suitable carbonate substrate provided the allylic oxidation product in 84% yield and

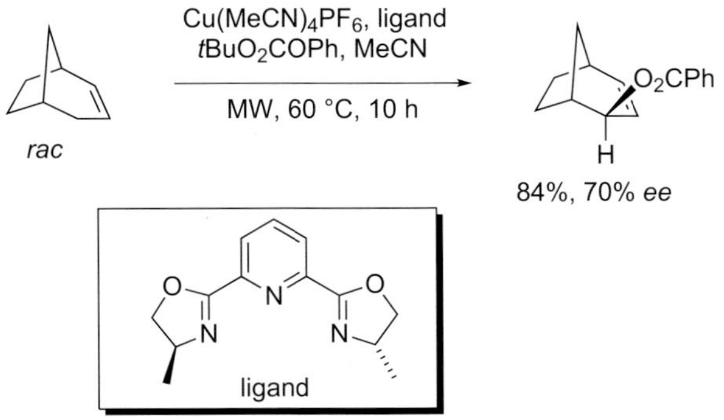

Scheme 6.40 Copper-catalyzed asymmetric allylic oxidation of bridged bicyclic alkenes.

with 70% *ee*. A control experiment using conventional heating at the same temperature required 2 days to provide the same product in 80% yield and with 66% *ee*.

Another industrially important oxidation reaction is the conversion of cyclohexene to adipic acid. The well-known Noyori method uses hydrogen peroxide, catalytic tungstate, and a phase-transfer catalyst to afford the clean oxidation of cyclohexene to adipic acid. Ondruschka and coworkers have demonstrated that a modified protocol employing microwave heating without solvent gave comparable yields of the desired product but at a much faster rate (Scheme 6.41) [102]. Optimum results were obtained using excess hydrogen peroxide as oxidant, 1 mol% of methyltrioctylammonium hydrogen sulfate as phase-transfer catalyst, and 1 mol% of sodium tungstate as a catalyst for 90 min under microwave reflux conditions (about 100 °C).

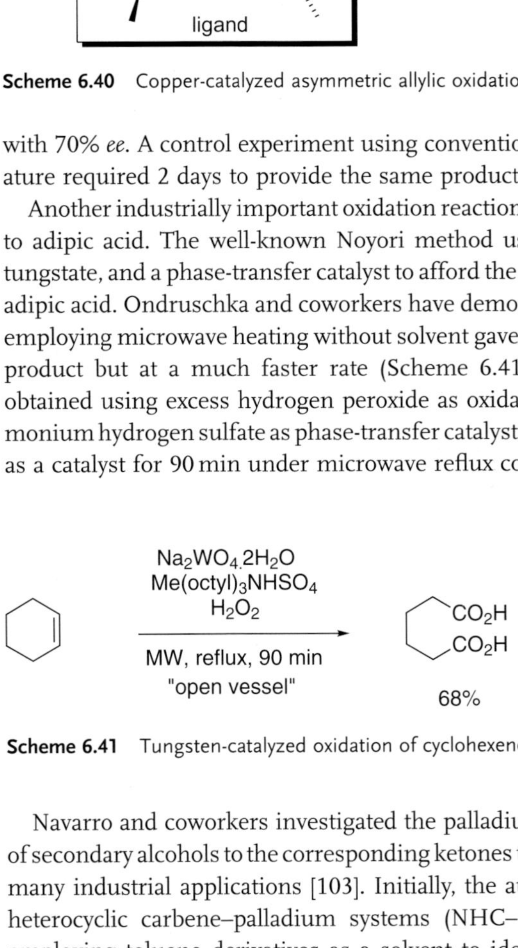

Scheme 6.41 Tungsten-catalyzed oxidation of cyclohexene.

Navarro and coworkers investigated the palladium-catalyzed anaerobic oxidation of secondary alcohols to the corresponding ketones that are of considerable interest in many industrial applications [103]. Initially, the authors screened a number of N-heterocyclic carbene–palladium systems (NHC–Pd) with alkaline *tert*-butoxides employing toluene derivatives as a solvent to identify the most suitable catalyst/oxidant combination. Under microwave conditions, commercially available (IMes)Pd(allyl)Cl together with sodium *tert*-butoxide in 2,4-dichlorotoluene as solvent/oxidant proved to be the most effective system. Gratifyingly, the complex loading could be reduced by one order of magnitude compared to thermal heating at milder conditions. The optimized protocol was applied toward a variety of secondary alcohols to validate the efficiency of the developed method (Scheme 6.42). Even

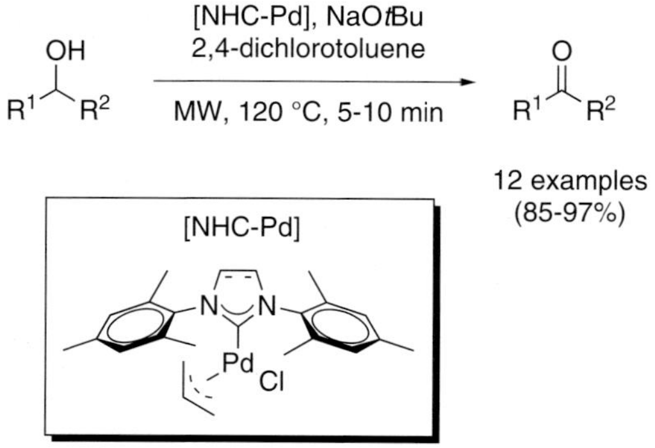

Scheme 6.42 Anaerobic oxidation of secondary alcohols.

substrates with rather challenging functionalities, such as thioethers or alkenes, were smoothly converted.

Rhodium and ruthenium-catalyzed hydrogen transfer-type oxidations of primary and secondary alcohols have been reported by Matsubara and coworkers (Scheme 6.43) [104]. Thus, secondary alcohols were converted into the corresponding ketones using 2 equiv of methyl acrylate as hydrogen acceptor and 5 mol% of bis (triphenylphosphine)rhodium(I) carbonyl chloride [$RhCl(CO)(PPh_3)_2$] as transition metal catalyst in a water/DMF solvent mixture. Microwave irradiation for 15 min to 140 °C provided the desired ketones in moderate to excellent yields, while without microwave irradiation only starting materials were recovered. Primary alcohols were not oxidized under these conditions, but required a ruthenium catalyst. Optimum conditions for the oxidation of primary alcohols utilized 2.5 mol% of tris(triphenylphosphine)ruthenium(II) dichloride [$RuCl_2(PPh_3)_3$] and 2 equiv of methyl vinyl ketone under solvent-free conditions (120 °C, 15 min).

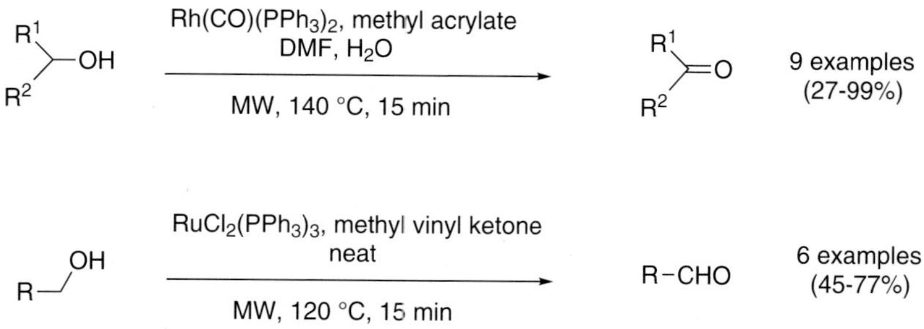

Scheme 6.43 Hydrogen transfer-type oxidations.

The oxidation of a thiazolidine derivative to the corresponding thiazole using activated manganese dioxide in dichloromethane at 100 °C is shown in Scheme 6.44. Further manipulation of this molecule led to dimethyl sulfomycinamate, a methanolysis product of the thiopeptide antibiotic sulfomycin I [105].

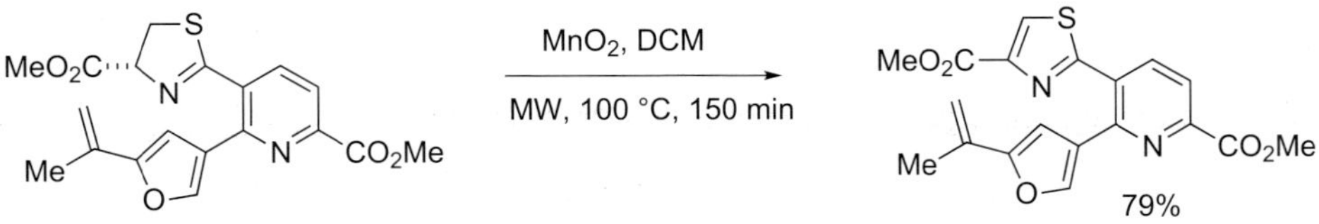

Scheme 6.44 Dehydrogenation of thiazolidines.

6.4 Reductions and Hydrogenations

A popular reduction process carried out under microwave conditions is catalytic transfer hydrogenation using hydrogen donors such as ammonium formate and a palladium catalyst like palladium-on-charcoal (Pd/C). Two transformations, involving the reduction of aromatic nitro groups are shown in Scheme 6.45. The reaction can be carried out either under sealed-vessel conditions (Scheme 6.45a) [106] or at atmospheric pressure under reflux (Scheme 6.45b) [107]. In the latter case, the reaction takes somewhat longer as the reaction temperature is limited by the boiling point of the solvent. Under sealed-vessel conditions, reactions are generally faster, but care must be taken not to generate too much pressure in the microwave reaction vessel due to the formation of ammonia, carbon dioxide, and hydrogen.

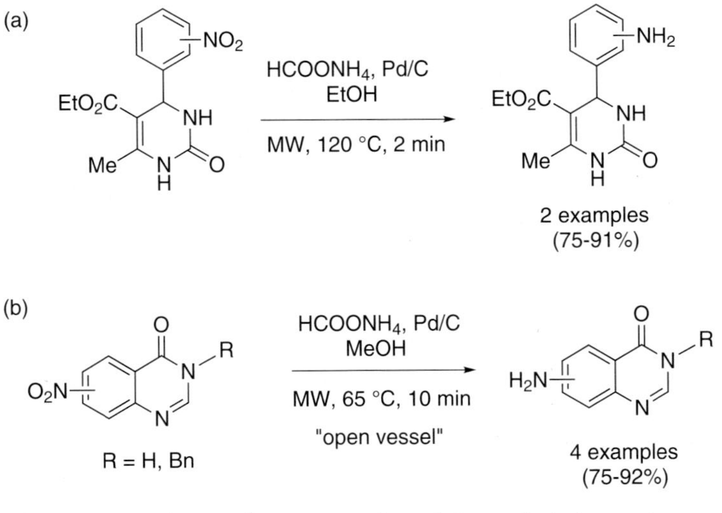

Scheme 6.45 Reduction of nitro groups via catalytic transfer hydrogenations.

Reductions of azide functionalities to amines with lipases suspended in organic media under microwave conditions have also been reported [108].

The same catalyst/hydrogen donor system was employed by Fustero *et al.* for the microwave-induced stereoselective reduction step of β-enamino esters to *cis*-β-amino esters (Scheme 6.46) [109]. Amino acids having both the β-amino and the acid functionality attached to an aliphatic ring have shown antibacterial activities, therefore, the interest in convenient synthesis of such compounds has considerably grown. In the optimized protocol, 0.2 equiv of Pd/C (10 wt%) and 10 equiv of ammonium formate were necessary. Microwave heating at 100 °C for 45 min furnished the desired *cis*-(±)-β-amino esters in high yields and with excellent stereoselectivity (*cis* : *trans* = >99 : 1). The efficient procedure can also be applied when a chiral moiety is present in the ester. Employing the (−)-8-phenylmenthol derivative of the β-enamino ester, phase-transfer catalysis hydrogenation afforded an 80 : 20 mixture of the *cis*-diastereoisomers in good yield. To complete the desired amino acid synthesis, the chiral amino esters were finally deprotected and oxidized following several classical steps.

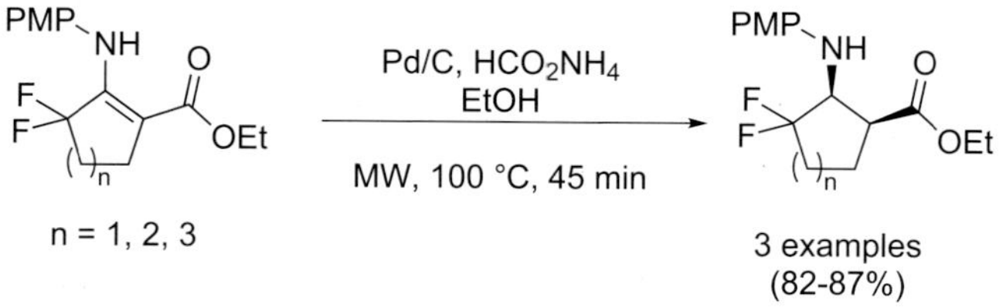

Scheme 6.46 Stereoselective reduction of β-enamino esters by catalytic transfer hydrogenation.

The group of Quinn reported on the reduction of heteroaromatic nitro groups by catalytic transfer hydrogenation. The resulting aryl amines are important synthetic targets and intermediates in the preparation of biologically relevant molecules [110]. The authors employed their previously elaborated microwave-mediated method for the hydrogenation of alkenes under catalytic hydrogen transfer conditions toward aromatic and heteroaromatic nitro compounds. Slight modifications and optimization steps led to a rapid standard procedure applicable toward numerous substrates using 1,4-cyclohexadiene as hydrogen donor and Pd/C as catalyst (Scheme 6.47). Chloro-substituted substrates experienced complete dechlorination under the optimized conditions, thus slight modifications of the protocols were required to keep the chloro-functionality in the product. Simply switching to 5 mol% of Pt/C (5 wt%) proved successful to obtain the desired aromatic chloro-compounds. In general, a very efficient method for the reduction of nitro groups to generate amines was elaborated that is tolerable to a variety of aromatic and heteroaromatic substrates.

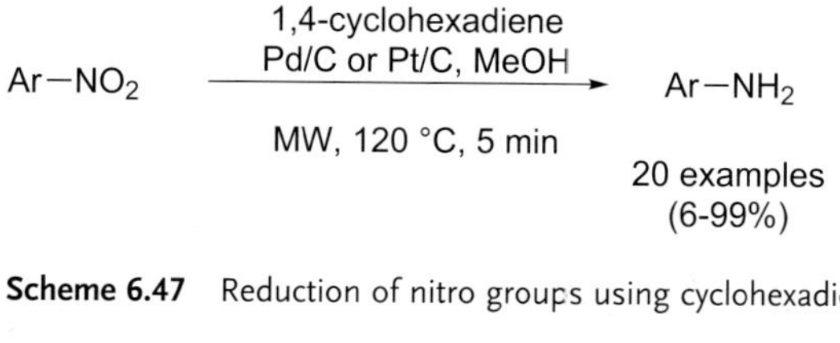

Scheme 6.47 Reduction of nitro groups using cyclohexadiene as hydrogen donor.

The group of Adolfsson has shown that ketones can be reduced to the corresponding alcohols via transfer hydrogenation in the absence of a transition metal catalyst when catalytic amounts of an alkali alkoxide are employed (Scheme 6.48) [111]. The reaction proceeded smoothly for a range of aromatic and aliphatic ketones using lithium isopropoxide (*i*-PrOLi) as catalyst and *i*-PrOH as hydrogen donor. Problematic ketone substrates, for example, where $R^1 = CN$ or acetylpyridines, that show coordination to the metal catalyst under traditional conditions can be converted to the alcohols in high yields. In addition, a scale-up from 0.5 to 82 mmol for acetophenone ($R^1 = H$, $R^2 = Me$) was conducted employing a multimode instrument under the same reaction conditions giving the alcohol in similar isolated yield.

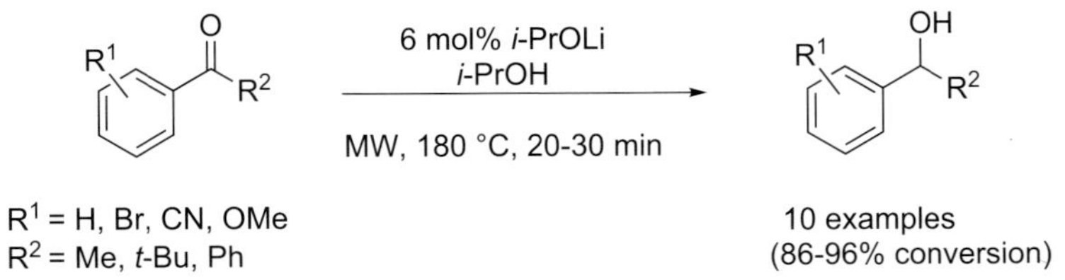

Scheme 6.48 Ketone reduction via transfer hydrogenation.

In this context, Ru-catalyzed transfer hydrogenations of 1,3-cycloalkanediones with *i*-PrOH as hydrogen donor have been described by Bäckvall and coworkers [112].

The group of Taddei has investigated the preparation of substituted piperidines via hydrogenation of the corresponding pyridine derivatives (Scheme 6.49) [113]. For introducing hydrogen into the microwave vial, a special gas-charging accessory for the dedicated microwave instrument was employed. Key to the success of reproducible hydrogenations was the prereduction of PtO_2 to Pt in acetic acid with hydrogen combined with microwave heating at 50 °C for 15 min. Subsequent addition of the pyridine building block, charging the vial again with hydrogen, and heating at 80 °C delivered the piperidine products. Importantly, the hydrogenation was as stereoselective as hydrogenations carried out at room temperature.

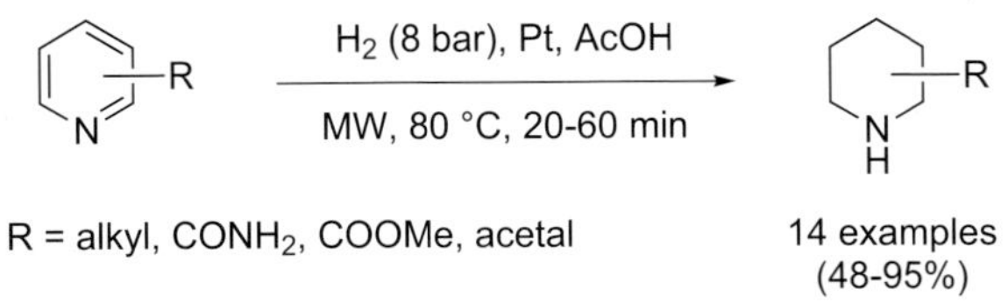

Scheme 6.49 Hydrogenation of pyridines.

Similar hydrogenation reactions under microwave conditions in a pressurized reactor were disclosed by Vanier and Kappe *et al.* [114].

In a 2006 publication, Gogoll and coworkers have developed the Raney/Ni-catalyzed reduction of thiolane **22** to *N*,*N'*-dialkyl-3,7-diazabicyclo[3.3.1]nonanes (bispidines) **23** (Scheme 6.50) [115]. Primary alcohols, which also serve as reagents and to which the reaction is limited, proved to be the best solvents. After initial thioacetal reduction and simultaneous debenzylation, the generated secondary amine is alkylated by the alcohol, which was converted to the corresponding carbonyl compound. Best reaction conditions to reach full conversion turned out to be 20 mass equiv of Raney/Ni, a temperature of 160 °C, and 55–120 min as reaction time (compared to 2 days under conventional reflux heating).

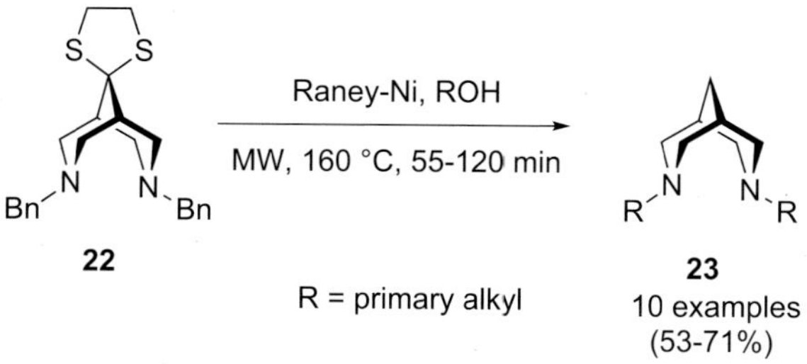

22 **23**

Scheme 6.50 Raney nickel reductions.

The groups of Calderone and Rapposelli disclosed the microwave-assisted reduction of spiromorpholone **24** with the borane–dimethyl sulfide complex $BH_3{\cdot}SMe_2$ in the course of the synthesis of conformationally restricted benzopyrans **25** by the introduction of a spirocyclic substituent at the C-4 position (Scheme 6.51) [116].

In the presence of a proline-based chiral diamine, ketones can be reduced in an asymmetric fashion using $BH_3{\cdot}SMe_2$ as reducing agent [117].

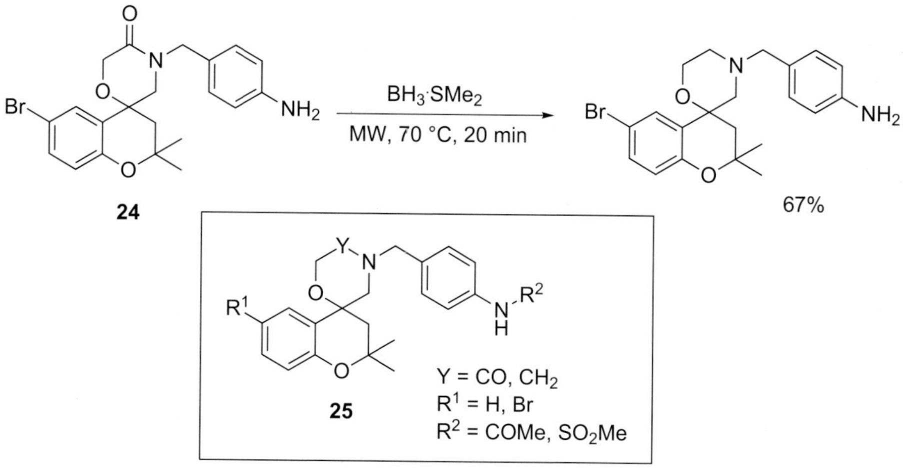

24 **25**

Scheme 6.51 Reduction of spiromorpholone.

In 2009, Campagne and coworkers described a convenient microwave method for the reduction of ketones and aldehydes. The developed protocol uses polymethylhydrosiloxane (PMHS) as reducing agent and inexpensive iron(III) chloride hexahydrate as Lewis acid, thus representing an environment-friendly setup [118]. In general, aromatic ketones gave higher yields than nonaromatic ketones or aldehydes (Scheme 6.52). Overall, the method proceeds highly chemoselective toward the ketone or aldehyde moiety, aromatic cores being present in the substrate remain unaffected.

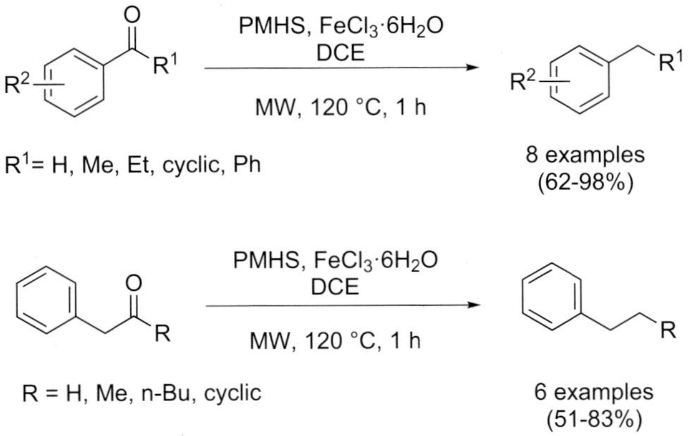

Scheme 6.52 Polymethylhydrosiloxane reduction of ketones and aldehydes.

Spencer *et al.* have reported on the reduction of nitroarenes to anilines mediated by $Mo(CO)_6$ and DBU (Scheme 6.53) [119]. The addition of DBU was essential for obtaining the anilines in high yields due to the facilitated liberation of CO from $Mo(CO)_6$ (33% versus 93% for nitrobenzene). The method proved to be chemoselective and tolerates a broad range of functional groups. By applying microwave heating to 150 °C, the reaction time could be reduced from several hours to 15–30 min compared to conventional heating.

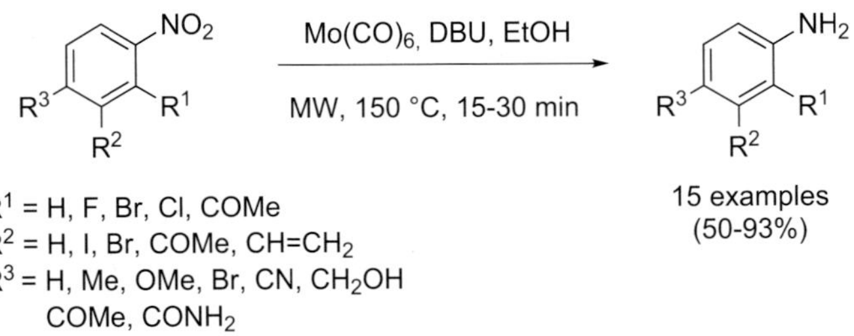

Scheme 6.53 Reduction of nitro compounds with $Mo(CO)_6$ and DBU.

In a related fashion, Alterman and coworkers have shown that benzylic alcohols can be deoxygenated using a combination of $Mo(CO)_6$ and Lawesson reagent under microwave conditions [120]. Nitro groups have also been reduced to amines using $SnCl_2$ or Fe/NH_4Cl as reagents [121].

Mata and coworkers presented a microwave-mediated reduction protocol for α,β-unsaturated alkenes involving ruthenium-based Grubb's catalyst [122]. The procedure was optimized utilizing benzyl crotonate as model substrate (Scheme 6.54) and then applied to several linear and cyclic unsaturated carbonyl compounds. The reaction was highly effective under microwave conditions, since full conversion was obtained after a few minutes and proved highly chemoselective as the benzyl ester moiety was not affected. Although the protocol is tolerable even for geminal and trisubstituted olefins, the authors identified two unexpected reaction outcomes. When fluorocinnamate was the substrate, immediate dehalogenation occurred; and when the terminal triple bond of benzyl 2-propynyl succinate was subject to reduction, a selective deprotection of the propargyl group was observed. This could be a result of decomposition of the catalyst forming a ruthenium hydride species under the applied conditions. Reductions of solid-supported olefins were also performed.

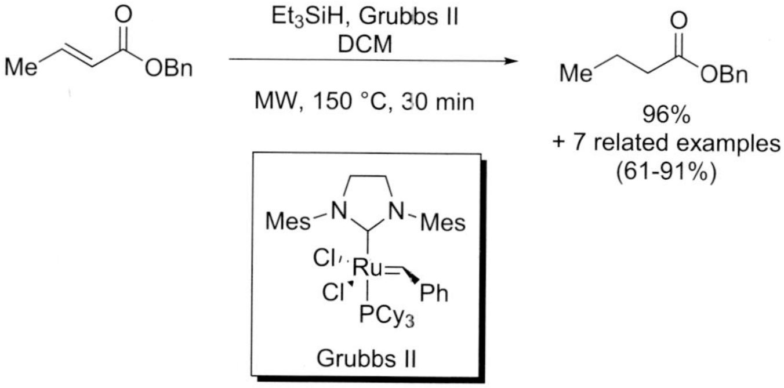

Scheme 6.54 Hydrogen-free reduction of α,β-unsaturated alkenes.

Lipshutz and Frieman have made investigations on [{(*R*)-(−)-DTBM-segphos}CuH] as a hydrosilylation reagent regarding stability and preparation (Scheme 6.55) [123]. This reagent proved to be very stable over several weeks at room temperature (no decrease in *ee* values) if prepared by using $Cu(OAc)_2{\cdot}H_2O$ as copper source and stored as stock solution in a bottle. Excellent *ee* values and good conversions were obtained in several asymmetric hydrosilylations at room temperature, but microwave heating could also be applied for the reduction of enone **26**. By using a substrate/catalyst ratio (S/C) of 1000 : 1, the reaction time could be reduced from several hours to only 10 min.

A modified Stryker's reagent for achiral conjugate additions of unsaturated substrates was recently reported by the same group [124].

An asymmetric hydrogenation step within the synthesis of the *C*-glycoside analog of *β*-D-galactosyl hydroxylysine (**28**) was disclosed by Kihlberg and coworkers (Scheme 6.56) [125]. Using Burk's catalyst and microwave heating (70 °C for 40 min)

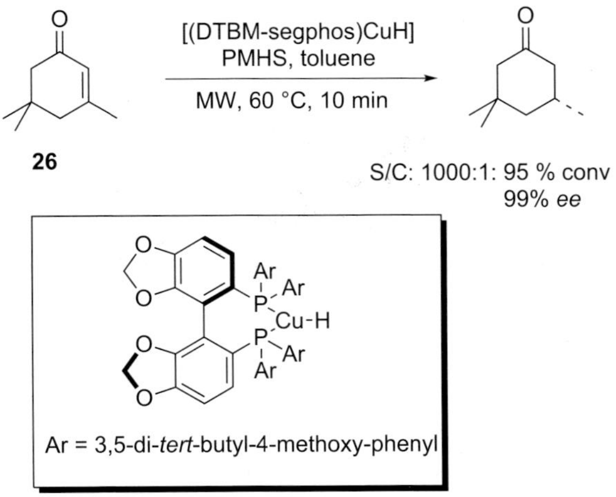

Scheme 6.55 "CuH in a bottle" for asymmetric hydrosilylations.

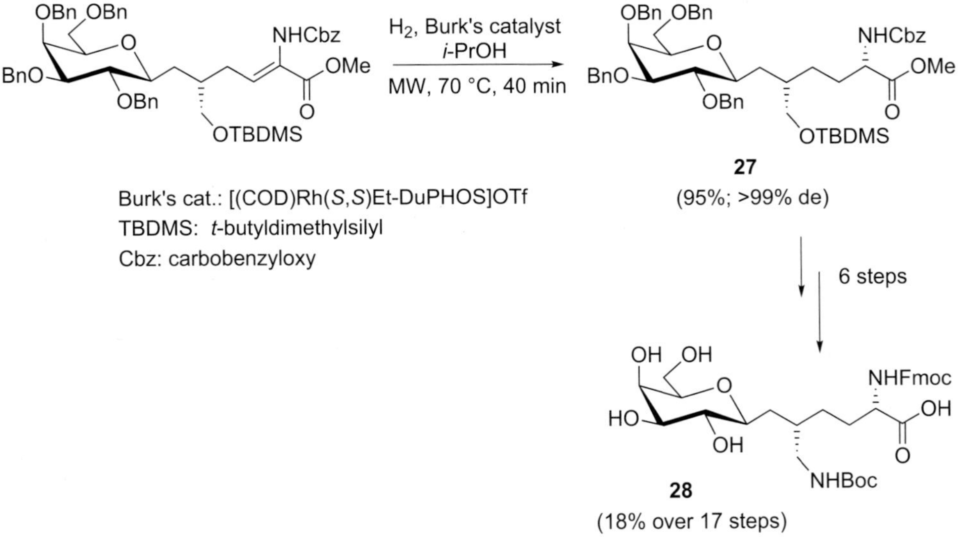

Scheme 6.56 Asymmetric hydrogenation using Burk's catalyst.

for the hydrogenation step, the corresponding product **27** was obtained in high yield and with excellent selectivity (>99% de), which could not be reached by conducting the reaction at room temperature and prolonged reaction time.

Wolff–Kishner carbonyl group reductions (Huang–Minlon modification) [126] and $LiBH_4$-mediated double-bond reductions [127] have both been reported to work efficiently in conjunction with microwave heating. Azides can be reduced selectively to primary amines using the $CeCl_3{\cdot}7H_2O/NaI$ system [128]. α-Haloketones have been reduced (dehalogenated) by Zn and NH_4Cl in ethanol [129].

6.5 Mitsunobu Reactions

A powerful stereochemical transformation is the Mitsunobu reaction. This reaction is very efficient for inverting the configuration of chiral secondary alcohols since a clean S_N2 process is generally observed (Mitsunobu inversion). Considering the fact that Mitsunobu chemistry is typically carried out at or below room temperature, high-temperature Mitsunobu reactions performed under microwave conditions would appear to have little chance of success. In 2001, it was established by the Kappe group that Mitsunobu chemistry can indeed be carried out using high-temperatures in the context of an enantioconvergent approach to the aggregation pheromones (*R*)- and (*S*)-sulcatol (Scheme 6.57a) [130]. While the conventional Mitsunobu protocol carried out at room temperature proved to be extremely sluggish, complete conversion of (*S*)-sulcatol to the (*R*)-acetate (S_N2 inversion) using essentially the standard Mitsunobu conditions (1.9 equiv of diisopropylazodicarboxylate and 2.3 equiv of triphenylphosphine) was achieved within 5 min at 180 °C under sealed-vessel microwave conditions. Despite the high reaction temperatures, no by-products could be identified in these Mitsunobu experiments, with enantiomeric purities of sulcatol (*R*)-acetate being >98% *ee* (80% isolated yield).

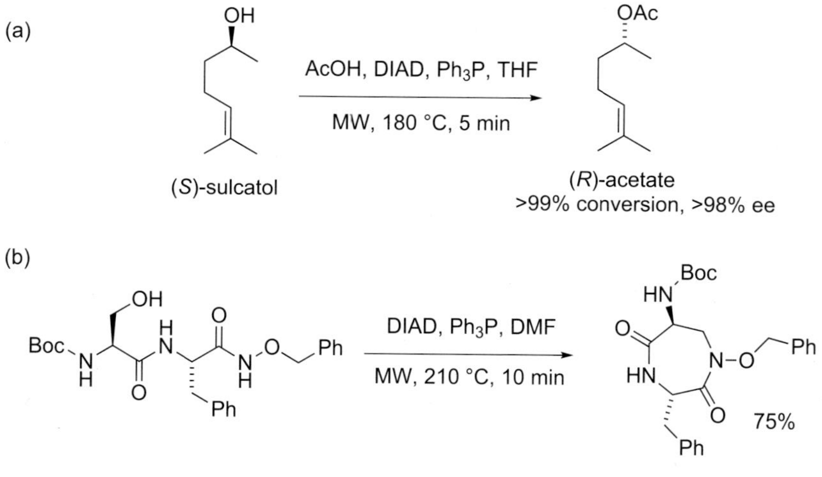

Scheme 6.57 Mitsunobu reactions.

An application of these rather unusual high-temperature Mitsunobu conditions for the preparation of conformationally constrained peptidomimetics based on the 1,4-diazepan-2,5-dione core was disclosed by the group of Taddei (Scheme 6.57b) [131]. Cyclization of a hydroxy hydroxamate dipeptide using the DIAD/Ph_3P microwave conditions (210 °C, 10 min) provided the desired 1,4-diazepan-2,5-dione in 75% isolated yield. Standard room-temperature conditions (*N*,*N*-dimethylformamide,

12 h) were significantly less efficient and gave only 46% of the desired compound. These transformations could also be carried out on a polystyrene resin.

Ghassemi and Fuchs have reported on a general method for the synthesis of a set of unsymmetric sulfamides in combination with an alternative method for a Boc-removal strategy (Scheme 6.58) [132]. The first step was a one-pot synthesis of Boc-protected sulfamides, followed by *N*-alkylation via Mitsunobu reaction to introduce more diversity. Silica-bonded phenylsulfonic acid (Si-TsOH) was used for Boc-removal, which was possible in 5 min at 100 °C for all examples. All synthesis steps have been performed using microwave heating. For releasing the sulfamides from Si-TsOH, NH_3 in MeOH was used.

Another microwave-mediated intramolecular S_N2 reaction forms one of the key steps in a recent catalytic asymmetric synthesis of the cinchona alkaloid quinine by Jacobsen and coworkers [133]. The strategy to construct the crucial quinuclidine core of the natural product relies on an intramolecular S_N2 reaction/epoxide ring opening (Scheme 6.59). After removal of the benzyl carbamate (Cbz) protecting group with diethyl aluminum chloride/thioanisol, microwave heating of the acetonitrile solution to 200 °C for 20 min provided a 68% isolated yield of the natural product as the final transformation in a 16-step total synthesis.

6.6
Glycosylation Reactions and Related Carbohydrate-Based Transformations

Glycosylation reactions involving oxazoline donors are generally rather slow and require prolonged reaction times due to the low reactivity of the donors. Oscarson and coworkers have reported the preparation of spacer-linked dimers of *N*-acetyllactosamine using microwave-assisted pyridinium triflate-promoted glycosylations with oxazoline donors (Scheme 6.60a) [134]. Using 2.2 equiv of each of the oxazoline donors and pyridinium triflate promoters, rapid and efficient coupling was achieved in DCM with four different diols. Employing 20 min microwave irradiation at 80 °C, moderate to high yields of the dimers were obtained, with yields increased by 12–15% over the conventional process. Fraser-Reid and coworkers have recently described related saccharide couplings employing *n*-pentenyl glycosyl donors and *N*-iodosuccinimide (NIS) as promoter in acetonitrile (Scheme 6.60b) [135].

Jensen and coworkers have described the microwave promoted synthesis of oligosaccharides (Scheme 6.61) [136]. Without the use of a strong Lewis acid, *O*-glycosylations could be accelerated from up to 31 h to 5–40 min by applying microwave heating. Glycosylation reactions have been performed with trichloroacetimidate (**30**) or DISAL (3,5-dinitrosalicylate, **29**) donors, respectively, in the latter case with α-selectivity for sterically hindered alcohols.

The group of Gervay-Hague has discovered that microwave heating can accelerate the α-glycosidation of per-*O*-silylated galactosyl iodide **31** with ceramide **32** (Scheme 6.62) [137]. If galactosyl iodide **31** is reacted with ceramide **32** under conventional heating or at room temperature for 48 h, low yields are obtained. The authors assume that the unsaturated side chain in **32** favors lipid packing that could

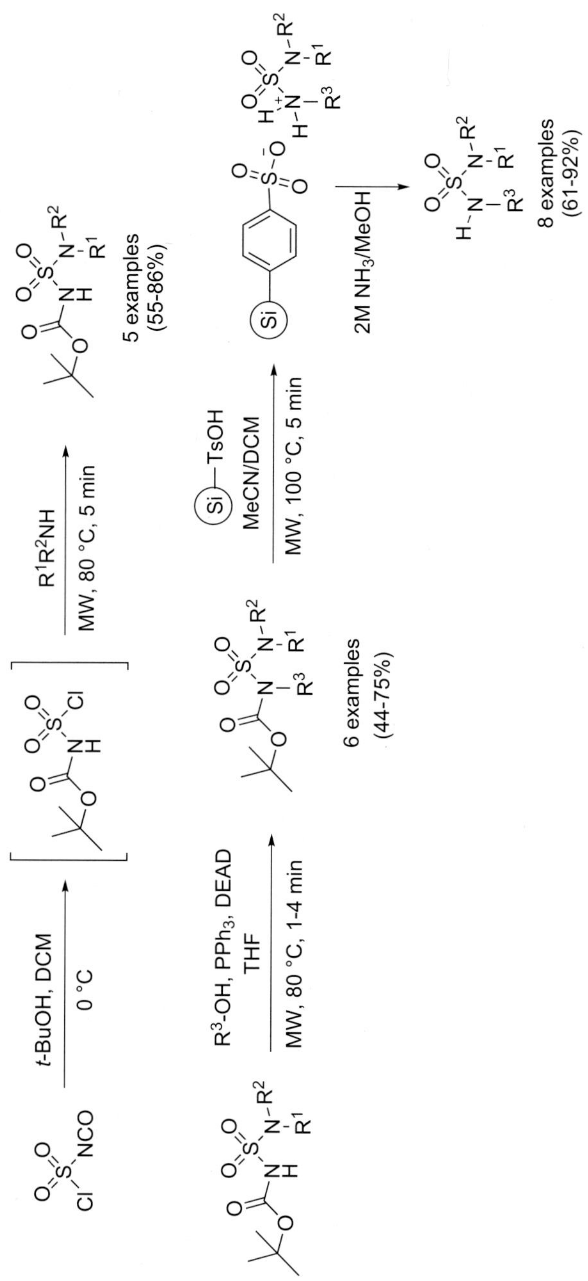

Scheme 6.58 Synthesis of unsymmetric sulfonamides.

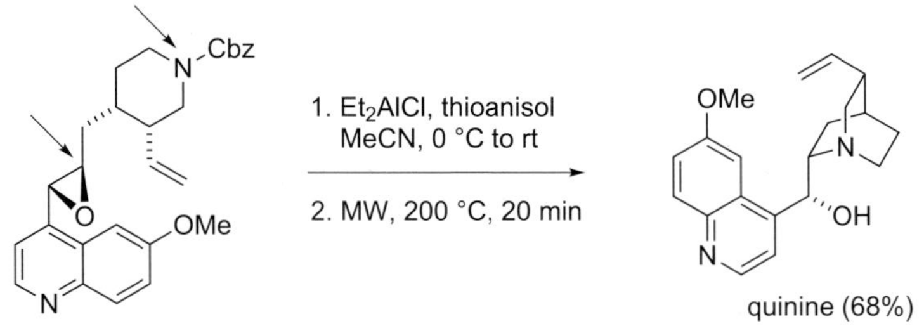

Scheme 6.59 Intramolecular S_N2 reaction in the total synthesis of quinine.

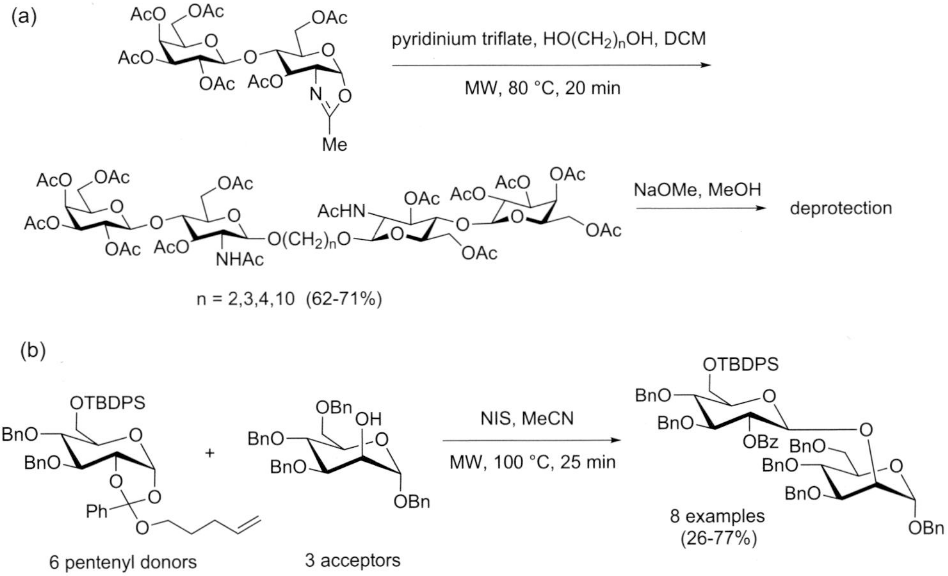

Scheme 6.60 Glycosylation reactions.

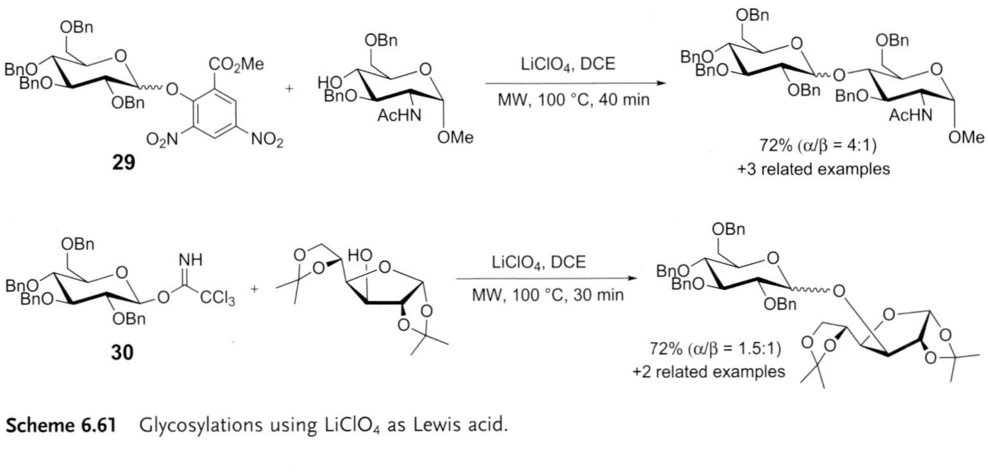

Scheme 6.61 Glycosylations using $LiClO_4$ as Lewis acid.

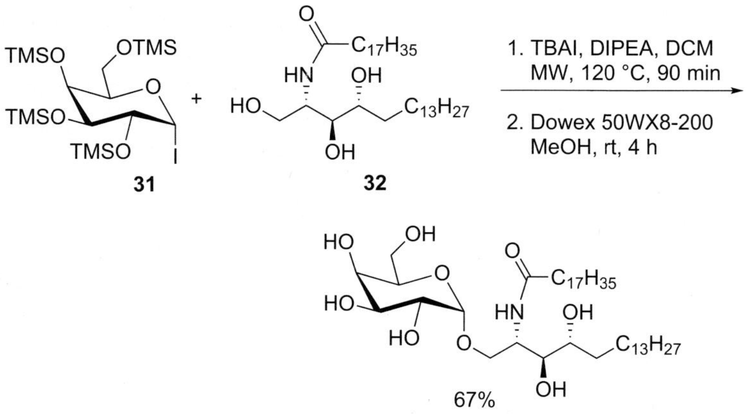

Scheme 6.62 Synthesis of α-linked glycolipids.

be overcome by applying microwave heating, thus obtaining the glycolipid product in higher yield. With this method that normally gives high yields when the reaction is performed at room temperature, only the α-isomer is obtained, unsaturated fatty acid side chains are tolerated, and multiple protection–deprotection steps are eliminated.

The glucosamine residues in heparin-like glycosaminoglycans have been found to exist as amines, acetamides, and *N*-sulfonates. To develop a completely general, modular synthesis of heparin, three degrees of orthogonal nitrogen protection are required. The synthesis of fully *N*-differentiated heparin oligosaccharides was demonstrated by Lohman and Seeberger (Scheme 6.63) [138]. One of the many synthetic steps involves the simultaneous installment of a *N*-diacetate and *O*-acetyl functionality in a trisaccharide building block. Microwave irradiation in isopropenyl acetate as solvent in the presence of *p*-toluenesulfonic acid (*p*-TsOH) at 90 °C for 5 h led to the desired product in 86% yield. This transformation could not be accomplished under a variety of thermal conditions, with only poor yields being achieved even after several days.

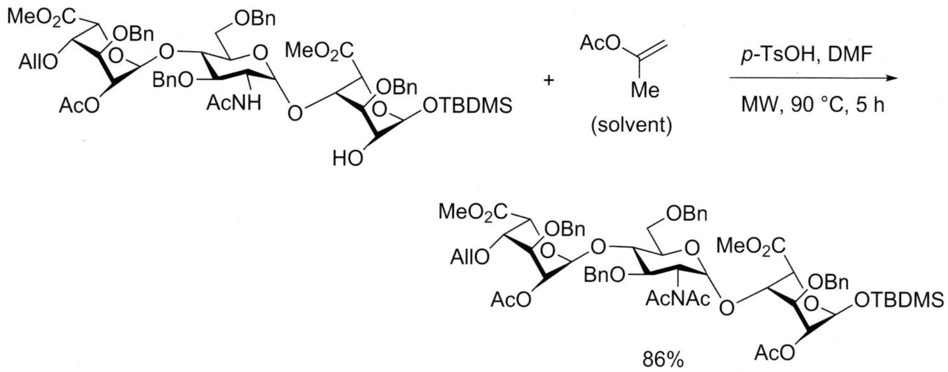

Scheme 6.63 Heparin oligosaccharide synthesis.

The reaction of modified carbasugars with heterocyclic bases for the synthesis of constrained carbanucleosides was studied in detail by the group of Sega (Scheme 6.64) [139]. Using conventional conditions employing sodium hydride or cesium carbonate as a base in DMF (120 °C, 48 h), the reaction between the two substrates shown in Scheme 6.64 was extremely sluggish and only provided modest yields of the desired carbanucleosides. Performing the reaction under microwave irradiation conditions (205 °C, 4 min) proved to be ineffective; only the starting materials were recovered. Switching to the ionic liquid 1-butyl-3-methylimidazolium tetrafluoroborate ($bmimBF_4$) as reaction medium also did not provide any product (cesium carbonate base, 120 °C, 48 h) either. Only by applying the combination of microwave heating and the use of the ionic liquid as reaction medium, could the transformation be successfully accomplished shown in Scheme 6.64 and provided a 64% isolated yield of the desired enantiomerically pure carbanucleoside after 4 min. Similar results were obtained with other heterocyclic bases.

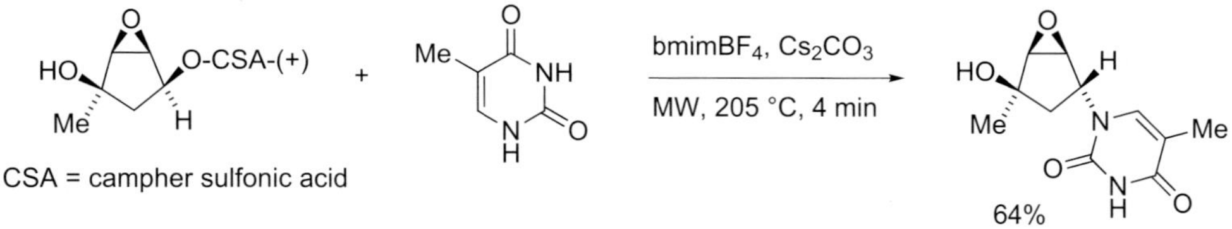

Scheme 6.64 Carbanucleoside synthesis in ionic liquids.

In view of the need for new carbohydrate functionalization methods that minimize the employment of protection groups, tin acetal-mediated regioselective functionalization methods are of great importance. A microwave-assisted tin-mediated, regioselective 3-*O*-alkylation of galactosides was presented by Pieters and coworkers (Scheme 6.65) [140]. Here, dibutylstannylene acetals were generated *in situ* by heating

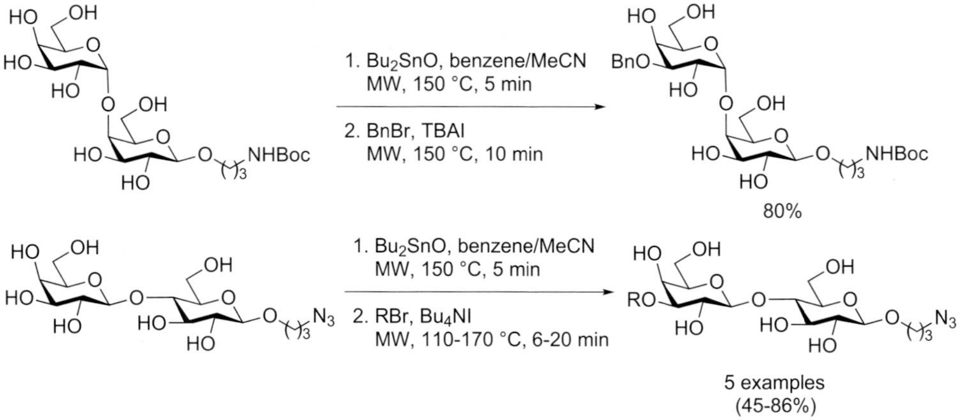

Scheme 6.65 Tin-mediated 3-*O*-alkylation of galactosides.

the unprotected carbohydrates with 1.1 equiv of dibutyltin oxide under microwave conditions at 150 °C for 5 min. After the addition of the alkylating reagent (5 equiv) and tetrabutylammonium iodide (TBAI) (2.5 equiv), the mixture was again irradiated at 110–170 °C for 6–20 min. Good yields of products were obtained with short overall reaction times.

A somewhat related microwave-promoted 5′-*O*-allylation of thymidine was described by the Zerrouki group (Scheme 6.66) [141]. While the classical method for the preparation of 5′-*O*-allylthymidine required various protection steps (four synthetic steps in total), the authors have attempted the direct allylation of thymidine under basic conditions. Employing sodium hydride as a base at room temperature in DMF resulted in the formation of perallylated compounds along with the desired monoallylated product (75% yield). The best result was achieved when both the deprotonation with sodium hydride (1.15 equiv) and the subsequent allylation (1.2 equiv of allyl bromide) were conducted under microwave irradiation for 2 min each. Under these conditions, a 97% of 5′-*O*-allylthymidine was isolated.

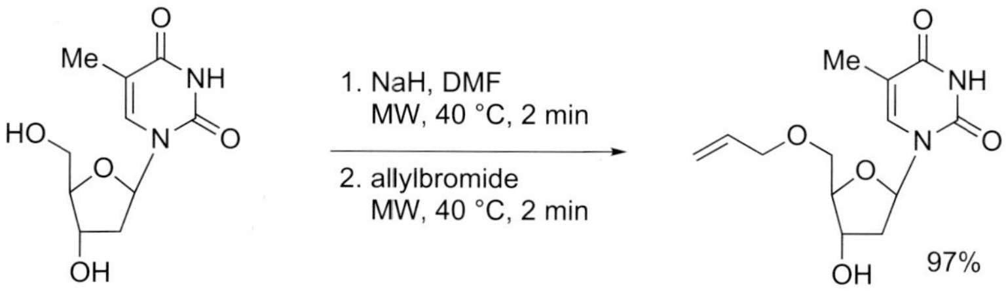

Scheme 6.66 Selective allylation of thymidine.

Bookser and Raffaele have reported on the synthesis of a nucleoside library via the Vorbrüggen glycosylation and subsequent deprotection with NH_3/MeOH (Scheme 6.67) [142]. They were successful in performing this reaction that is known to be a two-step synthesis (presilylation of the base and then reaction with trimethylsilyl triflate-activated sugar) in only one step by applying microwave heating at 130 °C for 5 min. An additional advantage of the elevated temperature is that otherwise insoluble silylated bases were solubilized under these conditions. For the library synthesis, 48 *N*-containing bases were reacted with *β*-D-ribofuranose **33**; of these, 32 yielded single isomers, 6 provided separable mixtures, and 7 afforded inseparable regioisomer mixtures. Altogether a set of 58 nucleosides was obtained in poor to good yields.

A rapid synthesis of carbasugars was reported by Pohl and coworkers starting from a protected D-glucose (Scheme 6.68) [143]. Microwave-enhanced iodination of the selectively protected glucose precursor took place in 1 min in toluene at 60 °C in the presence of iodine and triphenylphosphine. Subsequent base-promoted elimination of hydrogen iodide in the presence of 1,8-diazabicyclo[5.4.0]undec-7-ene (DBU) required 30 min (DMF, 80 °C), and the key Ferrier rearrangement induced by palladium dichloride was also completed within 5 min of microwave irradiation.

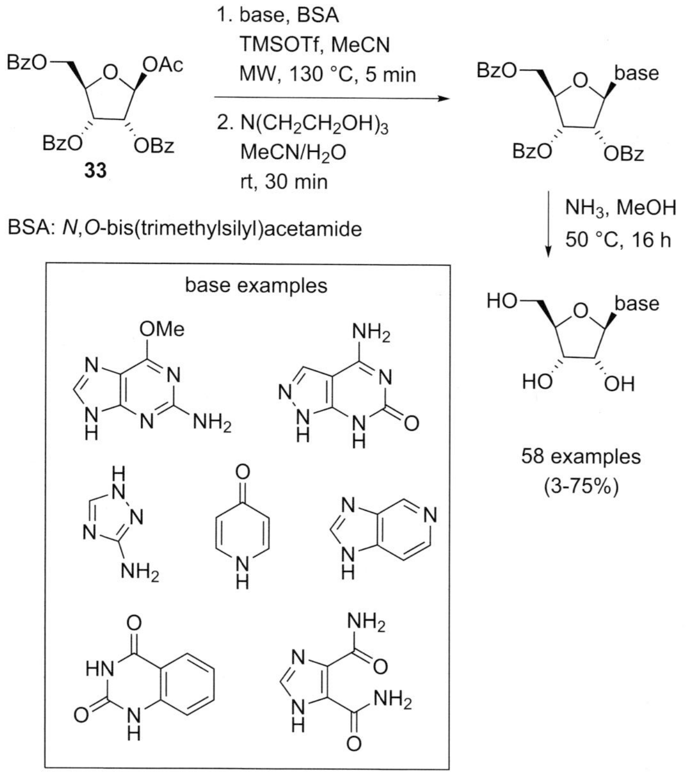

Scheme 6.67 One-step Vorbrüggen glycosylation.

The development of this series of microwave-assisted transformations significantly shortened the time to form the core carbocyclic structure from a protected glucose.

Bejugam and Flitsch have described the synthesis of glycosylamines from mono-, di-, and trisaccharides via direct microwave-assisted Kochetkov amination (Scheme 6.69) [144]. The reaction was found to be effective with just a fivefold excess (w/w) of ammonium carbonate with respect to the sugar, as compared to the

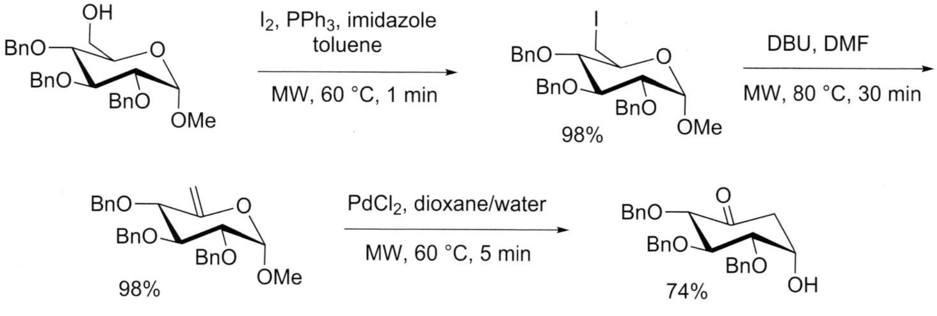

Scheme 6.68 Multistep synthesis of carbaglucose derivatives.

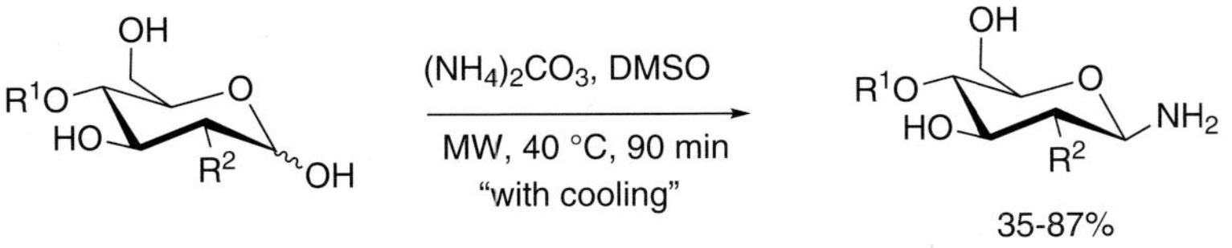

R^1 = H, Glc (β1-, Gal (β1-, GlcNAc (β1-, Glc (α1-4)Glc (α1-, Glc (α1
R^2 = OH, NHAc

Scheme 6.69 Synthesis of β-glycosylamines via Kochetkov amination.

40–50-fold excess needed under thermal conditions. All transformations were completed within 90 min in DMSO as solvent while maintaining the vessel temperature at an apparent 40 °C, using the "heating-while-cooling" technique.

Dondoni *et al.* have disclosed the synthesis of glycosyl amino acids where the carbohydrate and amino acid residues are linked via a pyridine ring (Scheme 6.70) [145]. The synthesis of those *C*-glycosylmethyl pyridylalanines was accomplished by a Hantzsch-type three-component cyclocondensation of *C*-glycosyl or *C*-galactosyl acetaldehydes (**34**), *β*-ketoester **35** that incorporates the amino acid residue, and enaminoester **36**. The purification was performed using three different polymer-bound scavengers in one step and one pot. Unreacted enamine was scavenged by sulfonic acid A-15, *β*-ketoester by the strongly basic anion exchange resin Ambersep, and aldehyde by the aminomethylated (AM) resin. Subsequent oxidation employing silica-supported PCC afforded the products in good yields. Attachment of the carbohydrate residue onto

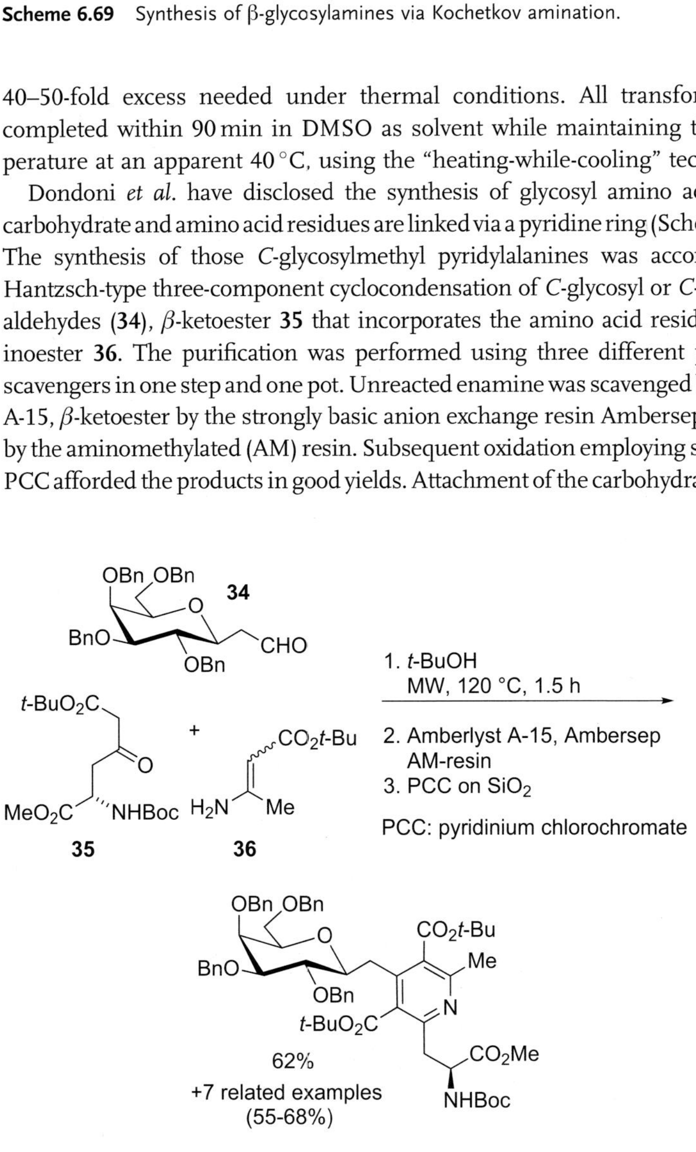

Scheme 6.70 Synthesis of carbon-linked glycosyl amino acids.

the β-ketoester building block and the amino acid residue onto the aldehyde building block, respectively, is also possible.

Li and Danishefsky have investigated the reaction of isonitriles with carboxylic acids to form *N*-formyl amides (Scheme 6.71) [146]. The reactions proceed well under microwave irradiation conditions at 150 °C for 30–45 min (depending on the substrates) possibly through a 1,3-*O*→*N*-acyl transfer, whereas at room temperature no reaction could be detected. When glycosyl-linked isonitriles are reacted with aspartate, the reactions occur in an anomerically specific way corresponding to the α- and β-*N*-linked glycosyl amino acids. Further transformations of the *N*-formyl amides to different amide types are also described.

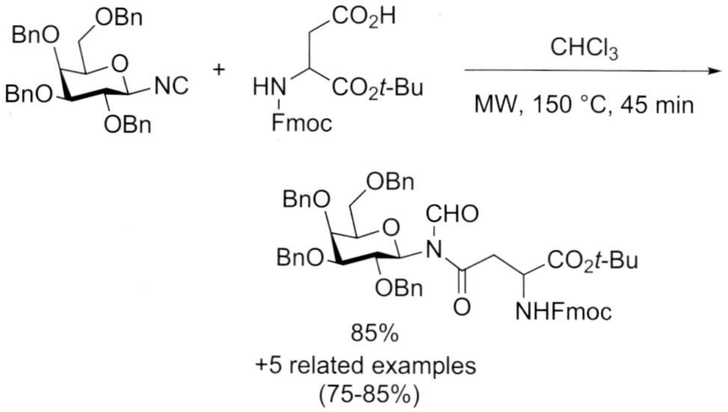

Scheme 6.71 Reaction of carbohydrate-derived isonitriles with carboxylic acids.

Other examples of microwave-assisted transformations related to carbohydrate chemistry involve glycosylation reactions catalyzed by $FeCl_3$ [147] or carried out in a solvent-free manner [148], Ferrier rearrangements [149], Fischer glycosylations [150], glycosylations of *N*-acetyl glycosamines [151], and acetal exchange-type glycosylations from methyl glycosides [152]. Microwave-assisted epoxide ring openings of 1,5:2,3-dianhydro-4,6-*O*-benzylidene-D-allitol with nucleobases have been reported [153]. Various rapid microwave-assisted protection and deprotection methods in the area of carbohydrate chemistry are known [134], and two general review articles on microwave-assisted carbohydrate chemistry were published in 2004 [154, 155].

6.7 Organocatalytic Transformations

Rodríguez and Bolm investigated thermal effects in the (*S*)-proline-catalyzed Mannich reaction (Scheme 6.72) [156]. By applying microwave heating with simultaneous cooling, reaction times and, most importantly, catalyst loadings could be reduced. Similar results were also achieved by heating in an oil bath for 23 h at the same temperature (45–50 °C). The corresponding reduced amino alcohols could be obtained in high yields and with excellent *ee* values.

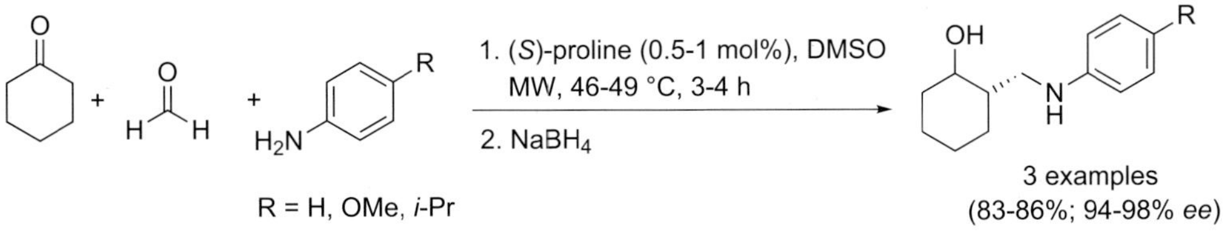

Scheme 6.72 Asymmetric Mannich reactions.

Mossé and Alexakis have reinvestigated asymmetric aldol- (Scheme 6.73a), Mannich- (Scheme 6.73b) and Diels–Alder reactions (Scheme 6.73c) applying organocatalysis under microwave conditions [157]. L-Proline (Scheme 6.73a), *N*-*i*Pr-2*R*,2′*R*-bipyrrolidine [(*R*,*R*)-*i*PBP] (**37**) (Scheme 6.73b), and McMillan imidazolidinone (**38**) (Scheme 6.73c) were used as organocatalysts. In all cases, the reaction times were reduced dramatically using microwave heating. In addition, for the aldol and Mannich reaction, catalyst loadings could be decreased maintaining the enantioselectivities achieved by conventional room-temperature experiments.

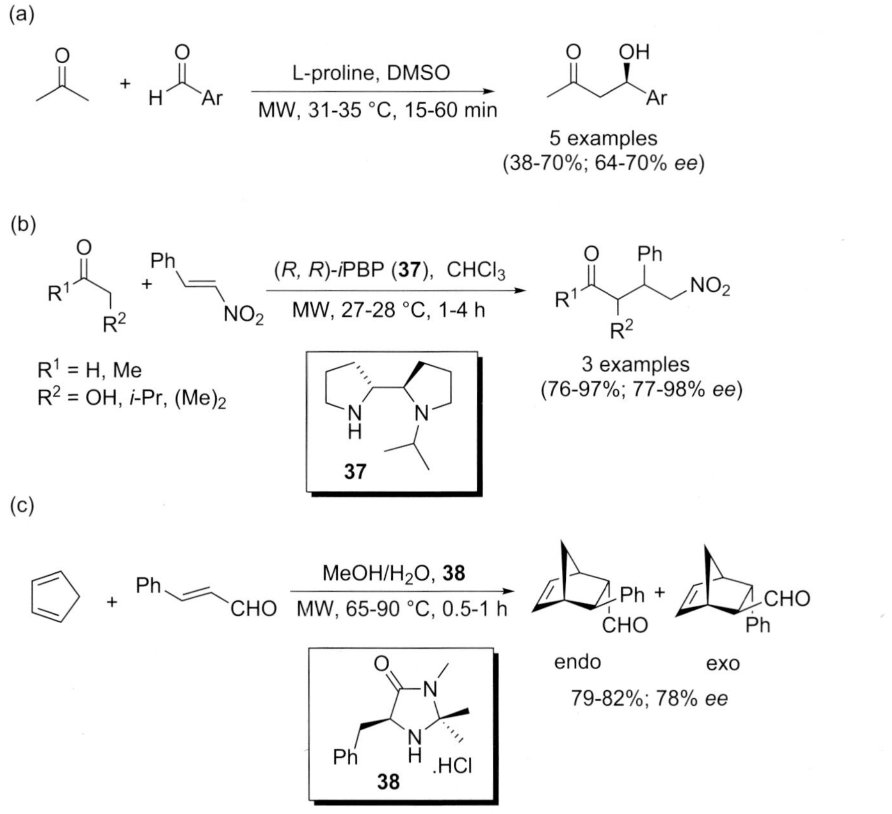

Scheme 6.73 Organocatalyzed asymmetric reactions.

In a related study, the group of Bräse has performed optimization studies toward solvent, time, temperature, and catalyst loading for the microwave-mediated organocatalytic asymmetric α-amination of disubstituted aldehydes with diethyl azodicarboxylate (Scheme 6.74) [158]. Optimum conditions in terms of both high yield and enantioselectivity proved to be MeCN as solvent, 50 mol% of L-proline as catalyst, a reaction temperature of 60 °C, and 30 min reaction time. Importantly, the reaction time could be decreased from several days to only 30 min by applying microwave heating compared to room-temperature conditions, with a concomitant increase in yield and enantioselectivity. In particular, improvements for otherwise unreactive starting materials bearing electron-withdrawing R-groups (e.g., F, NO_2, and CF_3) were observed.

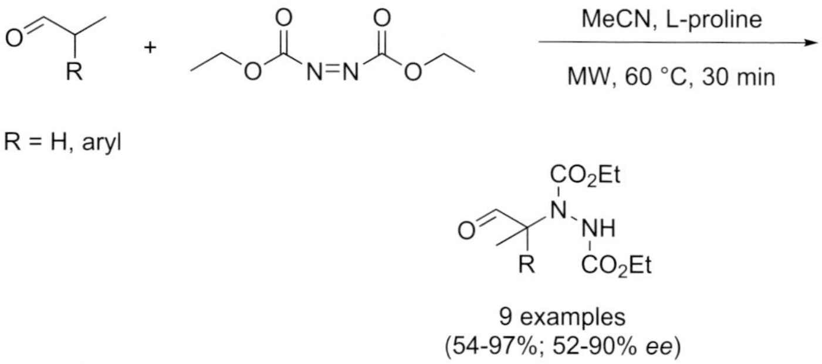

Scheme 6.74 Synthesis of α,α-disubstituted amino aldehydes.

For other examples of microwave-assisted organocatalytic transformations, see Refs [159, 160]. A detailed study on microwave effects in these reactions has been published [161].

6.8 Organometallic Transformations (Mg, Zn, and Ti)

Larhed and coworkers have described a safe, productive, and reproducible lab-scale protocol for fast generation of Grignard reagents from reluctant aryl chlorides and bromides under controlled microwave heating (Scheme 6.75) [162]. The Grignard reagents prepared in this way were used as precursors to promote a subsequent microwave-assisted Kumada coupling, toward the synthesis of novel HIV-1 protease inhibitor. The microwave-induced (150 °C for 1 h) Grignard formation in most of the aryl chlorides was initiated using only catalytic amounts of iodine as the sole activator.

Wunderlich and Knochel have reported on the direct zincation of highly functionalized aryl and heteroaryl derivatives using the complex base $(TMP)_2Zn{\cdot}2MgCl_2{\cdot}2LiCl$ (Scheme 6.76) [163]. With microwave heating, a tremendous time saving could be achieved–for example, for derivatives with $R^1 = CO_2Et$, $R^2 = R^3 = H$, $R^4 = Cl$, Br, the

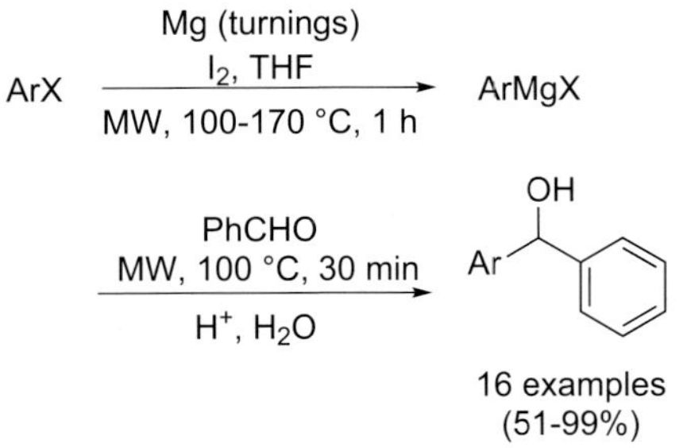

Scheme 6.75 Grignard reaction of aryl chlorides.

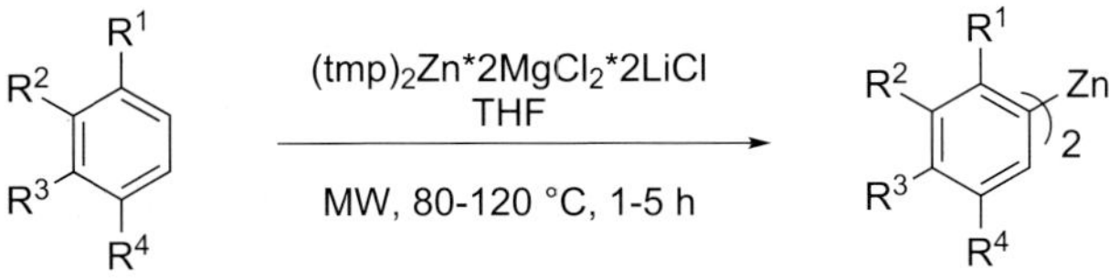

13 examples

R^1 = H, CN, CO_2Et, CO_2Me, $CONEt_2$
R^2 = H, F, COPh, CO_2Et
R^3 = H, CO_2Me
R^4 = H, Cl, Br, F, CN, CO_2Et

tmp: 2,2,6,6-tetramethylpiperidyl

Scheme 6.76 High-temperature zincation of functionalized (het)aromatics.

reaction time could be reduced from 110 h at 25 °C to 2 h at 80 °C. Interestingly, for substrates with $R^1 = CO_2Et$, $CONEt_2$, $R^2 = R^3 = R^4 = H$, microwave heating is crucial since under oil bath heating at the same temperature a very low yield is obtained (10–20% versus >90%). Importantly, sensitive functional groups are well tolerated under the reaction conditions. This zincation protocol was further successfully applied to heterocyclic systems such as pyridines, benzothiophene, or benzofuran. In a postderivatization step, the bis-zincated species were further transformed via Cu-mediated or Pd-catalyzed reactions.

More recently, the same group presented a similar protocol applying TMPZnCl·LiCl as efficient and selective base [164]. The benefit of microwave irradiation was demonstrated with the direct zincation of coumarin, which requires 7 days at room temperature, but reached full conversion after only 1 h microwave heating in THF at 80 °C. Also, weakly activated aromatics, even with sensitive functional groups, could be regio- and stereoselectively metallated under microwave irradiation at 160 °C for 2 h (Scheme 6.77). Again, the resulting zinc species readily underwent Cu-mediated allylation or acylation and also Negishi cross-couplings when quenched with corresponding electrophiles.

Moloney and coworkers have reported the Pd-catalyzed α-arylation of esters and amides with organozinc reagents in a one-pot fashion (Scheme 6.78) [165]. The

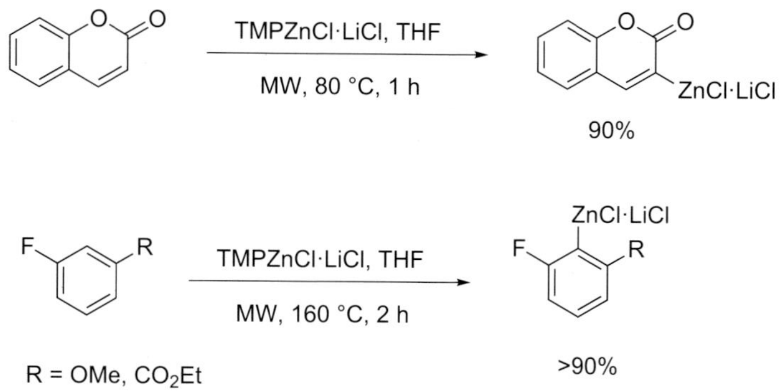

Scheme 6.77 Zincation of aromatic compounds.

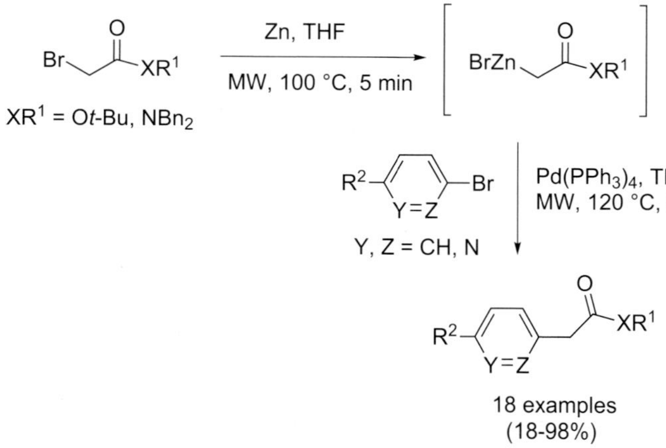

Scheme 6.78 α-Arylation of esters using Reformatsky reagents.

required Reformatsky reagents were readily prepared by microwave irradiation of the corresponding bromoacetate or bromoacetamide with Zn in THF for 5 min at 100 °C. Addition of this Reformatsky reagent to the coupling partner, an aryl bromide and the relevant catalyst/ligand in THF followed by further irradiation for 5 min at 120 °C provided the arylacetic esters or amides in good yields.

More recently, a reexamination of the reactivity of Reformatsky reagents in conjugate additions was performed by the same group. *t*-Butyl 2-bromopropionate **39a** and 2-bromopropionitrile **39b** were used as Reformatsky reagent precursors in the reaction with α,β-unsaturated malonate esters under both microwave and conventional conditions (Scheme 6.79) [166]. Compared to thermal heating, the protocol could be simplified when microwave irradiation is applied: the Reformatsky reagents are formed *in situ* when activated Zn, iodine, precursors **39a** and **39b**, respectively, and malonates in THF are microwave heated. Bromopropionitrile **39b**

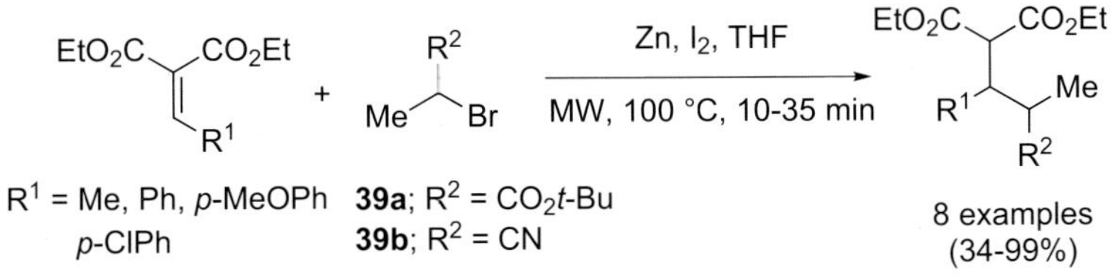

Scheme 6.79 Reformatsky conjugate additions.

showed increased reactivity compared to the thermal approach, whereas bromopropionate **39a** gave higher yields only when being activated ($R^1 = p$-ClPh).

Braga *et al.* have reported on the arylation of aromatic aldehydes via arylzinc addition using an aziridine-based ligand (Scheme 6.80) [167]. Employing microwave heating, the reactive arylzinc species are generated from aryl boronic acids and Et_2Zn in only 10 min followed by the addition of the aldehyde and ligand and further microwave heating for 5 min. The reaction time can be reduced from 1 day conventionally to only 15 min under microwave conditions. A key feature of this method is that both enantiomers of the product can be obtained selectively using only one enantiomer of the ligand by interchanging the functional groups of the boronic acids and aldehydes.

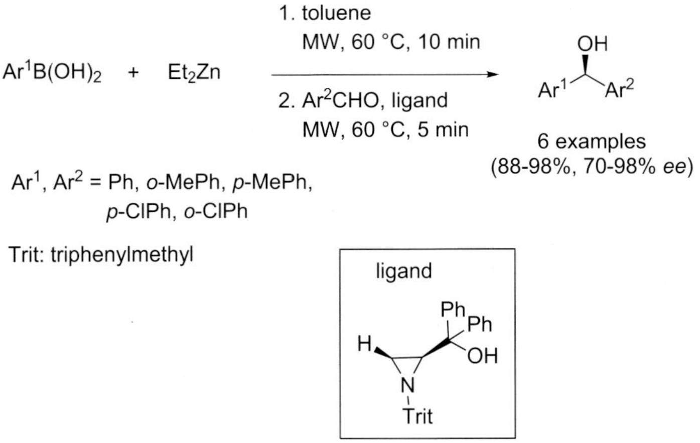

Scheme 6.80 Enantioselective arylation of aromatic aldehydes.

Enantioselective addition of aromatic aldehydes to Et_2Zn catalyzed by chiral aminonaphthols were described by Fülöp and coworkers [168]. Similar results were obtained by Espinet using Me_2Zn and chiral β-amino alcohols [169].

The synthesis of pyruvate-based enol ethers and enamines via the Petasis olefination has been demonstrated by Gallagher and coworkers (Scheme 6.81) [170]. The olefination of unsymmetrical oxalates is highly regioselective by employing the Petasis reagent and is dramatically improved when microwave heating is applied.

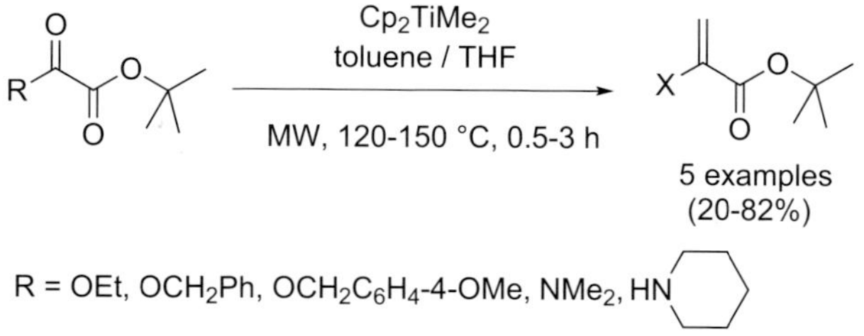

Scheme 6.81 Petasis olefination of unsymmetrical oxalates.

Shorter reaction times and higher yields are achieved by irradiating in toluene/THF for 0.5–3 h at 120–150 °C.

Other microwave-assisted Petasis olefinations have been described in the recent literature [171]. For examples of microwave-assisted Wittig olefinations, see Refs [172–174].

Stuhr-Hansen has reported on a microwave-induced high-yielding synthesis of alkenes by McMurry coupling of aldehydes and ketones with low-valent titanium (Scheme 6.82) [175]. For all the aldehydes and ketones including sulfur end-capped analogs studied, even at low-energy microwave conditions and within 10 min of microwave heating, complete conversion and above 80% yields of corresponding alkenes were realized. With microwave heating, the corresponding alkenes were produced much faster with higher yields compared to that with conventional heating, and in some cases, without damaging the sensitive *tert*-butyl sulfur-protecting groups.

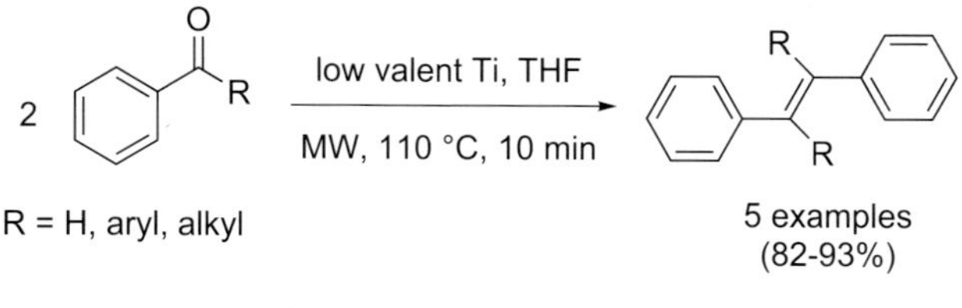

Scheme 6.82 Synthesis of alkenes by McMurry reaction.

6.9 Multicomponent Reactions

Luthman and coworkers have reported microwave-assisted Mannich reactions that employed paraformaldehyde as a source of formaldehyde, a secondary amine in the form of its hydrochloride salt, and a substituted acetophenone (Scheme 6.83) [176]. Optimized reaction conditions utilized equimolar amounts of reactants, dioxane as solvent, and microwave irradiation at 180 °C for 500 s to produce the desired β-aminoketones in moderate to good yields. Importantly, for several examples, the

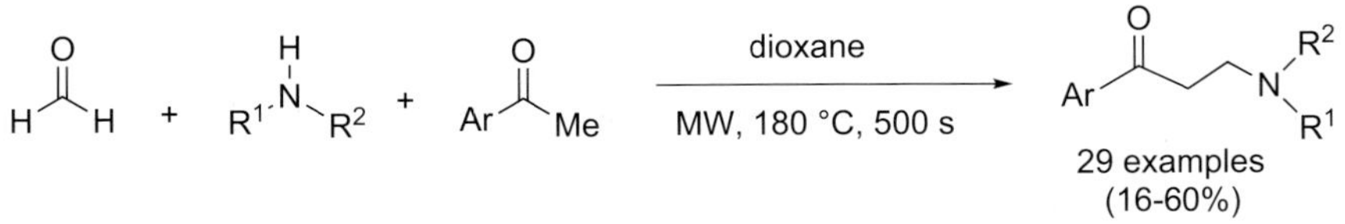

Scheme 6.83 Mannich reactions.

reaction was not only performed on a 2 mmol scale using a single-mode microwave reactor but was also run on a 40 mmol scale using a dedicated multimode instrument.

Leadbeater *et al.* have reported a different type of Mannich reaction that involved condensation of an aldehyde (1.5 equiv) with a secondary amine and a terminal acetylene in the presence of copper(I) chloride (10 mol%) to activate the terminal acetylene (Scheme 6.84) [177]. Optimum yields of the target propargylamines were obtained by microwave irradiation of the three building blocks with the catalyst in dioxane, doped with 1-(2-propyl)-3-methylimidazolium hexafluorophosphate ($pmimPF_6$) at 150 °C for 6–10 min. Tu and coworkers subsequently reported a diastereoselective modification of this method employing chiral amines such as (*S*)-proline methyl ester and changing the solvent to microwave-heated water [178]. A further variation of this three-component coupling was described by Varma and coworkers utilizing a solvent-free approach [179]. More recently, the group of Van der Eycken also reported on this A^3-coupling employing CuBr, toluene as solvent, and microwave heating at 100 °C for 25 min [180].

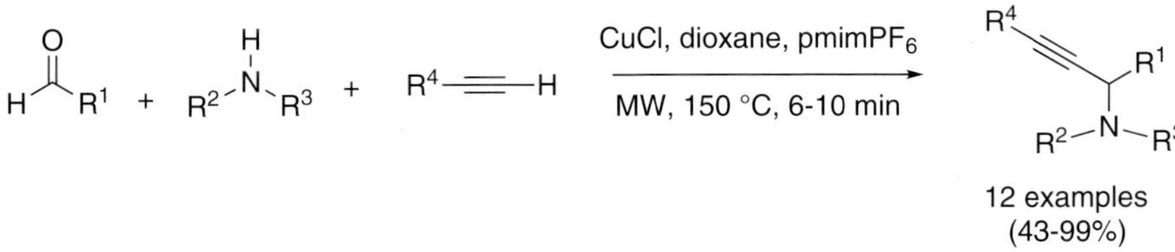

Scheme 6.84 Propargylamines via Mannich-type condensations.

A related work by the group of Van der Eycken extended the above-established method for the preparation of propargylamines toward the diversity-oriented synthesis of 3-benzazepines [181]. The target compounds were prepared in a two-step sequence, starting with the A^3-coupling providing the propargylamines, followed by a regio- and stereoselective intramolecular acetylene hydroarylation. Reoptimization of the A^3-coupling revealed slightly modified conditions, since a neat mixture of the amine, aldehyde, and alkyne in the presence of 15 mol% of Cu(I) iodide as the base heated in the microwave reactor at 90 °C for 30 min gave the best yields (Scheme 6.85). With similar reaction conditions (100 °C for 25 min), the corresponding propargylamines can also be generated via a tandem *anti*-Markovnikov hydroamination and alkyne addition, where two molecules of the alkyne are coupled to the amine. With these two routes, a small library of 12 different propargylamines has

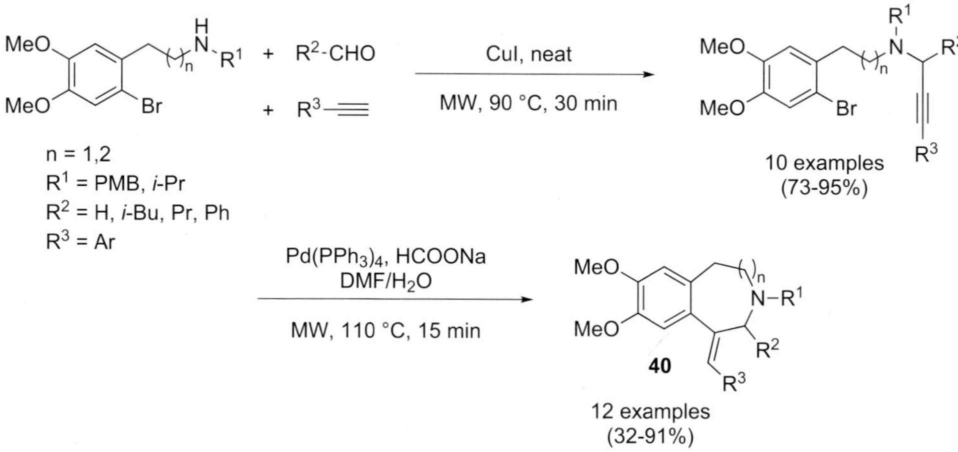

Scheme 6.85 Synthesis of 3-benzazepines from secondary propargylamines.

been prepared. The intermediates were further cyclized to corresponding 3-benzazepines **40** in a Pd-catalyzed process using sodium formate as reducing agent (Scheme 6.85). However, product compounds with eight-membered rings were obtained in only moderate yields.

A joint work by the group of Van der Eycken and Kappe dealt with diversity-oriented synthesis of dibenzoazocines and dibenzoazepines. Under microwave conditions, an intramolecular A^3-coupling reaction afforded the desired apogalanthamine analogs in generally high yield [182]. The presented approach allows the introduction of five points of diversity in this attractive scaffold. The required intermediate biaryl moiety was initially obtained via classical microwave-mediated Suzuki–Miyaura cross-coupling of suitable amines and electron-poor *o*-formyl arylboronic acids. These substrates were then reacted according to the previous optimized microwave method for A^3-couplings (Scheme 6.86). Interestingly, the products with a seven-membered ring were obtained as single diastereomers, whereas the eight-membered rings attached to a phenyl ring appeared as interconvertible 1 : 1 diastereomeric mixtures; when a thiophene ring is present, a diastereomeric ratio of 20 : 80 was observed.

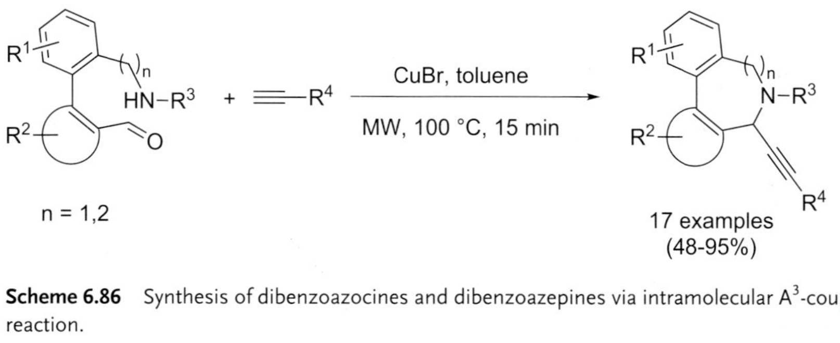

Scheme 6.86 Synthesis of dibenzoazocines and dibenzoazepines via intramolecular A^3-coupling reaction.

Another related work by the group of Van der Eycken described a further development of the above-described A^3-coupling (Scheme 6.87) [183]. Utilizing a ketone instead of the aldehyde in the so-called KA^2-coupling under solvent-free microwave conditions resulted in structurally attractive propargylamines. The convenient method overcomes the drawback that primary amines are difficult substrates in the more common A^3-couplings, which up to now limited the access to secondary propargylamines. A wide scope of substrates can be applied successfully under these conditions, although aliphatic alkynes have to be used in twofold excess and still provide only moderate yields. In general, sterical hindrance, especially on the ketone moiety, played an important role in terms of reaction outcome, since not only low yields were observed but also a mixture of diastereomers was isolated in that case.

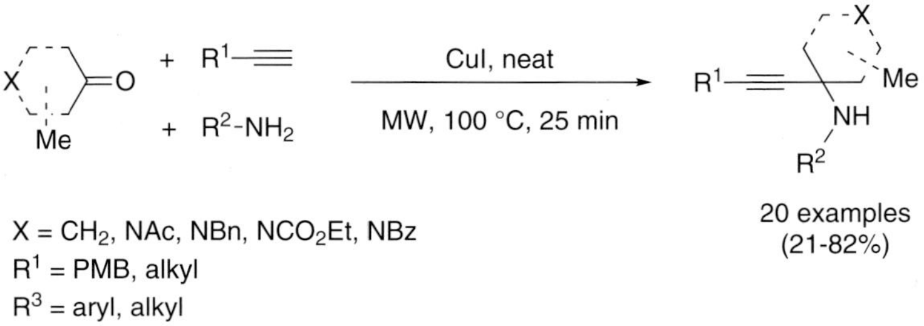

Scheme 6.87 Synthesis of secondary propargylamines by microwave-mediated KA^2-coupling.

Related to the Mannich three-component reaction is the Petasis or boronic-Mannich reaction, which involves the reaction between an aldehyde, an amine, and a boronic acid.

A high-speed microwave approach for the Petasis multicomponent reaction was published by Tye and coworkers [184]. Using a design of experiments (DoE) approach for reaction optimization, the best conditions for this transformation involved microwave heating of the components in DCM (1 M concentration) at 120 °C for 10 min (Scheme 6.88). These conditions were successfully applied to a range of Petasis reactions employing either glyoxalin acid or salicylaldehyde as the carbonyl component along with a number of aryl/heteroaryl boronic acids and amine components. The crude products were transformed to the corresponding methyl esters in order to simplify purification.

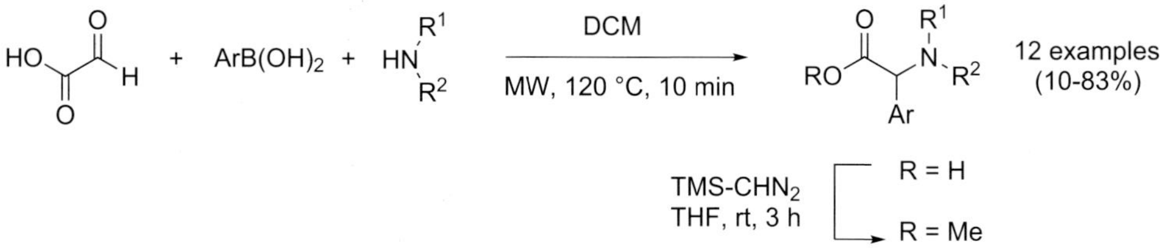

Scheme 6.88 The Petasis (boronic-Mannich) reaction.

Similar reaction conditions were described in a 2005 publication of Follmann *et al.* [185]. A wide range of electron-poor anilines and heterocyclic anilines, but only electron-rich boronic acids, can be employed to obtain the unnatural *N*-aryl-α-amino acids in reasonable yields.

Lamaty and coworkers presented a microwave-mediated solvent-free procedure for the Petasis reaction [186]. A variety of compounds are tolerated as coupling partners and in all cases high conversions were obtained with exclusive formation of the expected product. Although screening of reaction conditions determined a reaction temperature of 150 °C for 45 min as the optimum, the authors decided to apply a lower temperature (120 °C for 2 h) being more compatible with most organic structures (Scheme 6.89). Not using any solvent in this borono-Mannich type reaction simplified the workup of the final product, since immediate extraction with ethyl acetate and aqueous sodium hydroxide afforded the desired compounds in mostly high yields.

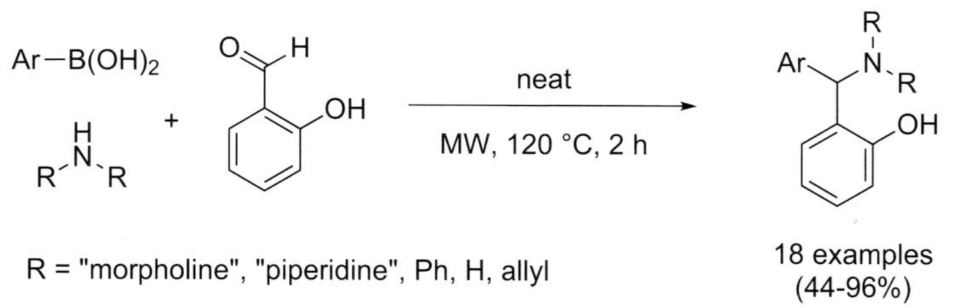

Scheme 6.89 Solvent-free microwave-assisted Petasis multicomponent reaction.

Another frequently used multicomponent reaction is the Kindler thioamide synthesis (the condensation of aldehyde, amine, and sulfur). Kappe and coworkers have described a microwave-assisted protocol that utilized a diverse selection of 13 aldehyde and 12 amine precursors in the construction of a representative 34-member library of substituted thioamides (Scheme 6.90) [187]. The three-component condensations of aldehydes, amines, and elemental sulfur were carried out using NMP as solvent, employing microwave flash heating at 110–180 °C for 2–20 min. A simple workup protocol allowed the isolation of synthetically valuable primary, secondary, and tertiary thioamide building blocks in 83% average yield and with >90% purity.

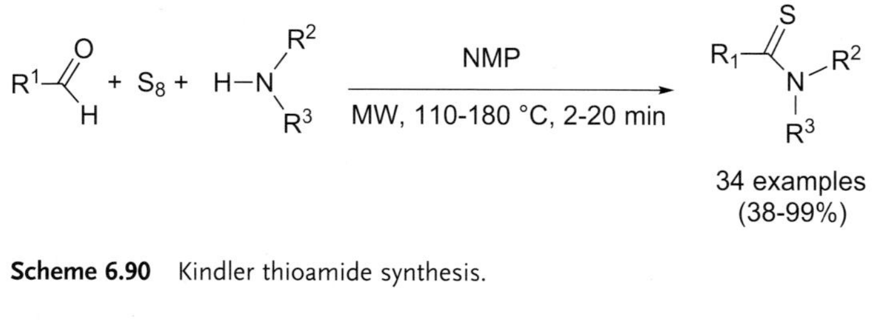

Scheme 6.90 Kindler thioamide synthesis.

The group of Beller has reported on the three-component reaction of amides, aldehydes, and dienophiles (AAD) to give 2-bromo-substituted 1-amido-2-cyclohexene derivatives **41** (Scheme 6.91) [188]. With this protocol, aromatic amides, sulfonamides, carbamates, and cyclic amides are well tolerated, the latter providing the products only in moderate yields (35–46%). Up to four stereogenic centers are generated in the AAD sequence and, in most cases, an all-*syn*-configuration is obtained. By applying α-bromo aldehydes, further scaffold decorations like cross-coupling or carbonylation reactions are possible at the 2-position.

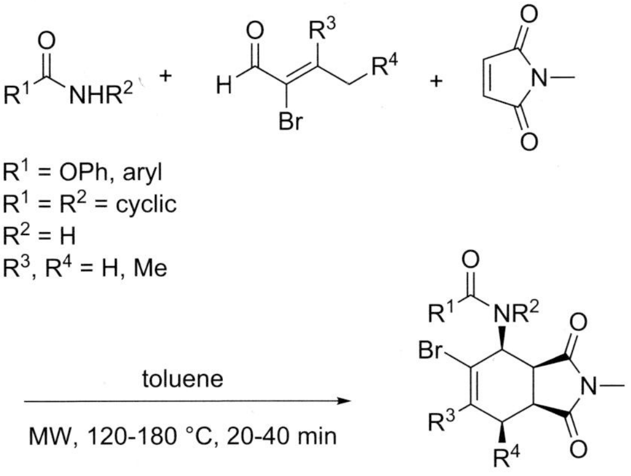

41
9 examples
(35-85%)

Scheme 6.91 Three-component synthesis of 1-amido-2-cyclohexenes.

A related and more general protocol for the AAD reaction was developed by the same group (Scheme 6.92) [189]. Optimization studies revealed that a solventless microwave protocol (150 °C, 20 min) gave a superior yield (90% versus 61%) for the reaction of acetamide, crotonaldehyde, and *N*-methyl-maleimide in a 50 times shorter reaction time compared to thermal heating at 110 °C. The addition of acetic anhydride (Ac_2O) as water-removing reagent proved to be beneficial for the reaction.

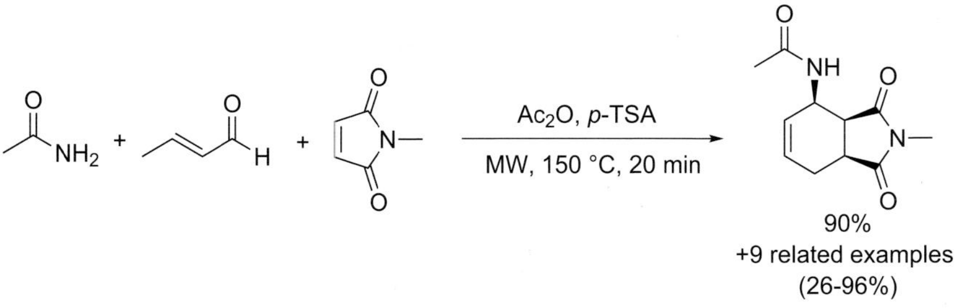

Scheme 6.92 Solvent-free synthesis of functionalized 1-amido-2-cyclohexenes.

All 10 cyclohexene derivatives are obtained as only one diastereomer, namely, as the all-*syn* product, due to the selective *endo*-addition of the dienophile.

A novel one-step three-component reaction for the synthesis of 3-aminoimidazo[1,2-*a*]pyridines has been demonstrated by Hulme and coworkers (Scheme 6.93) [190]. In this multicomponent reaction, an aminopyridine, aldehyde, and TMS-CN are reacted under microwave heating to give the desired products **42**. By using TMS-CN as nonclassical isonitrile equivalent, no further deprotection steps were necessary, making this method applicable for library synthesis.

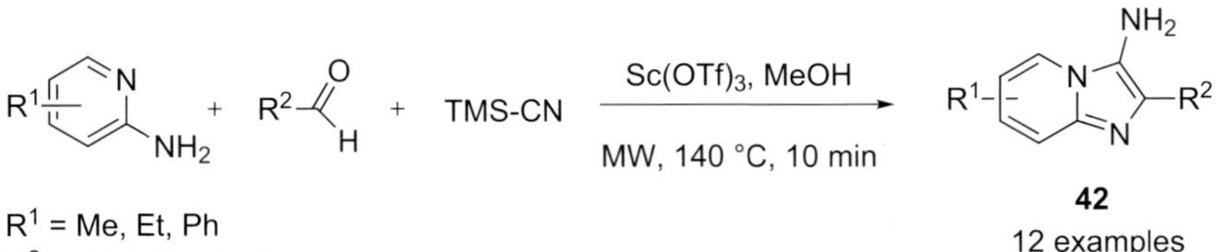

42
12 examples
(27-73%)

Scheme 6.93 Synthesis of 3-aminoimidazo[1,2-*a*]pyridines.

A closely related approach combining the Ugi-type chemistry shown in Scheme 6.93 with a Strecker synthesis leading to α-iminonitriles was disclosed by the same group in 2006 [191].

Using a related approach, DiMauro and Kennedy have developed a microwave protocol for the synthesis of a focused library of 3-amino-imidazopyridines **44** (Scheme 6.94) [192]. Starting from 2-aminopyridine-5-boronic acid pinacol ester, the imidazopyridine products **44** were obtained via an Ugi-type cyclization-leading to intermediate **43**-Suzuki coupling sequence in a one-pot procedure. Notably, the authors were not successful in performing the reaction sequence in one step nor can be the sequence reversed (first Suzuki then cyclization).

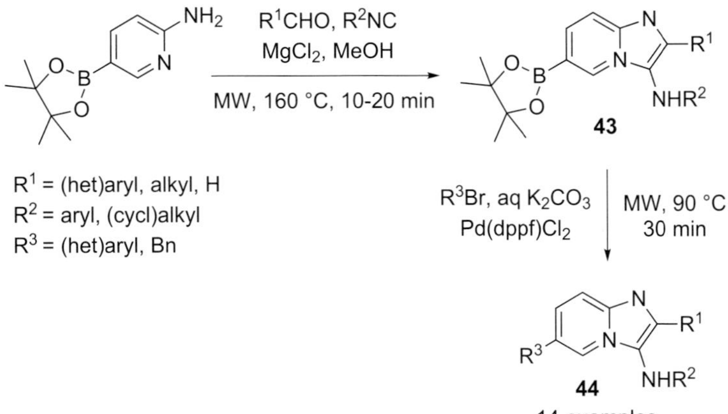

44
14 examples
(42-68%)

Scheme 6.94 One-pot cyclization/Suzuki coupling.

A 144-member library of furocoumarins via three-component reaction of 4-hydroxycoumarins, isocyanides, and arylaldehydes was prepared by Wu (Scheme 6.95) [193]. By applying a microwave protocol (DMF, 150 °C, 5 min), reduction in reaction times (24 h → 5 min), higher yields, and less by-products were observed compared to the conventional synthesis.

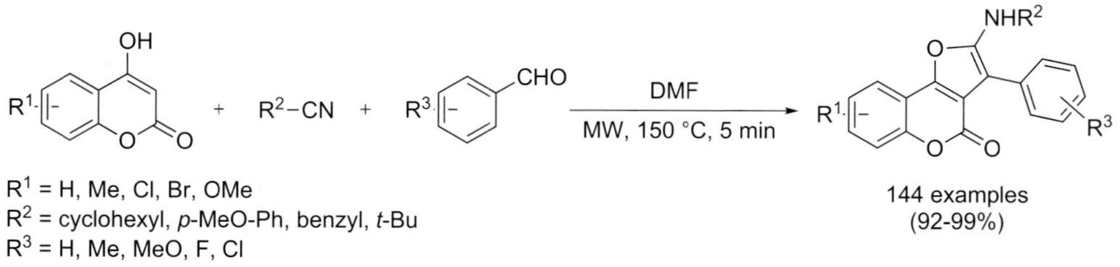

Scheme 6.95 Synthesis of furocoumarin libraries.

The synthesis of highly functionalized dibenz[*b,f*][1,4]oxazepin-11(10*H*)-ones **45** and dibenz[*b,f*][1,4]oxazepin-11(10*H*)-carboxamides **46**, respectively, has been disclosed by Dai and coworkers (Scheme 6.96) [194]. The oxazepine derivatives **45** and **46** were obtained via a one-pot/two-sequence Ugi reaction and intramolecular

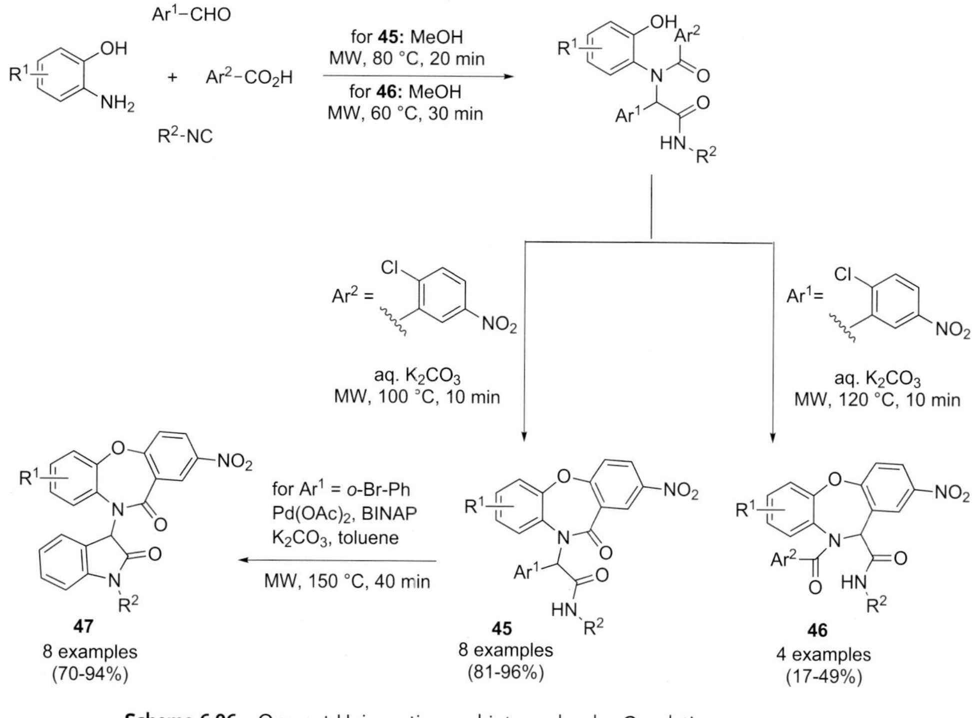

Scheme 6.96 One-pot Ugi reaction and intramolecular *O*-arylation.

O-arylation in good to excellent yields. It turned out that for the Ugi reaction in the synthesis of derivatives **46**, the acidity of the employed acid component played an important role, resulting in a 49% overall yield for $Ar^2 = o$-Br-Ph with a pK_a of 2.85 and decreased yields for less stronger acids. In addition, a postmodification of derivatives **45** was conducted via a Pd-catalyzed intramolecular *N*-arylation, which furnished derivatives **47** containing a 2-oxindole unit.

El Kaïm and coworkers have reported on the generation of quinoxaline scaffolds (Scheme 6.97) [195]. The reaction sequence consists of an Ugi four-component reaction with subsequent Smiles rearrangement leading to intermediate **48**. The Ugi–Smiles intermediate **48** further cyclizes to the quinoxaline product via an intramolecular Ullmann-type coupling. A higher yield of 62% could be obtained when the Ugi–Smiles intermediate is isolated prior to the Ullmann-type coupling step. By applying microwave heating for both steps, the reaction time can be reduced from 31 to only 1.8 h.

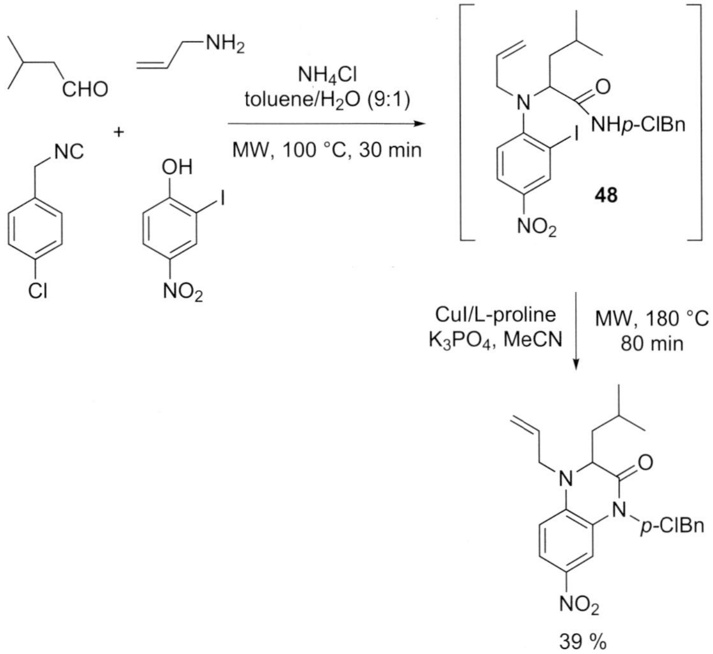

Scheme 6.97 Quinoxaline synthesis via an Ugi–Smiles sequence.

Tryptophan-derived diketopiperazines with variable side chains have been obtained by Ugi four-component reaction followed by a deprotection/cyclization sequence to generate the diketopiperazine fragment [196].

Hulme *et al.* described the synthesis of triazadibenzoazulenones [197]. The two-step microwave protocol starts with a common Ugi four-component coupling followed by a trifluoroacetic acid-mediated cyclization/deprotection sequence representing the first example of a postcondensation Ugi modification employing internal nucleophiles (Scheme 6.98). Microwave heating to 130 °C accomplished the double deprotection and tandem cyclization in only 20 min and the desired triazabenzoazulenes are obtained in acceptable overall yields. Although most substrates performed well in both steps, the use of *N*-methylated acid proved to be unsuccessful since the expected tetracyclic product was obtained only in traces.

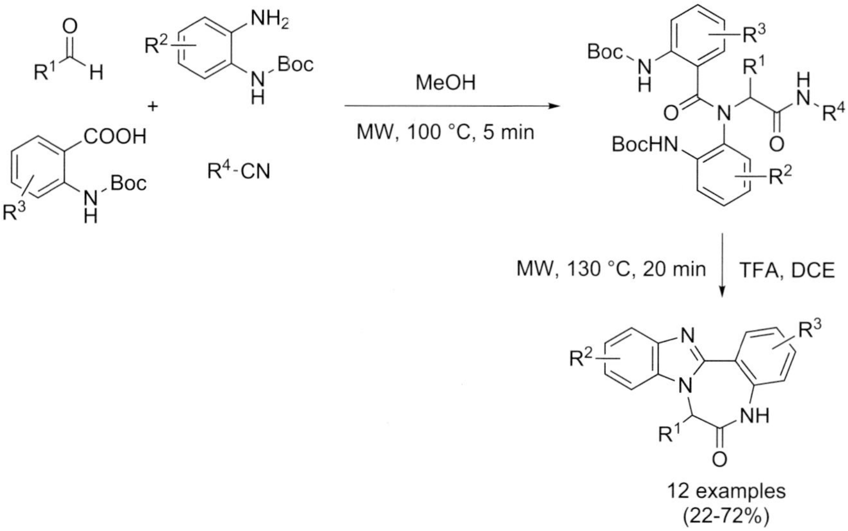

Scheme 6.98 Two-step preparation of triazadibenzoazulenones.

The group of Andreana has developed a one-pot two-step synthesis of regiochemically differentiated 1,4-benzodiazepin-3-ones (Scheme 6.99) [198]. The first step involves the Ugi four-component coupling reaction to form α-acylaminoamides that are subsequently converted to benzodiazepines by a reductive ($NO_2 \rightarrow NH_2$) aza-Michael cyclization employing Fe(0)/NH_4Cl in aqueous media as reducing agent. With bifunctional *o*-nitrobenzaldehyde **49**, exclusively C2, N4, C5-substituted benzodiazepines **50** are obtained; whereas with *o*-nitrobenzylamine **51**, the C2, N4-substituted derivatives **52** are formed. The reduction reaction gave higher yields in shorter times compared to reactions performed under oil bath heating. It has to be noted that at higher temperatures either decomposition or a 6-*exo*-aza-Michael cyclization to form 2,5-diketopiperazines occurred.

The same group has demonstrated that small-molecule diversity can be easily obtained in a single step starting from standard Ugi multicomponent reaction building blocks (Scheme 6.100) [199]. Depending on the solvent and sterics of the substituents, in particular of the amine and isocyanide precursors, three different scaffolds can be generated. When water is employed as protic solvent, either 2,5-diketopiperazines **53** via an aza-Michael reaction or 2-azaspiro[4.5]deca-6,9-diene-

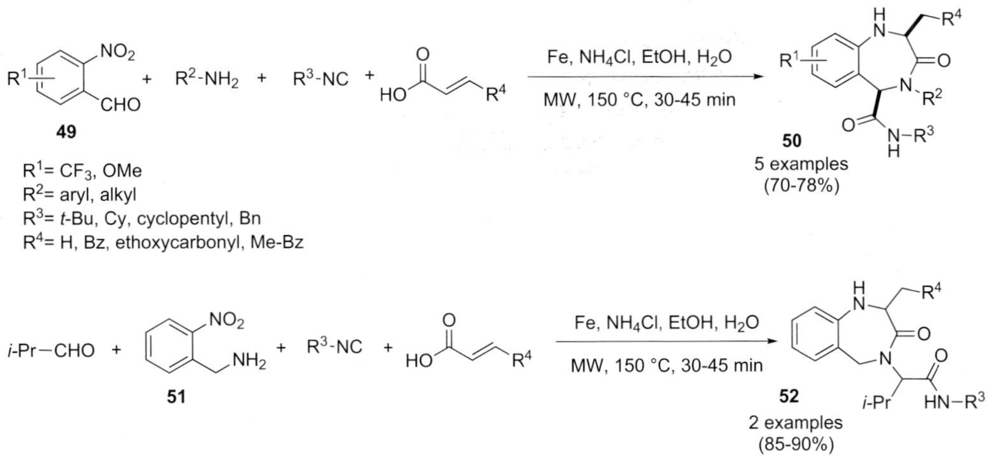

Scheme 6.99 One-pot synthesis of 1,4-benzodiazepin-3-ones.

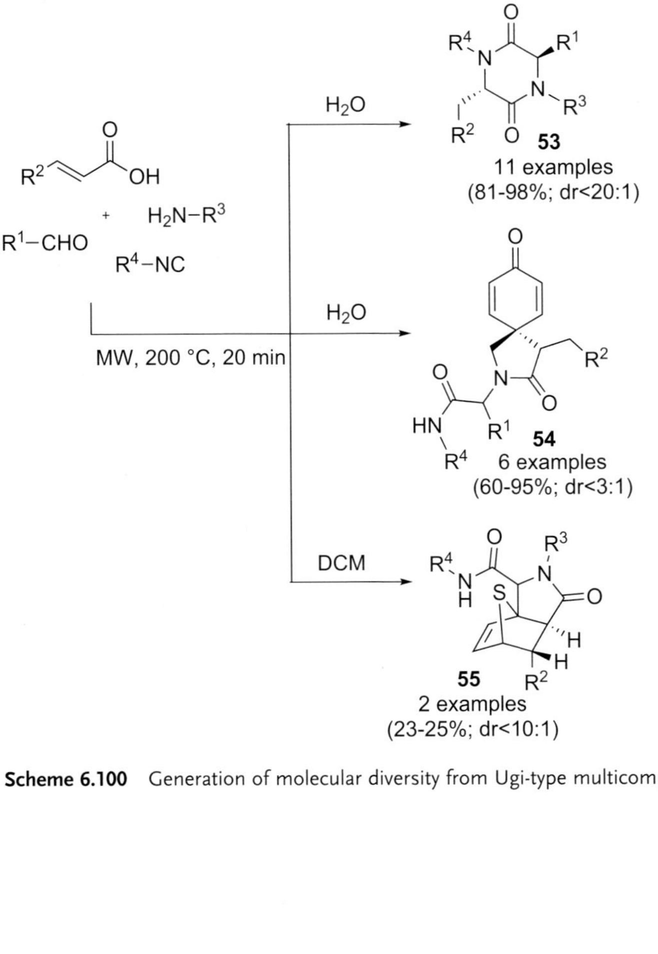

Scheme 6.100 Generation of molecular diversity from Ugi-type multicomponent reactions.

3,8-diones **54** from a 5-*exo*-Michael cyclization are obtained due to stabilization of zwitterionic intermediates by the protic solvent. When the reactions are performed in aprotic DCM, only the acyclic Ugi products, that are formed in the first reaction step prior to intramolecular cyclizations, could be isolated. If bulky R^4-substituents (*t*Bu, *i*Pr, and cyclic alkyl) are present, either product **54** or the acyclic Ugi products are accessible, depending on the amine substitution pattern. The tricyclic lactam product **55** was obtained via an intramolecular thiophene-derived Diels–Alder reaction when thiophene-2-carboxaldehyde in DCM was used. Interestingly, direct conversion to products **53–55** could be achieved only by microwave irradiation and not with conventional heating.

The group of Grimaud investigated the use of ammonia in Ugi-type couplings under microwave irradiation in order to establish a facile method to introduce a free NH functionality that is of considerable pharmacological interest in heterocyclic drug compounds [200]. The aim was to extend the scope of the earlier disclosed Ugi–Smiles coupling, utilizing electron-deficient phenols as acid surrogate, by the use of ammonia as the amine source. Under microwave irradiation, the use of aqueous ammonia (30% aqueous solution) led to satisfactory results tolerating various nitriles and nitrophenol derivatives albeit only aliphatic aldehydes can be used (Scheme 6.101a). More interestingly, the developed method could be effectively applied toward hydroxypyridines and hydroxypyrimidines to obtain the corresponding NH-heteroaryl compounds. Furthermore, in the related Ugi–Mumm coupling, where carboxylic acids were employed instead of the phenol derivatives, the method proved successful (Scheme 6.101b).

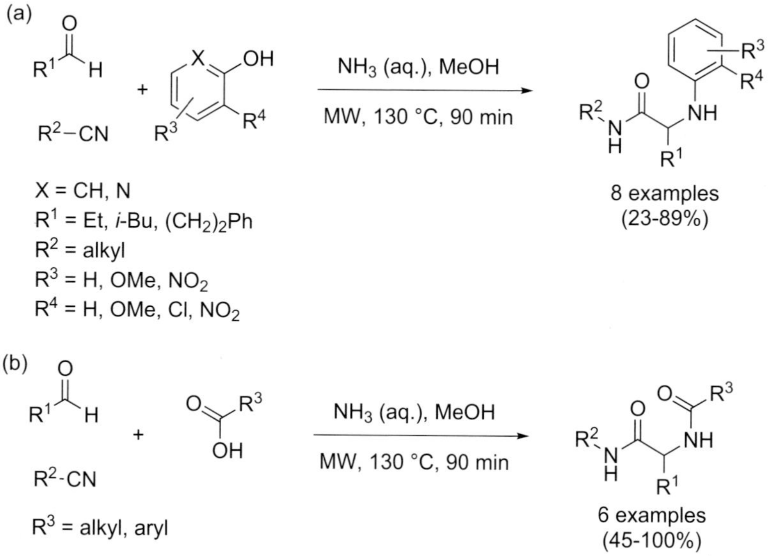

Scheme 6.101 Ugi couplings involving aqueous ammonia.

Miller and coworkers reported a one-pot protocol for the preparation of thiazolidin-4-ones by condensation of aromatic aldehydes, amines, and mercaptoacetic acid in ethanol (Scheme 6.102) [201]. The optimized procedure involved microwave irradiation of a mixture of amine hydrochloride, aldehyde, and mercaptoacetic acid (molar ratio 1 : 2 : 3) in the presence of 1.25 equiv of *N,N*-diisopropylethylamine (DIEA) base in ethanol at 120 °C for 30 min at atmospheric pressure.

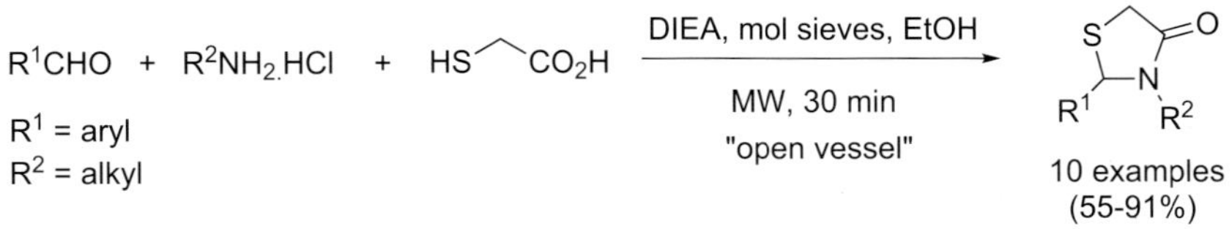

Scheme 6.102 4-Thiazolidinones via multicomponent chemistry.

Risitano *et al.* have reported on a synthesis of bridgehead aziridines in high yields and with excellent diastereocontrol by a three-component reaction of aldehyde, phenacyl chloride, and ammonium acetate in acetic acid (Scheme 6.103) [202]. As opposed to 2–3 h of traditional conductive heating, 5–10 min of microwave irradiation at 90 °C afforded the stereodefined bridgehead aziridines in 35–92% isolated yields.

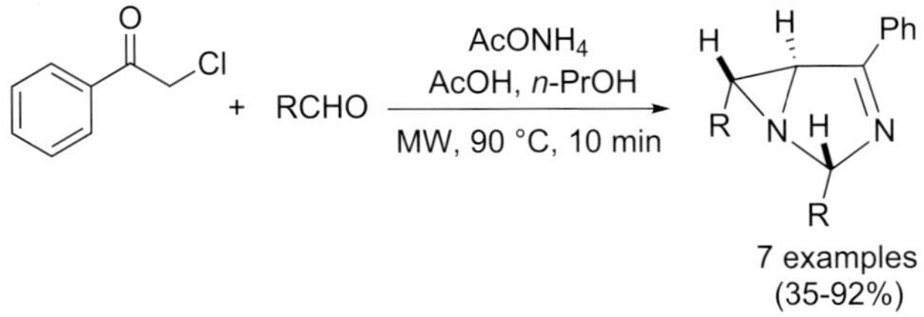

Scheme 6.103 Multicomponent synthesis of 1,3-diazabicyclo[3.1.0]hex-3-enes.

In 2009, Botta and coworkers reported on a one-pot multicomponent procedure to prepare 2,3-dihydropyrans [203]. The developed method can be explored with corresponding alkyne derivatives to generate appropriate carbohydrate intermediates on the route to direct linked *C-C*-1,3-disaccharides, which are biologically interesting scaffolds and important building blocks in carbohydrate chemistry. Alkyne, ethyl glyoxalate, and ethylvinyl ether are reacted together with 10 mol% of Grubbs catalyst II at 80 °C for two cycles of 10 min each. Extraction workup furnished the corresponding 2,3-dihydropyrans in an unexpected 2 : 1 *trans/cis* ratio in satisfactory yields (Scheme 6.104). The formation of the *trans*-isomer as major product indicates an unusual *exo*-attack in the occurring hetero Diels–Alder reaction. Common treatment with zinc(II) chloride in DCM affords quantitative conversion to the *trans*-isomer as sole product.

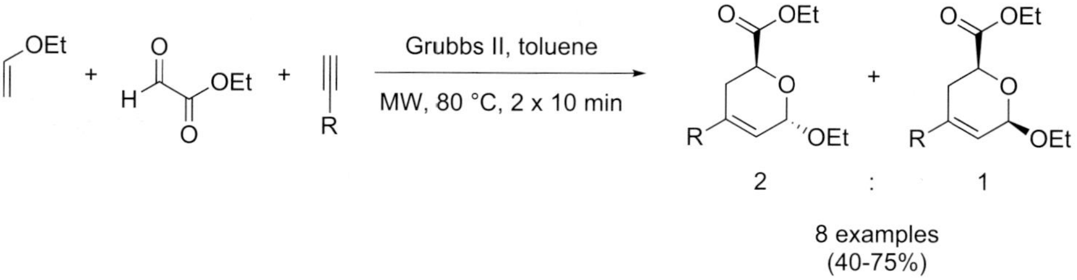

Scheme 6.104 One-pot synthesis of 2,3-dihydropyrans.

Tu *et al.* have reported on the one-pot four-component reaction for the synthesis of tetrahydroquinoline diones (Scheme 6.105) [204]. By reacting aromatic aldehydes, dimedone (**56**), Meldrum's acid (**57**), and amines under microwave conditions, a new series of *N*-functionalized quinolone products were obtained. Higher yields and shorter reaction times compared to conventional heating could be accomplished.

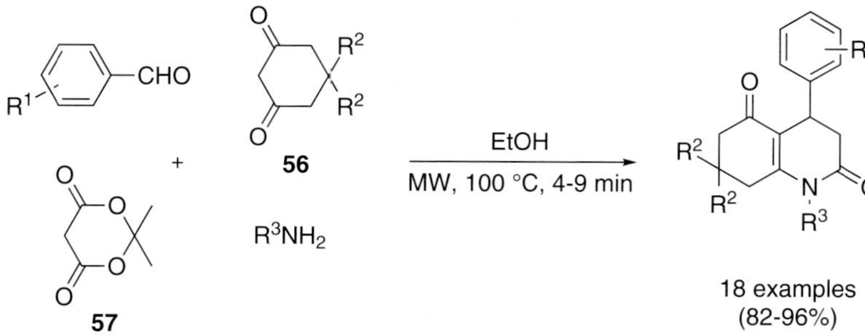

R^1 = H, 4-Cl, 2-Cl, 4-OMe, 4-F, 4-Br, 3-NO_2,
2,4-$(Cl)_2$, 4-Me, 3,4-$(MeO)_2$, hetaryl
R^2 = Me, H
R^3 = Me, cyclopropyl, cyclcpentyl, cyclohexyl, 4-MePh

Scheme 6.105 Synthesis of *N*-substituted tetrahydroquinoline diones.

In 2010, the same group reported on an unexpected route to novel azapodophyllotoxin derivatives via microwave-assisted multicomponent reaction employing aromatic aldehydes, tetronic acid (**58**), and dimedone (**56**) [205]. When performing the four-component reaction in aqueous ammonia under microwave irradiation, ^{1}H-NMR spectra revealed that an unexpected hydroxyl derivative **59** was formed rather than the desired unsaturated products. Interestingly, when aldehydes with electron-donating groups were employed, another unexpected transformation to products **60** occurred indicating that tetronic acid did not participate in the reaction process at all (Scheme 6.106). Although not the initially desired compounds have been formed, the presented method nevertheless provides rapid access to prominent heterocyclic scaffolds with biological significance.

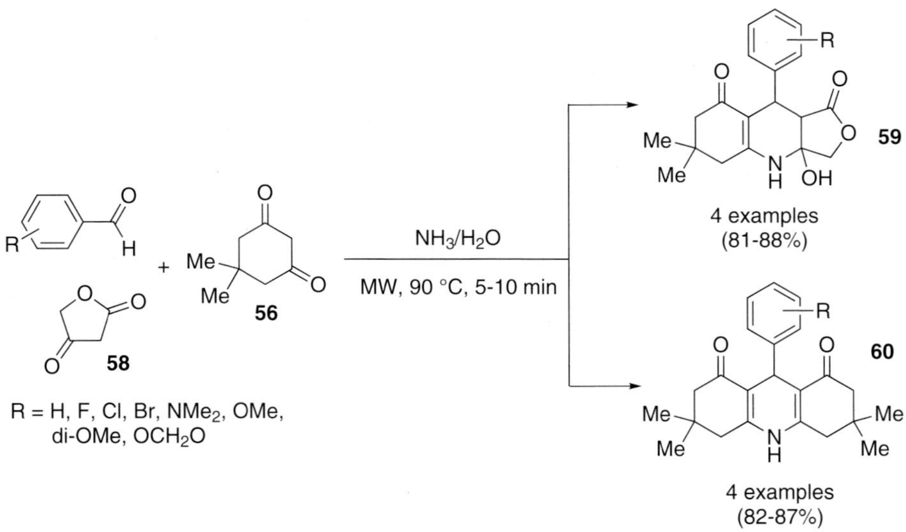

Scheme 6.106 Synthesis of azapodophyllotoxin derivatives.

The same group has described a number of related MCR approaches to heterocycles, involving Hantzsch-type condensations leading to dihydropyridin-2(1*H*9-ones [206], indeno[1,2-*b*]quinolines [207, 208], furo[3,4-*b*]quinolines [209], and pyrimido[1,2-*a*]quinolines [210]. The generation of 4-azafluorenone derivatives [211], furo[3′,4′:5,6]pyrido[2,3-*d*]pyrimidines [212], and aminopyridines [213] has also been disclosed by the same group using MCR strategies.

Tu and coworkers presented a four-component domino reaction for the synthesis of multifunctionalized quinazolines, which are important scaffolds for drug design [214]. The developed method employs aromatic aldehydes, cycloketones, and cyano amides and involves tandem formation of two different Knoevenagel intermediates, which in further sequences generate highly functionalized pyrido[3,4-*i*] quinazoline derivatives **61** (Scheme 6.107). Preliminary investigations revealed ethylene glycol as the best suited solvent and potassium carbonate as most effective base, acting as dehydrating agent and heterogeneous activator for the initial Knoevenagel condensation/[4 + 2] cycloaddition/intramolecular Michael addition/ nucleophilic reaction sequence. All building block combinations achieved full conversions within 12–24 min at 120 °C and furnished the desired compounds in high yields after simple workup. Importantly, all potential stereocenters can be well controlled in this multicomponent one-pot procedure, which makes it even more important for synthesis strategies.

In a subsequent study, the same authors applied this one-pot four-component domino reaction to prepare multifunctionalized tricyclo[6.2.2.0^{1,6}]dodecanes **62** (Scheme 6.107) [215]. Contrary to their findings of the above-mentioned examples, the application of aliphatic aldehydes directly leads to the tricyclic dodecanes instead of multifunctionalized quinazolines. Cs_2CO_3 proved to be more effective than K_2CO_3

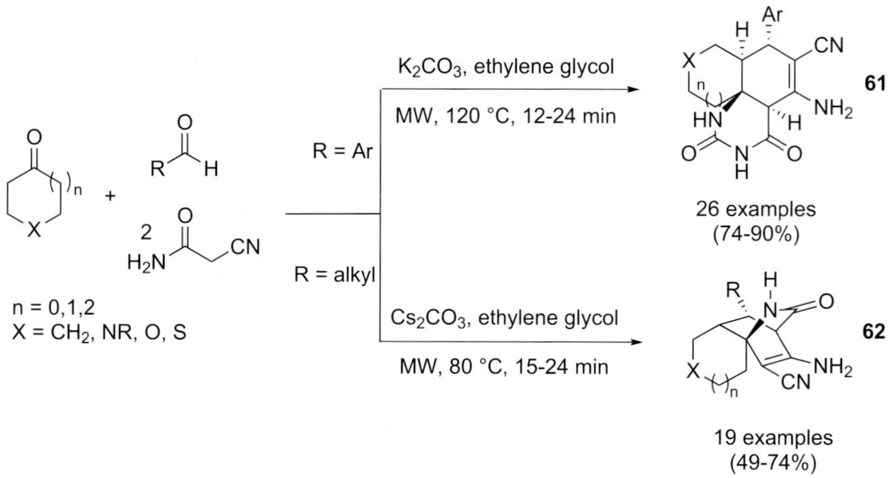

Scheme 6.107 Four-component domino synthesis of quinazolines and tricyclic dodecanes.

by acting as the base for the dehydration and decarboxylation step in this domino process.

Bremner and Organ have employed microwave-assisted continuous flow organic synthesis for the preparation of heterocyclic compounds via multicomponent reactions utilizing a self-designed capillary system as reactor (Scheme 6.108) [216]. Each of the three components are introduced into the microwave cavity through three separate leads, in equal concentrations and at the same flow rate. In this way, a set of tetrahydropyrazolo[3,4-*b*]quinolin-5(6*H*)-ones **63** and aminofuranes **64**, respectively,

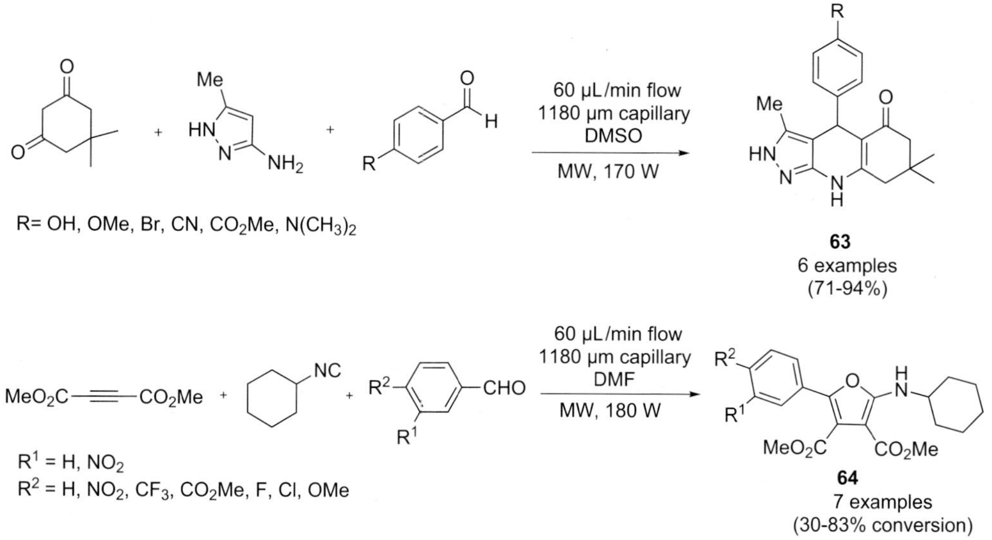

Scheme 6.108 Synthesis of heterocycles by multicomponent reactions.

could be accomplished in a few seconds and in very good conversions. Importantly, this flow system is an open system, that means, that high-boiling and good microwave-absorbing solvents like DMSO or DMF have to be used to reach higher temperatures and hence higher conversions.

The group of Coquerel presented an interesting approach toward the diastereoselective domino three-component synthesis of α-spiro-δ-lactams [217]. Under microwave irradiation, the Wolff rearrangement of cyclic 2-diazo-1,3-diketones in the presence of primary amines and α,β-unsaturated aldehydes directly lead to the corresponding spiro-compounds (Scheme 6.109). Numerous substrates can act as building blocks to successfully follow the amination/Wolff rearrangement/[2 + 4] cycloaddition sequence featuring an unprecedented reactivity of the *in situ* generated acylketenes as dienophiles in a 6π electrocyclic process. The synthesis can be performed either in a convenient one-pot multicomponent process or, for those substrates that suffer from undesired side reactions by the *in situ* generated water, in a step-wise two-pot reaction. Here, the corresponding intermediate 1-azadiene is generated from amine and aldehyde in 15 min under microwave irradiation and after removal of the volatiles reacts with the freshly added diazocompound in a second 15 min irradiation cycle to the final product in high yields.

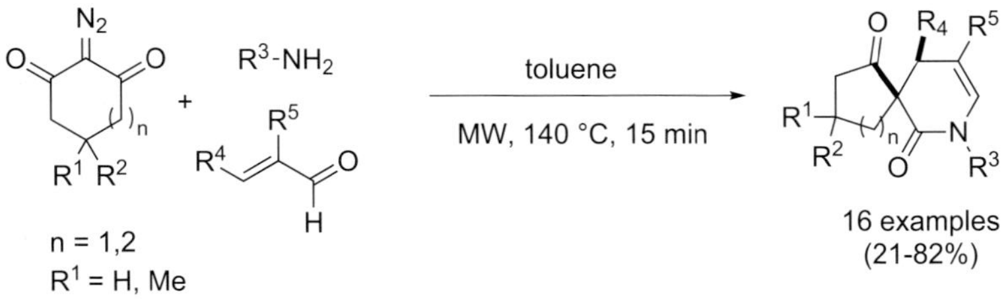

Scheme 6.109 Domino three-component synthesis of α-spiro-δ-lactams.

In a related earlier report of the same authors, the domino synthesis of oxazinones **65** and oxazinediones **66** via cyclic acylketenes is described [218]. Again, the microwave-mediated Wolff rearrangement initiates a three-component domino reaction involving primary amines and aryl as well as aliphatic aldehydes to generate the corresponding oxazinones (Scheme 6.110a). All compounds are obtained in good yields, no virtual preferences for aliphatic or aromatic aldehydes could be observed. In order to extend the scope, the developed method was also used in the synthesis of oxazinediones (Scheme 6.110b). Here, an isocyanate needs to be generated, which undergoes a hetero Diels–Alder reaction with the *in situ* formed acyl ketene. Gratifyingly, a 1 : 1 mixture of 2-diazo-5,5-dimethylcyclohexan-1,3-dione with simple acyl azides proved successful. Slight modifications of the protocol led to an initial microwave heating at 80 °C for 10 min to promote the Curtius rearrangement. A subsequent second irradiation cycle at 120 °C for another 5 min initiated the Wolff rearrangement and finally the Diels–Alder cycloaddition took place to furnish the target compounds in acceptable yields.

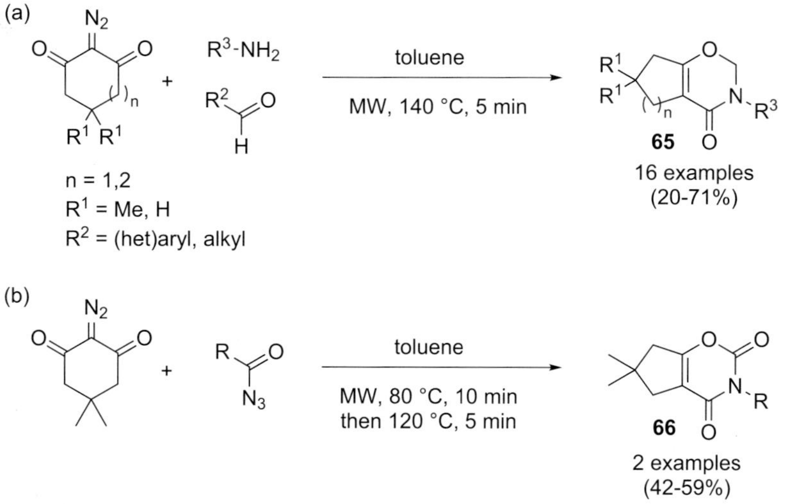

Scheme 6.110 Domino synthesis of oxazinones and oxazinediones.

Chebanov *et al.* have disclosed the microwave-assisted three-component reaction of 3-substituted 5-aminopyrazoles **67**, pyruvic acid or ethyl pyruvate (**68**), and aromatic aldehydes (Scheme 6.111) [219]. By changing the solvent from AcOH, which is used under conventional conditions, to an EtOH/HCl system, product isolation was facilitated and no further recrystallization was necessary. Compared to classical heating, an 8–12 fold decrease in reaction time was possible and higher yields could be obtained for the pyrazolopyridine carboxylic ethyl ester products **69** ($R^2 = Et$).

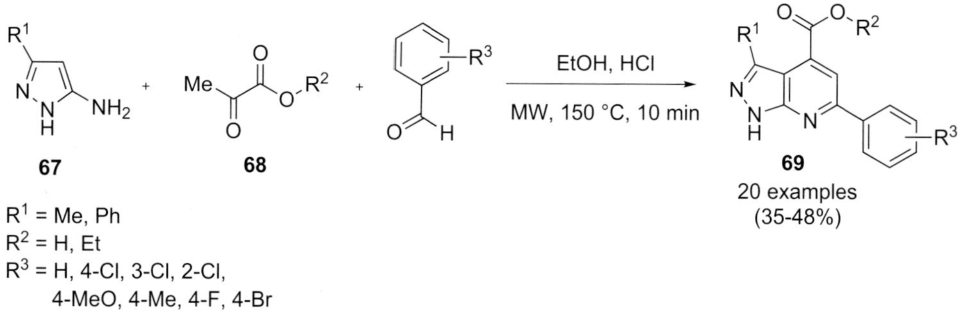

Scheme 6.111 Synthesis of pyrazolopyridines via a three-component approach.

The related three-component condensation reaction involving aminoazoles, aldehydes, and β-ketoamides leading to triazolopyrimidines has been described by the same group [220]. A series of polysubstituted (3′-indolyl)pyrazolo[3,4-*b*]pyridine and (3′-indolyl)benzo[*h*]quinoline derivatives were synthesized via one-pot multicomponent condensation of aldehydes, 3-cyanoacetyl indoles with 5-aminopyrazoles, or naphthylamines [221].

The three-component reaction of Meldrum's acid (**57**), aldehydes, and guanidine carbonate (**70**) for the synthesis of 2-amino-dihydropyrimidines was developed by Gorobets *et al.* (Scheme 6.112) [222]. The condensations were performed under both conventional (120–130 °C, 20–25 min) and microwave heating (150 °C, 5 min) showing slightly higher yields for most of the derivates applying conventional conditions. Importantly, to avoid high-pressure buildup in the closed microwave vials due to CO_2 formation, the reaction mixture was stirred for 10–15 min at room temperature before subjecting to microwave irradiation.

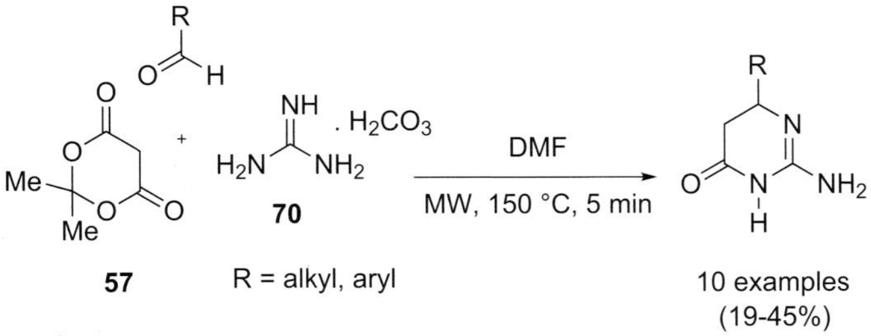

Scheme 6.112 Synthesis of 2-amino-dihydropyrimidin-4-ones.

Fang and Lam recently presented a rapid microwave method for the synthesis of 5-unsubstituted 3,4-dihydropyrimidin-2-ones [223]. The one-pot reaction involves oxaloacetic acid, aldehydes, and urea/thiourea derivatives to generate the potential biologically active heterocycles. After heating at 95 °C for 15 min, the desired dihydropyrimidinones were furnished in high yields (Scheme 6.113). Due to *in situ* decarboxylation after the cyclization step, the generated scaffold remains unsubstituted on the C-5 position. The short reaction time and the simple workup prove the entire procedure appropriate for library generation of the synthetically valuable compounds.

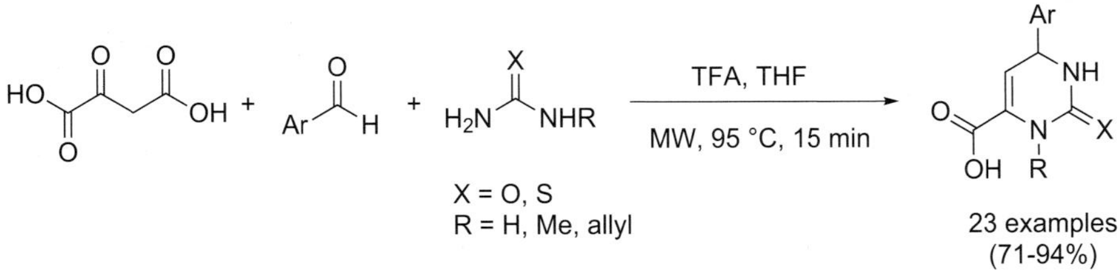

Scheme 6.113 Synthesis of 5-unsubstituted 3,4-dihydropyrimidinones.

In 2010, Van der Eycken and coworkers investigated the Biginelli-like three-component condensation utilizing 1,2,4-triazole and discovered an unexpected direction of the reaction when comparing microwave irradiation with thermal heating [224]. In the presence of an *ortho*-hydroxy functionality on the aromatic aldehyde, microwave heating for 30 min at 150 °C led to a novel polyheterocyclic scaffold **71** by forming an oxygen bridge (Scheme 6.114). In contrast, under classical thermal heating (40 °C for 16 h), a typical Biginelli-type product **72** is formed rather than the expected bicyclic pyrimidinone derivative. Various commercially available salicylic aldehyde derivatives give access to an array of functionalized polycyclic compounds.

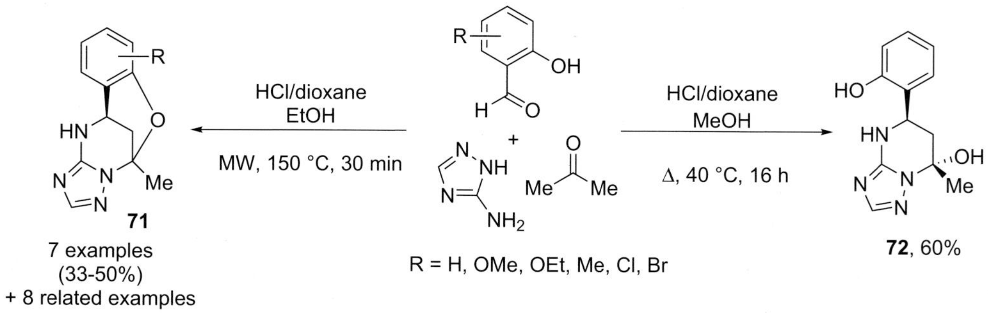

Scheme 6.114 Synthesis of oxygen-bridged heterocycles.

Yadav *et al.* have reported on a Biginelli-type reaction of unprotected aldoses, 2-methyl-2-phenyl-1,3-oxathiolan-5-one **73** as CH acidic building block, and urea/thiourea under acidic Montmorillonite K-10 catalysis and solvent-free conditions (Scheme 6.115) [225]. By employing the unprotected aldoses D-xylose ($n = 3$) and D-glucose ($n = 4$), a one-pot protocol was feasible avoiding protection and deprotection steps; furthermore, they induce asymmetry to the final products. Thiosugar-annulated

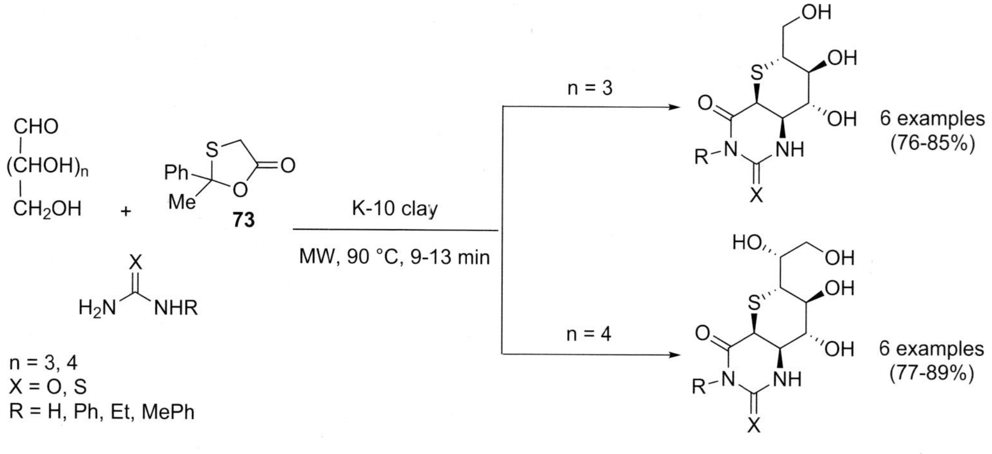

Scheme 6.115 Synthesis of thiosugar-annulated dihydropyrimidines.

dihydropyrimidines were obtained via intermolecular domino cyclocondensations with >95% diastereoselectivity in favor of the *cis*-isomer.

Yan *et al.* have developed a modified Kröhnke protocol for the synthesis of pyridines (Scheme 6.116) [226]. This typically two-step procedure could be performed in one pot by four-component cyclocondensation of a pyridinium salt, aromatic aldehydes, acetophenones, and ammonium acetate, leading to polysubstituted pyridines **74**. When acetophenone is replaced by cyclic ketones **75**, annulated pyridines **76** are obtained in very good yields. By applying microwave heating, less by-products are formed and only a simple filtration/recrystallization workup is required.

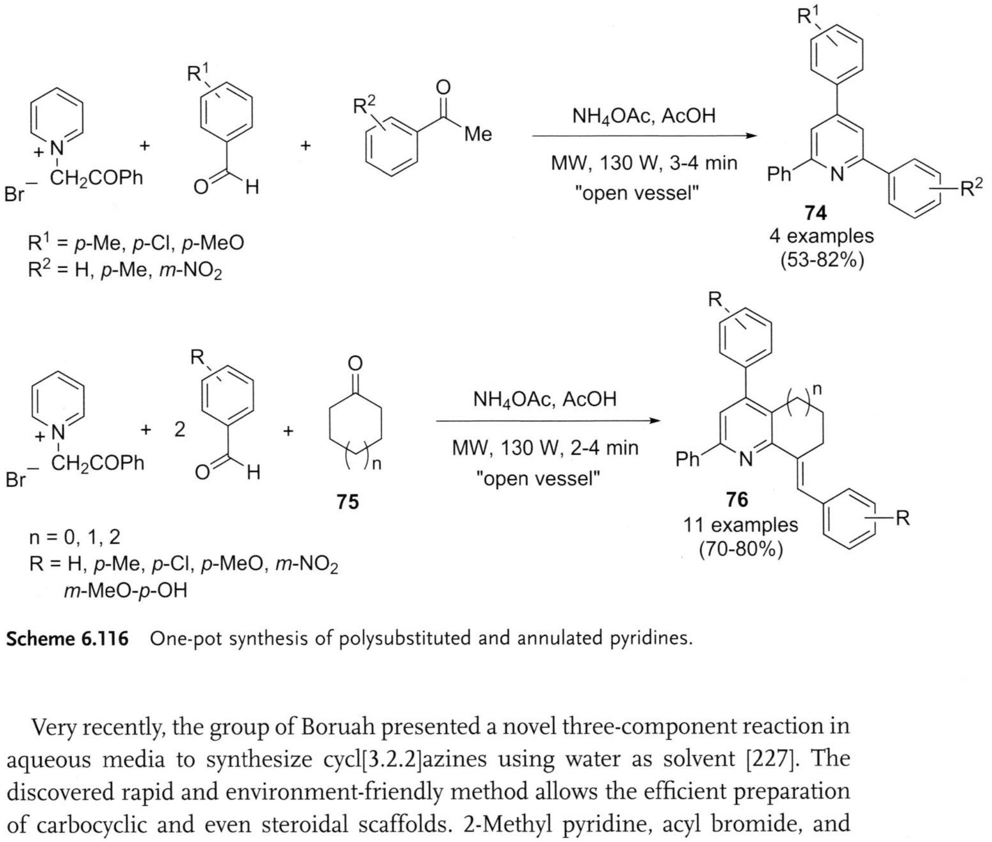

Scheme 6.116 One-pot synthesis of polysubstituted and annulated pyridines.

Very recently, the group of Boruah presented a novel three-component reaction in aqueous media to synthesize cycl[3.2.2]azines using water as solvent [227]. The discovered rapid and environment-friendly method allows the efficient preparation of carbocyclic and even steroidal scaffolds. 2-Methyl pyridine, acyl bromide, and alkyne are reacted together with potassium carbonate in water in an open vessel under microwave heating at 100 °C (Scheme 6.117). Within 2–5 min, the reaction was completed affording the desired compounds after extraction workup. Although high yields were generally obtained, the electron-deficient 4-nitrophenyl acyl bromide provided only poor yields (20–22%).

Several publications have reported on the use of terminal acetylenes as building blocks in MCRs leading to fused heterocycles [228–230]. Pyridine-3,5-dicarbonitrile

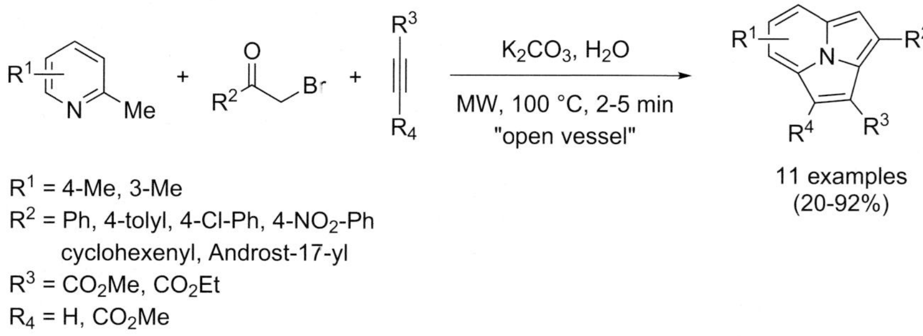

Scheme 6.117 Synthesis of cyc [3.2.2]azines in water.

libraries have been prepared in pseudo four-component condensations involving malononitrile as one of the building blocks [231]. Several other MCR strategies *en route* to *N*-heterocycles using malononitrile as building block have been reported [232, 233]. A three-component cascade reaction to yield 3-spirocyclopropanated β-lactames was reported by de Meijere and coworkers [234]. The three-component condensations of cyclic amino acid esters, carboxylic acids, and amines produced spiroimidazolinones in high yields [235]. The preparation of α-aminophosphonates using a three-component condensation approach has been described [236]. 2-Aminomorpholines have been obtained in a four-component approach starting from 1,2-amino alcohols, glyoxal, boronic acids, and amines [237]. For a microwave-assisted Gewald approach to 2-aminothiophenes, see Ref. [238]. One-pot condensation of 2-chlorobenzenthiols, chloroacetyl chloride, and primary amines leads to benzo[*b*][1,4]thiazin-3(4*H*)-ones [239].

6.10 Alkylation Reactions

A simple and atom economical synthesis of hydrogen halide salts of primary amines, directly from the corresponding halides, which avoids the production of significant amounts of secondary amine side products was described by researchers from Bristol-Myers Squibb [240]. Microwave irradiation of a variety of different alkyl halides or tosylates in a commercially available 7 M solution of ammonia in methanol at 100–130 °C for 15 min–2.5 h provided the corresponding primary amines directly as halide or sulfonate salts after evaporation of the solvent (Scheme 6.118a). All reactions gave 2–4% or less of the symmetrical secondary amine side product. A distinct advantage of using the high-temperature microwave approach (100 °C, 15 min) was evident for benzyl chlorides, where in some cases the corresponding reaction at room temperature (23 °C, 2 h) provided 20–25% of bis-alkylated secondary amine side product. The microwave method under otherwise identical conditions led to less than 4% of by-product.

Similar results were also observed by Ju and Varma in the synthesis of tertiary amines from the corresponding alkyl halides and primary or secondary amines

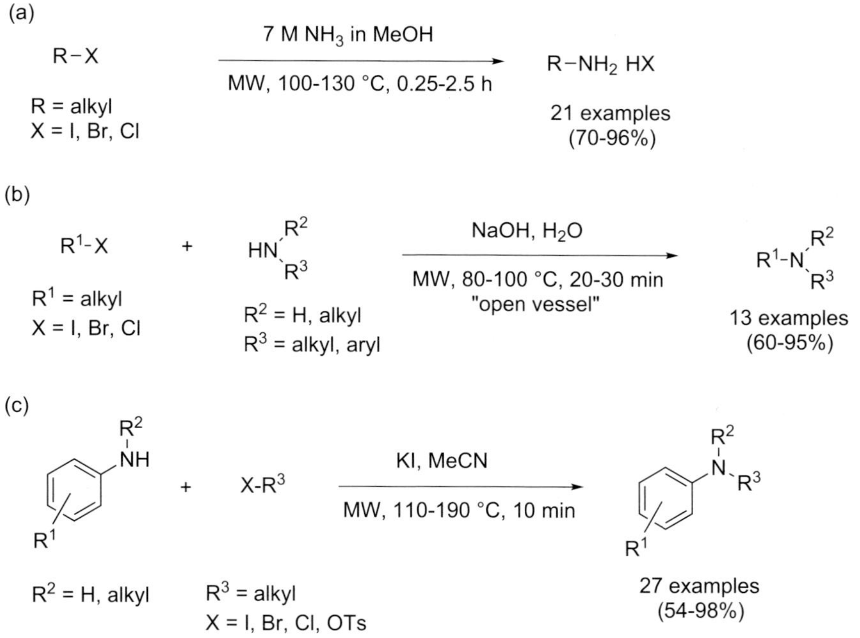

Scheme 6.118 Synthesis of primary and tertiary amines by *N*-alkylation reactions.

(Scheme 6.118b) [241]. Here, water was used as a solvent and 1.1 equiv of sodium hydroxide as a base.

The selective *N*-monoalkylation of anilines with alkyl halides and alkyl tosylates under microwave irradiation was described by Romera *et al.* (Scheme 6.118c) [242]. Using 10 mol% of potassium iodide as a catalyst and acetonitrile as solvent, many different *N*-alkylanilines were obtained in good yields with only minor quantities of dialkylation products being formed.

Lamberto *et al.* presented their investigations toward the microwave-assisted synthesis of viologens **78** [243]. These diquartenary 4,4′-bipyridine derivatives show attractive redox properties and have become common building blocks in the preparation of materials for liquid crystal displays. However, these interesting compounds suffer from extended reaction times (hours to days) under classical heating conditions. Starting from 4,4′-bipyridine, a number of symmetric and asymmetric viologens have been synthesized. Symmetric viologens were prepared by heating the bipyridine with 10–20 equiv of alkyl halide at 80–130 °C for 10 min. For asymmetric viologen derivatives **77**, monoalkylation proceeded at 60 °C within 20–60 min and employing 0.5 equiv of alkyl halide (Scheme 6.119). These derivatives were subsequently bis-alkylated to the asymmetric products by microwave heating with 10 equiv of alkyl halides for another 10 min at 130 °C.

A variety of related microwave-promoted *N*-alkylations involving more elaborate heterocyclic scaffolds are summarized in Scheme 6.120 [244–248].

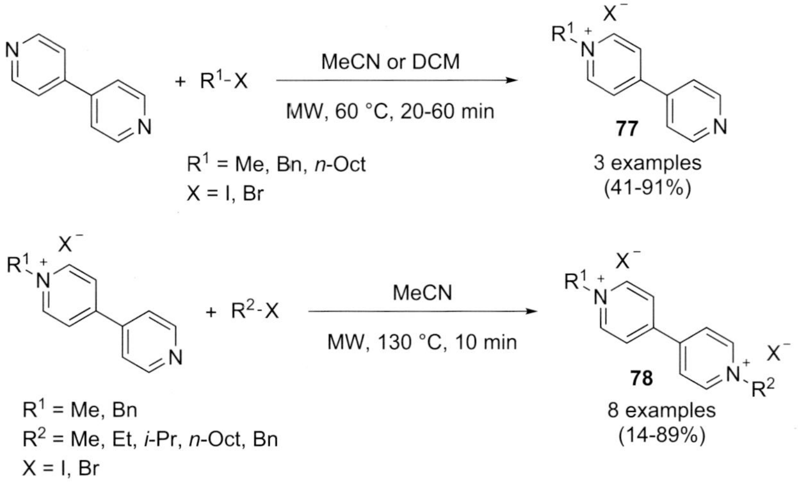

Scheme 6.119 Synthesis of viologens by *N*-alkylation.

For microwave-assisted *N*-allylations of 2(1*H*)-quinolones, see Ref. [249]. *S*-Alkylations of thiouracils with benzyl halides were reported [250, 251]. Camphor-derived chiral imidazolium ionic liquids have been synthesized by *N*-alkylation of imidazoles with appropriate camphor-derived alkyl halides [252]. Similarly, imidazolium salts have been prepared as *N*-heterocyclic carbene ligand precursors by microwave-assisted alkylation reactions [253].

Besides *N*-alkylation reactions, there are also reports in the literature concerning microwave-promoted *O*-alkylations. A mild method for the *O*-alkylation of phenols with alkyl bromides and chlorides has been developed by Wagner and coworkers (Scheme 6.121a) [254]. The protocol is applicable to substrates that are sensitive toward strong bases or hydrolysis or are difficult to extract from an aqueous phase. The procedure uses methanol as a solvent and potassium carbonate in stoichiometric amounts as a weak base. Optimum yields were obtained by heating the phenol with 1.2 equiv of the alkyl bromide (or 3 equiv of the less reactive chloride) to 100–140 °C for 15–30 min.

Organ *et al.* have used a similar protocol for the preparation of a key building block required in the synthesis of styrene-based nicotinic acetylcholine receptor (nAChR) antagonists (Scheme 6.121b) [255]. The authors have employed 4 equiv of each of the alkyl halides and potassium carbonates in a water/ethanol mixture to obtain 97% yield of the alkylated phenol. Subsequent one-pot *N*-alkylation and Suzuki coupling led to the desired nACHr antagonists (see Scheme 5.28) [255]. A procedure describing the *O*-alkylation of phenols utilizing polymer-supported bases has been described by Vasudevan *et al.* [246].

A rapid *O*-tosylation of a primary alcohol with tosyl chloride (1 equiv) in the presence of 4-(*N*,*N*-dimethylamino)pyridine (1 equiv) was reported by Botta and coworkers (Scheme 6.122) [256]. Microwave heating of the reaction mixture to 50 °C for 5 min provided the desired tosyl ester in 95% yield.

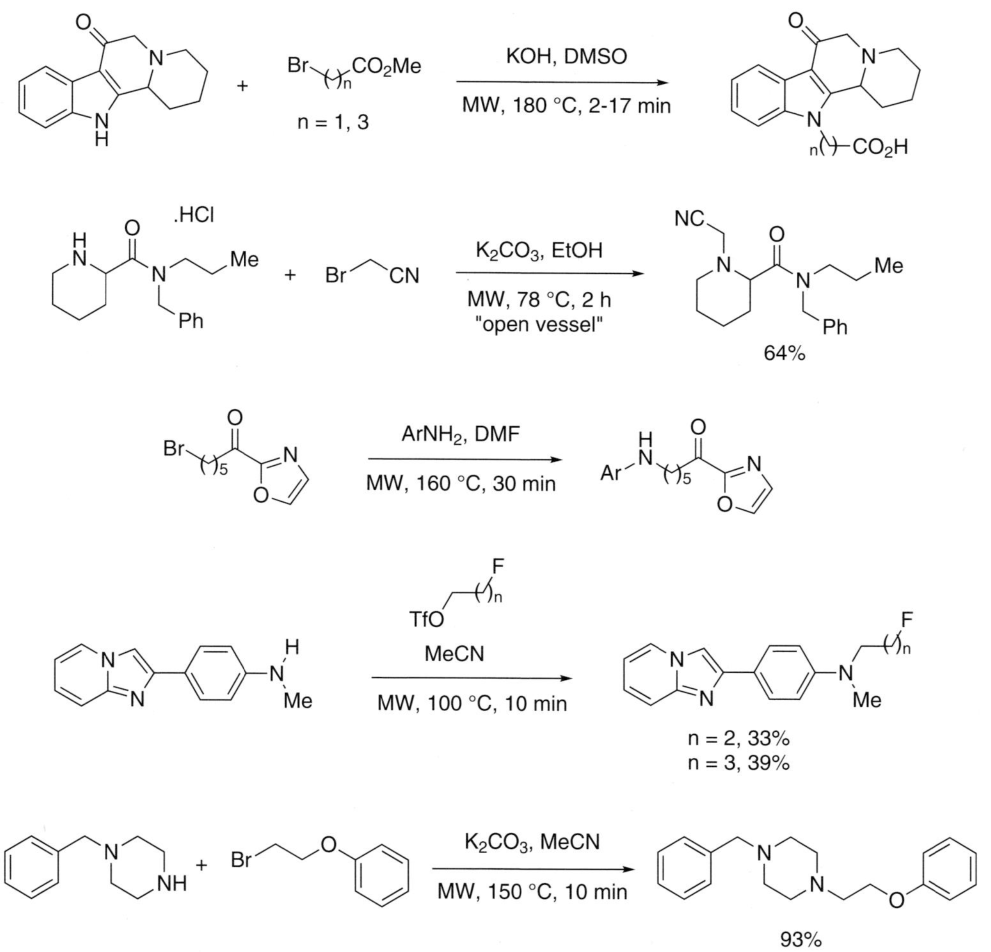

Scheme 6.120 Miscellaneous *N*-alkylation reactions.

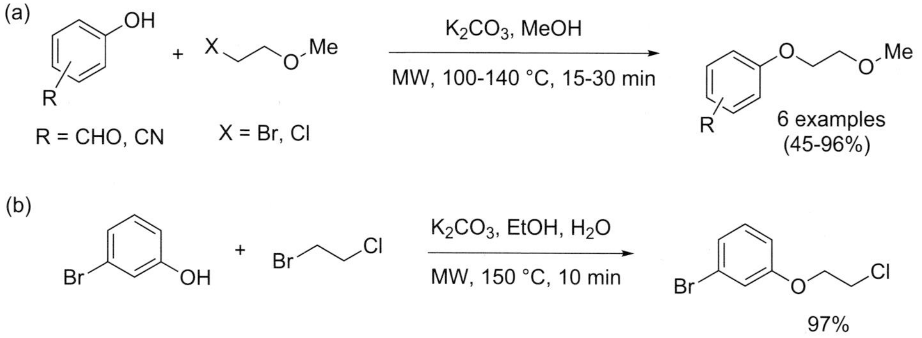

Scheme 6.121 *O*-Alkylations of phenols.

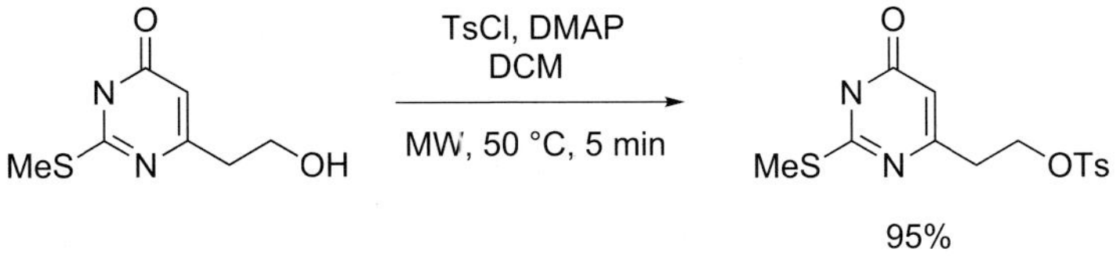

Scheme 6.122 *O*-Tosylation of alcohols.

Ollevier and Li have reported on the bismuth-catalyzed Sakurai reaction toward the synthesis of homoallylic alcohols (Scheme 6.123) [257]. Several aldehydes were successfully converted to the corresponding homoallylic alcohols by reaction with 1 equiv of allyltributylstannane and only 0.5 mol% of $Bi(OTf)_3$ as catalyst. By microwave irradiation at 160 °C for 5 min, the alcohols were obtained in good to excellent yields without any by-product formation.

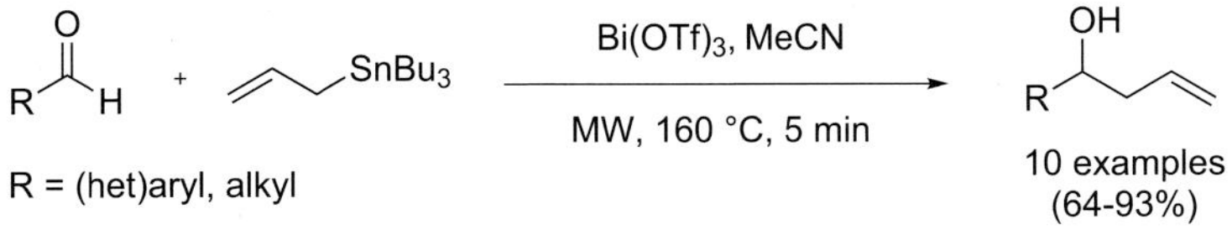

Scheme 6.123 Allylation of aldehydes with allylstannane.

In the course of the synthesis of (±)-baccatin III (**79**), a precursor of taxol, Takahashi and his group have developed a microwave-assisted protocol for the key ABC ring construction step (Scheme 6.124) [258]. The cyclization of protected cyanohydrin, which was synthesized in 36 steps completely with the aid of automated synthesizers, was accomplished by intramolecular alkylation in the presence of excess $LiN(TMS)_2$. The reaction time could be reduced from 10 h in refluxing dioxane employing thermal heating to 15 min at 145 °C using microwave irradiation.

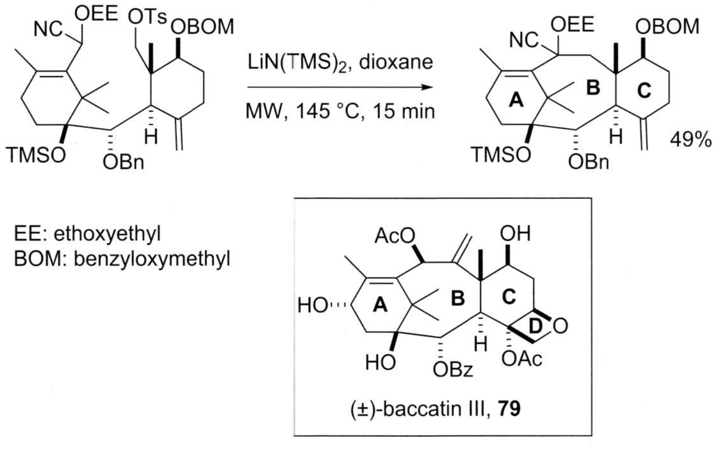

Scheme 6.124 Intramolecular alkylation in the synthesis of (±)-baccatin III.

$Yb(OTf)_3$-catalyzed etherifications of benzylic alcohols with allyl alcohol have been described by Handlon and Guo [259]. Ir-catalyzed monoalkylations of 1,3-dimethylbarbituric acid with benzyl alcohols were discussed by Grigg and coworkers [260]. Similarly, the alkylation of active methylene compounds with alkyl halides in ionic liquids [261] or under solvent-free conditions [262] has been described. The allylation of tetramic acids using allyl bromides in the presence of Et_3N has been reported [263].

6.11
Nucleophilic Aromatic Substitutions

An alternative to the palladium-catalyzed Buchwald–Hartwig and the related copper-catalyzed methods for C(aryl)−N, C(aryl)−O, and C(aryl)−S bond formations (see Section 5.2) are nucleophilic aromatic substitution (S_NAr) reactions. Such S_NAr reactions are notoriously difficult to perform and often require high temperatures and long reaction times. A number of publications report efficient nucleophilic aromatic substitutions driven by microwave heating involving either halogen-substituted aromatic or heteroaromatic systems. Scheme 6.125 summarizes some heteroaromatic systems and nucleophiles along with the reaction conditions that have been developed by Cherng in 2002 for microwave-assisted nucleophilic substitution reactions [264–266]. In general, the microwave-driven processes provide significantly higher yields of the desired products in much shorter reaction times. Typically, the substitutions were carried out neat or in high-boiling, strongly microwave-absorbing solvents such as NMP, hexamethyl phosphoramide (HMPA), or DMSO.

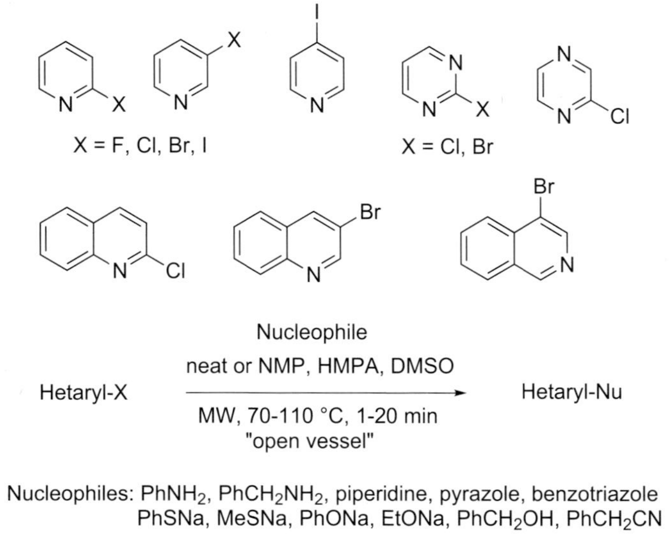

Scheme 6.125 Nucleophilic aromatic substitution reactions involving halo-substituted *N*-heteroaromatic ring systems.

In more recent work by other researchers, sealed-vessel microwave technology was utilized to access valuable medicinally relevant heterocyclic scaffolds or intermediates (Scheme 6.126) [267–272]. Additional examples not shown in Scheme 6.126 can be found in the literature [273–276]. Examples of nucleophilic aromatic substitutions in the preparation of chiral ligands for transition metal-catalyzed transformations are displayed in Scheme 6.127 [277, 278].

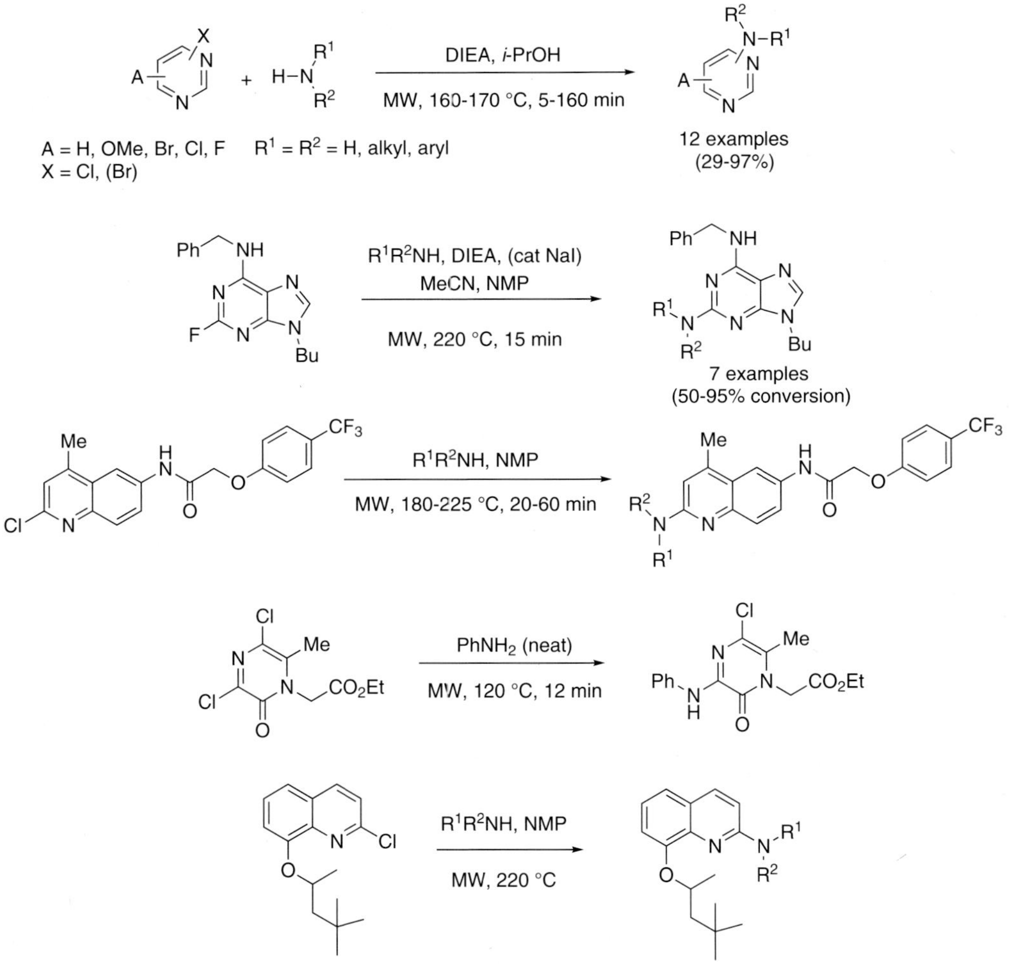

Scheme 6.126 Nucleophilic aromatic substitution reactions on medicinally relevant azaheterocyclic cores.

A pyridyl bis-*N*-heterocyclic carbene (NHC) ligand was prepared by Steel and Teasdale based on the nucleophilic aromatic substitution of dichloroisonicotinic amides with *N*-methylimidazole (Scheme 6.128) [279]. Microwave heating of the neat reagents at 140 °C for 10 min provided a 91% yield of the corresponding

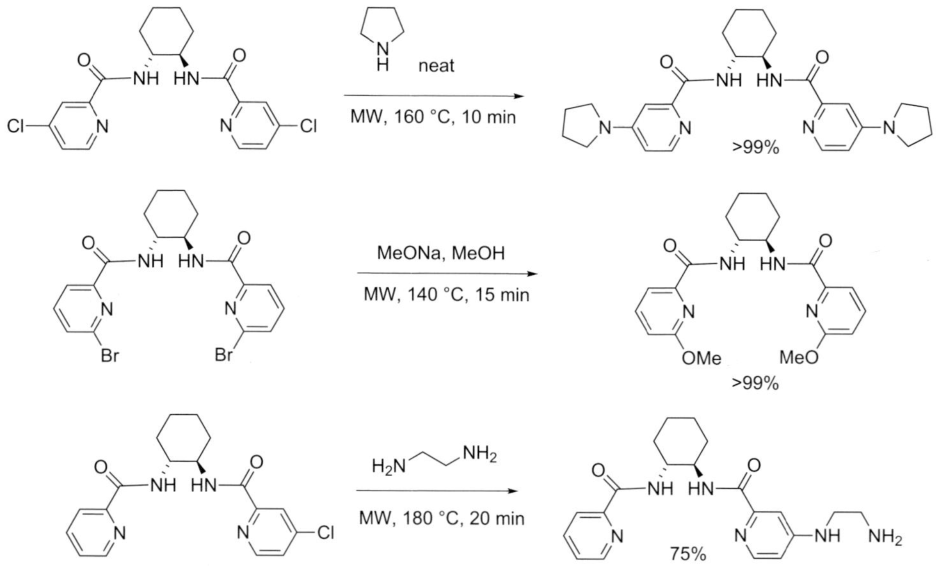

Scheme 6.127 Preparation of chiral ligands via nucleophilic aromatic substitution reactions.

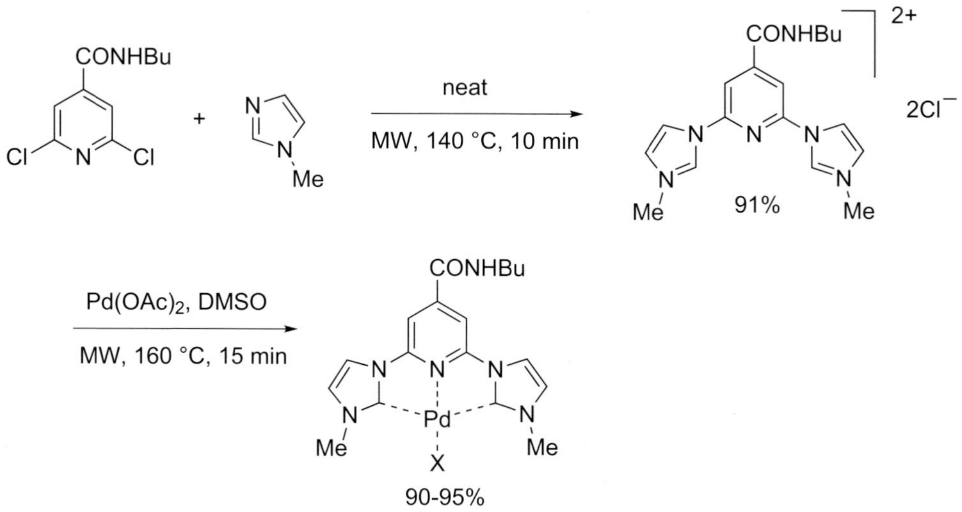

Scheme 6.128 Preparation of a pyridyl bis *N*-heterocyclic carbene palladium complex.

bis-hydrochloride salt, which subsequently was subjected to complexation by simple ligand exchange with palladium(II) acetate ($Pd(OAc)_2$) in degassed DMSO at 160 °C for 15 min. The resulting palladium complex proved useful in Suzuki and Heck reactions, and furthermore was immobilized on Tentagel resin to provide a long-lived and recyclable heterogeneous palladium catalyst.

In 2001, researchers from Wyeth-Ayerst described the preparation of RFI-641, a potent and selective inhibitor of the respiratory syncytial virus (RSV) [280]. The final key step in the synthesis involved the coupling of a diaminobiphenyl with 2 equiv of a chlorotriazine derivative under microwave irradiation (Scheme 6.129). The reaction was carried out in DMSO/phosphate buffer/sodium hydroxide in an open vessel at 105 °C. This protocol provided RFI-641 in 10% isolated yield.

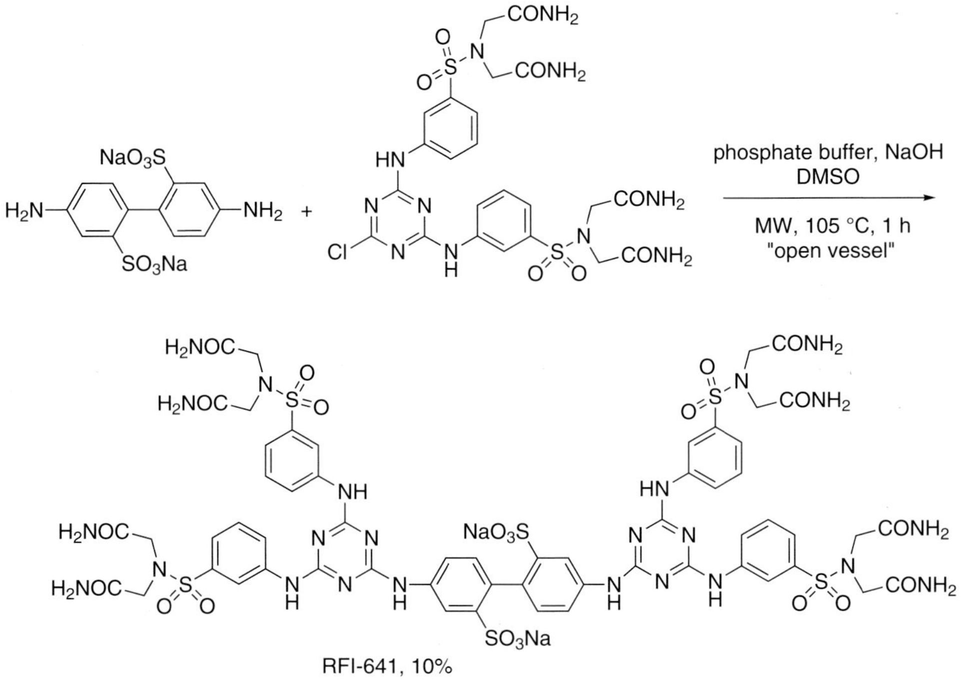

Scheme 6.129 Synthesis of the respiratory syncytial virus inhibitor RFI-641.

Nucleophilic aromatic substitution reactions are also possible on benzene rings if the aryl halide bears strongly electron-withdrawing substituents – such as nitro groups – in *ortho-* or *para*-position to the halide atom. Rapid substitution reactions can, therefore, be performed with nitro-substituted aryl fluorides such as 2,4-dinitrofluorobenzene (Sanger's reagent). In the example shown in Scheme 6.130 described by Cherng [281], amino acids are used as coupling partners using water as solvent. The presence of 2 equiv of sodium hydrogen carbonate was necessary in order to increase the nucleophilicity of the amino group. Typical reaction times under microwave irradiation were in the order of seconds, leading to *N*-arylated amino acids in excellent yields. Under conventional heating conditions (95 °C oil bath temperature), no product was detected within 1 min. The influence of the halogen atom on the efficiency of the substitution reaction was also investigated. As expected,

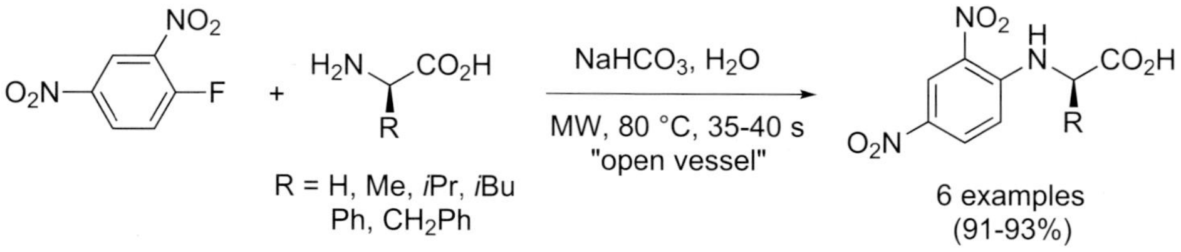

Scheme 6.130 Nucleophilic aromatic substitutions involving amino acids.

2,4-dinitrobromobenzene and 2,4-dinitrochlorobenzene reacted at a much slower rate and gave inferior yields compared to the fluoro analog (6–15 min, 48–64% yield).

The synthesis of *N*-substituted 5-nitroanthranilic acid derivatives was disclosed by Baqi and Müller (Scheme 6.131) [282]. The amination of 2-chloro-5-nitrobenzoic acid proceeded smoothly with a variety of aryl- and alkylamines under catalyst- and solvent-free microwave conditions. Aliphatic primary and secondary amines only needed 5 min to give the products in quantitative yields. In the absence of the NO_2 group in *para*-position or the carboxylate in *ortho*-position, respectively, no reaction occurred even under harsh reaction conditions (200 °C, 3 h).

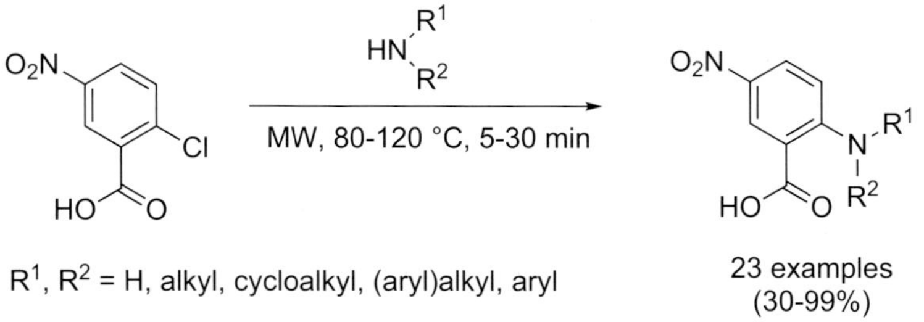

Scheme 6.131 Nucleophilic aromatic substitution for the generation of 5-nitroanthranilic acid derivatives.

Similar nucleophilic aromatic substitution reactions involving activated aryl halides are shown in Scheme 6.132 [283–285].

Liu and coworkers have developed a simple procedure for the synthesis of 2,6,9-substituted purines **81** (Scheme 6.133) [286]. 2-Chloro-6,9-substituted purines **80** are obtained via an one-pot two-step reaction of 2,6-chloropurine with anilines or amines in the first step, followed by *N*-alkylation or *N*-arylation at the 9-position with alkyl halides or phenyl boronic acid, respectively. In a subsequent reaction sequence, the generated 2-chloro-6,9-substituted purines **80** are further reacted with various amines and anilines under $NaBF_4$ catalysis to finally provide the 2,6,9-funtionalized purines **81** in low to excellent yields (13–99%).

A one-pot three-step conversion of aryl fluorides into phenols based on a consecutive nucleophilic aromatic substitution/isomerization/hydrolysis sequence was reported by Levin and Du (Scheme 6.134) [287]. The authors have discovered that 2-butyn-1-ol can function as a hydroxyl synthon via consecutive S_NAr displacement,

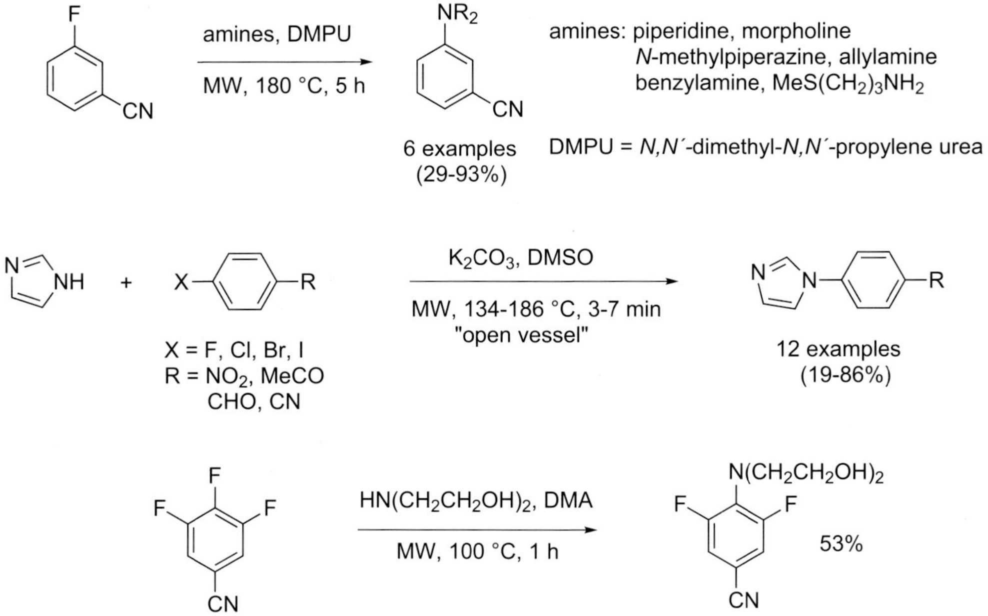

Scheme 6.132 Nucleophilic aromatic substitutions of activated aryl halides.

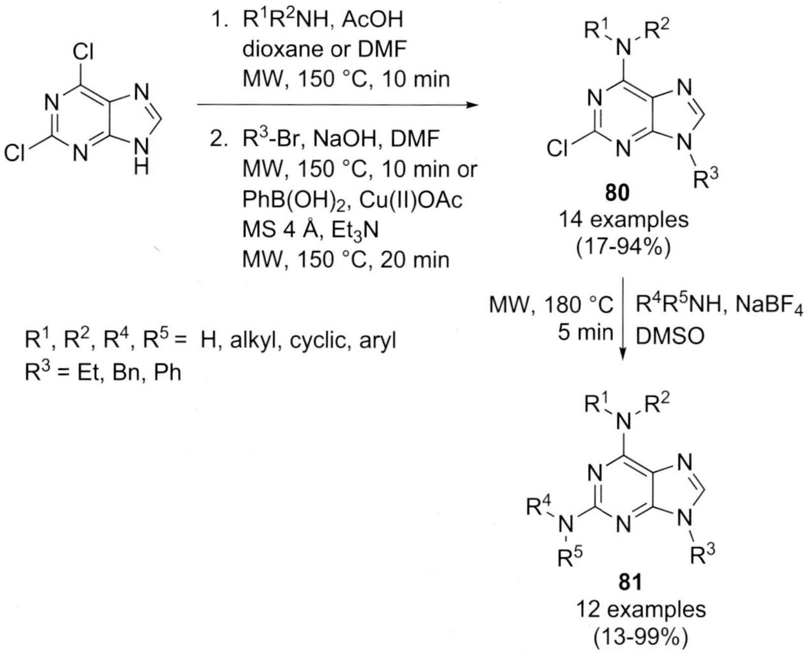

Scheme 6.133 Synthesis of 2,6,9-substituted purines via nucleophilic aromatic substitution.

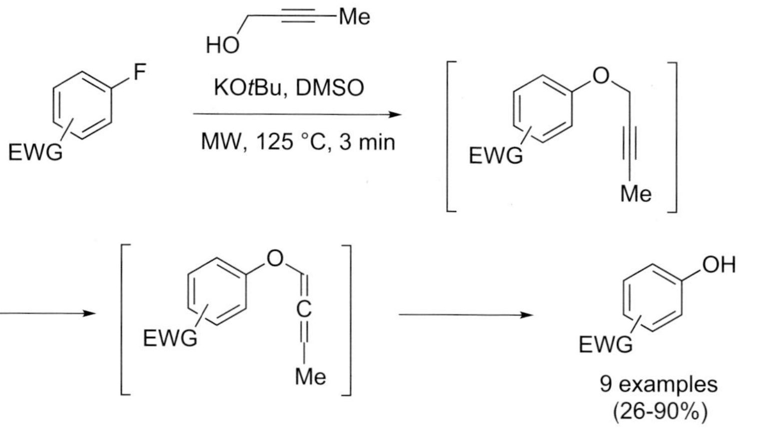

Scheme 6.134 One-pot conversion of aryl fluorides into phenols.

in situ isomerization to the allenyl ether, and subsequent hydrolysis to afford phenols rapidly and in good yields. In most cases, an excess of 2-butyn-1-ol (1–2 equiv) and potassium *tert*-butoxide (2–4 equiv) was required in order to achieve optimum yields.

In 2010, the group of Organ reported on the synthesis of bromo and fluoro benzofused sultams **82** [288]. These important cyclic sulfamides show versatile biological activity, especially benzothiaoxazepine-1,1-dioxide exhibits a manifold inhibition activity. β-Hydroxy-α-fluorobenzene sulfonamides underwent cyclization by an intramolecular S_NAr *O*-arylation route when subjected to microwave irradiation in DMF at 140 °C for 30 min in the presence of cesium carbonate to obtain the desired chiral benzothiaoxazepine-1,1-dioxides **82** in excellent yields (Scheme 6.135). For the scale-up synthesis of **82**, the authors developed a microwave-assisted continuous flow protocol employing capillaries in a homemade flow reactor applicable in a commercially available microwave reactor. Slight modifications of conditions were required to make the protocol prone for the flow approach: potassium *tert*-butoxide had to be used as base and irradiation at a maximum temperature of 80 °C furnished the desired chiral benzothiaoxazepine-1,1-dioxides in very good

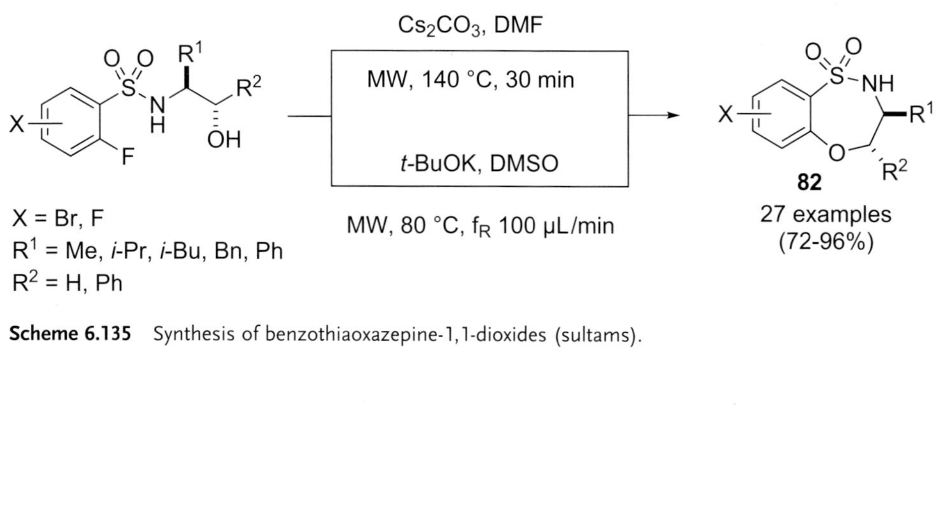

Scheme 6.135 Synthesis of benzothiaoxazepine-1,1-dioxides (sultams).

yields. With this flow protocol, a small library of 19 sultam derivatives could be generated in multigram scale. The preparation of 5–10 g product took approximately 2.5 h for each compound.

In a related work, Hanson and coworkers prepared benzothiaoxazepine-1,1-dioxides by an epoxide cascade protocol involving epoxide opening and S_NAr cyclization (Scheme 6.136a) [289]. These scaffolds can be further functionalized on the aryl fluoride position by a standard S_NAr reaction with nucleophilic species like phenols, thiols, amines, or sulfonamides under microwave irradiation at 150 °C in 30 min. Although the observed conversions were good, the need of large excess of the nucleophile remains undesirable in terms of purification. Thus, the authors developed a two-step one-pot synthesis-scavenging strategy employing oligomeric dichlorotriazine ($ODCT_{50}$). Conducting the scavenging process under microwave irradiation at 50 °C for only 30 min, furnished the products in low to good yields after filtration through silica SPE. This highly efficient time-saving process was then applied to generate a diverse library of benzothiaoxazepine-1,1-dioxides (Scheme 6.136b).

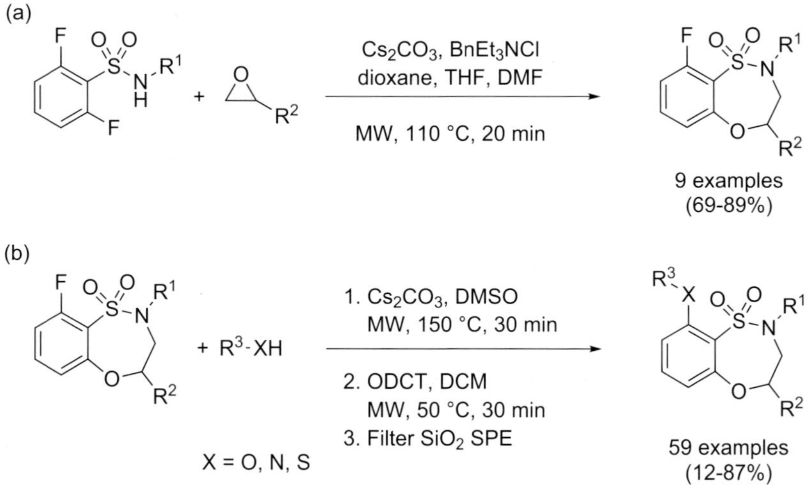

Scheme 6.136 Synthesis and two-step diversification/purification of benzothiaoxazepine-1,1-dioxides.

Microwave-assisted nucelophilic aromatic substitutions have also been reported on halouracils [290], polyfluoropyridine scaffolds [291], pentafluorophenyl substrates [292], dichloropyridines [293], 5-chloropyridazinones [294], 4-chloroquinolines [295], 6-chloro-purines [296], and 4-chloro-7*H*-pyrrolo[2,3-*d*]pyrimidines [297]. A tandem cross-coupling/nucleophilic aromatic substitution approach for the synthesis of carbazoles has been described by St. Jean *et al.* [298]. Direct high-temperature microwave-assisted aminations from a variety of aryl halides [299] or aryl triflates [300] under transition metal-free conditions have also been reported, probably involving a benzyne mechanism.

6.12
Ring-Opening Reactions

6.12.1
Cyclopropane and Cyclobutene Ring Openings

The fused cyclopropane 1-acetyl-4-phenyl-3-oxabicyclo[3.1.0]hexane shown in Scheme 6.137 is a key precursor for the synthesis of *endo,exo*-furofuranone derivatives, one of the largest subclass of lignans, which are known to possess varied biological activities. Brown and coworkers have found that while Lewis acid-assisted ring opening of the cyclopropane ring in MeOH at reflux temperature was extremely sluggish, the cyclopropane ring could effectively be opened at 120 °C by microwave irradiation (Scheme 6.137) [301, 302]. After screening a variety of conditions, best results were achieved by employing stoichiometric amounts of zinc triflate in MeOH. After 30 min at 120 °C, full consumption of the starting bicycle was achieved resulting in the formation of the desired methyl ether product and the corresponding enol ether.

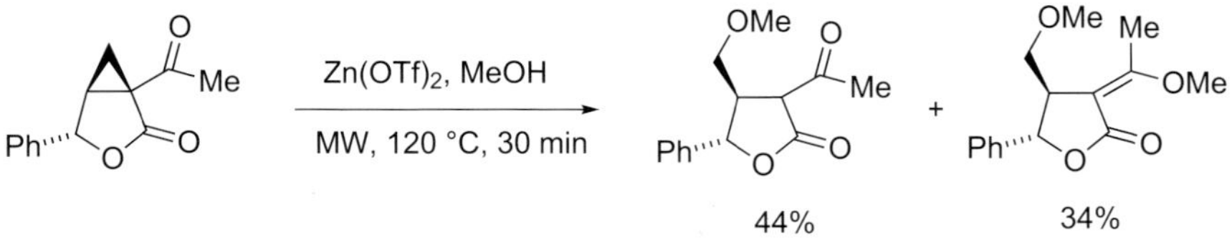

Scheme 6.137 Cyclopropane ring opening.

Baudoin and coworkers reported on the synthesis of tricyclic molecules by a microwave-enhanced electrocyclic ring opening and subsequent [4 + 2] cycloaddition [303]. A variety of initially prepared benzocyclobutenes (BCBs) could be transformed to the corresponding tricyclic compounds by reacting with a variety of dienophiles such as *N*-phenyl maleimide, maleic anhydride, and tetracyanoethylene (Scheme 6.138). Gratifyingly, the products were obtained as pure single

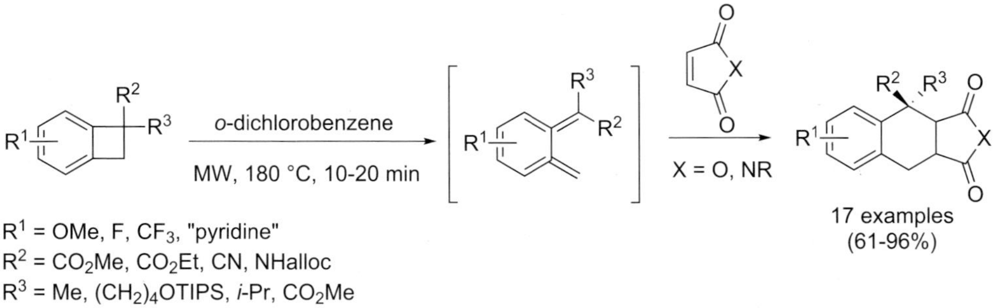

Scheme 6.138 Cycloadditions of 1,1-disubstituted benzocyclobutenes.

diastereomers in the same relative configuration reflecting a complete outward torquoselectivity of the 1-alkyl group in the BCB core. On the other hand, 1-monosubstituted BCBs could also be transformed with the same protocol, but the products were obtained as inseparable mixtures of the two diastereomers.

6.12.2 Aziridine Ring Openings

Aziridines with electron-withdrawing groups on the nitrogen atom readily undergo ring opening when treated with nucleophiles. The nucleophilic attack generally occurs at the least hindered carbon atom. Lake and Moberg have utilized the ring opening of chiral *N*-tosyl aziridines for the generation of C_3-symmetric tripodal tris (sulfonamide) ligands (Scheme 6.139) [304]. Exposing a 4.5 : 1 molar ratio of aziridine to ammonia in methanol to microwave irradiation at 160 °C for 45 min led to the clean formation of the tris(sulfonamide) ligand without any formation of unwanted mono(tosyl)amine by-products. The isolated product yield was 88% with 12% recovery of the aziridine starting material. For comparison purposes, ring opening at 50 °C required 4 days for completion and furnished only a 70% yield of the product.

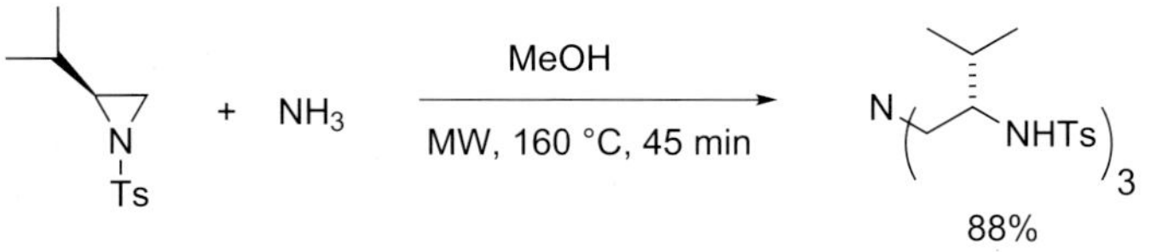

Scheme 6.139 Aziridine ring opening.

Pinho e Melo and coworkers investigated the synthesis of functionalized *N*-vinyl heterocycles under microwave irradiation [305]. Therefore, functionalized enamines bearing a strained, aziridine moiety were utilized to generate highly reactive azomethine ylides upon electrocyclic ring opening, which *in situ* take part in 1,3-dipolar cycloadditions to form biologically relevant five-membered heterocycles. As model vinyl aziridine, the authors employed ethyl *cis*-1-[(*E*)-4-(benzyloxy)-4-oxobut-2-en-2-yl]-3-phenylaziridine-2-carboxylate, which was successfully reacted under microwave conditions with several dipolarophiles to prepare functionalized *N*-heterocycles (Scheme 6.140a). All reactions proceeded stereo- and regioselectively with exception when *N*-phenylmaleimide was employed. In this case, both *exo*- and *endo*-adducts of the expected octahydropyrrolo[3,4-*c*]pyrrole derivatives were formed, with a distinct preference for the *exo*-adduct (Scheme 6.140b). In general, the simple and quick method has proven effective for the preparation of various heterocyclic moieties with attractive substitution patterns for pharmaceutical research.

Ring opening of (*S*)-*N*-tosyl-2-isopropylaziridine with suitable amines provides polydentate chiral amines [306]. α-Hydroxyamides have been obtained by ring opening of oxazolidin-2,4-diones with NaOMe [307].

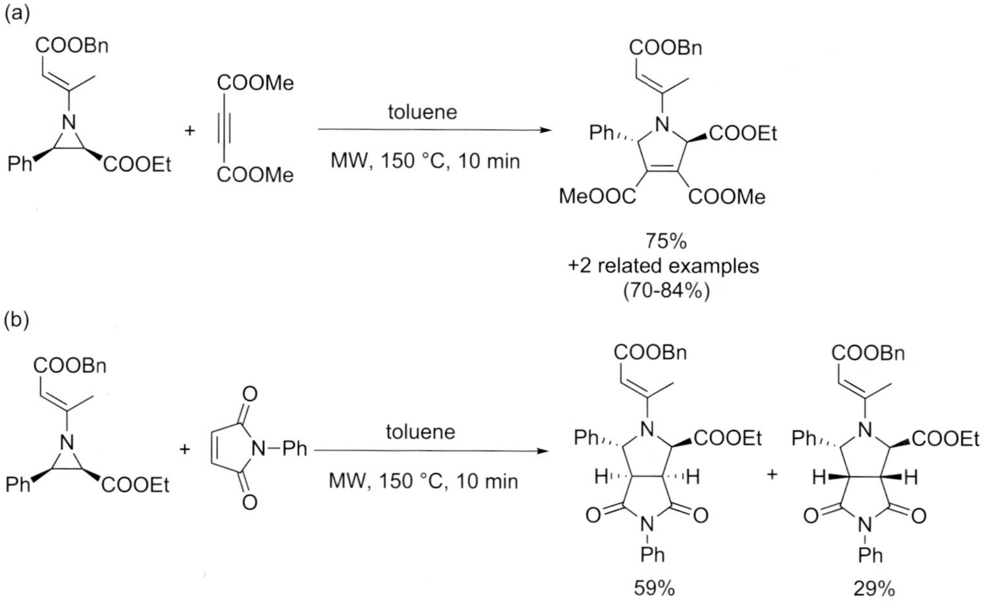

Scheme 6.140 Synthesis of heterocycles from *N*-vinylaziridine via electrocyclic ring opening.

Aryl pyranosides have been ring opened by nucleophiles in the presence of $Sc(OTf)_3$ [308]. Cyclic ureas have been ring opened under acid catalysis to produce enantiomerically pure (*S*)-2-(aminomethyl)-4-phenylbutanoic acids [309].

6.12.3
Epoxide Ring Openings

Several articles describe the ring opening of chiral epoxides under microwave irradiation conditions (see also Scheme 6.59). In the context of the preparation of novel β_2-adrenoceptor agonists related to formoterol and salmeterol, Fairhurst and a team from Novartis have described the synthesis of chiral ethanolamines by solvent-free microwave-assisted ring opening of a suitable chiral epoxide precursor with secondary benzylated amines (Scheme 6.141) [310]. The reaction occurred selectively at 110 °C at the least hindered position, yielding the desired ethanolamine derivatives in 52–82% isolated yield.

Lindsay and Pyne have described related ring-opening reactions of a chiral vinyl epoxide with ammonia or allylamine, as shown in Scheme 6.142 [311, 312]. In case of ammonia, the reaction was simply carried out in concentrated aqueous ammonia (28%) under sealed-vessel microwave conditions. After irradiation for 30 min at 110 °C, the amino alcohol was obtained in 98% yield [311]. In case of allylamine, MeCN was used as a solvent in the presence of 1 equiv of lithium triflate as Lewis acid. Again, a high yield of the expected amino alcohol was obtained (120 °C, 1 h, 97%

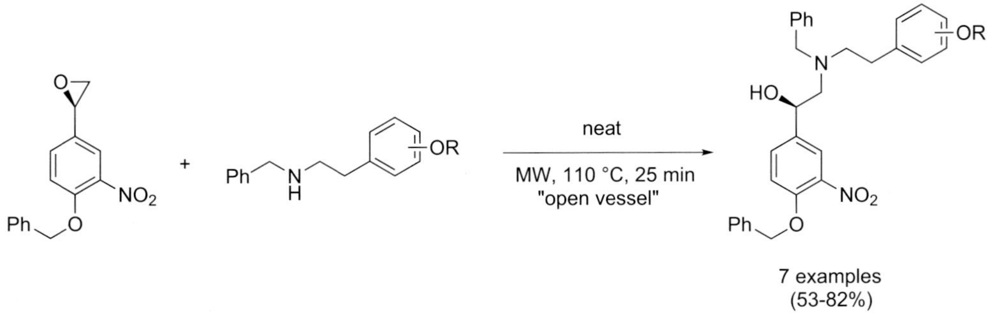

Scheme 6.141 Preparation of β_2-adrenoceptor agonists via epoxide ring opening.

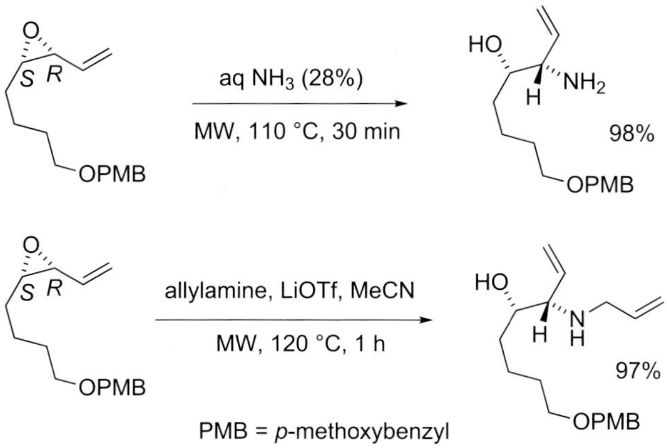

Scheme 6.142 Epoxide ring opening with amines.

yield) [312]. In both cases, clean S_N2 ring opening, with no evidence of other regio/stereoisomers, occurred.

A rather complex microwave-assisted ring opening of chiral difluorinated epoxycyclooctenones has been studied by Percy and coworkers (Scheme 6.143) [313]. The epoxide resisted conventional hydrolysis but reacted smoothly in basic aqueous medium (ammonia or *N*-methylimidazole) under microwave irradiation at 100 °C for

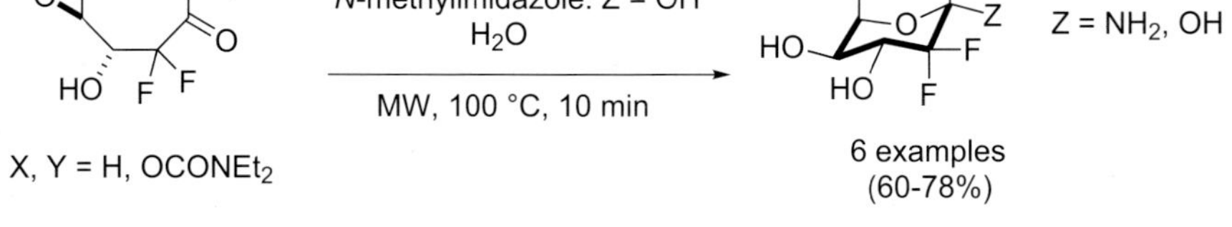

Scheme 6.143 Ring opening of chiral difluorinated epoxycyclooctenones.

10 min to afford unique hemiacetals and hemiaminals in good yield. Other nitrogen nucleophiles such as sodium azide or imidazole failed to trigger the reaction. The reaction with sodium hydroxide led to a much poorer conversion of the starting material.

Schirok from Bayer HealthCare has reported on a short and flexible synthesis of 1,3-substituted 7-azaindoles **84** starting from oxiranes **83** (Scheme 6.144) [314]. The reaction of epoxides **83** with a variety of substituted primary amines proceeds via epoxide opening (**A**)/cyclization (via S_NAr, **B**)/dehydration (**C**) sequence, which is accelerated by microwave irradiation. A higher temperature (200 °C) was required for the monochloro pyridine derivatives ($R^2 = H$) since they proved to be less reactive. As opposed to the conventional heating protocol, no addition of an acid was required for the dehydration step.

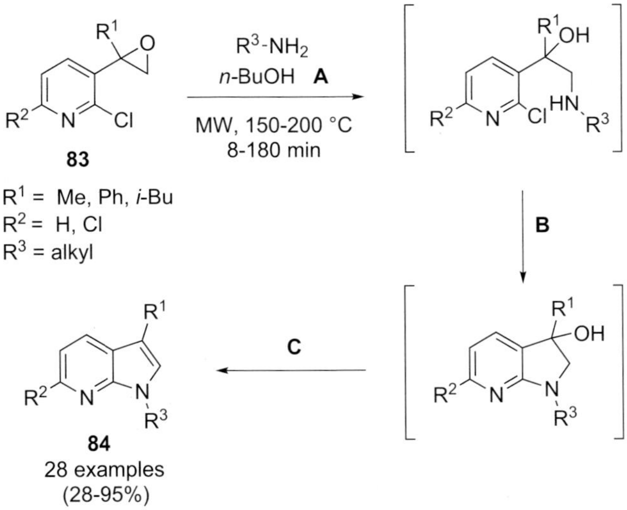

Scheme 6.144 Synthesis of 7-azaindoles.

In a 2005 publication, Pericàs and coworkers have described the investigation of fluoroepoxide **85** as a new, chiral resolution reagent (Scheme 6.145) [315]. By applying microwave heating, the regioselective ring opening of this reagent induced by α-chiral primary and secondary amines occurred in higher yields and with less by-products compared to that obtained by applying conventional heating. Because of the use of readily available enantiopure epoxide **85**, the resulting diastereomeric amine products could be very easily identified by ^{19}F, ^{1}H, ^{13}C NMR and by HPLC.

For closely related examples published by the same group, see Ref. [316]. Similar epoxide ring openings with amines have also been described by Flitsch and coworkers in the context of synthesizing 1-aminopropan-2-ols with antimalaria parasite activities [317]. The aminolysis of epoxides using microwave conditions has been reported by Lindsay and coworkers [318].

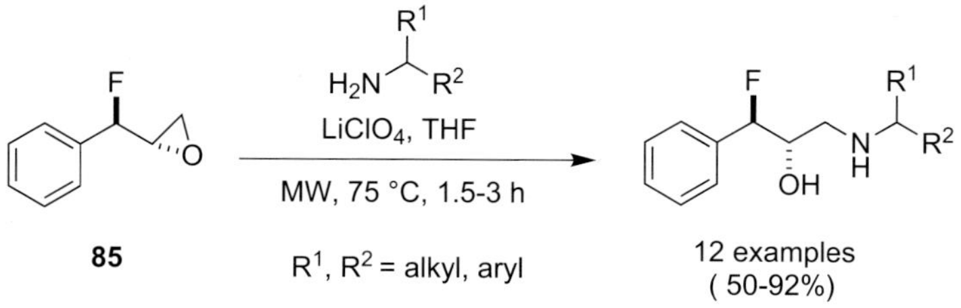

Scheme 6.145 Ring opening of epoxides.

Finally, several authors have reported on ring-opening transformations of epoxides with azides. Van Delft and coworkers have investigated the chelation-controlled ytterbium triflate-mediated azidolysis of a highly functionalized epoxycyclohexene derivative (Scheme 6.146a) [319]. The authors determined that ring opening of the desired 1,3-diazidocyclitol compound was best achieved by exposing a mixture of the starting epoxide, sodium azide, ytterbium triflate catalyst, triethylamine base, and 3 Å molecular sieves to microwave irradiation at 135 °C. Under these conditions, the desired bis-azide was obtained in 79% yield compared to a 49% product yield achieved at 80 °C using conventional heating for 4 days.

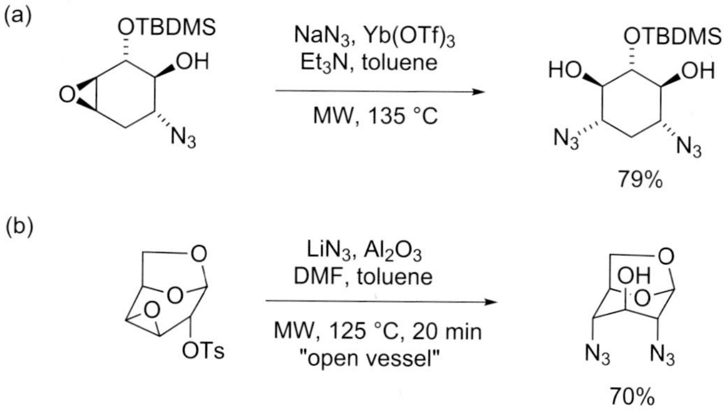

Scheme 6.146 Ring opening of epoxides with azide ions.

Related ring openings of levoglucosan-derived epoxides with 4 equiv of lithium azide in the presence of alumina were reported by Cleophax and coworkers in 2004 (Scheme 6.146b) [320].

The ring opening of cyclohexene epoxide with thiophenol has been described by Pironti and Colonna [321]. Ring opening of fused epoxides with ammonia provides cyclic β-amino alcohols in good yields [322]. For the ring opening of epoxides with *N*-arylpiperazines, see Ref. [323]. The ring opening of an epoxide ring by an internally displayed indole nitrogen was used in the total synthesis of balasubramide [324].

6.13
Addition and Elimination Reactions

6.13.1
Michael Additions

In 2002, Leadbeater and Torenius reported the base-catalyzed Michael addition of methyl acrylate to imidazole using ionic liquid-doped toluene as a reaction medium (Scheme 6.147a) [51]. A 75% product yield was obtained after 5 min of microwave irradiation to 200 °C employing equimolar amounts of Michael acceptor/donor and triethylamine base. As for the Diels–Alder reaction studied by the same group (see Scheme 6.23), pmimPF_6 was the utilized ionic liquid. Related microwave-promoted Michael additions studied by Jennings *et al.* involving indoles as heterocyclic amines are shown in Scheme 6.147b [244] and 6.147c [325]. Here, either lithium *bis*(trimethylsilyl)amide (LiHMDS) or KO*t*Bu in the presence of tetrabutylammonium iodide (TBAI) was employed as strongly basic reaction mediators.

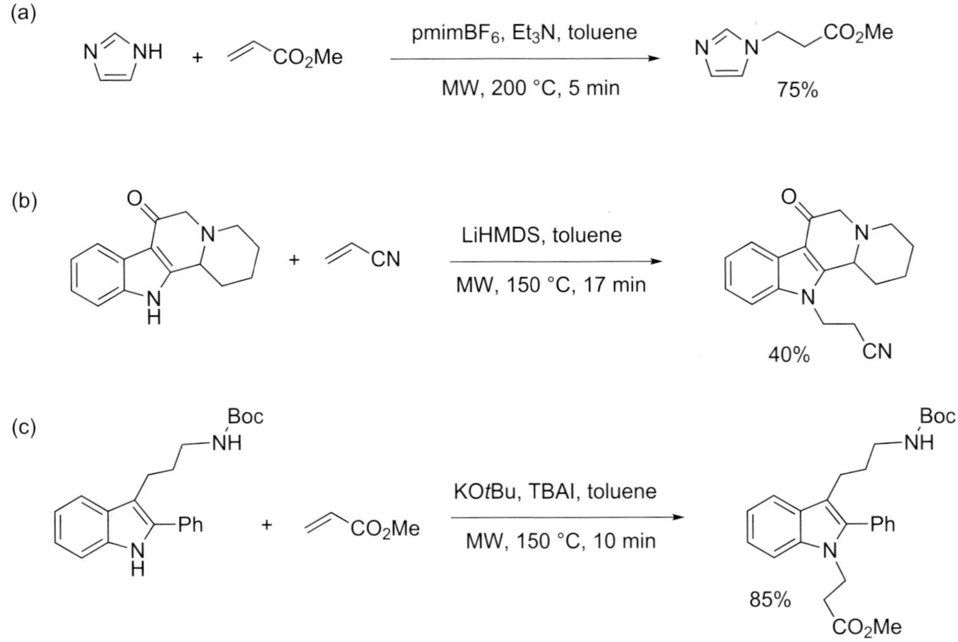

Scheme 6.147 Michael additions involving heterocyclic amines.

The Michael addition of phthalimide and saccharin to acrylic acid esters in the presence of a ZnO catalyst has been reported by Zare *et al.* [326].

An acid-catalyzed double Michael addition of water to the bridged bisdioxine moiety in a larger macrocyclic framework was described by Kollenz and coworkers

(Scheme 6.148) [327]. While conventional reaction conditions failed to provide any of the desired functionalized 2,4,6,8-tetraoxaadamantane products, microwave heating of the hydrophobic macrocyclic bisdioxine in a 1 : 1 mixture of 1,2-dichloroethane and acetic acid containing excess of concentrated hydrochloric acid at 170 °C for 40 min provided a 35% isolated yield of the desired oxaadamantane compound.

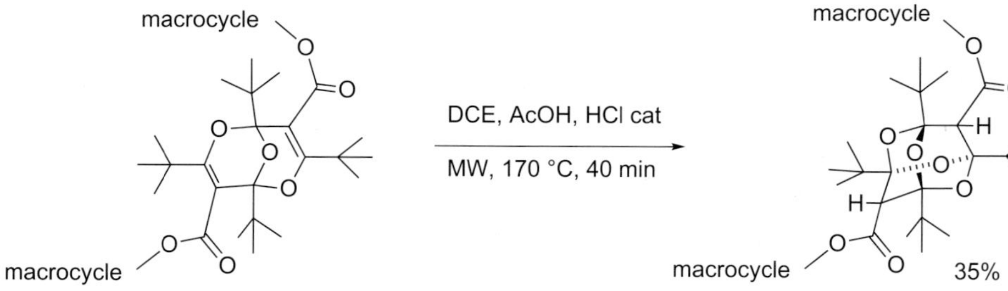

Scheme 6.148 Formation of 2,4,6,8-tetraoxaadamantanes via double Michael addition.

Enders *et al.* reported on organocatalytic quadruple domino reactions under microwave irradiation [328]. Employing acetaldehyde and several aromatic nitroalkenes, the authors investigated a comprehensive domino Michael addition/Henry condensation/Michael addition/aldol condensation cascade for the asymmetric synthesis of cyclohexene carbaldehydes (Scheme 6.149). Initial investigations revealed that the presence of water is crucial for the formation of the cyclic aldehydes. Nevertheless, rather long reaction times under microwave heating were required to obtain the desired products (5–12 h). The asymmetric cyclohexene carbaldehydes were obtained in moderate yields with high enantioselectivities (89–>99%), but with only low diastereomeric ratio (1.4–3.5 : 1).

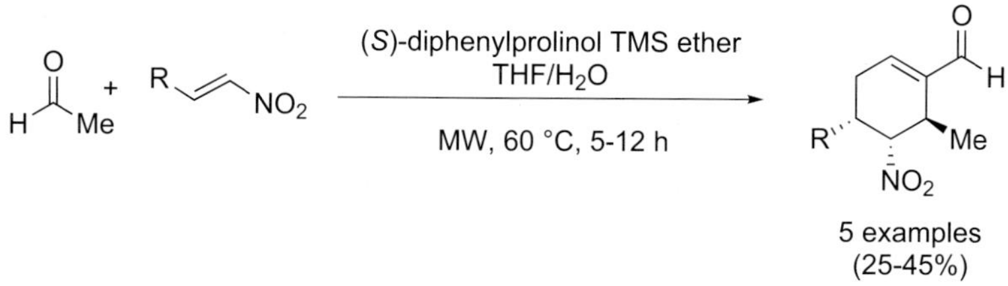

Scheme 6.149 Asymmetric synthesis of cyclohexene carbaldehydes.

A retro-Michael addition process, leading to an aminomethyl dihydrodipyridopyrazine analog, was described by Guillaumet and coworkers in the context of preparing DNA bisintercalators as antitumor agents (Scheme 6.150) [329]. Compared to 6 days of heating under conventional reflux conditions, the microwave-assisted reaction was completed in 9 h.

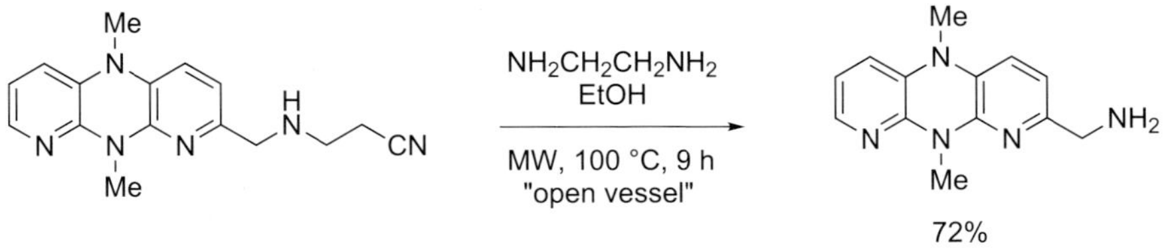

Scheme 6.150 Retro-Michael additions.

6.13.2
Addition to Alkynes

A direct addition of cycloethers to terminal alkynes was discovered by Zhang and Li (Scheme 6.151) [330]. Best results were obtained when the reactions were run without additional solvent and in the absence of additives such as transition metal catalysts, Lewis acids, or radical initiators. Typically, the cycloether was used in large excess (200 molar equiv) as solvent under sealed-vessel conditions. Employing 200 °C reaction temperature, moderate to good yields of the vinyl cycloether products (as mixtures of *cis*- and *trans*-isomers) were obtained. The reaction is proposed to follow a radical pathway.

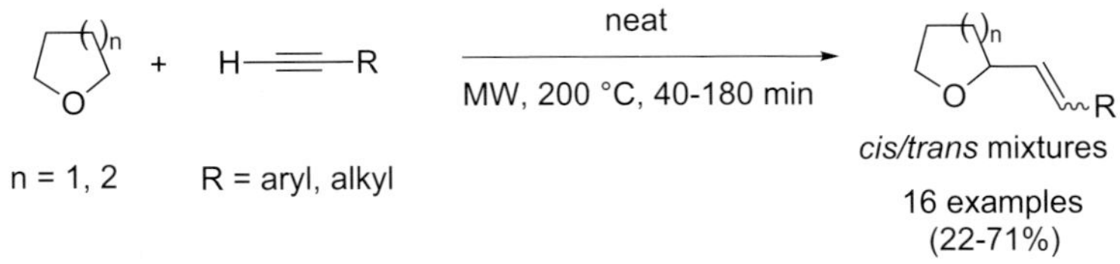

Scheme 6.151 Addition of cycloethers to terminal alkynes.

A hydrophosphination reaction of terminal alkynes with secondary phosphine–borane complexes has been developed by Mimeau and Gaumont as a general synthetic pathway to stereodefined vinylphosphine derivatives [331]. The regioselectivity of the reaction is controlled by the choice of activation method. While in the presence of a palladium catalyst the corresponding α-adducts are obtained (Markovnikov addition), thermal activation leads exclusively to the corresponding β-adducts (anti-Markovnikov addition). For the thermal activation process shown in Scheme 6.152, microwave heating was found to be the best method to achieve fast and efficient reactions. Typically, the secondary phosphine–borane was added to an excess of the alkyne and was then microwave heated under open-vessel conditions at 50–80 °C for 30–45 min to allow full conversion to the vinylphosphine products. As far as the stereochemistry is concerned, the *Z*-isomer was usually obtained as the major product, typically with a high degree of stereoselectivity (>95 : 5 for $R^1 = R^2 =$ phenyl).

The group of Stanovnik investigated the [2 + 2] cycloaddition of imidazolidine-2,4-diones to electron-poor acetylenes [332]. The utilized (5*Z*)-5-[(dimethylamino)

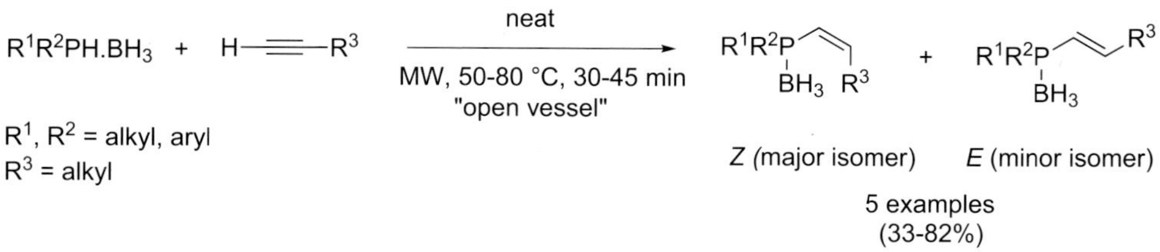

Scheme 6.152 Hydrophosphination of terminal alkynes.

methylene]imidazolidine-2,4-diones are important intermediates in the synthesis of natural products, thus modification of this scaffold is of considerable interest. Microwave irradiation of the reaction mixture at 80 °C for 1–2 h furnished the desired products (Scheme 6.153). In all cases, the corresponding (2(1′)*E*,3(4″)*E*)-2-[(dimethylamino)methylene]-3-(1-methyl-2,5-dioxoimidazolidin-4-ylidene)butanedioates **86** were formed via regiospecific [2 + 2] cycloaddition as the major product. However, depending on the substrate combination, the respective (2(1′)*E*,3(4″)*Z*)-isomers **87** were formed in varying ratio as a by-product. Interestingly, when ethyl 2-propynoate was employed, heating at 100 °C for 2 h was required to afford full conversion. Here, a (2(4′)*E*,3*E*)-isomer is obtained as the minor by-product. Furthermore, the *N*-unsubstituted imidazolidine-2,4-dione did not react as expected. Being insoluble in MeCN, the reaction was carried out in DMF, but resulted in dimethyl (2*E*)-2(2,5-dioxoimidazolidin-4-ylidene)butanedioate **88** due to hydrolyzation of the (dimethylamino)methylene group under the given reaction conditions (Scheme 6.153).

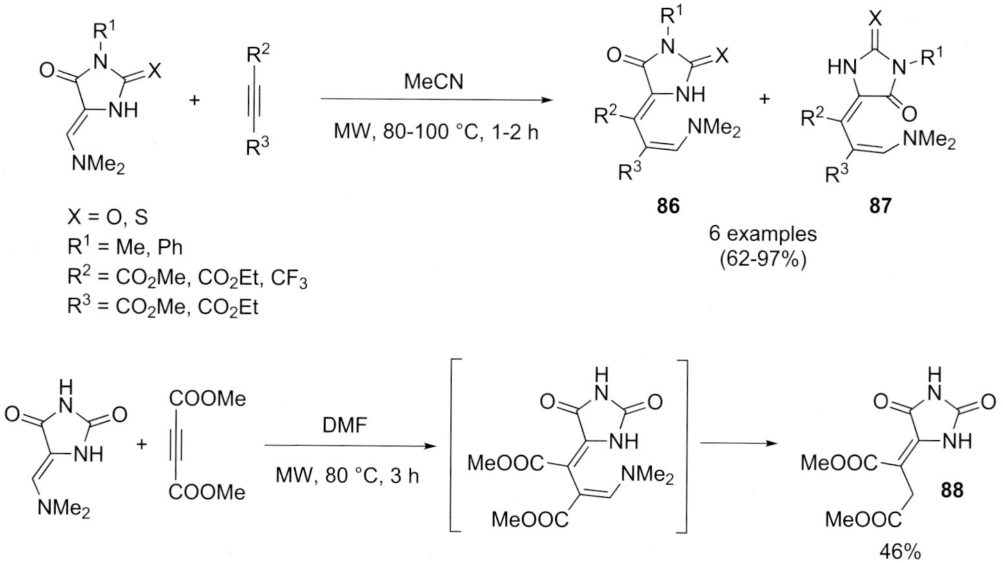

Scheme 6.153 [2 + 2] Cycloadditions of electron-poor acetylenes to imidazolidine-2,4 diones.

Wipf *et al.* have reported in 2004 that the hydrozirconation of alkynes with zirconocene hydrochloride (**89**) can be greatly accelerated by microwave irradiation (Scheme 6.154) [333]. A synthetically useful one-pot method for the preparation of allylic amides **90** was elaborated where an alkyne was first hydrozirconated by microwave irradiation, followed by rapid addition of imines in the presence of dimethyl zinc.

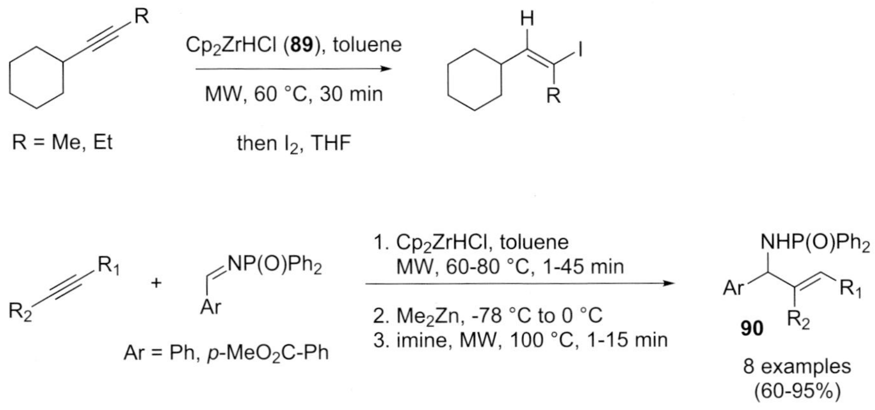

Scheme 6.154 Hydrozirconation–transmetalation–aldimine addition sequence.

Along similar lines, the Zr(IV)-catalyzed hydroboration of allyl propargyl esters was reported by Nelson and coworkers [334].

Stannylated allylsulfones can be prepared by hydrostannylation of propargyl sulfones using a Mo catalyst [335]. The same group has also reported the Mo-catalyzed hydrostannation of simple alkynes with tributyltin hydride [336, 337].

6.13.3
Addition to Alkenes

Similar to the addition of secondary phosphine–borane complexes to alkynes described in Scheme 6.152, the same hydrophosphination agents can also be added to alkenes, leading to alkylarylphosphines under fairly similar reaction conditions (Scheme 6.155) [338]. Again, the expected anti-Markovnikov addition products were exclusively obtained. In some cases, the additions also proceeded at room temperature, but required much longer reaction times (2 days). Treatment of the phosphine–borane complexes with a chiral olefin such as (−)-β-pinene led to chiral cyclohexene derivatives via a radical-initiated ring-opening mechanism. The same group has prepared ketophosphine–boranes by the neat addition of borane–phosphine complexes to activated olefins [339]. In a related work, Ackerman *et al.* have described microwave-assisted Lewis acid-mediated intermolecular hydroamination reactions of norbornene [340].

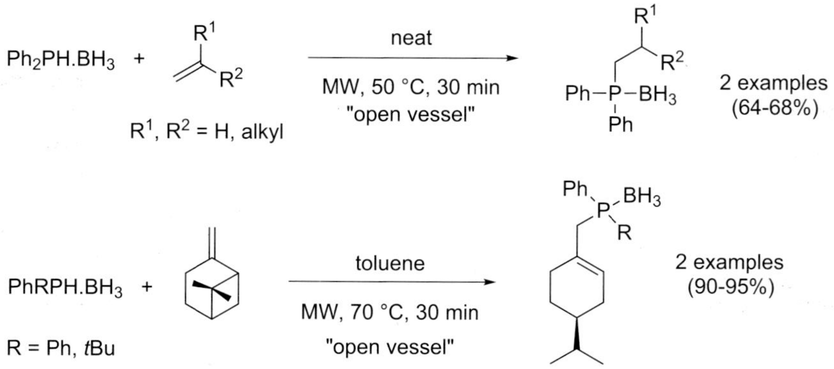

Scheme 6.155 Hydrophosphination of terminal alkenes.

Giner and Nájera examined the intermolecular hydroamination of alkenes [341]. Here, the authors screened several silver salts on their catalytic efficiency on the addition of 4-toluenesulfonamides to alkenes to identify silver(I) triflate (AgOTf) as the best suited catalyst. In the optimized protocol, the vial was charged in the dark under argon atmosphere with 4-toluensulfonamide, alkene, 5 mol% of AgOTf, and dry solvent. When cyclohexene was applied as the alkene, no solvent was added. The mixture was heated under air stream cooling at 90 °C for 30 min to afford the desired products in good yields (Scheme 6.156).

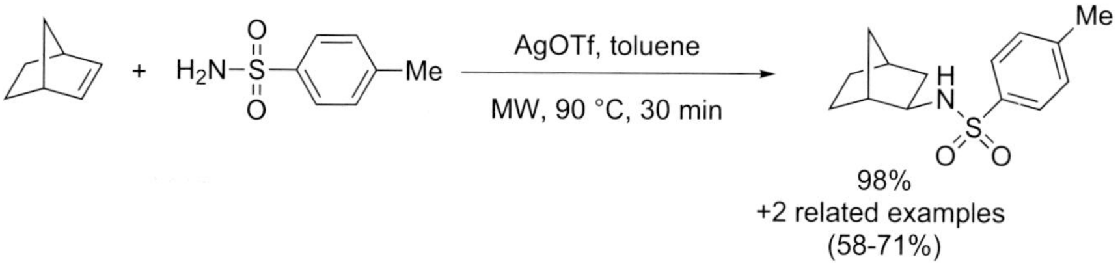

Scheme 6.156 Hydroamination of dienes with 4-toluenesulfonamide.

The microwave-assisted addition of sodium bisulfite to activated olefins has been reported by Crawley *et al.* [342]. α-Hydroxy-β-amino acids can be synthesized by conjugate addition of chiral amines to methacryloxy acrylates [343].

6.13.4
Addition to Nitriles

Bagley *et al.* have described the preparation of primary thioamides by treatment of nitriles with ammonium sulfide in methanol solution (Scheme 6.157) [344]. While the reaction with electron-deficient aromatic nitriles proceeds at room temperature, other aromatic and aliphatic nitriles require microwave heating at 80–130 °C for 15–30 min to furnish the thioamides in moderate to high yields. This protocol

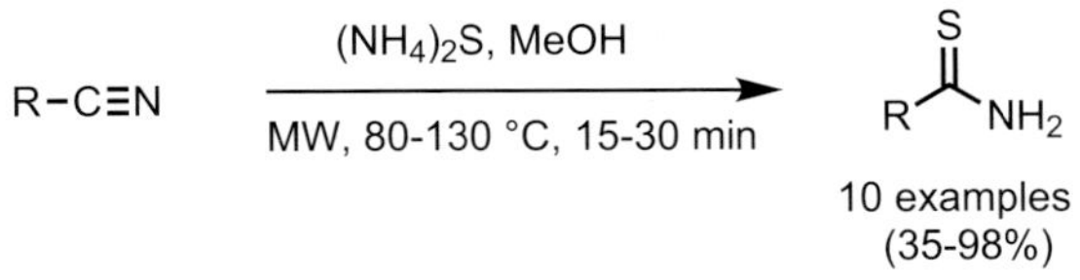

Scheme 6.157 Preparation of primary thioamides.

avoids the use of hydrogen sulfide gas under high pressure, proceeds in the absence of base, and provides thioamides usually without the need for chromatographic purification.

Similarly, amidines have been obtained by addition of ammonia to nitriles [345].

6.13.5 Elimination Reactions

A high-yielding concerted elimination process involving the conversion of *N*-sulfinyl aldimines into nitriles has been disclosed by Schenkel and Ellman (Scheme 6.158) [346]. The authors found that the self-condensation products derived from *N*-*tert*-butanesulfinyl aldimines readily undergo a concerted elimination of *tert*-butanesulfonic acid to provide the corresponding nitriles. The highest yields were obtained by heating an acetonitrile solution of the starting materials to 150 °C for 15 min. Incomplete conversion was observed at lower temperatures and significant amounts of undesired decomposition products were obtained at higher temperatures than 150 °C. These processes were applied to the synthesis of biologically important amine-containing compounds such as the seratonin 5-HT_4 agonist SC-53116. In a subsequent study, the same concept was applied to a one-pot method for the synthesis of nitriles starting from aldehydes and *tert*-butanesulfinamide [347].

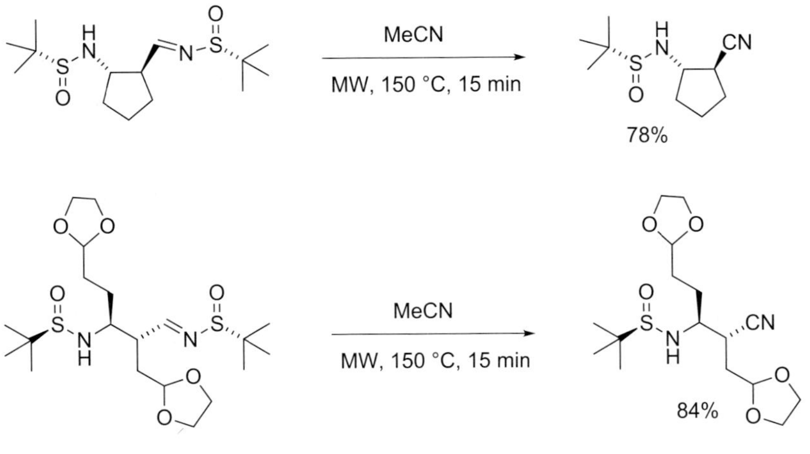

Scheme 6.158 Elimination of *tert*-butanesulfonic acid.

In the context of preparing analogs of chiral 1,2-dimethyl-3-(2-naphthyl)-3-hydroxy-pyrrrolidines, which are known nonpeptide antinociceptive agents, Collina *et al.* have reported the solvent-free dehydration of hydroxypyrrolidines to pyrrolines under microwave conditions (Scheme 6.159) [348]. In a typical experiment, the substrate was adsorbed on a large excess of anhydrous ferric(III) chloride on silica gel and then irradiated as powder under microwave conditions for 30 min at 150 °C. The microwave method takes place without racemization and provides higher yields in considerable shorter times than the conventionally heated process.

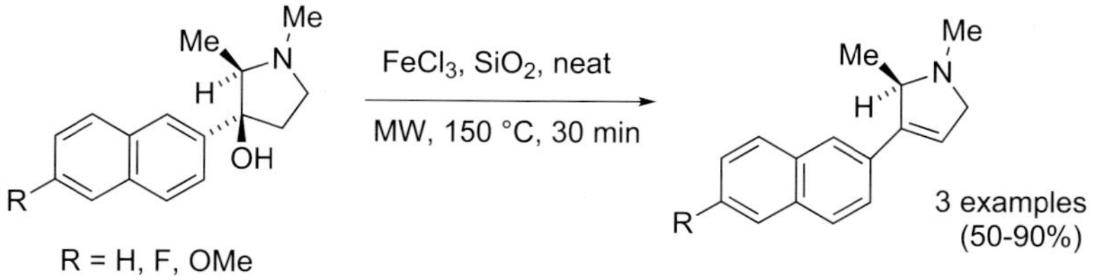

Scheme 6.159 Elimination of water from chiral pyrrolidines.

A publication by Porcheddu *et al.* describes a simple and efficient method for the synthesis of both aliphatic and aromatic isonitriles in high yields (Scheme 6.160) [349]. By employing 1.3–3.0 equiv of a cheap dehydration agent such as 2,4,6-trichloro [1,3,5]triazine (cyanuric chloride, TCT) aliphatic and aromatic formamides were transformed to their corresponding isonitriles in high yields at 50–100 °C within 3–10 min of controlled microwave irradiation.

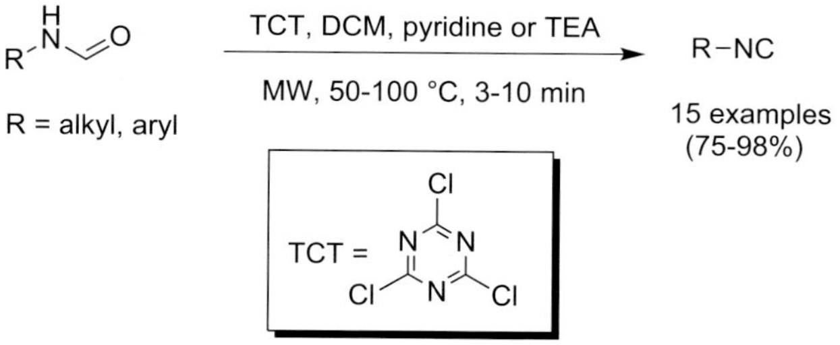

Scheme 6.160 Synthesis of isonitriles.

6.14 Substitution Reactions

A large variety of different electrophilic and nucleophilic substitution reactions performed by controlled microwave heating have been reported in the recent literature. Bose *et al.* have described electrophilic nitrations of electron-rich aromatic systems using diluted nitric acid as nitration agent (Scheme 6.161a) [350]. In their study, 4-hydroxycinnamic acid was treated with excess of ca. 15% nitric acid. On a 5 g scale, full nitration to the corresponding dinitrostyrene derivative was accomplished within 5 min at 80 °C. This nitro compound is a natural product with antifungal activity.

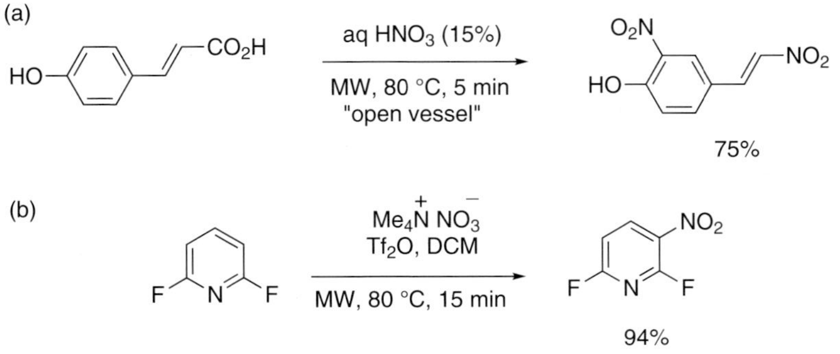

Scheme 6.161 Electrophilic aromatic nitrations.

A more elaborate electrophilic nitration system, involving a mixture of tetramethylammonium nitrate and trifluoromethane sulfonic (triflic) anhydride, to generate nitronium triflate *in situ*, was reported by Shackelford and a team from Pfizer as a general and mild nitration method (Scheme 6.161b) [351]. With this reagent, even unreactive heteroaromatics such as 2,6-difluoropyridine could be nitrated. Reactions were conducted under an inert gas in sealed vessels. The nitration reagent, nitronium triflate, was generated first by reacting tetramethylammonium nitrate and triflic anhydride at room temperature for at least 1.5 h before the addition of 2,6-difluoropyridine. The optimum microwave nitration conditions for full conversion utilized 1.5 equiv of the nitration reagent and irradiation of the dichloromethane solution at 80 °C for 15 min. This resulted in the formation of analytically pure product in 94% isolated yield. The reaction could be scaled from 3.44 mmol using a single-mode microwave reactor to 54.47 mmol in a multimode instrument providing about 9 g of material in nearly identical yields/purities.

Other electrophilic substitution reactions on aromatic and heteroaromatic systems are summarized in Scheme 6.162. Friedel–Crafts alkylation of *N,N*-dimethylaniline with squaric acid dichloride was accomplished by heating the two components in DCM to 120 °C in the absence of a Lewis acid catalyst to provide a 23% yield of the 2-aryl-1-chlorocyclobut-1-ene-3,4-dione product (Scheme 6.162a) [352]. Hydrolysis of the monochloride provided a 2-aryl-1-hydroxycyclobut-1-ene-3,4-dione, an inhibitor of protein tyrosine phosphatases [352]. Formylation of 4-chloro-3-nitrophenol with hexamethylenetetramine and trifluoroacetic acid (TFA) at 115 °C for 5 h furnished the corresponding benzaldehyde in 43% yield, which was further manipulated into a benzofuran derivative (Scheme 6.162b) [353]. 4-Chloro-5-bromopyrazolopyrimidine

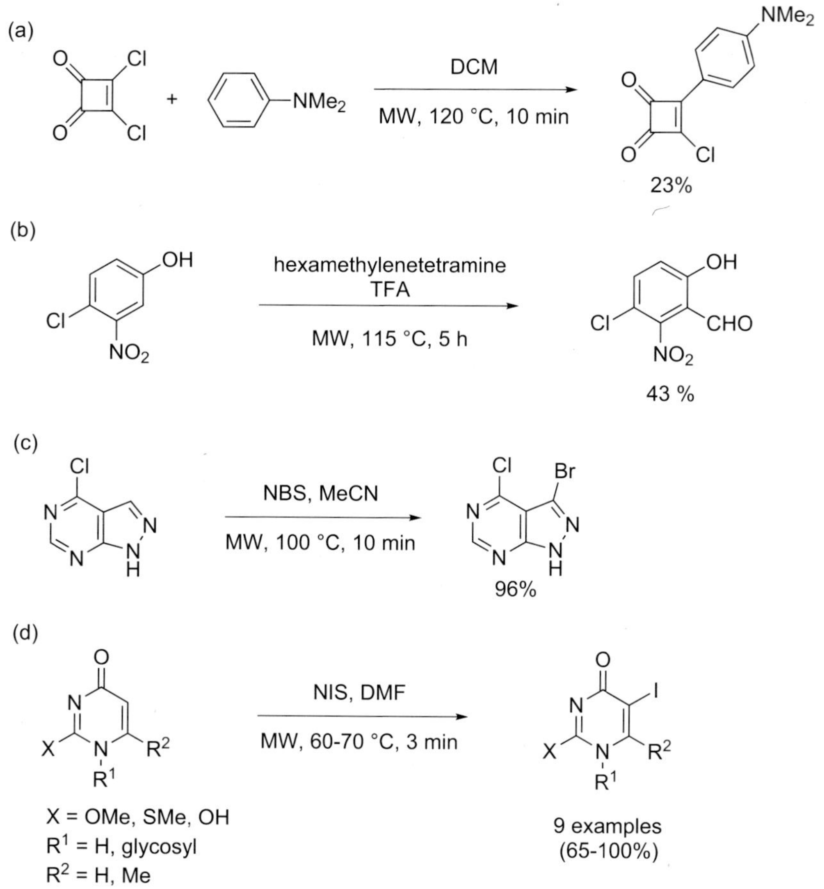

Scheme 6.162 Miscellaneous electrophilic aromatic substitutions.

is an important intermediate in the synthesis of decorated pyrazolopyrimidines with activity against multiple kinase subfamilies (see also Scheme 5.27) and can be prepared rapidly from 4-chloropyrazolopyrimidine and *N*-bromosuccinimide (NBS) by microwave irradiation in MeCN (Scheme 6.162c) [354]. Similarly, substituted pyrimidinones can be iodinated very effectively at the C-5 position with *N*-iodosuccinimide (NIS) in DMF (Scheme 6.162d) [312]. The latter transformation can also be applied on solid phase using a resin-bound pyrimidone [355].

In the context of developing a catalytic asymmetric total synthesis of quinine (see Scheme 6.59), Jacobsen and coworkers described the bromination of a 4-quinolinone analog with 1.5 equiv of dibromotriphenylphosphorane (Ph_3PBr_2) in MeCN at 170 °C (Scheme 6.163a) [133]. The desired 4-bromo-6-methoxyquinoline was obtained in 86% yield. In the context of synthesizing [^{18}F]-labeled radioligands as positron emission tomography (PET) imaging agents, Neumeyer and colleagues have reported the conversion of phenyltropane-derived tosylates into the corresponding fluoride by using tetrabutylammonium fluoride (TBAF) as fluoride source (Scheme 6.163b) [356]. Similar transformations have also been reported utilizing $Et_3N{\cdot}3HF$ as fluorination agent [357, 358]. Benzylic brominations with *N*-bromosuccinimide in the presence of a radical initiator were investigated by van Koten and coworkers [359].

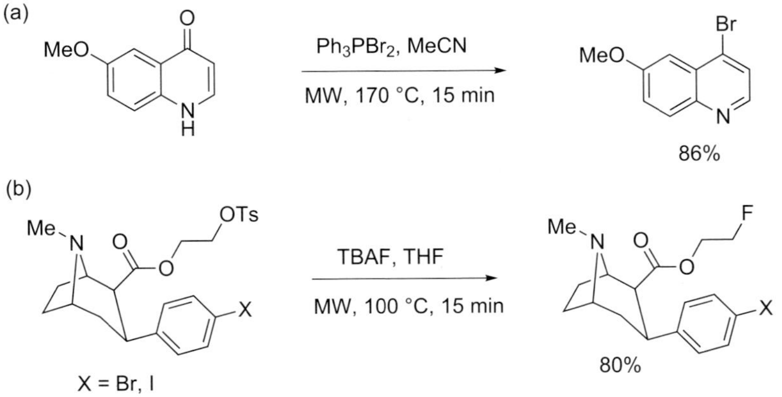

Scheme 6.163 Bromination and fluorination reactions.

A 2005 publication by Xiao and Shreeve describes the rapid electrophilic fluorination of 1,3-dicarbonyl compounds using Selectfluor as fluorinating agent (Scheme 6.164) [360]. By employing 1 equiv of Selectfluor under neutral reaction conditions, the 2-monofluorinated products were obtained. Treatment of 1,3-dicarbonyls with 3 equiv of Selectfluor in the presence of tetrabutylammonium hydroxide (TBAH) as base, difluorination can be achieved in a single reaction step, also resulting in high yields.

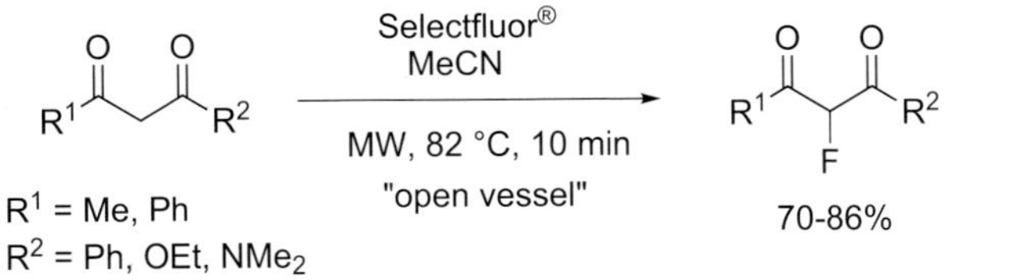

Scheme 6.164 Electrophilic fluorination of 1,3-dicarbonyl compounds.

Jones and coworkers reported on microwave-mediated fluorodenitrations and nitrodehalogenations to prepare labeled PET ligands [361]. The authors chose the synthesis of the ^{19}F-substituted CNS agent Haloperidol as model reaction of their developed method. The substrate was heated with anhydrous TBAF in DMSO at 160 °C for 10 min (2 min heat cycles with 2 min pause in between) to achieve 75% conversion (Scheme 6.165a). This rapid method also proved applicable toward the synthesis of the corresponding ^{18}F-labeled PET ligand compounds. Furthermore, the same principle was used for the synthesis of a fluoro analog of the acetyl cholinesterase inhibitor donepezil (**91**). Microwave heating at 130 °C for 5 min furnished the desired target in good yield (Scheme 6.165b). In addition to the fluorodenitration, the authors also investigated the introduction of the key nitro-leaving group into valuable arene compounds. Nitro de-iodinations were performed by heating a mixture of the aryl iodide, tetrabutylammonium nitrite, copper powder, and *N,N'*-dimethylethylenediamine in DMF at 110 °C (Scheme 6.165c). Various substrates could be functionalized with this method to obtain the desired compounds in less than 20 min.

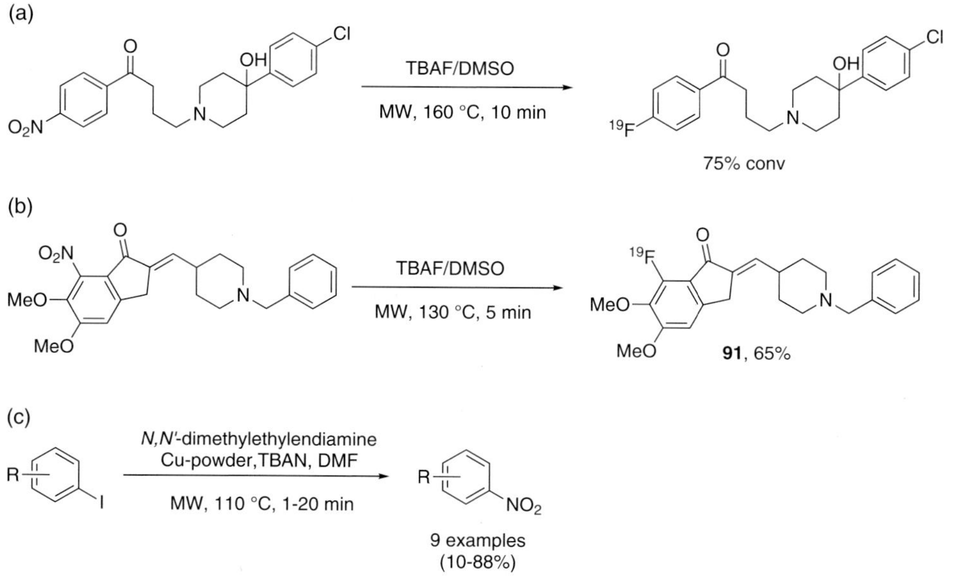

Scheme 6.165 Preparation of fluorolabeled PET ligands.

Deoxyfluorinations of amino alcohols without solvent using *N,N*-diethyl-α,α-difluorobenzylamine (DFBA) as reagent have been reported by Hara and coworkers [362]. Deoxyfluorinations of 3-deoxy-D-manno-2-octurosonic acid using *N,N*-diethyl-α,α-difluoro-(*m*-methylbenzyl)amine (DFMBA) as reagent have been studied by Fukase and coworkers [363]. Similar deoxyfluorinations of primary alcohols can also be achieved using TBAF·HF as reagent [364]. Applying $Et_3N \cdot 3HF$ as reagent, epoxides can be efficiently ring-opened to fluoroalcohols [365].

Rapid halide exchange reactions in aryl halides were investigated by Arvela and Leadbeater using nickel(II) halides as reagents (Scheme 6.166) [366]. The methodology can be used for the conversion of aryl chlorides to bromides, aryl iodides to bromides and chlorides, and aryl bromides to chlorides. The exchange reactions are fast for microwave heating (5 min) and can be performed without the need for exclusion of air and water in DMF as solvent. Typically, 2 equiv of nickel(II) halide is used in these transformations.

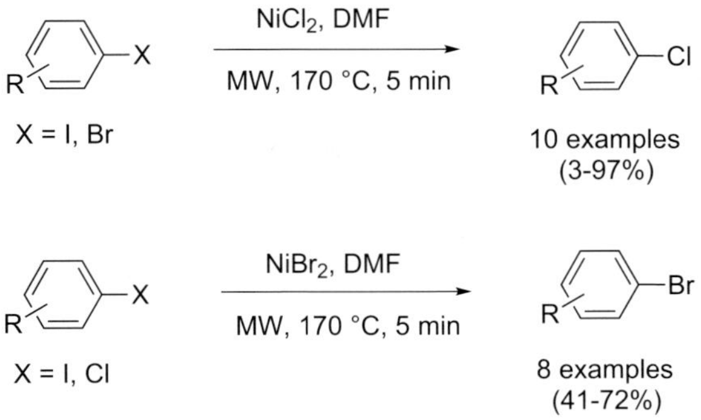

Scheme 6.166 Halide exchange reactions in aryl halides.

In 2009, Kappe and coworkers described an effective microwave-assisted fluorine–chlorine exchange [367]. Employing highly polar triethylamine trihydrofluoride (TREAT-HF) allows rapid heating to reaction temperatures beyond 200 °C accomplishing the desired transformation within a couple of minutes. The authors also highlight the use of a silicon carbide reaction vessel to avoid likely occurring corrosion of glass vials during the fluorination process. During optimization, utilizing dichloromethylbenzene, it became apparent that water present in the system favors hydrolysis of the substrate resulting in the formation of simple benzaldehyde as the major product instead of the desired difluoromethylbenzene. Distilling commercially available TREAT-HF prior to use significantly improved the results of the fluorination. The authors could also verify that TREAT-HF remarkably serves selective stepwise mono-, di-, and trifluorination of model trichloromethylthiobenzene by simply increasing the reaction temperature. Within only 5 min at 70 °C, 160 °C, or 250 °C, the corresponding products were obtained in high yields. With this set of information in hand, the authors turned the focus toward the fluorination of ethyl 3-dichloromethyl-1-methyl-1*H*-pyrazole-4-carboxylate. Heating at 250 °C for 5 min furnished the target difluorinated product **92** (Scheme 6.167).

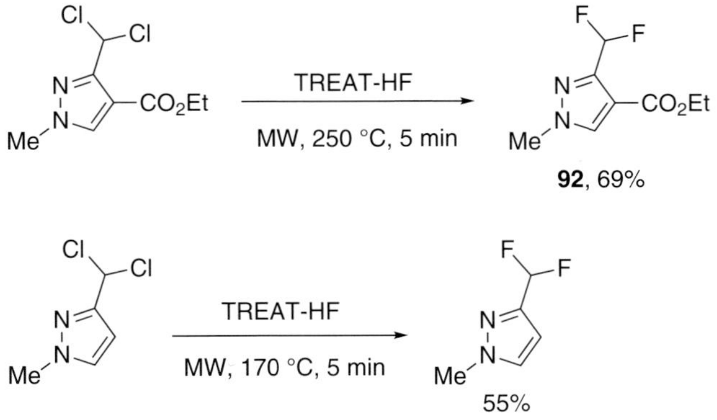

Scheme 6.167 Fluorination of dichloromethyl pyrazoles utilizing TREAT-HF.

The product heterocycle is a key precursor in the preparation of pyrazolyl carboxamides, which are industrially applied as active fungicidal ingredients. To demonstrate the efficiency of the protocol, the authors finally applied it toward a recently patented autoclave fluorination of 3-dichloromethylpyrazole. Utilizing microwave heating, the reaction was completed in only 5 min, whereas the patented process requires heating at 160 °C for 1 h.

Additional microwave-assisted substitution reactions, involving *N*-chlorinations [368], *O*-triflations [369], and *O*-sulfation [370], are shown in Scheme 6.168.

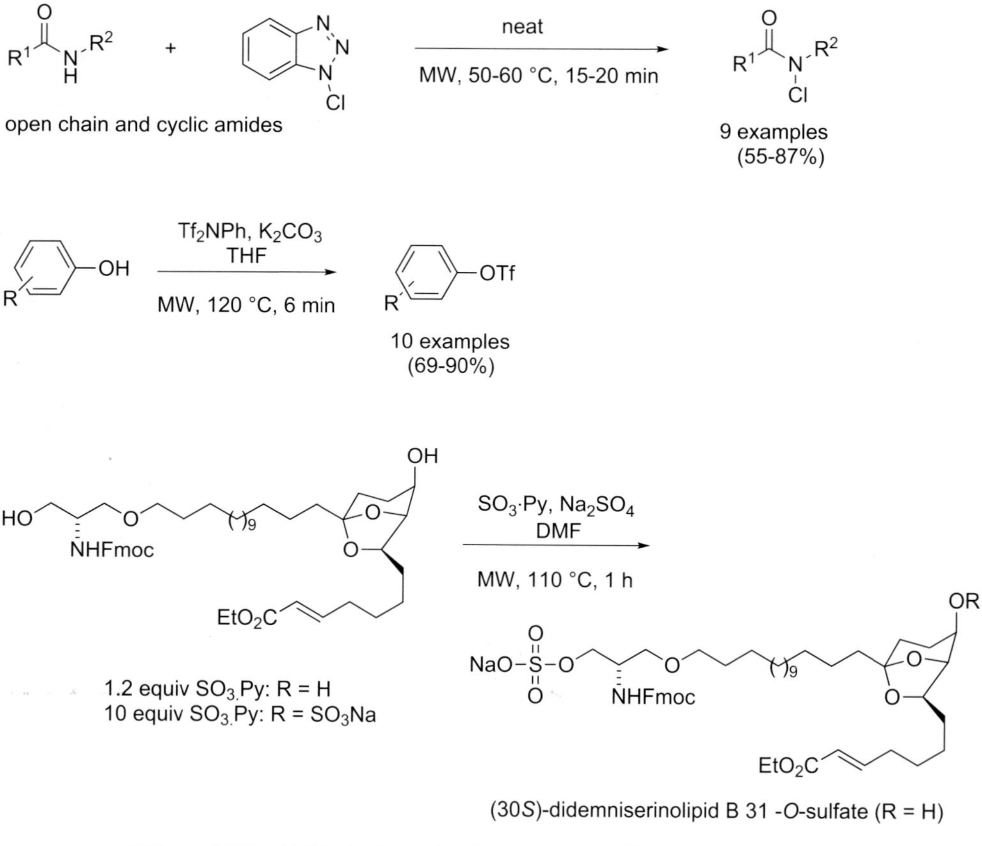

Scheme 6.168 *N*-Chlorinations, *O*-triflations, and *O*-sulfation reactions.

Ju and Varma have reported a double *N*-alkylation by alkyl dihalides of aniline derivatives for the synthesis of *N*-aryl azacycloalkanes (Scheme 6.169) [371]. By applying microwave irradiation at 120 °C for 20 min, it was possible to significantly improve the yields, compared to conventional heating, which is related to the elimination of side reactions. This "green chemistry" approach uses very mild basic aqueous conditions that not only tolerates a variety of functional groups but also

Scheme 6.169 Synthesis of *N*-aryl azacycloalkanes.

simplifies the isolation of the solid or liquid products by the occurring phase separation.

The same authors have subsequently reported on the direct syntheses of 4,5-dihydropyrazole, pyrazolidine, and 1,2-dihydrophthalazine derivatives by double alkylation of hydrazine compounds with alkyl dihalides or tosylates (Scheme 6.170) [372]. Using water as solvent, the reaction could be performed under microwave heating at 120 °C for 20 min without the need of a phase-transfer catalyst, making this protocol more environment-friendly.

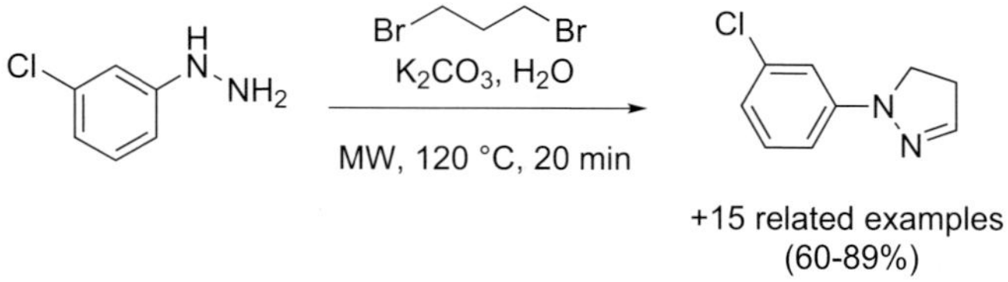

Scheme 6.170 Cyclocondensation of hydrazine derivatives with alkyl dihalides or ditosylates.

The same group has also presented the preparation of azides, thiocyanates, and sulfones in aqueous medium by nucleophilic displacement of halides or tosylates by the appropriate nucleophile [373]. Displacement of iodide with amino acid esters was used for the synthesis of $\psi[CH_2NH]$ amide bond surrogates [374].

Nucleophilic substitution of chiral mesylates by Cs phenoxylates in the presence of 18-crown-6 provided the corresponding aryl ether as a pure stereoisomer [375]. Purines have been used as nucleophiles in substitution reactions [376]. The cyanation of alkyl chlorides with NaCN in MeCN has been demonstrated [377]. For microwave-assisted Chichibabin reactions, see Ref. [378].

6.15
Enamine and Imine Formations

The formation of enamines from the corresponding enols in two series of complex thiazole-containing heterocyclic molecules has been reported by Bagley *et al.*

(Scheme 6.171) [105, 379]. Treatment of the enols with excess ammonium acetate in toluene at 120 °C for 30 min provided the expected products in good yields as single enamine tautomers. The enamines were then further manipulated into thiopeptide antibiotics again utilizing microwave heating in some of the reaction steps.

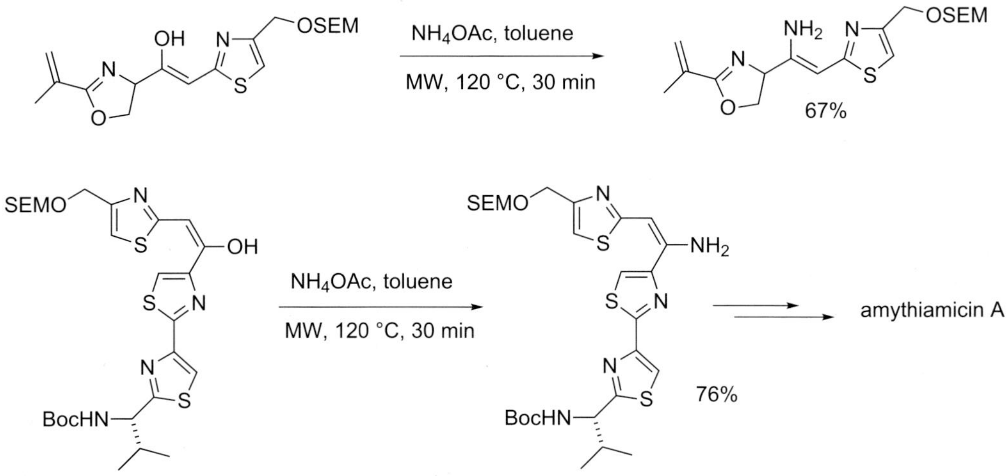

Scheme 6.171 Enamine formations.

Similarly, imines can be formed by condensation of ketones and suitable primary amines under microwave conditions (Scheme 6.172). Since the formation of imines is an equilibrium process, these transformations are best carried out under open-vessel conditions, allowing the formed water to be removed from the reaction mixture and therefore from the equilibrium (see Section 4.3). The groups of Loupy

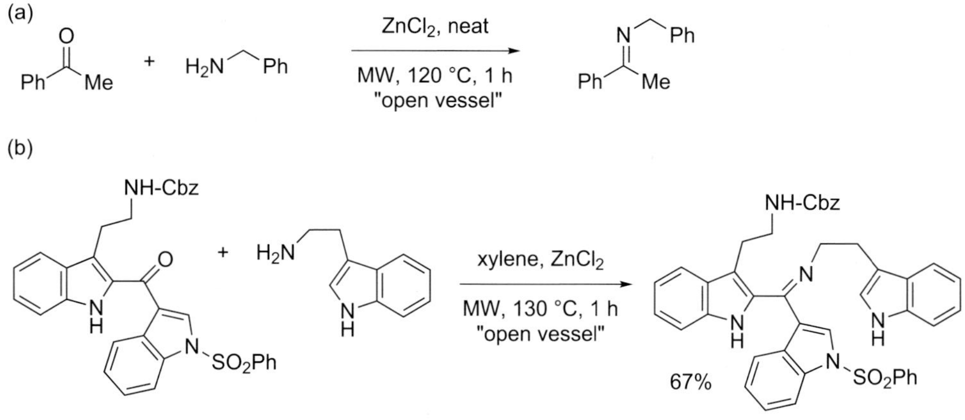

Scheme 6.172 Ketimines from ketones and primary amines.

(Scheme 6.172a) [380] and Langlois (Scheme 6.172b) [381] have independently described the efficient formation of ketimines from ketones and primary amines using anhydrous zinc(II) chloride (10–50 mol%) as catalyst. While in the example presented in Scheme 6.172a the reaction was performed in the absence of any solvent, *para*-xylene was used as a solvent in the second example (Scheme 6.172b) [381]. The formed ketimine from this study was subsequently used in the synthesis of the marine alkaloid bengacarboline.

6.16
Reductive Aminations

Öhberg and Westman were the first to report on reductive aminations under controlled microwave irradiation conditions. In a 2001 publication [382], the authors demonstrated a one-pot, three-step synthesis of thiohydantoins (Scheme 6.173). The first two steps consisted of a reductive amination involving 4-bromobenzaldehyde and an amino acid ester. The aldehyde was treated with the corresponding amino acid ester hydrochloride together with 1.1 equiv of triethylamine base in 1,2-dichloroethane (DCE) as solvent. After 5 min of irradiation at 140 °C, formation of the imine intermediate was complete. After the addition of 1.4 equiv of sodium triacetoxy borohydride, the reaction mixture was subsequently resubjected to microwave heating for another 9 min at 170 °C to give the desired *N*-benzylated amino acid esters. Addition of an isothiocyanate building block and triethylamine base (2 equiv each) followed by microwave heating to 170 °C for a further 5 min ultimately provided the target thiohydantoin products in acceptable overall yield after chromatographic purification.

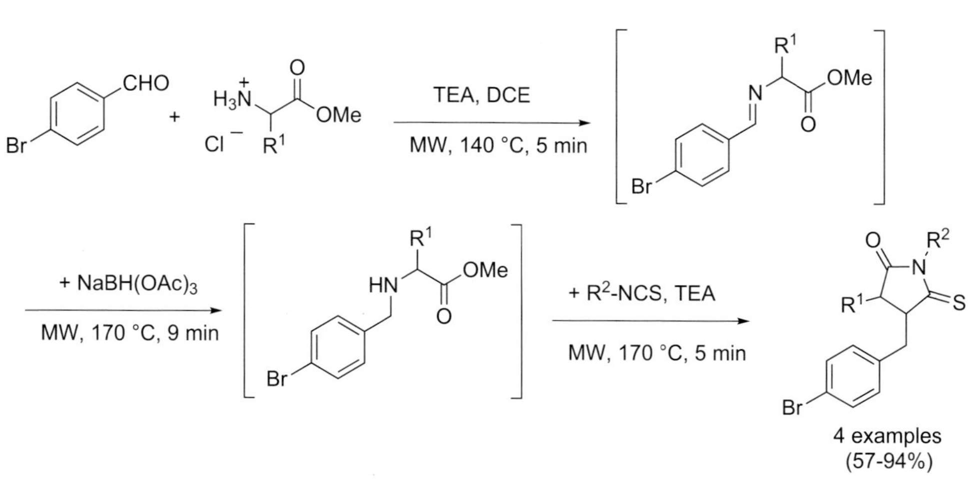

Scheme 6.173 Reductive amination and thiohydantoin synthesis.

A team from Medivir Ltd. has described a fast protocol for the direct reductive amination of aldehydes that uses dibutyltin dichloride as catalyst in the presence of phenylsilane as reductant (Scheme 6.174) [383]. Best results are obtained utilizing 10 mol% of dibutyltin dichloride as a catalyst and 2 equiv of phenylsilane as the reducing agent, employing THF as solvent and sealed-vessel microwave heating at 100 °C for 7 min. Suitable reactants for this novel protocol include aliphatic and aromatic aldehyde building blocks on the one hand and anilines and secondary and primary amines on the other. Workup typically involves extraction or cation exchange methods.

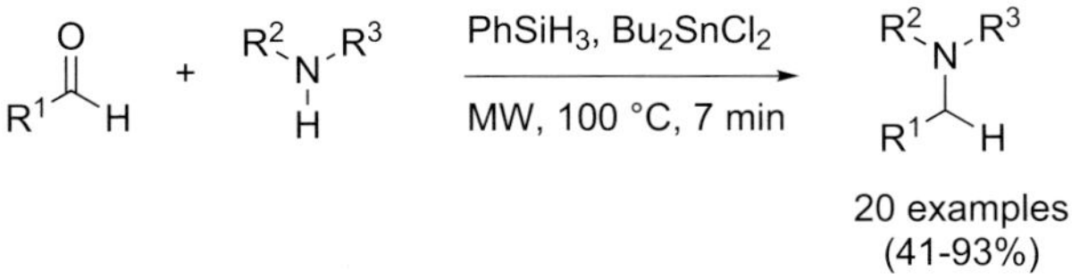

Scheme 6.174 Direct reductive amination of aldehydes.

Further applications of microwave-assisted reductive aminations are shown in Schemes 6.175 and 6.176. In the example shown in Scheme 6.175, Baran and Richter have utilized a reductive amination in their synthesis of the natural product (+)-hapalindole Q [384]. Employing 10 equiv of sodium cyanoborohydride and 40 equiv of ammonium acetate in a MeOH/THF mixture (150 °C, 2 min) allowed the preparation of the primary amine as a 6 : 1 mixture of diastereomers. Transformation to the isothiocyanate completed the total synthesis of (+)-hapalindole Q (Scheme 6.175).

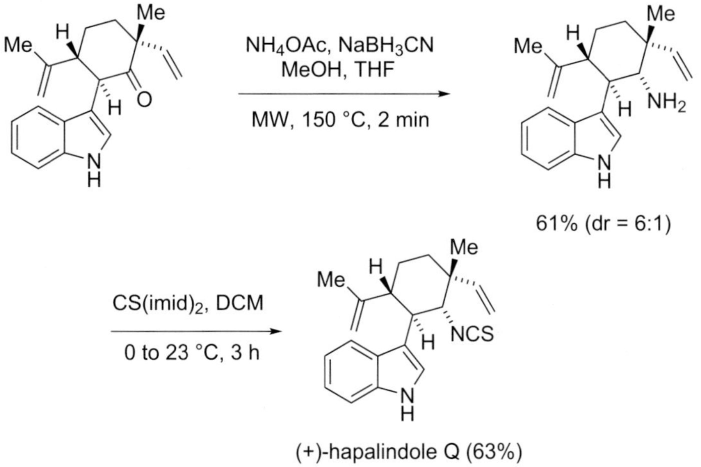

Scheme 6.175 Reductive amination in the synthesis of (+)-hapalindole Q.

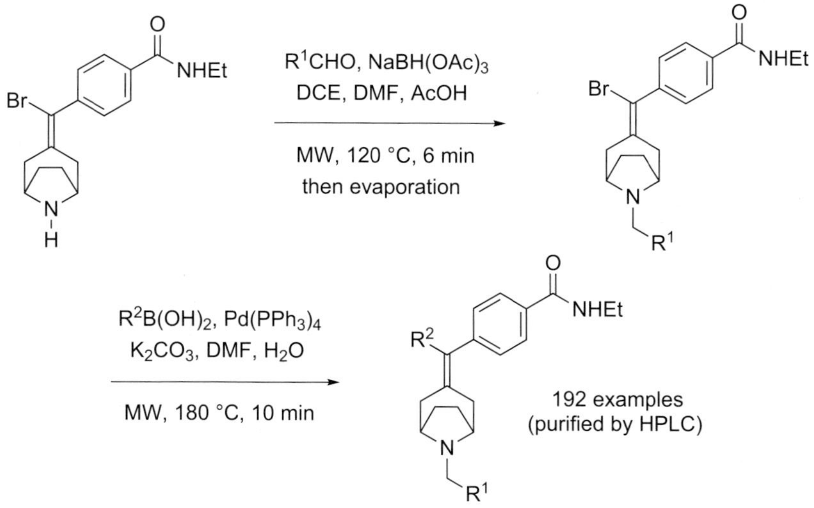

Scheme 6.176 Reductive amination and Suzuki couplings in the synthesis of tropanylidene opioid receptors.

In the chemistry described in Scheme 6.176, Coats and a group of researchers from Johnson and Johnson utilized successive reductive aminations and Suzuki cross-coupling reactions to prepare a 192-member library of tropanylidene benzamides [385]. This series of tropanylidene opioid agonists proved to be extremely tolerant with regard to structural variation while maintaining excellent opioid activity. For the solution-phase preparation of functionalized tropanylidenes, the authors simply dispensed DCE solutions of the bromo N–H precursor to a set of microwave vials, added the aldehydes (3 equiv) and a solution of sodium triacetoxy borohydride in dimethylformamide (2 equiv), and subjected the mixture to microwave irradiation for 6 min at 120 °C. Quenching the reductive amination with water and subsequent concentration allowed to directly perform a microwave-assisted Suzuki reaction on the crude products.

Santagada *et al.* have disclosed a reductive amination method for the generation of a reduced peptide bond by reaction of a protected amino acid aldehyde with an *N*-deprotected amino ester using sodium cyanoborohydride as reducing agent [386]. A related procedure for the preparation of *N*-alkylated glycine methyl esters from glycine methyl ester and aldehydes was described by the same group [387].

A special case of reductive amination was described by Kann and coworkers in the preparation of P-chirogenic *β*-amino phosphine borans starting from the corresponding P-chirogenic aldehydes, amines, and $NaBH(OAc)_3$ [388].

Microwave-promoted direct transformations of amines to ketones have been disclosed by Miyazawa *et al.* (Scheme 6.177) [389]. By using water as an oxygen source and catalytic amounts of Pd/C, the *retro*-reductive amination of mono- or

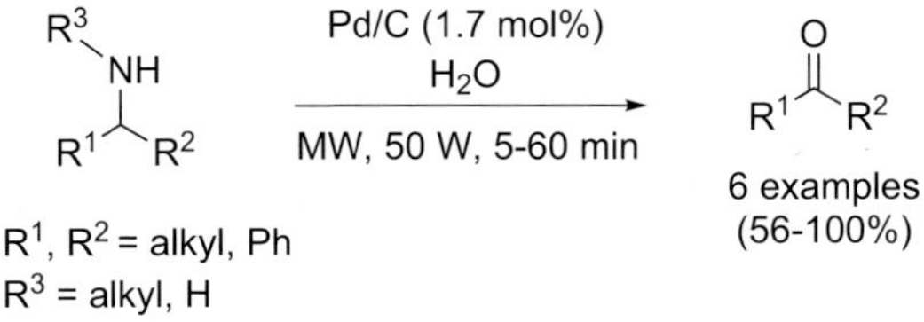

Scheme 6.177 Retro-reductive aminations.

di-*sec*-alkylamines to the corresponding ketones was performed very efficiently. This new and "green" method provides ketones with a high selectivity and faster rate compared to other methods that use late transition metal oxidants (e.g., $KMnO_4$, $Pb(OAc)_4$) in large amounts and strong bases such as *n*-butyllithium.

The reduction of an azide group with triphenylphosphine in THF by microwave heating to 130 °C for 5 min was described by Kihlberg and colleagues (Scheme 6.178) [390]. The use of diethyl 4-(hydrazinosulfonyl)-benzyl phosphonate as *in situ* diazene precursor for the reduction of trisubstituted *gem*-diiodoalkenes to terminal geminal diodides under microwave conditions has also been reported [391].

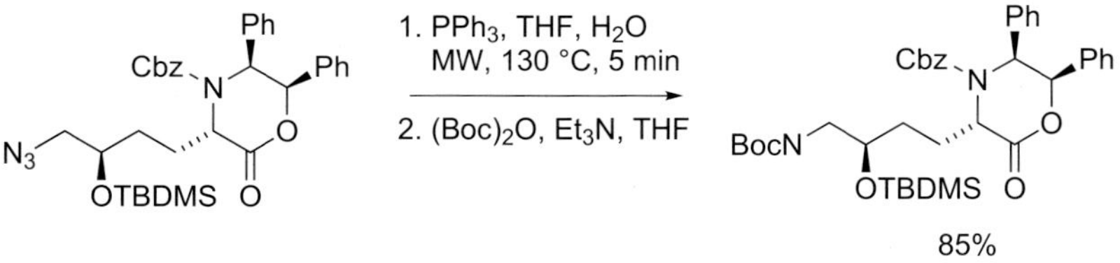

Scheme 6.178 Reduction of an azide group with triphenylphosphine.

6.17
Ester and Amide Formation

The preparation of 4-hydroxybenzoic acid esters (parabans) possessing antimicrobial activity via esterification of 4-hydroxybenzoic acid has been performed by Raghavan and coworkers (Scheme 6.179) [392]. Optimum results were obtained using alcohol (1-butanol) as solvent in the presence of catalytic amounts of zinc(II) chloride or *p*-toluenesulfonic acid (*p*-TsOH) under atmospheric conditions. After 5 min of

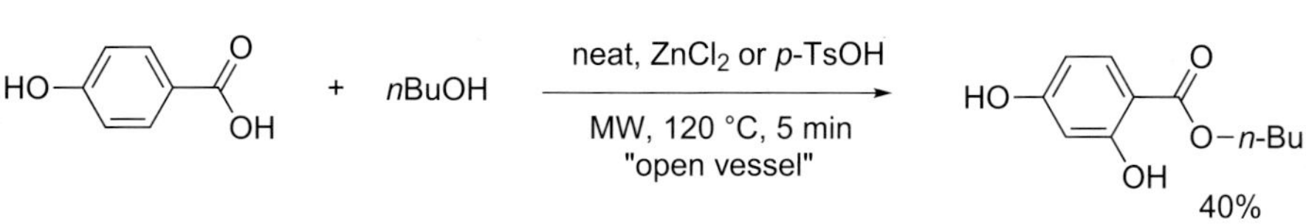

Scheme 6.179 Esterification of benzoic acid.

microwave irradiation to 120 °C, about 40% conversion to the ester was observed. Related studies on the synthesis of long-chain aliphatic esters were described by Mariani and coworkers [393].

In addition, ester bonds have been formed starting from benzotriazol-activated *N*-protected amino acids and terpene alcohols [394].

An unusual class of heterocycles are polyketide-derived macrodiolide natural products. Porco and coworkers have shown that stereochemically well-defined macrodiolides of this type can be obtained by cyclodimerization (transesterification) of nonracemic chiral hydroxy esters (Scheme 6.180) [395]. Preliminary experiments involving microwave irradiation demonstrated that exposing dilute solutions of the hydroxy ester (0.02 M) in chlorobenzene to sealed-vessel microwave irradiation conditions (200 °C, 7 min) in the presence of 10 mol% of a distannoxane transesterification catalyst led to a 60% isolated yield of the 16-membered macrodiolide heterocycle. Conventional reflux conditions in the same solvent (0.01 M of hydroxy ester) provided a 75% yield after 48 h at about 135 °C.

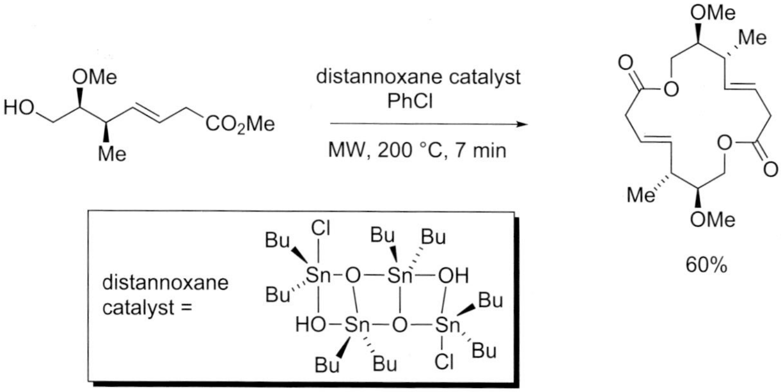

Scheme 6.180 Macrodiolide formation by cyclodimerization.

Carboxylic acid esters can be transesterified with methanol (and other alcohols) using $Sc(OTf)_3$ as a catalyst [396].

Colombo *et al.* generated a small library of aryl amides utilizing phosphortrichloride as dehydrating agent under microwave irradiation [397]. The authors employed anilines substituted with strong electron-withdrawing groups as preferred substrates for the condensation with various carboxylic acids. Applying the conditions described in Scheme 6.181, the authors generated a 46-membered library in sequential manner combined with a quick workup procedure utilizing solid-phase extraction on basic alumina. The protocol is tolerable to a variety of strong electron-withdrawing groups, although in some cases somewhat longer reaction times (15 min) were required to achieve full conversion. However, aminopyridyl substrates proved rather sluggish, conversion was incomplete, and therefore isolated yields were quite low. Finally, the

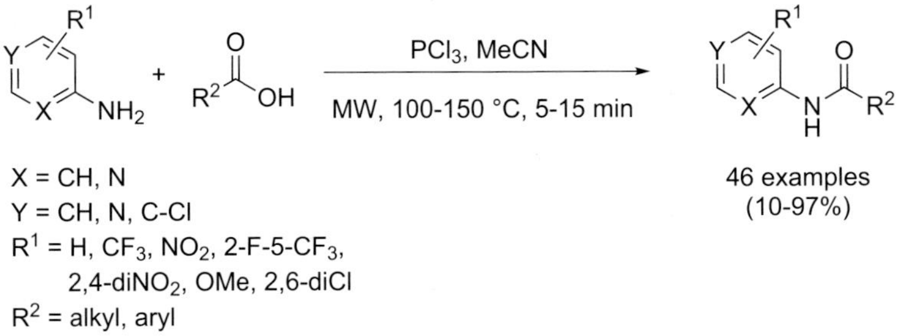

Scheme 6.181 Phosphorus-mediated synthesis of amides with substituted anilines.

method was also applied to anilines without electron-withdrawing groups. With the established standard protocol, the overcondensed product was generated in significant amount besides the desired amide due to the higher nucleophilicity. Slight modification of the procedure by simply reducing the reaction temperature to 100 °C or 120 °C, respectively, led to the pure amides in excellent yields.

Danishefsky and coworkers investigated the formation of amide bonds from isonitriles for the generation of tertiary amines *en route* to the total synthesis of cyclosporine A (**93**) [398]. A solution of azido valine in DCE was admixed with an equimolar amount of leucine isonitrile and microwave-heated at 155 °C for 40 min to furnish the corresponding *N*-formyl dipeptide (Scheme 6.182). After purification by

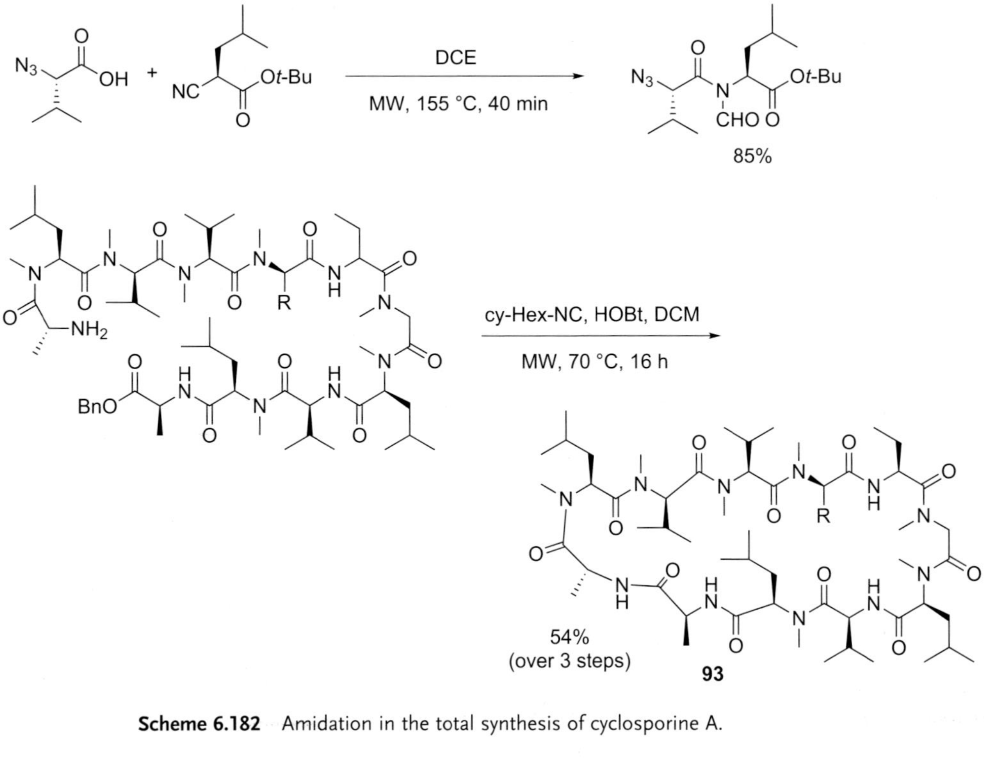

Scheme 6.182 Amidation in the total synthesis of cyclosporine A.

chromatography, the product could be used in an additional four-step process to achieve the required polyamide fragment. Furthermore, in the very last step of polylactamization in the cyclosporine synthesis, the aid of microwave irradiation proved beneficial. The present crude undecapeptide was heated with cyclohexylisonitrile in the presence of hydroxybenzotriazole (HOBt) in DCM under microwave irradiation at 70 °C for 16 h. Satisfyingly, the transformation proceeded very clean and in high conversion to furnish the desired cyclosporine A in good overall yield (Scheme 6.182).

Toma and coworkers have described the solvent-free synthesis of salicylanilides from phenyl salicylate or phenyl 4-methoxysalicylate and substituted anilines (Scheme 6.183) [399]. By exposing an equimolar mixture of the ester and the amine to microwave irradiation at 150–220 °C for 4–8 min under open-vessel conditions, good yields of the corresponding salicylanilides were obtained. This synthesis was carried out on a multigram scale (0.1 mol).

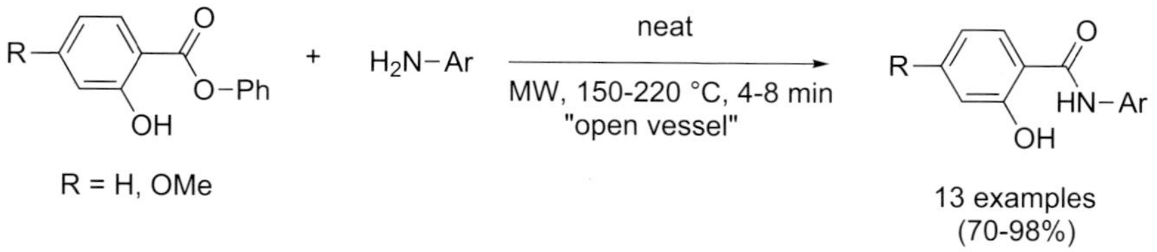

Scheme 6.183 Synthesis of salicylanilides.

Woodward and coworkers have demonstrated the direct coupling of cyclic secondary amines with esters employing the air-stable trimethylaluminum source DABAL-Me_3 for the preparation of tertiary amides (Scheme 6.184) [400]. Microwave heating at 130 °C for 5–16 min afforded the amides in good to excellent yields, with the longer reaction times necessary for more hindered amine–ester pairs. Weinreb amides can also be prepared in good yields by using a one-pot *in situ* deprotonation with NaH to liberate the free *N*-methoxy-*N*-methylamine followed by the coupling step.

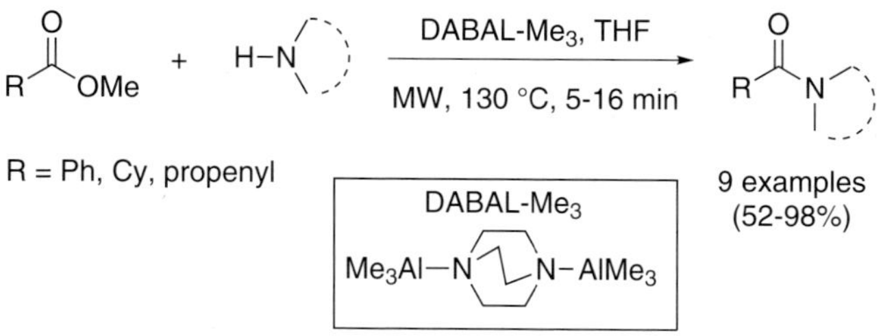

Scheme 6.184 Synthesis of tertiary amides from esters.

A double acylation reaction involving phosgene was used by Holzgrabe and Heller in their synthesis of diazepinone analogs of the muscarinic receptor antagonist

AFDX-384 (Scheme 6.185) [245]. Treatment of a solution of 6-oxo-5,11-dihydrobenzo[*e*]pyrido[3,2-*b*][1,4]diazepine and 2 equiv of *N,N*-diisopropylethylamine (DIEA) in dioxane with 1.75 equiv of phosgene at room temperature, followed by heating to 85 °C for 2 h, led to the carbonyl chloride that was transformed to the desired target structure by reaction with a suitable piperidine fragment (1.15 equiv) and DIEA base, again using microwave conditions (110 °C, 10 min). Subsequent *N*-debenzylation provided the desired AFDX-384 analog.

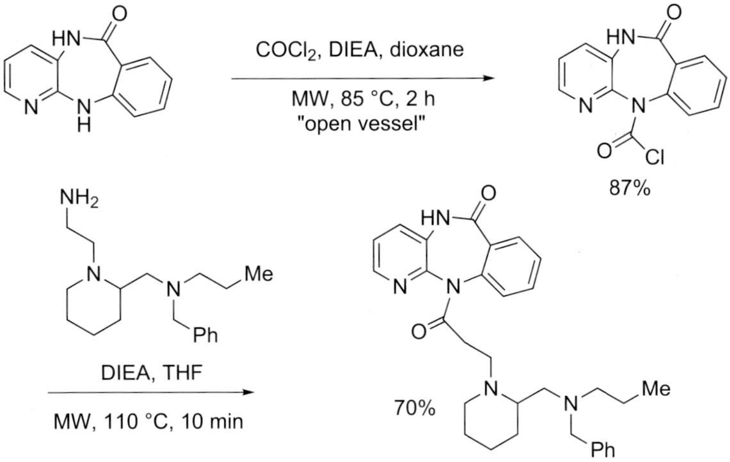

Scheme 6.185 Synthesis of muscarinic receptor antagonist analogs.

A high-throughput method for the monoacylation of 7-amino-5-aryl-6-cyanopyrido[2,3-*d*]pyrimidines with acid chlorides was reported by Nicewonger *et al.* (Scheme 6.186) [401]. Since incomplete conversions were achieved with either 1.5 or 3 equiv of the acid chloride, an optimized microwave protocol was elaborated that utilized 9 equiv of acid chloride at 230 °C using pyridine as a solvent. Under these conditions, however, mixtures of mono- and diacylated products were obtained. Treatment of the crude reaction mixture with an excess of macroporous Trisamine resin for 2 h allowed complete conversion of the diacylated product to the mono-acylated product while scavenging the excess acid chloride.

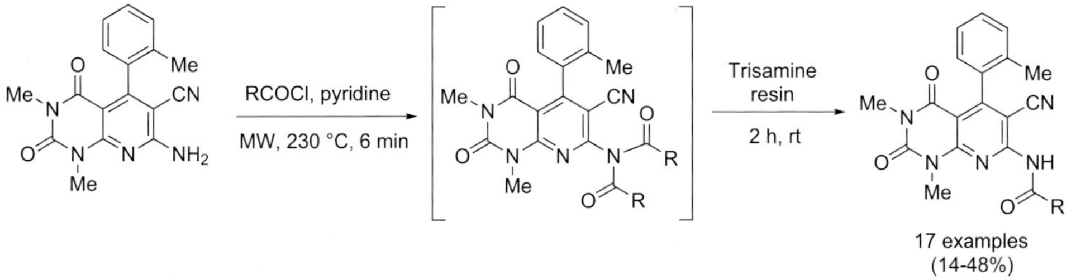

Scheme 6.186 Acylation of 7-amino-5-aryl-6-cyanopyrido[2,3-*d*]pyrimidines.

Miriyala and Williamson have described the synthesis of β-ketocarboxamides from primary and secondary amines and 2,2-dimethyl-2H,4H-1,3-dioxin-4-ones as reactive α-oxoketene precursor (Scheme 6.187) [402]. The experimental procedure involved heating a mixture of the dioxinone with 2–3 equiv of the amine without solvent in a sealed vessel under microwave irradiation to about 180 °C for 1–3 min. A small collection of 18 β-ketocarboxamides was prepared in very high yields using this protocol.

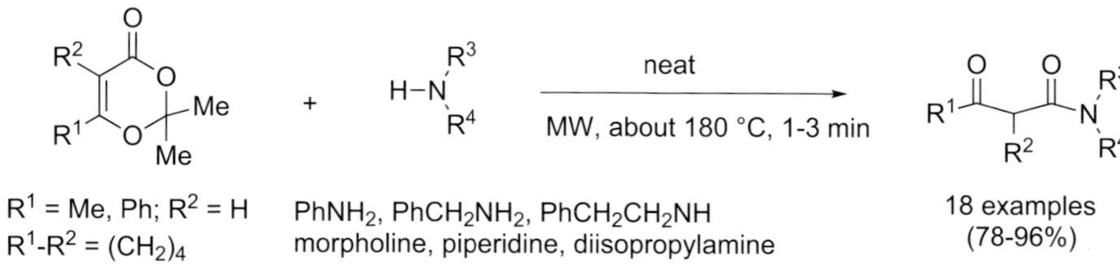

Scheme 6.187 Preparation of β-ketocarboxamides.

Benzoic acid amides have been obtained by direct, solvent-free amidation of benzoic acid with a variety of amines [403]. Similar results have been disclosed using acetic acid [404]. Benzoic acid amides have also been obtained by solvent-free Ritter reaction involving the reaction of benzylic alcohols with benzonitriles catalyzed by Nafion NR50 [405]. Salicylamides have been obtained via BCl_3-mediated coupling of phenols with isocyanates [406]. Succinimides can be prepared from succinic anhydrides and hydrazines using water as solvent [407] or by treatment with amines under solvent-free conditions [408]. Hydroxamic acids can be formed directly from esters and hydroxylamine under microwave conditions [409]. Sterically hindered amino acids such as *N*-benzylaminoisobutyric acid can be incorporated into dipeptides under microwave irradiation in the presence of zinc dust [410]. Similar results can be obtained with benzotriazol-activated *N*-protected amino acids [411, 412].

Caddick *et al.* have described the synthesis of functionalized sulfonamide via microwave-assisted displacement of pentafluorophenyl (PFP) sulfonate esters with amines (Scheme 6.188) [413]. Their ease of handling due to their higher crystallinity along with their shelf stability and their ability to react under aqueous reaction conditions makes them an attractive alternative to sulfonyl chlorides. The microwave-assisted reaction of alkyl PFP esters with amines is a facile process that proceeds

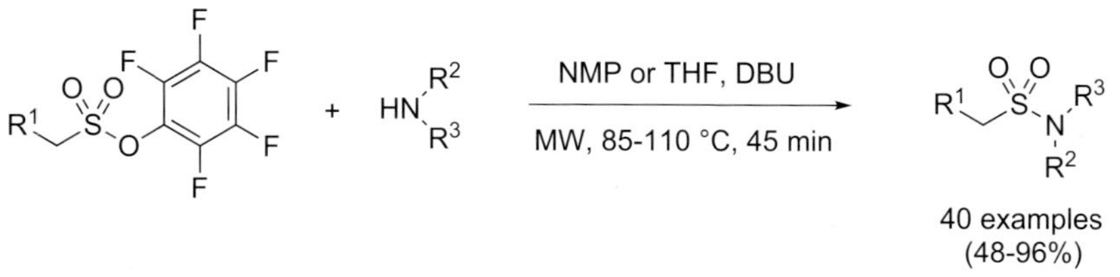

Scheme 6.188 Preparation of sulfonamides.

cleanly and in good yields with a number of different amines including primary, secondary, sterically hindered amines and anilines. Optimum conditions involved a reaction time of 45 min, 2 equiv of an amine base such as DBU, and a temperature of 85–110 °C using either 1-methyl-2-pyrrolidone or tetrahydrofuran as solvent.

Sulfonamides have also been obtained directly from sulfonic acids and amines using 2,4,6-trichloro-[1,3,5]triazine as dehydrating agent [414]. Alternatively, the treatment of sulfonylchlorides with *N*-arylpiperidines in the presence of Et_3N base has been reported [415].

6.18
Decarboxylation Reactions

In the context of the preparation of a library of pyrazole-based cyclooxygenase II (COX-II) inhibitors, the Organ group has described the microwave-assisted decarboxylation of a pyrazole carboxylic ester with 20% sulfuric acid (Scheme 6.189a) [255]. While the conventional protocol (reflux, 100 °C) required 96 h to provide a 86% yield, full conversion could be achieved within 5 min at 200 °C under microwave heating, leading to an 88% isolated product yield.

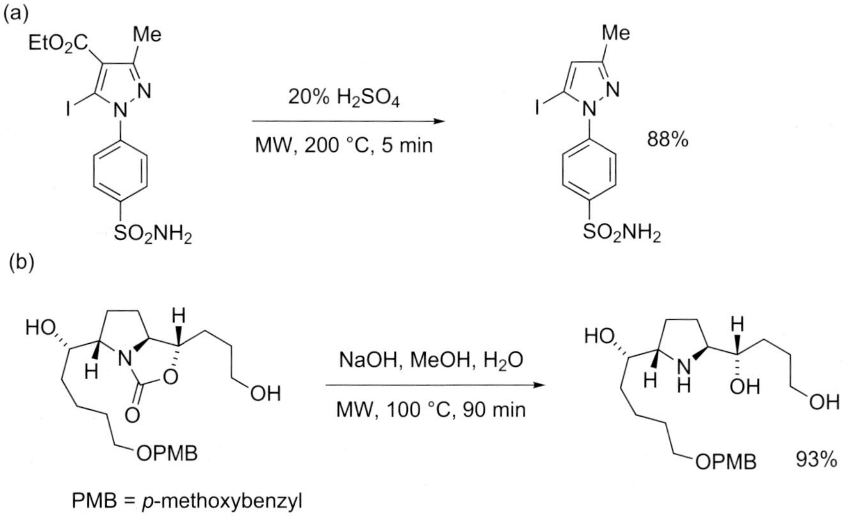

Scheme 6.189 Decarboxylation reactions.

Lindsay and Pyne have utilized microwave heating in a base-catalyzed cleavage of the oxazolodinone group during the total synthesis of the tricyclic core structure of the cromine alkaloids (Scheme 6.189b) [311].

Dealkoxycarbonylations of malonic esters have been described by the Curran and Moberg groups (Scheme 6.190). Zhang and Curran found that a variety of malonates and β-ketoesters could be rapidly dealkoxycarbonylated when heated under

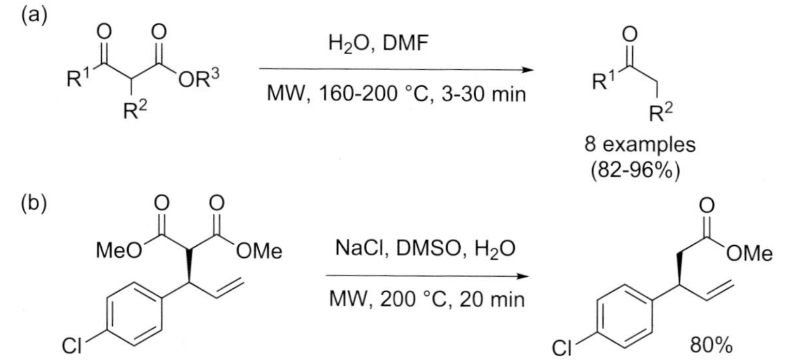

Scheme 6.190 Dealkoxycarboxylation reactions.

microwave conditions in wet DMF (2.4 equiv of water) to temperatures of 160–200 °C for 3–30 min (Scheme 6.190a) [416]. This novel transformation shows good generality for unsubstituted and monosubstituted malonates and ketoesters, while being ineffective for dialkylated analogs. The required time and temperature depended significantly on the substitution pattern, ranging from 3 min for unsubstituted to 30 min for substituted derivatives.

A similar dealkoxycarbonylation reaction utilizing the Krapcho conditions was used by Belda and Moberg in the synthesis of (*R*)-baclofen (Scheme 6.190b) from a chiral malonate precursor [277].

Related to the transformations described in Scheme 6.190 are the decomposition reactions of mono- and dialkylated Meldrum's acids highlighted in Scheme 6.191 [417].

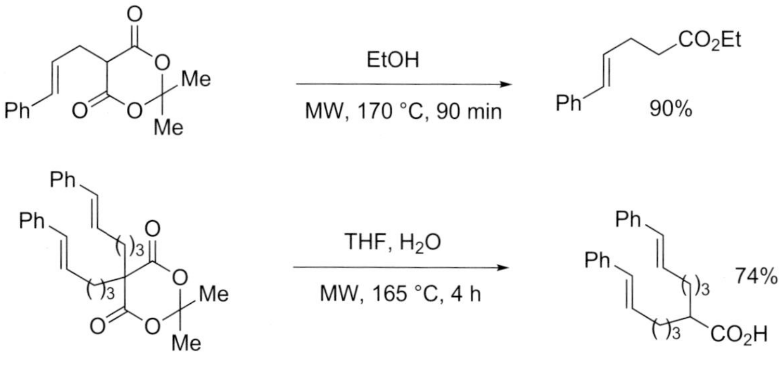

Scheme 6.191 Transesterification and decarboxylation reactions.

In 2009, Goossen *et al.* developed a simple microwave-mediated procedure for the protodecarboxylation of aromatic carboxylic acids [418]. Utilizing 4-methoxy- and 2-nitrobenzoic acid, a number of copper(I) catalysts and appropriate ligands were screened with various solvent mixtures. Five mol% of copper(I) oxide and 10 mol% of

1,10-phenanthroline were found to be the best catalyst/ligand system. After inertization of the reaction mixture, NMP/quinolie (3 : 1) was added. The mixture was heated under microwave irradiation at 190 °C for 15 min, and after cooling quenched with water (Scheme 6.192). The method proved tolerable to a variety of aryl and hetaryl substrates to furnish the compounds in high yields as determined by GC with *n*-tetradecane as internal standard. Isolated yields were somewhat lower as in most cases fractional distillation was required. Nevertheless, the efficient method avoids stoichiometric amounts of the heavy metal catalyst and provides a smooth practical approach since even first grade scale-up is possible in standard monomode microwave equipment.

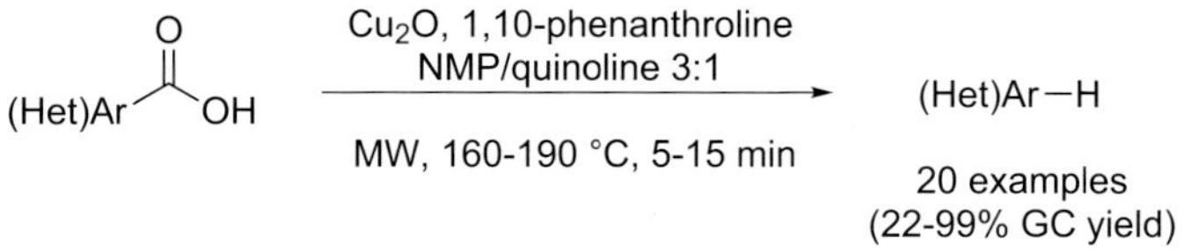

Scheme 6.192 Protodecarboxylation of aromatic carboxylic acids.

6.19
Free Radical Reactions

There are only a limited number of examples in the literature that involve radical chemistry under controlled microwave heating conditions [419]. Wetter and Studer have described radical carboaminoxylations of various nonactivated olefins and difficult radical cyclizations (Scheme 6.193) [420]. The thermally reversible homo-

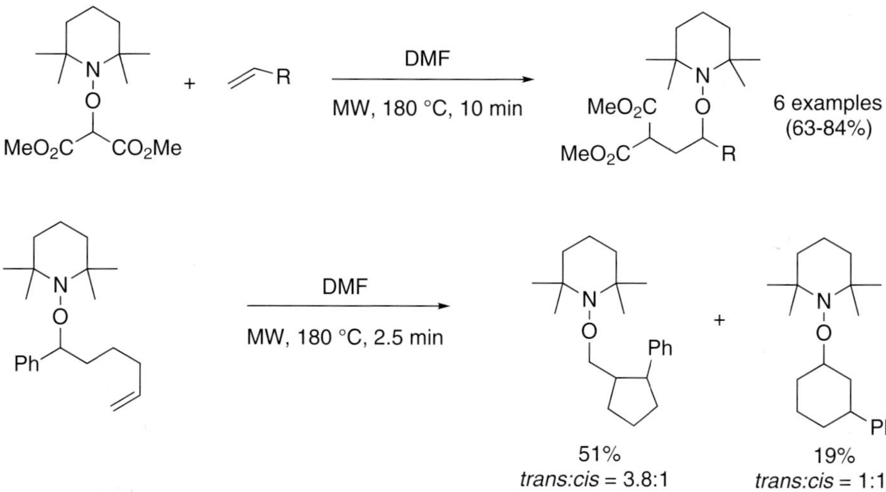

Scheme 6.193 Radical carboxaminations with malonyl radicals.

lysis of alkoxyamines generates the persistent radical 2,2,6,6-tetramethylpiperidinyl-1-ol (TEMPO) and a stabilized transient malonyl radical, which subsequently reacts with an olefin to give the corresponding carboaminoxylation product. Under conventional conditions (DMF, sealed tube, 135 °C), these radical addition processes take up to 3 days. Using sealed-vessel microwave heating at 180 °C, higher yields were obtained for all but one example when comparing microwave heating at 180 °C for 10 min and thermal heating at 135 °C for 3 days.

The same group disclosed a related free radical process, namely, an efficient one-pot sequence comprising a homolytic aromatic substitution followed by an ionic Horner–Wadsworth–Emmons olefination for the production of a small library of α,β-unsaturated oxindoles (Scheme 6.194) [421]. Suitable TEMPO-derived alkoxyamine precursors were exposed to microwave irradiation in DMF for 2 min to generate an oxindole intermediate via a radical reaction pathway (intramolecular homolytic aromatic substitution). After addition of potassium *tert*-butoxide base (1.2 equiv) and a suitable aromatic aldehyde (10–20 equiv), the mixture was exposed again to microwave irradiation at 180 °C for 6 min to provide the α,β-unsaturated oxindoles in moderate to high overall yields. A number of related oxindoles were also prepared via the same one-pot radical/ionic pathway (Scheme 6.194).

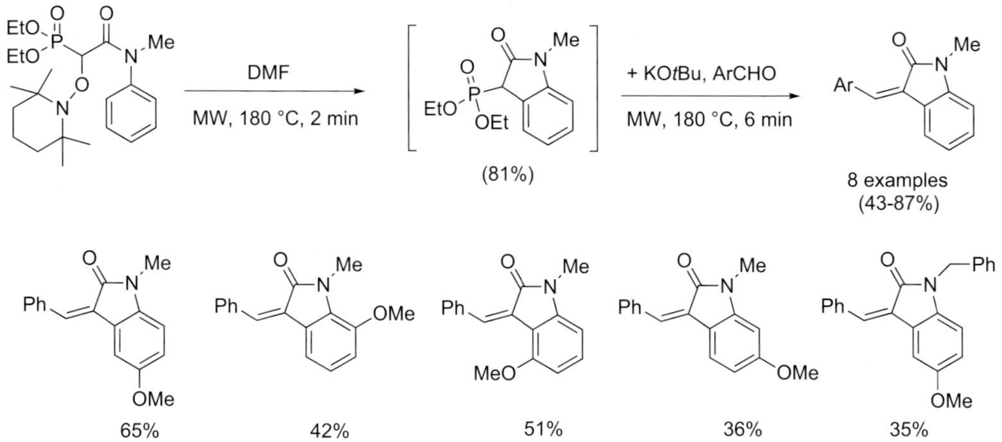

Scheme 6.194 One-pot homolytic aromatic substitution/Horner–Wadsworth–Emmons olefinations.

The synthesis of 2-arylindoles **96** via tandem radical cyclization of the corresponding acrylates **94** and subsequent oxidation was reported by the group of Reiser (Scheme 6.195) [422]. In the first step of the synthesis, acrylates **94** are transformed to 2-arylindolines **95** by a tributyltin hydride-mediated radical cyclization involving a 1,6-*H* transfer followed by a 5-*exo* ring closure under open-vessel microwave conditions. For the aromatization step, DDQ proved to be the oxidation reagent of

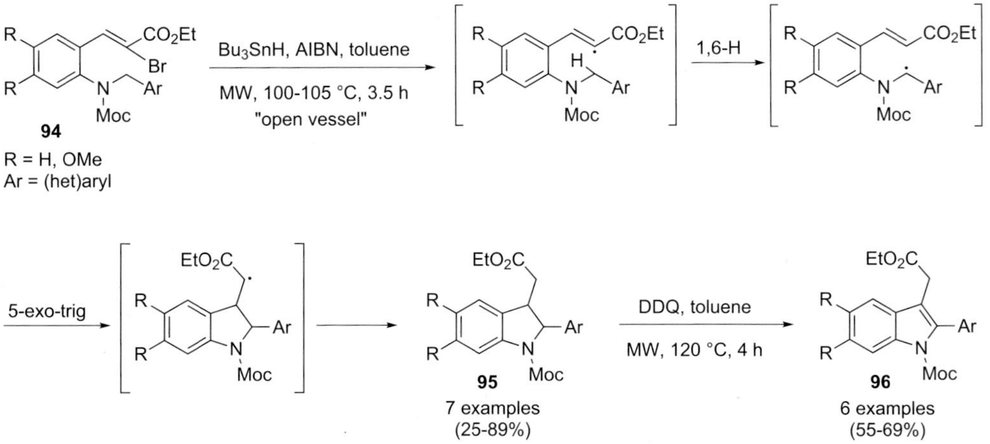

Scheme 6.195 Synthesis of 2-arylindoles via tandem radical cyclization.

choice and 2-arylindoles were obtained in good yields. For both procedures, higher yields and reduced reaction times could be achieved when applying microwave heating.

Walton and coworkers have reported on the use of *O*-phenyl oxime ethers as precursors for iminyl radicals *en route* for *N*-heterocycle synthesis (Scheme 6.196) [423]. Microwave irradiation was employed for the thermolysis of functionalized *O*-phenyl oximes **97** generating the iminyl radicals **98** followed by cyclization to the corresponding heterocycles. For the synthesis of dihydropyrroles **99**, toluene is used as solvent, whereas a non-H-atom donor solvent such as *t*-butylbenzene has to be employed for

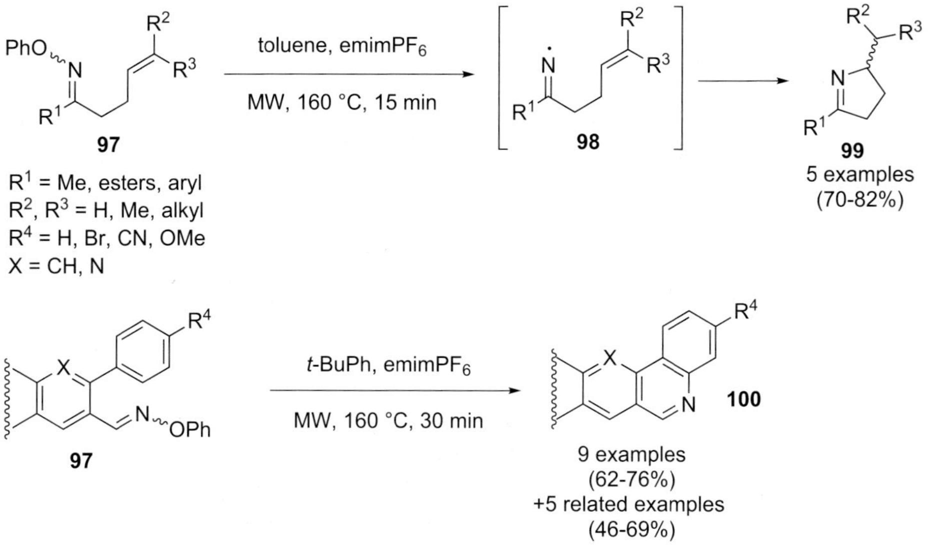

Scheme 6.196 Synthesis of *N*-heterocycles via iminyl radicals.

cyclizations onto aromatic radical acceptors for the synthesis of quinolines, phenanthridines (**100**, X = CH), benzonaphthiridines (**100**, X = N), and indolopyridines. For both protocols, the ionic liquid emimPF_6 needs to be used in order to reach the 160 °C reaction temperature. With this method, no initiator is necessary and in combination with microwave heating, cleaner product mixtures and shortened reaction times can be achieved. See also Ref. [424].

In 2004, Ericsson and Engman have reported on rapid radical group-transfer cyclizations of organotellurium compounds. They found that primary and secondary alkyl aryl tellurides, prepared by arenetellurolate ring opening of epoxides/*O*-allylation, underwent rapid (3–10 min) group-transfer cyclization to afford tetrahydrofuran derivatives in good yields when heated in a microwave cavity at 250 °C in ethylene glycol or at 180 °C in water (Scheme 6.197) [425]. To go to completion, similar transformations had previously required extended photolysis in refluxing benzene. This chemistry can be extended to primary alkyl aryl tellurides.

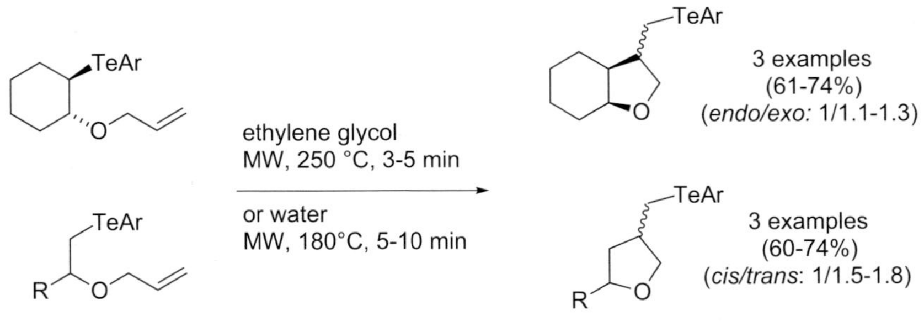

Scheme 6.197 Radical group transfer cyclizations.

Radical cyclizations of *N*-allyl-2-nitroanilines to yield substituted 1,2,3,4-tetrahydroquinoxalines have been described by Beifuss and coworkers [426, 427]. Radical addition reactions of diphenylphosphine sulfide to olefins have been reported [428]. A related publication describes the radical addition of phosphorous hydrides to bisolefines [429]. Bergmann cyclizations leading to poly-*N*-heterocycles assisted by microwave irradiation at high temperatures have been described [430]. Microwave irradiation of *N*-(alkoxy)thiazole-2(3*H*)-thiones generates alkoxy radicals that have been used by Hartung *et al.* for additions, β-fragmentations, and remote functionalizations [431].

Radical processes are, of course, of great importance in the field of polymer synthesis, and the applications of controlled microwave heating in this area are rapidly growing. The group of Ritter has described the direct preparation of (meth)acrylamide monomers from (meth)acrylic acid and an amine using microwave irradiation in a solvent-free environment. Irradiation of equimolar mixtures of the reagents in a sealed monomode reactor (no temperature control) for 30 min provided the desired amides in moderate to high yields and with good purity (Scheme 6.198) [432]. The addition of a radical initiator like 2,2′-azoisobutyronitrile

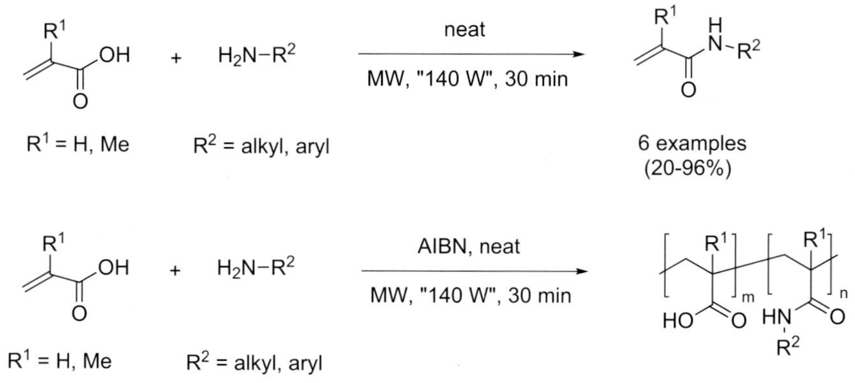

Scheme 6.198 Radical polymerizations.

(AIBN) to the starting mixture led directly to the corresponding poly(meth)acrylamides in a single step. Additional examples of microwave-assisted radical polymerization reactions have been described [433] and reviewed [434] by Schubert and coworkers.

6.20 Protection/Deprotection Chemistry

In carbohydrate chemistry, protection and deprotection reactions play a significant role. Apart from the examples described in Section 6.6, Oscarson and coworkers have summarized carbohydrate-based microwave-assisted protection group manipulations in a 2001 review article [134]. The simultaneous *O*-debenzylation and carbon–carbon double-bond reduction in a series of alkenoic acids by catalytic transfer hydrogenation was reported by the group of Pohl [435]. In the example shown in Scheme 6.199, treatment of the starting material in ethylene glycol with 10 equiv of ammonium formate as hydrogen donor and catalytic amounts of Pd/C led to complete debenzylation and C=C double-bond reduction within 5 min at 120°. Further deprotection of the *N*-methoxy-*N*-methyl (Weinreb) amide to the corresponding

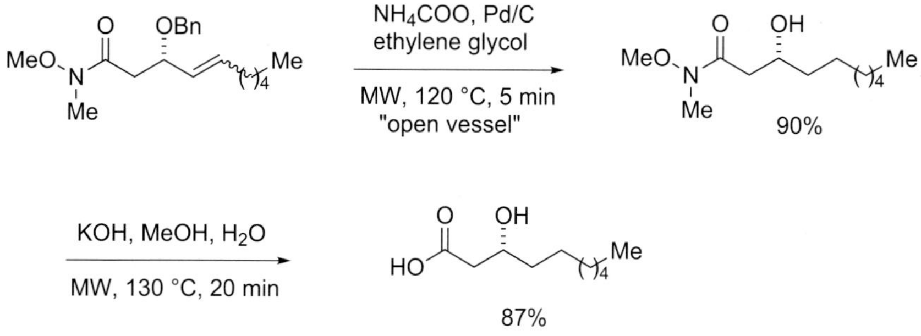

Scheme 6.199 Catalytic transfer hydrogenations.

acid (without elimination of the β-hydroxyl moiety) was accomplished by irradiation of the amide in a dilute potassium hydroxide/methanol/water mixture to 130 °C. After 20 min the desired acid was obtained in 87% isolated yield. The same transformation at room temperature required 4 days to reach completion.

In 2010, Trebanco and coworkers presented a facile microwave method for the challenging debenzylation of *N*-benzylamides utilizing trifluoromethane sulfonic acid (Scheme 6.200) [436]. The method tolerates various substrates such as alkyl-, aryl-, and heteroarylamides; even *N*-benzyltriazolones are converted almost quantitatively. However, *N*-benzylated pyrrolidinone and piperidinone required higher reaction temperatures (210 °C) to yield the expected products and *N*-benzylistatin remained unstable under the acidic conditions and underwent decomposition. Also, in terms of benzylated benzamides, the elaborated method proved sufficient for a set of functional moieties.

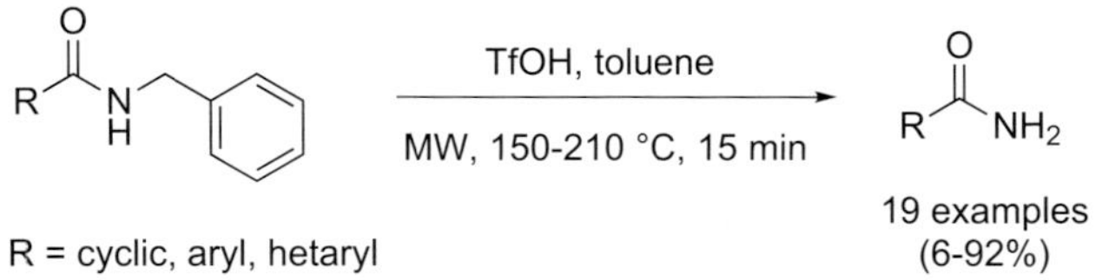

Scheme 6.200 Debenzylation of amides.

Baran *et al.* have reported an unusual deprotection of allyl esters in microwave-superheated water. The diallyl ester, structurally related to the sceptrin natural products (see Scheme 6.10), was cleanly deprotected at 200 °C within 5 min (Scheme 6.201) [25].

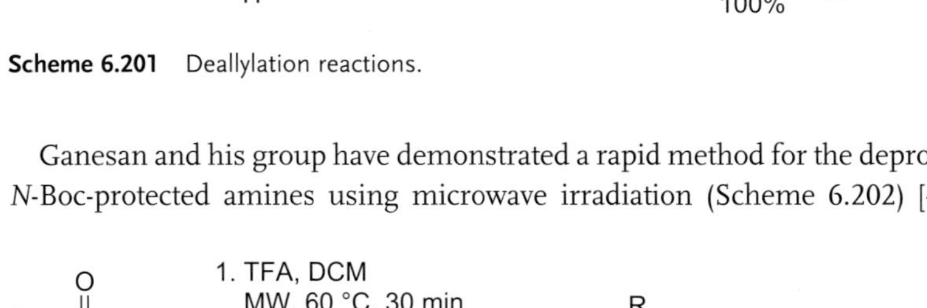

Scheme 6.201 Deallylation reactions.

Ganesan and his group have demonstrated a rapid method for the deprotection of *N*-Boc-protected amines using microwave irradiation (Scheme 6.202) [437]. The

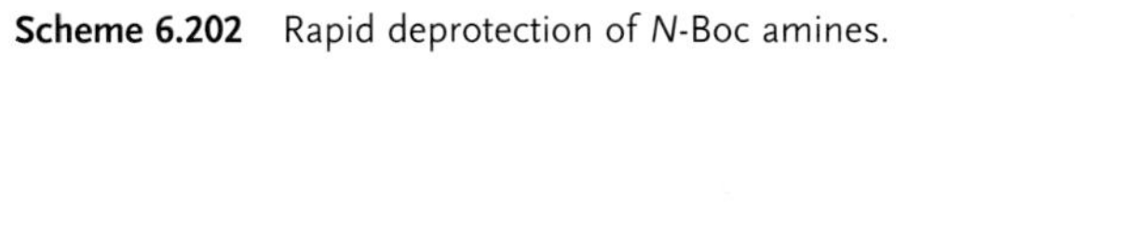

Scheme 6.202 Rapid deprotection of *N*-Boc amines.

procedure utilizes 5 equiv of trifluoroacetic acid (TFA) and DCM as solvent. The protected amines are heated at 60 °C for 30 min in the cleavage cocktail. The freebase amines can be rapidly isolated by scavenging the crude reaction mixture with basic Amberlyst A-21 ion exchange resin.

N-Boc deprotections of lactams were reported using SiO_2 as support under solvent-free conditions [438]. An aqueous *N*-Boc deprotection of a dipeptide was used in the synthesis of the Schöllkopf chiral auxiliary [439]. *N*-Boc deprotection of piperidine was reported by Holzgrabe and Heller [245].

Park and Chae developed an efficient demethylation of methyl aryl ethers utilizing an ionic liquid under microwave irradiation (Scheme 6.203) [440]. Since methyl-protected phenolic groups are widely used in multistep preparations of pharmaceutical agents and fine chemicals, versatile cleavage methods for deprotection are of considerable interest. Ionic liquids are known for beneficially assisting nucleophilic displacement reactions, which thus turned out to be attractive reagents for a microwave-mediated demethylation protocol. Neither an organic base nor any organic solvent was required for the demethylation when performed under microwave irradiation utilizing bmimBr as ionic liquid. Interestingly, the authors found that the power-controlled mode with simultaneous airflow cooling was essential for high yields. Only low power levels of <30 W, which led to temperatures below 200 °C, gave good results, while at higher levels decomposition of materials was observed. Also, the temperature-controlled mode at 220 °C target temperature led to tar-like materials instead of the desired compound. Only with substrates bearing a functional ester or aldehyde group, rather low yields were obtained. Nevertheless, a variety of structural interesting phenolic moieties could be generated with this microwave-assisted ionic liquid-mediated demethylation of methyl aryl ethers.

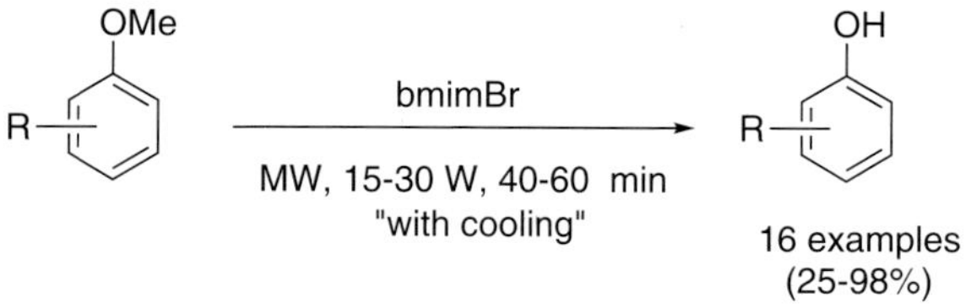

Scheme 6.203 Demethylation of methyl aryl ethers utilizing ionic liquids.

Other standard deprotection transformations carried out under microwave conditions, involving *N*-detosylations [441] and trimethylsilyl removal [442, 443], are summarized in Scheme 6.204.

The selective mono-decarboxylation of a chiral-substituted dimethylmalonate was achieved by Schaus and coworkers using microwave heating in DMSO containing small amounts of water [444]. In the carbohydrate field, a similar transformation was reported by Linker and coworkers using LiI in DMSO [445]. In the total synthesis of mersicarpine, the selective mono-decarboxylation of a dimethylmalonate unit was accomplished by Kerr employing LiCl in DMF [446].

Desai and coworkers have disclosed the sulfation of poly-hydroxyl scaffolds (Scheme 6.205) [447]. This otherwise troublesome reaction proceeded well under

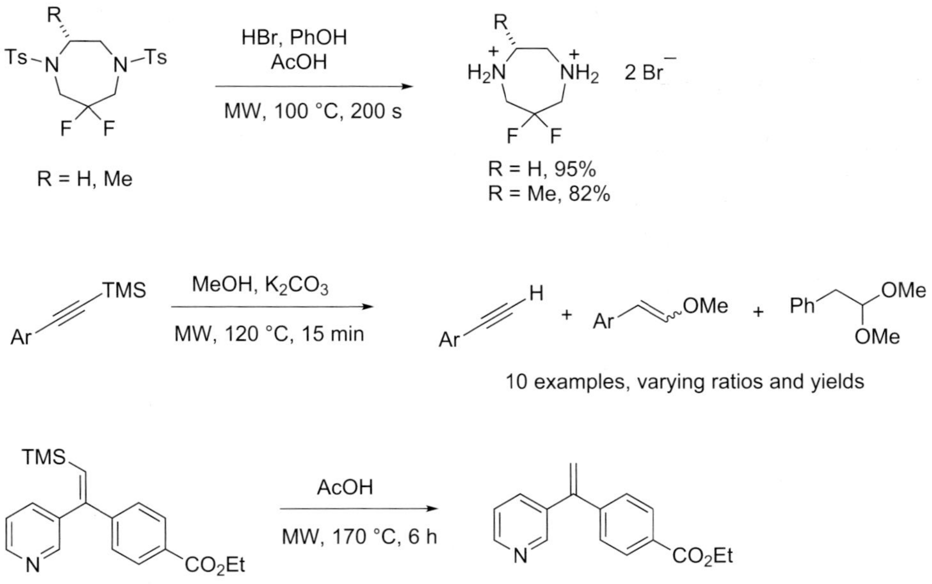

Scheme 6.204 Miscellaneous deprotection reactions.

microwave heating employing either $Me_3N{\cdot}SO_3$ (6–9 equiv/OH) and Et_3N (10 equiv/OH) as base or SO_3·py (6–12 equiv/OH) and pyridine (10 equiv/OH), respectively, giving the persulfated products in high yields and in short reaction times. With this method a range of functional groups are tolerated (amides, esters, and aldehydes),

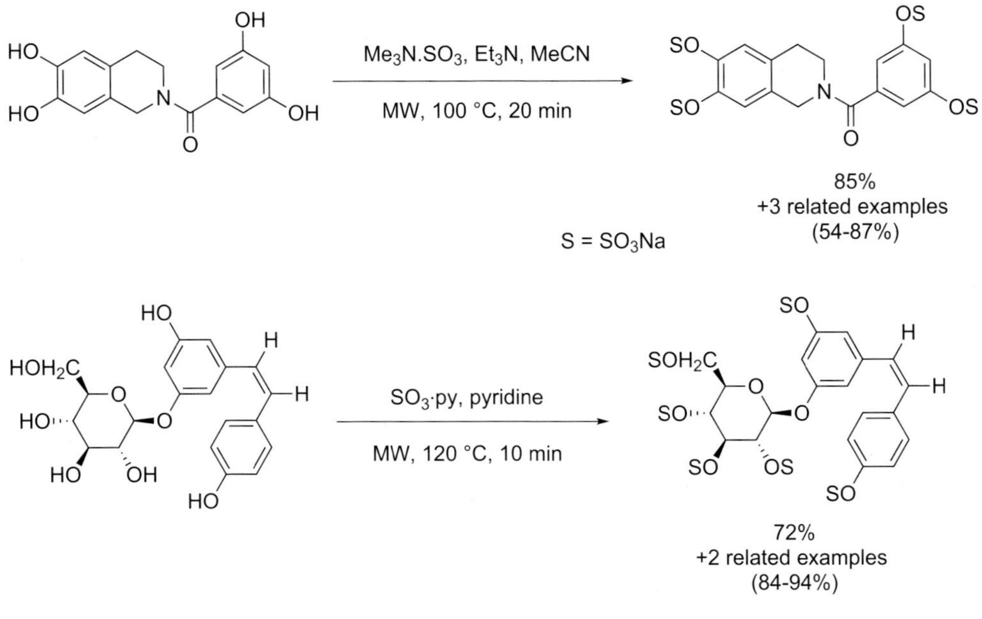

Scheme 6.205 Synthesis of highly sulfated organic scaffolds.

alcoholic and phenolic OH-groups are sulfated equally well, and the products are obtained with high purities.

Pyridine- and solvent-free acetylation of nucleosides employing Ac_2O in combination with molecular sieves has been reported by Sá and Meier [448]. The transprotection of aryl methyl ethers to aryl benzoates using benzoyl chloride in combination with hexabutylguanidinium chloride hydrochloride as reagent has been disclosed by Gras and coworkers [449]. Phenols can be rapidly protected with a 4-methoxybenzyl protection group by treatment with 4-methoxybenzyl chloride in DMF in the presence of K_2CO_3 applying microwave irradiation (or ultrasound) [450]. Acetal and ketal formation can be efficiently performed using $In(OTf)_3$ as a catalyst [451].

6.21
Preparation of Isotopically Labeled Compounds

The rapid synthesis of short-lived radiolabeled (e.g., ^{11}C, ^{18}F) substances used in positron emission tomography (PET) has been one of the first applications of single-mode microwave-assisted synthesis, and this area has been extensively reviewed [452]. A typical application of this technique is shown in Scheme 6.206. Considerable effort has been devoted to the design, synthesis, and pharmacological characterization of radiofluorinated derivatives of the 5-HT_{1A} receptor antagonist, WAY-100635, for the *in vivo* study of these receptors in human brain with PET. 6-[^{18}F]Fluoro-WAY-100635 can be efficiently synthesized in one step from the corresponding 6-nitro precursor using nucleophilic heteroaromatic fluorinations [453]. As radiofluorinating agent, the activated K[^{18}F]F-Kryptofix® 222 complex was employed (Kryptofix 222 = 4,7,13,16,21,24-hexaoxa-1,10-diazabicyclo[8.8.8]hexacosane). High incorporation yields were observed after 1 min of single-mode microwave irradiation of the nitro precursor in DMSO solution (no reaction temperature given). The same group has also described related nucleophilic aromatic substitutions on other substituted pyridine cores [454]. For other applications of microwave heating in the synthesis of fluorine-18-labeled materials, see Ref. [455].

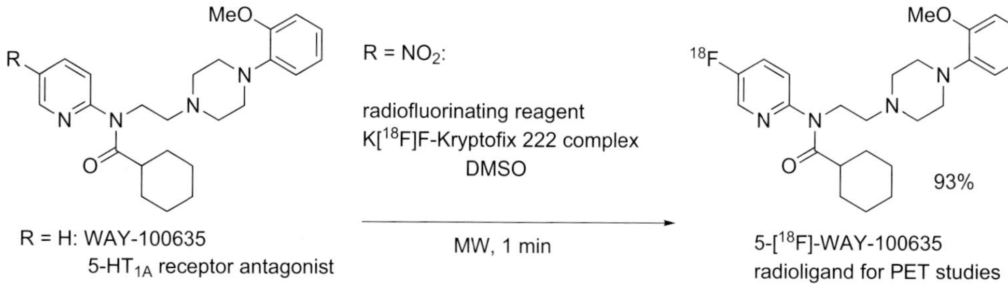

Scheme 6.206 Incorporation of ^{18}F by nucleophilic heteroaromatic substitution.

The incorporation of a fluorine-18 label can also be achieved by standard aliphatic nucleophilic substitution chemistry, as exemplified in Scheme 6.207. Here, the widely used reagent [^{18}F]β-fluoroethyl tosylate was utilized to prepare several important ^{18}F-labeled compounds [456].

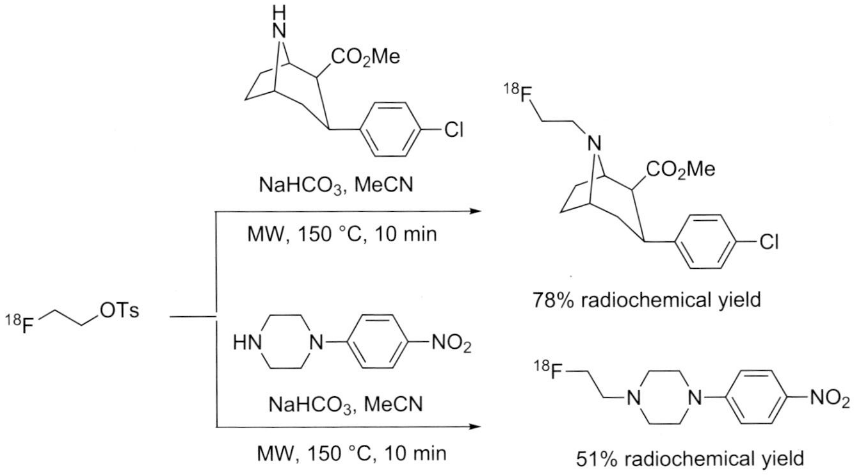

Scheme 6.207 Incorporation of ^{18}F by nucleophilic substitution.

The rapid synthesis of carbon-14-labeled [1-^{14}C]levulinic acid from simple building blocks has been demonstrated by Johansen and coworkers (Scheme 6.208) [457]. In all three of the synthetic steps, starting from bromo[1-^{14}C]acetic acid, microwave heating was used in order to accelerate the reactions, allowing the total preparation to proceed in less than 1 h. The labeled levulinic acid was subsequently transformed

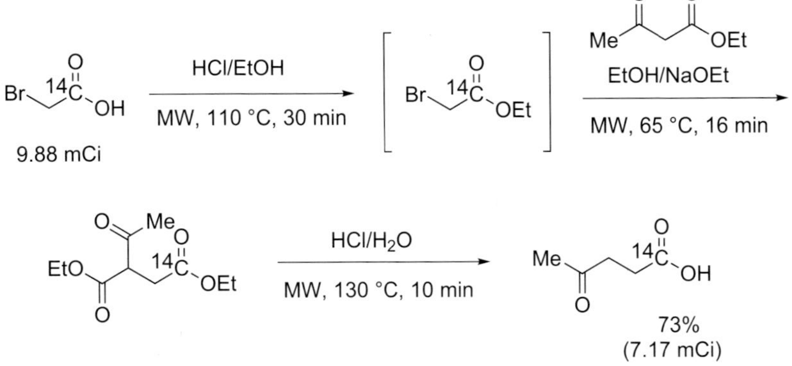

Scheme 6.208 Preparation of [1-^{14}C]levulinic acid.

into (5*Z*)-4-bromo-5-(bromomethylene)-2(5*H*)-furanone in a bromination/oxidation sequence (not shown), a potent quorum-sensing inhibitor.

The synthesis of ^{188}Re [458] and ^{99m}Tc [459] complexes as radiochemical labeling agents using microwave irradiation has been investigated by Park *et al.*

Apart from the preparation of radiotracers, microwave-assisted transformations have also been utilized for carrying out simple hydrogen–deuterium exchange reactions. In case of acetophenone, for example, simple treatment with deuterium oxide (D_2O) as solvent in the presence of molecular sieves at 180 °C for 30 min led to a complete incorporation of deuterium into the methyl group. The presence of the molecular sieves was found to be essential for the exchange of the acidic protons (Scheme 6.209a) [460]. In the same article, the authors also report the preparation of deuterium-labeled organophosphonium salts.

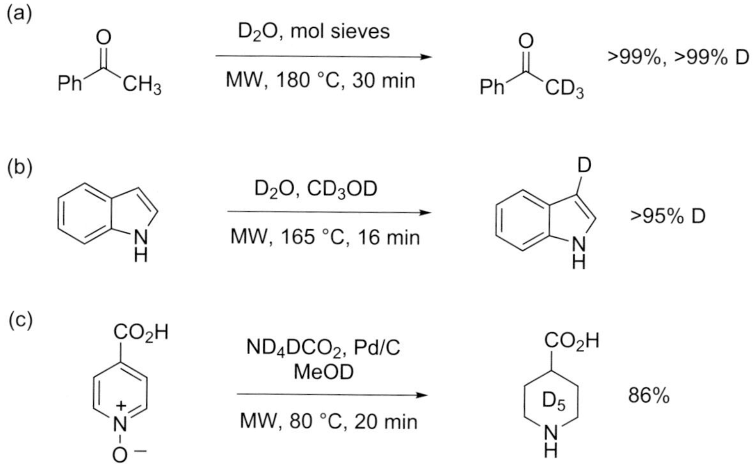

Scheme 6.209 Preparation of deuterium-labeled compounds.

Masjedizadeh and coworkers have recently described similar microwave-promoted hydrogen–deuterium exchange reactions in a series of heterocycles using mixtures of D_2O and deuteriomethanol (Scheme 6.209b) [461]. The rapid exchange method was applied to the deuteration of the antitumor antibiotic bleomycin A under catalyst-free conditions.

The application of deuterated ammonium formate as deuterium source in transfer deuteration reactions of aromatic heterocycles was reported by Derdau (Scheme 6.209c) [462]. By employing microwave irradiation, it is possible to accelerate the reaction from 12-18 h (50 °C, oil bath) to 20 min at 80 °C using deuteriomethanol as solvent and 10% Pd/C as a catalyst.

Deuterated anilines can be prepared by selective deuteration of anilines in *o*- and *p*-position using 1 equiv of concentrated HCl, see Ref. [463].

6.22 Miscellaneous Transformations

In the context of preparing potential inhibitors of dihydrofolate reductase (DHFR), the group of Organ has developed a rapid microwave-assisted method for the preparation of biguanide libraries (Scheme 6.210) [464]. Initial optimization work centered around the acid-catalyzed addition of amines to dicyandiamide. It was discovered that 150 °C was the optimum temperature for reaction rate and product recovery, as reactions heated beyond this point led to decomposition. While the use of hydrochloric acid as catalyst led to varying yields of product, evaluation of trimethylsilyl chloride (TMSCl) in MeCN as solvent led to improved results. Compared to the protic conditions, the reaction rate was greatly enhanced under TMSCl catalysis. The reaction already provided significant product recovery after 1 min and best results were typically obtained by exposing equimolar amounts of amine and dicyandiamide in acetonitrile (150 °C, 15 min) to 1.1 equiv of trimethylsilyl chloride. After addition of 3 equiv of 2-propanol and further microwave heating to 125 °C for 30 s, the desired biguanide products precipitated as hydrochloride salts. New, improved lead structures were discovered from screening the 60-member compound library prepared in this fashion.

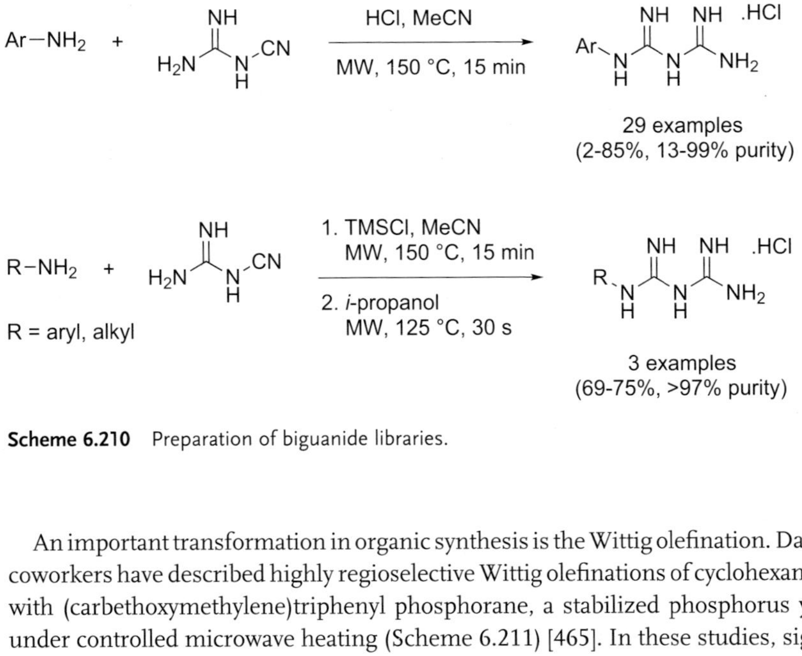

Scheme 6.210 Preparation of biguanide libraries.

An important transformation in organic synthesis is the Wittig olefination. Dai and coworkers have described highly regioselective Wittig olefinations of cyclohexanones with (carbethoxymethylene)triphenyl phosphorane, a stabilized phosphorus ylide, under controlled microwave heating (Scheme 6.211) [465]. In these studies, significant base and temperature effects were recorded. When the Wittig reaction was carried out in MeCN at 190 °C in the absence of a base, a very high selectivity for the *exo*-product was obtained. On the other hand, the same olefination carried out at 230 °C in DMF in the presence of 20 mol% of DBU as a strong base afforded the

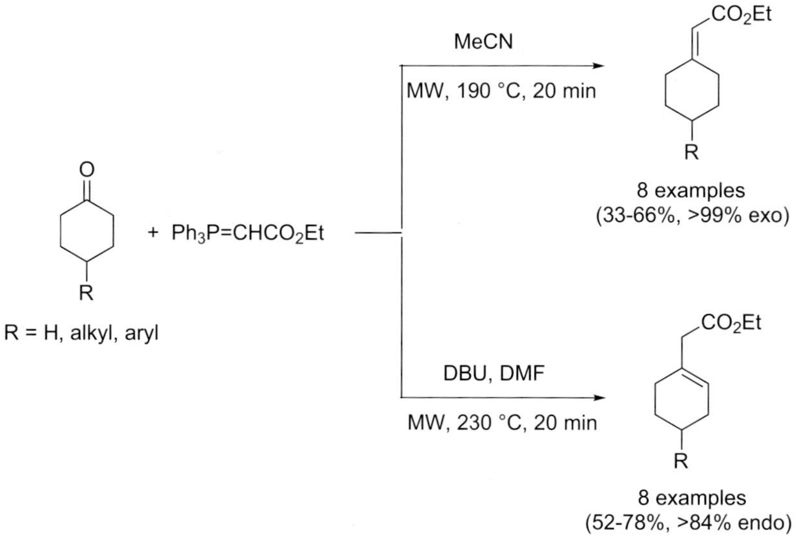

Scheme 6.211 Wittig olefinations.

thermodynamically more stable *endo*-products in >84 : 16 isomer ratio. In both cases, a threefold excess of the ketone was employed.

Another example of the Wittig reaction where the chemoselective synthesis of functionalized alkenes was investigated was reported by McNulty *et al.* (Scheme 6.212) [466]. Although the initial preparation of the required phosphonium salts utilizing air-stable triethylphosphane hydrobromide was performed neat under classical heating at 100 °C for 8 h, the subsequent Wittig olefination to afford the desired stilbenes could be successfully accelerated by microwave irradiation. Gratifyingly, the two-step process can be performed as a simplified one-pot procedure. The corresponding benzylic alcohol was heated with equimolar amounts of triethylphosphane hydrobromide in a microwave vial under conventional conditions. When cooled to room temperature, 1.1 equiv of potassium carbonate, water, and (het)aryl aldehyde were added. The sealed vial was finally heated under microwave irradiation at 75 °C for 30 min to furnish the desired *trans*-stilbenes in high yields and with excellent stereoselectivity upon simple filtration of the solid products (Scheme 6.212). The effective method with its mildly basic conditions can be applied toward a variety of aldehydes, even indole and pyrrole carboxaldehydes proceed smoothly without the necessity of NH protections. To extend the scope of the newly developed protocol, the

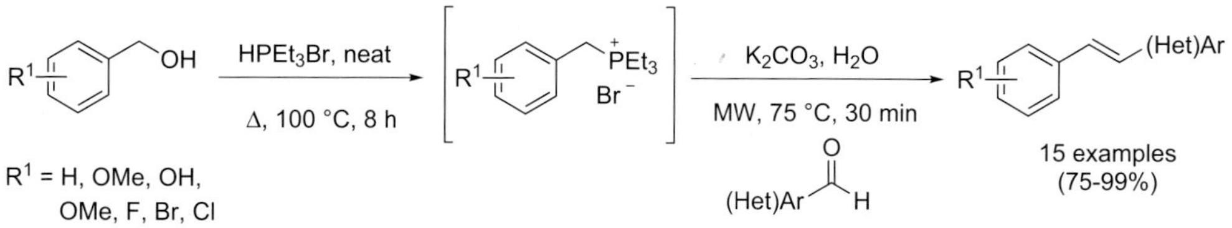

Scheme 6.212 One-pot synthesis of *trans*-stilbenes via Wittig olefination.

authors also added the intermediate ylides to ketones. However, slight modifications of the protocol have been required to achieve satisfactory results. When switching to sodium hydroxide or lithium hydroxide as a base and increasing the reaction time to 35 min, the desired compounds are generated in good yields.

In a 2010 publication, Triola and coworkers introduced a microwave-assisted synthesis of unsymmetrical disulfides [467]. Since up to now the synthesis of disulfides remained rather sluggish and time consuming, the authors investigated the effect of microwave irradiation on this process. A two-step deprotection–disulfide formation protocol proved effective for various thiols (Scheme 6.213). *S*-Trityl-protected Fmoc-penicillamine ($R^1 = Me$) or Fmoc-cysteine ($R^1 = H$), respectively, were deprotected utilizing 5% TFA and 3% triethylsilane in DCM under argon atmosphere. The residue was dissolved in a buffer containing ammonium acetate in MeCN/water (3 : 2). Ten equiv of the corresponding thiol and DMSO as oxidizing agent were added and the mixture was irradiated at 150 °C for 5 min. The method covers aromatic as well as aliphatic thiols. Even a biotin derivative worked satisfactory with the established microwave protocol. Most substrate combinations also work with only 5 equiv of thiol without reducing the yield. All examples proceeded quickly and no racemization was observed, thus proving the suitability of the method to selectively attach structurally demanding biomolecules.

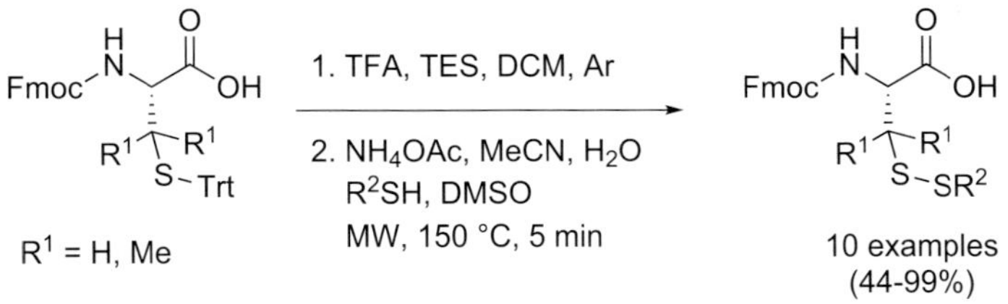

Scheme 6.213 Synthesis of unsymmetrical disulfides.

De Luca *et al.* presented a simple procedure for the preparation of *N*-monosubstituted ureas from amines employing potassium cyanate (Scheme 6.214) [468]. These important compounds occurring in natural products as well as in synthetic compounds show interesting biological activities and can be easily generated in acid aqueous media. Under microwave irradiation, the reaction time could be successfully reduced from 6–24 h at 60 °C under conventional heating to 1 h at 80 °C. Temperatures higher than 80 °C led to significantly decreased yields due to formation of unwanted by-products. However, the authors demonstrated the potential of the developed protocol by scaling up the reaction switching to a commercially available 80 mL microwave vessel. When aryl amines are employed, acetic acid instead of HCl furnished the best results. Furthermore, the protocol could be applied toward amino acids, which did not require any additional acidification at all. The chiral products precipitated in pure form after bringing the cooled mixture to pH 2, significant racemization was not obtained.

Kurth and coworkers investigated an effective route toward the synthesis of 2-substituted 1*H*-indazolones [469]. A variety of nucleophiles, such as amines,

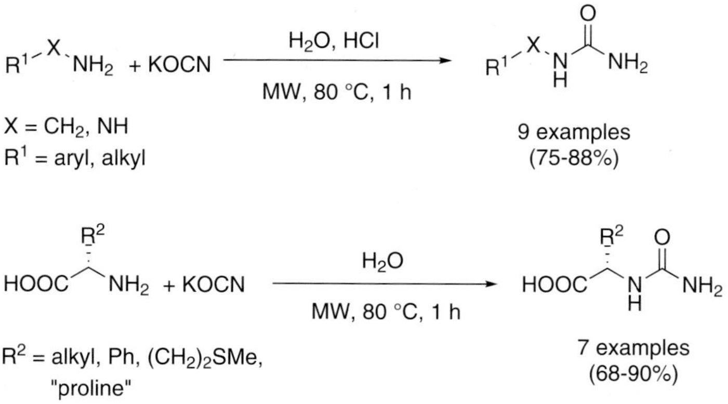

Scheme 6.214 Synthesis of *N*-monoalkylureas.

alkoxides, thiolates, cyanides, and iodides react with the corresponding indazoles to yield the target compounds. Three different indazole derivatives have been employed in nucleophilic ring-opening reactions with various substrates to generate the desired products **101** (Scheme 6.215). In terms of alkoxides and thiolates, the nucleophile was generated *in situ* in the reaction vial prior to adding the indazole and heating at 120 °C for 20 min. When potassium iodide was employed, an interesting result was obtained indicating a subsequent intramolecular *N*-alkylation after formation of the

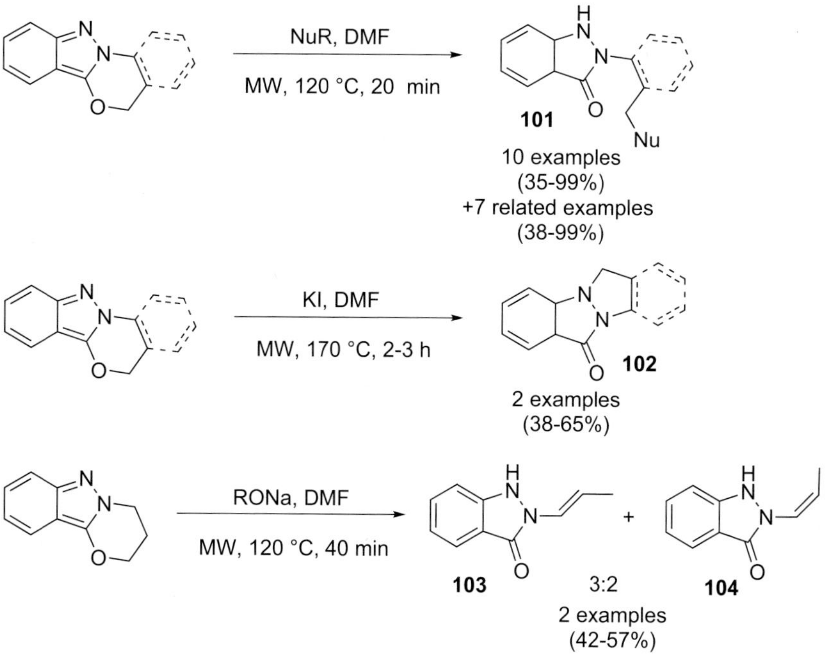

Scheme 6.215 Synthesis of 1*H*-indazolones.

intermediate potassium dihydroindazolide (ANRORC sequence: addition of nucleophile, ring opening, and ring closure) to generate the isomeric pyrazoloindazolones **102** (Scheme 6.215). However, when the general method was applied toward oxazolino[3,2-*b*]indazole, generally higher yields were obtained with all nucleophiles since the oxazine moiety proved more reactive toward the nucleophilic attack. Interestingly, when oxazino[3,2-*b*]indazole was reacted with alkoxides, elimination products **103** and **104** were obtained rather than the expected ether-containing target compounds.

Back in 2009, Zard and coworkers presented a route to biaryl-3-carboxylates esters involving a radical 1,2-aryl migration (Scheme 6.216) [470]. Usually, these potential pharmacophores are generated by transition metal-catalyzed couplings of two monoaryl units. Herein, the authors describe a novel two-step sequence starting with radical addition of a xanthate onto a unsaturated sulfone, followed by a one-pot isomerization of the resulting α,β-unsaturated ester, cyclization, elimination, and finally aerial oxidation. A number of required precursors **105** were prepared by radical addition of various xanthates to two different functionalized olefins. For the next step, microwave irradiation (150 °C, 10 min) was employed to enhance the elimination/cyclization step and to improve the aerial oxidation. To accomplish aromatization, the mixture was then left under air at room temperature for 2 days. Ester precursors featuring a cyclobutanone ring furnished only very low yields since opening of the arylcyclobutanone ring is obviously favored under the present conditions. In general, the authors established an interesting procedure giving access to diverse and complex bi-and tricyclic scaffolds.

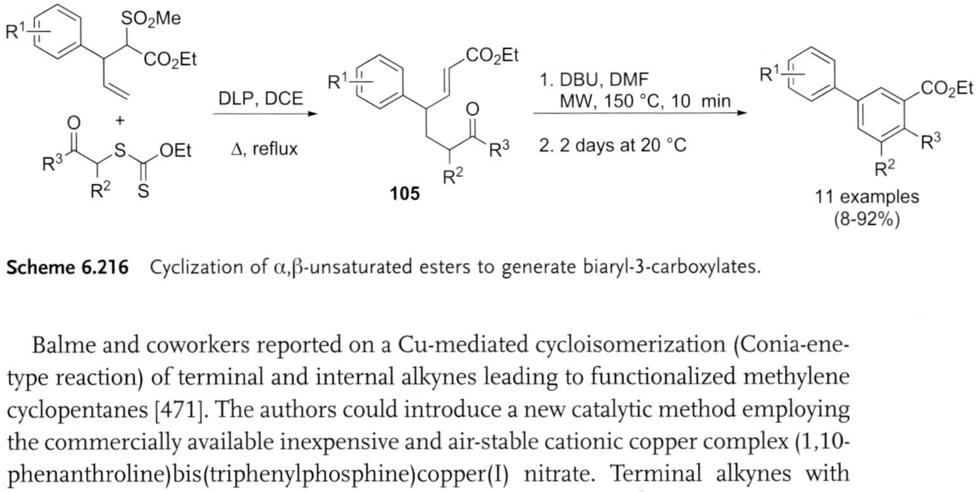

Scheme 6.216 Cyclization of α,β-unsaturated esters to generate biaryl-3-carboxylates.

Balme and coworkers reported on a Cu-mediated cycloisomerization (Conia-ene-type reaction) of terminal and internal alkynes leading to functionalized methylene cyclopentanes [471]. The authors could introduce a new catalytic method employing the commercially available inexpensive and air-stable cationic copper complex (1,10-phenanthroline)bis(triphenylphosphine)copper(I) nitrate. Terminal alkynes with various functional groups were used to optimize the reaction conditions, revealing that a low catalyst loading of 1 mol% is sufficient in the microwave-mediated process. Heating at 150 °C for 20–50 min afforded high yields for most of the substrates (Scheme 6.217). However, sulfone esters furnished predominantly the corresponding allylic sulfone by thermal 1,3-rearrangement of the initially formed target tertiary sulfone. Furthermore, internal alkynes could be successfully transformed with the

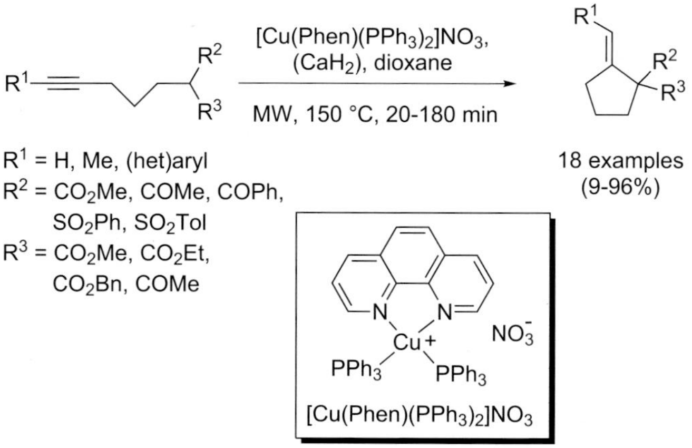

Scheme 6.217 Cu-catalyzed Conia-ene reaction of terminal and internal alkynes.

developed method. Slight modifications of the protocol by increasing the amount of copper catalyst (up to 20 mol%) and adding CaH_2 to enhance the cyclization led to satisfactory results. Depending on the substitution pattern, longer reaction times (up to 3 h) were required for satisfactory yields. In all cases, the unactivated internal alkynes yielded the corresponding 5-*exo* cyclized products with excellent stereoselectivity.

Hilmersson and coworkers have investigated the use of lanthanide(II) halides ($LnX_2 = SmBr_2$, SmI_2, YbI_2) in microwave-assisted reduction and reductive coupling reactions without the use of a co-solvent for a variety of functional groups such as α,β-unsaturated esters, aldehydes, ketones, imines, and alkyl halides (Scheme 6.218) [472]. A strong influence on the direction of the reactions was found to be the redox potential of the LnX_2 reagent, that is, $SmBr_2$ induced mainly the pinacol coupling of ketones (Scheme 6.218a) or the intramolecular reductive coupling product **107**, (Scheme 6.218b), respectively, while the weaker YbI_2 afforded mainly the reduction products **106**.

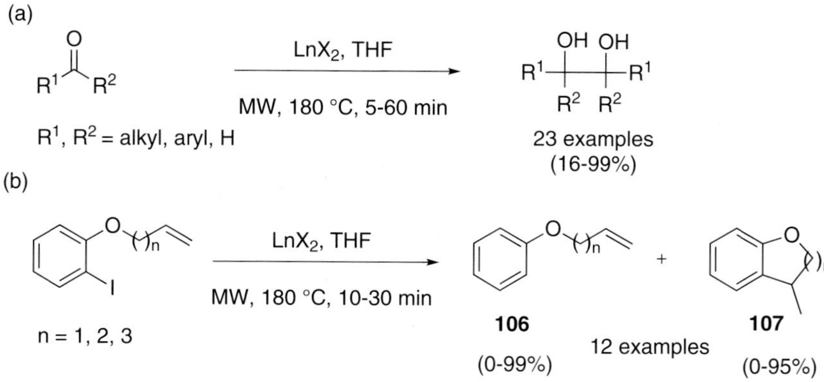

Scheme 6.218 Exploring $SmBr_2$-, SmI_2-, and YbI_2-mediated reactions.

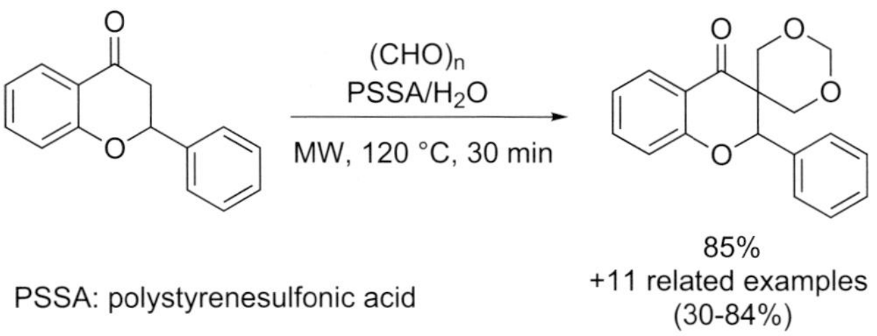

Scheme 6.219 Tandem bis-aldol reaction of ketones.

In order to attach a 1,3-dioxane moiety to ketones, Varma and Polshettiwar have performed tandem bis-aldol reactions of ketones with paraformaldehyde (Scheme 6.219) [473]. This one-pot protocol can be considered as eco-friendly, since polymer-supported PSSA and water as solvent are employed. Furthermore, phase separation under hot conditions occurred for most of the examples, thus facilitating the purification step (only decantation is needed).

Some other microwave-assisted reactions carried out under controlled conditions are summarized in Scheme 6.220 [474–476] and Scheme 6.221 [477–480].

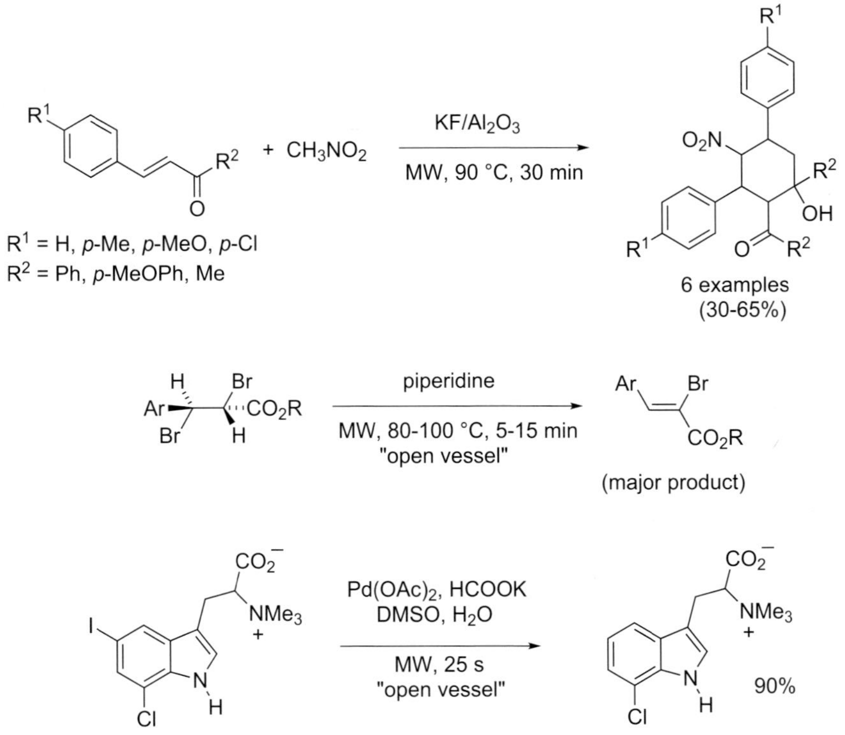

Scheme 6.220 Nitrocyclohexanol synthesis [474], α,β-dibromoesters transformations [475], and dehalogenation reactions [476].

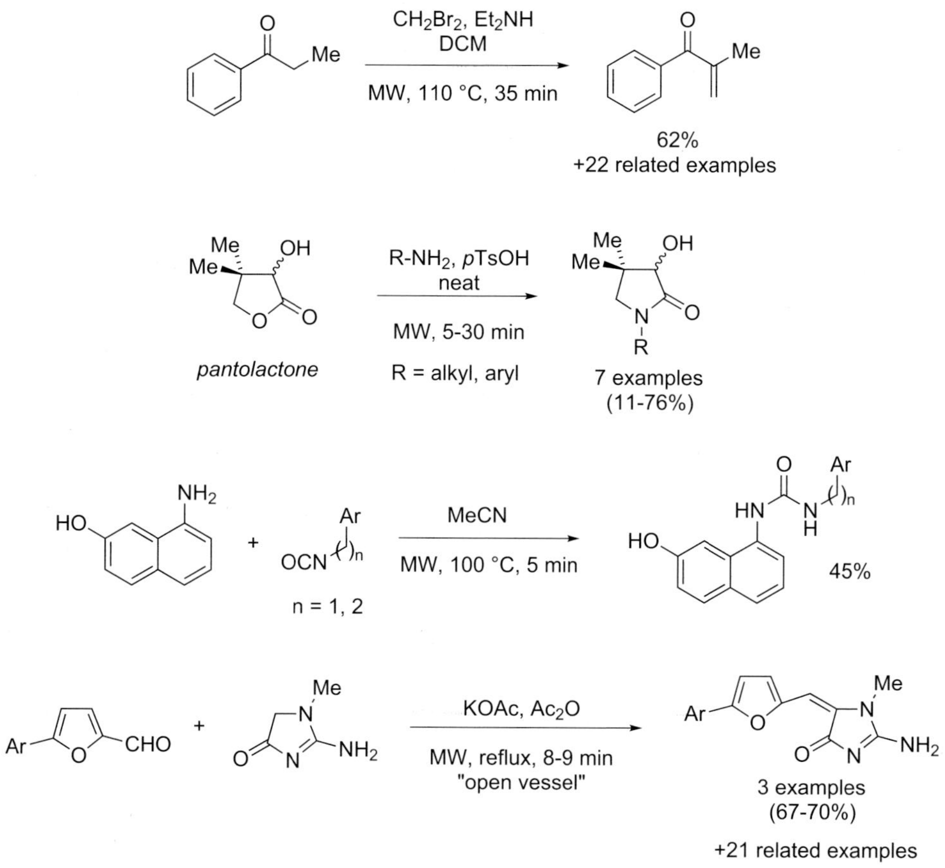

Scheme 6.221 α-Methylenation of ketones [477], lactam formation from lactones [478], urea formation [479], and Knoevenagel condensation [480].

In recent years, many more transformations performed under microwave heating were reported. Chiral *N*-Boc- and Cbz-protected (α-aminoacyl)benzotriazoles react with ethyl (triphenylphosphoranylidene)acetate to produce chiral phosphorus ylides [481]. A one-pot intramolecular aldol process to synthesize bicyclo [2.2.2]oct-5-en-2-ones has been developed in which a cyclic or acyclic ketone is reacted with a cyclic enone in the presence of strong acid [482]. $FeBr_3$-promoted oxidations of diarylalkynes to the corresponding diketones in DMSO as solvent have been reported [483]. For $Pb(OAc)_4$-promoted domino transformation leading to complex oxygen heterocycles, see Ref. [484]. Microwave-assisted Friedel–Crafts acylations catalyzed by BCl_3 have been discussed by Zhang and Zhang [485]. For the oxidation of Hantzsch-type dihydropyridines to pyridines with MnO_2, see Ref. [486]. Cyclic amidines can be obtained from the corresponding lactams and amines using $TiCl_4$ as a reagent [487]. The construction of substituted phenanthrenes via intramolecular condensation of 2′-methylbiphenyl-2-carbaldehydes using a mild base at 200 °C has been described [488]. CuBr-catalyzed homologation of alk-1-ynes with paraformaldehyde and *N,N*-dicyclohexylamine afforded the corresponding allenes [489]. Morita–Baylis–Hillman reactions have been performed in water/ionic liquid medium catalyzed by DABCO [490]. For an example of a microwave-assisted Grob fragmentation used in the synthesis of lamellarin alkaloids, see Ref. [491].

References

1 Baxendale, I.R., Lee, A.-I., and Ley, S.V. (2002) *Journal of the Chemical Society: Perkin Transactions*, 1850–1857.

2 Durand-Reville, T., Gobbi, L.B., Gray, B.L., Ley, S.V., and Scott, J.S. (2002) *Organic Letters*, **4**, 3847–3850.

3 Nordmann, G. and Buchwald, S.L. (2003) *Journal of the American Chemical Society*, **125**, 4978–4979.

4 Yamamoto, T., Wada, Y., Enokida, H., Fujimoto., M., Nakamura, K., and Yanagida, S. (2003) *Green Chemistry*, **5**, 690–692.

5 Bennett, C.J., Caldwell, S.T., McPhail, D.B., Morrice, P.C., Duthie, G.G., and Hartley, R.C. (2004) *Bioorganic & Medicinal Chemistry*, **12**, 2079–2098.

6 Cid, J.M., Romera, J.L., and Trabanco, A.A. (2004) *Tetrahedron Letters*, **45**, 1133–1136.

7 Trost, B.M. and Zhang, T. (2006) *Organic Letters*, **8**, 6007–6010.

8 Craig, D. and Gavin, D.H. (2006) *European Journal of Organic Chemistry*, 3558–3561.

9 Lin, Y.L., Cheng, J.Y., and Chu, Y.H. (2007) *Tetrahedon*, **63**, 10949–10957.

10 Gonda, J., Martinková, M., Zadrošová, A., Šoteková, M., Raschmanová, J., Čonka, P., Gajdošíková, E., and Kappe, C.O. (2007) *Tetrahedron Letters*, **48**, 6912–6915.

11 Schobert, R., Gordon, G.J., Mullen, G., and Stehle, R. (2004) *Tetrahedron Letters*, **45**, 1121–1124.

12 Farand, J.A., Denissova, I., and Barriault, L. (2004) *Heterocycles*, **62**, 735–748.

13 Sauer, E.L.O. and Barriault, L. (2004) *Journal of the American Chemical Society*, **126**, 8569–8575.

14 Sauer, E.L.O. and Barriault, L. (2004) *Organic Letters*, **6**, 3329–3332.

15 Morency, L. and Barriault, L. (2004) *Tetrahedron Letters*, **45**, 6105–6107.

16 McIntosh, C.E., Martínez, I., and Ovaska, T.V. (2004) *Synlett*, 2579–2581.

17 Ovaska, T.V. and Kyne, R.E. (2008) *Tetrahedron Letters*, **49**, 376–378.

18 Nicolaou, K.C., Lister, T., Denton, R.M., and Gelin, C.F. (2007) *Angewandte Chemie International Edition*, **46**, 7501–7505.

19 Pelc, M.J. and Zakarian, A. (2006) *Tetrahedron Letters*, **47**, 7519–7523.

20 Jacob, A.M. and Moody, C.J. (2005) *Tetrahedron Letters*, **46**, 8823–8825.

21 McErlean, C.S.P. and Moody, C.J. (2007) *The Journal of Organic Chemistry*, **72**, 10298–10301.

22 Quesada, E. and Taylor, R. (2005) *Synthesis*, 3193–3195.

23 Bremner, W.S. and Organ, M.G. (2008) *Journal of Combinatorial Chemistry*, **10**, 142–147.

24 Trost, B.M., Thiel, O.R., and Tsui, H.-C. (2003) *Journal of the American Chemical Society*, **125**, 13155–13164.

25 Baran, P.S., O'Malley, D.P., and Zografos, A.L. (2004) *Angewandte Chemie International Edition*, **43**, 2674–2677.

26 Braddock, D.C., Ahmad, S.M., and Douglas, G.T. (2004) *Tetrahedron Letters*, **45**, 6583–6587.

27 Steinhardt, S.E., Silverston, J.S., and Vanderwal, C.D. (2008) *Journal of the American Chemical Society*, **130**, 7560–7561.

28 Trost, B.M. and Gutierrez, A.C. (2007) *Organic Letters*, **9**, 1473–1476.

29 Métro, T.X., Pardo, D.G., and Cossy, J. (2007) *The Journal of Organic Chemistry*, **72**, 6556–6561.
30 Métro, T.X., Pardo, D.G., and Cossy, J. (2007) *Synlett*, 2888–2890.
31 Moseley, J.D., Sankey, R.F., Tang, O.N., and Gilday, J.P. (2006) *Tetrahedon*, **62**, 4685–4689.
32 Presset, M., Coquerel, Y., and Rodriguez, J. (2009) *The Journal of Organic Chemistry*, **74**, 415–418.
33 Sudrik, S.G., Chavan, S.P., Chandrakumar, K.R.S., Pal, S., Date, S.K., Chavan, S.P., and Sonawane, H.R. (2002) *The Journal of Organic Chemistry*, **67**, 1574–1579.
34 Shen, Z. and Dong, V.M. (2009) *Angewandte Chemie International Edition*, **48**, 784–786.
35 Serra-Muns, A., Guérinot, A., Reymond, S., and Cossy, J. (2010) *Chemical Communications*, **46**, 4178–4180.
36 Kesavan, S., Panek, J.S., and Porco, J.A. (2007) *Organic Letters*, **9**, 5203–5206.
37 Constant, S., Tortoioli, S., Müller, J., Linder, D., Buron, F., and Lacour, J. (2007) *Angewandte Chemie International Edition*, **46**, 8979–8982.
38 Pouwer, Re.H., Schill, H., Williams, C.M., and Bernhardt, P.V. (2007) *European Journal of Organic Chemistry*, 4699–4705.
39 Douelle, F., Tal, L., and Greaney, M.F. (2005) *Chemical Communications*, 660–662.
40 Yin, W., Ma, Y., Xu, J., and Zhao, Y. (2006) *The Journal of Organic Chemistry*, **71**, 4312–4315.
41 Närhi, K., Franzén, J., and Bäckvall, J.-E. (2006) *The Journal of Organic Chemistry*, **71**, 2914–2917.
42 Dieltiens, N. and Stevens, C.V. (2007) *Organic Letters*, **9**, 465–468.
43 Csutorás, C., Berényi, S., and Neumeyer, J.L. (2008) *Synthetic Communications*, **38**, 866–872.
44 Patil, B.S., Vasanthakumar, G.-R., and Babu, V.V.S. (2003) *The Journal of Organic Chemistry*, **68**, 7274–7280.
45 Das, S.K., Reddy, K.A., and Roy, J. (2003) *Synlett*, 1607–1610.
46 Trost, B.M. and Crawley, M.L. (2002) *Journal of the American Chemical Society*, **124**, 9328–9329.
47 Pinto, D.C.G.A., Silva, A.M.S., Almeida, L.M.P.M., Carrillo, J.R., Díaz-Ortiz, A., de la Hoz, A., and Cavaleiro, J.A.S. (2003) *Synlett*, 1415–1418.
48 Leonard, M.S., Carroll, P.J., and Joullié, M.M. (2004) *The Journal of Organic Chemistry*, **69**, 2526–2531.
49 Sarotti, A.M., Pisano, P.L., and Pellegrinet, S.C. (2010) *Organic & Biomolecular Chemistry*, **8**, 5069–5073.
50 Kissane, M., Lynch, D., Chopra, J., Lawrence, S.E., and Maguire, A.R. (2010) *Organic & Biomolecular Chemistry*, **8**, 5602–5613.
51 Leadbeater, N.E. and Torenius, H.M. (2002) *The Journal of Organic Chemistry*, **67**, 3145–3148.
52 Hong, B.-C., Shr, Y.-J., and Liao, J.-H. (2002) *Organic Letters*, **4**, 663–666.
53 Delgado, J.L., de la Cruz, P., Langa, F., Urbina, A., Casado, J., and López Navarrete, J.T. (2004) *Chemical Communications*, 1734–1735.
54 Li, J. and Grennberg, H. (2006) *Chemistry: A European Journal*, **12**, 3869–3875.
55 Brunetti, F.G., Herrero, M.A., Muñoz, J.deM., Giordani, S., Díaz-Ortiz, A., Filippone, S., Ruaro, G., Meneghetti, M., Prato, M., and Vázquez, E. (2007) *Journal of the American Chemical Society*, **129**, 14580–14581.
56 Chen, I.-H., Young, J.-N., and Yu, S.J. (2004) *Tetrahedron*, **60**, 11903–11909.
57 Eddolls, J.P., Iqbal, M., Roberts, S.M., and Santoro, M.G. (2004) *Tetrahedron*, **60**, 2539–2550.
58 Iqbal, M., Li, Y., and Evans, P. (2004) *Tetrahedron*, **60**, 2531–2538.
59 Petronijevic, F., Timmons, C., Cuzzupe, A., and Wipf, P. (2009) *Chemical Communications*, **45**, 104–106.
60 Van der Eycken, E., Appukkuttan, P., De Borggraeve, W., Dehaen, W., Dallinger, D., and Kappe, C.O. (2002) *The Journal of Organic Chemistry*, **67**, 7904–7909.
61 Kaval, N., Dehaen, W., Kappe, C.O., and Van der Eycken, E. (2004) *Organic & Biomolecular Chemistry*, **2**, 154–156.
62 Hughes, R.A., Thompson, S.P., Alcaraz, L., and Moody, C.J. (2004) *Chemical Communications*, 946–948.

63 Burland, P.A., Coisson, D., and Osborn, H.M.I. (2010) *The Journal of Organic Chemistry*, **75**, 7210–7218.

64 Frankowski, K.J., Hirt, E.E., Zeng, Y., Neuenswander, B., Fowler, D., Schoenen, F., and Aubé, J. (2007) *Journal of Combinatorial Chemistry*, **9**, 1188–1192.

65 Cook, S.P., Polara, A., and Danishefsky, S.J. (2006) *Journal of the American Chemical Society*, **128**, 16440–16441.

66 Tucker, C.J., Welker, M.E., Day, C.S., and Wright, M.W. (2004) *Organometallics*, **23**, 2257–2262.

67 Fuentes, A., Martínez-Palou, R., Jiménez-Vázquez, H.A., Delgado, F., Reyes, A., and Tamariz, J. (2005) *Monatshefte für Chemie*, **136**, 177–192.

68 Pinto, D.C.G.A., Silva, A.M.S., Brito, C.M., Sandulache, A., Carrillo, J.R., Prieto, P., Díaz-Ortiz, A., de la Hoz, A., and Cavaleiro, J.A.S. (2005) *European Journal of Organic Chemistry*, 2973–2986.

69 Mance, A.D. and Jakopčić, K. (2005) *Molecular Diversity*, **9**, 229–231.

70 Wu, J., Yu, H.a., Wang, Y., Xing, X., and Dai, W.-M. (2007) *Tetrahedron Letters*, **48**, 6543–6547.

71 Panunzio, M., Bandini, E., D'Aurizio, A., Millemaggi, A., and Xia, Z. (2007) *Synthesis*, 2060–2062.

72 Xing, X., Wu, J., and Dai, W.M. (2006) *Tetrahedron*, **62**, 11200–11206.

73 Wu, J., Sun, L., and Dai, W.M. (2006) *Tetrahedron*, **62**, 8360–8372.

74 Jiménez-Alonso, S., Estévez-Braun, A., Ravelo, Á.G., Zárate, R., and López, M. (2007) *Tetrahedron*, **63**, 3066–3074.

75 Kranjc, K. and Kočevar, M. (2006) *Collection of Czechoslovak Chemical Communication*, **71**, 667–678.

76 Sarotti, A.M., Joullié, M.M., Spanevello, R.A., and Suárez, A.G. (2006) *Organic Letters*, **8**, 5561–5564.

77 Gudipati, I.R., Sadasivam, D.V., and Birney, D.M. (2008) *Green Chemistry*, **10**, 275–277.

78 Saaby, S., Baxendale, I.R., and Ley, S.V. (2005) *Organic & Biomolecular Chemistry*, **3**, 3365–3368.

79 Young, D.D. and Deiters, A. (2007) *Angewandte Chemie International Edition*, **46**, 5187–5190.

80 Young, D.D., Sripada, L., and Deiters, A. (2007) *Journal of Combinatorial Chemistry*, **9**, 735–738.

81 Sripada, L., Teske, J.A., and Deiters, A. (2008) *Organic & Biomolecular Chemistry*, **6**, 263–265.

82 McIver, A., Young, D.D., and Deiters, A. (2008) *Chemical Communications*, 4750–4752.

83 Teske, J.A. and Deiters, A. (2008) *The Journal of Organic Chemistry*, **73**, 342–345.

84 Shanmugasundaram, M., Aguirre, A.L., Leyva, M., Quan, B., and Martinez, L.E. (2007) *Tetrahedron Letters*, **48**, 7698–7701.

85 Novák, P., Číhalová, S., Otmar, M., Hocek, M., and Kotora, M. (2008) *Tetrahedron*, **64**, 5200–5207.

86 Park, K.H., Choi, S.Y., Kim, S.Y., and Chung, Y.K. (2006) *Synlett*, 527–532.

87 Hrdina, R., Kadlčíková, A., Valterová, I., Hodačová, J., and Kotora, M. (2006) *Tetrahedron Asymmetry*, **17**, 3185–3191.

88 Zhou, Y., Porco, J.A., and Snyder, J.K. (2007) *Organic Letters*, **9**, 393–396.

89 Arrieta, A., Otaegui, D., Zubia, A., Cossío, F.P., Díaz-Ortiz, A., de la Hoz, A., Herrero, M.A., Prieto, P., Foces-Foces, C., Pizarro, J.L., and Arriortua, M.I. (2007) *The Journal of Organic Chemistry*, **72**, 4313–4322.

90 Bergner, I. and Opatz, T. (2007) *The Journal of Organic Chemistry*, **72**, 7083–7090.

91 Kim, Y., Kim, J., and Park, S.B. (2009) *Organic Letters*, **11**, 17–20.

92 Meng, L., Fettinger, J.C., and Kurth, M.J. (2007) *Organic Letters*, **9**, 5055–5058.

93 Hong, B.-C., Liu, K.-L., Tsai, C.-W., and Liao, J.-H. (2008) *Tetrahedron Letters*, **49**, 5480–5483.

94 Beryozkina, T., Appukkuttan, P., Mont, N., and Van der Eycken, E. (2006) *Organic Letters*, **8**, 487–490.

95 Katritzky, A.R., Singh, S.K., Meher, N.K., Doskocz, J., Suzuki, K., Jiang, R., Sommen, G.L., Ciaramitaro, D.A., and Steel, P.J. (2006) *ARKIVOC*, **v**, 43–62.

96 Perissutti, E., Frecentese, F., Fiorino, F., Severino, B., Cirillo, D., Santagada, V., and Caliendo, G. (2007) *Journal of Heterocyclic Chemistry*, **44**, 815–819.

97 Morrison, A.J., Paton, R.M., and Sharp, R.D. (2005) *Synthetic Communications*, **35**, 807–813.

98 McMillan, K.G., Tackett, M.N., Dawson, A., Fordyce, E., and Paton, R.M. (2006) *Carbohydrate Research*, **341**, 41–48.

99 Brummond, K.M. and Chen, D. (2005) *Organic Letters*, **7**, 3473–3475.

100 Dupau, P., Epple, R., Thomas, A.A., Fokin, V.V., and Sharpless, K.B. (2002) *Advanced Synthesis and Catalysis*, **344**, 421–433.

101 Clark, J.S., Clarke, M.-R., Clough, J., Blake, A.J., and Wilson, C. (2004) *Tetrahedron Letters*, **45**, 9447–9450.

102 Freitag, J., Nüchter, M., and Ondruschka, B. (2003) *Green Chemistry*, **5**, 291–295.

103 Landers, B., Berini, C., Wang, C., and Navarro, O. (2011) *The Journal of Organic Chemistry*, **76**, 1390–1397.

104 Takahashi, M., Oshima, K., and Matsubara, S. (2003) *Tetrahedron Letters*, **44**, 9201–9203.

105 Bagley, M.C., Dale, J.W., Xiong, X., and Bower, J. (2003) *Organic Letters*, **5**, 4421–4424.

106 Stiasni, N. and Kappe, C.O. (2002) *ARKIVOC*, **viii**, 71–79.

107 Alexandre, F.-R., Berecibar, A., Wrigglesworth, R., and Besson, T. (2003) *Tetrahedron Letters*, **44**, 4455–4458.

108 Mazumder, S., Laskar, D.D., Prajapati, D., and Roy, M.K. (2004) *Chemistry & Biodiversity*, **1**, 925–929.

109 Fustero, S., Sánchez-Roselló, M., Aceña, J.L., Fernández, B., Assensio, A., Sanz-Cervera, J.F., and del Pozo, C. (2009) *The Journal of Organic Chemistry*, **74**, 3414–3423.

110 Quinn, J.F., Bryant, C.E., Golden, K.C., and Gregg, B.T. (2010) *Tetrahedron Letters*, **51**, 786–789.

111 Ekström, J., Wettergren, J., and Adolfsson, H. (2007) *Advanced Synthesis and Catalysis*, **349**, 1609–1613.

112 Leijondahl, K., Fransson, A.B.L., and Bäckvall, J.-E. (2006) *The Journal of Organic Chemistry*, **71**, 8622–8625.

113 Piras, L., Genesio, E., Ghiron, C., and Taddei, M. (2008) *Synlett*, 1125–1128.

114 (a) Vanier, G.S. (2007) *Synlett*, 131–135; (b) Irfan, M., Fuchs, M., Glasnov, T.N., and Kappe, C.O. (2009) *Chemistry: A European Journal*, **15**, 11608–11618.

115 Toom, L., Grennberg, H., and Gogoll, A. (2006) *Synthesis*, 2064–2068.

116 Breschi, M.C., Calderone, V., Martelli, A., Minutolo, F., Rapposelli, S., Testai, L., Tonelli, F., and Balsamo, A. (2006) *Journal of Medicinal Chemistry*, **49**, 7600–7602.

117 Olivares-Romero, J.L. and Juaristi, E. (2008) *Tetrahedron*, **64**, 9992–9998.

118 Dal Zotto, C., Virieux, D., and Campagne, J.-M. (2009) *Synlett*, 276–278.

119 Spencer, J., Anjum, N., Patel, H., Rathnam, R.P., and Verma, J. (2007) *Synlett*, 2557–2558.

120 Wu, X., Mahalingam, A.K., and Alterman, M. (2005) *Tetrahedron Letters*, **46**, 1501–1504.

121 Hu, Y., Green, N., Gavrin, L.K., Janz, K., Kaila, N., Li, H.-Q., Thomason, Je.R., Cuozzo, J.W., Hall, J.P., Hsu, S., Nickerson-Nutter, C., Telliez, J.-Ba., Lin, L.-L., and Tam, S. (2006) *Bioorganic & Medicinal Chemistry Letters*, **16**, 6067–6072.

122 Poeylaut-Palena, A.A., Testero, S.A., and Mata, E.G. (2011) *Chemical Communications*, **47**, 1565–1567.

123 Lipshutz, B.H. and Frieman, B.A. (2005) *Angewandte Chemie International Edition*, **44**, 6345–6348.

124 Baker, B.A., Boskovic, Z.V., and Lipshutz, B.H. (2008) *Organic Letters*, **10**, 289–292.

125 Gustafsson, T., Hedenström, M., and Kihlberg, J. (2006) *The Journal of Organic Chemistry*, **71**, 1911–1919.

126 Stuhr-Hansen, N., Sørensen, J.K., Moth-Poulsen, K., Christensen, J.B., Bjørnholm, T., and Nielsen, M.B. (2005) *Tetrahedron*, **61**, 12288–12295.

127 Rakowitz, D., Maccari, R., Ottanà, R., and Vigorita, M.G. (2006) *Bioorganic & Medicinal Chemistry*, **14**, 567–574.

128 Bartoli, G., Di Antonio, G., Giovannini, R., Giuli, S., Lanari, S., Paoletti, M., and Marcantoni, E. (2008) *The Journal of Organic Chemistry*, **73**, 1919–1924.

129 Li, J., Ye, D., Liu, H., Luo, X., and Jiang, H. (2008) *Synthetic Communications*, **38**, 567–575.

130 Steinreiber, A., Stadler, A., Mayer, S.F., Faber, K., and Kappe, C.O. (2001) *Tetrahedron Letters*, **42**, 6283–6286.

131 Lampariello, L.R., Piras, D., Rodriquez, M., and Taddei, M. (2003) *The Journal of Organic Chemistry*, **68**, 7893–7895.

132 Ghassemi, S. and Fuchs, K. (2005) *Molecular Diversity*, **9**, 295–299.

133 Raheem, I.T., Goodman, S.N., and Jacobsen, E.N. (2004) *Journal of the American Chemical Society*, **126**, 706–707.

134 Söderberg, E., Westman, J., and Oscarson, S. (2001) *Journal of Carbohydrate Chemistry*, **20**, 397–410.

135 Mathew, F., Jayaprakash, K.N., Fraser-Reid, B., Mathew, J., and Scicinski, J. (2003) *Tetrahedron Letters*, **44**, 9051–9054.

136 Larsen, K., Worm-Leonhard, K., Olsen, P., Hoel, A., and Jensen, K.J. (2005) *Organic & Biomolecular Chemistry*, **3**, 3966–3970.

137 Du, W., Kulkarni, S.S., and Gervay-Hague, J. (2007) *Chemical Communications*, 2336–2338.

138 Lohman, G.J.S., and Seeberger, P. (2004) *The Journal of Organic Chemistry*, **69**, 4081–4093.

139 Paoli, M.L., Piccini, S., Rodriquez, M., and Sega, A. (2004) *The Journal of Organic Chemistry*, **69**, 2881–2883.

140 Ballell, L., Joosten, J.A.F., Ait el Maate, F., Liskamp, R.M.J., and Pieters, R.J. (2004) *Tetrahedron Letters*, **45**, 6685–6687.

141 Roy, V., Colombeau, L., Zerrouki, R., and Krausz, P. (2004) *Carbohydrate Research*, **339**, 1829–1831.

142 Bookser, B.C. and Raffaele, N.B. (2007) *The Journal of Organic Chemistry*, **72**, 173–179.

143 Ko, K.-S., Zea, C.J., and Pohl, N.L. (2004) *Journal of the American Chemical Society*, **126**, 13188–13189.

144 Bejugam, M. and Flitsch, S.L. (2004) *Organic Letters*, **6**, 4001–4004.

145 Dondoni, A., Massi, A., and Aldhoun, M. (2007) *The Journal of Organic Chemistry*, **72**, 7677–7687.

146 Li, X. and Danishefsky, S.J. (2008) *Journal of the American Chemical Society*, **130**, 5446–5448.

147 Seibel, J., Hillringhaus, L., and Moraru, R. (2005) *Carbohydrate Research*, **340**, 507–511.

148 Rat, S., Mathiron, D., Michaud, P., Kovensky, J., and Wadouachi, A. (2007) *Tetrahedron*, **63**, 12424–12428.

149 Lin, H.-C., Chang, C.-C., Chen, J.-Y., and Lin, C.-H. (2005) *Tetrahedron Asymmetry*, **16**, 297–301.

150 Bornaghi, L.F. and Poulsen, S.A. (2005) *Tetrahedron Letters*, **46**, 3485–3488.

151 Christensen, H., Steinicke Christiansen, M., Petersen, J., and Jensen, H.H. (2008) *Organic & Biomolecular Chemistry*, **6**, 3276–3283.

152 Yoshimura, Y., Shimizu, H., Hinou, H., and Nishimura, S.-I. (2005) *Tetrahedron Letters*, **46**, 4701–4705.

153 Abramov, M., Marchand, A., Calleja-Marchand, A., and Herdewijn, P. (2004) *Nucleosides, Nucleotides & Nucleic Acids*, **23**, 439–455.

154 Das, S.K. (2004) *Synlett*, 915–932.

155 Corsaro, A., Chiacchio, U., Pistara, V., and Romeo, G. (2004) *Current Organic Chemistry*, **8**, 511–538.

156 Rodríguez, B. and Bolm, C. (2006) *The Journal of Organic Chemistry*, **71**, 2888–2891.

157 Mossé, S. and Alexakis, A. (2006) *Organic Letters*, **8**, 3577–3580.

158 Baumann, T., Bächle, M., Hartmann, C., and Bräse, S. (2008) *European Journal of Organic Chemistry*, 2207–2212.

159 Westermann, B. and Neuhaus, C. (2005) *Angewandte Chemie International Edition*, **44**, 4077–4079.

160 Massi, A., Nuzzi, A., and Dondoni, A. (2007) *The Journal of Organic Chemistry*, **72**, 10279–10282.

161 Hosseini, M., Stiasni, N., Barbieri, V., and Kappe, C.O. (2007) *The Journal of Organic Chemistry*, **72**, 1417–1424.

162 Gold, H., Larhed, M., and Nilsson, P. (2005) *Synlett*, 1596–1600.

163 Wunderlich, S. and Knochel, P. (2008) *Organic Letters*, **10**, 4705–4707.

164 Mosrin, M., Monzon, G., Bresser, T., and Knochel, P. (2009) *Chemical Communications*, 5615–5617.

165 Bentz, E., Moloney, M.G., and Westaway, S.M. (2004) *Tetrahedron Letters*, **45**, 7395–7397.

166 Bentz, E., Moloney, M.G., and Westaway, S.M. (2007) *Synlett*, 733–736.

167 Braga, A.L., Paixão, M.W., Westermann, B., Schneider, P.H., and Wessjohann, L.A. (2008) *The Journal of Organic Chemistry*, **73**, 2879–2882.

168 Szatmári, I., Sillanpää, R., and Fülöp, F. (2008) *Tetrahedron Asymmetry*, **19**, 612–617.
169 Genov, M., Salas, G., and Espinet, P. (2008) *Journal of Organometallic Chemistry*, **693**, 2017–2020.
170 Cook, M.J., Fleming, D.W., and Gallagher, T. (2005) *Tetrahedron Letters*, **46**, 297–300.
171 Adriaenssens, L.V. and Hartley, R.C. (2007) *The Journal of Organic Chemistry*, **72**, 10287–10290.
172 Su, C.-R., Shen, Y.-C., Kuo, P.-C., Leu, Y.-L., Damu, A.G., Wang, Y.-H., and Wu, T.-S. (2006) *Bioorganic & Medicinal Chemistry Letters*, **16**, 6155–6160.
173 Duvall, J.R., Wu, F., and Snider, B.B. (2006) *The Journal of Organic Chemistry*, **71**, 8579–8590.
174 Morone, M., Beverina, L., Abbotto, A., Silvestri, F., Collini, E., Ferrante, C., Bozio, R., and Paganil, G.A. (2006) *Organic Letters*, **8**, 2719–2722.
175 Stuhr-Hansen, N. (2005) *Tetrahedron Letters*, **46**, 5491–5494.
176 Lehmann, F., Pilotti, Å., and Luthman, K. (2003) *Molecular Diversity*, **7**, 145–152.
177 Leadbeater, N.E., Torenius, H.M., and Tye, H. (2003) *Molecular Diversity*, **7**, 135–144.
178 Shi, L., Tu, Y.-Q., Wang, M., Zhang, F.-M., and Fan, C.-A. (2004) *Organic Letters*, **6**, 1001–1004.
179 Ju, Y., Li, C.-J., and Varma, R.S. (2004) *QSAR & Combinatorial Science*, **23**, 891–894.
180 Bariwal, J.B., Ermolat'ev, D.S., and Van der Eycken, E.V. (2010) *Chemistry: A European Journal*, **16**, 3281–3284.
181 Peshkov, V.A., Pereshivko, O.P., Donets, P.A., Mehta, V.P., and Van der Eycken, E.V. (2010) *European Journal of Chemistry*, **47**, 4861–4867.
182 Bariwal, J.B., Ermolat'ev, D.S., Glasnov, T.N., Van Hecke, K., Mehta, V.P., Van Meervelt, L., Kappe, C.O., and Van der Eycken, E.V. (2010) *Organic Letters*, **12**, 2774–2777.
183 Pereshivko, O.P., Peshkov, V.A., and Van der Eycken, E.V. (2010) *Organic Letters*, **12**, 2638–2641.
184 McLean, N.J., Tye, H., and Whittaker, M. (2004) *Tetrahedron Letters*, **45**, 993–995.
185 Follmann, M., Graul, F., Schäfer, T., Kopec, S., and Hamley, P. (2005) *Synlett*, 1009–1011.
186 Nun, P., Martinez, J., and Lamaty, F. (2010) *Synthesis*, 2063–2068.
187 Zbruyev, O.I., Stiasni, N., and Kappe, C.O. (2003) *Journal of Combinatorial Chemistry*, **5**, 145–148.
188 Hübner, S., Neumann, H., Michalik, D., Klaus, S., Strübing, D., Spannenberg, A., Jacobi von Wangelin, A., and Beller, M. (2007) *Synlett*, 1085–1090.
189 Strübing, D., Neumann, H., Jacobi von Wangelin, A., Klaus, S., Hübner, S., and Beller, M. (2006) *Tetrahedron*, **62**, 10962–10967.
190 Schwerkoske, J., Masquelin, T., Perun, T., and Hulme, C. (2005) *Tetrahedron Letters*, **46**, 8355–8357.
191 Masquelin, T., Bui, H., Brickley, B., Stephenson, G., Schwerkoske, J., and Hulme, C. (2006) *Tetrahedron Letters*, **47**, 2989–2991.
192 DiMauro, E.F. and Kennedy, J.M. (2007) *The Journal of Organic Chemistry*, **72**, 1013–1016.
193 Wu, J. (2006) *Chemistry Letters*, **35**, 118–119.
194 Xing, X., Wu, J., Luo, J., and Dai, W.M. (2006) *Synlett*, 2099–2103.
195 Oble, J., El Kaïm, L., Gizzi, M., and Grimaud, L. (2007) *Heterocycles*, **73**, 503–517.
196 Bohn Rhoden, C.R., Westermann, B., and Wessjohann, L. (2008) *Synthesis*, 2077–2082.
197 Hulme, C., Chappetta, S., Griffith, C., Lee, Y.-S., and Dietrich, J. (2009) *Tetrahedron Letters*, **50**, 1938–1942.
198 De Silva, R.A., Santra, S., and Andreana, P.R. (2008) *Organic Letters*, **10**, 4541–4544.
199 Santra, S. and Andreana, P.R. (2007) *Organic Letters*, **9**, 5035–5038.
200 Bartheleon, A., El Kaim, L., Gizzi, M., and Grimaud, L. (2010) *Synlett*, 2784–2788.
201 Gududuru, V., Nguyen, V., Dalton, J.T., and Miller, D.D. (2004) *Synlett*, 2357–2358.
202 Risitano, F., Grassi, G., Foti, F., and Moraci, S. (2005) *Synlett*, 1633–1635.

203 Castagnolo, S., Botta, L., and Botta, M. (2009) *Tetrahedron Letters*, **50**, 1526–1528.
204 Tu, S., Zhu, X., Zhang, J., Xu, J., Zhang, Y., Wang, Q., Jia, R., Jiang, B., Zhang, J., and Yao, C. (2006) *Bioorganic & Medicinal Chemistry Letters*, **16**, 2925–2928.
205 Shi, F., Ma, N., Zhang, Y., Zhang, G., Jiang, B., and Tu, S. (2010) *Synthetic Communications*, **40**, 235–241.
206 Tu, S., Zhu, X., Shi, F., Zhang, J., and Zhang, Y. (2007) *Journal of Heterocyclic Chemistry*, **44**, 837–844.
207 Tu, S.-J., Jiang, B., Jia, R.-H., Zhang, J.-Y., Zhang, Y., Yao, C.-S., and Shi, F. (2006) *Organic & Biomolecular Chemistry*, **4**, 3664–3668.
208 Tu, S.-J., Jiang, B., Zhang, J.-Y., Jia, R.-H., Zhang, Y., and Yao, C.-S. (2006) *Organic & Biomolecular Chemistry*, **4**, 3980–3985.
209 Tu, S., Zhang, Y., Jia, R., Jiang, B., Zhang, J., and Ji, S. (2006) *Tetrahedron Letters*, **47**, 6521–6525.
210 Tu, S., Li, C., Li, G., Cao, L., Shao, Q., Zhou, D., Jiang, B., Zhou, J., and Xia, M. (2007) *Journal of Combinatorial Chemistry*, **9**, 1144–1148.
211 Tu, S., Jiang, B., Jiang, H., Zhang, Y., Jia, R., Zhang, J., Shao, Q., Li, C., Zhou, D., and Cao, L. (2007) *Tetrahedron*, **63**, 5406–5414.
212 Tu, S.-J., Zhang, Y., Jiang, H., Jiang, B., Zhang, J.-Y., Jia, R.-H., and Shi, F. (2007) *European Journal of Organic Chemistry*, 1522–1528.
213 Tu, S., Jiang, B., Zhang, Y., Jia, R., Zhang, J., Yao, C., and Shi, F. (2007) *Organic & Biomolecular Chemistry*, **5**, 355–359.
214 Jiang, B., Tu, S.-J., Kaur, P., Wever, W., and Li, G. (2009) *Journal of the American Chemical Society*, **131**, 11660–11661.
215 Jiang, B., Li, C., Shi, F., Tu, S.-J., Kaur, P., Wever, W., and Li, G. (2010) *The Journal of Organic Chemistry*, **75**, 2962–2965.
216 Bremner, W.S. and Organ, M.G. (2006) *Journal of Combinatorial Chemistry*, **8**, 14–16.
217 Presset, M., Coquerel, Y., and Rodriguez, J. (2010) *Organic Letters*, **12**, 4212–4215.
218 Presset, M., Coquerel, Y., and Rodriguez, J. (2009) *Organic Letters*, **11**, 5706–5709.
219 Chebanov, V.A., Sakhno, Y.I., Desenko, S.M., Chernenko, V.N., Musatov, V.I., Shishkina, S.V., Shishkin, O.V., and Kappe, C.O. (2007) *Tetrahedron*, **63**, 1229–1242.
220 Chebanov, V.A., Muravyova, E.A., Desenko, S.M., Musatov, V.I., Knyazeva, I.V., Shishkina, S.V., Shishkin, O.V., and Kappe, C.O. (2006) *Journal of Combinatorial Chemistry*, **8**, 427–434.
221 Zhu, S.-L., Ji, S.-J., Zhao, K., and Liu, Y. (2008) *Tetrahedron Letters*, **49**, 2578–2582.
222 Ostras, K.S., Gorobets, N.Y., Desenko, S.M., and Musatov, V.I. (2006) *Molecular Diversity*, **10**, 483–489.
223 Fang, Z. and Lam, Y. (2011) *Tetrahedron*, **67**, 1294–1297.
224 Gorobets, N.Yu., Sedash, Y.V., Ostras, K.S., Zaremba, O.V., Shishkina, S.V., Baumer, V.N., Shishkin, O.V., Kovalenko, S.M., Desenko, S.M., and Van der Eycken, E.V. (2010) *Tetrahedron Letters*, **51**, 2095–2098.
225 Yadav, L.D.S., Awasthi, C., Rai, V.K., and Rai, A. (2007) *Tetrahedron Letters*, **48**, 4899–4902.
226 Yan, C.G., Cai, X.M., Wang, Q.F., Wang, T.Y., and Zheng, M. (2007) *Organic & Biomolecular Chemistry*, **5**, 945–951.
227 Gogol, S., Dutta, M., Gogol, J., and Boruah, R.C. (2011) *Tetrahedron Letters*, **52**, 813–816.
228 Nguyen, R.V. and Li, C.J. (2008) *Synlett*, 1897–1901.
229 Xiao, F., Chen, Y., Liu, Y., and Wang, J. (2008) *Tetrahedron*, **64**, 2755–2761.
230 Ohta, Y., Chiba, H., Oishi, S., Fujii, N., and Ohno, H. (2008) *Organic Letters*, **10**, 3535–3538.
231 Guo, K., Thompson, M.J., Reddy, T.R.K., Mutter, R., and Chen, B. (2007) *Tetrahedron*, **63**, 5300–5311.
232 Li, M., Zuo, Z., Wen, L., and Wang, S. (2008) *Journal of Combinatorial Chemistry*, **10**, 436–441.
233 Ravikumar Naik, T.R. and Bhojya Naik, H.S. (2008) *Molecular Diversity*, **12**, 139–142.
234 Zanobini, A., Brandi, A., and de Meijere, A. (2006) *European Journal of Organic Chemistry*, 1251–1255.

235 Ye, P., Sargent, K., Stewart, E., Liu, J.-F., Yohannes, D., and Yu, L. (2006) *The Journal of Organic Chemistry*, **71**, 3137–3140.

236 Yang, S., Gao, X.-W., Diao, C.-L., Song, B.-A., Jin, L.-H., Xu, G.-F., Zhang, G.-P., Wang, W., Hu, D.-Y., Xue, W., Zhou, X., and Lu, P. (2006) *Chinese Journal of Chemistry*, **24**, 1581–1588.

237 Régnier, T., Berrée, F., Lavastre, O., and Carboni, B. (2007) *Green Chemistry*, **9**, 125–126.

238 Sridhar, M., Rao, R.M., Baba, N.H.K., and Kumbhare, R.M. (2007) *Tetrahedron Letters*, **48**, 3171–3172.

239 Zuo, H., Li, Z.-B., Ren, F.-K., Falck, J.R., Lijuan, M., Ahn, C., and Shin, D.-S. (2008) *Tetrahedron*, **64**, 9669–9674.

240 Saulnier, M.G., Zimmermann, K., Struzynski, C.P., Sang, X., Velaparthi, U., Wittman, M., and Frennesson, D.B. (2004) *Tetrahedron Letters*, **45**, 397–399.

241 Ju, Y. and Varma, R.S. (2004) *Green Chemistry*, **6**, 219–221.

242 Romera, J.L., Cid, J.M., and Trabanco, A.A. (2004) *Tetrahedron Letters*, **45**, 8797–8800.

243 Lamberto, M., Rastede, E.E., Decker, J., and Raymo, F.M. (2010) *Tetrahedron Letters*, **51**, 5618–5620.

244 Jennings, L.D., Foreman, K.W., Rush, T.S., Tsao, D.H.H., Mosyak, L., Li, Y., Sukhdeo, M.N., Ding, W., Dushin, E.G., Kenny, C.H., Moghazeh, S.L., Petersen, P.J., Ruzin, A.V., Tuckman, M., and Sutherland, A.G. (2004) *Bioorganic & Medicinal Chemistry Letters*, **14**, 1427–1431.

245 Holzgrabe, U. and Heller, E. (2003) *Tetrahedron*, **59**, 781–787.

246 Vasudevan, A., Ji, Z., Frey, R.R., Wada, C.K., Steinman, D., Heyman, H.R., Guo, Y., Curtin, M.L., Guo, J., Li, J., Pease, L., Glaser, K.B., Marcotte, P.A., Bouska, J.J., Davidsen, S.K., and Michaelides, M.R. (2003) *Bioorganic & Medicinal Chemistry Letters*, **13**, 3909–3913.

247 Cai, L., Chin, F.T., Pike, V.W., Toyama, H., Liow, J.-S., Zoghbi, S.S., Modell, K., Briard, E., Shetty, H.U., Sinclair, K., Donohue, S., Tipre, D., Kung, M.-P., Dagostin, C., Widdowson, D.A., Green, M., Gao, W., Herman, M.M., Ichise, M., and Innis, R.B. (2004) *Journal of Medicinal Chemistry*, **47**, 2208–2218.

248 Alcázar, J., Diels, G., and Schoentjes, B. (2004) *QSAR & Combinatorial Science*, **23**, 906–910.

249 Kainmuller, E.K., Olle, E.P., and Bannwarth, W. (2006) *Chemical Communications*, 5459–5461.

250 Manetti, F., Esté, J.A., Clotet-Codina, I., Armand-Ugón, M., Maga, G., Crespan, E., Cancio, R., Mugnaini, C., Bernardini, C., Togninelli, A., Carmi, C., Alongi, M., Petricci, E., Massa, S., Corelli, F., and Botta, M. (2005) *Journal of Medicinal Chemistry*, **48**, 8000–8008.

251 Mugnaini, C., Alongi, M., Togninelli, A., Gevariya, H., Brizzi, A., Manetti, F., Bernardini, C., Angeli, L., Tafi, A., Bellucci, L., Corelli, F., Massa, S., Maga, G., Samuele, A., Facchini, M., Clotet-Codina, I., Armand-Ugón, M., Esté, J.A., and Botta, M. (2007) *Journal of Medicinal Chemistry*, **50**, 6580–6595.

252 Bica, K., Gmeiner, G., Reichel, C., Lendl, B., and Gaertner, P. (2007) *Synthesis*, 1333–1338.

253 Bon, R.S., de Kanter, F.J.J., Lutz, M., Spek, A.L., Jahnke, M.C., Hahn, F.E., Groen, M.B., and Orru, R.V.A. (2007) *Organometallics*, **26**, 3639–3650.

254 Sarju, J., Danks, T.N., and Wagner, G. (2004) *Tetrahedron Letters*, **45**, 7675–7677.

255 Organ, M.G., Mayer, S., Lepifre, F., N'Zemba, B., and Khatri, J. (2003) *Molecular Diversity*, **7**, 211–227.

256 Radi, M., Petricci, E., Maga, G., Corelli, F., and Botta, M. (2005) *Journal of Combinatorial Chemistry*, **7**, 117–122.

257 Ollevier, T. and Li, Z. (2007) *European Journal of Organic Chemistry*, 5665–5668.

258 Doi, T., Fuse, S., Miyamoto, S., Nakai, K., Sasuga, D., and Takahashi, T. (2006) *Chemistry: An Asian Journal*, **1**, 370–383.

259 Handlon, A. and Guo, Y. (2005) *Synlett*, 111–114.

260 Löfberg, C., Grigg, R., Keep, A., Derrick, A., Sridharan, V., and Kilner, C. (2006) *Chemical Communications*, 5000–5002.

261 Ranu, B.C., Banerjee, S., and Jana, R. (2007) *Tetrahedron*, **63**, 776–782.

262 Keglevich, G., Majrik, K., Vida, L., and Greiner, I. (2008) *Letters in Organic Chemistry*, **5**, 224–228.

263 Moloney, M. and Yaqoob, M. (2008) *Synlett*, 2107–2110.

264 Cherng, Y.-J. (2002) *Tetrahedron*, **58**, 887–890.

265 Cherng, Y.-J. (2002) *Tetrahedron*, **58**, 1125–1129.

266 Cherng, Y.-J. (2002) *Tetrahedron*, **58**, 4931–4935.

267 Luo, G., Chen, L., and Pointdexter, G.S. (2002) *Tetrahedron Letters*, **43**, 5739–5742.

268 Takvorian, A.G. and Combs, A.P. (2004) *Journal of Combinatorial Chemistry*, **6**, 171–174.

269 Arienzo, R., Clark, D.E., Cramp, S., Daly, S., Dyke, H.J., Lockey, P., Norman, D., Roach, A.G., Stuttle, K., Tomlinson, M., Wong, M., and Wren, S.P. (2004) *Bioorganic & Medicinal Chemistry Letters*, **14**, 4099–4102.

270 Hopkins, C.R., Neuenschwander, K., Scotese, A., Jackson, S., Nieduzak, T., Pauls, H., Liang, G., Sides, K., Cramer, D., Cairns, J., Maignand, S., and Mathieu, M. (2004) *Bioorganic & Medicinal Chemistry Letters*, **14**, 4819–4823.

271 Souers, A.J., Wodka, D., Gao, J., Lewis, J.C., Vasudevan, A., Gentles, R., Brodjian, S., Dayton, B., Ogiela, C.A., Fry, D., Hernandez, L.E., Marsh, K.C., Collins, C.A., and Kym, P.R. (2004) *Bioorganic & Medicinal Chemistry Letters*, **14**, 4873–4877.

272 Souers, A.J., Wodka, D., Gao, J., Lewis, J.C., Vasudevan, A., Brodjian, S., Dayton, B., Ogiela, C.A., Fry, D., Hernandez, L.E., Marsh, K.C., Collins, C.A., and Kym, P.R. (2004) *Bioorganic & Medicinal Chemistry Letters*, **14**, 4883–4886.

273 Priem, G., Anson, M.S., Macdonald, S.J.F., Pelotier, B., and Campbell, I.B. (2002) *Tetrahedron Letters*, **43**, 6001–6003.

274 Narayan, S., Seelhammer, T., and Gawley, R.E. (2004) *Tetrahedron Letters*, **45**, 757–759.

275 Boursereau, Y. and Coldham, I. (2004) *Bioorganic & Medicinal Chemistry Letters*, **14**, 5841–5844.

276 Loung, M., Loupy, A., Marque, S., and Petit, A. (2004) *Heterocycles*, **63**, 297–308.

277 Belda, O. and Moberg, C. (2002) *Synthesis*, 1601–1606.

278 Belda, O., Lundgren, S., and Moberg, C. (2003) *Organic Letters*, **5**, 2275–2278.

279 Steel, P.G. and Teasdale, C.W.T. (2004) *Tetrahedron Letters*, **45**, 8977–8980.

280 Nikitenko, A.A., Raifeld, Y.E., and Wang, T.Z. (2001) *Bioorganic & Medicinal Chemistry Letters*, **11**, 1041–1044.

281 Cherng, Y.-J. (2000) *Tetrahedron*, **56**, 8287–8289.

282 Baqi, Y. and Müller, C.E. (2007) *The Journal of Organic Chemistry*, **72**, 5908–5911.

283 Brown, G.R., Foubister, A.J., Roberts, C.A., Wells, S.L., and Wood, R. (2001) *Tetrahedron Letters*, **42**, 3917–3919.

284 Mečiarova, M., Podlesná, J., and Toma, S. (2004) *Monatshefte fur Chemie*, **135**, 419–423.

285 Niculescu-Duvaz, I., Scanlon, I., Niculescu-Duvaz, D., Friedlos, F., Martin, J., Marais, R., and Springer, C.J. (2004) *Journal of Medicinal Chemistry*, **47**, 2651–2658.

286 Huang, H., Liu, H., Chen, K., and Jiang, H. (2007) *Journal of Combinatorial Chemistry*, **9**, 197–199.

287 Levin, J.I. and Du, M.T. (2002) *Synthetic Communications*, **32**, 1401–1406.

288 Ullah, F., Samarakoon, T., Rolfe, A., Kurtz, R.D., Hanson, P.R., and Organ, M. (2010) *Chemistry: A European Journal*, **16**, 10959–10962.

289 Rolfe, A., Samarakoon, T.B., Klimberg, S.V., Brzozowski, M., Neuenswander, B., Lushington, G.H., and Hanson, P.R. (2010) *Journal of Combinatorial Chemistry*, **12**, 850–854.

290 Fang, W.-P., Cheng, Y.-T., Cheng, Y.-R., and Cherng, Y.-J. (2005) *Tetrahedron*, **61**, 3107–3113.

291 Sandford, G., Slater, R., Yufit, D.S., Howard, J.A.K., and Vong, A. (2005) *The Journal of Organic Chemistry*, **70**, 7208–7216.

292 Samaroo, D., Soll, C.E., Todaro, L.J., and Drain, C.M. (2006) *Organic Letters*, **8**, 4985–4988.

293 Larsson, A., Spjut, S., Kihlberg, J., and Almqvist, F. (2005) *Synthesis*, 2590–2596.

294 Cao, P., Qu, J., Burton, G., and Rivero, R.A. (2008) *The Journal of Organic Chemistry*, **73**, 7204–7208.

295 Gemma, S., Kukreja, G., Fattorusso, C., Persico, M., Romano, M.P., Altarelli, M., Savini, L., Campiani, G., Fattorusso, E., Basilico, N., Taramelli, D., Yardley, V., and Butini, S. (2006) *Bioorganic & Medicinal Chemistry Letters*, **16**, 5384–5388.

296 Qu, G.-R., Zhao, L., Wang, D.-C., Wu, J., and Guo, H.-M. (2008) *Green Chemistry*, **10**, 287–289.

297 Caldwell, J.J., Davies, T.G., Donald, A., McHardy, T., Rowlands, M.G., Aherne, G.W., Hunter, L.K., Taylor, K., Ruddle, R., Raynaud, F.I., Verdonk, M., Workman, P., Garrett, M.D., and Collins, I. (2008) *Journal of Medicinal Chemistry*, **51**, 2147–2157.

298 St. Jean, D.J., Poon, S.F., and Schwarzbach, J.L. (2007) *Organic Letters*, **9**, 4893–4896.

299 Shi, L., Wang, M., Fan, C.-A., Zhang, F.-M., and Tu, Y.-Q. (2003) *Organic Letters*, **5**, 3515–3517.

300 Xu, G. and Wang, Y.-G. (2004) *Organic Letters*, **6**, 985–987.

301 Swain, N.A., Brown, R.C.D., and Bruton, G. (2002) *Chemical Communications*, 2042–2043.

302 Swain, N.A., Brown, R.C.D., and Bruton, G. (2004) *The Journal of Organic Chemistry*, **69**, 122–129.

303 Chaumontet, M., Retailleau, P., and Baudoin, O. (2009) *The Journal of Organic Chemistry*, **74**, 1774–1776.

304 Lake, F. and Moberg, C. (2002) *European Journal of Organic Chemistry*, 3179–3188.

305 lopes, S.M.M., Beja, A.M., Silva, M.R., Paixão, J.A., Palacios, F., and Pinho e Melo, T.M.V.D. (2009) *Synthesis*, 2403–2407.

306 Pei, Y., Brule, E., and Moberg, C. (2006) *Organic & Biomolecular Chemistry*, **4**, 544–550.

307 Kurz, T. and Widyan, K. (2005) *Tetrahedron*, **61**, 7247–7251.

308 Qin, H.L., Lowe, J.T., and Panek, J.S. (2007) *Journal of the American Chemical Society*, **129**, 38–39.

309 Díaz-Sánchez, B.R., Iglesias-Arteaga, M.A., Melgar-Fernández, R., and Juaristi, E. (2007) *The Journal of Organic Chemistry*, **72**, 4822–4825.

310 Alikhani, V., Beer, D., Bentley, D., Bruce, I., Cuenoud, B.M., Fairhurst, R.A., Gedeck, P., Haberthuer, S., Hayden, C., Janus, D., Jordan, L., Lewis, C., Smithies, K., and Wissler, E. (2004) *Bioorganic & Medicinal Chemistry Letters*, **14**, 4705–4710.

311 Lindsay, K.B. and Pyne, S.G. (2004) *Synlett*, 799–781.

312 Lindsay, K.B. and Pyne, S.G. (2004) *Tetrahedron Letters*, **60**, 4173–4176.

313 Fawcett, J., Griffith, G.A., Percy, J.M., and Uneyama, E. (2004) *Organic Letters*, **6**, 1277–1280.

314 Schirok, H. (2006) *The Journal of Organic Chemistry*, **71**, 5538–5545.

315 Rodríguez-Escrich, S., Popa, D., Jimeno, C., Vidal-Ferran, A., and Pericàs, M.A. (2005) *Organic Letters*, **7**, 3829–3832.

316 Cattoen, X. and Pericas, M.A. (2007) *The Journal of Organic Chemistry*, **72**, 3253–3258.

317 Robin, A., Brown, F., Bahamontes-Rosa, N., Wu, B., Beitz, E., Kun, J.F.J., and Flitsch, S.L. (2007) *Journal of Medicinal Chemistry*, **50**, 4243–4249.

318 Desai, H., D'Souza, B.R., Foether, D., Johnson, B.F., and Lindsay, H.A. (2007) *Synthesis*, 902–910.

319 Busscher, G.F., Groothuys, S., de Gelder, R., Rutjes, F.R.J.T., and van Delft, F.L. (2004) *The Journal of Organic Chemistry*, **69**, 4477–4481.

320 Bailliez, V., Olesker, A., and Cleophax, J. (2004) *Tetrahedron*, **60**, 1079–1085.

321 Pironti, V. and Colonna, S. (2005) *Green Chemistry*, **7**, 43–45.

322 Xue, F. and Seto, C.T. (2006) *Bioorganic & Medicinal Chemistry*, **14**, 8467–8487.

323 Grundt, P., Prevatt, K.M., Cao, J., Taylor, M., Floresca, C.Z., Choi, J.-K., Jenkins, B.G., Luedtke, R.R., and Hauck Newman, A. (2007) *Journal of Medicinal Chemistry*, **50**, 4135–4146.

324 Johansen, M., Leduc, A., and Kerr, M. (2007) *Synlett*, 2593–2595.

325 Jennings, L.D., Foreman, K.W., Rush, T.S., Tsao, D.H.H., Mosyak, L., Kincaid, S.L., Sukhdeo, M.N., Sutherland, A.G., Ding, W., Kenny, C.H., Sabus, C.L., Liu, H., Dushin, E.G., Moghazeh, S.L., Labthavikul, P., Petersen, P.J., Tuckmand, M., and Ruzin, A.V. (2004) *Bioorganic & Medicinal Chemistry*, **12**, 5115–5131.

326 Zare, A., Hasaninejad, A., Khalafi-Nezhad, A., Moosavi Zare, A.R., Parhami, A., and Nejabat, G.R. (2007) *ARKIVOC*, **i**, 58–69.

327 Chebanov, V.A., Reidlinger, C., Kanaani, H., Wentrup, C., Kappe, C.O., and Kollenz, G. (2004) *Supramolecular Chemistry*, **16**, 121–127.

328 Enders, D., Krüll, R., and Bettray, W. (2010) *Synthesis*, 567–572.

329 Blanchard, S., Rodriguez, I., Tardy, C., Baldeyrou, B., Bailly, C., Colson, P., Houssier, C., Léonce, S., Kraus-Berthier, L., Pfeiffer, B., Renard, P., Pierré, A., Caubére, P., and Guillaumet, G. (2004) *Journal of Medicinal Chemistry*, **47**, 978–987.

330 Zhang, Y. and Li, C.-J. (2004) *Tetrahedron Letters*, **45**, 7581–7584.

331 Mimeau, D. and Gaumont, A.-C. (2003) *The Journal of Organic Chemistry*, **68**, 7016–7022.

332 Uršič, U., Grošelj, U., Meden, A., Svete, J., and Stanovnik, B. (2009) *Helvetica Chimica Acta*, **92**, 481–490.

333 Wipf, P., Janjic, J., and Stephenson, J.C.R. (2004) *Organic & Biomolecular Chemistry*, **2**, 443–445.

334 Stevens, B.D., Bungard, C.J., and Nelson, S.G. (2006) *The Journal of Organic Chemistry*, **71**, 6397–6402.

335 Kazmaier, U. and Wesquet, A. (2005) *Synlett*, 1271–1274.

336 Wesquet, A., Dörrenbächer, S., and Kazmaier, U. (2006) *Synlett*, 1105–1109.

337 Kazmaier, U., Dörrenbächer, S., Wesquet, A., Lucas, S., and Kummeter, M. (2007) *Synthesis*, 320–326.

338 Mimeau, D., Delacroix, O., and Gaumont, A.-C. (2003) *Chemical Communications*, 2928–2929.

339 Join, B., Delacroix, O., and Gaumont, A.C. (2005) *Synlett*, 1881–1884.

340 Ackermann, L., Kaspar, L.T., and Gschrei, C.J. (2004) *Organic Letters*, **6**, 2515–2518.

341 Giner, X. and Nájera, C. (2009) *Synlett*, 3211–3213.

342 Crawley, M.L., McLaughlin, E., Zhu, W., and Combs, A.P. (2005) *Organic Letters*, **7**, 5067–5069.

343 Herrera, R., Jiménez-Vázquez, H.A., and Tamariz, J. (2005) *ARKIVOC*, **vi**, 233–249.

344 Bagley, M.C., Chapaneri, K., Glover, C., and Merritt, E.A. (2004) *Synlett*, 2615–2617.

345 Hammerland, L.G., Johansson, M., Malmström, J., Mattsson, J.P., Minidis, A.B.E., Nilsson, K., Peterson, A., Wensbo, D., Wållberg, A., and Österlund, K. (2006) *Bioorganic & Medicinal Chemistry Letters*, **16**, 2467–2469.

346 Schenkel, L.B. and Ellman, J.A. (2004) *Organic Letters*, **6**, 3621–3624.

347 Tanuwidjaja, J., Peltier, H.M., Lewis, J.C., Schenkel, L.B., and Ellman, J.A. (2007) *Synthesis*, 3385–3389.

348 Collina, S., Loddo, G., Barbieri, A., Linati, L., Alcaro, S., Chimenti, P., and Azzolina, O. (2004) *Tetrahedron Asymmetry*, **15**, 3601–3608.

349 Porcheddu, A., Giacomelli, G., and Salaris, M. (2005) *The Journal of Organic Chemistry*, **70**, 2361–2363.

350 Bose, A.K., Ganguly, S.N., Manhas, M.S., Srirajan, V., Bhattacharjee, A., Rumthao, S., and Sharma, A.H. (2004) *Tetrahedron Letters*, **45**, 1179–1181.

351 Shackelford, S.A., Anderson, M.B., Christie, L.C., Goetzen, T., Guzman, M.C., Hananel, M.A., Kornreich, W.D., Li, H., Pathak, V.P., Rabinovich, A.K., Rajapakse, R.J., Truesdale, L.K., Tsank, S.M., and Vazir, H.N. (2003) *The Journal of Organic Chemistry*, **68**, 267–275.

352 Xie, J., Comeau, A.B., and Seto, C.T. (2004) *Organic Letters*, **6**, 83–86.

353 Plé, P.A., Green, T.P., Hennequin, L.F., Curwen, J., Fennell, M., Allen, J., Lambert-van der Brempt, C., and Costello, G. (2004) *Journal of Medicinal Chemistry*, **47**, 871–887.

354 Wu, T.Y.H., Schultz, P.G., and Ding, S. (2003) *Organic Letters*, **5**, 3587–3590.

355 Paolini, L., Petricci, E., Corelli, F., and Botta, M. (2003) *Synthesis*, 1039–1042.

356 Peng, X., Zhang, A., Kula, N.S., Baldessarini, R.J., and Neumeyer, J.L. (2004) *Bioorganic & Medicinal Chemistry Letters*, **14**, 5635–5639.

357 Inagaki, T., Fukuhara, T., and Hara, S. (2003) *Synthesis*, 1157–1159.

358 Kobayashi, S., Yoneda, A., Fukuhara, T., and Hara, S. (2004) *Tetrahedron Letters*, **45**, 1287–1289.

359 Amijs, C.H.M., van Klink, G.P.M., and van Koten, G. (2003) *Green Chemistry*, **5**, 470–474.

360 Xiao, J.-C. and Shreeve, J.M. (2005) *Journal of Fluorine Chemistry*, **126**, 473–476.

361 LaBeaume, P., Placzek, M., Daniels, M., Kendrick, I., Ng, P., McNeel, M., Afrouze, R., Alexander, A., Thomas, R., Kallmerten, A.E., and Jones, G.B. (2010) *Tetrahedron Letters*, **51**, 1905–1909.

362 Nomoto, T., Fukuhara, T., and Hara, S. (2006) *Synlett*, 1744–1746.

363 Fujimoto, Y., Iwata, M., Imakita, N., Shimoyama, A., Suda, Y., Kusumoto, S., and Fukase, K. (2007) *Tetrahedron Letters*, **48**, 6577–6581.

364 Le, H.P. and Müller, C.E. (2006) *Bioorganic & Medicinal Chemistry Letters*, **16**, 6139–6142.

365 Akiyama, Y., Hiramatsu, C., Fukuhara, T., and Hara, S. (2006) *Journal of Fluorine Chemistry*, **127**, 920–923.

366 Arvela, R.K. and Leadbeater, N.E. (2003) *Synlett*, 1145–1148.

367 Kremsner, J.M., Rack, M., Pilger, C., and Kappe, C.O. (2009) *Tetrahedron Letters*, **50**, 3665–3668.

368 Katritzky, A.R., Majumder, S., and Jain, R. (2003) *ARKIVOC*, **xiii**, 74–79.

369 Bengtson, A., Hallberg, A., and Larhed, M. (2002) *Organic Letters*, **4**, 1231–1233.

370 Kiyota, H., Dixon, D.J., Luscombe, C.K., Hettstedt, S., and Ley, S.V. (2002) *Organic Letters*, **4**, 3223–3226.

371 Ju, Y. and Varma, R.S. (2005) *Organic Letters*, **7**, 2409–2411.

372 Ju, Y. and Varma, R.S. (2005) *Tetrahedron Letters*, **46**, 6011–6014.

373 Ju, Y., Kumar, D., and Varma, R.S. (2006) *The Journal of Organic Chemistry*, **71**, 6697–6700.

374 Campiglia, P., Aquino, C., Bertamino, A., Sala, M., Gomez-Monterrey, I.M., Novellino, E., and Grieco, P. (2008) *Tetrahedron Letters*, **49**, 731–734.

375 Curti, C., Zanardi, F., Battistini, L., Sartori, A., Rassu, G., Pinna, L., and Casiraghi, G. (2006) *The Journal of Organic Chemistry*, **71**, 8552–8558.

376 Conejo-García, A., Núñez, M.C., Marchal, J.A., Rodríguez-Serrano, F., Aránega, A., Gallo, M.A., Espinosa, A., and Campos, J.M. (2008) *European Journal of Medicinal Chemistry*, **43**, 1742–1748.

377 Morini, G., Comini, M., Rivara, M., Rivara, S., Lorenzi, S., Bordi, F., Mor, M., Flammini, L., Bertoni, S., Ballabeni, V., Barocelli, E., and Plazzi, P.V. (2006) *Bioorganic & Medicinal Chemistry Letters*, **16**, 4063–4067.

378 Palmer, A.M., Grobbel, B., Brehm, C., Zimmermann, P.J., Buhr, W., Feth, M.P., Holst, H.C., and Simon, W.A. (2007) *Bioorganic & Medicinal Chemistry*, **15**, 7647–7660.

379 Bagley, M.C., Dale, J.W., Jenkins, R.L., and Bower, J. (2004) *Chemical Communications*, 102–103.

380 Vo-Thanh, G., Lahrache, H., Loupy, A., Kim, I.-J., Chang, D.-H., and Jun, C.-H. (2004) *Tetrahedron*, **60**, 5539–5543.

381 Pouilhés, A., Langlois, Y., and Chiaroni, A. (2003) *Synlett*, 1488–1490.

382 Öhberg, L. and Westman, J. (2001) *Synlett*, 1893–1896.

383 Kangasmetsä, J.J. and Johnson, T. (2005) *Organic Letters*, **7**, 5653–5655.

384 Baran, P.S. and Richter, J.M. (2004) *Journal of the American Chemical Society*, **126**, 7450–7451.

385 Coats, S.J., Schulz, M.J., Carson, J.R., Codd, E.E., Hlasta, D.J., Pitis, P.M., Stone, D.J., Zhang, S.-P., Colburn, R.W., and Dax, S.L. (2004) *Bioorganic & Medicinal Chemistry Letters*, **14**, 5493–5498.

386 Santagada, V., Frecentese, F., Fiorino, F., Cirillo, D., Perissutti, E., Severino, B., Terracciano, S., and Caliendo, G. (2004) *QSAR & Combinatorial Science*, **23**, 899–901.

387 Santagada, V., Frecentese, F., Perissutti, E., Fiorino, F., Severino, B., Cirillo, D., Terracciano, S., and

Caliendo, G. (2005) *Journal of Combinatorial Chemistry*, **7**, 618–621.
388 Johansson, M.J., Andersson, K.H.O., and Kann, N. (2008) *The Journal of Organic Chemistry*, **73**, 4458–4463.
389 Miyazawa, A., Tanaka, K., Sakakura, T., Tashiro, M., Tashiro, H., Prakash, G.K.S., and Olah, G.A. (2005) *Chemical Communications*, 2104–2106.
390 Gustafsson, T., Schou, M., Almqvist, F., and Kihlberg, J. (2004) *The Journal of Organic Chemistry*, **69**, 8694–8701.
391 Cloarec, J.-M. and Charette, A.B. (2004) *Organic Letters*, **6**, 4731–4734.
392 Liao, X., Raghavan, G.S.V., and Yaylayan, V.A. (2002) *Tetrahedron Letters*, **43**, 45–48.
393 Villa, C., Mariani, E., Loupy, A., Grippo, C., Grossi, G.C., and Bargagna, A. (2003) *Green Chemistry*, **5**, 623–626.
394 Katritzky, A.R. and Angrish, P. (2006) *Synthesis*, 4135–4142.
395 Su, Q., Beeler, A.B., Lobkovsky, E., Porco, J.A., and Panek, J.S. (2003) *Organic Letters*, **5**, 2149–2152.
396 Remme, N., Koschek, K., and Schneider, C. (2007) *Synlett*, 491–493.
397 Colombo, M., Bossolo, S., and Aramini, A. (2009) *Journal of Combinatorial Chemistry*, **11**, 335–337.
398 Wu, X., Stockdill, J.L., Wang, P., and Danishefsky, S.D. (2010) *Journal of the American Chemical Society*, **132**, 4098–4100.
399 Veverková, E., Mečiarová, M., Toma, Š., and Balko, J. (2003) *Monatshefte für Chemie*, **134**, 1215–1219.
400 Glynn, D., Bernier, D., and Woodward, S. (2008) *Tetrahedron Letters*, **49**, 5687–5688.
401 Nicewonger, R., Fowke, A., Nguyen, K., Ditto, L., Caserta, J., Harris, M., and Baldino, C.M. (2003) *Molecular Diversity*, **7**, 247–252.
402 Miriyala, B. and Williamson, J.S. (2003) *Tetrahedron Letters*, **44**, 7957–7959.
403 Gelens, E., Smeets, L., Sliedregt, L.A.J.M., van Steen, B.J., Kruse, C.G., Leurs, R., and Orru, R.V.A. (2005) *Tetrahedron Letters*, **46**, 3751–3754.
404 Ferroud, C., Godart, M., Ung, S., Borderies, H., and Guy, A. (2008) *Tetrahedron Letters*, **49**, 3004–3008.
405 Polshettiwar, V. and Varma, R.S. (2008) *Tetrahedron Letters*, **49**, 2661–2664.
406 Zhang, L. and Zhang, J.Y. (2005) *Journal of Combinatorial Chemistry*, **7**, 622–626.
407 Martelanc, M., Kranjc, K., Polanc, S., and Kočevar, M. (2005) *Green Chemistry*, **7**, 737–741.
408 Hren, J., Kranjc, K., Polanc, S., and Kočevar, M. (2008) *Synthesis*, 452–458.
409 Massaro, A., Mordini, A., Reginato, G., Russo, F., and Taddei, M. (2007) *Synthesis*, 3201–3204.
410 Cianci, J., Baell, J.B., and Harvey, A.J. (2007) *Tetrahedron Letters*, **48**, 5973–5975.
411 Katritzky, A.R., Todadze, E., Angrish, P., and Draghici, B. (2007) *The Journal of Organic Chemistry*, **72**, 5794–5801.
412 Katritzky, A.R., Khashab, N.M., Yoshioka, M., Haase, D.N., Wilson, K.R., Johnson, J.V., Chung, A., and Haskell-Luevano, C. (2007) *Chemical Biology and Drug Design*, **70**, 465–468.
413 Caddick, S., Wilden, J.D., Bush, H.D., and Judd, D.B. (2004) *QSAR & Combinatorial Science*, **23**, 902–905.
414 De Luca, L. and Giacomelli, G. (2008) *The Journal of Organic Chemistry*, **73**, 3967–3969.
415 Flores Toque, H.A., Priviero, F.B.M., Teixeira, C.E., Perissutti, E., Fiorino, F., Severino, B., Frecentese, F., Lorenzetti, R., Baracat, J.S., Santagada, V., Caliendo, G., Antunes, E.N., and De Nucci, G. (2008) *Journal of Medicinal Chemistry*, **51**, 2807–2815.
416 Curran, D.P. and Zhang, Q. (2003) *Advanced Synthesis and Catalysis*, **345**, 329–332.
417 Bélanger, G., Lévesque, F., Pâquet, J., and Barbe, G. (2005) *The Journal of Organic Chemistry*, **69**, 291–296.
418 Goossen, L.J., Manjolinho, F., Khan, B.A., and Rodriguez, N. (2009) *The Journal of Organic Chemistry*, **74**, 2620–2623.
419 Olofsson, K., Kim, S.-Y., Larhed, M., Curran, D.P., and Hallberg, A. (1999) *The Journal of Organic Chemistry*, **64**, 4539–4541.
420 Wetter, C. and Studer, A. (2004) *Chemical Communications*, 174–175.
421 Teichert, A., Jantos, K., Harms, K., and Studer, A. (2004) *Organic Letters*, **6**, 3477–3480.

422 Prediger, I., Weiss, T., and Reiser, O. (2008) *Synthesis*, 2191–2198.

423 Portela-Cubillo, F., Scott, J.S., and Walton, J.C. (2008) *The Journal of Organic Chemistry*, **73**, 5558–5565.

424 Portela-Cubillo, F., Scott, J.S., and Walton, J.C. (2007) *Chemical Communications*, 4041–4043.

425 Ericsson, C. and Engman, L. (2004) *The Journal of Organic Chemistry*, **69**, 5143–5146.

426 Merişor, E., Conrad, J., Klaiber, I., Mika, S., and Beifuss, U. (2007) *Angewandte Chemie International Edition*, **46**, 3353–3355.

427 Merişor, E., Conrad, J., Mika, S., and Beifuss, U. (2007) *Synlett*, 2033–2036.

428 Parsons, A., Sharpe, D., and Taylor, P. (2005) *Synlett*, 2981–2983.

429 Jessop, C.M., Parsons, A.F., Routledge, A., and Irvine, D.J. (2006) *European Journal of Organic Chemistry*, 1547–1554.

430 Du, Y., Creighton, C.J., Yan, Z., Gauthier, D.A., Dahl, J.P., Zhao, B., Belkowski, S.M., and Reitz, A.B. (2005) *Bioorganic & Medicinal Chemistry*, **13**, 5936–5948.

431 Hartung, J., Daniel, K., Gottwald, T., Groß, A., and Schneiders, N. (2006) *Organic & Biomolecular Chemistry*, **4**, 2313–2322.

432 Goretzki, C., Krlej, A., Steffens, C., and Ritter, H. (2004) *Macromolecular Rapid Communications*, **25**, 513–516.

433 Zhang, H. and Schubert, U.S. (2004) *Macromolecular Rapid Communications*, **25**, 1225–1230.

434 Wiesbrock, F., Hoogenboom, R., and Schubert, U.S. (2004) *Macromolecular Rapid Communications*, **25**, 1739–1764.

435 Jaipuri, F.A., Jofre, M.F., Schwarz, K.A., and Pohl, N.L. (2004) *Tetrahedron Letters*, **45**, 4149–4152.

436 Rombouts, F., Franken, D., Martinez-Lamenca, C., Braeken, M., Zavattaro, C., Chen, J., and Trebanco, A.A. (2010) *Tetrahedron Letters*, **51**, 4815–4818.

437 Srinivasan, N., Yurek-George, A., and Ganesan, A. (2005) *Molecular Diversity*, **9**, 291–293.

438 Murray, A.J., Parsons, P.J., and Hitchcock, P. (2007) *Tetrahedron*, **63**, 6485–6492.

439 Carlsson, A.-C., Jam, F., Tullberg, M., Pilotti, Å., Ioannidis, P., Luthman, K., and Grøtli, M. (2006) *Tetrahedron Letters*, **47**, 5199–5201.

440 Park, J. and Chae, J. (2010) *Synlett*, 1651–1656.

441 Wellner, E., Sandin, H., and Pääkkönen, L. (2002) *Synthesis*, 223–226.

442 Wettergren, J. and Minidis, A.B.E. (2003) *Tetrahedron Letters*, **44**, 7611–7612.

443 Belema, M., Nguyen, V.N., and Zusi, F.C. (2004) *Tetrahedron Letters*, **45**, 1693–1697.

444 Bode, C.M., Ting, A., and Schaus, S.E. (2006) *Tetrahedron*, **62**, 11499–11505.

445 Yin, J., Spindler, J., and Linker, T. (2007) *Chemical Communications*, 2712–2713.

446 Magolan, J., Carson, C.A., and Kerr, M.A. (2008) *Organic Letters*, **10**, 1437–1440.

447 Raghuraman, A., Riaz, M., Hindle, M., and Desai, U.R. (2007) *Tetrahedron Letters*, **48**, 6754–6758.

448 Sá, M.M. and Meier, L. (2006) *Synlett*, 3474–3478.

449 Marette, C., Larrouquet, C., Tisnès, P., Deloye, J.-B., and Gras, E. (2006) *Tetrahedron Letters*, **47**, 6947–6950.

450 Luzzio, F.A. and Chen, J. (2008) *The Journal of Organic Chemistry*, **73**, 5621–5624.

451 Gregg, B.T., Golden, K.C., and Quinn, J.F. (2008) *Tetrahedron*, **64**, 3287–3295.

452 (a) Stone-Elander, S., Elander, N., Thorell, J.-O., and Fredriksson, A. (2007) *Ernst Schering Research Foundation Workshop*, **64**, 243–269; (b) Jones, J.R. and Lu, S.Y. (2006) *Microwaves in Organic Synthesis*, 2nd edn (ed. A. Loupy), Wiley-VCH Verlag GmbH, Weinheim, pp. 820–859; (c) Jones, J.R. and Lu, S.Y. (2002) *Microwaves in Organic Synthesis* (ed. A. Loupy), Wiley-VCH Verlag GmbH, Weinheim, pp. 435–462; (d) Stone-Elander, S. and Elander, N. (2002) *Journal of Labelled Compounds & Radiopharmaceuticals*, **45**, 715–746; (e) Elander, N., Jones, J.R., Lu, S.-Y., and Stone-Elander, S. (2000) *Chemical Society Reviews*, **29**, 239–250.

453 Karramkam, M., Hinnen, F., Berrehouma, M., Hlavacek, C., Vaufrey, F., Halladin, C., McCarron, J.A., Pike, V.W., and Dollé, F. (2003) *Bioorganic & Medicinal Chemistry*, **11**, 2769–2782.

454 Karramkam, M., Hinnen, F., Vaufrey, F., and Dollé, F. (2003) *Journal of Labelled Compounds & Radiopharmaceuticals*, **46**, 979–992.

455 Wüst, F. and Kniess, T. (2003) *Journal of Labelled Compounds & Radiopharmaceuticals*, **46**, 699–713.

456 Lu, S.-Y., Chin, F.T., McCarron, J.A., and Pike, V.W. (2004) *Journal of Labelled Compounds & Radiopharmaceuticals*, **47**, 289–297.

457 Persson, T., Johansen, S.K., Martiny, L., Giskov, M., and Nielsen, J. (2004) *Journal of Labelled Compounds & Radiopharmaceuticals*, **47**, 627–634.

458 Park, S.H., Gwon, H.J., and Park, K.B. (2004) *Chemistry Letters*, **33**, 1278–1279.

459 Park, S.H., Gwon, H.J., Park, J.S., and Park, K.B. (2004) *QSAR & Combinatorial Science*, **23**, 868–874.

460 Yamamoto, M., Oshima, K., and Matsubara, S. (2004) *Chemistry Letters*, **33**, 846–847.

461 de Keczer, S.A., Lane, T.S., and Masjedizadeh, M.R. (2004) *Journal of Labelled Compounds & Radiopharmaceuticals*, **47**, 733–740.

462 Derdau, V. (2004) *Tetrahedron Letters*, **45**, 8889–8893.

463 Martins, A. and Lautens, M. (2008) *Organic Letters*, **10**, 4351–4353.

464 Mayer, S., Daigle, D.M., Brown, E.D., Khatri, J., and Organ, M.G. (2004) *Journal of Combinatorial Chemistry*, **6**, 776–782.

465 Wu, J., Wu, H., Wei, S., and Dai, W.-M. (2004) *Tetrahedron Letters*, **45**, 4401–4404.

466 McNulty, J., Das, P., and McLeod, D. (2010) *Chemistry: A European Journal*, **16**, 6756–6760.

467 Görmer, K., Waldmann, H., and Triola, G. (2010) *The Journal of Organic Chemistry*, **75**, 1811–1813.

468 De Luca, L., Porchedddu, A., Giacomelli, G., and Murgia, I. (2010) *Synlett*, 2439–3442.

469 Donald, M.B., Conrad, W.E., Oakdale, J.S., Butler, J.D., Haddadin, M.J., and Kurth, M.J. (2010) *Organic Letters*, **12**, 2524–2527.

470 Quiclet-Sire, B., Revol, G., and Zard, S.Z. (2009) *Organic Letters*, **11**, 2832–2835.

471 Montel, S., Bouyssi, D., and Balme, G. (2010) *Advanced Synthesis and Catalysis*, **352**, 2315–2320.

472 Dahlén, A., Prasad, E., Flowers, R.A., II, and Hilmersson, G. (2005) *Chemistry: A European Journal*, **11**, 3279–3284.

473 Polshettiwar, V. and Varma, R.S. (2007) *The Journal of Organic Chemistry*, **74**, 7420–7422.

474 Correc, O., Guillou, K., Hamelin, J., Paquin, L., Texier-Boullet, F., and Toupet, L. (2004) *Tetrahedron Letters*, **45**, 391–395.

475 Hamelin, J., Soudi, A., and Benhaoua, H. (2003) *Synthesis*, 2185–2188.

476 Borrelli, F., Campagnuolo, C., Capasso, R., Fattorusso, E., and Taglialatela-Scafati, O. (2004) *European Journal of Organic Chemistry*, 3227–3232.

477 Hon, Y.-S., Hsu, T.-R., Chen, C.-Y., Lin, Y.-H., Chang, F.-J., Hsieh, C.-H., and Szu, P.-H. (2003) *Tetrahedron*, **59**, 1509–1520.

478 Barrios, I., Camps, P., Comes-Franchini, M., Muñoz-Torrero, D., Ricci, A., and Sánchez, L. (2003) *Tetrahedron Letters*, **59**, 1971–1979.

479 McDonnell, M.E., Zhang, S.-P., Nasser, N., Dubin, A.E., and Dax, S.L. (2004) *Bioorganic & Medicinal Chemistry Letters*, **14**, 531–534.

480 Rábarova, E., Koiš, P., Lácová, M., and Krutošíková, A. (2004) *ARKIVOC*, **i**, 110–112.

481 Katritzky, A.R., Vincek, A.S., and Suzuki, K. (2005) *ARKIVOC*, **v**, 116–126.

482 Jung, M.E. and Maderna, A. (2005) *Tetrahedron Letters*, **46**, 5057–5061.

483 Giraud, A., Provot, O., Peyrat, J.-F., Alami, M., and Brion, J.-D. (2006) *Tetrahedron*, **62**, 7667–7673.

484 Corbu, A., Gauron, G., Castro, J.M., Dakir, M., and Arseniyadis, S. (2008) *Tetrahedron Asymmetry*, **19**, 1730–1743.

485 Zhang, L. and Zhang, J.Y. (2006) *Journal of Combinatorial Chemistry*, **8**, 361–367.

486 Bagley, M. and Lubinu, M.C. (2002) *Synthesis*, 1283–1288.

487 Hellal, M., Bihel, F., Mongeot, A., and Bourguignon, J.-J. (2006) *Organic & Biomolecular Chemistry*, **4**, 3142–3146.

488 Monsieurs, K., Rombouts, G., Tapolcsányi, P., Mátyus, P., and Maes, B.U.W. (2006) *Synlett*, 3225–3230.

489 Nakamura, H., Sugiishi, T., and Tanaka, Y. (2008) *Tetrahedron Letters*, **49**, 7230–7233.

490 de Souza, R.O.M.A., de Souza, A.L.F., Fernandez, T.L., Silva, A.C., Pereira, V.L.P., Esteves, P.M., Vasconcellos, M.L.A.A., and Antunes, O.A.C. (2008) *Letters in Organic Chemistry*, **5**, 379–382.

491 Liermann, J.C. and Opatz, T. (2008) *The Journal of Organic Chemistry*, **73**, 4526–4531.

7
Literature Survey Part C: Heterocycle Synthesis

Heterocyclic scaffolds form the core of many pharmaceutically relevant substances. Not surprisingly, therefore, many publications in the area of microwave-assisted organic synthesis from both academia and industry deal with the preparation of heterocycles [1]. In this chapter, the description of heterocycle synthesis is structured by ring size and the number of heteroatoms in the ring.

7.1
Three-Membered Heterocycles with One Heteroatom

The vanadium-catalyzed epoxidation of hindered homoallylic alcohols has been described by Prieto and coworkers [2]. The reaction times for the epoxidation in a series of *cis*- and *trans*-2-methyl alkenoles were significantly reduced from 6–10 days to less than 3 h using open-vessel microwave irradiation. Standard conditions for the one-pot oxidation of the alkenes utilized a bis(acetylacetonate)oxovanadium(IV)/ *tert*-butyl hydroperoxide reagent mixture (1.4 mol% catalyst, 1.1 equiv peroxide oxidant) in toluene solution. For the examples shown in Scheme 7.1, moderate to high chemical yields and diastereoselectivities were observed. This process, being carried out under open-vessel conditions (see Section 4.3), was fully scalable and could be performed on a 30 g scale [2].

More recently, Yanada and coworkers reported on the preparation of aziridine-fused 2-azabicyclo[2.2.1]hept-5-en-3-ones by cycloaddition of azides to the bicycloheptene core [3].

7.2
Four-Membered Heterocycles with One Heteroatom

Linder and Podlech studied the microwave-assisted decomposition of diazoketones derived from α-amino acids [4]. In the presence of imines, the initially formed ketene intermediates reacted spontaneously via [2 + 2] cycloaddition to form β-lactams with a *trans*-substitution pattern at positions C-3 and C-4 (Scheme 7.2) [4]. In order to avoid

Microwaves in Organic and Medicinal Chemistry, Second Edition.
C. Oliver Kappe, Alexander Stadler, and Doris Dallinger.

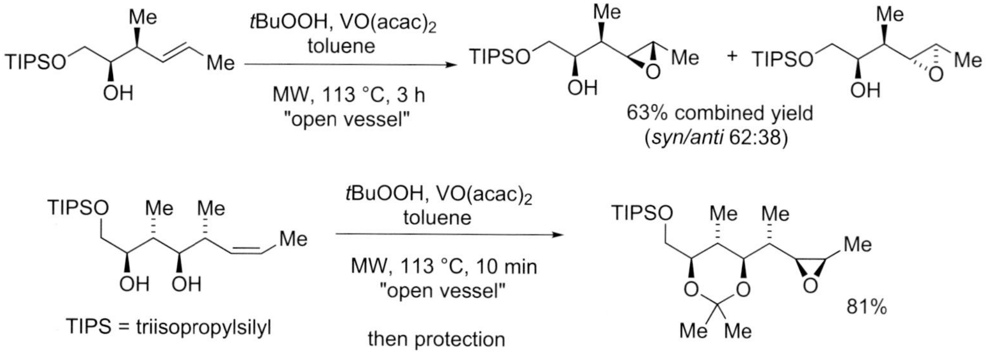

Scheme 7.1 Epoxidation of homoallylic alcohols.

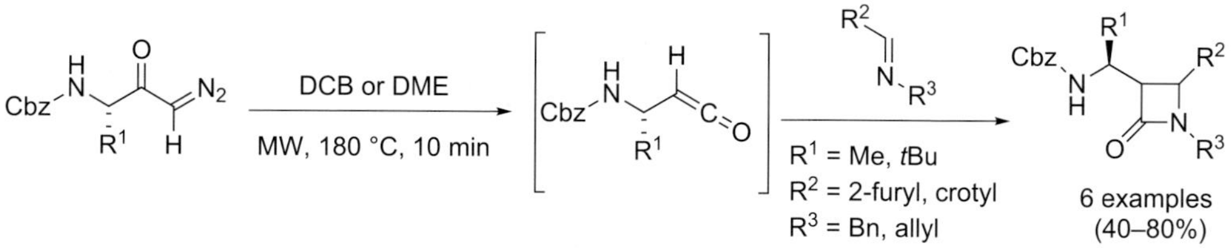

Scheme 7.2 Formation of β-lactams.

the use of the high-boiling solvent 1,2-dichlorobenzene, most transformations were carried out in 1,2-dimethoxyethane under sealed vessel conditions. Solvent-free protocols where the substrates were adsorbed on inorganic alumina support only led to the corresponding homologated β-amino acids. Obviously, traces of water present on the support trapped the intermediate ketene.

In more recent studies by the group of Xu, various β-lactams have been prepared by [2 + 2] cycloaddition of ketenes with imines under microwave conditions [5, 6].

7.3 Five-Membered Heterocycles with One Heteroatom

7.3.1 Pyrroles

Pyrrole is one of the most prominent heterocycles, having been known for more than 150 years, and it is the structural skeleton of several natural products, synthetic pharmaceuticals, and electrically conducting materials. A simple access to the pyrrole ring system involves the conversion of cyclic anhydrides into five-membered imides. Mortoni *et al.* have described the conversion of 2-methylquinoline-3,4-dicarboxylic acid anhydride to a quinoline-3,4-dicarboxoximide library by treatment of the anhydride with a diverse set of primary amines under microwave conditions (Scheme 7.3) [7].

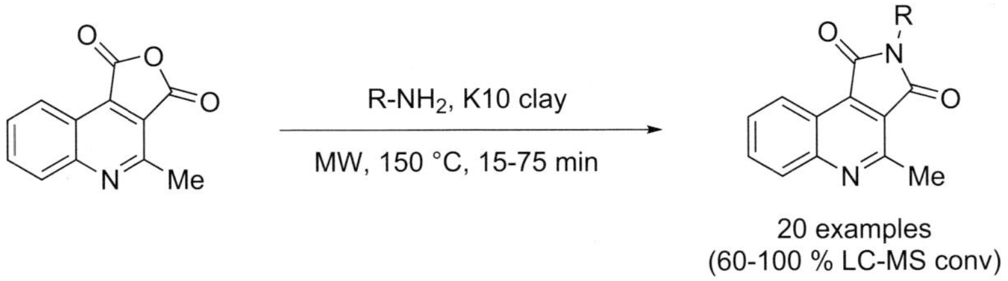

Scheme 7.3 Formation of imides from anhydrides.

The authors have studied a range of different conditions including "dry media" protocols (see Section 4.1), whereby the starting materials were adsorbed on an inorganic support and then were irradiated by microwaves. Best results were achieved with wet montmorillonite K10 clay, which afforded complete conversions, also with amines showing low reactivity on other supports, such as silica or alumina. Optimum yields were obtained when a mixture of the anhydride and the support was grounded in a mortar until a homogeneous powder was formed. After addition of 1 equiv of primary amine, the material was irradiated at 150 °C for 15–75 min, leading to high product yields. Alternatively, similar high yields were also achieved using toluene as solvent, albeit longer reaction times were typically required.

One of the most common approaches to pyrrole synthesis is the Paal–Knorr reaction in which 1,4-dicarbonyl compounds are converted to pyrroles via acid-mediated dehydrative cyclization in the presence of a primary amine. The group of Taddei has reported a microwave-assisted variation of the Paal–Knorr procedure in which a small array of tetrasubstituted pyrroles was obtained (Scheme 7.4) [8]. The pyrroles were effectively synthesized by heating a solution of the corresponding 1,4-dicarbonyl compound in the presence of 5 equiv of the primary amine in acetic acid at 180 °C for 3 min. The same result was obtained heating an identical mixture under open-vessel microwave conditions (reflux) for 5 min. Interestingly, the authors were not able to achieve meaningful product yields when carrying out the same transformation in an oil bath. A small array of tetrasubstituted pyrroles was prepared using this method.

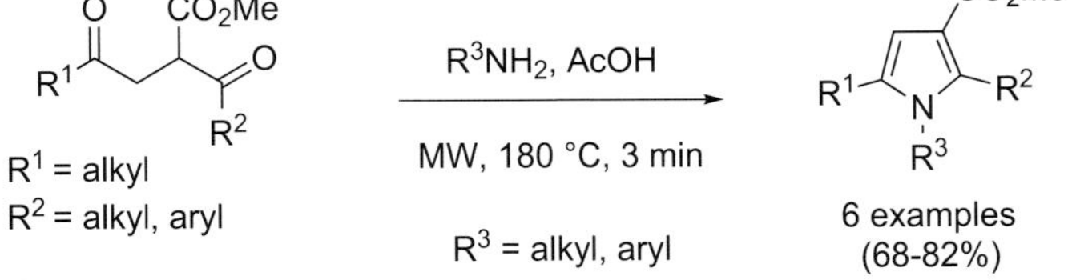

Scheme 7.4 Paal–Knorr pyrrole synthesis.

The group of Ley from the University of Cambridge has reported on the synthesis of substituted pyrroles from the corresponding 1,4-dicarbonyl compounds via the Paal–Knorr reaction (Scheme 7.5) [9]. Magnesium nitride (Mg_3N_2) was used as source of ammonia that is generated *in situ* by the reaction of Mg_3N_2 with protic

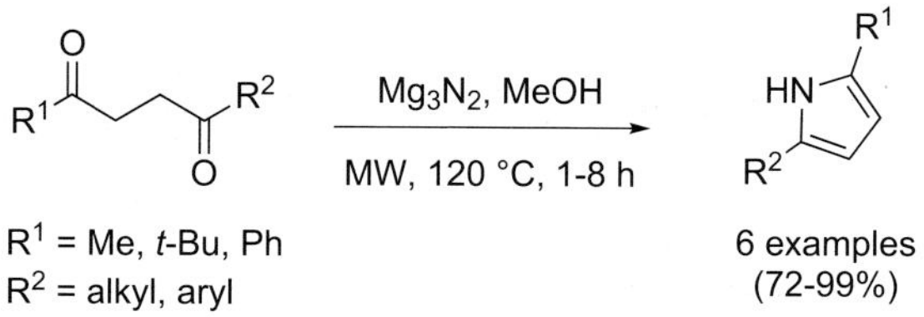

Scheme 7.5 Synthesis of pyrroles using magnesium nitride as ammonia source.

solvents like MeOH. Reaction times were typically 1 h, the only exception being the derivative with $R^1 = R^2 = t$-Bu, where 8 h irradiation was necessary to achieve complete conversion.

Paal–Knorr reactions involving a specific set of 1,4-dicarbonyl compounds and amines under microwave conditions have been reported by Werner and Brummond [10]. The same strategy has also been used to prepare *N*-phenylpyrroles using aniline [11]. 3,4-Disubstituted pyrroles have also been obtained by Piloty–Robinson synthesis [12]. For an alternative microwave-assisted pyrrole synthesis, see Ref. [13].

Wilson *et al.* presented a simplified Paal–Knorr condensation in water under microwave irradiation to prepare a set of *N*-aryl pyrroles [14]. Whereas most established methods employ acid catalysts, this novel approach provides a catalyst-free procedure for the synthesis of these important structural motifs. When appropriate aryl amine or sulfonamide together with 1.3 equiv 2,5-dimethoxytetrahydrofuran in water (0.64 M) was microwave heated at 150 °C for 30 min, the corresponding *N*-aryl pyrroles and *N*-aryl sulfonyl pyrroles were obtained in excellent yields after simple filtration (Scheme 7.6). Approximately 50-fold scale-up with this inexpensive, straightforward method was possible without changing of conditions by switching to a commercially available 80 mL microwave vessel. Thus, rapid preparation of significant gram amounts of valuable heterocycles is easily accessible.

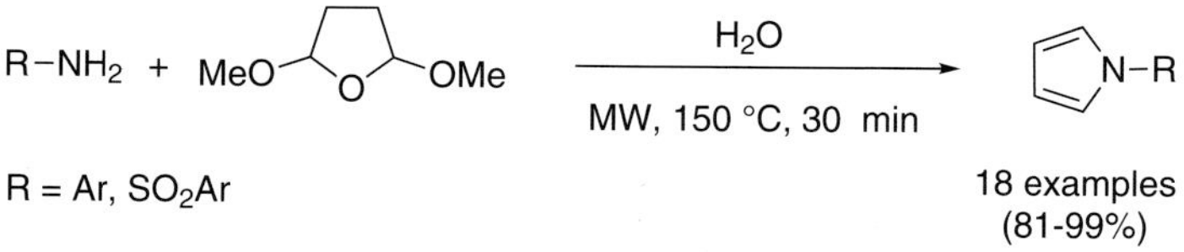

Scheme 7.6 Microwave-assisted synthesis of *N*-aryl pyrroles in water.

A different approach toward highly substituted pyrroles involving a one-pot sila-Stetter/Paal–Knorr strategy was realized by Bharadwaj and Scheidt (Scheme 7.7) [15]. In this multicomponent synthesis, catalyzed by a thiazolium salt, an acyl anion conjugate addition reaction of an acylsilane (sila-Stetter) is coupled *in situ* with the conventional Paal–Knorr approach. Employing microwave conditions at 160 °C for 15 min, the acylsilane is combined with the α,β-unsaturated ketone in the presence of 20 mol% of the thiazolium salt catalyst and 30 mol% of 1,8-diazabicyclo[5.4.0]undec-7-ene (DBU) base in addition to 4 equiv of 2-propanol. This sequence is followed by

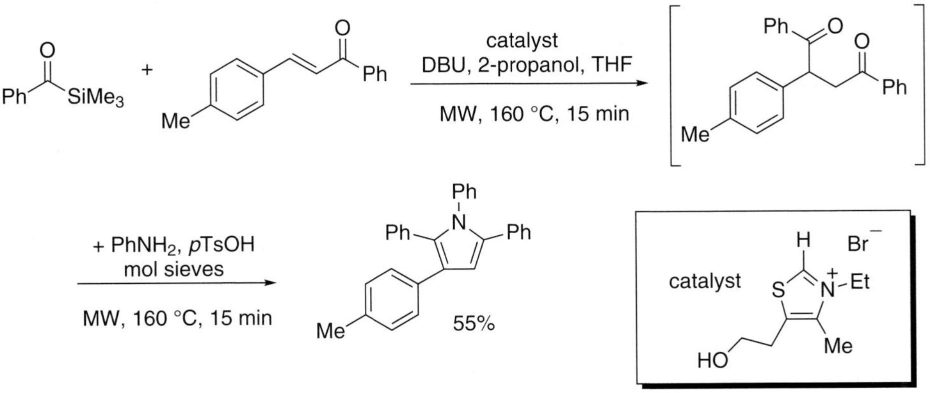

Scheme 7.7 Sila-Stetter/Paal–Knorr pyrrole synthesis.

the addition of aniline and *p*-toluenesulfonic acid (*p*-TsOH), and a second 15 min heating cycle at 160 °C smoothly provided the desired pyrrole in 55% yield. This streamlined approach generated the target heterocycle in 3% of the time required using conventional heating (30 min versus 16 h) [15].

Tejedor *et al.* have utilized the combination of two domino processes for a microwave-promoted synthesis of tetrasubstituted pyrroles [16]. The protocol combines two coupled domino processes: the triethylamine-catalyzed synthesis of enol-protected propargylic alcohols and their sequential transformation into pyrroles via a spontaneous rearrangement from 1,3-oxazolidines (Scheme 7.8). Overall, these two linked and coupled domino processes build up two carbon–carbon bonds, two carbon–nitrogen bonds, and an aromatic ring in a regioselective and efficient manner. The tetrasubstituted pyrroles could be directly synthesized from the enol-protected propargylic alcohols and the primary amines by microwave irradiation (domestic oven) of a silica gel-adsorbed mixture of both components. The process is general for the amine and tolerates a range of functionalities in the aldehyde, leading to the desired diverse pyrrole structures in moderate to good overall yields.

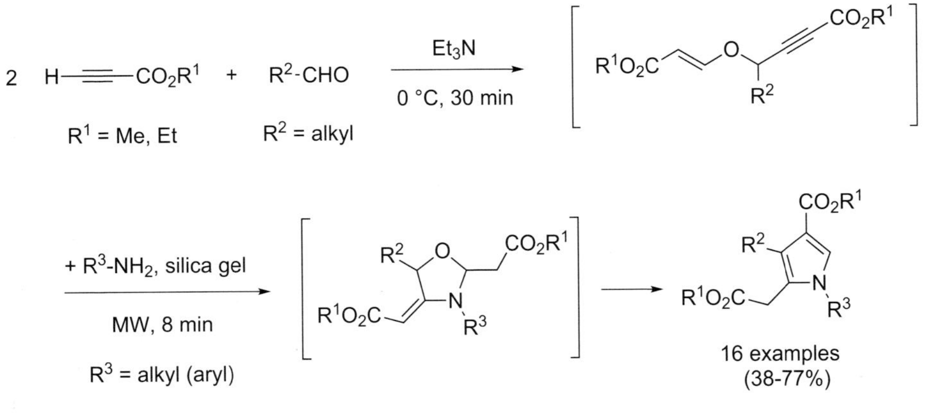

Scheme 7.8 Pyrroles via coupled domino processes.

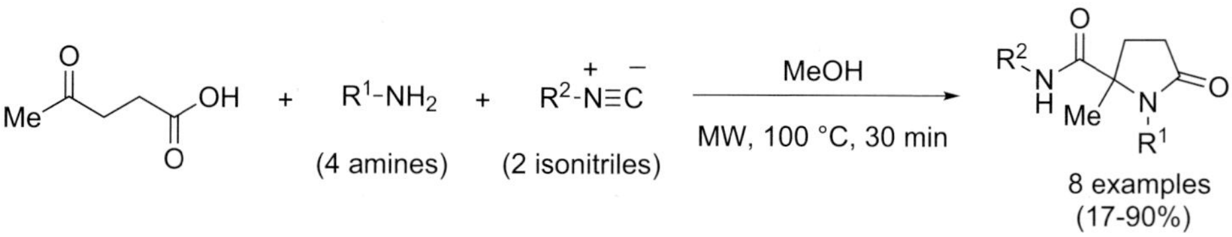

Scheme 7.9 Ugi-type three-component condensation reactions.

Another method to prepare pyrrole rings is via Ugi-type three-component condensation (Scheme 7.9). In the protocol published by Tye and Whittaker [17], levulinic acid was reacted with two different isonitriles and four amine building blocks (1.5 equiv) to provide a set of eight pyrrole derivatives. While the previously published protocol at room temperature required up to 48 h reaction time and provided only moderate product yields, the microwave method (100 °C, 30 min) optimized by a design of experiments (DoE) approach led to high yields of the desired lactams for most of the examples studied.

In addition to cyclocondensation reactions of the Paal–Knorr type, cycloaddition processes play a prominent role in the construction of pyrrole rings. For the preparation of pyrroles, 1,3-dipolar cycloadditions of azomethine ylides with olefin dipolarophiles are very important. The group of de la Hoz has studied the microwave-induced thermal isomerization of imines, derived from α-amino esters, to azomethine ylides (Scheme 7.10) [18]. In the presence of equimolar amounts of β-nitrostyrenes, three isomeric pyrrolidines (nitroproline esters) were obtained under solvent-free conditions in 81–86% yield within 10–15 min at 110–120 °C in the [3 + 2] cycloaddition process. Interestingly, using classical heating in an oil bath (toluene reflux, 24 h) only two of the three isomers were observed.

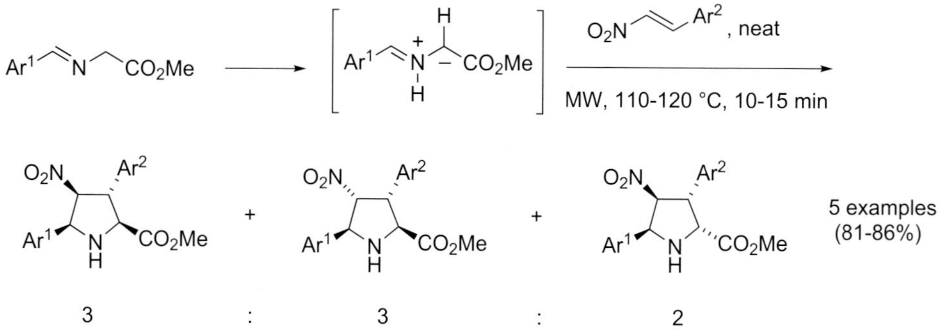

Scheme 7.10 Azomethine ylide–olefin [3 + 2] cycloadditions.

An intramolecular variation of the same reaction principle was used by Bashiardes *et al.* to synthesize fused pyrrolidine and pyrrole derivatives (Scheme 7.11) [19]. The condensation of *O*-allylic and *O*-propargylic salicylaldehydes with α-amino esters was carried out either in the absence of a solvent or – if both components were solids – in minimal amounts of xylene. All reactions performed under microwave conditions

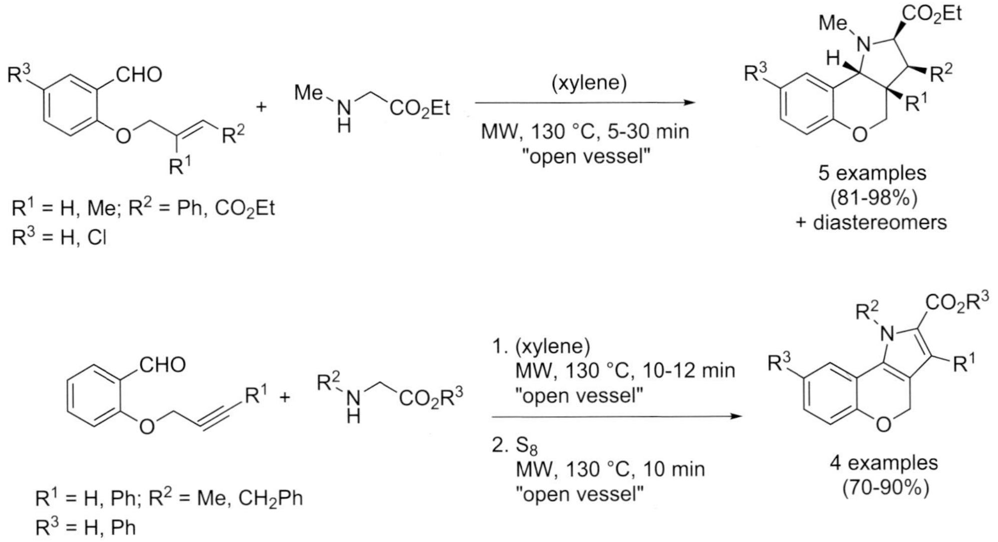

Scheme 7.11 Intramolecular azomethine ylide–olefin/acetylene [3 + 2] cycloadditions.

were completed rapidly after a few minutes, and typically provided higher yields compared to the corresponding thermal protocols. In the case of intramolecular olefin cycloadditions, mixtures of hexahydrochromeno[4,3-*b*]pyrrole diastereoisomers were obtained, whereas transformations involving acetylene tethers provided directly chromeno[4,3-*b*]pyrroles after *in situ* oxidation by elemental sulfur (Scheme 7.11). Independent work by Pospíšil and Potáček described very similar transformations under strictly solvent-free conditions [20].

In a joint work, the groups of Bonchio, Prato, Maggini, and Fontana investigated the functionalization of carbon nanostructures in ionic liquids under microwave irradiation [21]. In their 2009 report, the cycloaddition of azomethine ylides toward [60]fullerenes was described, screening for the best-suited ionic liquid. This particular functionalization led to fulleropyrrolidines, which are highly attractive nanomaterials in solar energy conversion as well as in molecular medicine. In a typical procedure, [60]fullerene was dispersed in [omim][BF_4]/*o*-DCB (1 : 3) together with 2 equiv sarcosine and 4 equiv aldehyde and the mixture was microwave heated for 10 min to 100 °C bulk temperature under simultaneous cooling with a constant power output of 12 W to obtain the desired monopyrrolidinofullerenes in acceptable yields (Scheme 7.12). This smooth and effective method was expanded to various aldehydes including fluoro-tagged species and could be successfully applied toward the functionalization of single-walled nanotubes (SWNTs) as well [21].

The synthesis of indolizinones via a thermally induced cycloisomerization of tertiary propargylic alcohols was described by Kim *et al.* (Scheme 7.13) [22]. Heating the starting materials for 20–30 min in EtOH to 150 °C without the addition of a catalyst afforded the indolizinones in excellent yields.

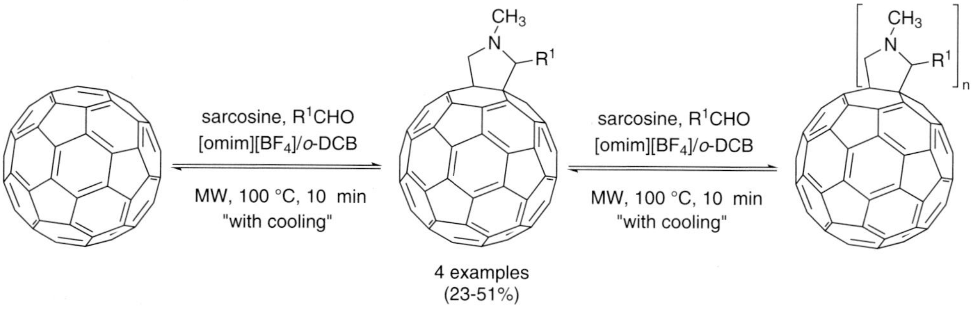

Scheme 7.12 Azomethine ylide cycloadditions to [60]fullerene under microwave irradiation.

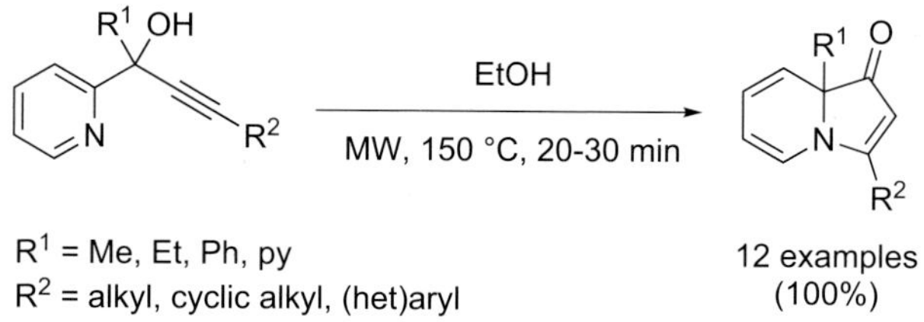

Scheme 7.13 Catalyst-free cycloisomerization approach to indolizinones.

Already in 2001, Sarko and coworkers disclosed the synthesis of an 800-membered solution-phase library of substituted prolines based on multicomponent chemistry (Scheme 7.14) [23]. The process involved microwave irradiation of an α-amino ester with 1.1 equiv of an aldehyde in 1,2-dichloroethane or *N*,*N*-dimethylformamide to 180 °C for 2 min. After cooling, 0.8 equiv of a maleimide dipolarophile was added to the solution of the imine and the mixture was subsequently resubjected to microwave irradiation at 180 °C for another 5 min. This produced the desired products in good yields and purities as determined by HPLC, after scavenging excess aldehyde with polymer-supported sulfonyl hydrazide resin. Analysis by LC–MS of each compound verified purity and identity, thus indicating that a high-quality library had been produced.

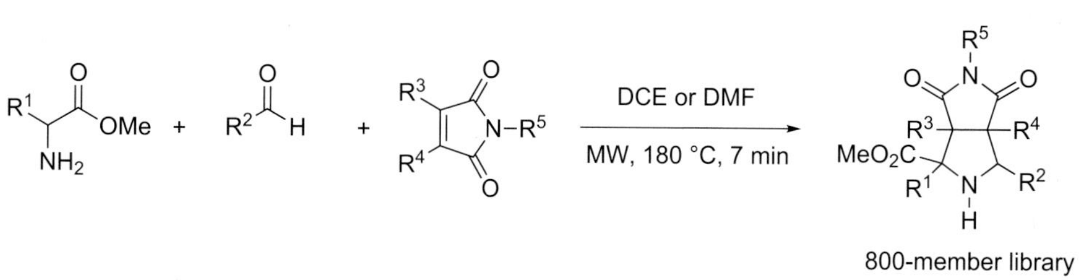

Scheme 7.14 Fused prolines via azomethine ylide–maleimide [3 + 2] cycloadditions.

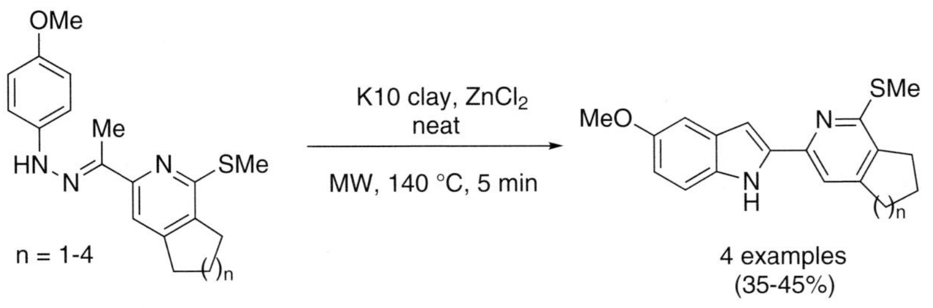

Scheme 7.15 Clay-catalyzed Fischer indole synthesis.

A classical method to synthesize indoles is via the Fischer indolization involving the cyclization of aryl hydrazones in the presence of strong acids. Lipińska has reported the generation of 2-heteroaryl-5-methoxyindoles using a solvent-free, clay-catalyzed protocol, starting from the corresponding *in situ* generated aryl hydrazones (Scheme 7.15) [24]. The microwave-induced Fischer indole synthesis was performed on montmorillonite K10 clay modified with zinc chloride, providing the desired indoles within a relatively short time frame. The 2-substitued indole products were subsequently transformed into 9-methoxyindolo[2,3-*a*]quinolizine alkaloids of the sempervirine type.

Another microwave-assisted Fischer indole synthesis using $ZnCl_2$ in triethylene glycol as reaction medium has been reported by Lipińska in 2006 [25, 26]. Indoles have also been obtained by reaction of (2-aminobenzyl)triphenylphosphonium bromide with aromatic or α,β-unsaturated aldehydes [27], and by Pd-catalyzed intramolecular cyclization of *N*-vinyl-2-chloroanilines [28]. Isoindoles have been obtained by Hughes and coworkers utilizing *o*-substituted aryl oximes under microwave irradiation in the presence of Ac_2O that undergo a sp^3 C–H activated cyclization [29].

The Leimgruber–Batcho synthesis is another widely used method for preparing indole-containing structures. The reaction depends on the acidity of a methyl group positioned adjacent to an aromatic nitro group. The direct condensation reaction with *N,N*-dimethylformamide dimethyl acetal (DMFDMA) under acid catalysis facilitates the introduction of the future indole α-carbon as the enamine. Subsequent catalytic reduction of the nitro group leads to spontaneous cyclization and formation of an indole derivative. Under conventional conditions, the first condensation reaction usually requires overnight heating in *N,N*-dimethylformamide that often leads to less than optimal yields. Ley and coworkers have described microwave-assisted Leimgruber–Batcho reactions in the presence of 2 mol% of anhydrous copper(I) iodide in *N,N*-dimethylformamide/DMFDMA mixtures as solvent (Scheme 7.16a) [30]. Typically, irradiation times from 10 min to 10 h at 180 °C led to acceptable yields of the (hetero)aromatic enamines. The nitro intermediates were then subjected to catalytic hydrogenation using 10 mol% palladium-on-charcoal (Pd/C) in methanol, or to transfer hydrogenation using an encapsulated nanoparticulate palladium catalyst (10 mol%), formic acid and triethylamine (5 equiv each) as the reducing agents. For example, *trans*-2[β-(dimethylamino)vinyl]nitronaphthalene was smoothly converted to the corresponding 1*H*-benz[g]indole within 2 h at 120 °C using microwave irradiation in ethyl acetate solvent (Scheme 7.16b) [30]. Other

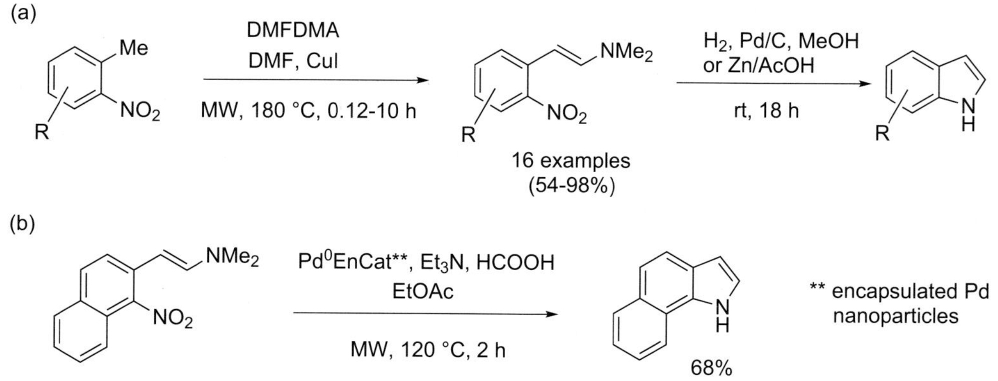

Scheme 7.16 Leimgruber–Batcho indole synthesis.

microwave-assisted transformations involving immobilized palladium catalysts are described in Section 8.6.

The group of Valdés and Barluenga presented a microwave-assisted method for the regioselective synthesis of indoles from imines and difunctionalized arenes [31]. The protocol was elaborated under the recently introduced "on-water" methodology referring to microwave procedures in aqueous suspension when reagents are virtually insoluble in water. In the typical procedure, a reaction vial was charged with equimolar amounts of imine and 1,2-dihaloarene, 8 mol% Xphos, 2 mol% $Pd_2(dba)_3$, 0.125 equiv TBAB, and aqueous potassium hydroxide (3 N). Microwave heating at 140–160 °C for a period of 30–60 min furnished the desired indoles in good to high yields (Scheme 7.17). The conditions were applied to a set of imines and *o*-dihalobenzene derivatives as well as *o*-chlorosulfanes, which proceed with total regioselectivity introducing suitable substituents on the pharmacologically attractive scaffold.

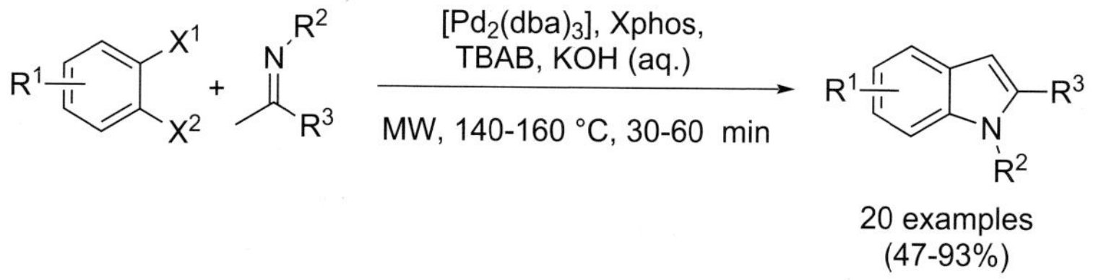

Scheme 7.17 Microwave-promoted on-water synthesis of indoles.

In 2009, the group of Laufer at the University of Tuebingen introduced another easy indole preparation by microwave-assisted Hemetsberger–Knittel synthesis [32]. The authors focused on the synthesis of substituted indole-2-carboxylates that are of interest in the design of kinase inhibitors. The employed Hemetsberger–Knittel approach involves condensation of an aryl aldehyde with azidoacetate to give corresponding α-azidocinnamates. These intermediates furnish the desired indoles upon heating in good yields. The substrates in *n*-hexane are microwave heated at

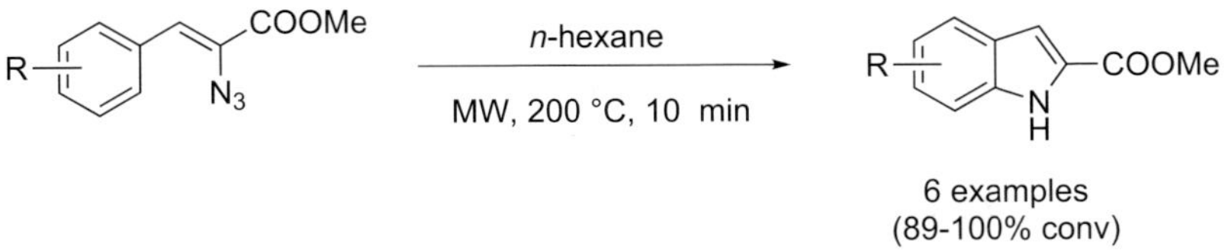

Scheme 7.18 Microwave-promoted indole synthesis by intramolecular cyclization of α-azidocinnamates.

200 °C for 10 min (Scheme 7.18) and furnish the corresponding indoles in excellent yields by crystallization upon cooling. This simple method shows potential to be adapted for ring closure of all previously described α-azidocinnamates leading to various aromatic *N*-heterocycles.

7.3.2
Furans

As in the case of the Paal–Knorr pyrrole synthesis described in Scheme 7.4 by Taddei and coworkers [8], similar reaction conditions were used by the same authors to cyclize 1,4-dicarbonyl compounds into furans (Scheme 7.19). Therefore, heating a solution of the 1,4-dicarbonyl compounds in ethanol/water in the presence of catalytic amounts of hydrochloric acid at 140 °C for 3 min provided excellent yields of the corresponding trisubstituted furan derivatives.

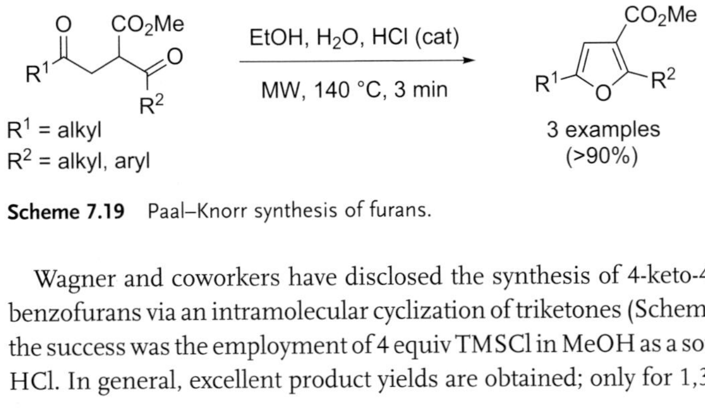

Scheme 7.19 Paal–Knorr synthesis of furans.

Wagner and coworkers have disclosed the synthesis of 4-keto-4,5,6,7-tetrahydrobenzofurans via an intramolecular cyclization of triketones (Scheme 7.20) [33]. Key to the success was the employment of 4 equiv TMSCl in MeOH as a source of anhydrous HCl. In general, excellent product yields are obtained; only for 1,3-dicarbonyls with

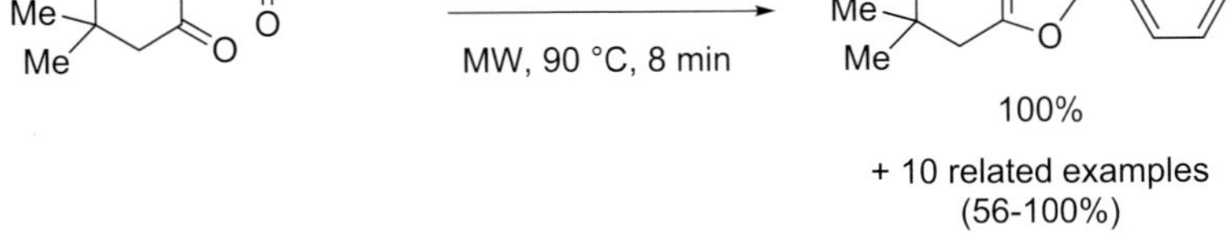

Scheme 7.20 Synthesis of 4-keto-4,5,6,7-tetrahydrobenzofurans.

an amide functionality, a lower yield (56%) of the corresponding oxazole is achieved. Whereas acyclic and seven-membered triketones furnished the products in high yields, five-membered triketones were completely unreactive and only starting material was recovered.

A closely related strategy developed by Mulzer uses HCl in EtOH to form a furan ring from a 1,4-dicarbonyl component as part of the total synthesis of providencin [34]. Benzofurans have been obtained in one pot from *o*-bromophenols and ketones using Pd-catalyzed enolate arylation chemistry [35].

A more complex two-step furan synthesis was described by Jen and coworkers in the context of preparing 2,5-dihydrofuran derivatives as electron acceptors for highly nonlinear optical (NLO) chromophores (Scheme 7.21) [36]. In the first step, α-hydroxyketones are condensed with equimolar quantities of CH-acidic cyano precursors in the presence of 10 mol% of sodium ethoxide in ethanol to furnish reactive 2-imino-2,5-dihydrofuran derivatives. Those intermediates can then be further condensed with a second CH-acidic carbonyl or cyano compound (1.2 equiv) under similar reaction conditions to provide the desired 2-methylene-2,5-dihydrofurans bearing strongly electron-withdrawing groups at the *exo*-methylene functionality. By again using microwave heating, these 2,5-dihydrofuran derivatives can be easily coupled with aromatic, heteroaromatic, and polyene conjugated bridges through their acidic methyl terminals at C-4 [36]. The properties of the resulting NLO chromophores can therefore be tuned very rapidly utilizing high-speed microwave synthesis.

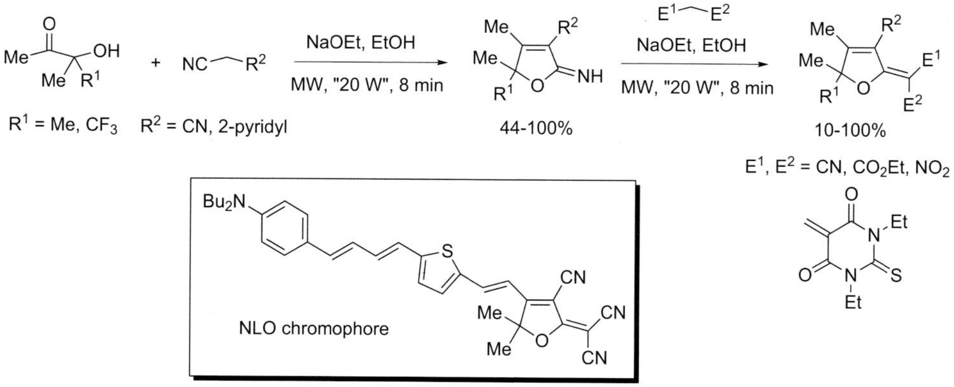

Scheme 7.21 Preparation of 2,5-dihydrofurans.

Furo[3,4-*c*]pyrolediones are important intermediates for the synthesis of diketopyrrolopyrrole (DPP) pigments. Smith and coworkers have described the preparation of several different 3,6-diaryl-substituted furo[3,4-*c*]pyrrole-1,4-diones by microwave-assisted cyclization of readily available 4-aroyl-4,5-dihydro-5-oxo-2-arylpyrrole-3-carboxylates (Scheme 7.22) [37]. While conventional heating in Dowtherm® A at 230–240 °C for 64 h provided only moderate product yields, microwave irradiation of the neat starting material to 200–270 °C for 10–15 min provided significantly increased product yields.

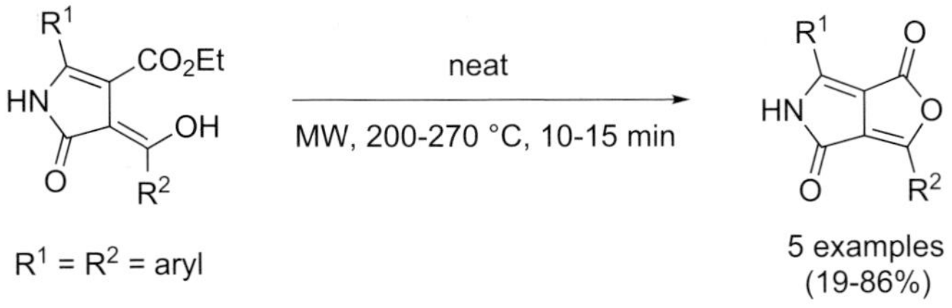

Scheme 7.22 Preparation of furo[3,4-*c*]pyrolediones.

7.3.3
Thiophenes

In the context of preparing benzothienyloxy phenylpropanamines as inhibitors of serotonin and norepinephrine uptake, a group from Eli Lilly and Company has developed a two-step synthesis of benzo[*b*]thiophenes (Scheme 7.23) [38]. Thus, a 2-mercapto-3-phenylpropenoic acid derivative was cyclized with iodine in 1,2-dimethoxyethane at 120 °C to give 5-fluoro-4-methoxybenzothiophene-2-carboxylic acid in 67% yield. Decarboxylation under strongly basic conditions involving 1,8-diazabicyclo[5.4.0]undec-7-ene as base in *N,N*-dimethylacetamide (DMA) as solvent at 200 °C led to the desired benzo[*b*]thiophene intermediate in moderate yield, which was subsequently further manipulated into the benzothienyloxy phenylpropanamine target structures [38]. A subsequent publication by Eli Lilly and Company disclosed an improved protocol for the decarboxylation of a related benzo[*b*]thiophene-2-carboxylic acid providing the final product in 93% yield on a 170 mmol scale using either the DBU/DMA conditions or the traditional quinoline base as solvent [39].

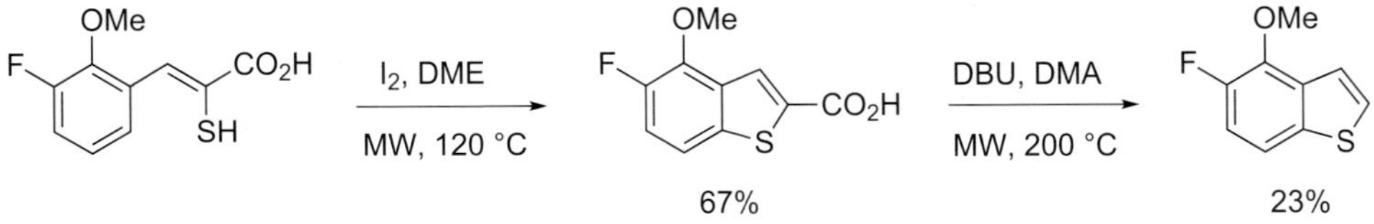

Scheme 7.23 Preparation of benzo[*b*]thiophenes.

7.4
Five-Membered Heterocycles with Two Heteroatoms

7.4.1
Pyrazoles

The groups of Giacomelli and Taddei have developed a rapid solution-phase protocol for the synthesis of 1,4,5-trisubstituted pyrazole libraries (Scheme 7.24) [40]. The transformations involved the cyclization of a monosubstituted hydrazine with an enamino-β-ketoester derived from a β-ketoester and *N,N*-dimethylformamide

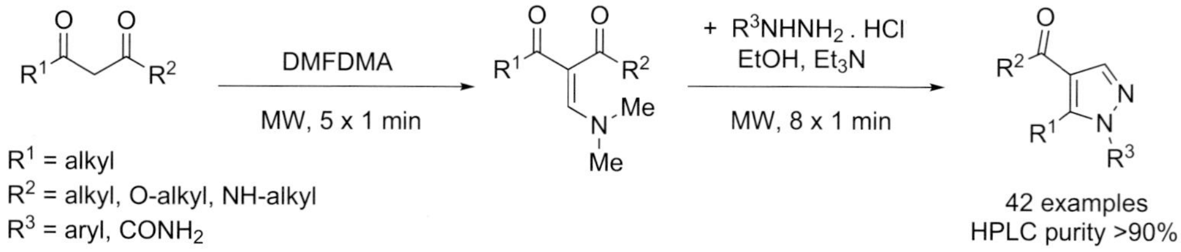

Scheme 7.24 Synthesis of 1,4,5-trisubstituted pyrazoles.

dimethyl acetal. The sites for molecular diversity in this approach are the substituents on the hydrazine (R^3) and on the starting β-ketoester (R^1, R^2). Subjecting a solution of the β-ketoester in DMFDMA as solvent to 5 min of microwave irradiation (domestic oven) led to full and clean conversion to the corresponding enamine. After evaporation of excess DMFDMA, ethanol was added to the crude reaction mixture followed by 1 equiv of the hydrazine hydrochloride and 1.5 equiv of triethylamine base. Further microwave irradiation for 8 min provided – after purification by filtration through a short silica gel column – the desired pyrazoles in >90% purity.

A somewhat related approach was followed by Molteni *et al.* who have described the three-component, one-pot synthesis of fused pyrazoles by reacting cyclic 1,3-diketones with DMFDMA and a suitable bidentate nucleophile, such as a hydrazine derivative (Scheme 4.9) [41]. Again, the reaction proceeds via initial formation of an enaminoketone as the key intermediate from the 1,3-diketone and DMFDMA precursors, followed by a tandem addition–elimination/cyclodehydration step. The details of this reaction, carried out in superheated water as solvent, were already described in Section 4.5.1.

Humphries and Finefield at Pfizer have described the two-step synthesis of pyrazoles using microwave heating for both steps (Scheme 7.25) [42]. In the first step, the enol ketone precursor **1** was prepared from aryl methyl ketones and ethyl

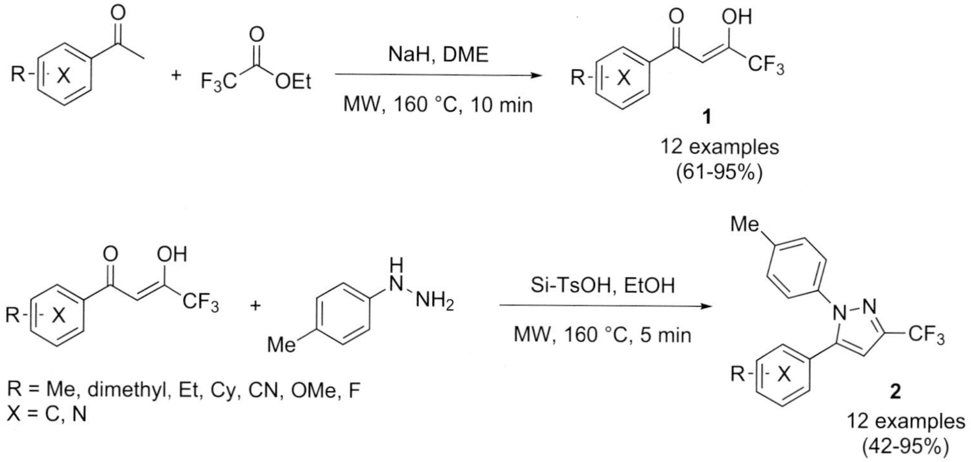

Scheme 7.25 Synthesis of 1,5-diarylpyrazoles.

trifluoroacetate. For the second step, enole ketones were reacted with 4-methylphenyl hydrazine and silica-supported *p*-toluenesulfonic acid (Si-TsOH) to give the corresponding pyrazoles **2** in good to excellent yields. Compared to standard *p*-TsOH, higher yields could be achieved using Si-TsOH, and additionally the reaction workup was simplified.

For an application of this strategy to the generation of analogs of the pyrazole COX-2 inhibitor celecoxib, see Ref. [43].

In a related fashion, Vernier and coworkers reported on the condensation of hydrazines with ω-cyanoacetophenones to produce 5-aminopyrazoles [44]. A multistep microwave approach for the synthesis of *N*-pyrazole ureas including the preparation of the p38α inhibitor BIRB 796 was disclosed by Bagley *et al.* in 2006 [45]. Alternative pyrazole synthesis involving hydrazine building blocks has been reported elsewhere [46–49].

As shown in Scheme 7.26, hydrazines also serve as building blocks for the preparation of medicinally interesting 3,5-diaryl-5-alkyl-4,5-dihydropyrazoles. As reported by Cox *et al.*, β-alkyl chalcones rapidly add hydrazine hydrate (2 equiv) within 30 min under microwave conditions at 150 °C to form *N*1-unsubstituted 4,5-dihydropyrazoles [50]. These unstable intermediates react efficiently with a number of electrophiles to form stable *N*1-acyl dihydropyrazoles. The current methodology allows the incorporation of many substitution patterns not available from the previously published approaches. Substituted hydrazines provide similar dihydropyrazole products, although the yields using aryl hydrazines are low. Microwave heating is not a requirement for these condensation reactions, as refluxing overnight in ethanol provided similar results; however, the shorter reaction times and the ability to easily perform multiple reactions in parallel with an automated handling system make the microwave route a more attractive option [50].

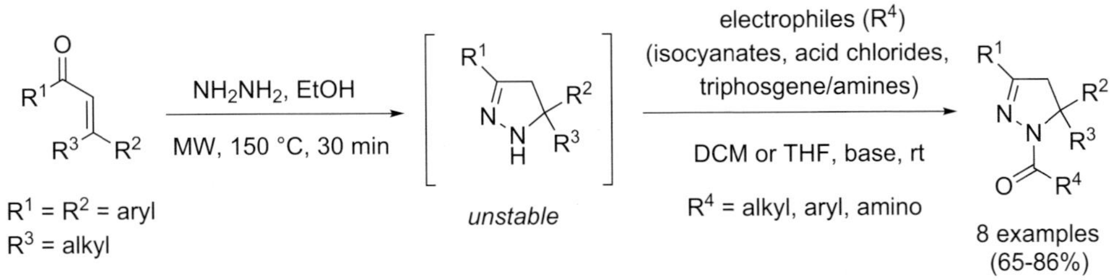

Scheme 7.26 4,5-Dihydropyrazole synthesis.

Kappe and coworkers recently presented a rapid microwave-assisted protocol to synthesize the canonical transient receptor potential channel (TRPC) inhibitor Pyr3, which could be implicated in diverse biological functions [51]. Beneficially, the initial two steps can be effectively conducted in a one-pot procedure, not requiring isolation of the pyrazole condensation product. The final amidation step was successfully performed utilizing the crude reduced aniline after simple filtration. In the optimized procedure, a suspension of 4-nitrophenylhydrazine hydrochloride and 1.05

equiv ethyl 2-(ethoxymethylene)-4,4,4-trifluoro-3-oxobutyrate in ethanol was microwave heated for 2 min at 160 °C. After cooling to 50 °C, 2 equiv cyclohexene and 1 mol% Pd/C (10% w/w) were added and the sealed vial was heated for additional 2 min at 160 °C in the microwave reactor. The cooled mixture was concentrated *in vacuo*, the residue was dissolved in acetonitrile, and then 1.1 equiv 2,3,3-trichloroacrylic acid and 1.1 equiv phosphorus trichloride were added. This mixture was subjected to microwave irradiation at 150 °C for another 5 min to furnish the desired Pyr3 in good overall yield (Scheme 7.27). The authors explored this effective method by investigating the preparation of other pyrazole-based NFAT transcription regulators as well. Satisfactory results were obtained after only minor modifications of the three-step microwave protocol [51].

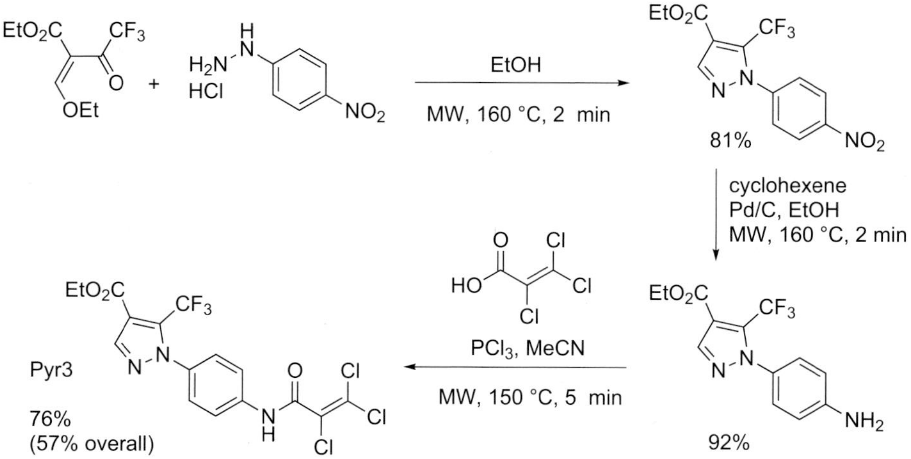

Scheme 7.27 Microwave-assisted three-step synthesis of TRPC inhibitor Pyr3.

An alternative strategy to generate 4,5-dihydropyrazoles is to perform 1,3-dipolar cycloaddition reactions of nitrile imines and olefins. Langa and coworkers have reported the preparation of pyrazolino[60]fullerenes by nitrile imine–fullerene [3 + 2] cycloadditions (Scheme 7.28a) [52]. The required nitrile imines were generated *in situ* from the corresponding hydrazones in chloroform. After evaporation of solvent and addition of C_{60}, triethylamine, and toluene, subsequent irradiation by microwaves under open-vessel conditions for 25 min provided the desired isoindazolylpyrazolino[60]fullerene dyads in moderate yield. Similarly, the microwave-assisted sidewall functionalization of single-walled carbon nanotubes via Diels–Alder cycloaddition with *ortho*-quinodimethane was demonstrated by the same group (Scheme 7.28b) [53]. The required *ortho*-quinodimethane was generated *in situ* from 4,5-benzo-1,2-oxathiin-2-oxide (sultin) by refluxing in 1,2-dichlorobenzene under open-vessel microwave irradiation conditions. In the presence of ester-functionalized SWNTs, cycloaddition takes place within 45 min. Conventional refluxing in 1,2-dichlorobenzene for 3 days leads only to a low degree of conversion.

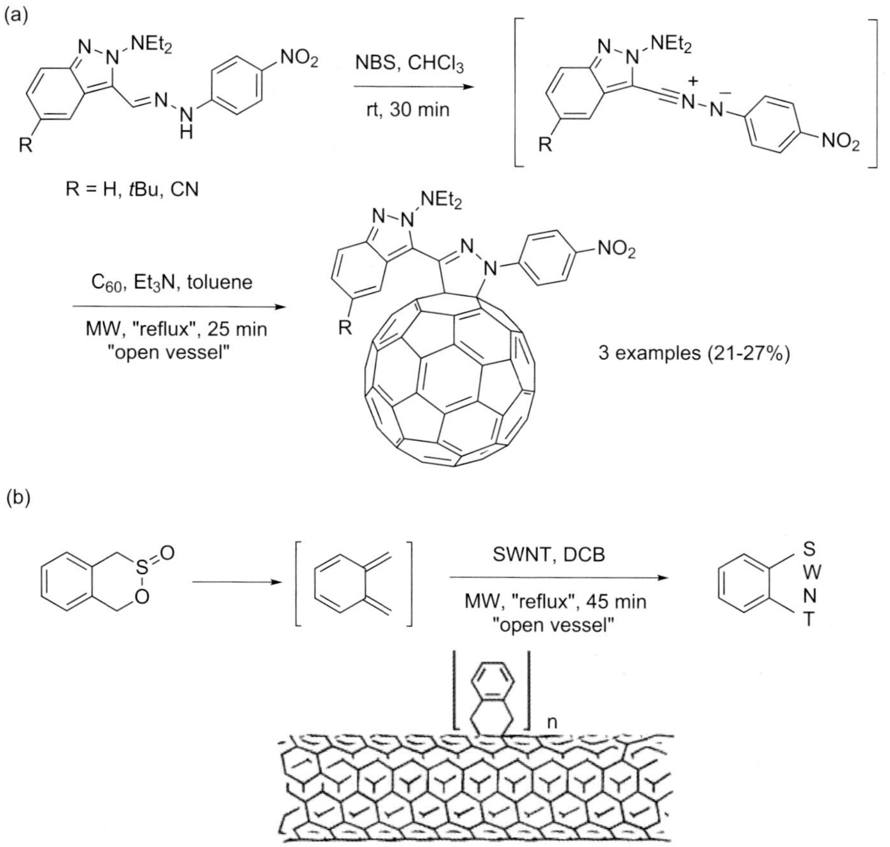

Scheme 7.28 Functionalization of fullerene and single-walled carbon nanotubes via cycloaddition chemistry.

7.4.2 Imidazoles

A simple, high-yielding synthesis of 2,4,5-trisubstituted imidazoles from 1,2-diketones and aldehydes in the presence of ammonium acetate was reported by Wolkenberg *et al.* (Scheme 7.29) [54]. Utilizing microwave irradiation (180 °C), alkyl-, aryl-, and heteroaryl-substituted imidazoles were formed in very high yields ranging from

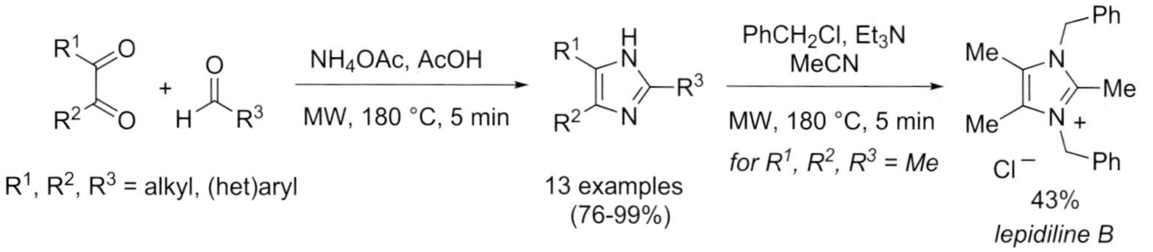

Scheme 7.29 Preparation of imidazoles from 1,2-diketones.

76 to 99% within 5 min by condensing 1,2-diketones with aldehydes in the presence of 10 equiv of ammonium acetate in acetic acid. Further microwave-assisted alkylation of 2,4,5-trimethylimidazole with benzyl chloride in the presence of a base led to the alkaloid lepidiline B in 43% overall yield. Lepidiline B with its symmetrical imidazolium structure exhibits micromolar cytotoxicity against several human cancer cell lines. The microwave-assisted two-step synthesis was evaluated, optimized, and completed within 2 h. Notably, the preparation of the intermediate 2,4,5-trimethylimidazole was technically simpler, faster, and higher yielding than previous routes [54].

A closely related protocol for the synthesis of imidazoles was independently investigated by Sparks and Combs (Scheme 7.30) [55]. Here, the authors have employed readily available unsymmetrical keto-oximes as building blocks, initially leading to *N*-hydroxyimidazoles. Diaryl keto-oximes were condensed with various aldehydes (1.1 equiv) in the presence of 4 equiv of ammonium acetate under microwave conditions at 160 °C. Under these conditions, the *N*-hydroxyimidazoles were formed in high yields and were subsequently quantitatively reduced with titanium trichloride (120 °C, 5 min) to imidazoles. The authors have uncovered that at higher reaction temperatures (200 °C) treatment of the keto-oximes with aldehydes and ammonium acetate led directly to the desired imidazoles by *in situ* cleavage of the thermolabile N–O bond. Utilizing these optimized conditions, a diverse set of 2,4,5-tri(hetero)arylimidazoles was prepared.

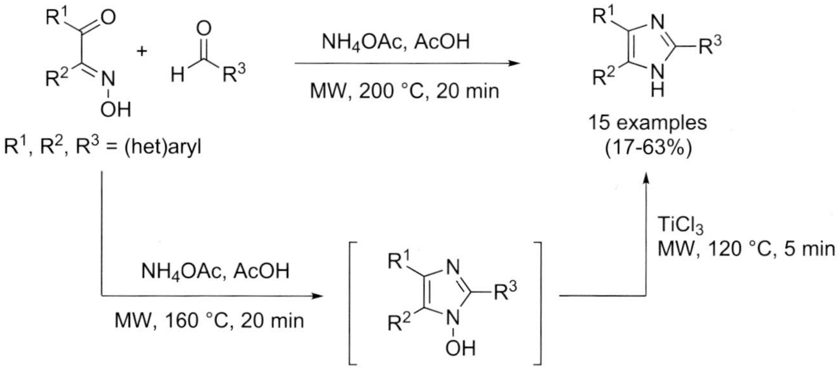

Scheme 7.30 Preparation of imidazoles from keto-oximes.

The synthesis of mono- and disubstituted 2-aminoimidazoles by the reaction of diverse α-haloketones with *N*-acetylguanidine was reported by Lam and coworkers (Scheme 7.31) [56]. In the first reaction step, the corresponding imidazol-2-acetamides were obtained in excellent yields. Further deacylation with H_2SO_4 and $MeOH/H_2O$ (1 : 1) or EtOH, respectively, and subsequent formation of the free amine with 5 M KOH in MeOH furnished the 2-aminoimidazole products, again in high yields. The overall reaction time could be cut down to only 20 min compared to several days under conventional room-temperature conditions.

Prati and coworkers established a one-pot synthesis of imidazole-4-carboxylates. The diversely functionalized heterocycles were achieved by microwave-assisted

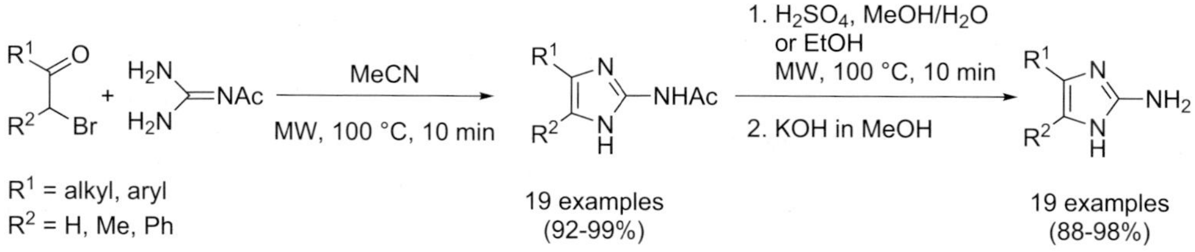

Scheme 7.31 Synthesis of substituted 2-aminoimidazoles.

1,5-electrocyclization of azomethine ylides [57]. The required intermediates were initially prepared by treatment of 1,2-diaza-1,3-dienes with primary amines and then subsequently subjected to microwave irradiation in the presence of corresponding aldehydes to initiate the cyclization process (Scheme 7.32). Access to 2-unsubstituted derivatives is given by utilizing paraformaldehyde. In a representative procedure, 1,2-diaza-1,3-diene is dissolved in acetonitrile, treated with 1.05 equiv amine and stirred at room temperature until decoloration of the solution. Then, 2 equiv of the aldehyde is added and the sealed vessel is subjected to microwave irradiation at 150 °C for 20 min to afford the desired imidazoles. The simple protocol proved successful for a variety of substrates, revealing that best yields were achieved with an alkyl side chain on the 1,2-diaza-1,3-diene [57].

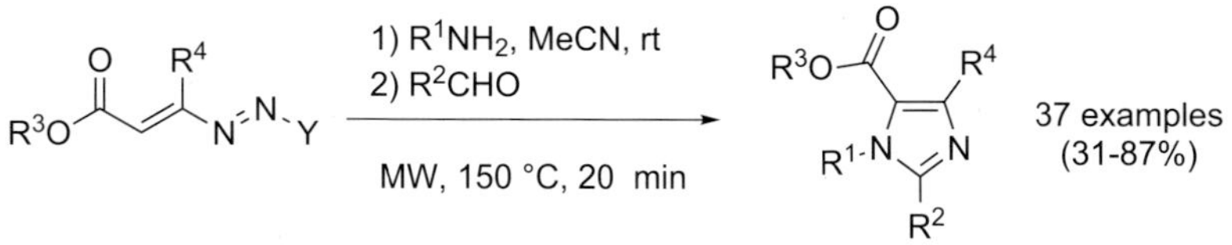

Scheme 7.32 One-pot synthesis of 3*H*-imidazole-4-carboxylates.

Multicomponent reactions (MCRs) are of increasing importance in organic and medicinal chemistry. The Ugi four-component condensation in which an amine, an aldehyde or ketone, a carboxylic acid, and an isocyanide combine to yield an α-acylamino amide is particularly interesting due to the wide range of products obtainable through variation of the starting materials. The reaction of heterocyclic amidines with aldehydes and isocyanides, using 5 mol% of scandium triflate as a catalyst in an Ugi-type three-component condensation (Scheme 7.33), generally requires extended reaction times of up to 72 h at room temperature for the generation

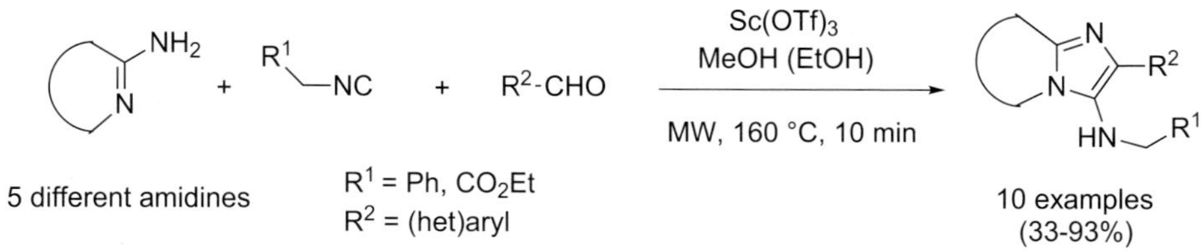

Scheme 7.33 Ugi-type three-component condensation reactions.

of the desired fused 3-aminoimidazoles. Tye and coworkers have demonstrated that this process can be speeded up significantly by performing the reaction under microwave conditions [58]. A reaction time of 10 min at 160 °C using methanol as solvent (in some cases ethanol was employed) produced similar yields of products as the same process at room temperature, albeit at a fraction of the time.

A different multicomponent route to imidazoles was described by the group of O'Shea, involving the diversity-tolerant three-component condensation of an aldehyde, a 2-oxo-thioacetamide, and an alkyl bromide (5 equiv) in the presence of ammonium acetate (Scheme 7.34) [59]. This allowed the preparation of a 24-membered 4(5)-alkylthio-1*H*-imidazole demonstration library from 21 different aldehydes, 12 alkyl bromides, and two 2-oxo-thioacetamides. The library was synthesized in a parallel format using a custom-built reaction vessel. Alkylthioimidazoles have been shown to be potential acyl-CoA/cholesterol acyltransferase inhibitors, analgesic agents, and angiotensin II receptor antagonists.

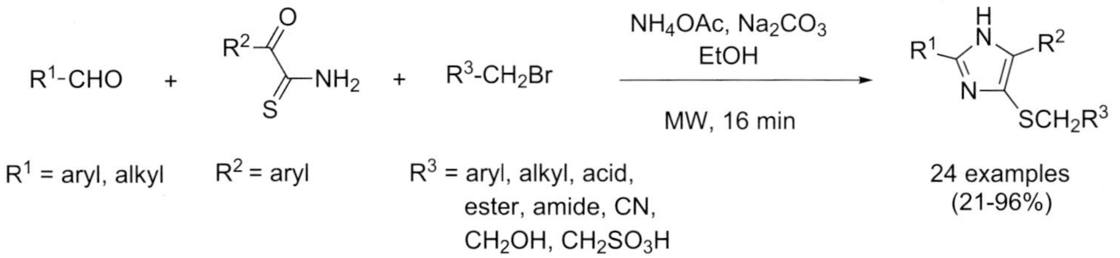

Scheme 7.34 Three-component condensation reactions leading to 4-alkylthioimidazoles.

The solvent-free preparation of 1,2,3-trisubstituted imidazolidin-4-ones from aldehydes and *N*-substituted α-amino acid amides was reported by Pospíšil and Potáček (Scheme 7.35) [60]. The general procedure simply involved heating equimolar mixtures of the aldehyde and amine building blocks under open-vessel microwave irradiation for 5 min to 200 °C. After cooling to room temperature, the imidazolidin-4-one products were purified by flash chromatography.

Kim and Varma have described the preparation of a range of cyclic ureas from diamines and urea [61]. In the example highlighted in Scheme 7.36, ethylenediamine and urea were condensed in the presence of 7.3 mol% of zinc(II) oxide in *N*,*N*-dimethylformamide as solvent at 120 °C to furnish imidazolidine-2-one in 95% isolated yield. Key to the success of this method is to perform the reaction

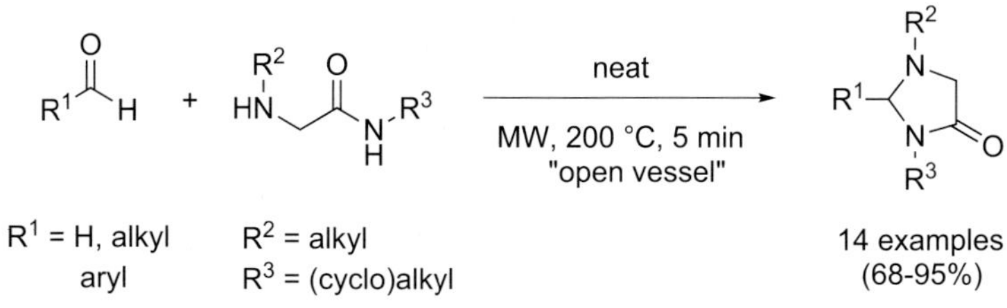

Scheme 7.35 Synthesis of imidazolidin-4-ones.

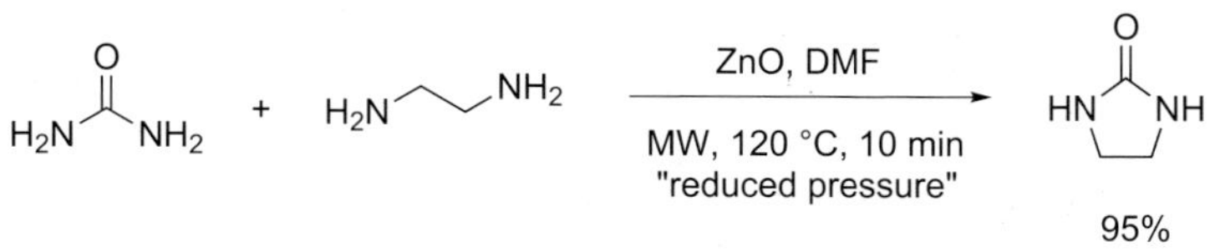

Scheme 7.36 Synthesis of imidazolidinones.

under reduced pressure in order to remove the formed ammonia from the reaction mixture. This method was extended to a variety of diamines and amino alcohols [61].

Merriman *et al.* have reported the cyclization of *N*-acyl-1,2-diaryl-1,2-ethanediamine derivatives, obtained from a solid-phase approach, to 4,5-diarylimidazolines by treatment with trimethylsilyl polyphosphate (TMS-PP) in dichloromethane solution (Scheme 7.37) [62]. Best results were obtained by microwave heating at 140 °C for 8 min. A library of 38 compounds was prepared by this method, leading to a novel and potent family of $P2X_7$ receptor antagonists.

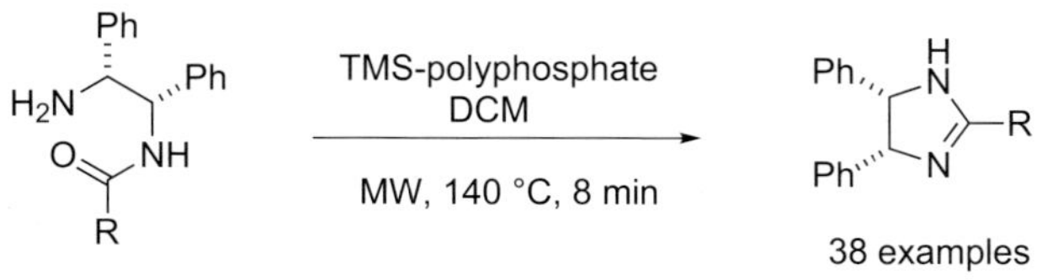

Scheme 7.37 Synthesis of 4,5-diarylimidazolines.

Berteina-Raboin and coworkers were successful in the synthesis of 2,3,6-trisubstituted imidazo[1,2-*a*]pyridine derivatives via a one-pot, three-step cyclization/Suzuki coupling/Pd-catalyzed heteroarylation sequence (Scheme 7.38) [63]. In previous optimization studies, the best conditions for a one-pot, two-step Suzuki reaction/heteroarylation starting from 6-bromoimidazo[1,2-*a*]pyridine were evaluated. $Pd(OAc)_2/PPh_3$ proved to be superior to $Pd(PPh_3)_4$ as catalyst system for the reaction sequence; in addition, it is essential to perform the Suzuki coupling as the

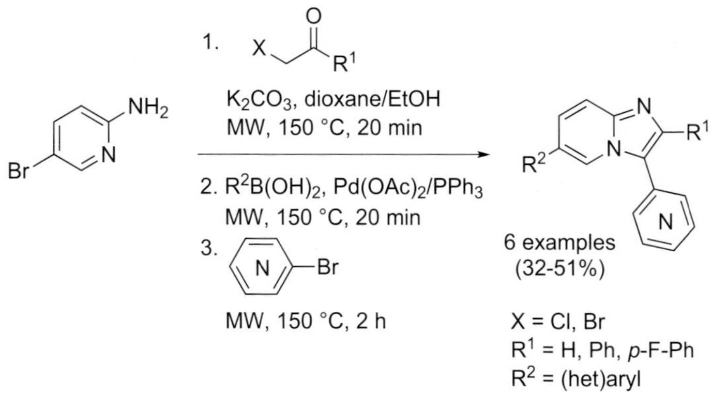

Scheme 7.38 Synthesis of polysubstituted imidazo[1,2-*a*]pyridines.

first step. In the three-step sequence, 2-amino-5-bromopyridine was reacted with α-halogenocarbonyls to 6-bromoimidazo[1,2-α]pyridine derivatives followed by the Suzuki coupling and heteroarylation in the same pot, giving the products in somewhat lower yields compared to the two-step sequence.

Using a somewhat related strategy, the group of Van der Eycken prepared aminoimidazoles of types **4** and **6** (Scheme 7.39) [64]. By a cyclocondensation sequence of 2-aminopyrimidines with α-bromo carbonyl compounds at 150 °C, imidazopyrimidinium salts **3** are obtained as intermediates that undergo cleavage using hydrazine hydrate furnishing 1-substituted 2-aminoimidazoles **4**. However, when the cyclocondensation step is performed at 80 °C, 2-hydroxy-2,3-dihydroimidazopyrimidinium salts **5** are formed that can be subsequently cleaved with hydrazine hydrate to give the 1-unsubstituted 2-aminoimidazoles **6**. The cleavage step for the formation of products **6** is proposed to proceed via a Dimroth-type rearrangement with an *in situ* generation of 2-amino-5-hydroxyimidazolidine, see also Ref. [65].

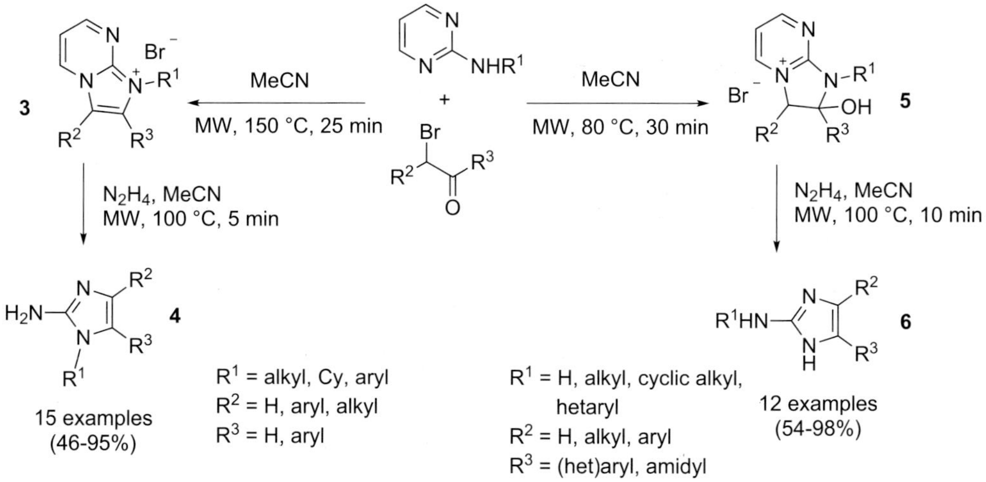

Scheme 7.39 Synthesis of 2-aminoimidazoles from 2-aminopyrimidines.

Comparison studies of thermal versus microwave heating in the synthesis of 2,4-disubstituted 5-aminoimidazoles **9** have been conducted by Lam and coworkers from the National University of Singapore (Scheme 7.40) [66]. Higher yields and considerable rate enhancements (up to about 130 times for the overall reaction) could be achieved by applying microwave heating for all three synthetic steps. The synthesis of 5-aminoimidazole scaffolds **9** was performed in parallel employing a multivessel rotor system, making the protocol amenable for high-throughput synthesis.

1,3-Diarylimidazolinium chlorides, useful as precursors to *N*-heterocyclic carbene ligands, can be synthesized efficiently by condensation of *N,N′*-diarylethylenediamine dihydrochlorides with triethyl orthoformate as demonstrated by Delaude and coworkers [67]. Similar results have been achieved by the group of Yu using carboxylic acids instead of orthoformate in the presence of pyridine [68].

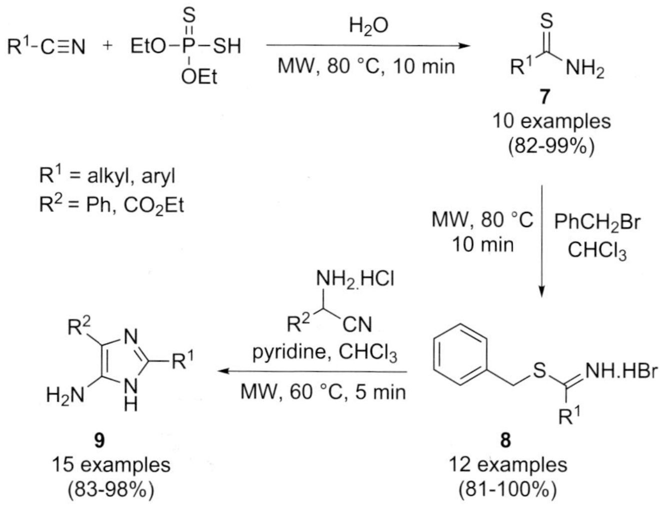

Scheme 7.40 Parallel synthesis of 2,4-disubstitued 5-aminoimidazoles.

2-Substituted benzimidazoles have been obtained earlier by $SnCl_2$-mediated reduction/condensation of *o*-nitroanilines with carboxylic acids [69], or by the condensation of aromatic 1,2-diamines with aromatic aldehydes in the presence of SiO_2 [70]. Similar approaches were used for the generation of xanthines by the group of Stanovnik [71].

Three different microwave-assisted synthetic routes to benzimidazole derivatives are summarized in Scheme 7.41, involving either the condensation of 1,2-phenylenediamines with either carboxylic acids (Scheme 7.41a and b) [72, 73], or 2 equiv of aldehydes (Scheme 7.41c) [74], or by cyclization of *N*-acylated diaminopyrimidines mediated by a strong base (Scheme 7.41d and e) [75, 76].

7.4.3
Isoxazoles

One obvious synthetic route to isoxazoles and dihydroisoxazoles is by [3 + 2] cycloaddition of nitrile oxides with acetylenes or olefins, respectively. In the example elaborated by Giacomelli *et al.* shown in Scheme 7.42, nitroalkanes were converted *in situ* to nitrile oxides by 1.25 equiv of the reagent 4-(4,6-dimethoxy[1,3,5]triazin-2-yl)-4-methylmorpholinium chloride (DMTMM) and 10 mol% of *N*,*N*-dimethylaminopyridine (DMAP) as catalyst [77]. In the presence of an olefin or acetylene dipolarophile (5 equiv), the generated nitrile oxide 1,3-dipoles undergo cycloaddition with the double or triple bond, respectively, in order to furnish 4,5-dihydroisoxazoles or isoxazoles. Here, open-vessel microwave conditions were chosen and full conversion with very high isolated yields of products was achieved within 3 min at 80 °C. The reactions could also be carried out utilizing a resin-bound acetylene [77].

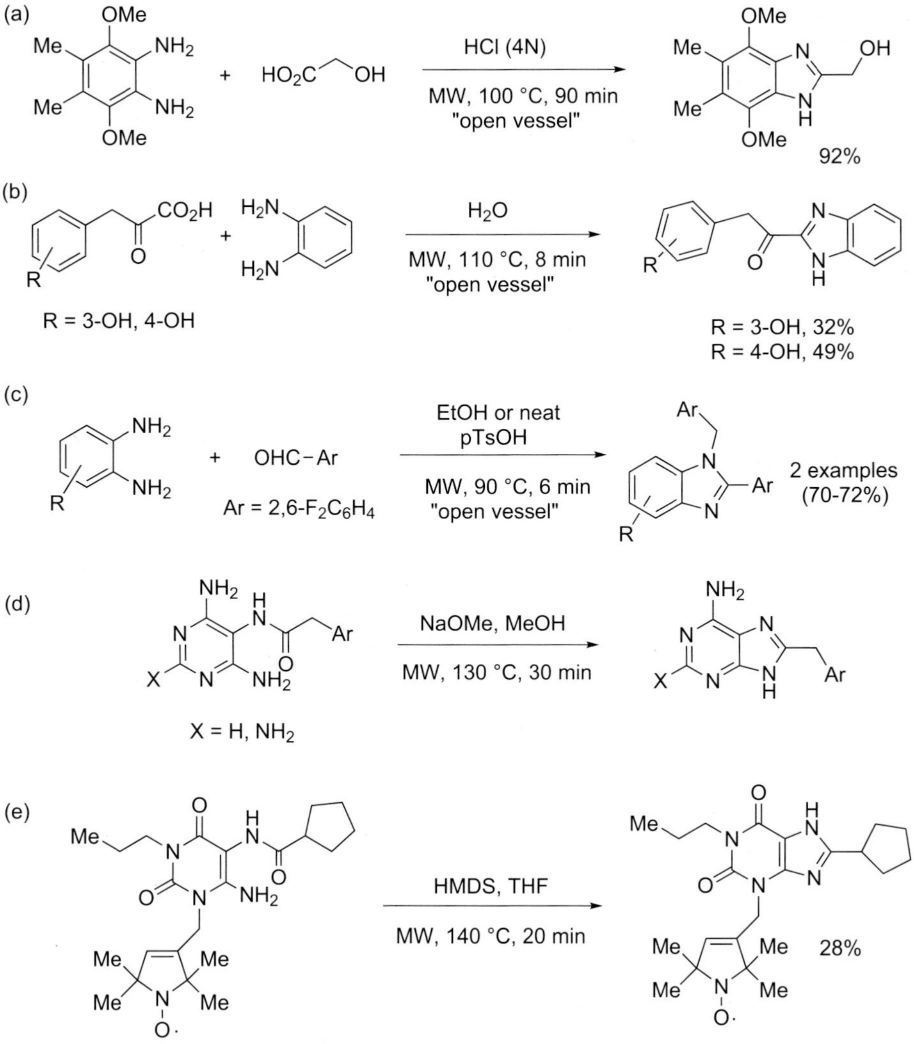

Scheme 7.41 Synthesis of benzimidazole derivatives.

Related to the nitrile oxide cycloadditions described in Scheme 7.42 are the 1,3-dipolar cycloaddition reactions of nitrones with olefins leading to isoxazolidines. The group of Comes-Franchini has described cycloadditions of *Z*-α-phenyl-*N*-methylnitrone with allylic fluorides, leading to enantiopure fluorine-containing isoxazolidines and amino polyols (Scheme 7.43) [78]. The reactions were carried out under solvent-free conditions in the presence of 5 mol% of either scandium(III) or indium(III) triflate. In the racemic series, an optimized 74% yield of an *exo*/*endo* mixture of cycloadducts was obtained within 15 min at 100 °C. In the case of the enantiopure allyl fluoride, a similar product distribution was achieved after 25 min at 100 °C. Reduction of the isoxazolidine cycloadducts with lithium aluminum

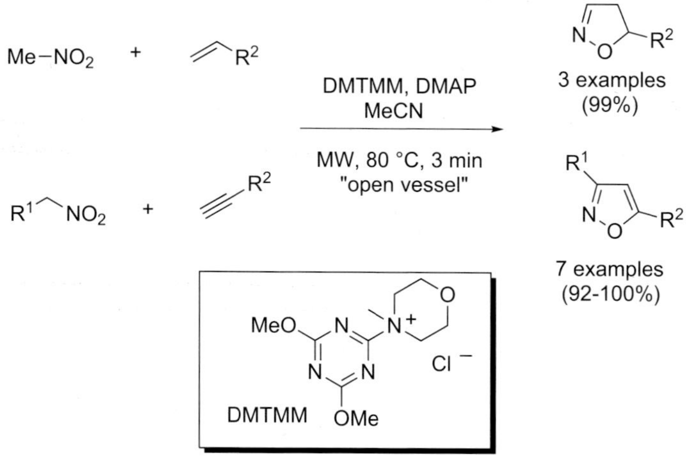

Scheme 7.42 Nitrile oxide cycloaddition reactions.

hydride provided fluorinated enantiopure polyols with four stereocenters of pharmaceutical interest.

A remarkable switch in selectivity was observed by Wagner and coworkers in the cycloaddition of nitrones to free and coordinated *E*-cinnamonitrile (Scheme 7.44) [79]. While the reaction of the nitrone with free cinnamonitrile occurs exclusively at the C=C bond furnishing isoxazolidine-4-carbonitriles (Scheme 7.44a), cycloaddition of the same nitrone to transition metal-coordinated cinnamonitrile occurs at the nitrile C≡N bond, leading to 1,2,4-oxadiazoline complexes, from which the heterocyclic ligand could be released and isolated in high yield (Scheme 7.44b). Microwave irradiation in dichloromethane solution (100 °C) enhances the reaction rates of both transformations considerably, without changing their regioselectivity with respect to the thermal reactions. The two nitrile ligands in complexes of the type [MCl_2(cinnamonitrile)$_2$] (M = platinum or palladium) are significantly different in reactivity.

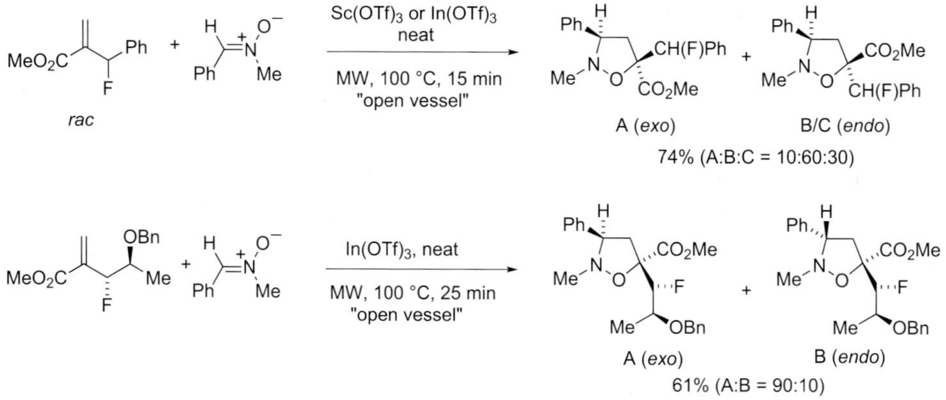

Scheme 7.43 Nitrone–allyl fluoride cycloaddition reactions.

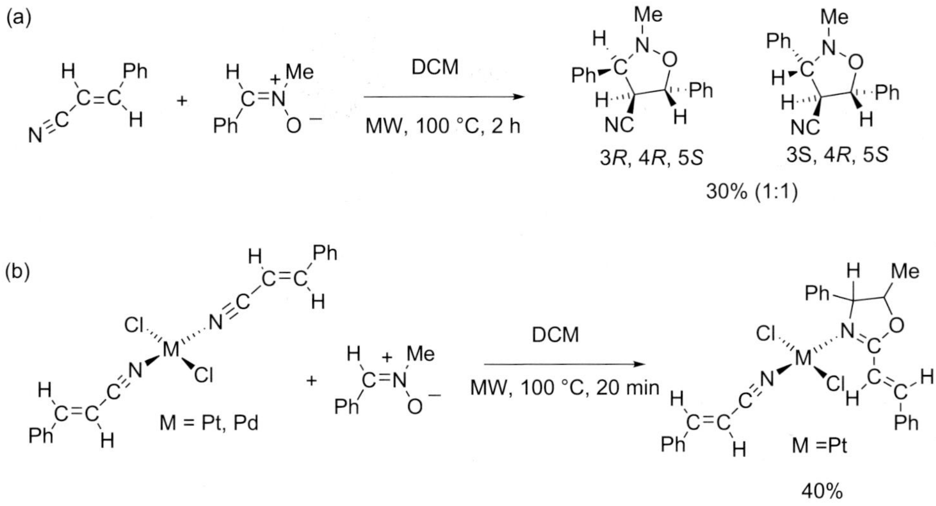

Scheme 7.44 Nitrone–cinnamonitrile cycloaddition reactions.

Thus, short-term microwave irradiation allows the selective synthesis of the mono-cycloaddition product (M = platinum), even in the presence of an excess of nitrone. Using longer reaction times, this complex can be further transformed into the biscycloaddition product. Related work has been published by the same authors in 2004 [80].

7.4.4
Oxazoles

A simple two-step synthesis of 5*H*-alkyl-2-phenyl-oxazole-4-ones has been reported by Trost *et al.* (Scheme 7.45) [81]. α-Bromo acid halides were condensed with benzamide in the presence of pyridine base at 60 °C to form the corresponding imides. Microwave irradiation of the imide intermediates in *N*,*N*-dimethylacetamide containing sodium fluoride at 180 °C for 10 min provided the desired 5*H*-alkyl-2-phenyl-oxazole-4-ones (oxalactams) in 44–82% yields. This class of heterocycles served as excellent precursors for the asymmetric synthesis of α-hydroxycarboxylic acid derivatives [81].

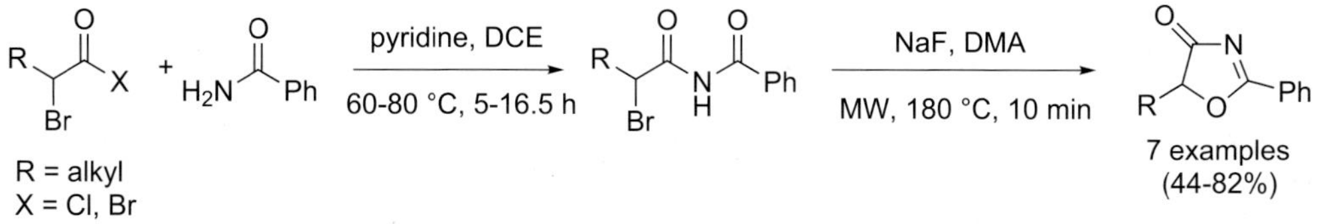

Scheme 7.45 Preparation of 5*H*-alkyl-2-phenyl-oxazole-4-ones.

The group of Wipf has described a one-pot oxazole synthesis via base-catalyzed condensation of oximes and acid chlorides at 180 °C [82]. Benzoxazoles were

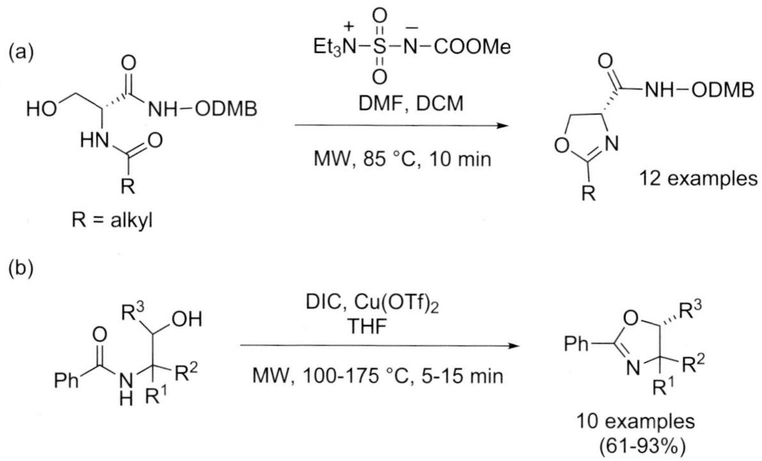

Scheme 7.46 Preparation of 2-oxazolines.

obtained by acid-catalyzed cyclization of *N*-acetylated *o*-aminophenols at 200 °C in a study of Huxley [83].

An alternative procedure for the synthesis of aliphatic 2-substituted oxazoline hydroxamates was described by Pirrung *et al.* in the context of preparing inhibitors of *E. coli* LpxC zinc amidase [84]. As shown in Scheme 7.46a, the protocol utilized the cyclization of suitable amides, formed *in situ* by acylation of a serine-derived *O*-2,4-dimethoxybenzyl (DMB)-protected hydroxamate. The cyclization was best performed employing 1.25 equiv of Burgess reagent at 85 °C for 10 min. Owing to the use of Burgess reagent and the formation of *N*-acylated by-products, chromatographic purification of the intermediate *O*-protected oxazoline was required. The purified oxazoline hydroxamates were immediately deprotected with 2% trifluoroacetic acid (TFA) in hexafluoroisopropanol. Linclau and coworkers have reported a related strategy where (chiral) *N*-(β-hydroxy)amides were cyclized with 1 equiv of diisopropylcarbodiimide (DIC) in the presence of 5 mol% of copper(II) triflate in tetrahydrofuran (THF) (Scheme 7.46b) [85]. Microwave heating for 5 min to 150 °C provided good to excellent yields for most of the desired 2-oxazolines.

As demonstrated by the Katritzky group, simple oxazolines can be obtained by treatment of β-amino alcohols with readily available *N*-acylbenzotriazoles (Scheme 7.47a) [86]. Accordingly, microwave irradiation of a solution of 2-amino-2-methyl-1-propanol with 0.5 equiv of an *N*-acylbenzotriazole in chloroform at 80 °C for 10 min produced the desired oxazolines along with the uncyclized intermediates, *N*-(2-hydroxy-1,1-dimethylethyl)amides. Addition of 3 equiv of thionyl chloride to the reaction mixture and subsequent further microwave irradiation for 2 min (80 °C) resulted in complete conversion of the uncyclized intermediates to the oxazoline product. An analogous protocol was used to generate the corresponding thiazolines (Scheme 7.47b) [86]. Marrero-Terrero and coworkers have reported a similar preparation of 4,4-disubstituted 2-oxazolines by solvent-free condensation of β-amino alcohols with carboxylic acids using zinc oxide as inorganic support [87].

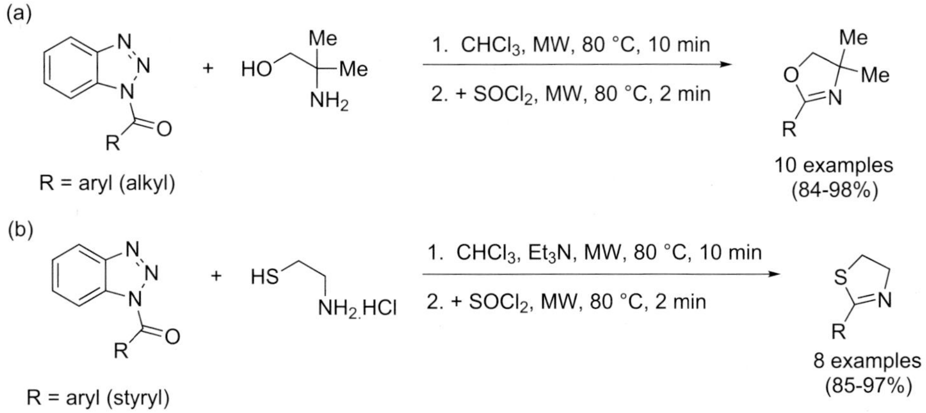

Scheme 7.47 Preparation of oxazolines.

The condensation of enantiomerically pure amino alcohols (derived from amino acids) with aldehydes to furnish 1,3-oxazolidines was studied by Kuhnert and Danks (Scheme 7.48) [88]. Under solvent-free conditions, microwave irradiation of equimolar mixtures of the amino alcohol and the aldehyde for less than 3 min provided high isolated yields of 1,3-oxazolidines in excellent diastereoselectivity. In the case of (−)-ephedrine, prolonged microwave irradiation (3 min) produced quantitative conversions and high diastereoselectivities. For shorter irradiation times (80 s), mixtures of the two diastereomers were obtained with moderate conversions. Apparently, the microwave conditions are suitable to drive the equilibrium between the two diastereomers toward the thermodynamically more stable *syn,syn* isomer, which was confirmed in a separate control experiment [88]. Similar results were obtained by Holzgrabe and coworkers in related systems using chloroform as solvent under microwave conditions [89].

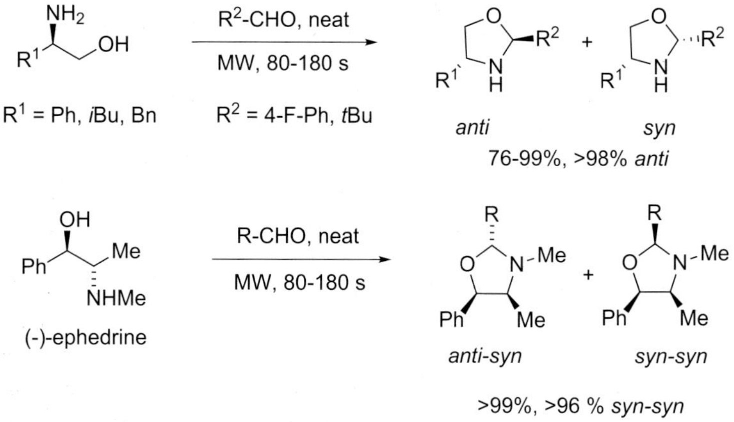

Scheme 7.48 Preparation of 1,3-oxazolidines.

For the synthesis of benzoxazoles, Player and coworkers have developed a simple method that involves microwave heating of a 2-aminophenol with 1.1 equiv of an acid chloride in dioxane or xylene (Scheme 7.49) [90]. Best results were obtained when the reaction mixture was exposed to microwave irradiation at 210–250 °C for 10–15 min. The addition of base or Lewis acids was not necessary. By choosing a set of six 2-aminophenols and eight acid chlorides, a library of 48 benzoxazoles was prepared in good to excellent yields.

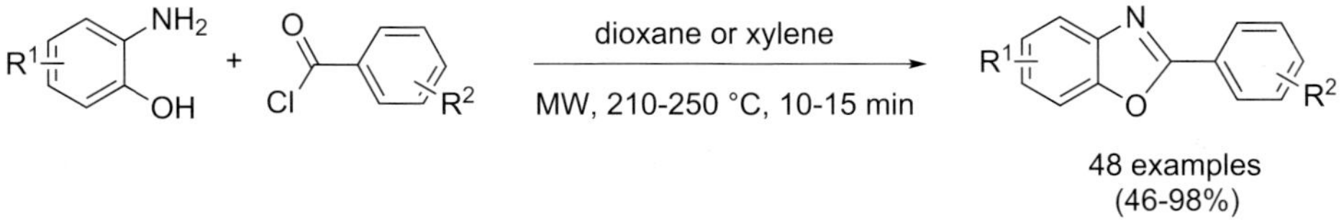

Scheme 7.49 Preparation of benzoxazoles.

A somewhat similar method for the preparation of 1,3-oxazolo[4,5-*d*]pyridazinones has been described by Ivachtchenko and coworkers (Scheme 7.50) [91]. Here, the key intermediate 5-amino-4-hydroxy-3(2*H*)-pyridazinone was treated with 1.5 equiv of a carboxylic acid in 1-methyl-2-pyrrolidone in the presence of polyphosphoric acid (PPA). The desired oxazolopyridazinones were obtained in good yields by microwave heating to 230 °C for 15–20 min. Reaction of the same precursor with 1,1′-carbonyldiimidazole (CDI) in dioxane at 170 °C furnished the 1,3-oxazolo[4,5-*d*]pyridazine-2(3*H*)-7(6*H*)-dione in 42% yield. Here, the conventional reaction conditions (dioxane, reflux) failed and did not provide any isolable product.

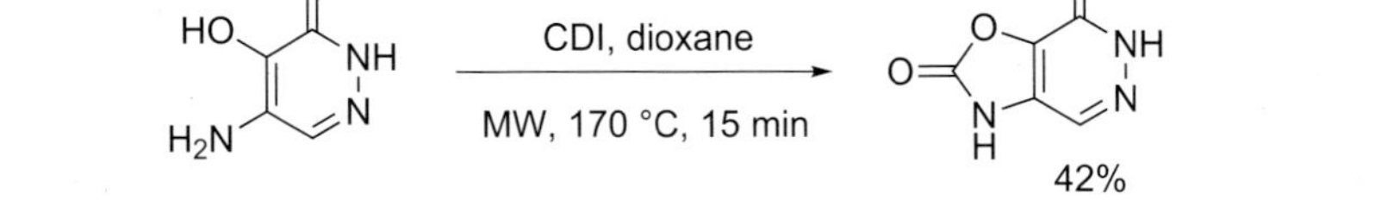

Scheme 7.50 Preparation of oxazolopyridazinones.

As reported by various groups, isoxazolidinyl analogs can be prepared via a cycloaddition approach using nitrones and allyl alcohol [92, 93] or acrylate building blocks [94].

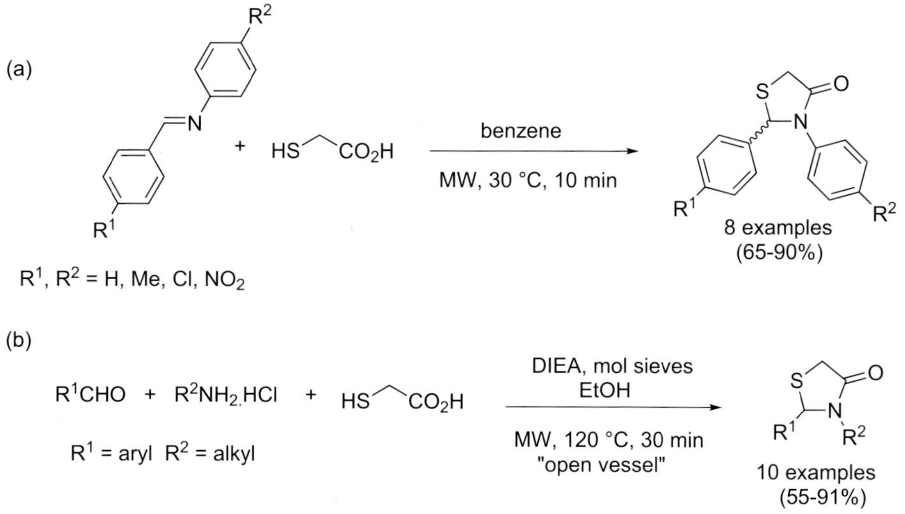

Scheme 7.51 Preparation of thiazolidin-4-ones.

7.4.5
Thiazoles

The group of Bolognese has disclosed the synthesis of thiazolidin-4-ones by condensation of benzylideneanilines and mercaptoacetic acid (Scheme 7.51a) [95]. The authors have found that microwave heating of an equimolar mixture of the two components in benzene at 30 °C for only 10 min provides excellent yields of the thiazolidin-4-one heterocycles. Surprisingly, when the same transformation was carried out at reflux temperature (80 °C), much longer reaction times (2 h) were required and the products were obtained in significantly lower yields (25–69%).

More recently, Miller and coworkers reported a one-pot protocol for the preparation of thiazolidin-4-ones by condensation of aromatic aldehydes, amines, and mercaptoacetic acid in ethanol (Scheme 7.51b) [96]. The optimized procedure involved microwave irradiation of a mixture of amine hydrochloride, aldehyde, and mercaptoacetic acid (molar ratio 1 : 2 : 3) in the presence of 1.25 equiv of *N,N*-diisopropylethylamine (DIEA) base in ethanol at 120 °C for 30 min at atmospheric pressure.

In a recent report, Shimizu and Gaonkar introduced a simple, high-yielding microwave-assisted method for the synthesis of rosiglitazone. This compound belongs to the thiazolidinedione drug class showing antihyperglycemic activities in curing diabetes mellitus type II [97]. The multistep route starts with the preparation of thiazolidine-2,4-dione, which is a key intermediate in the synthesis of several pharmacologically active heterocycles. The initially generated 2-imino-thiazolidin-4-one was microwave heated at 140 °C for 10 min to furnish the desired thiazolidine-2,4-dione **10** in 90% yield. The second required building block was achieved by condensing 2-chloropyridine with 2-(*N*-methylamino)ethanol. Heating under microwave irradiation without any solvent at 140 °C for 20 min gave the desired 2-(methyl(pyridin-2-yl)amino)ethanol **11** in 92% yield. In the next step, the freshly prepared **11** was coupled to 4-fluorobenzaldehyde in aqueous toluene, employing 3 equiv potassium hydroxide powder and 10 mol% tetrabutylammonium hydrogensulfate (TBAHS) as phase-transfer catalyst. Microwave heating at moderate 85 °C for 20 min afforded 4-[2-(methyl(pyridin-2-yl)amino)ethoxy]benzaldehyde **12** in 90% yield. Subsequently, this product was subjected to a Knoevenagel condensation with the previously prepared **10**. Both compounds were suspended with silica gel in toluene containing catalytic amounts of piperidine and acetic acid and microwave heated at 130 °C for 20 min to give (*Z*)-5-{4-[2-methyl(pyridin-2-yl)amino)ethoxy]benzylidene}thiazolidine-2,4-dione **13** in 93% yield (Scheme 7.52). The desired rosiglitazone could be isolated in excellent 95% yield (73% over four steps) after magnesium-catalyzed reduction. The high-yielding protocol proves suitable for generating a library of various pharmaceutically attractive thiazolidines [97].

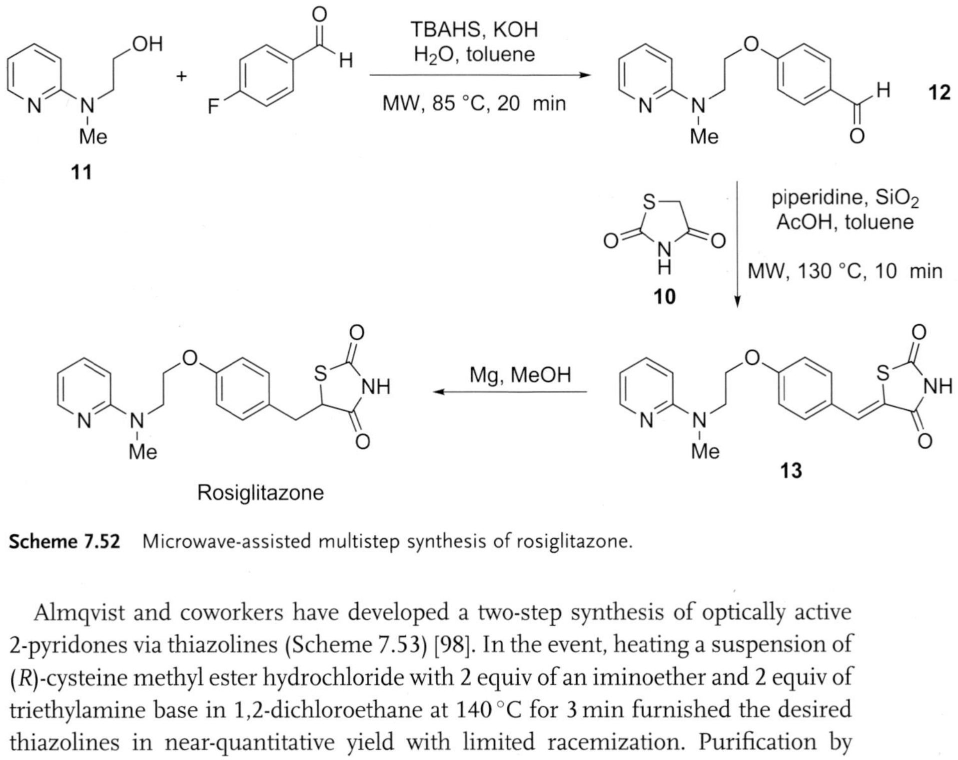

Scheme 7.52 Microwave-assisted multistep synthesis of rosiglitazone.

Almqvist and coworkers have developed a two-step synthesis of optically active 2-pyridones via thiazolines (Scheme 7.53) [98]. In the event, heating a suspension of (*R*)-cysteine methyl ester hydrochloride with 2 equiv of an iminoether and 2 equiv of triethylamine base in 1,2-dichloroethane at 140 °C for 3 min furnished the desired thiazolines in near-quantitative yield with limited racemization. Purification by filtration through a short silica gel column and concentration of the filtrate gave a crude product that was directly carried on in the next step. Thus, after addition of a presaturated hydrochloric acid 1,2-dichloroethane solution containing 1.5 equiv of an acyl Meldrum's acid derivative, the resulting solution was again heated under microwave conditions to 140 °C for 2 min. Applying this two-step protocol, a small array of six bicyclic 2-pyridinones was prepared in 70–96% yield in only 5 min of

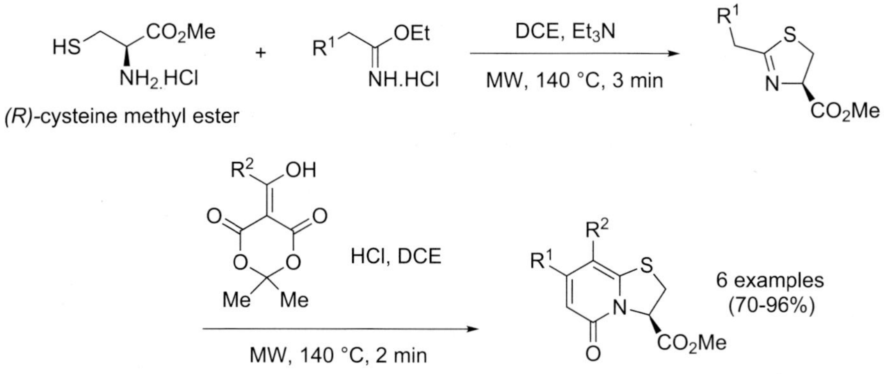

Scheme 7.53 Preparation of thiazolines and 2-pyridones.

microwave irradiation time. The optical purity of the heterocyclic products proved to be slightly lower (78–88% *ee*) compared to the conventional protocol (75–97% *ee*), but the total reaction time was reduced from 2 days to 5 min.

The same group subsequently described the rapid functionalization of the bicyclic 2-pyridone ring, in particular various aminomethylation pathways (Scheme 7.54) [99]. Primary aminomethylene substituents were introduced via cyanodehalogenation followed by a borane dimethyl sulfide reduction of the formed nitrile. To incorporate tertiary aminomethylene substituents in the 2-pyridone framework, a microwave-assisted Mannich reaction using preformed iminium salts proved to be effective.

Gallagher and coworkers have elaborated a rather complex strategy for the synthesis of fused thiazolines of the penem type (bicyclic β-lactams) based on the

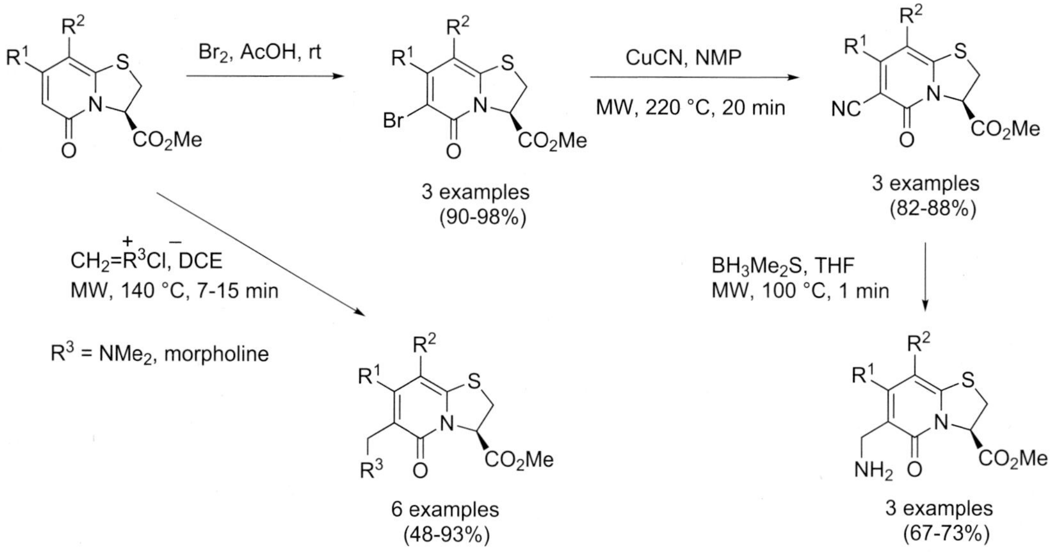

Scheme 7.54 Functionalization of 2-pyridones.

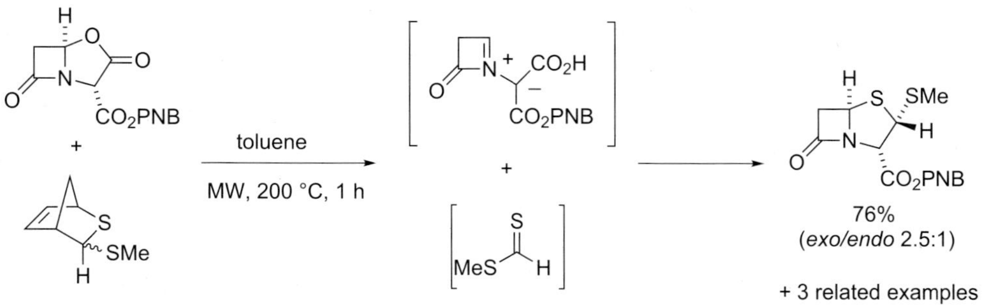

Scheme 7.55 Synthesis of penem derivatives by azomethine ylide–thione cycloadditions.

generation of β-lactam-based azomethine ylides [100]. In the example shown in Scheme 7.55, an oxazolidinone precursor reacts via sequential ring cleavage to give an azomethine ylide dipole, which then undergoes spontaneous cycloaddition with an *in situ* generated *S*-alkyl dithioformate dipolarophile to produce the target penem. The thermal process did require 2 days to go to completion and this only achieved a very modest 19% yield of the cycloadduct. Under microwave irradiation conditions utilizing toluene as solvent at 200 °C, a 76% isolated yield of the cycloadduct was obtained within 1 h. Somewhat lower yields were obtained when the transformation was carried out in ionic liquid-doped toluene or under open-vessel conditions.

Alternative methods for the microwave-assisted preparation of thiazole derivatives are summarized in Scheme 7.56.

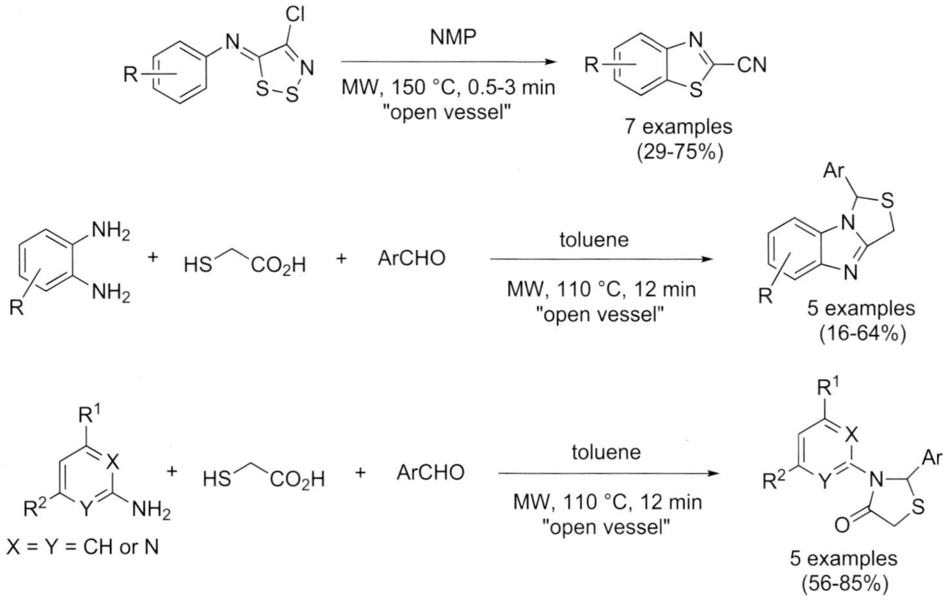

Scheme 7.56 Synthesis of benzothiazoles [101, 102], thiazolobenzimidazoles [74], and heteroaryl-substituted thiazolidinones [74].

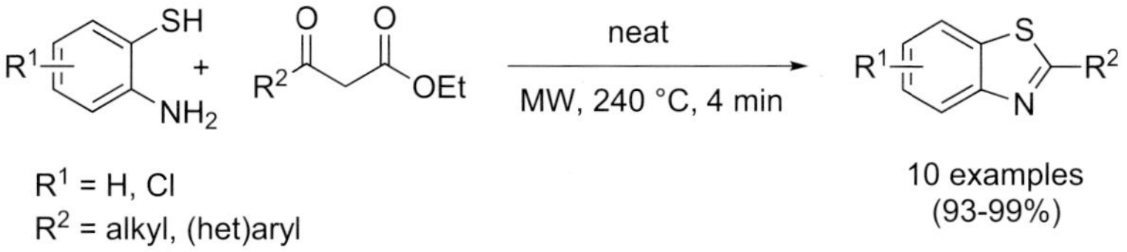

Scheme 7.57 One-pot solvent-free synthesis of 2-aryl- and 2-alkylbenzothiazoles.

A recent publication by Biehl and coworkers described an improved and eco-friendly synthesis of 2-alkyl- and aryl-substituted benzothiazoles (Scheme 7.57) [103]. As opposed to conventional heating, under controlled microwave heating the neat reaction of aromatic and aliphatic β-ketoesters and *o*-aminothiophenols proceeded faster and in higher yields. It was also possible to scale up selected reactions to give multigram quantities with similar high yields.

For an alternative Mn(III)-promoted benzothiazole synthesis starting from thioformanilides, see Ref. [104].

Endothiopeptides have been converted directly to thiazoles in the presence of 1.5 equiv of TMSCl in MeCN [105]. For examples of Hantzsch thiazole syntheses, see Refs. [106, 107].

The group of Botta at the University of Siena developed a one-pot two-step protocol for the microwave synthesis of functionalized rhodanine derivatives [108]. These 2-thiazolidine-4-one scaffolds are considered privileged scaffolds to constitute compounds with a variety of pharmacological activities. Especially, the herein generated 5-alkylidene derivatives show significant DDX3 inhibitory activity in context of HIV replication. To overcome the isolation and purification problems of already presented multicomponent approaches, the authors elaborated an efficient one-pot protocol for preparing the desired compounds by combining the Holberg method and a common Knoevenagel condensation under microwave irradiation. After preliminary screening of the reaction conditions, the optimized one-pot protocol was established as follows. A 10 mL microwave vial was charged with 1 equiv bis(carboxymethyl) trithiocarbonate in 1 mL 1,2-dimethoxyethane, 1 equiv triethylamine and 1 equiv corresponding amine. The mixture was irradiated at 90 °C for 10 min. Then 1 equiv aldehyde was added and the mixture was microwave heated at 110 °C for additional 5 min to furnish the desired rhodanines in moderate to good yields (Scheme 7.58). This rapid, sequential two-step protocol gives an easy access to numerous biologically valuable sulfur-containing heterocycles with a high level of structural diversity.

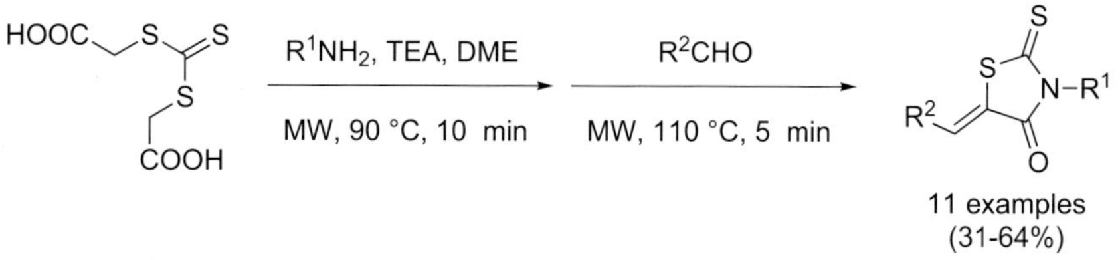

Scheme 7.58 One-pot two-step microwave synthesis of 3-substituted 5-arylidene rhodanines.

7.5 Five-Membered Heterocycles with Three Heteroatoms

7.5.1 1,2,3-Triazoles

The 1,3-dipolar cycloaddition of azides to alkynes is a versatile route to 1,2,3-triazoles. Different combinations of substituents on the azide and on the alkyne allow the preparation of diverse *N*-substituted 1,2,3-triazoles. Katritzky and Singh have described the synthesis of *C*-carbamoyl-1,2,3-triazoles by microwave-induced cycloaddition of benzyl azides to acetylenic amides (Scheme 7.59) [109]. Employing equimolar mixtures of the azide and acetylene under solvent-free conditions, the authors were able to achieve good to excellent isolated product yields by microwave heating to 55–85 °C for 30 min. In general, the triazole products were obtained as mixtures of regioisomers. Control experiments at 55–60 °C carried out under thermal (oil bath) conditions did not show any conversion even after 24 h. In the case of bis-azides, besides the anticipated monocycloadducts, in some cases the bis-triazoles could also be obtained (not shown) [110]. Other microwave-assisted 1,3-dipolar cycloaddition reactions of azides with symmetrical acetylenes have been studied by Savin *et al.* [111].

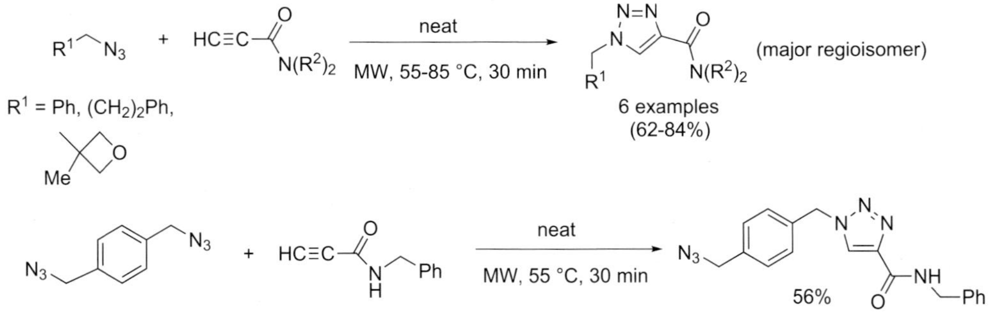

Scheme 7.59 1,3-Dipolar cycloaddition of benzyl azides with acetylenic amides.

A significant advance in this area has been the development of a copper(I)-catalyzed variation of this classical 1,3-dipolar cycloaddition process. For terminal acetylenes, this method leads to a regiospecific coupling of the two reaction partners. The Kappe group has exploited a microwave-assisted version of this copper(I)-catalyzed azide–acetylene ligation process ("click chemistry") for the preparation of 6-(1,2,3-triazol-1-yl)-dihydropyrimidones (Scheme 7.60) [112]. Here, a suitable heterocyclic azide intermediate (obtained by microwave-assisted azidation) was treated with phenylacetylene in *N,N*-dimethylformamide employing 2 mol% of copper(II) sulfate/sodium ascorbate as catalyst precursor. After completion of the cycloaddition process, the triazole product could be precipitated in high yield (73%) and purity by addition to ice/water. For the model reaction displayed in Scheme 7.60, full

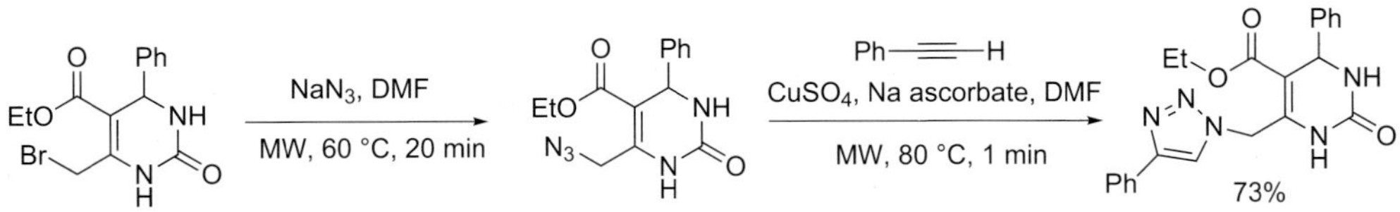

Scheme 7.60 Copper(I)-catalyzed azide–acetylene ligation.

conversion at room temperature required 1 h. By carrying out the same reaction utilizing controlled microwave heating at 80 °C, complete conversion was achieved within 1 min. A library of 27 6-(1,2,3-triazol-1-yl)-dihydropyrimidones was prepared with four points of diversity.

For certain substrates, Fokin, Van der Eycken, and coworkers subsequently discovered that the azidation and ligation step can be carried out in a one-pot fashion, thereby simplifying the overall protocol (Scheme 7.61) [113]. This procedure eliminates the need to handle organic azides, as they are generated *in situ*. Other applications of microwave-assisted copper(I)-catalyzed azide–acetylene ligations ("click chemistry") have been reported [114].

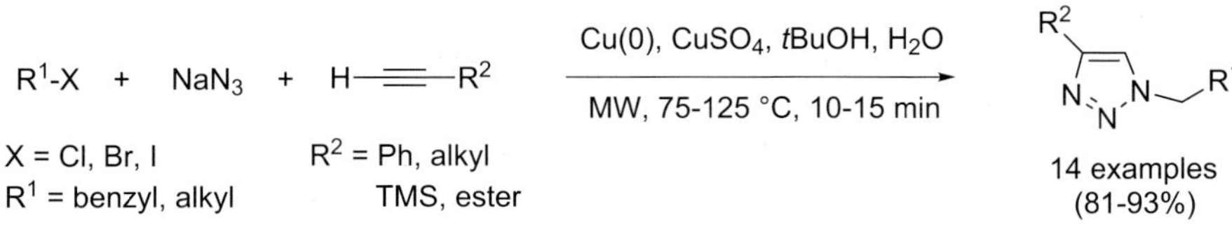

Scheme 7.61 Copper(I)-catalyzed azide–acetylene ligation in one-pot fashion.

Lam and Gao disclosed a study on triazolo[4,5-*b*]pyridin-5-ones, which have been obtained by [3 + 2] cycloaddition of NaN_3 to 5-benzenesulfonyl-3,4-dihydro-1*H*-pyridin-2-one in DMF [115].

7.5.2
1,2,4-Triazoles

A recent publication by Yeung *et al.* describes a convenient and efficient one-step, base-catalyzed synthesis of 3,5-disubstituted 1,2,4-triazoles by the condensation of a nitrile and a hydrazide (Scheme 7.62) [116]. Under the reaction conditions, a diverse range of functionality and heterocycles are tolerated. The reactivity of the nitrile partner is relatively insensitive to electronic effects.

Scheme 7.62 One-step synthesis of 3,5-disubstituted 1,2,4-triazoles.

The condensation of hydrazides with 3,6-dichloropyridazines furnishes 1,2,4-triazolo[4,3-*b*]pyridazines via a similar mechanistic pathway [117].

7.5.3
1,2,4-Oxadiazoles

The 1,2,4-oxadiazole class of heterocycles has been shown to possess a variety of CNS (central nervous system)-related activities. A group from ArQule and Pfizer has reported the rapid synthesis of 1,2,4-oxadiazoles based on the coupling of amidoximes with carboxylic acids in the presence of *O*-benzotriazol-1-yl-*N*,*N*,*N′*,*N′*-tetramethyluronium hexafluorophosphate (HBTU) (Scheme 7.63) [118]. The authors have made extensive use of statistical DoE methods in order to optimize the protocol. The optimum conditions utilized equimolar amounts of carboxylic acid, amidoxime, HBTU, and 2.35 equiv of *N*,*N*-diisopropylethylamine base in *N*,*N*-dimethylformamide as the solvent. Employing microwave irradiation of this mixture for 2 min at 191 °C provided >90% conversion for the majority of the 30 library compounds studied. Similar results were achieved by Santagada *et al.* in an independent investigation of the same general transformation [119].

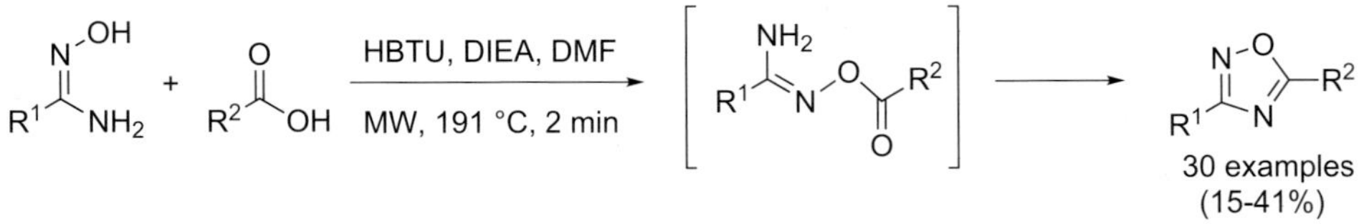

Scheme 7.63 Synthesis of 1,2,4-oxadiazoles.

Adib *et al.* have disclosed a three-component, one-pot reaction of nitriles, where the amidoxim generated *in situ* further reacted with aldehydes to give 1,2,4-oxadiazoles (Scheme 7.64) [120]. First the nitrile and hydroxylamine were reacted at 100 °C for 1 min, and then addition of aldehyde and irradiation at 150 °C for 3 min under solvent-free conditions gave oxadiazoles in excellent yields.

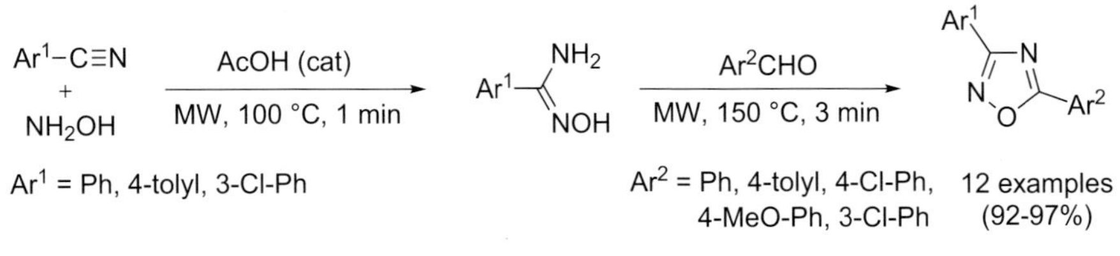

Scheme 7.64 One-pot synthesis of 1,2,4-oxadiazoles.

A related approach to 3,5-diamino-1,2,4-oxadiazoles has been described by Kurz *et al.* [121].

7.5.4
1,3,4-Oxadiazoles

In the context of preparing potential inhibitors of histone deacetylase, Vasudevan and a team from Abbott have described the cyclization of 1,2-diacyl hydrazides to 1,3,4-oxadiazoles with Burgess reagent under microwave conditions (150 °C, 15 min) (Scheme 7.65a) [122]. A different approach was chosen by Natero *et al.* who prepared 2-chloromethyl-1,3,4-oxadiazoles by treatment of acyl hydrazides with 1-chloro-2,2,2-trimethoxyethane (Scheme 7.65b) [123]. Here, the reagent was used as solvent under microwave heating to 160 °C for 5 min.

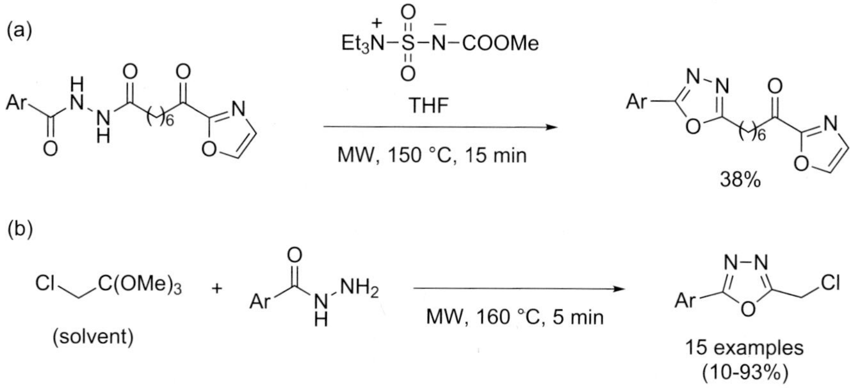

Scheme 7.65 Synthesis of 1,3,4-oxadiazoles.

Bolm and García Mancheño have obtained 1,3,4-oxadiazoles by treatment of tetrazoles with acetic anhydride at high temperatures [124]. Heterocyclic sulfonylureas such as [1,2,5]thiadiazolo[3,4-*b*]pyrazines can be obtained by condensation of the corresponding *o*-diaminohetarenes with sulfamide as demonstrated by the group of Albericio [125].

7.5.5
1,3,2-Diazaphospholidines

A method for the preparation of 1,3,2-diazaphospholidine heterocycles was described by Deng and Chen (Scheme 7.66) [126]. The authors have found that

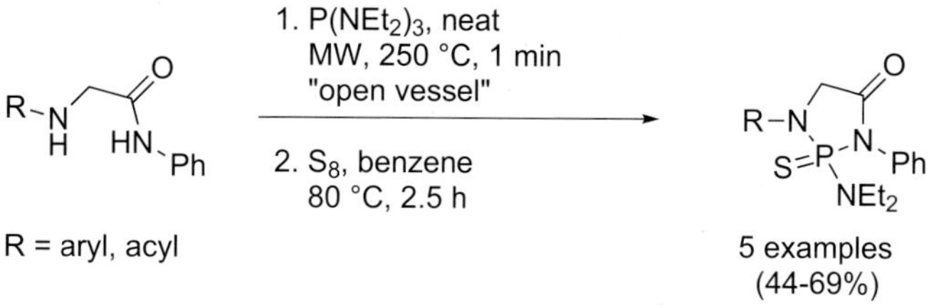

Scheme 7.66 Synthesis of 1,3,2-diazaphospholidin-4-ones.

hindered 1,2-diamino substrates such as α-amino acid amides can be treated with tris(diethylamino)phosphine as reagent/solvent under open-vessel microwave conditions at 250 °C for 1 min to furnish a trivalent phosphorus intermediate. Subsequent thiation of this intermediate with elemental sulfur in refluxing benzene provided the target 1,3,2-diazaphospholidin-4-ones in good overall yields.

7.6
Five-Membered Heterocycles with Four Heteroatoms

In 2004, three research groups have independently described the synthesis of (hetero)aryl tetrazoles by cycloaddition reactions of the corresponding aromatic nitriles with the reagent system trimethylsilyl azide/dibutyltin oxide (TMS-N_3/DBTO) (Scheme 7.67). Lukyanov and coworkers have used various nicotinonitriles as starting materials (Scheme 7.67a) [127] and have obtained moderate yields of tetrazole products by employing 4 equiv of trimethylsilyl azide and 0.3 equiv of dibutyltin oxide in anhydrous dioxane at 140 °C for 8 h. Under these conditions, full conversion was not achieved and 15–63% of the starting nitriles were recovered. Schulz *et al.* have reported the preparation of aryltetrazole boronate esters using similar reaction conditions (Scheme 7.67b) [128]. While the cycloaddition proved to be problematic with the free boronic acids, high conversions were achieved with boronic acids protected as pinacol esters (prepared *in situ*). Typically, the nitrile was reacted with 2 equiv of trimethylsilyl azide and 0.1 equiv of dibutyltin oxide in 1,2-dimethoxyethane (DME) at 150 °C for 10 min. After addition of more TMS-N_3/DBTO

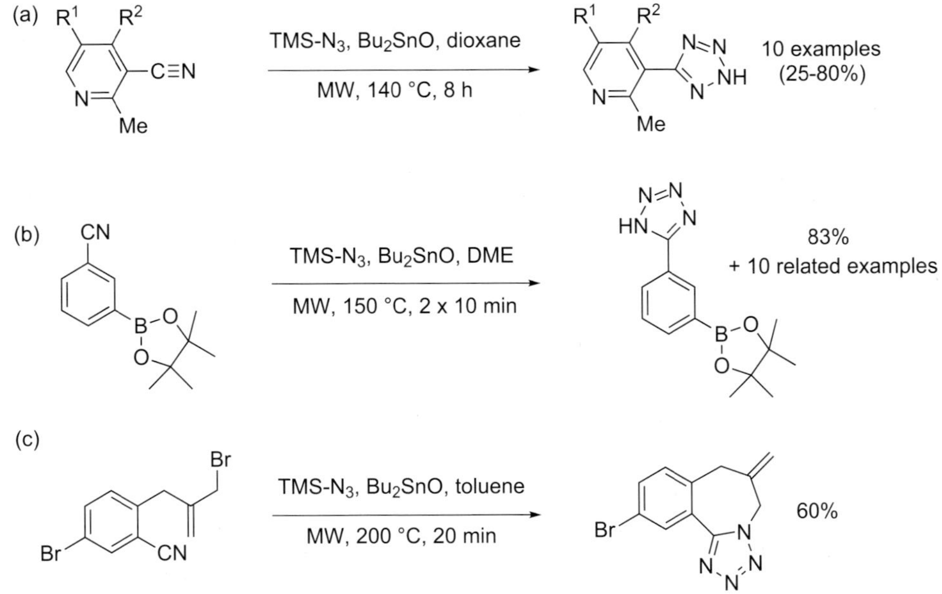

Scheme 7.67 Synthesis of tetrazoles.

reagent mixture, the vessel was reheated for 10 min at 150 °C to complete the reaction. This provided a high isolated yield of the aryltetrazole boronates that were subsequently used in microwave-assisted Suzuki couplings with aryl bromides. In a more complex variation of the same reaction, the group of Frejd has described the synthesis of fused tetrazole derivatives via a tandem cycloaddition and *N*-allylation reaction sequence (Scheme 7.67c) [129]. Here, 5 equiv of trimethylsilyl azide and 1 equiv of dibutyltin oxide were employed (toluene, 200 °C, 20 min).

5-Substituted tetrazoles have been prepared by Shie and Fang in a one-pot fashion by *in situ* generation of nitriles from aldehydes and ammonia in THF at room temperature, followed by cycloaddition with $NaN_3/ZnBr_2$ [130]. For related examples presented by the group of Richman, see Ref. [131].

7.7 Six-Membered Heterocycles with One Heteroatom

7.7.1 Piperidines

De Kimpe and coworkers investigated the microwave-assisted synthesis of stereodefined piperidines and their transformation into conformationally constrained products [132]. Novel 4-chloro-2-[1-(arylmethyl)aziridine-2-ylmethyl]-2-phenylbutyronitriles with appropriate leaving groups have been explored in 6-*exo-tet* ring closures toward intermediate bicyclic aziridinium salts to generate the desired piperidines (Scheme 7.68a). In a representative procedure, a 1 M solution of 2-[1-benzylaziridine-

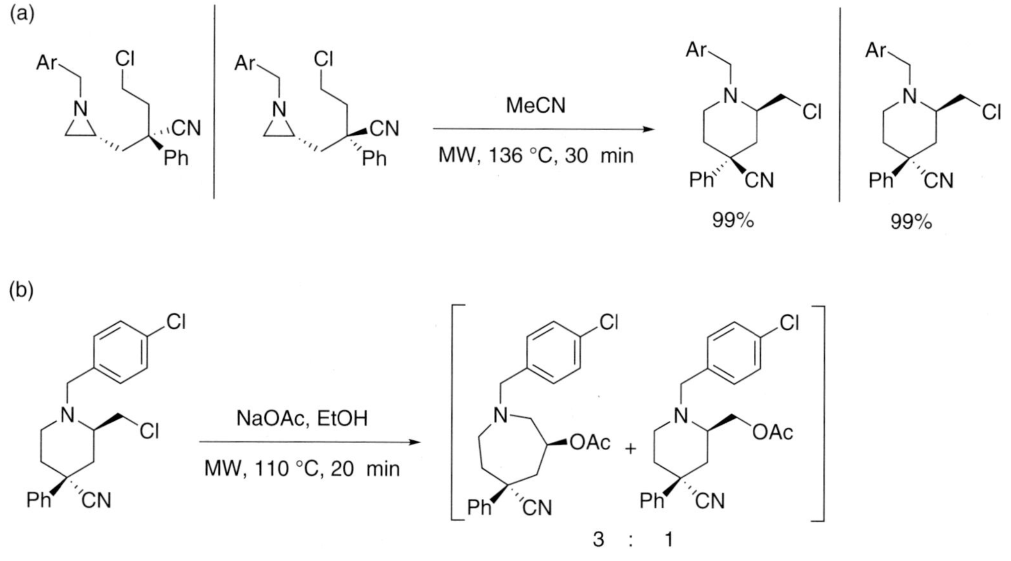

Scheme 7.68 Microwave-assisted synthesis and transformation of 2-chloromethyl-4-phenylpiperidine-4-carbonitriles.

2-ylmethyl]-4-chloro-2-phenylbutyronitrile in acetonitrile was placed in a sealed 80 mL microwave vessel and subjected under stirring to microwave heating at 136 °C for 30 min to furnish corresponding 1-benzyl-2-chloromethyl-4-phenylpiperidine-4-carbonitrile. The *cis*-piperidine can be transferred in a microwave-assisted two-step procedure to ammonium *cis*-1-(4-chlorobenzyl)-2-hydroxymethyl-4-phenylpiperidine-4-carboxylate in good yield. When the *trans*-isomer piperidine was utilized, microwave heating at 150 °C for 100 min directly leads to the *trans*-1-(4-chlorobenzyl)-2-hydroxymethyl-4-phenylpiperidine-4-carboxylic acid in high yield. Both resulting products represent a new class of biologically relevant constrained γ-amino acids.

In another microwave-assisted example, the *trans*-piperidine carbonitrile can be transferred in an unusual ring enlargement to a corresponding azepane (Scheme 7.68b). A solution of *trans*-1-(4-chlorobenzyl)-4-phenylpiperidine-4-carbonitrile in ethanol was admixed with 2 equiv NaOAc and subjected to microwave irradiation at 110 °C for 20 min to afford *trans*-6-acetoxy-1-(4-chlorobenzyl)-4-phenylazepan-4-carbonitrile together with *trans*-2-acetoxymethyl-1-(4-chlorobenzyl)-4-phenylpiperidine-4-carbonitrile in an approximately 3 : 1 ratio. Further treatment with lithium hydroxide in refluxing methanol gives the corresponding hydroxylated derivatives, demonstrating the synthetic flexibility of these valuable piperidine building blocks [132].

7.7.2
Pyridines

The Bohlmann–Rahtz synthesis of trisubstituted pyridines from β-aminocrotonates and an ethynyl ketone has found application in the preparation of a variety of heterocycles containing the substituted pyridine motif. Bagley *et al.* have developed a microwave-assisted modification of this one-pot heteroannulation method that is best conducted in dimethyl sulfoxide at 170 °C for 20 min, providing the desired pyridines in 24–94% yield (Scheme 7.69) [133, 134]. Typically, 2 equiv of the β-aminocrotonates was employed.

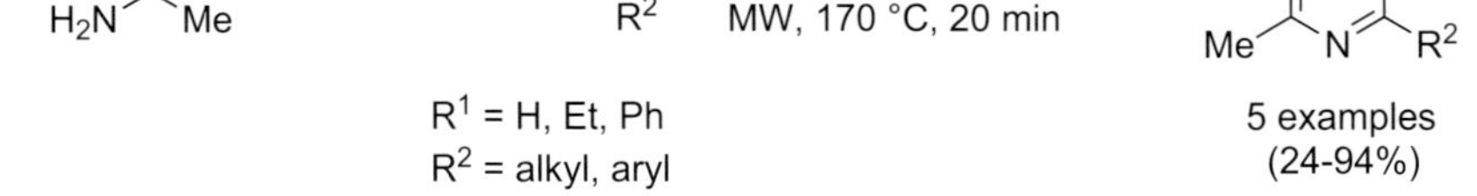

Scheme 7.69 The Bohlmann–Rahtz synthesis of pyridines.

The group of Tejedor presented their investigations on the diversity-oriented domino synthesis of nicotinic acid derivatives [135]. These functionalized pyridines constitute a variety of important biologically and pharmaceutically relevant molecules. The established microwave protocol deals with a simple one-pot domino reaction of propargyl vinyl ethers with primary amines to furnish substituted nicotinates with valuable diversity points. In a typical procedure, a solution of propargyl vinyl ether with 1.1 equiv methoxyamine hydrochloride and 55 mol% sodium acetate in ethanol or isopropanol was microwave heated at 100 °C for 60 min to furnish the desired pyridine derivatives (Scheme 7.70). The developed method tolerates a diverse range of substitution patterns, which could be introduced in the initial preparation sequence of the propargyl ethers, with aromatic substituents showing higher efficiency. Thus, the authors introduced a practical approach to rapidly generate libraries of valuable scaffolds for drug discovery.

Scheme 7.70 Microwave-assisted diversity-oriented domino synthesis of nicotine acid derivatives.

Another well-known method for the preparation of heterocycles is the Hantzsch dihydropyridine synthesis. In 2001, Öhberg and Westman presented a microwave-assisted Hantzsch dihydropyridine synthesis that allowed the rapid preparation of heterocycles of this type in a multicomponent, one-pot fashion (Scheme 7.71a) [136]. Thus, suitable aliphatic or (hetero)aromatic aldehydes were reacted with 5 equiv of a β-ketoester and 4 equiv of concentrated aqueous ammonia that was used both as a reagent and as a solvent. Best yields were obtained by exposing the reaction mixture to microwave heating at 140–150 °C for 10–15 min. In order to prepare a diverse set of products, six different aldehydes and four different β-ketoesters or 1,3-dicarbonyl substrates were used. All 24 compounds were formed in moderate to good yield. A somewhat related multicomponent transformation for the synthesis of

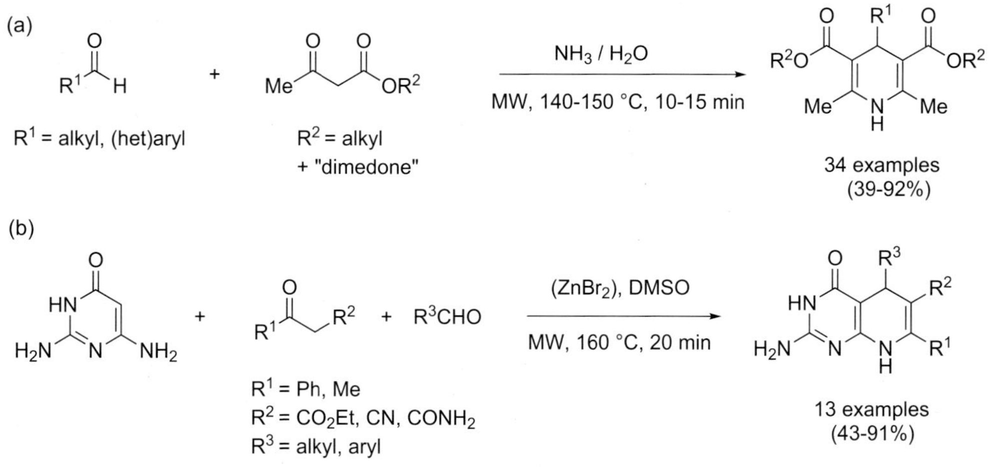

Scheme 7.71 Hantzsch dihydropyridine synthesis.

5-deaza-5,8-dihydropterins (pyrido[2,3-*d*]pyrimidines) was described by Bagley and Singh (Scheme 7.71b) [137]. Heating of 2,6-diaminopyrimidin-4-one with an aldehyde (2 equiv) and a suitable CH-acidic carbonyl compound (2 equiv) in dimethyl sulfoxide in the presence of 20 mol% of zinc bromide as Lewis acid at 160 °C for 20 min provided good yields of the desired fused dihydropyridines.

In a recent report, Albrecht *et al.* presented a microwave protocol to synthesize pharmacologically important quinolinone derivatives from anilines and various ester-containing building blocks [138]. In a representative one-pot procedure, *o*-anisidine together with 1.1 equiv dimethyl acetylene dicarboxylate and diphenyl ether was microwave heated in a sealed 10 mL tube at 120 °C for 5 min and for additional 15 min at 250 °C to obtain the expected methyl 8-methoxy-4-oxo-1,4-dihydroquinoline-2-carboxylate (Scheme 7.72a). Utilizing diethyl methoxymethylene malonate or ethyl 2-oxocyclohexanecarboxylate as condensation substrate afforded the desired quinolinone derivatives in acceptable yield as well. When trying to scale up the reaction, the authors found that open-vessel conditions were required in order to allow the volatile side products being removed from the reaction equilibrium. Nevertheless, the observed yields significantly dropped when 10-fold scale-up was performed. To further explore the developed method, also the synthesis of corresponding 4-chloroquinolones was performed. In a one-pot two-step sequence, the initially prepared *N*-functionalized *o*-anisidine precursor was microwave heated at 120 °C for 5 min, then phosphoroxytrichloride was added and the mixture was heated for another 5 min at 150 °C. The resulting quinoline can be employed in further scaffold decoration reactions.

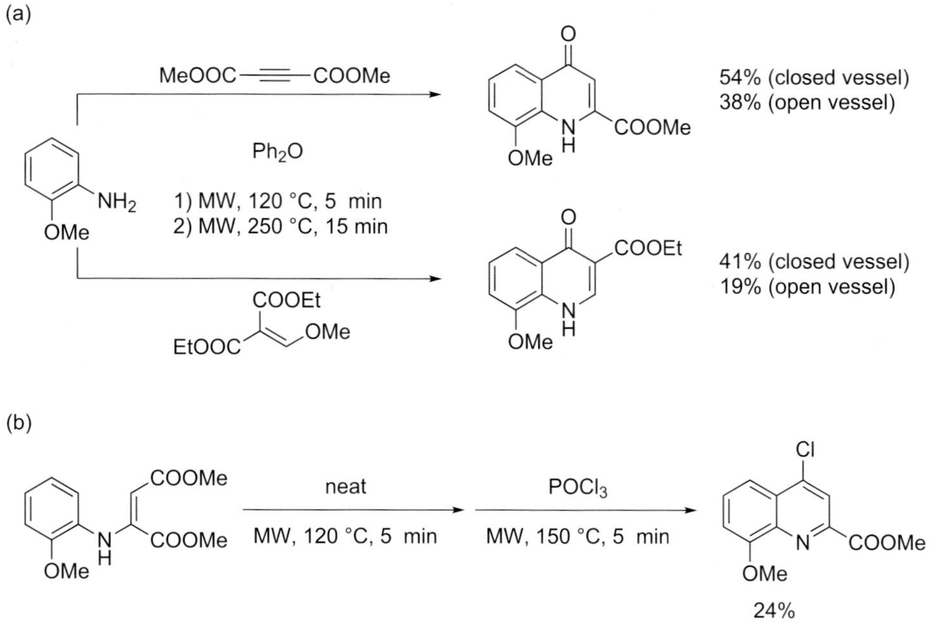

Scheme 7.72 Microwave-assisted one-pot synthesis of quinolinones.

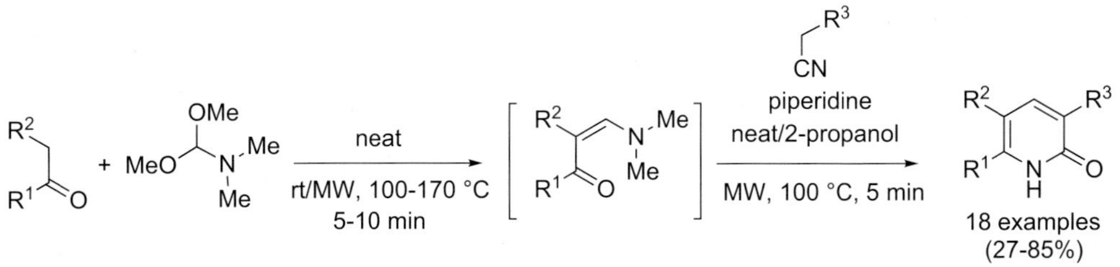

Scheme 7.73 Three-component synthesis of fused pyridones.

The Kappe group has described a multicomponent, one-pot, two-step pathway to 3,5,6-substituted 2-pyridones (Scheme 7.73) [139]. In the first step, equimolar mixtures of a CH-acidic carbonyl compound and *N,N*-dimethylformamide dimethyl acetal were reacted to form the corresponding enamines, either at room temperature or under microwave conditions. After addition of a methylene active nitrile (1 equiv), 2-propanol as solvent, and catalytic amounts of piperidine base, the reaction mixture was heated to 100 °C for 5 min under microwave conditions. In most cases, the desired heterocyclic product precipitated directly after cooling of the reaction mixture and could be collected by filtration.

An alternative method to access 3-bromo-2(1*H*)-pyridinones involving the NBS-mediated cyclization of dienamino esters was reported by Vounatsos in 2006 [140].

In Scheme 7.74, the multistep synthesis of 2,3-dihydro-4-pyridones is highlighted [141]. The pathway described by Panunzio *et al.* starts from a dioxin-4-one precursor that is reacted with 2 equiv of benzyl alcohol under solvent-free microwave conditions to furnish the corresponding β-diketo benzyl esters. Subsequent treatment with 1 equiv of *N,N*-dimethylformamide dimethyl acetal, again under solvent-free conditions, produces an enamine, which is then cyclized with an amine building block (1.1 equiv) to produce the desired 4-pyridinone products. All microwave protocols were conducted under open-vessel conditions using power control.

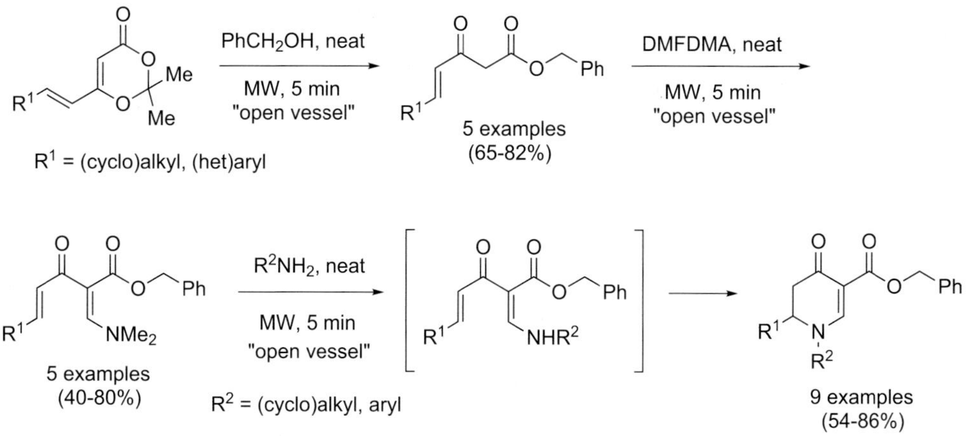

Scheme 7.74 Multistep preparation of 2,3-dihydropyridin-4-ones.

Many of the traditional condensation reactions leading to heterocycles require high temperatures and conventional reaction conditions very often involve heating of the reactants in an oil, metal, or sand bath for many hours or even days. One example published by Kappe and coworkers is illustrated in Scheme 7.75a, namely, the formation of 4-hydroxyquinolin-2-(1*H*)-ones from anilines and malonic esters [142]. The corresponding conventional thermal protocol involves heating of the two components in equimolar amounts in an oil bath at 220–300 °C for several hours (without solvent), whereas similar high yields can be obtained by microwave heating at 250 °C for 10 min. Here, it was essential to use open-vessel technology, since 2 equiv of a volatile by-product (ethanol) are formed that under normal (atmospheric pressure) conditions is simply distilled off and therefore removed from the equilibrium (see Section 4.3). Similar results were previously obtained by Lange *et al.* (Scheme 7.75b) [143].

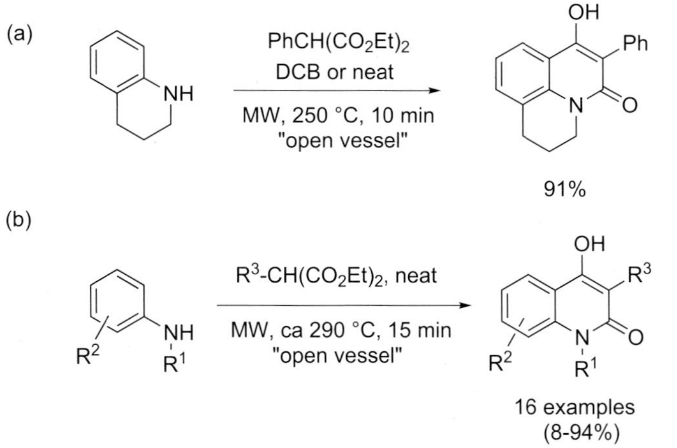

Scheme 7.75 Formation of 4-hydroxyquinolin-2-(1*H*)-ones from anilines and malonic esters.

The synthesis of polysubstituted quinolines applying the Friedländer condensation was disclosed by the group of Wang (Scheme 7.76) [144]. By conducting the reaction under solvent-free conditions with *p*-toluenesulfonic acid as catalyst, the quinoline products could be obtained in excellent yields through a simple workup

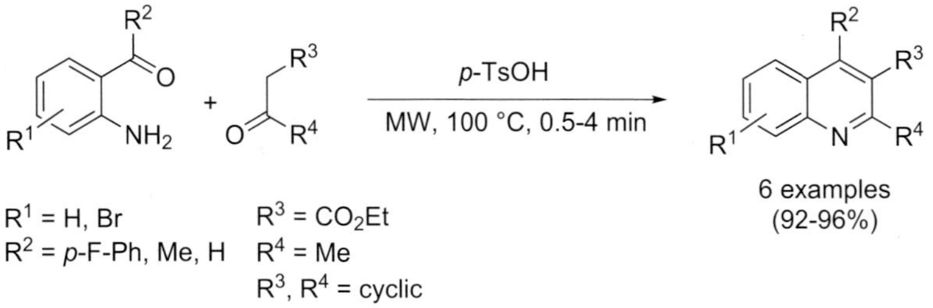

Scheme 7.76 Quinolines via Friedländer annulation.

procedure. Identical results were achieved performing the reactions under conventional heating at the same temperature, but with prolonged reaction times.

Alternatively, quinolines have been prepared by condensation of 2-aminoaryl ketones with phenylacetylene in the presence of $In(OTf)_3$ [145].

The group of Tu has reported on the synthesis of a series of pyrido[2,3-*d*] pyrimidine-4,7-diones (Scheme 7.77) [146]. By reacting isoxazoles **14** with 2,6-diaminopyridinone **15** or naphthalene-2-amine, pyridopyrimidines **16** and **17** with a hydroxyiminoethyl substituent at the 6-position are obtained, whereas 6-benzamide functionalized products **19** are generated by the reaction of oxazolones **18** with 2,6-diaminopyridinone. The authors propose a Michael addition, cyclization, and ring-opening reaction sequence, where the acetic acid acts as both solvent and catalyst.

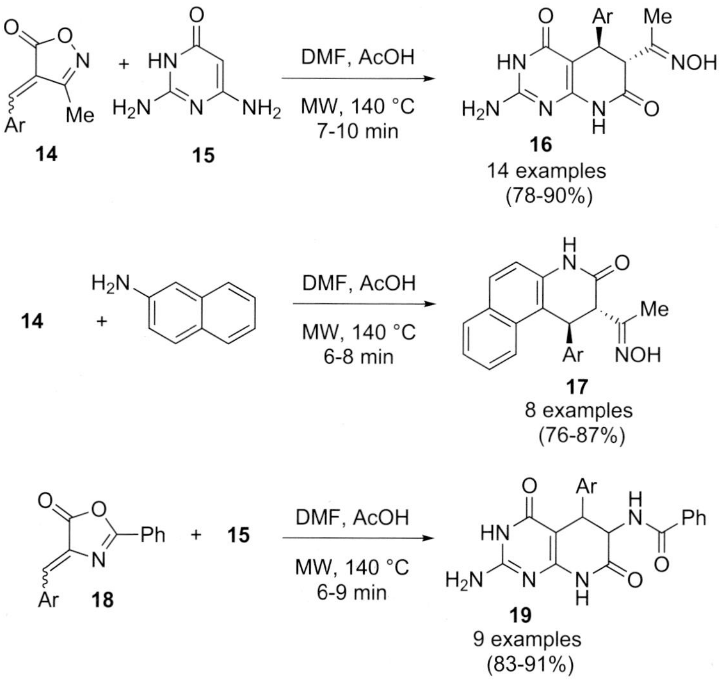

Scheme 7.77 Synthesis of pyrido[2,3-*d*]pyrimidinedione derivatives.

An efficient multistep synthesis of biologically active 4-aryl-3-alkenyl-substituted quinolin-2-(1*H*)-ones has been demonstrated by the Kappe group (Scheme 7.78) [147]. All the synthetic transformations (six steps) required for the synthesis of the desired target quinolin-2-(1*H*)-ones have been carried out under controlled microwave heating. The key steps in the synthesis include a sequential Pd-catalyzed Suzuki/ Heck scaffold decoration method to allow introduction of substituents late in the synthetic scheme and facilitate preparation of 2(1*H*)-quinolinone analogs with a greater diversity for potentially generating combinatorial libraries.

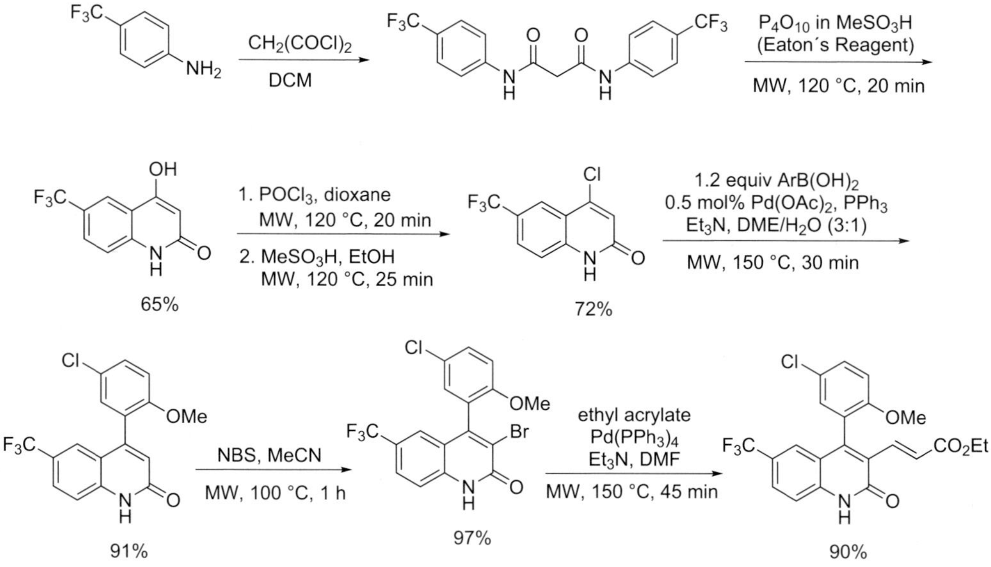

Scheme 7.78 Multistep synthesis of functionalized 4-arylquinolin-2(1*H*)-ones.

Related cyclizations of activated malonic esters and anilines leading to 4-hydroxy-2(1*H*)-quinolones have been described by Rivkin [148]. The same approach was also used for the preparation of the corresponding aza analogs starting from α-aminoazines [149], or bis(pentafluorophenyl)imidodicarbonate [150].

The group of Gorobets reported on a rapid one-pot three-step synthesis of pharmaceutically attractive 3-quinolinecarbonitriles [151]. The reaction sequence involves 1,3-cyclohexanediones and dimethylformamide dimethyl acetal, forming corresponding enamines, which react with *N*-substituted cyanoacetamides to construct the desired 2-pyridone ring via its piperidinium salts to be cyclized under microwave irradiation (Scheme 7.79). Microwave heating at 120 °C for 10 min furnished the desired products in good yield and high purity. This facile procedure has proven effective for a variety of substrates to enlarge the diversity of synthetically available potentially active 2-pyridone derivatives [151].

The Pictet–Spengler reaction is another important reaction principle for the generation of (fused) pyridine systems. Several groups have independently reported on microwave-assisted Lewis acid or ionic liquid-mediated Pictet–Spengler cyclizations involving iminium ion intermediates. Srinivasan and Ganesan have described a Pictet–Spengler approach to the tetrahydro-β-carboline ring system, utilizing the one-pot condensation of tryptophan methyl ester or tryptamine with aliphatic and aromatic aldehydes (Scheme 7.80) [152]. The most active catalytic system, in particular for transformations involving tryptophan, was found to be a combination of 10 mol% of ytterbium(III) triflate and 50 mol% of the ionic liquid 1-butyl-3-methylimidazolium chloroaluminate salt ([bmim]Cl-$AlCl_3$). Employing 1.2 equiv of the aldehyde building block, very high yields of products were obtained under microwave irradiation in dichloromethane at 100–120 °C for 30–60 min.

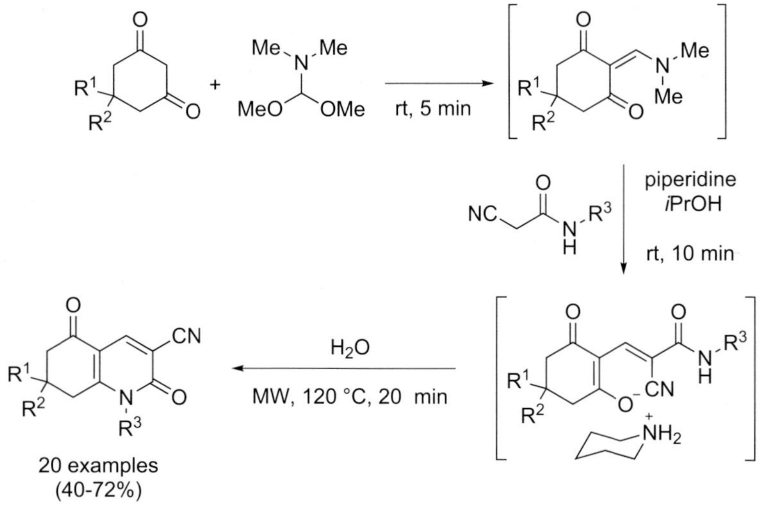

Scheme 7.79 Three-step one-pot microwave-assisted synthesis of 2,5-dioxo-1,2,5,6,7,8-hexahydro-3-quinolinecarbonitriles.

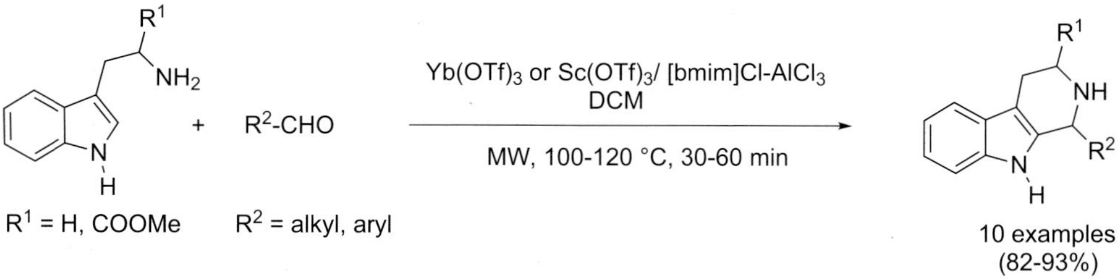

Scheme 7.80 Lewis acid-catalyzed Pictet–Spengler reactions.

In a closely related approach, Yen and Chu have reported the preparation of tetrahydro-β-carbolinediketopiperazines employing a three-step Pictet–Spengler, Schotten–Baumann, and intramolecular ester amidation step (Scheme 7.81) [153]. Throughout the synthesis, the ionic liquid 1-butyl-2,3-dimethylimidazolium hexafluorophosphate (bdmimPF_6) was employed. In a typical experiment, (*S*)-tryptophan methyl ester was dissolved in a 1 : 1 mixture of the ionic liquid and tetrahydrofuran containing 10% (v/v) of trifluoroacetic acid. After 25 s of microwave irradiation at 60 °C, the mixture was evaporated to dryness and after addition of 5 equiv of *N*,*N*-diisopropylethylamine and (*S*)-proline acid chloride in bdmimPF_6/THF was stirred at room temperature for 3 min. Finally, after *in situ* deprotection with 20% piperidine (v/v) in bdmimPF_6/THF at 60 °C for 1 min in the microwave reactor, the target compounds were obtained as mixtures of diastereomers.

Grieco and coworkers independently described the same type of Pictet–Spengler cyclization reactions involving tryptophan methyl ester and aldehydes, albeit using methanol as solvent and hydrochloric acid as a catalyst (microwave irradiation, 50 °C, 20–50 min) [154]. Moderate to good product yields were obtained.

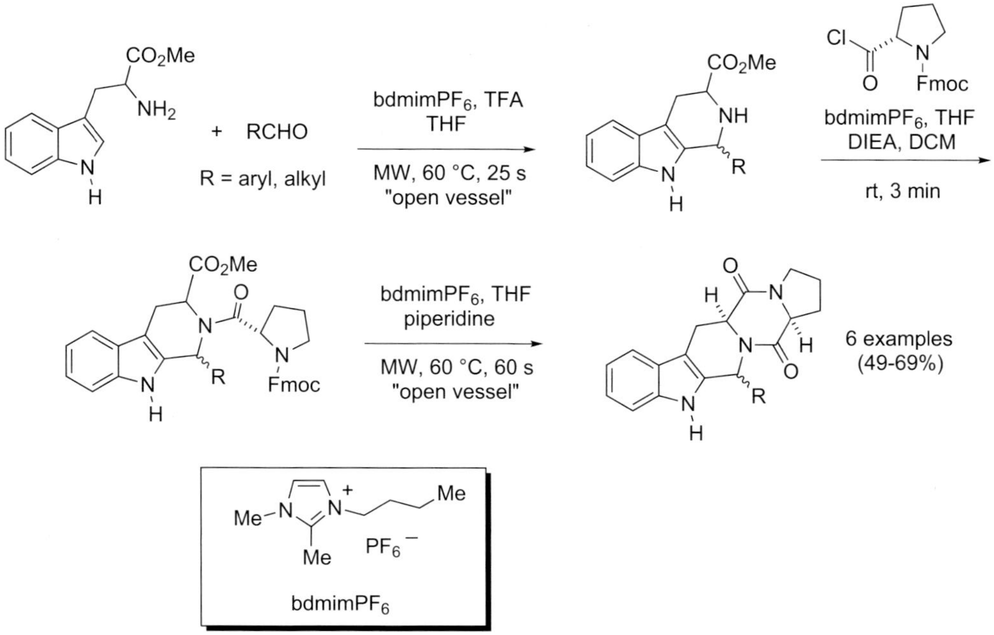

Scheme 7.81 Ionic liquid-mediated preparation of tetrahydro-β-carbolinediketopiperazines.

Yen and Chu subsequently also disclosed a related Pictet–Spengler reaction involving tryptophan and ketones for the preparation of 1,1-disubstituted indole alkaloids [155]. In the approach shown in Scheme 7.82, tryptophan was reacted with numerous ketones (12 equiv) in toluene in the presence of 10 mol% of trifluoroacetic acid catalyst. Using microwave irradiation at 60 °C under open-vessel conditions, the desired products were obtained in high yields. Compared to transformations carried out at room temperature, reaction times were reduced typically from days to minutes.

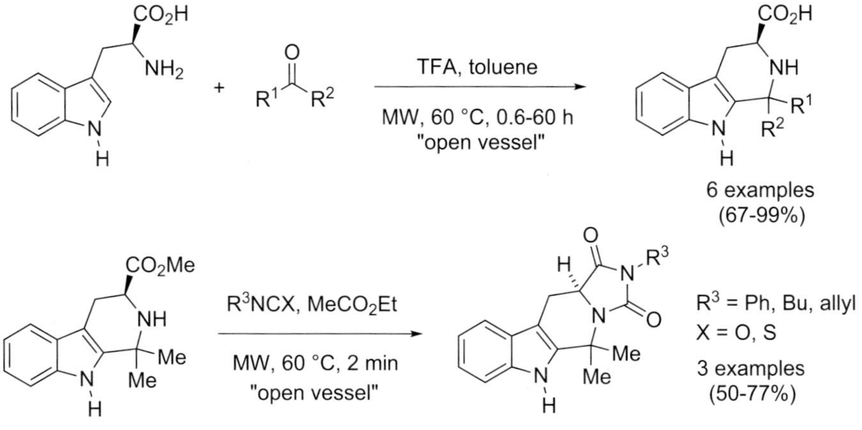

Scheme 7.82 Trifluoroacetic acid-mediated Pictet–Spengler reactions.

Subsequent treatment with isocyanates or isothiocyanates led to tetrahydro-β-carboline hydantoins.

Different types of N-acyliminium ion-based cyclizations that are assisted by microwave irradiation are highlighted in Scheme 7.83 [156].

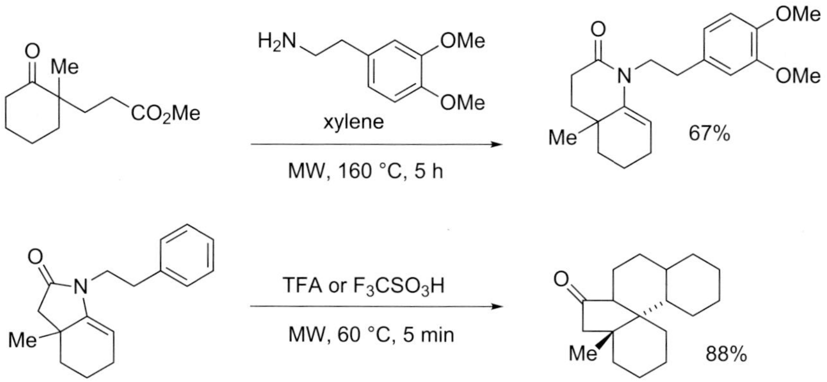

Scheme 7.83 N-Acyliminium ion-based cyclizations.

Preparation of fused pyridines by Pictet–Spengler heterocyclization using DMSO at high temperatures as a formaldehyde source has been reported by the group of Besson [157], whereas Almqvist and coworkers studied the condensation of thiazolidine precursors with Meldrum's acid derivatives [158]. A one-pot synthesis of 2,3-dihydro-1*H*-pyrrolo[3,2-*c*]quinolines from substituted 2-iodoanilines and 2,3-dihydro-1*H*-pyrrole using Pd catalysis was reported by Hu in 2008 [159]. For additional preparations of pyridine and quinoline heterocycles, see Refs. [160–162].

As reported by Padwa and coworkers, exposing a suitable nitrofuran precursor to microwave irradiation in 1-methyl-2-pyrrolidone (NMP) in the presence of 2,6-lutidine catalyst provided 1,4-dihydro-2*H*-benzo[4,5]furo[2,3-*c*]pyridin-3-one as the major product in 36% yield (Scheme 7.84) [163]. In contrast, under thermal conditions, removal of the *tert*-butyl group was observed as the major reaction product.

The Skraup cyclization is another reaction principle that provides a rapid entry into the quinoline moiety. Theoclitou and Robinson have published the preparation of a 44-member library based on the 2,2,4-substituted 1,2-dihydroquinoline scaffold, using the Lewis acid-catalyzed cyclization of substituted anilines or amino heterocycles with the corresponding ketones (Scheme 7.85) [164]. Best results were obtained using 10 mol% of scandium(III) triflate as a catalyst in acetonitrile as solvent at 140–150 °C. A variety of anilines and unsymmetrical ketones (5 equiv) can be used to expand the scope of the substituents, particularly five-membered ring heterocycles, to yield a variety of fused dihydropyridines.

The Friedländer reaction is the acid- or base-catalyzed condensation of an *ortho*-acylaniline with an enolizable aldehyde or ketone. Hénichart and coworkers have described microwave-assisted Friedländer reactions for the synthesis of indolizino [1,2-*b*]quinolines, the constitutive heterocyclic core of camptothecin-type antitumor

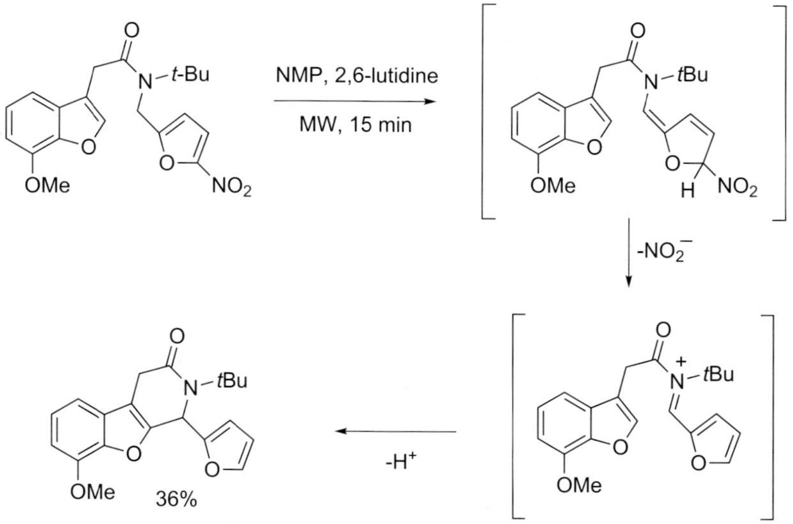

Scheme 7.84 *N*-Acyliminium ion-based cyclizations leading to fused pyridones.

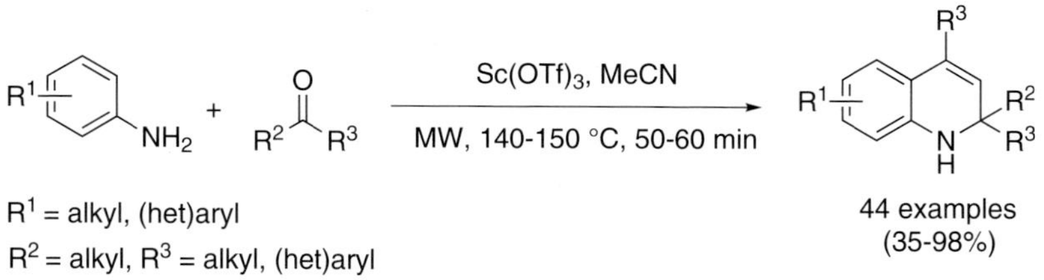

Scheme 7.85 Skraup dihydroquinoline synthesis.

agents (Scheme 7.86) [165]. The process involved the condensation of *ortho*-aminobenzaldehydes (or imines) with tetrahydroindolizinediones to form the quinoline structures. Employing 1.25 equiv of the aldehyde or imine component in acetic acid as solvent provided the desired target compounds in 57–91% yield within 15 min. These transformations were carried out under open-vessel conditions at the reflux temperature of the acetic acid solvent.

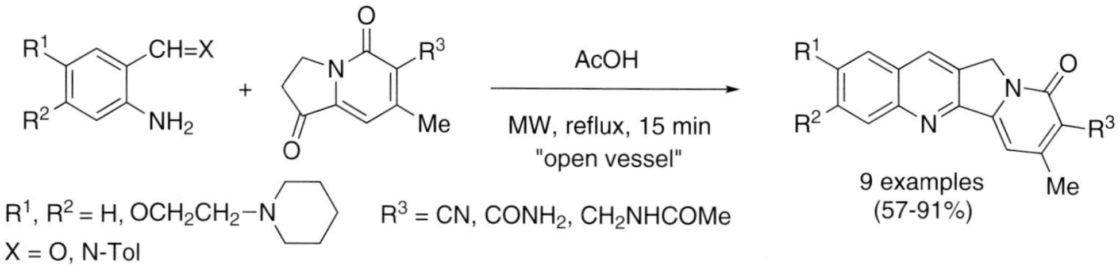

Scheme 7.86 Friedländer synthesis of indolizino[1,2-*b*]quinolines.

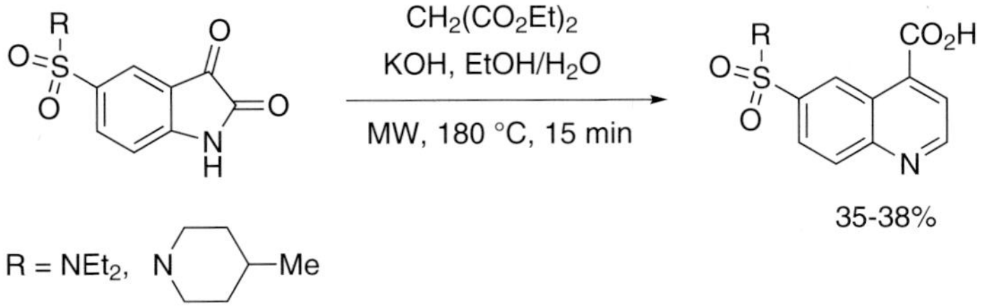

Scheme 7.87 Synthesis of 6-sulfamoylquinoline-4-carboxylic acids under Pfitzinger conditions.

The Pfitzinger reaction of isatins with α-methylene carbonyl compounds is widely used for the synthesis of substituted quinoline-4-carboxylic acids. Ivachtchenko *et al.* have recently reported on the Pfitzinger reaction in a series of 5-sulfamoylisatins (Scheme 7.87) [166]. Treatment of 5-sulfamoylisatins with diethyl malonate under basic conditions (ethanol/water) surprisingly led to 6-sulfamoylquinoline-4-carboxylic acids, instead of the anticipated Pfitzinger products, namely, the 2-oxo-1,2-dihydroquinoline-4-carboxylic acids (not shown). A careful mechanistic investigation involving ^{13}C-labeled ethanol demonstrated that the key step in this process is the reaction of *in situ* generated acetaldehyde (from the solvent) with the hydrolytically cleaved isatin ring. Moderate yields of the quinoline products were isolated after 15 min of microwave irradiation of the isatin precursors in 2.5 N potassium hydroxide solution at 180 °C.

The generation of a library of 2-aminoquinoline derivatives was disclosed by Wilson *et al.* (Scheme 7.88) [167]. The process involved microwave irradiation of the secondary amine and aldehyde components to form an enamine (1,2-dichloroethane, 180 °C, 3 min), followed by addition of the resulting crude enamine to a 2-azidobenzophenone derivative (0.8 equiv) and further microwave heating for 7 min at the same temperature.

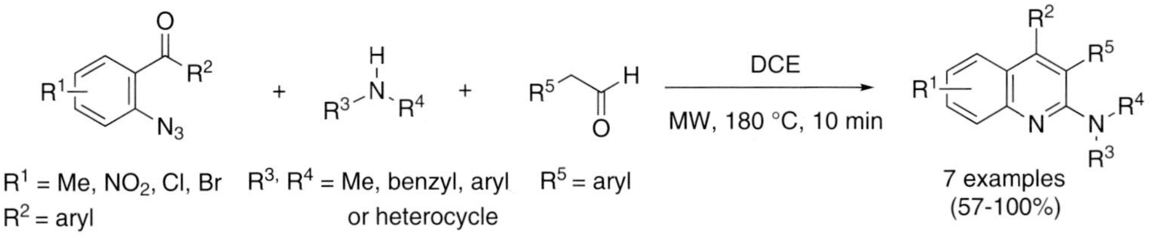

Scheme 7.88 Synthesis of a 2-aminoquinoline library via three-component reaction.

A general hetero-Diels–Alder cycloaddition of fulvenes with azadienes to furnish tetrahydro[1]pyridines has been described by Hong *et al.* (Scheme 7.89) [168]. In the event, a solution of the azadiene and fulvene (1.2 equiv) precursors in chlorobenzene was heated under open-vessel microwave irradiation for 30 min at 125 °C to provide the target compounds in excellent yields and in exclusive regio- and diastereoselectivity. Performing the reactions under conventional conditions or under microwave irradiation in different solvents provided significantly reduced yields.

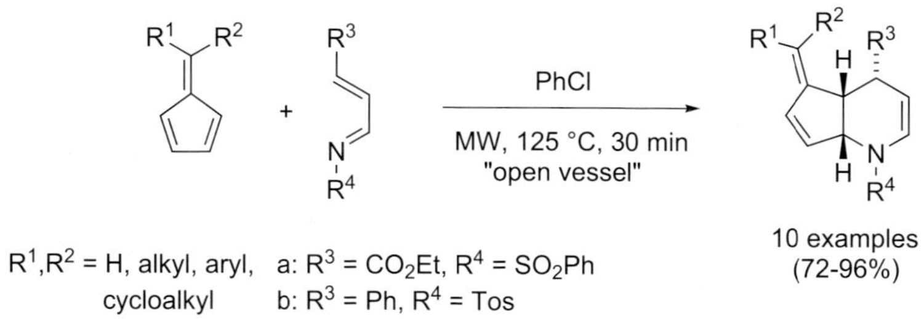

Scheme 7.89 Azadiene Diels–Alder cycloaddition of fulvenes.

A multicomponent assembly of pyrido-fused tetrahydroquinolines was accomplished by Lavilla and coworkers in a one-pot process from the interaction of dihydroazines, aldehydes, and anilines (Scheme 7.90) [169]. The reactions were conducted with 20 mol% of scandium(III) triflate as a catalyst in dry acetonitrile in the presence of 4 Å molecular sieves employing equimolar amounts of building blocks. This protocol provided the cycloadducts shown in Scheme 7.90 in 80% yield as a 2 : 1 mixture of diastereoisomers using microwave irradiation at 80 °C for 5 min. The same reaction at room temperature requires 12 h to reach completion.

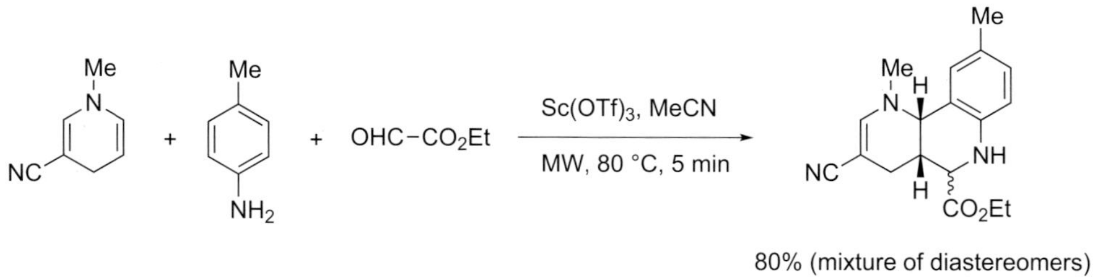

Scheme 7.90 Multicomponent synthesis of pyrido-fused tetrahydroquinolines.

Moody and coworkers have employed a "biomimetic" hetero-Diels–Alder–aromatization sequence for the construction of the 2,3-dithiazolepyridine core unit in amythiamicin D and related thiopeptide antibiotics (Scheme 7.91a) [170]. The key cycloaddition reaction between the azadiene and enamine component was carried out by microwave irradiation at 120 °C for 12 h and gave the required 2,3,6-tris(thiazolyl)pyridine intermediate in a moderate 33% yield. Coupling of the remaining building blocks then completed the first total synthesis of the thiopeptide antibiotic amythiamicin D. In previous work, the authors established the concept of this biomimetic approach in a series of model structures (Scheme 7.91b) [171]. Here, performing the reaction in 1,2-dichlorobenzene typically gave the best results.

7.7.3 Pyrans

Both natural and unnatural compounds with a 2*H*,5*H*-pyrano[4,3-*b*]pyran-5-one skeleton are of interest to medicinal chemistry. Several natural products like the

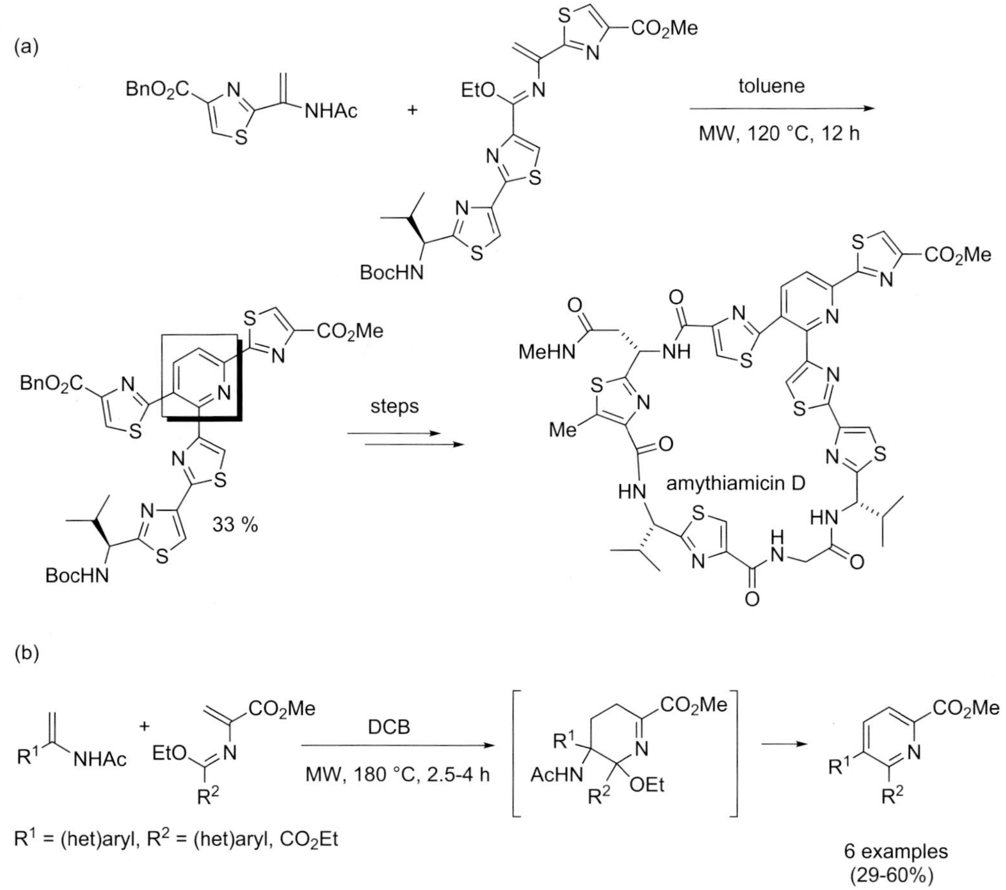

Scheme 7.91 Biomimetic hetero-Diels–Alder–aromatization sequences.

pyripyropenes incorporate this bicyclic ring system. The group of Beifuss has described an efficient microwave-promoted domino synthesis of the 2*H*,5*H*-pyrano[4,3-*b*]pyran-5-one skeleton by condensation of α,β-unsaturated aldehydes with 4-hydroxy-6-methyl-2*H*-pyran-2-one (Scheme 7.92) [172]. It is assumed that in the presence of an amino acid catalyst a Knoevenagel condensation occurs first, which is then followed by a 6π-electron electrocyclization to the pyran ring. While the conventional thermal protocol required up to 25 h of reaction time (refluxing ethyl acetate), the microwave method employing 1.1 equiv of the pyran-2-one and 0.5 equiv each of β-alanine catalyst and calcium sulfate as dehydrating agent could be run much more efficiently. Best yields for a variety of diverse aldehyde building blocks were obtained at 110 °C within 10–90 min, although in some cases somewhat higher yields were achieved by the conventional thermal method (80 °C).

The sodium bromide-catalyzed three-component cyclocondensation of aryl aldehydes, CH-acidic nitriles, and dimedone under solvent-free conditions was studied

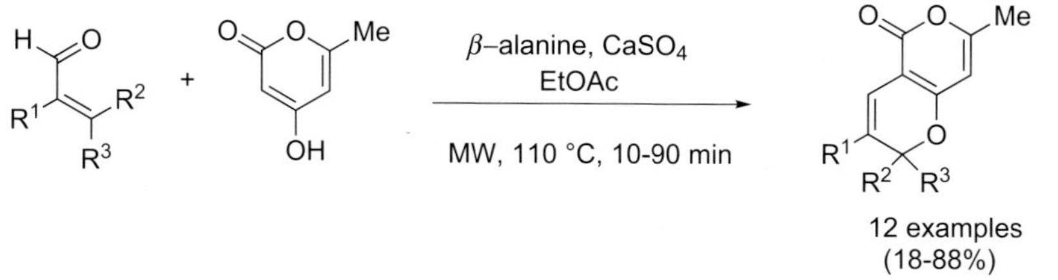

Scheme 7.92 Domino Knoevenagel condensation/6π-electron electrocyclization for the synthesis of 2*H*,5*H*-pyrano[4,3-*b*]pyran-5-ones.

by Devi and Bhuyan (Scheme 7.93) [173]. Utilizing equimolar amounts of building blocks and 20 mol% of sodium bromide as catalyst, microwave irradiation for 10 min at 70 °C produced the anticipated tetrahydrobenzo[*b*]pyrans in good to excellent yields.

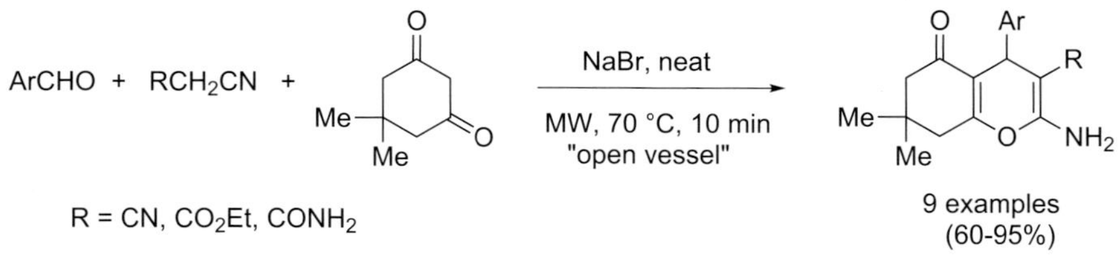

Scheme 7.93 Multicomponent condensation for the synthesis of tetrahydrobenzo[*b*]pyrans.

Utilizing the microwave protocol shown in Scheme 7.94, Kabalka and Mereddy were able to synthesize a variety of functionalized flavones and chromones in very high yields by the cyclization of 1-(2-hydroxyaryl)-1,3-propanediones [174].

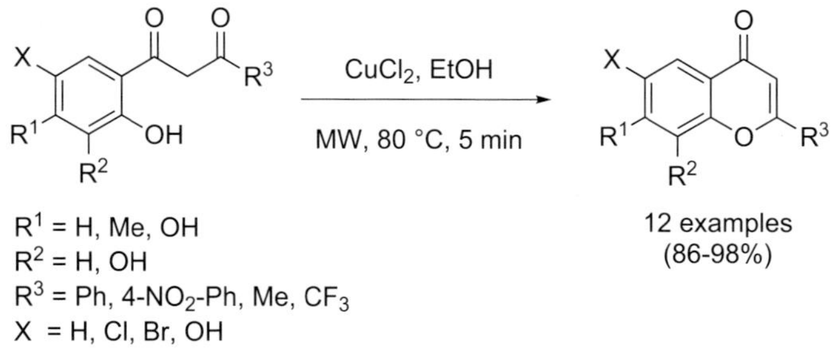

Scheme 7.94 Synthesis of flavones and chromones.

A related flavone synthesis by Seijas employs solvent-free conditions [175]. Coumarin libraries have been generated from activated phenols and β-ketoesters using a similar approach [176]. Fused α-pyrones have been obtained by Lewis acid-catalyzed regiocontrolled 6-*endo-dig* cyclization of 2-(2-arylethynyl)heteroaryl esters [177].

A pyran ring is formed in the intramolecular Diels–Alder cycloaddition of olefin-tethered enantiopure (1*S*,2*R*)-1,2-dihydroxycyclohexa-3,5-diene-1-carboxylic acid derivatives (derived from the biodihydroxylation of benzoic acid). For the three cases displayed in Scheme 7.95, Mihovilovic *et al.* found that moderate to high yields of the desired cycloadducts could be obtained by exposing a solution of the precursor to microwave irradiation at 135–210 °C for extended periods of time [178]. In all cases, the yields obtained in the microwave protocol were higher than in the conventional run (refluxing toluene, 3–7 days).

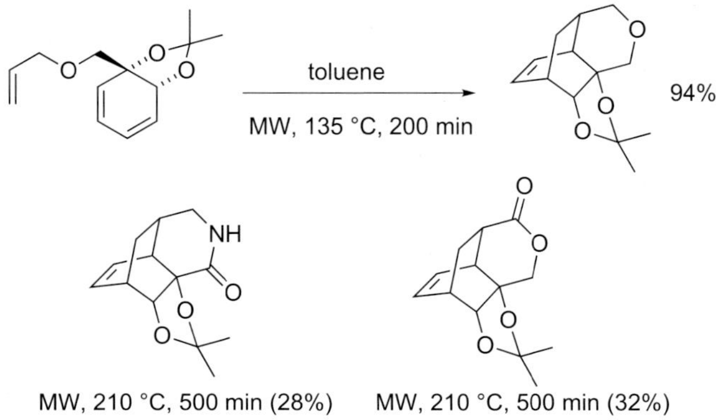

Scheme 7.95 Intramolecular Diels–Alder cyclization of biodihydroxylated benzoic acid derivatives.

In 2009, Brimble and coworkers investigated the synthesis of 6,6-bisbenzannulated spiroketals by double intramolecular hetero-Michael addition (DIHMA) [179]. Spiroketals are present in numerous bioactive natural products, and therefore considered privileged scaffolds in drug discovery; especially, the benzannulated derivatives of the rubromycin family have attracted considerable interest. The authors proposed a mechanism for the formation of the desired spiro compounds by DIHMA of dihydroxy ynone intermediates generated from corresponding benzopyrones. Microwave irradiation proved beneficial to achieve this transformation (Scheme 7.96). In the optimized procedure, a solution of benzopyrone in dichloromethane was added to 216 equiv dry ground potassium carbonate. The solvent was removed *in vacuo* and the residue was subjected to microwave irradiation at 120 °C for 30 min to furnish the desired 2,2′-spirobi[chroman]-4-ones in good yield. This efficient method allows the facile generation of a focused library of bisbenzannulated spiroketals for biological evaluation [179].

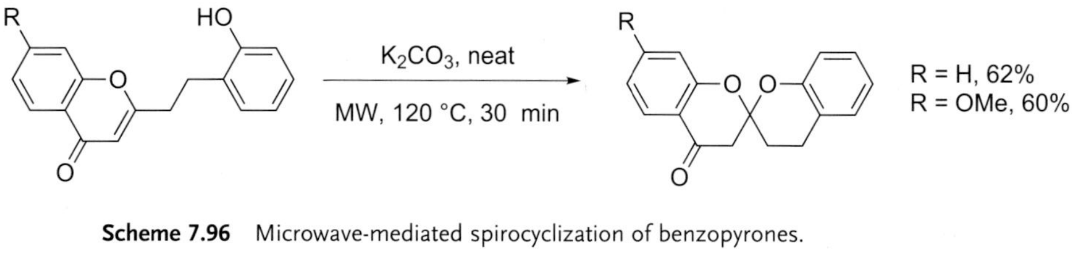

Scheme 7.96 Microwave-mediated spirocyclization of benzopyrones.

7.8
Six-Membered Heterocycles with Two Heteroatoms

7.8.1
Pyrimidines

2-Substituted pyrimidines can be obtained in a tandem oxidation–heteroannulation fashion from propargylic alcohols and amidines. As demonstrated by Bagley *et al.*, both benzamidine and acetamidine react with 1-phenyl-2-propyn-1-ol in the presence of either manganese dioxide or *ortho*-iodoxybenzoic acid (IBX) as *in situ* oxidation agent to provide the corresponding pyrimidines (Scheme 7.97a) [180]. Up to 84% yield was obtained exposing the alcohol, the amidine hydrochloride salt (1.2 equiv), and sodium carbonate as base (2.4 equiv) to an excess of the oxidant in acetonitrile at 120 °C for 40 min. In subsequent work, the same group has expanded on this protocol and has shown that a more diverse set of substituted pyrimidines can be obtained using similar reaction conditions by directly starting from the corresponding alkynones (Scheme 7.97b) [181, 182]. This method also allows the preparation of 2-amino-substituted pyrimidines using guanidines as starting materials.

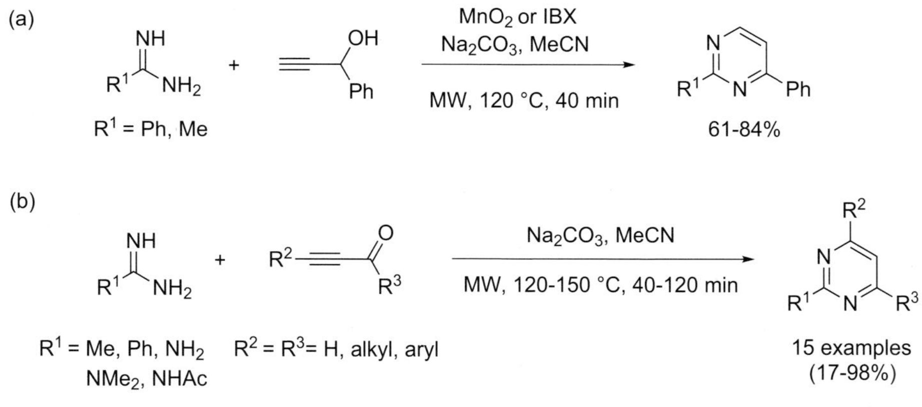

Scheme 7.97 Synthesis of 2,4,6-trisubstituted pyrimidines from amidine and alkyne building blocks.

In a related, though more recent work, the Bagley group introduced an effective procedure for the synthesis of solvatochromic 2,4,6-triarylpyrimidines from propargylic alcohols utilizing barium manganate as oxidant [183]. The target compounds have attracted interest due to their fluorescence properties, which make them potential intermediates or precursors in the development of optoelectronic devices. The authors could demonstrate that barium manganate was superior in a microwave-assisted tandem oxidation/cyclocondensation than the usually applied manganese dioxide. In a typical procedure, propargylic alcohol together with 1 equiv benzamidine and 3 equiv barium manganate suspended in ethanol/acetic acid (5 : 1) was irradiated in a sealed microwave vial at 150 °C for 45 min to afford the desired

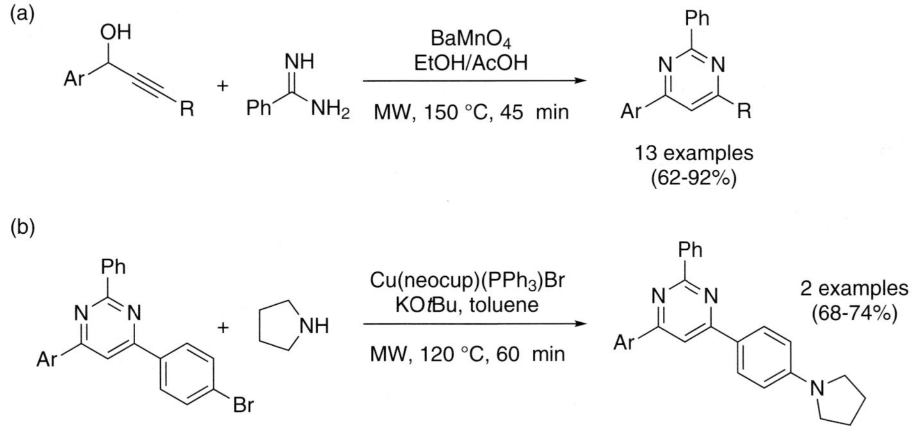

Scheme 7.98 Microwave-assisted *in situ* tandem oxidation–heteroannulation synthesis of 2,4,5-triarylpyrimidines and library diversification.

2,4,6-triarylpyrimidines in high yields (Scheme 7.98a). The generated scaffolds can also be used to broaden the electronic profile of the π-extended pyrimidines as was shown by copper-mediated *N*-arylation of initially prepared bromo derivatives (Scheme 7.98b). Overall, a variety of highly fluorescent compounds could be prepared in a facile two-step microwave protocol.

An important multicomponent transformation for the synthesis of dihydropyrimidines is the Biginelli reaction, involving the acid-catalyzed condensation of aldehydes, CH-acidic carbonyl components, and urea-type building blocks (Scheme 7.99) [184, 185]. Under conventional conditions, this MCR typically requires several hours of heating under reflux conditions, whereas microwave-assisted protocols can be completed within 10–20 min providing improved product yields.

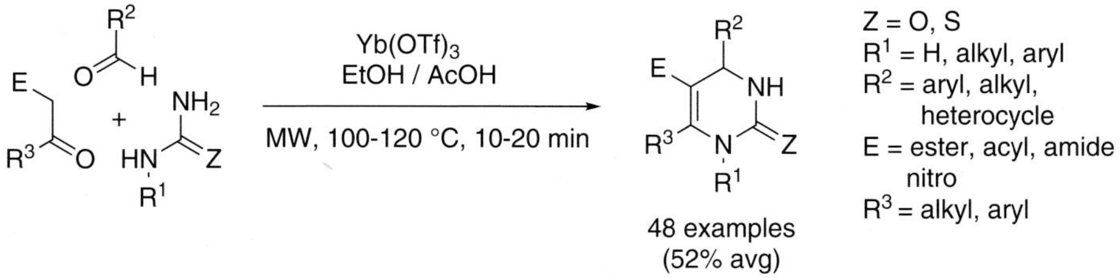

Scheme 7.99 Biginelli three-component synthesis of dihydropyrimidines.

A multistep microwave-assisted route to 2-amino-3,4,5,6-tetrahydropyrimidines was disclosed by Wellner and coworkers (Scheme 7.100) [186]. The construction of the central basic scaffold was achieved solely by the application of microwave-assisted chemistry, without any need of activating agents or protecting group manipulations. The initially required diaminoaryl ether building block was synthesized by nucleophilic aromatic substitution from fluorobenzenes (1.1 equiv) and 1,3-diaminopropane-2-ol. With sodium hydride (1.2 equiv) as nonnucleophilic base

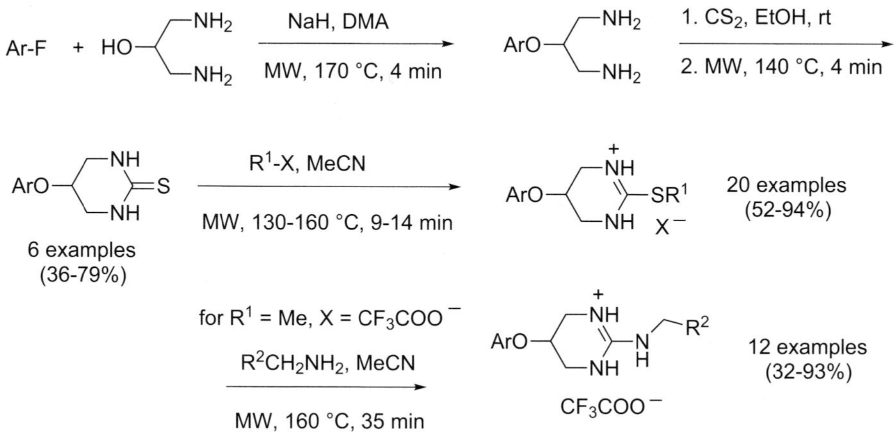

Scheme 7.100 Preparation of cyclic isothioureas and guanidines.

and *N,N*-dimethylacetamide as solvent, moderate to good yields of the diaminoaryl ethers were obtained under microwave conditions (170 °C, 4 min). Addition of carbon disulfide at room temperature led to spontaneous formation of a dithiocarbonate intermediate that was transformed by thermolytic cleavage and hydrogen sulfide extrusion (140 °C, 4 min) to the cyclic thioureas that crystallized spontaneously and could be filtered off without any need for further purification. Further *S*-alkylation with alkyl halides (2 equiv) in acetonitrile (130–160 °C, 9–14 min) provided the isothiouronium salts, which could be further reacted – after transforming the halide to trifluoroacetic acid salts – with primary amines (1.1 equiv) again using acetonitrile as solvent and applying microwave conditions (160 °C, 35 min).

Another multistep protocol that initially involves the formation of fused pyrimidines (quinazolines) was described by Besson and coworkers in the context of synthesizing 8*H*-quinazolino[4,3-*b*]quinazolin-8-ones via double Niementowski condensation reactions (Scheme 7.101) [187]. In the first step of the sequence, an anthranilic acid was condensed with formamide (5 equiv) under open-vessel microwave conditions (Niementowski condensation). Subsequent chlorination with excess $POCl_3$, again using open-vessel conditions, produced the anticipated 4-chloroquinazoline derivatives, which were subsequently condensed again with anthranilic acids in acetic acid to produce the tetracyclic 8*H*-quinazolino[4,3-*b*]quinazolin-8-one target structures. The final condensation reactions were completed within 20 min under open-vessel reflux conditions, but not surprisingly could also be performed within 10 min using sealed vessel heating at 130 °C.

The same authors have described a related Niementowski condensation for the preparation of 3*H*-nitroquinazolin-4-ones. Subsequent manipulation of this structure led to 8*H*-thiazolo[5,4-*f*]quinazolin-9-ones via a series of open-vessel microwave-assisted transformations that are highlighted in Scheme 7.102 [188].

Movassaghi and Hill from the MIT have developed a protocol for the synthesis of quinazolines via electrophilic activation of secondary amides utilizing 2-chloropyridine (2-ClPyr) in combination with Tf_2O and subsequent reaction with weakly

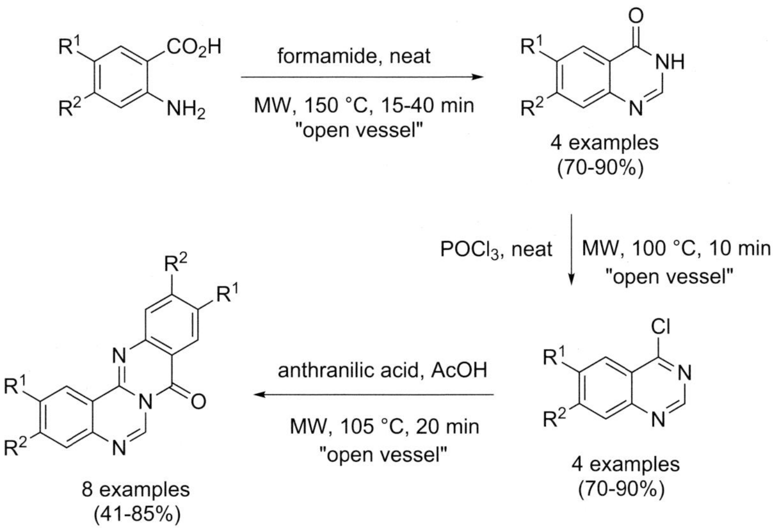

Scheme 7.101 Formation of 8*H*-quinazolino[4,3-*b*]quinazolin-8-ones via double Niementowski condensation.

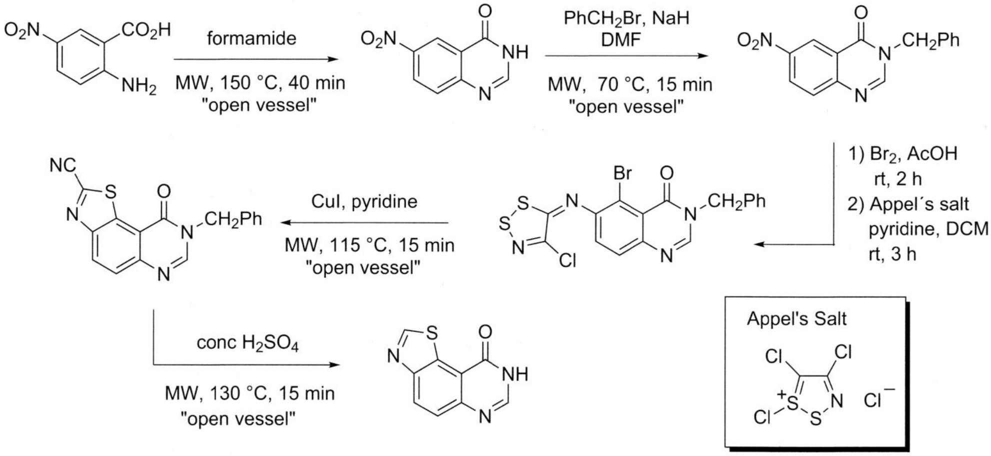

Scheme 7.102 Formation of 8*H*-thiazolo[5,4-*f*]quinazolin-9-ones.

nucleophilic nitriles (Scheme 7.103) [189]. Whereas electron-rich *N*-vinyl and *N*-aryl amides afforded the corresponding pyrimidine derivatives at room temperature, microwave heating to 140 °C was necessary for less reactive substrates. In addition, one example was presented where a primary amide was employed instead of the nitrile under the same microwave conditions giving the quinazoline ($R^1 = Ph$, $R^2 = H$, $R^3 = OMe$, $R^4 = {}^{c}C_6H_{11}$) in 74% yield. For alternative synthetic approaches to quinazolines, see Refs. [190, 191].

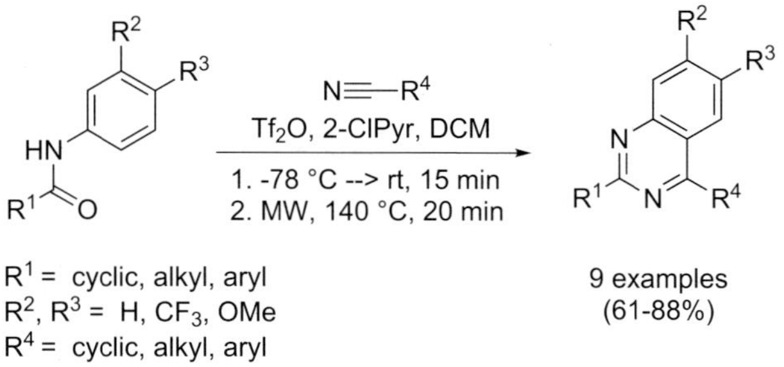

Scheme 7.103 Single-step synthesis of pyrimidine derivatives.

In a subsequent publication, the same authors reported a comparative study of oil bath and microwave heating for the synthesis of quinazolines **20** and pyridine **21** (Scheme 7.104) [192]. Microwave irradiation proved to be beneficial for electron-deficient and sterically hindered amides or for weak nucleophiles. Improved yields could be achieved with microwave heating at 140 °C for 20 min compared to oil bath heating under otherwise identical conditions. At lower temperatures of 75 °C, no difference in the reaction outcome was observed; however, 30–70% lower yields were obtained.

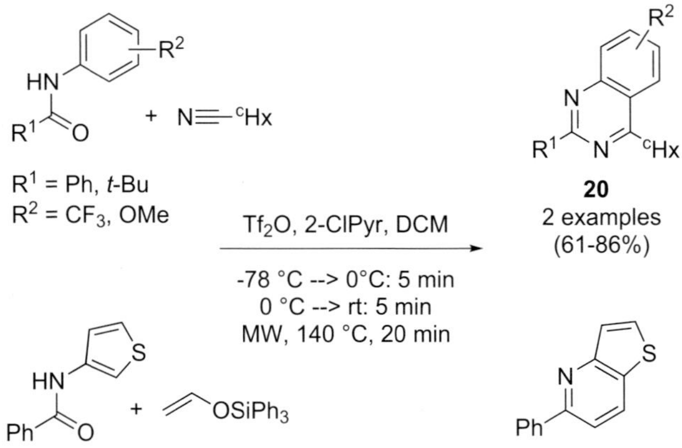

Scheme 7.104 Single-step synthesis of azaheterocycles.

The group of Liu at Shanghai Institute for Biological Sciences developed an environmentally benign method for the synthesis of quinazoline derivatives in water [193]. In the optimized protocol, a mixture of 2-iodobenzoic acid with 1.5 equiv corresponding amidine hydrochloride in water was stirred at room temperature under nitrogen atmosphere for 10 min. Then 2 equiv cesium carbonate as a base were added and the mixture stirred for additional 5 min. Finally, 10 equiv iron (III) chloride were added and the sealed vial was microwave heated at 120 °C for 30 min (Scheme 7.105). The general procedure can be slightly modified by adding

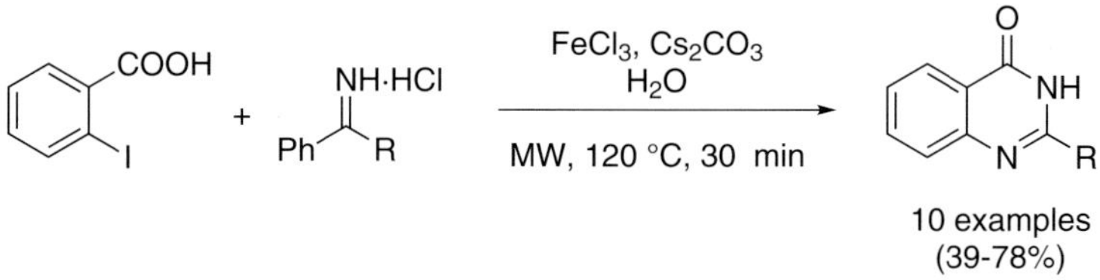

Scheme 7.105 Iron-catalyzed synthesis of quinazolinone derivatives under microwave conditions.

ligands, switching catalyst, and/or changing the solvent to maximize yields for individual substrates.

A one-pot preparation of pyrrolo[1,2-*a*]quinazoline libraries with three points of diversification by condensation of α-cyanoketones and 2-hydrazinobenzoic acids was developed by Hulme and coworkers (Scheme 7.106) [194]. The protocol simply involved heating a solution of both building blocks in equimolar amounts in acetic acid to 150 °C for 5 min. In many cases, the final products precipitated directly out of the reaction mixture. In such cases, simple washing with diethyl ether yielded the products in >95% purity. A 63-member library was prepared by employing seven α-cyanoketones and nine 2-hydrazinobenzoic acids.

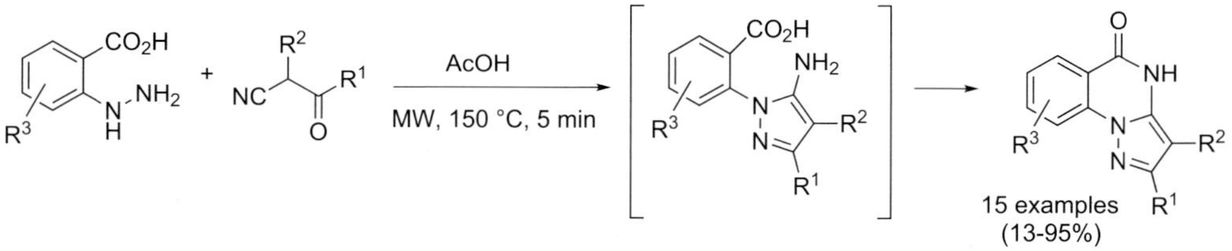

Scheme 7.106 Formation of pyrrolo[1,2-*a*]quinazolines.

A microwave-assisted rapid and highly selective method for the synthesis of new pyrazolo[1,5-*a*]pyrimidines has been demonstrated by Ming *et al.* (Scheme 7.107) [195]. The synthesis of 12 new pyrazolo[1,5-*a*]pyrimidines was performed by the reaction of 5-amino-1*H*-pyrazoles with a variety of enaminones under microwave irradiation. Under controlled microwave heating, the synthesis has shown the same

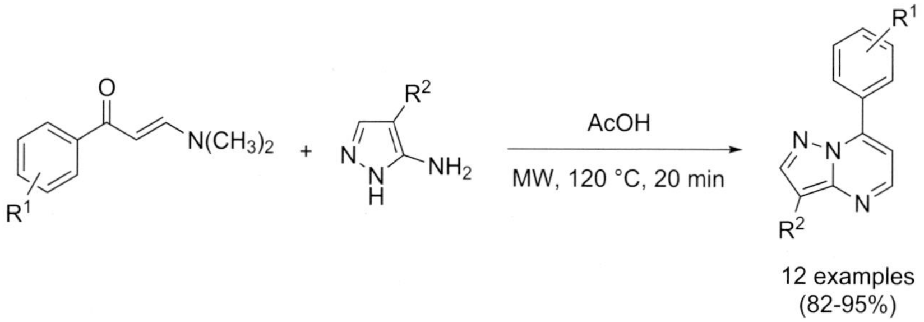

Scheme 7.107 Synthesis of pyrazolo[1,5-*a*]pyrimidines.

high regioselectivity as in case of conventional heating; however, the yields with microwave heating were improved by 10–20% and the reaction rates enhanced up to 50 times. For a related approach to pyrazolo[1,5-*a*]pyrimidines starting from 1,3-dicarbonyl compounds and 5-amino-1*H*-pyrazoles, see Refs. [48, 196].

A small library of 2,4(1*H*,3*H*)-quinazolinediones (X = O) and 2-thioxoquinazolines (X = S) was prepared via the reaction of substituted methyl anthranilate with diverse iso(thio)cyanates without the addition of catalyst, ligand, or base by Liu and coworkers (Scheme 7.108) [197]. Aryl-substituted iso(thio)cyanates proved to give higher product yields compared to aliphatic ones. In addition, the products could be easily isolated by precipitation and filtration due to the employed DMSO/H_2O solvent mixture.

Scheme 7.108 Synthesis of quinazolinediones and thioxoquinazolines.

For alternative syntheses of quinazolin-4-ones from anthranilic acid, amines, and carboxylic acid derivatives, see Refs. [198–201].

Pyrimidines have been synthesized in one step from ketones and formamidine using *p*-TsOH as a catalyst [202], and from enones and benzamidines [203]. Related approaches starting from ureas [204], enaminomethylene malonates [205], and benzamidines have also been disclosed [206]. Pyrazolo[1,5-*a*]pyrimidines and 1,2,4-triazolo[1,5-*a*]pyrimidines were synthesized as novel carbohybrids by condensation of 2-*C*-formyl glycals with 3-aminopyrazoles and 3-amino-1,2,4-triazoles [207].

A simple, efficient, and high-yielding synthesis of quinazolin-4-ylamines and thieno[3,2-*d*]pyridin-4-ylamines based on the condensation of appropriately functionalized *N*′-(2-cyanophenyl)-*N*,*N*-dimethylformamidines and primary amines has been reported by Han and coworkers (Scheme 7.109) [208]. Optimization of the reaction parameters resulted in the use of acetonitrile/acetic acid as solvent mixture and 1.2 equiv of the corresponding amine. In general, microwave heating to 160 °C for 10 min provided excellent product yields.

The group of Borrell has described a variety of different microwave-assisted approaches to biologically active pyrido[2,3-*d*]pyrimidin-7(8*H*)-ones [209–211]. In the multicomponent method described in Scheme 7.110a, the target structures were obtained in a one-pot fashion by cyclocondensation of α,β-unsaturated esters, amidines or guanidines, and CH-acidic nitriles (malononitrile or ethyl cyanoacetate) [209, 210]. Employing sodium methoxide as base in methanol as solvent, low to

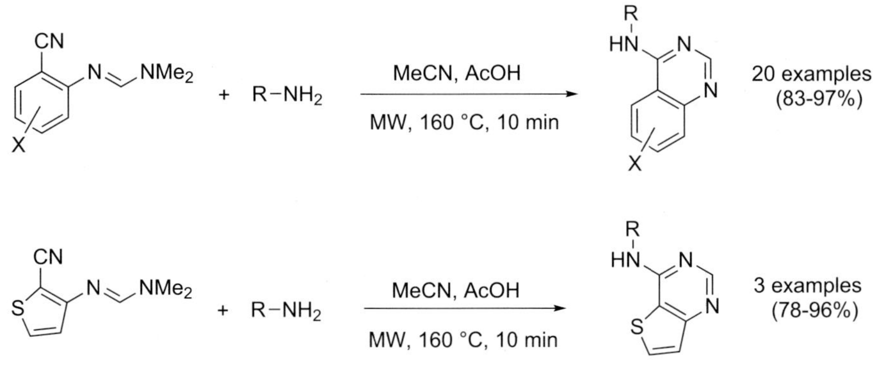

Scheme 7.109 Formation of aminoquinazoline and thieno[3,2-*d*]pyrimidine derivatives.

excellent yields were obtained by microwave heating to 100–140 °C for 10 min. Typically, the resulting pyrido[2,3-*d*]pyrimidin-7(8*H*)-ones possessing four diversity centers crystallized directly from the reaction mixture in high purity. In a modification of this strategy (Scheme 7.110b) [211], the same authors subsequently disclosed a stepwise protocol that allowed the isolation of 2-methoxy-6-oxo-1,4,5,6-tetrahydropyridin-3-carbonitrile intermediates under conventional reaction conditions, which were subsequently transformed to 2-aminopyridones by treatment

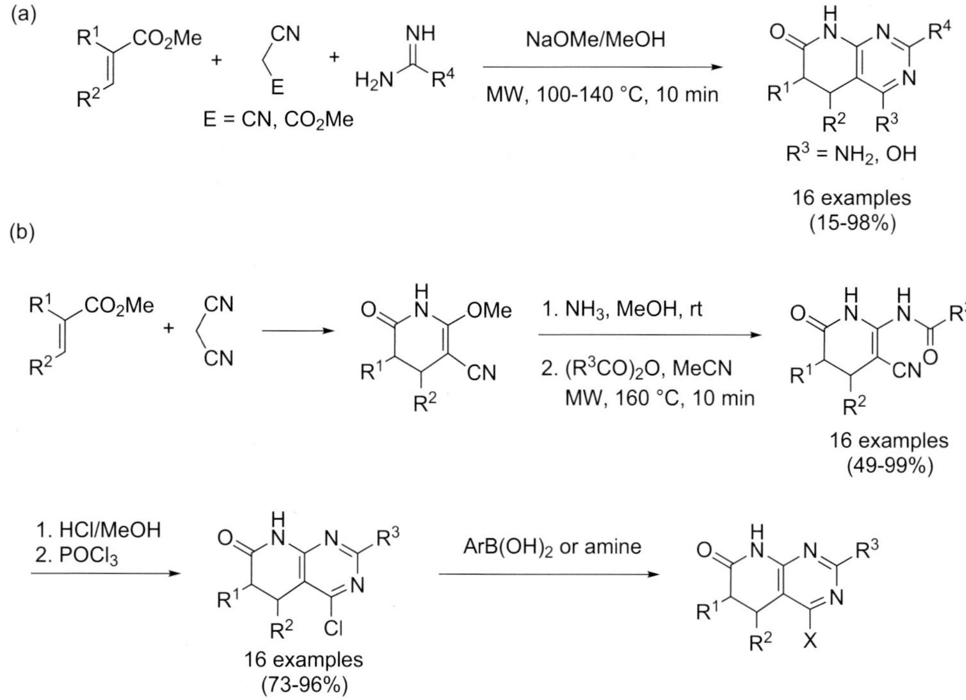

Scheme 7.110 Formation of pyrido[2,3-*d*]pyrimidines.

with ammonia at room temperature. Microwave-assisted acylation of the amino group with carboxylic acid anhydrides (R^3) in acetonitrile at 160 °C for 10 min furnished 2-acylaminopyridones that were subsequently cyclized with hydrochloric acid in methanol to the corresponding 4-oxopyrido[2,3-*d*]pyrimidines in almost quantitative yield. Further incorporation of a chloro substituent by treatment with phosphorus oxychloride provided an ideal substrate for further decoration of the pyrido[2,3-*d*]pyrimidine scaffold using Suzuki chemistry, or nucleophilic displacement reactions with amines. For both types of transformations, rapid microwave-assisted protocols were utilized, allowing access to a diverse set of pyrido[2,3-*d*] pyrimidines with four diversity centers.

Several other microwave-assisted approaches to fused pyrimidines are summarized in Scheme 7.111 [212–214].

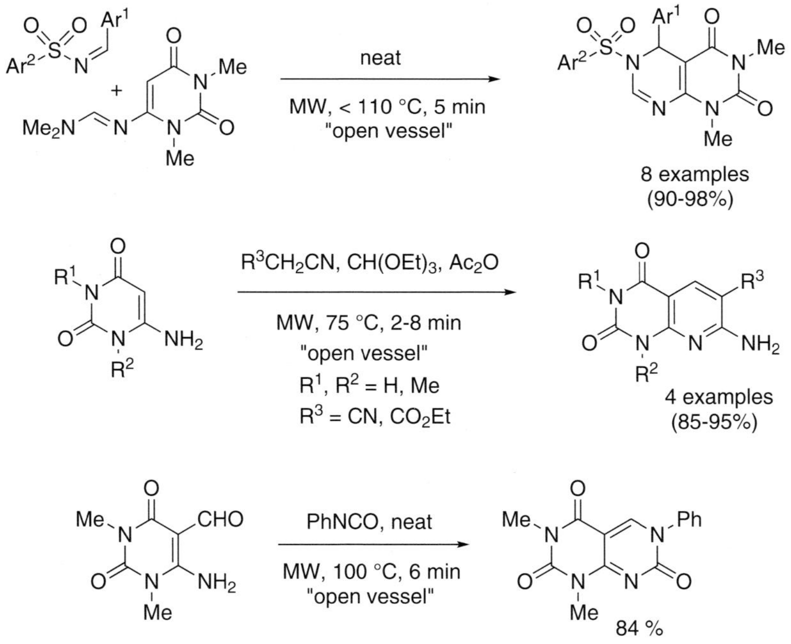

Scheme 7.111 Formation of bicyclic pyrimidines.

In 2009, the group of Kappe presented an efficient parallel microwave synthesis of 2-styrylquinazolin-4(3*H*)-ones, which are important heterocycles in naturally occurring alkaloids exhibiting a variety of biological activities (Scheme 7.112) [215]. Initially, a rapid catalyst-free one-pot microwave procedure was elaborated converting commercially available anthranilic acids into quinazolin-4(3*H*)-ones. These intermediates were subsequently transformed into the corresponding 2-styryl derivatives by condensation with aromatic aldehydes. The availability of numerous anthranilic acids and aromatic aldehydes makes this procedure susceptible for rapid library generation of this attractive scaffold. Compared to conventional sequential methods,

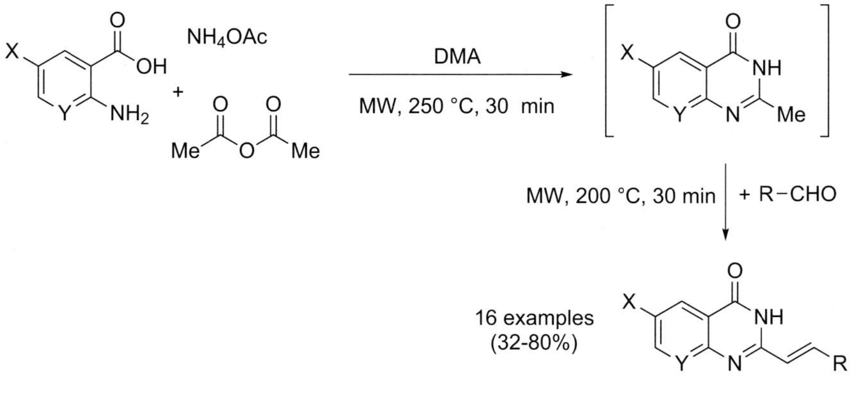

Scheme 7.112 Two-step one-pot microwave-assisted synthesis of 2-styrylquinazolines.

the parallel approach considerably increased the efficiency. The applied high-throughput reactor for efficient parallel approach enables performing 80 reactions in parallel on an approximately 100 mg scale.

Various other examples in this chapter have already highlighted how *N,N*-dimethylformamide dimethyl acetal can be utilized efficiently as a synthon for the construction of heterocyclic rings (see Schemes 7.16, 7.24, 7.73, and 7.74). Westman *et al.* have described a two-step method for the generation of a large variety of different heterocyclic scaffolds, which is based on the initial formation of alkylaminopropenones and alkylaminopropenoates from *N,N*-dimethylformamide diethyl acetal (DMFDEA) and the corresponding CH-acidic carbonyl compounds (Scheme 7.113) [216]. Treatment of the CH-acidic carbonyl compound with 1.5–2.5 equiv of DMFDEA in *N,N*-dimethylformamide at 180 °C resulted in full conversion to the enamine synthons within 5 min. The enamines were obtained in 53–93% yield (based on LC–MS analysis) and were used without further purification in the next step.

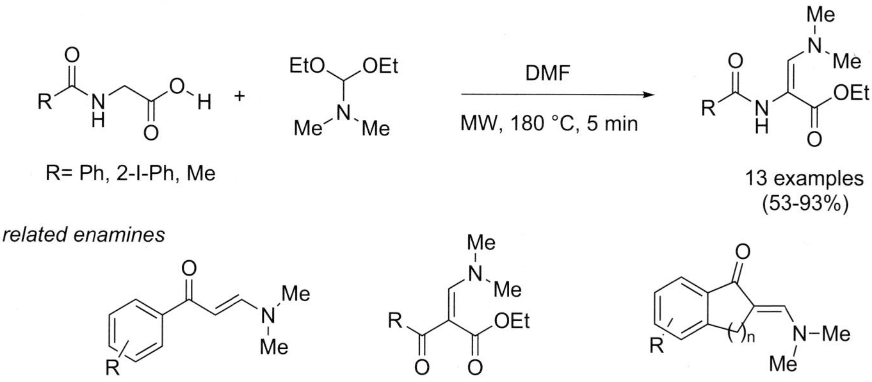

Scheme 7.113 Formation of alkylaminopropenones and alkylaminopropenoates.

For the preparation of the anticipated heterocyclic library compounds (Scheme 7.114), solutions of the prepared enamine synthons were split and diluted

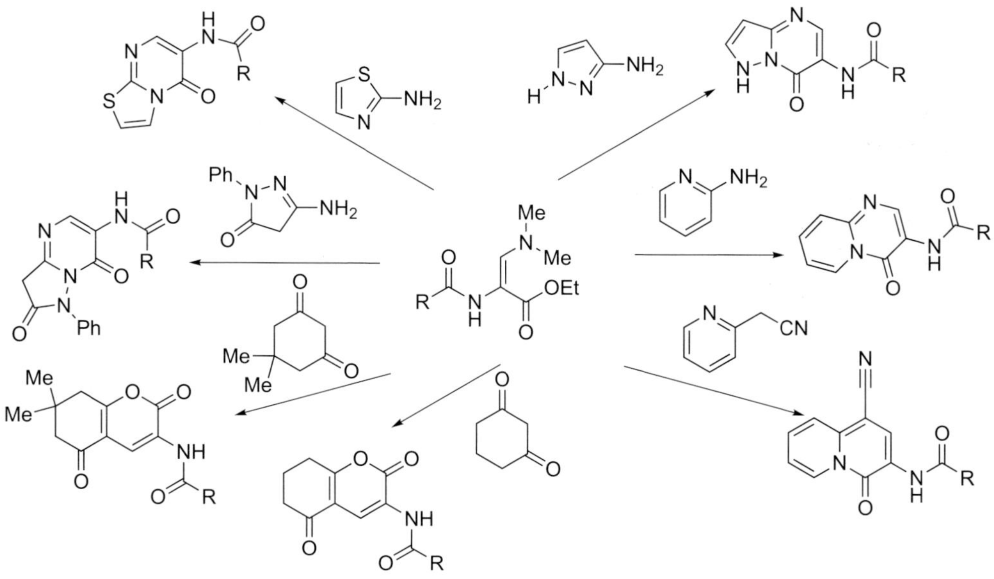

Scheme 7.114 Reaction of enamine synthons with dinucleophiles.

with an appropriate solvent and 1.2 equiv of a dinucleophile (hydrazine, hydroxylamine, amidines; see Scheme 7.114 for more complex building blocks). Subsequent resubjection to microwave conditions in acetic acid/DMF mixtures to 180 °C for a further 5 min furnished the desired target compounds in moderate to good overall yields [216]. More than 100 compounds were prepared using this strategy, and examples are highlighted in Scheme 7.114.

In a subsequent article, the same group has described a cascade sequence involving (triphenylphosphoranylidene)ethenone ($PPh_3C{=}C{=}O$) as a versatile building block for the construction of heterocycles and other unsaturated amides (Scheme 7.115) [217]. Typically, the phosphoranylidene reagent was reacted with a carbonyl compound possessing an adjacent hydroxyl, amine, or thiol group in α or β position to form five- or six-membered heterocycles (Scheme 7.115b). The reactions were performed at 180 °C for 5–8 min in 1,2-dichloroethane (DCE) and resulted in 54–97% product yields (LC–MS). Alternatively, the phosphoranylidene reagent (1.5 equiv) could also be used in a multicomponent reaction with an aldehyde and an amine to form α,β-unsaturated amides under similar reaction conditions (Scheme 7.115a) [217].

7.8.2
Pyrazines

The group of Larhed introduced a rapid microwave method for the synthesis of *N*-1, *C*-6 functionalized 3,5-dichloro-2(1*H*)-pyrazinones [218]. These important potential drug scaffolds can be obtained either via a one-pot two-step protocol generating the

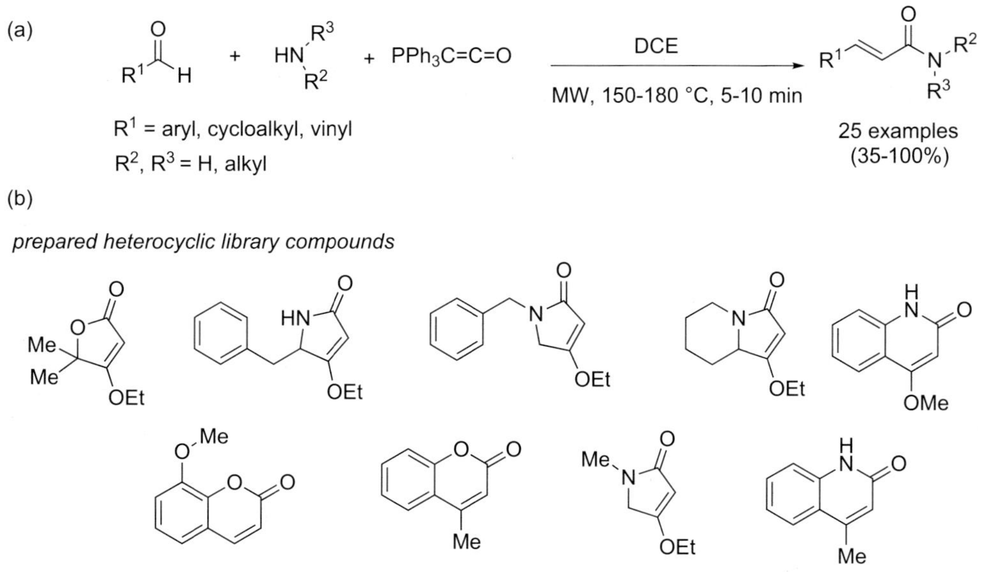

Scheme 7.115 Reactions of the (triphenylphosphoranylidene)ethenone reagent.

essential α-aminonitrile intermediates *in situ* (Scheme 7.116a) or in an even more facile one-step protocol starting from α-aminonitriles (Scheme 7.116b). The two-step approach involves a primary amine, a nitrile source, and an aldehyde to form the corresponding α-aminonitrile in a Strecker-type reaction. To initiate the cyclization, the intermediate is treated with HCl gas and oxalyl chloride to furnish the desired heterocyclic scaffold. Especially, the usually rather slow ring closure can be significantly enhanced when applying microwave irradiation for 10 min. The desired compounds were obtained in moderate to high yields after purification by flash chromatography. In case of weakly nucleophilic anilines, it was beneficial to isolate and purify the formed intermediate α-aminonitrile rather than performing the one-pot two-step sequence. The generated compounds can readily be used in further scaffold decorations and transformation as demonstrated by the substitution of the *C*-3 chloro functionality by a Boc-protected primary amino group under microwave conditions (Scheme 7.116c). This newly prepared pyrazinone scaffold shows beneficial alignment of CO and NH groups and could be successfully employed as a β-strand inducer in hepatitis C virus NS3 protease inhibitors [218].

ortho-Quinodimethane derivatives are reactive dienes and can be generated *in situ* by a number of methods. The inter- and intramolecular Diels–Alder reactions of these compounds form the basis of the synthesis of a wide range of target molecules. Díaz-Ortiz *et al.* have described the generation of a pyrazine *ortho*-quinodimethane derivative that cycloadds to nonactivated dienophiles under microwave conditions (Scheme 7.117) [219]. Heating of 2,3-bis(dibromomethyl)pyrazine with 5 equiv of sodium iodide in the presence of a small amount of *N*,*N*-dimethylformamide as

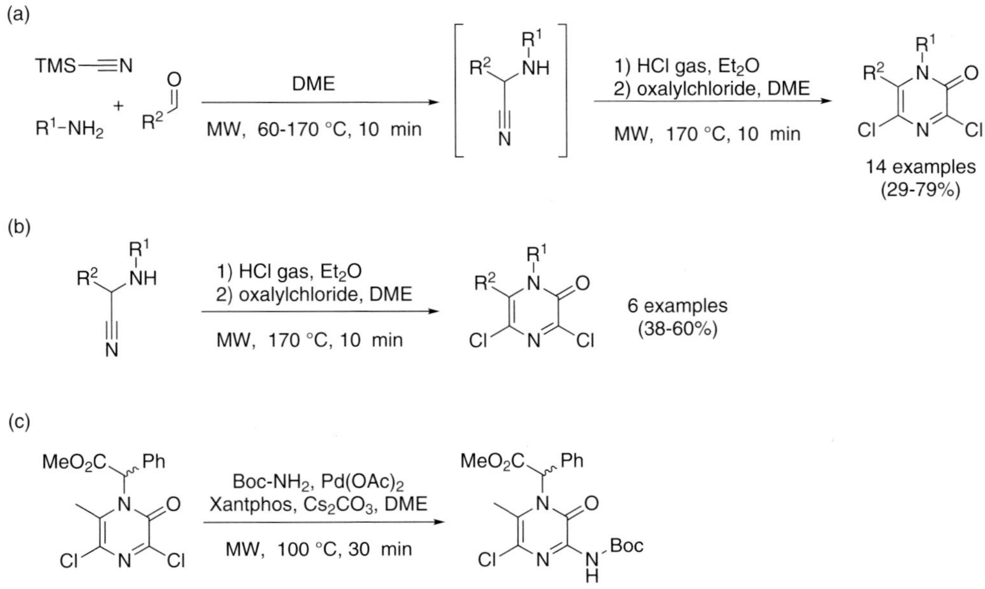

Scheme 7.116 Microwave-assisted synthesis of *N*-1, *C*-6 disubstituted 3,5-dichloro-2(1*H*)-pyrazinones.

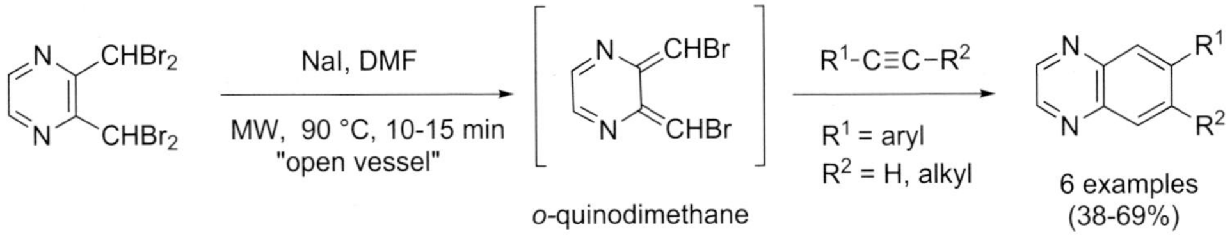

Scheme 7.117 Diels–Alder cycloadditions of pyrazine *ortho*-quinodimethane with dienophiles.

solvent and 3.6 equiv of an alkyne or enamine (not shown) dienophile resulted in the formation of quinoxaline cycloadducts within 10–15 min at 90 °C in 38–69% yield. A higher reaction temperature led to decomposition of the *ortho*-quinodimethane intermediate before reaction with the dienophile and to a corresponding decrease in the yield.

The manganese(IV) dioxide-catalyzed synthesis of quinoxalines has been demonstrated by Kim *et al.* from Seoul National University (Scheme 7.118) [220]. The quinoxalines were obtained from a variety of α-hydroxyketones by trapping with aromatic or aliphatic 1,2-diamines under solventless microwave conditions for 1 min at 70 °C. By applying the same protocol, also benzimidazoles from aldehydes and 1,2-diaminobenzene were prepared.

Kamila and Biehl have reported on the synthesis of potentially biologically active quinoxalin-2-ones (Scheme 7.119) [221]. Symmetrical 1,2-diaminobenzenes were

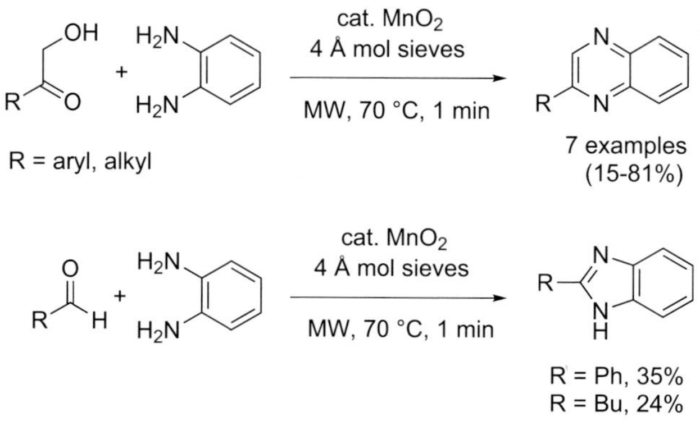

Scheme 7.118 Synthesis of quinoxalines.

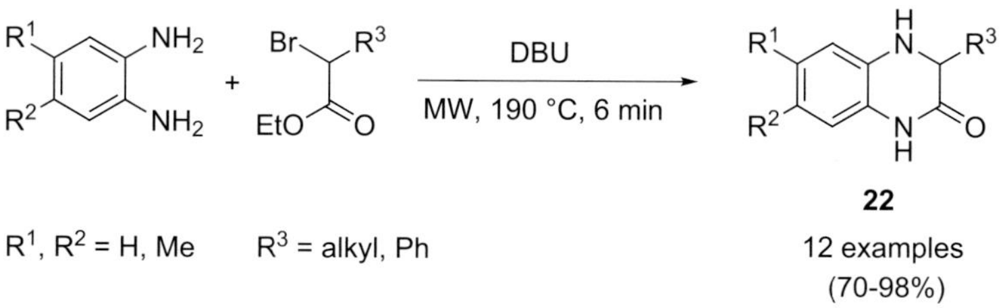

Scheme 7.119 Synthesis of quinoxalinones.

reacted with ethyl 2-bromoalkyl/phenyl acetates and DBU as base under solvent-free microwave conditions to give quinoxalinones **22** in good to excellent yields.

A different approach to quinoxalines and heterocyclic pyrazines was described by the Lindsley group, involving the cyclocondensation of 1,2-diketones and aryl/heteroaryl 1,2-diamines (Scheme 7.120) [222]. Optimized reaction conditions required heating an equimolar mixture of the diketone and diamine component for 5 min at 160 °C in

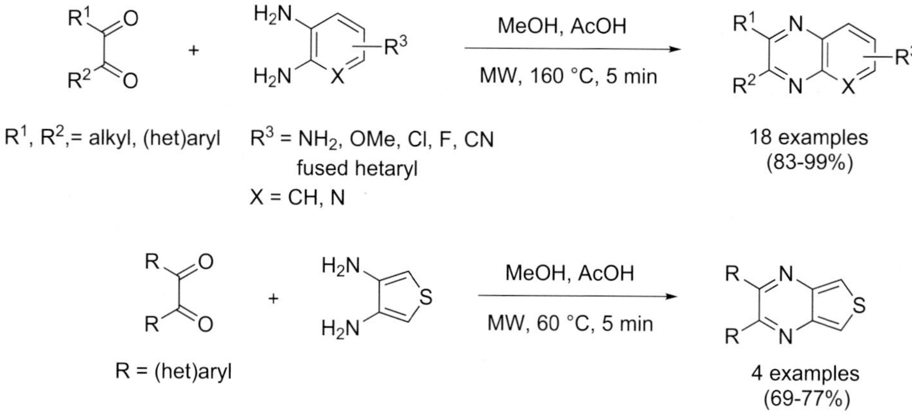

Scheme 7.120 Synthesis of quinoxalines and heterocyclic pyrazines.

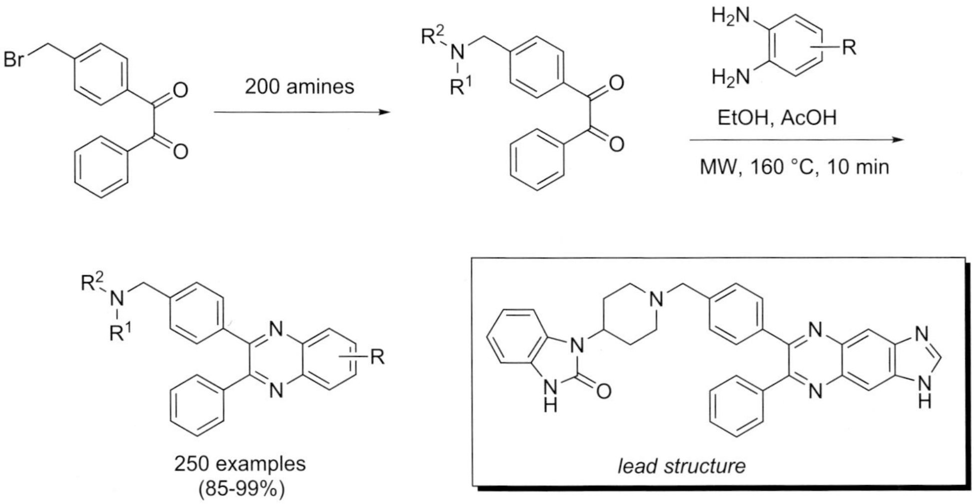

Scheme 7.121 Synthesis of quinoxalines as allosteric Akt kinase inhibitors.

a 9 : 1 methanol/acetic acid solvent mixture to deliver the substituted quinoxalines in excellent yields. This approach could easily be extended with equal success to heteroaryl pyrazines such as pyrido[2,3-*b*]pyrazines and thieno[3,4-*b*]pyrazines. The same group has employed 1,2-diketone building blocks for the preparation of other heterocyclic structures (see Schemes 7.29, 7.134, and 7.135).

An application of this synthetic strategy by the same group led to the development of a series of potent and selective allosteric Akt (protein kinase B/PKB) kinase inhibitors that induced apoptosis in tumor cells and inhibited Akt phosphorylation *in vivo* (Scheme 7.121) [223].

The phosphorus-promoted coupling of unprotected amino acids toward the one-step synthesis of 2,5-diketopiperazines was disclosed by the group of Bräse (Scheme 7.122) [224]. Symmetrical and unsymmetrical 2,5-diketopiperazines are generated by the condensation of amino acids with methyl dichlorophosphite in toluene. The ionic liquid 1,3-dimethylimidazolium dimethyl phosphate needs to be

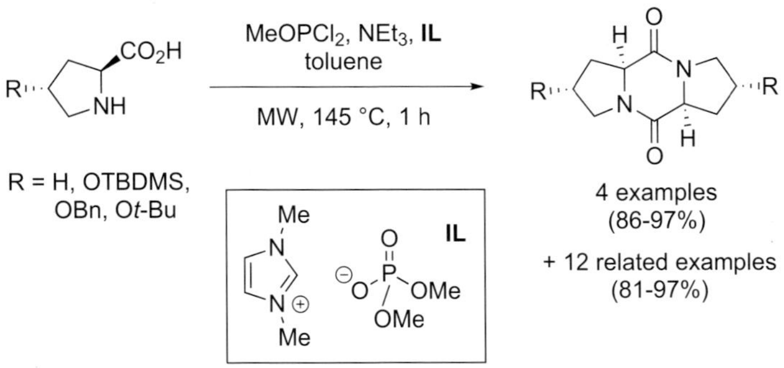

Scheme 7.122 Synthesis of symmetrical and unsymmetrical 2,5-diketopiperazines.

used as a heating aid in order to reach 145 °C. Unsymmetrical derivatives are obtained by the combination of proline, sarcosine, indoline, and octahydroindole-carboxylic acids without the formation of any by-products (1.2 equiv excess of one amino acid is crucial in this case). Advantages of this protocol are excellent yields, complete retention of the stereochemistry, tolerance of base-stable protecting groups, and a simple workup procedure.

7.8.3 Pyridazines

Polysubstituted pyridazines have been synthesized by cyclization of 1,4-dicarbonyl compounds with hydrazine followed by oxidation with DDQ as reported by the group of Taddei (Scheme 7.123) [225]. Fused and polycyclic pyridazine analogs have been generated by cycloaddition chemistry [226, 227]. 3,6-Di(pyridin-2-yl)pyridazines can be obtained by cycloaddition of suitable dienophiles to 3,6-disubstituted tetrazines in a cycloaddition–cycloreversion pathway [228].

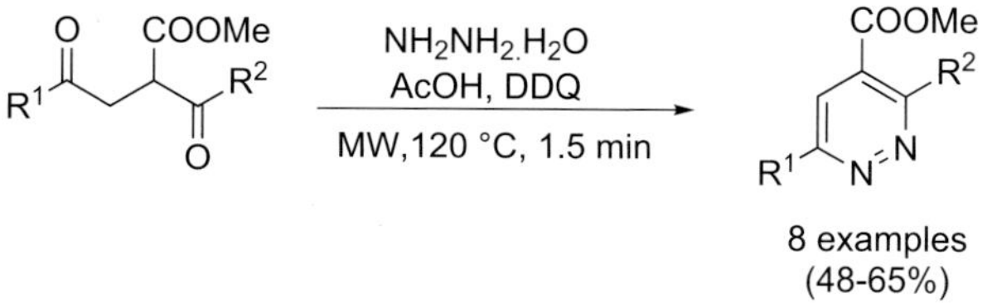

Scheme 7.123 Microwave-assisted synthesis of polysubstituted pyridazines.

7.8.4 Oxazines

In the context of synthesizing libraries of thiophene derivatives of potential therapeutic interest, Rault and coworkers have described the preparation of thieno[3,2-*d*][1,3]oxazine-2,4-diones (thiaisatoic anhydride) using a microwave approach (Scheme 7.124) [229]. The synthesis of the thiaisatoic anhydrides started from the corresponding amino esters. Alkaline hydrolysis was performed in 7 N potassium

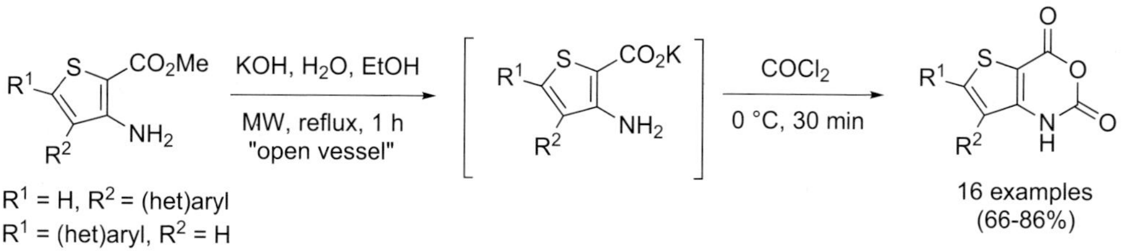

Scheme 7.124 Synthesis of thiaisatoic anhydrides.

hydroxide (water/ethanol 1 : 3) under microwave irradiation for 1 h. After cooling the solution to 0 °C, gaseous phosgene was bubbled through the solution for 30 min with stirring, leading to precipitation of the desired thiaisatoic anhydride analogs in 66–86% yield. The transformations were carried out on a 15 g scale.

The solvent-free synthesis of substituted spiroindolinonaphth[2,1-*b*][1,4]oxazines through condensation of 2-methylene-1,3,3-trimethylindoline derivatives with 1-nitroso-2-naphthol under microwave irradiation was described by Fedorova and coworkers (Scheme 7.125) [230]. In a typical reaction, the equimolar mixture of the two starting materials was irradiated to 65–110 °C for 15 min to produce the desired spiroindolinonaphth[2,1-*b*][1,4]oxazines, useful as photochromic compounds. In a related procedure, addition of morpholine to the reaction mixture led to the formation of the corresponding 6′-amino-functionalized spiroindolinonaphth[2,1-*b*][1,4]oxazines, exhibiting a strong hypsochromic color shift [230].

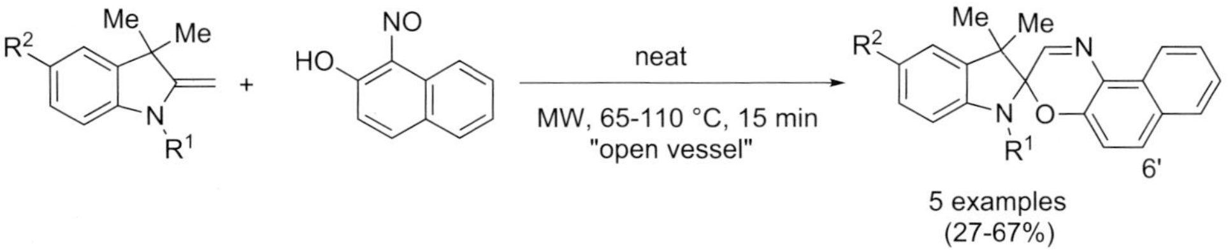

Scheme 7.125 Synthesis of spiroindolinonaphth[2,1-*b*][1,4]oxazines.

The group of Caliendo has described the multistep synthesis of benzoxazine libraries, in the search for compounds with vasorelaxant activity related to cromakalim [231]. As highlighted in Scheme 7.126, all of the required synthetic manipulations were carried out under microwave irradiation conditions. In all cases, a reduction in time compared to the corresponding thermal protocols was reported.

A 2005 publication by Dai *et al.* describes a regioselective one-pot synthesis of 2-alkyl-3,4-dihydro-3-oxo-2*H*-1,4-benzoxazines (Scheme 7.127) [232]. The use of a base such as DBU is critical for achieving a regioselective *O*-alkylation of 2-aminophenols with 2-bromoalkanoates to give an acyclic intermediate, which subsequently undergoes an intermolecular amidation reaction to furnish the desired 2-alkyl-3,4-dihydro-3-oxo-2*H*-1,4-benzoxazines. The desired products possessing alkyl, aryl, halogen, nitro, or sulfonyl substituents and additional ring structures have been prepared in 13–82% yield. Microwave heating was necessary to induce annulation reaction, for acyclic intermediates bearing an electron-withdrawing group. For a different approach to 3,4-dihydro-3-oxo-2*H*-1,4-benzoxazines, see Ref. [233].

The same product class was synthesized by the Dai group using a slightly different approach (Scheme 7.128) [234]. The first step was a reductive *N*-alkylation of substituted 2-aminophenols with aromatic aldehydes to obtain the *N*-alkylated anilines **23**. Regioselective *O*-alkylation of **23** with 2-bromoalkanoates furnished acyclic intermediates that showed spontaneous cyclization to 1,4-benzoxazines **24**. Higher temperatures (180–200 °C) for the second step were necessary when electron-withdrawing groups were present at the C-6 position ($R^1 = NO_2$, Cl).

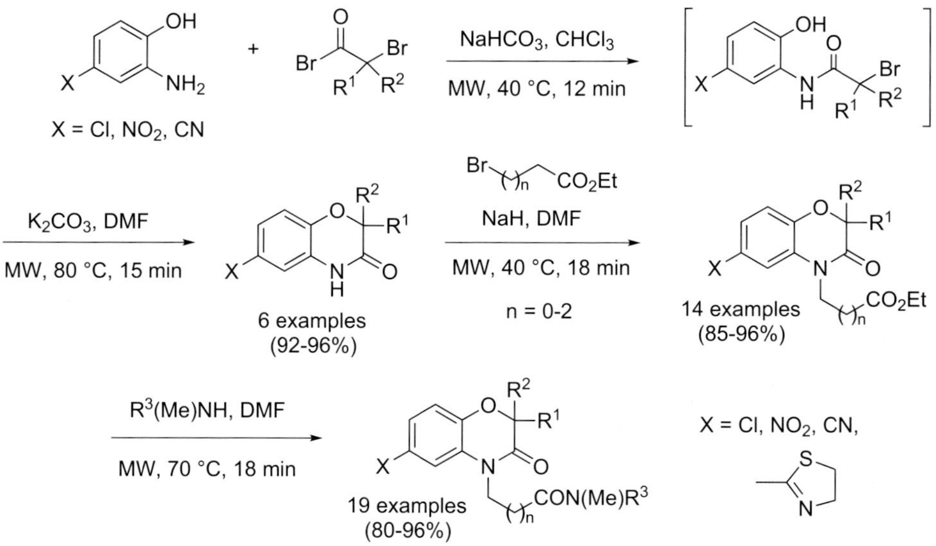

Scheme 7.126 Multistep synthesis of substituted benzoxazine derivatives.

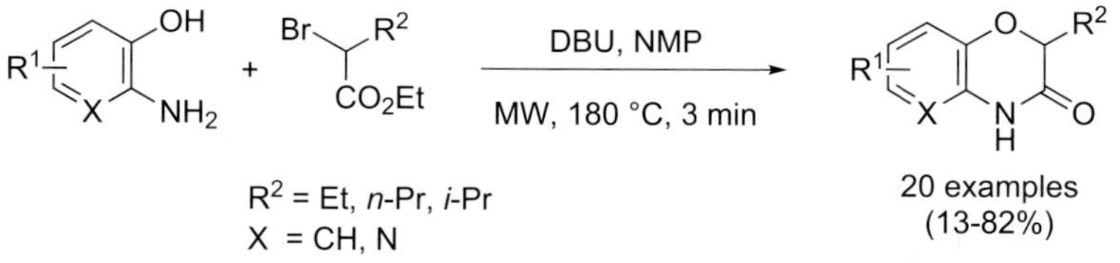

Scheme 7.127 Regioselective synthesis of 3,4-dihydro-3-oxo-2*H*-1,4-benzoxazines.

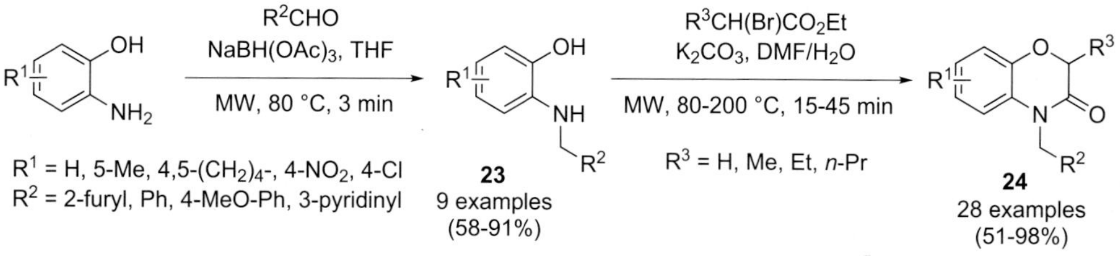

Scheme 7.128 Multistep synthesis of 1,4-benzoxazine scaffolds.

Highly functionalized benzoxazine derivatives **26** were obtained in a different approach by the same group via a one-pot Ugi four-component reaction of 2-aminophenols, aromatic aldehydes, α-bromoalkanoic acids, and isocyanides and subsequent base-catalyzed *O*-alkylation of **25** (Scheme 7.129) [235]. By additional post-modification of scaffolds **27** through CuI-catalyzed intramolecular amidations, a novel class of heterocyclic derivatives could be achieved.

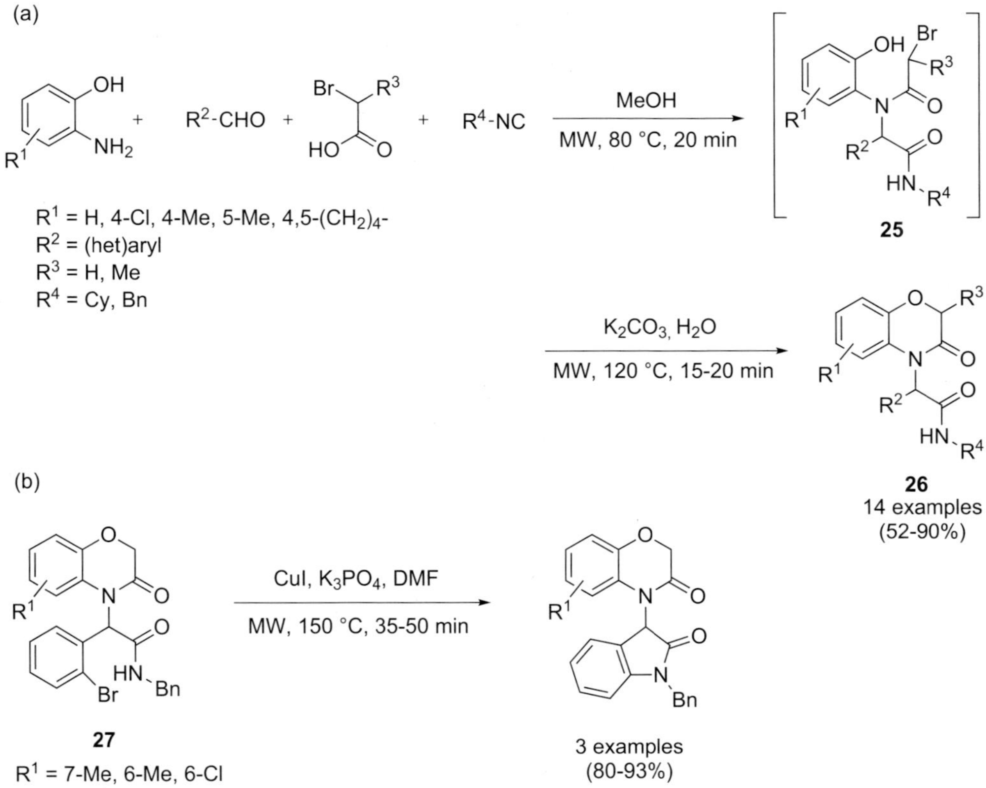

Scheme 7.129 One-pot Ugi four-component reaction and intramolecular *O*-alkylation.

7.8.5 Thiazines

Phenothiazines are well known as intermediates for pharmaceuticals, and are also active as insecticides as well as antioxidants. These compounds are usually prepared by thiation of diphenylamines with elemental sulfur. In this context, the group of Toma has elaborated the synthesis of 3-phthalimidophenothiazine as shown in Scheme 7.130 [236]. Using a variety of different high-boiling solvents under conventional thermal reflux conditions provided low isolated yields of the desired product. The highest conversion and isolated product yield (55%) was achieved by

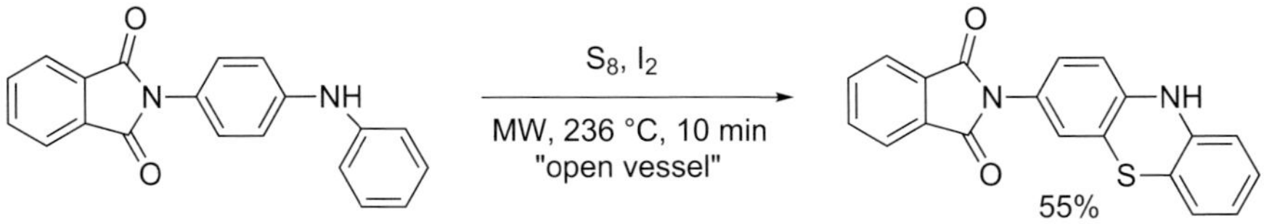

Scheme 7.130 Synthesis of phenothiazines.

microwave irradiation of a mixture of the starting *N*-(4-phenylaminophenyl)phthalimide with 5 equiv of elemental sulfur and 6.25 mol% of iodine as catalyst to 236 °C for 10–20 min. This reaction was run on a 15 g scale, and subsequent deprotection with hydrazine hydrate in ethanol provided 3-aminophenothiazine.

7.9
Six-Membered Heterocycles with Three Heteroatoms

Lee and Rana reported on the preparation of a 20-member library of 4,6-diamino-2,2-dimethyl-1,2-dihydro-1-phenyl-1,3,5-triazines, which are established inhibitors of dihydrofolate reductase (DHFR). The authors have utilized a one-pot three-component method, involving the microwave-assisted condensation of *N*-cyanoguanidine, acetone, and an aromatic amine (Scheme 7.131) [237]. The most suitable conditions involved heating a solution of the aniline with 1.1 equiv of *N*-cyanoguanidine (dicyandiamide) in acetone as solvent containing 1 equiv of hydrochloric acid to 90 °C for 35 min. After leaving the solution at 4 °C overnight, the desired triazine products precipitated as hydrochloride salts from the solution.

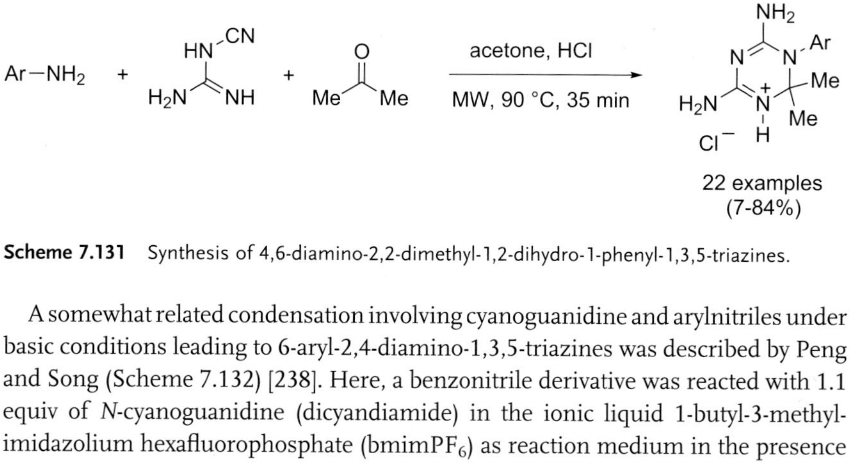

Scheme 7.131 Synthesis of 4,6-diamino-2,2-dimethyl-1,2-dihydro-1-phenyl-1,3,5-triazines.

A somewhat related condensation involving cyanoguanidine and arylnitriles under basic conditions leading to 6-aryl-2,4-diamino-1,3,5-triazines was described by Peng and Song (Scheme 7.132) [238]. Here, a benzonitrile derivative was reacted with 1.1 equiv of *N*-cyanoguanidine (dicyandiamide) in the ionic liquid 1-butyl-3-methylimidazolium hexafluorophosphate (bmimPF_6) as reaction medium in the presence

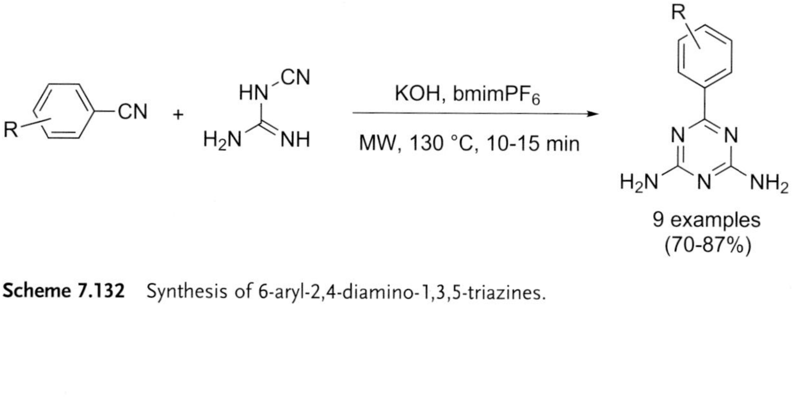

Scheme 7.132 Synthesis of 6-aryl-2,4-diamino-1,3,5-triazines.

of 20 mol% of powdered potassium hydroxide as catalyst. Highest yields were obtained at 130 °C and 10–15 min of irradiation time. The ionic liquid could be recycled at least five times.

Recently, Chen *et al.* developed a high-yielding microwave-assisted synthesis of multisubstituted 1,3,5-triazines (Scheme 7.133) [239]. Especially, *N*-4,6-disubstituted 1,3,5-triazine-4,6-diamine analogs have proven relevant scaffolds in chemotherapeutic agents and several other pharmacologically attractive compounds. The required biguanide precursors **28** were prepared by microwave heating of an equimolar mixture of dicyandiamide, arylamine, and concentrated hydrochloric acid in dioxane. Irradiation of the mixture at 90 °C under open-vessel conditions for 15 min led to precipitation of the corresponding arylbiguanide hydrochlorides upon cooling. In the next step, **28** was suspended in tetrahydrofuran together with 1.5 equiv sodium methoxide and 3 equiv ethyl ester and subjected to microwave heating under open-vessel conditions for 20 min at 70 °C to furnish the intermediate triazines **29** in good yield. In the final transformation step, **29** together with 1.1 equiv arylmethyl bromide and 1.5 equiv sodium *tert*-butoxide in dioxane was microwave heated under open-vessel conditions at 90 °C for 15 min to obtain the desired 2-(arylmethyl)amino-4-arylamino-6-alkyl-1,3,5-triazines **30** in high overall yield. The simple and efficient method can also be used for parallel synthesis to give rapid access to numerous biologically active derivatives.

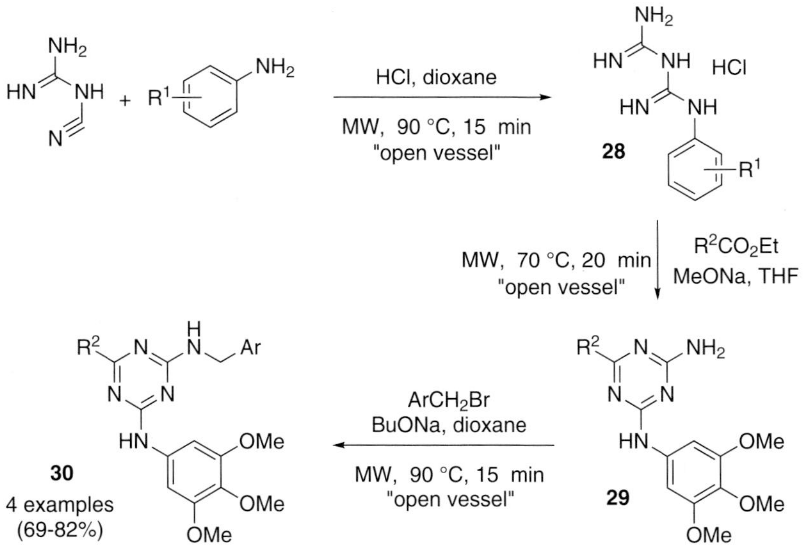

Scheme 7.133 Microwave-assisted synthesis of 2-(arylmethyl)amino-4-arylamino-6-alkyl-1,3,5-triazines.

In addition to the previous syntheses involving 1,2-diketones as building blocks discussed above, Zhao *et al.* have described the preparation of 1,2,4-triazine libraries by condensation of the 1,2-diketones with acyl hydrazides and ammonium acetate (Scheme 7.134) [240]. Microwave heating of an equimolar mixture of the two starting

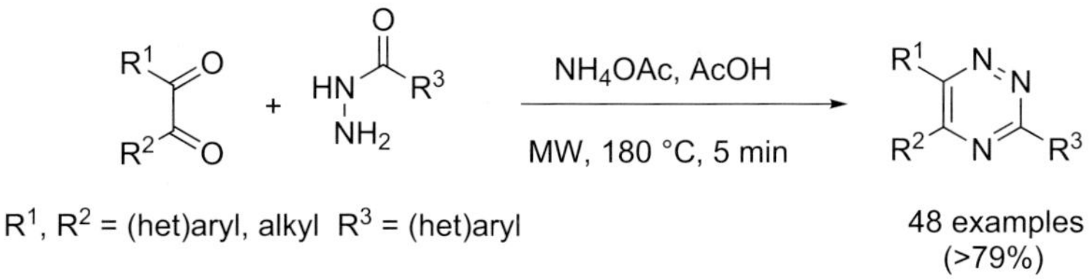

Scheme 7.134 Synthesis of 3,5,6-trisubstituted-1,2,4-triazines.

materials with 10 equiv of ammonium acetate in acetic acid to 180 °C for 5 min resulted in the formation of the anticipated 1,2,4-triazine products in excellent yield for all of the 48 examples studied. For most of the library compounds, the desired products precipitated directly from the reaction mixture upon cooling. The conventional protocols involving refluxing acetic acid conditions provided much lower product yields and required reaction times of 6–24 h.

In Scheme 7.135, an application of this triazine synthesis by the same authors toward the preparation of the basic canthine tetracylcic alkaloid skeleton is highlighted [241]. Here a suitable "indole-tethered" acyl hydrazide was prepared in quantitative yield in a rapid microwave protocol "on demand," as the hydrazide was shown to slowly decompose at room temperature. Treatment of the acyl hydrazide with 1,2-diketones under the standard conditions reported in Scheme 7.134 (180 °C, 5 min) provided the anticipated triazine-tethered indole products, along with a tetracyclic canthine analog formed by intramolecular inverse electron demand hetero-Diels–Alder cycloaddition. By increasing the reaction temperature to 220 °C and prolonging the reaction time to 40 min, the desired canthine derivatives could be obtained directly in moderate to good yields.

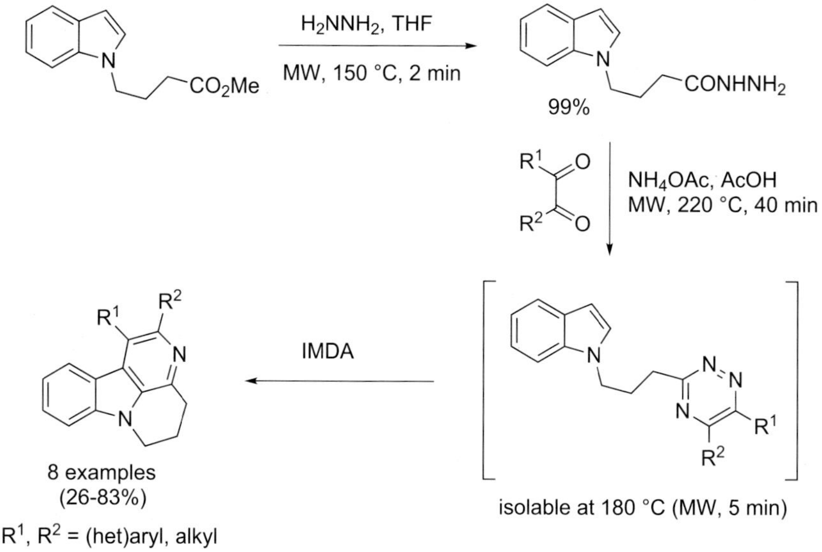

Scheme 7.135 One-pot preparation of the tetracylcic canthine alkaloid skeleton.

An even more complex pathway involving inverse electron demand Diels–Alder reactions between imidazoles and 1,2,4-triazines, linked by a tri- or tetramethylene tether from the imidazole N-1 position to the triazine C-3, produced 1,2,3,4-tetrahydro-1,5-naphthyridines or 2,3,4,5-tetrahydro-1*H*-pyrido[3,2-*b*]azepines, respectively (Scheme 7.136). The sequence carried out by Snyder and coworkers is believed to proceed by cycloaddition with subsequent loss of nitrogen, followed by a stepwise loss of a nitrile [242]. Here, the best conditions for the cycloaddition were found to involve microwave irradiation to 210 °C for 20 min in 1,2-dichlorobenzene in the presence of 15 equiv of ammonium acetate (trimethylene tether). The yields under these conditions were comparable to results obtained in refluxing diphenyl ether under thermal heating (259 °C, 80 min, 89% yield). Ammonium acetate is believed to act as an energy-transfer reagent in these reactions.

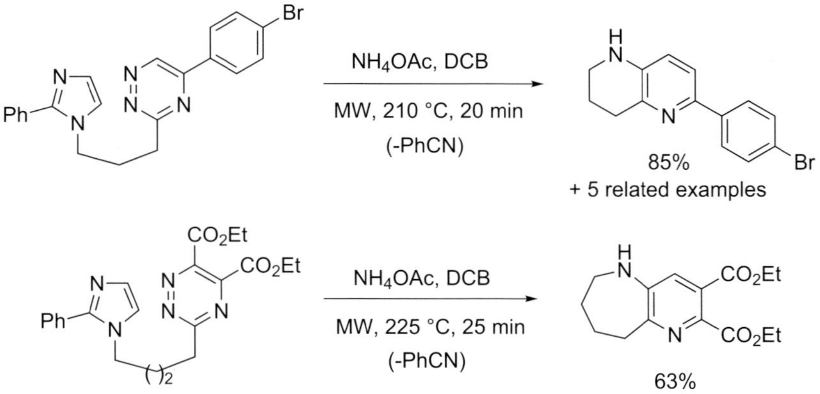

Scheme 7.136 Cycloaddition of polymethylene-tethered imidazole/triazine pairs.

7.10
Larger Heterocyclic and Polycyclic Ring Systems

The preparation of 5,11-dihydrobenzo[*e*]pyrido[3,2-*b*][1,4]diazepin-6-one by cyclocondensation of 3-amino-2-chloropyridine and 2-aminobenzoate in the presence of the strong base potassium *tert*-butoxide was described by Holzgrabe and Heller (Scheme 7.137) [243]. Microwave heating of an equimolar mixture of the two starting materials with 3 equiv of base in dry dioxane under an argon atmosphere at 100 °C for

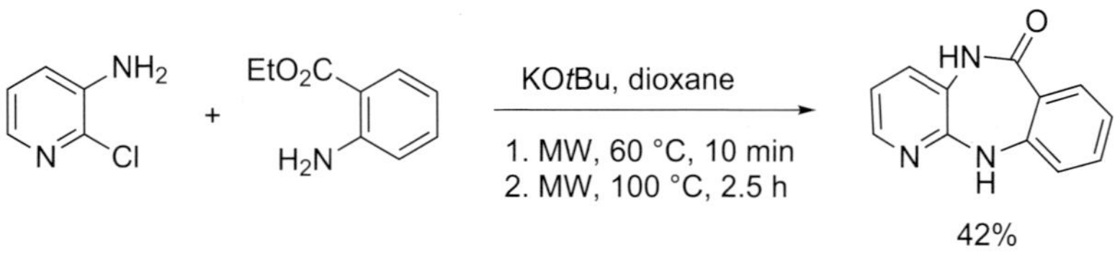

Scheme 7.137 Preparation of 5,11-dihydrobenzo[*e*]pyrido[3,2-*b*][1,4]diazepin-6-one.

2.5 h provided a 42% yield of the tricyclic product, which was subsequently used as the starting material for the synthesis of diazepinone analogs of the muscarinic receptor antagonist AFDX-384 [243].

Vasudevan *et al.* have used an intramolecular transamidation to generate a collection of bicyclic fused azepinones wherein the substitution patterns of up to six positions could be varied, four of these in a stereospecific manner [244]. In the example highlighted in Scheme 7.138, the β-lactam precursor (generated via imine formation and subsequent [2 + 2] cycloaddition of a bifunctional amine component) is heated in *N,N*-dimethylformamide solution at 200 °C for 40 min to undergo ring opening/ring expansion to furnish the desired 1,4-diazepin-5-one in 70% yield. This methodology has been used to synthesize a library of 120 1,4-diazepin-5-ones and other related ring systems.

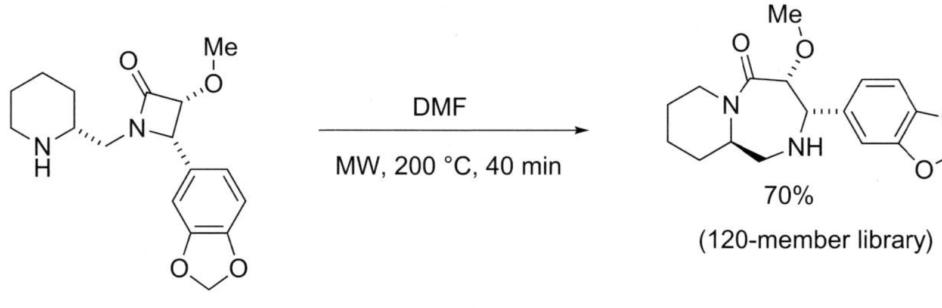

Scheme 7.138 Preparation of bicyclic fused azepinones via intramolecular transamidation.

The one-pot total synthesis of the quinazolinobenzodiazepine alkaloid sclerotigenin via a novel microwave-assisted domino reaction was reported by Liu *et al.* (Scheme 7.139) [245]. The alkaloid was obtained in 55% yield by condensation of 2 equiv of anthranilic acid with *N*-Boc-glycine in the presence of triphenyl phosphite using pyridine as solvent and microwave heating at 230 °C for 20 min. Related products were obtained employing substituted anthranilic acids and other amino acids as starting materials ($R^1 \neq H$). The authors also developed a three-component, one-pot sequential modification that allowed the preparation of related alkaloids having different substituents on the aromatic rings, starting from two different anthranilic acid derivatives.

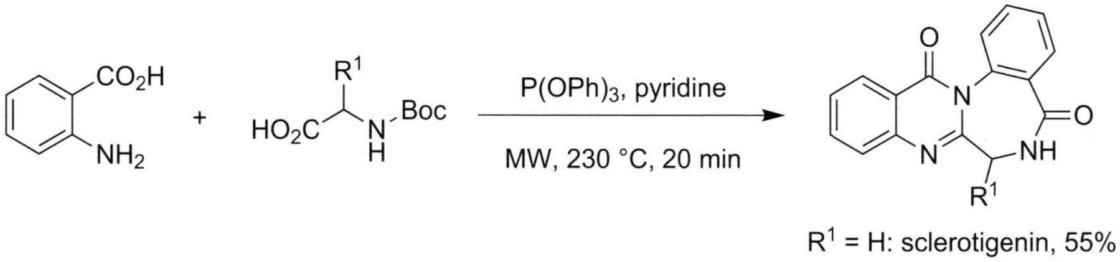

Scheme 7.139 Synthesis of quinazolinobenzodiazepines.

The groups of Fujii and Ohno from Kyoto University have developed a Cu(I)-catalyzed domino three-component coupling/indole formation/*N*-arylation protocol

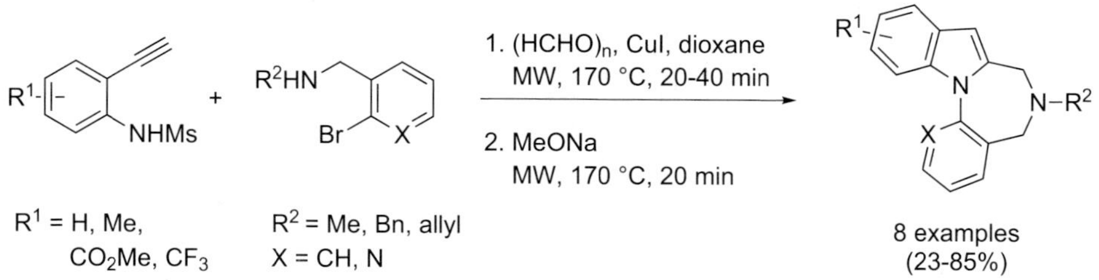

Scheme 7.140 Synthesis of indole-fused 1,4-diazepines.

for the synthesis of indole-fused 1,4-diazepines (Scheme 7.140) [246]. The first step toward the indole-fused tetracyclic scaffolds is a Mannich-type reaction of 2-ethynylanilines, *o*-bromobenzylamines, and paraformaldehyde, followed by the indole formation, subsequent deprotection of the *N*-mesyl group by addition of MeONa, and the final *N*-arylation. In addition to the fused benzodiazepines, heterocycle-fused 1,4-diazepines with a pyridine or thiophene moiety were also prepared.

The group of Waldmann at the Max Planck Institute in Dortmund investigated a silver-catalyzed one-pot cascade synthesis of diverse alkaloid ring systems [247]. In a representative procedure, arylalkyne was dissolved in dry ethanol and admixed with 1.05 equiv aniline derivative and 10 mol% 2,6-lutidine. Then 10 mol% silver triflate was added and the mixture was subjected to microwave irradiation in a sealed vial at 150 °C for 45 min to afford the target compounds **31** in good yields (Scheme 7.141a). The reaction was most effective with electron-rich substrates to furnish tetracyclic indoloisoquinolones, which are claimed to show antitumor activity. Beneficially, the microwave protocol also tolerates indole- and furan-derived acetylenic aldehydes, leading to complex benzoindolizines **32** after a cascade reaction of only 8 min microwave heating followed by treatment of the crude mixture with formic acid at room temperature to induce decarboxylation (Scheme 7.141b). Furthermore, when anilines with a substituted ethyl malonate moiety were employed with acetylenic benzaldehydes under the optimized microwave protocol, the corresponding benzazepino[2,1-*a*]isoquinolines **33** were formed in single diastereomers in high yield (Scheme 7.141c). Finally, the authors explored their developed procedure toward the synthesis of the pentacyclic natural product targets fascaplysin and homofascaplysin C to demonstrate the efficiency of the method to access biologically active structurally diverse alkaloids [247].

The group of Van der Eycken investigated the synthesis of indoloazocines **35** via intramolecular alkyne carbocyclization. These eight-membered nitrogen heterocycles fused to the indole core represent an interesting alkaloid-type scaffold [248]. The target compounds were simply achieved by Lewis acid-catalyzed cyclization of appropriate amides bearing an alkyne moiety (Scheme 7.142). The required precursors **34** were synthesized from tryptamides and 3-substituted 2-propynoic acids. The intramolecular cyclization was achieved by microwave heating of a mixture of propynoic acid amides with 5 mol% mercury(II) triflate in dichloromethane. A reaction time of 15–20 min at 80 °C proved sufficient for most substrates to obtain the

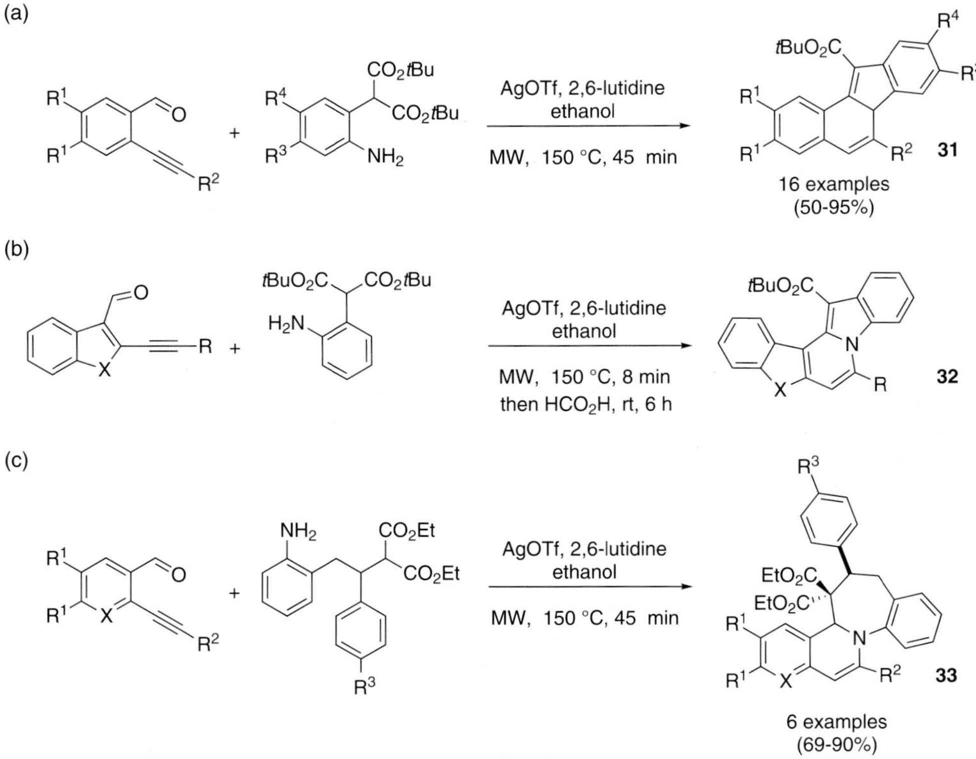

Scheme 7.141 Silver-catalyzed cascade synthesis of alkaloid ring systems.

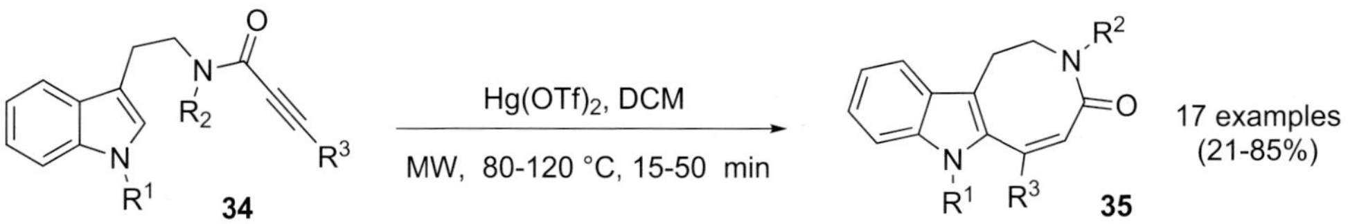

Scheme 7.142 Generation of indoloazocines by intramolecular alkyne carbocyclization.

target compounds in good yield. However, sterically challenging amides required up to 50 min at slightly enhanced reaction temperatures of 100 or 120 °C. In these cases, also the amount of catalyst has to be increased up to 30 mol% to achieve full conversion. The bulky *tert*-butyl group on the alkyne moiety totally inhibited the cyclization [248]. In general, the authors introduced a facile procedure for the preparation of indoloazocines to explore the diversity of alkaloid analogs.

Wipf and coworkers described a dipolar cycloaddition–*retro*-Mannich domino reaction to prepare pyrrolo[1,3]diazepines, which are of interest due to their inhibitory effects on HIV-1 protease [249]. The authors developed a simple microwave-accelerated synthesis of piperidyl 4-arylthio-3-oxazolin-5-ones **36** and their *in situ* conversion to corresponding diazepines **37**. In a typical procedure, ethyl cyanoformate and diethylamine were added at 0 °C to a solution of thiophenol in dry dichloromethane. After stirring under nitrogen atmosphere for 2 h, freshly prepared stock solution of titanium(IV) chloride and $BF_3 \cdot EtO_2$ (1 : 2) in dichloromethane was added dropwise, subsequently followed by initially prepared 1-benzyl-3-piperidone. The mixture was heated in a sealed microwave vial at 80 °C for 15 min to afford the desired intermediate **36** in acceptable yield. Adding **36** to a solution of 2 equiv acetylene derivative in chlorobenzene and microwave heating the mixture at 150 °C for 10 min furnished the desired targets **37** in moderate to good yields (Scheme 7.143). However, in most cases both applicable regioisomers were observed as an inseparable mixture. Furthermore, the reaction scope is limited to moderately to highly activated and thermally stable dipolarophiles. Nevertheless, the prepared diazepines are susceptible for further modifications and transformations to generate several valuable macrocycles [249].

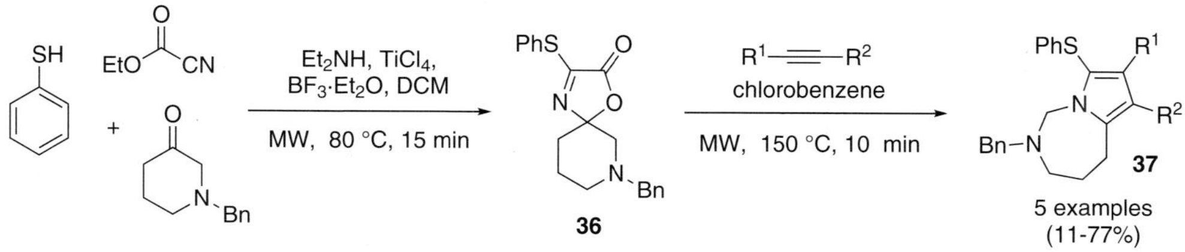

Scheme 7.143 Domino reactions of piperidyl 4-arylthio-3-oxazolin-5-ones with 1,3-dipolarophiles.

The construction of a diazaadamantane skeleton under microwave conditions was explored by Ivachtchenko *et al.* (Scheme 7.144) [250]. Cleavage of seminatural tetrahydro(−)-cytisine by acidic methanol provided the corresponding free diamine ester, which was carried on directly without purification by addition of 1,1′-carbonyldiimidazole in a water/methanol mixture at room temperature to an *N*-acyl-imidazole intermediate (not shown) that was cyclized to the target diazaadamantane system by microwave irradiation at 130 °C for 20 min. The reaction under microwave conditions was higher yielding and led to fewer side products as compared to the thermal run.

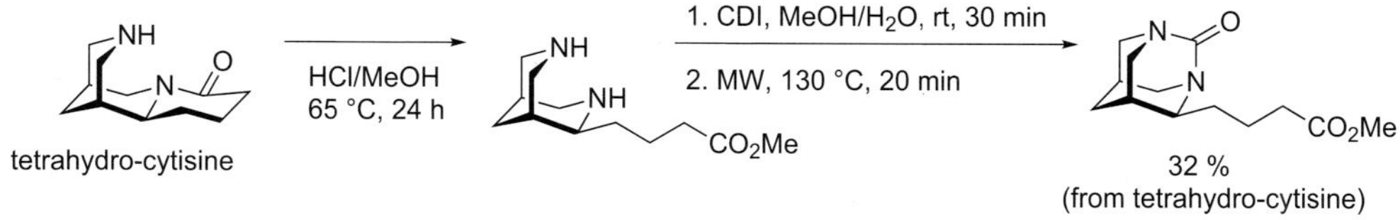

Scheme 7.144 Preparation of diazaadamantane skeletons.

Corroles are porphyrin analogs that lack one *meso* carbon bridge. Collman and Decréau have shown that the classical Gross synthesis of corroles from aldehydes and pyrroles adsorbed on an inorganic solid support such as alumina can be efficiently carried out under microwave irradiation (Scheme 7.145) [251]. Compared to con-

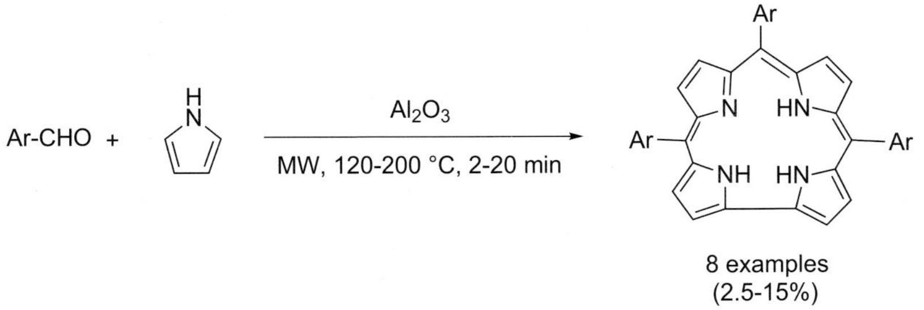

Scheme 7.145 Preparation of corroles from aromatic aldehydes and pyrroles.

ventional heating, the microwave technique afforded an increase in corrole yields of 20–50% and led to noticeable cleaner reactions. The general conditions involved mixing the aldehyde component with 3 equiv of pyrrole and oven-dried basic alumina, before the mixture was heated by microwave irradiation to 120–200 °C for 2–20 min.

The preparation of metallophthalocyanines under solvent-free conditions from 1,2-phthalonitrile or phthalic anhydride and urea in the presence of metal templates was reported by Bogdal and coworkers (Scheme 7.146) [252]. Among the many conditions screened by the authors, the use of either copper(II) or cobalt(II) chloride in the presence of small amounts of water provided the best results.

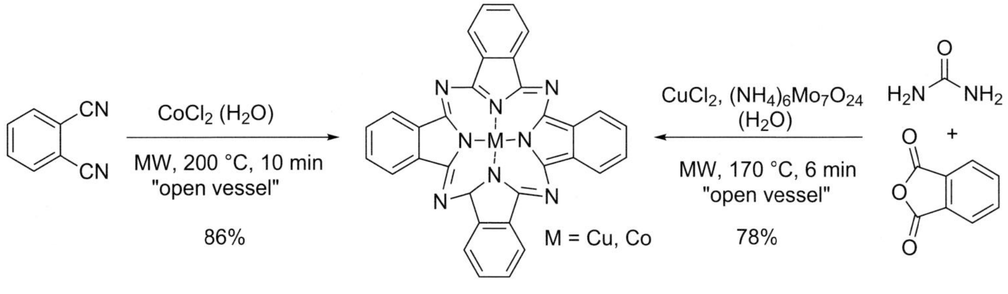

Scheme 7.146 Preparation of metallophthalocyanines.

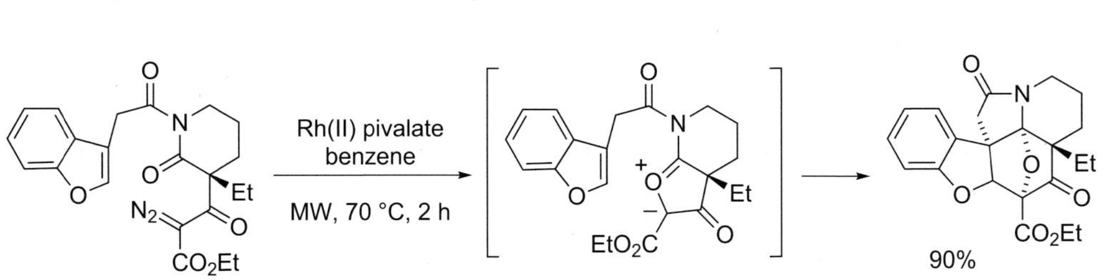

Scheme 7.147 Intramolecular [3 + 2] cycloaddition reactions of push–pull dipoles across heteroaromatic π-systems.

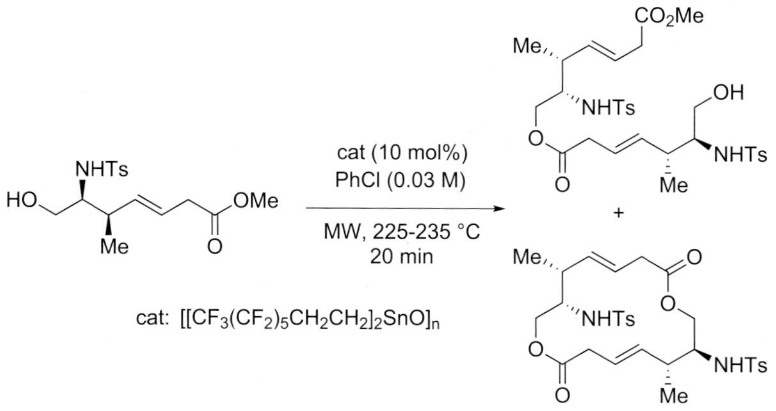

Scheme 7.148 Synthesis of complex macrodiolides by cyclodimerization of hydroxy esters.

Mejía-Oneto and Padwa have explored intramolecular [3 + 2] cycloaddition reactions of push–pull dipoles across heteroaromatic π-systems triggered by microwave irradiation [253]. The push–pull dipoles were generated from the rhodium(II)-catalyzed reaction of a diazo imide precursor containing a tethered heteroaromatic ring. In the examples shown in Scheme 7.147, microwave heating a solution of the diazo imide precursor in dry benzene in the presence of a catalytic amount of

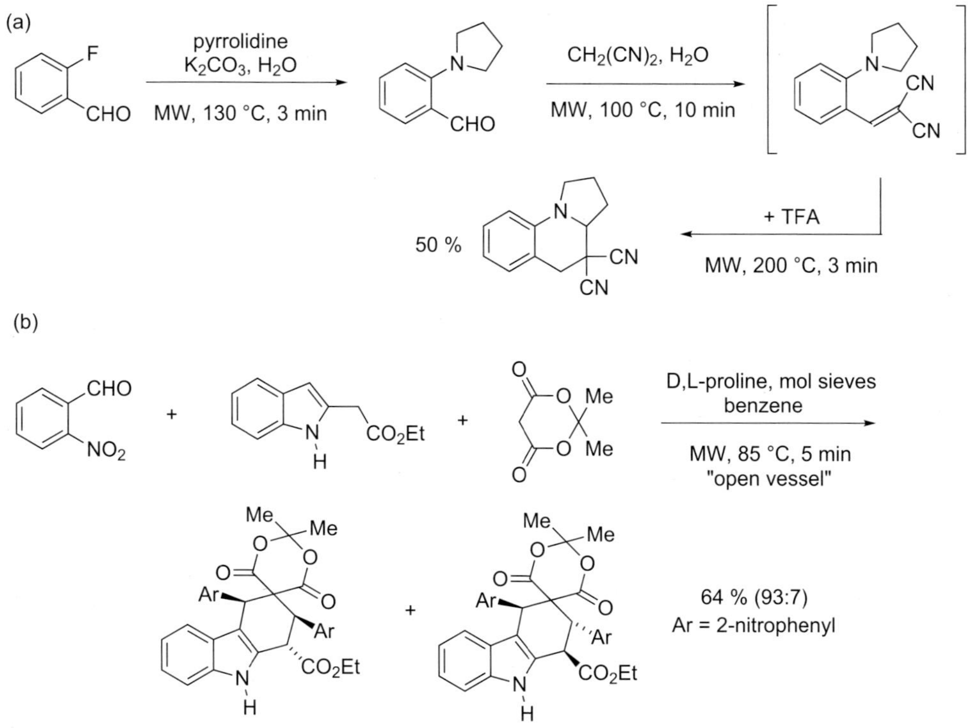

Scheme 7.149 Preparation of polycyclic and spiroheterocyclic structures.

rhodium(II) pivalate and 4 Å molecular sieves for 2 h at 70 °C produced a transient cyclic carbonyl ylide dipole that spontaneously underwent cycloaddition across the tethered benzofuran π-system to form a pentacyclic structure related to alkaloids of the vindoline type.

The synthesis of complex macrodiolides involving microwave-accelerated transesterification of chiral, nonracemic, hydroxy esters has been demonstrated by Porco and coworkers at Boston University (Scheme 7.148) [254]. Dramatic accelerations in cyclodimerizations of monomeric units containing ethers and amino acid subunits were seen under microwave heating with a fluorous tin oxide catalyst, affording optimal yields. Cyclodimerizations studied under microwave heating indicated an initial dependence of macrodiolide formation on microwave power. Use of a rapid cyclodimerization process under microwave incubation has been demonstrated to prepare a 127-member library of diverse macrodiolides with high levels of macrocyclic ring, stereochemical, and functional diversity.

Further reaction pathways leading to polycyclic structures and spiroheterocycles are summarized in Scheme 7.149 [255, 256]. Other synthetic transformations leading to larger heterocyclic or polycyclic ring systems have been described by several authors; see Refs. [257–269].

References

1 Van der Eycken, E. and Kappe, C.O. (eds) (2006) *Microwave-Assisted Synthesis of Heterocycles*, Springer, Berlin.

2 Torres, G., Torres, W., and Prieto, J.A. (2004) *Tetrahedron*, **60**, 10245–10251.

3 Ishikura, M., Hasunuma, M., Yamada, K., and Yanada, R. (2006) *Heterocycles*, **68**, 2253–2257.

4 Linder, M.R. and Podlech, J. (2001) *Organic Letters*, **3**, 1849–1851.

5 Jiao, L., Liang, Y., and Xu, J. (2006) *Journal of the American Chemical Society*, **128**, 6060–6069.

6 Hu, L., Wang, Y., Li, B., Du, D.-M., and Xu, J. (2007) *Tetrahedron*, **63**, 9387–9392.

7 Mortoni, A., Martinelli, M., Piarulli, U., Regalia, N., and Gagliardi, S. (2004) *Tetrahedron Letters*, **45**, 6623–6627.

8 Minetto, G., Raveglia, L.F., and Taddei, M. (2004) *Organic Letters*, **6**, 389–392.

9 Veitch, G.E., Bridgwood, K.L., Rands-Trevor, K., and Ley, S.V. (2008) *Synlett*, 2597–2600.

10 Werner, S., Iyer, P.S., Fodor, M.D., Coleman, C.M., Twining, L.A., Mitasev, B., and Brummond, K.M. (2006) *Journal of Combinatorial Chemistry*, **8**, 368–380.

11 Mattson, A.E., Bharadwaj, A.R., Zuhl, A.M., and Scheidt, K.A. (2006) *The Journal of Organic Chemistry*, **71**, 5715–5724.

12 Milgram, B.C., Eskildsen, K., Richter, S.M., Scheidt, W.R., and Scheidt, K.A. (2007) *The Journal of Organic Chemistry*, **72**, 3941–3944.

13 Gupton, J.T., Banner, E.J., Scharf, A.B., Norwood, B.K., Kanters, R.P.F., Dominey, R.N., Hempel, J.E., Kharlamova, A., Bluhn-Chertudi, I., Hickenboth, C.R., Little, B.A., Sartin, M.D., Coppock, M.B., Krumpe, K.E., Burnham, B.S., Holt, H., Du, K.X., Keertikar, K.M., Diebes, A., and Ghassemi, S. (2006) *Tetrahedron*, **62**, 8243–8255.

14 Wilson, M.A., Filzen, G., and Weimaker, G.S. (2009) *Tetrahedron Letters*, **50**, 4807–4809.

15 Bharadwaj, A.R. and Scheidt, K.A. (2004) *Organic Letters*, **6**, 2465–2468.

16 Tejedor, D., González-Cruz, D., García-Tellado, F., Marrero-Tellado, J.J.,

and López Rodríguez, M. (2004) *Journal of the American Chemical Society*, **126**, 8390–8391.
17 Tye, H. and Whittaker, M. (2004) *Organic and Biomolecular Chemistry*, **2**, 813–815.
18 Díaz-Ortiz, A., de la Hoz, A., Herrero, M.A., Prieto, P., Sánchez-Migallón, A., Cossio, F.P., Arrieta, A., Vivanco, S., and Foces-Foces, C. (2003) *Molecular Diversity*, **7**, 175–180.
19 Bashiardes, G., Safir, I., Mohamed, A.S., Barbot, F., and Laduranty, J. (2003) *Organic Letters*, **5**, 4915–4918.
20 Pospíšil, J. and Potáček, M. (2004) *European Journal of Organic Chemistry*, 710–716.
21 Guryanov, I., Toma, F.M., López, A.M., Carraro, M., Da Ros, T., Angelini, G., D'Aurizio, E., Fontana, A., Maggini, M., Prato, M., and Bonchio, M. (2009) *Chemistry – A European Journal*, **15**, 12837–12845.
22 Kim, I., Choi, J., Lee, S., and Lee, G.H. (2008) *Synlett*, 2334–2338.
23 Wilson, N.S., Sarko, C.R., and Roth, G.P. (2001) *Tetrahedron Letters*, **42**, 8939–8941.
24 Lipińska, T.M. (2004) *Tetrahedron Letters*, **45**, 8831–8834.
25 Lipińska, T.M. and Czarnocki, S.J. (2006) *Organic Letters*, **8**, 367–370.
26 Lipińska, T.M. (2006) *Tetrahedron*, **62**, 5736–5747.
27 Kraus, G.A. and Guo, H. (2008) *Organic Letters*, **10**, 3061–3063.
28 Fuwa, H. and Sasaki, M. (2009) *The Journal of Organic Chemistry*, **74**, 212–221.
29 Savarin, C.G., Grise, C., Murry, J.A., Reamer, R.A., and Hughes, D.L. (2007) *Organic Letters*, **9**, 981–983.
30 Siu, J., Baxendale, I.R., and Ley, S.V. (2004) *Organic and Biomolecular Chemistry*, **2**, 160–167.
31 Barluenga, J., Jiménez-Aquino, A., Aznar, F., and Valdés, C. (2010) *Chemistry – A European Journal*, **16**, 11707–11711.
32 Lehmann, F., Holm, M., and Laufer, S. (2009) *Tetrahedron Letters*, **50**, 1708–1709.
33 Goncalves, S., Wagner, A., and Mioskowski, C. (2009) *Tetrahedron Letters*, **50**, 274–276.
34 Schweizer, E., Gaich, T., Brecker, L., and Mulzer, J. (2007) *Synthesis*, 3807–3814.
35 Eidamshaus, C. and Burch, J.D. (2008) *Organic Letters*, **10**, 4211–4214.
36 Liu, S., Haller, M.A., Ma, H., Dalton, L.R., Jang, S.-H., and Jen, A.K.-Y. (2003) *Advanced Materials*, **15**, 603–607.
37 Morton, C.J.H., Riggs, R.L., Smith, D.M., Westwood, N.J., Lightfoot, P., and Slawin, A.M.Z. (2005) *Tetrahedron*, **61**, 727–738.
38 Boot, J.R., Brace, G., Delatour, C.L., Dezutter, N., Fairhurst, J., Findlay, J., Gallagher, P.T., Hoes, I., Mahadevan, S., Mitchell, S.N., Rathmell, R.E., Richards, S.J., Simmonds, R.G., Wallace, L., and Whatton, M.A. (2004) *Bioorganic & Medicinal Chemistry Letters*, **14**, 5395–5399.
39 Allen, D., Callaghan, O., Cordier, F.L., Dobson, D.R., Harris, J.R., Hotten, T.M., Owton, W.M., Rathmell, R.E., and Wood, V.A. (2004) *Tetrahedron Letters*, **45**, 9645–9647.
40 Giacomelli, G., Porcheddu, A., Salaris, M., and Taddei, M. (2003) *European Journal of Organic Chemistry*, 537–541.
41 Molteni, V., Hamilton, M.M., Mao, L., Crane, C.M., Termin, A.P., and Wilson, D.M. (2002) *Synthesis*, 1669–1674.
42 Humphries, P.S. and Finefield, J.M. (2006) *Tetrahedron Letters*, **47**, 2443–2446.
43 Ahlström, M.M., Ridderström, M., Zamora, I., and Luthman, K. (2007) *Journal of Medicinal Chemistry*, **50**, 4444–4452.
44 Pinkerton, A.B., Huang, D., Cube, R.V., Hutchinson, J.H., Struthers, M., Ayala, J.M., Vicario, P.P., Patel, S.R., Wisniewski, T., DeMartino, J.A., and Vernier, J.-M. (2007) *Bioorganic & Medicinal Chemistry Letters*, **17**, 807–813.
45 Bagley, M.C., Davis, T., Dix, M.C., Widdowson, C.S., and Kipling, D. (2006) *Organic and Biomolecular Chemistry*, **4**, 4158–4164.
46 Bagley, M., Lubinu, M.C., and Mason, C. (2007) *Synlett*, 704–708.
47 Borisov, A.V., Gorobets, N.Y., Yermolayev, S.A., Zhuravel, I.O., Kovalenko, S.M., and Dosenko, S.M. (2007) *Journal of Combinatorial Chemistry*, **9**, 909–911.

48 Daniels, R.N., Kim, K., Lebois, E.P., Muchalski, H., Hughes, M., and Lindsley, C.W. (2008) *Tetrahedron Letters*, **49**, 305–310.

49 Quiroga, J., Trilleras, J., Insuasty, B., Abonia, R., Nogueras, M., Marchal, A., and Cobo, J. (2008) *Tetrahedron Letters*, **49**, 3257–3259.

50 Cox, C.D., Breslin, M.J., and Mariano, B.J. (2004) *Tetrahedron Letters*, **45**, 1489–1493.

51 Glasnov, T., Groschner, K., and Kappe, C.O. (2009) *ChemMedChem*, **4**, 1816–1818.

52 Delgado, J.L., de la Cruz, P., López-Arza, V., Langa, F., Kimball, D.B., Haley, M.M., Araki, Y., and Ito, O. (2004) *The Journal of Organic Chemistry*, **69**, 2661–2668.

53 Delgado, J.L., de la Cruz, P., Langa, F., Urbina, A., Casado, J., and López Navarrete, J.T. (2004) *Chemical Communications*, 1734–1735.

54 Wolkenberg, S.E., Wisnoski, D.D., Leister, W.H., Wang, Y., Zhao, Z., and Lindsley, C.W. (2004) *Organic Letters*, **6**, 1453–141456.

55 Sparks, R.B. and Combs, A.P. (2004) *Organic Letters*, **6**, 2473–2475.

56 Soh, C.H., Chui, W.K., and Lam, Y. (2008) *Journal of Combinatorial Chemistry*, **10**, 118–122.

57 Preti, L., Attanasi, O.A., Caselli, E., Avi, G., Ori, C., Davoli, P., Felluga, F., and Prati, F. (2010) *European Journal of Organic Chemistry*, 4312–4320.

58 Ireland, S.M., Tye, H., and Whittaker, M. (2003) *Tetrahedron Letters*, **44**, 4369–4371.

59 Coleman, C.M., MacElroy, J.M.D., Gallagher, J.F., and O'Shea, D.F. (2002) *Journal of Combinatorial Chemistry*, **4**, 87–93.

60 Pospíšil, J. and Potáček, M. (2004) *Heterocycles*, **63**, 1165–1173.

61 Kim, Y.J. and Varma, R.S. (2004) *Tetrahedron Letters*, **45**, 7205–7208.

62 Merriman, G.H., Ma, L., Shum, P., McGarry, D., Volz, F., Sabol, J.S., Gross, A., Zhao, Z., Rampe, D., Wang, L., Wirtz-Brugger, F., Harris, B.A., and Macdonald, D. (2005) *Bioorganic & Medicinal Chemistry Letters*, **15**, 435–438.

63 Koubachi, J., El Kazzouli, S., Berteina-Raboin, S., Mouaddib, A., and Guillaumet, G. (2007) *The Journal of Organic Chemistry*, **72**, 7650–7655.

64 Ermolat'ev, D.S. and Van der Eycken, E.V. (2008) *The Journal of Organic Chemistry*, **73**, 6691–6697.

65 Ermolat'ev, D.S., Babaev, E.V., and Van der Eycken, E.V. (2006) *Organic Letters*, **8**, 5781–5784.

66 Soh, C., Chui, W.K., and Lam, Y. (2006) *Journal of Combinatorial Chemistry*, **8**, 464–468.

67 Aidouni, A., Demonceau, A., and Delaude, L. (2006) *Synlett*, 493–495.

68 Lin, S.-Y., Isome, Y., Stewart, E., Liu, J.-F., Yohannes, D., and Yu, L. (2006) *Tetrahedron Letters*, **47**, 2883–2886.

69 VanVliet, D.S., Gillespie, P., and Scicinski, J.J. (2005) *Tetrahedron Letters*, **46**, 6741–6743.

70 Al-Dweik, M., Zahra, J., Khanfar, M., El-Abdelah, M.M., and Zeller, K.-P. (2009) *Monatshefte für Chemie*, **140**, 221–228.

71 Ursic, U., Groselj, U., Meden, A., Svete, J., and Stanovnik, B. (2008) *Tetrahedron Letters*, **49**, 3775–3778.

72 Boufatah, N., Gellis, A., Maldonado, J., and Vanelle, P. (2004) *Tetrahedron*, **60**, 9131–9137.

73 Ferro, S., Rao, A., Zappalà, M., Chimirri, A., Barreca, M.L., Witvrouw, M., Debyser, Z., and Monforte, P. (2004) *Heterocycles*, **63**, 2727–2734.

74 Rao, A., Chimirri, A., Ferro, S., Monforte, A.M., Monforte, P., and Zappalá, M. (2004) *ARKIVOC*, **v**, 147–155.

75 Dymock, B., Barril, X., Beswick, M., Collier, A., Davies, N., Drysdale, M., Fink, A., Fromont, C., Hubbard, R.E., Massey, A., Surgenor, A., and Wright, L. (2004) *Bioorganic & Medicinal Chemistry Letters*, **14**, 325–328.

76 Ilaš, J., Pečar, S., Hockemeyer, J., Euler, H., Kirfel, A., and Müller, C.E. (2005) *Journal of Medicinal Chemistry*, **48**, 2108–2114.

77 Giacomelli, G., De Luca, L., and Porcheddu, A. (2003) *Tetrahedron*, **59**, 5437–5440.

78 Bernardi, L., Bonini, B.F., Comes-Franchini, M., Fochi, M., Folegatti, M., Grilli, S., Mazzanti, A., and Ricci, A. (2004) *Tetrahedron: Asymmetry*, **15**, 245–250.
79 Desai, B., Danks, T.N., and Wagner, G. (2003) *Dalton Transactions*, 2544–2549.
80 Desai, B., Danks, T.N., and Wagner, G. (2004) *Dalton Transactions*, 166–171.
81 Trost, B.M., Dogra, K., and Franzini, M. (2004) *Journal of the American Chemical Society*, **126**, 1944–1945.
82 Wipf, P., Fletcher, J.M., and Scarone, L. (2005) *Tetrahedron Letters*, **46**, 5463–5466.
83 Huxley, A. (2006) *Synlett*, 2658–2660.
84 Pirrung, M.C., Tumey, L.N., McClerren, A.L., and Raetz, C.R.H. (2003) *Journal of the American Chemical Society*, **125**, 1575–1586.
85 Crosignani, S., Young, A.C., and Linclau, B. (2004) *Tetrahedron Letters*, **45**, 9611–9615.
86 Katritzky, A.R., Cai, C., Suzuki, K., and Singh, S.K. (2004) *The Journal of Organic Chemistry*, **69**, 811–814.
87 García-Tellado, F., Loupy, A., Petit, A., and Marrero-Terrero, A.L. (2003) *European Journal of Organic Chemistry*, 4387–4391.
88 Kuhnert, N. and Danks, T.N. (2001) *Green Chemistry*, **3**, 68–70.
89 Diwischek, F., Heller, E., and Holzgrabe, U. (2003) *Monatshefte für Chemie*, **134**, 1105–1111.
90 Pottorf, R.S., Chadha, N.K., Katkevics, M., Ozola, V., Suna, E., Ghane, H., Regberg, T., and Player, M.R. (2003) *Tetrahedron Letters*, **44**, 175–178.
91 Frolov, E.B., Lakner, F.J., Khvat, A.V., and Ivachtchenko, A.V. (2004) *Tetrahedron Letters*, **45**, 4693–4696.
92 Chiacchio, U., Rescifina, A., Saita, M.G., Iannazzo, D., Romeo, G., Mates, J.A., Tejero, T., and Merino, P. (2005) *The Journal of Organic Chemistry*, **70**, 8991–9001.
93 Rescifina, A., Chiacchio, M.A., Corsaro, A., De Clercq, E., Iannazzo, D., Mastino, A., Piperno, A., Romeo, G., Romeo, R., and Valveri, V. (2006) *Journal of Medicinal Chemistry*, **49**, 709–715.
94 Yong, S.R., Ung, A.T., Pyne, S.G., Skelton, B.W., and White, A.H. (2007) *Tetrahedron*, **63**, 5579–5586.
95 Bolognese, A., Correale, G., Manfra, M., Lavecchia, A., Novellino, E., and Barone, V. (2004) *Organic and Biomolecular Chemistry*, **2**, 2809–2813.
96 Gududuru, V., Nguyen, V., Dalton, J.T., and Miller, D.D. (2004) *Synlett*, 2357–2358.
97 Gaonkar, S.L. and Shimizu, H. (2010) *Tetrahedron*, **66**, 3314–3317.
98 Emtenäs, H., Taflin, C., and Almqvist, F. (2003) *Molecular Diversity*, **7**, 165–169.
99 Pemberton, N., Åberg, V., Almstedt, H., Westermark, A., and Almqvist, F. (2004) *The Journal of Organic Chemistry*, **69**, 7830–7835.
100 Sanchez, S., Bateson, J.H., O'Hanlon, P.J., and Gallagher, T. (2004) *Organic Letters*, **6**, 2781–2783.
101 Frére, S., Thiéry, V., Bailly, C., and Besson, T. (2003) *Tetrahedron*, **59**, 773–779.
102 Frére, S., Thiéry, V., and Besson, T. (2003) *Synthetic Communications*, **33**, 3789–3798.
103 Kamila, S., Zhang, H., and Biehl, E.R. (2005) *Heterocycles*, **65**, 2119–2126.
104 Mu, X.-J., Zou, J.-P., Zeng, R.-S., and Wu, J.-C. (2005) *Tetrahedron Letters*, **46**, 4345–4347.
105 Kazmaier, U. and Ackermann, S. (2005) *Organic and Biomolecular Chemistry*, **3**, 3184–3187.
106 Merritt, E. and Bagley, M. (2007) *Synthesis*, 3535–3541.
107 Yuan, C., St. Jean, D.J., Liu, Q., Jr., Cai, L., Li, A., Han, N., Moniz, G., Askew, B., Hungate, R.W., Johansson, L., Tedenborg, L., Pyring, D., Williams, M., Hale, C., Chen, M., Cupples, R., Zhang, J., Jordan, S., Bartberger, M.D., and Sun, Y. (2007) *Bioorganic & Medicinal Chemistry Letters*, **17** 6056–6061.
108 Radi, M., Botta, L., Casaluce, G., Bernardini, M., and Botta, M. (2010) *Journal of Combinatorial Chemistry*, **12**, 200–205.
109 Katritzky, A.R. and Singh, S.K. (2002) *The Journal of Organic Chemistry*, **67**, 9077–9079.
110 Katritzky, A.R., Zhang, Y., Singh, S.K., and Steel, P.J. (2003) *ARKIVOC*, **xv**, 47–64.

111 Savin, K.A., Robertson, M., Gernert, D., Green, S., Hembre, E.J., and Bishop, J. (2003) *Molecular Diversity*, **7**, 171–174.

112 Khanetskyy, B., Dallinger, D., and Kappe, C.O. (2004) *Journal of Combinatorial Chemistry*, **7**, 884–892.

113 Appukkuttan, P., Dehaen, W., Fokin, V.V., and Van der Eycken, E. (2004) *Organic Letters*, **7**, 4223–4225.

114 (a) Ermolat'ev, D., Dehaen, W., and Van der Eycken, E. (2004) *The QSAR & Combinatorial Science*, **24**, 915–918; (b) Kappe, C.O. and Van der Eycken, E. (2010) *Chemical Society Reviews*, **39**, 1280–1290.

115 Gao, Y. and Lam, Y. (2008) *Journal of Combinatorial Chemistry*, **10**, 327–332.

116 Yeung, K.S., Farkas, M.E., Kadow, J.F., and Meanwell, N.A. (2005) *Tetrahedron Letters*, **46**, 3429–3432.

117 Aldrich, L.N., Lebois, E.P., Lewis, L.M., Nalywajko, N.T., Niswender, C.M., Weaver, C.D., Conn, P.J., and Lindsley, C.W. (2009) *Tetrahedron Letters*, **50**, 212–215.

118 Evans, M.D., Ring, J., Schoen, A., Bell, A., Edwards, P., Berthelot, D., Nicewonger, R., and Baldino, C.M. (2003) *Tetrahedron Letters*, **44**, 9337–9341.

119 Santagada, V., Frecentese, F., Perissutti, E., Cirillo, D., Terracciano, S., and Caliendo, G. (2004) *Bioorganic & Medicinal Chemistry Letters*, **14**, 4491–4493.

120 Adib, M., Jahromi, A.H., Tavoosi, N., Mahdavi, M., and Bijandazeh, H.R. (2006) *Tetrahedron Letters*, **47**, 2965–2967.

121 Kurz, T., Lolak, N., and Geffken, D. (2007) *Tetrahedron Letters*, **48**, 2733–2735.

122 Vasudevan, A., Ji, Z., Frey, R.R., Wada, C.K., Steinman, D., Heyman, H.R., Guo, Y., Curtin, M.L., Guo, J., Li, J., Pease, L., Glaser, K.B., Marcotte, P.A., Bouska, J.J., Davidsen, S.K., and Michaelides, M.R. (2003) *Bioorganic & Medicinal Chemistry Letters*, **13**, 3909–3913.

123 Natero, R., Koltun, D.O., and Zablocki, J.A. (2004) *Synthetic Communications*, **34**, 2523–2529.

124 García Mancheño, O. and Bolm, C. (2007) *Organic Letters*, **9**, 2951–2954.

125 Sevilla, S., Forns, P., Fernàndez, J.C., de la Figuera, N., Eastwood, P., and Albericio, F. (2006) *Tetrahedron Letters*, **47**, 8603–8606.

126 Deng, S.-L. and Chen, R.-Y. (2004) *Monatshefte fur Chemie*, **135**, 1113–1119.

127 Bliznets, I.V., Vasil'ev, A.A., Shorshnev, S.V., Stepanov, A.E., and Lukyanov, S.M. (2004) *Tetrahedron Letters*, **45**, 2571–2573.

128 Schulz, M.J., Coats, S.J., and Hlasta, D.J. (2004) *Organic Letters*, **6**, 3265–3268.

129 Ek, F., Manner, S., Wistrand, L.-G., and Frejd, T. (2004) *The Journal of Organic Chemistry*, **69**, 1346–1352.

130 Shie, J.J. and Fang, J.M. (2007) *The Journal of Organic Chemistry*, **72**, 3141–3144.

131 Semple, G., Skinner, P.J., Gharbaoui, T., Shin, Y.-J., Jung, J.-K., Cherrier, M.C., Webb, P.J., Tamura, S.Y., Boatman, P.D., Sage, C.R., Schrader, T.O., Chen, R., Colletti, S.L., Tata, J.R., Waters, M.G., Cheng, K., Taggart, A.K., Cai, T.-Q., Carballo-Jane, E., Behan, D.P., Connolly, D.T., and Richman, J.G. (2008) *Journal of Medicinal Chemistry*, **51** 5101–5108.

132 Vervisch, K., D'hooge, M., Törnroos, K.W., and De Kimpe, N. (2010) *The Journal of Organic Chemistry*, **75**, 7734–7744.

133 Bagley, M.C., Lunn, R., and Xiong, X. (2002) *Tetrahedron Letters*, **43**, 8331–8334.

134 Bagley, M.C. and Xiong, X. (2004) *Organic Letters*, **6**, 3401–3404.

135 Tejedor, D., Méndez-Abt, G., and García-Tellado, F. (2010) *European Journal of Organic Chemistry*, 6582–6587.

136 Öhberg, L. and Westman, J. (2001) *Synlett*, 1296–1298.

137 Bagley, M.C. and Singh, N. (2002) *Synlett*, 1718–1720.

138 Albrecht, M., Osetska, O., Rantanen, T., Fröhlich, R., and Bolm, C. (2010) *Synlett*, 1081–1084.

139 Gorobets, Yu.N., Yousefi, B., and Kappe, C.O. (2004) *Tetrahedron*, **60**, 8633–8644.

140 Adams, J., Hardin, A., and Vounatsos, F. (2006) *The Journal of Organic Chemistry*, **71**, 9895–9898.

141 Panunzio, M., Lentini, M.A., Campana, E., Martelli, G., Tamanini, E., and Vicennati, P. (2004) *Synthetic Communications*, **34**, 345–359.

142 Stadler, A., Pichler, S., Horeis, G., and Kappe, C.O. (2002) *Tetrahedron*, **58**, 3177–3183.

143 Lange, J.H.M., Verveer, P.C., Osnabrug, S.J.M., and Visser, G.M. (2001) *Tetrahedron Letters*, **42**, 1367–1369.

144 Jia, C.S., Zhang, Z., Tu, S.J., and Wang, G.W. (2006) *Organic and Biomolecular Chemistry*, **4**, 104–110.

145 Lekhok, K., Prajapati, D., and Boruah, R. (2008) *Synlett*, 655–658.

146 Tu, S., Zhang, J., Jia, R., Jiang, B., Zhang, Y., and Jiang, H. (2007) *Organic and Biomolecular Chemistry*, **5**, 1450–1453.

147 Glasnov, T.N., Stadlbauer, W., and Kappe, C.O. (2005) *The Journal of Organic Chemistry*, **70**, 3864–3870.

148 Rivkin, A. and Adams, B. (2006) *Tetrahedron Letters*, **47**, 2395–2398.

149 Chichetti, S.M., Ahearn, S.P., Adams, B., and Rivkin, A. (2007) *Tetrahedron Letters*, **48**, 8250–8252.

150 Chichetti, S.M., Ahearn, S.P., and Rivkin, A. (2008) *Tetrahedron Letters*, **49**, 6081–6083.

151 Yermolayev, S.A., Gorobets, Yu.N., and Desenko, S.M. (2009) *Journal of Combinatorial Chemistry*, **11**, 44–46.

152 Srinivasan, N. and Ganesan, A. (2003) *Chemical Communications*, 916–917.

153 Yen, Y.-H. and Chu, Y.-H. (2004) *Tetrahedron Letters*, **45**, 8137–8140.

154 Campiglia, P., Gomez-Monterrey, I., Lama, T., Novellino, E., and Grieco, P. (2004) *Molecular Diversity*, **8**, 427–430.

155 Kuo, F.-M., Tseng, M.-C., Yen, Y.-H., and Chu, Y.-H. (2004) *Tetrahedron*, **60**, 12079–12088.

156 Padwa, A., Lee, H.I., Rashatasakhon, P., and Rose, M. (2004) *The Journal of Organic Chemistry*, **69**, 8209–8218.

157 Mésangeau, C., Yous, S., Pérès, B., Lesieur, D., and Besson, T. (2005) *Tetrahedron Letters*, **46**, 2465–2468.

158 Aberg, V., Norman, F., Chorell, E., Westermark, A., Olofsson, A., Sauer-Eriksson, A.E., and Almqvist, F. (2005) *Organic and Biomolecular Chemistry*, **3**, 2817–2823.

159 Tomaszewski, M.J., Whalley, A., and Hu, Y.J. (2008) *Tetrahedron Letters*, **49**, 3172–3175.

160 Duvelleroy, D., Perrio, C., Parisel, O., and Lasne, M.-C. (2005) *Organic and Biomolecular Chemistry*, **3**, 3794–3804.

161 Fernandez, S.Y., Raw, S.A., and Taylor, R.J.K. (2005) *The Journal of Organic Chemistry*, **70**, 10086–10095.

162 Lee, H.K., Cao, H., and Rana, T.M. (2005) *Journal of Combinatorial Chemistry*, **7**, 279–284.

163 Crawford, K.R., Bur, S.K., Straub, C.S., and Padwa, A. (2003) *Organic Letters*, **5**, 3337–3340.

164 Theoclitou, M.-E. and Robinson, L.A. (2002) *Tetrahedron Letters*, **43**, 3907–3910.

165 Perzyna, A., Houssin, R., Barby, D., and Hénichart, J.-P. (2002) *Synlett*, 2077–2079.

166 Ivachtchenko, A.V., Khvat, A.V., Kobak, V.V., Kysil, V.M., and Williams, C.T. (2004) *Tetrahedron Letters*, **45**, 5473–5476.

167 Wilson, N.S., Sarko, C.R., and Roth, G.P. (2002) *Tetrahedron Letters*, **43**, 581–583.

168 Hong, B.-C., Wu, J.-L., Gupta, A.K., Hallur, M.S., and Liao, J.-H. (2004) *Organic Letters*, **6**, 3453–3456.

169 Carranco, I., Díaz, J.L., Jiménez, O., Vendrell, M., Albericio, F., Royo, M., and Lavilla, R. (2005) *Journal of Combinatorial Chemistry*, **7**, 33–41.

170 Hughes, R.A., Thomson, S.P., Alcaraz, L., and Moody, C.J. (2004) *Chemical Communications*, 946–948.

171 Moody, C.J., Hughes, R.A., Thompson, S.P., and Alcaraz, L. (2002) *Chemical Communications*, 1760–1761.

172 Leutbecher, H., Conrad, J., Klaiber, I., and Beifuss, U. (2004) *The QSAR & Combinatorial Science*, **23**, 895–898.

173 Devi, I., and Bhuyan, P.J. (2004) *Tetrahedron Letters*, **45**, 8625–8627.

174 Kabalka, G.W. and Mereddy, A.R. (2005) *Tetrahedron Letters*, **46**, 6315–6317.

175 Seijas, J.A., Vazquez-Tato, M.P., and Carballido-Reboredo, R. (2005) *The Journal of Organic Chemistry*, **70**, 2855–2858.

176 Katkevičs, M., Kontijevskis, A., Mutule, I., and Suna, E. (2007) *Chemistry of Heterocyclic Compounds*, **43**, 151–159.

177 Hellal, M., Bourguignon, J.J., and Bihel, F.J.J. (2008) *Tetrahedron Letters*, **49**, 62–65.
178 Mihovilovic, M.D., Leisch, H.G., and Mereiter, K. (2004) *Tetrahedron Letters*, **45**, 7087–7090.
179 Choi, P.J., Rathwell, D.C.K., and Brimble, M.A. (2009) *Tetrahedron Letters*, **50**, 3245–3248.
180 Bagley, M.C., Hughes, D.D., Sabo, H.M., Taylor, P.H., and Xiong, X. (2003) *Synlett*, 1443–1446.
181 Bagley, M.C., Hughes, D.D., and Taylor, P.H. (2003) *Synlett*, 259–261.
182 Bagley, M.C., Hughes, D.D., Lubinu, M.C., Merritt, E.A., Taylor, P.H., and Tomkinson, N.C.O. (2004) *The QSAR & Combinatorial Science*, **23**, 859–867.
183 Bagley, M.C., Lin, Z., and Pope, S.J.A. (2009) *Tetrahedron Letters*, **50**, 6818–6822.
184 Stadler, A. and Kappe, C.O. (2001) *Journal of Combinatorial Chemistry*, **3**, 624–630.
185 Kappe, C.O. and Stadler, A. (2003) *Combinatorial Chemistry, Part B* (eds B.B. Bunin and G. Morales), Elsevier Science, pp. 197–223.
186 Sandin, H., Swanstein, M.-L., and Wellner, E. (2004) *The Journal of Organic Chemistry*, **69**, 1571–1580.
187 Alexandre, F.-R., Berecibar, A., Wrigglesworth, R., and Besson, T. (2003) *Tetrahedron*, **59**, 1413–1419.
188 Alexandre, F.-R., Berecibar, A., and Besson, T. (2002) *Tetrahedron Letters*, **43**, 3911–3913.
189 Movassaghi, M. and Hill, M.D. (2006) *Journal of the American Chemical Society*, **128**, 14254–14255.
190 Chilin, A., Marzaro, G., Zanatta, S., and Giuotto, A. (2007) *Tetrahedron Letters*, **48**, 3229–3231.
191 Ferrini, S., Ponticelli, F., and Taddei, M. (2007) *Organic Letters*, **9**, 69–72.
192 Hill, M.D. and Movassaghi, M. (2008) *Tetrahedron Letters*, **49**, 4286–4288.
193 Zhang, X., Ye, D., Sun, H., Guo, D., Wang, J., Huang, H., Zhang, X., Jiang, H., and Liu, H. (2009) *Green Chemistry*, **11**, 1881–1888.
194 Vasquez, T.E., Nixey, T., Chenera, B., Gore, V., Bartberger, M., Sun, Y., and Hulme, C. (2003) *Molecular Diversity*, **7**, 161–164.
195 Ming, L., Shuwen, W., Lirong, W., Huazheng, Y., and Xiuli, Z. (2005) *Journal of Heterocyclic Chemistry*, **42**, 925–930.
196 Shaaban, M.R. (2008) *Journal of Fluorine Chemistry*, **129**, 1156–1161.
197 Li, Z., Huang, H., Sun, H., Liang, H., and Liu, H. (2008) *Journal of Combinatorial Chemistry*, **10**, 484–486.
198 Liu, J.-F., Lee, J., Dalton, A.M., Bi, G., Yu, L., Baldino, C.M., McElroy, E., and Brown, M. (2005) *Tetrahedron Letters*, **46**, 1241–1244.
199 Liu, J.-F., Ye, P., Sprague, K., Sargent, K., Yohannes, D., Baldino, C.M., Wilson, C.J., and Ng, S.-C. (2005) *Organic Letters*, **7**, 3363–3366.
200 Kostakis, I.K., Elomri, A., Seguin, E., Iannelli, M., and Besson, T. (2007) *Tetrahedron Letters*, **48**, 6609–6613.
201 Kalusa, A., Chessum, N., and Jones, K. (2008) *Tetrahedron Letters*, **49**, 5840–5842.
202 Tyagarajan, S. and Chakravarty, P.K. (2005) *Tetrahedron Letters*, **46**, 7889–7891.
203 Nie, A., Wang, J., and Huang, Z. (2006) *Journal of Combinatorial Chemistry*, **8**, 646–648.
204 Devi, I. and Bhuyan, P.J. (2005) *Tetrahedron Letters*, **46**, 5727–5729.
205 Avalos, M., Babiano, R., Cintas, P., Hursthouse, M.B., Jiménez, J.L., Lerma, E., Light, M.E., and Palacios, J.C. (2006) *Tetrahedron Letters*, **47**, 1989–1992.
206 Yoburn, J.C. and Baskaran, S. (2005) *Organic Letters*, **7**, 3801–3803.
207 Sagar, R. and Park, S.B. (2008) *The Journal of Organic Chemistry*, **73**, 3270–3273.
208 Yoon, D.S., Han, Y., Stark, T.M., Haber, J.C., Gregg, B.T., and Stankovich, S.B. (2004) *Organic Letters*, **6**, 4775–4778.
209 Mont, N., Teixidó, J., Borrell, J.I., and Kappe, C.O. (2003) *Tetrahedron Letters*, **44**, 5385–5388.
210 Mont, N., Teixidó, J., Borrell, J.I., and Kappe, C.O. (2003) *Molecular Diversity*, **7**, 153–159.
211 Mont, N., Fernández-Megido, L., Teixidó, J., Kappe, C.O., and Borrell, J.I. (2004) *The QSAR & Combinatorial Science*, **23**, 836–849.
212 Gohain, M., Prajapati, D., Gogoi, B.J., and Sandhu, J.S. (2004) *Synlett*, 1179–1182.

213 Devi, I. and Bhuyan, P.J. (2004) *Synlett*, 283–286.
214 Devi, I., Borah, H.N., and Bhuyan, P.J. (2004) *Tetrahedron Letters*, **45**, 2405–2408.
215 Baghbanzadeh, M., Molnar, M., Damm, M., Reidlinger, C., Dabiri, M., and Kappe, C.O. (2009) *Journal of Combinatorial Chemistry*, **11**, 676–684.
216 Westman, J., Lundin, R., Stalberg, J., Ostbye, M., Franzen, A., and Hurynowicz, A. (2002) *Combinatorial Chemistry & High Throughput Screening*, **5**, 565–570.
217 Westman, J. and Orrling, K. (2002) *Combinatorial Chemistry & High Throughput Screening*, **5**, 571–574.
218 Gising, J., Örtqvist, P., Sandström, A., and Larhed, M. (2009) *Organic and Biomolecular Chemistry*, **7**, 2809–2815.
219 Díaz-Ortiz, A., de la Hoz, A., Moreno, A., Prieto, P., León, R., and Herrero, M.A. (2002) *Synlett*, 2037–2038.
220 Kim, S.Y., Park, K.H., and Chung, Y.K. (2005) *Chemical Communications*, 1321–1323.
221 Kamila, S. and Biehl, E.R. (2006) *Heterocycles*, **68**, 1931–1939.
222 Zhao, Z., Wisnoski, D.D., Wolkenberg, S.E., Leister, W.H., Wang, Y., and Lindsley, C.W. (2004) *Tetrahedron Letters*, **45**, 4873–4876.
223 Lindsley, C.W., Zhao, Z., Leister, W.H., Robinson, R.G., Barnett, S.F., Defeo-Jones, D., Jones, R.E., Hartman, G.D., Huff, J.R., Huber, H.E., and Duggan, M.E. (2005) *Bioorganic & Medicinal Chemistry Letters*, **15**, 761–764.
224 Jainta, M., Nieger, M., and Bräse, S. (2008) *European Journal of Organic Chemistry*, 5418–5424.
225 Minetto, G., Lampariello, L.R., and Taddei, M. (2005) *Synlett*, 2743–2746.
226 Caprosu, M.D., Butnariu, R.M., and Mangalagiu, I.I. (2005) *Heterocycles*, **65**, 1871–1879.
227 Zbancioc, G.N., Caprosu, M.C., Moldoveanu, C.C., and Mangalagiu, I.I. (2005) *ARKIVOC*, **x**, 189–198.
228 Hoogenboom, R., Moore, B.C., and Schubert, U.S. (2006) *The Journal of Organic Chemistry*, **71**, 4903–4909.
229 Le Foulon, F.-X., Braud, E., Fabis, F., Lancelot, J.-C., and Rault, S. (2003) *Tetrahedron*, **59**, 10051–10057.
230 Koshkin, A.V., Fedorova, O.A., Lokshin, V., Guglielmetti, R., Hamelin, J., Texier-Boullet, F., and Gromov, S.P. (2004) *Synthetic Communications*, **34**, 315–322.
231 Caliendo, G., Perissutti, E., Santagada, V., Fiorino, F., Severino, B., Cirillo, D., d'Emmanuele di Villa Bianca, R., Lippolis, L., Pinto, A., and Sorrentino, R. (2004) *European Journal of Medicinal Chemistry*, **39**, 815–826.
232 Dai, W.M., Wang, X., and Ma, C. (2005) *Tetrahedron*, **61**, 6879–6885.
233 Yuan, Y., Liu, G., Li, L., Wang, Z., and Wang, L. (2007) *Journal of Combinatorial Chemistry*, **9**, 158–170.
234 Feng, G., Wu, J., and Dai, W.M. (2006) *Tetrahedron*, **62**, 4635–4642.
235 Xing, X., Wu, J., Feng, G., and Dai, W.M. (2006) *Tetrahedron*, **62**, 6774–6781.
236 Mečiarova, M., Toma, S., and Magdolen, P. (2003) *Synthetic Communications*, **33**, 3059–3054.
237 Lee, H.-K. and Rana, T.M. (2004) *Journal of Combinatorial Chemistry*, **6**, 504–508.
238 Peng, Y. and Song, G. (2004) *Tetrahedron Letters*, **45**, 5313–5316.
239 Chen, H., Dao, P., Laporte, A., and Garbay, C. (2010) *Tetrahedron Letters*, **51**, 3174–3176.
240 Zhao, Z., Leister, W.H., Strauss, K.A., Wisnoski, D.D., and Lindsley, C.W. (2003) *Tetrahedron Letters*, **44**, 1123–1128.
241 Lindsley, C.W., Wisnoski, D.D., Wang, Y., Leister, W.H., and Zhao, Z. (2003) *Tetrahedron Letters*, **44**, 4495–4498.
242 Lahue, B.R., Lo, S.-M., Wan., Z.-K., Woo, G.H.C., and Snyder, J.K. (2004) *The Journal of Organic Chemistry*, **69**, 7171–7182.
243 Holzgrabe, U. and Heller, E. (2003) *Tetrahedron*, **59**, 781–787.
244 Vasudevan, A., Villamil, C.I., and Djuric, S.W. (2004) *Organic Letters*, **6**, 3361–3364.
245 Liu, J.F., Kaselj, M., Isome, Y., Chapnick, J., Zhang, B., Bi, G., Yohannes, D., Yu, L., and Baldino, C.M. (2005) *The Journal of Organic Chemistry*, **70**, 10488–10493.

246 Ohta, Y., Chiba, H., Oishi, S., Fujii, N., and Ohno, H. (2008) *Organic Letters*, **10**, 3535–3538.
247 Waldmann, H., Eberhardt, L., Wittstein, K., and Kumar, K. (2010) *Chemical Communications*, **46**, 4622–4624.
248 Donets, P.A., Van Hecke, K., Meervelt, L.V., and Van der Eycken, E.V. (2009) *Organic Letters*, **11**, 3618–3621.
249 Liang, M., Saiz, C., Pizzo, C., and Wipf, P. (2009) *Tetrahedron Letters*, **50**, 6810–6813.
250 Ivachtchenko, A., Khvat, A., Tkachenko, S.E., Sandulenko, Y.B., and Vvedensky, V.Y. (2004) *Tetrahedron Letters*, **45**, 6733–6736.
251 Collman, J.P. and Decréau, R.A. (2003) *Tetrahedron Letters*, **44**, 1207–1210.
252 Burczyk, A., Loupy, A., Bogdal, D., and Petit, A. (2005) *Tetrahedron*, **61**, 179–188.
253 Mejía-Oneto, J.M. and Padwa, A. (2004) *Organic Letters*, **6**, 3241–3244.
254 Beeler, A.B., Acquilano, D.E., Su, Q., Yan, F., Roth, B.L., Panek, J.S., and Porco, J.A. (2005) *Journal of Combinatorial Chemistry*, **7**, 673–681.
255 Kaval, N., Dehaen, W., Mátyus, P., and Van der Eycken, E.V. (2004) *Green Chemistry*, **6**, 125–127.
256 Cochard, F., Laronze, M., Sigaut, P., Sapi, J., and Laronze, J.-Y. (2004) *Tetrahedron Letters*, **45**, 1703–1707.
257 Abbiati, G., Arcadi, A., Bellinazzi, A., Beccalli, E., Rossi, E., and Zanzola, S. (2005) *The Journal of Organic Chemistry*, **70**, 4088–4095.
258 Vvedensky, V.Y., Ivanov, Y.V., Kysil, V., Williams, C., Tkachenko, S., Kiselyov, A., Khvat, A.V., and Ivachtchenko, A.V. (2005) *Tetrahedron Letters*, **46**, 3953–3956.
259 Kaval, N., Halasz-Dajka, B., Vo-Thanh, G., Dehaen, W., Van der Eycken, J., Mátyus, P., Loupy, A., and Van Der Eycken, E. (2005) *Tetrahedron*, **61**, 9052–9057.
260 Lebrini, M., Bentiss, F., Vezin, H., Wignacourt, J.P., Roussel, P., and Lagrenee, M. (2005) *Heterocycles*, **65**, 2847–2860.
261 Buckley, B.R., Boxhall, J.Y., Bulman Page, P.C., Chan, Y., Elsegood, M.R.J., Heaney, H., Holmes, K.E., McIldowie, M.J., McKee, V., McGrath, M.J., Mocerino, M., Poulton, A.M., Sampler, E.P., Skelton, B.W., and Whiteuckley, A.H. (2006) *European Journal of Organic Chemistry*, 5117–5134.
262 Koizumi, H., Itoh, Y., and Ichikawa, T. (2006) *Chemistry Letters*, **35**, 1350–1351.
263 Giuseppone, N., Schmitt, J.L., and Lehn, J.M. (2006) *Journal of the American Chemical Society*, **128**, 16748–16763.
264 Miljanic, O.S., Dichtel, W.R., Mortezaei, S., and Stoddart, J.F. (2006) *Organic Letters*, **8**, 4835–4838.
265 Su, S. and Porco, J.A. (2007) *Organic Letters*, **9**, 4983–4986.
266 Cho, T.J., Shreiner, C.D., Hwang, S.-H., Moorefield, C.N., Courneya, B., Godínez, L.A., Manríquez, J., Jeong, K.-U., Cheng, S.Z.D., and Newkome, G.R. (2007) *Chemical Communications*, 4456–4458.
267 Dogutan, D.K., Zaidi, S.H.H., Thamyongkit, P., and Lindsey, J.S. (2007) *The Journal of Organic Chemistry*, **72**, 7701–7714.
268 Chandrasekharam, M., Rao, C.S., Singh, S.P., Kantam, M.L., Reddy, M.R., Reddy, P.Y., and Toru, T. (2007) *Tetrahedron Letters*, **48**, 2627–2630.
269 Elghamry, I. and Tietze, L.F. (2008) *Tetrahedron Letters*, **49**, 3972–3975.

8
Literature Survey Part D: Combinatorial Chemistry and High-Throughput Organic Synthesis

8.1
Solid-Phase Organic Synthesis

8.1.1
Peptide Synthesis and Related Examples

One of the first dedicated applications of microwaves toward solid-phase chemistry was the synthesis of small peptide molecules, presented by Wang and coworkers [1]. As a preliminary test, the authors coupled Fmoc-Ile and Fmoc-Val with Gly-preloaded Wang resin using the corresponding symmetric anhydrides (Scheme 8.1). The reactions were carried out rapidly in a modified domestic microwave oven with mountings in the sidewall for sufficient inert gas supply. Furthermore, a dedicated custom-made solid-phase reaction vessel was employed under atmospheric pressure conditions.

The microwave protocol increased the reaction rate at least two- to threefold, as conversion was only 60–80% within 6 min under conventional heating. This improved coupling efficiency was duplicated with numerous amino acid derivatives and further two peptide fragments were coupled with the Gly-Wang resin. These couplings were completed within 2 min as determined by a quantitative ninhydrin assay.

For a more representative investigation, the authors synthesized a fragment of the acyl carrier protein ($^{65-74}$ACP) using preformed active esters in *N,N*-dimethylformamide (DMF) [1]. Each coupling step included only 4 min of irradiation in a modified domestic oven with an average coupling yield of >99% (Scheme 8.2).

Importantly, it was demonstrated that no significant racemization occurred in the peptide formation. Furthermore, the complete coupling of difficult sequence peptides could be accomplished within a few minutes and it was determined that under microwave irradiation conditions peptide fragments have higher reactivity than single amino acid derivatives. However, the exact reaction temperature during the irradiation period was not determined.

Microwaves in Organic and Medicinal Chemistry, Second Edition.
C. Oliver Kappe, Alexander Stadler, and Doris Dallinger.

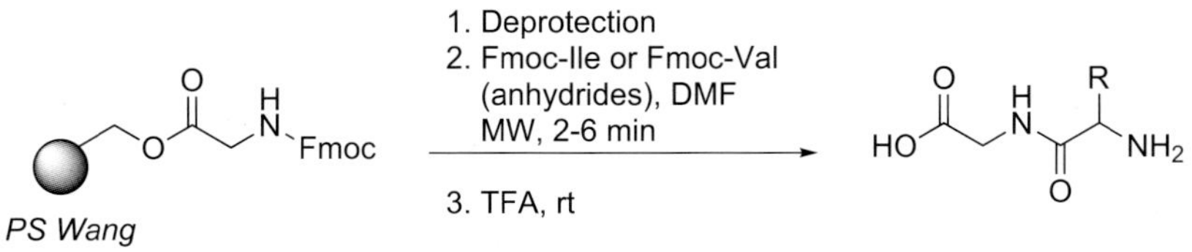

Scheme 8.1 Rapid solid-phase peptide coupling.

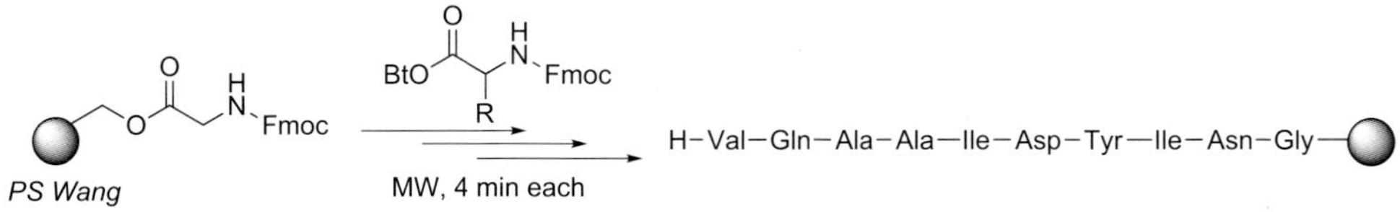

Scheme 8.2 Synthesis of $^{65-74}$ACP employing stepwise coupling of amino acid esters.

In a related study, microwave irradiation has been applied to the coupling of sterically hindered amino acids, leading to di- and tripeptides (Scheme 8.3) [2]. Erdélyi and Gogoll investigated a variety of common coupling reagents, for example, benzotriazol-1-yl-oxytripyrrolidinophosphonium hexafluorophosphate (PyBOP), 2-(7-aza-1*H*-benzotriazol-1-yl)-1,1,3,3-tetramethyluronium hexafluoride (HATU), *O*-(benzotriazol-1-yl)-*N*,*N*,*N*′,*N*′-tetramethyluronium tetrafluoroborate (TBTU), and Mukaiyama reagent, for peptide synthesis. Under controlled microwave conditions with IR temperature measurement, coupling of the amino acids via the corresponding anhydrides or *N*-hydroxybenzotriazole (HOBt)-activated esters was completed within 20 min without the need of double or triple coupling steps as in conventional protocols. The azobenzotriazole derivatives showed increased coupling efficiency up to 110 °C. Above this temperature, decomposition of the reagents was indicated by the color change of the reaction mixtures. However, no degradation of the solid support was observed. Furthermore, both LC–MS and ^{1}H NMR confirmed the absence of racemization during the high-temperature treatment in the presence of *N*,*N*-diisopropylethylamine (DIEA) as base.

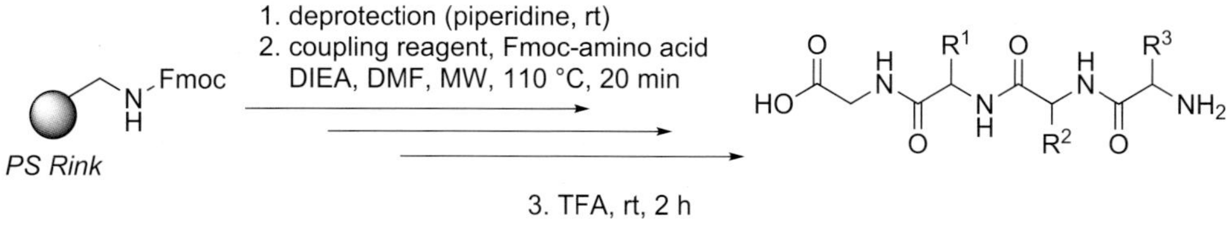

Scheme 8.3 Efficient tripeptide synthesis employing various coupling reagents.

Somfai and coworkers reported on the efficient coupling of a set of carboxylic acids suitable as potential scaffolds for peptide synthesis to a polymer-bound hydrazide linker [3]. Indole-like scaffolds were selected for this small library synthesis as these structures are found in numerous natural products with interesting activities. The best results have been obtained using 2-(7-aza-1*H*-benzotriazol-1-yl)-1,1,3,3-

tetramethyluronium hexafluoride and N,N-diisopropylethylamine in N,N-dimethylformamide as a solvent. Heating the reaction mixtures at 180 °C for 10 min achieved the desired products in high yields (Scheme 8.4). In this application, no Fmoc protection of the indole nitrogen is required.

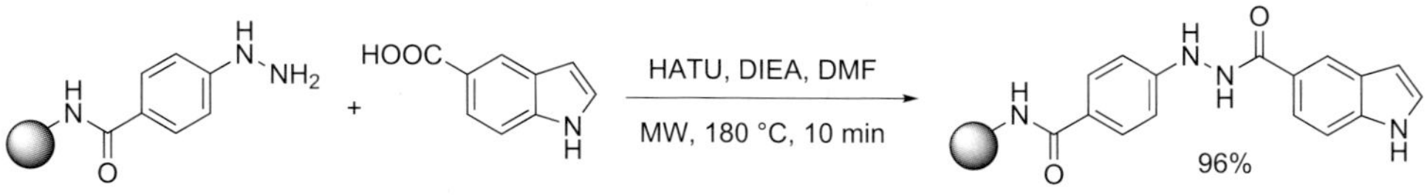

Scheme 8.4 Coupling of indolyl acid to polystyrene-bound hydrazide linker.

A notable increase in the overall speed of peptoid synthesis was reported by Kodadek and coworkers [4]. Therein, the authors present a multistep protocol for the generation of various peptoids employing a domestic microwave oven (Scheme 8.5). Reaction times were drastically reduced, requiring less than 1 min for the coupling of each residue. With this protocol, nine different primary amines were used to generate different 9-residue homo-oligomers, one 20-residue homo-oligomer, and one 9-residue hetero-oligomer. These transformations were performed in N,N-dimethylformamide as a solvent, employing N,N′-diisopropylcarbodiimide (DIC) as activating agent in two 15 s runs with manual stirring between the irradiation steps. In this case, the temperature of the solutions did not exceed 35 °C as determined by inserting a thermometer after the second 15 s run. For comparative purposes, these couplings were carried out also by conventional thermal heating at 37 °C, leading to similar results. Both methods, however, provided better yields and purities than conventional room-temperature couplings. In general, this protocol allows a convenient method for high-throughput peptoid synthesis, as microwave acceleration reduces the overall production time. For example, a nine-residue peptoid can be synthesized in 3 h, compared to 20–32 h employing the standard protocol.

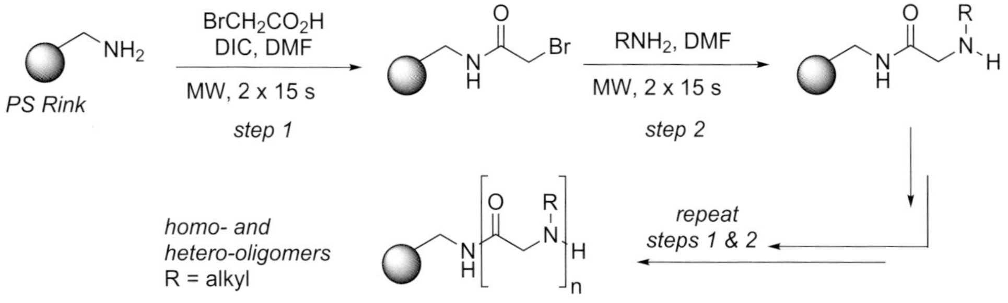

Scheme 8.5 Construction of peptoid sequences. Cleavage was carried out using trifluoroacetic acid at room temperature.

In another application, the group of Berteina-Raboin demonstrated the solid-supported synthesis of the indole core of melatonin analogs under microwave irradiation (Scheme 8.6) [5]. Coupling of the benzoic acid derivative to the Rink

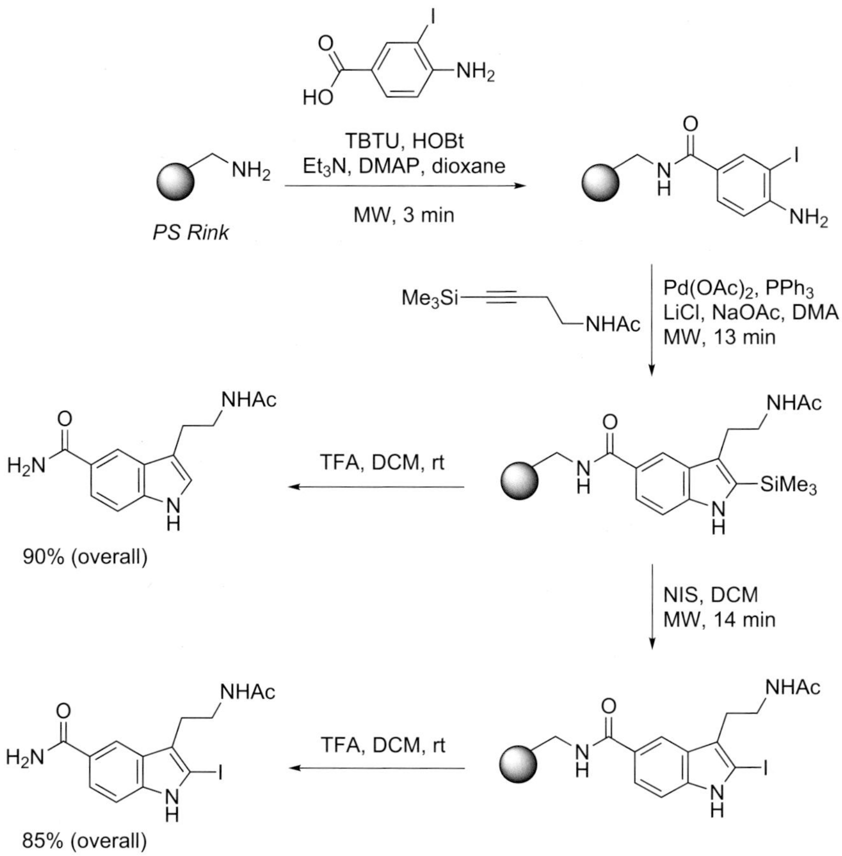

Scheme 8.6 Solid-phase synthesis of 5-carboxamido-*N*-acetyltryptamines.

amide resin was achieved by using the peptide coupling reagents *N*-hydroxybenzotriazole and *O*-(benzotriazol-1-yl)-*N*,*N*,*N*′,*N*′-tetramethyluronium tetrafluoroborate. Subsequent indole ring formation was carried out by palladium-catalyzed coupling directly on the solid phase. Synthesis of the corresponding iodo compound was accomplished by treatment of the polymer-bound indole derivative with *N*-iodosuccinimide (NIS). Finally, the desired compounds were released from the resin by using conventional trifluoroacetic acid (TFA)/dichloromethane cleavage cocktails at room temperature. Carrying out these reactions under microwave conditions led to a substantial increase in yields and a significant reduction in the reaction time, compared to conventional thermal heating, where each individual step takes 24–48 h. Due to the limited thermal stability of the solid support, the temperature during the reactions was kept below 140 °C, which was achieved by strict power control.

Helen E. Blackwell and coworkers from the University of Wisconsin-Madison have described the synthesis of *N*-acyl-L-homoserine lactones (AHLs) via a

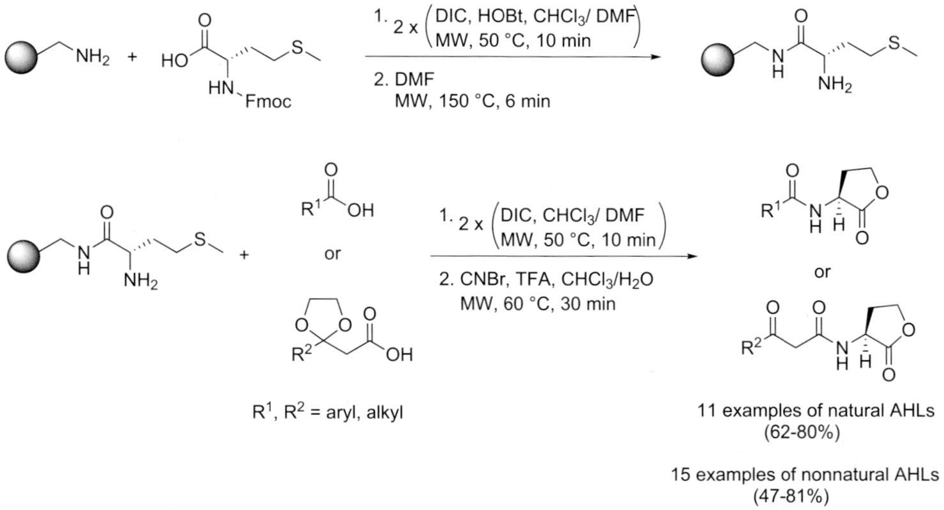

Scheme 8.7 Solid-phase synthesis of natural and nonnatural AHLs.

microwave-assisted solid-phase route (Scheme 8.7) [6]. Both natural and nonnatural AHLs could be obtained in high purities and short reaction times. Among the two libraries, a set of nonnative AHLs has been found to have high inhibitory affinity for bacterial quorum sensing [7].

Murray and Gellman from the same university have evaluated the effects of microwave heating on the solid-phase synthesis of high-purity β-peptides (Scheme 8.8) [8]. They showed that by microwave irradiation using an automated microwave peptide synthesizer, 14-helical β-peptides, especially those containing the structure-promoting residue *trans*-2-aminocyclohexanecarboxylic acid (ACHC), can be obtained in a purity superior to that of both conventional heating and standard solid-phase peptide synthesis (SPPS). For shorter peptides (i.e., the hexamer), microwave heating and conventional heating gave the same results.

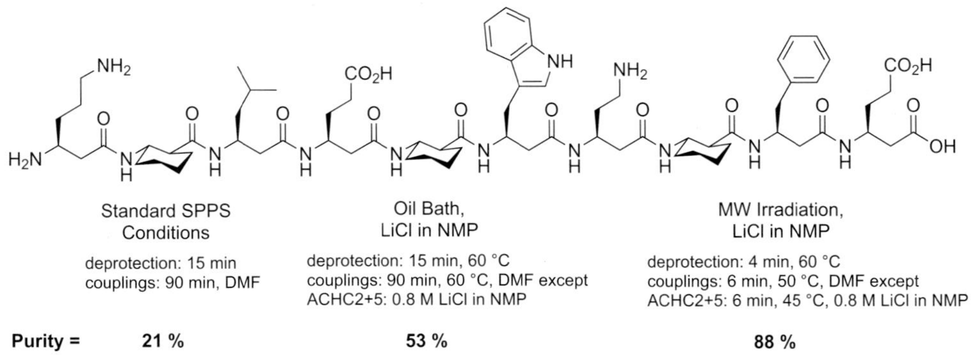

Scheme 8.8 Synthesis of 14-helical β-peptides.

Microwave technology was also used in the synthesis of difficult α-peptide sequences [9–11], for the preparation of glycopeptides [12], peptoids [13], α-peptides [14], and pseudopeptides [15], and for the generation of *N*-glycosyl amino acids (in solution phase) [16]. Furthermore, the issue of racemization and aspartamide formation was extensively investigated by Collins [17]. A detailed discussion of the benefits of microwave-assisted peptide synthesis has been presented in Ref. [10]. Further information can also be found in Section 4.7.2.1.

Carotti and coworkers have developed a multistep solid-phase synthesis of imatinib that acts as a selective tyrosine kinase inhibitor (Scheme 8.9) [18]. By applying microwave heating in five steps of the synthesis (preparation of linker **1**, nucleophilic substitution, reduction of the nitro group, formation of guanidine, and final cyclization), the entire process could be accelerated. Key steps were the guanylation of aniline **2**, where a higher yield and purity of product **3** could be obtained under microwave irradiation, and the final cyclization to resin-bound imatinib, where the reaction time could be reduced from 20 h to 50 min. In addition, resin stability was ensured due to the shorter reaction time.

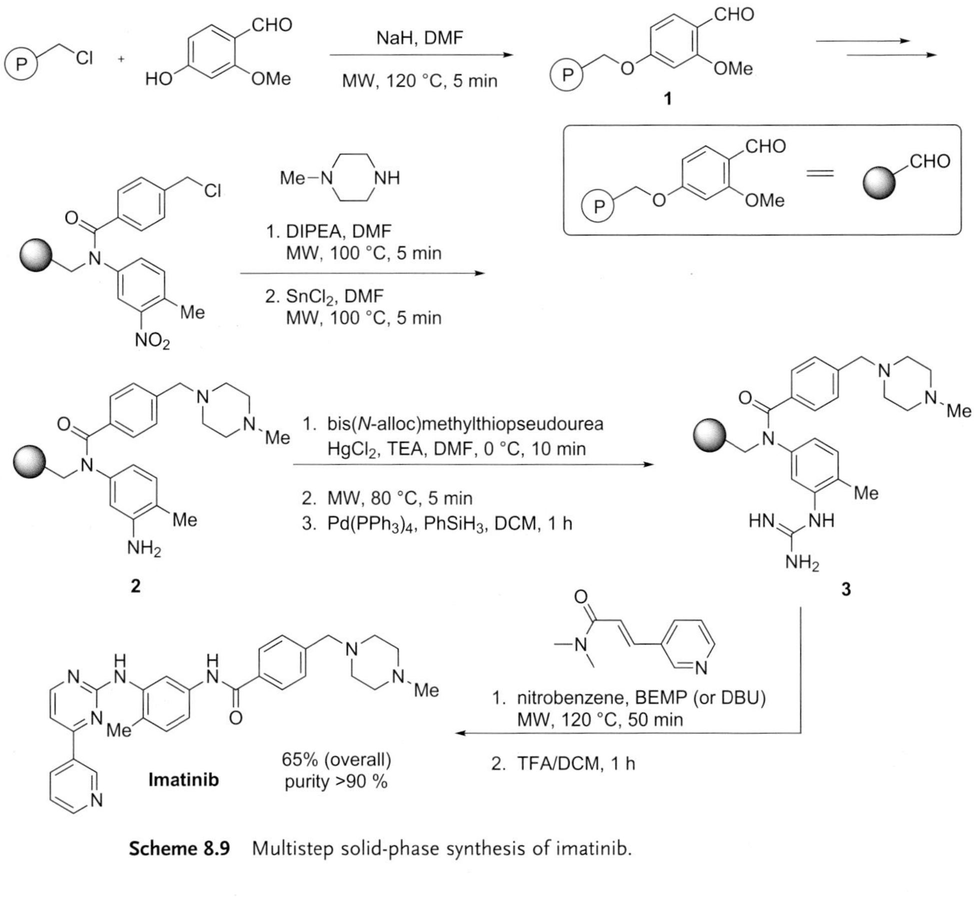

Scheme 8.9 Multistep solid-phase synthesis of imatinib.

8.1.2
Resin Functionalization

The functionalization of commercially available standard solid supports is of particular interest for combinatorial purposes to enable a broad range of reactions to be studied. Since these transformations usually require long reaction times under conventional thermal conditions, it was obvious to combine microwave chemistry with the art of resin functionalization.

As a suitable model reaction, the coupling of various substituted carboxylic acids to polymer resins has been investigated by Stadler and Kappe [19]. The resulting polymer-bound esters served as useful building blocks in a variety of further solid-phase transformations. In a preliminary experiment, benzoic acid was attached to Merrifield resin under microwave conditions within 5 min. This functionalization was additionally used to determine the effect of microwave irradiation on the cleavage of substrates from polymer supports (see Section 8.1.8). The benzoic acid was quantitatively coupled within 5 min via its cesium salt utilizing standard glassware under atmospheric reflux conditions at 200 °C.

In a more extended study, 34 substituted carboxylic acids were coupled to chlorinated Wang resin, employing an identical reaction protocol (Scheme 8.10). In a majority of cases, the microwave-mediated conversion reached at least 85% after 3–15 min at 200 °C. These microwave conditions represented a significant rate enhancement, in contrast to the conventional protocol, which took 24–48 h. The microwave protocol has additional benefits in comparison to the conventional method, as the amounts of acid and base equivalents can be reduced and potassium iodide as an additive can be eliminated from the reaction protocol [19]. High loadings of the resin-bound esters could be obtained very rapidly; even sterically demanding acids were coupled successfully. Importantly, in all the examples given, the loadings accomplished after 15 min of microwave irradiation were actually higher than those achieved using the thermally heated protocols.

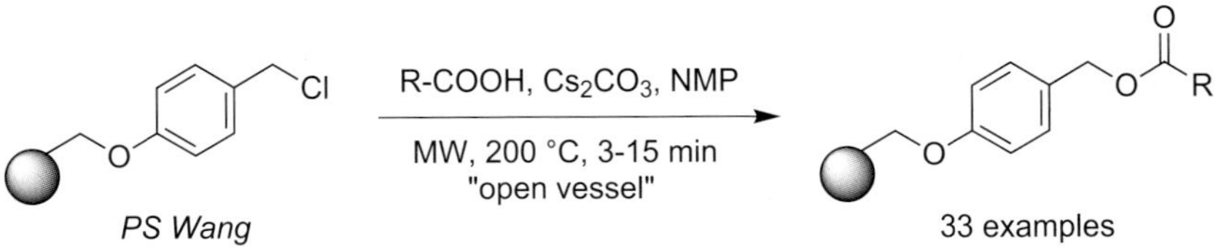

Scheme 8.10 Resin functionalization with carboxylic acids using chlorinated Wang resin as the polymer support.

In a related study by the same authors, the effect of microwave irradiation on carbodiimide-mediated esterifications on solid support has been investigated, employing benzoic acid [20]. Activation of the carboxylic acid was carried out using *N,N′*-diisopropylcarbodiimide via the *O*-acyl isourea or the symmetrical anhydride protocol (Scheme 8.11).

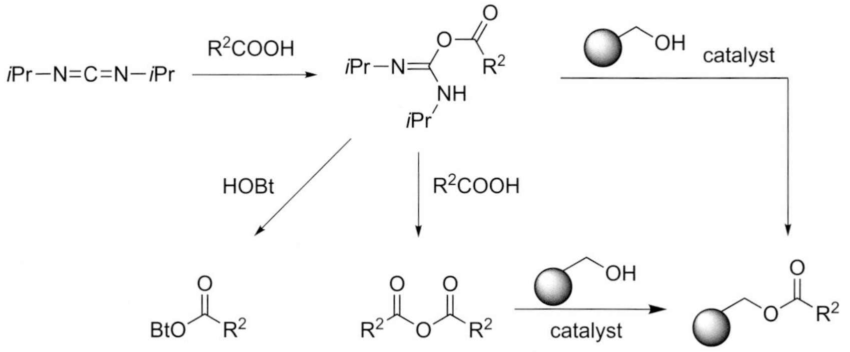

Scheme 8.11 Carbodiimide-mediated pathways for esterification reactions.

The isourea protocol was carried out in a dichloromethane–*N,N*-dimethylformamide mixture in sealed vessels, whereas the anhydride reactions were carried out in 1-methyl-2-pyrrolidinone (NMP) under atmospheric pressure. The isourea protocol showed some deficiencies, as complete conversion could not be obtained, due to unexpected side reactions at higher temperatures. The anhydride protocol was superior to this method, as it could be carried out quantitatively at 200 °C within 10 min under open-vessel conditions without the need for high-pressure vessels.

Song and coworkers have reported the synthesis of a series of functionalized Merrifield resins [21]. Utilizing a modified domestic microwave oven with a reflux condenser, the reaction rates were dramatically enhanced over the conventional methods, as high conversions were achieved within 25 min (Scheme 8.12). Since the choice of solvent was crucial in this procedure, the authors used a solvent mixture to balance between the aspects of good resin swelling properties and high microwave absorption efficiency. These microwave-mediated pathways provide convenient

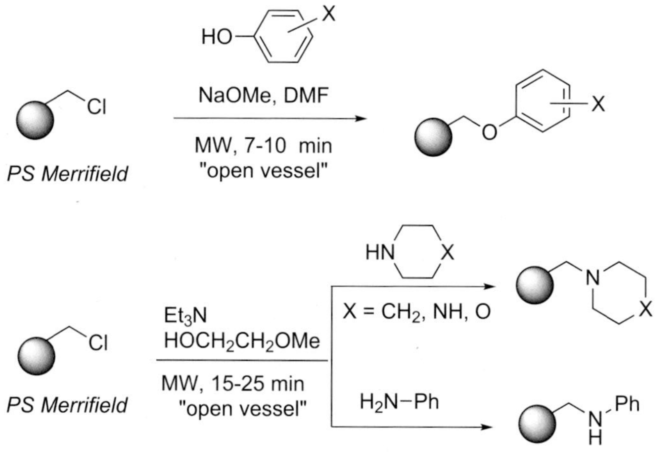

Scheme 8.12 Efficient preparation of functionalized resins for several solid-phase applications.

methods for rapid and efficient solid-phase synthesis, using PS-Merrifield resin as either a support or a scavenger (see Section 8.7).

More recently, the group of Gong developed a traceless, solid-phase synthesis of 2,4,6-trisubstituted thiazolo[4,5-*d*]pyrimidine-5,7-dione derivatives via urea formation by a microwave-mediated reaction of a thiazole amino ester resin with isocyanate [22]. The sequence started with standard Merrifield resin, which was treated with dipotassium cyanodithioimidocarbonate at room temperature. The so-prepared solid support was further modified with 2-bromoethylacetate to obtain the required thiazole amino ester resin that was swollen in DMSO before admixing with isocyanate and diisopropylethylamine and microwave heating at 150 °C for 30 min (Scheme 8.13). After cooling, the resin was filtered, washed with water, *N*,*N*-dimethylformamide, methanol, and dichloromethane, and finally dried in a vacuum oven to yield the desired thiazolourea resin. Three more synthetic steps at room temperature provide the final sulfone-bound thiazolo[4,5-*d*]pyrimidine-5,7-dione resin, which furnished in a desulfonative substitution reaction with appropriate amines the desired 2,4,6-trisubstituted thiazolo[4,5-*d*]pyrimidine-5,7-dione derivative in acceptable yield over all six steps [22]. The reaction works best if aryl isocyanates are employed; alkyl isocyanates led to somewhat lower yields. The authors employed the novel method toward the generation of a 48-member library of valuable thiazolo[4,5-*d*]pyrimidine-5,7-diones.

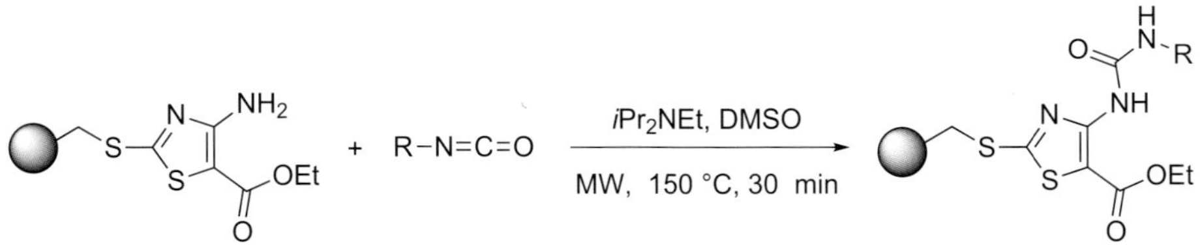

Scheme 8.13 Microwave-mediated modification of thiazole amino ester resin with isocyanate.

In a 2010 report, the same authors presented a traceless solid-phase synthesis of thiazolo[4,5-*b*]pyridine derivatives. The key modification of the solid support to generate the corresponding thiazolopyridine resin was performed utilizing a microwave-mediated Friedländer protocol [23]. Initially, conventional Merrifield resin was modified at room temperature with dipotassium cyanodithioimidocarbonate. The resulting solid-supported cyanocarbonimidodithioate was swollen in DMF and treated with α-bromoacetophenone in the presence of triethylamine at 80 °C to generate the corresponding thiazole resin in a Thorpe–Ziegler cyclization. In the following microwave-mediated Friedländer reaction, the thiazole resin was admixed with 3 equiv ketone and 3 equiv aluminum chloride in acetonitrile. The mixture was microwave heated at 150 °C for 15 min to afford the desired thiazolopyridine resin (Scheme 8.14). To release the pyridine derivatives from the resin in traceless manner, a two-step cleavage process, involving *m*-chloroperbenzoic acid and subsequent treatment with an appropriate amine in the presence of triethylamine, was employed.

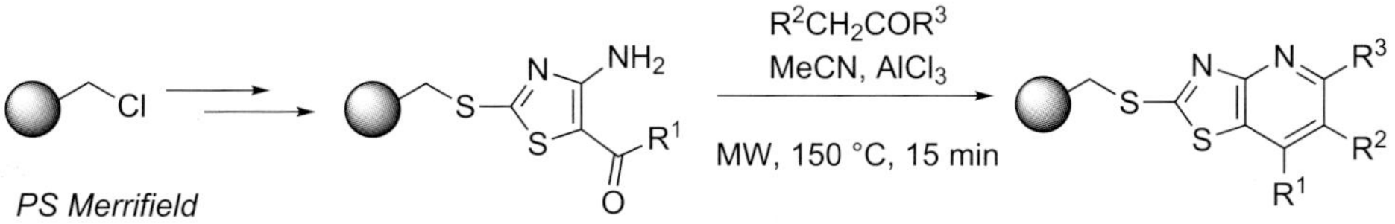

Scheme 8.14 Microwave-mediated solid-phase synthesis of thiazolo[4,5-*b*]pyridines.

Purifying the residue by column chromatography furnished 2-aminated thiazolo[4,5-*b*]pyridines in acceptable overall yields (five steps). The authors used this novel protocol to generate a 50-member library of these structurally important heterocycles. By varying the applied α-bromoacetophenones, ketones, and amines, various functionalities can be introduced to provide pharmacologically attractive diversity of the fused thiazole ring system [23].

An interesting approach of resin functionalization was presented by the group of Yaylayan, describing microwave-assisted PEGylation of Merrifield resin [24]. Treating commercially available polystyrene Merrifield resin with polyethylene glycol (PEG 200) at 170 °C for only 2 min afforded the corresponding hybrid polymer combining advantages of both insoluble and soluble polymers (Scheme 8.15). The Merrifield resin was suspended in excess PEG, which acts as the solvent and at the same time prevents cross-linking of the Merrifield resin. The mixture was irradiated in a pulsed sequence for a total irradiation time of 2 min. The product was simply purified by subsequent washings with water, 10% hydrochloric acid, and methanol. However, it turned out that the yield decreased with increasing molecular weight of the PEG and also with increasing chlorine loading of the Merrifield resins.

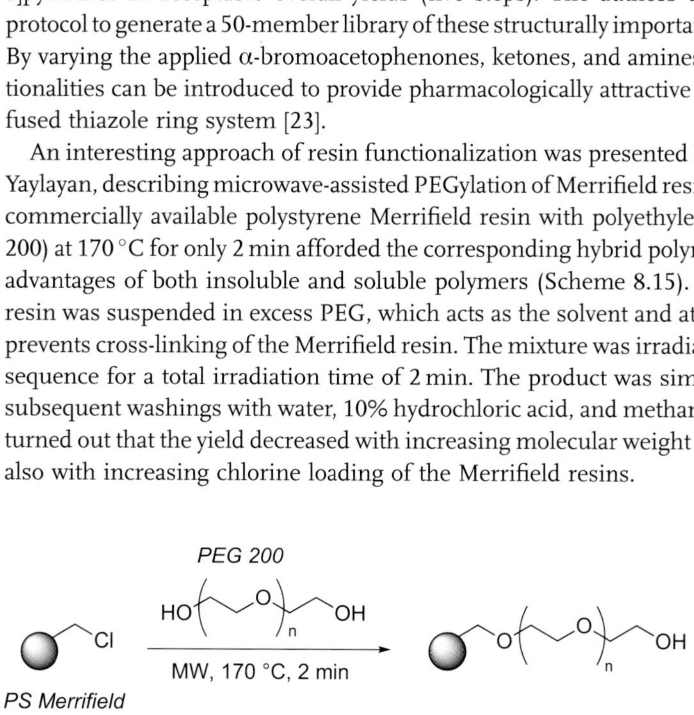

Scheme 8.15 PEGylation of Merrifield resin.

Westman and Lundin described the solid-phase synthesis of aminopropenones and aminopropenoates [25]. Two different three-step methods for the preparation of these heterocycles have been developed. The first method involved formation of the respective ester from *N*-protected glycine derivatives and Merrifield resin (Scheme 8.16a), while the second method employed a different approach utilizing simple aqueous methylamine solution for functionalization of the solid support (Scheme 8.16b). The desired heterocycles were obtained by treatment of the generated polymer-bound benzylamine with the corresponding acetophenones according to similar conditions shown in Scheme 8.16a, utilizing 5 equiv of *N*,*N*-dimethylformamide diethyl acetal (DMFDEA) as reagent. The final step in the synthesis of the pyridopyrimidinones (Scheme 8.16a) involved the release of the

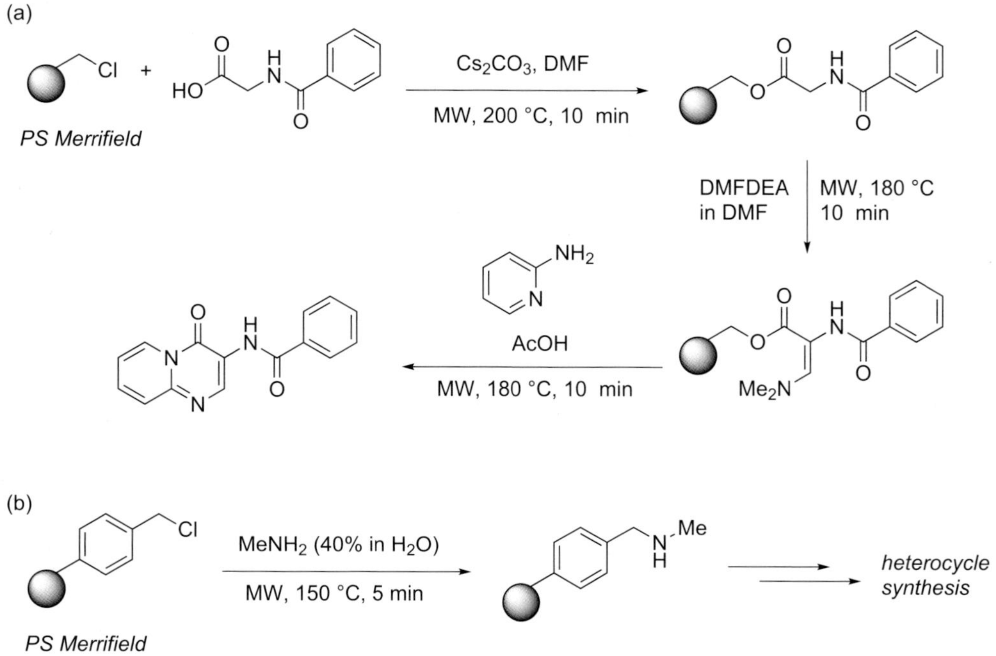

Scheme 8.16 Reaction strategies for the polymer-supported synthesis of dialkylaminopropenones.

products from the solid support by intramolecular cyclization, whereupon the pure products were obtained.

In a dedicated combinatorial approach, Strohmeier and Kappe reported the rapid parallel synthesis of polymer-bound enones [26]. This approach involved a two-step protocol utilizing initial high-speed acetoacetylation of Wang resin with a selection of common β-ketoesters (Scheme 8.17) and subsequent microwave-mediated Knoevenagel condensations with a set of 13 different aldehydes (see Scheme 8.45).

These transesterifications are believed to proceed by the initial formation of a highly reactive α-oxoketene intermediate, with the elimination of the alcohol component of the acetoacetic ester being the limiting factor. Subsequent trapping of the ketene intermediate affords the transacetoacetylated products. Acetoacetylations were performed successfully within 1–10 min under these microwave conditions at 170 °C. Furthermore, the acetoacetylated products can be obtained in a

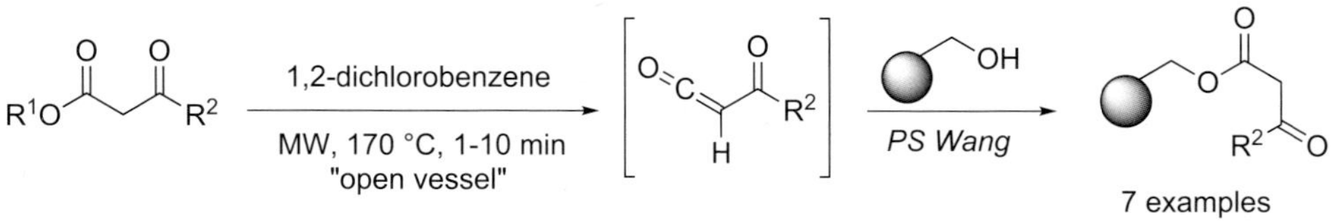

Scheme 8.17 Acetoacetylation reactions.

parallel fashion in a single 10 min run employing a multivessel rotor system. It is worth mentioning that these transesterifications need to be carried out under open-vessel conditions, so that the alcohol by-product can be removed from the reaction mixture (see Section 4.3).

The group of Yil-Kauhaluoma at Helsinki University reported on a new method for the synthesis of *N*-unsubstituted pyrazoles on solid support [27]. In this approach, the solid support acts as protecting group for the utilized amino acids. Initially, several amino acid methyl esters were attached to formyl-functionalized AMEBA resin by standard reductive amination, followed by hydrolysis and *N*-nitrosation. The three-step procedure is susceptible also to parallel processing to increase diversity in an efficient way. The following one-pot cycloaddition was carried out under microwave irradiation when the modified resin in toluene was treated with 5 equiv of acetylene esters in the presence of acetic acid anhydride as water removing agent for 30 min at 150 °C (Scheme 8.18). Via *in situ* formed mesoionic sydnones, which are immediately acting as azomethine-imine-type dipoles, a 1,3-cycloaddition takes place to afford polymer-bound pyrazoles. Traceless cleavage utilizing trifluoroacetic acid furnished the desired pyrazole scaffolds in acceptable yields and high purity. Generally, alkynes bearing electron-withdrawing groups proved to be most effective. In case of terminal alkynes, the corresponding 3-carboxylates were formed as the major product in inseparable mixtures of 3- and 4-carboxylate

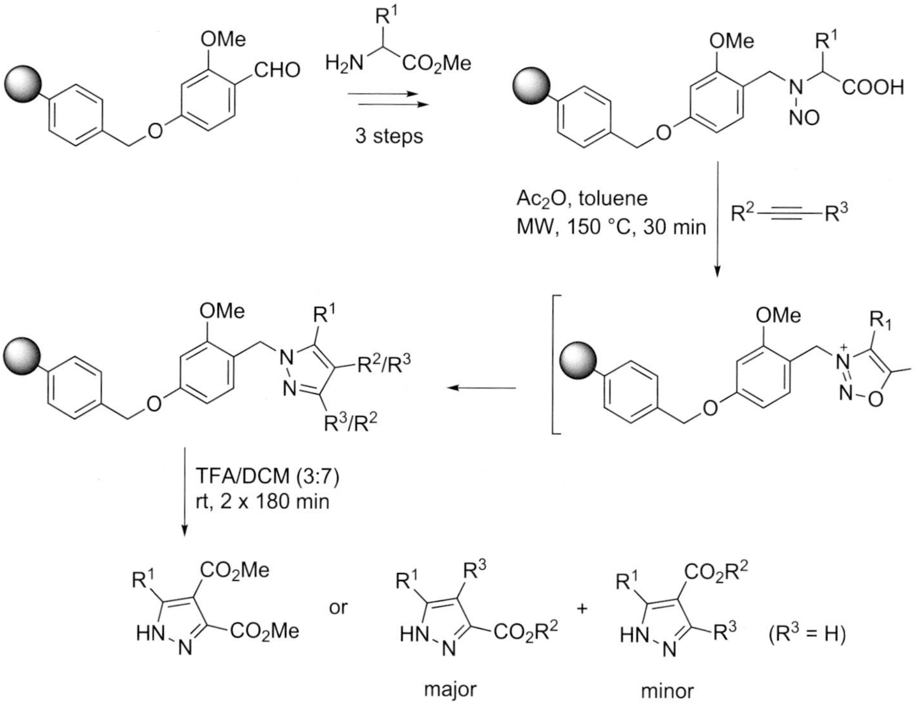

Scheme 8.18 Microwave-assisted solid-supported synthesis of *N*-unsubstituted pyrazoles.

regioisomers [27]. However, the novel method allows introducing a variety of substituents into the pyrazole scaffold to provide attractive building blocks for further transformations.

More recently, Qin *et al.* described a general solid-phase synthesis of 1,2,5-trisubstituted imidazolidin-4-ones by microwave-assisted condensation of solid-supported α-amino amides with aldehydes [28]. The developed procedure allows a facile diversification of the core scaffold at N-1, C-2, and C-5 positions. For the solid-supported sequence, the authors utilized acylated and further modified aminomethyl-terminated Argogel. The initially generated resin was then again acylated with an Fmoc-protected α-amino acid to provide the required α-amino amide resin after subsequent deprotection at room temperature. This appropriate solid support was treated with aromatic aldehydes under microwave heating at 180 °C for 10 min to furnish the desired resin-bound diastereomeric imidazolidin-4-one in acceptable overall yield (Scheme 8.19). The so-prepared intermediate could be further modified via *N*-derivatizations, which allows even for sole acetylation of the *cis*-isomers according to the chosen reaction conditions, which can be explained by steric hindrance [28]. Finally, release of the generated 1,2,5-trisubstituted imidazolin-4-ones can be achieved by simple photolysis yielding the target compounds in excellent purity.

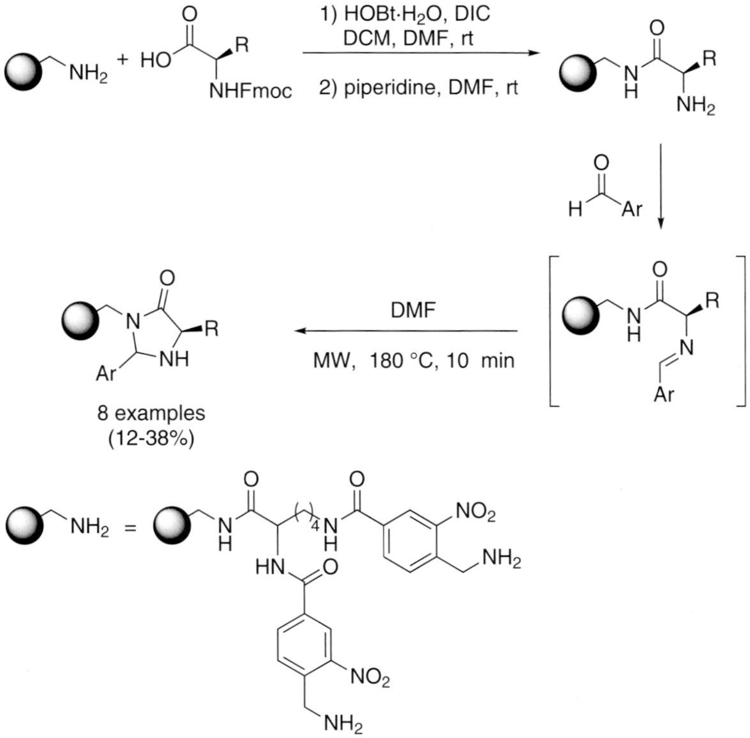

Scheme 8.19 Microwave-assisted synthesis of diastereomeric imidazolidin-4-one.

The group of Kusumoto presented the solid-phase synthesis of indol-2-ones via radical cyclization [29]. A set of commercially available 2-bromoanilines and acryloyl chloride derivatives have been employed to generate a 40-member library of corresponding 2-oxindoles, which are known to be important cores for several drug candidate compounds (Scheme 8.20). *N,N*-Dimethylformamide has been found to be the best solvent for the radical cyclization, inducing a reagent concentration effect of the solid support. Interestingly, the radical-induced cyclization with tributyltin hydride (Bu_3SnH) and 2,2′-azobisisobutyronitrile (AIBN) did not proceed at all under thermal conditions and also the microwave-mediated solution-phase cyclization remained ineffective. However, after preparing the corresponding polymer-bound acrylamides in a multistep sequence, the radical cyclization of these intermediates could be performed utilizing a dedicated single-mode microwave reactor. After conventional cleavage from the applied Rink amide resin using a trifluoroacetic acid/dichloromethane mixture, the desired compounds were obtained in moderate yields and purity.

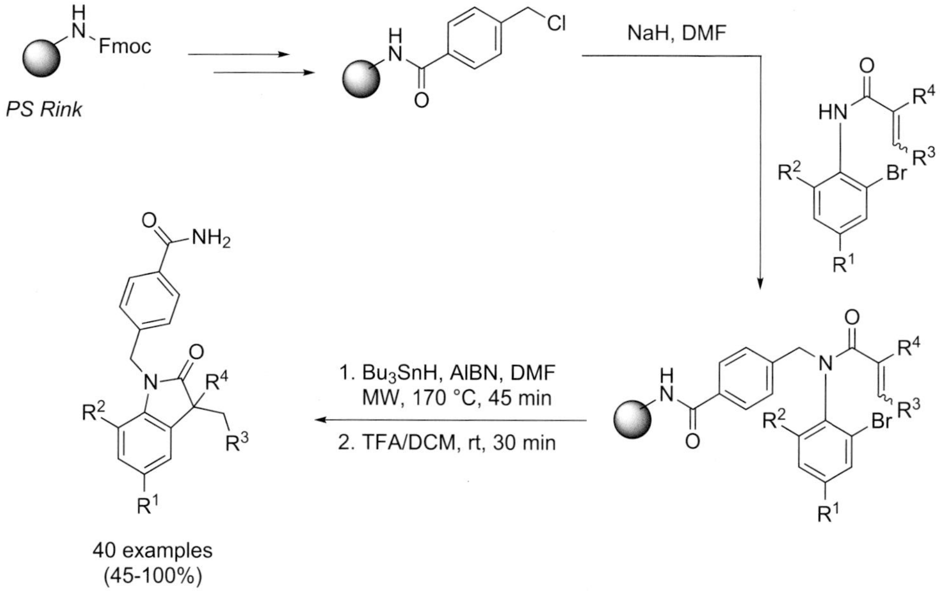

Scheme 8.20 Solid-phase synthesis of indol-2-ones via radical cyclization.

8.1.3
Transition Metal Catalysis

Palladium-catalyzed cross-coupling reactions are one of the cornerstones in modern organic synthesis. Combination of both microwave irradiation and solid-phase synthesis to such chemical transformations is therefore of great interest. One of the first publications dealing with such reactions was presented by the group of

Hallberg in 1996 [30]. Therein, the authors investigated the effect of microwave irradiation toward Suzuki- and Stille-type cross-coupling reactions on solid phase.

The reactions were carried out in sealed Pyrex tubes, employing a prototype single-mode microwave cavity. The reagents were added to the resin-bound aryl halide under a nitrogen atmosphere and irradiated for the time periods indicated. Rather short reaction times provided almost quantitative conversions, as well as minimal degradation of the solid support (Scheme 8.21).

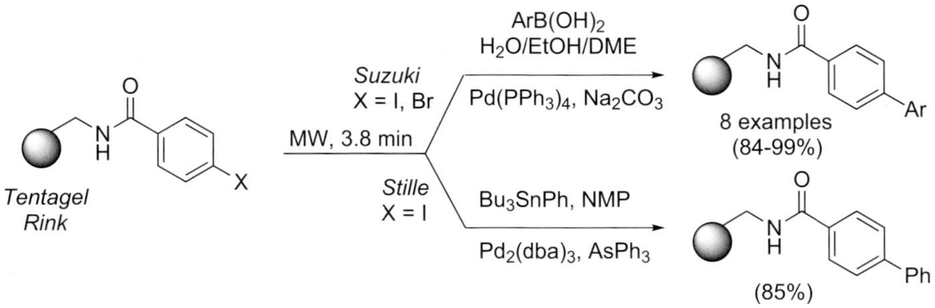

Scheme 8.21 Palladium-catalyzed cross-coupling reactions.

In a related study, the same group investigated molybdenum-catalyzed alkylations in solution and on solid phase [31], demonstrating that microwave irradiation could also be applied to highly enantioselective reactions (Scheme 8.22). For these examples, commercially available and stable molybdenum hexacarbonyl [$Mo(CO)_6$] was used to generate the catalytic system *in situ*. In contrast to the solution-phase protocol, the conversion rates for the solid-phase examples were rather poor. However, the enantioselectivity was excellent (>99% *ee*) for both the solution- and solid-phase reactions.

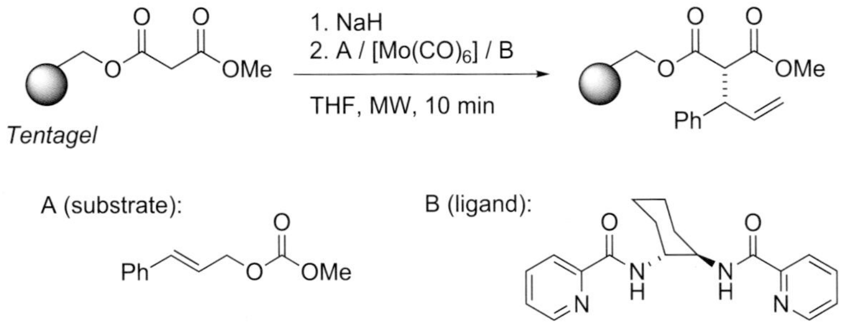

Scheme 8.22 Solid-phase molybdenum-catalyzed allylic alkylation.

In addition, further studies by Hallberg and Alterman described the microwave-promoted preparation of tetrazoles employing organonitriles [32]. After establishing a solution-phase protocol, this protocol was also adapted for solid-phase examples (Scheme 8.23). Full conversion to the corresponding nitriles was achieved after very

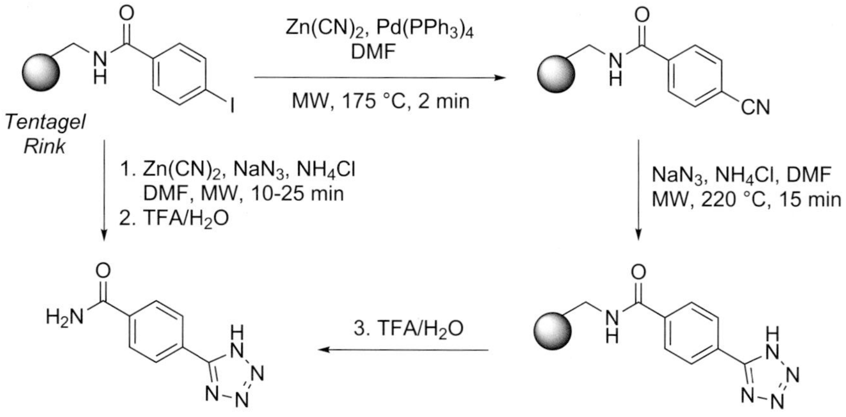

Scheme 8.23 Tetrazole synthesis on solid phase.

short reaction times, at a maximum temperature of 175 °C. Subsequently, the nitriles were treated with sodium azide to form the desired tetrazoles at 220 °C for 15 min. Despite the rather high temperatures, only negligible decomposition of the solid support was observed. It is worth noting that the formation of tetrazoles could easily be carried out as a one-pot reaction in good yields, eliminating the need for tetrakis (triphenylphosphine)palladium catalyst in the reaction mixture. Furthermore, reaction times were drastically reduced by using microwave heating; comparable yields were achieved after 3–96 h using conventional thermal heating.

In a more recent study, Wannberg and Larhed reported solid-supported aminocarbonylations employing molybdenum hexacarbonyl as solid source of carbon monoxide [33]. The carbon monoxide is liberated smoothly at the reaction temperature by addition of the strong base 1,8-diazabicyclo[5.4.0]undec-7-ene (DBU). In this transformation, 5 equiv of $Mo(CO)_6$ were used and immediately before the irradiation a 30-fold excess of DBU was added to the reaction mixture. Cleavage with a conventional mixture of trifluoroacetic acid and dichloromethane furnished the desired sulfamoylbenzamide in good yields (Scheme 8.24).

Combs *et al.* presented a study on the solid-phase synthesis of oxazolidinone antimicrobials via microwave-mediated Suzuki coupling [34]. A valuable oxazolidinone scaffold was coupled to Bal resin (PS-PEG resin with a 4-formyl-3,5-dimethoxyphenoxy linker) to afford the corresponding resin-bound secondary amine (Scheme 8.25). After subsequent acylation, the resulting intermediate was transformed by microwave-assisted Suzuki coupling to the corresponding biaryl compound. Cleavage with trifluoroacetic acid/dichloromethane yielded the desired target structures.

Another palladium-catalyzed carbon–carbon coupling, which could efficiently be accelerated by microwave heating, is the Sonogashira reaction as demonstrated by Erdélyi and Gogoll [35]. Aryl bromides and iodides have been coupled efficiently with various terminal acetylene derivatives (Scheme 8.26). In a set of preliminary experiments, it has been found that 5 mol% of the palladium catalyst and

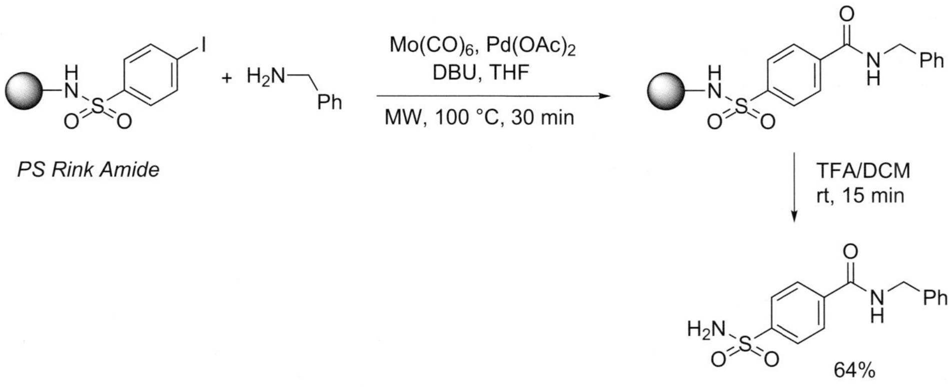

Scheme 8.24 Aminocarbonylations on solid phase.

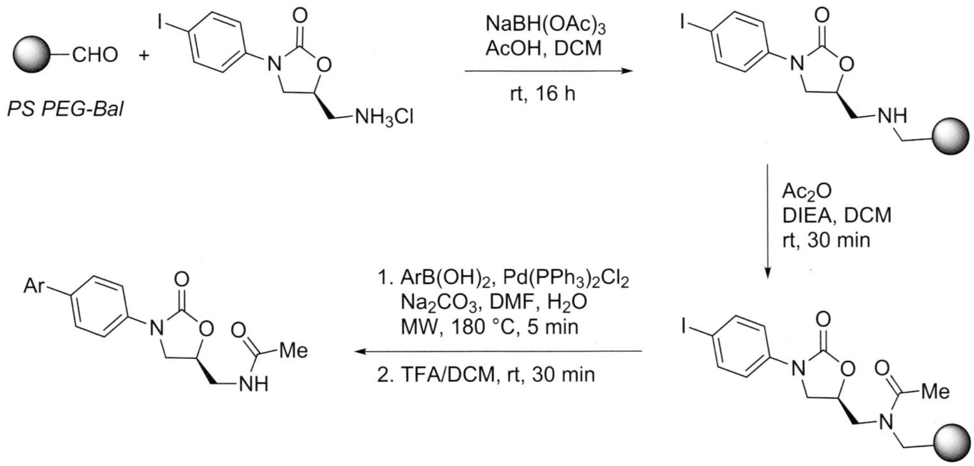

Scheme 8.25 Synthesis of biaryl oxazolidinones.

10 mol% of copper(I) iodide as co-catalyst in a mixture of diethylamine and *N,N*-dimethylformamide gave the best results. Depending on the substrates used, different irradiation times were required to achieve full conversion. The reactions were performed in a dedicated single-mode instrument utilizing a modified reaction vessel for simplified resin handling.

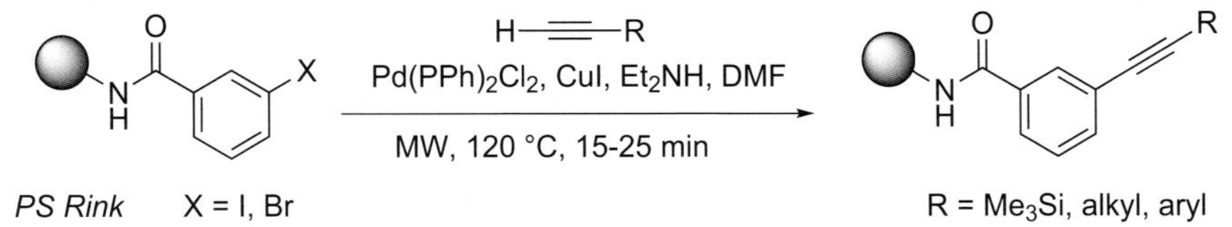

Scheme 8.26 Sonogashira couplings on solid phase.

In 2004, Walla and Kappe disclosed the first polymer-supported cross-coupling reaction involving aryl chlorides [36]. The authors performed rapid Negishi couplings utilizing organozinc reagents to prepare biaryl compounds from various aryl halides. As catalytic system, the highly reactive but air-stable tri-*tert*-butylphosphonium tetrafluoroborate and tris(dibenzylideneacetone)dipalladium(0) have been employed (Scheme 8.27). Subsequent rapid cleavage under microwave conditions furnished the desired biaryl carboxylic acids in high yield and excellent purity.

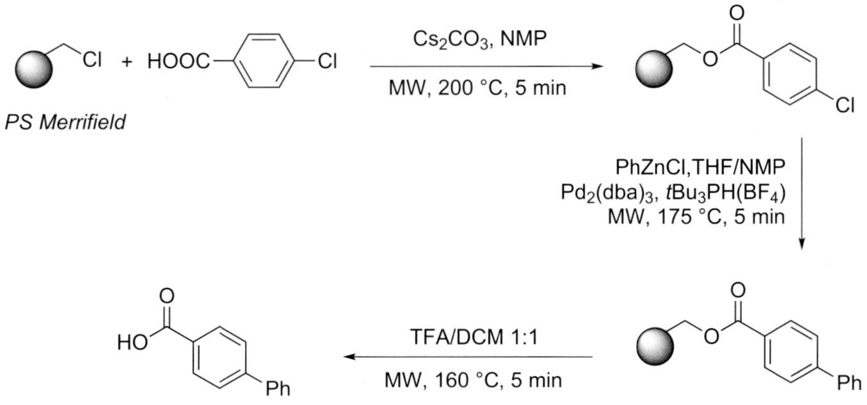

Scheme 8.27 Negishi coupling on polymer support.

The first examples of microwave-mediated solid-phase carbon–nitrogen cross-coupling reactions were reported in 1999 by the group of Combs [37], using a boronic acid and a copper(II) catalyst. The reactions were carried out in a domestic microwave oven at full power for 3 × 10 s. After five cycles of heating with the addition of fresh reagents, none of the remaining starting benzimidazole amide could be detected after cleavage from the solid support (Scheme 8.28). This represented a reduction in the reaction time from 48 h under conventional heating at 80 °C to less than 5 min by microwave heating. However, it should be noted that both possible *N*-arylated regioisomeric products were obtained with this microwave-heated procedure. To assess the versatility of this reaction on the solid support, several heterocyclic carboxylic acids were coupled to the PS-PEG resin (PAL linker). Applying the microwave conditions furnished the desired products in good yields and excellent purities (not shown).

A 2003 publication by Weigand and Pelka disclosed a polymer-bound Buchwald–Hartwig amination [38]. Activated, electron-deficient aryl halides have been coupled with conventional PS Rink resin under microwave irradiation. Subsequent acidic cleavage afforded the desired aryl amines in moderate to good yields (Scheme 8.29). Commercially available Fmoc-protected Rink amide resin was suspended in 20% piperidine/*N*,*N*-dimethylformamide at room temperature for 30 min to achieve deprotection. After washing and drying, the resin was placed in a silylated microwave vessel and suspended under argon atmosphere in dimethoxyethane

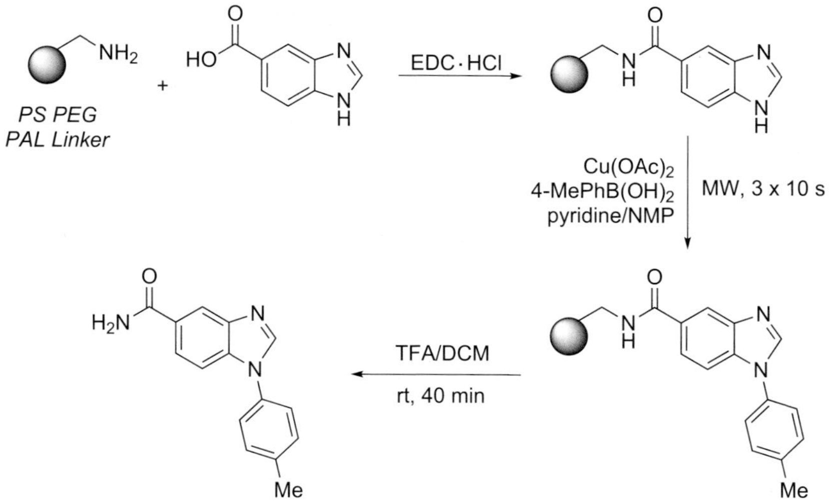

Scheme 8.28 Copper(II)-mediated *N*-arylation of polymer-bound benzimidazole.

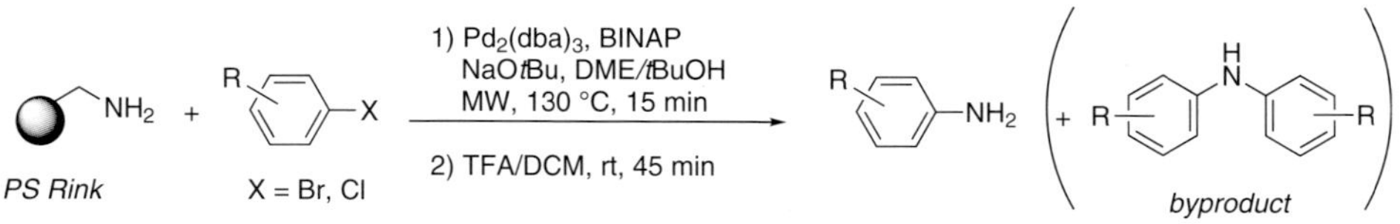

Scheme 8.29 Palladium-catalyzed Buchwald–Hartwig amination.

(DME)/*tert*-butanol. After approximately 10 min, 10 equiv of the aryl halide was added and the mixture was stirred under argon for an additional 10 min. Subsequently, 0.2 equiv of the palladium catalyst, 0.3 equiv of 2,2′-bis(diphenylphosphino)-1,1′-binaphthyl (BINAP) ligand, and 10 equiv of sodium *tert*-butoxide (NaO*t*Bu) base were added and the sealed vessel submitted to microwave irradiation at 130 °C for 15 min. Finally, after several washings, the product was cleaved employing conventional 5% trifluoroacetic acid in dichloromethane. After purification, the desired anilines were obtained in good yields and excellent purities.

Transition metal catalysis on solid support can also be applied toward indole formation, as shown by Dai *et al.* [39]. The authors reported a palladium- or copper-catalyzed procedure for the generation of a small indole library (Scheme 8.30), representing the first example of a solid-phase synthesis of 5-arylsulfamoyl-substituted indole derivatives. The most crucial step was the cyclization of the key polymer-bound sulfonamide intermediates. Whereas the best results for the copper-mediated cyclization were achieved using 1-methyl-2-pyrrolidinone as a solvent, the palladium-catalyzed variation required tetrahydrofuran in order to achieve comparable results. Both procedures afforded the desired indoles in good yields and excellent purities [39].

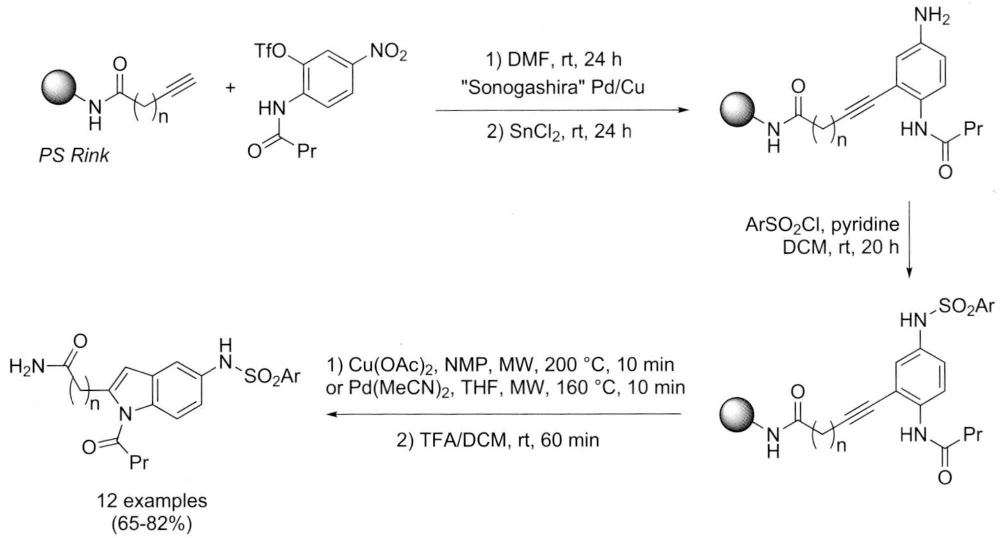

Scheme 8.30 Transition metal-catalyzed indole formation on solid phase.

The group of Van der Eycken described the decoration of polymer-bound 2(1*H*)-pyrazinone scaffolds performing various transition metal-catalyzed transformations [40]. The readily prepared pyrazinone has been specifically decorated at the C-3 position, employing microwave-mediated Suzuki, Stille, Sonogashira, and Ullmann protocols (Scheme 8.31) to introduce additional diversity. In all cases, smooth cleavage of the desired products was achieved by treatment of the resin with a 1 : 2 mixture of trifluoroacetic acid in dichloromethane upon 20 min of microwave irradiation at 120 °C.

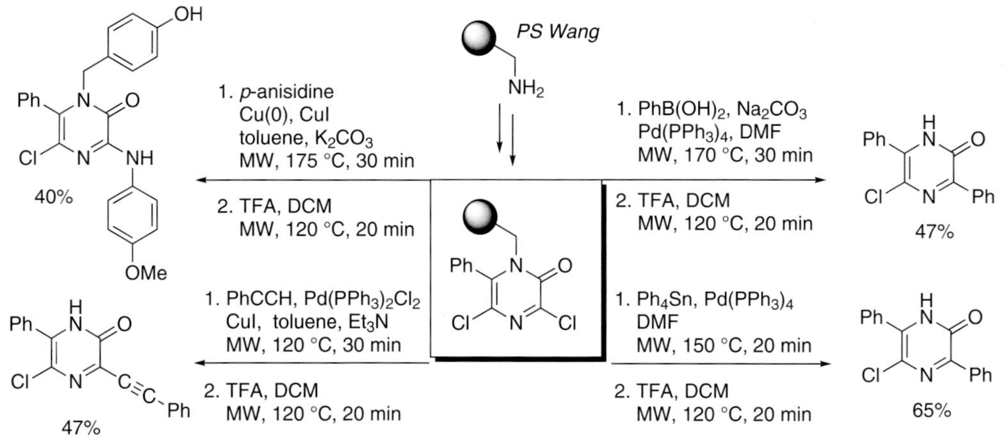

Scheme 8.31 Scaffold decoration of polymer-bound pyrazinone scaffolds.

8.1.4
Substitution Reactions

Another interesting field is that of microwave-mediated substitution reactions on solid phase. In this context, an innovative study was presented by Wenschuh and coworkers [41]. Therein, the authors described the synthesis of trisamino- and amino-oxy-1,3,5-triazines on cellulose and polypropylene membranes, applying SPOT synthesis techniques. This research demonstrates the usefulness of planar solid supports in addition to granulated polystyrene or PEG resins in microwave-assisted solid-phase organic synthesis (SPOS). The development of the SPOT synthesis protocols allows the rapid generation of highly diverse spatially addressed single compounds under mild conditions. This SPOT synthesis protocol required the investigation of suitable planar polymeric supports bearing an orthogonal ester-free linker system, which were cleavable under dry conditions.

Several functionalized membranes could be synthesized by conventional methods at room temperature. In contrast, microwave heating was employed for both the synthesis of the triazine membrane and the practical generation of an 8000-member library of triazines, bound to an amino-functionalized cellulose membrane (Scheme 8.32).

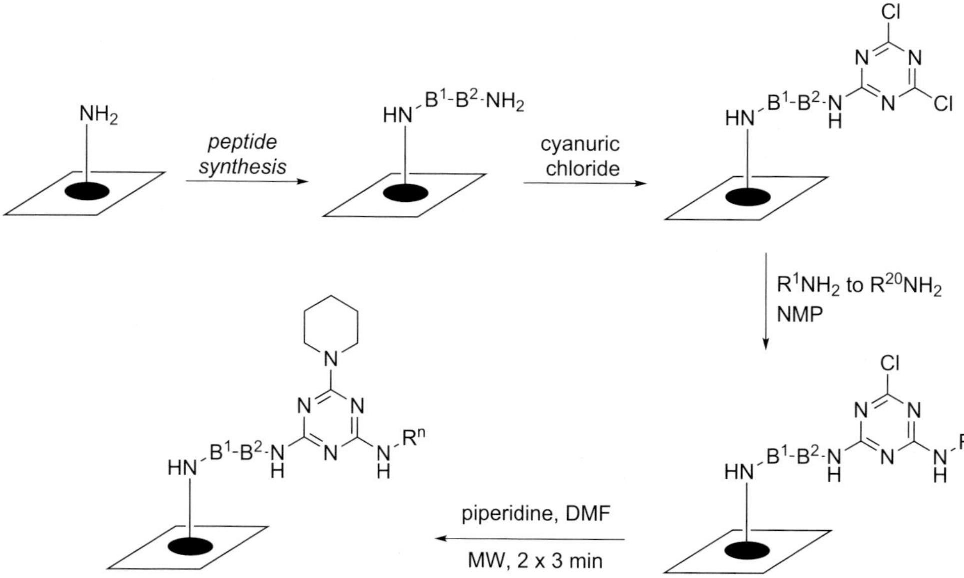

Scheme 8.32 Library generation on a cellulose membrane employing the SPOT technique.

For the preparation of the triazine membranes, the entire solid support (cellulose or polypropylene membrane) was treated with a 5 M solution of the corresponding amine in 1-methyl-2-pyrrolidinone and a 1 M solution of cesium phenolate in dimethyl sulfoxide (2 μL each at one spot) and subsequently heated using a domestic microwave oven for 3 min. After washing the support successively with *N*,*N*-dimethylformamide, methanol, and dichloromethane, the membrane was air dried. All possible 400 dipeptides composed of the 20 proteinogenic L-amino acids (B^1 and B^2) were synthesized in 20 replicas, as illustrated in Scheme 8.32. Subsequently, cyanuric chloride was attached to each dipeptide, followed by selective substitution of one of the two chlorine atoms by 20 different amines. The second chlorine atom was ultimately replaced by piperidine under microwave irradiation conditions. Thereafter, the resulting library of functionalized dipeptides was tested directly on the cellulose sheet, for binding in the murine IgG mab Tab2 assay. Cleavage of the generated compounds could be achieved under mild conditions by treatment with trifluoroacetic acid vapor, leaving the compounds adsorbed on the polymeric support. Since the synthetic conditions described could be applied to the parallel assembly of 8000 cellulose-bound triazines, the method could be of high potential for the parallel screening of small molecule compound libraries.

A similar approach has been described by the same authors for the synthesis of related cyclic peptidomimetics [42]. A set of 10 nucleophiles was employed for the substitution of the chlorine atom of the cyclic triazinyl peptide bound to the cellulose membrane. Due to the above observed rate enhancement effects for nucleophilic substitution of the solid-supported monochloro triazines, these reactions were carried out rapidly by microwave heating. All products were obtained in high purities, enabling the systematic modification of the molecular properties of the cyclic peptidomimetics.

In general, this method was tested on the cyclization of peptides of various chain lengths. Following the SPOT method described above [41], the *N*-terminal amino acids were attached to a photolinker-modified cellulose membrane (Scheme 8.33). 2,4,6-Trichloro-[1,3,5]-triazine was linked to the free *N*-terminus of the peptides, subsequently followed by deprotection of the Lys side chain. Cyclization was achieved by nucleophilic attack of the free amino group at the triazine moiety under basic conditions. Finally, the peptides were cleaved from the solid support by dry-state UV irradiation. For examples utilizing microwave heating, the remaining chlorine functionality on the triazine was substituted by a set of 10 different nucleophiles immediately before the cleavage step [42].

A more recent study applied this so-called SPOT synthesis for the preparation of pyrimidines [43]. The group of Blackwell described primarily the appropriate support modification of commercially available cellulose sheets (Scheme 8.34). The initial introduction of the amine spacer was achieved within 15 min utilizing microwave irradiation compared to 6 h conventional heating. The attachment of the acid-cleavable Wang-type linker was performed by classic methods at ambient temperature.

The readily prepared support was then used for dihydropyrimidine and chalcone synthesis in a Claisen–Schmidt condensation as described in Section 4.7.2.1 (Scheme 4.24) [43].

This new robust support/linker system for SPOT synthesis is compatible with a range of organic reactions and highly applicable for microwave conditions as demonstrated by the same group to construct deazalumazine dyes [44], a small

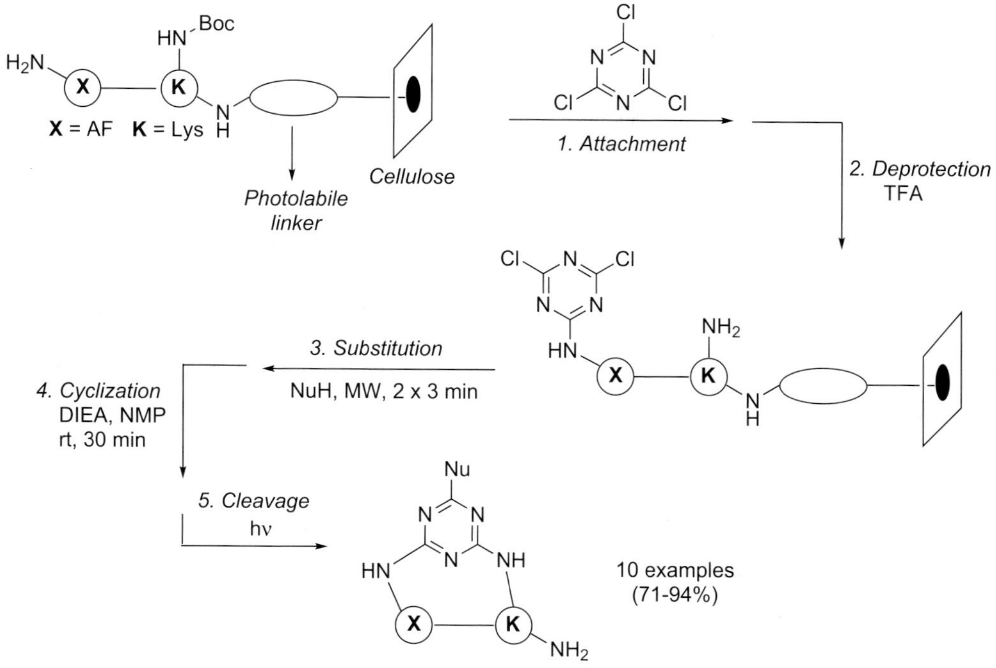

Scheme 8.33 Microwave-mediated synthesis and UV-promoted cleavage of cyclic triazines on cellulose membrane.

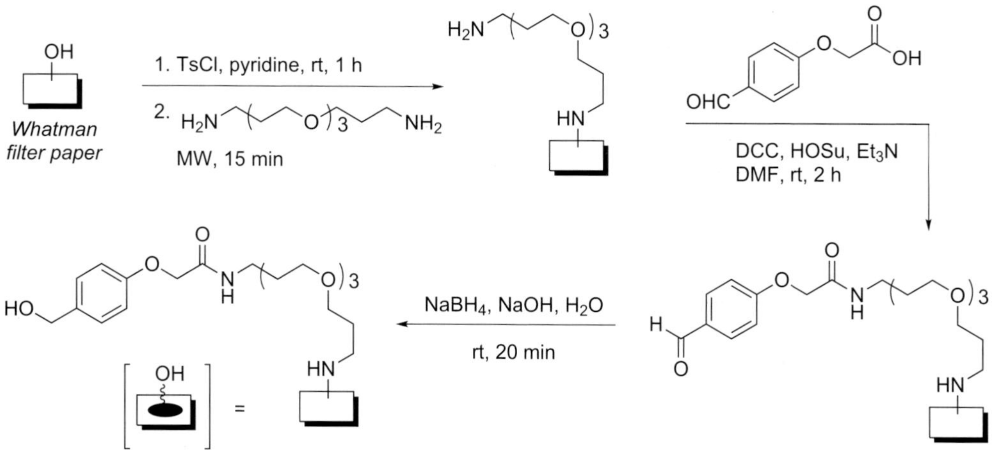

Scheme 8.34 Support modification for SPOT synthesis of pyrimidines.

molecule array based on the Ugi four-component reaction [45], and a selection of diketopiperazines again based on Ugi chemistry [46].

In 2004, a SPOS approach in DNA synthesis was presented, involving a 10-step synthesis of a phosphoramidite building block of 1′-aminomethylthymidine starting from 2-deoxyribose [47]. Such oligonucleotides with chemical modifications are routinely immobilized on surfaces to generate DNA microarrays. To facilitate detection, the oligonucleotides are usually labeled with dyes or radioactive elements. Grünefeld and Richert have described the microwave-mediated deprotection of a corresponding *N*-allyloxycarbonyl (alloc)-protected nucleoside and acylation with the residue of pyrene-1-yl-butanolic acid (PyBA) (Scheme 8.35). The removal of the alloc protecting group was achieved on a support (controlled pore glass, cpg) under microwave conditions (80 °C, 10 min) to ensure full conversion as this step has proven to proceed sluggish and rather difficult under conventional conditions.

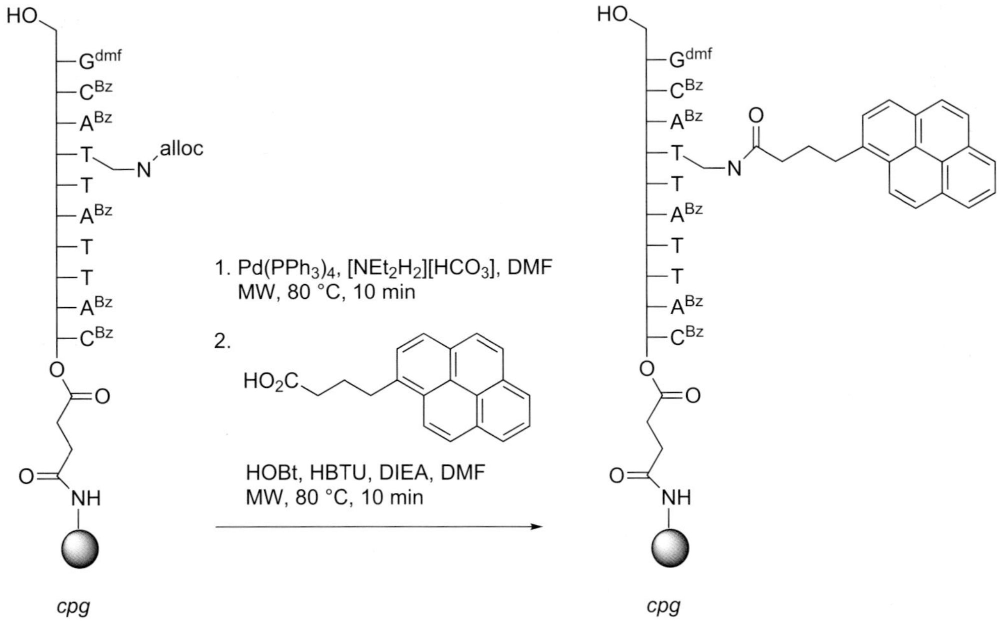

Scheme 8.35 Deprotection of oligonucleotides on controlled pore glass.

The readily prepared immobilized phosphoramidite could be used to efficiently synthesize oligodeoxyribonucleotides with modified thymidine residues. Whereas the effect of microwave irradiation on the deprotection by exposing the strand to tetrakis-triphenylphosphine palladium(0) and diethylammonium bicarbonate was small using dichloromethane as solvent, complete removal of the alloc group was achieved in *N*,*N*-dimethylformamide within 10 min at 80 °C (Scheme 8.35). After the reaction, the solid-supported product was washed with *N*,*N*-dimethylformamide and dichloromethane and dried before being subjected to acylation. The coupling of the PyBA was carried out under microwave irradiation at 80 °C for 10 min employing a mixture of 1 equiv of PyBA, 1 equiv of 1-hydroxybenzotriazole, and 0.9 equiv *O*-benzotriazol-1-yl-*N*,*N*,*N*′,*N*′-tetramethyluronium hexafluorophosphate (HBTU) together with 4.86 equiv of *N*,*N*-diisopropylethylamine in *N*,*N*-dimethylformamide.

The progress of deprotection and acylation was verified by recording MALDI-TOF spectra of the crude products. In this particular case, microwave irradiation not only accelerated the palladium(0)-catalyzed deprotection and the following amide formation, but also led to cleaner reactions. Apparently, the microwave heating breaks up the inaccessibility of the reaction sites, so that the intermolecular reaction sites prevail and competing reactions are suppressed [47].

In a related study, Turnbull and coworkers described the attachment of carbohydrates to amino-derived glass slides. They found significant rate enhancement when performing this step under microwave irradiation compared to classic heating [48]. This method should be an efficient aid for the construction of functional carbohydrate array systems.

The group of Grieco presented a method to efficiently perform macrocyclizations on solid phase (Scheme 8.36) [49]. Preparation of the macrocyclic peptides requires several standard transformations, which are not described in detail herein. The final intramolecular nucleophilic aromatic substitution step was carried out under microwave irradiation at 50 °C. The cyclization product was obtained in good yield after 10 min reaction time and subsequent cleavage, whereas under conventional heating the process required 16 h furnishing somewhat lower yields.

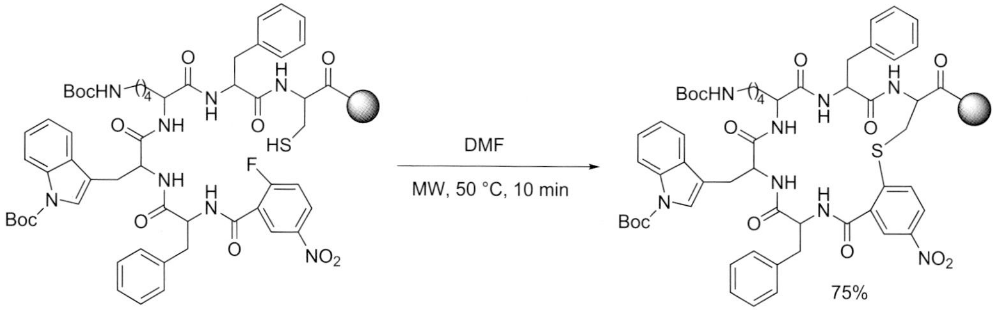

Scheme 8.36 Macrocyclization on solid phase.

In a related study by the same group, the synthesis of cyclic peptides containing unnatural thioether side chain bridges has been reported [50]. Instead of cyclization as described above, a microwave-mediated thioalkylation was performed employing the corresponding alkyl iodides and *N,N*-diisopropylethylamine as catalyst (Scheme 8.37). The prepared precursor can be successfully cyclized following removal of the Fmoc and OFm protecting groups. Acidic cleavage utilizing trifluoroacetic acid furnished the desired products in good yield, whereas only moderate yields were achieved after conventional heating for several hours.

A different approach toward cyclic peptides was presented by Leatherbarrow and coworkers employing ring-closing metathesis (RCM) on solid support [51]. The authors reported on the synthesis of conformationally strained cyclic peptides of the Bowman–Birk inhibitor type, which are naturally occurring serine protease inhibitors containing nine-residue disulfide-constrained loops. Leatherbarrow and his group demonstrated that the disulfide could be replaced by an all-carbon link

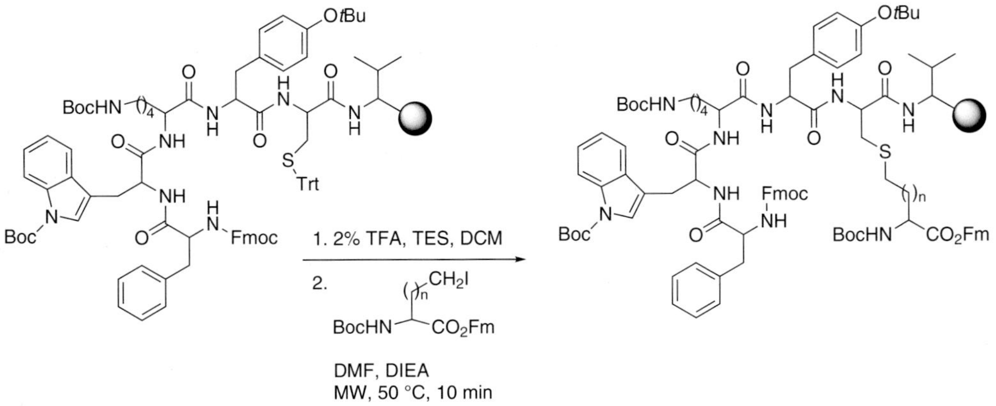

Scheme 8.37 Microwave-mediated thioalkylation to prepare precursors for macrocyclic peptide synthesis.

employing RCM of a linear dienic peptide under microwave irradiation (Scheme 8.38). The reaction proceeded efficiently utilizing Grubbs second-generation ruthenium benzylidene catalyst in dichloromethane applying four cycles of microwave irradiation. Final cleavage was carried out by treatment with 95% trifluoroacetic acid, and the desired cyclic peptides were obtained in rather poor yields and as mixture of *E*/*Z* isomers.

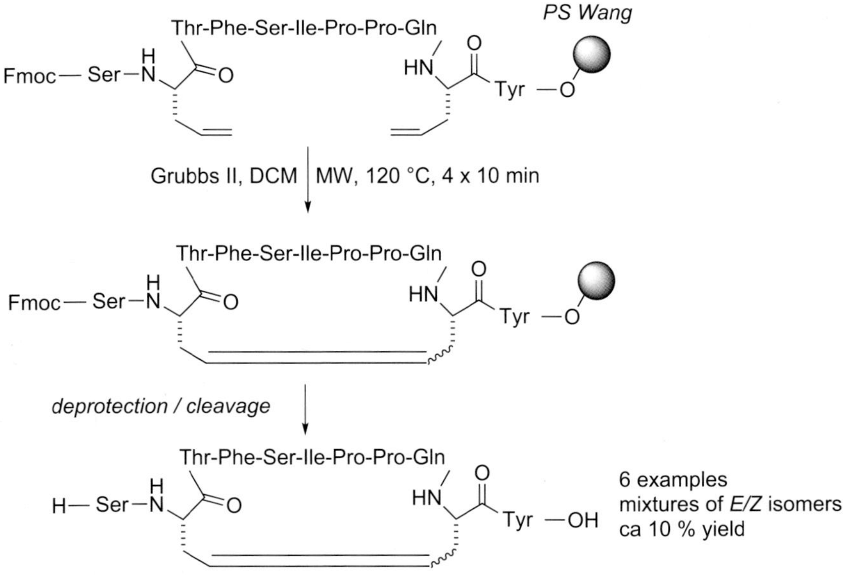

Scheme 8.38 Cyclic peptides via ring-closing metathesis.

The first report on the use of cellulose beads as support for microwave-assisted SPOS was presented by the group of Taddei [52]. The authors developed a library generation of pyrazoles and isoxazoles, which are pharmacologically important heterocyclic scaffolds via *in situ* generation of polymer-bound enamines. The synthesis was performed using commercially available amino cellulose (Perloza VT-100) under mild conditions. Cellulose shows good swelling properties in the used polar solvents, is biodegradable, and could furthermore be recycled for follow-up reactions. The utilized cellulose beads contained aminoaryl ethyl sulfone groups in flexible chains (Scheme 8.39). Initially, the solid support was treated with an excess of formyl imidazole and the corresponding β-keto compounds to generate cellulose-bound enaminones in a one-pot Bredereck-type condensation. The reaction was catalyzed by (±)-campher-10-sulfonic acid (CSA) and carried out under microwave irradiation in an open vessel in order to allow the formed methanol to be removed from the reaction equilibrium [52].

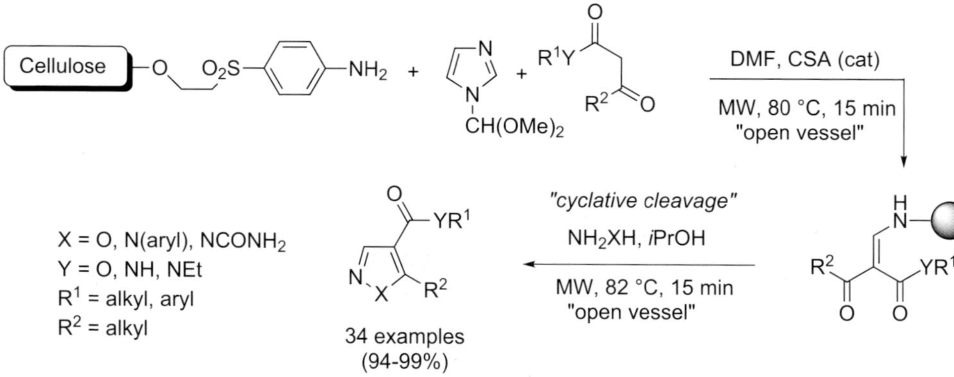

Scheme 8.39 Generation of pyrazole and oxazole libraries on cellulose beads.

Subsequent cyclization and cleavage from the support was achieved by microwave heating in 2-propanol with several hydrazines or hydroxylamines to afford the desired heterocyclic targets. The cellulose-bound aniline could be recycled by simple washings with methanol and diethyl ether and drying in vacuum and reused up to 10 times without loss of efficiency or decreasing purity of the resulting compounds.

The microwave-assisted solid-phase synthesis of purines has been reported by the group of Al-Obeidi [53]. The heterocyclic scaffold was first attached to the acid-sensitive methoxybenzaldehyde (AMEBA)-linked polystyrene via an aromatic nucleophilic substitution reaction by conventional heating in 1-methyl-2-pyrrolidinone in the presence of *N,N*-diisopropylethylamine. The key aromatic nucleophilic substitution of the iodine with primary and secondary amines was conducted by microwave heating for 30 min at 200 °C in NMP (Scheme 8.40) [53]. The resin was subsequently washed with tetrahydrofuran and methanol and, after drying, the products were cleaved from the solid support using trifluoroacetic acid/water at 60 °C.

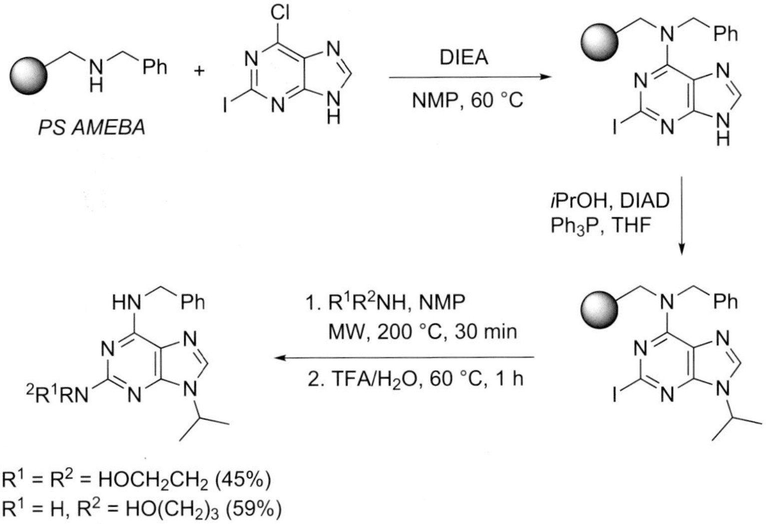

Scheme 8.40 Purine synthesis on solid support.

8.1.5
Multicomponent Chemistry

Multicomponent reactions (MCRs), where three or more components build a single product, have received considerable interest for several years. Since most of these reactions tolerate a wide range of building block combinations, these types of reactions are frequently applied for combinatorial purposes.

The first solid-phase application toward the Ugi four-component condensation generating an 18-member acylamino amide library was presented in 1999 by Nielsen and Hoel [54]. The authors described a library generation utilizing amino-functionalized PEG–polystyrene (Tentagel S RAM) as the solid support (Scheme 8.41). A set of three aldehydes, three carboxylic acids, and two isonitriles was used for the generation of the 18-member acylamino amide library.

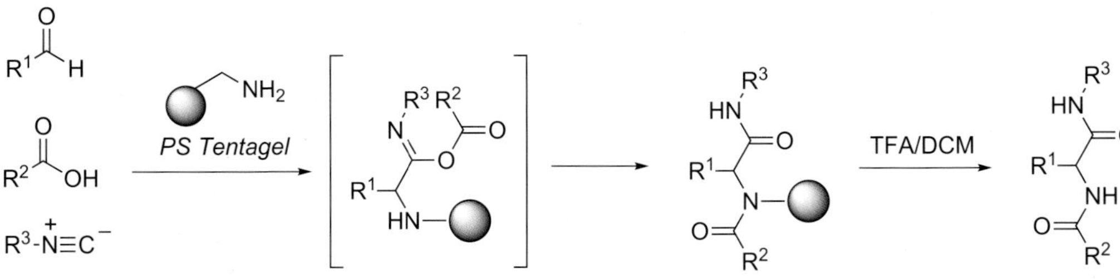

Scheme 8.41 Solid-supported Ugi condensation.

In a typical procedure, the PS-Tentagel Fmoc-protected amino resin was deprotected using 20% piperidine in *N,N*-dimethylformamide. The resin was swollen in a mixture of a 1 M solution of the corresponding aldehyde in dichloromethane and a 1 M solution of the corresponding carboxylic acid in methanol. After 30 min, a 1 M solution of the corresponding isonitrile in dichloromethane was added to the preswollen resin mixture. The vial was flushed with nitrogen, subsequently sealed, and irradiated for 5 min. Subsequent cleavage of the resins with trifluoroacetic acid/dichloromethane afforded the products in high purities but varying yields, after simple evaporation of the solvent.

Additional examples of the Ugi four-component condensation, performed on a Rink amide resin, have been reported by the group of Bradley [55].

Henkel reported on another method for microwave-assisted heterocycle synthesis leading to a small set of imidazole derivatives [56]. These pharmaceutically important scaffolds were synthesized utilizing polymer-bound 3-*N,N*-(dimethylamino)isocyanoacrylate. This polymer support was easily prepared by treatment of (4-bromomethylphenoxy)methyl polystyrene with a twofold excess of the corresponding isocyanoacrylate potassium salt in *N,N*-dimethylformamide (Scheme 8.42). The obtained intermediate was subsequently treated with *N,N*-dimethylformamide diethyl acetal in a mixture of tetrahydrofuran and ethanol to generate the desired polymer-bound substrate.

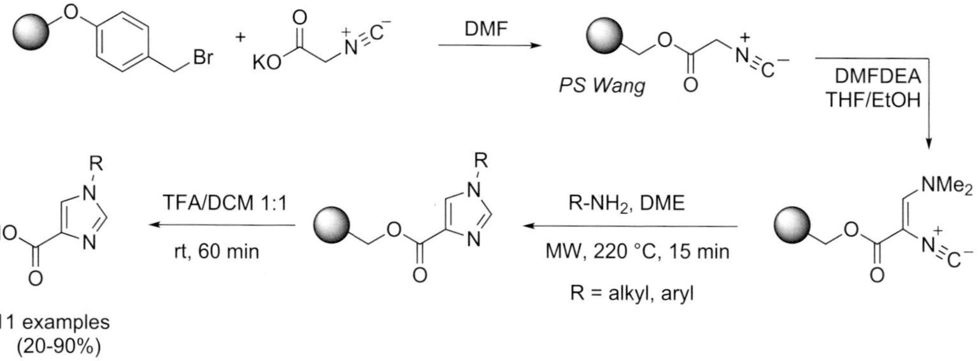

Scheme 8.42 Polymer-bound synthesis of 1-substituted 4-imidazolecarboxylic acids.

The best results for the imidazole synthesis were obtained by microwave-assisted reaction of an eightfold excess of the polymer-supported isonitrile suspended in 1,2-dimethoxyethane with the corresponding amines. Cleavage with 50% trifluoroacetic acid in dichloromethane afforded the desired heterocyclic scaffolds in moderate yields.

In a related study, the group of Giacomelli has synthesized a set of 1-alkyl-4-imidazolecarboxylates by applying a "catch and release" strategy (Scheme 8.43) [57]. The resin-bound isocyanoacrylate **4** was prepared under microwave conditions in only 20 min, compared to 36 h of oil bath heating. For the next step, resin **4** was reacted with a set of amines under open-vessel microwave conditions to give imidazolecarboxylates **5** in excellent yields and high purities. Furthermore, the solid support could be recycled and used successfully for three to four cycles.

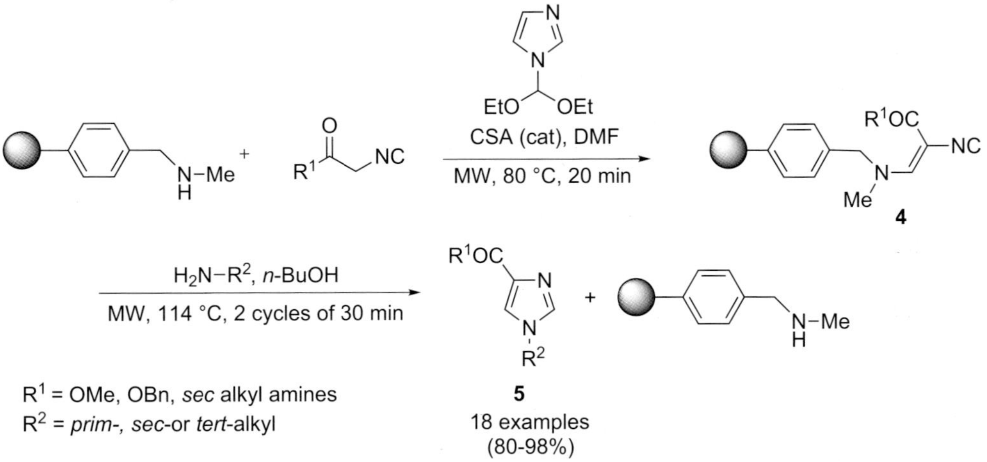

Scheme 8.43 Synthesis of 1-alkyl-4-imidazolecarboxylates employing catch and release strategy.

Another interesting multicomponent reaction is the Gewald synthesis, leading to 2-acyl-3-aminothiophenes, which are of current interest since they are commercially used as dyes and conducting polymers and have shown extensive potential for pharmaceutical purposes. Earlier reports of the classical Gewald synthesis have described the rather long reaction times by conventional heating and laborious purification of the resulting thiophenes.

In view of these issues, Frutos Hoener *et al.* investigated a "one-pot" microwave-assisted Gewald synthesis on solid support [58]. The reactions were carried out in less than 1 h employing commercially available cyanoacetic acid Wang resin as the solid support. This solid-phase "one-pot" two-step microwave-promoted process constitutes an efficient route to 2-acylaminothiophenes (Scheme 8.44), which requires no filtration in between the two reaction steps. In total, 12 diverse aldehydes, ketones, and acylating agents have been employed to generate the desired thiophene products in high yields and generally good purities.

8.1.6
Condensation Reactions

As discussed in Section 8.1.4, polymer-bound acetoacetates can be used as precursors for the solid-phase synthesis of enones [26]. For these Knoevenagel condensations, the crucial step is to initiate enolization of the CH-acidic component. In general, enolization can be initiated with a variety of catalysts (e.g., piperidine, piperidinium acetate, ethylenediamine diacetate); however, for the microwave-assisted procedure, piperidinium acetate was found to be the catalyst of choice, if temperatures were kept below 130 °C. At higher reaction temperatures, significant cleavage of material was obtained from the resin.

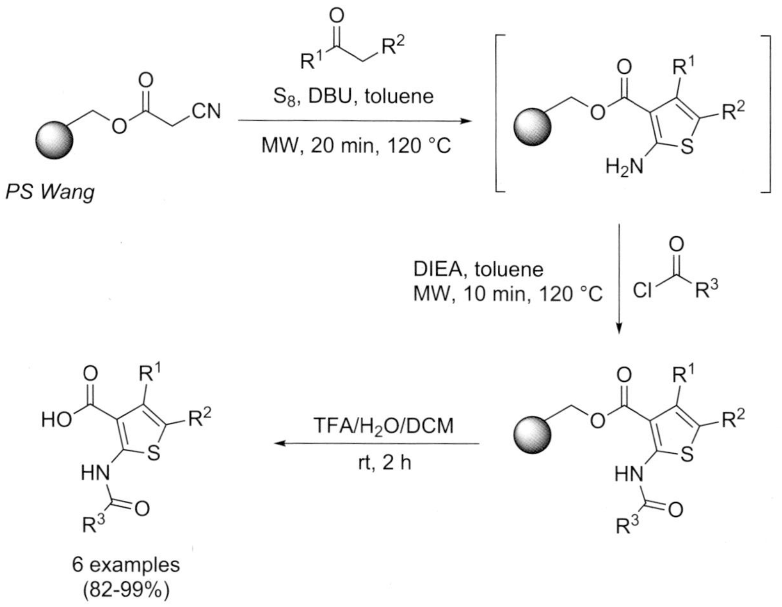

Scheme 8.44 One-pot Gewald synthesis on polymeric support.

To ensure complete conversion for all examples of a 21-member library, irradiation times of 30–60 min were used (Scheme 8.45), employing a multivessel rotor system for parallel microwave-assisted synthesis.

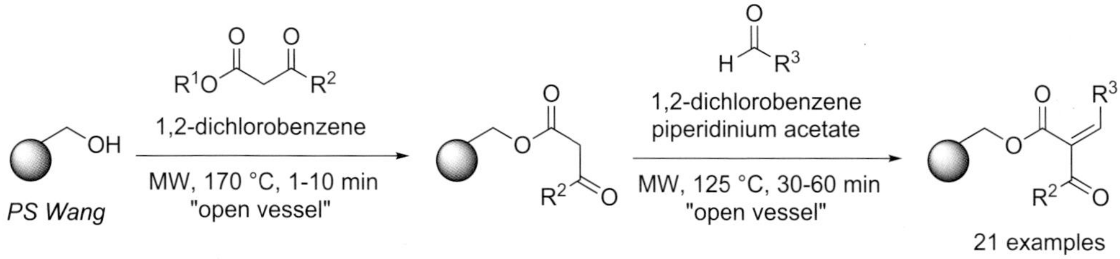

Scheme 8.45 Parallel synthesis of polymer-bound enones.

Microwave-assisted Knoevenagel reactions have also been utilized in the preparation of resin-bound nitroalkenes [59]. The generation of various resin-bound nitroalkenes was described employing resin-bound nitroacetic acid, which was condensed with a different variety of aldehydes under microwave conditions (Scheme 8.46).

In order to demonstrate the potential of these resin-bound products for combinatorial applications, the readily prepared nitroalkenes were subsequently employed in Diels–Alder reactions with 2,3-dimethylbutadiene [59]. In addition, the resin-bound nitroalkenes were also used in a "one-pot" three-component tandem [4 + 2]/[3 + 2] reaction with ethyl vinyl ether and styrene.

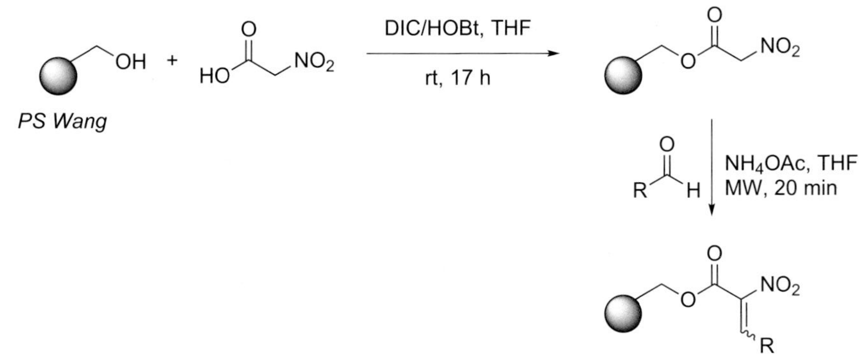

Scheme 8.46 Generation of polymer-bound nitroalkenes via Knoevenagel condensation.

The group of Janda presented a microwave-mediated oxazole synthesis utilizing β-ketoesters bound to a novel polymeric resin [60]. The desired polymer support has been prepared by transesterification reactions by *tert*-butyl β-ketoesters and hydroxybutyl-functionalized *JandaJel* resin and subsequent standard diazo transfer. The resulting α-diazo β-ketoesters have been employed for the synthesis of an array of oxazoles (Scheme 8.47).

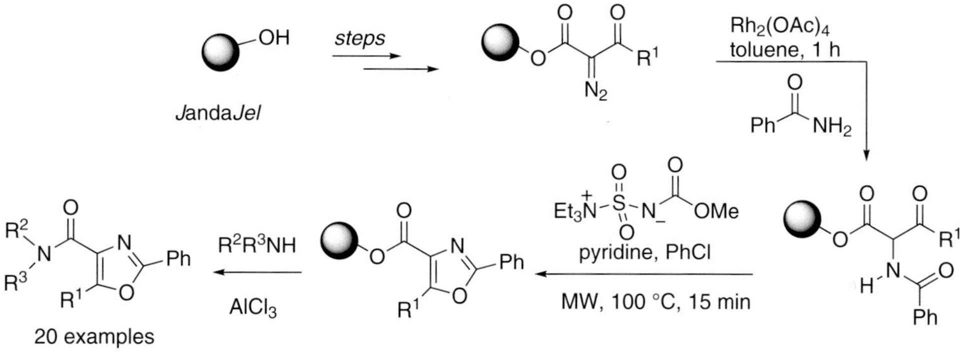

Scheme 8.47 Oxazole synthesis on functionalized *JandaJel*.

For the microwave-assisted polymer-bound cyclization, the use of 3 equiv of Burgess reagent and 20 equiv of pyridine in chlorobenzene turned out to be superior. Due to the thermal sensitivity of the Burgess reagent, the temperature was kept rather low, but irradiation for 15 min at 100 °C furnished satisfactory results for a range of oxazoles [60]. Cleavage from the solid support was achieved by a diversity introducing amidation leading to the corresponding oxazole amides in reasonable yields.

8.1.7
Rearrangements

Microwave-assisted rearrangement reactions on solid phase have not been discussed in the literature very often. So far, only two examples describing Claisen rearrangements have been reported [61, 62]. In the first study, Merrifield resin-bound *O*-allyl salicylic esters were rapidly rearranged to the corresponding *ortho*-allyl salicylic esters employing microwave heating [61] (Scheme 8.48). Acid-mediated cleavage of these resin-bound ester products afforded the corresponding *ortho*-allyl salicylic acids.

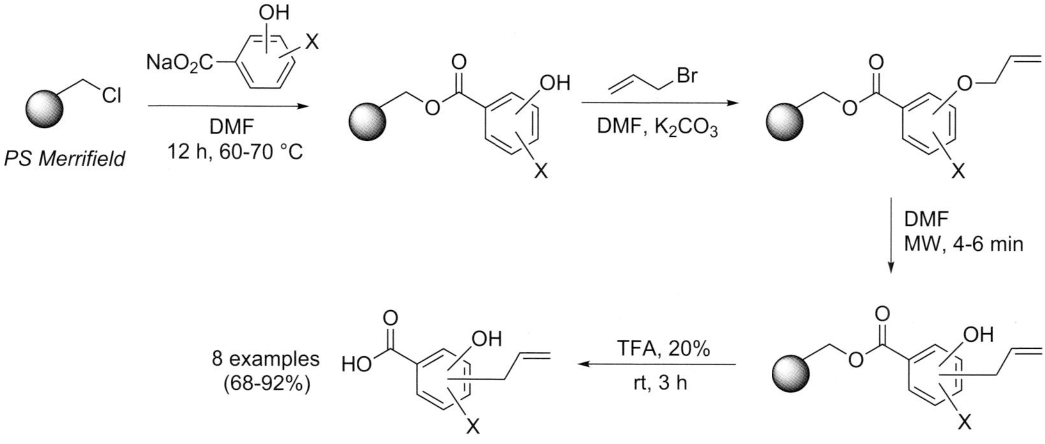

Scheme 8.48 Claisen rearrangement on solid phase.

The resin-bound salicylic esters were suspended in *N,N*-dimethylformamide and placed in an Erlenmeyer flask within a domestic microwave cavity. After microwave irradiation for 4–6 min (1 min cycles), the reaction mixture was allowed to cool to ambient temperature and the resin was collected by filtration and washed with methanol and dichloromethane. The desired compounds were subsequently cleaved with trifluoroacetic acid in dichloromethane. Removal of the solvent by evaporation gave the corresponding acid products in high yields. Compared to conventional thermal heating (*N,N*-dimethylformamide, 140 °C), reaction times could be drastically reduced from 10–16 h to a few minutes using microwave flash heating and in addition higher yields of products were obtained.

A more recent report investigated a microwave-assisted tetronate synthesis. For this purpose, Schobert and Jagusch performed domino addition/Wittig olefinations of polymer-bound α-hydroxy esters with the cumulated phosphorus ylide $Ph_3P{=}C{=}C{=}O$ [62]. Employing α-hydroxy allyl esters can give polymer-supported allyl tetronates or Claisen-rearranged tetronic acids, depending on the reaction conditions.

The desired immobilized α-hydroxy esters can be obtained by ring opening of the corresponding glycidyl esters by OH-, NH-, or SH-terminal polystyrenes of the Merrifield or Wang type (Scheme 8.49). Applying microwave irradiation (85 °C, 30 min), lithium perchlorate showed the highest efficiency for this ring opening. The following tandem Wittig reaction was carried out in tetrahydrofuran under microwave irradiation, employing catalytic amounts of benzoic acid. The formation of the polymer-bound tetronates was complete within 20 min irradiation at 80 °C [62]. Quantitative cleavage of the tetronates was achieved by treatment with 50% trifluoroacetic acid in dichloromethane at room temperature for 2 h.

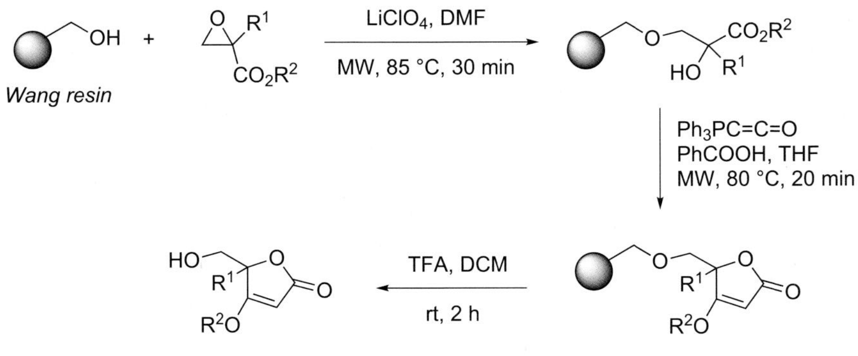

Scheme 8.49 Tetronate synthesis via domino addition/Wittig olefination.

Allyl esters ($R^2 = CH_2C(R^3){=}CH_2$) could be transformed to the corresponding 3-allyltetronic acids by maintaining microwave irradiation at 120 °C for 60 min [62]. However, cleavage of the tetronic acids under the above-mentioned conditions remained somewhat troublesome. To obtain satisfying amounts of product, the polymer-bound intermediates had to be protected at the O-4 position, in order to accomplish cleavage with a 9: 1 mixture of TFA in DCM (Scheme 8.50).

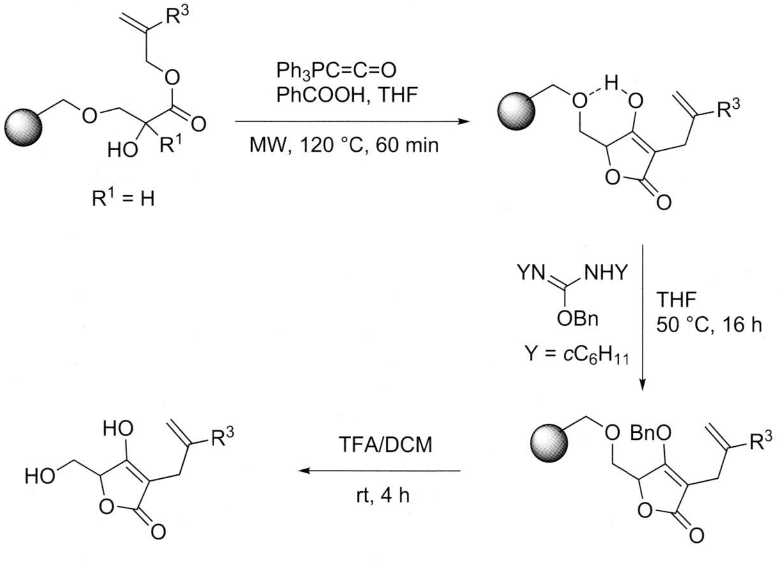

Scheme 8.50 Release of polymer-bound tetronic acids.

8.1.8
Cleavage Reactions

One of the key steps in combinatorial solid-phase synthesis is clearly the cleavage of the desired product from the solid support. A variety of cleavage protocols have been investigated, depending on the nature of the employed linker. It would appear that a complete microwave-assisted protocol including attachment of the starting material to the solid support, scaffold preparation, scaffold decoration, and cleavage of the resin-bound product would be desirable.

A protocol for the microwave-assisted acid-mediated resin cleavage was presented by Stadler and Kappe [19]. Several resin-bound carboxylic acids (see Scheme 8.10) were cleaved from traditional non-acid-sensitive Merrifield resin employing 50% trifluoroacetic acid in dichloromethane under microwave heating (Scheme 8.51).

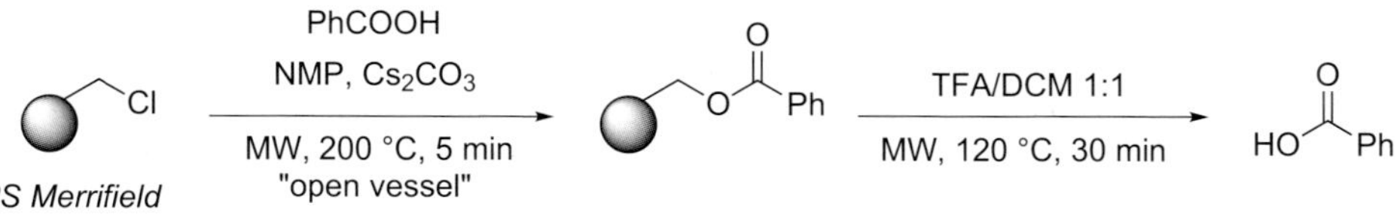

Scheme 8.51 Acidic cleavage under elevated pressure.

In general, acidolysis of the Merrifield linker requires acids with a high ionizing power, such as hydrogen fluoride, trifluoromethanesulfonic acid, or hydrogen bromide/acetic acid. Therefore, under conventional conditions, cleavage does not take place with trifluoroacetic acid. Employing microwave irradiation allows performing these cleavages at elevated pressure/temperature using sealed vessels. Thus, the resin-bound ester and the trifluoroacetic acid/dichloromethane mixture were placed inside a 100 mL PFA reactor and heated with stirring for 30 min at 120 °C. Evaporation of the filtrate to dryness furnished the recovered benzoic acid in quantitative yield and excellent purity. No degradation of the polymer support could be detected, albeit the reaction conditions were rather harsh for solid-phase chemistry.

A different protocol was presented by Glass and Combs, elaborating the Kenner safety catch principle for the generation of amide libraries [63, 64]. In another application of microwave-assisted resin cleavage, *N*-benzoylated alanine attached to 4-sulfamylbutyryl resin was cleaved (after activation of the linker with bromoacetonitrile employing Kenner's safety catch principle) with a variety of amines (Scheme 8.52). By using microwave heating, even cleavage with normally unreactive aniline could be accomplished within 15 min at about 140 °C.

The microwave approach was used for the parallel synthesis of an 880-member library utilizing 96-well plates, employing 10 different amino acids coupled to the 4-sulfambutyryl resin, each bearing a different acyl group, and using 88 diverse amines in the cleavage step. Sets of four plates were placed in a domestic microwave oven and were heated at a temperature of about 80 °C for 60 s. After the plates had cooled down, the solutions from the wells were drained into a collection microtiter plate and were combined with the dimethyl sulfoxide resin washings from the respective wells to afford 10 mM solutions of the products in dimethyl sulfoxide for biological screening. Existing temperature gradients between wells of the microtiter plates did not represent a significant issue for this type of chemistry.

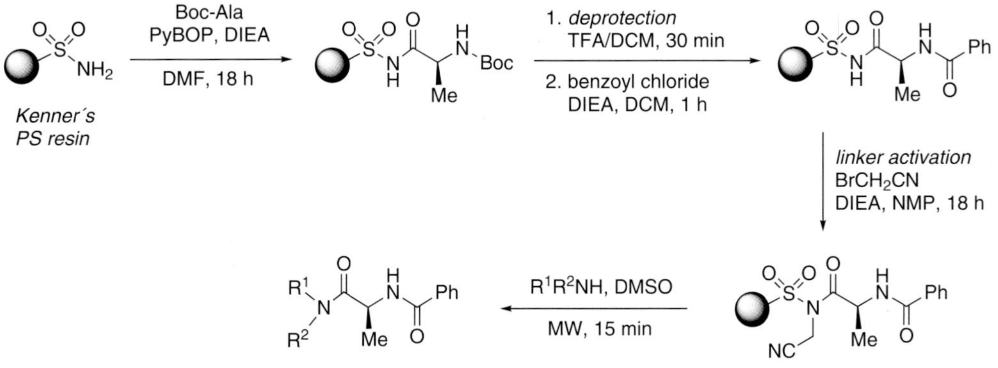

Scheme 8.52 Solid-supported amide synthesis employing the "safety catch" principle.

In closely related work, similar solid-phase chemistry was employed by the same research group to also prepare biaryl urea compound libraries via microwave-assisted Suzuki couplings, followed by cleavage from the resin with amines [64].

A 2004 study discussed the microwave-assisted parallel synthesis of di- and trisubstituted ureas utilizing dedicated 96-well plates [65]. In a typical procedure, modification of the utilized Marshall resin was achieved by treatment with *p*-nitrophenyl chloroformate and *N*-methylmorpholine (NMM) in dichloromethane at low temperatures. The resulting resin was further modified by attaching various amines to obtain a set of polymer-bound carbamates (Scheme 8.53).

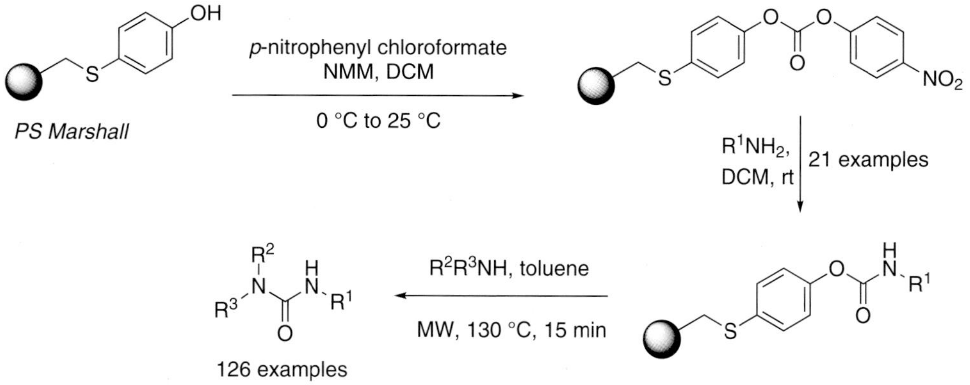

Scheme 8.53 Parallel synthesis of substituted ureas from thiophenoxy carbamate resins.

The immobilized carbamates (40 μmol) were admixed with 10 μmol of different primary or secondary amines dissolved in 400 μL anhydrous toluene and irradiated in a multimode microwave instrument, generating a ramp to reach 130 °C within 45 min and holding this temperature for an additional 15 min. After cooling, the resins were filtered and the filtrates were evaporated to achieve the desired substituted ureas in good purity and reasonable yields. Anilines reacted rather sluggish and 2-substituted benzyl carbamates afforded somewhat poorer results.

In a study by the Kappe group, multidirectional cyclative cleavage transformations leading to bicyclic dihydropyrimidinones have been employed [66]. This approach required the synthesis of 4-chloroacetoacetate resin **6** as the key starting material, which was prepared by microwave-assisted acetoacetylation of commercially available hydroxymethyl polystyrene resin under open-vessel conditions. The precursor **6** was subsequently treated with urea and various aldehydes in a Biginelli-type multicomponent reaction, leading to the corresponding resin-bound dihydropyrimidinones **7** (Scheme 8.54). The desired furo[3,4-*d*]pyrimidine-2,5-dione scaffold **8** was obtained in high purity by a novel protocol for cyclative release under microwave irradiation at 150 °C for 10 min.

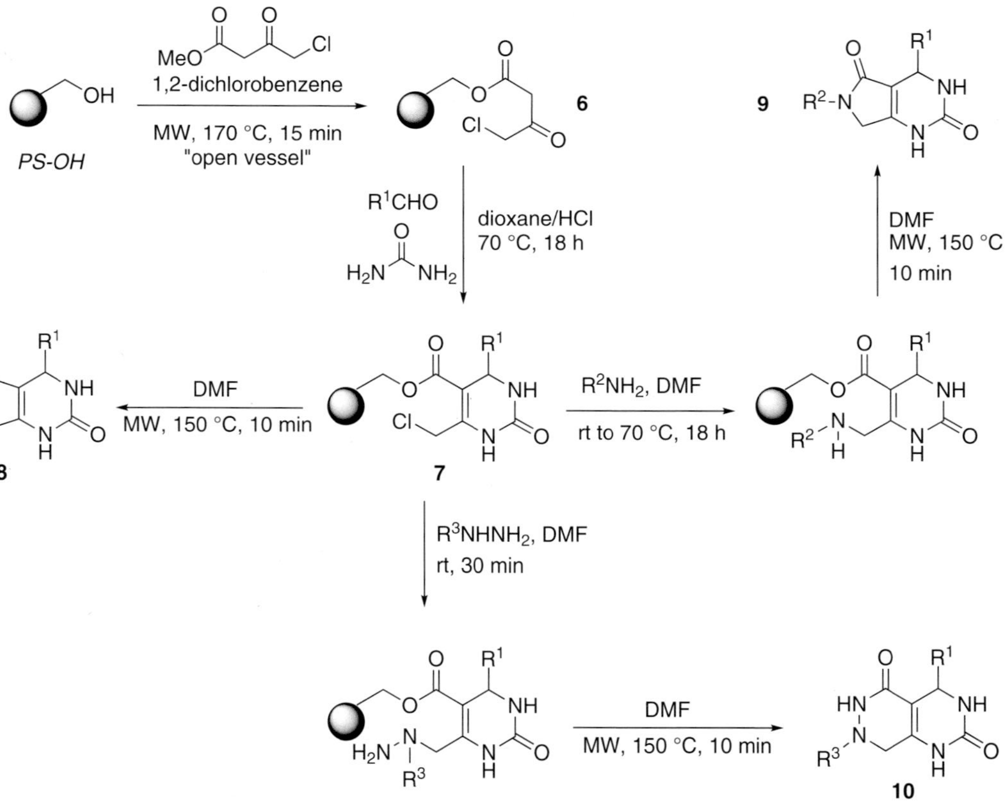

Scheme 8.54 Preparation of bicyclic dihydropyrimidinones employing cyclative cleavage.

Alternatively, pyrrolo[3,4-*d*]pyrimidine-2,5-diones **9** were synthesized using **7**, which was first treated with a representative set of primary amines to substitute the chlorine. Subsequent cyclative cleavage was carried out as described previously, leading to the corresponding products **9** in high purity but moderate yield.

The synthesis of pyrimido[4,5-*d*]pyridazine-2,5-diones **10** was carried out in a similar manner, employing several hydrazines (R^3 = H, Me, Ph) for the nucleophilic substitution of **7**, prior to cyclative cleavage. Due to the high nucleophilicity of the hydrazines, reaction times for the substitution step could be reduced to 30 min. In the case of phenylhydrazine, concomitant cyclization could not be avoided, which led to very low overall yields of the isolated products.

In an earlier report presented from Kurth and coworkers, the microwave-mediated intramolecular carbanilide cyclization to hydantoins was described (Scheme 8.55) [67]. Since the hydantoin moiety imparts a broad range of biological activities, several protocols involving both reactions in solution and on solid phase have been investigated. Reaction studies were carried out employing a single-mode microwave cavity with a variety of several base and solvent combinations; barium hydroxide in *N,N*-dimethylformamide proved to be the best combination for this cyclization reaction. Under these solution-phase conditions, carbanilides could be converted in high yields to the corresponding hydantoins within 2–7.5 min. With the appropriate solid-supported protocol, the carbanilide cyclization would act as a method for resin release of the hydantoins; reaction times could be drastically reduced to several minutes compared to 48 h under thermal heating.

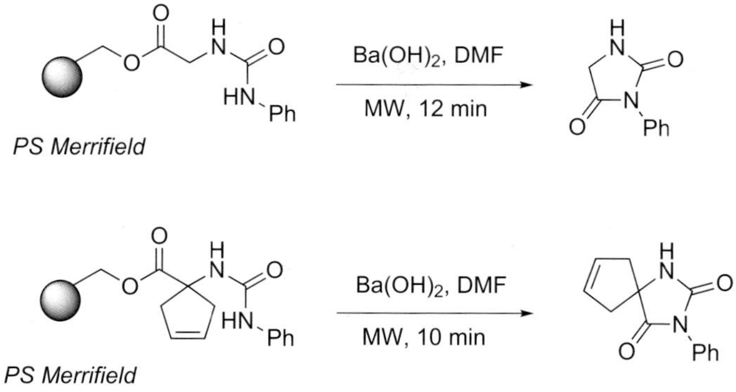

Scheme 8.55 Intramolecular carbanilide cyclization on solid support.

For this solid-phase approach, conventional *i*PrOCH_2-functionalized polystyrene resin (Merrifield linker) was employed. After attachment of the corresponding substrate, the resin was preswollen in a solution of barium(II) hydroxide in *N,N*-dimethylformamide within an appropriate sealed microwave vial. The vial was heated in the microwave cavity for 5 × 2 min cycles (overall 10 min) with the reaction mixture being allowed to cool to room temperature in between irradiation cycles (Scheme 8.55), leading to comparatively modest isolated yields of hydantoins.

A study by the group of Chassaing discussed the traceless solid-phase preparation of phthalimides by cyclative cleavage from conventional Wang resin (Scheme 8.55) [68]. In order to employ the optimum conditions for the cyclative cleavage step, *ortho*-phthalic acid was chosen as the model compound to be attached to the polystyrene resin by an efficient Mitsunobu protocol.

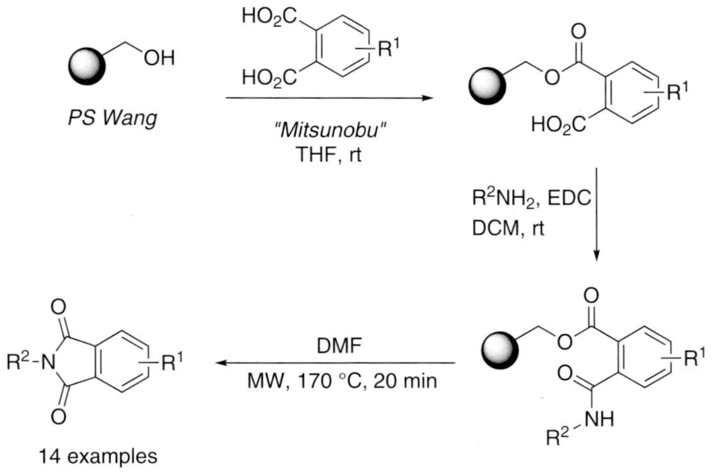

Scheme 8.56 Polymer-supported phthalimide synthesis.

For the cyclative cleavage step, it turned out that aprotic conditions were definitely superior to the use of protic media. Thus, employing *N,N*-dimethylformamide as a solvent at somewhat elevated temperatures furnished the desired compounds in high yields and excellent purities. With the optimized conditions in hand, different phthalic acids and various amines were employed to prepare a set of phthalimides (Scheme 8.56). However, the origin of the amines showed an effect on the outcome of the reaction. Benzyl derivatives furnished somewhat lower yields, probably due to the reduced activities of these amines. Aromatic amines could not be included in the study as auto-induced ring closure occurred during the conversion of the polymer-bound phthalic acid.

8.1.9
Miscellaneous

Several other reaction types on solid support have also been investigated utilizing microwave heating. For instance, in an early report Yu *et al.* monitored the addition of resin-bound amines to isocyanates employing on-bead FTIR measurements in order to investigate the differences in reaction progress under microwave heating and thermal conditions [69].

The corresponding isocyanates were added to the respective resin-bound amine suspended in dichloromethane within an open glass tube. The resulting reaction mixtures were each irradiated in a single-mode microwave cavity for 2 min intervals (no temperature measurement given) (Scheme 8.57). After each step, samples were collected for on-bead FTIR analysis. Within 12 min (six cycles of irradiation), every reaction was completed. Acidic cleavage of the polymer-bound ureas furnished the corresponding hydrouracils.

A very interesting approach toward solid-supported synthesis under microwave heating was introduced by Chandrasekhar *et al.* [70]. The authors developed a

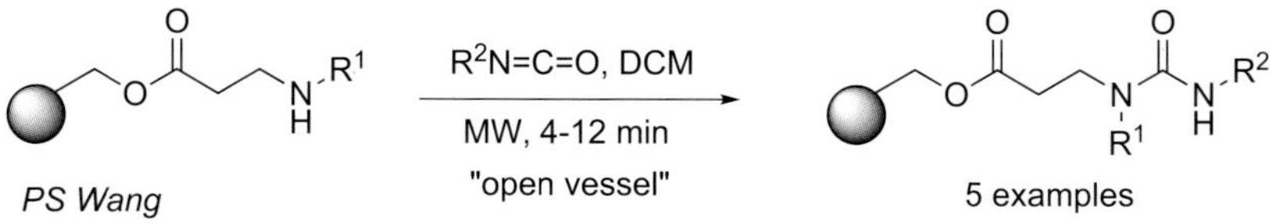

Scheme 8.57 Addition of isocyanates to resin-bound amines.

synthesis of *N*-alkyl imides on solid phase under solvent-free conditions employing tantalum(V) chloride-doped silica gel as a Lewis acid catalyst (Scheme 8.58). Surprisingly, in this rather unusual method, dry and unswollen polystyrene resin is involved. The reaction was carried out employing a domestic microwave oven, performing 1 min irradiation cycles with thorough agitation after each step. Within 5–7 min, the reaction was completed and the corresponding *N*-alkyl imide product was obtained in good yield, after subsequent cleavage from the polystyrene resin–silica gel mixture employing trifluoroacetic acid/dichloromethane. In addition, employing two resin-bound amines and three different anhydrides in this solid-phase protocol, a set of six cyclic imides was synthesized in good yields.

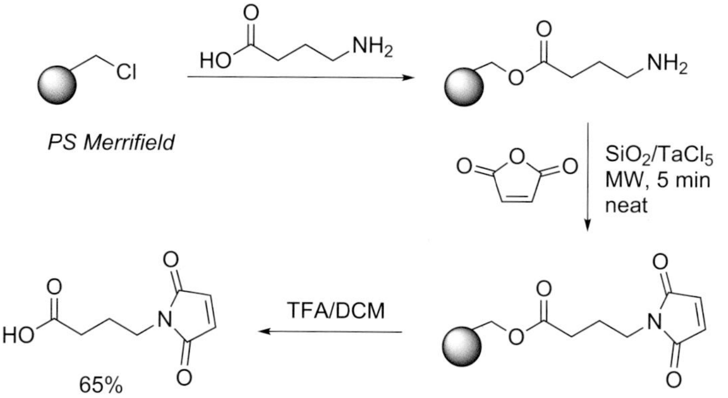

Scheme 8.58 Solvent-free preparation of 4-(1,3-2,3-dihydro-1*H*-2-isoindolyl)butanoic acid.

Weik and Rademann presented the use of phosphoranes as polymer-bound acylation equivalents [71]. The authors have chosen a norstatine isostere as synthetic target and employed the classical polymer-bound triphenylphosphine for their studies (Scheme 8.59). Initial alkylation of the polymer-supported reagent (PSR) was achieved with bromoacetonitrile under microwave irradiation. Simple treatment with triethylamine transformed the polymer-bound phosphonium salt into the corresponding stable phosphorane, which could be efficiently coupled with various protected amino acids. In this acylation step, the exclusion of water was crucial.

Best results have been achieved employing *N*-(3-dimethylaminopropyl)-*N*′-ethylcarbodiimide hydrochloride (EDC) as coupling agent. After Fmoc deprotection with piperidine in *N,N*-dimethylformamide, additional diversity could be introduced by acylation of the liberated amine position. Finally, the acyl cyano phosphoranes could

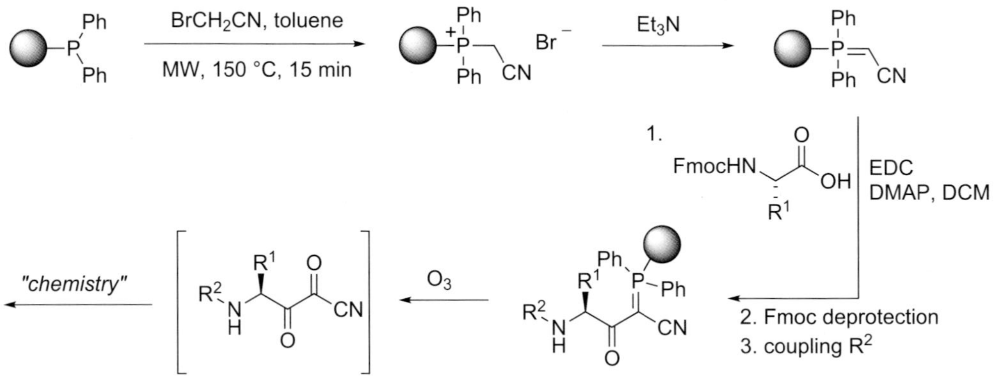

Scheme 8.59 Phosphoranes as polymer-bound acylanion equivalents.

be efficiently cleaved by ozonolysis at −78 °C or utilizing freshly distilled 3,3-dimethyloxirane at room temperature [71]. The released compounds are highly activated electrophiles that can be converted *in situ* with appropriate nucleophiles.

In an early study of the Hallberg group, microwave irradiation was employed in the accelerated solid-phase synthesis of aryl triflates [72]. In addition to the solution-phase protocol, the authors also elaborated a rapid and convenient solid-phase procedure (Scheme 8.60). The use of *N*-phenyl triflimide as a triflating agent in microwave-mediated protocols is an appropriate choice, since it is a stable, crystalline agent, which often results in improved selectivity. Applying the commercially available chlorotrityl linker as a solid support allows mild cleavage conditions to be employed to obtain the desired aryl triflates in good yields.

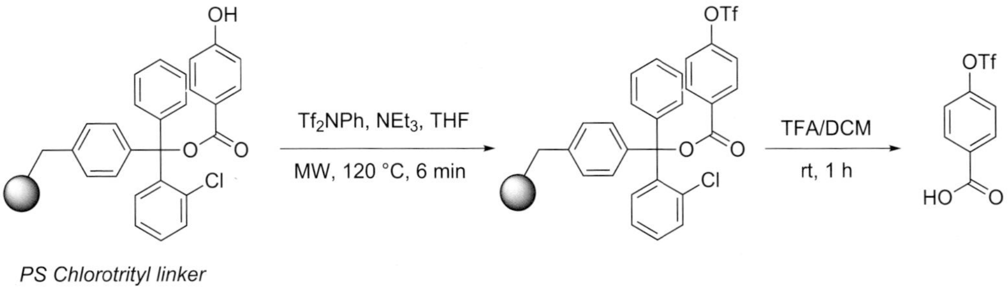

Scheme 8.60 Solid-phase triflating procedures.

In addition to the ionic liquid-mediated procedure in solution, Leadbeater *et al.* also presented a solid-phase protocol for a one-pot Mannich reaction employing the above-mentioned chlorotrityl linker [73]. In this approach, *p*-chlorobenzaldehyde and phenylacetylene were condensed with readily prepared immobilized piperazines (Scheme 8.61).

The substrates were admixed with 50 mol% of copper(I) chloride and small amounts of 1-(2-propyl)-3-methylimidazolium hexafluorophosphate (pmimPF_6) in dioxane. The mixture was heated to 110 °C within 2 min and kept at this reaction

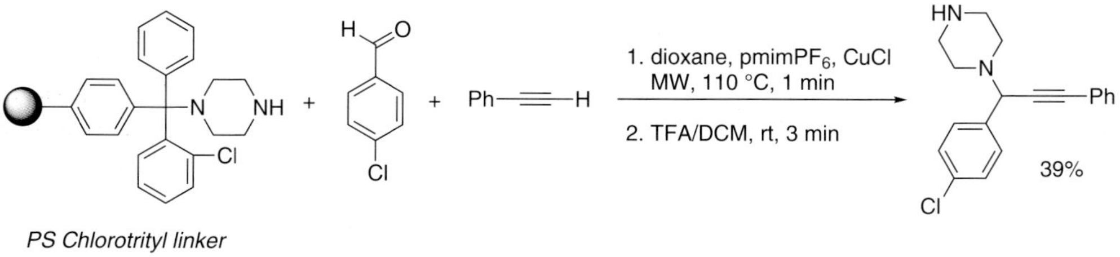

Scheme 8.61 One-pot Mannich reaction utilizing polymer-supported piperazine.

temperature for an additional minute. After cooling to room temperature, the product was rapidly released from the polymer support employing 20% trifluoroacetic acid in dichloromethane furnishing the corresponding bis-TFA salt in moderate yield.

The group of Botta demonstrated the feasibility of their microwave-assisted iodination protocol toward a polymer-supported substrate [74]. A corresponding pyrimidinone attached to conventional Merrifield polystyrene resin was suspended in *N,N*-dimethylformamide, treated with 2 equiv of *N*-iodosuccinimide, and subjected to microwave irradiation for 3 min. Treatment of the polymer-bound intermediate with Oxone® released the desired 5-iodouracil in almost quantitative yield (Scheme 8.62).

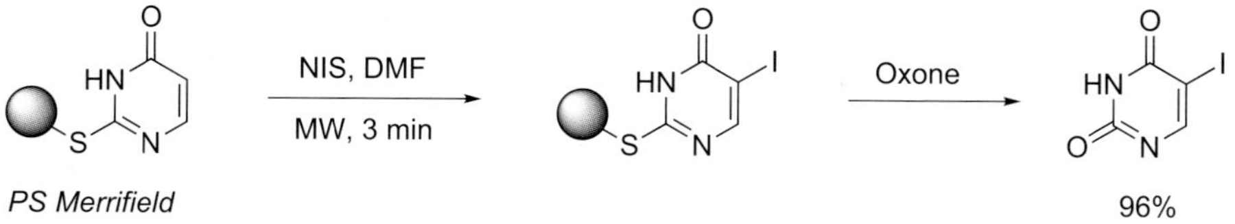

Scheme 8.62 Iodination of polymer-bound pyrimidinone.

Van der Eycken and coworkers presented a study describing the microwave-assisted solid-phase Diels–Alder cycloaddition reaction of 2(1*H*)-pyrazinones with dienophiles [75]. After fragmentation of the resin-bound primary cycloadduct formed from Diels–Alder reaction of the 2(1*H*)-pyrazinone with an acetylenic dienophile, separation of the resulting pyridines from the pyridinone by-products was achieved by applying a traceless linking concept, whereby the pyridinones remained on the solid support with concomitant release of the pyridine products into solution (Scheme 8.63).

For this approach, a novel, tailor-made, and readily available linker, derived from inexpensive syringaldehyde, was developed [75]. The novel linker was produced by cesium carbonate-activated coupling of commercially available syringaldehyde to the Merrifield resin under microwave heating conditions. Subsequently, the aldehyde moiety was reduced at room temperature within 12 h and the benzylic position was finally brominated by treatment with a large excess of thionyl bromide (10 equiv) leading to the desired polymeric support (Scheme 8.64).

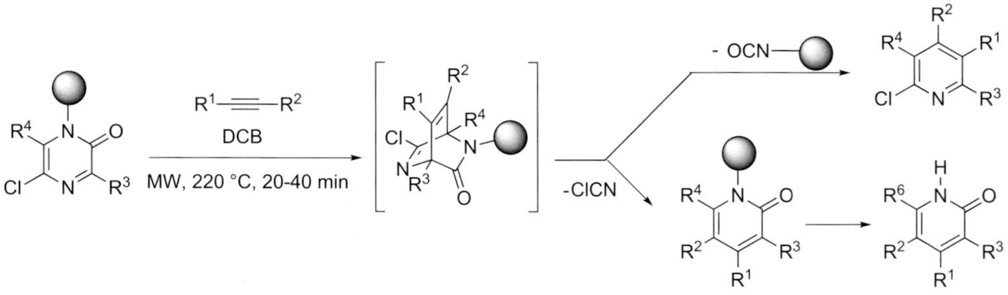

Scheme 8.63 General reaction sequence for 2(1*H*)-pyrazinone Diels–Alder cycloadditions with acetylenic dienophiles.

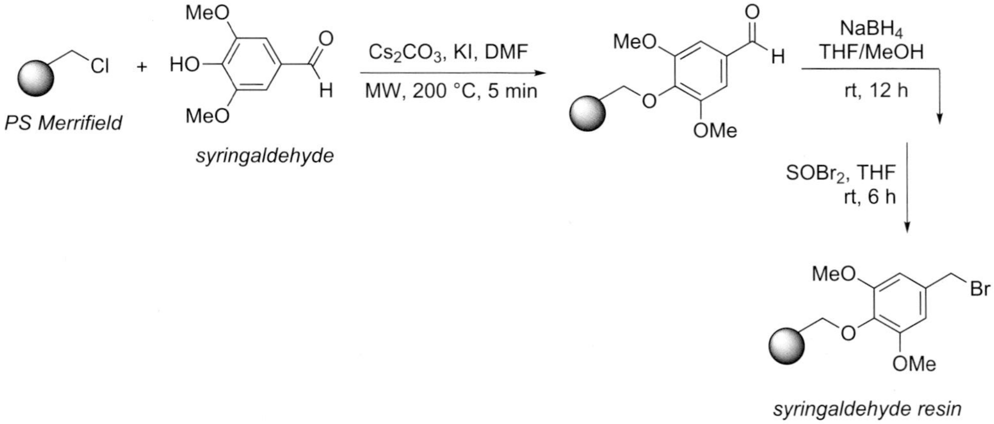

Scheme 8.64 Preparation of brominated syringaldehyde resin.

For the development of an appropriate cleavage strategy from the novel syringaldehyde resin, the authors adapted a preliminary elaborated solution-phase model study on intramolecular Diels–Alder reactions for the solid-phase procedure (Scheme 8.65). The resulting pyridines could be easily separated from the polymer-bound by-products by employing a simple filtration step and subsequent evaporation of the solvent. The remaining resins were each washed and dried. After drying, the resins were each treated with trifluoroacetic acid/dichloromethane cleavage solutions under microwave irradiation to obtain the corresponding pyridinones.

Utilizing the novel syringaldehyde resin, a smooth release from the support could be performed upon microwave heating of a suspension of the resin-bound pyridinones in trifluoroacetic acid/dichloromethane (5 : 95) at 120 °C for only 10 min. The very mild cleavage conditions for this new linker, as well as its stability toward various reaction conditions and its easy accessibility, make it highly suitable for ongoing pyrazinone chemistry.

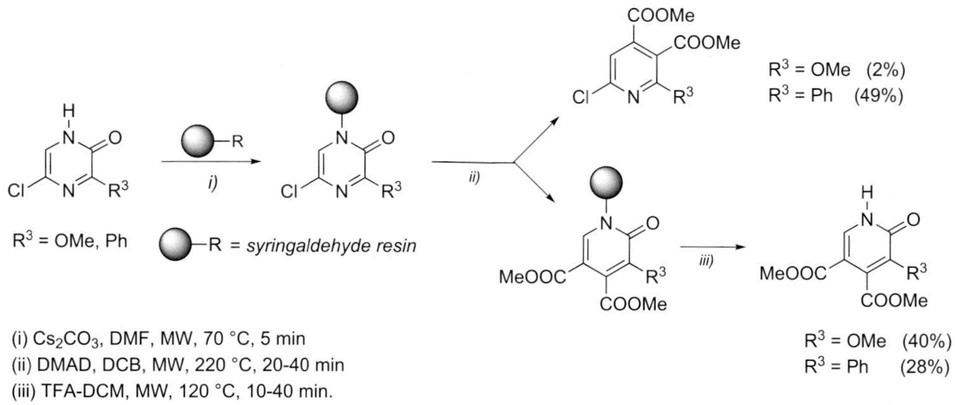

Scheme 8.65 Intermolecular 2(1*H*)-pyrazinone Diels–Alder reactions on solid support.

Lam and coworkers investigated a versatile microwave-mediated solid-supported approach to prepare various bi- and tricyclic heteroannulated 1,3-oxazin-6-ones. Those heterocyclic scaffolds are of considerable interest due to their widespread pharmacological and biological activities [76]. Starting from conventionally available Wang resin, several requisite ester linkers could be prepared to allow traceless cleavage with concomitant formation of the 1,3-oxazin-6-one ring. To generate the various heteroannulated oxazinones, different resin-bound heterocycles (imidazoles, pyrazoles **11**, pyrroles **12**, and thiophenes **13**) were prepared. These intermediates were treated under identical microwave conditions for further acylation and subsequent traceless cleavage (Scheme 8.66). The resin-bound imidazoles were formed by reacting Wang resin with bromoacetyl chloride in the presence of 4-dimethylaminopyridine (DMAP). The resulting solid support was treated at room temperature with imidamides and potassium *tert*-butoxide in *tert*-butanol to initiate cyclization to the desired polymer-bound imidazole. To generate the corresponding resin-bound intermediates **11**, **12**, and **13**, the key solid support was prepared by esterification of Wang resin with cyanoacetic acid. When treating this intermediate with trimethyl orthoformate in the presence of acetic anhydride and further condensation with hydrazines, **11** was obtained. On the other hand, when the intermediate cyanated resin was treated with α-aminoketones in THF under microwave irradiation, the solid-supported pyrroles **12** were formed. Finally, the resin-bound thiophene scaffolds **13** were obtained by a modified microwave-mediated Gewald synthesis when condensing the cyanated resin with cycloketones and elemental sulfur in the presence of morpholine (Scheme 8.66).

From this point, the resin-bound heterocycles **11–13** have been treated identically. The respective resin in acetonitrile was admixed with furanoyl chloride and triethylamine in the presence of 4-dimethylaminopyridine and heated under microwave irradiation at 120 °C for 20 min. After this acylation, the resin was subsequently treated with hydrazine monohydrate in *N*,*N*-dimethylformamide under microwave heating at 100 °C for 20 min. Finally, cyclative cleavage from the resin was performed by suspending the resin in acetonitrile with dibromotetrachloroethane, triphenylphosphine, and *N*,*N*-diisopropylethylamine and heating under microwave

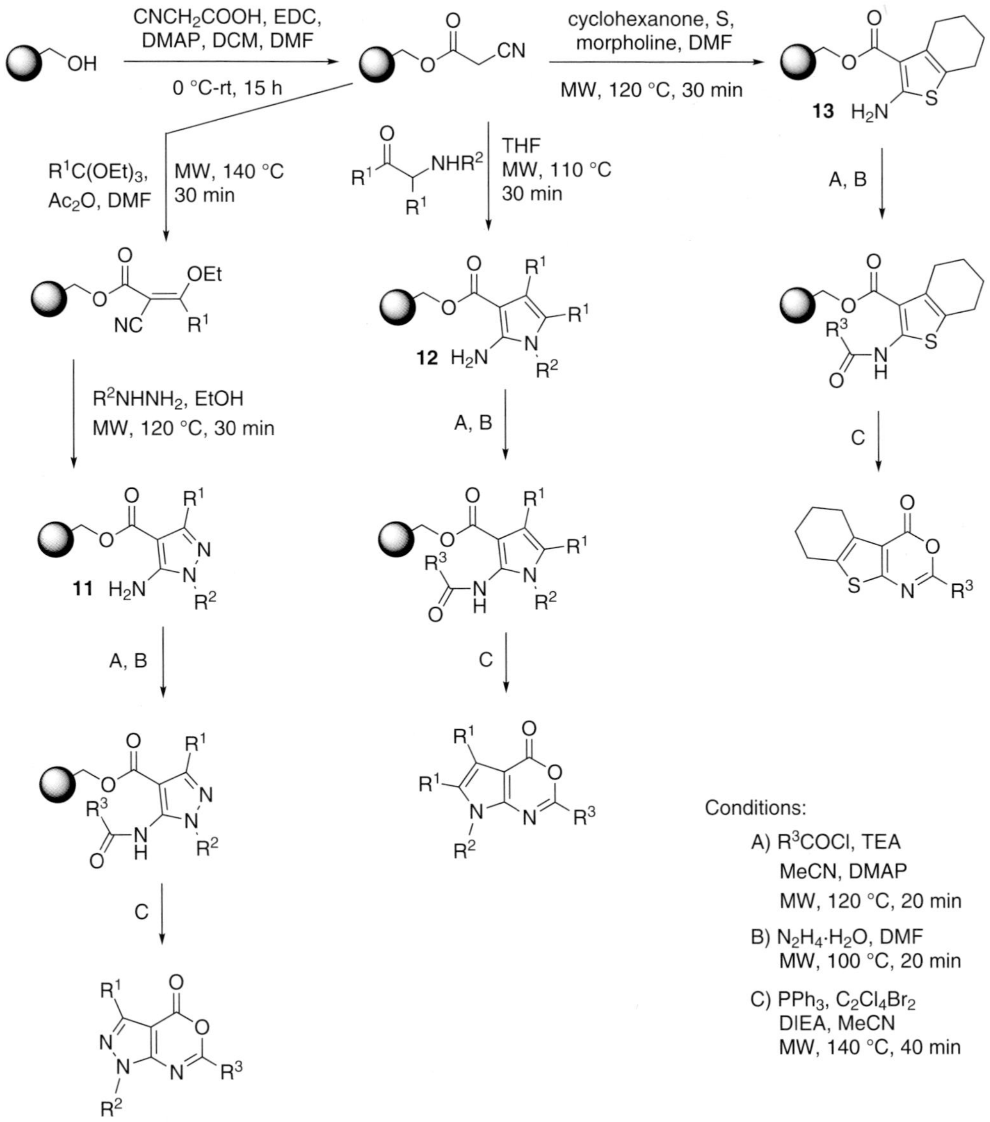

Scheme 8.66 Solid-phase synthesis of heteroannulated 1,3-oxazin-6-ones.

irradiation at 140 °C for 40 min. This procedure furnished the different bi- and tricyclic scaffolds with several points of diversity in good yields [76].

8.2
Soluble Polymer-Supported Synthesis

Besides solid-phase organic synthesis involving insoluble cross-linked polymer supports, chemistry on soluble polymer matrices, sometimes called liquid-phase organic synthesis, has emerged as a viable alternative. Problems associated with the heterogeneous nature of the ensuing chemistry and on-bead spectroscopic characterization in SPOS have led to the development of soluble polymers as alternative matrices for combinatorial library production (see also Section 4.7.2.2).

Several microwave-assisted protocols for soluble polymer-supported syntheses have been described. Among the first examples of so-called liquid-phase synthesis have been aqueous Suzuki couplings. Schotten and coworkers presented the use of PEG-bound aryl halides and sulfonates in these palladium-catalyzed cross-couplings [77]. The authors demonstrated that no additional phase-transfer catalyst (PTC) is needed when the PEG-bound electrophiles are coupled with appropriate aryl boronic acids. The polymer-bound substrates were coupled with 1.2 equiv of the boronic acids in water under short-term microwave irradiation utilizing sealed vessels in a domestic microwave oven (Scheme 8.67). Workup involved precipitation of the polymer-bound biaryl from a suitable organic solvent with diethyl ether. Water and insoluble impurities need to be removed prior to precipitation in order to achieve high recoveries of the products.

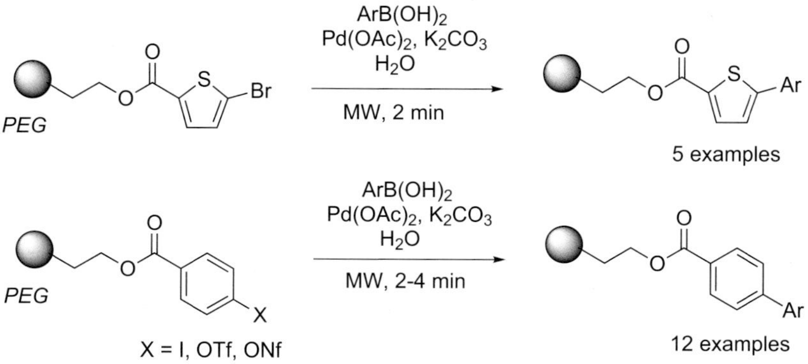

Scheme 8.67 PEG-supported Suzuki couplings.

Another palladium-catalyzed coupling reaction performed successfully on soluble polymers is the Sonogashira coupling. Xia and Wang presented an approach, where the utilized PEG 4000 acts simultaneously as polymeric support, solvent, and phase-transfer catalyst in both steps, coupling and hydrolysis [78]. PEG-bound 4-iodobenzoic acid was readily reacted with several terminal alkynes under rapid microwave conditions (Scheme 8.68). Attachment of the 4-iodobenzoic acid was achieved by transesterification using *N,N'*-dicyclohexylcarbodiimide (DCC) and *N,N*-dimethylaminopyridine in dichloromethane. Cleavage of the coupling products from the PEG support was carried out efficiently by simple saponification under microwave irradiation and subsequent acidification. The process time for this step was drastically reduced from 8 h at 50 °C (classical heating) to 2 min in an open beaker in a domestic microwave oven.

Another example, where PEG played the role of polymeric support, solvent, and PTC, was presented by the group of Lamaty [79]. In this study, a Schiff base-protected

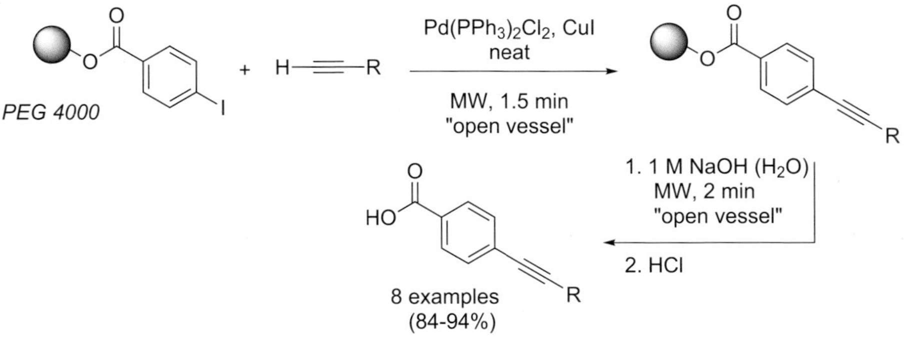

Scheme 8.68 Sonogashira couplings on PEG support.

glycine was reacted neat with various electrophiles (RX) under microwave irradiation (Scheme 8.69). After alkylation, the corresponding amino esters were released from the polymer support by transesterification employing methanol in the presence of triethylamine.

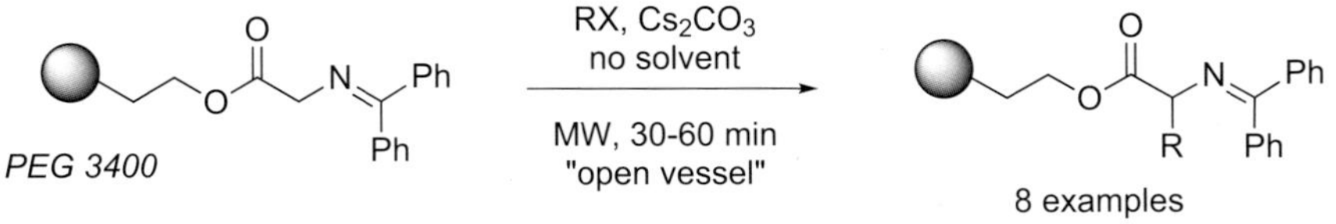

Scheme 8.69 Microwave-assisted alkylations on PEG.

A method for microwave-assisted transesterifications has been described earlier by Vanden Eynde and Rutot [80]. The authors have investigated the microwave-mediated derivatization of poly(styrene-*co*-allyl alcohol) as key step for the polymer-assisted synthesis of heterocycles. Several β-ketoesters were employed in this procedure and multigram quantities of products were obtained when exposing neat mixtures of the reagents in open vessels to microwave irradiation utilizing a domestic microwave oven (Scheme 8.70). The successful derivatization of the polymer was confirmed by IR, ^{1}H NMR, and ^{13}C NMR spectra. The prepared soluble supports were used for the preparation of various bicyclic heterocycles like pyrazolopyridinediones or coumarins.

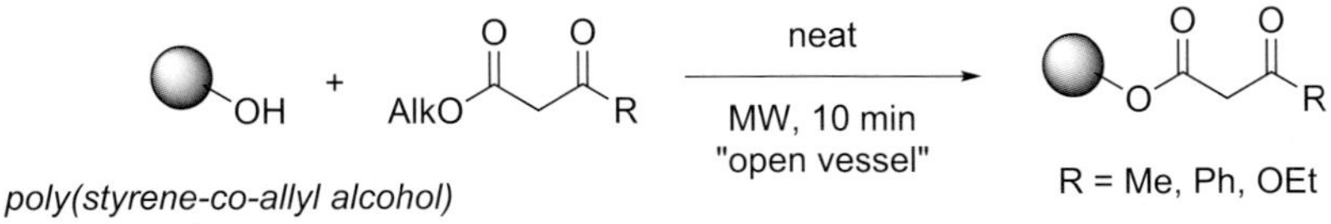

Scheme 8.70 Transesterification of poly(styrene-*co*-allyl alcohol).

Another procedure for the preparation of valuable heterocyclic scaffolds has been presented by Xia and Wang, that is, the Biginelli condensation on PEG support [81, 82].

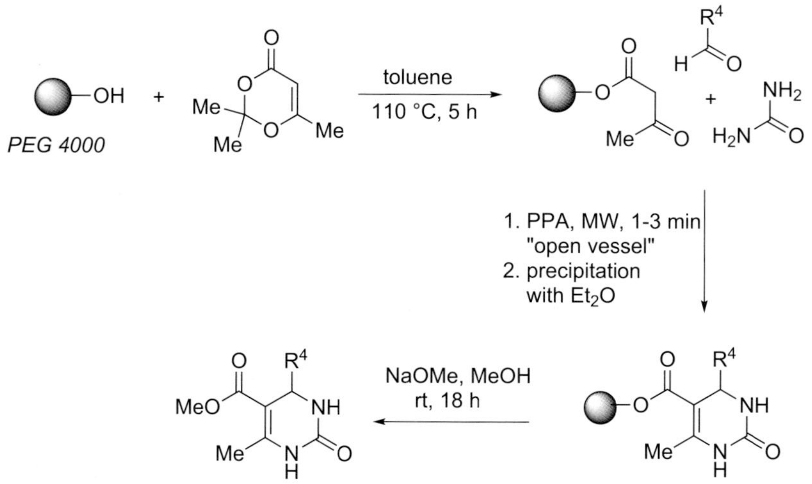

Scheme 8.71 One-pot Biginelli cyclocondensation on PEG support.

Polymer-bound acetoacetate was prepared by reacting commercially available PEG 4000 with 2,2,6-trimethyl-4*H*-1,3-dioxin-4-one in refluxing toluene (Scheme 8.71). In a representative procedure, the polymer-bound acetoacetate was suspended with 2 equiv each of urea and the corresponding aldehyde, and two to three drops of polyphosphoric acid (PPA) as a catalyst in a glass beaker. The open beaker was placed in a large container containing alumina as heat sink in a domestic microwave oven [82]. During microwave heating, the PEG-bound substrate melted, ensuring a homogeneous reaction mixture. After the reaction, diethyl ether was added to achieve precipitation of the polymer-bound products. The desired compounds were released by treatment with sodium methoxide (NaOMe) in methanol at room temperature. All dihydropyrimidines were obtained in high yields, and purification was achieved by recrystallization from ethanol.

A related support frequently used for liquid-phase synthesis is methoxy-polyethylene glycol (MeO-PEG). The group of Taddei presented a general procedure for the microwave-assisted synthesis of organic molecules on MeO-PEG [83]. The use of MeO-PEG under microwave conditions in open vessels simplified the process of polymer-supported synthesis (Scheme 8.72) as the polymer-bound products precipitate while cooling after removal from the microwave oven. In addition, cleavage could be performed rapidly under microwave irradiation employing a 1 : 1 mixture of trifluoroacetic acid and 2-propanol furnishing the desired nicotinic acid as corresponding pyridinium trifluoroacetate.

Wu and Sun presented a versatile procedure for the liquid-phase synthesis of 1,2,3,4-tetrahydro-β-carbolines [84]. After successful esterification of the utilized MeO-PEG-OH with Fmoc-protected tryptophan, a one-pot cyclocondensation with various ketones and aldehydes was performed under microwave irradiation (Scheme 8.73). The desired products were released from the soluble support in good yields and high purity. The interest in this particular scaffold is due to the

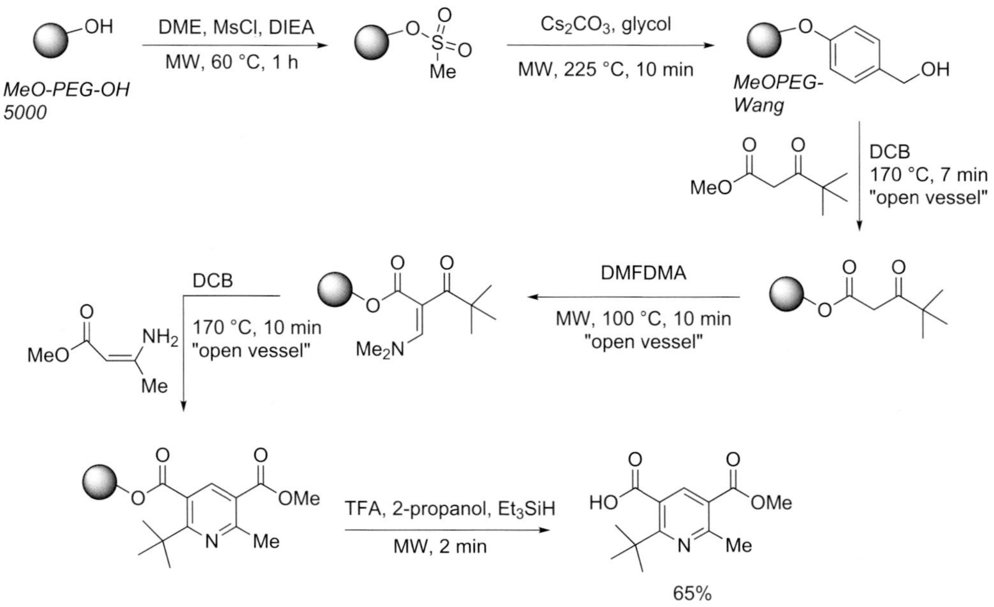

Scheme 8.72 Liquid-phase organic synthesis utilizing MeO-PEG.

knowledge of the 1,2,3,4-tetrahydro-β-carboline pharmacophore being an important structural element in several natural alkaloids, and the template possessing multiple sites for combinatorial modifications. The microwave-assisted liquid-phase protocol furnished purer products than homogeneous protocols and product isolation/purification was definitely simplified.

Generation of the desired polymer-bound tryptophan was carried out rapidly under microwave irradiation employing a classic esterification protocol using *N,N′*-dicyclohexylcarbodiimide and catalytic amounts of *N,N*-dimethylaminopyridine followed by subsequent Fmoc deprotection (Scheme 8.73). The cyclocondensations with

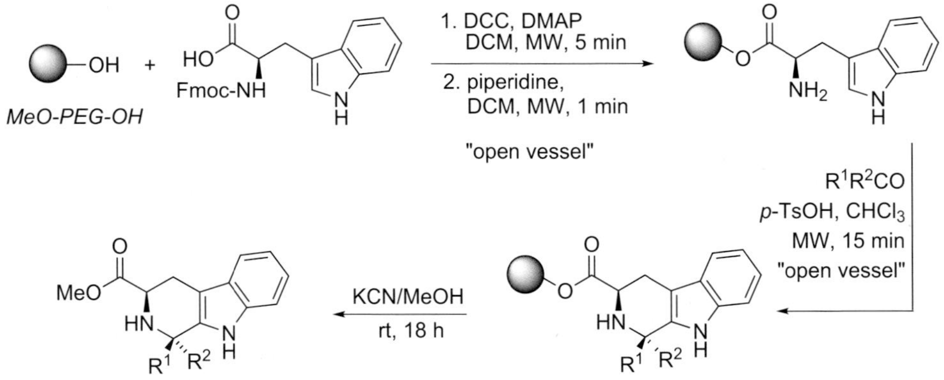

Scheme 8.73 Liquid-phase preparation of 1,2,3,4-tetrahydro-β-carbolines.

various carbonyl compounds were performed with catalytic amounts of *p*-toluenesulfonic acid (*p*-TsOH) within 15 min of microwave irradiation to obtain quantitative conversion as monitored by ^{1}H NMR. Finally, treatment of the polymer-bound heterocycles with 1 mol% potassium cyanide (KCN) in methanol furnished the desired target structures in high yields. The crude products were found to be enriched with the corresponding *trans*-isomers.

In a related study by the same group, the microwave-assisted liquid-phase synthesis of benzopiperazinones (quinoxalin-2-ones) was described [85]. The [6,6]-ring quinoxalinone system has also been reported to show promising pharmaceutical properties, and Sun and coworkers have employed a hybrid strategy using both combinatorial and microwave syntheses to generate a quinoxalin-2-one library from readily available building blocks. Thus, 4-fluoro-3-nitrobenzoic acid was coupled efficiently with a commercially available polyethylene glycol (PEG 6000). Various primary amines were then condensed to the readily prepared immobilized *ortho*-fluoronitrobenzene via nucleophilic *ipso*-fluoro displacement. Following reduction of the resulting *ortho*-nitroanilines furnished the corresponding *ortho*-phenylenediamines, which were subsequently reacted with chloroacetyl chloride to form polymer-bound 1,2,3,4-tetrahydroquinoxalin-2-ones (Scheme 8.74). Each step was performed within a few minutes under microwave irradiation in a dedicated

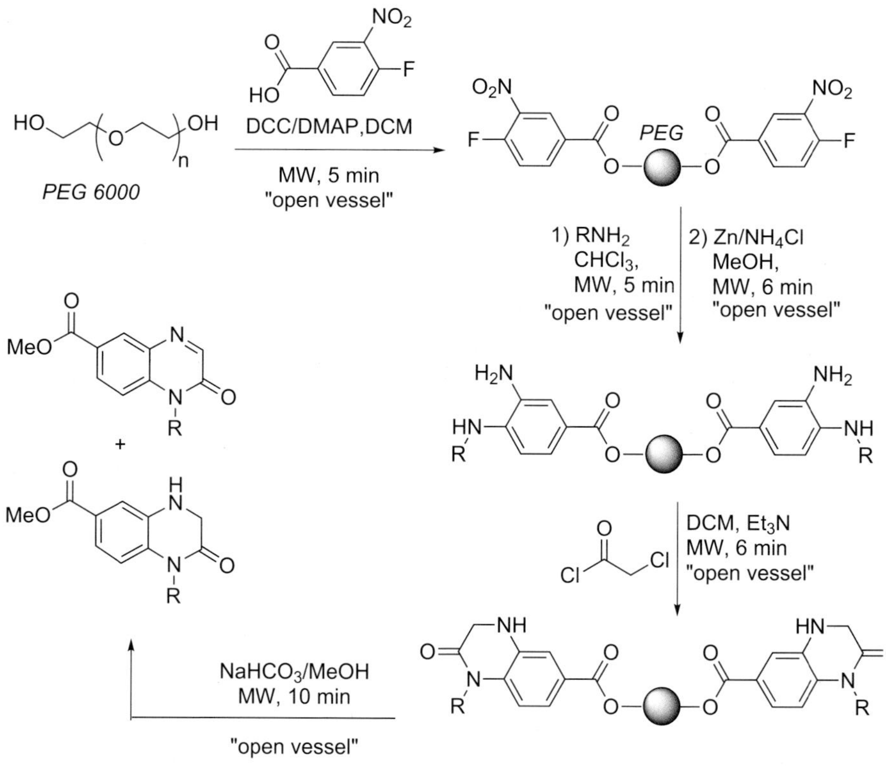

Scheme 8.74 Liquid-phase synthesis of quinoxalin-2-ones.

single-mode reactor. The PEG-bound intermediates were precipitated in ethanol and removed by filtration, leaving the by-products in the ethanolic phase. Finally, cleavage of the target molecules was achieved within 10 min under microwave irradiation employing a methanolic solution of sodium bicarbonate ($NaHCO_3$), furnishing both the desired dihydroquinoxalinones and the corresponding oxidized compounds in varying ratios. The dihydro products were fully converted to the oxidized form upon storage within several days [85].

Utilizing the identical aryl fluoride linker on conventional MeO-PEG polymer, the same authors also presented a microwave-accelerated liquid-phase synthesis of benzimidazoles (Scheme 8.75) [86]. This bicyclic pharmacophore is an important and valuable structural element in medicinal chemistry, showing a broad spectrum of pharmacological activities, such as antihistaminic, antiparasitic, and antiviral effects.

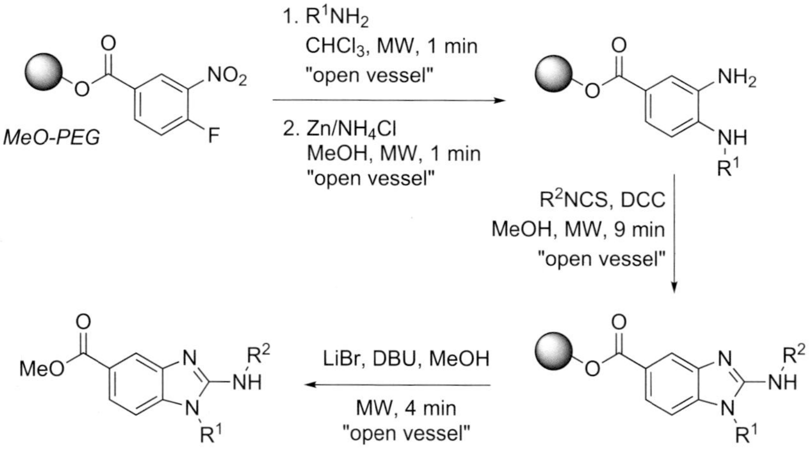

Scheme 8.75 Liquid-phase synthesis of benzimidazoles.

Following the strategy described above with *ipso*-fluoro displacement and subsequent reduction, the resulting *ortho*-phenylenediamines were treated with several aromatic isothiocyanates in the presence of *N,N'*-dicyclohexylcarbodiimide to form the corresponding PEG-bound benzimidazoles. Cleavage of the desired compounds was readily achieved by microwave-mediated transesterification (methanol) with lithium bromide and 1,8-diazabicyclo[5.4.0]undec-7-ene. MeO-PEG-OH was separated by precipitation and filtration to obtain the crude products in high yields and good purities [86].

In a closely related study, Tung and Sun discussed the microwave-assisted liquid-phase synthesis of chiral quinoxalines [87]. Various L-α-amino acid methyl ester hydrochlorides were coupled to MeO-PEG-bound *ortho*-fluoronitrobenzene by the already described *ipso*-fluoro displacement. Reduction under microwave irradiation resulted in spontaneous synchronous intramolecular cyclization to the corresponding 1,2,3,4-tetrahydroquinoxalin-2-ones (Scheme 8.76). Retention of the chiral

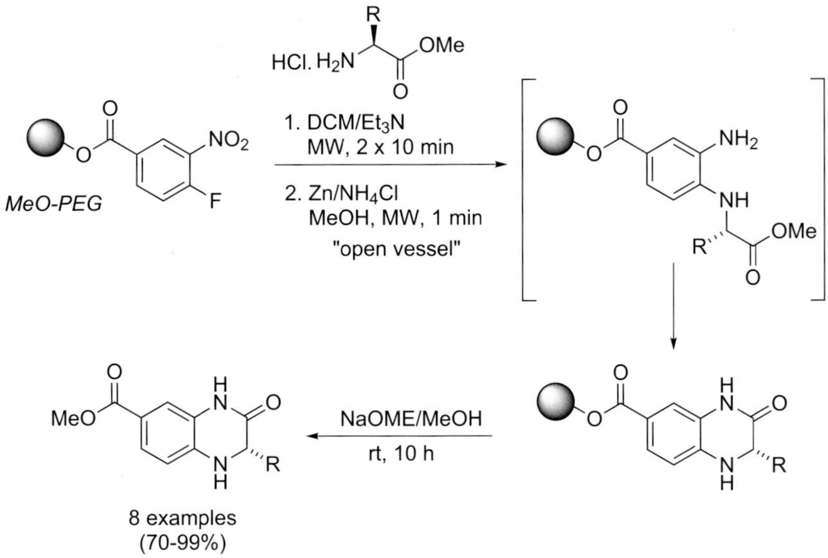

Scheme 8.76 Liquid-phase synthesis of chiral quinoxalinones.

moiety could not be monitored during reaction, but after release of the desired products it was found that about 10% of the product underwent racemization.

A related variation of this method led to the generation of bis-benzimidazoles [88, 89]. The versatile immobilized *ortho*-phenylenediamine template was prepared as described above in several microwave-mediated steps. Additional *N*-acylation was performed utilizing the initially used 4-fluoro-3-nitrobenzoic acid at room temperature exclusively at the primary aromatic amine moiety (Scheme 8.77). Various amines have been used to introduce diversity by nucleophilic aromatic substitution. Cyclization to the polymer-bound benzimidazole was achieved by refluxing several hours in a mixture of trifluoroacetic acid and chloroform. Individual steps at ambient temperature for selective reduction, cyclization with several aldehydes, and final detachment from the polymer support were necessary to obtain the desired bis-benzimidazoles. A set of 13 examples has been prepared in high yields and good purities [88].

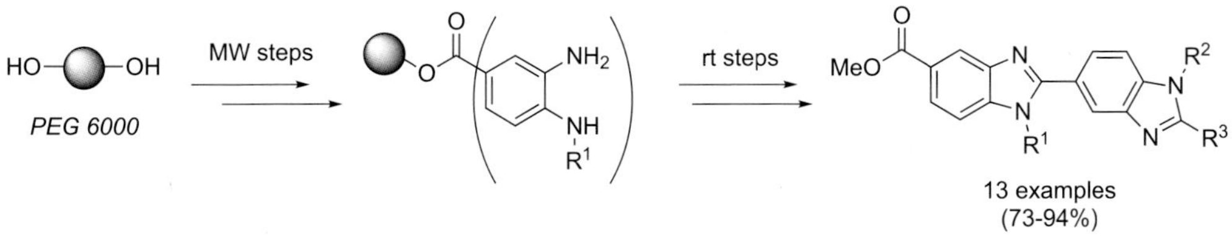

Scheme 8.77 Liquid-phase synthesis of bis-benzimidazoles.

The cyclization to the desired head-to-tail linked bis-benzimidazoles can also be performed utilizing aryl or alkyl isothiocyanates with *N*,*N′*-dicyclohexylcarbodi-imide [89].

Upon completion, the formed insoluble *N,N'*-dicyclohexylthiourea has to be removed by filtration and the desired PEG-bound products were precipitated by addition of diethyl ether to the reaction mixture. Results were more or less identical to the cyclizations with the above-mentioned aldehydes.

In a closely related study by the same group, mercury chloride was utilized as a catalyst to perform cyclization to the benzimidazoles [90].

In a further application of the bis-hydroxylated polymer support PEG 6000, the group of Sun presented several reports on the microwave-accelerated liquid-phase synthesis of thiohydantoins [91, 92] (Scheme 8.78). Whereas efficient coupling of Fmoc-protected amino acids under DCC/DMAP activation succeeded within 14 min in a dedicated single-mode instrument under open-vessel conditions, no conversion at all could be obtained after the same period of conventional heating (refluxing dichloromethane) [91]. After conventional deprotection employing 10% piperidine in dichloromethane at ambient temperature, various isothiocyanates were introduced under microwave irradiation. In order to drive the transformation to completion, 3 equiv of the isothiocyanates had to be used. The resulting PEG-bound thiourea derivatives were successfully released from the support by cyclative cleavage under mild basic conditions employing potassium carbonate. This traceless cleavage protocol ensured that only the desired compounds were released from the soluble support. Removal of the polymer by precipitation and filtration afforded the heterocyclic compounds in high yields and excellent purities.

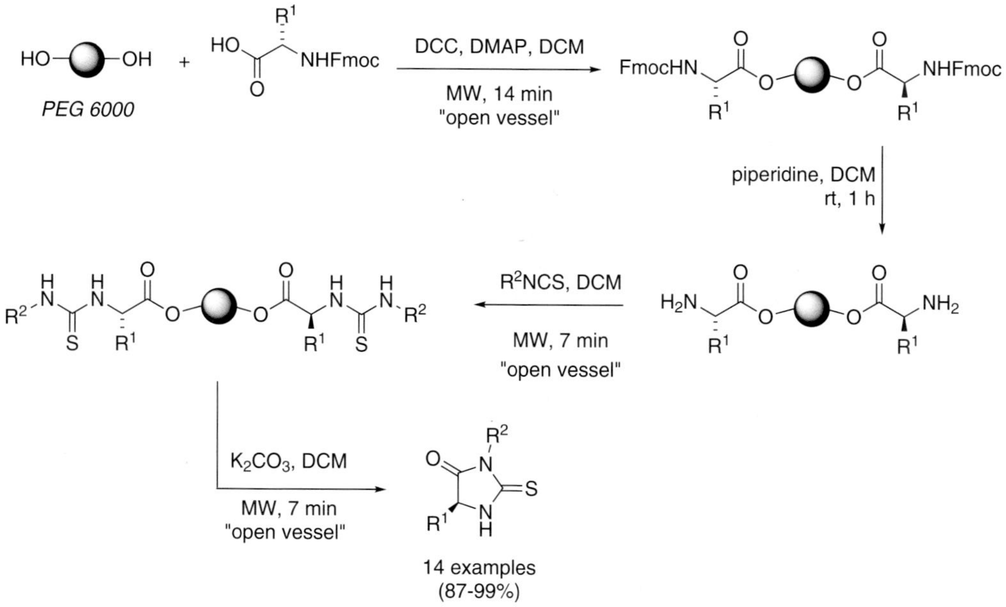

Scheme 8.78 Liquid-phase synthesis of thiohydantoins.

In addition, the authors chose 3-chloropropionyl chloride as the immobilized building block to achieve a ring expansion approach, which led to the generation of a

14-member library of the corresponding thioxotetrahydropyrimidinones [92, 93]. The initially prepared polymer-bound chloropropionyl ester was efficiently transformed into the corresponding diamines by transamination utilizing several primary amines. These diamine intermediates could also be obtained by treatment of the pure polymeric support with acryloyl chloride and subsequent addition of the corresponding amines (Scheme 8.79).

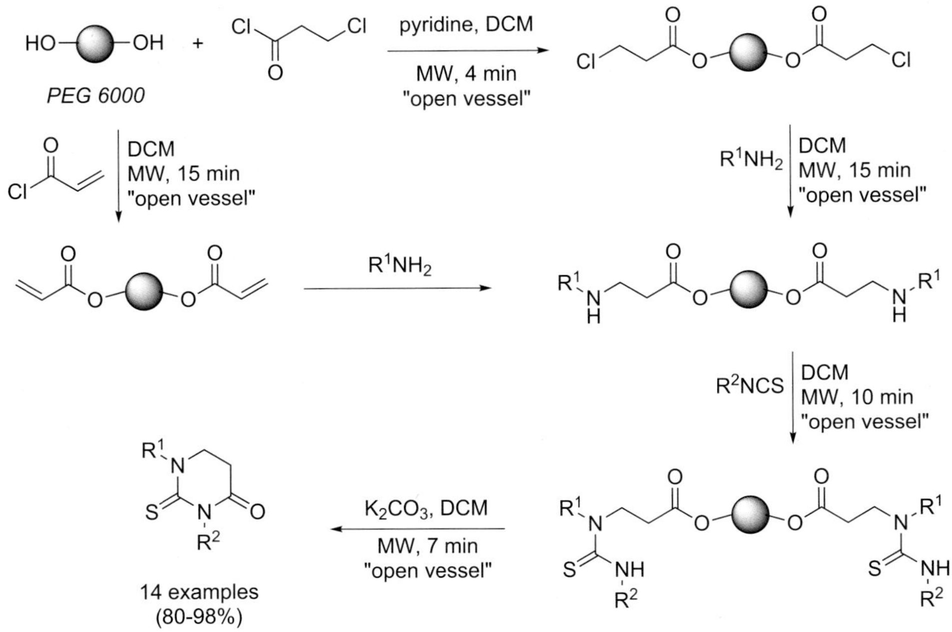

Scheme 8.79 Liquid-phase synthesis of 1,3-disubstituted thioxotetrahydropyrimidinones.

In analogy to the thiohydantoin synthesis, the PEG-bound diamines have been treated with various alkyl and aryl isothiocyanates and after applying traceless cyclative cleavage the desired thioxotetrahydropyrimidinones were obtained in excellent yields. Contrary to the synthesis of thiohydantoins, purification of the products was more complicated if the excess of isothiocyanates was higher than 2.2 equiv [92].

In a more recent work, Sun and coworkers investigated also the synthesis of thiohydantoin-fused tetrahydro-β-carbolines (Scheme 8.80) [94]. A soluble polymer-supported strategy using the PEG resin combined with a traceless cyclative cleavage step was applied to achieve products **14** in high stereoselectivity. By employing microwave heating, the thermodynamically stable *trans* diastereomers were obtained preferentially.

A series of related articles by the same group highlight the efficiency of using PEG-derived soluble supports for heterocyclic synthesis [95].

Another related study from the Sun laboratory described the synthesis of hydantoins utilizing acryloyl chloride to prepare a suitable polymer support [96].

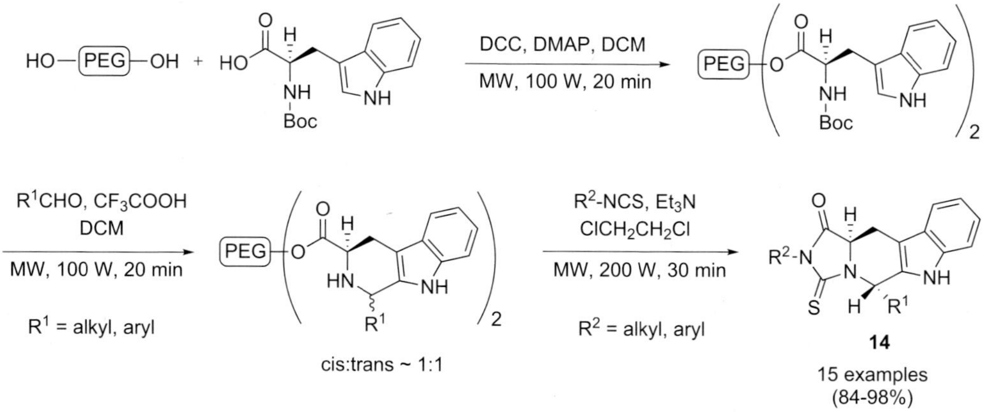

Scheme 8.80 Synthesis of tetracyclic β-carbolines.

All steps were carried out under reflux conditions in a dedicated microwave instrument utilizing 50 mL round-bottom flasks.

The soluble polymer support was dissolved in dichloromethane and treated with 3 equiv of chloroacetyl chloride for 10 min under microwave irradiation. The following nucleophilic substitution utilizing 4 equiv of various primary amines was carried out in *N,N*-dimethylformamide as solvent. The resulting PEG-bound amines were reacted with 3 equiv of aryl or alkyl isothiocyanates in dichloromethane to furnish the polymer-bound urea derivatives after 5 min of microwave irradiation (Scheme 8.81). After each step, the intermediates were purified by simple precipitation employing diethyl ether and filtration to remove by-products and unreacted

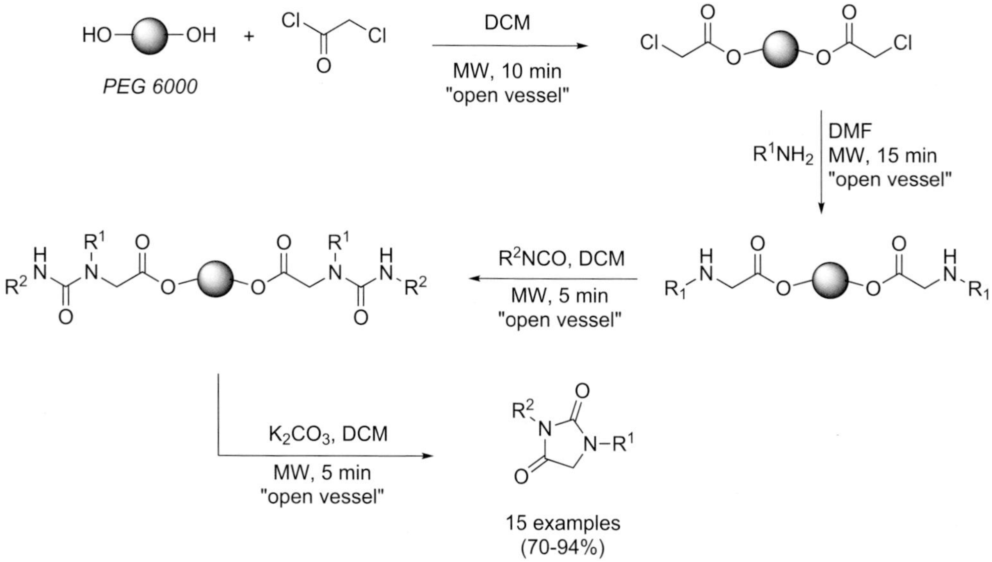

Scheme 8.81 Liquid-phase synthesis of hydantoins.

substrates. Finally, traceless release of the desired compounds by cyclative cleavage was achieved under mild basic conditions within 5 min of microwave irradiation. The 1,3-disubstituted hydantoins were obtained in varying yields but high purity.

In recent work, the Sun laboratory reported on a microwave-mediated traceless polymer-supported synthesis of 1,3,5-oxadiazinones utilizing soluble PEG-based polymers [97]. The novel approach involved preparation of polymer-bound bis-Boc-guanidines that underwent intramolecular cyclization for cleavage to yield the desired oxadiazinones in high yields (Scheme 8.82). The guanidines were chosen as appropriate precursors due to their relevance in pharmaceutically active heterocyclic moieties. Commercially available PEG 6000 was modified stepwise under microwave conditions to generate the corresponding PEG-bound piperazine **15**. In the crucial next step, the guanidine moiety was introduced by reaction with **15** in the presence of 3 equiv triethylamine under microwave irradiation at 150 W for 7 min. The desired PEG-bound Boc-protected intermediate **16** was obtained in high yields, which was further reacted with appropriate amines under additional 7 min of microwave irradiation, leading to PEG-bound oxadiazinone **17** in high yields. In the final step of the sequence, the desired products **18** were released from the polymer support

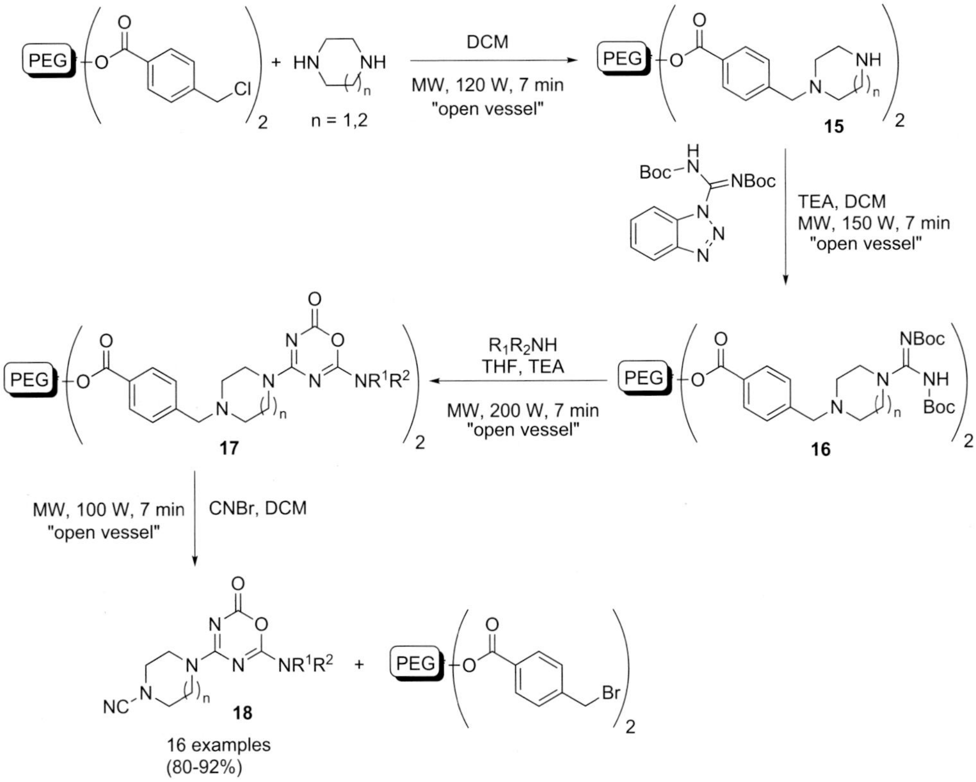

Scheme 8.82 Cyclization of bis-Boc-guanidines to generate oxadiazinones.

via cyclative cleavage with simultaneous cyanation again under microwave heating at 100 W for 7 min (Scheme 8.82). Simple filtration after precipitation with cold diethyl ether furnished also the bromo derivative of the initial functionalized PEG, which could be successfully recycled [97]. The novel approach worked for piperazines and 1,4-diazepanes to modify the polymer support and tolerates a variety of cyclic and aliphatic secondary amines for the cyclization step to provide remarkable diversity for the biologically interesting heterocyclic scaffold. Microwave irradiation proved to be highly beneficial for the entire sequence reducing the reaction times to only a few minutes for each step whereas under classic thermal heating several hours were required to complete each individual transformation.

A cycloalkane-based thermomorphic system for performing Pd-catalyzed cross-coupling reactions was developed by Chiba [98]. The use of substrates with cycloalkane-soluble tags facilitates separation of the desired products and the homogeneous Pd catalyst via simple liquid–liquid extraction, thereby eliminating the need for a catalyst removal process.

8.3 Fluorous-Phase Organic Synthesis

Fluorous-phase organic synthesis is a separation and purification technique for organic synthesis and process development (see also Section 4.7.2.3). This technique involves the use of an organic molecule, which is rendered soluble in fluorocarbon solvents by attachment of a suitable fluorocarbon group ("fluorous tag"). These hydrophobic perfluorinated hydrocarbon compounds are immiscible with water and many common organic solvents. At the desired stage of the synthesis, the fluorous label is cleaved and the product is rendered "organic" again [99].

One early example of microwave-assisted fluorous synthesis was presented by the group of Hallberg and involves palladium-catalyzed Stille couplings of fluorous tin reagents with aryl halides or triflates (Scheme 8.83) [100]. While the comparable thermal process required 1 day for completion, the microwave-heated reactions (sealed vessels in monomode reactors) were completed within 2 min, with the additional benefit of reduced homocoupling of the tin reagent. The desired biaryl products were isolated in good yields and purities after a three-phase extraction. Similar results were also achieved by the same group utilizing so-called F-21 fluorous tags ($CH_2CH_2C_{10}F_{21}$) on the tin reagent [101].

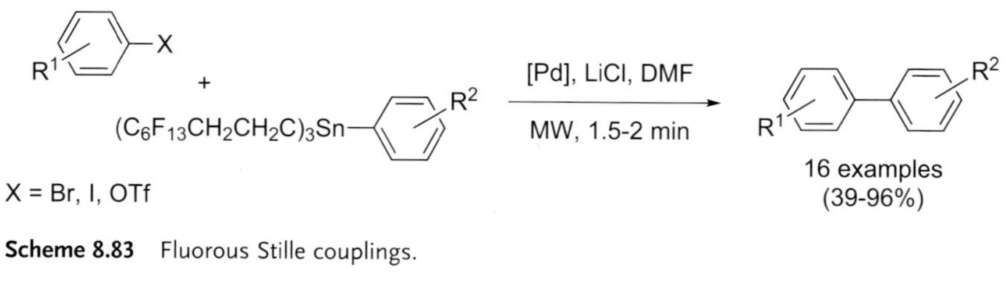

Scheme 8.83 Fluorous Stille couplings.

In an additional application of fluorous chemistry, Hallberg's group has reported radical-mediated cyclizations utilizing benzotrifluoride (BTF) as solvent under microwave irradiation [101]. In the presence of 2,2′-azobisisobutyronitrile as radical initiator, the aryl iodide shown in Scheme 8.84 smoothly underwent microwave-mediated cyclization to the corresponding indole derivative in high isolated yield. The ability to promote highly fluorous reactions with microwave heating deserves special attention. With these highly fluorous tin reagents, microwave irradiation is more than an expedient means of reducing reaction times. Reactions conducted under traditional heating either did not work at all or did not work nearly as well. The advantage of microwave heating may be the rapid coalescence of the organic and fluorous phases to form a homogeneous solution [101].

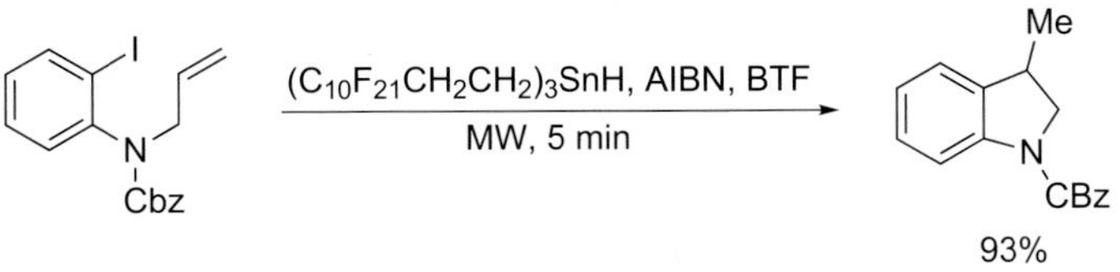

Scheme 8.84 Radical cyclizations using fluorous tin reagents.

Zhang *et al.* reported on a palladium-catalyzed carbon–sulfur cross-coupling of aryl perfluoroalkoxysulfonates with thiols (Scheme 8.85) [102]. The fluorous substrates were obtained from commercially available phenols by treatment with perfluoro-octanesulfonyl fluoride ($C_8F_{17}SO_2F$) under basic conditions. Various thiols were reacted with a slight excess of the perfluorinated sulfonates in a microwave-mediated Suzuki-type reaction employing 10 mol% of [1,1′-bis(diphenylphosphino)ferrocene] dichloropalladium(II) ($PdCl_2dppf$) as catalyst to furnish the corresponding aryl sulfides. Purification of the products after aqueous workup was achieved by fluorous solid-phase extraction (F-SPE).

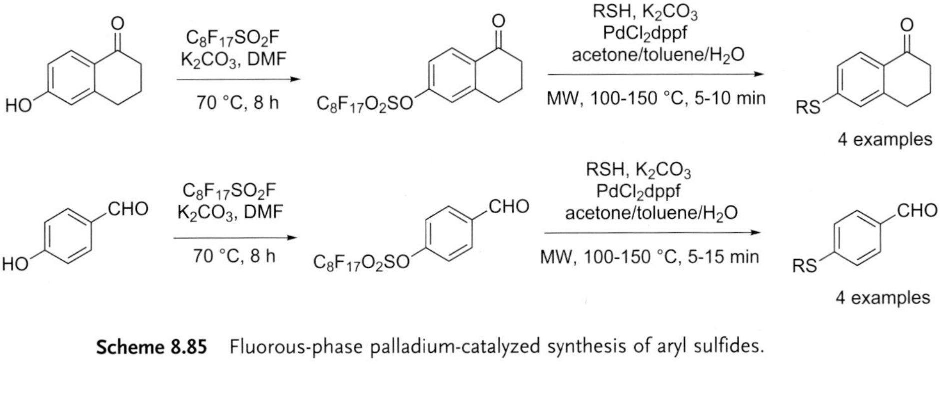

Scheme 8.85 Fluorous-phase palladium-catalyzed synthesis of aryl sulfides.

In another study from the same laboratories, Zhang and Nagashima described a parallel synthesis approach toward an *N*-alkylated dihydropteridinone library utilizing fluorous amino acids [103]. The fluorous substrates were readily prepared from

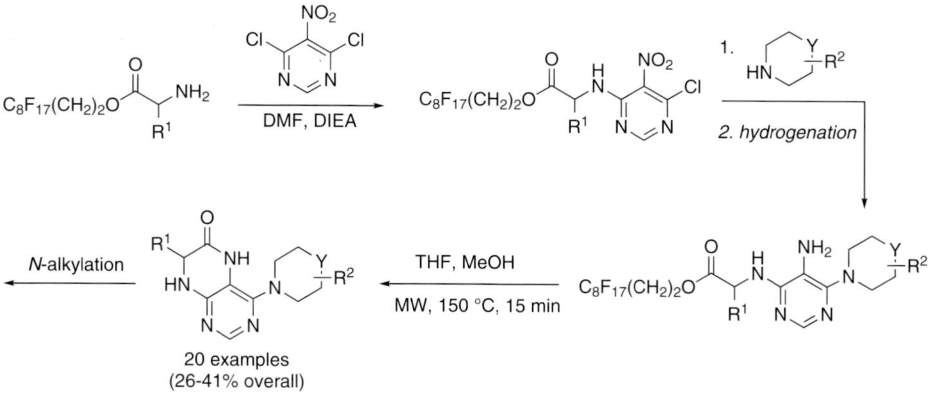

Scheme 8.86 Generation of dihydropteridinones via cyclative cleavage from a fluorous tag.

corresponding *N*-Boc-amino acids employing 3-(perfluorooctyl)propanol as the fluorinating agent, followed by deprotection under acidic conditions to afford the corresponding TFA salts of the amino acids. Four of these TFA salts were used without further purification in parallel displacement reactions utilizing 1.5 equiv of 4,6-dichloro-5-nitropyrimidine (Scheme 8.86). After addition of *N,N*-diisopropylethylamine, these exothermic reactions were completed within 10 min. Subsequently, a second displacement was carried out by addition of 3 equiv of five different cyclic amines. The resulting 20 intermediates were purified by fluorous solid-phase extraction and subjected to hydrogenation with platinum-on-charcoal for 12 h.

As only small amounts of the hydrogenation products underwent spontaneous cyclization, the compounds were heated in a sealed vessel under microwave irradiation. Due to the low solubility of the cyclized products at room temperature, they precipitated while cooling and could be easily collected by filtration. The desired dihydropteridinones were obtained in high purity and could be readily used for generating a more diverse library by subsequent *N*-alkylation [103].

The same group additionally presented further investigations on fluorous-phase palladium-catalyzed carbon–carbon couplings [104]. Fluorinated aryl octylsulfonates were reacted in slight excess with aryl boronic acids under microwave conditions to efficiently achieve the corresponding biaryls in a Suzuki-type cross-coupling (Scheme 8.87).

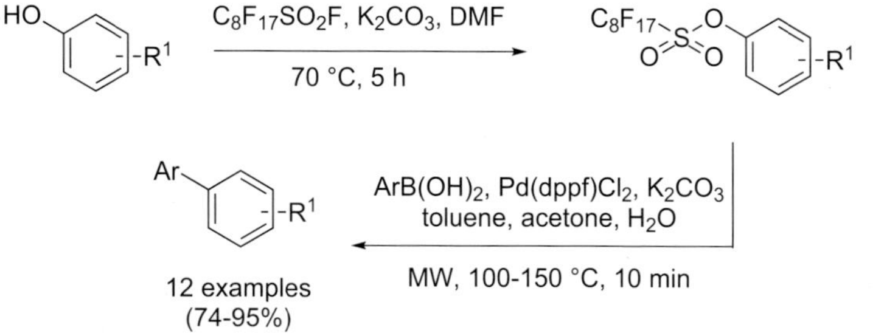

Scheme 8.87 Fluorous-phase Suzuki-type couplings.

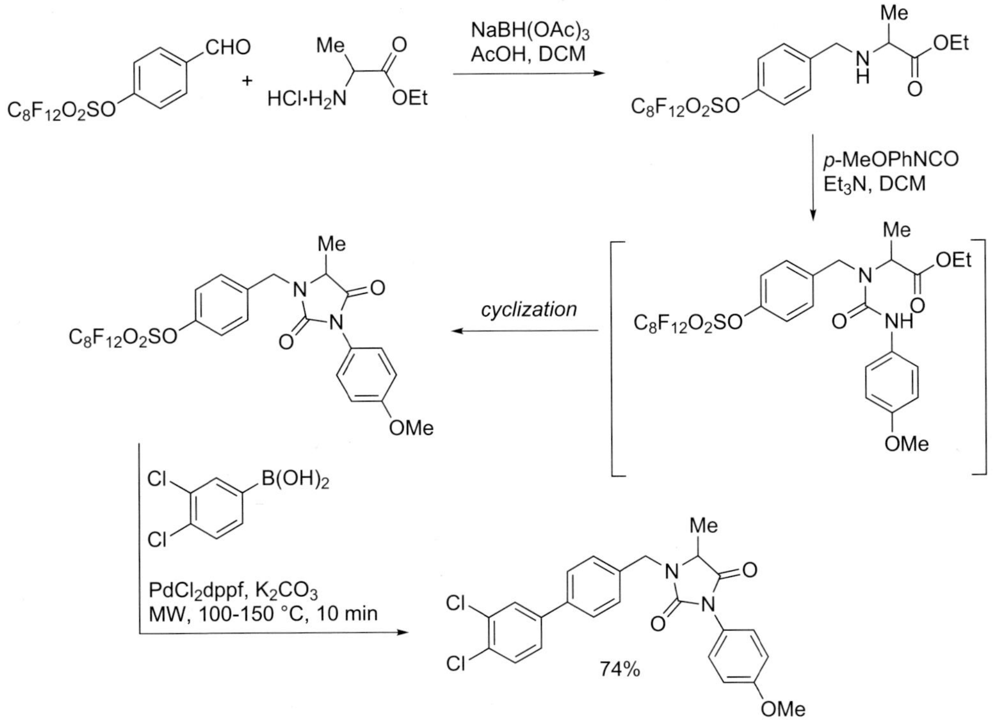

Scheme 8.88 Fluorous-phase synthesis of *N,N'*-disubstituted hydantoins.

The described fluorous tag strategy was also applied toward the synthesis of biaryl-substituted hydantoins (Scheme 8.88) [105]. 4-Hydroxybenzaldehyde was converted into the corresponding perfluorinated species, which underwent a reductive amination. The resulting amine was treated with an isocyanate to produce the fluorous-tagged urea, which spontaneously cyclized to form the corresponding hydantoin. Finally, the fluorous tag was detached by a Suzuki-type carbon–carbon bond formation to furnish the desired target structure in good yield.

In a related approach from the same laboratory, the perfluorooctylsulfonyl tag was furthermore employed in a traceless strategy for the deoxygenation of phenols (Scheme 8.89) [105]. These reactions were carried out in the solvent mixture toluene/acetone/water 4 : 4 : 1, utilizing 5 equiv of formic acid and potassium carbonate/[1,1′-bis(diphenylphosphino)ferrocene]dichloropalladium(II) as catalytic system. After 20 min of irradiation, the reaction mixture was subjected to fluorous

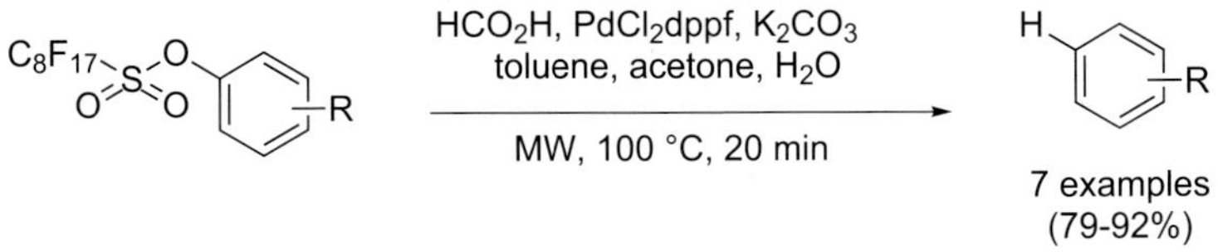

Scheme 8.89 Fluorous-phase traceless deoxygenation of phenols.

solid-phase extraction to afford the desired products in high yields. The application of this new traceless fluorous tag has in addition been applied for the synthesis of pyrimidines and hydantoins.

In 2005, the same group reported on a solution-phase synthesis of biaryl-substituted proline analogs by a three-component 1,3-dipolar cycloaddition and a Suzuki coupling (Scheme 8.90) [106]. Use of controlled microwave irradiation in both steps and perfluoroalkylsulfonyl-protected hydroxybenzaldehydes has enhanced both the reaction and separation processes for the synthesis of bicyclic proline analogs.

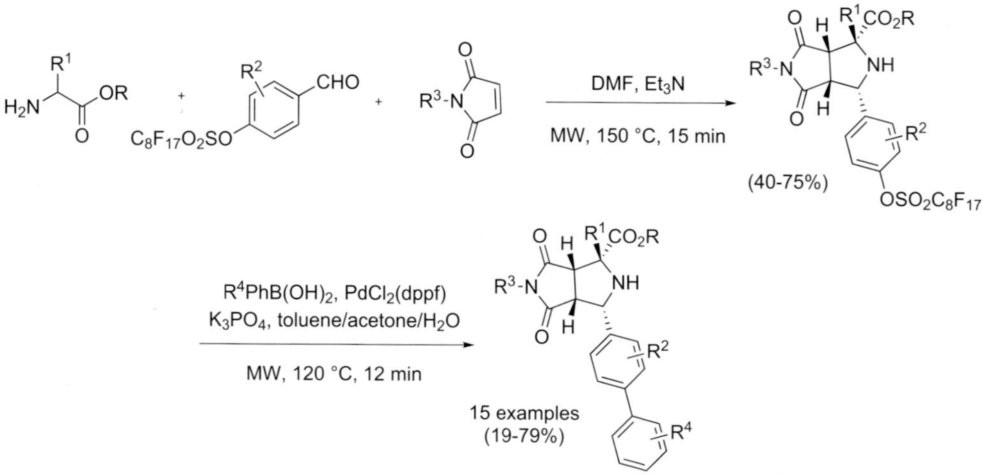

Scheme 8.90 Fluorous synthesis of biaryl-substituted proline analogs.

Fluorous tags can also be introduced as acid-labile protecting groups in the synthesis of carboxamides and sulfonamides [107]. Ladlow and coworkers presented the parallel generation of an 18-member library of biaryl carboxamides utilizing perfluorooctane-tagged iodopropane ($C_8F_{17}(CH_2)_3I$) to prepare the corresponding fluorous substrates (Scheme 8.91). Removal of the fluorous tag was achieved by

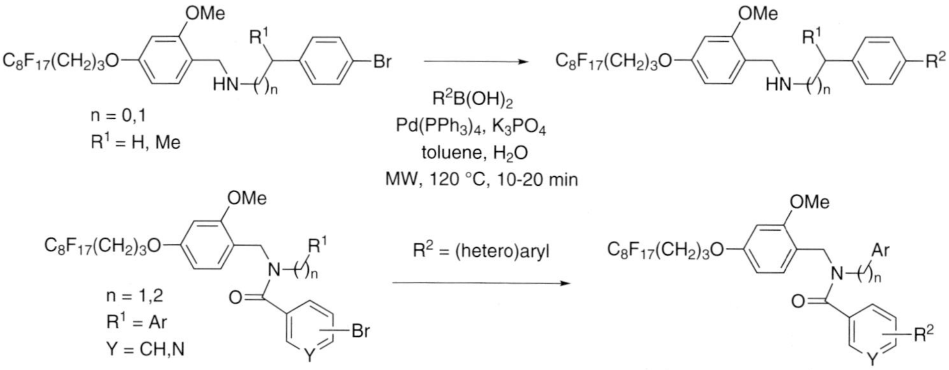

Scheme 8.91 Suzuki reactions utilizing fluorous-tagged acid-labile protecting groups.

treatment with a mixture of trifluoroacetic acid, triethylsilane, and water (90 : 5 : 5). Purification by subsequent SPE afforded the desired products in good yields and excellent purities.

Furthermore, multicomponent reactions can also be performed under fluorous-phase conditions as shown for the Ugi four-component reaction [108]. To improve the efficiency of a recently reported Ugi/de-Boc/cyclization strategy, Zhang and Tempest introduced a fluorous Boc group for amine protection and carried out the Ugi multicomponent condensation under microwave irradiation (Scheme 8.92). The desired fluorous condensation products were easily separated by fluorous solid-phase extraction and deprotected by treatment with trifluoroacetic acid/tetrahydrofuran under microwave irradiation. The resulting quinoxalinones were purified by a second F-SPE to furnish the products in excellent purity. This methodology was also applied toward a benzimidazole synthesis, employing benzoic acid as a substrate.

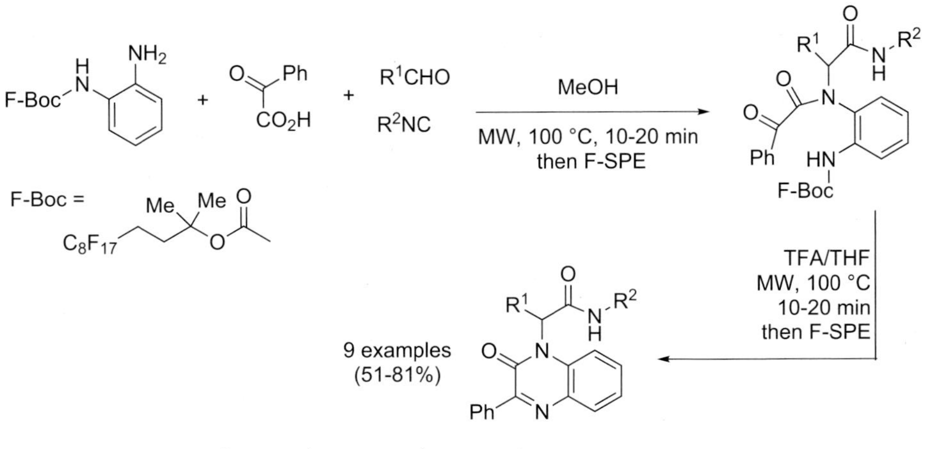

Scheme 8.92 Fluorous-phase quinoxalinone synthesis.

In another related study, Lu and Zhang described the microwave-assisted fluorous-phase synthesis of a pyridine/pyrazine library employing a three-component condensation reaction [109]. Imidazo[1,2-*a*]pyridines/pyrazines are biologically interesting compounds, showing several activities such as antifungal, antibacterial, and benzodiazepine receptor antagonistic properties. As multicomponent reactions are powerful tools to generate sets of highly diverse molecules, this technique was applied toward the fluorous synthesis of the desired pyridines/pyrazines under microwave irradiation.

For this purpose, perfluorooctanesulfonyl-tagged benzaldehydes were reacted with 1.1 equiv of 2-aminopyridines (or 2-aminopyrazines), 1.2 equiv of isonitriles, and catalytic amounts of scandium(III) triflate [$Sc(OTf)_3$] under microwave irradiation in a mixture of dichloromethane and methanol (Scheme 8.93). A ramp time of 2 min was employed to achieve the adjusted temperature, and then the reaction mixture was maintained at the final temperature for an additional 10 min. The fluorous tag is a multifunctional tool in this reaction, protecting the phenol

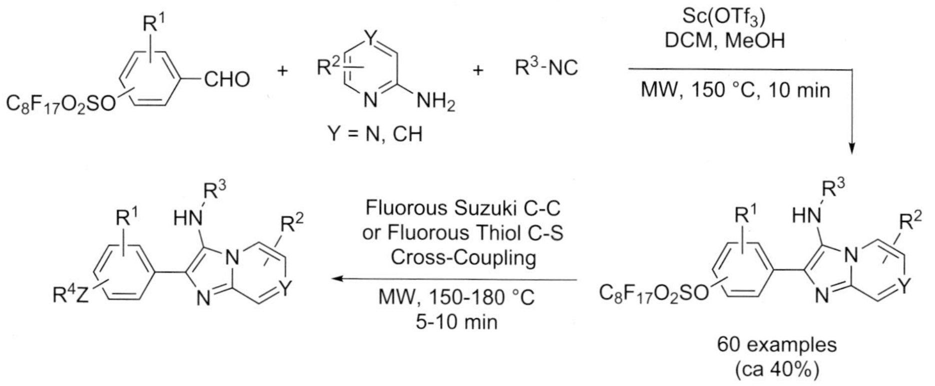

Scheme 8.93 Fluorous synthesis of imidazo[1,2-*a*]pyridine/pyrazine derivatives.

in the condensation step, being the phase tag for purification, and serving as activating group in the following cross-coupling reactions. After cooling to room temperature, the resulting fluorous-tagged condensation products were purified by simple solid-phase extraction or recrystallization before being subjected to various palladium-catalyzed cross-coupling reactions [109]. However, yields for both the multicomponent reaction and the cross-coupling were rather low. Interestingly, the established microwave-optimized protocols have subsequently been converted to conventional heating in order to apply for quick parallel library synthesis.

In 2009, Yan and coworkers established a 1,4-benzodiazepine-2,5-dione library by microwave-assisted fluorous synthesis [110]. The authors employed perfluoro-octanesulfonyl-protected 4-hydroxybenzaldehydes in an Ugi four-component reaction to form 46-member library of fluorous condensed products. In a postcondensation reaction, 30 representative intermediates were cyclized in parallel manner using 10% acetyl chloride in methanol to the fluorous benzodiazepinediones that were obtained in generally good to high yields. Finally, 20 of those benzodiazepinediones were subjected to a combined cleavage/arylation by microwave-assisted Suzuki coupling (Scheme 8.94). A standard microwave vial was charged with 1 mmol fluorous compound, 0.9 equiv boronic acid, 4 mol% Pd(dppf)Cl_2, and 2 equiv potassium carbonate in acetone/toluene/water (4 : 4 : 1) as a solvent. The mixture was heated under microwave irradiation at 150 °C for 20 min. With eight different boronic acids employed, additional diversity was introduced into the heterocyclic scaffolds and in total 36 of the desired biarylated 1,4-benzodiazepine-2,5-diones were obtained as diastereomers in moderate to good yield after standard fluorous solid-phase extraction. The variety of applicable building blocks demonstrates the general feasibility of this versatile reaction sequence to generate pharmaceutically attractive moieties.

In a related, though more recent, work, the same group explored their developed method to various amines and cyanides. Furthermore, the authors enhanced the method by performing even the cyclization step under microwave conditions [111]. The Ugi four-component cyclocondensation under basic conditions was performed utilizing five Boc-protected anthranilic acids, two fluorous benzaldehydes, five

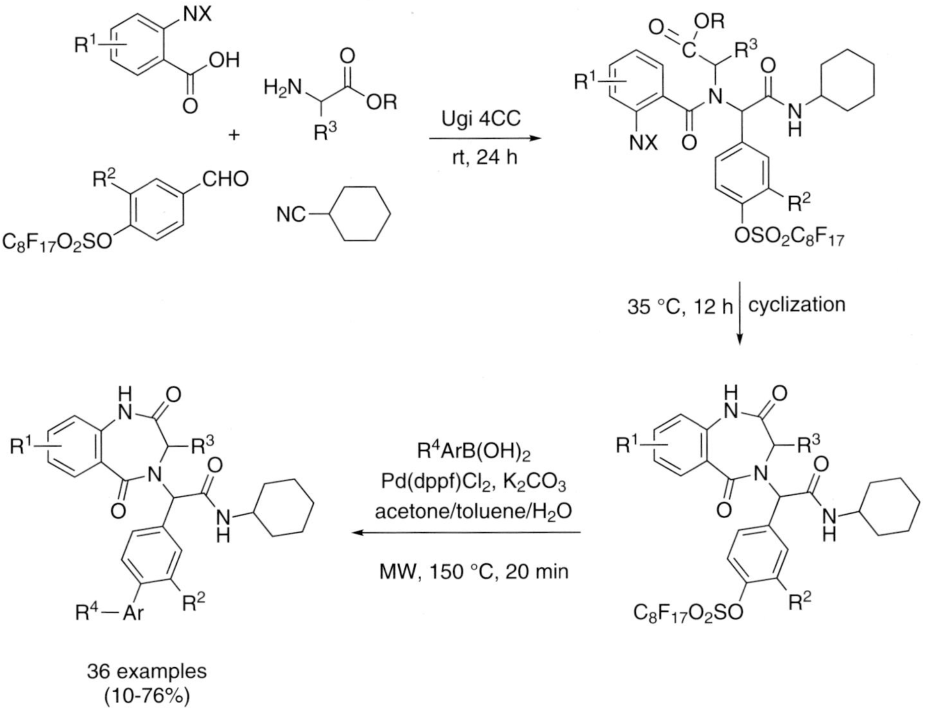

Scheme 8.94 Fluorous linker cleavage by microwave-assisted Suzuki coupling.

amines, and two different cyanides. The intermediate Ugi compounds were efficiently cyclized under microwave conditions by dissolving them in 10 equiv methanol/trifluoroacetic acid (1 : 1) and heating at 150 °C for 20 min (Scheme 8.95). After the reaction, the mixtures were neutralized with 1 N aqueous NaOH and 16 corresponding fluorous benzodiazepines were isolated by F-SPE. These compounds underwent cleavage from the fluorous tag by microwave-mediated Suzuki coupling under identical conditions as cited above to furnish a library of 31 1,4-benzodiazepine-2,5-diones in moderate to good yields [111].

Regioselective Heck vinylations of enamides have been studied by the Hallberg group under controlled microwave irradiation using a palladium catalyst and fluorous bidentate 1,3-bis(diphenylphosphino)propane (F-dppp) ligands [112]. The Heck reaction gave identical yields of the products when using nonfluorous and fluorous ligands, although the regioselectivity was slightly lower employing the fluorous ligand. Utilizing microwave heating showed enhanced reaction rates over classical heating. Vinylations of both cyclic and acyclic enamides with vinyl triflates were accelerated under microwave heating using fluorous ligands (F-dppp) and palladium(II) acetate as catalyst (Scheme 8.96).

Fluorous ligands introduce an ease of purification to completely remove the tagged phosphine ligand, palladium catalyst complexed ligand, and oxidized ligand by direct fluorous solid-phase separation before product isolation. Similarly, another example

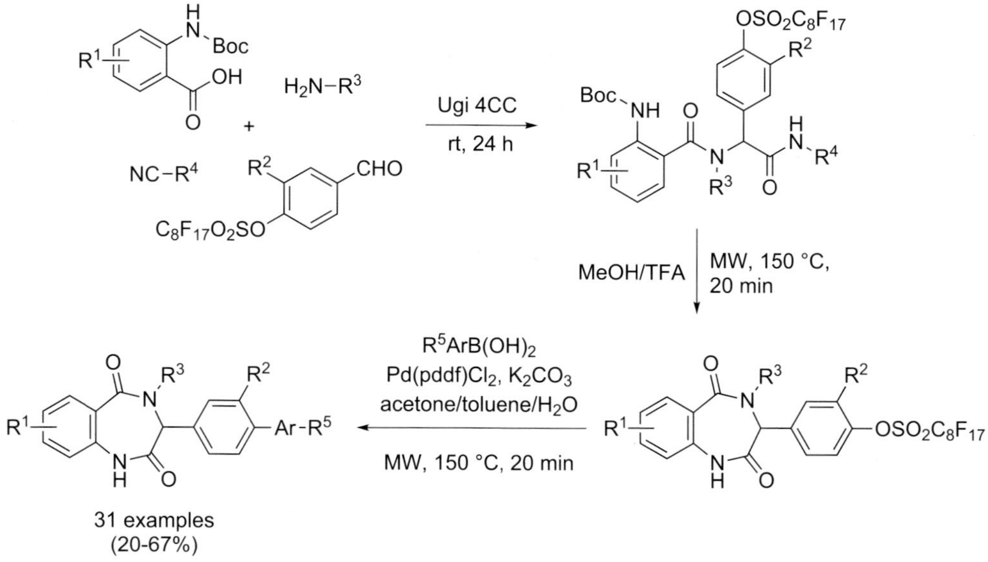

Scheme 8.95 Microwave-assisted fluorous synthesis of benzodiazepinediones.

of fluorous palladium-catalyzed microwave-induced synthesis of aryl sulfides has been reported where the product purification is aided by fluorous solid-phase extraction [113].

In a 2004 report, Larhed and coworkers have presented microwave-mediated fluorous reaction conditions for palladium-catalyzed aminocarbonylations [114]. A set of aryl halides was reacted with carbonyl hydrazides and molybdenum

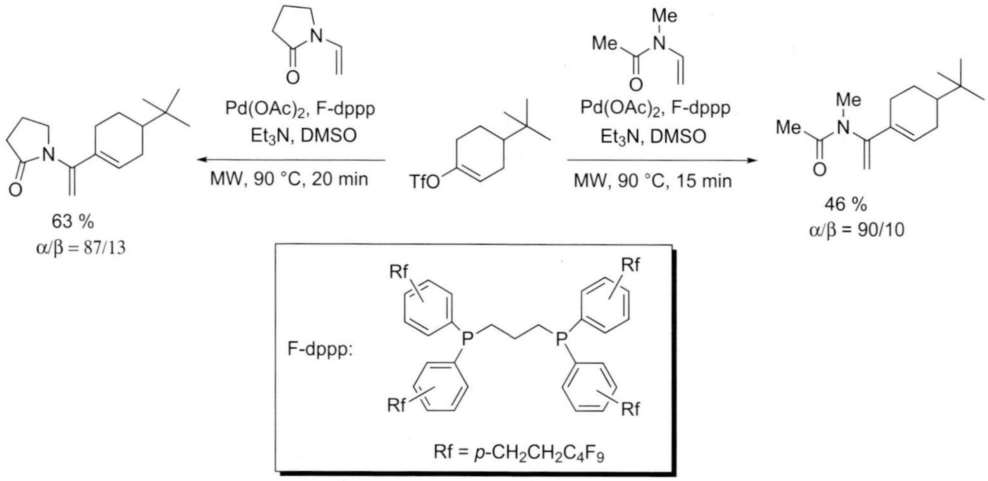

Scheme 8.96 Vinylation of enamides employing fluorous ligands.

hexacarbonyl [$Mo(CO)_6$] as source of carbon monoxide, employing fluorous triphenylphosphine (F-TPP) as ligand and the perfluorocarbon liquid FC-84 as perfluorinated solvent (Scheme 8.97).

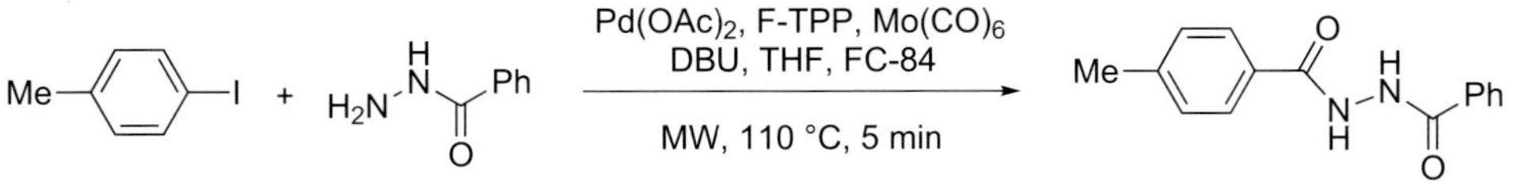

Scheme 8.97 Fluorous hydrazidocarbonylation.

The authors demonstrated the recyclability of the fluorous reagents without significant loss of efficiency for the model reaction shown in Scheme 8.97. After each reaction, the organic layer was separated and the perfluorinated liquid was subjected to the next mixture. Performing six cycles of the reaction afforded the corresponding product in 64–79% yield (Figure 8.1).

An interesting approach involving the use of a fluorous dienophile as diene scavenger under microwave conditions has been investigated by Werner and Curran [115]. The classical Diels–Alder-type cycloaddition of diphenylbutadiene with maleic anhydride as dienophile was accelerated under microwave heating at 160 °C for 10 min. A structurally related fluorous dienophile scavenged the excess diene (1.5 equiv) under the same microwave conditions and aided the purification of the final cycloaddition product (Scheme 8.98). The desired compounds were isolated in 79–93% yield and 90–93% purity after fluorous solid-phase extraction. Subsequent elution of the F-SPE column with diethyl ether additionally furnished the fluorous Diels–Alder adduct.

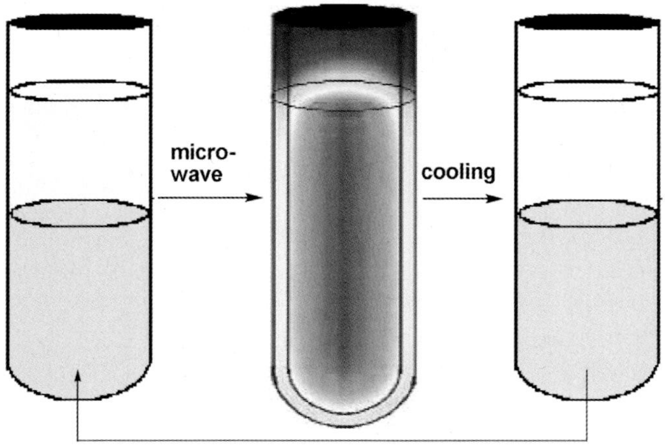

Figure 8.1 Recycling of fluorous ligand in aminocarbonylations. Reproduced with permission from Ref. [114].

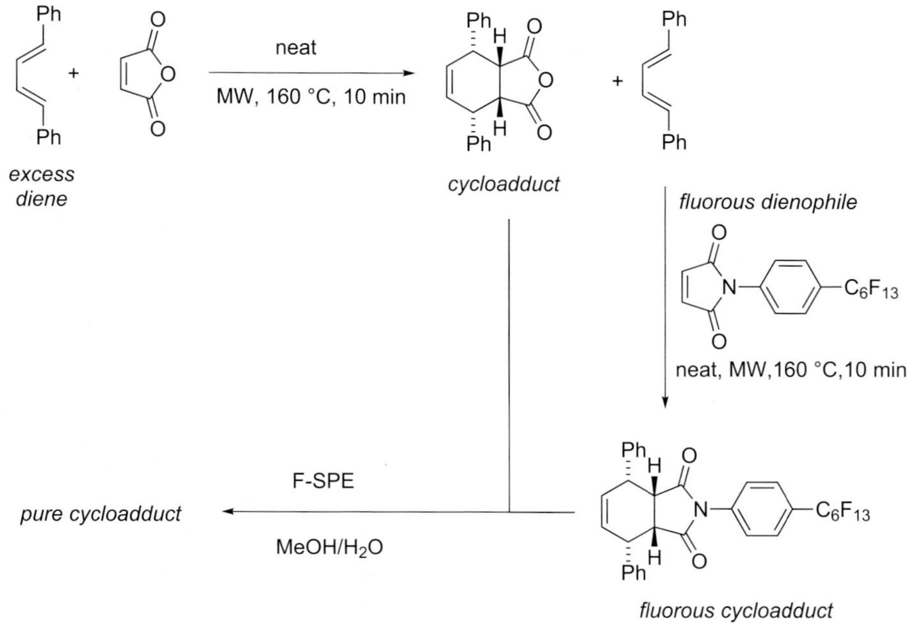

Scheme 8.98 Diene scavenging utilizing a fluorous dienophile.

8.4
Grafted Ionic Liquid-Phase-Supported Synthesis

Ionic liquids are particularly applicable for use in microwave-mediated liquid-phase reactions as they efficiently couple with microwaves resulting in very rapid heating profiles (see Section 4.5.2). Furthermore, ionic liquids are immiscible with a wide range of organic solvents and provide an alternative non-aqueous two-phase system. An interesting approach involves attachment of ionic liquid components toward the organic substrates.

The very first report on the use of ionic liquids as soluble supports was presented by Fraga-Dubreuil and Bazureau in 2001 [116]. The reactivity of the microwave-induced solvent-free Knoevenagel condensation of a formyl group on the ionic liquid (IL) phase with malonate derivatives ($E^1CH_2E^2$) catalyzed by 2 mol% of piperidine has been studied (Scheme 8.99). The progress of the reaction could be monitored easily by ^1H and ^{13}C NMR spectroscopy, and the final products could be cleaved from the IL phase and extracted in high yield without additional need for purification. More importantly, the reaction times are short (15–60 min) and it was possible to reuse the ionic liquid in another cycle of the synthesis. This methodology was also instrumental to provide grafted aldimines in short microwave irradiation times. This enables to regioselectively introduce amines onto the IL phase as precursors for some conventional regioselective 1,3-cycloadditions with imidates for the preparation of diversely substituted imidazoles on the ionic liquid phase (not shown).

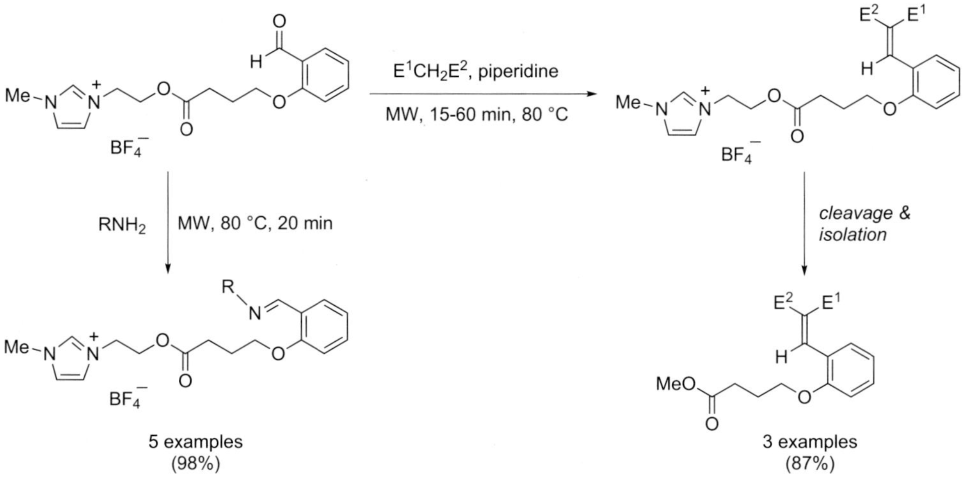

Scheme 8.99 Ionic liquid-mediated Knoevenagel condensation.

This technique was extended by grafting the ionic liquid moiety to polyethylene glycol units prior to attaching the benzaldehyde, as presented by the same group in a more recent study [117]. The corresponding imidazoles were treated with several ω-chlorinated alcohols to efficiently form the desired ionic liquids under microwave irradiation (Scheme 8.100). Since not only the length of the alkyl spacer attached to the imidazolium cation influences drastically the physical properties (viscosity, hydrophobicity) of the resulting polyethylene glycol ionic liquid phases (PEG-ILPs) but also the counterion, a series of novel ionic liquids has been prepared by simple anion metathesis exchange. The synthesized imidazolium chlorides were treated with inorganic salts to achieve PEG-ILPs with tetrafluoroborate (BF_4^-), hexafluorophosphate (PF_6^-), and bis[(trifluoromethyl)sulfonyl]amide (NTf_2^-) anions. The prepared compounds showed varying viscosity and miscibility with organic solvents and could be efficiently used for microwave-assisted reactions.

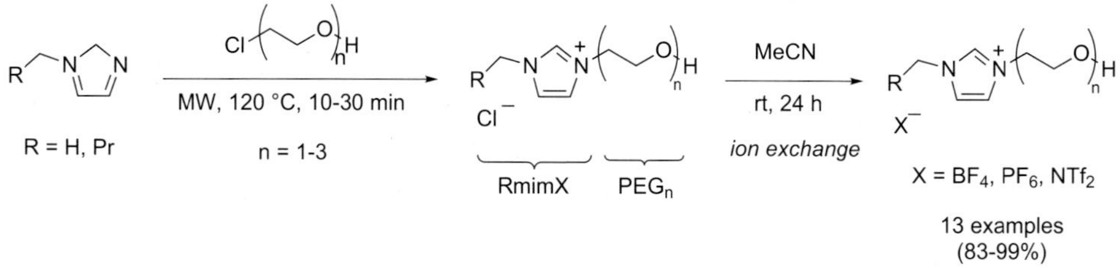

Scheme 8.100 Preparation of polyethylene glycol ionic liquid phases.

In subsequent work, it has been demonstrated that this ionic liquid phase-bound organic synthesis methodology is also applicable toward the formation of several heterocyclic scaffolds. Fraga-Dubreuil and Bazureau described the generation of a small 4-thiazolidinone library utilizing their previously described ionic liquid phase (see above) in a one-pot three-component condensation [118]. Equimolar amounts of the ILP, a primary amine, and mercaptoacetic acid were irradiated at 100 °C to afford – via an initial imine formation – the desired ILP-grafted thiazolidinones (Scheme 8.101). The cyclization step was monitored by ^{1}H NMR and required a rather long time compared to other microwave-mediated transformations. Cleavage of the products was achieved by amide formation utilizing either primary amines or pyrrolidine (not shown). This ester aminolysis usually requires rather harsh conditions and was performed within 10–20 min at 100 °C under microwave irradiation employing potassium *tert*-butoxide base as a catalyst. The use of β-substituted primary amines in the cleavage step required increased temperatures of 150 °C.

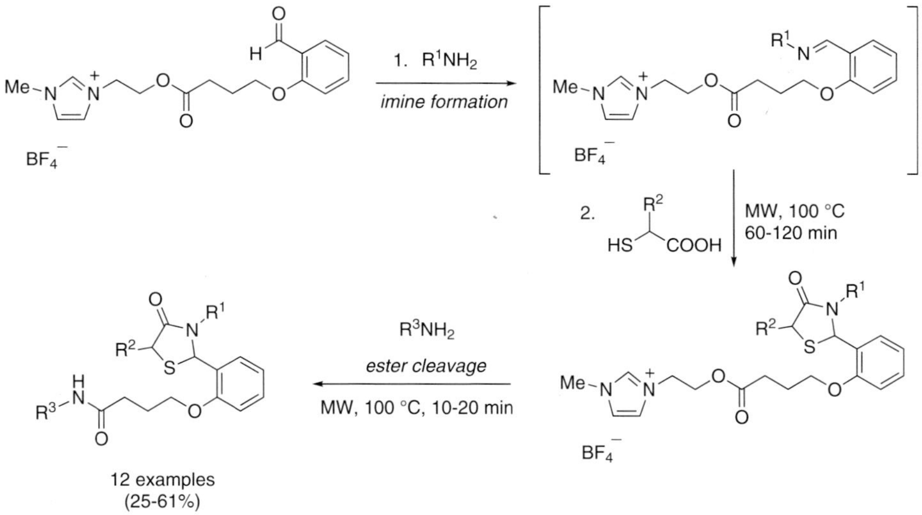

Scheme 8.101 Generation of a 4-thiazolidinone library utilizing grafted ionic liquids.

In a related study, the group of Bazureau applied the use of the polyethylene glycol-grafted ionic liquid phases also toward the preparation of 2-thioxotetrahydro-pyrimidinones [119]. After the initial formation of acrylate-bound ILPs utilizing acryloyl chloride in refluxing dichloromethane, several primary amines were attached in a Michael addition-type reaction at ambient temperature. Subsequent treatment of the ILP-bound β-amido esters with alkyl isothiocyanates in dry acetonitrile gave the final intermediate (Scheme 8.102). Finally, cyclative cleavage was achieved under microwave irradiation employing 2 equiv of diethylamine. The resulting mixtures were extracted with chloroform and purified after evaporation by flash chromatography to afford the desired 2-thioxotetrahydropyrimidin-4-(1*H*)-ones.

These comprehensive studies by Bazureau and coworkers have demonstrated the usefulness of specifically designed ionic liquids as supports in organic reactions such as the Biginelli or Hantzsch synthesis [120] as well as the synthesis of

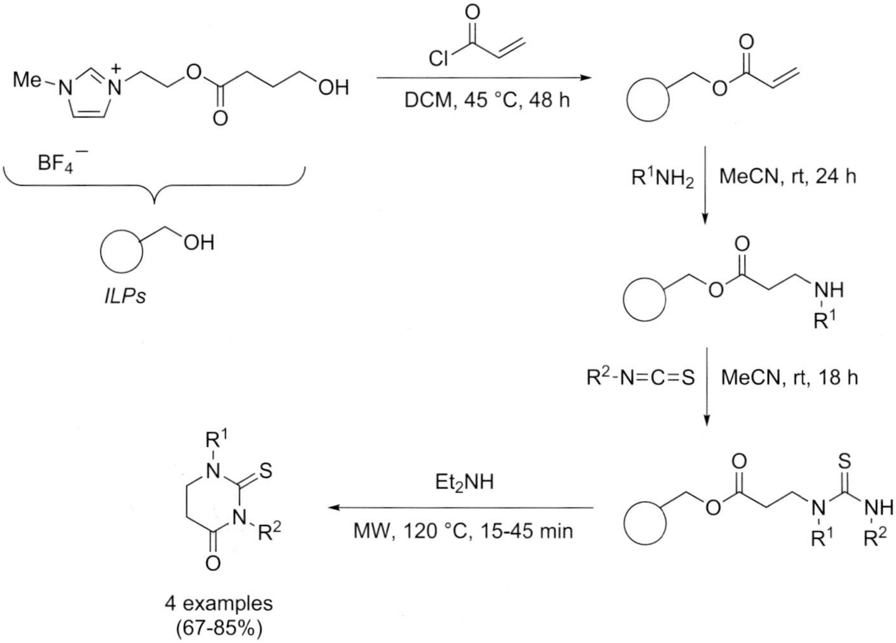

Scheme 8.102 Preparation of 2-thioxotetrahydropyrimidin-4-(1*H*)-ones by ionic liquid-phase organic synthesis.

polyhydroquinolines [121]. Other researchers have described similar ionic liquid-based approaches for the Gewald synthesis of 2-aminothiazole [122].

The group of Buijsman presented the microwave-mediated preparation of a different *N*-imidazolium-based ionic analog of the well-known AMEBA solid support (Scheme 8.103). With this soluble support, a set of various sulfonamides and amides was prepared and the use of this novel linker in the synthesis of a potent analog of the antiplatelet drug tirofiban was presented [123].

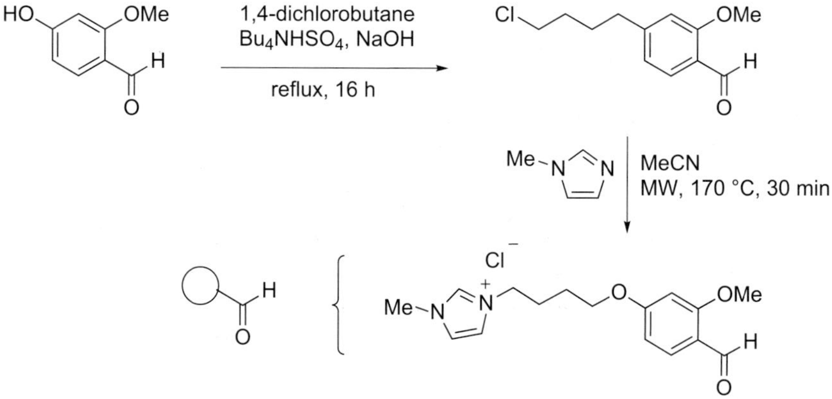

Scheme 8.103 Preparation of an *N*-imidazolium-based soluble AMEBA linker.

8.5
Polymer-Supported Reagents

Apart from traditional solid-phase organic synthesis, the use of polymer-supported reagents has gained increasing attention from practitioners in the field of combinatorial chemistry [124, 125]. The use of PSRs combines the benefits of SPOS with the advantages of solution-phase synthesis (see also Section 4.7.2.4).

The polymer material involved in the development of polymer supports for immobilization of catalysts and/or reagents is generally a functionalized polystyrene/divinylbenzene copolymer. Cross-linking with divinylbenzene renders mechanical strength to the polymer support making it resistant against detrimental conditions like vigorous stirring. Chemical functionality is introduced onto the polymer support by physical adsorption or by chemical bonding. These cross-linked poly(styrene-*co*-divinylbenzene) resins (1–2% cross-linking) are stable with high loading capacity ($>1\,\text{mmol}\,\text{g}^{-1}$) and high swelling properties and are compatible with a variety of nonprotic solvents.

Microwave-assisted organic transformations using polymer-supported reagents and/or a catalyst (see also Sections 4.7.2.4 and 8.6) have been widely exploited in the chemical community. Above all, the considerations on characteristics of polymeric material as support for catalysts and/or reagents, the ease of separation after the reaction, recyclability, and possibility of reinstating its catalytic activity after reaction make their application increasingly popular.

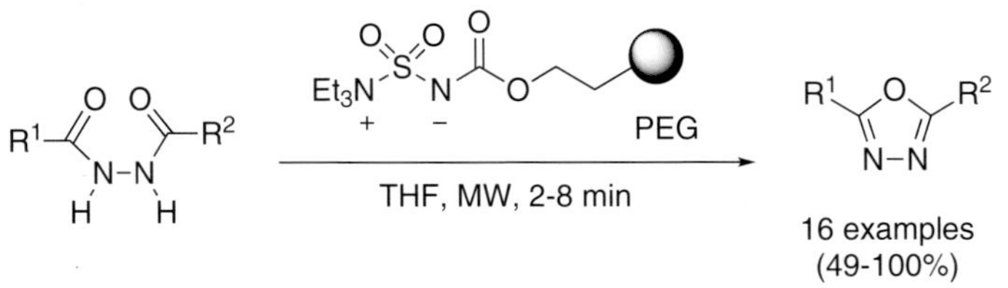

Scheme 8.104 Synthesis of 1,3,4-oxodiazoles utilizing polymer-bound Burgess reagent.

A microwave-induced synthesis of 1,3,4-oxadiazoles has been described by Brain *et al.* employing a soluble polymer-supported Burgess reagent (Scheme 8.104) [126]. The use of this supported Burgess reagent proposes a mild procedure for the cyclodehydration of 1,2-diacylhydrazines. Applying soluble PEG-supported Burgess reagent [127], only 40% conversion was obtained after 3 h of conventional reflux in tetrahydrofuran. However, under microwave heating the reaction reached completion within 2–8 min. The cyclodehydration was performed on 1,2-diacylhydrazines with a variety of substituted aromatic- or heteroaromatic rings with the soluble polymer-supported Burgess reagent. The corresponding 1,3,4-oxadiazoles were obtained in varying yields but generally high purity within 2 min of microwave irradiation.

In more recent work, Brain and Brunton applied a different polystyrene-supported dehydrating agent (Figure 8.2a) for the synthesis of 1,3,4-oxadiazoles under thermal

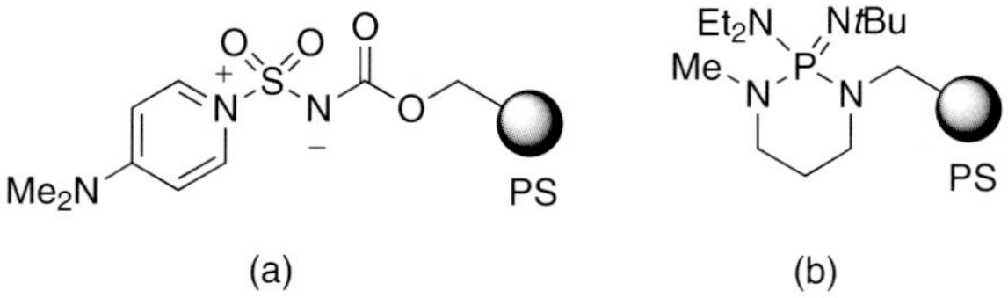

Figure 8.2 Polystyrene-bound dehydrating agent (a) and polymer-supported phosphazene base (b) utilized for oxadiazole synthesis.

and microwave conditions [128]. Addition of a homogeneous base such as guanidine improved the cyclodehydration reaction and led to completion under thermal and even more significantly under microwave conditions. However, this required subsequent purification steps for the removal of the base for the clean and quantitative isolation of the desired products. On the other hand, use of a polymer-supported phosphazene base (Figure 8.2b) circumvents the need for additional purification of the reaction mixture for product isolation. Microwave heating proved advantageous over the thermal protocol with a dramatic reduction in reaction times together with high yields and purity of the corresponding 1,3,4-oxadiazoles.

An expeditious synthesis of 1,2,4-oxadiazoles has been demonstrated by Wang *et al.* at Abbott Laboratories (Scheme 8.105) [129]. The synthesis of 1,2,4-oxadiazoles from a variety of readily available carboxylic acids and amidoximes using solid-phase reagents and catalysts has been devised in two approaches with high yields and simplification of purification process. Microwave technology was utilized for a rapid optimization of reaction conditions with reduction in reaction times from hours to minutes.

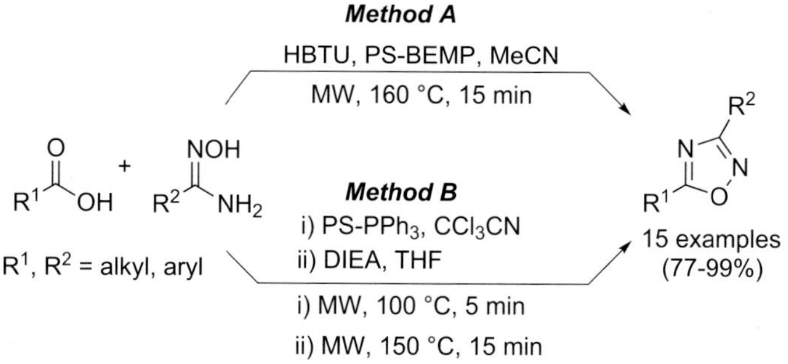

Scheme 8.105 Synthesis of 1,2,4-oxadiazoles using polymer-supported reagents.

In a related work, the same group has reported on the generation of 1,3,4-oxadiazoles via condensation of acid hydrazides and carboxylic acids employing trichloroacetonitrile in combination with polymer-supported PPh_3 as reagents [130]. The same reagent combination was also used in the synthesis of triazolopyridines by the same authors [131].

Ley and coworkers have reported on a symbiotic combination of solid-supported reagents and microwave-assisted organic synthesis to rapidly generate compound libraries of 5-substituted 2-amino-1,3,4-oxadiazoles and their corresponding thiadiazole analogs (Scheme 8.106) [132]. A one-pot preparation of the 2-aminosulfonylated analogs through a three-component coupling of an acylhydrazine, an isocyanate, and sulfonyl chloride promoted by a polymer-supported phosphazene base (PS-BEMP) under microwave heating is also described.

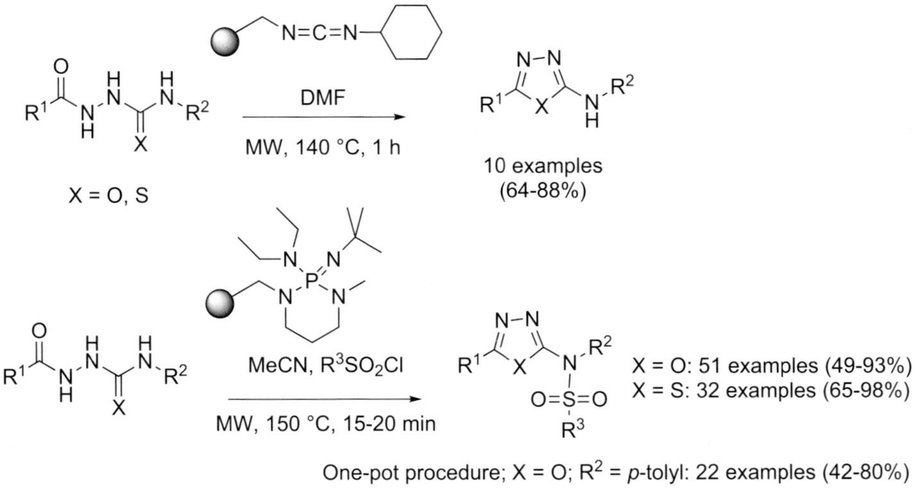

Scheme 8.106 Preparation of 1,3,4-oxadiazoles using polymer-supported reagents.

Earlier, the Ley group has studied the microwave-assisted thionation of amides using a polymer-supported thionating reagent [133]. The polymer-supported amino thiophosphate serves as a convenient substitute to its homogeneous analog for the microwave-induced rapid conversion of amides to thioamides. Under microwave conditions, the reaction is completed within 15 min as opposed to 30 h by conventional reflux in toluene (Scheme 8.107). The reaction has been studied for a range of secondary and tertiary amides and GC-MS monitoring showed that it proceeded almost quantitatively. More importantly, the use of the ionic liquid 1-ethyl-3-methylimidazolium hexafluorophosphate (emimPF_6) was described herein for the first time to dope unpolar solvents such as toluene to assist the heating under microwave irradiations (see Section 4.5.2).

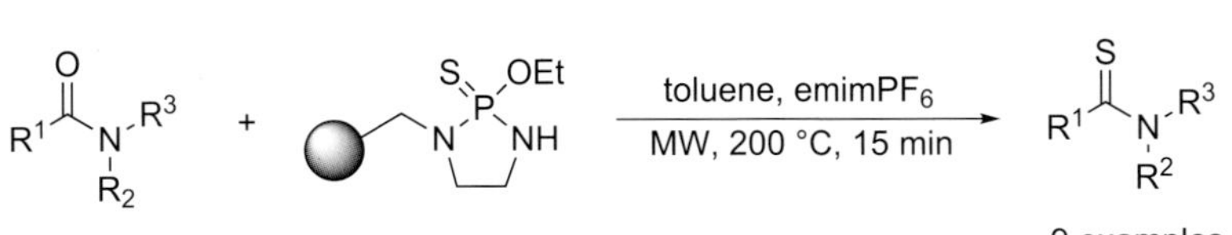

Scheme 8.107 Thionation of amides utilizing polymer-bound aminothiophosphate.

A microwave-induced one-pot synthesis of olefins by Wittig olefination has been reported by Westman using a polymer-supported triphenylphosphine [134].

The conventional Wittig reaction often requires long reaction times and most of all in solution-phase synthesis the triphenylphosphine unavoidably forms the triphenylphosphine oxide by-product in solution. On the other hand, a similar reaction protocol using a solid-supported triphenylphosphine has been developed. Microwave heating first speeds up the actual chemistry and the products can be obtained free from organophosphorus contamination by simple filtration. The preparation of the actual Wittig reagent with *in situ* formation of the corresponding ylide has been exemplified in high yields within minutes as opposed to several days of using conventional methods.

The olefinations using a solid-supported triphenylphosphine were achieved within 5 min of microwave heating at 150 °C using potassium carbonate as the base and methanol as the solvent (Scheme 8.108). A variety of aldehydes and organic halides were studied with the solid-supported phosphine, yielding the corresponding olefins in excellent purity.

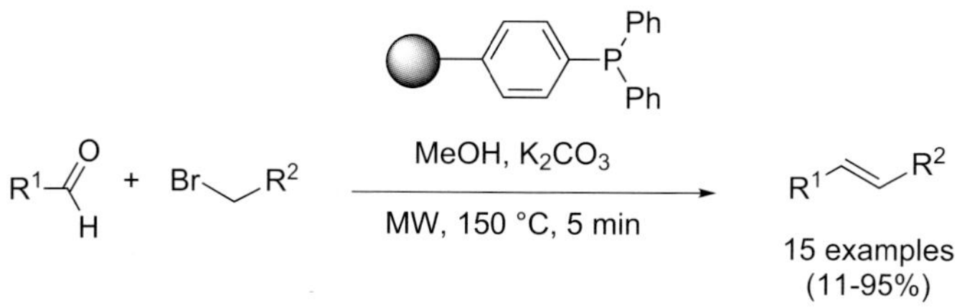

Scheme 8.108 Wittig olefinations using resin-bound triphenylphosphine.

Desai and Danks reported on rapid solid-phase transfer hydrogenation utilizing a polymer-supported hydrogen donor (ammonium formate derivative) for the reduction of electron-deficient alkenes in the presence of Wilkinson's catalyst [RhCl$(PPh_3)_3$] under microwave heating [135]. Conventional hydrogen donors such as ammonium formate sublime at high temperatures and also release toxic gases like ammonia. Use of the polymer-supported formate on the other hand overcomes these shortcomings and it could be easily used for automated robotic synthesis.

In the present work, the utilized formate, immobilized on Amberlite® IRA-938, is admixed with the corresponding substrate in a minimum amount of dimethyl sulfoxide and the reaction mixture is irradiated in a sealed vessel for 30 s (Scheme 8.109). After cooling, the mixture was diluted with dichloromethane, washed with water, and dried. Evaporation of the solvent furnished the successfully hydrogenated compounds in high yields.

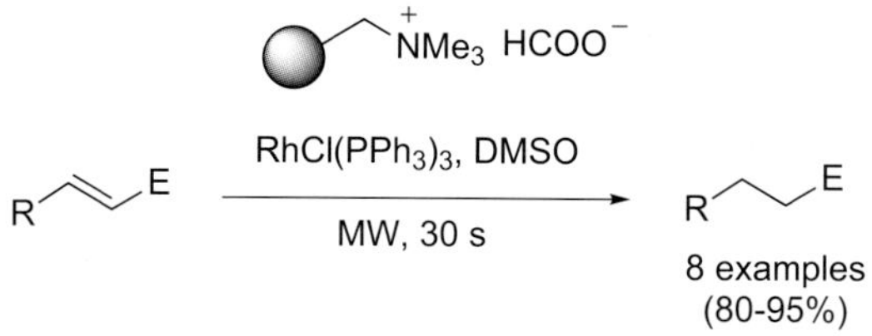

Scheme 8.109 Transfer hydrogenations using a resin-bound formate.

Other authors have described the application of a recyclable polymer-supported bromine chloride resin for a regio- and chemoselective bromomethoxylation of various substituted alkenes under microwave conditions [136]. The utilized immobilized brominating reagent (perbromide resin, 1.6 mmol g^{-1} Br) was generated from a commercially available chloride exchange resin by simple bromination at room temperature (Scheme 8.110). The short-term microwave irradiations were carried out in a modified domestic oven equipped with a mounted reflux condenser in refluxing methanol.

Scheme 8.110 Regio- and chemoselective bromomethoxylation.

However, the bromomethoxylation was observed only across the isolated double bonds whereas the conjugated double bonds remained unaffected to the reaction conditions. The crude products were easily recovered by filtration and evaporation of the solvent. The polymer-supported bromine resin could be recycled and regenerated by successive washings with methanol, acetonitrile, and chloroform and then passing bromine in tetrachloromethane through it.

The conversion of isothiocyanates to isonitriles has been studied by Ley and Taylor under microwave conditions using a polymer-supported [1,3,2]oxaphospholidine [137]. The use of 3-methyl-2-phenyl[1,3,2]oxaphospholidine in solution is less appreciated [138] due to the associated toxicity and instability of the phosphorus-derived reagent, as well as the need to isolate the products from a complex reaction mixture by vacuum distillation. This drawback has been resolved by attaching the active [1,3,2]oxaphospholidine on a polymer matrix.

Commercial Merrifield resin was treated with aminoethanol and the resulting precursor was condensed with bis(diethylamino)phosphine in anhydrous toluene to generate the desired polymer-supported reagent. When using the novel solid-supported reagent, the conversions of the isothiocyanates could be afforded in a parallel fashion under microwave irradiation conditions, and the corresponding isonitriles were isolated with high purity (Scheme 8.111). This approach avoided any exposure to the highly toxic isocyanides during isolation and also limited the toxicity by the use of an immobilized phosphorus-derived reagent.

To demonstrate the feasibility for library generation, the produced isocyanides were subjected to a Ugi three-component condensation with various primary amines and carboxybenzaldehyde. The resulting 2-isoindolinone derivatives were obtained in high to excellent yields.

The synthesis of highly substituted benzoxazoles employing solid-supported reagents was developed by the group of Botta (Scheme 8.112) [139]. In the first

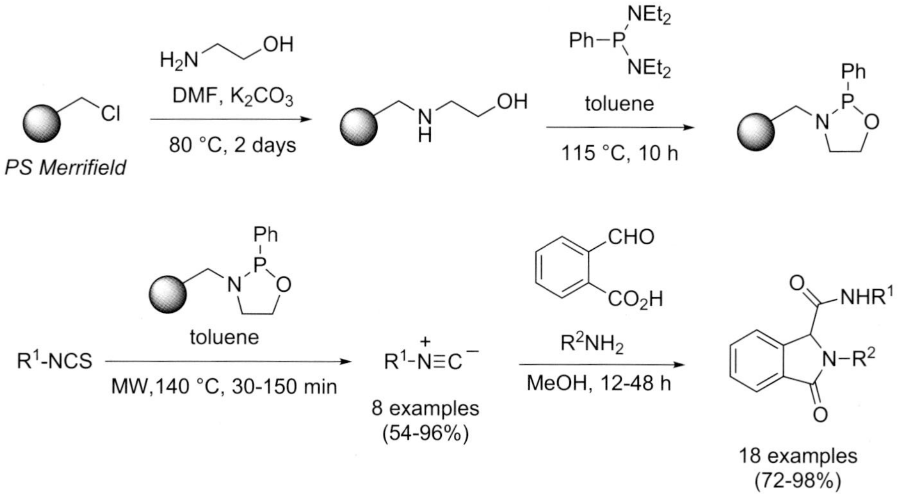

Scheme 8.111 Conversion of isothiocyanates to isonitriles utilizing polymer-supported [1,3,2] oxazaphospholidine.

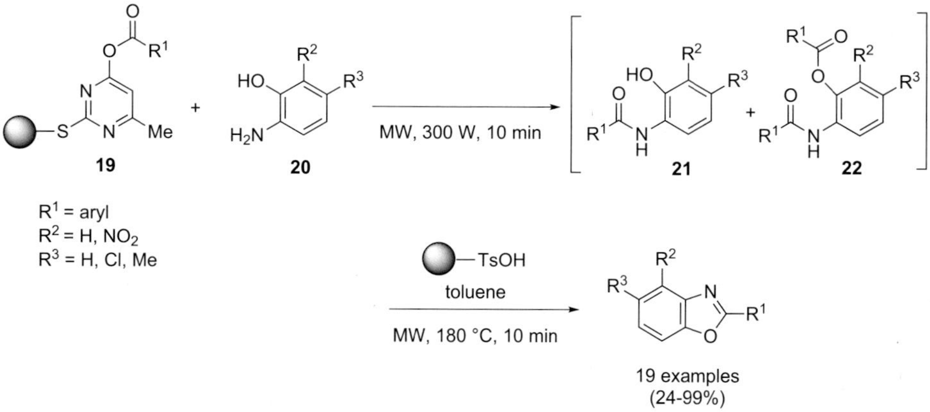

Scheme 8.112 One-pot two-step synthesis of benzoxazoles.

step, solid-supported reagents **19** were reacted neat with 2-aminophenols **20** to give the uncyclized mono- and diacylated intermediates **21** and **22**. Subsequent addition of toluene and polymer-bound *p*-toluenesulfonic acid to catalyze the cyclodehydration and irradiation at 180 °C for 10 min gave the benzoxazole derivatives in moderate to excellent yields. Importantly, the products are obtained in high purity only by final filtration from the polymer-supported reagents. By combining a parallel synthesizer for evaporation and filtration/washing steps and microwave heating, the small library was produced in short time and renders the protocol amenable for automation.

Bradley and coworkers discussed the solid-phase-mediated synthesis of isonitriles from formamides, utilizing polystyrene-bound sulfonyl chloride as a suitable supported reagent [140]. This commercially available polymer-supported reagent offers an efficient method of isonitrile generation under microwave irradiation, employing a simple filtration and acidic workup. Six formamide derivatives were converted into the corresponding isonitriles in good purities utilizing 3 equiv of the polymer-supported reagent (Scheme 8.113). However, increasing the degree of substitution of the formamide has a decreasing effect on the purity of the desired products. The isonitriles were isolated by filtering off the resin, pouring the filtrate into water, and extraction with dichloromethane. Evaporation of the solvent furnished the desired products in approximately 70% yield. Furthermore, the sulfonyl chloride resin could be quantitatively regenerated by treatment of the formed sulfonic acid resin with phosphorus pentachloride (PCl_5) in *N,N*-dimethylformamide at room temperature. Solid-phase-mediated isonitrile synthesis under microwave irradiation led to much faster reactions, therefore allowing rapid access to this important class of compounds, amenable for a broad range of subsequent syntheses.

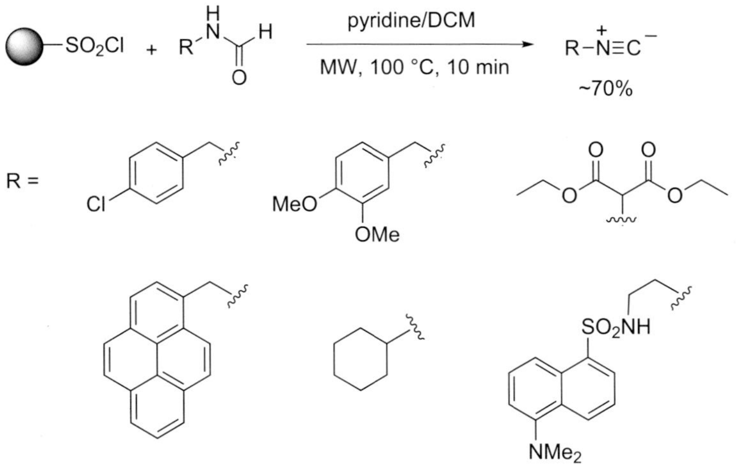

Scheme 8.113 Solid-phase-mediated isonitrile synthesis.

The group of Moser described the use of a polymer-bound borohydride to perform reductive aminations of tetrameric isoquinolines (Scheme 8.114) [141]. These tetrameric isoquinolines, serving as lead compounds in the research for antibacterial distamycin A analogs, were prepared from the corresponding isoquinoline, imidazole, and pyrrole building blocks by standard amide bond formations. The final derivatization by reductive amination was efficiently accelerated by microwave irradiation in the presence of Merrifield resin-supported cyanoborohydride.

The group of Linclau demonstrated the effective *O*-alkylation of carboxylic acids using a polymer-supported *O*-methylisourea reagent [142]. Microwave heating afforded complete esterifications within 15–20 min, requiring only 2 equiv of the polymer-bound *O*-methylisourea (Scheme 8.115). Furthermore, the polymer-supported reagent simplified purification as the pure methyl esters were furnished by evaporation after a simple filtration step.

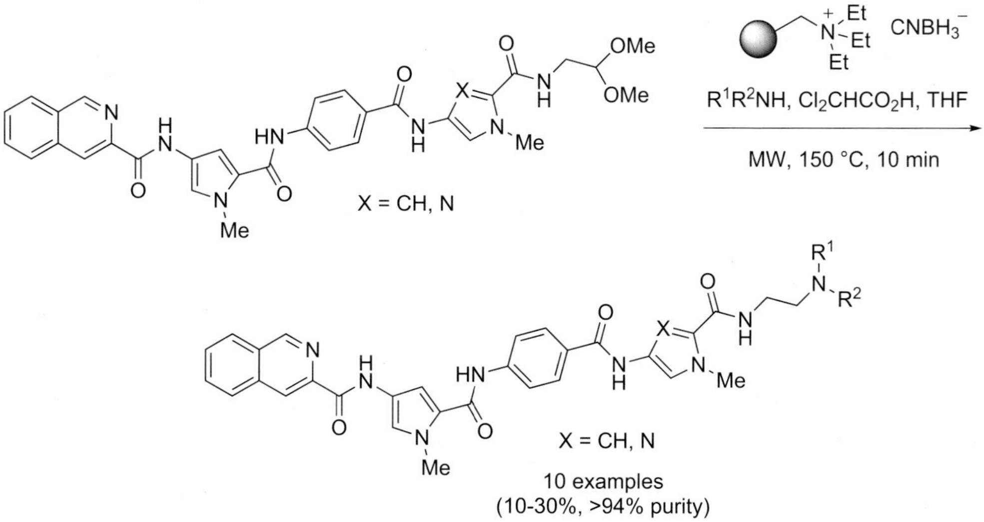

Scheme 8.114 Reductive amination of tetrameric isoquinolines.

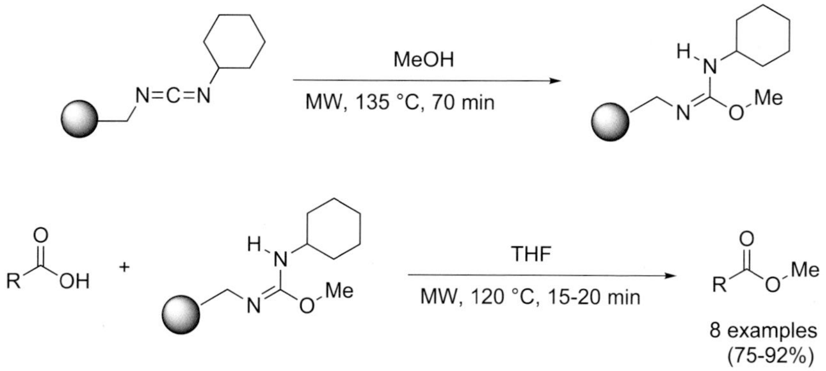

Scheme 8.115 Polymer-supported *O*-alkylation of carboxylic acids.

Sauer *et al.* presented the use of polystyrene-bound carbodiimide for convenient and rapid amide synthesis [143]. An equimolar mixture of 1-methylindole-3-carboxylate, corresponding amine, and 1-hydroxybenzotriazole in 1-methyl-2-pyrrolidinone was admixed with 2 equiv of the polymer-bound carbodiimide and irradiated for 5 min at 100 °C (Scheme 8.116). After cooling, the mixture was diluted with methanol and subjected to solid-phase extraction utilizing silica carbonate. Evaporation of the filtrate furnished the desired compounds in excellent purity.

A detailed investigation by Turner and coworkers describes the preparation and application of solid-supported cyclohexane-1,3-dione (CHD) as a so-called "capture and release" reagent for amide synthesis, as well as its use as a novel scavenger resin [144]. Within this report, a three-step synthesis of polymer-bound cyclohexane-1,3-dione (CHD resin, Scheme 8.117) from cheap and readily available starting

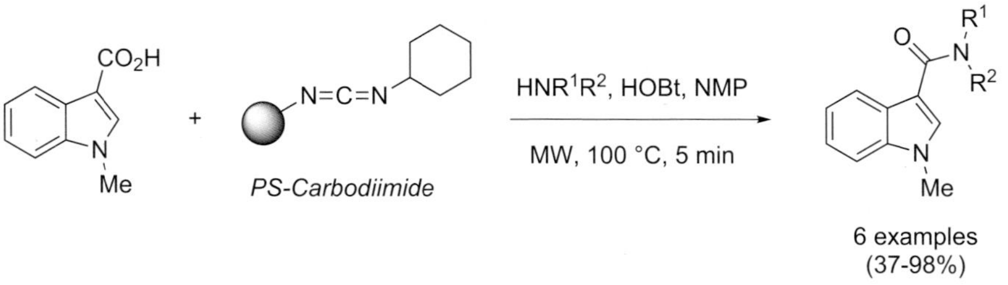

Scheme 8.116 Amide synthesis utilizing polymer-bound carbodiimide.

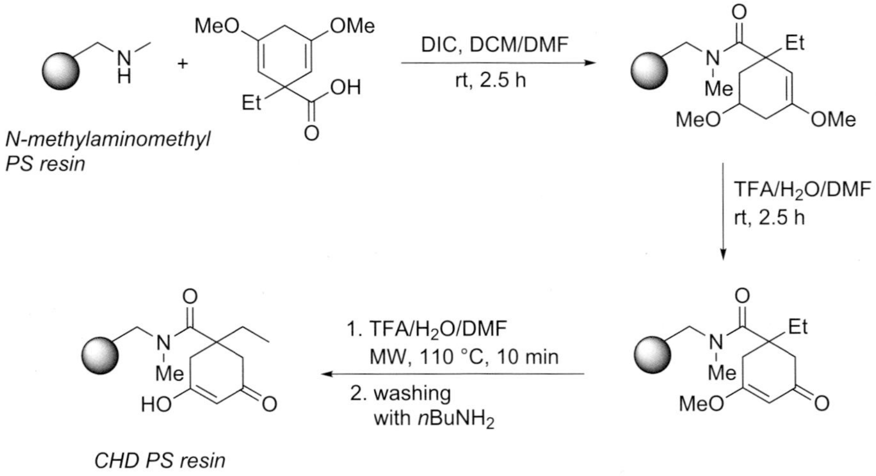

Scheme 8.117 Preparation of resin-bound cyclohexan-1,3-dione.

materials is described. The key step in this reaction is the microwave-assisted complete hydrolysis of the 3-methoxy cyclohexen-1-one resin to the desired CHD resin.

This novel resin-bound CHD derivative was then utilized in the preparation of an amide library under microwave irradiation. The enol ester formed by the reaction of the starting resin-bound CHD with an acyl or aroyl chloride upon treatment with amines leads to the corresponding amide, regenerating the CHD. This demonstrates the feasibility of the CHD resin as a "capture and release" reagent for the synthesis of amides. The "resin capture–release" methodology [145] aids in the removal of impurities and facilitates product purification.

The synthesis of amides using the resin-bound CHD enol ester was accelerated under microwave heating in 30 min, which is otherwise afforded after overnight stirring at room temperature (Scheme 8.118), as described by the group of Turner [144]. Interestingly, kinetic monitoring has shown that the release of the amide into solution is apparently accelerated by microwave heating. Also, the presence of an excess of amine was necessary for complete release of the amide. Aromatic enol esters generally gave higher yields and purities of their corresponding amides than

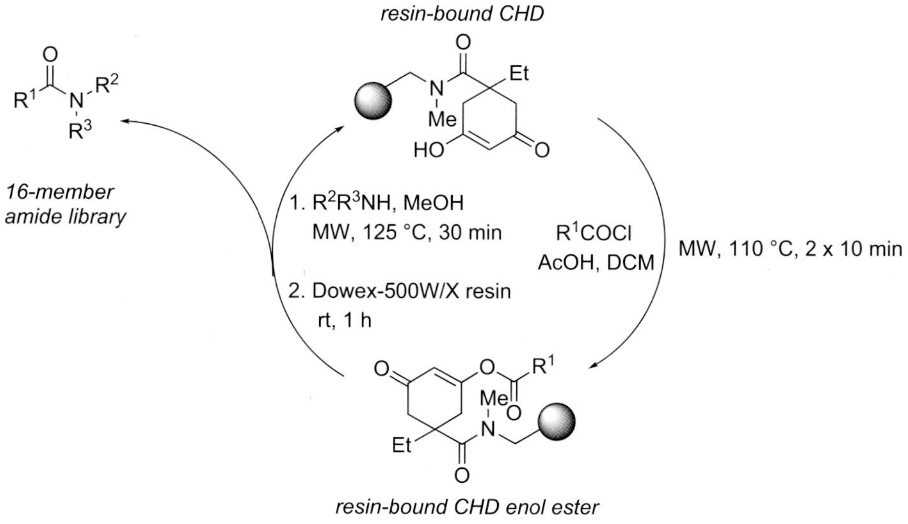

Scheme 8.118 Generation of an amide library utilizing resin capture–release methodology.

the aliphatic enol esters, whereas lower yields for aniline were explained by the reduced amine nucleophilicity.

In order to explore the potential of CHD resins as scavenger materials, 1-ethyl-3,5-dimethoxycyclohexa-2,5-dienecarboxylic acid was anchored to a commercially available trisamine resin (Scheme 8.119), yielding a high-loading cyclohexane-1,3-dione scavenger resin (CHD-SR) [144].

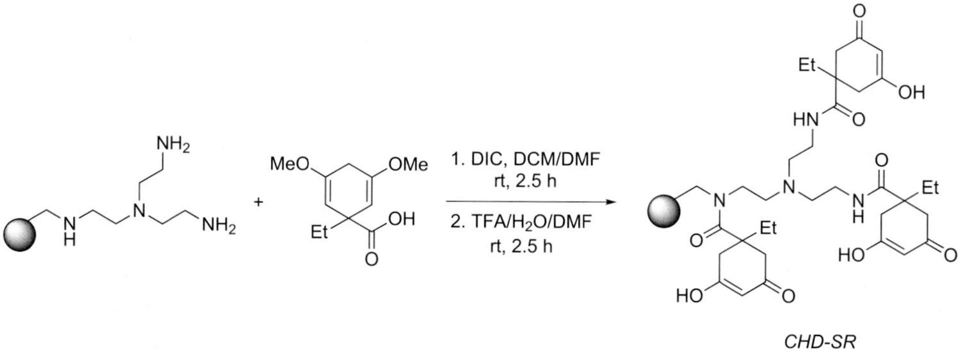

Scheme 8.119 Preparation of high-loading cyclohexane-1,3-dione scavenger resin.

Within the same study, the scavenging ability of this novel resin as an allyl cation scavenger was demonstrated in the palladium-catalyzed *O*-alloc deprotection of the *O*-alloc benzyl alcohol [144]. Benzyl alcohol was obtained in high yield with only small traces of by-product, thereby eliminating the need for further purification.

Another example of the "capture and release" strategy was recently presented by Porcheddu *et al.*, utilizing polystyrene-bound piperazine for the generation of a

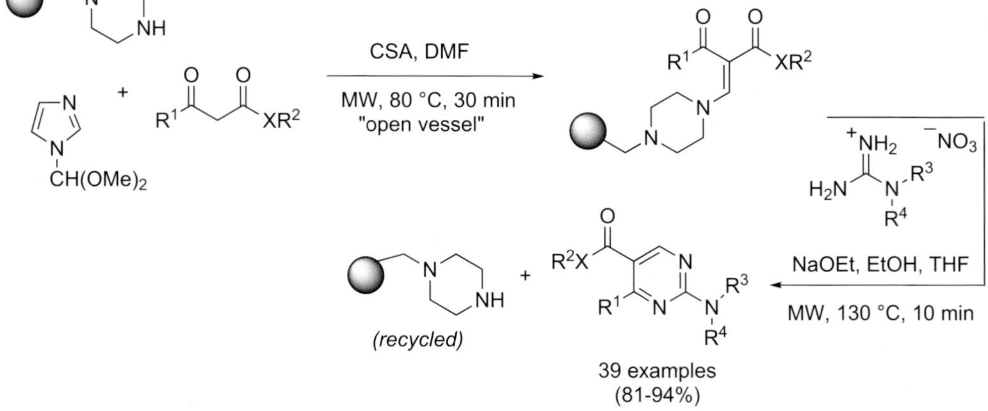

Scheme 8.120 Generation of a pyrimidine library via capture and release strategy.

pyrimidine library [146]. The authors have disclosed a novel synthetic route in order to access these valuable heterocycles involving polymer-bound enaminones, which are condensed with guanidines (Scheme 8.120). The utilized enaminones were readily prepared by condensation of 6 equiv of *N*-formylimidazole diacetal and 3 equiv of β-keto compounds to the polymer-bound piperazine in the presence of 10 mol% of camphorsulfonic acid. The mixture was heated in an open flask in a dedicated single-mode microwave reactor for 30 min to achieve full conversion of the substrates. After cooling, the functionalized resin was collected by filtration and subjected to the releasing cyclization. This step was performed by treating the polymer-bound enaminone with 1 equiv of the corresponding guanidine, which was liberated from its nitrate immediately prior to reaction. The mixture was irradiated in a sealed flask for 10 min at 130 °C to achieve the desired 2,4,5-trisubstituted pyrimidines. After the reaction, the piperazine resin could be collected by filtration and reused after several washings with ethanol. The products were isolated from the filtrate after aqueous workup by evaporation in high yields and excellent purity.

A recyclable solid-supported reagent has been developed for acylations by the group of Botta [147]. The solid-supported acylating agent could be easily applied in a parallel synthesis of the corresponding amides starting from amines (Scheme 8.121). The solid-supported reagent could be removed by simple filtration after the reaction, the product isolated by evaporation, and the supported reagent reused at least twice before any decrease in the reaction yield. The 4-acyloxypyrimidines were prepared

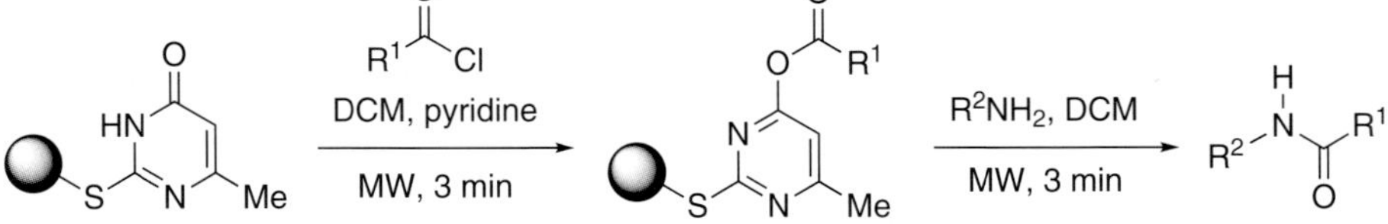

Scheme 8.121 Amide synthesis utilizing reusable polymer-supported acylation reagent.

utilizing a domestic microwave oven with different acyl chlorides on a solid-supported pyrimidine linker. The solid-supported acyloxypyrimidines prepared in this way were used as solid-supported reagents for the rapid and selective acylation of amines under microwave irradiation conditions (Scheme 8.121). The employed solid-supported pyrimidine could be recycled for several reaction runs.

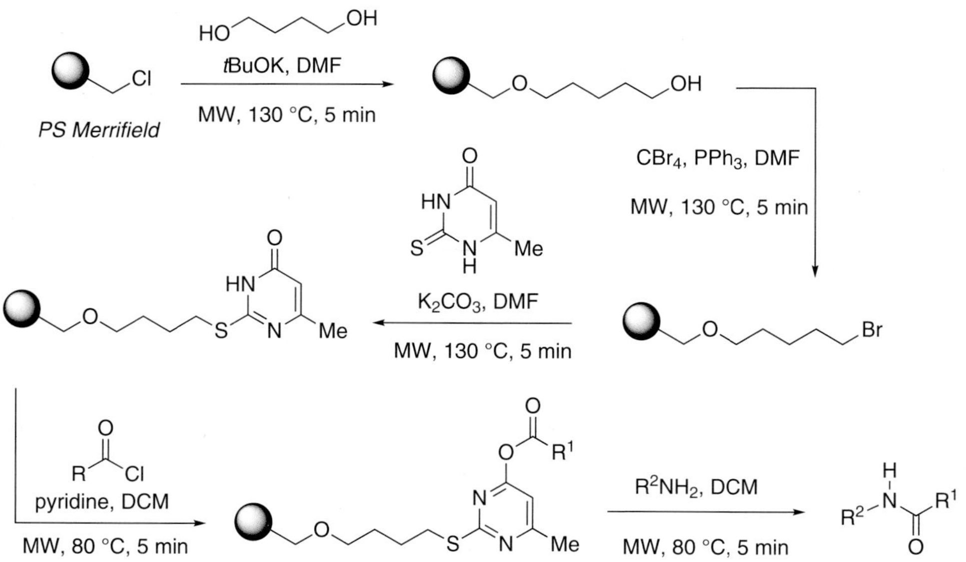

Scheme 8.122 Synthesis of a novel solid-supported reagent and its use for acylation.

In a related and more recent study by the same group, syntheses of various solid-supported acylating agents are presented for microwave-mediated transformation of amines, alcohols, phenols, and thiophenols [148]. In a microwave-mediated procedure, Merrifield resin was first modified by attaching 1,4-butanediol to introduce a spacer unit. Bromination and subsequent reaction with commercially available 6-methyl-2-thiouracil followed by treatment with corresponding acyl chloride afforded the desired polymer-bound pyrimidines (Scheme 8.122). The acylating ability of this supported reagent has been proven by reaction with benzylamine. Furthermore, the utilized thiouracil building block can be anchored to different solid supports to furnish various polymer-supported acylating agents [148].

Similarly, a solid-supported imide has been reported as an acylating reagent under microwave conditions by Nicewonger *et al.* [149]. The starting imide was immobilized on aminomethyl polystyrene and in this case benzoyl chloride was chosen to prepare the corresponding acylating reagent (Scheme 8.123). Primary amines and piperazines were acylated smoothly at room temperature; however, the more hindered secondary amines required more time and heat, whereas anilines could not be acylated at all. Above all, the resin-bound acylating agent was shown to be recyclable after washing with *N,N*-dimethylformamide and reactivation using microwave conditions.

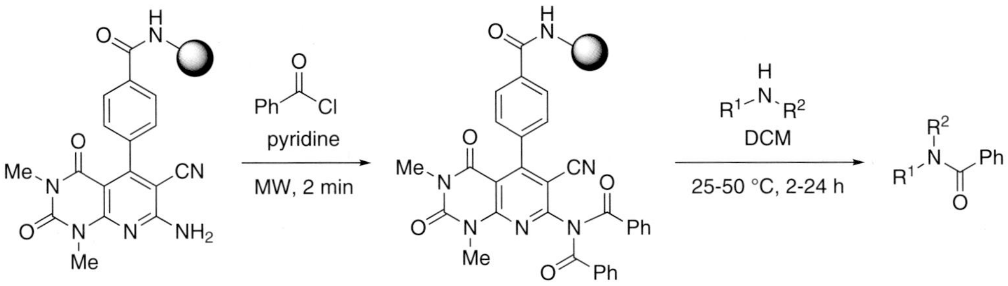

Scheme 8.123 Recyclable polymer-supported imides for amide synthesis.

The group of Swinnen presented the use of a polymer-supported Mukaiyama reagent for the synthesis of amides [150]. The utilized polymer-bound *N*-alkyl-2-chloro pyridinium triflate was initially synthesized from commercially available polystyrene Wang resin. To prove its effectiveness in even difficult coupling reactions, it was used in a microwave-accelerated amide synthesis from sterically hindered pivalic acid (Scheme 8.124). The mixture was subjected to microwave irradiation at 100 °C for 10 min. Employing 3 equiv of acid furnished the desired product in 80% yield. Remaining starting amine was removed by filtering the mixture on a sulfonic acid-derivatized SPE column.

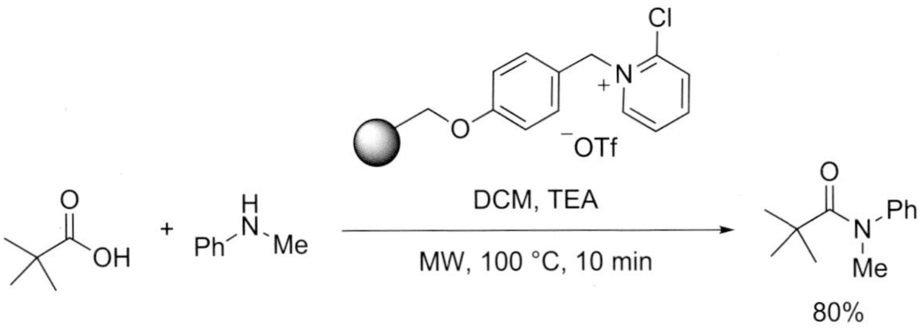

Scheme 8.124 Polymer-supported Mukaiyama reagent for amide synthesis.

As demonstrated by the Ley group, microwave heating has been applied in accelerating several slow intermediate reactions in the total synthesis of the natural product (+)-plicamine (see Scheme 4.31) as well as a number of related spirocyclic templates [151–153]. The use of microwave heating in combination with polymer-bound reagents, catalysts, and scavengers was beneficial in gaining rapid access to several key intermediates for the synthesis of the target molecule.

Lindner and coworkers have reported on the nucleophilic halide/azide exchange of preactivated silica gels **23** (Scheme 8.125) [154]. By applying microwave heating, the reaction time could be reduced from 48 h at 80 °C to only 5 min at 125 °C. For the azidopropyl-modified silica support **24a** an 87% and for the 11-azidoundecyl silica **24b** even quantitative conversion was obtained. The azido-modified silica supports were then subjected to click chemistry with alkyne-functionalized cinchona

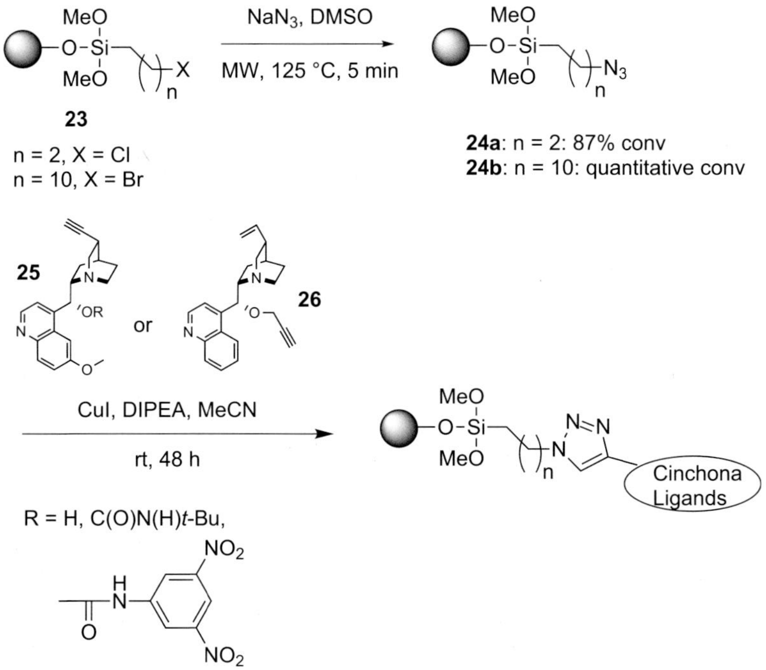

Scheme 8.125 Synthesis of azido-modified silica gels.

alkaloids **25** and **26**, respectively, in order to produce an effective chiral stationary phase for HPLC enantiomer separation.

Esterification reactions using a Mukaiyama-type supported reagent were described by the Taddei group [155]. A similar supported reagent was used by Crosignani for the preparation of a 2-oxazoline library [156]. In a related study, a polymer-supported carbodiimide was employed to prepare a library of imidazothiazol-3-ones and imidazothiazine-4-ones [157], in the generation of a 3,4-dehydroproline amide discovery library [158], and for the synthesis of carboxamide libraries containing oxepines and pyrans [159]. Resin-bound isocyanates were prepared as amine scavengers by Bradley [160].

8.6 Polymer-Supported Catalysts

Catalysts immobilized on polymeric support have an important additional advantage over conventional homogeneous catalysts, in that the spent catalyst can be removed after the reaction by simple filtration, as it was already demonstrated for polymer-supported reagents in Section 8.5. In many cases, the catalyst system could be regenerated and recycled several times without any significant loss of activity. Furthermore, transition metal catalysts immobilized on polymer resins have significant benefits in reducing metal contamination of the reaction mixture and the product.

8.6.1
Catalysts on Polymeric Support

In a novel approach to improve the synthesis of valuable Biginelli compounds with different substitution patterns, the Kappe group has described the selective protection of dihydropyrimidinones [161]. This procedure was successfully catalyzed by a polymer-supported *N,N*-dimethylaminopyridine (PS-DMAP) (Scheme 8.126). This microwave-induced selective *N*3-acylation was applied to diversely substituted dihydropyrimidinone scaffolds with 30–97% isolated yield within 10–20 min of microwave heating. The methodology involving a polymer-supported DMAP is preferred in comparison with the solution-phase protocol due to ease of purification and workup and the possibility of adaptation to a high-throughput format. The purification of the final products was improved by microwave-induced scavenging techniques (see also Section 8.7).

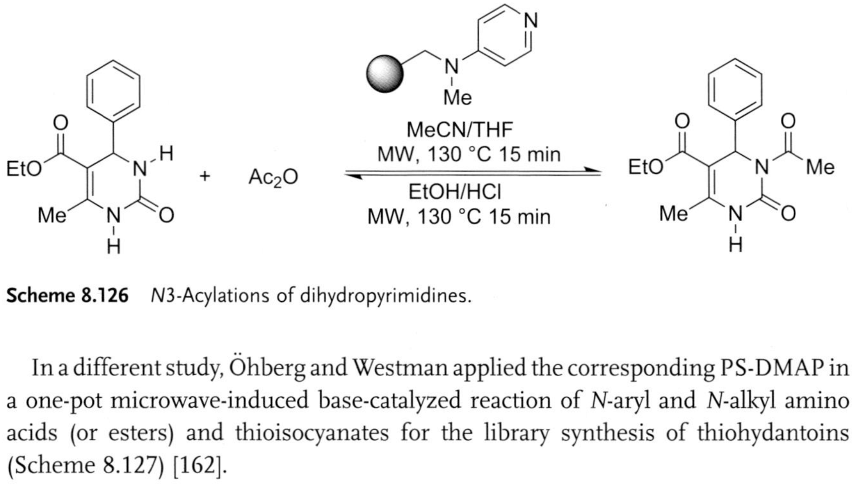

Scheme 8.126 *N*3-Acylations of dihydropyrimidines.

In a different study, Öhberg and Westman applied the corresponding PS-DMAP in a one-pot microwave-induced base-catalyzed reaction of *N*-aryl and *N*-alkyl amino acids (or esters) and thioisocyanates for the library synthesis of thiohydantoins (Scheme 8.127) [162].

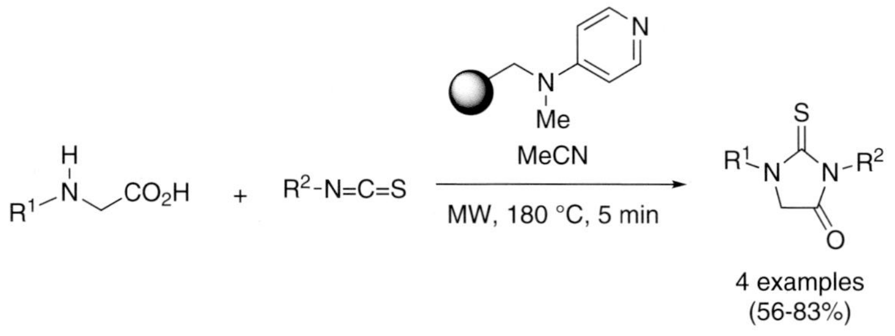

Scheme 8.127 PS-DMAP-catalyzed thiohydantoin synthesis.

Thiohydantoins are of interest due to their ease of preparation and range of biological properties associated with this heterocyclic ring system. PS-DMAP as the base gave slightly lower yields as opposed to the use of triethylamine, but resulted in a cleaner reaction mixture and easier purification procedure. Cyclization of a number of *N*-substituted amino acids and thioisocyanates gave satisfactory yields within 5 min of microwave heating.

In 2009, Pericás and coworkers employed for the first time a solid-supported organocatalyst for stereoselective Mannich reactions [163]. The authors found that (2*S*,4*R*)-hydroxyproline connected to the polymeric backbone via a 1,2,3-triazole linker gave the best results. As a model reaction, to demonstrate the catalyst efficiency, the highly stereoselective Mannich reaction of aldehydes or ketones with *N*-(*p*-methoxyphenyl) (NPMP) ethyl glyoxylate imine was chosen. In an optimized microwave protocol, the initially prepared modified Merrifield resin and the imine were suspended in a microwave vial with *N*,*N*-dimethylformamide and admixed with cyclic ketone. The mixture was irradiated under open-vessel conditions with 1 W for 3–4 h (Scheme 8.128). The temperature never exceeded 33 °C during this period as monitored with an immersing fiber-optic probe. After the reaction, the resin was filtered, and the filtrate was diluted with water and extracted with hexanes. After washing with water and drying, the desired products were obtained in high yields upon removal of the solvent under vacuum. The polymer-supported catalyst could easily be recovered and recycled. In three consecutive runs, the catalytic activity remained virtually unaffected [163].

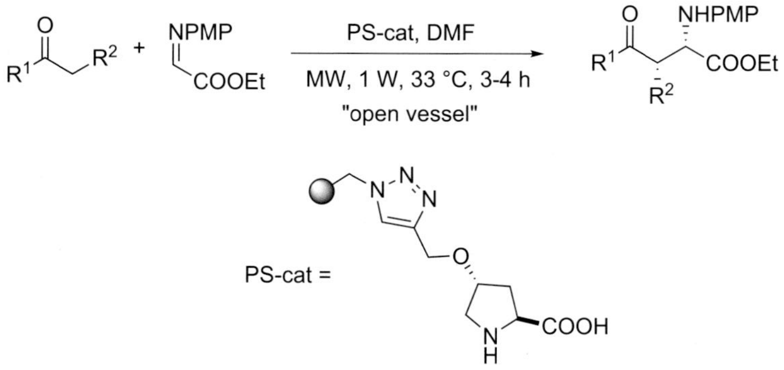

Scheme 8.128 Microwave-mediated Mannich reaction employing a polymer-supported proline derivative.

Since metal catalysts are widely employed in organic transformations but are sometimes difficult to withdraw from the reaction mixtures, the use of the corresponding polymer-bound species in microwave-mediated reactions to simplify the workup is of certain interest. Ring-closing metathesis can be efficiently applied for the preparation of several small, medium, and macrocyclic ring systems. Microwave heating provides a much desirable acceleration of such reactions using ruthenium-based catalysts. Ruthenium-based Grubbs catalysts have been preferred in many studies involving ring-closing olefin metathesis due to their tolerance to a variety of common organic functional groups and low sensitivity to air and moisture.

In this context, the group of Kiddle presented a rapid microwave-accelerated ring-closing metathesis reaction of diethyl diallylmalonate in the ionic liquid 1-butyl-3-methylimidazolium tetrafluoroborate ($bmimBF_4$) employing both homogeneous and immobilized Grubbs catalyst in a sealed pressure tube utilizing a domestic microwave oven (Scheme 8.129) [164]. The authors observed that the polymer-bound Grubbs catalyst showed significantly lower conversion (40%) as opposed to the homogeneous analog (80–100%). Similarly, notable accelerations in ring-closing metathesis have also been realized by Organ *et al.* using polymer-supported second-generation Grubbs catalyst under microwave heating together with the advantages of simple purification procedures [165].

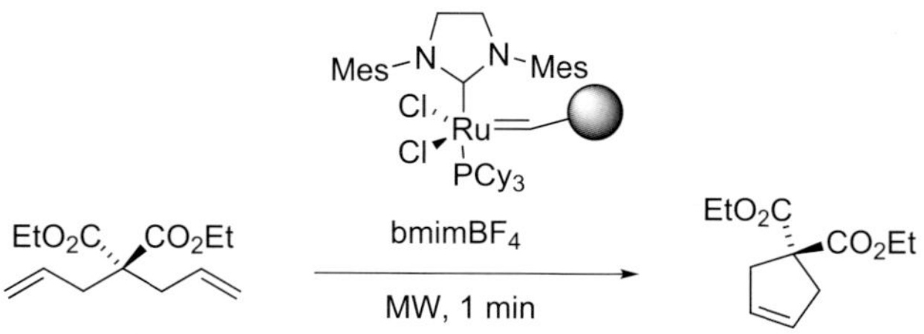

Scheme 8.129 Ring-closing metathesis utilizing ionic liquids.

The group of Ley successfully employed the commercially available encapsulated palladium species Pd-EnCat TPP30 as catalyst in microwave-mediated copper-free Sonogashira coupling reactions [166]. This easy-handling and reusable catalyst leads to more sustainable chemical practices in method development. A variety of aryl halides were reacted with a broad range of terminal acetylenes to generate an array of Sonogashira products (Scheme 8.130). In a typical procedure, a microwave vial was charged with Pd-EnCat TPP30 (0.01 mmol Pd), 0.3 mmol DBU, 0.25 mmol aryl halide, 0.3 mmol acetylene derivative, and acetonitrile. The vial was sealed and irradiated under simultaneous cooling at constant power of 70 W at 100 °C for 10 min. After the reaction, the mixture was diluted with diethyl ether and filtered to recover the catalyst. The filtrate was concentrated to obtain the coupling products in high to excellent yield. Some substrate combinations required a slightly higher reaction temperature of 120 °C as well as enhanced reaction times up to 60 min [166]. However, unprotected 4-bromoaniline and aryl chlorides in general showed no conversion at all. Furthermore, the catalyst could be reused several times without

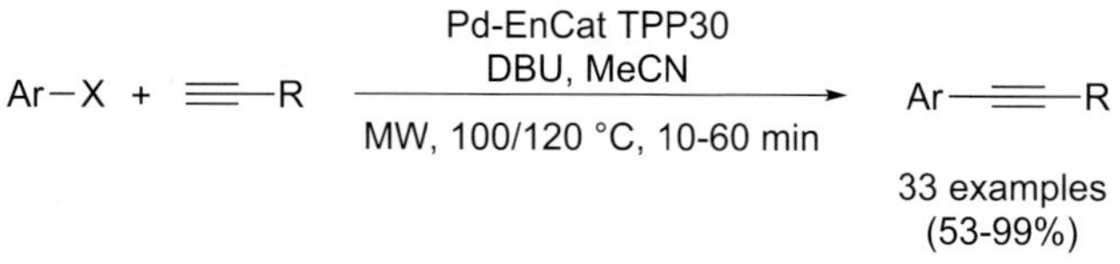

Scheme 8.130 Microwave-mediated Sonogashira couplings employing encapsulated palladium catalyst TPP30.

the loss of efficiency. Even after the sixth cycle, the conversion of the model reaction remained still identical to the initial run.

Asymmetric catalysis provides access to several synthetically important compounds, and immobilized catalysts together with solid-supported chiral ligands have been equally instrumental. Chiral ligands immobilized on a solid support provide the advantage of rapidly removing them after reaction while retaining their activity for further applications [167].

The microwave-assisted molybdenum-catalyzed allylic allylation has been studied using a free and polymer-supported bis-pyridylamide ligand [168]. However, the microwave-assisted catalytic reaction of 3-phenylprop-2-enyl methyl carbonate with dimethyl malonate ($CH_2(COOMe)_2$) in the presence of *N,O*-bis(trimethylsilyl)acetamide (BSA) and a polymer-bound bis-pyridyl ligand was rather slow.

Even so, the microwave-induced reaction with a polymer-supported bis-pyridyl ligand (Scheme 8.131) was also slow presumably because the insoluble resin-bound ligand subsequently renders the catalyst heterogeneous. However, when the experiment was performed with the twofold concentration of the reagents, after 30 min of microwave irradiation at 160 °C the reaction is complete and the product showed a branch-to-linear ratio of 35 : 1 and an enantiomeric excess of 97%. The polymer-supported ligand has obvious advantages over its unsupported analog because it could be reused, after sufficient washing and vacuum drying, for at least seven times without any loss of activity.

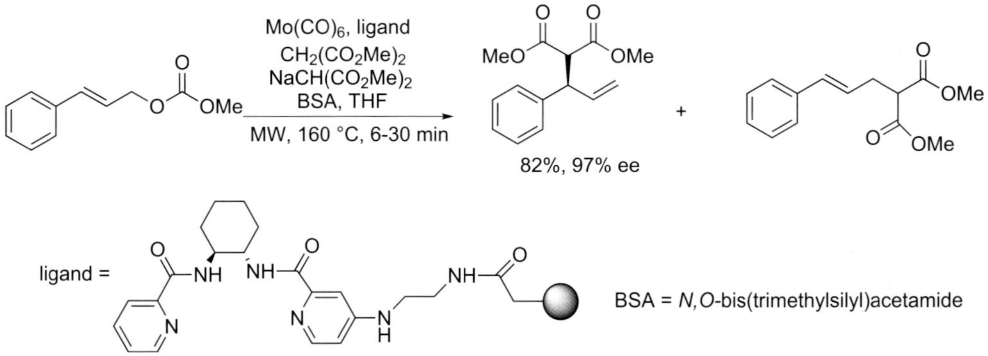

Scheme 8.131 Molybdenum-catalyzed allylic allylation.

The presented polymer-supported bis-pyridyl ligand has also been applied in the microwave-assisted asymmetric allylic alkylation [168], a key step in the enantioselective synthesis of (*R*)-baclofen (Scheme 8.132), as reported by Moberg and coworkers. The (*R*)-enantiomer is a useful agonist of the $GABA_B$ (γ-aminobutyric acid) receptor, and the racemic form is used as a muscle relaxant (antispasmodic). Under microwave heating, the enantioselectivity could be improved to 89% when using toluene as the solvent [168].

In a more recent study, Wang and coworkers discussed microwave-assisted Suzuki couplings employing a reusable polymer-supported palladium complex [169].

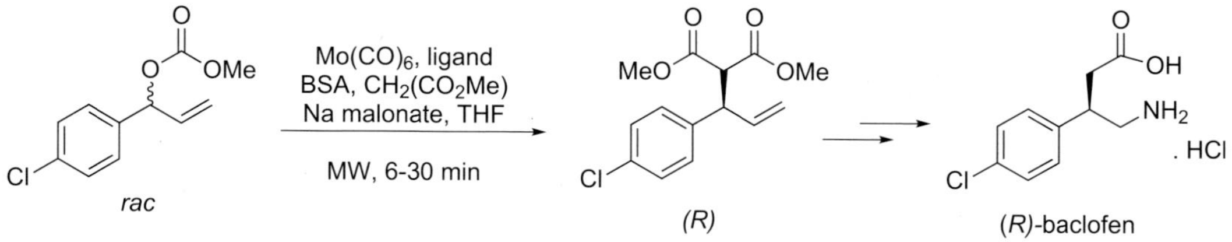

Scheme 8.132 Enantioselective synthesis of (*R*)-baclofen.

The supported catalyst was prepared from commercial Merrifield polystyrene resin under ultrasound sonication. In a typical procedure for the biaryl synthesis, 1 mmol of the corresponding aryl bromide together with 1.1 equiv of phenylboronic acid, 2.5 equiv of potassium carbonate, and 10 mg of the polystyrene-bound palladium catalyst (1.125 mmol Pd g^{-1}, 1 mol%) was admixed with 10 mL of toluene and 1 mL of water in a round-bottom flask and irradiated for 10 min in a domestic microwave oven (Scheme 8.133). After cooling, the mixture was filtered and the catalyst extracted with toluene and dried. Thus, the recycled polymer-bound catalyst can be reused five times without loss of efficiency. To isolate the desired biaryl compounds, the water layer of the filtrate was separated, and the organic phase was washed with water and dried over anhydrous magnesium sulfate. After filtration and evaporation of the solvent, the crystalline biaryls were achieved in high yields.

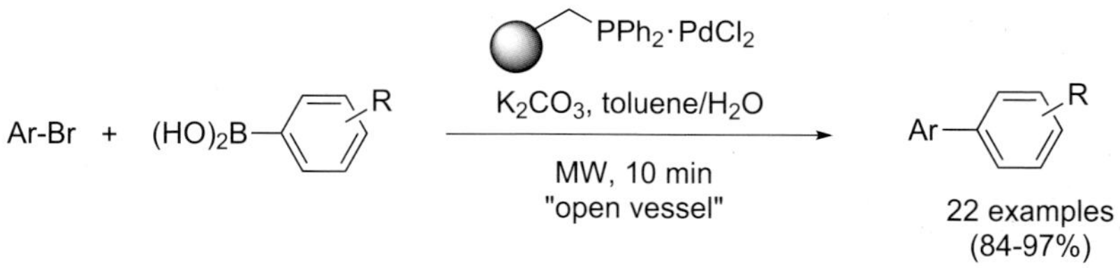

Scheme 8.133 Suzuki coupling utilizing a polymer-supported palladium catalyst.

More recently, the group of Kirschning [170] as well as a team from Abbott Laboratories [171] have reported very efficient Suzuki coupling reactions involving immobilized Pd catalysts (Scheme 8.134). The methods can be used to couple a range of aryl halides and triflates with boronic acids in solvents such as water or ethanol. The Kirschning group has used a novel type of solid Pd(II) precatalyst that can easily be prepared from 4-pyridinealdoxime and Na_2PdCl_4 (catalyst **27**), whereas the Abbott researchers employed polyethylene-supported FibreCat Pd catalysts (catalysts **28–30**) under similar conditions.

The same types of FibreCat Pd catalysts were also employed in Buchwald–Hartwig aminations [172] and in intramolecular Heck cyclizations [173]. More recently, Wang reported the use of an immobilized Pd catalyst for performing cross-couplings of sodium tetraphenylborate with aryl halides in water [174].

Srivastava and Collibee employed polymer-supported triphenylphosphine in palladium-catalyzed cyanations [175]. Commercially available resin-bound

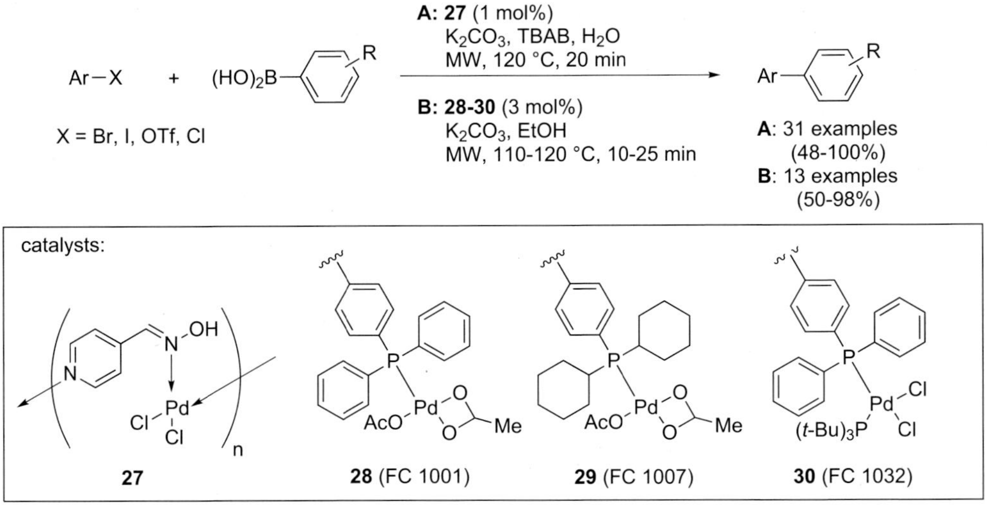

Scheme 8.134 Suzuki couplings utilizing various immobilized Pd catalysts.

triphenylphosphine was admixed with palladium(II) acetate in *N,N*-dimethylformamide in order to generate the heterogeneous catalytic system. Under nitrogen atmosphere in a sealed microwave reaction vessel, the mixture was stirred for 2 h to achieve complete formation of the active palladium–phosphine complex. The septum was removed and equimolar amounts of zinc(II) cyanide and the corresponding aryl halide were added. After purging with nitrogen and resealing, the vessel was transferred to the microwave reactor and irradiated at 140 °C for 30–50 min (Scheme 8.135). Finally, the resin was removed by filtration and evaporation of the solvent furnished the desired benzonitriles in high yields and excellent purities.

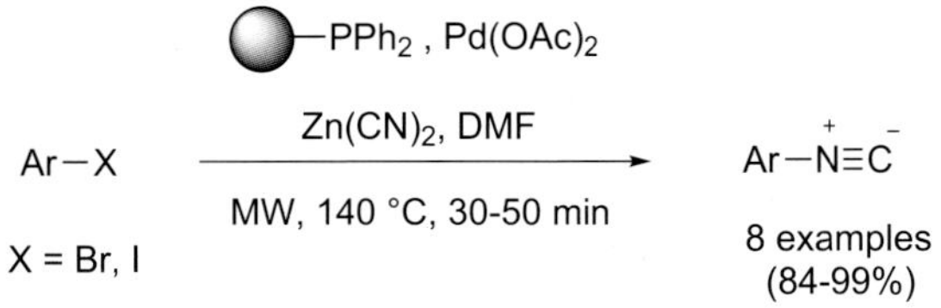

Scheme 8.135 Palladium-catalyzed cyanations utilizing polymer-bound triphenylphosphine.

Bergbreiter and Furyk presented an oligo(ethylene glycol)-bound palladium(II) complex for microwave-mediated Heck couplings of various aryl halides [176]. This novel catalyst was prepared in a multistep procedure from polyethylene glycol monomethyl ether via its methanesulfonyl intermediate, potassium thioacetate, and α,α′-dichloro-*m*-xylene. Final complexation was achieved by treatment of the 1,3-bis(MeOPEG$_{350}$thiomethyl)benzene with a solution of dichloro(bisbenzonitrile)palladium(II) [$Pd(PhCN)_2Cl_2$] in acetonitrile. The Heck reactions were carried out in *N,N*-dimethylacetamide (DMA) utilizing 4 equiv of alkene and either 2 equiv of triethylamine or 3 equiv of potassium carbonate as a base. The catalyst (0.01 mol%) was added as a DMA solution immediately before subjecting the vessel to microwave irradiation. Heating at 150 °C for 10–60 min furnished the desired cinnamate products in moderate to good yields (Scheme 8.136). The catalyst can be recycled performing the reactions in a 10% aqueous DMA/heptane mixture (1: 2) with slight loss of efficiency in up to four cycles [176].

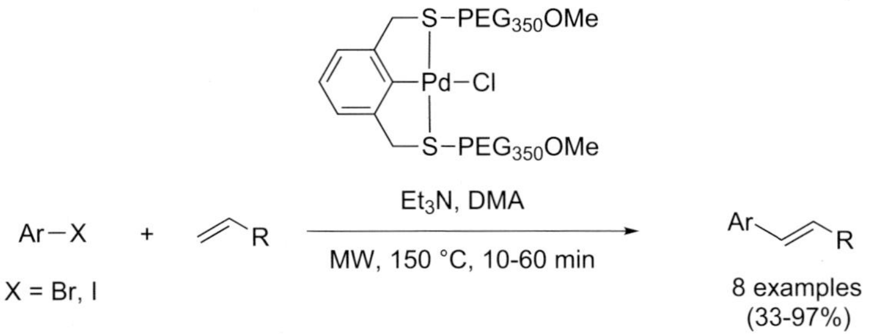

Scheme 8.136 Heck couplings utilizing an oligo(ethylene glycol)-bound SCS-palladium(II) complex as catalyst.

An application involving polystyrene-immobilized aluminum(III) chloride for a ketone–ketone rearrangement was presented by Gopalakrishnan *et al.* [177].

Kirschning and coworkers have developed the Ru catalyst **31** for various metathesis reactions (Scheme 8.137) [178]. The Ru complex is noncovalently (ion exchange) immobilized on glass–polymer composite Raschig rings that have the advantage of easy removal and reactivation after the reaction. By applying microwave heating, the reaction could be enhanced from 2 h to 4 min but with slightly reduced isolated yields. A similar type of Raschig ring Pd catalyst was developed by the same group for Suzuki cross-couplings [179].

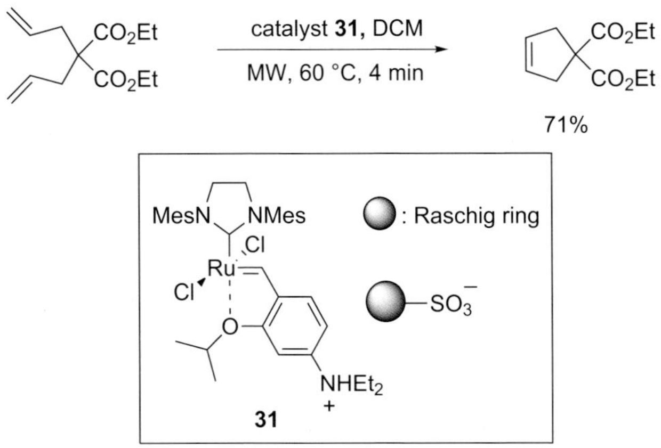

Scheme 8.137 Ru catalyst on Raschig rings for olefin metathesis.

8.6.2 Silica-Grafted Catalysts

Monoliths comprising cross-linked organic media with a well-defined porosity have emerged as useful supports in immobilizing catalysts. These supports offer advantages of lower back pressure, enhanced diffusional mass transfer, and ease of synthesis with many possibilities of introducing structural variety. The group of Buchmeiser presented the use of poly-(*N,N*-dipyrid-2-yl-7-oxanorborn-2-en-5-yl-carbamino·$PdCl_2$)-grafted monolith supports for catalyzed Heck reactions [180]. Equally useful are silica-based materials as supports in slurry reactions under conventional as well as microwave conditions. Norborn-2-ene surface-functionalized silica has been used to graft *N,N*-dipyrid-2-yl-norbor-2-ene-5-ylcarbamide monomer, and the tentacles of poly-(*N,N*-dipyrid-2-ylnorbor-2-ene-5-ylcarbamide) are generated by controlled polymerization (Figure 8.3).

The catalyst generated by palladium loading was utilized in microwave-assisted Heck reactions of iodoarenes with styrene that showed quantitative conversions to the corresponding C–C coupling products. The material could be easily removed by filtration and showed only minor leaching of Pd (<2.5%) into the reaction mixture. These catalysts were successfully applied in the coupling of selected aryl iodides and aryl bromides under flow-through conditions in cartridges [180].

8.6.3 Catalysts Immobilized on Glass

Solution-phase organic coupling reactions in the presence of palladium catalysts are among the most intensively studied organic transformations. Most of these reactions incorporate the susceptibility of the palladium catalysts to be poisoned, together with a disadvantage of making it necessary to involve relatively large amounts (1–5 mol%) for appreciable conversions. Organopalladium complexes in the form of homogeneous catalysts containing phosphine ligands provide alternatives to overcome these disadvantages; however, their separation and recovery after the reaction become difficult. Palladium metal on porous glass tubing has been investigated as a catalyst in reactions conducted continuously or batchwise. In an early report, Strauss and

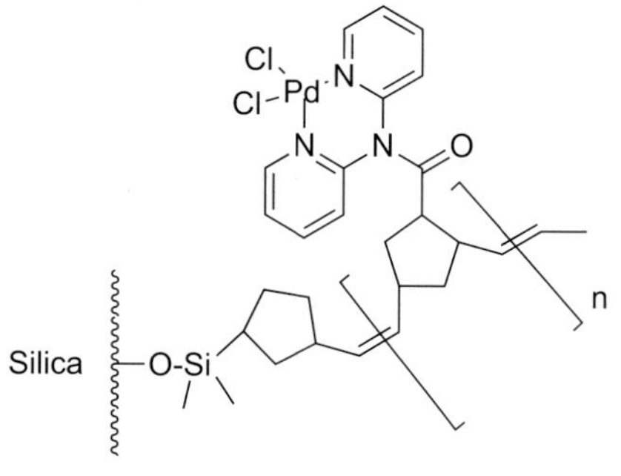

Figure 8.3 Silica-grafted palladium catalysts.

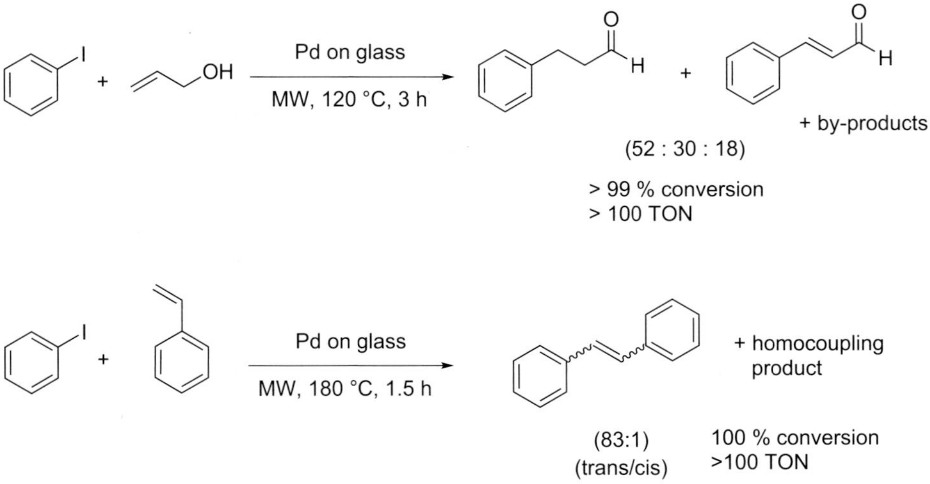

Scheme 8.138 Heck coupling utilizing palladium on porous glass.

coworkers described the Heck coupling of iodobenzene with allyl alcohol or styrene (Scheme 8.138) under microwave conditions in a pressure-tight batch reactor [181]. Although a reduced molar ratio of the catalyst (0.02 mol% palladium) was employed, >99% conversions with high regioselectivity were observed.

In the same report, the group of Strauss furthermore presented the effective use of palladium on porous glass to achieve quantitative conversions for the coupling of phenylacetylene with iodobenzene and 4-bromobenzaldehyde. In addition, satisfactory results were obtained for the coupling of phenylacetylene with 4-bromoacetophenone and 2-bromopyridine [181].

Stadler and Kappe reported on a palladium-doped microwave process vial, wherein palladium deposited on the inner glass surface was applied as a catalyst in heterogeneous C–P couplings leading to triphenylphosphines under microwave conditions (Scheme 8.139) [182]. The catalytic activity under microwave heating is, however, reported somewhat lower (85% yield) than that using palladium-on-charcoal (Pd/C) (98% yield) under identical reaction conditions (190 °C for 3 min). The vials with doped palladium could effectively be reused several times without significant loss of catalytic activity. In addition, the necessity of catalyst filtration and additional reaction workup is eliminated, simplifying the overall synthetic process.

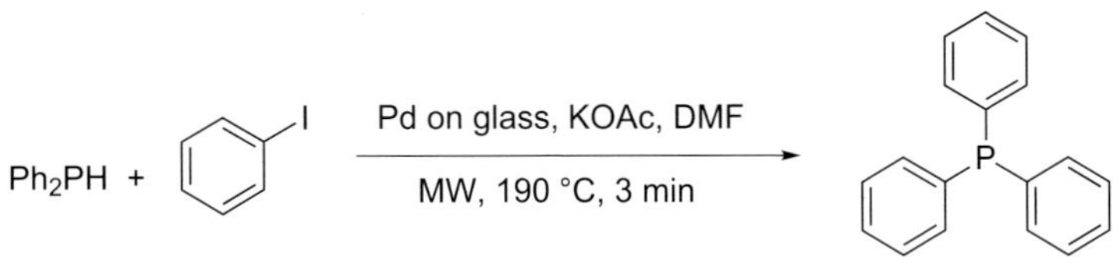

Scheme 8.139 Synthesis of triphenylphosphine by C–P coupling.

The concept of performing microwave chemistry (e.g., Suzuki cross-couplings) in Pd-coated microcapillaries has been introduced by Organ [183]. A variety of other transformations have been reported by the same group using metal films on glass capillaries in flow format [184]. A similar approach was chosen by Ondruschka applying Pd on porous glass for Suzuki and Heck chemistry [185].

8.6.4
Catalysts Immobilized on Carbon

Carbon is inert in nature and has a high surface area, making it highly suitable as a support for catalysts. The surface characteristics and porosity of carbon could be easily tailored for different applications. Acid treatment is often applied to modify the surface chemistry of carbon for specific applications. Typically, active metal species are immobilized on carbon for application in catalysis.

Palladium-on-charcoal (Pd/C) is a catalyst heavily employed in catalytic transfer hydrogenations (CTH). Microwave-assisted open-vessel reduction of double bonds and hydrogenolysis of several functional groups have been studied safely and rapidly by Bose and coworkers, using 10% Pd/C catalyst and ammonium formate as the hydrogen donor [186]. In addition, the microwave-assisted selective hydrogenolysis of only the *O*-benzyl (Bn) group while retaining the *N*-benzyl group together with the reduction of the unsaturated ester to give a saturated side chain in a β-lactam (see Scheme 8.140a) has been reported. In a stereoselective preparation of β-lactam synthons using microwave-assisted catalytic transfer hydrogenation conditions, unsaturation in a sugar moiety was successfully removed without disturbing the β-lactam ring.

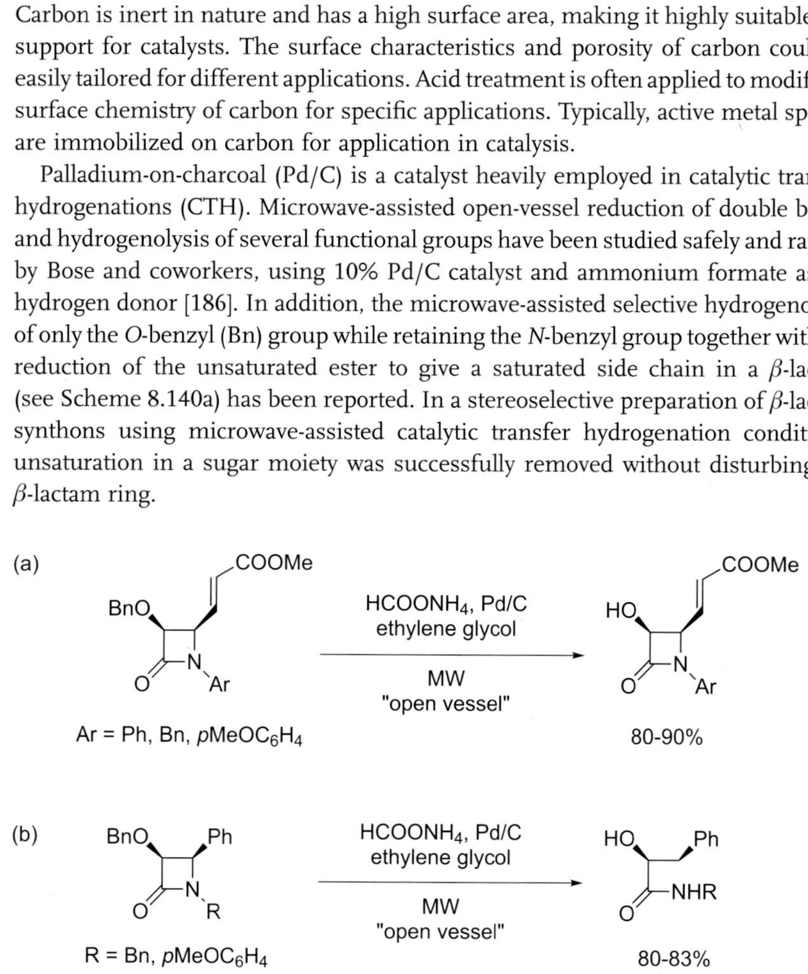

Scheme 8.140 Hydrogenolysis reactions catalyzed by palladium-on-charcoal.

In addition, microwave-assisted Pd/C-catalyzed transfer hydrogenolysis has been documented to cause rapid scission of 4-phenyl-2-azetidinones (Scheme 8.140b). The hydrogenolysis conditions selectively deprotect the benzyloxy group at the C-3 position to a hydroxy group with high yields within a few minutes. Microwave-assisted reduction of phenyl hydrazone was also carried out using 10% Pd/C in the presence of ammonium formate to give the corresponding amine within 4 min in 92% yield.

Related examples of microwave-assisted transformations involving palladium-on-charcoal are reported in Chapters 5–7. The use of graphite as support in microwave synthesis has been discussed in Section 4.1.

8.6.5
Miscellaneous

The group of Choudary reported on a novel layered double hydroxide-supported nanopalladium catalyst [LDH-Pd(0)] with superior activity for C−C coupling reactions compared to other supported catalysts, ranging from acidic to weakly basic Pd/C, Pd/SiO_2, Pd/Al_2O_3, and resin-($PdCl_4^{2-}$) [187]. In comparison to the homogeneous palladium(II) chloride catalyst, the LDH-Pd(0) catalytic system showed higher activity and selectivity with excellent yields and high turnover frequencies (TOFs) in the microwave-assisted Heck olefination of electron-poor and electron-rich chloroarenes in nonaqueous ionic liquids (Scheme 8.141).

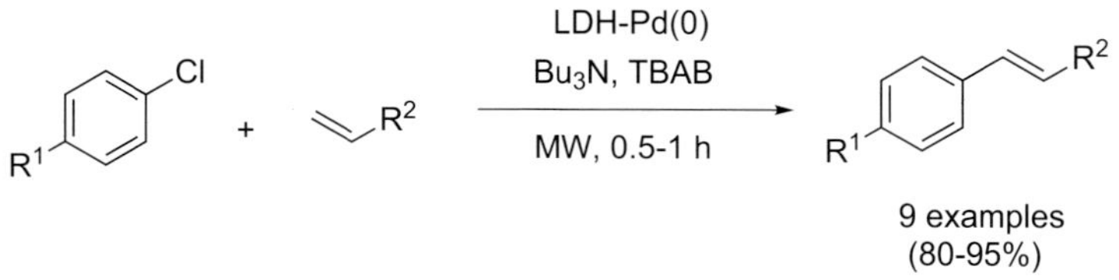

Scheme 8.141 Nanopalladium catalyst for Heck olefinations.

Under microwave heating, the Heck olefinations were achieved in 30–60 min as opposed to 10–40 h of conventional heating. The recyclable heterogeneous LDH-Pd(0) catalytic system circumvents the need to use expensive and air-sensitive basic phosphines as ligands in the palladium-catalyzed coupling of chloroarenes. This novel Mg–Al layered double hydroxide (LDH) support in the catalytic system stabilizes the nanopalladium particles and also provides adequate electron density to the anchored palladium(0) species and facilitates the oxidative addition of the deactivated electron-rich chloroarenes.

Some early examples involving microwave-assisted solvent-free Sonogashira couplings using palladium powder doped on alumina/potassium fluoride as catalyst have been described by Kabalka *et al.* [188]. In addition, this novel catalytic system has been used in the microwave-assisted solvent-free Sonogashira coupling–cyclization of *ortho*-iodophenol with terminal alkynes and similarly of *ortho*-ethynylphenols with aromatic iodides to generate 2-substituted benzo[*b*]furans (Scheme 8.142) [188]. All experiments within this report were carried out in a domestic microwave oven, utilizing septum-sealed round-bottom flasks. The alumina in the reaction mixture

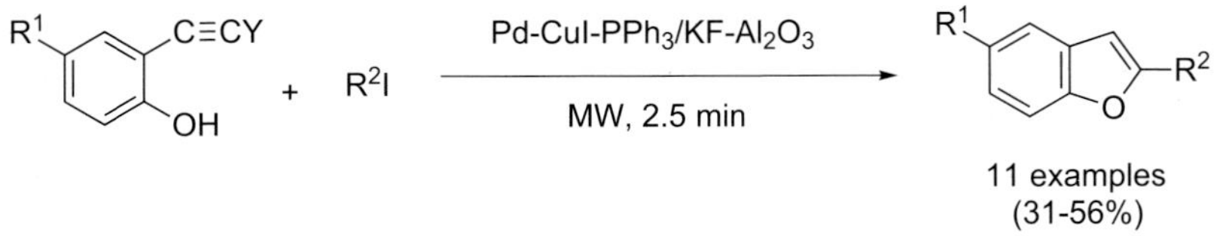

Scheme 8.142 Solvent-free Sonogashira coupling–cyclization.

acts as a temperature moderator in order to prevent the reagents from extensively reacting with the metal catalyst. After the reaction, the mixture was suspended with hexane, filtered, and the products were isolated by evaporation of the solvent and finally purified by flash chromatography.

The group of Ley reported on using palladium-doped perovskites as recyclable and reusable catalysts for Suzuki couplings [189]. The microwave-mediated cross-couplings of phenylboronic acid with aryl halides were achieved within 1 h utilizing the supported catalyst (0.25 mol% palladium) in aqueous 2-propanol (Scheme 8.143). The addition of water was crucial as transformations in nonaqueous mixtures remained ineffective.

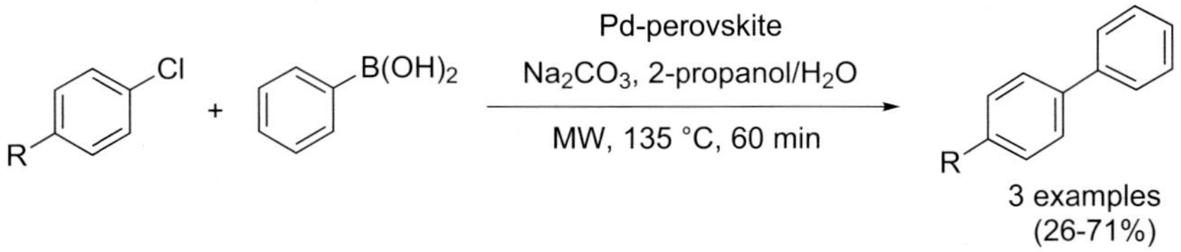

Scheme 8.143 Suzuki couplings utilizing palladium-doped perovskite as catalyst.

Similar Suzuki couplings have been performed by Hu and coworkers utilizing a poly(dicyclohexylcarbodiimide)/palladium nanoparticle composite [190]. This PDHC-Pd catalyst showed remarkable activity and stability under microwave irradiation. Almost quantitative conversion (95% isolated yield) was obtained after 40 min of microwave heating of a mixture of iodobenzene with phenylboronic acid in dioxane. Reusing the immobilized catalyzed showed no significant loss of efficiency, as the fifth cycle still furnished 90% isolated yield of the desired biphenyl.

More recently, Cu- and Pd-containing perovskites were shown to be useful catalysts for Ullmann and Sonogashira couplings [191]. A palladium catalyst based on modified starch has been shown to be active in Sonogashira cross-couplings [192]. Similarly, the use of Pd-*N*-heterocyclic carbene ligands grafted onto silica in C–C coupling chemistry has been demonstrated [193].

A polymer-supported BEMP base was used for the ionic immobilization of carboxylic acid derivative possessing a Suzuki-active bromide functionality. The Suzuki reaction was carried out under microwave conditions directly on the ionic complex [194].

More examples of microwave-assisted transformations involving immobilized catalysts are described in Ref. [195].

8.7 Polymer-Supported Scavengers

The use of microwave heating in most examples of the syntheses discussed within this and the preceding chapter has the common purpose of speeding up the performed chemical transformation, in some cases to obtain a desired selectivity and often to improve the conversion and yield of the desired products over the conventional approach. Moreover, the isolation of a clean and homogeneous product becomes an integral part of any synthesis; thus, microwave heating has also been instrumental in improving many purification techniques. This final section will account specific examples for the use of microwave irradiation in combination with several functionalized polymers for scavenging and purification techniques.

The Kappe group has described the microwave-induced *N*3-acylation of the dihydropyrimidine (DHPM) Biginelli scaffold [161] (see Scheme 4.29) using various anhydrides. The process involved purification of the reaction mixture by a microwave-assisted scavenging technique. Volatile acylating agents such as acetic anhydride could be simply removed by evaporation while other nonvolatile anhydrides such as benzoic anhydride (Bz_2O) required an elaborate workup suitable for a high-throughput format. Several scavenging reagents possessing amino functionalities with different loadings were applied to sequester the excess of benzoic anhydride (Bz_2O) from the reaction mixture [161, 196].

In order to compare the effectiveness, the scavenging was studied under conventional (room temperature) and microwave conditions (Figure 8.4). At room temperature, the polystyrene-based diamine (3.0 equiv amine functionality) required 1–2 h for complete sequestration of excess anhydride. In contrast, under sealed vessel microwave heating, this was accelerated to completion within 5–10 min.

Comparable results were achieved with the silica-based diamine (3.0 equiv amine functionality). At room temperature, the excess anhydride was scavenged within 4 h. On the other hand, microwave heating at 100 °C in a sealed vessel significantly reduced the quenching time to 5–10 min. Scavenging using the Stratosphere Plugs,

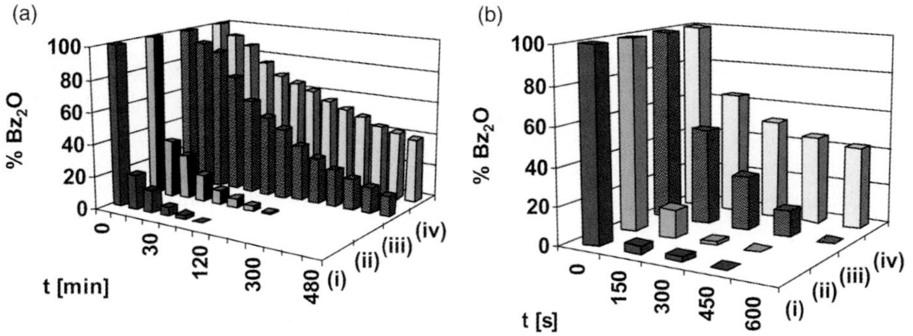

Figure 8.4 Scavenging efficiency of polymer-supported scavengers ((i) polystyrene, (ii) functionalized silica, (iii) plugs, and (iv) lanterns) at room temperature (a) and under microwave heating (b). Reproduced with permission from Ref. [161].

bearing a diethylenetriaminomethyl polystyrene function, and Synphase Lanterns, having an aminomethyl polystyrene function, proved somewhat less effective.

A parallel solution-phase asymmetric synthesis of α-branched amines has been reported by Ellman and coworkers based on stereoselective addition of organomagnesium reagents to enantiomerically pure *tert*-butanesulfinyl imines [197]. Microwave heating was utilized in two of the steps of the synthesis of asymmetric amines, both for the imine formation and for the resin capture (Scheme 8.144).

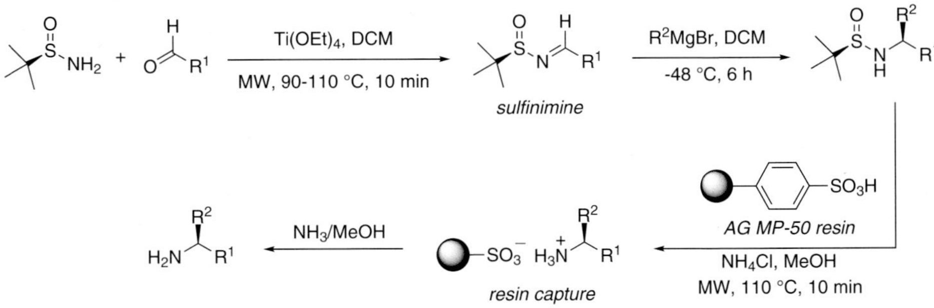

Scheme 8.144 Asymmetric synthesis of α-branched amines.

The initial condensation step in the synthesis of the sulfinimine substrates was catalyzed by titanium(IV) ethoxide used as Lewis acid under microwave heating. The excess of the Lewis acid was scavenged using a large amount of a support-bound diethanolamine (not shown in Scheme 8.144). Contrary to room-temperature conditions, the imine was formed within 10 min of microwave irradiation and often in quantitative yields. After the Grignard addition step, microwave irradiation was utilized for acidic alcoholysis of the sulfinimide using macroporous sulfonic acid resin (AG MP-50). However, under conventional reflux conditions, only partial sulfinyl group cleavage in methanol was observed. On the other hand, in some cases under microwave heating at 110 °C, there was complete consumption within 10 min. The microwave-assisted resin capture of amines allowed the preparation of a range of α-phenylethylamines and diphenylmethylamine derivatives in analytically pure form, in good overall yields, and with high enantiomeric purity. The "resin capture methodology" afforded the amine hydrochlorides without requiring chromatography or crystallization of any intermediates or products.

The group of Messeguer presented the use of a high-loading polystyrene Wang aldehyde resin for the scavenging of excess amines in the preparation of piperazinium derivatives [198]. In the initial step, chloroacetyl chloride was reacted with 50% excess of primary amine at 0 °C for 30 min. After filtration and evaporation, the residue was dissolved in dioxane, suspended with 3.5 equiv of Wang aldehyde resin, and irradiated in a domestic microwave oven for 20–40 min, applying 4 min intervals. The resin was filtered off and the resulting chloroacetamide was treated with 3 equiv of the corresponding primary diamines (Scheme 8.145). After the reaction, the solution was again suspended with the aldehyde resin and subjected to the

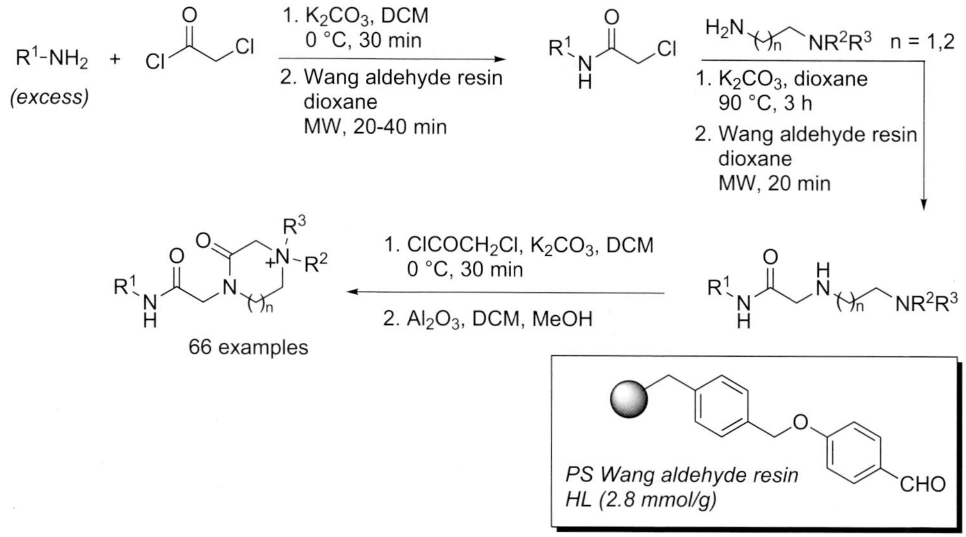

Scheme 8.145 Efficient amine scavenging utilizing a polystyrene aldehyde scavenger.

microwave-assisted scavenging (domestic microwave oven). After five cycles of irradiation for 4 min, the resin was filtered off and the filtrate evaporated to obtain the corresponding glycinamides. Further modification and cyclization steps furnished the desired heterocyclic scaffolds.

Porco and Lei described the use of polymer-bound anthracene as effective dienophile scavenger [199]. The utilized scavenging agent can easily be prepared

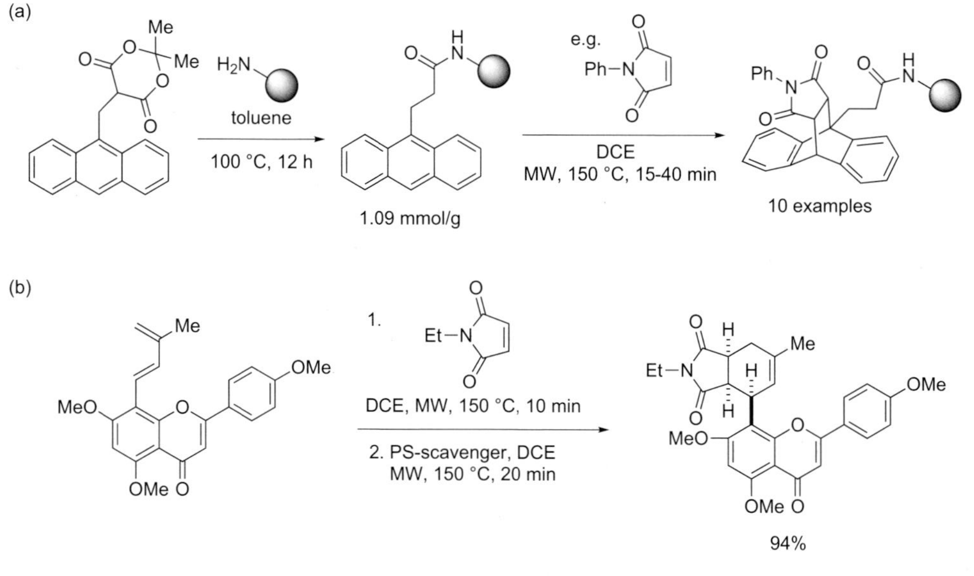

Scheme 8.146 Preparation and application of a novel polystyrene-bound anthracene as dienophile scavenger.

by treatment of the commercially available corresponding Meldrum's acid derivative with aminomethyl polystyrene resin (Scheme 8.146a). The reactivity of this novel polymer-bound scavenger was examined for a series of 10 different dienophiles such as *N*-phenylmaleimide. Under microwave heating, 2–3 equiv of the resin were sufficient to achieve effective scavenging of reactive dienophiles in less than 30 min. Rather unreactive derivatives, such as 1,4-naphthoquinone, required 40 min of irradiation in 1,2-dichloroethane (DCE).

To demonstrate the effectiveness of this scavenger, it was successfully applied in the synthesis of eight different flavonoid Diels–Alder cycloadducts (Scheme 8.146b). The two-step microwave-mediated procedure furnished the desired compounds in high yields and excellent purities.

Finally, extensive work by the Ley group has demonstrated the usefulness of tagged reagents and scavengers in various transformations when applied in conjunction with microwave heating [200].

References

1 Yu, H.-M., Chen, S.-T., and Wang, K.-T. (1992) *The Journal of Organic Chemistry*, **57**, 4781–4784.

2 Erdélyi, M. and Gogoll, A. (2002) *Synthesis*, 1592–1596.

3 Lindquist, C., Tedebark, U., Ersoy, O., and Somfai, P. (2003) *Synthetic Communications*, **33**, 2257–2262.

4 Olivos, H.J., Alluri, P.G., Reddy, M.M., Salony, D., and Kodadek, T. (2002) *Organic Letters*, **4**, 4057–4059.

5 Finaru, A., Berthault, A., Besson, T., Guillaumet, G., and Berteina-Raboin, S. (2002) *Organic Letters*, **4**, 2613–2615.

6 Geske, G.D., Wezeman, R.J., Siegel, A.P., and Blackwell, H.E. (2005) *Journal of the American Chemical Society*, **127**, 12762–12763.

7 Geske, G.D., O'Neill, J.C., Miller, D.M., Mattmann, M.E., and Blackwell, H.E. (2007) *Journal of the American Chemical Society*, **129**, 13613–13625.

8 Murray, J.K. and Gellman, S.H. (2005) *Organic Letters*, **7**, 1517–1520.

9 (a) Rizzolo, F., Sabatino, G., Chelli, M., Rovero, P., and Papini, A.M. (2007) *International Journal of Peptide Research and Therapeutics*, **13**, 203–208; (b) Chen, P.H., Tseng, Y.H., Mou, Y., Tsai, Y.-L., Guo, S.-M., Huang, S.-J., Yu, S.S.-F., and Chan, J.C.C. (2008) *Journal of the American Chemical Society*, **130**, 2862–2868; (c) Coantic, S., Subra, G., and Martinez, J. (2008) *International Journal of Peptide Research and Therapeutics*, **14**, 143–147.

10 Bacsa, B., Horváti, K., Bősze, S., Andreae, F., and Kappe, C.O. (2008) *The Journal of Organic Chemistry*, **73**, 7532–7542.

11 Bacsa, B., Bösze, S., and Kappe, C.O. (2010) *The Journal of Organic Chemistry*, **75**, 2103–2106.

12 Matsushita, T., Hinou, H., Kurogochi, M., Shimizu, H., and Nichimura, S.-I. (2005) *Organic Letters*, **7**, 877–880.

13 Díaz-Mochón, J.J., Fara, M.A., Sanchez-Martin, R.M., and Bradley, M. (2008) *Tetrahedron Letters*, **49**, 923–926.

14 Fara, M.A., Díaz-Mochón, J.J., and Bradley, M. (2006) *Tetrahedron Letters*, **47**, 1011–1014.

15 Park, M.S., Oh, H.S., Cho, H., and Lee, K.H. (2007) *Tetrahedron Letters*, **48**, 1053–1057.

16 Paolini, I., Nuti, F., de la Cruz Pozo-Carrero, M., Barbetti, F., Kolesinska, B., Kaminski, Z.J., Chelli, M., and Papini, A.M. (2007) *Tetrahedron Letters*, **48**, 2901–2904.

17 Palasek, S.A., Cox, Z.J., and Collins, M.J. (2007) *Journal of Peptide Science*, **13**, 143–148.

18 Leonetti, F., Capaldi, C., and Carotti, A. (2007) *Tetrahedron Letters*, **48**, 3455–3458.

19 Stadler, A. and Kappe, C.O. (2001) *European Journal of Organic Chemistry*, 919–925.

20 Stadler, A. and Kappe, C.O. (2001) *Tetrahedron*, **57**, 3915–3920.

21 Yang, H., Peng, Y., Song, G., and Qian, X. (2001) *Tetrahedron Letters*, **42**, 9043–9046.

22 Lee, T., Park, J.-H., Lee, D.-H., and Gong, Y.-D. (2009) *Journal of Combinatorial Chemistry*, **11**, 495–499.

23 Lee, T., Lee, D., Lee, I.Y., and Gong, Y.-D. (2010) *Journal of Combinatorial Chemistry*, **10**, 95–99.

24 Yaylayan, V.A., Siu, M., Bélangere, J.M.R., and Paré, J.R.J. (2002) *Tetrahedron Letters*, **43**, 9023–9025.

25 Westman, J. and Lundin, R. (2003) *Synthesis*, 1025–1030.

26 Strohmeier, G.A. and Kappe, C.O. (2002) *Journal of Combinatorial Chemistry*, **4**, 154–161.

27 Harju, K., Vesterinen, J., and Yil-Kauhaluoma, J. (2009) *Organic Letters*, **11**, 2219–2221.

28 Qin, L.-Y., Cole, A.G., Metzger, A., O'Brien, L., Sun, X., Wu, J., Xu, Y., Xu, K., Zhang, Y., and Henderson, I. (2009) *Tetrahedron Letters*, **50**, 419–422.

29 Akamatsu, H., Fukase, K., and Kusumoto, S. (2004) *Synlett*, 1049–1053.

30 Larhed, M., Lindeberg, G., and Hallberg, A. (1996) *Tetrahedron Letters*, **37**, 8219–8222.

31 Kaiser, N.-F.K., Bremberg, U., Larhed, M., Moberg, C., and Hallberg, A. (2000) *Angewandte Chemie, International Edition*, **39**, 3596–3598.

32 Alterman, M. and Hallberg, A. (2000) *The Journal of Organic Chemistry*, **65**, 7984–7989.

33 Wannberg, J. and Larhed, M. (2003) *The Journal of Organic Chemistry*, **68**, 5750–5753.

34 Combs, A.P., Glass, B.M., and Jackson, S.A. (2003) *Method Enzymology*, **369**, 223–231.

35 Erdélyi, M. and Gogoll, A. (2003) *The Journal of Organic Chemistry*, **68**, 6431–6434.

36 Walla, P. and Kappe, C.O. (2004) *Chemical Communications*, 564–565.

37 Combs, A.P., Saubern, S., Rafalski, M., and Lam, P.Y.S. (1999) *Tetrahedron Letters*, **40**, 1623–1626.

38 Weigand, K. and Pelka, S. (2003) *Molecular Diversity*, **7**, 181–184.

39 Dai, W.-M., Guo, D.-S., Sun, L.-P., and Huang, X.-H. (2003) *Organic Letters*, **5**, 2919–2922.

40 Kaval, N., Dehaen, W., and Van der Eycken, E. (2005) *Journal of Combinatorial Chemistry*, **7**, 90–95.

41 Scharn, D., Wenschuh, H., Reineke, U., Schneider-Mergener, J., and Germeroth, L. (2000) *Journal of Combinatorial Chemistry*, **2**, 361–369.

42 Scharn, D., Germeroth, L., Schneider-Mergener, J., and Wenschuh, H. (2001) *The Journal of Organic Chemistry*, **66**, 507–513.

43 Bowman, M.D., Jeske, R.C., and Blackwell, H.E. (2004) *Organic Letters*, **6**, 2019–2022.

44 Bowman, M.D., Jacobson, M.M., and Blackwell, H.E. (2006) *Organic Letters*, **8**, 1645–1648.

45 Lin, Q., O'Neil, J.C., and Blackwell, H.E. (2005) *Organic Letters*, **7**, 4455–4458.

46 Lin, Q. and Blackwell, H.E. (2006) *Chemical Communications*, 2884–2886.

47 Grünefeld, P. and Richert, C. (2004) *The Journal of Organic Chemistry*, **69**, 7543–7555.

48 Yates, E.A., Jones, M.O., Clarke, C.E., Powell, A.K., Johnson, S.R., Porch, A., Edwards, P.P., and Turnbull, J.E. (2003) *Journal of Materials Chemistry*, **13**, 2061–2063.

49 Grieco, P., Campiglia, P., Gomez-Monterry, I., Lama, T., and Novellino, E. (2003) *Synlett*, 2216–2218.

50 Campiglia, P., Gomez-Monterry, I., Longobardo, L., Lama, T., Novellino, E., and Grieco, P. (2004) *Tetrahedron Letters*, **45**, 1453–1456.

51 Miles, S.M., Leatherbarrow, R.J., Marsden, S.P., and Coates, W.J. (2004) *Organic and Biomolecular Chemistry*, **2**, 281–283.

52 De Luca, L., Giacomelli, G., Porcheddu, A., Salaris, M., and

Taddei, M. (2003) *Journal of Combinatorial Chemistry*, **5**, 465–471.

53 Austin, R.E., Okonya, J.F., Bond, D.R.S., and Al-Obeidi, F. (2002) *Tetrahedron Letters*, **43**, 6169–6171.

54 Hoel, A.M.L. and Nielsen, J. (1999) *Tetrahedron Letters*, **40**, 3941–3944.

55 Portal, C., Launay, D., Merritt, A., and Bradley, M. (2005) *Journal of Combinatorial Chemistry*, **7**, 554–560.

56 Henkel, B. (2004) *Tetrahedron Letters*, **45**, 2219–2221.

57 De Luca, L., Giacomelli, G., and Porcheddu, A. (2005) *Journal of Combinatorial Chemistry*, **7**, 905–908.

58 Frutos Hoener, A.P., Henkel, B., and Gauvin, J.-C. (2003) *Synlett*, 63–66.

59 Kuster, G.J. and Scheeren, H.W. (2000) *Tetrahedron Letters*, **41**, 515–519.

60 Clapham, B., -H- Lee, S., Koch, G., Zimmermann, J., and Janda, K.D. (2002) *Tetrahedron Letters*, **43**, 5407–5410.

61 Kumar, H.M.S., Anjaneyulu, S., and Reddy, B.V.S. (2000) *Synlett*, 1129–1130.

62 Schobert, R. and Jagusch, C. (2003) *Tetrahedron Letters*, **44**, 6449–6451.

63 Glass, B.M. and Combs, A.P. (2001) Chapter 4.6: Case study 4–6: rapid parallel synthesis utilizing microwave irradiation, in *High-Throughput Synthesis. Principles and Practices* (ed. I. Sucholeiki), Marcel Dekker, Inc., New York, pp. 123–128.

64 Glass, B.M. and Combs, A.P. (2001) Rapid parallel synthesis utilizing microwave irradiation (Article E0027), in *5th International Electronic Conference on Synthetic Organic Chemistry* (eds C.O. Kappe, P. Merino, A. Marzinzik, H. Wennemers, T. Wirth, J.J. Vanden Eynde, and S.-K. Lin), CD-ROM edition, MDPI, Basel, Switzerland.

65 Martinez-Teipel, B., Green, R.C., and Dolle, R.E. (2004) *The QSAR & Combinatorial Science*, **23**, 854–858.

66 Pérez, R., Beryozkina, T., Zbruyev, O.I., Haas, W., and Kappe, C.O. (2002) *Journal of Combinatorial Chemistry*, **4**, 501–510.

67 Gong, Y.-D., Sohn, H.-Y., and Kurth, M.J. (1998) *The Journal of Organic Chemistry*, **63**, 4854–4856.

68 Martin, B., Sekijic, H., and Chassaing, C. (2003) *Organic Letters*, **5**, 1851–1853.

69 Yu, A.-M., Zhang, Z.-P., Yang, H.-Z., Zhang, C.-X., and Liu, Z. (1999) *Synthetic Communications*, **29**, 1595–1599.

70 Chandrasekhar, S., Padmaja, M.B., and Raza, A. (1999) *Synlett*, 1597–1599.

71 Weik, S. and Rademann, J. (2003) *Angewandte Chemie, International Edition*, **42**, 2491–2494.

72 Bengtsson, A., Hallberg, A., and Larhed, M. (2002) *Organic Letters*, **4**, 1231–1233.

73 Leadbeater, N.E., Torenius, H.M., and Tye, H. (2003) *Molecular Diversity*, **7**, 135–144.

74 Paolini, L., Petricci, E., Corelli, F., and Botta, M. (2003) *Synlett*, 1039–1042.

75 Kaval, N., Van der Eycken, J., Caroen, J., Dehaen, W., Strohmeier, G.A., Kappe, C.O., and Van der Eycken, E. (2003) *Journal of Combinatorial Chemistry*, **5**, 560–568.

76 Che, J., Raghavendra, M.S., and Lam, Y. (2009) *Journal of Combinatorial Chemistry*, **11**, 378–384.

77 Blettner, C.G., König, W.A., Stenzel, W., and Schotten, T. (1999) *The Journal of Organic Chemistry*, **64**, 3885–3890.

78 Xia, M. and Wang, Y.-G. (2002) *Journal of Chemical Research S*, 173–175.

79 Sauvagnat, B., Lamaty, F., Lazaro, R., and Martinez, J. (2000) *Tetrahedron Letters*, **41**, 6371–6375.

80 Vanden Eynde, J.J. and Rutot, D. (1999) *Tetrahedron*, **55**, 2687–2694.

81 Xia, M. and Wang, Y.-G. (2002) *Tetrahedron Letters*, **43**, 7703–7705.

82 Xia, M. and Wang, Y.-G. (2003) *Synthesis*, 262–266.

83 Porcheddu, A., Ruda, G.F., Sega, A., and Taddei, M. (2003) *European Journal of Organic Chemistry*, 907–912.

84 Wu, C.-Y. and Sun, C.-M. (2002) *Synlett*, 1709–1711.

85 Chang, W.-J., Yeh, W.-B., and Sun, C.-M. (2003) *Synlett*, 1688–1692.

86 Bendale, P.M. and Sun, C.-M. (2002) *Journal of Combinatorial Chemistry*, **4**, 359–361.

87 Tung, C.-L. and Sun, C.-M. (2004) *Tetrahedron Letters*, **45**, 1159–1162.

88 Lin, M.-J. and Sun, C.-M. (2004) *Synlett*, 663–666.

89 Yeh, W.-B., Lin, M.-J., and Sun, C.-M. (2004) *Combinatorial Chemistry & High Throughput Screening*, **7**, 251–255.

90 Su, Y.-S., Lin, M.-J., and Sun, M.-C. (2005) *Tetrahedron Letters*, **46**, 177–180.

91 Lin, M.-J. and Sun, C.-M. (2003) *Tetrahedron Letters*, **44**, 8739–8742.

92 Yeh, W.-B., Lin, M.-J., Lee, M.-J., and Sun, C.-M. (2003) *Molecular Diversity*, **7**, 185–197.

93 Yeh, W.-B. and Sun, C.-M. (2004) *Journal of Combinatorial Chemistry*, **6**, 279–282.

94 Chang, W.J., Kulkarni, M.V., and Sun, C.M. (2006) *Journal of Combinatorial Chemistry*, **8**, 141–144.

95 Lin, M.J. and Sun, C.M. (2006) *Journal of Combinatorial Chemistry*, **8**, 455–458.

96 Lee, M.-J. and Sun, C.-M. (2004) *Tetrahedron Letters*, **45**, 437–440.

97 Yellol, G.S., Chung, T.-W., and Sun, C.-M. (2010) *Chemical Communications*, **46**, 9170–9172.

98 Kim, S., Yamamoto, K., Hayashi, K., and Chiba, K. (2008) *Tetrahedron*, **64**, 2855–2863.

99 (a) Zhang, W. (2006) *Topics in Current Chemistry*, **266**, 145–166; (b) Zhang, W. (2004) *Chemical Reviews*, **104**, 2531–2556; (c) Olofsson, K., and Larhed, M. (2004) Chapter 10.19, in *Handbook of Fluorous Chemistry* (eds J.A. Gladysz, D.P. Curran, and I.T. Horvath), Wiley-VCH Verlag GmbH, Weinheim, pp. 359–365.

100 Larhed, M., Hoshino, M., Hadida, S., Curran, D.P., and Hallberg, A. (1997) *The Journal of Organic Chemistry*, **62**, 5583–5587.

101 Olofsson, K., Kim, S.-Y., Larhed, M., Curran, D.P., and Hallberg, A. (1999) *The Journal of Organic Chemistry*, **64**, 4539–4541.

102 Zhang, W., Lu, Y., and Chen, C.H.-T. (2003) *Molecular Diversity*, **7**, 199–202.

103 Nagashima, T. and Zhang, W. (2004) *Journal of Combinatorial Chemistry*, **6**, 942–949.

104 Zhang, W., Chen, C.H.-T., Lu, Y., and Nagashima, T. (2004) *Organic Letters*, **6**, 1473–1476.

105 Zhang, W., Nagashima, T., Lu, Y., and Chen, C.H.-T. (2004) *Tetrahedron Letters*, **45**, 4611–4613.

106 Zhang, W. and Chen, C.H.T. (2005) *Tetrahedron Letters*, **46**, 1807–1810.

107 Villard, A.-L., Warrington, B., and Ladlow, M. (2004) *Journal of Combinatorial Chemistry*, **6**, 611–622.

108 Zhang, W. and Tempest, P. (2004) *Tetrahedron Letters*, **45**, 6757–6760.

109 Lu, Y. and Zhang, W. (2004) *The QSAR & Combinatorial Science*, **23**, 827–835.

110 Liu, A., Zhou, H., Su, G., Zhang, W., and Yan, B. (2009) *Journal of Combinatorial Chemistry*, **11**, 1083–1093.

111 Zhou, H., Zhang, W., and Yan, B. (2010) *Journal of Combinatorial Chemistry*, **12**, 206–214.

112 Vallin, K.S.A., Zhang, Q., Larhed, M., Curran, D.P., and Hallberg, A. (2003) *The Journal of Organic Chemistry*, **68**, 6639–6645.

113 Zhang, W., Lu, Y., and Chen, C.H.-T. (2003) *Molecular Diversity*, **7**, 199–202.

114 Herrero, M.A., Wannberg, J., and Larhed, M. (2004) *Synlett*, 2335–2338.

115 Werner, S. and Curran, D.P. (2003) *Organic Letters*, **5**, 3293–3296.

116 Fraga-Dubreuil, J. and Bazureau, J.P. (2001) *Tetrahedron Letters*, **42**, 6097–6100.

117 Fraga-Dubreuil, J., Famelart, M.-H., and Bazureau, J.P. (2002) *Organic Process Research & Development*, **6**, 374–378.

118 Fraga-Dubreuil, J. and Bazureau, J.P. (2003) *Tetrahedron*, **59**, 6121–6130.

119 Hakkou, H., Vanden Eynde, J.J., Hemelin, J., and Bazureau, J.P. (2004) *Tetrahedron*, **60**, 3745–3753.

120 Legeay, J.C., Vanden Eynde, J.J., and Bazureau, J.P. (2005) *Tetrahedron*, **61**, 12386–12397.

121 Legeay, J.C., Goujon, J.Y., Vanden Eynde, J.J., Toupet, L., and Bazureau, J.P. (2006) *Journal of Combinatorial Chemistry*, **8**, 829–833.

122 Hu, Y., Wei, P., Huang, H., Han, S.-Q., and Ouyang, P.-K. (2006) *Heterocycles*, **68**, 375–380.

123 De Kort, M., Tuin, A.W., Kuiper, S., Overkleeft, H.S., Van der Marel, G.A., and Buijsman, R.C. (2004) *Tetrahedron Letters*, **45**, 2171–2175.

124 Jinni, L. and Toy, P.H. (2009) *Chemical Reviews*, **109**, 815–838.

125 (a) Kirschning, A., Monenschein, H., and Wittenberg, R. (2001) *Angewandte Chemie, International Edition*, **40**, 650–679; (b) Tzschucke, C.C., Markert, C., Bannwarth, W., Roller, S., Hebel, A., and Haag, R. (2002) *Angewandte Chemie, International Edition*, **41**, 3964–4000.

126 Brain, C., Paul, J.M., Loong, Y., and Oakley, P.J. (1999) *Tetrahedron Letters*, **40**, 3275–3278.

127 Wipf, P. and Venkatraman, S. (1996) *Tetrahedron Letters*, **37**, 4659–4662.

128 Brain, C.T. and Brunton, S.A. (2001) *Synlett*, 382–384.

129 Wang, Y., Miller, R.L., Sauer, D.R., and Djuric, S.W. (2005) *Organic Letters*, **7**, 925–928.

130 Wang, Y., Sauer, D.R., and Djuric, S.W. (2006) *Tetrahedron Letters*, **47**, 105–108.

131 Wang, Y., Sarris, K., Sauer, D.R., and Djuric, S.W. (2007) *Tetrahedron Letters*, **48**, 2237–2240.

132 Baxendale, I.R., Ley, S.V., and Martinelli, M. (2005) *Tetrahedron*, **61**, 5323–5349.

133 Ley, S.V., Leach, A.G., and Storer, R.I. (2001) *Journal of the Chemical Society, Perkin Transactions 1*, 358–361.

134 Westman, J. (2001) *Organic Letters*, **3**, 3745–3747.

135 Desai, B. and Danks, T.N. (2001) *Tetrahedron Letters*, **42**, 5963–5965.

136 Gopalkrishnan, G., Kasinath, V., Pradeep Singh, N.D., Santhana Krishnan, V.P., Anand Soloman, K., and Rajan, S.S. (2002) *Molecules*, **7**, 412–419.

137 Ley, S.V. and Taylor, S.J. (2002) *Bioorganic & Medicinal Chemistry Letters*, **12**, 1813–1816.

138 Mukaiyama, T. and Yokota, Y. (1965) *Bulletin of the Chemical Society of Japan*, **38**, 858.

139 Radi, M., Saletti, S., and Botta, M. (2008) *Tetrahedron Letters*, **49**, 4464–4466.

140 Launay, D., Booth, S., Clemens, I., Merritt, A., and Bradley, M. (2002) *Tetrahedron Letters*, **43**, 7201–7203.

141 Hu, W., Bürli, R.W., Kaizerman, J.A., Johnson, K.W., Gross, M.I., Iwamoto, M., Jones, P., Lofland, D., Difuntorum, S., Chen, H., Bozdogan, B., Appelbaum, P.C., and Moser, H.E. (2004) *Journal of Medicinal Chemistry*, **47**, 4352–4355.

142 (a) Crosignani, S., White, P.D., and Linclau, B. (2002) *Organic Letters*, **4**, 2961–2963; (b) Crosignani, S., Launay, D., Linclau, B., and Bradley, M. (2003) *Molecular Diversity*, **7**, 203–210; (c) Crosignani, S., White, P.D., and Linclau, B. (2004) *The Journal of Organic Chemistry*, **69**, 5897–5905.

143 Sauer, D.R., Kalvin, D., and Phelan, K.M. (2003) *Organic Letters*, **5**, 4721–4724.

144 Humphrey, C.E., Easson, M.A.M., Tierney, J.P., and Turner, N.J. (2003) *Organic Letters*, **5**, 849–852.

145 Kirschning, A., Monenschein, H., and Wittenberg, R. (2000) *Chemistry – A European Journal*, **6**, 4445–4450.

146 Porcheddu, A., Giacomelli, G., De Luca, L., and Ruda, A.M. (2004) *Journal of Combinatorial Chemistry*, **6**, 105–111.

147 Petricci, E., Botta, M., Corelli, F., and Mugnaini, C. (2002) *Tetrahedron Letters*, **43**, 6507–6509.

148 Petricci, E., Mugnaini, C., Radi, M., Corelli, F., and Botta, M. (2004) *The Journal of Organic Chemistry*, **69**, 7880–7887.

149 Nicewonger, R.B., Ditto, L., Kerr, D., and Varady, L. (2002) *Bioorganic & Medicinal Chemistry Letters*, **12**, 1799–1802.

150 Crosignani, S., Gonzalez, J., and Swinnen, D. (2004) *Organic Letters*, **6**, 4579–4582.

151 Baxendale, I.R., Ley, S.V., and Piutti, C. (2002) *Angewandte Chemie, International Edition*, **41**, 2194–2197.

152 Baxendale, I.R., Ley, S.V., Nessi, M., and Piutti, C. (2002) *Tetrahedron*, **58**, 6285–6304.

153 Baxendale, I.R., Lee, A.-L., and Ley, S.V. (2005) Chapter 6, in *Microwave-Assisted Organic Synthesis* (eds P. Lidström and J.P. Tierney), Blackwell Publishing, Oxford.

154 Kacprzak, K.M., Maier, N.M., and Lindner, W. (2006) *Tetrahedron Letters*, **47**, 8721–8726.

155 Donati, D., Morelli, C., and Taddei, M. (2005) *Tetrahedron Letters*, **46**, 2817–2819.

156 Crosignani, S. and Swinnen, D. (2005) *Journal of Combinatorial Chemistry*, **7**, 688–696.

157 Le Bas, M.D.H. and O'Shea, D.F. (2005) *Journal of Combinatorial Chemistry*, **7**, 947–951.

158 Werner, S., Kasi, D., and Brummond, K.M. (2007) *Journal of Combinatorial Chemistry*, **9**, 677–683.

159 Mao, S., Probst, D., Werner, S., Chen, J., Xie, X., and Brummond, K.M. (2008) *Journal of Combinatorial Chemistry*, **10**, 235–246.

160 Galaffu, N. and Bradley, M. (2005) *Tetrahedron Letters*, **46**, 859–861.

161 Dallinger, D., Gorobets, N.Yu., and Kappe, C.O. (2003) *Molecular Diversity*, **7**, 229–245.

162 Öhberg, L. and Westman, J. (2001) *Synlett*, 1893–1896.

163 Alza, E., Rodríguez-Escrich, C., Sayalero, S., Bastero, A., and Pericás, M.A. (2009) *Chemistry – A European Journal*, **15**, 10167–10172.

164 Mayo, K.G., Nearhoof, E.H., and Kiddle, J.J. (2002) *Organic Letters*, **4**, 1567–1570.

165 Organ, M.G., Mayer, S., Lepifre, F., N'Zemba, B., and Khatri, J. (2003) *Molecular Diversity*, **7**, 211–227.

166 Sedelmeier, J., Ley, S.V., Lange, H., and Baxendale, I.R. (2009) *European Journal of Organic Chemistry*, 4412–4420.

167 Jacobsen, E.N., Pfaltz, A., and Yamamoto, H. (1999) *Comprehensive Asymmetric Catalysis*, Springer, Berlin.

168 Belda, O., Lundgren, S., and Moberg, C. (2003) *Organic Letters*, **5**, 2275–2278.

169 Bai, L., Zhang, Y., and Wang, J.X. (2004) *The QSAR & Combinatorial Science*, **23**, 875–882.

170 Solodenko, W., Schön, U., Messinger, J., Glenschert, A., and Kirschning, A. (2004) *Synlett*, 1699–1702.

171 Wang, Y. and Sauer, D.R. (2004) *Organic Letters*, **6**, 2793–2796.

172 Christensen, H., Kiil, S., Dam-Johansen, K., and Nielsen, O. (2007) *Organic Process Research & Development*, **11**, 956–965.

173 Ribelin, T.P., Judd, A.S., Akritopoulou-Zanze, I., Henry, R.F., Cross, J.L., Whittern, D.N., and Djuric, S.W. (2007) *Organic Letters*, **9**, 5119–5122.

174 Bai, L. and Wang, J.X. (2008) *Advanced Synthesis and Catalysis*, **350**, 315–320.

175 Srivastava, R.R. and Collibee, S.E. (2004) *Tetrahedron Letters*, 8895–8897.

176 Bergbreiter, D.E. and Furyk, S. (2004) *Green Chemistry*, **6**, 280–285.

177 Gopalakrishnan, G., Kasinath, V., and Pradeep Singh, N.D. (2002) *Organic Letters*, **4**, 781–782.

178 Michrowska, A., Mennecke, K., Kunz, U., Kirschning, A., and Grela, K. (2006) *Journal of the American Chemical Society*, **128**, 13261–13267.

179 Dawood, K.M. and Kirschning, A. (2005) *Tetrahedron*, **61**, 12121–12130.

180 Buchmeiser, M.R., Lubbad, S., Mayr, M., and Wurst, K. (2003) *Inorganica Chimica Acta*, **345**, 145–153.

181 Li, J., Mau, A.W.-H., and Strauss, C.R. (1997) *Chemical Communications*, 1275–1276.

182 Stadler, A. and Kappe, C.O. (2002) *Organic Letters*, **4**, 3541–3443.

183 Shore, G., Morin, S., and Organ, M.G. (2006) *Angewandte Chemie, International Edition*, **45**, 2761–2766.

184 Shore, G., Morin, S., Mallik, D., and Organ, M.G. (2008) *Chemistry – A European Journal*, **14**, 1351–1356.

185 Schmöger, C., Szuppa, T., Tied, A., Schneider, F., Stolle, A., and Ondruschka, B. (2008) *ChemSusChem*, **1**, 339–347.

186 Banik, B.K., Barakat, K.J., Wagle, D.R., Manhas, M.S., and Bose, A.K. (1999) *The Journal of Organic Chemistry*, **64**, 5746–5753.

187 Choudary, B.M., Madhi, S., Chowdhari, N.S., Kantan, M.L., and Sreedhar, B. (2002) *Journal of the American Chemical Society*, **124**, 14127–14136.

188 Kabalka, G.W., Wang, L., and Pagni, R.M. (2001) *Tetrahedron*, **57**, 8017–8028.

189 Smith, M.D., Stepan, A.F., Ramarao, C., Brennan, P.E., and Ley, S.V. (2003) *Chemical Communications*, 2652–2653.

190 Liu, Y., Chemtong, C., and Hu, J. (2004) *Chemical Communications*, 398–399.

191 Lohmann, S., Andrews, S.P., Burke, B.J., Smith, M.D., Attfield, J.P., Tanaka, H.,

Kaneko, K., and Ley, S.V. (2005) *Synlett*, 1291–1295.

192 Gronnow, M.J., Luque, R., Macquarrie, D.J., and Clark, J.H. (2005) *Green Chemistry*, **7**, 552–557.

193 Polshettiwar, V. and Varma, R.S. (2008) *Tetrahedron*, **64**, 4637–4643.

194 Vedantham, P., Guerra, J.M., Schoenen, F., Huang, M., Gor, P.J., Georg, G.I., Wang, J.L., Neuenswander, B., Lushington, G.H., Mitscher, L.A., Ye, Q.-Z., and Hanson, P.R. (2008) *Journal of Combinatorial Chemistry*, **10**, 185–194.

195 Desai, B. and Kappe, C.O. (2004) *Topics in Current Chemistry*, vol. 242 (ed. A. Kirschning), Springer, Berlin, pp. 177–207.

196 Dallinger, D., Gorobets, N.Yu., and Kappe, C.O. (2003) *Organic Letters*, **5**, 1205–1208.

197 Mukade, T., Dragoli, D.R., and Ellman, J.A. (2003) *Journal of Combinatorial Chemistry*, **5**, 590–596.

198 Masip, I., Ferrándiz-Huertas, C., García-Martínez, C., Ferragut, J.A., Ferrer-Montiel, A., and Messeguer, A. (2004) *Journal of Combinatorial Chemistry*, **6**, 135–141.

199 Lei, X. and Porco, J.A. (2004) *Organic Letters*, **6**, 795–798.

200 Siu, J., Baxendale, I.R., Lewthwaite, R.A., and Ley, S.V. (2005) *Organic and Biomolecular Chemistry*, **3**, 3140–3160.

Index

Microwaves in Organic and Medicinal Chemistry, Second Edition.
C. Oliver Kappe, Alexander Stadler, and Doris Dallinger.

b

C

i

n

q

r

t